Lecture Notes in Computer Science 3038

Commenced Publication in 1973
Founding and Former Series Editors:
Gerhard Goos, Juris Hartmanis, and Jan van Leeuwen

AF165948

Editorial Board

Takeo Kanade
  Carnegie Mellon University, Pittsburgh, PA, USA
Josef Kittler
  University of Surrey, Guildford, UK
Jon M. Kleinberg
  Cornell University, Ithaca, NY, USA
Friedemann Mattern
  ETH Zurich, Switzerland
John C. Mitchell
  Stanford University, CA, USA
Moni Naor
  Weizmann Institute of Science, Rehovot, Israel
Oscar Nierstrasz
  University of Bern, Switzerland
C. Pandu Rangan
  Indian Institute of Technology, Madras, India
Bernhard Steffen
  University of Dortmund, Germany
Madhu Sudan
  Massachusetts Institute of Technology, MA, USA
Demetri Terzopoulos
  New York University, NY, USA
Doug Tygar
  University of California, Berkeley, CA, USA
Moshe Y. Vardi
  Rice University, Houston, TX, USA
Gerhard Weikum
  Max-Planck Institute of Computer Science, Saarbruecken, Germany

Springer-Verlag Berlin Heidelberg GmbH

Marian Bubak Geert Dick van Albada
Peter M.A. Sloot Jack J. Dongarra (Eds.)

Computational Science - ICCS 2004

4th International Conference
Kraków, Poland, June 6-9, 2004
Proceedings, Part III

 Springer

Volume Editors

Marian Bubak
AGH University of Science and Technology
Institute of Computer Science and Academic Computer Center CYFRONET
Mickiewicza 30, 30-059 Kraków, Poland
E-mail: bubak@uci.agh.edu.pl

Geert Dick van Albada
Peter M.A. Sloot
University of Amsterdam, Informatics Institute, Section Computational Science
Kruislaan 403, 1098 SJ Amsterdam, The Netherlands
E-mail: {dick,sloot}@science.uva.nl

Jack J. Dongarra
University of Tennessee, Computer Science Department
1122 Volunteer Blvd, Knoxville, TN 37996-3450, USA
E-mail: dongarra@cs.utk.edu

Library of Congress Control Number: Applied for

CR Subject Classification (1998): D, F, G, H, I, J, C.2-3

ISSN 0302-9743
ISBN 978-3-540-22116-6   ISBN 978-3-540-24688-6 (eBook)
DOI 10.1007/978-3-540-24688-6

This work is subject to copyright. All rights are reserved, whether the whole or part of the material is
concerned, specifically the rights of translation, reprinting, re-use of illustrations, recitation, broadcasting,
reproduction on microfilms or in any other way, and storage in data banks. Duplication of this publication
or parts thereof is permitted only under the provisions of the German Copyright Law of September 9, 1965,
in its current version, and permission for use must always be obtained from Springer-Verlag. Violations are
liable to prosecution under the German Copyright Law.

springeronline.com

© Springer-Verlag Berlin Heidelberg 2004
Originally published by Springer-Verlag Berlin Heidelberg New York in 2004.

Typesetting: Camera-ready by author, data conversion by PTP-Berlin, Protago-TeX-Production GmbH
Printed on acid-free paper   SPIN: 11009337   06/3142   5 4 3 2 1 0

Preface

The International Conference on Computational Science (ICCS 2004) held in Kraków, Poland, June 6–9, 2004, was a follow-up to the highly successful ICCS 2003 held at two locations, in Melbourne, Australia and St. Petersburg, Russia; ICCS 2002 in Amsterdam, The Netherlands; and ICCS 2001 in San Francisco, USA.

As computational science is still evolving in its quest for subjects of investigation and efficient methods, ICCS 2004 was devised as a forum for scientists from mathematics and computer science, as the basic computing disciplines and application areas, interested in advanced computational methods for physics, chemistry, life sciences, engineering, arts and humanities, as well as computer system vendors and software developers. The main objective of this conference was to discuss problems and solutions in all areas, to identify new issues, to shape future directions of research, and to help users apply various advanced computational techniques. The event harvested recent developments in computational grids and next generation computing systems, tools, advanced numerical methods, data-driven systems, and novel application fields, such as complex systems, finance, econo-physics and population evolution.

Keynote lectures were delivered by David Abramson and Alexander V. Bogdanov, *From ICCS 2003 to ICCS 2004 – Personal Overview of Recent Advances in Computational Science*; Iain Duff, *Combining Direct and Iterative Methods for the Solution of Large Sparse Systems in Different Application Areas*; Chris Johnson, *Computational Multi-field Visualization*; John G. Michopoulos, *On the Pathology of High Performance Computing*; David De Roure, *Semantic Grid*; and Vaidy Sunderam, *True Grid: What Makes a Grid Special and Different?* In addition, three invited lectures were delivered by representatives of leading computer system vendors, namely: Frank Baetke from Hewlett Packard, Eng Lim Goh from SGI, and David Harper from the Intel Corporation.

Four tutorials extended the program of the conference: Paweł Płaszczak and Krzysztof Wilk, *Practical Introduction to Grid and Grid Services*; Grzegorz Młynarczyk, *Software Engineering Methods for Computational Science*; the *CrossGrid Tutorial* by the CYFRONET CG team; and the Intel tutorial.

We would like to thank all keynote, invited and tutorial speakers for their interesting and inspiring talks.

Aside of plenary lectures, the conference included 12 parallel oral sessions and 3 poster sessions. Ever since the first meeting in San Francisco, ICCS has attracted an increasing number of more researchers involved in the challenging field of computational science. For ICCS 2004, we received 489 contributions for the main track and 534 contributions for 41 originally-proposed workshops. Of these submissions, 117 were accepted for oral presentations and 117 for posters in the main track, while 328 papers were accepted for presentations at 30 workshops. This selection was possible thanks to the hard work of the Program

Committee members and 477 reviewers. The author index contains 1395 names, and almost 560 persons from 44 countries and all continents attended the conference: 337 participants from Europe, 129 from Asia, 62 from North America, 13 from South America, 11 from Australia, and 2 from Africa.

The ICCS 2004 proceedings consists of four volumes, the first two volumes, LNCS 3036 and 3037 contain the contributions presented in the main track, while volumes 3038 and 3039 contain the papers accepted for the workshops. Parts I and III are mostly related to pure computer science, while Parts II and IV are related to various computational research areas. For the first time, the ICCS proceedings are also available on CD. We would like to thank Springer-Verlag for their fruitful collaboration. During the conference the best papers from the main track and workshops as well as the best posters were nominated and presented on the ICCS 2004 Website. We hope that the ICCS 2004 proceedings will serve as a major intellectual resource for computational science researchers, pushing back the boundaries of this field. A number of papers will also be published as special issues of selected journals.

We owe thanks to all workshop organizers and members of the Program Committee for their diligent work, which ensured the very high quality of the event. We also wish to specifically acknowledge the collaboration of the following colleagues who organized their workshops for the third time: Nicoletta Del Buono (New Numerical Methods) Andres Iglesias (Computer Graphics), Dieter Kranzlmueller (Tools for Program Development and Analysis), Youngsong Mun (Modeling and Simulation in Supercomputing and Telecommunications).

We would like to express our gratitude to Prof. Ryszard Tadeusiewicz, Rector of the AGH University of Science and Technology, as well as to Prof. Marian Noga, Prof. Kazimierz Jeleń, Dr. Jan Kulka and Prof. Krzysztof Zieliński, for their personal involvement. We are indebted to all the members of the Local Organizing Committee for their enthusiastic work towards the success of ICCS 2004, and to numerous colleagues from ACC CYFRONET AGH and the Institute of Computer Science for their help in editing the proceedings and organizing the event. We very much appreciate the help of the Computer Science and Computational Physics students during the conference. We owe thanks to the ICCS 2004 sponsors: Hewlett-Packard, Intel, IBM, SGI and ATM, SUN Microsystems, Polish Airlines LOT, ACC CYFRONET AGH, the Institute of Computer Science AGH, the Polish Ministry for Scientific Research and Information Technology, and Springer-Verlag for their generous support.

We wholeheartedly invite you to once again visit the ICCS 2004 Website (http://www.cyfronet.krakow.pl/iccs2004/), to recall the atmosphere of those June days in Kraków.

June 2004   Marian Bubak, Scientific Chair 2004
on behalf of the co-editors:
G. Dick van Albada
Peter M.A. Sloot
Jack J. Dongarra

Organization

ICCS 2004 was organized by the Academic Computer Centre CYFRONET AGH University of Science and Technology (Kraków, Poland) in cooperation with the Institute of Computer Science AGH, the University of Amsterdam (The Netherlands) and the University of Tennessee (USA).

All the members of the Local Organizing Committee are the staff members of CYFRONET and/or ICS. The conference took place at the premises of the Faculty of Physics and Nuclear Techniques AGH and at the Institute of Computer Science AGH.

Conference Chairs

Scientific Chair – Marian Bubak (Institute of Computer Science and ACC CYFRONET AGH, Poland)
Workshop Chair – Dick van Albada (University of Amsterdam, The Netherlands)
Overall Chair – Peter M.A. Sloot (University of Amsterdam, The Netherlands)
Overall Co-chair – Jack Dongarra (University of Tennessee, USA)

Local Organizing Committee

Marian Noga
Marian Bubak
Zofia Mosurska
Maria Stawiarska
Milena Zając
Mietek Pilipczuk
Karol Frańczak
Aleksander Kusznir

Program Committee

Jemal Abawajy (Carleton University, Canada)
David Abramson (Monash University, Australia)
Dick van Albada (University of Amsterdam, The Netherlands)
Vassil Alexandrov (University of Reading, UK)
Srinivas Aluru (Iowa State University, USA)
David A. Bader (University of New Mexico, USA)

J.A. Rod Blais (University of Calgary, Canada)
Alexander Bogdanov (Institute for High Performance Computing and Information Systems, Russia)
Peter Brezany (University of Vienna, Austria)
Marian Bubak (Institute of Computer Science and CYFRONET AGH, Poland)
Rajkumar Buyya (University of Melbourne, Australia)
Bastien Chopard (University of Geneva, Switzerland)
Paul Coddington (University of Adelaide, Australia)
Toni Cortes (Universitat Politècnica de Catalunya, Spain)
Yiannis Cotronis (University of Athens, Greece)
Jose C. Cunha (New University of Lisbon, Portugal)
Brian D'Auriol (University of Texas at El Paso, USA)
Federic Desprez (INRIA, France)
Tom Dhaene (University of Antwerp, Belgium)
Hassan Diab (American University of Beirut, Lebanon)
Beniamino Di Martino (Second University of Naples, Italy)
Jack Dongarra (University of Tennessee, USA)
Robert A. Evarestov (SPbSU, Russia)
Marina Gavrilova (University of Calgary, Canada)
Michael Gerndt (Technical University of Munich, Germany)
Yuriy Gorbachev (Institute for High Performance Computing and Information Systems, Russia)
Andrzej Goscinski (Deakin University, Australia)
Ladislav Hluchy (Slovak Academy of Sciences, Slovakia)
Alfons Hoekstra (University of Amsterdam, The Netherlands)
Hai Jin (Huazhong University of Science and Technology, ROC)
Peter Kacsuk (MTA SZTAKI Research Institute, Hungary)
Jacek Kitowski (AGH University of Science and Technology, Poland)
Dieter Kranzlmüller (Johannes Kepler University Linz, Austria)
Domenico Laforenza (Italian National Research Council, Italy)
Antonio Lagana (Università di Perugia, Italy)
Francis Lau (University of Hong Kong, ROC)
Bogdan Lesyng (ICM Warszawa, Poland)
Thomas Ludwig (Ruprecht-Karls-Universität Heidelberg, Germany)
Emilio Luque (Universitat Autònoma de Barcelona, Spain)
Michael Mascagni (Florida State University, USA)
Edward Moreno (Euripides Foundation of Marilia, Brazil)
Jiri Nedoma (Institute of Computer Science AS CR, Czech Republic)
Genri Norman (Russian Academy of Sciences, Russia)
Stephan Olariu (Old Dominion University, USA)
Salvatore Orlando (University of Venice, Italy)
Marcin Paprzycki (Oklahoma State University, USA)
Ron Perrott (Queen's University of Belfast, UK)
Richard Ramaroson (ONERA, France)
Rosemary Renaut (Arizona State University, USA)

Alistair Rendell (Australian National University, Australia)
Paul Roe (Queensland University of Technology, Australia)
Hong Shen (Japan Advanced Institute of Science and Technology, Japan)
Dale Shires (U.S. Army Research Laboratory, USA)
Peter M.A. Sloot (University of Amsterdam, The Netherlands)
Gunther Stuer (University of Antwerp, Belgium)
Vaidy Sunderam (Emory University, USA)
Boleslaw Szymanski (Rensselaer Polytechnic Institute, USA)
Ryszard Tadeusiewicz (AGH University of Science and Technology, Poland)
Pavel Tvrdik (Czech Technical University, Czech Republic)
Putchong Uthayopas (Kasetsart University, Thailand)
Jesus Vigo-Aguiar (University of Salamanca, Spain)
Jens Volkert (University of Linz, Austria)
Koichi Wada (University of Tsukuba, Japan)
Jerzy Wasniewski (Technical University of Denmark, Denmark)
Greg Watson (Los Alamos National Laboratory, USA)
Jan Węglarz (Poznań University of Technology, Poland)
Roland Wismüller (LRR-TUM, Germany)
Roman Wyrzykowski (Technical University of Częstochowa, Poland)
Jinchao Xu (Pennsylvania State University, USA)
Yong Xue (Chinese Academy of Sciences, ROC)
Xiaodong Zhang (College of William and Mary, USA)
Alexander Zhmakin (Soft-Impact Ltd, Russia)
Krzysztof Zieliński (Institute of Computer Science and CYFRONET AGH, Poland)
Zahari Zlatev (National Environmental Research Institute, Denmark)
Albert Zomaya (University of Sydney, Australia)
Elena Zudilova (University of Amsterdam, The Netherlands)

Reviewers

Abawajy, J.H.	Aluru, S.	Balogh, Z.
Abe, S.	Anglano, C.	Bang, Y.C.
Abramson, D.	Archibald, R.	Baraglia, R.
Adali, S.	Arenas, A.	Barron, J.
Adcock, M.	Astalos, J.	Baumgartner, F.
Adriaansen, T.	Ayani, R.	Becakaert, P.
Ahn, G.	Ayyub, S.	Belleman, R.G.
Ahn, S.J.	Babik, M.	Bentes, C.
Albada, G.D. van	Bader, D.A.	Bernardo Filho, O.
Albuquerque, P.	Bajaj, C.	Beyls, K.
Alda, W.	Baker, M.	Blais, J.A.R.
Alexandrov, V.	Baliś, B.	Boada, I.
Alt, M.	Balk, I.	Bode, A.

Bogdanov, A.
Bollapragada, R.
Boukhanovsky, A.
Brandes, T.
Brezany, P.
Britanak, V.
Bronsvoort, W.
Brunst, H.
Bubak, M.
Budinska, I.
Buono, N. Del
Buyya, R.
Cai, W.
Cai, Y.
Cannataro, M.
Carbonell, N.
Carle, G.
Caron, E.
Carothers, C.
Castiello, C.
Chan, P.
Chassin-de-
  Kergommeaux, J.
Chaudet, C.
Chaves, J.C.
Chen, L.
Chen, Z.
Cheng, B.
Cheng, X.
Cheung, B.W.L.
Chin, S.
Cho, H.
Choi, Y.S.
Choo, H.S.
Chopard, B.
Chuang, J.H.
Chung, R.
Chung, S.T.
Coddington, P.
Coeurjolly, D.
Congiusta, A.
Coppola, M.
Corral, A.
Cortes, T.
Cotronis, Y.

Cramer, H.S.M.
Cunha, J.C.
Danilowicz, C.
D'Auriol, B.
Degtyarev, A.
Denazis, S.
Derntl, M.
Desprez, F.
Devendeville, L.
Dew, R.
Dhaene, T.
Dhoedt, B.
D'Hollander, E.
Diab, H.
Dokken, T.
Dongarra, J.
Donnelly, D.
Donnelly, W.
Dorogovtsev, S.
Duda, J.
Dudek-Dyduch, E.
Dufourd, J.F.
Dumitriu, L.
Duplaga, M.
Dupuis, A.
Dzwinel, W.
Embrechts, M.J.
Emiris, I.
Emrich, S.J.
Enticott, C.
Evangelos, F.
Evarestov, R.A.
Fagni, T.
Faik, J.
Fang, W.J.
Farin, G.
Fernandez, M.
Filho, B.O.
Fisher-Gewirtzman, D.
Floros, E.
Fogel, J.
Foukia, N.
Frankovic, B.
Fuehrlinger, K.
Funika, W.

Gabriel, E.
Gagliardi, F.
Galis, A.
Galvez, A.
Gao, X.S.
Garstecki, L.
Gatial, E.
Gava, F.
Gavidia, D.P.
Gavras, A.
Gavrilova, M.
Gelb, A.
Gerasimov, V.
Gerndt, M.
Getov, V.
Geusebroek, J.M.
Giang, T.
Gilbert, M.
Glasner, C.
Gobbert, M.K.
Gonzalez-Vega, L.
Gorbachev, Y.E.
Goscinski, A.M.
Goscinski, W.
Gourhant, Y.
Gualandris, A.
Guo, H.
Ha, R.
Habala, O.
Habib, A.
Halada, L.
Hawick, K.
He, K.
Heinzlreiter, P.
Heyfitch, V.
Hisley, D.M.
Hluchy, L.
Ho, R.S.C.
Ho, T.
Hobbs, M.
Hoekstra, A.
Hoffmann, C.
Holena, M.
Hong, C.S.
Hong, I.

Hong, S.
Horan, P.
Hu, S.M.
Huh, E.N.
Hutchins, M.
Huynh, J.
Hwang, I.S.
Hwang, J.
Iacono, M.
Iglesias, A.
Ingram, D.
Jakulin, A.
Janciak, I.
Janecek, J.
Janglova, D.
Janicki, A.
Jin, H.
Jost, G.
Juhola, A.
Kacsuk, P.
Kalousis, A.
Kalyanaraman, A.
Kang, M.G.
Karagiorgos, G.
Karaivanova, A.
Karl, W.
Karypis, G.
Katarzyniak, R.
Kelley, T.
Kelly, W.
Kennedy, E.
Kereku, E.
Kergommeaux, J.C. De
Kim, B.
Kim, C.H.
Kim, D.S.
Kim, D.Y.
Kim, M.
Kim, M.J.
Kim, T.W.
Kitowski, J.
Klein, C.
Ko, P.
Kokoszka, P.
Kolingerova, I.

Kommineni, J.
Korczak, J.J.
Korkhov, V.
Kou, G.
Kouniakis, C.
Kranzlmüller, D.
Krzhizhianovskaya, V.V.
Kuo, T.W.
Kurka, G.
Kurniawan, D.
Kurzyniec, D.
Laclavik, M.
Laforenza, D.
Lagan, A.
Lagana, A.
Lamehamedi, H.
Larrabeiti, D.
Latt, J.
Lau, F.
Lee, H.G.
Lee, M.
Lee, S.
Lee, S.S.
Lee, S.Y.
Lefevre, L.
Leone, P.
Lesyng, B.
Leszczynski, J.
Leymann, F.
Li, T.
Lindner, P.
Logan, B.
Lopes, G.P.
Lorencz, R.
Low, M.Y.H.
Ludwig, T.
Luethi, J.
Lukac, R.
Luksch, P.
Luque, E.
Mairandres, M.
Malawski, M.
Malony, A.
Malyshkin, V.E.
Maniatty, W.A.

Marconi, S.
Mareev, V.
Margalef, T.
Marrone, S.
Martino, B. Di
Marzolla, M.
Mascagni, M.
Mayer, M.
Medeiros, P.
Meer, H. De
Meyer, N.
Miller, B.
Miyaji, C.
Modave, F.
Mohr, B.
Monterde, J.
Moore, S.
Moreno, E.
Moscato, F.
Mourelle, L.M.
Mueller, M.S.
Mun, Y.
Na, W.S.
Nagel, W.E.
Nanni, M.
Narayanan, M.
Nasri, A.
Nau, B.
Nedjah, N.
Nedoma, J.
Negoita, C.
Neumann, L.
Nguyen, G.T.
Nguyen, N.T.
Norman, G.
Olariu, S.
Orlando, S.
Orley, S.
Otero, C.
Owen, J.
Palus, H.
Paprzycki, M.
Park, N.J.
Patten, C.
Peachey, T.C.

Peluso, R.
Peng, Y.
Perales, F.
Perrott, R.
Petit, F.
Petit, G.H.
Pfluger, P.
Philippe, L.
Platen, E.
Plemenos, D.
Pllana, S.
Polak, M.
Polak, N.
Politi, T.
Pooley, D.
Popov, E.V.
Puppin, D.
Qut, P.R.
Rachev, S.
Rajko, S.
Rak, M.
Ramaroson, R.
Ras, I.
Rathmayer, S.
Raz, D.
Recio, T.
Reichel, L.
Renaut, R.
Rendell, A.
Richta, K.
Robert, Y.
Rodgers, G.
Rodionov, A.S.
Roe, P.
Ronsse, M.
Ruder, K.S.
Ruede, U.
Rycerz, K.
Sanchez-Reyes, J.
Sarfraz, M.
Sbert, M.
Scarpa, M.
Schabanel, N.
Scharf, E.
Scharinger, J.

Schaubschlaeger, C.
Schmidt, A.
Scholz, S.B.
Schreiber, A.
Seal, S.K.
Seinstra, F.J.
Seron, F.
Serrat, J.
Shamonin, D.P.
Sheldon, F.
Shen, H.
Shende, S.
Shentu, Z.
Shi, Y.
Shin, H.Y.
Shires, D.
Shoshmina, I.
Shrikhande, N.
Silvestri, C.
Silvestri, F.
Simeoni, M.
Simo, B.
Simonov, N.
Siu, P.
Slizik, P.
Slominski, L.
Sloot, P.M.A.
Slota, R.
Smetek, M.
Smith, G.
Smolka, B.
Sneeuw, N.
Snoek, C.
Sobaniec, C.
Sobecki, J.
Sofroniou, M.
Sole, R.
Soofi, M.
Sosnov, A.
Sourin, A.
Spaletta, G.
Spiegl, E.
Stapor, K.
Stuer, G.
Suarez Rivero, J.P.

Sunderam, V.
Suzuki, H.
Szatzschneider, W.
Szczepanski, M.
Szirmay-Kalos, L.
Szymanski, B.
Tadeusiewicz, R.
Tadic, B.
Talia, D.
Tan, G.
Taylor, S.J.E.
Teixeira, J.C.
Telelis, O.A.
Teo, Y.M
Teresco, J.
Teyssiere, G.
Thalmann, D.
Theodoropoulos, G.
Theoharis, T.
Thurner, S.
Tirado-Ramos, A.
Tisserand, A.
Toda, K.
Tonellotto, N.
Torelli, L.
Torenvliet, L.
Tran, V.D.
Truong, H.L.
Tsang, K.
Tse, K.L.
Tvrdik, P.
Tzevelekas, L.
Uthayopas, P.
Valencia, P.
Vassilakis, C.
Vaughan, F.
Vazquez, P.P.
Venticinque, S.
Vigo-Aguiar, J.
Vivien, F.
Volkert, J.
Wada, K.
Walter, M.
Wasniewski, J.
Wasserbauer, A.

Watson, G.	Xiao, Y.	Zhang, J.W.
Wawrzyniak, D.	Xu, J.	Zhang, N.X.L.
Weglarz, J.	Xue, Y.	Zhang, X.
Weidendorfer, J.	Yahyapour, R.	Zhao, L.
Weispfenning, W.	Yan, N.	Zhmakin, A.I.
Wendelborn, A.L.	Yang, K.	Zhu, W.Z.
Weron, R.	Yener, B.	Zieliński, K.
Wismüller, R.	Yoo, S.M.	Zlatev, Z.
Wojciechowski, K.	Yu, J.H.	Zomaya, A.
Wolf, F.	Yu, Z.C.H.	Zudilova, E.V.
Worring, M.	Zara, J.	
Wyrzykowski, R.	Zatevakhin, M.A.	

Workshops Organizers

Programming Grids and Metasystems

V. Sunderam (Emory University, USA)
D. Kurzyniec (Emory University, USA)
V. Getov (University of Westminster, UK)
M. Malawski (Institute of Computer Science and CYFRONET AGH, Poland)

Active and Programmable Grids Architectures and Components

C. Anglano (Università del Piemonte Orientale, Italy)
F. Baumgartner (University of Bern, Switzerland)
G. Carle (Tubingen University, Germany)
X. Cheng (Institute of Computing Technology, Chinese Academy of Science, ROC)
K. Chen (Institut Galilée, Université Paris 13, France)
S. Denazis (Hitachi Europe, France)
B. Dhoedt (University of Gent, Belgium)
W. Donnelly (Waterford Institute of Technology, Ireland)
A. Galis (University College London, UK)
A. Gavras (Eurescom, Germany)
F. Gagliardi (CERN, Switzerland)
Y. Gourhant (France Telecom, France)
M. Gilbert (European Microsoft Innovation Center, Microsoft Corporation, Germany)
A. Juhola (VTT, Finland)
C. Klein (Siemens, Germany)
D. Larrabeiti (University Carlos III, Spain)
L. Lefevre (INRIA, France)
F. Leymann (IBM, Germany)
H. de Meer (University of Passau, Germany)
G. H. Petit (Alcatel, Belgium)

J. Serrat (Universitat Politècnica de Catalunya, Spain)
E. Scharf (QMUL, UK)
K. Skala (Ruder Boskoviç Institute, Croatia)
N. Shrikhande (European Microsoft Innovation Center, Microsoft
Corporation, Germany)
M. Solarski (FhG FOKUS, Germany)
D. Raz (Technion Institute of Technology, Israel)
K. Zieliński (AGH University of Science and Technology, Poland)
R. Yahyapour (University Dortmund, Germany)
K. Yang (University of Essex, UK)

Next Generation Computing

E.-N. John Huh (Seoul Women's University, Korea)

Practical Aspects of High-Level Parallel Programming (PAPP 2004)

F. Loulergue (Laboratory of Algorithms, Complexity and Logic,
University of Paris Val de Marne, France)

Parallel Input/Output Management Techniques (PIOMT 2004)

J. H. Abawajy (Carleton University, School of Computer Science, Canada)

OpenMP for Large Scale Applications

B. Chapman (University of Houston, USA)

Tools for Program Development and Analysis in Computational Science

D. Kranzlmüller (Johannes Kepler University Linz, Austria)
R. Wismüller (TU München, Germany)
A. Bode (Technische Universität München, Germany)
J. Volkert (Johannes Kepler University Linz, Austria)

Modern Technologies for Web-Based Adaptive Systems

N. Thanh Nguyen (Wrocław University of Technology, Poland)
J. Sobecki (Wrocław University of Technology, Poland)

Agent Day 2004 – Intelligent Agents in Computing Systems

E. Nawarecki (AGH University of Science and Technology, Poland)
K. Cetnarowicz (AGH University of Science and Technology, Poland)
G. Dobrowolski (AGH University of Science and Technology, Poland)
R. Schaefer (Jagiellonian University, Poland)
S. Ambroszkiewicz (Polish Academy of Sciences, Warsaw, Poland)
A. Koukam (Université de Belfort-Montbeliard, France)
V. Srovnal (VSB Technical University of Ostrava, Czech Republic)
C. Cotta (Universidad de Málaga, Spain)
S. Raczynski (Universidad Panamericana, Mexico)

Dynamic Data Driven Application Systems

F. Darema (NSF/CISE, USA)

HLA-Based Distributed Simulation on the Grid

S. J. Turner (Nanyang Technological University, Singapore)

Interactive Visualisation and Interaction Technologies

E. Zudilova (University of Amsterdam, The Netherlands)
T. Adriaansen (CSIRO, ICT Centre, Australia)

Computational Modeling of Transport on Networks

B. Tadic (Jozef Stefan Institute, Slovenia)
S. Thurner (Universität Wien, Austria)

Modeling and Simulation in Supercomputing and Telecommunications

Y. Mun (Soongsil University, Korea)

QoS Routing

H. Choo (Sungkyunkwan University, Korea)

Evolvable Hardware

N. Nedjah (State University of Rio de Janeiro, Brazil)
L. de Macedo Mourelle (State University of Rio de Janeiro, Brazil)

Advanced Methods of Digital Image Processing

B. Smolka (Silesian University of Technology, Laboratory of Multimedia Communication, Poland)

Computer Graphics and Geometric Modelling (CGGM 2004)

A. Iglesias Prieto (University of Cantabria, Spain)

Computer Algebra Systems and Applications (CASA 2004)

A. Iglesias Prieto (University of Cantabria, Spain)
A. Galvez (University of Cantabria, Spain)

New Numerical Methods for DEs: Applications to Linear Algebra, Control and Engineering

N. Del Buono (University of Bari, Italy)
L. Lopez (University of Bari, Italy)

Parallel Monte Carlo Algorithms for Diverse Applications in a Distributed Setting

V. N. Alexandrov (University of Reading, UK)
A. Karaivanova (Bulgarian Academy of Sciences, Bulgaria)
I. Dimov (Bulgarian Academy of Sciences, Bulgaria)

Modelling and Simulation of Multi-physics Multi-scale Systems

V. Krzhizhanovskaya (University of Amsterdam, The Netherlands)
B. Chopard (University of Geneva, CUI, Switzerland)
Y. Gorbachev (St. Petersburg State Polytechnical University, Russia)

Gene, Genome and Population Evolution

S. Cebrat (University of Wrocław, Poland)
D. Stauffer (Cologne University, Germany)
A. Maksymowicz (AGH University of Science and Technology, Poland)

Computational Methods in Finance and Insurance

A. Janicki (University of Wrocław, Poland)
J.J. Korczak (University Louis Pasteur, Strasbourg, France)

Computational Economics and Finance

X. Deng (City University of Hong Kong, Hong Kong)
S. Wang (Chinese Academy of Sciences, ROC)
Y. Shi (University of Nebraska at Omaha, USA)

GeoComputation

Y. Xue (Chinese Academy of Sciences, ROC)
C. Yarotsos (University of Athens, Greece)

Simulation and Modeling of 3D Integrated Circuits

I. Balk (R3Logic Inc., USA)

Computational Modeling and Simulation on Biomechanical Engineering

Y.H. Kim (Kyung Hee University, Korea)

Information Technologies Enhancing Health Care Delivery

M. Duplaga (Jagiellonian University Medical College, Poland)
D. Ingram (University College London, UK)
K. Zieliński (AGH University of Science and Technology, Poland)

Computing in Science and Engineering Academic Programs

D. Donnelly (Siena College, USA)

Sponsoring Institutions

Hewlett-Packard
Intel
SGI
ATM
SUN Microsystems
IBM
Polish Airlines LOT
ACC CYFRONET AGH
Institute of Computer Science AGH
Polish Ministry of Scientific Research and Information Technology
Springer-Verlag

Table of Contents – Part III

Workshop on Programming Grids and Metasystems

Workshop on First International Workshop on Active and Programmable Grids Architectures and Components

Workshop on Next Generation Computing

Workshop on OpenMP for Large Scale Applications

Workshop on Tools for Program Development and Analysis in Computational Science

Workshop on Modern Technologies for Web-Based Adaptive Systems

Workshop on Agent Day 2004 – Intelligent Agents in Computing Systems

Workshop on Dynamic Data Driven Applications Systems

Workshop on HLA-Based Distributed Simulation on the Grid

Workshop on Interactive Visualisation and Interaction Technologies

Workshop on Computational Modeling of Transport on Networks

Workshop on Modeling and Simulation in Supercomputing and Telecommunications

Workshop on QoS Routing

Workshop on Evolvable Hardware

Table of Contents – Part I

Track on Models and Algorithms

Track on Data Mining and Data Bases

Track on Networking

Poster Papers

Table of Contents – Part II

Track on Numerical Algorithms

Track on Finite Element Method

Track on Neural Networks

Track on Applications

Poster Papers

Table of Contents – Part IV

Workshop on Advanced Methods of Digital Image Processing

Workshop on Computer Graphics and Geometric Modelling (CGGM 2004)

Workshop on Computer Algebra Systems and Applications (CASA 2004)

Workshop on New Numerical Methods for DEs: Applications to Linear Algebra, Control and Engineering

Workshop on Parallel Monte Carlo Algorithms for Diverse Applications in a Distributed Setting

Workshop on Modelling and Simulation of Multi-physics Multi-scale Systems

Workshop on Gene, Genome, and Population Evolution

Workshop on Computational Methods in Finance and Insurance

Workshop on Computational Economics and Finance

Workshop on GeoComputation

Workshop on Simulation and Modeling of 3D Integrated Circuits

Workshop on Computational Modeling and Simulation on Biomechanical Engineering

Workshop on Information Technologies Enhancing Health Care Delivery

Workshop on Computing in Science and Engineering Academic Programs

High-Performance Parallel and Distributed Scientific Computing with the Common Component Architecture*

David E. Bernholdt

Oak Ridge National Laboratory**
Oak Ridge, TN 37831-6016 USA
bernholdtde@ornl.gov

Abstract. [1] In the scientific computing community, parallel and, increasingly, distributed computing are both important paradigms for the development of large-scale simulation software. The ability to bridge seamlessly between these two paradigms is a valuable characteristic for programming models in this general domain. The Common Component Architecture (CCA)[2] is a software component model specially designed for the needs of the scientific community, including support for both high-performance parallel and distributed computing.

The Common Component Architecture (CCA) provides a means for software developers to manage the complexity of large-scale scientific simulations and to move toward a *plug-and-play* environment for high-performance computing. In the scientific computing context, component models also promote collaboration using independently developed software, thereby allowing particular individuals or groups to focus on the aspects of greatest interest to them. The CCA is being applied within an increasing range of disciplines, including combustion research, mesoscale storm prediction, global climate simulation, and computational chemistry, as well as connecting to instruments and sensors.

In this talk, I will introduce the basic concepts behind component-based software engineering in general, and the common component architecture in particular. I will emphasize the mechanisms by which the CCA provides for both high-performance parallel computing and distributed computing, and how it integrates with several popular distributed computing environments. Finally, I will offer examples of several applications using the CCA in parallel and distributed contexts.

---

* Research supported by the Office of Mathematical, Information and Computational Sciences (MICS) of the U.S. Dept. of Energy, Scientific Discovery through Advanced Computing (SciDAC) program.
** Oak Ridge National Laboratory is managed by UT-Battelle, LLC for the US Dept. of Energy under contract DE-AC-05-00OR22725.

[1] This is the abstract of an invited talk for the Programming Grids and Metasystems Workshop.
[2] For more information, please see http://www.cca-forum.org

Multiparadigm Model Oriented to Development of Grid Systems

Jorge Luis Victória Barbosa[1], Cristiano André da Costa[1],
Adenauer Corrêa Yamin[2], and Cláudio Fernando Resin Geyer[3]

[1] Informatics Department, University of Vale do Rio dos Sinos
São Leopoldo, RS, Brazil
{barbosa,cac}@exatas.unisinos.br
[2] Informatics Department, Catholic University of Pelotas (UCPel)
Pelotas, RS, Brazil
adenauer@ucpel.tche.br
[3] Informatics Institute, Federal University of Rio Grande do Sul (UFRGS)
Porto Alegre, RS, Brazil
geyer@inf.ufrgs.br

Abstract. Multiparadigm approach integrates programming language paradigms. We propose *Holoparadigm* (*Holo*) as a multiparadigm model oriented to development of grid systems. Holo uses a logic blackboard (called *history*) to implement a coordination mechanism. The programs are organized in levels using abstract entities called *beings*. First, we describe the principal concepts of the Holoparadigm. After, the principles of a language based on the Holoparadigm are presented. Besides, we propose the *Grid Holo* (*GHolo*), a platform to support the multi-domain heterogeneous distributed computing of programs developed in Holo. GHolo is based on object mobility and blackboards. This distributed model can be fully implemented on Java platform.

Keywords: Multiparadigm, Mobility, Blackboard and Grid Systems.

1 Introduction

Several programming language paradigms were developed to make computer programming more effective. There is no ideal solution since each one has advantages and disadvantages. Multiparadigm approaches mix two or more basic paradigms trying to get a more powerful and general solution and to overcome the specific limitations of each paradigm taking advantage of its most useful characteristics. Several multiparadigm languages and environments have been proposed as for example [3, 12, 18, 20, 23]. Each paradigm has sources of implicit parallelism, for example, AND parallelism and OR parallelism in logic programming [4, 27]. Another example is object-oriented paradigm that allows the exploitation of inter-object parallelism and intra-object parallelism [9, 21]. The multiparadigm approach integrates paradigms. So, it also integrates their parallelism sources. In this context, interest in automatic exploitation of parallelism in multiparadigm software has emerged. The development of distributed software using multiparadigm models has

M. Bubak et al. (Eds.): ICCS 2004, LNCS 3038, pp. 2–9, 2004.
© Springer-Verlag Berlin Heidelberg 2004

received attention of the scientific community [9, 13, 14, 21, 25] with some systems considering mobility, heterogeneous hardware and cluster architectures.

In this paper we propose *Holoparadigm* (*Holo*) as a multiparadigm model oriented to development of grid systems. A logic blackboard (called *history*) implements the coordination mechanism and a new programming entity (called *being*) organizes the several encapsulated levels of beings and histories (multi-domains). A new multiparadigm language (*Hololanguage*) implements the main ideas introduced by Holoparadigm. Besides, we propose a platform to support the distributed execution of programs developed in Holo. This platform is called *Grid Holo* (*GHolo*). GHolo has a heterogeneous network as physical execution environment and is based on object mobility, blackboards and multi-domain organization (tree of beings). A prototype was implemented using Java [17] and special libraries to support mobility (Voyager [28]) and blackboards (Jada [10]).

The paper is organized in five sections. Section two presents the Holoparadigm and the Hololanguage. In section three is proposed the Grid Holo. Section four describes related works. Finally, section five draws some conclusions and presents directions for future works.

2 Holoparadigm and Hololanguage

Being is the main Holoparadigm abstraction. There are two kinds of beings: *elementary being* (atomic being without composition levels) and *composed being* (being composed by other beings). An elementary being (figure 1a) is organized in three parts: *interface*, *behavior* and *history*. The interface describes the possible interactions between beings. The behavior contains actions, which implement functionalities. The history is a shared storage space in a being. A composed being (figure 1b) has the same organization, but may be composed by others beings (*component beings*).

Each being has its history. The history is encapsulated in the being. In composed being, the history is shared by component beings. Several levels of encapsulated history can possibly exist. A being uses the history in a specific composition level. For example, figure 1c shows two levels of encapsulated history in a being with three composition levels. Behavior and interface parts are omitted for simplicity.

Automatic distribution is one of the main Holoparadigm goals. Figure 2 exemplifies a possible distribution of the being presented in the figure 1b. Besides that, the figure presents the mobility in Holo. The being is distributed in two nodes of the distributed architecture. The history of a distributed being is called *distributed history*. This kind of history can be implemented using DSM techniques [24] or distributed shared spaces [2, 10, 11].

Mobility [16] is the dislocation capacity of a being. In Holo, there are two kinds of mobility: *logical mobility* (being is moved when crosses one or more borders of beings) and *physical mobility* (dislocation between nodes of distributed architectures). Figure 2 exemplifies two possible mobilities in the being initially presented in the figure 1b. After the dislocation, the moveable being is unable to contact the history of the source being (figure 2, mobility A). However, now the being is able to use the history of the destiny being. Here, physical mobility only occurs if the source and destiny beings are in different nodes of the distributed architecture (it is the case in

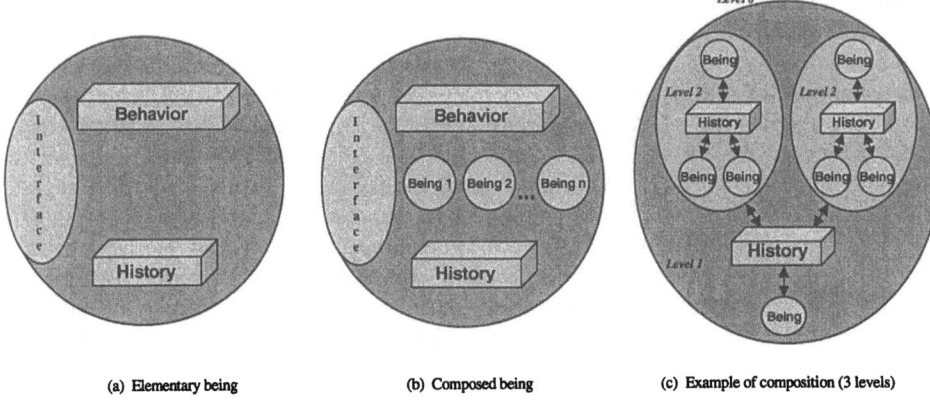

(a) Elementary being   (b) Composed being   (c) Example of composition (3 levels)

Fig. 1. Being organization

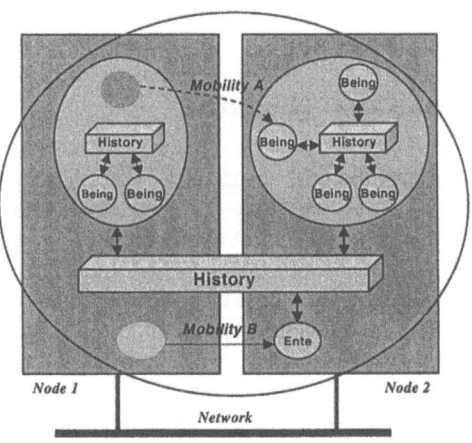

Fig. 2. Distributed being and mobility

our example). Logical and physical mobilities are independent. Occurrence of one does not imply in the occurrence of the other. For example, the mobility B in the figure 2 is a physical mobility without logical mobility. In this example, the moved being does not change its history view (supported by the blackboard). This kind of situation could happen if the execution environment aims to speedup execution through locality exploitation.

The coordination model used in Holo is based on the *blackboard architecture* [22] (figure 3a). A blackboard is composed by a common data area (blackboard) shared by a collection of programming entities called *knowledge sources* (KSs). Control is implicit in the blackboard access operations. The read and write operations in the blackboard are used to communication and synchronization between KSs. This kind of control is called *implicit invocation*. A composed being architecture is similar to the blackboard architecture, since several components share a common data area. In Holo, KSs are beings and the blackboard is the history. Since blackboard implicit

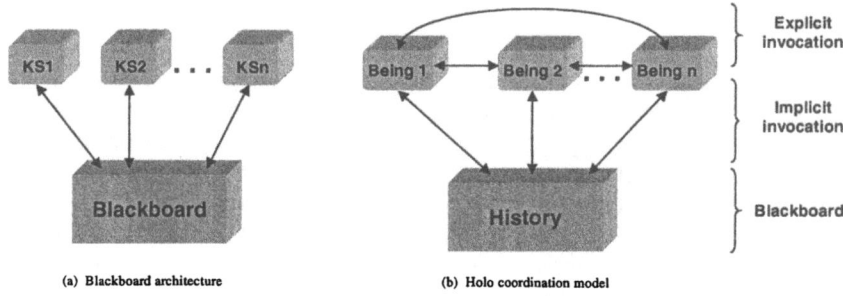

Fig. 3. Holo coordination model

invocation has several limitations, explicit invocation was introduced in the coordination model. Explicit invocation is any direct call between entities. So, the beings influence the others using the history, but can change information directly too.

Figure 3b shows the Holo coordination model. History is a logic blackboard, i.e., the information stored is a group of logic terms. Interaction with the history uses two kinds of Linda-like [8] operations: *affirmation* and *question*. An affirmation puts terms in the history, like asserts in Prolog databases. Moreover, a question permits to consult terms from the history. A consult does a search in the database using unification of terms. A question is *blocking* or *non-blocking*. A blocking question only returns when a unifying term is found. Therefore, blocking questions synchronize beings using the implicit invocation. In a non-blocking question, if a unifying term is not found, the question immediately fails. Besides that, a question is *destructive* or *non-destructive*. A destructive question retracts the unifying term. The non-destructive one does not remove it.

Hololanguage (so-called Holo) is a programming language that implements the concepts of Holoparadigm. A program is composed by descriptions of beings. The language supports logical mobility and concurrency between actions of a being. Besides that, the Hololanguage permits both kinds of blackboard interaction proposed by the Holoparadigm. Five kinds of actions are supported: (1) *logic action* (LA) is a logic predicate; (2) *imperative action* (IA) is a group of imperative commands; (3) *modular logic action* (MLA) contains several logic actions encapsulated in a module; (4) *modular imperative action* (MIA) encapsulates several imperative actions; (5) *multiparadigm action* (MA) integrates logic and imperative actions.

Actions are composed using an *Action Composition Graph* (ACG). Following the Peter Wegner's opinion [29] about the impossibility of mixing logic and imperative behaviors, we have created the *Action Invocation Graph* (AIG). This graph determines the possible order of action calls during a program execution. MAs, IAs and MIAs call any action. LAs and MLAs only call LAs and MLAs. Therefore, there are two regions of actions during an execution, namely, imperative and logic regions.

If an execution flow goes in the logic region, the only way to return to the imperative region is finishing the flow (returning the results asked from the imperative region). This methodology eliminates many problems, which emerge when logic and imperative commands are mixed (for example, distributed backtracking [5]). We believe AIG is an important contribution to the discussion presented by Wegner [29, 30].

3   Grid Holo

Holo to grid systems is called Grid Holo (GHolo). The being multi-level organization (see figure 1c) is adequate to modelling heterogeneous multi-domain distributed systems. Holo was created to support implicit distribution, i.e., automatic exploitation of distribution using mechanisms provided by basic software (compiler and execution environment). Holoparadigm abstractions are hardware independent. However, the model is dedicated to distributed architectures. When the hardware is distributed, there are two main characteristics to be considered: (1) *mobility support*: it is necessary to implement the physical mobility treatment when there is a move to a being located in another node; (2) *dynamic and hierarchical history support*: distribution involves data sharing between beings in different nodes (distributed history). There are several levels of history (hierarchical history). Besides that, access history is adapted during the execution to support the mobility (dynamic history).

GHolo is the software layer that supports the grid distributed computing of programs in Holo. It creates support to physical mobility and dynamic/hierarchical history in a grid. GHolo project is based on a structure called *Tree of Beings* (HoloTree, see figure 4). This structure is used to organize a being during its execution. The tree organizes the beings in levels. A being only can access the history of the composed being to which it belongs. This is equivalent to access the history of being localized in the superior level. A logical mobility is implemented moving a leaf (elementary being) or a tree branch (composed being) from the *source being* to the *destiny being*. The Jada spaces [10] are used to support the change of context. After the mobility, the being moved has direct access to the destiny being's space.

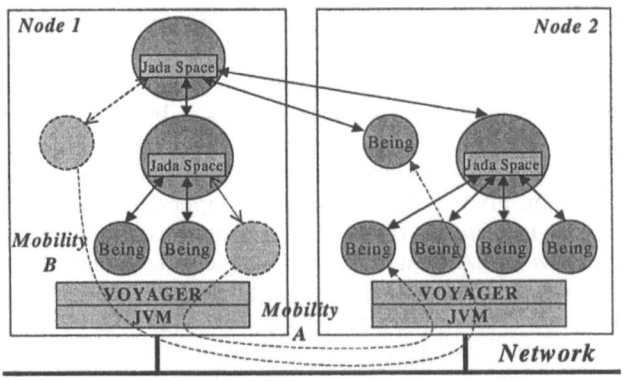

Fig. 4. Grid Holo (distributed HoloTree)

The figure 4 presents the GHolo architecture to the being initially shown in the figure 2. The figure shows the HoloTree distributed in two nodes. The changes to both mobilities of figure 2 are demonstrated. Each being is implemented using an object and a Jada space (*history*). GHolo initialization involves the creation of a *Voyager-enabled program* [28] (Voyager environment) in each node that will be used. Since Voyager executes on the Java Virtual Machine each node will also have a running JVM. During the initialization, a *Table Environment (TE)* indicates the nodes that will be used by GHolo. During a program execution, if a logical mobility results in a physical mobility, it executes a *moveTo* operation in Voyager (mobility A, figures 2 and 4). When a physical mobility is realized without logical mobility (mobility B, figures 2 and 4), the tree of beings does not change, but a *moveTo* operation is realized. This kind of mobility does not have any kind of relation with the program. It is a decision of the environment to support a specific functionality (load balancing, tolerance fault, etc). GHolo does not support this kind of decision yet, but there is on going work in this subject [31].

4 Related Works

There are other multiparadigm implementations over distributed environment. I^+ model [21] supports the distribution of objects, which implement methods using functions (functional classes) and predicates (logic classes). The implementation is based on the translation of functional classes into Lazy ML (LML) modules and translation of logic classes into Prolog modules. The distributed architecture is a network of Unix workstations using 4.3 BSD sockets to implement message passing. The runtime environment was initially implemented using C language, Quintus Prolog and LML. In a second phase, programs were only translated into C language. I^+ does not focus mobility. In addition, none kind of shared space is supported between objects.

Ciampolini et al [9] have proposed DLO, a system to create distributed logic objects. This proposal is based on previous works (Shared Prolog [7], ESP [11] and ETA [2]). The implementation is based on the translation of DLO programs into clauses of a concurrent logic language called Rose [6]. The support to Rose execution is implemented on a MIMD distributed memory parallel architecture (transputer-based Meiko Computing Surface). The runtime environment consists of a parallel abstract machine that is an extension of the WAM [1]. This proposal does not support mobility and it is not applied on a network of workstations. DLO does not support levels of spaces.

Oz multiparadigm language [20] is used to create a distributed platform called Mozart [25]. Oz uses a constraint store similar to a blackboard and supports the use of several paradigm styles [15, 20]. Besides, Mozart has special support to mobility of objects [13] and distributed treatment of logic variables [14]. Mozart distributed architecture is a network of workstations providing standard protocols such as TCP/IP. The runtime environment is composed by four software layers [25]: Oz centralized engine (Oz virtual machine [19]), language graph layer (distributed algorithms to decide when to do a local operation or a network communication), memory management layer (shared communication space and distributed garbage collection) and reliable message layer (transfer of byte sequences between nodes). The physical mobility supported by Mozart is completely transparent, i. e., the system decides when to move an object. None kind of logical mobility is used. The shared spaced supported by Mozart is monotonic and stores constraints, while Holo being's history is a non-monotonic logic blackboard that stores logic tuples (terms). In addition, Mozart does not provide levels of encapsulated contexts composed by objects accessing a shared space.

Tarau has proposed Jinni [26], a logic programming language that supports concurrency, mobility and distributed logic blackboards. Jinni is implemented using BinProlog [5] a multi-threaded Prolog system with ability to generate C/C++ code. Besides that, it has special support to Java, such as a translator allowing packaging of Jinni programs as Java classes. Jinni is not a multiparadigm platform. In addition, Jinni does not work with logical mobility and with levels of encapsulated blackboards.

So, Oz and Jinni have a kind of mobility. In addition, they can be executed over network of workstations. However, we believe that the support to hierarchy of spaces as proposed by Holo is an innovation.

5 Conclusion

We have proposed the use of Holo to multi-domain heterogeneous systems. Holoparadigm concepts estimulate the grid programming. Besides, we proposed the Grid Holo (GHolo), namely, an environment to grid computing that automatically manages the tree of beings distribution.

One important aspect of Holo is the coordination model, which simplifies the management of mobility. For this coordination, the Holo model uses a logic blackboard while GHolo proposes the use of spaces to implement it. Another important concept is the autonomous management of mobility. Holo model does not deal with physical distribution so mobility is always at logic level, i.e., between beings. GHolo execution environment can define what kind of mobility is necessary: a logical or a physical one. A logical mobility requires changes in history sharing, while physical also involves objects mobility.

Future works will improve our proposal. One ongoing work [31] aims to propose a dynamic scheduling of distributed objects, which can be directly used in GHolo. Optimizations over initial execution kernel are also under development.

References

1. Ait-Kaci, H. Warren's Abstract Machine – A Tutorial Reconstruction. MIT Press, 1991.
2. Ambriola, V.; Cignoni, G. A.; Semini; L. A Proposal to Merge Multiple Tuple Spaces, Object Orientation and Logic Programming. Computer Languages, Elmsford, v.22, n.2/3, p.79-93, July/October 1996.
3. Apt, R. et al. Alma-0: An Imperative Language that Supports Declarative Programming. ACM Transactions on Programming Languages and Systems, New York, v.20, September 1998.
4. Barbosa, J. L. V.; Vargas, P. K.; Geyer, C. GRANLOG: An Integrated Granularity Analysis Model for Parallel Logic Programming. Workshop on Parallelism and Implementation Technology (constraint) Logic Programming, London, 2000.
5. Bosschere, K.; Tarau, P. Blackboard-based Extensions in Prolog. Software – Practice and Experience, v.26, n.1, p.49-69, January 1996.
6. Brogi, A. AND-parallelism without shared variables. Seventh International Conference on Logic Programming. MIT Press, p.306-324, 1990.

7. Brogi, A.; Ciancarini, P. The Concurrent Language, Shared Prolog. ACM Transaction on Programming Languages and Systems. New York, v.13, n.1, p.99-123, January 1991.
8. Carriero, N.; Gelernter, D. Linda in context. Communications of the ACM, v.32, n.4, p.444-458, 1989.
9. Ciampolini, A.; Lamma, E.; Stefanelli, C; Mello, P. Distributed Logic Objects. Computer Languages, v.22, n.4, p.237-258, December 1996.
10. Ciancarini, P.; Rossi, D. JADA: A Coordination Toolkit for Java. http://www.cs.unibo.it/~rossi/jada, 2003.
11. Ciancarini, P. Distributed Programming with Logic Tuple Spaces. New Generating Computing, Berlin, v.12, n.3, p.251-283, 1994.
12. Hanus, M. The Integration of Functions into Logic Programming from Theory to Practice. Journal of Logic Programming, New York, v.19/20, p.583-628, May/July 1994.
13. Haridi, S. et al. Programming Languages for Distributed Applications. New Generating Computing, v.16, n.3, p.223-261, 1998.
14. Haridi, S. et al. Efficient Logic Variables for Distributed Computing. ACM Transactions on Programming Languages and Systems, v. 21, n.3, p.569-626, May 1999.
15. Henz, M. Objects in Oz. Saarbrüchen: Universität des Saarlandes, May 1997. (PhD Thesis)
16. IEEE Transactions on Software Engineering, v.24, n.5, May 1998. (Special Issue on Mobility)
17. Java. http://www.sun.com/java, 2003
18. Lee, J. H. M.; Pun, P. K. C. Object Logic Integration: A Multiparadigm Design Methodology and a Programming Language. Computer Languages, v.23, n.1, p.25-42, April 1997.
19. Meh, M.; Scheidhauer, R.; Schulte, C. An Abstract Machine for OZ. Seventh International Symposium on Programming Languages, Implementations, Logics and Programs (PLIP'95), Springer-Verlag, LNCS, September 1995.
20. Muller, M.; Muller, T.; Roy, P. V. Multiparadigm Programming in Oz. Visions for the Future of Logic Programming: Laying the Foundations for a Modern Successor of Prolog, 1995.
21. Ng, K. W.; Luk, C. K. I+: A Multiparadigm Language for Object-Oriented Declarative Programming. Computer Languages, v.21, n.2, p. 81-100, July 1995.
22. Nii, H. P. Blackboard systems: the blackboard model of problem solving and the evolution of blackboard architectures. AI Magazine, v.7, n.2, p.38-53, 1986.
23. Pineda, A.; Hermenegildo, M. O'CIAO: An Object Oriented Programming Model Using CIAO Prolog. Technical report CLIP 5/99.0 , Facultad de Informática, UMP, July 1999.
24. Proceedings of the IEEE, v.87, n.3, march 1999. (Special Issue on Distributed DSM)
25. Roy, P. V. et al. Mobile Objects in Distributed Oz. ACM Transactions on Programming Languages and Systems, v.19, n.5, p.804-851, September 1997.
26. Tarau, P. Jinni: Intelligent Mobile Agent Programming at the Intersection of Java and Prolog. PAAM'9, The Practical Applications Company, 1999.
27. Vargas, P. K.; Barbosa, J. L. V.; Ferrari, D.; Geyer, C. F. R.; Chassin, J. Distributed OR Scheduling with Granularity Information. XII Symposium on Computer Architecture and High Performance Computing, Brazil, 2000.
28. Voyager. http://www.recursionsw.com/products/voyager/voyager.asp, 2003
29. Wegner, P. Tradeoffs between Reasoning and Modeling. In: Agha, G.; Wegner, P.; Yonezawa, A. (eds.). Research Direction in Concurrent Object-Oriented Programming. Mit Press, p.22-41, 1993.
30. Wegner, P. Why interaction is more powerful than algorithms. Communications of the ACM, v. 40, n. 5, p.80-91, May 1997.
31. Yamin, A. C. ExEHDA: Execution Environment for High Distriubted Applications. PPGC/UFRGS, 2001. (PHD proposal)

The Effect of the 2nd Generation Clusters: Changes in the Parallel Programming Paradigms

Jari Porras, Pentti Huttunen, and Jouni Ikonen

Lappeenranta University of Technology, Lappeenranta, FIN-53100, Finland
{Jari.Porras,Pentti.Huttunen,Jouni.Ikonen}@lut.fi

Abstract. Programming paradigms for networks of symmetric multi-processor (SMP) workstation (2nd generation of clusters) are discussed and a new paradigm is introduced. The SMP cluster environments are explored in regard to their advantages and drawbacks with a special focus on memory architectures and communication. The new programming paradigm provides a solution to write efficient parallel applications for the 2nd generation of clusters. The paradigm aims at improving the overlap of computation and communication and the locality of communication operations. The preliminary results with large message sizes indicate improvements in excess of 30% over traditional MPI implementations.

1 Introduction

The main goal of utilizing a parallel computing environment is, most likely, the need to increase the performance of time-critical applications. There are numerous factors that impact the overall performance of a parallel application such as the number of processors, the memory architecture, and communication between processors to name a few. In order to maximize the performance of a parallel system all factors have to be considered and an optimal balance must be found.

This paper discusses the latest trends in parallel computing environments. The study shows that a new environment has emerged. This new environment is a successor of a network of workstations where each workstation consists of one processor. The 2nd generation of clusters is composed of workstations that have two or more processors. Furthermore, such systems generally pose a more heterogeneous execution environment due to various hardware and software components found in them. The new parallel environment requires changes in traditional programming paradigm or entirely new programming paradigms to efficiently utilize the available resources. This paper discusses requirements for programming paradigms for 2nd generation clusters, and describes an implementation of a communication library (MPIT) that supports the special features of SMP clusters, such as multilevel memory architecture.

M. Bubak et al. (Eds.): ICCS 2004, LNCS 3038, pp. 10–17, 2004.
© Springer-Verlag Berlin Heidelberg 2004

2   Developments in Parallel Environments

Parallel environments have developed significantly during the past few years as processor and network technologies have evolved. Presently, the environments can be divided into the following categories:

2.1   Symmetric Multiprocessors (SMP)

In SMP machines the processors are connected through an internal network in such a way that all processors have equal access to the local (global) memory. Communication between processors occurs through this memory. Therefore, the communication is expedient, but can suffer from congestion on the network. Current multiprocessor PCs represent this category of parallel systems.

2.2   Clusters or Networks of Workstations (NOW)

Clusters are environments where processors are physically separated into different workstations and the workstations are connected through a network. The popularity of these environments is often contributed to the significant improvement in the performance-to-cost ratio compared to proprietary parallel systems. However, these environments introduce additional complexity to programming and running of parallel applications. This is a result of the distributed memory architecture and the external communication network. Together, these two characteristics introduce latencies to memory access and to communication between processors. For the purpose of this paper, the networks of single processor workstations (such as Beowulf clusters) are denoted as 1st generation of the clusters.

2.3   Clusters of SMP Machines

Due to the low cost of multiprocessor workstations, clusters of SMP workstations have seen a considerably increased in number [3]. The workstations are connected in a similar way than in the traditional cluster environments but each node is a symmetric multiprocessor workstation with two or more processors. Therefore, the communication can occur internally or externally depending on the location of source and destination processors. An SMP cluster is denoted as a 2nd generation cluster.

3   Changes in the Parallel Programming Paradigms

Programming paradigms for parallel environments commonly consists of a communication library to allow processors to exchange messages. In fact, most popular programming paradigms, such as MPI [10] and PVM [5], are merely message passing libraries. The communication introduces overhead to parallel applications, and thus must be carefully implemented to minimize its contribution.

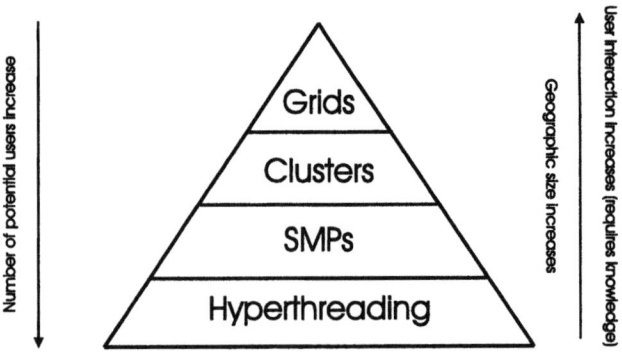

Fig. 1. The four levels of programmability in parallel systems.

There are various programming paradigms to write parallel applications based on the environment in which the applications are run. Fig. 1 depicts the paradigms and their main characteristics.

At the lowest level, parallelism occurs within a processor through the Hyperthreading technology [9]. Hyperthreading is the simplest approach from the programmer's point of view as the parallelization is automatically performed by the processors. The communication is handled implicitly through the shared memory (or even registers). The utilization of user-level threads in an SMP workstation represents the second level in Fig. 1. The programmer must be familiar with the thread-programming paradigm to efficiently implement a parallel application. Since the processors are still within a single workstation in an SMP environment, the communication between the processors occurs via the shared memory. In the 1st generation of clusters the memory architecture is distributed. Therefore, the environment requires messages to be transferred from one workstation to another with a help of a communication library such as MPI [10] or PVM [5]. The 2nd generation of clusters combine the SMP and traditional cluster environments by incorporating shared and distributed memory environments. Thus, a new paradigm is required in order to fully utilize the available resources and to obtain maximum performance. Finally, the top level in Fig. 1 illustrates an environment that integrates (SMP) clusters to each other to form grids. The grids add another level of message passing as communication between clusters is necessitated.

4   New 2-Level Communication Paradigm

The 2-level programming paradigm introduced in this section is not novel per se, since several studies have been conducted on 2-level communication paradigms:

1. Multithreaded and thread-safe MPI implementations have been proposed in [1], [2], [11], [12], [14], [15]
2. Hybrid models with MPI and OpenMP have been studied in [4], [6], [7], [8], [13]

The programming paradigm, MPIT, developed by the authors is a communication library built on top of MPI. The library provides functions to handle MPI operations in a thread-safe manner. MPIT considers both intra and inter workstation communication by using shared memory and message passing according to the resources available. The POSIX threads are used to run instances of the code on processors in a workstation. The intra workstation communication among threads is handled through the shared memory. The message passing is based on the MPI library and used only in inter workstation communication. The determination of whether shared memory or message passing is utilized is done automatically within the library.

An MPIT application operates in a single process on a workstation regardless of the number of processors. This process is created when the MPI environment is initialized through a 3rd party software (such as mpirun). The MPI environment must be initialized prior to calling the MPIT initialization function. The initialization of the MPIT library consists of the creation of the communication and the worker threads and local message buffers. The communication thread is responsible for handling all incoming and outgoing messages from a process (i.e. workstation), whereas worker threads execute the actual code.

In addition to the initialization and termination functions, the MPIT library provides a set of communication routines. The routines are very similar to the standard MPI communication operations. As a matter of fact, the only difference is that the MPIT routines have an additional argument, the target thread id. The target thread identifier allows for messages to be sent to a specific thread on a workstation rather than merely to the workstation. However, if the thread identifier is not defined, the message is sent to the workstation and processed by any of the worker threads on that workstation. Analogous to MPI, there are blocking and non-blocking send and receive operations available.

Also, MPIT provides various thread manipulation routines. These routines allow the programmer to control the number of worker threads on a workstation and the execution of the threads. With the help of the thread manipulation routines, the programmer is capable of adjusting the load imposed by the MPIT application on the workstation; more threads can be created, if the load on the workstation is low, whereas existing threads can be terminated to lower the impact of the MPIT application on the workstation's resources. The remaining thread manipulation routines are related to synchronization. There are barrier synchronization routines as well as routines to control the execution of a single thread.

Aspects like scheduling and load balancing can be addressed by this approach. The communication thread can be instructed to perform scheduling and load balancing operations to dynamically adjust loads on workstations or to retrieve new work for the workstation. This feature provides automatic scheduling functionality that occurs simultaneously with the execution of the worker threads. Therefore, the proposed programming paradigm supports not only the overlap of communication and computation but also the overlap of scheduling/load balancing and computation. The support for the scheduling/load balancing and computation overlap is very important in heterogeneous and non-dedicated 2nd generation clusters due to the necessity of frequent scheduling and load balancing operations.

The MPIT programming paradigm also provides a mechanism to assign priorities to the threads executed in a workstation. With this priority scheme a preference may be given either to the communication thread or the worker threads. Thus, it provides the means for the paradigm to adapt to the needs of the application and to the requirements set by the underlying network. For example, in an application where a large number of messages need to be exchange, the communication thread should be assigned with a higher priority than the worker threads. This allows for the communication operations to occur practically instantly minimizing the overhead introduced by the communication. A reversed priority scheme is applicable when there is no need for constant communication.

5   Performance

The MPIT library was tested in a cluster of 4 dual processor workstations. Table 1 shows the configuration of the workstations.

Table 1. The configuration of dual processor workstations used in the performance analysis

CPUs	Memory	Network	OS	MPI
2 x Pentium III 800MHz	1GB	Myrinet 2Gb/s	Red Hat 2.4.24	MPICH-GM 1.2.5..10

In order to estimate the performance of the MPIT library, a new benchmark was implemented. This benchmark makes it possible to evaluate the MPIT performance with various processor topologies, messaging schemes, and message types and sizes. The benchmark configuration for the preliminary results shown in this paper was as follows: the benchmark had 2 phases, computation and communication. During the computation phase each processor spent 1 second in a busy-wait loop, after which the process sent and received a message. The processor topology was a ring, in which each processor send a message to its right neighbor and received a message from its left neighbor.

The performance of MPI and MPIT were compared with 2 cluster configurations. The results shown in Fig. 2 illustrate three different cases of where one processors of all 4 workstations were used. In the dedicated system no other user processes were run, whereas in two other scenarioes 1 and 3 CPUs on different workstations were executing other user processes. As all the processes were distributed among the workstations all communication operations involved the Myrinet network.

The results clearly indicate that a simple MPI application is faster when the message size is below 16KB. However, the MPIT application provides up to 30%

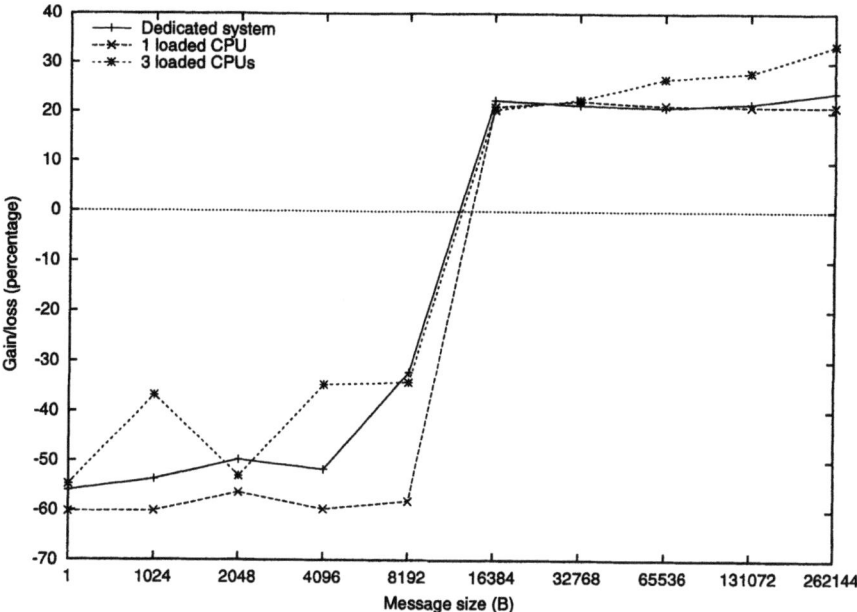

Fig. 2. Performance gain/loss of the MPIT library with 4 processes (1 process per workstation).

increase in the performance with message sizes greater than 16KB. The significant diffences with small message sizes is explained by the short message transfer times, whereas their impact becomes more apparent with the large message sizes. In addition, the current implementation of the MPIT library incurs unnecessary overhead when a message is sent synchronously to a remote process. This is due to a fact the communication thread is involved in the transmission even though it is not mandatory. This issue will be solved in the next release of the software. The performance improvement over the MPI application can be contributed to the overlap of computation and communication. The overlap allows processors to continue their execution immediately after initiating a communication operation. This also leads to a more synchronized execution of the processes which minimizes wait times during the communication phases.

The second test included running the benchmark on all available 8 processors on the test cluster. The test was run only in the dedicated mode where no other user processes were running. The results are shown in Fig. 3. The figure also entails the results from the dedicated 4 processor test for comparison purposes. Again, the MPI application has the advantage over the MPIT library with message sizes smaller than 16KB. With larger message sizes, the MPIT library has significant better performance than MPI. Furthermore, the MPIT performance has improved slightly compared to the 4 processor test.

These preliminary results indicate that MPIT has the potential to provide superior performance than MPI. However, at this point, MPI outperforms MPIT

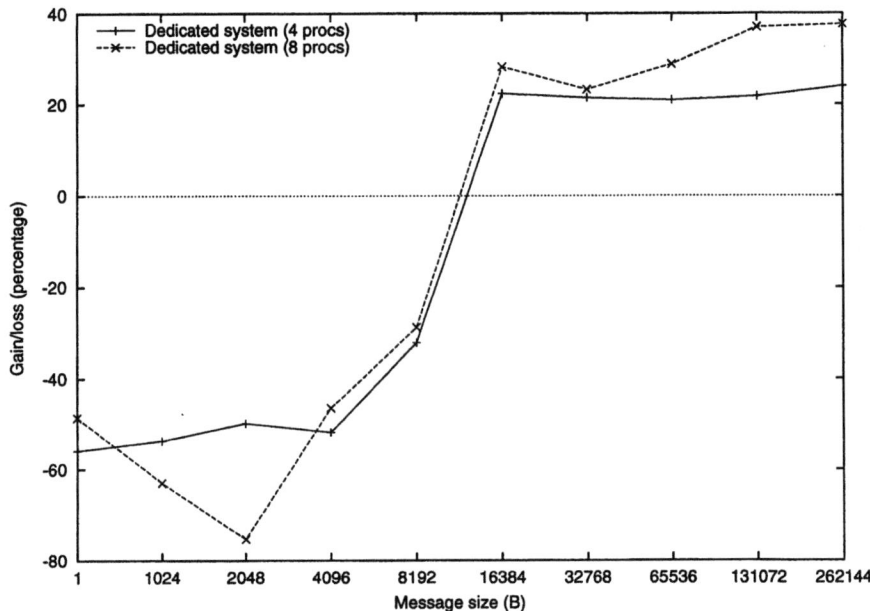

Fig. 3. Performance gain/loss of the MPIT library with 4 and 8 processes (1 process per workstation) in dedicated systems.

with small message sizes. As earlier discusses, the next release of the MPIT code will include an improved transmission mechanism for small messages, which should allow MPIT to excel regardless of the message size.

6   Conclusions

In this paper programming paradigms for cluster of SMP workstations were discussed. Also, a new programming paradigm, implemented as a thread-safe communication library on top of MPI, was introduced. The library, MPIT, optimizes the communication within and between the workstations of the cluster. It is important to realize that MPIT does not require any changes to the actual MPI implementation.

The performance results obtained show that an implementation like MPIT is required order to take full advantage of the 2nd generation clusters. Although, the performance of MPIT was inferior to MPI with small message sizes, it outpermed MPI with large message sizes (16KB and up). The new release of the MPIT library will include a new design for the transmission of small messages. Overall, the preliminary results are very encouraging as nearly 40% improvements over MPI were observed on a cluster of dual processor workstations.

In addition to the new communication scheme, MPIT includes features to perform automatic scheduling and load balancing, and to prioritize execution of communication and worker threads. In fact, as MPIT encompasses all these feature it could be extended for use in GRID environments where communication, load balancing, and scheduling are even further more complex issues than in 2nd generation clusters.

References

1. Bader, D.A., Jájá, J.: SIMPLE: A Methodology for Programming High Performance Algorithms on Clusters of Symmetric Multiprocessors (SMPs). Journal of Parallel and Distributed Computing **58** (1999) 92–108
2. Chowdappa, A.K., Skjellum, A., Doss, N.E.: Thread-Safe Message Passing with P4 and MPI. Technical Report TR-CS-941025, Computer Science Department and NSF Engineering Research Center, Mississippi State University (1994)
3. Clusters@TOP500 list. http://clusters.top500.org
4. Djomehri, M.J., Jin, H.H.: Hybrid MPI+OpenMP Programming of a Overset CFD Solver and Performance Investigations. NAS Technical Report NAS-02-002 (2002)
5. Geist A. et. al.: PVM: Parallel Virtual Machine. MIT press, 1994.
6. He, Y., Ding, C.H.Q.: MPI and OpenMP Paradigms on Clusters of SMP Architectures: the Vacancy Tracking Algorithm for Multi-Dimensional Array.
7. Hu, Y., Lu, H., Cox, A. and Zwaenepoel, W.: OpenMP for Networks of SMPs. Journal of Parallel and Distributed Computing **60** (2000) 1512–1530
8. Kee, Y-S., Kim, J-S., Ha, S.: ParADE: An OpenMP Programming Environment for SMP Cluster Systems. Proceedings of Supercomputing 2003 (2003)
9. Leng, T., Ali, R., Hsieh, J., Mashayekhi, V., Rooholamini, R.: An Empirical Study of Hyper-Threading in High Performance Computing Clusters. Proceedings of LCI International Conference on Linux Clusters: The HPC Revolution 2002 (2002)
10. Pancheco, P.: Parallel Programming with MPI. Morgan Kaufmann (1997).
11. Protopopov, B. Skjellum A.: A multithreaded message passing interface (MPI) Architecture: Performance and Program Issues. Journal of Parallel and Distributed Computing **61** (2001) 449–466
12. Rauber, T., Runger, G., Trautmann, S.: A Distributed Hierarchical Programming Model for Heterogeneous Cluster of SMPs. Proceedings of the International Parallel and Distributed Processing Symposium (2003) 381–392
13. Smith, L., Bull, M.: Development of mixed mode MPI / OpenMP Applications. Scientific Programming **9** (2001) 83–98
14. Tang, H., Yang, T.: Optimizing Threaded MPI Execution on SMP Clusters. Proceedings of Supercomputing 2001 (2001) 381–392
15. Tang, H., Shen, K., Yang, T.: Compile/Run-time Support for Threaded MPI Execution on Multiprogrammed Shared Memory Machines. Proceedings of Programming Principles of Parallel Processing (1999) 107–118

JavaSymphony, a Programming Model for the Grid*

Alexandru Jugravu[1] and Thomas Fahringer[2]

[1] University of Vienna, Institute for Software Science, Liechtensteinstr. 22,
A-1090 Wien, Austria
[2] University of Innsbruck, Institute for Software Science, Technikerstr. 25/7,
A-6020 Innsbruck, Austria

Abstract. In previous work, JavaSymphony has been introduced as a
high level programming model for performance-oriented distributed and
parallel Java programs. We have extended JavaSymphony to simplify
the development of Grid applications written in Java, allowing the programmer to control parallelism, load balancing, and locality at a high
level of abstraction. In this paper, we introduce new features to support
the widely popular workflow paradigm. The overall idea is to provide a
high level programming paradigm that shields the programmer from low
level implementation details such as RMI, sockets, and threads. Experiments will be shown to demonstrate the usefulness of JavaSymphony as
a programming paradigm for the Grid.

1 Introduction

Numerous research projects have introduced class libraries or language extensions for Java to enable parallel and distributed high-level programming. However, most approaches tend towards automatic management of locality, parallelism and load balancing which is almost entirely under the control of the runtime system. Automatic load balancing and data migration can easily lead to
performance degradation. JavaSymphony, on the other hand, is a programming
paradigm for wide classes of heterogeneous systems that allows the programmer
to control locality, parallelism and load balancing at a high level of abstraction.

In previous work, we describe the JavaSymphony programming paradigm[1],
implementation[2] and we evaluate the usefulness of our system for several parallel and distributed applications[3]. In this paper we describe important JavaSymphony mechanisms for controlling parallelism, load balancing, and locality, which
are crucial for the Grid. Moreover, we introduce new features to simplify the development of Grid workflow applications at a high level, including a user interface
to build the graphical representation of a workflow process, an XML-based description language for workflow applications, library support for the definition of

---

* This research is partially supported by the Austrian Science Fund as part of Aurora
Project under contract SFBF1104.

M. Bubak et al. (Eds.): ICCS 2004, LNCS 3038, pp. 18–25, 2004.
© Springer-Verlag Berlin Heidelberg 2004

the workflow and its components, and a specialized scheduler for workflow applications. We describe the components of our workflow model, which includes activities, links, conditional branches, and loops.

We also demonstrate these features for an existing JavaSymphony application, which is suitable to be modelled by using a workflow.

The rest of this paper is organized as follows. The next section discusses related work. In Section 3 we briefly introduce the main features of the JavaSymphony programming paradigm, showing also how they are related to Grid computing. Section 4 presents the new framework for developing JavaSymphony workflow applications, and illustrates an experiment to translates an existing application into a workflow application using our tool. Finally, some concluding remarks are made and future work is outlined in Section 5.

2   Related Work

Java's platform independent bytecode can be executed securely on many platforms, making Java an attractive basis for portable Grid computing. Java offers language elements to express thread based parallelism and synchronization.

Therefore, its potential for developing parallel and distributed applications was noticed long before the Grid computing turn out to be relevant for computer scientists. Many projects tried to use the huge potential of the Internet by supplying Java-based systems for web-computing (e.g. Bayanihan[4], Javelin[5], JaWS[6]). Such systems are commonly based on volunteer computing and/or on a three-tier architecture with hosts, brokers and clients.

Other systems are oriented towards cluster computing (e.g JavaParty[7], ProActive[8]) and offer high-level APIs for distributing Java objects over a computing resources of a cluster of workstations or a dedicated SMP cluster.

The Grid community cannot ignore the advantages of using Java for Grid computing: portability, easy deployment of Java's bytecode, component architecture provided through JavaBeans, a wide variety of class libraries that include additional functionality such as secure socket communication or complex message passing, etc. Moreover, Java Commodity Grid (CoG) Kit[9] allows access to the services provided by the Globus toolkit(http://www.globus.org).

3   JavaSymphony Preliminaries

JavaSymphony is a 100% Java library that enables the programmer to specify and control locality, parallelism, and load balancing at a high level of abstraction, without dealing with error-prone and low-level details (e.g. create and handle remote proxies for Java/RMI or socket communication). The developers of distributed application can use the JavaSymphony API to write distributed Java applications. A typical JavaSymphony application registers with the JavaSymphony Runtime System (JRS), allocates resources (machines where JRS is active), distributes code as Java objects among the resources, and invokes methods

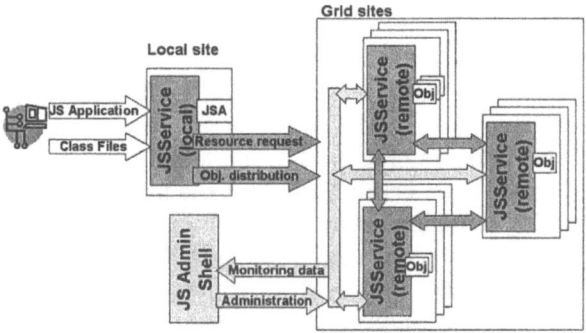

Fig. 1. JavaSymphony on the Grid

of these objects. The objects may also communicate with each other through remote method invocation. Finally the application un-registers and is removed
from the JRS's list of active applications.

Dynamic Virtual Distributed Architectures (called VAs) allow the programmer to define the structure and the characteristics of a heterogeneous network of computing resources (e.g. CPU type, speed, or machine configuration),
and to control mapping, load balancing, migration of objects and code placement.
JavaSymphony also offers possibilities to specify, control and query the properties of the computing resources. We call these properties *constraints*, which can
be static (e.g. operating system name, available memory/disc space, java version,
etc.) or dynamic (e.g. free memory, load, free disk space, network bandwidth and
latency, etc.)

JavaSymphony remote objects (JS objects) are used to distribute objects onto virtual architectures. Under JavaSymphony one can distribute any
type of Java objects, by encapsulated them in JS Objects. The programmer is
not forced to implement specific interfaces and to encode special methods. Thus
any Java object can be used in parallel applications without adding a single line
of code. The mapping of the JS objects can be explicit, on specific level-1 VAs
(single resources), or controlled by constraints associated with the resources.
JavaSymphony applications use the distributed objects by remotely invoking
their methods.

Other programming features offered by JavaSymphony include a variety
of remote method invocation types (synchronous, asynchronous and one-sided);
(un)lock mechanism for VAs and JS objects; high level API to access a variety of
static or dynamic system parameters, e.g. machine name, user name, JVM version or CPU load, idle time, available memory; selective remote class-loading;
automatic and user-controlled mapping of objects; conversion from Java conventional objects to JS objects for remote access; single-threaded versus multi-
threaded JS objects; object migration, automatic or user-controlled; distributed
event mechanism; synchronization mechanisms and persistent objects. More details about the JavaSymphony programming constructs can be found in [1,3].

The **JavaSymphony Runtime System (JRS)** is implemented as an agent based system, with agents running on each computing resource to be used by JavaSymphony applications (see [1,2]). We intend to implement the JS agents as Grid services, running on various grid resources (see Figure 1). The JS applications correspond to Grid distributed applications. The resources are requested using VAs defined in the JS application. JS constraints provide support for managing, querying and monitoring the parameters of the resources. The JS objects play the role of Grid submitted Java jobs. In addition, the distributed components of an application may freely communicate each with other, a feature which is currently not supported or hard to implement for typical Grid applications. Other features like migration, events, synchronization may be extremely useful for collaborative Grid applications.

4   Grid Workflow Applications under JavaSymphony

We define a Grid Workflow application as a set of one or more linked activities, which collectively realize a common goal. Information (files, messages, parameters, etc.) is passed from one participant to another for action, according to a set of procedural rules and the whole process is using a Grid infrastructure.

JavaSymphony programming paradigm is flexible enough to allow the implementation of a large range of distributed applications, including workflow applications. However, the developer usually has to manage the resources, build Java objects, and control the mapping of these objects onto resources. For better performance, the developer also must incorporate a scheduling strategy adapted to his particular applications. All these issues require a significant programming effort.

Many distributed applications follow a well-defined pattern, and therefore many of the above-mentioned programming issues could be automated. We are particularly interested in automatic resource allocation and scheduling. In recent times, workflow applications became quite popular in Grid community and many research and industry groups proposed standards to model and develop workflow applications and built workflow definition languages or schedulers for workflow applications[10,11,12,13]. On the other hand the workflow applications may support automatic resource discovery, allocation and scheduling.

Motivated by these aspects, we have considered to support the developing of workflow application in JavaSymphony. This support consists of:

A graphical user interface for building the structure of a workflow, specify constraints associated with the resources or characteristics of the components (called activities) of the workflow.

A specification language for describing the workflow and its components. We use an XML based language and we intend to keep it simple. A JavaSymphony workflow application will be automatically generated using the description of the workflow and the code for its activities.

Library support for workflow applications consists of a set of classes used to standardize the workflows and their activities in JavaSymphony.

A specialized scheduler which automatically finds resources (as VAs), maps components onto resources and runs the distributed computation, according to the rules defined for the workflow.

We have fully implemented the specification language and the GUI. Our GUI uses the same graphical engine as Teuta[14]. We are currently working on the library support and the scheduler for JavaSymphony workflow applications.

4.1   Workflow Model

Our tool is used to build the graphical representation of a workflow as a graph. We have followed the terminology and specifications proposed by the Workflow Management Coalition[15] and used several types of basic workflow elements:

Activities. The activities perform the computational parts of a workflow. They are represented as vertices of the associated graph. Each activity will be associated with a Java class that extends the *JSActivity* abstract class provided by JavaSymphony library. Instances of the associated classes will be placed onto computing resources where they will perform specific computation.

Dummy Activities. The dummy activities represent a special type of activities. They are supposed to perform evaluation of complex conditional expressions that may influence the scheduling of the workflow, but they require minimal computing power. They will not be placed onto distributed computing resources, but will run locally within the scheduler.

Links. Between the vertices of the associated graph, there may be directed edges, which represent the dependency relation (dataflow or control flow dependency) between the activities. Sending data from one activity to another implies that there is also a control flow dependency between the two. Therefore we may consider that all links are associated with dataflow dependencies.

Initial and Final States. Each workflow has an entry and an exit point, which we call **initial state**, respectively **final state**. These states are not associated with computation. They are also used to mark subworkflows of a workflow. Each subworkflow has a unique initial state and a unique final state.

Conditional branches. The execution plan of a workflow is dynamically changed using conditional branches. In our model, a conditional branch has one entry (link from an activity to the conditional branch) and two or more exits. The successors of the conditional branch correspond to entry points of subworkflows. When the execution reaches the conditional branch, a single successor will be chosen and the rest of them will be omitted in the execution.

Loops. For consistency reasons, the loops in JavaSymphony may be associated only with entire (sub)workflows units, with a single entry (initial state) and a single exit (final state) point. For a (sub)workflow which has a loop associated with it, the entire sequence of activities is executed repeatedly a fixed number of times (for-loops) or until an associated condition is satisfied (until-loops).

4.2   DES Encryption/Decryption Application

In the following, we show how we have used our tool to transform an existing JavaSymphony application into a workflow application. This application

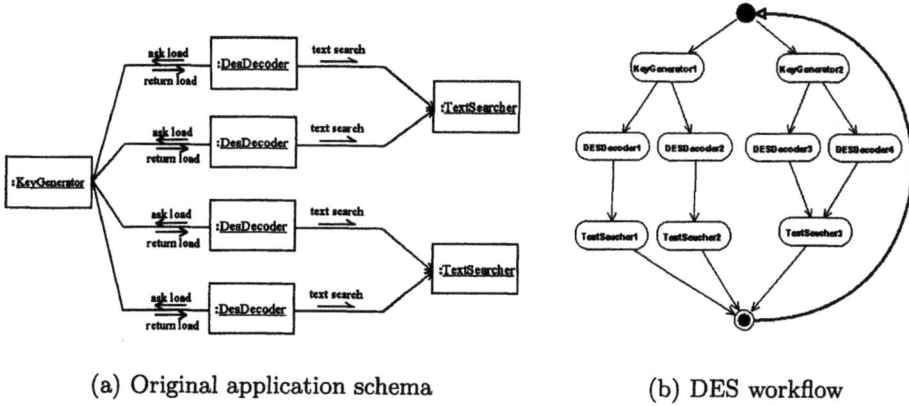

(a) Original application schema   (b) DES workflow

Fig. 2. JavaSymphony DES decoding algorithm

implements DES encryption/decryption algorithm[16]. In previous work[1], we have studied the performance of this application in comparison with two other Java-based programming paradigms for concurrent and distributed applications: JavaParty[7] and ProActive[8].

DES encryption/decryption algorithm[16] uses a key of 56 bits, which is extended with another 8 parity bits. DES tries to detect the key that has been used to encrypt a message using DES, based on a "brute-force" approach (every possible key is tested). The assumption is that we know a string that must appear in the encrypted message.

Figure 2(a) shows our original design of the DES decoding algorithm. The *DesDecoder* objects process a space of possible keys, which are provided by one or several *KeyGenerator* object(s). A *DesDecoder* acquires the keys from the *Key-Generator* through a synchronous method invocation. The *KeyGenerator* keeps track of the keys that have already been generated. After the *DesDecoders* have decoded a message by using their asigned keys, a one-sided method invocation is used to transfer the decoded messages to a *TextSearcher* object, which validates it by searching the known string in the messages.

The most important component of the application is the *DESDecoder* class. One *DESDecoder* plays an active role: it requests new data from a *KeyGenerator* as a *DESJob*, applies the decryption algorithm and sends (buffers of) processed data further to a *TextSearcher*, which (in)validates it.

4.3   DES Encryption/Decryption as a Workflow Application

The DES JavaSymphony application described above is not implemented as a workflow application. There are three types of functional components, which we used, but the data exchanged does not follow a "flow", as for workflow applications. There is a bi-directional communication between *KeyGenerator* and *DESDecoder(s)* (i.e. request and response), and respectively unidirectional between *DESDecoder(s)* and *TextSearcher(s)* (send data). However we can modify the functional components to build a workflow application.

The activities of the workflow are associated with the three types of objects: *KeyGenerator, DESDecoder* and *TextSearcher*. There is workflow precedence between *KeyGenerator(s)* and *DESDecoders*, respectively between *DESDecoders* and *TextSearchers* (see Figure 2(b)). Dataflow precedence overlaps the workflow precedence: There is unidirectional communication from *KeyGenerator* to *DES-Decoders* and from *DESDecoders* to *TextSearchers*. The implementation for the three classes has to be changed accordingly.

The graphical representation of the workflow is shown in Figure 2(b) and has been built by using our tool. In our example, there are 2 *KeyGenerator* activities, 4 *DESDecoder* activities, and 3 *TextSearcher* activities represented as vertices. The arrows between vertices represent the links of the workflow (data and control flow). There is no need for conditional branches or dummy activities. The process is repetitive and therefore a loop is associated with the whole workflow.

In addition, the developer has to implement the classes for the activities such that they implement a specific JavaSymphony interface for activities. Using the GUI, the developer associates the activities with these classes. He may also specify input parameters, resource constraints (for the VA generation), and activity characteristics relevant for the scheduling (e.g. estimated computation load, minimum/average/maximum execution time for the activity, priority etc.). For the data links between activities, one may specify link constraints (e.g. bandwidth or latency constraints) or communication characteristics (e.g. estimated communication load), which are relevant for the scheduling.

The GUI produces a file, which describes the workflow (structure, constraints, characteristics of the workflow elements, etc.) in a XML-based workflow specification language . The scheduler will use the workflow description file to map activities of the workflow onto computing resources, to enact these activities and to control communication between them. The scheduler is currently under development.

5   Conclusions and Future Work

JavaSymphony is a system designed to simplify the development of parallel and distributed Java applications that use heterogeneous computing resources ranging from small-scale cluster computing to large scale Grid computing.

In contrast with most existing work, JavaSymphony allows the programmer to explicitly control locality of data, parallelism, and load balancing based on dynamic virtual distributed architectures (VAs). These architectures impose a virtual hierarchy on a distributed system of physical computing nodes.

JavaSymphony is implemented as a collection of Java classes and runs on any standard compliant Java virtual machine. No modifications to the Java language are made and no pre-processors or special compilers are required.

In this paper we have presented a framework for developing Grid workflow applications in JavaSymphony. We consider that JavaSymphony offers a very suitable programming paradigm for developing Java-based Grid applications. We plan to further investigate the applicability of JavaSymphony in the context of Grid computing and implement an automatic scheduler for JavaSymphony workflow applications.

References

1. Jugravu, A., Fahringer, T.: Javasymphony: A new programming paradigm to control and to synchronize locality,parallelism, and load balancing for parallel and distributed computing. to appear in Concurency and Computation, Practice and Experience (2003)
2. Jugravu, A., Fahringer, T.: On the implementation of JavaSymphony. In: HIPS 2003, Nice, France, IEEE (2003)
3. Fahringer, T., Jugravu, A., Martino, B.D., Venticinque, S., Moritsch, H.: On the Evaluation of JavaSymphony for Cluster Applications. In: IEEE International Conference on Cluster Computing (Cluster2002), Chicago USA (2002)
4. Sarmenta, L.F.G., Hirano, S., Ward, S.A.: Towards Bayanihan: Building an extensible framework for volunteer computing using Java. In ACM, ed.: ACM 1998 Workshop on Java for High-Performance Network Computing, New York, NY, USA, ACM Press (1998)
5. Alan, M.N.: Javelin 2.0: Java-based parallel computing on the internet (2001)
6. Lalis, S., Karipidis, A.: Jaws: An open market-based framework for distributed computing over the internet (2000)
7. Haumacher, B., Moschny, T., Reuter, J., Tichy, W.F.: Transparent distributed threads for java. In: 5th International Workshop on Java for Parallel and Distributed Computing ,Nice, France, April 22-26, IEEE Computer Society (2003)
8. Baude, F., Caromel, D., Morel, M.: From distributed objects to hierarchical grid components. In: International Symposium on Distributed Objects and Applications (DOA), Catania, Sicily, Italy, 3-7 November, Springer Verlag, Lecture Notes in Computer Science, LNCS (2003)
9. Laszewski, G., Foster, I., Gawor, J., Lane, P.: A java commodity grid kit. Concurrency and Computation: Practice and Experience **13** (2001)
10. Andrews, T., Curbera, F., Dholakia, H., Goland, Y., Klein, J., Leymann, F., Liu, K., Roller, D., Smith, D., Systems, S., Thatte, S., Trickovic, I., Weerawarana, S.: Business process execution language for web services (bpel4ws). Specification version 1.1, Microsoft, BEA, and IBM (2003)
11. Erwin, D.W., Snelling, D.F.: UNICORE: A Grid computing environment. Lecture Notes in Computer Science **2150** (2001)
12. The Condor Team: (Dagman (directed acyclic graph manager)) http://www.cs.wisc.edu/condor/dagman/.
13. Krishnan, S., Wagstrom, P., von Laszewski, G.: GSFL : A Workflow Framework for Grid Services. Technical Report, The Globus Project (2002)
14. Fahringer, T., Pllana, S., Testori, J.: Teuta: Tool Support for Performance Modeling of Distributed and Parallel Applications. In: International Conference on Computational Science. Tools for Program Development and Analysis in Computational Science., Krakow, Poland, Springer-Verlag (2004)
15. WfMC: Workflow Management Coalition: http://www.wfmc.org/ (2003)
16. Wiener, M.J.: Efficient DES key search, technical report TR-244, Carleton University. In: William Stallings, Practical Cryptography for Data Internetworks. IEEE Computer Society Press (1996)

Adaptation of Legacy Software to Grid Services

Bartosz Baliś[1,2], Marian Bubak[1,2], and Michał Węgiel[1]

[1] Institute of Computer Science, AGH, al. Mickiewicza 30, 30-059 Kraków, Poland
[2] Academic Computer Centre – CYFRONET, Nawojki 11, 30-950 Kraków, Poland
{balis,bubak}@uci.agh.edu.pl, mwegiel@student.uci.agh.edu.pl

Abstract. Adaptation of legacy software to grid services environment is gradually gaining in significance both in academic and commercial settings but presently no comprehensive framework addressing this area is available and the scope of research work covering this field is still unsatisfactory. The main contribution of this paper is the proposal of a versatile architecture designed to facilitate the process of transition to grid services platform. We provide thorough analysis of the presented solution and confront it with fundamental grid requirements. In addition, the results of performance evaluation of a prototype implementation are demonstrated.

Keywords: Legacy software, grid services, design patterns

1 Introduction

In this paper we intend to propose a versatile framework enabling for semi-automated migration from legacy software to grid services environment [1,8]. A detailed analysis of the presented solution can be found in [2] which contains rationale supporting our design decisions and justification for rejection of other approaches. Here, we outline the most recent stage of evolution of our concept and demonstrate the results of performance evaluation of its prototype implementation.

The need for a framework enabling for cost-effective adaptation of legacy software to grid services platform is a widely recognized issue. Nonetheless, it is still addressed inadequately as compared to its significance. We have summarized the related research, which originates from both academia and industry, in [2]. Recent experiences in this area were presented at the GGF9 Workshop [3]. Currently no comprehensive framework facilitating migration to grid services environment is available and, as discussed in [2], existing approaches [4,5] possess numerous limitations.

2 System Structure

In the proposed architecture, we can distinguish three main components: **back-end host**, **hosting environment** and service **client**. As depicted in Fig. 1, they

M. Bubak et al. (Eds.): ICCS 2004, LNCS 3038, pp. 26–33, 2004.
© Springer-Verlag Berlin Heidelberg 2004

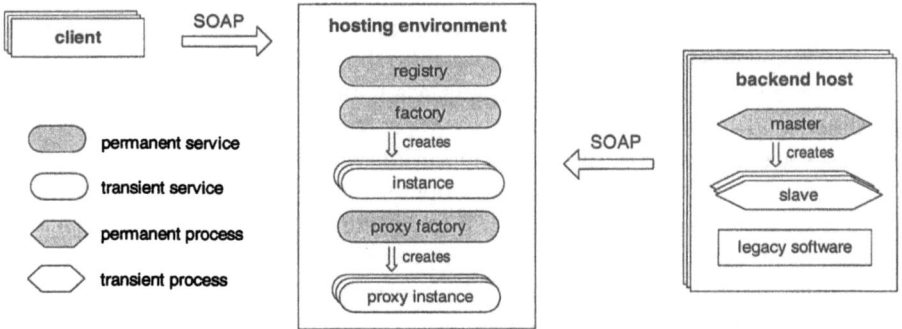

Fig. 1. System architecture

are potentially located on separate machines. The presented diagram reflects the configuration for a single legacy system which is exposed as a grid service.

Communication between components takes place by means of SOAP messages. Data is exchanged by remote method invocations performed on services deployed within the hosting environment.

2.1   Backend Host

Backend host represents the machine on which legacy software is installed and executed. In order to enhance performance and scalability as well as improve reliability and fault tolerance, we may decide to deploy several redundant copies of a single legacy application on different computing nodes, possibly in various geographic locations. For this reason, in the context of a particular service, we can end up having multiple backend hosts constituting dynamic pool of available processing resources.

We propose the following approach to maintenance and utilization of such a conglomerate.

- Backend hosts do not fulfill the function of network servers. Instead, they are devised to act as clients. Legacy applications are not required to be directly accessible from the outside.
- We employ registration model. Backend hosts are expected to volunteer to participate in computations and offer their capabilities of serving client requests.

Each backend host comprises two flavors of processes: **master** and **slave** ones. The master process is a permanent entity responsible for host registration and creation of slave processes. Whenever the load of a particular backend host in comparison with its resources allows it to serve a new client, the master process is obliged to report this fact by calling the registry service. Along with this invocation, the estimated processing capability (in the sense of a certain metric) and validity timestamp are provided. The call blocks until the given time expires or one of the pending clients is assigned. In the latter case, a new slave process is spawned in order to take over further processing. There is only one master process per host. Slave processes are transient – as clients come and go, the number of concurrently running slave processes changes respectively. These processes are in charge of direct collaboration with legacy software on behalf of the clients that they represent. This cooperation can assume various forms and range from local library usage to communication over proprietary protocols. The primary activity performed by each slave process is intercepting subsequent client invocations, translating them into legacy interface and delivering the obtained results. Slave processes communicate with the service container by means of blocking method invocations performed on the proxy instance.

2.2   Hosting Environment

The hosting environment contains a collection of grid services that jointly fulfill the role of shielding clients from unmediated interaction with backend hosts. For each legacy system, there are three permanent services deployed: **registry**, **factory** and **proxy factory**. Depending on the number of simultaneously served clients, we have varying number of transient services, two types of which exist: **instance** and **proxy instance**.

Access to all services is authorized and restricted to subjects holding adequate identities. In case of internal services, such as registry, proxy factory and proxy instance, permission is granted on the basis of host certificates. For the remaining services, namely factory and instance, we employ user certificates. Clients are not eligible to use internal services. Moreover, both types of instances can be accessed exclusively by their owners.

Registry is a service that controls the mapping between clients and backend hosts, which offered their participation in request processing. It is obliged to assign consecutively appearing clients to the chosen backend hosts. For the purpose of selection, registry maintains priority queue to which pending master processes are inserted. The ordering criterion is processing capability advertised during each registration.

The remaining services form two pairs, each consisting of factory and the corresponding instance. We discriminate two types of factories and instances: ordinary and proxy ones. The former are designated for clients whereas the latter are auxiliary entities providing mediation facilities. Proxy entities are logically separated because they serve internal purposes and should be transparent to the clients.

3   System Operation

Let us now turn our attention to the scenarios that take place during typical client-service interaction. From the client's perspective, the following sequence of actions normally is performed: (1) new service is created, (2) certain methods are invoked, (3) the service is destroyed. On the internal system side, each of the above-mentioned steps involves a number of activities. They are schematically presented in Fig. 2 as a time sequence diagram. The prerequisite for all depicted scenarios is an operative environment with permanent services up and running. Moreover, before any client interaction takes place, at least one master process should be registered in the context of the considered service. It is essential since otherwise no actual processing would be feasible.

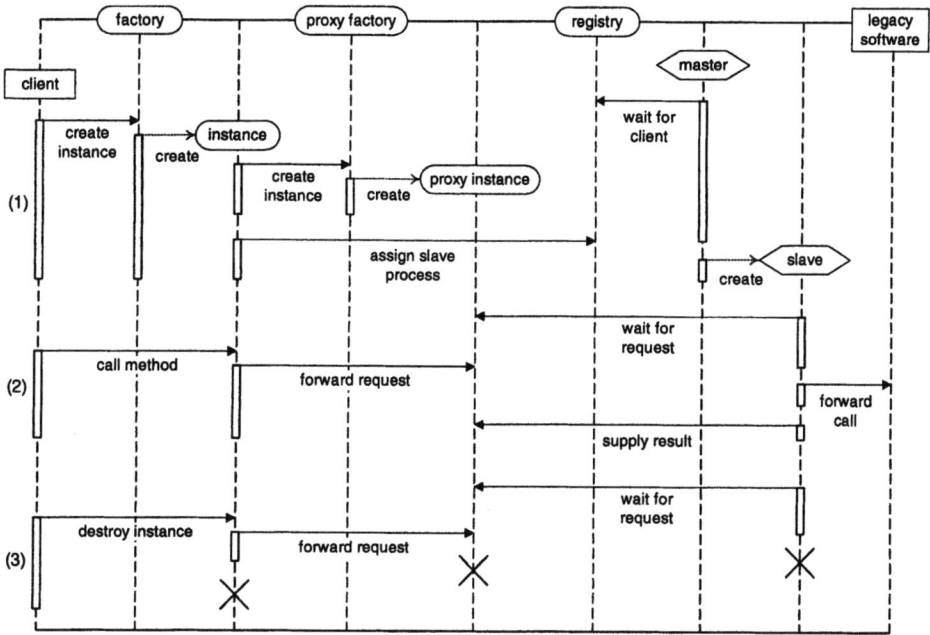

Fig. 2. Scenarios: (1) service construction (2) method invocation (3) service destruction

All scenarios are triggered by clients. Procedures accompanying service construction and destruction are executed by container's callback methods connected with lifetime management.

3.1   Service Construction

The scenario of service construction encompasses several actions. In the first place, proxy factory is contacted and the corresponding proxy instance is created. Following this, registry service is invoked. As a result of this call, one of the registered master processes is selected and provided with the location of the newly deployed proxy instance. The chosen master process spawns a new slave process which in turn starts to wait for client requests using the created proxy instance.

3.2   Method Invocation

After a successful initialization, three transient entities are created: instance, proxy instance and slave process. Since they cannot be shared, we can treat them as a private property of a particular client. Whenever a client invokes a method, its complete description, together with passed parameters, is forwarded to the proxy instance. The request is then further propagated to slave process. The call blocked on waiting for client action returns and slave process translates the obtained invocation to its legacy equivalent. Next, legacy processing is started. As soon as it is finished, the obtained results are supplied to the proxy instance via a separate asynchronous method invocation. This in turn causes that control is transferred back to service instance and after that to the client from which it originated.

Effective scenario of method invocation is critical to the system performance due to the frequency with which it occurs. The communication between instance and proxy instance takes place within the borders of a local container so it should not pose a bottleneck. Apart from this, we have two internal remote invocations per client call.

3.3   Service Destruction

Grid services support explicit and soft state mode of destruction. The former takes place on client demand whereas the latter is executed automatically when instance lifetime expires. The scenario of service removal is relatively simple. The request of destruction is forwarded to proxy instance. Once it is delivered, it is intercepted by the slave process, which terminates its execution. In the next steps, proxy instance and instance are deleted by the container.

4   System Features

Having examined individual components and scenarios comprising the proposed architecture, we will now characterize it from a wider perspective and confront its properties with requirements that have to be satisfied in a grid environment.

4.1   Security

We can point out two major aspects concerning the security of our infrastructure.

- There is no need to introduce open incoming ports on backend hosts.
- It is possible to obtain the identity of machines with which we cooperate and verify that the processing is delegated only to trusted nodes.

We owe both these advantages to the fact that backend hosts act as clients rather than servers. As we stated earlier, we perform authentication and authorization procedures. If needed, communication integrity and privacy can be ensured by means of digital signatures and encryption. In our solution security configuration can be thought of as two lists of identities. First for clients that are entitled to use our service, and second for hosts that are permitted to perform processing for our service. In consequence, maintenance of security policy should not involve excessive administrative effort.

4.2 Scalability

The combination of several factors enables to achieve high scalability of our architecture. Since actual processing is always delegated to backend hosts, computations can be heavily distributed across a large number of nodes. Furthermore, service instances residing in the container do not incur large resource consumption. Their activity is in fact reduced to message forwarding. Moreover, thanks to registration model we earn automatic load balancing. Backend hosts decide about client acceptance themselves and volunteer only when they can guarantee that processing can be finished within reasonable amount of time. This brings high responsiveness to unexpected changes in utilization. In addition, master processes advertise temporal capabilities of machines on which they are executed. This can be used for resource reservation, which is essential for the assurance of the quality of service.

4.3 Fault Tolerance

The proposed solution offers a high degree of resilience to component failures. The most important reason for this is the fact that processes are not bound to any specific endpoint. Being clients, they can arbitrarily change their location between subsequent method calls without any serious repercussions. Thus, process migration is supported. In case of legacy systems which maintain internal state or work in a transactional manner, process migration has to be disabled or automatic repetition of all preformed method invocations should be allowed.

Immunity to sudden changes of configuration is ensured by the eager registration model. At any point in time, it is not necessary to poll backend hosts in order to determine operative nodes. We have up-to-date system image sustained continuously.

4.4 Portability

Unquestionably, an important advantage of our architecture is the fact that we make no assumptions as regards programming language or platform on which both backend host and hosting environment are based. Our solution is versatile enough to accommodate a variety of legacy systems and container implementations. Furthermore, legacy software can remain in the same location where it was initially installed. There is no necessity of moving programs between machines or changing their configuration.

5   Performance Evaluation

In order to estimate the communication overhead introduced in the proposed framework, we developed a simple benchmark – echo grid service exposing a single method which repeated the string passed as its parameter. There were two functionally equivalent implementations of the above service. Measurements were performed on the client side and embraced solely method invocation scenario. No security mechanism was employed. In consequence, the results reflect the overhead introduced merely by our core architecture. The influence of different security modes on efficiency of Globus I/O library is quantified in [6].

We performed two types of tests: bandwidth- and latency-oriented. The former used large messages whereas the latter transmitted small amount of data.

The experiment was carried out on a single-processor desktop machine running Linux operating system. It simultaneously played the role of the hosting environment, backend host and service client. As for software configuration, Globus Toolkit 3.0 [9] together with gSOAP 2.1 [10] were used. We performed time measurement for message payload being 1 and 10^5 bytes. Number of method calls ranged from 1 to 256 and 8192 for bandwidth and latency measurement, respectively.

The obtained measurement results are presented in Fig. 3. In order to enhance clarity, logarithmic scale (base 2) is used. We observe a linear dependence between the number of method calls and the time needed for their execution. Thereby, the introduced overhead can be computed as a quotient of directional coefficients of linear approximations of individual functions.

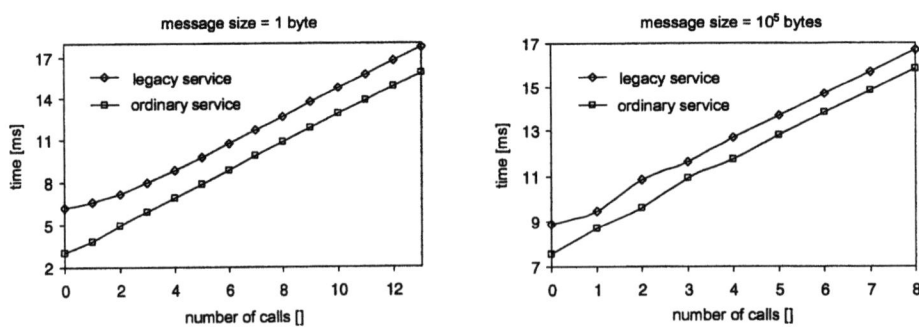

Fig. 3. Results of performance evaluation

In case of transmitting small SOAP messages, the obtained ratio was 3.65 whereas for relatively large ones we got the proportion equal to 1.79. This phenomenon can be explained if we take into account that the impact of communication overhead diminishes along with the message enlargement.

On the basis of the conducted experiments, we can draw a conclusion that in our architecture, in average case, we should expect approximately threefold decreased communication efficiency.

6   Concluding Remarks

Implementation of initial version of the presented framework took place while porting grid monitoring system, OCM-G [7], to grid services environment. This project served as a proof of the concept of the overall solution. Since then the architecture has undergone a number of refinements. Its current form is to a large extent stable. Our work is now focused on developing additional test cases in order to evaluate the proposed architecture under a variety of circumstances. The anticipated direction of project evolution is implementation of tools facilitating the application of our framework. This includes, among others, utilities allowing for automatic interface and code generation.

References

1. Foster, I., Kesselman, C., Nick, J., Tuecke, S.: The Physiology of the Grid. http://www.globus.org/research/papers/ogsa.pdf
2. Baliś, B., Bubak, M., Węgiel, M.: A Framework for Migration from Legacy Software to Grid Services. To be published in Proc. Third Cracow Grid Workshop, Cracow, Poland 2003.
3. Contributions to GGF9 Workshop *Designing and Building Grid Services*. http://www.gridforum.org
4. Huang, Y., Taylor, I., Walker, D., Davies, R.: Wrapping Legacy Codes for Grid-Based Applications. In Proc. International Workshop on Java for Parallel and Distributed Computing, Nice, France, 2003.
5. Kuebler, D., Eibach, W.: Adapting Legacy Applications as Web Services. http://www-106.ibm.com/developerworks/webservices/library/ws-legacy
6. Baliś, B., Bubak, M., Rząsa, W., Szepieniec, T., Wismüller, R.: Two Aspects of Security Solution for Distributed Systems in the Grid on the Example of OCM-G. To be published in Proc. Third Cracow Grid Workshop, Cracow, Poland 2003.
7. Baliś, B., Bubak M., Funika, W., Szczepieniec, T., Wismüller, R., Monitoring Grid Applications with Grid-enabled OMIS Monitor. In Proc. First European Across Grids Conference, Santiago de Compostela, Spain, February 2003. To appear.
8. Open Grid Services Infrastructure Specification. http://www.gridforum.org/ogsi-wg/
9. Globus Project homepage: http://www.globus.org
10. gSOAP Project homepage: http://www.cs.fsu.edu/~engelen/soap.html

Grid Service Registry for Workflow Composition Framework

Marian Bubak[1,2], Tomasz Gubała[2,3], Michał Kapałka[1], Maciej Malawski[1,2], and Katarzyna Rycerz[1,2]

[1] Institute of Computer Science, AGH, al. Mickiewicza 30, 30-059 Kraków, Poland
[2] Academic Computer Centre – CYFRONET, Nawojki 11, 30-950 Kraków, Poland
[3] Faculty of Sciences, Section of Computational Science, University of Amsterdam
Kruislaan 403, 1098 SJ Amsterdam, The Netherlands
{bubak,malawski,kzajac}@uci.agh.edu.pl
T.Gubala@cyfronet.krakow.pl, kapalka@icslab.agh.edu.pl

Abstract. The system presented in this paper supports the user in composing the flow of distributed application from existing Grid services. The flow composition system builds workflows on an abstract level with semantic and syntactic description of services available in a Grid services registry. This paper presents concepts of an overall system architecture and it focuses on one of the two main modules of the system – the distributed Grid service registry.

Keywords: Grid workflow, workflow composition, distributed registry, ontologies, Grid programming

1   Motivation

The future Grid is often described as a geographically-distributed set of services deployed on different sites by some service providers. These *Grid services*, as described in the Open Grid Service Architecture (OGSA) specifications [7,12], should allow the user to do computations, access some resources, get information or use external scientific devices. Usually each of them will provide only a part of functionality required by the user, so using many of them, connected in a workflow, will be required quite often. It is one of the new approaches to programming Grid applications. Unfortunately, as the number of available services grows, the task of manual workflow composition becomes very difficult and time-consuming. A mature flow composition system is needed to support the user and thus make the task feasible. This kind of system would also help in fast Grid application development (prototyping) and in finding services appropriate for completing a partial solution prepared by the user. It would also allow for automatic, on-demand creation of workflows for specified tasks or rebuilding old workflows when they cannot be run on the Grid anymore due to frequent changes of the environment. Two approaches to mapping an abstract workflow defined in terms of application components onto a set of available Grid resources are presented in [5]. In the Web Services community, the most popular registry system

M. Bubak et al. (Eds.): ICCS 2004, LNCS 3038, pp. 34–41, 2004.
© Springer-Verlag Berlin Heidelberg 2004

is the Universal Description, Discovery and Integration standard [13]. The implementation of a distributed UDDI registry, within a full-fledged P2P environment, supported by sophisticated registry ontologies, can be found in METEOR-S [9, 4]. The core specification of the UDDI standard (already in version 3.0) also introduces a new, distributed registry architecture. A simpler, but still very flexible and efficient solution for sharing any kind of information stored in XML documents can be found in JXTA Search [14]. Another example of this approach is the myGrid project [10] where the improved UDDI system has been successfully introduced in a Grid OGSA environment. To the best of our knowledge, there is still no mature distributed system for semi-automatic workflow composition on a Grid. As a proof-of-concept and for feasibility studies we have developed a prototype of such a system – an Application Flow Composer (AFC) [2]. It is based on the Common Component Architecture (CCA) and builds flows comprised of CCA components. We perceive, however, that syntactic information about components (the types of their ports) is not sufficient for efficient workflow composition. The other disadvantage of the system is its centralization, leading to lack of scalability and fault tolerance. Therefore, we have decided to design a new, Grid-enabled distributed system. It is intended to be used to compose workflows from Grid services, utilizing their semantic description.

2 Overall Structure of the System

The new system consists of two main elements: a flow composer and a distributed registry (see Fig. 1). Both the flow composer and the registry are distributed and should run in a heterogeneous, geographically-spread Grid environment. They can communicate in a standard way (e.g. via SOAP) or, if the highest efficiency is required, via a dedicated, fast protocol. This distributed topology makes both subsystems more scalable and fault-tolerant.

The roles of the flow composer and the registry can be explained with the following scenario. The user who wants to build a new Grid application consisting of Grid services, describes the initial conditions of the application workflow and sends them to flow composition unit. The flow composer divides this initial workflow document into smaller parts and tries to find appropriate solutions (in terms of Grid services) for each of them. To achieve this, it has to ask the registry for information about services that conform to a given semantic or syntactic description. After it finds all the necessary components, it combines results from all subtasks and builds the final workflow description that will be returned to the user. When multiple solutions can be applied, more than one final document is prepared. The user should choose the one that is the most appropriate. Sometimes the user's help is required during the composition process. The system might ask the user questions in order to ease and speed up the decision-making process, especially regarding semantic matchmaking which is always difficult for an artificial entity.

The flow composer architecture can be seen as a distributed set of agents, composing workflow on a user request. We use the well-known term *agent* for

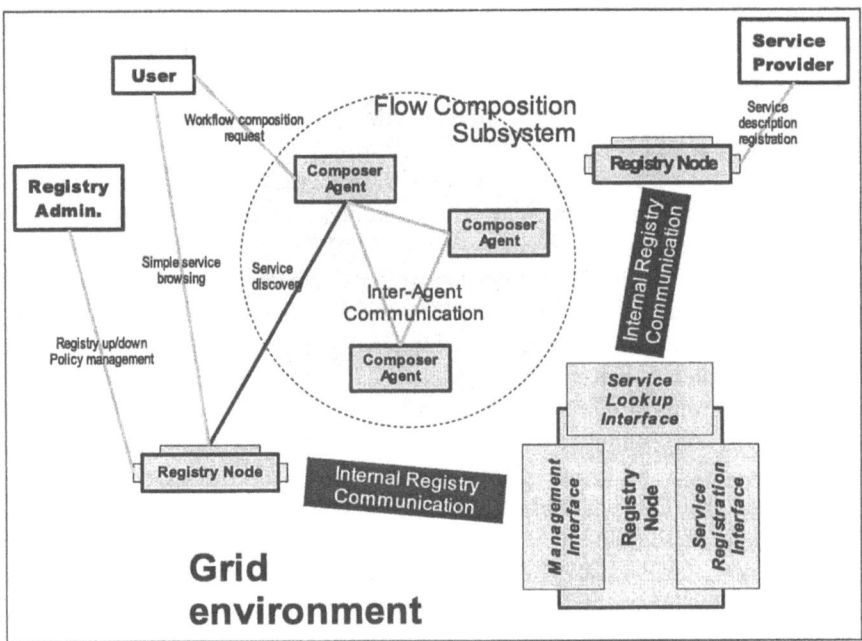

Fig. 1. General view of the system architecture

describing the components of the flow composer, because each of them operates on behalf of the user, contains its own rules of quasi-autonomous decision making and uses resources (i.e. descriptions of Grid services). This agent-based approach has several advantages. It makes the solution scalable and efficient as the agents can spread on many resources and work together on one composition problem. It also imparts some fault-tolerance on the system as a failure of one agent shouldn't stop the whole process. A more detailed description of the flow composer subsystem and the workflow composition process is given in [3].

The role of the distributed service registry in the entire framework is crucial. It is responsible for efficient delivery of data describing every available Grid service for the use of the flow composer. In this paper we describe our approach to building the registry together with requirements analysis and design.

3   Distributed Registry

3.1   Requirements

The registry should enable publication and searching for information about components of a workflow. It should be as universal as possible in order to reduce the need for multiple registries within one Grid environment. However, as we are not able to make a registry capable of storing each kind of information, we will concentrate on Grid services and their descriptions, aiming at flexibility and extensibility. The registry should meet the following requirements which stem from demands of Grid computing and the flow composition system:

- be distributed and scalable,
- be efficient, at least as far as searching for data is concerned (the speed of updating and inserting new data into the registry isn't so crucial),
- allow for some redundancy of stored data (i.e. replication), thus lacking single points of failures,
- do not depend too much on the format of the information that can be stored, so new data formats can be introduced or the current one extended,
- be as fault-tolerant as possible,
- use technologies that are Grid-enabled,
- be simple to use.

3.2   Contents of the Registry

We want the registry to store information about Grid services. However, a simple syntactic description is not sufficient and should be extended so that every Grid service is described at least by:

- its unique identifier (within the whole Grid),
- the syntactic description of its ports and messages (e.g. WSDL),
- semantic information about the service itself and the ports, messages, tasks, data transformations, etc. it provides,
- the domain (or domains) it belongs to.

This complete description of each Grid service will be called a Grid Service Information Document (GSID). We do not specify its format, however, we assume that it will be an XML document, which should guarantee portability, flexibility and extensibility. It will also allow for using many pre-existing technological solutions, such as parsers, XML databases, etc. The *domain* field enclosed in every GSID is used by the registry's internal data distribution algorithm. All information in the registry is classified according to its domain, which can be, for example, a scientific domain (like biology, mathematics, etc). It can be a separate field, provided explicitly during service registration, or a function of the service's semantic description. It plays a very important role and thus it should be selected carefully taking into account the following requirements:

- it should be itself a simple data structure, e.g. the name of an ontology,
- it must be a part of an uncomplicated structure (like a set of ontology trees) with not too many disjoint parts (e.g. trees), to enable efficient data distribution,
- the number of domains shouldn't be too small, as it could limit the number of registry nodes, since these two quantities are strongly correlated.

The user will be able get a list of all most general domains (ie. domains that have no parents) or a list of all direct subdomains of a specific domain from the registry. This will assure that at least for the "domain" field the ontologies

will be well known to users. The list of domains can change as users add more
ontologies, however the uniqueness and global consistency of domain names in
the whole registry is still guaranteed.

The syntactical description of a Grid service is relatively simple. We can
apply Grid/web service standards, like WSDL. The semantical description seems
to be much more complicated. We have decided to use ontologies to express
semantics, but the choice of notation is still to be made and many problems (e.g.
whether to use shared or local ontologies) still have to be solved. Fortunately,
many projects are exploring the areas of ontologies and semantic description of
web/Grid services – these include OWL-S and the myGrid project [10,11].

3.3   Architecture of the Registry

The registry should be highly distributed and scalable. To achieve that, its architecture is a structure of equal and independent registry nodes with a given
communication topology. By *equal* we mean that every node works in the same
way and presents to the user the same interfaces. Therefore, the complicated
structure of the registry is hidden behind the interfaces and all GSID documents
can be accessed by sending requests to an arbitrary registry node regardless of
their distribution within the whole registry.

As long as the efficiency is not taken into account, it doesn't matter for a user
which node he or she will use to access the registry through one of its interfaces.
The node that receives the user's request is responsible for its proper redirection
to other nodes of the registry when necessary. The algorithm of request redirection (routing) is under development. The problem is that no node can store
complete information about data distribution within the whole registry as this
solution wouldn't be too scalable. On the other hand, remembering only the
domains of the nearest neighbors may not be sufficient and broadcasting user's
requests throughout all the nodes should be strictly limited.

Every registry node will be responsible for storing GSID documents related
to some specific domains (see Fig. 2). The choice of nodes responsible for given
domains can be done both manually, by a system administrator, and automatically, if that is allowed by the registry's configuration. The main goal of such an
approach is to store information very close to the places where the corresponding Grid services are deployed and/or most frequently used. It is reasonable to
assume that every domain can be represented by at most one registry node as
this solves problems of data coherency. However, data redundancy is required
(see Sect. 3.1) so we have to introduce data replication between nodes, and, as
efficiency remains essential, cache mechanisms should be added to the registry
– this will require additional copying of data between many nodes. However, all
data can be kept coherent and up-to-date in this kind of complicated structure
and the concept of a single designated node for each domain, as well as other
advanced techniques may be applied for that purpose. The number of registry
nodes and the communication topology are dynamic so new nodes and inter-
connecting links can be added and removed without restarting. Of course, when

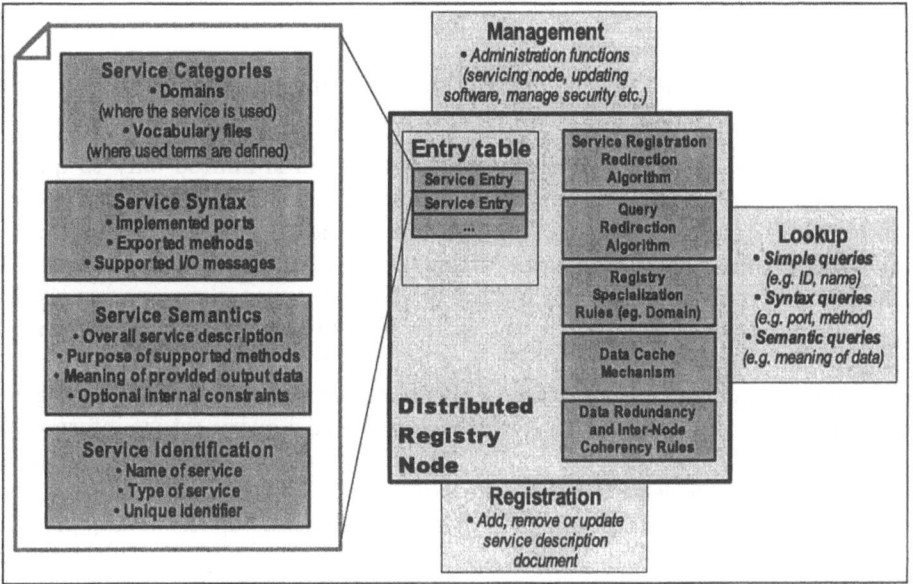

Fig. 2. Details of a single registry node

the number of nodes changes, some data has to be transferred and some time required to propagate all the required information throughout the whole registry.

3.4   Registry Interfaces

Each registry node will present the following interfaces to the outside world: a registration interface, a management interface and a lookup interface.

The registration interface is responsible for publishing GSID documents and unregistering them when needed. The registration requester may be a user (e.g. a developer of a Grid service) or any other entity that can do it automatically (e.g. searching for services and then publishing some of them). The registration scenario can be the following: the user wants to register Grid service S in domain D (placed somewhere in one of the domain trees). If the domain doesn't exist, a domain registration request has to be sent by a privileged entity. After that, service S can be registered by sending a registration request to any node of the registry. This node will take care of redirecting the request to the right place (the node responsible for domain D) at which service S will be registered.

The lookup interface allows users to search for data in the registry. An essential step is the specification of the type of queries the registry should support. As data lookup should be very fast, we'd rather not put any complicated ontological reasoning machine inside the registry. On the other hand, we cannot confine users to using only these queries that are related to syntactic information (this would be highly insufficient e.g. for our flow composer). Thus, the following types of queries should be supported by the registry:

- search by ID,
- search by syntactic information,
- search by semantic service description but using only names of ontologies and their attributes, not ontological reasoning.

In the last case, the entity that is asking a question is limited only to queries like: "find all services that have ports which are described by one of the following ontologies: O_1, O_2, O_3". The registry does not care about ontological relations; it only compares names of ontologies (strings). However, these relations might be introduced here by the entity asking the question, by properly preparing the list of ontology names that is put in the query (e.g. O_2 and O_3 may be subclasses of O_1). This approach should be efficient and flexible.

In order to make the migration of clients closer to data possible (e.g. agents of the flow composer) the registry will, in response to a query, return information on where exactly (i.e. on which node) the requested GSID document has been found. It will enable selecting the node which stores the data the user is looking for and asking next questions directly there, thus speeding up the lookup process.

4 Implementation Choices

It's now too early to make detailed implementation choices; we thus prefer to point out several requirements. First of all, as we would like the system to be portable, the use of a highly portable programming language (e.g. Java or C#) is desirable. Secondly, as the system has to be distributed, scalable and, at the same time, efficient, a non-hierarchical peer-to-peer paradigm with a lightweight underlying platform and fast communication protocols should be used. This will also allow for automatic deployment of system components in containers on nodes. For the first prototype of the system we are going to use Globus Toolkit 3.0 [8] as a hosting environment. For ontological reasoning we might use the Jena2 framework [15].

The registry should be conformable to a well-known standard to make it more universal and portable. This can be, for example, the third version of the UDDI specification that already supports distributed registries and data replication [13]. As the registry is going to be itself a Grid application, it seems obvious that it has to use a platform that is Grid-enabled. This could be one of the existing peer-to-peer environments or Grid middleware packages. We are considering using the JXTA Search system as a base and extending it with the required functionality. For an internal registry document database we investigate the possibility of using XML-dedicated database management systems, like Apache Xindice [1]

5 Conclusion and Future Works

This paper presents the design of a workflow composition system, with emphasis on the registry component (service). The general concept of a distributed registry is proposed basing on nodes specialized according to data domains, with mechanisms of data replication and caching. The main interfaces of the registry are designed to meet the requirements of the flow composition system. While the main functionality of the registry system is defined, there are still many issues that need to be solved, e.g. the choice of data replication mechanisms, query routing among registry nodes and implementation technologies. Our current work is focused on these issues and on implementing the proof-of-concept prototype. The following algorithms and related protocols still have to be developed:

- request redirection algorithm,
- data replication algorithm,
- request and response caching, fault tolerance, backup servers.

Acknowledgements. This work was partly funded by the European Commission, Project IST-2001-32243, CrossGrid, Project IST-2001-34808, GRID-START and the Polish State Committee for Scientific Research, SPUB-M 112/E-356/SPB/5.PR UE/DZ 224/2002-2004. We are grateful to Piotr Nowakowski for his comments.

References

1. Apache Xindice, The Apache XML Project, http://xml.apache.org/xindice
2. Bubak, M., Górka, K., Gubała, T., Malawski, M., Zając, K.: Component-based System for Grid Application Workflow Composition, in: Dongarra, J., et al. (Eds.): 10th European PVM/MPI Users' Group Meeting, Venice, Italy, September 29 - October 2, 2003, Proceedings. LNCS 2840, Springer 2003, pp. 611-618
3. Bubak, M., Gubała, T., Kapałka, M., Malawski, M., Rycerz, K.: Design of Distributed Grid Workflow Composition System. Cracow Grid Workshop 2003, November 2003, CYFRONET-AGH, Cracow
   http://www.cyfronet.krakow.pl/cgw03
4. Cardoso, J. and Sheth, A.: Semantic e-Workflow Composition. Journal of Intelligent Information Systems (JIIS), Vol. **12**, 191-225, 2003
5. Deelman, E., et al.: Mapping Abstract Complex Workflows onto Grid Environment. Journal of Grid Computing **1**, 25-39, 2003
6. Expert Group Report: Next Generation Grid(s), European Grid Research 2005-2010, June 2003
7. Foster, I., et al.: The Physiology of the Grid: An Open Grid Services Architecture for Distributed Systems Integration. OGSA WG, Global Grid Forum, June 2002
8. Globus Toolkit 3.0, http://www.globus.org/ogsa
9. METEOR-S Project: http://lsdis.cs.uga.edu/proj/meteor/SWP.htm
10. myGrid Project, http://mygrid.man.ac.uk/
11. OWL-S http://www.daml.org/services/owl-s/1.0/
12. Tuecke, S., et al.: Open Grid Services Infrastructure (OGSI) version 1.0, OGSI WG, Global Grid Forum, June 2003
13. UDDI 3.0 http://www.oasis-open.org/committees/uddi-spec/doc/tcspecs.htm
14. JXTA Search web page, http://search.jxta.org/
15. Jena – A Semantic Web Framework for Java, http://jena.sourceforge.net/

A-GWL: Abstract Grid Workflow Language*

Thomas Fahringer[1], Sabri Pllana[2], and Alex Villazon[1]

[1] Institute for Computer Science, University of Innsbruck
Technikerstraße 25/7, 6020 Innsbruck, Austria
{Thomas.Fahringer,Alex.Villazon}@uibk.ac.at
[2] Institute for Software Science, University of Vienna
Liechtensteinstraße 22, 1090 Vienna, Austria
pllana@par.univie.ac.at

Abstract. Grid workflow applications are emerging as one of the most interesting application classes for the Grid. In this paper we present A-GWL, a novel Grid workflow language to describe the workflow of Grid applications at a high level of abstraction. A-GWL has been designed to allow the user to concentrate on describing scientific Grid applications. The user is shielded from details of the underlying Grid infrastructure. A-GWL is XML-based which defines a graph of activities that refers to computational tasks or user interactions. Activities are connected by control- and data-flow links. We have defined A-GWL to support the user in orchestrating Grid workflow applications through a rich set of constructs including sequence of activities, sub-activities, control-flow mechanisms (sequential flow, exclusive choice, and sequential loops), data-flow mechanisms (input/output ports), and data repositories. Moreover, our work differs from most existing Grid workflow languages by advanced workflow constructs such as parallel execution of activities with pre- and post-conditions, parallel loops, event-based synchronization mechanisms, and property-based selection of activities. In addition, the user can specify high-level constraints and properties for activities and data-flow links.

1 Introduction

In the past years extensive experience has been gained with single site applications and parameter studies for the Grid. For some time, Grid workflow applications are emerging as an important new alternative to develop truly distributed applications for the Grid. Workflow Grid applications can be seen as a collection of activities (mostly computational tasks) that are processed in some order. Usually both control- and data-flow relationships are shown within a workflow. Although workflow applications have been extensively studied in areas such as business process modeling [12,5] and web services (BPEL4WS [2], XLANG [10], WSFL [7], SWFL [4]), this programming model is relatively new in the Grid computing area.

---

* The work described in this paper is supported in part by the Austrian Science Fund as part of Aurora Project under contract SFBF1104.

M. Bubak et al. (Eds.): ICCS 2004, LNCS 3038, pp. 42–49, 2004.
© Springer-Verlag Berlin Heidelberg 2004

Low-level XML-based workflow languages, like Grid Service Flow Language (GSFL) [6] and Grid Workflow [3], have been proposed to describe Grid workflow applications. However, these languages are missing some important control flow constructs such as branches, loops, split and join.

In the GridPhyN Project [1], Grid workflow applications are constructed by using partial/full abstract descriptions of "components", which are executable programs that can be located on several resources. Their abstract workflow is limited to acyclic graphs. Moreover, a reduced data-flow model (no customization is allowed) is supported, and there is no advanced parallel execution of components.

In the GridLab Triana Project [11], workflows are experiment-based: Each unit of work is executed for a specific experiment. A workflow (task graph) defines the order in which experiments are executed. Simple control-flow constructs can be used to define parallel processing, conditional execution (*if*, *switch*) and iterations (simple sequential *loops* and *while*).

In this paper we describe the Abstract Grid Workflow Language (A-GWL) which allows the user to compose scientific workflow applications in an intuitive way. A-GWL is an XML-based language for describing Grid workflow applications at a high level of abstraction without dealing with implementation details. A-GWL does not consider the implementation of activities, how input data is delivered to activities, how activities are invoked or terminated, etc. A-GWL has been carefully crafted to include the most essential workflow constructs including activities, sequence of activities, sub-activities, control-flow mechanisms, data-flow mechanisms, and data repositories. Moreover, we introduce advanced workflow constructs such as parallel activities, parallel loops with pre- and post-conditions, synchronization mechanism, and event based selection of activities. We support advanced data-flow constructs such as direct communication and broadcast communication. In addition, the user can specify high-level constraints on activities and on data-flow in order to support the underlying workflow enactment machine to optimize the execution of workflow applications.

The paper is organized as follows. Section 2 depicts the workflow application development process, from the abstract and concrete level to the execution of the workflow application (*composition, reification* and *execution*). Section 3 introduces the basic concepts of A-GWL as well as advanced constructs for control- and data-flow. Finally, Section 4 concludes the paper.

2   Overview

In this section we give an overview of our Grid workflow application development process from abstract representation, to the actual execution on the Grid.

Figure 1.a depicts the complete process of developing a Grid workflow application. Figure 1.b shows the flow of three fundamental activities *Composition, Reification* and *Execution*. The dashed lines indicate the iterative process of workflow application development.

Composition. The user composes the Grid workflow by using the Unified Modeling Language (UML) [8]. Alternatively, a High-level Textual Representa-

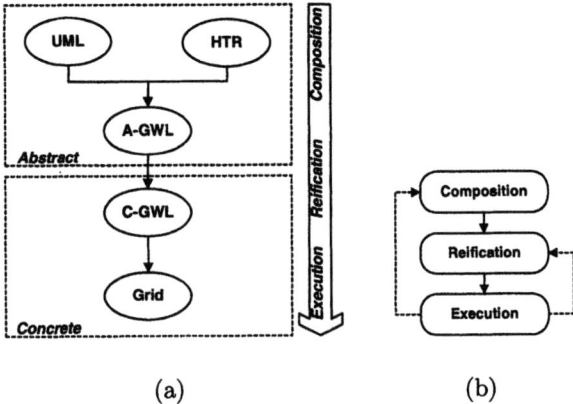

Fig. 1. The development of Grid workflow applications. Acronyms: UML - Unified Modeling Language, HTR - High-level Textual Representation, A-GWL - Abstract Grid Workflow Language, C-GWL - Concrete Grid Workflow Language.

tion (HTR) may be used for the workflow composition. Our tool Teuta [9] supports the UML based workflow composition, and the automatic conversion to Abstract Grid Workflow Language (A-GWL). A-GWL is an XML based language for describing the Grid workflow at a high level of abstraction. At this level activities correspond mostly to computational entities and the specification of its input and output data. There is no notion of how input data is actually delivered to activities, or how activities are implemented, invoked and terminated. A-GWL contains all the information specified by the user during the workflow composition.

Reification. The software infrastructure transforms A-GWL to the Concrete Grid Workflow Language (C-GWL). C-GWL contains the information specified by the user and the additional information provided by the software infrastructure (i.e. the actual mapping to Grid resources), which is needed for the execution of the workflow. It describes the realization of A-GWL by specifying how activities are actually implemented (e.g. Grid services), how they are deployed, invoked, and terminated. C-GWL also describes how data is transferred and in what form. C-GWL is an XML based language as well. Currently, we are in the process of the development of specification for C-GWL.

Execution. C-GWL is interpreted by the underlying workflow enactment machine to construct and execute the full Grid workflow application.

3   A-GWL Constructs

In A-GWL, an activity is a generic *unit of work*. At this level, it is not seen how the activity is implemented at the concrete level. An activity may be mapped to Grid services, software components or web services under C-GWL. A *workflow instance* is the mapping of an A-GWL workflow to a C-GWL workflow that is actually executed on the Grid. Different workflow instances can run concurrently. The user can specify constraints that are used to tune the execution of the workflow application. An activity can be a basic *computation*, a *sequence* of activities, or a composed *sub-activity*.

The composition of the application is done by specifying both control-flow and data-flow between activities. A-GWL supports basic control-flow (sequential flow, exclusive choice and sequential loops) and advanced control-flow constructs such as parallel execution of activities, parallel loops with pre- and post-synchronizations and selection of activities.

Basic data-flow is specified by connecting input and output ports between activities. The user can associate constraints with data-flow, which will be mapped, e.g., to concrete mechanisms that utilise secure communication or advanced QoS for data-transfer. In addition to basic data-flow between activities, A-GWL supports advanced data-flow between activities and repositories. In the following, we describe the most essential A-GWL elements in more detail.

3.1   Basic Elements

ACTIVITY: An activity in A-GWL is presented as a *black-box* with input/output ports, constraints, and properties. Input and output ports are used to specify the data-flow. Control-flow is made through explicit links among activities and by using different control-flow constructs (e.g. sequences, loops, etc).

The user specifies data-flow by connecting input and output ports among activities which may not necessarily be directly connected with control flow links.

Additional information about activities can be provided through constraints and properties. Constraints provide additional information that should be honored by the underlying workflow enactment engine when executing the Grid workflow application. Examples for constraints are as follows: execute activity on Grid machines running under a specific operating system, with a minimum amount of free dynamic memory, with more than some pre-defined idle time value, deployment requirements, etc. Properties describe information that can be useful for the workflow enactment engine for instance to schedule applications. Performance information (estimated execution times or floating point operations) is frequently indicated via properties.

The XML representation[1] for the ACTIVITY construct is given as follows,

```
<activity name="name">
  <input name="name"(repository="name" query=expression)? />*
  <output name="name" (repository="name")? /> *
  (<constraints> constraints+ </constraints>)?
  (<properties> properties+ </properties>)?
  <control-in name="name"/> *
  <control-out name="name"/> *
</activity>
```

Every *activity*, *input*, *output*, *control-in* and *control-out* have a unique *name* in a workflow application. Input and output ports can be used to specify data-flow between activities and/or repositories (see Section 3.4 and 3.5).

---

[1]   We use the notation convention from [2] for the informal syntax of XML grammar.

A-GWL supports the notion of hierarchical decomposition of activities. An activity containing other activities (sub-workflow), is also an activity. In that case, all the input ports of the incoming activities of the sub-workflow are mapped to the input ports of the composed activity, and the output ports of all the outgoing activities are mapped to the output ports of the composed activity. Analogously (re)connections are handled for the control-flow.

3.2   Basic Control-Flow

A-GWL supports basic control-flow among activities as in conventional programming languages. The control-flow constructs are: the standard sequential flow (SEQUENCE), exclusive choice (SWITCH, IF-THEN-ELSE), and sequential loops (FOR, WHILE and DO-UNTIL). A-GWL can be used to specify multiple exits from loops. The BREAK construct can be used to exit the execution of a loop construct. We also support an EXIT construct to terminate the execution of the entire workflow instance.

3.3   Advanced Control-Flow

In addition to basic control-flow constructs that are commonly found in workflow languages, A-GWL supports advanced control-flow elements that are highly useful for scientific Grid workflow applications. These constructs provide the user with enough abstraction to compose advanced and complex Grid workflow application and specify concurrency in a very natural way (parallel execution of activities, parallel loops).

PARALLEL: The PARALLEL construct is used to execute activities concurrently. The XML representation of the PARALLEL construct is as follows,

```
<parallel syncCondition=IdentifierSet>
  (<fork preCondition=expression > activity </fork>)+
  (<default> activity </default>) ?
</parallel>
```

Before executing the activities specified in `<fork>`, the *preCondition* is evaluated for all the activities. Only those that evaluate to true, are executed concurrently. If no *preCondition* is specified, all the activities are executed in parallel. The user can also specify a synchronizing post-condition (*syncCondition*) in order to wait for specific activities at the end of the PARALLEL construct, and then continue the control-flow, e.g. this feature can be used to select the *fastest* activity. Other examples of post-conditions include "wait for all activities to finish", "wait for one", and so on.

If no *syncCondition* is specified, the PARALLEL construct will wait for all activities to be finished. By default, the same input is passed to each parallel activity. However, different input sources can also be specified, e.g. from other activities or from repositories (see Section 3.5 for advanced data-flow).

Finally, the optional `<default>` activity inside the PARALLEL construct is used to specify an activity that is executed if all the pre-conditions of all activities of a parallel construct evaluate to false.

PARALLEL-FOR: The PARALLEL-FOR construct can be used to specify parallel for-loops. The XML representation for PARALLEL-FOR construct is,

```
<parallel-for index="i..j">
  <body> activity </body>
  <gathering> activity </gathering>
</parallel-for>
```

A PARALLEL-FOR consists of two parts: *body* and *gathering*. The body describes the activities of a loop. A PARALLEL-FOR defines that each of its iterations can be executed in parallel. It does not specify any parallelism within a loop iteration. Parallelism within a loop iteration can be achieved through the parallel construct. Each activity inside of a `parallel-for` loop will receive the loop *index* as an implicit input value. If there is a need for other input values they need to be explicitly defined within the body for each activity. Once all concurrent running activities have finished, the *gathering* part will be executed. Gathering is a special activity provided by the user that utilises the output of all the concurrent activities, and potentially produces some combined output.

The PARALLEL-FOR construct may be used to model parameter studies on the Grid. The index can be used to access different input data for all concurrent running activities. In the gathering phase, all results can be e.g. visualised or combined to create a final result of the parameter study.

SELECT: A-GWL introduces the SELECT construct which selects one activity from a set of activities based on their *properties* and a specific *criteria*. Activity properties may define, for instance, information about performance characteristics, quality of service parameters, etc. The selection criteria is an expression defined over properties to determine minimum, maximum, average, threshold values, etc. The XML representation for the SELECT construct is:

```
<select property="aProperty" criteria="aCriteria">
  activity+
</select>
```

The SELECT contruct is frequently used to determine an activity with the longest or shortest predicted execution time.

3.4   Basic Data-Flow

As described in Section 3.1, the input and output ports of activities are used to model basic data-flow. If the activity refers by name to the output port of another activity, a data-flow link is established. Data-flow can be expressed among arbitrary activities not necessarily connected by any control flow link at all. For example, a data-flow link between activities A and B is simply established by defining the output of A as `<output name="oA">`, and the input of B as `<input name="oA">`. If more than one activity declares oA as a desired input, then the output will be sent to all those that *subscribed* to it. This can be seen as some kind of simple *notification* or *broadcasting* mechanism.

3.5   Advanced Data-Flow

REPOSITORY: In addition to the basic data-flow among activities, A-GWL supports data-flow between activities and special entities called *repositories*, which are abstractions for data containers. They are used to model, for instance, saving intermediate results or querying data resources without bothering about how a repository is actually implemented (file servers, databases, etc.).

A repository is not an activity. It has a unique name. The insertion and retrieval of information is made by using the repository name without specifying any explicit input/output ports, i.e. data-flow between activities and repositories is specified by using the input/output ports of activities and the name of the repository. If an activity should store its output in a repository, then it should specify the name of the repository and attribute of the <output> tag (see Section 3.1). For instance, an activity A that wants to store its output (named oA) in a repository R, must simply define the following output: <output name="oA" repository="R"/>.

Enhanced Data-Flow: In addition to basic data-flow among activities, A-GWL extends the basic data-flow links with optional user-defined *constraints* and *properties* which can be used by the underlying workflow enactment engine to tune the data-transfer mechanism and behavior. Constraints again imply requirements that should be fulfilled by the enactment machine whereas properties is simply extra information. The user may have experience about the kind of output and input that is required by the activities. He thus may define constraints such as "store locally", "use high-speed network", "use-cache", "use-replica", "secure transfer", etc. by specifying specific mnemonics. Input and output data types, estimated data volumes, etc. can be described through properties.

The extension for the specification of constraints and properties are associated with the **input** and **output** ports of activities (here we show only the output port). The XML representation is,

```
<output name="name" (repository="name") ? >
  (<constraints> constraints+ </constraints>) ?
  (<properties> properties+ </properties>) ?
</output>
```

3.6   Synchronization with Events

A-GWL supports synchronization of activities based on *events*. An illustrative example is the A-GWL construct FIRST.

FIRST: This construct selects one activity from a set of activities based on the occurence of a specific event. Every activity A_i in a set of activities $\{A_1, .., A_N\}$ is associated with an event e_i in a set of events $\{e_1, .., e_N\}$. FIRST construct waits for one of the events from the set of events to occure. The first event that occures determines which activity is selected for execution. The XML representation for the FIRST construct is,

```
<first>
  (<event id="eventId">
      activity
   </event>) +
</first>
```

4 Conclusions and Future Work

In this paper we have described the abstract Grid workflow language (A-GWL) to simplify the specification of Grid workflow applications at a high level of abstraction without being distracted by low level details (e.g. how to start and stop tasks, how to transfer input and output data, what communication protocol to use, etc.). A-GWL is XML-based and has been carefully crafted to cover the most essential workflow constructs including sequence of activities, sub-activities, control-flow/data-flow/event mechanisms, and data repositories.

We have developed a UML interface to A-GWL called Teuta [9], in order to graphically orchestrate Grid workflow applications. We are currently in the process to develop a Grid workflow enactment engine to execute Grid workflow applications based on A-GWL.

References

1. GriPhyN: Grid Physics Network Project. www.griphyn.org.
2. T. Andrews, F. Curbera, H. Dholakia, Y. Goland, J. Klein, F. Leymann, K. Liu, D. Roller, D. Smith, S. Thatte, I. Trickovic, and S. Weerawarana. Business Process Execution Language for Web Services. Version 1.1, BEA, IBM, Microsoft, SAP, and Siebel, May 2003.
3. H. Bivens. Grid Workflow. Sandia National Laboratories, http://vir.sandia.gov/~hpbiven/, April 2001.
4. Y. Huang. SWFL: Service Workflow Language. Technical report, Welsh e-Science Centre-Cadiff University, 2003. http://www.wesc.ac.uk/projects/swfl/.
5. Business Process Management Initiative. Business Process Modelling Language. www.bpmi.org/bmpi-downloads/BPML-SPEC-1.0.zip, June 2002.
6. S. Krishnan, P. Wagstrom, and G. Laszewski. GSFL : A Workflow Framework for Grid Services. Preprint ANL/MCS-P980-0802, Argonne National Laboratory, August 2002.
7. F. Leymann. Web Services Flow Language (WSFL 1.0). Technical report, IBM Software Group, May 2001.
8. OMG. Unified Modeling Language Specification. http://www.omg.org, March 2003.
9. S. Pllana, T. Fahringer, J. Testori, S. Benkner, and I. Brandic. Towards an UML Based Graphical Representation of Grid Workflow Applications. In *The 2nd European Across Grids Conference*, Nicosia, Cyprus, January 2004. Springer-Verlag.
10. S. Thatte. XLANG: Web services for Business Process Design. Technical report, Microsoft Corporation, 2001.
11. Cadiff University. The Triana Project. http://trianacode.org/triana/.
12. The Workflow Management Coalition. http://www.wfmc.org/.

Design of Departmental Metacomputing ML

Frédéric Gava

LACL, University Paris XII, Créteil, France
gava@univ-paris12.fr

Abstract. Bulk Synchronous Parallel ML or BSML is a functional data-parallel language for programming bulk synchronous parallel (BSP) algorithms. The execution time can be estimated and dead-locks and indeterminism are avoided. For large scale applications, more than one parallel machine is needed. We consider here the design and cost-model of a BSML-like language devoted to the programming of such applications: Departmental Metacomputing ML or DMML.

Introduction. Bulk-Synchronous Parallel ML (BSML) is an extension of ML for programming Bulk Synchronous Parallel (BSP) algorithms as functional programs. BSP computing [4] is a parallel programming model introduced by Valiant to offer a high degree of abstraction like PRAM models. Such algorithms offer portable, predictable and scalable performances on a wide variety of architectures. BSML expresses them with a small set of primitives taken from the *confluent* BSλ-calculus. Those operations are implemented as a parallel library (http://bsmllib.free.fr) for the functional programming language Objective Caml (http://www.ocaml.org).

In recent years there has been a trend towards using *a set* of parallel machines in particular SMP clusters for these kinds of problems. Programming this kind of *supercomputers* is still difficult and libraries, languages and formal cost models are needed. Computing with these "cluster of clusters" is usually called *departmental metacomputing*.

BSML is not suited for departmental metacomputing, in particular to the heterogeneous nature of computing resources and networks. In fact the BSP model itself is not well suited to these two-tiered architectures. This paper describes our first work on the design of a model and a functional programming language for departmental metacomputing.

Bulk Synchronous Parallelism. A BSP computer contains a set of processor-memory pairs, a communication network allowing inter-processor delivery of messages and a global synchronization unit which executes collective requests for a synchronization barrier. Its performance is characterized by 3 parameters expressed as multiples of the local processing speed: p is the number of processor-memory pairs, l is the time required for a global synchronization and g is the time for collectively delivering a 1-relation (communication phase where every processor receives/sends at most one word). The network can deliver an

M. Bubak et al. (Eds.): ICCS 2004, LNCS 3038, pp. 50–53, 2004.
© Springer-Verlag Berlin Heidelberg 2004

h-relation in time gh for any arity h. A BSP program is executed as a sequence of *super-steps*, each one divided into three successive disjoint phases. In the first phase each processor uses its local data to perform sequential computations and to request data transfers to/from other nodes. In the second phase the network delivers the requested data transfers and in the third phase a global synchronization barrier occurs, making the transferred data available for the next super-step. The execution time of a super-step s is thus the sum of the maximal local processing time, of the data delivery time and of the global synchronization time:

$$\text{Time}(s) = \max_{i:processor} w_i^{(s)} + \max_{i:processor} h_i^{(s)} * g + l$$

where $w_i^{(s)} =$ local processing time and h_i is the number of words transmitted (or received) on processor i during super-step s. The execution time of a BSP program is therefore the sum of each super-steps.

There are some arguments against using BSP for metacomputing. First the global synchronization barrier is claimed to be expensive and the BSP model too restrictive. For example, divide-and-conquer parallel algorithms are a class of algorithms which seem to be difficult to write using the BSP model. This problem can be overcome at the programming language level without modifying the BSP model. The main problem is that this model does no take into account the different capacity of the parallel machines and of the networks. So bulk synchronous parallelism does not seem to be suitable for metacomputing.

The global synchronization barrier could be remove. [3] introduces MPM which is a model directly inspired by the BSP model. It proposes to replace the notion of super-step by the notion of m-step defined as: at each m-step, each process performs a sequential computation phase then a communication phase. During this communication phase the processes exchange the data they need for the next m-step. The model uses the set of "incoming partners" for each process at each m-step. Execution time for a program is thus bounded the number of m-steps of the program. The MPM model takes into account that a process only synchronizes with each of its incoming partners and is therefore more accurate.

The heterogeneous nature of networks has been investigated. [2] proposes a two-levels Hierarchical BSP Model for a supercomputing without subset synchronization and two levels of communications. A BSP^2 computer consists of a number of *uniformly* BSP units, connected by a communication network. The execution of a BSP^2 program proceeds in **hyper-steps** separated by global synchronizations. On each hyper-step each BSP unit performs a complete BSP computation (some super-steps) and communicates some data with other BSP units. However, the authors noted that none of the algorithms they have analyzed showed any significant benefit from this approach and the experiments do not follow the model. The failure of the BSP^2 model comes from two main reasons: first the BSP units are generally different in practice and second the time required for the synchronization of all the BSP units is too expensive.

DMM: Departmental Metacomputing Model. To preserve the work made on BSP algorithms and to deal with the different architectures of each parallel computer we propose a two-tiered model: the BSP model for each parallel unit and the MPM model for coordinating this heterogeneous set of BSP units. Thus, a DMM computer is characterized by the followings parameters: P the number of parallel computers, L the latency of the global communication network, G the time to exchange one word between two units, $\mathcal{P} = \{p_0, \ldots, p_{P-1}\}$ the list of the number of processes for each BSP unit, $\mathcal{L} = \{l_0, \ldots, l_{P-1}\}$ the times required for a global synchronization for each BSP unit and $\mathcal{G} = \{g_0, \ldots, g_{P-1}\}$ the time for collectively delivering a 1-relation on each BSP unit. We supposed that $\forall i\,(l_i \ll L)$ and $(g_i \ll G)$.

We proposes to replace the notion of hyper-step by the notion of d-step defined as: at each d-step, each BSP unit performs a BSP computation phase then a communication phase to deal with the values of other BSP units. Each processor within a unit parallel accesses to the outer world and so its messages passed on the local network, then on the global network and then to the destination unit local network. $\Omega_{d,i}$ is defined as the set of j such as the BSP unit j received messages from BSP unit i during d-step d. $\Phi_{d,i}$ is inductively defined as:

$$
\begin{cases}
\Phi_{1,i} = \max_{j \in \Omega_{1,i}} (W_{(1,j)}, W_{(1,i)}) + \sum_{j \in \Omega_{1,i}} ((g_i + G + g_j) \times h^j_{1,i}) + (g_i \times h^1_i + l_i) + L) \\
\Phi_{d,i} = \max_{j \in \Omega_{d,i}} (\Phi_{d-1,j} + W_{(1,j)}, \Phi_{d-1,i} + W_{(1,i)}) + \sum_{j \in \Omega_{d,i}} ((g_i + G + g_j) \times h^j_{d,i}) \\
\qquad\qquad\qquad\qquad\qquad\qquad\qquad\qquad\qquad\qquad\qquad + (g_i \times h^d_i + l_i) + L)
\end{cases}
$$

where $h^j_{d,i}$ denotes the number of words received by BSP unit i from the BSP unit j ("incoming partner") during the d-step d for $i \in \{0, \ldots, P-1\}$ and h^d_i is the maximum number of words exchanged within the BSP unit i during the d-step d and $W_{(d,j)}$ is the sum of the super-steps of a BSP units j.

Thus $\Psi = \max\{\Phi_{R,j}/j \in \{0, 1, \ldots, P-1\}\}$ bounds execution time for a complete program, where R is the number of d-steps of the program. The DMM model takes into account that a BSP unit only synchronizes with each of its incoming partner and is therefore more accurate than the BSP^2 model. Moreover more algorithms for irregular problems could be analyzed efficiently.

Departmental Metacomputing ML. Rather than a full Departmental Metacomputing ML (DMML) language we provide a library for Objective Caml. It is based on the elements given in figure 1. It gives access to the DMM parameters of the underling architecture: dm_p() gives the static number of BSP units, (dm_bsp_p i) the static number of processes of the BSP unit i. There are also two abstract polymorphic types: 'a par (resp. 'a dpar) which represents the type of p_i-wide parallel vectors (resp. P-wide departmental vectors) objects of type 'a, one per process (resp. per BSP unit).

The DMML parallel constructs operates on parallel (resp. departmental) vectors. Those vectors are created by mkpar and mkdpt so that (mkpar f) stores (f i) on process i and (mkdpt f) stores (f i) on BSP units i. Asynchronous phases are programmed with mkdpt, mkpar, apply and applydpt:

```
apply (mkpar f) (mkpar e) stores (f i)(e i) on process i
applydpt(mkdpt f)(mkdpt e) stores (f i)(e i) on BSP units i.
```

```
dm_p: unit->int                          dm_bsp_p: int->int
mkpar: (int -> 'a) -> 'a par             mkdpt: (int -> 'a) -> 'a dpar
apply: ('a -> 'b) par -> 'a par -> 'b par
applydpt: ('a -> 'b) dpar -> 'a dpar -> 'b dpar
put: (int->'a option) par -> (int->'a option) par}
get: (int->int->bool)par dpar->'a par dpar->(int->int->'a option)par dpar
```

Fig. 1. DMML Library

The communication phases are expressed by put and get. Consider:
get(mkdpt(fun i->mkpar(fun j->$f_{i,j}$)))(mkdpt(fun a->mkpar(fun b->$v_{a,b}$)))
For a process j of the BSP unit i, to receive a value $v_{a,b}$ from the process b of the BSP unit a (it is an incoming partner), the function $f_{i,j}$ at process j of the BSP unit i must be such that ($f_{i,j}$ a b) evaluates to true. To receive no value, ($f_{i,j}$ a b) must evaluate to false. Our expression evaluates to a departmental vector containing parallel vector of functions $f'_{i,j}$ of delivered messages on every process of every BSP unit. At process j of the BSP unit i, ($f'_{i,j}$ a b) evaluates to None if process j of the BSP unit i received no message from process b of the BSP unit a or evaluates to Some $v_{a,b}$ if received a value of the process b of the BSP unit a (it is an incoming partner).

Conclusions and Future Works. The BSP model has been used for the design of great variety of parallel algorithms. It also allowed the production of reliable and portable codes with predictable efficiency. However, additional complexity introduced by metacomputing forces a revision of the model. We have considered how to extend the BSP model hierarchically using the MPM model and we have also described a new functional parallel library which follows this new model. Several directions will be followed for future work: parallel implementation of DMML as a library using Madeleine [1], design and validation of new algorithms for this model, comparison with other hierarchical models [5].

Acknowledgments. This work is supported by a grant from the French Ministry of Research and the ACI Grid program, under the project CARAML.

References

1. O. Aumage, L. Bougé, and al. Madeleine II: A Portable and Efficient Communication Library. *Parallel Computing*, 28(4):607–626, 2002.
2. J. M. R. Martin and A. Tiskin. BSP modelling a two-tiered parallel architectures. In B. M. Cook, editor, *WoTUG'99*, pages 47–55, 1999.
3. J. L. Roda and al. Predicting the execution time of Message Passing Models. *Concurrency: Practice and Experience*, 11(9):461–477, 1999.
4. D. B. Skillicorn, J. M. D. Hill, and W. F. McColl. Questions and Answers about BSP. *Scientific Programming*, 6(3), 1997.
5. T. L. Williams and R. J. Parsons. The Heterogeneous BSP Model. *LNCS*, volume 2000 ,page 112–118, Springer Verlag, 2000.

A Grid-Enabled Scene Rendering Application

M. Caballer, V. Hernández, and J.E. Román

D. de Sistemas Informáticos y Computación, Universidad Politécnica de Valencia,
Camino de Vera, s/n, E-46022 Valencia, Spain.
{micafer,vhernand,jroman}@dsic.upv.es

Abstract. The work presented in this paper represents a step forward
in the wide adoption of grid technologies in more conventional commercial applications. The application treated herein is a global illumination
scene rendering engine in which grid tools have been used to execute
the computation remotely, as well as to incorporate certain interesting
characteristics. This engine can be incorporated, for instance, in an electronic commerce application in order to add visualization capabilities to
a conventional virtual store. The rendering engine implements a parallel
version of the radiosity illumination algorithm, providing high quality
and allowing a fast response for moderately complex scenes.

1   Introduction

This paper presents a remote scene rendering service, implemented with standard grid tools. These tools provide some desirable features such as uniformity,
transparency, reliability, ubiquity and security [4] which can be very important
in some applications but are lacking in more conventional Internet technologies.
In particular, Globus Toolkit 2 [3] has been selected for the middleware layer,
since it is currently one of the most widely accepted grid frameworks.

This work is a follow-up of the EU-funded research project VRE-Commerce
(see *http://www.grycap.upv.es/vrecommerce*), whose main objective was the integration of fast synthetic image generation in electronic commerce applications,
with special focus on the furniture and ceramics sectors. The application enables
the user to obtain realistic renders of customized environments in which products from the virtual store have been placed. These images are generated with
radiosity techniques [6], in particular with a parallel implementation so that the
response time is bearable for an e-commerce setting (see [2] for details).

2   Prototype Implementation

The complete operation of the application is illustrated in Figure 1. The user
accesses the virtual store located in the web server and makes the image generation request (step 1). A CGI program (linked with the Globus libraries) located
in the web server is in charge of managing this request on behalf of the user and
having it processed by the grid. The required steps are described next. In this

M. Bubak et al. (Eds.): ICCS 2004, LNCS 3038, pp. 54–57, 2004.
© Springer-Verlag Berlin Heidelberg 2004

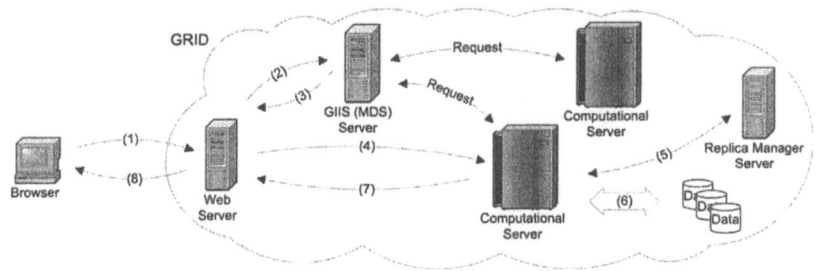

Fig. 1. Scheme of the rendering service.

process, several different Globus components participate, which may be placed in different physical locations.

Discovery of Grid Resources. First, the Monitoring and Discovery Service (MDS) included in Globus is used to obtain information about the resources available in the grid. The CGI program uses the information provided by MDS in order to select the most suitable resource. The decision is based on the number of free nodes. This information is also used to select the number of nodes to allocate for the radiosity program. If there are not enough free resources, the program will suggest the user to wait and try later (steps 2 and 3).

Job Submission. Once the computational server has been chosen, the CGI program launches the radiosity calculation via the Globus Resource Allocation Manager (GRAM). Before this, it is necessary to authenticate within the grid by means of a *proxy*, containing the certificate and public key of the grid user. Once authenticated, the job is submitted to the remote resource (step 4).

Update of Geometrical Data Files. The radiosity program needs a set of files containing the geometrical models of furniture pieces or the tiles textures in the case of ceramics. These files belong to the different manufacturers that offer products in the application. The files are cached locally in each computational server. The grid scheme adds the possibility to update the master copies of the files by the manufacturers, as well as to add new ones. The program analyzes the scene looking for the furniture or ceramic pieces used, and then queries the Replica Manager for their location (step 5). Then the dates of remote and local files are compared so that download is only done if necessary (step 6).

Return of Results. The CGI program uses the identifier returned by the GRAM service to detect the job finalization. When the job is finished the program uses the I/O functions provided by the GASS (Global Access to Secondary Storage) service to send the image generated by the radiosity algorithm back to the web server (step 7). Using these functions the CGI program can obtain the image with no need of modification of the radiosity code. Finally, the web server will display the image in the client's web browser (step 8).

Fig. 2. Some of the images used for validation and timing.

3   Testing and Results

Several tests have been carried out to verify the operation and performance of the whole system. Four test cases have been used to measure times in order to assess the overhead incurred by the Globus components. The chosen scenes are representative examples of the scenes that can be designed with the application. The generated images are illustrated in Figure 2.

The measured time begins when the user clicks the "Generate Image" button and finishes when the system starts to send the resulting image (the transfer time of the image is not taken into account because it depends on the type of Internet connection). The times considered for comparison have been the time consumed by the radiosity program and the accumulated time corresponding to the grid functions (which includes steps 1 – 5 and 7 in figure 1). The time obtained for each of the test cases is shown in Table 1, along with the initial number of polygons and patches at the end of execution, as well as the number of lights (to give an indication of the size of the problem).

The time added by the grid operations is slightly larger than 10 seconds, which is reasonable considering the response time requirement of 1 – 2 minutes established initially in the VRE-Commerce project. Although the time consumed by Globus functions represents a non-negligible percentage, it is constant, and, therefore, it will be less significant in more complex scenes. More details about the tests can be found in [1].

Table 1. Characteristics of the test scenes and measured times (in seconds).

Scene	Polygons	Patches	Lights	Radiosity	Globus	Total
Ex. 1	2426	28562	2	46	12	59
Ex. 2	5160	28740	2	56	12	68
Ex. 3	4908	24372	2	60	11	71
Ex. 4	1374	20118	2	17	13	30

4 Conclusions and Future Work

Grid tools pursue a number of properties (ubiquity, security, etc.) that are very important for distributed applications. Globus Toolkit 2 provides all the functionality required for our application. The benefits obtained by applying grid technologies have been the following in this case:

- Scalability: it is possible to add computational resources easily, so that the application can automatically use them.
- Fault tolerance: if any computational server fails, the system can redirect the request to the active ones.
- Security: the user must be authenticated to access the system, and all network data is encrypted. Although in this application there are no confidential data, security may be a key issue in other commercial applications.
- Load balancing: the system selects the most suitable grid resource.
- Data management: the system enables easy update of the furniture data and ceramic textures.

The system operation has been tested successfully in different situations, showing a robust behavior against possible failures. The execution overheads, owing to Globus components, are reasonable for the application.

With respect to future development of the system, different working lines are being considered. On one hand, after the release of Globus Toolkit 3, the logical extension would be to migrate to this new platform to take advantage of its new possibilities, adapting the application to the Open Grid Services Architecture defined in [5]. Among other things, this involves standardizing the way information is exchanged between the client application and the grid service. This opens the door to providing service to other applications such as standalone room design tools or, more generally, anyone requiring high-quality, fast rendering. For certain applications such as film making, this would entail harnessing greater amount of computing power and using geographically distributed resources. Finally, a desirable feature is to represent resulting renders with a 3-D modeling language (such as VRML or X3D) to allow for interactivity.

References

1. Caballer M., Hernández V., Román J.E.: A grid-enabled scene rendering application. Technical Report DSIC-II/04/04, Universidad Politécnica de Valencia (2004)
2. Caballer M., Guerrero D., Hernández V., Román J.E.: A parallel rendering algorithm based on hierarchical radiosity. Lecture Notes in Computer Science **2565** (2003) 523–536
3. Foster I., Kesselman C.: Globus: A metacomputing infrastructure toolkit. Internat. J. Supercomputer Appl. **11** (1997) 115–128
4. Foster I., Kesselman C.: The Grid: Blueprint for a New Computing Infrastructure. Morgan Kaufmann (1998)
5. Foster I., Kesselman C., Nick J., Tuecke S.: The physiology of the grid: An open grid services architecture for distributed systems integration (2002)
6. Sillion F., Puech C.: Radiosity and Global Illumination. Morgan Kaufmann (1994)

Rule-Based Visualization in a Computational Steering Collaboratory

Lian Jiang, Hua Liu, Manish Parashar, and Deborah Silver

Dept of Electrical and Computer Engineering,
Rutgers University, Piscataway, NJ 08854, USA
{lianjian,marialiu,silver,parashar}@caip.rutgers.edu

Abstract. In this paper, we introduce the concept of rule-based visualization for a computational steering collaboratory and show how these rules can be used to steer the behaviors of visualization subsystems. Feature-based visualization allows users to extract regions of interests, and then visualize, track and quantify the evolution of these features. Rules define high level user policies and are used to automatically select and tune the appropriate visualization technique based on application requirements and available computing/network resources. Such an automated management of the visualization subsystem can significantly improve the effectiveness of computational steering collaboratories in wide area Grid environments.

1 Introduction

A computational steering collaboratory is an environment in which geographically distributed scientists can collaboratively investigate complex and multidisciplinary simulations using online monitoring, remote visualization and computational steering techniques. Computational steering in such an environment not only shortens the period between changes to parameters and the viewing of the results, but also enables a what-if analysis which makes cause-effect relationships more evident [1].

The ability to flexibly manipulate the visualization subsystem in a computational steering collaboratory is important for both computational steering and multi-user collaboration as visualization is typically the basis for interactive monitoring and steering. For example, scientists often tend to look at the isosurface of a scalar field. These are specified using thresholds and are generally manipulated interactively using the visualization subsystem. However, for large-scale long-running simulations it may not be feasible to download an entire dataset or even one timestep of the dataset to a visualization platform. Therefore, visualization routines are co-located at the simulation platform. However, this can prevent them from being interactive and thresholds have to be selected a priori, which may not be most effective. Using rule-based control, scientists can enable the visualization subsystem to automatically pick the appropriate threshold. In this paper we present such a rule based visualization system. Rules are decoupled from the system and can be externally injected into a rule base. This

M. Bubak et al. (Eds.): ICCS 2004, LNCS 3038, pp. 58–65, 2004.
© Springer-Verlag Berlin Heidelberg 2004

allows scientists to conveniently add, delete, modify, disable, and enable rules that will control visualization behavior. Scientists can manipulate rules not only before the simulation, but also during the simulation. For example, while the simulation is running the scientist may specify a rule such as "if the number of the extracted regions is greater than 100, then increase the threshold by 5". The visualization subsystem will automatically adjust the threshold when the rule conditions evaluate to true. However, if the scientist knows from monitoring the first 200 timesteps that the value of x in the second rule is not appropriate and should be changed, she can modify the rule during the simulation and the modified rule will be applied to the rest of the simulation.

Rules can also be used to support collaborative visualization in heterogeneous environments where the collaborators' resources and display capabilities may differ. Some visualization techniques are more computation intensive than others, or may require more powerful graphic capabilities to display. For example, rendering an isosurface with millions of polygons may be too compute/network intensive for a thin client such as a PDA. Either the polygons can be reduced using straightforward triangle decimation techniques, or a more abstract feature-based representation can be displayed [2]. Such automated adaptations can be simply achieved using a rule such as "if there are more than 10k triangles, then display a higher level abstraction".

The rule based visualization subsystem presented in this paper builds on Discover, which is a computational collaboratory for interactive grid applications and provides the infrastructure for enabling rules to be dynamically composed and securely injected into the application, and executed at runtime so as to enable it to autonomically adapt and optimize its behavior [3]. In this paper, we integrate the rule mechanism into a feature-based visualization subsystem and demonstrate how this can be used to improve monitoring, steering and collaboration.

2   The Discover Computational Collaboratory

Discover is a virtual, interactive and collaborative PSE that enables geographically distributed scientists and engineers to collaboratively monitor, and control high performance parallel/distributed applications using web-based portals [3]. As shown in Figure 1, Discover provides a 3-tier architecture composed of detachable thin-clients at the front-end, a network of web servers in the middle, and the Distributed Interactive Object Substrate (DIOS++) [4] at the back-end.

DIOS++ enables rule based autonomic adaptation and control of distributed scientific applications. It is composed of 3 key components: (1) autonomic objects that extend computational objects with sensors (to monitor the state of an object), actuators (to modify the state of an object), access policies (to control accesses to sensors and actuators) and rule agents (to enable rule-based autonomic self-management), (2) mechanisms for dynamically and securely composing, deploying, modifying and deleting rules, and (3) a hierarchical control network that is dynamically configured to enable runtime accesses to and management

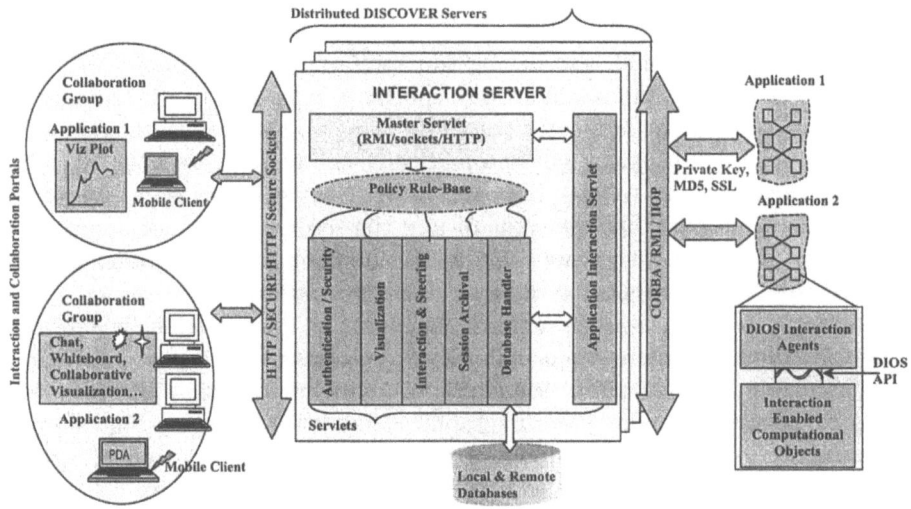

Fig. 1. Architectural Schematic of the DISCOVER Computational Collaboratory

of the autonomic objects and their sensors, actuators, access policies and rules. The rule engine responds to the users' actions (e.g. rule creation, deletion, modification, activation, deactivation) by decomposing and distributing the rules to corresponding rule agents, collecting rule execution results from the rule agents and reporting to users. The rule engine also controls and coordinates rule agents. Rules are evaluated and executed in parallel by the distributed rule agents. Priority and dynamic locking mechanisms are used at each rule agent to solve rule conflicts. The Discover collaboratory allows users to collaboratively interact with, interrogate, control, and steer remote simulations using web based pervasive portals, and has been used to enable remote feature extraction and tracking for the 3D Ritchmyer Meshkof compressible turbulence kernel (RM3D) [5,6].

3   Feature Extraction and Feature Tracking

Feature-based visualization allows scientists to extract regions of interests, and then visualize, track and quantify the evolution of these features. The first step in feature-based framework is defining the feature of interests. In [2], features are defined as connected regions which satisfy some threshold since this is a basic definition for most visualization routines (such as isosurfaces or volume rendering).

Features are tracked from one timestep to the next to capture how the features evolve [2]. Feature events can be classified into the following categories: continuation (a feature continues from one timestep t_i to the next t_{i+1}), creation (a new feature appears in t_{i+1}), dissipation (a feature in t_i does not appear in t_{i+1}), bifurcation (a feature in t_i splits into one or more features in t_{i+1}) and amalgamation (two or more features from t_i merge into one in t_{i+1}). Feature tracking allows events to be catalogued and enables complex queries to be performed on the dataset. Queries include data-mining type exploration, such as "how many new regions appear in timestep t_i?", "in which timesteps do large regions merge?", or "where does a large region break up?". The framework of feature based visualization is shown in Figure 2. First, the features (shown as the isosurfaces in this figure) are extracted from the time-varying data. Then the evolution of these features are tracked and the features are abstracted for different levels, one example of which shown in this figure is ellipsoid. Finally, the quantification, such as volume, mass, centroid, of the features is computed. The tracking, abstraction and quantification results can be used for enhanced visualization and event querying in a heterogeneous collaboratory.

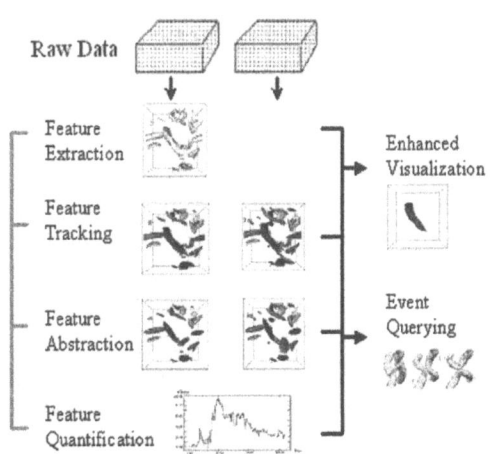

Fig. 2. Feature based visualization

After the features are extracted, the feature attributes, such as isosurface, mass and volume, can be computed. The features can also be abstracted using a simpler shape. One such reduced representation is an ellipsoid that provides an iconic abstraction to blob-like regions. An ellipsoid can capture the essential shape and orientation of the region with a limited number of parameters (center + axes) as opposed to an isosurface which may contain thousands of triangles. For huge datasets which may contain hundreds or thousands of evolving regions, and ellipsoid may provide a suitable first abstraction. Other types of abstractions include more line-like regions like skeletons [7] and vortex cores [8], or for vector fields, critical points and critical curves [9]. For scalar fields, contour trees [10] can also provide an abstraction. A contour tree is a graph which describes the topological changes of the isosurface with respect to the threshold.

An example of ellipsoid abstraction is shown in Figure 2. The scientist can choose between displaying ellipsoids or isosurfaces depending on the requirements and computing/network resources.

In addition to the standard feature extraction and tracking implemented within AVS, in [5], a distributed AMR version of feature extraction and tracking is described. AMR (Adaptive Mesh Refinement) is a technique used in computational simulations to concentrate grid points in areas where the errors are large. This results in a set of nested grids with varying resolutions. In distributed AMR grids, features can span multiple refinement levels and multiple processors; therefore, tracking must be performed across time, across levels, and across processors. The distributed algorithm [5], called *Ftrack*, is implemented within Discover and run in-situ, i.e., together with the simulation so that the data does not have to be copied or stored for visualization processing.

4   Rule Definitions

Ftrack, the feature-based visualization system integrated with Discover [3], is capable of self-management based on runtime user-defined rules. The rules are categorized into: (1) *Steering rules* apply intra-function management (e.g. changing the runtime behaviors of the visualization functions by dynamically altering their parameters). (2) *Configuration rules* apply inter-function management (e.g. organizing the visualization functions by selecting the appropriate functions to be executed). Examples of rules and the related sensors, actuators applicable to *Ftrack* are presented below:

Steering Rule: The rule for choosing an appropriate threshold mentioned in Section 1 is an example of steering rule.

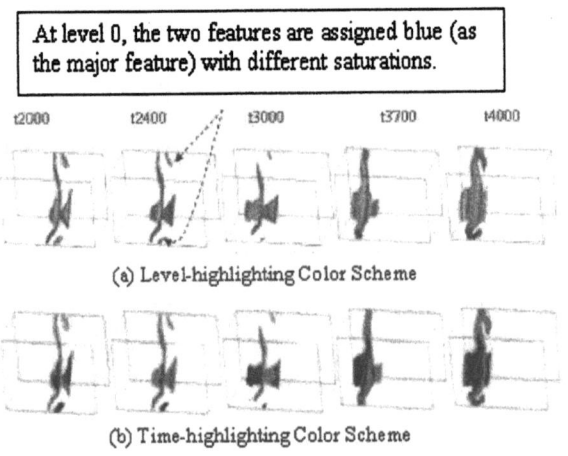

(a) Level-highlighting Color Scheme

(b) Time-highlighting Color Scheme

Fig. 3. Two color schemes used in AMR tracking for RM3D: (a) a level-highlighting color scheme is used. (b) a time-highlighting color scheme is used.

Another example given here is a rule for changing the color schemes. The level-highlighting color scheme gives each level a different hue and assign different saturation to each feature, as shown in figure 3(a). On the other hand, the time-highlighting color scheme assigns a different hue to the feature in different timestep. If a feature spans several levels, then each part of the feature falling in a different level is assigned a different saturation of the feature's hue, as shown in figure 3(b).

Rules can be used to select the color schemes. For example, when the number of features is very small (below 3), there is no need to highlight time-tracking. A scientist may prefer to see the tracking in refinement levels. For this purpose, *Ftrack* exposes one sensor *getAverageNumberofFeatures()* and two actuators *useLevelHighlightingScheme()* and *useTimeHighlightingScheme()*. A corresponding rule could be:

```
IF getAverageNumberofFeatures()<3 THEN useLevelHighlightingScheme()
                                  ELSE useTimeHighlightingScheme()
```

Configuration Rule: Scientists may choose to vary the visualization techniques based upon the computing/network resources available at the time of visualization. For example, when the number of grid cells in the dataset exceeds a threshold, a scientist at a thin client may want to display ellipsoids instead of isosurfaces. Thus, *Ftrack* should expose a sensor getNumCells() (to get the number of the cells in the input volume) and four actuators: *enableEllipsoidFitting(),* *disableEllipsoidFitting(), enableIsosurface(), disableIsosurface().* Corresponding rules could be:

```
Rule 1: IF getNumCells()>10K
        THEN {enableEllipsoidFitting(); disableIsosurface()}
        ELSE {disableEllipsoidFitting(); enableIsosurface()}
```

The scientist can modify this rule with operations like "change 10K to 50K" or "switch THEN and ELSE statements". If a scientist is working on a PDA, which typically has poor graphics resolution as well as limited memory capacity, the rule specified could be:

```
Rule 2: IF isPDA()=TRUE THEN {enableEllipsoidFitting();
                              disableIsosurface()}
```

5   Rule-Based Visualization Using Discover

The visualization subsystem, *Ftrack*, which performs the distributed feature extraction and tracking algorithms, is designed as a DIOS++ object. As shown in Figure 4, the *Ftrack* object consists of three actors, each managing a part of the visualization task:

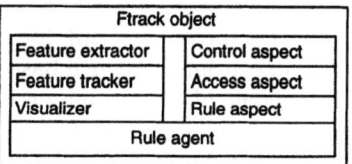

Fig. 4. Ftrack Object Structure

Feature extractor: this actor extracts interesting features, and computes the geometry attributes(e.g. isosurface) and quantification attributes (e.g. volume, mass, tensor, etc.) for each feature. Furthermore, it can calculate global statistics such as the average number of features, the variation of feature numbers, etc.

Feature tracker: this actor tracks the extracted features. The input is the feature information coming from feature extractor and the user steering commands from portals or rule operations from the rule engine. The output is the tracking information which can be represented as a set of graphs. A graph can be a feature tree [5] describing the feature correlation tracked on different levels, or a direct acyclic graph (DAG) [2] describing the feature correlation tracked on different timesteps.

Visualizer: this actor includes various visualization modules supported by the feature extractor and feature tracker. It utilizes the results coming from the other two actors to create visualizations of the features (for example, an isosurface rendering displaying quantification) and sends the image or data to the Discover portals.

The rule agent (RA), which is dynamically generated by the rule engine, is shared by the three actors. The rule engine and the RA form the interface between the actors and rules. Control aspects in the rule handler define sensors and actuators for the three actors and allow the state of the visualization subsystem to be externally monitored and controlled. The access aspect controls access to these sensors/actuators based on user privileges and capabilities. The rule aspect contains the rules that are used to automatically monitor, adapt and control the object. The rule-based visualization process is summarized below. Detail of the operation and evaluation of rule-based control in DIOS++/Discover can be found in [4].

Initialization and Interaction: During initialization, the application uses the DIOS++ APIs to register its objects, export their aspects, interfaces and access policies to the local computational nodes, which will further export the information to the Gateway. The Gateway then updates its registry. Since the rule engine is co-located with Gateway, it has access to the Gateway's registry. The Gateway interacts with the external access environment and coordinates accesses to the application's sensor/actuators, policies and rules.

At runtime the Gateway may receive incoming interaction or rule requests from users. The Gateway first checks the user's privileges based on the user's role, and refuses any invalid access. It then forwards valid interaction requests to *Ftrack* and forwards valid rule requests to the rule engine. Finally, the responses to the user's requests or from rules executions are combined, collated and forwarded to the user.

Rule Deployment and Execution: The Gateway transfers valid rules to the rule engine. The rule engine dynamically creates rule agents for *Ftrack* and other objects if they do not already exist. It then composes a script for each agent that defines the rule agent's lifetime and rule execution sequence based on rule priorities. For example, the script for the rule agent of *Ftrack* may specify that this agent will terminate itself when it has no rules, or that Rule 1 of the Configuration Rules (see Section 4) has higher priority than Rule 2.

While typical rule execution is straightforward (actions are issued when their required conditions are fulfilled), the application dynamics and user interactions make things unpredictable. As a result, rule conflicts must be detected at runtime. In DIOS++, rule conflicts are detected at runtime and are handled by simply disabling the conflicting rules with lower priorities. This is done by locking the required sensors/actuators. For example, configuration rules Rule 1 and Rule 2 conflict if *getNumCells()* is less than 10K (*disableEllipsoidFitting()* and *enableIsosurface()* should be used) while *isPDA()* is TRUE (*enableEllipsoidFitting()* and *disableIsosurface()* should be used). Assuming Rule 1 has higher priority, the script will inform the rule agent to fire Rule 1 first. After Rule 1 is executed, interfaces *enableEllipsoidFitting()* and *disableIsosurface()* are locked during the period when getNumCells() is less than 10K. When Rule 2 is issued, it cannot be executed as its required interfaces are locked. The two interfaces will be unlocked when *getNumCells()* becomes greater than 10K. By modifying the rules in the rule base through the Discover portal, the scientist can promote the priority of Rule 2 to be higher than that of Rule 1 so that the visualization subsystem always displays ellipsoids if *isPDA()* is TRUE.

The rule agent at *Ftrack* will continue to exist according to the lifetime specified in its script. If the scientist modifies the rules, for example, promoting the priority of Rule 2, the rule engine can dynamically modify the script to change the behavior of the rule agent.

6   Conclusion

This paper presented a rule based visualization system that improves the flexibility of visualization in a WAN-based computational steering collaboratory. Rules can be used to steer in-situ visualization and aid in data mining. In a heterogeneous environment, rules can help the scientists specify the type of interaction and visualization desired based on system capabilities.

This work was done at the Vizlab and TASSL, Rutgers University. This material is based upon work supported by the National Science Foundation under Grant Nos. 0082634, 9984357, 0103674, and 0120934. Any opinions, findings, and conclusions or recommendations expressed in this material are those of the author(s) and do not necessarily reflect the views of the National Science Foundation.

References

1. Mulder, J.D., Wijk, J.J.v., Liere, R.v.: A survey of computional steering environments. Future Generation Computer Systems (1999)
2. Silver, D., Wang, X.: Tracking and visualizing turbulent 3d features. IEEE Trans. on Visualizatin and Computer Graphics (1997)
3. Mann, V., Matossian, V., Muralidhar, R., Parashar, M.: Discover: An enviroment for web-based interaction and steering of high-performance scientific applications. Concurrency-Practice and experience (2000)
4. Liu, H., Parashar, M.: Dios++: A framework for rule-based autonomic management of distributed scientific applications. In: Proc. of the 9th International Euro-Par Conference (Euro-Par 2003). (2003)
5. Chen, J., Silver, D., Jiang, L.: The feature tree: Visualizing feature tracking in distributed amr datasets. In: IEEE Symposium on Parallel and Large-Data Visualization and Graphics. (2003)
6. Chen, J., Silver, D., Parashar, M.: Real-time feature extraction and tracking in a computational steering environment. In: Proc. of Advanced Simulations Technologies Conference (ASTC). (2003)
7. Reinders, F., Jacobson, M.E.D., Post, F.H.: Skeleton graph generation for feature shape description. Data Visualization (2000)
8. Banks, D., Singer, B.: A predictor-corrector technique for visualizing unsteady flow. IEEE Trans. Visualization and Computer Graphics (1995)
9. Helman, J., Hesselink, L.: Representation and display of vector field topology in fluid flow data sets. IEEE Computer (1989)
10. Sural, S., Qian, G., Pramanik, S.: Segmentation and histogram generation using the hsv color space for image retrieval. In: Int. Conf. on Image Processing. (2002)

Placement of File Replicas in Data Grid Environments

J.H. Abawajy

Carleton University, School of Computer Science, Ottawa, Ontario,
Canada, K1S 5B6
abawjem@scs.carleton.ca

Abstract. In this paper, we address the problem of file replica placement in Data Grids given a certain traffic pattern. We propose a new file replica placement algorithm and compare its performance with a standard replica placement algorithm using simulation. The results show that file replication improve the performance of the data access but the gains depend on several factors including where the file replicas are located, burstness of the request arrival, packet loses and file sizes.

1 Introduction

Grid computing is a type of parallel and distributed system that enables sharing, selection, and aggregation of geographically distributed resources. One class of grid computing and the focus of this paper is Data Grids that provide geographically distributed storage resources to large computational problems that require evaluating and mining large amounts of data [5],[8]. In a multi-user and multi-owned systems such as Grids, the resource managers (RMS) are expected to provide good resource utilization and application response time. There is an added problem of managing several petabytes of data in Data Grid environments with high availability and access optimization being some of the key challenges to be supported. This is because, in addition to the high latency of wide-area network, Data Grid environments are prone to node and link failures [2], [5].

One way of coping with some of these problems is to distribute multiple copies of a file across different sites in the system. It has been shown that file replication can improve the performance of the applications [8], [6], [5], [12]. There is a fair amount of work on file replication in Grid environments. However, most of the work done so far relates to creating the underlying infrastructure for replication and mechanisms for creating/deleting replicas [12]. We believe that, in order to obtain the maximum possible gains from file replication, a strategic placement of the file replicas in the system is the key. To this end, we propose a replica placement service called Proportional Share Replication (PSR) algorithm. PSR places file replicas in sites such that failures are masked and network load is minimized. Moreover, it distributes the load of data requests as evenly as possible over all the replicas.

The rest of the paper is organized as follows. Section 4 presents the proposed replica placement algorithm. This section also discusses a baseline policy used

M. Bubak et al. (Eds.): ICCS 2004, LNCS 3038, pp. 66–73, 2004.
© Springer-Verlag Berlin Heidelberg 2004

to compare the performances of PSR. Section 5 discusses the performance evaluation framework and the results of the proposed policy. Conclusion and future directions are presented in Section 6.

2   System Model

In this paper, we use a hierarchical Data Grid model (see Figure 1), which is one of the most common architectures in current use [11], [5], [6] [2]. The sites are classified into four regions: local sites, regional sites, national sites and international sites [5]. The connectivity bandwidth and delays between different levels of the sites is shown on Figure 1. Since the the load on the network is shared, the bandwidth can vary unpredictably in practice. Each Data Grid user is assigned a weight relative to the percentage of overall traffic in the system. Also, each site has an associated weight that represents the rate of data request expected to traverse it. Each higher level node is assigned a weight representing the sum of the traffic from its children, and the edge distance represents the one-way link transfer delay. The cumulative cost from a child node to a parent node is the sum of the edge distances.

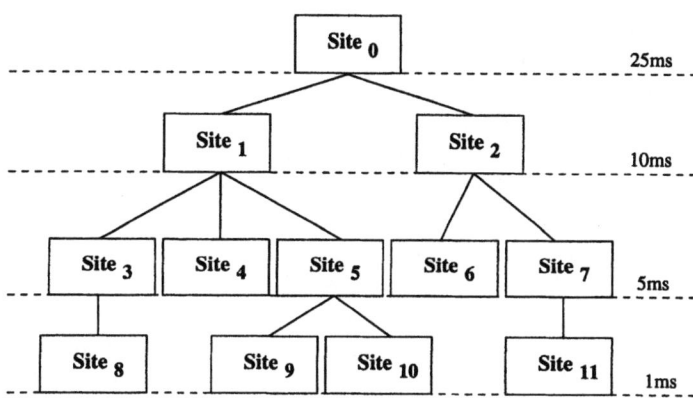

Fig. 1. Hierarchical Data Grid system model

We assume that once created, data is mostly read-only [5]. Also, as in [6] and [2], we assume that initially, only one copy (i.e., the master copy) of the files exists at the root site ($Site_0$) while all requests for file are generated at the leaves. Hence, in the hierarchical Data Grid system model, request for files travels upward from the leave node and any node along the path of the request in the hierarchy that has the requested file services the request. In other word, when a node v on level k requires a file, it sends a request to its parent node u at level $k+1$. Upon receiving the request, if u does not have the requested file, it forewords the request to its parent in turn. This process recursively continues up the hierarchy until a node that has the requested file is encountered or the request reaches the root node.

In the absence of file replicas, the root node services all data requests. Note that when there are more than one replicas of a file in the system, depending where the replica is located in the hierarchy, a site may be blocked from servicing requests for file. For example, if there are file replicas in $Site_0$, $Site_1$ and $Site_2$, the file replica at $Site_0$ cannot be used as all requests for file can be serviced by $Site_1$ and $Site_0$, respectively. Therefore, it is important to place file replicas at appropriate locations in the hierarchical Data Grid systems to reap the benefits of replication.

3   Problem Statement

The goal of replication is to allow researchers to use computing power and data storage with a reasonable response time rather than access all the data at the root. Therefore, given the network topology and data request patterns shown in Figure 1, the goal is to determine an optimal placement of a set of f file replicas on the n sites such that data access delays are reduced, data access availability is improved and the load of data request within the system both on the network and site level are balanced.

In this paper, we assume that both f and n are known. Note that there are many ways we can determine the number of replicas (i.e., f). For example, we can determined it based on the popularity of the file as in [9]. It has been noted that the access pattern in Data Grid environments shows considerable repetitiveness and locality [3], [5]. Therefore, we can use the intensity of access patterns from a given locations. Alternatively, the number of replicas can be based on a cost model that evaluates data access costs and performance gains of creating each replica. The estimation of costs and gains is based on factors such as run-time accumulated read/write statistics, response time, and bandwidth.

Broadly speaking, file replication makes identical files available at multiple locations within the Data Grid environments. The overall file replication problem includes making the following decisions: (1) when file replicas are created and deleted; (2) how many file replicas are created; (3) where these replicas should be placed in the system; and (4) how data consistency is enforced. Data consistency problem dictates that multiple copies of file must appear as a single logical data object. This problem has been address in [13] and [3]. As most of data in Data Grids are read-only [3],[5], data consistency is not a serious problem. An economic model for replica creation and deletion is discussed in [2]. Also, a scheme that is based on a cost model that evaluates data access costs and performance gains of creating each replica is discussed in [12] whereas replication of a file is done based on the popularity of a file in [9] and [14].

Although, data replication is one of the major optimization techniques for promoting high data availability, low bandwidth consumption, increased fault tolerance, and improved scalability, the problem of file replica placement as defined in this section has not been well studied for large-scale grid environments.

The common thread among existing work is that they focus on which replica should be used and not where should the file replicas are placed. Our work differs from these studies in that we are interested in determining the best location for the file replica. Therefore, the work described in this paper will help build on a high level tool above the replica management software. This tool will make replica placement decisions to increase the overall efficiency in terms of availability, response time and fault-tolerance of the Data Grid.

4   Replica Placement Service

Replica placement service is one components of the data grid architecture that decides where in the system a file replicas should be placed. In this section, we discuss a new and a baseline file replication strategies.

4.1   Proportional Share Replica Location Policy

The proportional share replica (PSR) policy works as shown in Figure 2. The main idea underlying the PSR policy is that each file replica should service approximately equal number of request rates in the system. The objective is to place the replicas on a set of sites systematically selected such that file access parallelism is increased while the access costs are decreased. For the Data Grid architecture shown in Figure 1, PSR selects $Site_1$ as the location of the new replica when the number of replica is one. However, when the number of replicas is set to two, the algorithm selects $Site_5$ and $Site_2$ as optimal location for the two replicas.

Note that collectively the sites will be able to offload about $\frac{1}{root\ copy + f}$ data requests from the root node. Also note that none of the nodes hinder any other node where a replica of the file is located from receiving its share of file requests. Note also that a given request for file reaches the root node only if there is no node along the path traveled by the request.

4.2   Affinity Replica Location Policy

The affinity replica placement algorithm replicates data on or near the client machines where the file is accessed most. In other words, a copy of the file will be placed near the client hat generates access traffic the most. In Figure 1, for example, when the number of replica is one, the algorithm selects $Site_6$ as the placement of the new replica. Most of existing replica management systems employs a variation of this type of replication scheme [1], [2], [12]. This algorithm is similar to the cascading replica placement algorithm discussed in [6].

5   Performance Analysis

In this section, we compare the proportional and affinity file replica placement schemes discussed in previous section on simulated Data Grid environments. We

1. Select a set of V sites, $V \in \{Site_1, \ldots, Site_n\}$ such that each site in V meets the following conditions:
   a) does not have f already.
   b) does not block any site where a replica of the file exits from servicing its share of file requests.
2. Compute an ideal load distribution over the replicas as follows:

$$load = \frac{1}{root\ copy + f} \times R \qquad (1)$$

   where R is the total number of data access requests in the system.
3. Place replicas as follows:
   REPEAT
   a) Select a site $v \in V$ that is able to service a set of requests slightly greater than or equal to the ideal load.
   b) Place a replica of the file in v.
   UNTIL all replicas are placed.

Fig. 2. Pseudo-code of the proportional policy

used ns2 network simulator but modified it such that we capture grid specific components in a similar manner to [12], [1], [2]. We also added some extra nodes whose sole purpose is to generate background traffic at a specific rate such that we model a shared interconnection network, as is the real grid environment. In all experiments reported in this paper, we run a total of 30 simulations. As in [5], [12], [1], [2], we used the mean response time as a chief performance metric.

5.1   Experimental Setup

The grid topology, number of sites, connectivity bandwidth and the number of users is as shown in Figure 1. Each request is a 4-tuple field with the arrival time, a source node, a server node and file size. The request arrival process is modeled as exponential distribution with mean of 30 requests per time period. The file size is drawn from a hybrid distribution, with a log-normal body, and a Pareto tail. The median transfer size is 5MB. The largest transfer is 53MB, while the smallest is 3MB bytes. The source and server nodes are generated at random based on the desired request rate for each user. The default server for all requests is the root node.

5.2   Results and Discussions

Figure 3 shows the performance of the proportional and affinity replication policies as a function of the number of replicas. We observe that replication do improve the mean response time as shown by both policies. The main observation is that the proportional algorithm performs better than the affinity algorithm.

In proportional policy, the tendency is to choose sites that serve multiple clients and offload traffic from the busiest network links whereas this is not the case in affinity algorithm. Also the secondary influence in proportional algorithm appears to favor the placement of file replicas in sub-trees that reflect the highest traffic generating clients. This observation is not surprising considering that this sub-tree generates most of the load on upstream links. The Affinity policy cannot process requests as quickly as it is receiving them. When the node is unable to handle the load of requests it is receiving, the size of its message queue begins to increase. Therefore, balancing the load among the replicas is important as it is done in proportional policy.

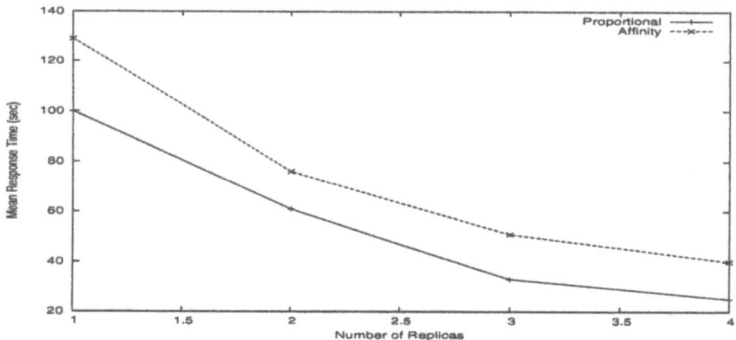

Fig. 3. Performance of the replication policies

It is important to investigate the sensitivity of the replication policy to the network congestion and packet loss factors. This can be modeled by varying the request arrival rate while keeping the file size to 20MB. Figure 4 shows the mean response time (y-axis) of the three policies as a function of the request arrival rate (x-axis). Note that 120 request per second produces an average of 80% utilization on the busiest link in the network. At this level, we observed that about 0.2% of packets were lost when all the requests are serviced from the master copy (i.e., NoReplica case), which required retransmission of the lost packets. This partially explains the performance of the NoReplica case. In contrast, the packet loses in Affinity is 0.064819% while that of Proportional is 0.03511% which is far less than both cases. The conclusion is that network related factors such as the number of packet losses have significant impact on the performance of the replica locations.

6   Conclusions and Future Direction

Many scientific and engineering applications perform large amount of data analysis on increasingly large datasets. There is a need for efficient tools and middleware infrastructures that allow secure access to massive amounts of data, to

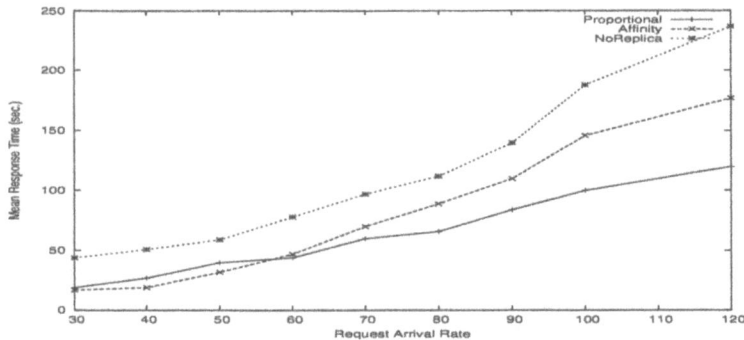

Fig. 4. Sensitivity of the replication policies to arrival rate

move and replicate data and to manage coherence of data in Data Grid environments [8]. In this paper, we addressed one of these data management problems, which is the file replication problem while focusing on the problems of strategic placement of the replicas with the objectives of increased availability of the data and improved response time while distributing load equally. We proposed a new replica placement algorithm that distributes the request to all replicas in such a way that data request processing parallelism is obtained. To the our knowledge, no replication approach balances the load of data requests within the system both on the network and host levels as well as to improve reliability. Our results show that:

1. Replication improves the data access performance of the overall system as measured by the response time. This observation concurs with results in the literature.
2. Many factors including network related issues such as the rate of packet losses affect the performance of replication approaches.
3. The placement of replicas in the system matters and it depends on several factors including the rate of data request arrival and the size of files.
4. Balancing the data request among the replicas is important in improving the performance.

The replica selection by individual users may dependent on a number of factors, including the security policies of the sites which contains these replicas, queuing and network delays, ease of access and so forth. These issues will be addressed in future. The simplifying read only assumption makes it possible to initially address the following question: What are efficient algorithms for creating and disseminating replicas in a Data Grid. We are currently working with both read and write aspects of the file replication strategy discussed.

References

1. David, W. B., Cameron, D. G. , Capozza, L., Millar, A. P., Stocklinger, K., Zini, F.: Simulation of Dynamic Grid Replication Strategies in Optorsim. In Proceedings of 3rd Int'l IEEE Workshop on Grid Computing (2002) 46-57
2. David, W. B.: Evaluation of an Economy-Based File Replication Strategy for a Data Grid, International Workshop on Agent based Cluster and Grid Computing (2003) 120-126
3. Stockinger, H., Samar, A., Allcock, B., Foster, I., Holtman, K., Tierney, B.: File and Object Replication in Data Grids. In 10th IEEE Symposium on High Performance and Dis-tributed Computing (2001) 305-314
4. Li, B., Colin, M., Deng, X., Sohraby, K.: The Optimal Placement of Web Proxies in the Internet. In Proceedings of IEEE INFOCOM (1999) 1282-1290
5. Hoschek, W., Janez, F. J., Samar, A., Stockinger, H., Stockinger, K.: Data Management in an International Data Grid Project. In Proceedings of GRID Workshop (2000) 77-90
6. Ranganathana, K., Foster, I.: Identifying Dynamic Replication Strategies for a High Performance Data Grid. In Proceedings of the International Grid Computing Workshop. (2001) 75-86
7. EU Data Grid Project. http://www.eu-datagrid.org.
8. Chervenak, A., Foster, I. , Kesselman, C., Salisbury, C., Tuecke, S.: The Data Grid: Towards an architecture for the distributed management and analysis of large scientific datasets. Journal of Network and Computer Applications (2000) 187-200
9. Allocck, W., Foster, I., Nefedova, V., Chervnak, A., Deelman, E., Kesselman, C. Lee, J., Sim, A., Shoshani, A., Tierney, B., Drach, B., Williams, D:. High-Performance Remote Access to Climate Simulation Data: A Challenge Problem for Data Grid Technologies (2001) 46-46
10. Particle Physics Data Grid (PPDG); http://www.ppdg.net.
11. Grid Physics Network (GriphyN); http://www.griphyn.org.
12. Lamehamedi, H., Szymanski, B., Shentu, Z., Deelman, E.: Data Replication Strategies in Grid Environments. In Proceedings of 5th International Conference on Algorithms and Architecture for Parallel Processing (2002) 378-383
13. Deris, M.M., Abawajy J.H., Suzuri, H.M.: An Efficient Replicated Data Access Approach for Large-Scale Distributed Systems. To appear in IEEE International Sympo-sium on Cluster Computing and the Grid (2004), April 19-22, Chicago, Illinois, U.S.A.
14. Abawajy J.H.: An Integrated Resource Scheduling Approach on Cluster Computing Systems. In the Proceedings of the IEEE International Parallel and Distributed Processing Symposium (2003) 251-256
15. Ranganathan, K., Iamnitchi, A., Foste, I.T.: Improving Data Availability through Dynamic Model-Driven Replication in Large Peer-to-Peer Communities. In 2nd IEEE/ACM International Symposium on Cluster Computing and the Grid (2002). 376-381
16. Holtman, K.: CMS Data Grid System Overview and Requirements. The Compact Muon Solenoid (CMS) Experiment Note 2001/037, CERN, Switzerland (2001)

Generating Reliable Conformance Test Suites for Parallel and Distributed Languages, Libraries, and APIs

Łukasz Garstecki[1,2]

[1] Faculty of Electronics, Telecommunications and Informatics,*
Gdańsk University of Technology, ul. Narutowicza 11/12, 80-952 Gdańsk, Poland
galu@eti.pg.gda.pl
[2] Laboratoire Informatique et Distribution**, ENSIMAG - antenne de Montbonnot,
ZIRST, 51, avenue Jean Kuntzmann, 38-330 Montbonnot Saint Martin, France
Lukasz.Garstecki@imag.fr

Abstract. This paper outlines a new methodology for generating Conformance Test Suites (CTS) for parallel and distributed programming languages, libraries and APIs. The author has started his research in the field of conformance testing for parallel data-driven language Athapascan, invented a methodology for designing and analyzing CTSs called Consecutive Confinements Method (CoCoM), developed a tool called CTS Designer, which implements standard ISO/IEC 13210 and the CoCoM methodology, and finally created a CTS for the crucial part of the Athapascan language. Although CoCoM was originally meant for parallel and distributed software, it is not limited only to it and can be also used for a variety of other languages, libraries and APIs.

Keywords: Software testing, parallel data-driven languages.

1 Introduction

Conformance testing is a black-box testing and is aimed at demonstrating that an implementation, called Implementation Under Test (IUT), fulfills its specification. Conformance testing is very well known for example in the domain of protocols and communication systems, where IUT can be relatively easy represented by a Finite State Machine. Conformance testing of programming languages, libraries and APIs requires however different solutions and the only one standard directly connected to this issue found by the author is ISO/IEC 13210 [1]. Although dedicated to POSIX, this standard is quite general and can be applied to different programming languages, libraries and APIs. In the paper we utilize the notions provided by it. The specification of an IUT is called a *base standard* and basic entities in the base standard, such as functions, constants, header files, etc.

---

* Funded in part by the State Committee for Scientific Research (KBN) under Grant 4-T11C-004-22.
** Laboratory is funded by CNRS, INRIA, INPG and UJF.

M. Bubak et al. (Eds.): ICCS 2004, LNCS 3038, pp. 74–81, 2004.
© Springer-Verlag Berlin Heidelberg 2004

are called *elements*. Each element has a number of *conformance requirements* it must satisfy. The specification for testing a requirement is called an *assertion*, and is required to be true in a *conforming implementation*. Additionally, we will provide a notion of a *potential error* of an assertion, which defines potential failures in this assertion. Standard ISO/IEC 13210 specifies how to identify elements in a base standard, and how to retrieve requirements and assertions for them, but unfortunately does not specify how to design and implement conformance tests. According to the knowledge of the author, there exist only two tools that are based on this standard[1] (apart from CTS Designer developed by the author): TET [2] developed by The Open Group and DejaGnu [3] developed by Free Software Foundation. Although these tools are well established and very useful, and moreover there exists even a special version of TET meant for distributed software, the author revealed a serious problem that can appear in Conformance Test Suites (CTSs), and which is not considered by these tools and by any solution known to the author. The problem lies in relations between tests in a CTS, and can result in an unreliable CTS, i.e. a CTS that can pass for faulty IUTs. The problem presents in Sect. 2, while Sect. 3 specifies precisely conditions under which the problem appears. Next, Sect. 4 is presenting a methodology called *Consecutive Confinements Method* (CoCoM), which solves the problem. Finally, Sect. 5 contains conclusions and plans for the future work.

The CoCoM methodology, presented in the paper, can be seen as a complement of existing solutions, or in other words as a framework, where different methodologies and tools could be used. However, the author developed its own prototype tool called *CTS Designer*, based on standard ISO/IEC 13210 and methodology CoCoM, and finally created a CTS for the crucial part of a parallel data-driven language called Athapascan [4] using his tool.

For the sake of simplicity and clarity we will refer throughout the paper to programming languages, libraries and APIs shortly as "programming languages", but the presented methodology applies to any of them equally. Also wherever we use the term "conformance testing", we mean conformance testing for programming languages, libraries and APIs.

2   Unreliable Conformance Test Suites

As we have written at the beginning, conformance testing is black box testing, so very often the only one way to test an element is to use another elements under tests. This takes place especially in a case of parallel and distributed programming languages, where usually the only one way to test a function sending a message is to use another function that is receiving a message, since in conformance testing we cannot access a source code of an IUT. The problem is, that afterward very often to test the receiving function, we will use the sending function, that we have tested just before. And when we use a function in a test, we

---

[1] Actually there are based on the former version of this standard, namely ISO/IEC 13210-1994, IEEE 1003.3-1991.

implicitly assume that its implementation is valid. So using words of mathematics, first we prove A, assuming that B is correct, and then we prove B assuming that A is correct. Below, we present a small example showing what can happen, if we do not care about this problem. Elements used in this example are: *write* (denoted as W), *read* (R) and *access* (A). These are three methods taken from Athapascan [4] that relatively write, read and access[2] a shared variable. For the sake of clarity let us assume that the shared variable is an integer, denoted as I. Now, let us assume that our system consists only of two elements: W and R, and let us assume that W always writes one more (+1), and R always reads one less (-1). The situation is presented in System 1 in Fig. 1.

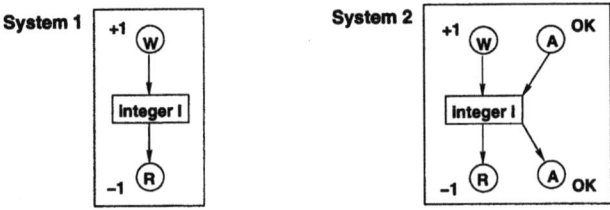

Fig. 1. Simple systems with read, write and access

Now, we can pose a question: *Does System 1 contain a functional error?* It can be sure that there is no failure, because failures occur when provided services are not conforming to their specifications [5], which is not the case here. May be we can say that there is a defect, however it is also not so sure because it could be a conscious decision of system designers, because for example storing zero value could be very costly or even impossible (especially in mechanic, not computer systems). However, regardless of whether we consider that this is a defect or not, we do not have any observable error here, because provided services will be always conforming to their specifications. Now, let us slightly change our system and let us add element A and let us assume that it works correctly. The situation is presented in System 2 in Fig. 1.

Now, our system will obviously contain failures, because for example if we accessed integer I previously written by W, we would obtain a wrong value of I. That means that the same implementations of elements W and R can be now evaluated as non-conforming to their specifications. "Can be" because by observing only responses of this system, we are not able to say which elements are faulty, because the same symptoms would appear, if for example element A always subtracted 1 when writing, and added 1 when reading. Luckily, in conformance testing, the principal question is whether **all** elements conform to their specification or not. In other words, we don't have to find which elements are faulty, but whether at least one element is faulty and no matter which.

---

[2] That means they can read it and write it.

So, in these two simple examples we could see that the same implementations of elements W and R can be once evaluated as conforming to their specification and the other time as non-conforming. This pushes us toward a very important conclusion, namely that conformance is relative, i.e. conformance of an element to its specification, depends not only on the element itself, but also on the *context of this element*, i.e. other elements either from IUT or out of IUT. If we do not care about the context of an element, we can obtain for example a CTS, presented in Fig. 2, which would pass for the faulty implementation of System 2 presented in Fig. 1.

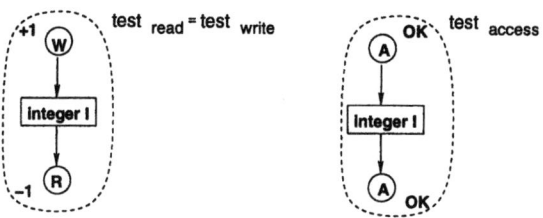

Fig. 2. A simple CTS for System 2 in Fig. 1, passing for an invalid IUT

In the next section we will say when the context of an element has to be considered and which elements from this context should be taken into account.

3   Dynamic Correctness of a Test

As we have written before, very often we can test a given element from IUT only by means of other elements, which are under test too. But, if an element is under test, we cannot be sure whether this element is valid or not (i.e. whether its implementation is valid or not) unless we do appropriate tests. Meanwhile, traditionally we implicitly assume that elements used in a test are valid. This pushes us to introducing two new terms: *static* and *dynamic correctness* of a test. Test is *statically correct*, if that test (namely its source code) is correct (i.e. it passes for a good implementation of an element under test and fails for an implementation that contains a potential error, for which the test was designed) under assumptions that all elements used in the test are valid. Test is *dynamically correct*, if the running test (namely its executable) is valid. In other words, by static correctness of a test we mean traditional correctness of a test. In turn, to prove that a test is dynamically correct, we should prove that it is *statically correct* and all elements used in this test are valid[3]. But to

---

[3] It is also worth mentioning here, that elements could be hardware too. For example, network with a very low performance could give untrue results, so it is a good idea to do performance tests of the network before we start conformance testing of parallel and distributed software [6].

prove that elements used in a given test are valid, we need dynamically correct tests that will test them. In that way, we can build a directed graph $G(V, E)$ of dependencies between tests, where tests are represented by nodes and directed edge E_{ij} indicates that test T_i needs test T_j to prove its dynamic correctness. We will say that test t_i *directly depends on* test t_j. Moreover, if there exist tests $t_k, t_{k+1}, \ldots, t_{k+n}$, where $n > 0$, such that $\forall_{l=\{k,\ldots,k+n-1\}} t_l$ depends on t_{l+1}, we will say that test t_k *indirectly depends on* test t_{k+n}. Finally, if t_i depends on t_j and t_j depends on t_i, we will say that test t_i *has direct cyclic dependence with* test t_j. We can also say that if t_i indirectly depends on t_j and t_j indirectly depends on t_i, then test t_i *has indirect cyclic dependence with* test t_j. In the farther part of this paper we will shortly say: *depends on* instead of *indirectly depends on* and *cyclic dependence* instead of *indirect cyclic dependence*. In the directed graph $G(V, E)$ cyclic dependence manifests as a cycle.

However, errors of elements used in a test do not affect the dynamic correctness of the test always in the same way. We should distinguish three different relations between error b_i of element e_i, for which test t_i is designed, and error b_j of an element e_j used in test t_i.

1. *masking error* – error b_j can *mask* b_i, i.e. test t_i can pass for an invalid implementation of element e_i.
2. *triggering error* – no masking error, but error b_j can *trigger* b_i, i.e. test t_i can fail for a valid implementation of element e_i.
3. *no relation* – error b_j has no influence on the dynamic correctness of test t_i.

By analyzing different cases one by one, we could find that the overall test result of a CTS can be wrong only if it contains at least two tests t_i and t_j which have cyclic dependence, and errors that are to be detected by these two tests can mask each other. We will say then that tests t_i and t_j have *strong cyclic dependence*. Notice for example, that in Fig. 2 non-conforming IUT would be evaluated as "conforming", because there is strong cyclic dependence between tests $test_{write}$ and $test_{read}$. In the rest of situations, even if a CTS contains dynamically incorrect tests (what means that at least one of elements is not valid), then single tests can return incorrect test results, but we could show that at least one test from the CTS will fail, and thus the whole CTS will fail, because in conformance testing IUT conforms to its base standard only if all tests pass.

The obvious conclusion from our considerations is that we should get rid of strong cyclic dependencies if we want a reliable CTS. The easiest solution is to find another test scenarios using another elements. However it can happen that we have already used all possible elements, that are able to test a given element e_i, and in each of those tests we will have strong cyclic dependencies. In such a situation, we should run all these tests. If all these tests pass, that means that either there is no error in element e_i or that there are errors, but these errors are not observable. In such a situation we will say that the tests for element e_i are *dynamically p-correct* (*p* comes here from *partially*). Naturally, if some other tests depend on dynamically p-correct tests, they can not be supposed to be dynamically correct. Examples of dynamically p-correct tests can be tests *write-read* for W and R in System 1 presented in Fig. 1. P-correct tests can frequently take place especially in black box testing of parallel and distributed software, where very often the only one way to check sent or received messages, or to examine the state of shared variables is to use another elements from the IUT.

4 Consecutive Confinements Method

Now, we are ready to give the algorithm for constructing a reliable CTS, i.e. a CTS which will pass for valid IUTs, i.e. implementations that fulfill all requirements from their specification, and will fail for invalid implementations, i.e implementations that contain at least one potential error defined for at least one assertion of at least one element, assuming that this error is observable.

Algorithm 1 *Design a conformance test suite for a base standard*

1. *Find all elements in the base standard and retrieve assertions (from a base standard) for them, using standard ISO/IEC 13210 [1].*
2. *Identify implicit preconditions in assertions.*
3. *For each assertion, enumerate potential errors.*
4. *For each potential error find at least one test scenario.*
5. *For each test scenario in the CTS make analysis using Algorithm 2.*
6. *Take any test t (for potential error b of assertion a of element e) from the CTS such that t has strong cyclic dependence with at least one another test in the CTS. If such a test does not exist, then finish.*
7. *If possible, write a new version of test t for potential error b of element e, using different elements (than in the previous versions of the test) that are able to test element e. Otherwise, i.e. if no more tests can be found, include all tests found for error b into the CTS, and mark these tests as dynamically p-correct.*
8. *Go to Step 5.*

For the sake of clarity, we have skipped some implementation, but important details. For example, each created test t for a given potential error shall be kept in a memory, even if t is dynamically incorrect and was replaced by another dynamically correct test t', because dynamic correctness is not a feature, which is constant (in contrary to static correctness). For example, it can happen later, that another test t_z will be added to CTS, such that t' will have a strong cyclic dependence with t_z, and thus t' will become dynamically incorrect. In such a situation, we will have to look for another test, alternative to t'. Then, first of all we have to know that test t was already designed, and moreover if it happens that no other alternative tests can be found, both tests t and t' will be necessary.

In Step 4 and 7 many different methodologies and tools for designing test scenarios can be used. This is why we wrote that CoCoM can be seen as a framework. Algorithm 2 for making analysis of relations between a given test t and all other tests in a CTS, which is used in Step 5, is presented below. In Step 8 we come back to Step 5, where we will remake the analysis for all test scenarios. We have to do it, because after adding a new test to a CTS, theoretically each test in a CTS can begin to have strong cyclic dependence with the newly added test (directly or indirectly). However, if we store the state of the whole analysis in memory, then only analysis for the newly added test must be done manually, while the analysis for all other tests can be easily done automatically (using the information provided earlier in Step 1 of Algorithm 2), like it is for example in the case of the prototype tool CTS Designer, developed by the author.

Algorithm 1 gives the answer for the question how to combine different elements into tests to obtain a reliable CTS, i.e. when it is enough to create only one test scenario to detect a given potential error of an element, and when we have to create more test scenarios using different elements from IUT. Below we present Algorithm 2 for the analysis of relations between tests.

Algorithm 2 *Make analysis of relations between test t and all tests in CTS*

1. *For each element e_x used in test t do*
   a) *Describe briefly a function of element e_x in test t.*
   b) *For each assertion a_x of element e_x do*
      i. *If assertion a_x is not used in test t, then go to Step 1b.*
      ii. *For each potential error b_x of assertion a_x do*
         A. *If error b_x cannot mask error b, then go to Step 1(b)ii.*
         B. *For each test t_x for potential error b_x, mark that test t depends on test t_x.*
2. *Find cycles (direct and indirect) in graph of dependencies between tests.*

So, as we can see, by the *consecutive confinements* we can narrow the group of tests, with which a given test t can have strong cyclic dependence. That is why the methodology is called Consecutive Confinements Method – shortly CoCoM.

Step 1a was provided to facilitate further decisions in Step 1b, whether a given assertion a_x is used in test t or not. Steps 1(b)iiB and 2 can be naturally done automatically. The rest of points must be done manually if no supporting formalisms are used[4]. However, it is worth mentioning that we can also reduce the amount of manual analysis by excluding from analysis cases, where cycles can not happen. For example, if element e_x (used in test t for element e) does not have any test, which uses element e (directly of indirectly), there is no point in analyzing relations between test t and assertions of element e_x, because cycles will not appear. In the similar way, we can also reduce analysis at the level of potential errors. It is also worth mentioning that Step 2 can be done "in fly", i.e. new cycles can be searched after any changes in the state of analysis made in Step 1. This takes place for example in CTS Designer.

5   Conclusions and Future Work

The paper presented a problem because of which a CTS can be unreliable, i.e. it can pass for invalid IUTs. The problem can appear if elements use themselves

---

[4] Notice, that CoCoM does not put any constraints on notations used for expressing assertions and errors, so they can vary from the natural language to formalisms that allow for automatic generation of tests.

to test each other. The paper precisely specified situations where CTSs can be unreliable and proposed a methodology for generation reliable CTSs, called CoCoM. The methodology also gives clear criteria when a given element can be tested by only one another element (if it is necessary), and when it requires to be tested by a group of different elements, what allows for optimizations of a CTS. The methodology was proved to be useful for parallel and distributed languages by designing a CTS for the crucial part of parallel data-driven language Athapascan by means of the prototype tool CTS Designer, developed by the author on the basis of standard ISO/IEC 13210 and the CoCoM methodology. CTS Designer works currently under MS Windows, and generates CTSs ready to execute on Linux. The results of the executed CTS can be viewed as XHTML pages. In the nearest future, the author plans to make experiments with other parallel and distributed languages, libraries and APIs, like PVM and MPI.

Acknowledgments. I wish to thank Bogdan Wiszniewski from Gdańsk University of Technology and Jean Louis Roch and Jacques Chassin de Kargommeaux from L'Institut National Polytechnique de Grenoble for their helpful suggestions and comments on my work, and their cooperation in conformance testing of Athapascan language.

References

1. ISO/IEC: ISO/IEC 13210:1999(E), IEEE Std 2003, Information Technology — Requirements and Guidelines for Test Methods Specifications and Test Method Implementations for Measuring Conformance to POSIX Standards. (1999)
2. The Open Group http://tetworks.opengroup.org/documents/docs33.html: Test Environment Tool – TETware User Guide. Revision 1.4 edn. (2003)
3. Savoye, R.: Deja Gnu – The GNU Testing Framework. Free Software Foundation, http://www.gnu.org/software/dejagnu/. 1.4.3 edn. (2002)
4. Roch, J.L., Galilée, F., Doreille, M., Cavalheiro, G., Maillard, N., Revire, R., Defrenne, A.: Athapascan : API for Asynchronous Parallel Programming, http://www-id.imag.fr/ Logiciels/ath1/manual. (2002)
5. Laprie, J.C.: Dependability: Basic Concepts and Terminology. Springer-Verlag (1991)
6. Garstecki, Ł., Kaczmarek, P., Krawczyk, H., Wiszniewski, B.: Conformace testing of parallel languages. In Kacsuk, P., Kranzlmüller, D., Németh, Z., Volkert, J., eds.: Distributed and Parallel Systems Cluster and Grid Computing. Volume 706 of The Kluwer International Series in Engineering and Computer Science., Kluwer Academic Publishers (2002) 173–181
7. Garstecki, Ł., Kaczmarek, P., de Kergommeaux, J.C., Krawczyk, H., Wiszniewski, B.: Testing for conformance of parallel programming pattern languages. In: Lecture Notes in Computer Science, Springer Verlag (2001) 323–330
8. Garstecki, Ł.: Generation of conformance test suites for parallel and distributed languages and apis. In: Eleventh Euromicro Conference on Parallel, Distributed and Network-Based Processing, IEEE (2003) 308–315

A Concept of Replicated
Remote Method Invocation*

Jerzy Brzezinski and Cezary Sobaniec

Institute of Computing Science
Poznan University of Technology, Poland
{Jerzy.Brzezinski,Cezary.Sobaniec}@cs.put.poznan.pl

Abstract. In this paper we present a new infrastructure for building distributed applications that communicate through remote objects. The objects are available by the use of the Java RMI, enhanced by replication. Shared objects may be arbitrary complex with references to other objects allowing nested invocations. The replication process may be controlled and tuned by careful design of shared objects granularity, explicit synchronization operations, and replication domains definitions.

1   Introduction

Replication is one of the key concepts in distributed systems, introduced mainly for increasing data availability and performance. It consists in maintaining multiple copies of data items (objects) on different servers. However, replication introduces some serious problems like consistency. Centralized systems have only one version of every object, and its state at any time reflects *the last* write operation on that object. In a distributed system the notion of *last* is not so obvious because of the lack of a common clock. Every system using replication must therefore use some kind of *consistency protocol* that will arrange communication between replicating servers. Consistency protocol is a practical implementation of a *consistency model* that defines guarantees provided by the system regarding operation execution order.

In this paper we present a concept of a system for sharing objects in a distributed environment that supports replication. The fact of replica existence must be known by the user, and the system expects some cooperation in order to maintain the replicas in consistent state. We use Java RMI as the basic operation model, because it is a widely accepted solution, and allows tight integration of remote objects with application code.

Java Remote Method Invocation [1] is an extension of the Java programming language and its Java Object Model into Distributed Object Model [2]. It allows applications to call methods of remote objects in the same manner as they call methods of local objects. Essentially it is a kind of *remote procedure call* known

---

* This work was supported in part by the State Committee for Scientific Research (KBN), Poland, under grant KBN 7 T11C 036 21

M. Bubak et al. (Eds.): ICCS 2004, LNCS 3038, pp. 82–89, 2004.
© Springer-Verlag Berlin Heidelberg 2004

from such products as Sun RPC. The unique feature of RMI is that it seamlessly integrates with Java programming language hiding the fact of true object location. On the client side the application uses an automatically generated *stub* that implements the interface of the shared object. The stub intercepts all method invocations, transmits the invocation with arguments to the server side, receives the results, and finally returns them to the application. On the server side there is a *skeleton* that receives a method invocation from a client, unpacks it, and issues an appropriate method of a locally available object.

Related work. CORBA and DCOM are example technology standards that define a distributed object model which allows invocations on remote objects in a very similar way local methods are invoked. Because of their emphasize on interoperability they are not so tightly integrated with a programming language as the Java RMI. What is more important: they do not support replication.

NinjaRMI [3] is a ground-up reimplementation of Java RMI with three important enhancements: UDP unicast, UDP multicast, and cryptography support for reliable, unicast RMI. The system optimizes communication, and is a good base to build a replication system on top of it.

JGroup [4] uses a group communication approach. A specialized *group reference* maintains references to all replicas and uses a reliable unicast when communicating replicas. Clients invoke methods of the group reference which causes an iteration process through replicas until a working one is found. JGroup increases reliability but does not solve the problem of consistency.

Manta [5] generalizes standard RMI by providing a *Group Method Invocation*. Methods can be invoked on a local copy, on a remote object, or on an explicit group of objects. GMI may be therefore treated as an implementation tool for building a replication system.

Aroma [6] intercepts remote method calls and propagates them to other replicas. Method invocations are transfered using underlying group communication with reliable, totally ordered multicast protocol to provide strong replica consistency. One of the most important drawbacks of this approach is a consequence of the most important advantage: full transparency. Intercepting, decoding, and distributing method invocations is a costly process.

Highly efficient replication was used for a long time in *Distributed Shared Memory* (DSM) systems. They usually relax memory consistency model in order to achieve a better performance. However, DSM systems are usually memory-oriented instead of object-oriented. Objects in DSM systems are usually simple encapsulations of data, and do not allow to build complex object structures.

Replicated Remote Method Invocation (RRMI) is an efficient object replication infrastructure that will increase the availability of shared remote objects. The system seamlessly integrates with the Java object model thanks to the RMI concept, and, on the other hand, provides efficient replication using consistency protocols adapted from DSM systems. The shared object model does not restrict possible complexity of object structures, allowing the developer to create objects that reference other objects, with methods that call methods of other objects.

We give the developer tools for optimizing the replication management such as: explicit synchronization and *replication domains*.

2   Design of RRMI

Replication strategy. Object replication in general is a complex task resulting from potential object complexity. Our previous experience [7] concerning object replication was limited to simple read-write objects, where objects may be treated as variables wrapped with objects that provide simple access methods for these variables. General objects provide methods that modify their state in an arbitrary way. The first problem that arises while replicating such objects is distinction between methods: which methods modify the state, and which are read-only methods. The problem may be generalized to the problem of identification of commuting methods, i.e. methods that may be safely executed concurrently, possibly on different replicas. A simple example is a queue object that can execute concurrently two modifying methods *add* and *get* unless it is empty. In RRMI the user annotates the shared object interface with statements indicating whether the method is a modifying one or not. The statements are then interpreted by a stub compiler to optimize method invocations.

Replication requires some kind of consistency management in order to hide replicas divergence from the user. An important part of consistency management is updating policy. Objects can be updated by means of *state transfer* or *operation transfer*. State transfer is usually used by DSM systems which use a flat page of memory as the basic sharing structure. In such systems update consists in transferring the whole updated page, or in computing so called *diffs* [8] that efficiently represent changes applied to memory page. Tracking changes made on objects, and especially object structures, is not that simple, and causes considerable computing overhead. The alternative is to transfer the whole object state every time it changes, which results in unacceptable communication overhead. For these reasons we have chosen *operation transfer* as the basic replica update policy.

Usually replication techniques are divided into two distinctive groups: *passive replication* and *active replication* [9]. The classification is somehow related to replica update policy but it emphasizes another aspect of replication: the organization aspect. In passive replication the processing takes place only at a single server, called the primary replica, and then is transfered (using *state transfer*) to other replicas. In active replication all replicas are contacted directly by the client, and are instructed to execute a given processing. Our approach is a mixture of these two approaches. Clients of RRMI contact only a single server (as in passive replication), the server registers appropriate method invocations on shared objects, and *later* sends the invocation history to other replicas where the methods are re-executed. This procedure assumes obviously a deterministic processing on replicas (as in active replication). Our system tries to be as *lazy* as possible: updates are sent to the nodes that really need them, usually after explicit requests.

State transfer systems are easy to implement because they require keeping only the newest version of all objects in the system with some kind of timestamps indicating its version. When a replica needs to be updated, the current version is located, transfered, and replaces the old one, regardless of the state of the replica. This is not the case in advanced state transfer systems (i.e. using diffs), nor operation transfer systems. They need to apply all changes made on an object by all servers, and apply them in order. In case of object replication system that permits references between objects and therefore nested calls, it is also necessary to apply the changes in causal order. We register and reconstruct the causal order of method invocations by the use of vector clocks which are also used by the consistency manager (see Sect. 3).

Shared object structure. Objects are very different entities. They may be as small as single variables, or as large as databases. This raises important problems when designing a system that provides replication facility of such objects. Every object registered and maintained by the replication infrastructure needs some additional data structures that keep track of its replicas consistency. The best performance can be achieved when the size of maintenance structures is small compared to the size of shared objects. The size of a shared object therefore should be as large as possible. However, this limits the possible concurrency level of applications accessing shared objects. In an extreme case the system would handle a single shared object with permanent access conflicts on replicas that are constantly inconsistent. In order to minimize possible number of conflicts and to increase concurrency, the shared data space should be divided into small, possibly autonomous entities. This claim is however in conflict with the previous one. In RRMI we leave the decision of object granularity to the user. He is responsible for designing shared objects in a manner that will be the best possible trade-off between concurrency and maintenance overhead. The system does not have enough knowledge about the application characteristics to make the decision in a better way.

Construction of more advanced objects very often requires the use of other objects because they are inherently complex. For example, one object may consist of a couple of other objects. Our system does not require explicit registration of all internal objects within the shared space. It is only necessary to register the top-level, interface objects. The whole internal structure will be replicated along with the main object. This concept is similar to *clouds* of objects used in Manta [5]. However, our objects may still have references to other shared objects which is not the case in Manta clouds. We use regular Java references to point to internal objects, and remote references to point to other shared objects. The kind of reference therefore semantically defines boundaries of replication entities.

Nested invocations. Objects are not isolated structures. Usually they are in relations with other objects which is expressed in programming languages as *references* to other objects. References allow invocations of other objects' methods, that is *nested invocations*. Pure active replication systems that permit nested invocations face the problem of repeated calls when the nested object is called several times by its callee that is also replicated. The system should either detect such repeated calls and silently ignore them, or avoid them by ignoring some calls at their source. RRMI solve the problem in a slightly different manner. Because of the replication strategy, the objects are called directly by the client only once. Such calls are then transmitted to replicas, but the re-execution is explicit, therefore may be treated differently — it is just not replicated. The only method invocations that are replicated come directly from the user. Whenever a nested method is called by a shared object, it is executed only on locally available replica of that object. Such nested calls are not registered in the history, and are not replicated. This assumes that every shared object reachable through remote references is available locally.

Naming. In standard RMI, objects, before they can be accessed, are *bound* to a name within a *registry*. Registry is a simple naming service that holds an array of objects' stubs, and sends them to clients when they issue *lookup* operation. The stub internally contains enough information to locate the server that provides the object implementation, i.e. its address and port number. The standard implementation does not assume replication, so every object is registered only once, at the server that provides its implementation. In an environment supporting replication the registry is aware of object replicas and returns stubs pointing to *the nearest* server. "Near" may be defined differently: based on the network delay, latencies, server load, and other factors.

The process of object creation and registration needs to be done atomically in order to avoid conflicting name resolutions at different servers. These operations are integrated within the consistency management. Successful object creation guarantees that there was no other object with a given name before that creation.

Remote references may be passed by value between shared objects as method parameters. When such a reference is transfered to another server, its representation is updated to point to that server, if the server holds a replica of that object. This is necessary to properly handle nested calls: it was assumed that replicas of objects that are referenced are available locally, at the same server.

Replication domains. Objects in RRMI are allowed to call methods of other objects because they are expected to be available locally. In the worst case, when there is some kind of link between every object in the system, the assumption will require total replication of all objects. Obviously this does not lead to optimal efficiency, because some replicas will be updated even when they are not accessed by anyone. We introduce a new concept of *replication domains* [10] in order to alleviate the problem. A replication domain essentially is a set of shared objects that may reference each other, and a set of servers that maintain their replicas. There are no references between objects from different domains. Objects from different domains may be replicated on a different set of servers. The concept reflects different requirements of concurrent applications that run on top of RRMI. Applications have different expectations concerning objects availability, and their consistency management. The programmer is responsible for defining replication domains and objects' associations to domains.

Transport layer. One of the most important factors that influence the availability of remote objects is the communication infrastructure that is used to access the objects or their replicas. We use two different transport mechanisms to communicate between processes. Remote invocations coming from clients are transfered using TCP connections. However, in contrary to the standard RMI implementation, we establish a single connection between a client and a server, and keep it open between consecutive calls. Java RMI uses a new TCP connection every time a method is called, which may result in a poor efficiency if the methods are issued frequently. On the other hand, a client is usually located at the same host as the server being contacted by the client, or on a host that is located close to such a server (e.g. within the same LAN). As a consequence, the connection overhead introduced by TCP protocol remains low, and its reliability gives the client precise monitoring of the link status.

Communication between servers maintaining objects' replicas is carried out differently. The servers are located on different nodes, usually distant. In order to facilitate and optimize the communication we have developed a new message passing library dedicated for the Java environment. It is called JMPL — *Java Message Passing Library*. The library was used in our previous project Jash [7]. RRMI uses the second revision of JMPL which is substantially rewritten and enhanced. Currently the library is much more flexible, supports dynamic reconfiguration, uses new features of the Java language, and is ready to work in an unreliable environment. JMPL uses UDP as the main transport protocol, but provides typical features of the TCP protocol like: reliable communication, ordering, flow control, message fragmentation, or windowing.

3   Consistency Maintenance

Replication introduces the problem of consistency when replicas are allowed to change. The problem becomes more complicated when modifications can be done concurrently at different replicas, i.e. in a multi-master model. Concurrent conflicting writes should not be allowed, or, in an optimistic approach [11], detected and resolved.

Our approach in RRMI system at this stage of development is close to pessimistic replication. It means that it is not possible to submit concurrent, conflicting object updates. Such updates would require a roll-back mechanism that is necessary to reschedule non-commutative operations in the case of a conflict. However, RRMI tries to postpone update propagations to maximum extent in order to minimize delays when submitting modifications to a shared object.

Traditionally, one of the most important goals of systems supporting replication in a distributed system was to hide the fact of replication from the user, i.e. to make the replication as transparent as possible. Although possible, this transparency comes at a cost. Usually it requires a kind of *strong consistency* where all replicas are updated or invalidated at the same time. Such protocols tend to be highly inefficient, and are only useful in systems with a high read/write operation ratio. Other models have been developed and they relax somehow consistency constrains at the cost of user awareness of the replication. In RRMI we have used *entry consistency* (EC) [12] as the basic consistency model. Entry consistency is a kind of *relaxed consistency* that uses special synchronization operations on *locks*. These operations are used explicitly by the user to instruct the underlying replication managers when and where a given portion of data should be updated. There are two synchronization operations: *acquire* and *release*, that indicate blocks of code when a replica should be updated. EC additionally distinguishes between acquires in *shared* mode and acquires in *exclusive* mode, in order to make possible concurrent reads of objects. Another unique feature of EC compared to *release consistency* (RC), is that it requires explicit association of shared objects to locks. In RC the association is implicit and may be deduced by analyzing the code. RRMI is a public system that is designed to be used by different applications at the same time. For this reason, the associations of locks to shared objects should be explicit, and available for applications communicating only with the system.

The consistency management layer intercepts method invocations in stubs, especially in server stubs. The user is expected to select a lock for every shared object during object creation. Several objects may be synchronized using a single lock. Consider the following example code:

```
Lock l = new Lock("mylock");
SharedObject x = new SharedObject("myobject1", l);
SharedObject y = new SharedObject("myobject2", l);
l.acquire(Lock.RW);
x.method1(argument1);
y.method2(argument2);
l.release();
```

Here, the lock was explicitly acquired. The same processing may be also done using implicit synchronization. The server stub will detect that the appropriate lock has not been acquired, and will acquire it before calling the method, and release after the method returns. This model is close to strong consistency, and is attractive for its full transparency. However, it does not allow the user to optimize the processing in order to decrease the number of synchronization operations. In the example, there is no synchronization operations between method1() and method2(), neither explicit nor implicit. Explicit synchronization may be necessary in deadlock prevention. Another possible usage of this technique is an atomic update of a set of objects, an operation not possible with strong consistency without implementing additional synchronization mechanisms.

Processing history. Modifications of objects are not sent directly to other servers right after their submission. RRMI uses operation transfer which requires gathering information about the processing. This is done by maintaining *histories* of processing at every server. Servers register all top level local method invocations submitted by clients, and incorporate parts of histories from other servers sent during synchronization operations (namely acquires). The histories are applied at servers in causal order. Causal order is preserved thanks to vector clocks that represent number of synchronization operations (acquires) issued at servers.

4 Conclusions

In this paper we have presented a new infrastructure for building distributed applications that communicate through remote objects. The construction of the system is currently being finalized, and we expect to get more practical experience and carry out some performance evaluation tests. Future work will be concentrated on introducing other consistency models, especially those which are more client-oriented, and on increasing dependability of the system through object recovery in case of server crashes.

References

1. Sun Microsystems: Java Remote Method Invocation Specification. (1998)
2. Wollrath, A., Riggs, R., Waldo, J.: A distributed object model for the Java system. In: 2nd Conference on Object-Oriented Technologies & Systems (COOTS), USENIX Association (1996) 219–232
3. Gribble, S.D., Welsh, M., Brewer, E.A., Culler, D.: The MultiSpace: An evolutionary platform for infrastructural services. In: In Proc. of the 16th USENIX Annual Technical Conference, Monterey, CA, USA (1999) 157–170
4. Montresor, A.: The JGroup distributed object model. In: Proc. of the Intl. Working Conference on Distributed Applications and Interoperable Systems, Helsinki, Finland (1999) 389–402
5. Maassen, J., Kielmann, T., Bal, H.E.: Parallel application experience with replicated method invocation. *Accepted for publication in* Concurrency and Computation: Practice and Experience (2001)
6. Narasimhan, N., Moser, L.E., Melliar-Smith, P.M.: Transparent consistent replication of Java RMI objects. In: Proc. of the International Symposium on Distributed Objects and Applications (DOA'00), Antwerp, Belgium (2000)
7. Brzeziński, J., Sobaniec, C., Szychowiak, M., Wawrzyniak, D.: Jash: A platform for sharing data in distributed Java applications. In: Proc. of the Int'l Conf. on Parallel and Distributed Processing Techniques and Applications (PDPTA '98), Las Vegas, USA (1998) 1430–1437
8. Amza, C., Cox, A.L., Dwarkadas, S., Keleher, P., Lu, H., Rajamony, R., Yu, W., Zwaenepoel, W.: Treadmarks: Shared memory computing on networks of workstations. IEEE Computer **29** (1996) 18–28
9. Pedone, F., Wiesmann, M., Schiper, A., Kemme, B., Alonso, G.: Understanding replication in databases and distributed systems. In: Proc. of the 20th International Conference on Distributed Computing Systems (ICDCS'00), Taipei, Taiwan (2000) 464–474
10. Sobaniec, C., Wawrzyniak, D.: Using domains to support replication in a distributed shared object system. In: Proc. of Int'l Conf. on Parallel and Distributed Computing and Networks (PDCN 2004), Innsbruck, Austria (2004)
11. Saito, Y., Shapiro, M.: Optimistic replication. Technical Report MSR-TR-2003-60, Microsoft Research (2003)
12. Bershad, B.N., Zekauskas, M.J., Sawdon, W.A.: The Midway distributed shared memory system. In: Proc. of the 38th IEEE Int'l Computer Conf. (COMPCON Spring'93). (1993) 528–537

Discovery of Web Services with a P2P Network

Florian Forster and Hermann De Meer

University of Passau
Faculty of Mathematics and Computer Science
Innstrasse 33
94032 Passau
{forsterf,demeer}@fmi.uni-passau.de

Abstract. In the concept of Web Services, Universal Description, Discovery and Integration is still the weakest part. As a central instance, it does not easily scale to a growing number of users and lacks acceptance by the industry. In Peer-to-Peer Networks, which are highly popular, discovery of resources is one of the strongest parts. A central registry is not required when integrating Web Services in a Peer-to-Peer network. Each Web Service is responsible for itself and the Peer-to-Peer Network provides the framework for discovery, publication and registration of Web Services. This paper shows, how both technologies fit together and gives details on both structure and design of the Peer-to-Peer network. This results in a more feasible solution than the central Universal Description, Discovery and Integration infrastructure and provides an easy way for registration, publishing and discovery of Web Services.

1 Introduction

Web Services (WS) and Peer-to-Peer (P2P) networks have evolved to popular technologies over the last few years. P2P was mostly pushed by users of popular software like the MP3-sharing program Napster[1] or the distributed analysis program SETI@Home [14]. WS are a concept coming from global business players like Microsoft, IBM or Sun.

P2P focuses on a de-centralized architecture of equal peers for the purpose of sharing resources of the participating peers, whereas WS are using a Client/Server approach, and have the intention of Business-to-Business (B2B) or Business-to-Costumer (B2C) integration.

WS use the markup language XML [1] for communication via a standardized protocol called SOAP [3]. Description of the services is done by the Web Service Description Language (WSDL [4]) . Publishing and discovery relies on the concept of Universal Description, Discovery and Integration (UDDI [2]). The goal of WS is to provide a common interface for Internet applications enabling everybody who implements software to rely on certain standards. With the use of the Web Service Flow Language (WSFL [15]), virtual organizations can be

---

[1] http://www.napster.org

M. Bubak et al. (Eds.): ICCS 2004, LNCS 3038, pp. 90–97, 2004.
© Springer-Verlag Berlin Heidelberg 2004

built up by making use of different WS. Discovery of adequate business partners plays an important role in the creation of virtual organizations.

P2P networks focus on the possibility for users to share their resources. Each peer who joins the network has to register itself, and the provided resources. Hybrid P2P networks use a central instance for registration and discovery, whereas in pure P2P networks this is done by active announcement to the network. A lot of research and development effort is put in discovery algorithms for P2P networks, which resulted in various strategies and P2P network topologies.

P2P networks enforce the sharing of resources, and WS took the first step in the creation of virtual organizations by the invention of WSFL. Creating a P2P network of WS provides an easy way of sharing resources for organizations. In this paper, the P2P network organizes the discovery and sharing of resources, whereas the WS provide and describe the resources shared. This combination also fits with the concept of *The Grid*[2]. The first step will be to show, how discovery, publishing and registration of WS can be done in a P2P network. This paper shows how to distribute information regarding WS among the nodes in a P2P network, thus making the centralized component of the UDDI concept obsolete.

After giving some references to related work in section 2, section 3 provides a brief overview to the concepts of WS and P2P. Section 4 provides a proposal for a P2P architecture replacing UDDI and gives detailed information about publishing (Section 4.3), discovery (Section 4.4) and registration (Section 4.2) in the network. Finally, section 5 summarizes the ideas.

2   Related Work

The basic idea that WS could benefit from P2P networks was mentioned in [5], [6], [7],[12]. A modified Gnutella Client, capable of searching for WSDL files has been implemented by the ProSa Discovery Project[8]. In [9], a general purpose query language was proposed, allowing complex searches suitable for WS. [10] proposed the *Web Service Discovery Architecture* which provides a basic framework for discovery of WS with a P2P network by evaluating different technologies and standards. Using JXTA as an implementation framework resulted in a basic P2P network for WS in [11]. This approach also focused on using WSDL, meaning that complex search queries are not possible in this network. The METEOR-S System [17] offers a scalable infrastructure of registries for semantic publication and discovery of WS. The Web Service Inspection Language [16] is a complementary concept to UDDI which does not need a central infrastructure and therefore is not taken into account in this paper.

---

[2] http://www.grid.org

3 Motivation

3.1 Web Services

WS basically provide a framework for interoperability between applications, and thus sharing of resources or information, over the Internet. Therefore, the concept of WS defines the roles provider, consumer and registry. Obviously, provider and consumer are absolutely necessary for a WS, but the registry entity is optional. Provider and consumer could contact each other without using the registry.

Compared with WSDL and SOAP, the UDDI standard still is in a relatively immature stage of development. Although IBM, Ariba, SAP and Microsoft are offering an infrastructure for a central UDDI registry, this entity is not widely used. As for the moment, most business still rely on direct contact to their partners. One reason my be the susceptibility of UDDI to security threats. Business simply do not seem to trust the UDDI registries and thus do not use them. Furthermore, the registries tend to be outdated because the update process of the entries in the registry is still often done manually. In the traditional approach, UDDI was designed as a single server, which leads to have a single point of failure for discovery of Web Services. IBM, Ariba and Microsoft tried to solve this by creating several servers, replicating each other. This approach still faces feasibility problems like many Client/Server Architectures. For WS, being one of the future technologies, it is very likely that there will be a huge number of WS in a few years.

That is why an infrastructure for UDDI registries has to be created for WS very fast. It is very unlikely that the growth of the UDDI infrastructure satisfies the upcoming requirements of WS. The capacity of the UDDI infrastructure would enforce an upper bound to the number of enlisted WS, and thus limit the growth of WS. Nevertheless, a global UDDI registry could help to create new B2B or B2C relationships by providing detailed information about a business and their services, allowing complex search queries.

3.2 P2P

P2P is a concept for sharing resources, mostly files, amongst peers participating in a network. P2P is divided into the concept of hybrid and pure P2P. Whereas in hybrid P2P Systems, a central entity, which provides a registry and helps in the discovery process, is available, pure P2P Systems consist of equal peers only. By joining the network, a new peer automatically registers itself to the network, either by signing up at a central entity or by announcing its presence to the network. P2P networks rely on the availability of a good discovery mechanism. Due to the fact that peers join or leave the network often, peers cannot establish direct contact without discovering one another first. Peers mostly provide very simple information during the registration process. Because of this, discovery of information has to stick to simple search queries.

Compared to UDDI, P2P has two important advantages. First, it is widely accepted among users. Second, the registration to a P2P network is done automatically and is very simple in nature. In the next section, a P2P architecture for discovery and publication of WS is presented.

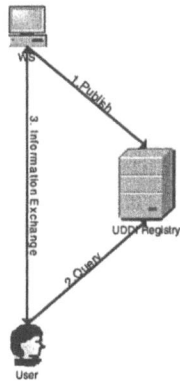

Fig. 1. Service Discovery with UDDI

4   Web Service Publishing and Discovery with P2P

4.1   Discovery and Publishing in WS and P2P

Discovery and publishing in WS is done by using a central entity. A provider of a WS submits several description files to a UDDI registry. The process of publishing is clearly defined by the *UDDI Publishing API*. Consumers searching for services send a query to the UDDI registry. The functionality of the search is defined by the *UDDI Query API*. Due to the fact that UDDI stores many description files, complex search queries can be implemented in the client software. After retrieving information from the UDDI registry, the consumer can contact the provider of the WS and get the requested information (see Fig. 1).

Discovery and publishing in a P2P network can be done in various ways. In a hybrid P2P network, it is very similar to the process of WS. A peer joining the network registers itself at a central instance, which manages all the resources in the P2P network. Search queries are sent to the central instance, which replies with several sources providing the requested information. In a pure P2P network there is no such central instance. Search queries are either broadcasted to a limited neighbourhood or sent to a specific set of peers. The usage of a central instance guarantees a success rate of 100%, meaning that if a ressource is present in the network, then it will be registered at the central instance and thus be found by queries. Guaranteeing a 100% success rate in a pure P2P network results in heavy traffic. Each search query has to be sent to an overwhelming part of the network to give the success assurance of 100%. But a hybrid P2P network depends on the central instance. If the central instance fails, no more discovery is possible within the network, whereas the pure P2P network does not have to face this problem due its distributed design.

As P2P networks are widely accepted and have proved to be feasible even for a large number of users, they could be an appropriate replacement for the UDDI infrastructure. The P2P network has to provide an easy way of registration and publishing of information. Furthermore, it has to give a nearly 100% assurance

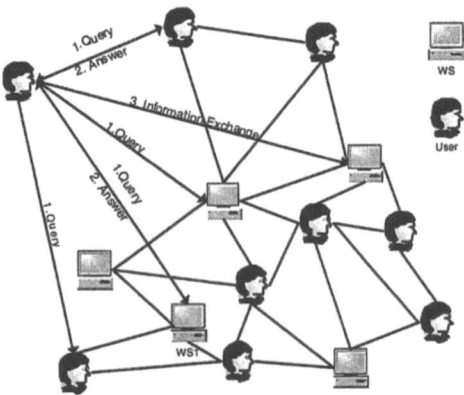

Fig. 2. Service Discovery with P2P

that a WS is found in the network by using appropriate queries, and has to provide the possibility for complex search queries, in order to be suitable as a replacement for UDDI. In this paper every WS is its own registry, similar to peers in pure P2P networks. These hosts are the essential part of the P2P network replacing UDDI. The P2P architecture consists of two different layers. On the first layer there are the peers providing information about WS (WS-Peer), whereas the second layer consists of consumers (User-Peer) joining the network for discovery purpose. Publishing of information is done by simply joining the P2P network. An algorithm like the one in Gnutella [13] is used for discovery. For better performance, more sophisticated algorithms for search have to be implemented in the future. A peer searching for a specific WS, sends its query to the network. The query is then forwarded in the network until its TTL[3] is reached. In case of a match to the query, a peer replies to the requester. After that, the requester can contact the peer providing the requested information directly (see Fig. 2). Details on publishing, discovery and registration are given in the next sections.

4.2   Registration of a Peer

For joining the network, a peer has to know the address of a WS-Peer for boot-strapping purposes. The joining peer is provided with a mixed set of addresses of User-Peers and WS-Peers which becomes the starting set for its neighbourhood. In case of a WS-Peer, only other WS-Peers are considered as neighbours. Each of the neighbours is contacted and the time-stamp of the response is stored along with its address. If a peer forwards a search query or gets responses to a search query, it adds the addresses of the peers involved in this process to its neigh-bourhood along with the time-stamp of the query received last. In case of a full neighbourhood set, the oldest peer is deleted from the set. For the sake of discov-ery, at least one of the peers in the neighbourhood has to be a WS-Peer. If none of the User-Peers has contact to the WS-Peer layer no WS at all can be found.

---

[3] Time to Live

Furthermore, a WS-Peer announces the provided information about its own WS, meaning **key,name** and **description** for each local file to its neighbourhood on a regular base. Each peer in the neighbourhood then adds the information to its cache (see section 4.4). Due to the dynamic nature of the neighbourhood, over time, more peers will have information pointing directly to the specific WS-Peer and thus increasing the chance for this WS-peer to be found with a search query.

As seen, the network consists of two layers. User-Peers are connected to each other and to WS-Peers, whereas WS-Peers are only connected to WS-Peers. By making sure that every User-Peer has at least one WS-Peer in its neighbourhood, a search query reaches the WS-Peer-layer. Due to the announcement of information and the dynamics of the neighbourhood, information regarding WS is spread along the caches in the P2P network. User-Peers involved in a search are only forwarding queries or, in case of a local match, provide a direct link to the WS-Peer. As shown, registering peers is as easy as it is in pure P2P networks

4.3   Publishing Information

For the complete description of a WS, UDDI relies on several models. These models are used to provide detailed information about business. For publishing its information a provider joins the P2P network and provides the files containing these models to the network. The uniqueness of **BusinessKey, ServiceKey** and **BindingKey**, each identifying a specific model, is a problem as there is no central instance assuring the uniqueness of the key. Usage of a hash-function mapping the static IP address of a provider to an unique identifier is sophisticated enough for this purpose. The **tModelKey** can be generated by using a hash-function on the description of the *tModel*. Besides the adjustment of the creation of an unique identifier for the description files, no further changes have to be made to the UDDI standard for the description files.

UDDI encourages WS covering the same topic to use the same *tModel* for providing a common interface to clients. Because there is no central instance for the storage of such *tModels* the idea of similar services using the same model has to be solved. For example, if a new company is planing to provide a weather service via a WS on its own, it has to query the P2P network for an appropriate *tModel* first. If it does not find one it can create its own and introduce it to the P2P network. If there is already an appropriate *tModel* in the network, the *bindingTemplate* has to contain a reference to this remote file.

Each participating company in the P2P network is now able to publish their information in the network. Remote references to models are solved by querying the P2P network for the unique ID and then storage of the results in a local cache for further requests. There is no additional effort for WS providers, compared to UDDI, by using a P2P network. The same files have to be created for using a P2P network as a replacement for UDDI.

4.4   Finding and Caching of Information

Searching for a business makes use of the function *find_business*. Each peer receiving the query parses the local files, using a SAX or DOM parser, and searches the elements **businessKey, name** and **description** for a match to the query. After checking the local repository it checks its local cache. The cache is split into four sections. There is a cache for business, services, bindings and tModels. Each cache contains a complex structure which includes, in the case of the business-cache, the elements **businessKey, name, description** and a reference to the source of this file. In the case of a match in the local repository, the peer sends the information **businessKey, name, description** to the requester. In the case of a match in the cache, it sends an answer which redirects the requester to the original source. The requester then caches the information itself and contacts the original source getting requested information. If the original source is not available, the requester informs the peer of the old entry in the cache. In case of using one of the *get*-functions the requester would first use the proper *find*-function with the unique key as parameter. This search is going to result in only one match and the requester is then able to request the file containing detailed information from the providing peer directly. Finding and getting information regarding services, bindings or tModels works the same way.

This strategy enables the peers to fill their local caches, making them able to reply to search queries. The cache is kept up-to-date by the method described in the last paragraph. Although each peer is able to answer queries, the complete file is only available at the WS provider, which limits the possibilities of abuse. Due to the fact that peers can search their local repository, complex search queries are possible within the network. After showing that the P2P network is able to provide the same functionality as the *UDDI Query API*, the next section provides details on the registration process for peers.

5   Conclusion

WS still lack an easy way of registration and discovery. Even though much effort is put into the concept of UDDI, UDDI is still not accepted amongst providers of WS. This is because of the fact that the registration process in UDDI is complicated and updates are still often done manually. Furthermore, as a central instance for registration, publishing and discovery, UDDI may face feasibility problems in the future, as the number of WS is very likely to grow fast in the near future. This paper showed a way on how to distribute the information regarding a WS in a P2P network by providing two architectural layers. For this purpose techniques and algorithms of P2P networks were used. Under the new scheme, the description files for WS remain at the host providing the WS. This host participates in a P2P network and registers its service within the network. Consumers searching for WS also join the network. By using SAX or DOM parsers for local files, complex search queries are still possible for the clients. This results in a P2P architecture suitable for replacing the central UDDI registry, helping to create virtual organizations for the purpose of resource sharing. The success rate for discovery depends on the algorithm used. Further work has to be done for finding an appropriate one.

References

1. Tim Bray et al. (2000): Extensible Markup Language (XML) 1.0 (Second Edition) in *W3C Recommendation 6 October 2000*, W3C, http://www.w3.org/TR/REC-xml

2. Ariba, IBM, Microsoft (2000): UDDI Technical White Paper, UDDI Project, http://www.uddi.org/pubs/Iru_UDDI_Technical_White_Paper.pdf

3. Martin Gudgin et al. (2003): SOAP Version 1.2 Part 1: Messaging Framework in *W3C Recommendation 24 June 2003*, W3C, http://www.w3c.org/TR/2003/REC-soap12-part1-20030624/

4. Erik Christensen et. al (2001): Web Services Description Language (WSDL) 1.1 in *W3C Note 15 March 2001*, W3C, http://www.w3.org/TR/wsdl

5. Remigiusz Wojciechowski, Christof Weinhardt (2002): Web Services und Peer-to-Peer Netzwerke in *Schoder, Fischbach, Teichmann (2002): Peer-to-Peer*, Springer Verlag, 99–117

6. Jeff Schneider (2001): Convergence of Peer and Web Services, http://www.openp2p.com/pub/a/p2p/2001/07/20/convergence.html

7. Nathan Torkington (2001): Why you should care about P2P and Web Services at *Yet Another Perl Conference*, http://prometheus.frii.com/~gnat/yapc/2001-ws-p2p/

8. Olaf Gregor (2003): ProSa-Discovery-Project - Discovery von WebServices mit Gnutella, http://vsis-www.informatik.uni-hamburg.de/~1gregor/

9. Wolfgang Hoschek (2002): A Unified Peer-to-Peer Database Framework and its Application for Scalable Service Discovery, in *Proc. of the Int. IEEE/ACM Workshop on Grid Computing, Baltimore, USA, November 2002*, Springer Verlag, 126–144

10. Wolfgang Hoschek (2002): Peer-to-Peer Grid Databases for Web Service Discovery, in *Grid Computing: Making the Global Infrastructure a Reality (to appear), Editors: Fran Berman, Geoffrey Fox and Tony Hey*, Wiley Press.

11. Ludwig Mittermeier and Roy Oberhauser (2002): Ad-hoc-Web-Services durch P2P-Technologien, in *JavaSpektrum 04/2002*, SIGS Verlag, 28–39

12. Mike P. Papazoglou, Bend J. Kraemer, Jian Yang (2003): Leveraging Web-Services and Peer-to-Peer Networks, in *Johann Eder and Michele Missikoff (2003): Lecture Notes in Computer Science Vol. 2681*, Springer, 485–501

13. Gene Kan (2001): Gnutella in *Andy Oram (2001): Peer-to-Peer: Harnessing the Power of Disruptive Technologies*, O'Reilly 94–122

14. David Anderson (2001): Seti@Home in *Andy Oram (2001): Peer-to-Peer: Harnessing the Power of Disruptive Technologies*, O'Reilly, 67–76

15. Frank Leymann (2001): Web Services Flow Language 1.0, IBM Document, http://www-3.ibm.com/software/solutions/webservices/pdf/WSFL.pdf

16. Keith Ballinger et al. (2001): Web Services Inspection Language 1.0, http://www-106.ibm.com/developerworks/webservices/library/ws-wsilspec.html

17. Verma, K. et al. (2003): METEOR'S WSDI: A Scalable Infrastructure of Registries for Semantic Publication and Discovery of Web Services in *Journal of Information Technology and Management* (to appear).

Achieving Load Balancing in Structured Peer-to-Peer Grids

Carles Pairot[1], Pedro García[1], Antonio F. Gómez Skarmeta[2], and Rubén Mondéjar[1]

[1] Department of Computer Science and Mathematics, Universitat Rovira i Virgili
Avinguda dels Països Catalans 26, 43007 Tarragona, Spain
{cpairot,pgarcia}@etse.urv.es
[2] Department of Computer Engineering, Universidad de Murcia
Apartado 4021, 30001 Murcia, Spain
skarmeta@fcu.um.es

Abstract. In this paper, we introduce the concept of a structured peer-to-peer grid and present our contribution to this new world by means of DERMI, a wide-area decentralized object middleware. In addition, we focus on the design and implementation of a load balancing facility by using the functionalities provided by this middleware. We present two different approaches to achieve load balancing in our system: a completely decentralized solution by using the *anycall* abstraction, and a distributed interceptor-based one. Each of them can be used in a wide variety of scenarios, depending on our needs.

1 Introduction

Grid and Peer-to-Peer (P2P) Networks are two concepts which may seem different at first sight. There exist no exact definitions that allow us to compare their similarities and differences in an unambiguous way. Nevertheless, the general idea seems to split both worlds into those cases where infrastructure is used to allow seamless access to supercomputers and their datasets (Grids), and those which enable ad hoc communities of low-end clients to advertise and access the files on the communal computers (P2P).

Trying to merge the best of both worlds, the concept of a P2P Grid was devised [1]. A P2P Grid contains a set of services including the services of Grids and P2P networks and support naturally environments having features of both limiting cases. In P2P Grids' architecture, Web services play a very important role. In addition, there is an event service which links these Web services and other resources together. In P2P Grids, everything is a resource, and they are exposed directly to users and to other services. An important feature is that these entities are built as distributed objects which are constructed as services allowing their properties and methods to be accessed by a message-based protocol.

Normally, these P2P Grids have been built on top of unstructured P2P networks. This kind of networks at the extreme case are exemplified by the Gnutella-like systems, where all participants join the network in an anarchic way, and messages are flooded all over the network (normally messages are only sent a few hops away from

M. Bubak et al. (Eds.): ICCS 2004, LNCS 3038, pp. 98–105, 2004.
© Springer-Verlag Berlin Heidelberg 2004

the origin, thus being extremely inefficient when searching for a precise resource). Other alternative techniques to flooding include the use of Random Walkers, or expanding ring techniques. However, they do not elegantly solve the problem of resource location in a deterministic time slot, being it impossible to know how long it will take to find a concrete resource, if it is found.

Many research efforts have lately been directed towards building structured P2P networks, also called overlay routing networks, which are the base for constructing distributed hash tables (DHTs). These systems provide a self-organizing substrate ideally suited for large-scale P2P applications. A wide variety of decentralized services can be built on top of these overlays. Examples include application-level multicast, network storage, and content distribution. Being structured overlays, applications can locate any resource in a small number of network hops (typically $O(log\ n)$), and they require small-sized per-node routing tables.

Relating P2P Grids with structured peer-to-peer overlay networks, we have developed a structured P2P Grid middleware platform called DERMI [4]. DERMI is a distributed event-based object middleware built on top of a DHT-based P2P overlay network. It is a first step towards building a complete P2P Grid middleware platform, providing many services. DERMI provides a solid building block for the development of wide-area distributed applications. DERMI implements many of the functionalities required for a P2P Grid, including an event service which glues the distributed objects representing resources.

In this paper we introduce the term of structured P2P Grid, which we define as a P2P Grid whose network substrate is based on a structured DHT approach. This means that the P2P Grid itself is built on top of a Key-Based Routing (KBR) infrastructure. The main advantage of this approach is that these kinds of Grids can benefit from better message routing and its inherent locality properties.

The main contribution of this paper is describing two load balancing mechanisms, which we have designed by using the different services provided by DERMI.

The paper is structured as follows: we start in Section 2 with a brief description of DERMI and the main services it provides. Section 3 describes the mechanisms we have designed to achieve load-balancing using DERMI. Next, we discuss the contributions of DERMI to the P2P Grid world and continue exploring existent work related to our system. Finally, we present a summary and an outline of future research.

2 DERMI Overview

DERMI is a distributed event-based object middleware built on top of a decentralized DHT-based P2P overlay network. It models method calls as events and subscriptions under an underlying publish/subscribe layer also built on top of the overlay. We have built a prototype implementation using Pastry [5] as the routing P2P substrate and Scribe [2] as its message oriented middleware (MOM) infrastructure. However, any other DHT-based overlay platform like Tapestry, Chord or Kademlia could have also been chosen to achieve our objective.

DERMI mimics Java RMI programming paradigm and provides a *dermic* tool to generate both stubs and skeletons for our remote objects, which transparently manage object publications/subscriptions and its inherent notifications. Indeed, it provides many features already found in Java RMI such as remote exception handling, pass by value and by reference, and dynamic class loading.

Nevertheless, DERMI is quite different from RMI. Naturally, our solution is aimed at wide-area environments, whereas RMI can suffer from bottlenecks due to its centralized client/server nature. Besides, another existent difference is in the communication layer located between stubs and skeletons. While in conventional RMI, a TCP socket is established between the caller (stub) and the callee (skeleton), DERMI stubs and skeletons both use the event service by making subscriptions and sending notifications in order to communicate the method calls and their results.

DERMI provides a set of features, including object mobility, replication, caching, and discovery services. Further, it provides the typical synchronous and asynchronous remote procedure calls. However, two new invocation abstractions are provided, namely *anycall* and *manycall*.

The anycall abstraction benefits from the anycast primitive provided by the event service (Scribe). Therefore, an *anycall* means sending an anycast notification to the group, which will make the sender's closest member in the network to answer, as long as it satisfies a condition. This feature can be very useful for having object replicas of any service. If a client wants to have access to this service, it will *anycall* to the object replicas' group and one of them (the client does not care which one) will answer.

The manycall abstraction is a variation of the *anycall*. It basically works by sending a *manycast* message to the members of a group, i.e. the message is sent to several group members. It is routed further until enough satisfying members are found to satisfy a global condition.

Moreover, DERMI implements a decentralized object location service which allows us to locate object references in order to be able to execute remote procedure calls on them. Object handles are inserted in the DHT using as their key a hash on their object identifier. To look up an object we simply hash its object UID and subsequently obtain its handle. At this point we can already call its methods.

Finally, DERMI provides distributed interception facilities. Distributed interception allows us to apply connection-oriented programming concepts in a distributed setting. With this service, it is possible to reconnect and locate type-compatible interceptors at runtime in a distributed application. In order not to change the subscriptions of both interceptor skeletons and intercepted remote objects each time a new interceptor is added or removed, we extended our event service's classes (Scribe) so as to natively support this feature. This service is very useful for very dynamic Aspect Oriented Programming (AOP) environments.

For more information on DERMI's design and implementation of these services, please refer to [4].

3 DERMI's Load Balancing Mechanisms

Load balancing can be defined as *distributing processing and communication activity evenly across a computer network so that no single device is overwhelmed*. Load balancing is especially important for networks where it is difficult to predict the number of requests that will be issued to a server. For instance, busy web sites typically employ two or more web servers in a load balancing scheme. If one server starts to get swamped, requests are forwarded to another server with more capacity.

In distributed objects technology, the problem is the same: a resource which provides a service can become overwhelmed by many clients. In such case, this resource is clearly a centralized hot spot, becoming a single point of failure. To avoid such hot spots, replication mechanisms are taken into account by making the object server replicated. Nevertheless, this is not the solution, if there is no awareness of which replica to choose depending on its current load. Therefore, apart from having multiple replicas of the same service, we need a mechanism to choose one or another according to certain parameters which may include server load or network bandwidth, to name the most common ones.

As an example, we could imagine a service like a video broadcast. If we expect our video server to be joined by an important amount of clients, we should consider replicating it. However, this does not solve the problem yet. We require some mechanism to decide which replica is to be assigned to each client that connects to our video broadcast service, depending on certain criteria (in this case, network bandwidth, or server's node stress). Naturally, this assignment process must be made transparent to the client.

Having described the problem, and keeping in mind the services DERMI offers, we present the load balancing facilities we have built on top of our middleware as follows.

In DERMI we can use two different methodologies to achieve load balancing. Before describing them both, we outline the common background they share. Firstly we must have our object server replicated among other nodes. Notice that DERMI provides us with mechanisms to do this, which are transparent to the developer. By doing so we ensure that we have all the needed replicas joined into a multicast group, and that any changes within the object server will be easily propagated to their replicas by sending event messages to the group (which will be received by all members, in this case all replicas). Once we have the object server's replicas joined into a multicast group, we can opt for two load balancing options, which are described as follows.

3.1 Decentralized Load Balancing Management

Decentralized load balancing management refers to using the *anycall* primitive to achieve load balancing purposes. The working principle for this methodology is totally decentralized, meaning that we do not have any centralized authority that knows the load of each replica.

Instead, we will be efficiently asking for the client's nearest replica to serve our request. If the replica feels it is not overwhelmed, it will serve us. On the contrary, it will answer that our request cannot be served. In the latter case, and by following the *anycall* behaviour, another object replica will be contacted and so on.

One scenario where this load balancing scheme would fit very well is for example in a distributed simulation of some kind. Imagine we have a group of servers waiting for data chunks to analyze. These chunks could be provided to them by using the anycall approach: the chunk would be delivered to the closest server, which would check if it has already reached its maximum number of concurrent data chunks to analyze. If this is the case, it delegates processing to another server, and so on; else it starts the chunk's analysis. A diagram showing this process can be seen on Fig. 1.

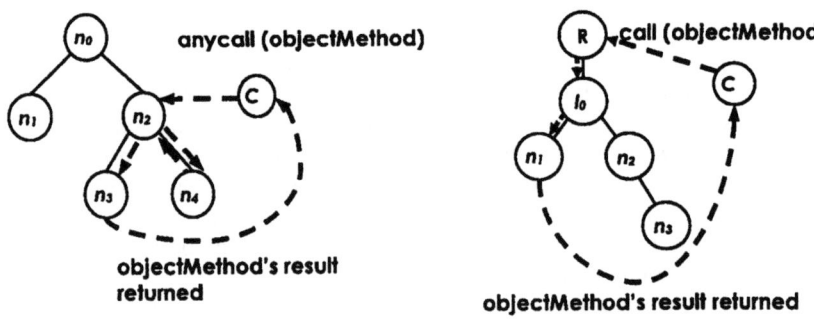

Fig. 1. Decentralized and Interceptor-based load balancing example

Notice this load balancing system is only valid when dealing with nodes that do not suffer network congestion. If nodes themselves would become overloaded in terms of network messages, they would be saturated and even unable to respond to our anycalls. This would prevent our decentralized load balancing mechanism to work properly.

3.2   Interceptor-Based Load Balancing Management

To deal with possible network congested nodes, we have another alternative to achieve load balancing: Interceptor-based load balancing management. In this case, we make use of our distributed interception mechanism. We will have a special interceptor node, which will be aware of each server replica load. This idea is similar to the *agent* concept found in NetSolve [1]. Particularly, we add an interceptor object to the multicast group of the object server's replicas. This means that when any client wishes to call a server's method, such call will be transformed into an event that will be sent to the multicast group itself, being automatically intercepted by the distributed interceptor.

The interceptor will, in turn, be aware of the load of each replica and make a direct call to the replica whose load is adequate to solve the client's request. All object replicas will inform periodically about their possible congestion state to the interceptor, by means of sending special events to the group. These events will be captured and stored by the interceptor itself, thus making it aware of the overall replicas' status. Notice the interceptor will in fact know about each replica's capacity for serving requests, and will decide which will serve the current client's RPC. Naturally, this approach can incur in being the interceptor itself a single point of failure: if the interceptor node fails, load balancing fails as well. As explained in [4], we can use replication mechanisms for the interceptor itself, thus assuring correct load balancing even in case of failure.

Clearly, if any of the replicas becomes overloaded in terms of network messages, the interceptor would be aware of this fact, and it would thus route no more messages to that node. Notice in such case, load balancing would continue to work properly, in contrast to the problem we have explained in the decentralized alternative. However, in this case, we have an only entity which controls all the load balancing process. If we wish to distribute this task among all servers, we should use the first approach (decentralized).

One scenario where this kind of load balancing scheme proves to be useful is for the video broadcast example explained before. Clearly, the broadcast of a video stream could overwhelm a server node's network adapter, and it would thus be unable to respond to anycalls (if we used the decentralized load balancing system). Nevertheless, if we use the interceptor approach, the interceptor itself knows the state of all available servers and tells the client which the most appropriate for him to connect is. A scheme for this procedure is shown in Fig. 1.

4 DERMI and P2P Grids

In the introduction we have defined the term structured peer-to-peer grid and have stated DERMI as a middleware platform built on top of such architecture. However, does DERMI provide the majority of services that a P2P Grid should offer? The answer is mainly yes. However, it leaves out a few features that P2P Grids do clearly support. One of these is web services. Nevertheless, the implementation of web services in DERMI is relatively easy. Stub and skeleton generation can be modified so as their public interfaces are defined and described using XML; and they can, for example use SOAP as their access protocol, thus achieving the transition from pure distributed objects to web services. We consider DERMI's services (*anycall/manycall, decentralized object location, and distributed interception*) to be an interesting and innovative approach to the P2P Grid world. This kind of grids will take advantage of DHT properties such as locality awareness and a better routing infrastructure.

5 Related Work

Existent RPC-based solutions in the Grid world are typically built using the GridRPC API [6]. This API was designed to address one of the factors that hindered widespread acceptance of Grid computing – the lack of a standardized, portable and simple programming interface. Since the GridRPC interface does not dictate the implementation details of the servers that execute the procedure call, there exist different implementations of the GridRPC API, each having the ability to communicate with one or more Grid computing systems.

Two well-known implementations of GridRPC are: one which lies on top of NetSolve [1], and Ninf-G [1], which is a full reimplementation of Ninf on top of the Globus Toolkit [8].

In NetSolve, there exist 3 main entities: the *client*, the *server* and the *agent*. Clients and servers model the typical behaviour in RPC systems, whereas agents maintain a list of all available servers and perform resource selection for all client requests as well as ensuring load balancing of the servers. Notice the interceptor approach used in DERMI to solve load-balancing issues resembles of a NetSolve agent. Further, fault detection and recovery is managed in a way that is transparent to the user. The agent keeps track of the status of all available servers so that in case of a problem, the agent can choose a new server to handle the problem. In DERMI, if there exist several replicas of an object server. We do not need to have an agent object which monitors the status of these servers (in case we use decentralized load balancing management). If a server goes down, the others will be unaffected and so will be the clients' requests. Our object replicas belong to a multicast group and clients *anycall* to that group, meaning that any (or the most adequate, depending on certain criteria) of the remaining object servers will answer.

Ninf-G is a GridRPC system built on top of the Globus Toolkit. Globus provides Grid's lower-level features such as authentication, authorization, secure communication, directory services, and many others. Nevertheless, the Globus Toolkit alone is insufficient for programming the Grid at higher-level layers. Ninf-G is a full reimplementation of Ninf. This system provides a mechanism to *gridify* an application or a library by using the Ninf IDL language to generate the necessary stubs. The executable is registered into Globus' MDS, and Globus-I/O is used for communication between client and server.

In such GridRPC systems, no efficient one-to-many asynchronous calls can be made. They can surely be emulated by means of *n* sequential calls incurring in a penalty overhead. In DERMI we have the notion of a communication bus that allows us to have efficient one-to-many asynchronous calls by using the underlying event service.

Both DERMI and Ninf-G have a directory service. DERMI's directory service is totally decentralized; whereas in Ninf-G MDS is used. MDS consists of two layers of LDAP servers and Grid Index Information Service (GIIS), which manages project-wide information; and Grid Resource Information Service (GRIS) which is responsible for site local information.

The latest version of the Globus Toolkit (3.0) introduced the concept of *Grid Services*. There is a clear resemblance between these Grid Services and Web Services in P2P Grids. However, by definition a service is a *Grid Service* if and only if it adheres to the OGSI specification. In DERMI we could transform our pure distributed objects into web services, in the way we have described in section 4. In addition, we believe that the OGSI specification and the Grid community could benefit from the DHT properties such as locality and even the services DERMI offers such as *anycall* and the *manycall* abstractions.

6 Conclusions and Future Work

Peer-to-peer Grids have until now been constructed on top of unstructured P2P networks. This kind of grids presents very interesting features including web services and an event service, among others. However, we believe that they can benefit from the efficient routing primitives available in structured P2P networks built on DHT-based architectures.

We have defined the term of structured P2P Grid term as a P2P Grid whose P2P network substrate is based on a DHT topology. We have presented a structured P2P Grid in the form of DERMI, an event-based object middleware built on top of a DHT. We have also introduced two mechanisms to achieve load balancing in our architecture, which we believe that can be a solid building block for the development of future applications on top of P2P Grids.

As future work, we are on the way to compare both described load balancing approaches by means of simulation results. By doing so we expect to provide exact data on when to use the decentralized method, or the interceptor-based one. Additionally, we are currently researching on adding support for shared sessions and artifacts, which mean the first step in creating a collaboration framework to provide wide-area CSCW services.

Acknowledgements. This work has been partially funded by the Spanish Ministry of Science and Technology through project TIC-2003-09288-C02-00.

References

1. Berman, F., Fox, G.C., and Hey, A.J.G., *Grid Computing: Making the Global Infrastructure a Reality*, John Wiley & Sons, Ltd, England, 2003.
2. Castro, M., Druschel, P., et al, "Scalable Application-level Anycast for Highly Dynamic Groups", *Proc. of NGC'03*, September 2003.
3. DERMI Website, http://ants.etse.urv.es/DERMI
4. Pairot, C., García, P., Gómez Skarmeta, A.F., "DERMI: A Decentralized Peer-to-Peer Event-Based Object Middleware", to appear in the *Proceedings of ICDCS'04*, Tokyo, Japan.
5. Rowstron, A., and Druschel, P., "Pastry: Scalable, decentralized object location and routing for large-scale peer-to-peer systems", *IFIP/ACM International Conference on Distributed Systems Platforms (Middleware)*, pp. 329-350, November 2001.
6. Seymour, K., Nakada, H., et al, "GridRPC: A Remote Procedure Call API for Grid Computing", *Proc of GRID 2002*, Springer, Baltimore, MD, USA, pp. 274-278.
7. Terpstra, W., Behnel, S., et al, "A Peer-to-peer Approach to Content-Based Publish/Subscribe", *Proc. of DEBS'03*, 2003.
8. The Globus Toolkit, http://www.globus.org.

A Conceptual Model for Grid-Adaptivity of HPC Applications and Its Logical Implementation Using Components Technology

A. Machì and S. Lombardo

ICAR/CNR Department of Palermo
{machi, s.lombardo}@pa.icar.cnr.it

Abstract. Today grid middleware is complex to be used, the development of grid-aware applications is error-prone and the Virtual Organization grid paradigm contrasts with traditional High Performance Computing (HPC) optimisation strategies relying on resource stability and known cost models. The authors analyse several aspects of grid adaptivity, and identify 4 roles: the active resource/execution manager, the proactive resource administrator, the reactive quality-service coordinator, the passive resource coordinator. They present a hierarchical model for a component-based grid-software infrastructure in which the resource administrator and the resource coordinator roles are assigned to grid middleware and the quality-service role to HPC skeletons. Roles interactions through interfaces are described for a component based infrastructure implementation. The resource administrator mimics functionalities of components containers of service-oriented architectures. The resource coordinator manages the life cycle of sets of processes over a pool of grid resources. It offers to upper infrastructure layers a Virtual Private Grid façade, simulating a processor cluster facility.

1 Introduction

The computational Grid paradigm defines a flexible, secure, coordinated large-scale resource-sharing model. Its focus is on large-scale problem-solving in dynamic, multi-institutional Virtual Organizations [1].

High performance computing has been, instead, traditionally oriented to performance optimisation of proprietary resources on local or wide area networks. Optimisation exploits knowledge of management policies at any level (computational models, resource connection patterns, cost models of the processor interaction graphs). In particular, the structured parallel programming approach has embodied such knowledge into patterns for the management of set of processes described by notable Directed Graphs, called *skeletons* and *parmods* [2-3]. Skeletons are automatically coded by parallel compilers to keep parallel efficiency and software portability high, while maintaining parallel programming difficulty low.

Code developed with such a structured approach for environments mapped on static networks of resources, managed with stable policies of exclusive resource allocation or partitioning, is inefficient on wide-area networks of dynamically discoverable and shareable resources. Resource unreliability, intrinsic in the Virtual Organization model, deceives any forecast based on cost models.

M. Bubak et al. (Eds.): ICCS 2004, LNCS 3038, pp. 106–113, 2004.
© Springer-Verlag Berlin Heidelberg 2004

One approach to such a problem is to develop self-adaptive parallel coordination patterns, where some node in the application process graph maintains awareness of past grid nodes performance statistics and of grid resources present status to optimally adapt process coordination. For instance, grid-awareness can be used to steer load balancing by means of optimal mapping of virtual processes over physical nodes according to their effective performance [4]. Otherwise, it can be used for substituting faulty nodes in a process graph or to redistribute data stream processing among workers of a farm when any worker is not honouring its performance contract [5].

Programming adaptivity for each coordination pattern is a complex and error-prone activity and makes porting of legacy (structured) parallel code hard.

We propose to adopt a hierarchical programming approach in which grid adaptivity is distributed among various layers of the software environment, playing different roles: the execution environment (active resource/execution manager), the application coordination middleware (proactive resource administrator), the application component layer (reactive quality-service coordinator) and the platform management middleware (passive platform coordinator). Clear separation of roles allows independent implementation of layers, as well as easy maintenance and run-time substitution of grid infrastructure components.

Specifically, adaptivity is factorised into the following tasks: a) discovery and reservation of grid nodes and services and definition of application virtual process graph (task assigned to the execution environment); b) optimal mapping of application process graph (assigned to the application coordination middleware); c) load balancing for nodes of the actual instance of the process graph (assigned to application components); d) monitoring of graph physical process set and (re) configuration of their ports (assigned to the lower middleware layer).

Two patterns are defined: the *Virtual Private Grid (VPG)*, a passive facade pattern which hides management of physical nodes and processes on the grid, and the *Grid-aware Component Administrator*, a reactive pattern which hides actual management of application components constrained to honour Quality of Service criteria.

The remainder of this paper is organized as follows. Section 2 sketches the hierarchical adaptivity model. Section 3 describes the functionalities of logical components implementing the model. Sections 4 and 5 expose current status of implementation of the runtime support layer prototype and research perspectives.

2   A Hierarchical Conceptual Model for Grid Adaptivity

Let's represent an HPC application, at conceptual level, as a Directed Graph (*DVG*) describing interaction of Virtual processors. Let's also represent it, at implementation level, with the Direct Graph (*DPG*) of the Physical processing units that implement *DVG*. Each graph node represents an application functionality or resource (a data or event source/sink, a computation, a control or a service). Each arc describes a data/event transfer or a path. *DVG* nodes are assigned weights representing their expected processing load, while arcs are labelled with weights representing expected data throughput. *DPG* nodes are assigned weights representing their effective available power while nodes are labelled according to the effective bandwidth available. After virtual process coding and *DVG* graph composition, application life-cycle can be factored into five main phases: *discovery* and reservation of a meaningful

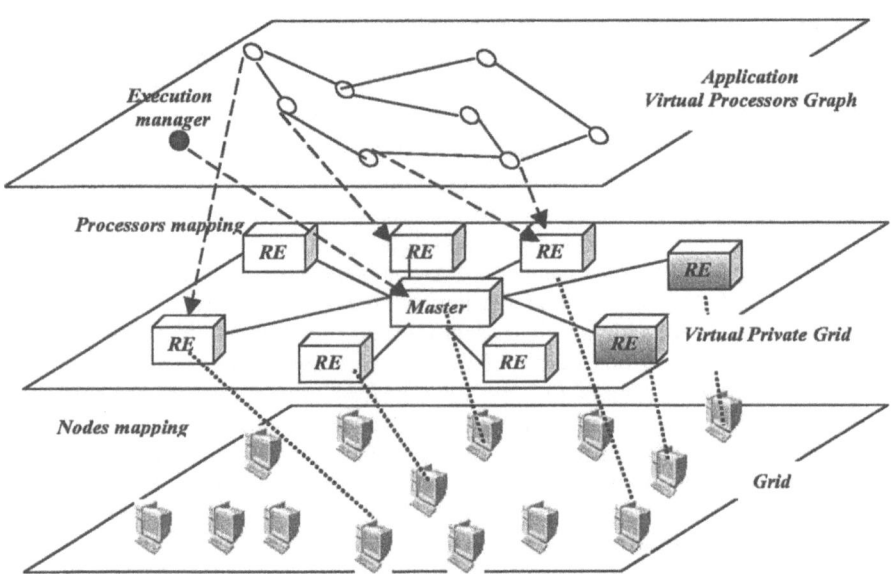

Fig. 1. Hierarchical representation of an HPC application over the grid. *Upper layer*: HPC application as a coordination of Virtual Processors *DVG* and an external Process Manager. *Intermediate layer*: Virtual Private Grid of computational resources (*Master* + *Remote Engines*) administrated by middleware on behalf of application to support adaptivity: *DPG* + spare nodes (dashed). *Lower level*: actual temporary mapping of VPG nodes over the grid.

set of proper physical resources, *mapping* of a *DVG* into a *DPG*, *initialisation* of *DPG*, *activation* plus execution of *DVG* processes on *DPG*, and *release* of allocated resources at completion.

HPC practice requires *DVG* execution to satisfy Quality of Service constraints often expressed as *performance contracts* to be honoured by the platform which enables *DVG* operation.

Adaptivity of the HPC application is required to cope with *events* affecting the composition of *DVG* resources set or even the behaviour of any of its elements:

– Fault of any processing node (graph disconnection)
– Fault of any connection between nodes (graph disconnection).
– Insertion of an additional resource (sub graph join)
– Redirection of a link to an external service (leaf node cut+leaf node join)
– Change of node effective power or link bandwidth (node/arc weight change)

Other events correspond to the occurrence of overall processing states:

– Performance contract violation (insufficient *DVG* node weights or non-optimal *DVG* to *DPG* mapping)
– Parallel inefficiency (excessive *DVG* node weights, or sub optimal *DVG* mapping).

Adaptation to such events is possible only if proper constraints are satisfied:

– Graph disconnection is a catastrophic event whose recovery from checkpoints is possible only if a mechanism for periodical backup of application status is expressly provided by application or run-time support [6].

- Adaptation to worker node fault or to node join is possible in master-slave implementations of skeletons like a *map* or a *farm* if a mechanism is provided to force any graph node to perform synchronous re-initialisation of the communication environment.
- Adaptation to dynamic redirection of a link to a leaf node implementing a service requires asynchronously forcing of the node to close its channel to the service and to open a new one.
- Adaptation to variance of worker node effective power or link bandwidth is possible for stream-parallel skeletons if knowledge of worker sub graph weights is expressly used for workload partitioning in the implementation template [7].
- Finally, adaptation to global performance states requires access to resource management privileges, normally reserved to a coordination process external to the process graph.

Each adaptivity case requires a different level of activity and of grid-awareness. It can be modelled using different actors playing hierarchically cooperative roles.

At topmost hierarchy level we find (re)selection of proper resources (nodes, services and *DVG*s). It requires grid discovery ability, detailed grid-awareness, reservation privileges and an adequate policy to coordinate resource provision and application Quality of Service. These tasks define the role of an *active resource/execution manager*.

At intermediate level we lay adaptive management of available grid resources for optimal execution of a defined application graph of processes. Taking advantage of self optimisation capability embodied in parallel skeleton templates, grid-adaptivity may be factorised in two roles: optimal administration of a pool of resources on behalf of a quality-application, and optimal administration of the set of resources assigned to a single application. The first role requires definition of each application quality in terms of a performance contract, selection of optimal subset for *DVG* to *DPG* mapping, monitoring of *DPG* performance and a policy for *DPG* reconfiguration. These tasks define a *proactive resource administrator* role, driven by a moderately complex ontology. The second role mimics load (re) balancing of physical processes over a cluster of virtually privates inhomogeneous resources labelled with their effective quality indexes plus partial reconfiguration of the processor graph in the event of resource modifications. These tasks define a *reactive quality-service coordinator* role, as the one implemented in some parallel skeletons.

At lowest level we lay effective monitoring of resource status, support for *DVG* to *DPG* mapping and re-mapping, detection and registration of events requiring attention and possible adaptation, tasks executable by a passive *resource coordinator*.

The hierarchical role model for HPC grid-adaptivity may be mapped to a component-based grid software infrastructure. The resource administrator and the resource coordinator roles are assigned to grid middleware while the quality-service coordinator role is assigned to skeletons. The *resource administrator* mimics functionalities of components containers of service-oriented architectures. The *resource coordinator* manages the life cycle of sets of processes on top of a pool of grid resources and offers to the upper layers a *Virtual Private Grid* facade simulating a processor cluster facility.

Next section describes a logical view of the software infrastructure as cooperation among software components playing model roles.

3 Functional Model of Middleware Infrastructure

Software component technology is a young programming-paradigm, even though its definition is quite old. Its aim is to enable the development of applications by composing existing software elements in an easy way. Among various definitions of the component concept, we report Szyperski's one [8]: *"A software component is a unit of composition with contractually specified interfaces and explicit context dependencies only. A software component can be deployed independently and is subject to composition by third parties"*.

The Globus Project has proposed the OGSA model for implementing and coordinating services over the grid. A component architecture for adaptive grid programming compliant with OGSA model been defined in [1]. A component architecture focusing on HPC grid programming is presently being developed by the italian *Grid.it* project [9]. The details of the architecture are a topic of current research In the working model [10] components expose their functionalities through a series of interfaces that differ for the interaction paradigm: (Remote Procedure Calls, streams, events, configuration). Interface signature together with implementation technology and communication protocol defines a *port type*. Components with same or compatible *port-types* can be connected together. In the framework of the *Grid.it* project, we exploit this component architecture to implement a graceful distribution of adaptivity roles, events and actions. Figure 2 shows the components implementing the architecture and their interactions. Each component is represented as an UML-package and its interfaces as UML-classes [11].

Passive roles provide slave functionalities through *provide-ports* (factory, service-provide and config), active roles use them via Remote Procedure Calls *use-ports* (discovery, system, service-invoke) as in CCA compliant frameworks [15]. Event ports of the run time support provide to the reactive application component an event bus for meaningful events registration and notification, to enable its reactive role.

The Execution Environment uses services exposed by Grid services, Component administrator and Application components.

The Component Administrator component exposes the following interfaces:

- **component_factory**: an extension of the factory design pattern [12] to the domain of distributed computing. It has the same goal of the OGSA Factory Port Type, even though it differs in some details. Factory services include submission of a Virtual Process Graph with its Quality-of-Service profile and VPG hardware resources creation and modification.
- **service_provide:** it exposes a set of functionalities about the status of submitted applications.
- **component_config:** modification of leaf nodes of *DVG* (external services binding).

The Component Administrator uses services of VPG master service-provide port to:

- *deploy* a set of processes (*DVG*) with related libraries on the VPG;
- *start* a set of processes (*DPG*) on the VPG;
- *retrieve* information about the status of managed hosts (nodes of the VPG) and about life status of started processes.

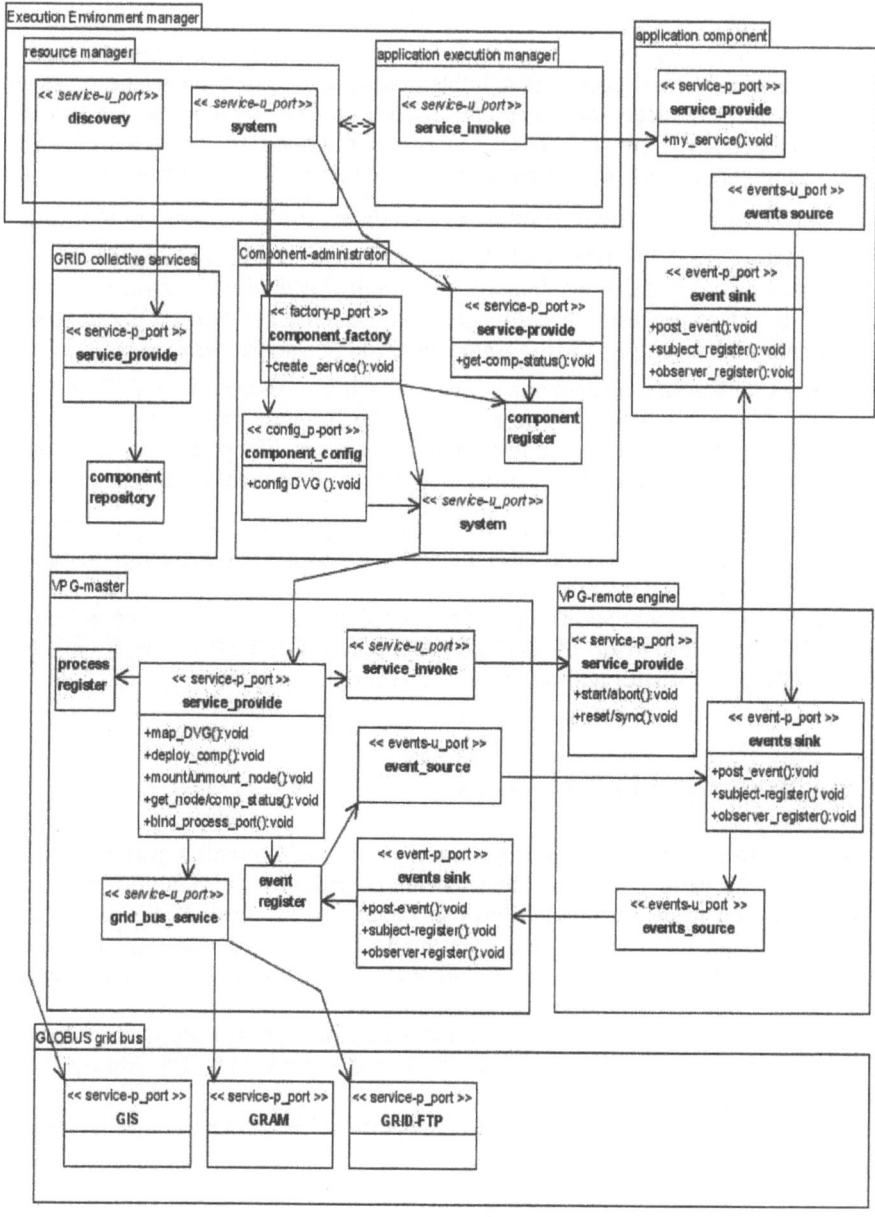

Fig. 2. Grid Layered Components Framework

- *retrieve* information about the proceeding status of *DPG* processes to detect Quality-of-Service violations;
- *notify* self-configuration requests to application-components.

4 Current Implementation of the VPG RunTime Support

A VPG Runtime Support prototype is presently being developed as a research activity of Project *Grid.it*. Actual component implementation is based on usage of several design patterns [13]: acceptor-connector, reactor, proxy, wrapper and adapter. A platform independent SDK for these patterns is provided by open-source object-oriented framework ACE [14], which enables code portability of the run-time system.

The *Virtual Private Grid* pattern is implemented by the following two components:

1. **VPG-Master**: the VPG front-end. It administers hosts by exposing methods for administrating node facilities (*mount, un-mount, keep alive, get-status*) and for controlling set of processes (*deploy, start, kill, delete, get-status*). It exposes this functionality by accepting XML-commands through a socket service-provide port.

2. **VPG-Remote Engine**: a daemon running on each host mounted on VPG as a slave for VPG-Master requests. It implements the remote run-time environment, administering, under master control, local processes lifecycle (*run, kill, status, clean*) and redirects events between VPG master and Application components.

The Master communicates with each Remote Engine in two ways: by ports-connection to invoke control of process lifecycle, and by event notification to delivery component to component event messages.

Grid nodes management and file transfer is implemented over the Globus Toolkit2 services: GRAM (for start-up of the Remote Engine), GridFTP for deploying *DPG* processes, GSI protocols for authorization and secure file transfer.

5 Conclusions and Future Work

The paper proposes a hierarchical programming approach in which grid adaptivity is distributed among various levels of the software environment, playing different roles. The approach constitutes a novel enterprise model for the grid. It allows a Virtual Organization to maintain application coordination while committing grid resources administration and coordination to middleware components or external services constrained by Quality of Service requirements.

Description of a resource coordinator middleware prototype has been presented. The development of a Component Administrator, aimed to provide a HPC library service is in progress [16].

Mapping of model hierarchical roles to OGSA model is a topic of current research.

Acknowledgements. Work supported by Italian MIUR Project FIRB-Grid.it, Work-package 8: Programming Environments and Workpackage 9 Scientific Libraries.

References

1. F.Berman, G.C. Fox, A.J.G.Hey: Grid Computing. Making the Global Infrastructure a Reality. Wiley 2003
2. J. Darlington, Y. Guo, H. W. To, J. Yang, *Parallel skeletons for structured composition*, In Proc. of the 5th ACM/SIGPLAN Symposium on Principles and Practice of Parallel Programming, Santa Barbara, California, July 1995, *SIGPLAN Notices* 30(8),19-28.G.
3. M. Vanneschi: The programming model of ASSIST, an environment for parallel and distributed portable applications. Parallel Computing 28(12): 1709-1732 (2002)
4. Z. Chen, J.Dongarra, P. Luszczek, and K. Roche "Self Adapting Software for Num. Linear Algebra and LAPACK for Clusters
    `www.cs.utk.edu/~luszczek/articles/lfc-parcomp.pdf`
5. C. Ferrari, Concettina Guerra, G. Canotti "A grid-aware approach to protein structure comparison" Journal of Parallel and Distributed Computing Vol 63 Issue 7-8 pp. 728-737.
6. P.D'Ambra, M.Danelutto, D. di Serafino, M.Lapegna:"Integrating MPI-based numerical software into an advanced parallel computing environment".Proc. of the 11th EUROMICRO Conf. on Parallel, Distributed and Netwok-based Processing, IEEE Publ., 2003, pp.283-291.
7. A. Machì, F. Collura "Skeleton di componenti paralleli riconfigurabili su griglia computazionale map e farm ".TR ICAR-PA-12-03 - Dec 2003.
8. Szypeski, C., Component Software: Beyond Object-Oriented Programming, Addison-Wesley, 1998.
9. M. Vanneschi "Grid.it : a National Italian Project on Enabling Platforms for High-performance Computational Grids" GGF Intern. Grid Summer School on Grid Computing Vico Equense Italy July 2003
    `www.dma.unina.it/~murli/SummerSchool/session-14.htm.`
10. [M. Aldinucci, M. Coppola, M. Danelutto, M. Vanneschi, C. Zoccolo "Grid.it component model "Project Grid.it WP8 Deliverable, Jan 2004.
11. J. Rumbaugh, I. Jacobson, G. Booch "The Unified Modeling Language Reference Manual" Addison Wesley 1998.
12. E. Gamma, R. Helm, R. Joyhnson, J. Vlissides "Design Patterns . Elements of Reusable Object-Oriented Software". Addison-Wesley.
13. Douglas C. Schmidt, Michael Stal, Hans Rohnert and Frank Buschmann "Pattern-Oriented Software Architecture: Patterns for Concurrent and Networked Objects" Wiley & Sons in 2000, ISBN 0-471-60695-2.
14. The Adaptive Communication Environment http://www.cs.wustl.edu/~schmidt/ACE.html
15. R. Armstrom, D. Gannon, K. Keahey, S.Kohn, L.McInnes, S. Parker, B. Smolinsk. "Toward a common component architecture for high-performance scientific computing". In Conference on High Performance Distributed Computing, 1999
16. S. Lombardo, A. Machì "A model for a component based grid-aware scientific library service." TR ICAR-PA-01-04 Jan 2004.

Global Discovery Service for JMX Architecture

Jacek Midura, Kazimierz Balos, and Krzysztof Zielinski

Deapartment of Computer Science
AGH University of Science & Technology
Al. Mickiewicza 30, 30-059 Krakow, Poland
{kbalos,kz}@agh.edu.pl

Abstract. Distributed computing in wide area networks requires special approach to solve the problem of communication between secure areas. Since there is no possibility to establish connection from the outside to secure region and no possibility to discover hidden resources (i.e. grids), special mechanisms are required to do this job. Article presents a new approach to solve these two problems by introduction of agent oriented design for establishing connections and resources discovery. Moreover, presented architecture stands for resources access transparency and is suitable for resources management in grids, especially using JMX (Java Management Extension) technology.

Keywords: Discovery service, resources management, firewall, JMX, MBean

1 Introduction

JMX (Java Management Extension) [1] specifies a framework for distributed resources management. It offers general architecture suitable for construction of monitoring systems and management applications. The spectrum of its applicability covers among others grids systems [4,5], active networks, and mobile systems. Full exploitation of JMX functionality is influenced by security policy and communication limitation imposed on Internet by firewalls.

Monitoring and management of distributed managed resources over Internet requires their dynamic discovery and binding to a management application at the first place. The built into JMX implementations standard discovery services do not work over firewalls. This is a basic constraint that reduces a scope of JMX applications over public networks.

The main goal of this paper is to present Global Discovery Service (GDS) for JMX architecture. This service is called global because it has been designed to work over firewalls. The proposed service could be used not only for resource discovery but offers also mechanisms to take control over resources hidden by firewalls. In the design and implementation phases every measures have been taken to make GDS scalable and lightweight.

The paper is structured as follows. In Section 2 major requirements of Global Discovery Service have been discussed. In Section 3 GDS concept and architecture have been described. GDS implementation details have been presented in Section 4. The paper is ended with conclusions.

M. Bubak et al. (Eds.): ICCS 2004, LNCS 3038, pp. 114–118, 2004.
© Springer-Verlag Berlin Heidelberg 2004

2 Requirements of Global Discovery Service

The JMX architecture described in [1,2] is fully operational in a wide area network which allows communication with RMI protocol via firewalls. The discovery services exploit broadcast communication that imposes even more harder to fulfill requirement on routers configuration over the public network.

Dynamic discovery can be divided into two categories: passive discovery and active discovery. In the first case client notifies the server that its state has changed. In second scenario there are special objects: *Discovery Client* on the client side and *Discovery Responder* on the agent side. *DiscoveryResponder* notices the request sent via multicast and then replies using unicast transmission.

Discovery mechanism is usually supported by the heartbeat mechanism, which is used for connection failure detection. To monitor the connection, the connector client sends periodic heartbeats (ping requests) to the connector server that acknowledges them by sending a reply (ping responses). If heartbeat message is lost, the connector retries until either the connection is reestablished or the number of retries has been exhausted and connection is released

The typical configuration of Internet firewalls allows communication via HTTP or HTTPS protocol under the condition that this communication is initiated from a secure region to a public network. This asymmetry creates the additional difficulties not only in JMX discovery services activity but also in standard management operations invocation.

Most of very important nowadays applications represent collection of systems connected via public network. The typical example are grids that represent collection of clusters connected into one system via Internet. Another example are mobile systems that should be localized despite its connectivity to secure part of the system. Though these problems could be resolved by VPN establishment over the public network, this solution breaks our assumption that JMX based applications operate over standard Internet.

During the design phase of GDS the following major requirements have been taken into account:

- the service should provide possibility of management and monitoring resources through MBeans located in the secure region, hidden by firewall, using only HTTP/HTTPS connections.,
- it should provide discovery of resources hidden in the secure area,
- the service should also provide mechanism supporting localization of resources in a network topology, especially provide information about agents' neighborhood,
- the implementation should be lightweight and scalable and should be fully compliant with services already defined within JMX framework.

3 GDS Concept and Architecture

During the design phase of GDS a few problems have to be resolved step by step. The process of obtaining a final solution is repeated in this section to justify the role of the proposed system components.

The most important problem concerns the management application which isn't able to establish connection to MBeans Server located in a secure region. The reason is that communication must be initiated from the secure region. To solve this problem a new element called *Join Agent* has been introduced. Function of this agent is initiation of communication connection with Management Application from secure region on one side and with MBean Server on the other. To discover MBean Server in the secure region Join Agent may use the standard JMX Discovery Services. This solution has still three disadvantages:

- It is impossible to discover of MBean Servers by Management Application,
- It is very difficult to chose a good strategy when Join Agent should establish connection to Management Application – there is no information when this application has been started.
- It is necessary to inform Join Agent in some way where Management Application has been started.

To eliminate these drawbacks the second element called *Mediator Agent* has been introduced. It is placed outside the secure region (or in the same region as Management Application) and by assumption is active all the time. The system activity sequence in this case is as follows: first Mediator Agent is started, next in any time MBean Server via Join Agent could establish communication with Mediator Agent. The Management Application (one or more) may also open communication with Mediator Agent obtaining this way access to all already connected MBean Servers. The proposed solution has been depicted in Fig. 1. It is free of drawbacks of the two elements solution and makes possible:

- discovery of MBean Servers connected to Mediator Agents,
- establishing communication at any time because Mediator Agents is always active so the Join Agents can access it,
- easy finding of Mediator Agents by Join Agents because localization of it could be fixed under well know address.

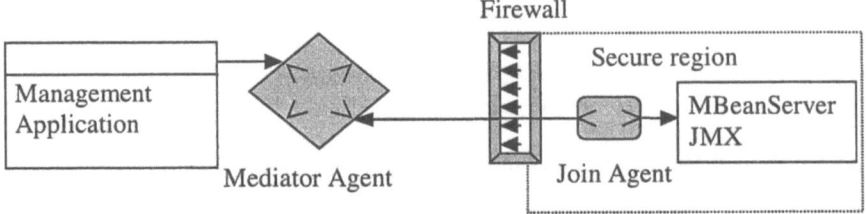

Fig. 1. GDS system elements

To finally clarify the concept of GDS it is necessary to explain how the operation invocation could be forwarded between Mediator Agent to Join Agent by the connection opened in the opposite direction. JMX notification mechanism is employed for this purpose. Join Agents is implemented as MBean which is registered with Mediator Agents. Mediator forwards operations' invocations encapsulated in a notification events. Internal structure of the Mediator Agents may be summarized as follows:

- allows connection of Join Agents and Management Applications – connector servers have to be installed for that purpose,

- forwards management operations' invocations to suitable Join Agents,
- makes possible searching for connected agents with Discovery Service.

The last interesting feature of presented system is its ability to display neighborhood relations between discovered agents. To achieve this goal there is performed special discovery procedure with TTL initially set to 1 and increased subsequently.

4 GDS Implementation

Concept of global discovery mechanism inspired the GDS prototype development, which is presented in this section. GDS implementation was split into three modules: management application, Join Agent and Mediator Agent. All elements of GDS infrastructure excluding management application are designed either as MBean Servers or MBeans, what makes them consistent with infrastructure of existing management systems. Such an approach makes management methods invocations of all MBeans transparent to client, which should bother neither with complexity of communication through secure regions nor with the discovery issues [9].

MBean Server with monitored resources and Join Agent is connected to Mediator Agent, which dynamically creates another MBean Server, called MBean Server Proxy. MBean Server Proxy is special implementation of JMX *MBeanServer* interface, and is equipped only with required connectors and mechanisms for forwarding methods and returning results from remote MBean Server. Having this special implementation of *MBeanServer* interface, there is no possibility to distinguish between real MBean Server and MBean Server Proxy by client application, what is the main goal and advantage of presented GDS [9].

Fig. 2 shows the main window of management application connected to remote agents and managing hidden resources.

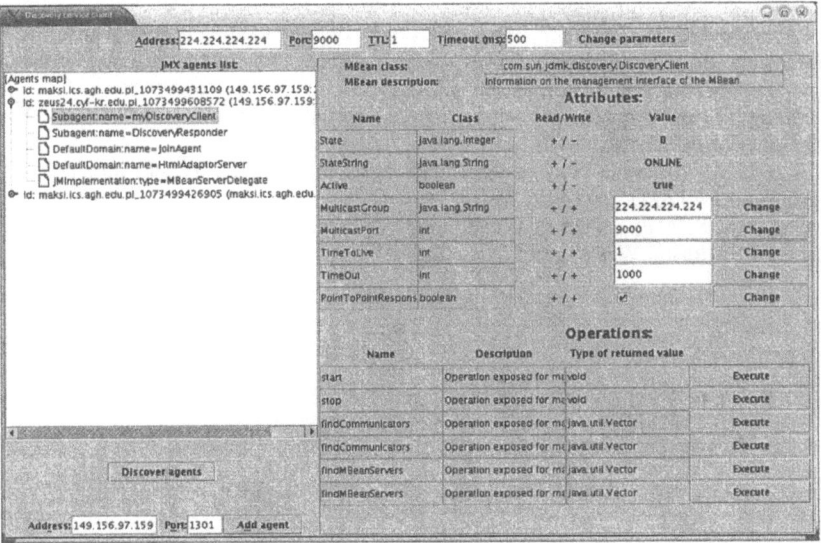

Fig. 2. Screenshot of management application

5 Conclusions

Presented architecture and its first implementation prove that dynamic resources discovery is not reserved only for local area networks, but can be fully operational in wide area networks, as it is in the Internet. GDS provides not only the global discovery utility functions, but makes discovered resources fully transparent for management applications.

Plans for the future GDS evolution should involve moving from JDMK to JMX reference implementation in order to create GDS package not limited by proprietary license. In some cases traffic encryption should be also considered, what probably can be achieved using standard secure communicators found in Sun's implementation of JMX Remote API [2].

References

1. Sun Microsystems: Java Management Extension Reference Implementation (JMX), http://java.sun.com/products/JavaManagement/
2. Sun Microsystems: JMX Remote API Specification (JMX Remote API), http://developer.java.sun.com/developer/earlyAccess/jmx/
3. Sun Microsystems: Java Dynamic Management Kit Specification v. 4.2 (JDMK), http://java.sun.com/products/jdmk/, JDMK42_Tutorial.pdf, 121-128
4. The Open Grid Services Infrastructure Working Group (OGSI-WG): OGSI Specification, http://www.ggf.org/ogsi-wg
5. The Open Grid Services Architecture Working Group (OGSA-WG): Globus Tutorial, www.globus.org/ogsa/
6. M. Sylvain, G. Leduc: A Dynamic Neighborhood Discovery Protocol for Active Overlay Networks, IWAN2003 Conference, Kyoto Japan (10-12 Jan. 2003)
7. Object Management Group (OMG): Notification Service Specification. New Edition, v.1.0 (2000)
8. A. Laurentowski, K. Zieliński: Experiences from Implementation and Evaluation of Event Processing and Distribution of Notifications in an Object Monitoring Service, The Computer Journal Vol. 47 No. 2 (2003) 1-16
9. J. Midura: Monitoring system of distributed computer resources. M.A. thesis, Kraków (2003, in polish) 31-32, 50-70
10. L. Bizon, M. Rozenau: Web services application in computer system integration. M.A. thesis, Kraków (2003, in polish) 33-37
11. R. Stallings: SNMP, SNMP v2.0 and CMIP. A Practical Guide to Network Management Standards, Addison Wesley (1993)

Towards a Grid Applicable Parallel Architecture Machine

Karolj Skala and Zorislav Sojat

Ruder Boskovi, Institute, Center for Informatics and Computing
Bijenicka 54, HR-10000 Zagreb, Croatia.
{skala,sojat}@irb.hr

Abstract. An approach towards defining a flexible, commonly programmable, efficient multi-user Virtual Infrastructural Machine (VIM) is proposed. A VIM is a software infrastructure which enables programming a Grid infrastructure scalable at the lowest possible level. This shall be attained by a Virtual Machine with inherent parallel execution, executing compiled and interpreted computer languages on a very complex p-code level. Based on this interactive deformalized (i.e. natural language like) human-Grid interaction languages should be developed, enabling parallel programming to be done by inherently parallel model description of the given problem.

1 Introduction

The development of the e-science, which, predictably, will evolve towards the e-*society*, justifies the necessary efforts towards better understanding of computer implementation, through active Grids, of naturally distributed parallelism of the environment we live in.

It is a fact that new technologies, clusters and specifically the Grids, are in the execution of programmed tasks quite different from classical single- or multi-processor computing environments by using the parallel architecture of a heap of hugely independent computers, connected through different communication links.

This poses a great many problems in the development of a viable Grid system. The main issues lie in developing such a structure (i.e. software, including middleware issues), infrastructure (i.e. hardware, including low level software and firmware) and applications (including high level portals etc.) which can effectively utilize the extremely different execution environments[1], and actually enable a seamless integration of different kinds of Grid-enabled equipment, including inter alias scientific instruments, experimentation and production tools, mobile units of all kinds etc. into one Grid system.

To enable effective utilization of the multiplicity and variety of Grid resources, in this paper we propose a direction for further scientific and practical exploration towards a human usable viable Grid system, i.e. a specific set of, as we see it, necessary steps towards defining a Grid Applicable Parallel Architecture Machine (GAPAM).

---

[1] I.e. cross-Grid, cross-platform and cross-cluster execution on computers of different makes, operating systems, performances and other parameters.

M. Bubak et al. (Eds.): ICCS 2004, LNCS 3038, pp. 119–123, 2004.
© Springer-Verlag Berlin Heidelberg 2004

2 The Grid Software Infrastructure

Actually the approach taken is an attempt to define the software infrastructure of a complex programmable system which has to be *flexible*, *easily programmable* in commonly known programming languages, and *efficient* in resource utilization and execution speed. A further requirement is for it to be multi-user, where all user privacy and group activities are supported. It shall be scalable, to encompass the whole range of Grid-enabled equipment, from simple user Grid Terminals, up to clusters of huge computing or data power, including fast and large data experiment, collection and visualisation equipment.

A major *flexibility* prerequisite is that the final system should not be an Operating System (OS), but a Virtual Infrastructural Machine (VIM), which should be applicable either in hardware, firmware, as a kernel, or as a layer, daemon or just an application. Specific programmable devices, like visualization equipment, measurement equipment, experimental equipment, Grid Terminals and other similar equipment, as well as different scientific, technical and production tools, have also to be taken into account.

We state that most of the problems mentioned may be much easier to solve, and some of the inherent heterogeneity problems would be solved in advance, by construction of a simple and consistent high performance flexible software infrastructure as a Virtual Infrastructural Machine (VIM). A VIM is actually a Virtual Machine (VM), defined in such a way that its basic states are distributed Grid states, i.e. its basic design is *infrastructural*.

This can be obtained by providing a lowest programmable common denominator, in the form of a Virtual Infrastructural Parallel Architecture Machine, so a viable Grid system could be written in programming languages translated into the interpretative, therefore easily reconfigurable, VIM p-code, executed in a closed interpretative software (or even hardware) driven machine. Such a Grid Applicable Parallel Architecture Machine (GAPAM) would tackle the particular sensitivity of all of the resources on a Grid to security issues in a strictly controlled way, as the execution of Grid software is done in a closed module fashion of a Virtual Machine.

Actually, we are advocating a goal of developing a system easily usable by not only computer experts and programmers, but also a wide circle of scientists and other common users. To attain such a goal, it is not only necessary to define, as mentioned above, a GAPAM, but also to develop, based on this VIM, a human interaction interface, in the form of interactive parallel model description language(s).

3 Approach

To attain this overall goal, we should primarily explore the inherent parallelism principles of most complex problems, through the prism of their immanent natural distributed parallelism, and to find effective and efficient methods of implementation of interactive programming systems which are able to automatically execute inherently parallel described models on parallel architectures, by a sequence of coordinated articulation and dearticulation of the model.

On one side there is the goal of producing such a VIM which allows proper articulation of a formal programming language into elements which enable loosely synchronized parallel execution. This will ease the strain of programming parallel applications.

On the other side, our goal should be producing targeted dearticulated and deformalized, in other words natural-language like, model definition languages, which will, through an interactive programming environment, facilitate the use of complex active Grid technologies, specifically enabling their widespread use by non-specially educated users.

The existing parallel programming languages (e.g. Occam, Orca etc.) are primarily based on explicit parallelism rules, i.e. the human user has to explicitly state the possibilities of parallel execution of certain sub-modules, i.e. sub-algorithms.

It is obvious from general knowledge on programming languages that the underlining principles of some of the "lateral" and less popular languages are not derivatives of the single instruction stream thought process. By this we primarily mean the object-oriented languages like Smalltalk, inherently interactive languages like Forth, Prolog, Lisp, and inherently parallel languages like APL, J and similar.

Fitting languages are primarily those that enable conceptual parallelism of natural process description, independent of their present day single-stream implementations. Those are then the languages that shall be used as the basis for their further dearticulation and deformalization.

The systematic and inter-compatible articulation[2] of the implementation principles of a particular programming language in parallel computing means finding those articulations of a specific programming principle which are elegantly and efficiently applicable onto programming of co-work of a set of identical or different computers (like in a cluster or a Grid).

The dearticulation of a programming language is actually the production of a "superstructure" above the programming system. This so called "superstructure" dearticulated from the programming language is effectively such a language which is specifically targeted for the description of particular scientific (or other user) models in a way as to be linguistically as close to the user as possible, and by this also more usable. By dearticulating the language, a level of interaction easier acceptable by a human is obtained.

A major principle leading to proper dearticulation, which is, per se, in the end very user-specific, is interactivity. To augment the measure of acceptability and possibility of such interactive programming, specifically by the scientists, which in their work, in modern science, always use formalized models of specific objects or areas of their scientific endeavours, special targeted dearticulation shall be used. By such dearticulation GAPAM languages would be adapted to specific targeted user communities.

The main means of attaining such dearticulation is the definition of a "superstructure" of the language by using denotational synonyms, and the exploration of the possibilities of deformalisation of such dearticulated languages. In this sense *deformalisation* is to be regarded as development of synonyms, easy and partially automatic definition of abstracts, i.e. abstraction layers, and a non-too-formal

---

[2] The articulation of a language is extracting of a lover level of symbols, articulating sentences into words, words into morphemes etc. Language articulation on any level may be syntactic (like phonetics), or semantic (like semiotics).

grammar, which enables interactive, step by step, and recursive formalisation of human input, by leading to the creation of a model description language in a form of a knowledge base for specific user or user groups and types.

The method to be used for the given purpose and goals is primarily consisting of proper *reformalization*, and consequently proper articulation of those model-definition languages, deformalized and dearticulated in the described way, into strictly formal forms of chosen programming languages. The mentioned model-definition languages, due to their deformalized and dearticulated nature may not strictly be called programming languages.

The articulation of those languages into formal forms shall be done in such a way that it is reasonably easy, and certainly feasible, to have a high quality implementation of such model-definition languages on clusters and Grids.

The implementation of the whole system shall, consequently, be done by the principle of imbedding, where a lower level executional virtual machine, i.e. the Virtual Infrastructural Machine (VIM), is used to implement a higher level interpretative Virtual Machine (VM), on a multilevel principle, where each successive higher level VM implements a specific formal language or deformalized dearticulation of such a language towards a specifically targeted human language.

As any dearticulated and deformalized language, the same as any natural (i.e. non-formal) language, allows (and necessitates) polysignificance and unobvious ness, the interactivity of the system must support guided significance definition of each specific model definition. In such a way the reformalization of the dearticulated deformalized language is attained in an interactive way.

The user programs written in such a way would be multiply understandable and executable. The multiplicity of the understanding and execution possibilities of such user programs is essential for the attainment of better human-machine interaction principles. They would be understandable on the scientist's dearticulated language, i.e. the community that uses this particular target language for their model descriptions will be able to understand the notation and notions used in such an executable model. As such the dearticulated and deformalized language may be quite close to common scientific jargon used in the specific field; such a model description can be easily readable and understandable for all humans knowledgeable in the particular field. On the other hand, the model description programmes will be, after their first articulation, and the interactive reformalization, understandable on the chosen underlying programming language. This has a consequence of transferability to other machines outside the described Virtual Infrastructural Machine system, as well as the understandability to other humans which do not know the specific science field dearticulated language, but do know the formal programming language.

Such model description programmes will be executable in a classical manner on a single sequential execution machine, as well as on fast parallel architectures, specifically on clusters and Grids. These parallel implementations are attained by the previously explored implementation articulation through the programming language imbedding system.

4 Conclusion and the Vision of Future

In this paper we tried to give a vision of an approach towards higher human interactivity with future Grid systems by defining several steps we see necessary regarding the conception of a system able to accept problem definitions in dearticulated and deformalized, natural-like, model description languages, enabling non-technical users to start using full power of the Grids.

Our vision of the future is a society where integrated Grids allow easy programming and interactive solving of user problems in real time by execution of particular models described by the model description near natural languages on an amount of machines chosen for the most effective execution, but possibly highly distributed even for basic VIM parallel complex instructions. A common user approaches the Grid as approaching a companion, a highly developed tool, which understands a great deal of the users near-natural language communication. It is such communication between the human and the machine that enables the Grid to start being, as W. Ross Ashby termed it, an *Intelligence Amplifier*.

And finally, regarding the dearticulation and deformalization, to cite A. N. Whitehead:

"By relieving the brain of all unnecessary work, a good notation sets it free to concentrate on more advanced problems, and in effect increases the mental power of the race."

References

1. Raymond Greenlaw, Lawrence Snyder, "Achieving Speedups for APL on an SIMD Distributed Memory Machine", International Journal of Parallel Programming, 1990.
2. Charles Antony Richard Hoare, "Communicating Sequential Processes", Prentice Hall International, 1985.
3. Kenneth E. Iverson, "Concrete Math Companion", ISI, 1995.
4. Siniša Marin, Mihajlo Ristić, Zorislav Šojat, "An Implementation of a Novel Method for Concurrent Process Control in Robot Programming", ISRAM '90, Burnaby, BC, 1990.
5. Robin Milner, "Computing in Space", CARS, 2003.
6. Karolj Skala, "e-Science", Ruđer, Vol. 3, No. 7/8, Institute Ruđer Bošković, Zagreb, 2002, pp. 11-14.
7. Karolj Skala, Zorislav Šojat; Grid for Scientific and Economic devolopment of Croatia, 4th CARNet Users Conference - CUC 2002, September 25-27, 2002, Zagreb, Croatia
8. Zorislav Šojat, "An Approach to an Active Grammar of (Non-Human) Languages", 27. Linguistisches Kolloquium, Münster, 1992.
9. Zorislav Šojat, "An Operating System Based on Device Distributed Intelligence", 1ᵃ Orwellian Symposium, Baden Baden, 1984.
10. Zorislav Šojat, "Nanoračunarstvo i prirodno distribuirani paralelizam" ('Nanocomputing and naturally distributed parallelism'), Ruđer, Vol. 3, No. 7/8, Institute Ruđer Bošković, Zagreb, 2002, pp. 20-22.
11. W. M. Waite, "Implementing Software for Non-Numeric Applications", Prentice Hall, New York, 1973.
12. Krzysztof Zielinski, ed., "Grid System Components", AGridnet Consortium, draft, December 2002.

A XKMS-Based Security Framework for Mobile Grid into the XML Web Services

Namje Park, Kiyoung Moon, Jongsu Jang, and Sungwon Sohn

Information Security Research division
Electronics Telecommunications research Institute (ETRI)
161 Gajeong-Dong, Yuseong-Gu, Daejeon, 305-350, Korea
{namjepark,kymoon,jsjang,swsohn}@etri.re.kr

Abstract. Mobile Grid, or Wireless Grid Services refers to value-added Grid Service by processing in Mobile environment. Besides Mobile Internet the traditional Internet computing is experiencing a conceptual shift from Client-Server model to Grid and Peer-to-Peer computing models. As these trends, Mobile Internet and the Grid, are likely to find each other the resource constraints that Wireless devices pose today affect the level of interoperability between them. The goal of this paper is to investigate how well the most limited Wireless devices can make use of Grid Security Services. This paper describes a novel security approach on Mobile Grid Services to validate certificate based on current Mobile XML Web Services Platform environment using XKMS (XML Key Management Specification) and SAML (Security Assertion Markup Language), XACML (extensible Access Control Markup Language) in XML Security mechanism.

1 Introduction

Grid Computing emerges as a technology for coordinated large-scale resource sharing and problem solving among many autonomous groups. In Grid's resource model, the resource sharing relationships among virtual organizations are dynamic. However, Grid requires a stable quality of service provided by virtual organizations and the changing of sharing relationship can never happen frequently. This model works for a conventional distributed environment but is challenged in the highly variational wireless mobile environment[3].

Besides Mobile Internet the traditional Internet computing is experiencing a conceptual shift from Client-Server model to Grid and Peer-to-Peer computing models. As these trends, Mobile Internet and the Grid, are likely to find each other the resource constraints that Wireless devices pose today affect the level of interoperability between them[2].

Grid is the umbrella that covers many of today's distributed computing technologies. Grid technology attempts to support flexible, secure, coordinated information sharing among dynamic collections of individuals, institutions, and resources. This includes data sharing but also access to computers, software and devices required by computation and data-rich collaborative problem solving. So far the use of Grid services has required a modern workstation, specialized software

M. Bubak et al. (Eds.): ICCS 2004, LNCS 3038, pp. 124–132, 2004.
© Springer-Verlag Berlin Heidelberg 2004

installed locally and expert intervention. In the future these requirements should diminish considerably. One reason is the emergence of Grid Portals as gateways to the Grid. Another reason is the 'Web Service' boom in the industry. The use of XML as a network protocol and an integration tool will ensure that future Grid peer could be a simple wireless device[2,3].

Furthermore, open Mobile Grid service infrastructure will extend use of the Grid technology or services up to business area using Web Services technology. Therefore differential resource access is a necessary operation for users to share their resources securely and willingly. Therefore, this paper describes a novel security approach on open Mobile Grid service to validate certificate based on current Mobile Grid environment using XKMS (XML Key Management Specification) and SAML (Security Assertion Markup Language), XACML (eXtensible Access Control Markup Language) in XML(eXtensible Markup Language) security mechanism.

This paper is organized as follows. First we investigate related work on Grid and mobile web service, XML web service security. Then we propose a design of security system platform for open mobile Grid service and explain experimented XKMS model for certificate validation service. Finally, we explain function of system and then we conclude this paper.

2 Mobile XML Web Services

A mobile XML web service can feature one of the following architectures: wireless portal network, wireless extended Internet, or wireless ad hoc network.

In a wireless portal network, the wireless information devices connect to the Internet backend services through portal entry points. The portal creates a "walled garden" and controls access to Internet contents. Wireless portal networks support widely deployed thin-client wireless technology, such as WAP (Wireless Application Protocol). The portal receives the message, checks the user's privilege, and then translates the request to a SOAP (Simple Object Access Protocol) message or an XML-RPC call to an appropriate partner web service. The web service replies, and the portal translate the response back to a WML (Wireless Markup Language) document. The portal sends the WML document back to the wireless device for display. In this way, the portal works as a proxy for wireless users. The portal operator provides user authorization and management services. Many partner vendors can provide real application web services under the ASP (Application Service Provider) model.

Wireless extended Internet is the wired Internet's expansion to wireless devices. Wireless information devices can have their own IP addresses (through Internet Protocol 6) and full network functionalities. Those devices usually run smart, fat clients that interact with multiple backend services simultaneously and store/process application data on the device. Smart devices support sophisticated user interfaces, offline processing, and automatic transactions. They can also implement flexible, application-specific security policies. Like the Internet itself, the wireless extended Internet architecture is decentralized and eliminates any single point of failure. However, as you will see later, centralized web services hubs are still required to support advanced security schemes and user interfaces. Unlike the portal architecture, the hubs themselves can be decentralized. Different vendors can provide similar hub

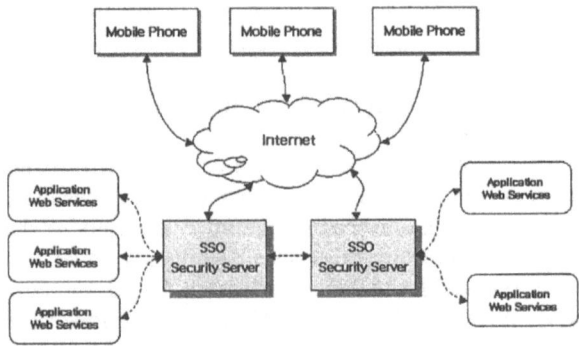

Fig. 1. Mobile Web Services Architecture

services that can interoperate with each other. Fig. 1 shows a topography diagram for such networks.

The extended wireless Internet architectures blended with decentralized hub web services will provide the foundation for future wireless web services applications, an approach we focus on throughout this article. Since most of the supporting technologies are just emerging, many challenges prevail.

The wireless ad hoc networks allow wireless devices to become servers to peers. Wireless peers can provide content, network traffic routing, and many other services. The ad hoc network truly leverages wireless networks' dynamic nature. However, because wireless peer-to-peer technology is still embryonic, its many performance and security issues must be solved before it can be widely used.

3   Security Framework for Providing Secure Open Mobile Grid

Web services can be used to provide mobile security solutions by standardizing and integrating leading security solutions using XML messaging. XML messaging is referred to as the leading choice for a wireless communication protocol and there are security protocols for mobile applications based upon it. Among them are the follows. SAML is a protocol to transport authentication and authorization information in an XML message. It could be used to provide single sign on web services. XML signatures define how to digitally sign part or all of an XML document to guarantee data integrity. The public key distributed with XML signatures can be wrapped in XKMS formats. XML encryption allows applications to encrypt part or all of an XML document using references to pre-agreed symmetric keys. The WS-Security, endorsed by IBM and Microsoft, is a complete solution to provide security to web services. It is based on XML signatures, XML encryption, and an authentication and authorization scheme similar to SAML. When a mobile device client requests access to a back-end application, it sends authentication information to the issuing authority. The issuing authority can then send a positive or negative authentication assertion depending upon the credentials presented by the mobile device client. While the user still has a session with the mobile applications, the issuing authority can use the earlier reference to send an authentication assertion stating that the user was, in fact, authenticated by a

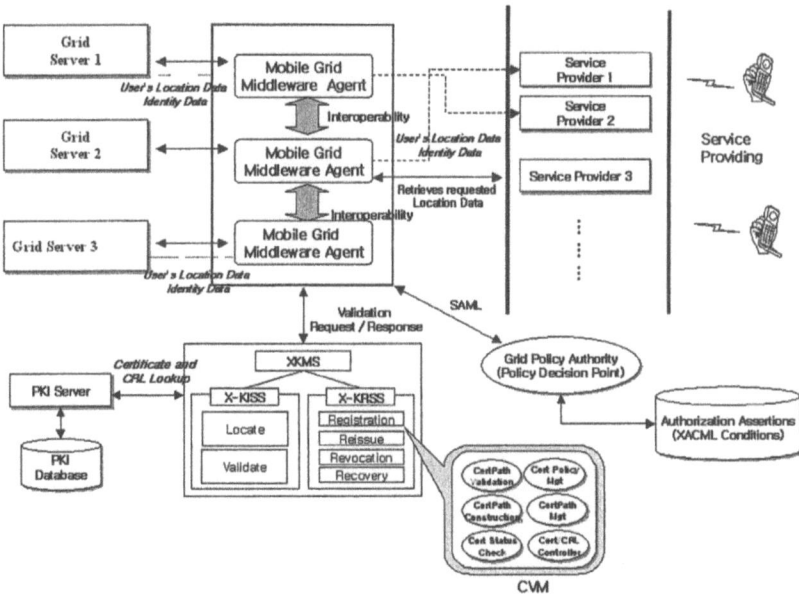

Fig. 2. Security Service Model for Open Mobile Grid Middleware

particular method at a specific time. As mentioned earlier, location-based authentication can be done at regular time intervals, which means that the issuing authority gives out location-based assertions periodically as long as the user credentials make for a positive authentication.

CVM (Certificate Validation Module) in XKMS system perform path validation on a certificate chain according to the local policy and with local PKI (Public Key Infrastructure) facilities, such as certificate revocation (CRLs) or through an OCSP (Online Certificates Status Protocol). In the CVM, a number of protocols (OCSP, SCVP, and LDAP) are used for the service of certificate validation. For processing the XML client request, certificate validation service from OCSP, LDAP (Lightweight Directory Access Protocol), SCVP (Simple Certificate Validation Protocol) protocols in XKMS based on PKI are used. The XKMS client generates an 'XKMS validate' request. This is essentially asking the XKMS server to go and find out the status of the server's certificate. The XKMS server receives this request and performs a series of validation tasks e.g. X.509 certificate path validation. Certificate status is determined. XKMS server replies to client application with status of the server's certificate and application acts accordingly. Using the OCSP protocol, the CVM obtained certificate status information from other OCSP responders or other CVMs. Using the LDAP protocol, the CVM fetched CRL (Certificate Revocation List) from the repository. And CA (Certificate Authority) database connection protocol (CVMP;CVM Protocol) is used for the purpose of that the server obtains real-time certificate status information from CAs. The client uses OCSP and SCVP. With XKMS, all of these functions are performed by the XKMS server component. Thus, there is no need for LDAP, OCSP and other registration functionality in the client application itself.

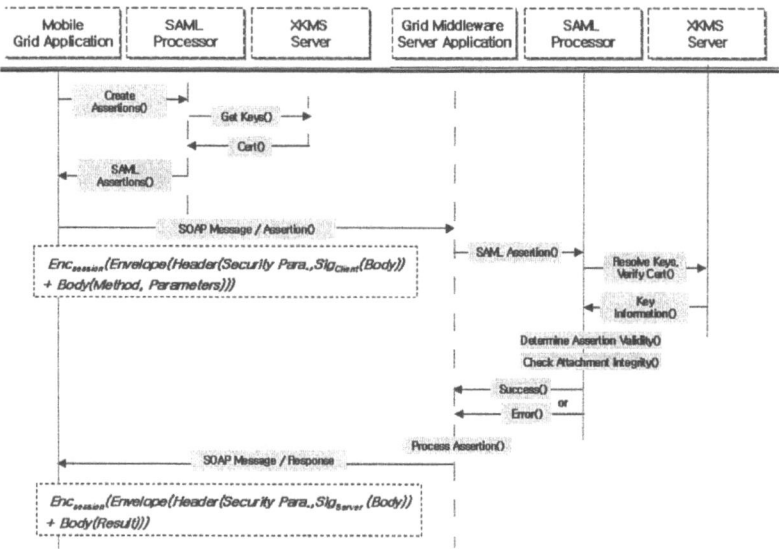

Fig. 3. Security Protocol for Secure Open Mobile Grid Service

4 Mobile Grid Application Security Protocol

Three types of principals are involved in our protocol: Mobile Grid application (server/client), SAML processor, and XKMS server (including PKI). Proposed invocation process for secure Mobile Grid security service consists of two parts: initialization protocol and invocation protocol. The initialization protocol is prerequisite for invoking Grid web services securely. Through the initialization protocol, all principals in our protocol set up security environments for their web services, as shown in fig. 3. The flow of setting up security environments is as follows.

The client first registers its information for using web services, and then gets its id/password that will be used for verifying its identity when it calls web services via secure channel. Then, the client gets SAML assertions and installs security module to configure its security environments and to make a secure SOAP message. It then generates a key pair for digital signature, and registers its public key to a CA.

The client creates a SOAP message, containing authentication information, method information, and XML signature, XML encrypts it, and then sends it to a server. The message is in following form: $Enc_{session}(Envelope$ $(Header(SecurityParameters,$ $Sig_{client}(Body)) + Body(Method, Parameters))))$, where $Sig_x(y)$ denotes the result of applying x's private key function (that is, the signature generation function) to y. The protocol shown in fig. 4 shows the use of end-to-end bulk encryption[12]. The security handlers in server receive the message, decrypt it, and translate it by referencing security parameters in the SOAP header. To verify the validity of the SOAP message and authenticity of the client, the server first examines the validity of the client's public key using XKMS. If the public key is valid, the server receives it from CA and verifies the signature. The server invokes web services after completion

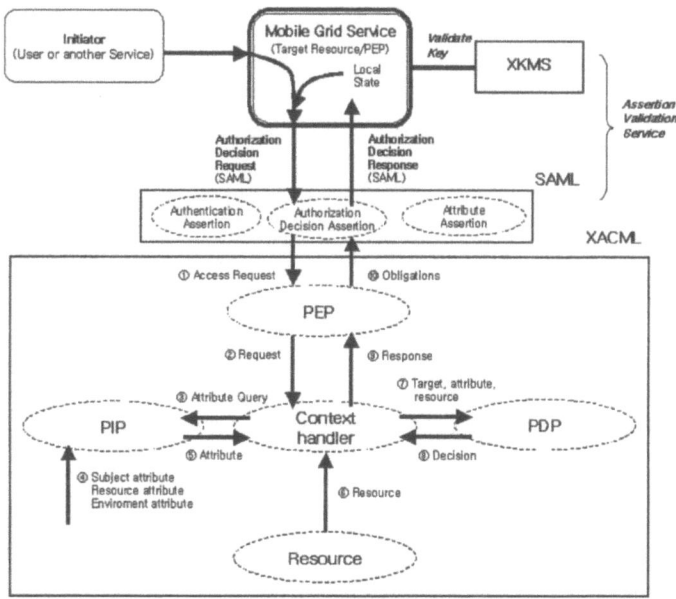

Fig. 4. Security Message Flow using XKMS in Open Mobile Grid Middleware

of examining the security of the SOAP message. It creates a SOAP message, which contains result, signature, and other security parameters. Then, it encrypts the message using a session key and sends it back to the client. Lastly, the client examines the validity of the SOAP message and server, and then receives the result[14,15].

In current Grid service, there is no mechanism of differential resource access. To establish such a security system we are seeking, a standardized policy mechanism is required. We employ the XACML specification to establish the resource policy mechanism that assigns differential policy to each resource (or service). SAML also has the policy mechanism while XACML provides very flexible policy mechanism enough to apply to any resource type. For our implementing model, SAML provides a standardized method to exchange the authentication and authorization information securely by creating assertions from output of XKMS (e.g. assertion validation service in XKMS). XACML replaces the policy part of SAML as shown in fig 4.

Once the three assertions are created and sent to the protected resource, there is no more verification of the authentication and authorization at the visiting site. This, SSO (Single Sign-On), is a main contribution of SAML in distributed security systems.

Fig. 4 shows the flow of SAML and XACML integration for differential resource access. Once assertions are done from secure identification of the PKI trusted service, send the access request to the policy enforcement point (PEP) server (or agent) and send to the context handler. Context handler parses the attribute query and sends it to PIP (policy information point) agent. The PIP gathers subject, resource and environment attributes from local policy file, and the context handler gives the required target resource value, attribute and resource value to PDP (policy decision point) agent. Finally, the PDP decides access possibility and send context handler so *that PEP agent allow or deny the request*[10,13].

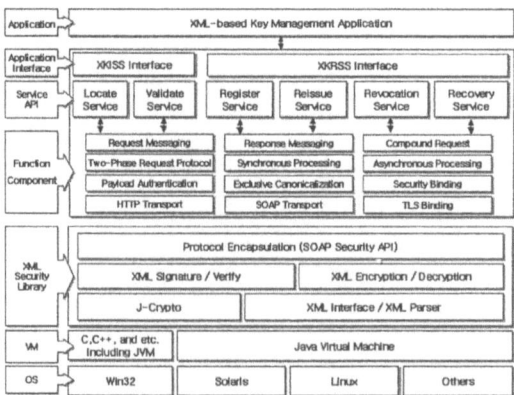

Fig. 5. Package Library Architecture of XKMS Server System based on CAPI

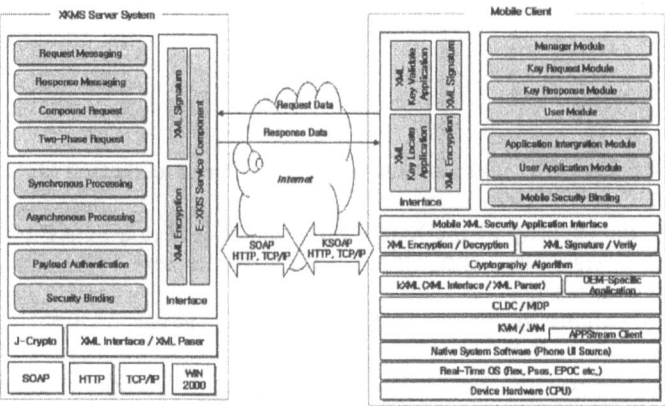

Fig. 6. Testbed Configuration of XKMS Component for Open Grid

5 Design of Mobile Grid Application Key Management Using XKMS

XKMS has been implemented based on the design described in previous section. Package library architecture of XKMS based on CAPI (Cryptographic Application Programming Interface) is illustrated in fig. 5.

Components of the XKMS are XML security library, service components API, application program. Although XKMS service component is intended to support XML applications, it can also be used in order environments where the same management and deployment benefits are achievable. XKMS has been implemented in Java and it runs on JDK (Java Development Kit) ver. 1.3 or more.

The figure for representing Testbed architecture of XKMS service component is as follows fig. 6. We use Testbed system of windows PC environment to simulate the processing of various service protocols. The protocols have been tested on pentium 3 and pentium 4 PCs. It has been tested on windows 2000 server, windows XP.

Java 2, Micro Edition (J2ME) is a set of technologies and specifications developed for small devices like smart cards, pagers, mobile phones, and set-top boxes. J2ME uses subset of Java 2, Standard Edition (J2SE) components, like smaller virtual machines and leaner APIs. J2ME has categorized wireless devices and their capabilities into profiles: MIDP, PDA and Personal. MIDP and PDA profiles are targeted for handhelds and Personal profile for networked consumer electronic and embedded devices. As the technology progresses in quantum leaps any strict categorization is under threat to become obsolete. It is already seen that J2ME Personal profile are being used in high-end PDAs such as PocketPCs and Mobile Communicators. We will concentrate on the most limited category of wireless J2ME devices that use Mobile Information Device Profile (MIDP). Applications that these devices understand are Midlets. Typically maximum size of a midlet varies from 30-50kbs and user can download four to six applications to his mobile phone. Midlet is a JAR-archive conforming to the Midlet content specification[2].

The XKMS server is composed server service component of XKMS platform package. And the message format is based on Specification of W3C (World Wide Web Consortium).

6 Conclusion

Mobile Grid services are so attractive that they can cover all walks of life. However, current Grid is growing slower than expected. Many problems like accuracy, privacy, security, customer requirement have to be addressed. It should be understood that there is no single universal solution to Grid.

We propose a novel security approach on open Grid to validate certificate based on current Grid security environment using XKMS and SAML, XACML in XML security. This service model allows a client to offload certificate handling to the server and enable to provide central administration of XKMS polices. In order to obtain timely certificate status information, the server uses several methods such as CRL, OCSP etc. Our approach will be a model for the future security system that offers security of open Grid security.

References

1. XML Key Management Specification Version 2.0 (W3C Working Draft), April 2003.
2. Miika Tuisku, Wireless Java-enabled MIDP devices as peers in Grid infrastructure, Helsinki Institute of Physics, CERN.
3. Ye Wen, Mobile Grid Major area examination, University of California, 2002.
4. E. Faldella and M.Prandini, A Novel Approach to On-Line Status Authentication of Public Key Certificates, in Proc. the 16th Annual Computer Security Applications Conference, Dec 2000.
5. Y. Elley, A. Anderson, S. Hanna, S. Mullan, R. Perlman and S. Proctor, Building Certification Paths: Forward vs. Reverse, Proc. the Network and Distributed System Security Symposium Conference, 2001.
6. M. Naor and K. Nissim, Certificate Revocation and Certificate Update, IEEE Journal on Selected Areas in Communications, 18 (4) (2000).
7. Jonghyuk Roh et. Al., Certificate Path Construction and Validation in CVS, KICS-Korea IT Forum, 2002.

8. M. Prandini, Efficient Certificate Status Handling within PKIs: an Application to Public Administration Services, in Proc. the 15th Annual Computer Security Applications Conference, 1999.
9. Donald E, Eastlake, Kitty Niles, Secure XML, Pearson addsion wesley, 2003.
10. Namje Park et. Al., Certificate Validation Scheme of Open Grid Service Usage XKMS, GCC 2003, 2003.
11. Euinam Huh, Jihye Kim, Hyeju Kim, Kiyoung Moon, Policy based on Grid security infrastructure implementation for dirrerential resource access, ISOC 2003, 2003.
12. Yuichi Nakamur, et. Al., Toward the Integration of web services security on enterprise environments, IEEE SAINT '02, 2002.
13. Diana Berbecaru, Antonio Lioy, Towards Simplifying PKI Implementation: Client-Server based Validation of Public Key Certificates, IEEE ISSPIT 2002, pp.277-281.
14. Jose L. Munoz et. Al., Using OCSP to Secure Certificate-Using transactions in M-Commerce. LNCS 2846 (2003) 280-292.
15. Sungmin Lee et. Al., TY*SecureWS:An integrated Web Service Security Solution based on java, LNCS 2738 (2003) 186-195.

A Proposal of Policy-Based System Architecture for Grid Services Management

Edgar Magaña, Epifanio Salamanca, and Joan Serrat

Universitat Politècnica de Catalunya
Network Management Group
Jordi Girona, 1-3 D4-213
Barcelona, Spain
{emagana,epi}@nmg.upc.es, serrat@tsc.upc.es

Abstract. Communication technologies are evolving along with Grid Computing. The trend of Open Grid Services Architecture (OGSA) is that "everything" is represented as a service, which will be provided to users without any kind of distinctions related to network technology, operative platform, administrative domain, etc. As a result of this trend, there has been a significant management requirement increase in order to cope with the yet elusive rapid and autonomous service creation, deployment, activation and management. This paper describes a proposal to cope with the above mentioned management requirements. Our policy-based management approach enables the establishment, configuration and administration of services on a Grid network that uses the Globus Architecture for Reservation and Allocation (GARA) as an interface for reservation and provisioning of grid resources.

1 Introduction

The Grid concept was introduced to denote a proposed form of distributed computing that involves coordinating and sharing computing, application, data, storage, or network resources across dynamic and geographically dispersed organizations. Grid technologies promise to change the way organizations tackle complex computational problems. Nevertheless, Grid computing is an evolving area of computing, where standards and technology are still being developed to enable this new paradigm. The sharing of resources inherent in Grid networks introduces challenging resource management problems due to the fact that many applications need to meet stringent end-to-end performance requirements across multiple computational resources (memory, processor, storage, etc). Furthermore, they should also guarantee fault tolerance and network level Quality-of-Service (QoS) parameters as well as arbitrate conflicting demands.

In order to cope with these resource management challenges, there is a first solution developed for Globus that is called Globus Resource Management Architecture (GRMA) [1]. This solution addresses the wide QoS problem of providing dedicated access to collections of computers in heterogeneous distributed system. But this architecture does not tackle two important issues, namely the

M. Bubak et al. (Eds.): ICCS 2004, LNCS 3038, pp. 133–140, 2004.
© Springer-Verlag Berlin Heidelberg 2004

reservation in advance and the resources heterogeneity. Thereby a new solution for resources management was needed. To this aim the Globus Architecture for Reservation and Allocation (GARA) was proposed. GARA extends GRMA in two major ways: The first one introduces the generic resource object, which encompasses network flows, memory blocks, disk blocks and processes. The second one introduces the reservation as a first class entity in the resource manager architecture [1]. Academy and industry have shown special interest on this technology during the last years, thereby several research projects are working on new proposals for improving the management of the grid such as MANTRIP [8] or RNTL e-Toile project [9]. Yang et al [10] presented a management middleware that involves active networks technology and policy-based management. People from Globus are still working on better solutions for management, such as the work presented in [11] related to obtain QoS using the GARA system.

We claim that there is a need for a major entity in charge of providing high-level management to allow quick and autonomous deployment, activation and reservation of services and resources. Additional relevant properties for such entity are fault tolerance and interdomain management support. This paper proposes the fulfillment of the above mentioned goals by means of a policy-based architecture that exploits the facilities offered by the Globus Architecture for Reservation and Allocation (GARA).

The paper is structured as follows. Section 2 analyses the most important management system requirements for the emerging grid services. Section 3 is an overview of the GARA system, which is a fundamental background to understand Section 4 that is related to present the policy-based management proposal and its capability to cater with the requirements mentioned in the second section. Section 5 presents an example scenario. Finally we conclude the paper with some remarks in Section 6.

2   Management System Requirements for Grid Services

Grid Computing is affected by continuous innovations and updates due to the huge interest generated among the industry and the scientific community. For this reason a new grid generation is emerging, where schema conversion technologies via common meta-models and ontologies are required to allow data to be moved between storage resources and be shared between tools with different data format requirements. Grid services that automate this process and build on-line libraries for conversion and mapping tools will be of use in a variety of scientific, social and business domains.

At the same time, there are other innovations with a significant impact on Grid Computing, like the creation of mechanisms to handle a large number of heterogeneous resources (computing, storage, networks, services, applications, etc.), the adequate use of data from a wide variety of sources and technical domains, the necessity of designing and implementing new techniques to allow for the interoperability between different data sources is constantly growing.

One of the key challenges will be the co-ordination and orchestration of resources to solve a particular problem or perform a business process. The current generation of grid architecture relies heavily on the program designer or the user to express their requirements in terms of resource usage. Such requirements are usually hardcoded in a program using low-level primitives. But the grid needs to handle resources in a more dynamic way. In other words, grid applications will require the co-ordination and orchestration of grid elements at run time.

Two of the most important grid features are management autonomy and fault-tolerance. This requires redundant resources (computation, data base, network), and also an autonomous and self-regulatory model that ensures the proper working of the management architecture by itself. Maintenance and Scalability are crucial; solutions should be able to support Grids consisting of billions of nodes. This requirement immediately stresses the importance of finding an appropriate trade-off between control systems exerted by users and administrators.

Current solutions for accomplishing these grid management requirements have limitations, particularly related to high-bandwidth, network reconfiguration, fault-tolerance, reliability, scalability, flexibility and persistence. Our approach goes one step further offering the possibility to handle the grid management requirements above mentioned by means of an extension of the Policy-Based System designed and implemented as part of the IST FAIN project [3].

3   Globus Architecture for Reservation and Allocation (GARA)

The most well known technology available to reserve and allocate low-level resources such as memory, processing or storage is the Globus Architecture for Reservation and Allocation (GARA) [2]. As mentioned in the introduction, GARA enlarges the Globus resource management architecture basically in two ways: the first one is to generate an architecture with the ability of co-reserving resources and the second one is to generate a generic resource object in order to manage heterogeneous resources.

The GARA architecture allows applications appropriate access to end-to-end Quality of Service (QoS). To do so, it provides a mechanism for making QoS reservation for different types of resources, including disks, computers and networks. GARA provides management for separately administered resources. In Figure 1 we show the GARA's architecture, which consists of three main components, the information service, local resource managers and co-allocation/reservation agents. The information service allows applications to discover resource properties such as current and future availability. Local resource managers have been implemented for a variety of resource types, this explains the use of term "resource manager" rather than the more specific "bandwidth broker", although each one implements reservation, control and monitoring operations for a specific resource. The reservation and allocation agents compute the resource requirements and send the request to the Globus Resource Allocation

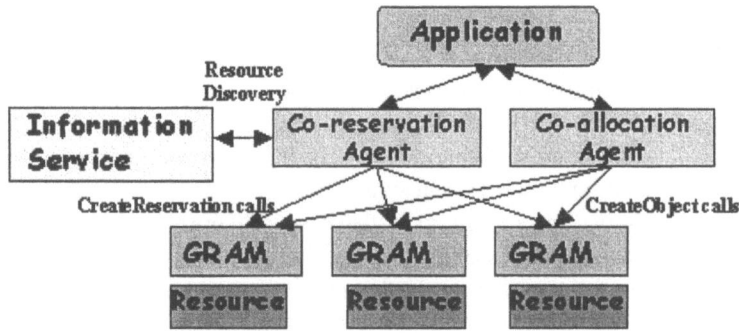

Fig. 1. PBM system with GARA

Manager (GRAM). The GRAM takes the request, authenticates it and if successful forwards it to the local scheduler in order to allocate or serve the resource and finally returns the job handle to GARA. The support for heterogeneous resources is provided by a generic resource object that encapsulates network flows, memory blocks, disk blocks, and even processes.

The support for advanced reservation is allowed by separating the reservation from allocation. For immediate resource allocation, the requested allocation is performed at the time of reservation, but in the case of advance reservation only a reservation handle is returned and the resources need to be reserved at the service start time. A new entity that is called Co-reservation Agent provides this advanced functionality. Its function is similar to the co-allocation agent one, except that after calculating the resource requirement for advanced reservation, it does not allocate but simply reserves the resources. GARA provides a simple, uniform interface for making advance reservations for any type of supported resource. The API provides functions for creating, modifying, binding, claiming, canceling, and monitoring reservations. Although it is as uniform as possible, different resources require different parameters. To accommodate these different needs within a single API, the *create function call* accepts a Resource Specification Language (RSL) string to specify the parameters for single reservation.

4   Policy-Based Management Proposal

Policy-Based Networking has attracted significant industry interest in recent years. Presently, it is promoted by the Distributed Management Task Force (DMTF) [5] and is standardized within the Internet Engineering Task Force (IETF) Policy working group [4]. Policy-Based Network Management (PBNM) offers a more flexible, customizable management solution allowing each network node to be configured on the fly, for a specific application tailored for a consumer.

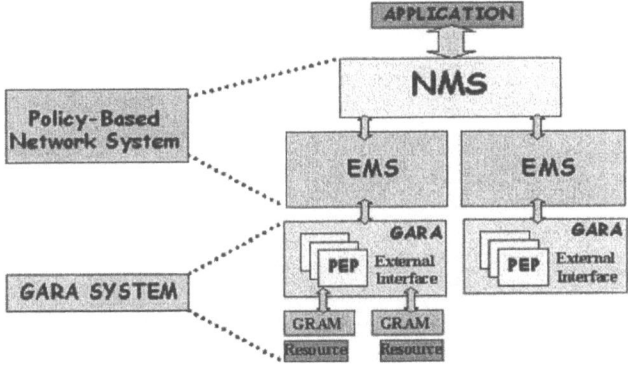

Fig. 2. PBM system with GARA

The Policy-Based Management (PBM)[1] system, presented in this paper, is designed as a hierarchically distributed architecture. Formerly designed to manage Active Networks [3], extensions in terms of management capabilities, have laid an efficient and robust PBM system, which supports co-existence of different management strategies thus facilitating customisation, interoperability with different vendor's equipment, and dynamic extensibility of its functionality to support new services and their deployment as well as new resources on the network.

In comparison with previous traditional network management approaches, PBM offers a more flexible and customisable management solution allowing each managed element (i.e., switch, router, scientific instruments, etc) to be configured on the fly to support customer tailored services. The PBM architecture is able to address a number of domain issues that can be translated to fixed configuration settings. For example, QoS issues often needs complex interactions between relevant network management components. These complex interactions can be easily implemented in the current PBM architecture. Moreover, according to the policy framework, policies are defined or modified by an administrative tool and the intervention of the administrator is just required for special activities. On the other hand is possible to define new types of policies and to extend the system as much as necessary to manage new target elements, without extending the core information model. The proposed PBM system consists of two levels: the Network Management System (NMS) and the Element Management System (EMS), as we show in Figure 2. The NMS is the entry point of the management architecture; therefore it offers the necessary GUIs for controlling the entire architecture. In fact, it is the receiver of policies, resulting from Service Level Agreements between various categories of users. The enforcement of these SLAs requires reconfiguration of the network, which is carried out by means of policies sent to the NMS. These policies, called network-level policies, are processed by the NMS Policy Decision Points (PDPs), which decide when

---

[1] From here and forward, we will express the Policy-Based Management System just as PBM for simplifying the notation.

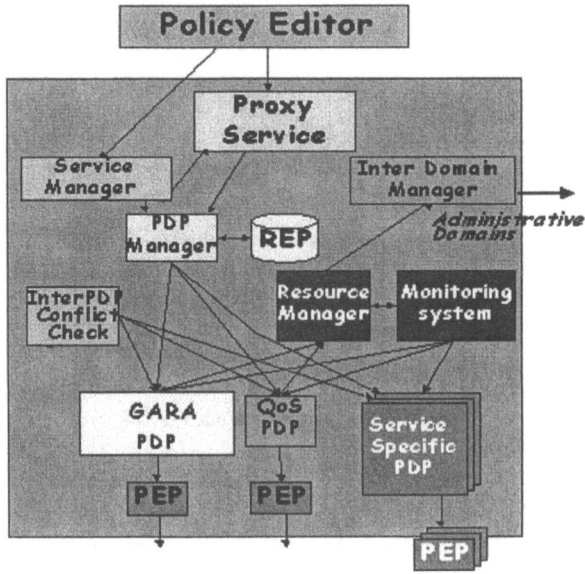

Fig. 3. Policy-Based Management Components

policies can be enforced. When enforced, they are delivered to the NMS Policy Enforcement Points (PEPs) that map them to element level policies, which are, sent to the EMSs. EMS PDPs follow a similar procedure at the element level. Finally, the node PEPs execute the enforcement actions on the managed target elements. The PBM system interaction between components is show in Figure 3, whilst an in depth description is provided in [7]. The flexibility and scalability of our PBM system proposal makes easy the integration process with GARA. With this goal in mind we have added element-level PEPs within the GARA system, as shown in Figure 3, in order to enforce the element-level policies sent by the EMS PDPs which are derived from polices sent by the NMS. The use of open interfaces in GARA allows all PEPs across the network to share the same view of grid resources, making them platform independent. The new PEPs will contain the necessary code to adapt the parameters from high-level policies to resource-level commands.

5   Case Example

The functionality of the entire system is explained in this section. We propose a configuration service scenario because it is one of the simplest cases and therefore it will ease the understanding of the functionalities and specially the management framework advantages. Figure 4 shows the above mentioned example where we distinguish the two main entries to the management framework. The first one comes directly from the application and the second one is from the network administrator Policy GUI. For this example we consider the second one only. Let's

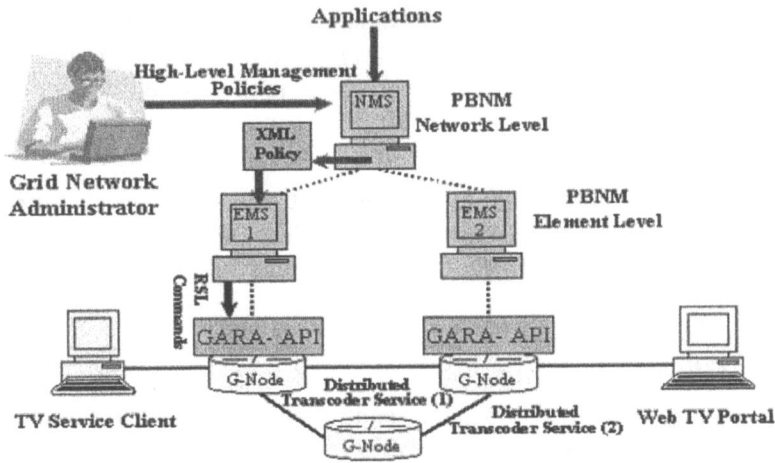

Fig. 4. Example scenario

assume that a TV service client is receiving a data video stream from a Web TV portal (server). As this server offers the video in coded format, clients are using a Distributed Transcoder Service (DTS) on G-Nodes (routers with the Globus Toolkit installed). At a given time, and due to other network requests, the G-Node's resources where our client is attached become unable to continue offering the decoding task with the appropriate quality constraints. The PBM is alerted on that fact by its monitoring system and therefore it starts a resource search task. As soon as the PBM discovers resources available on another G-Node, it generates the necessary resource-level (RSL) commands through GARA's PEPs and contacts with the appropriated interface in order to reserve and allocate the resources for the second DTS (Distributed Transcoder Service (2) in the figure). Once the PBM receives the notification of a correct policy execution, a re-configuration of resources task will be issued on the first G-Node in order to de-allocate all unused resources in that node to be offered to other applications or services. Note that finally the service is running on both G-Nodes without any service interruption as well as without the need of any manual change by the network administrator. The detailed interactions between components inside the PBM system can be found in [6].

6   Conclusions

Our proposal consists of an architecture taking advantage of the synergy obtained by coupling policy based management technology and GARA to obtain a full-featured management framework, which fulfills the most important requirements for managing grid services. The scenario illustrates the benefits of the proposed framework in terms of automation, scalability, flexibility and fault tolerance. The proposal is suitable for giving support for dynamic, reconfigurable on-demand, secure and highly customizable computing storage and networking environments. The next step will be to develop further grid services and the corresponding GARA's PEPs as well as carrying out the appropriate performance testing.

Acknowledgements. Part of the work presented in this paper is based on results obtained in the context of the EU IST Project FAIN (IST-10561), partially funded by the Commission of the European Union.

References

1. K. Czajkowski, I. Foster, N. Karonis, C. Kesselman, S. Martin, W. Smith, and S. Tuecke. *A resource management architecture for metacomputing systems.* In The 4th Workshop on Job Scheduling Strategies for Parallel Processing, 1998.
2. I. Foster, C. Kesselman, C. Lee, R. Lindell, K. Nahrstedt, and A. Roy. *A Distributed Resource Management Architecture that Supports Advance Reservation and Co-Allocation.* In the International Workshop on Quality of Service, June 1999.
3. EU IST FAIN Project Web Site http://www.ist-fain.org/
4. IETF Policy Web Site http://www.ietf.org/html.charters/policy-charter.html
5. Distributed Management Task Force Web Site http://www.dmtf.org/
6. E. Magaña, E. Salamanca, J. Vivero, A. Galis, B. Mathieu, Y. Carlinet, O. Koufopavlou, C. Tsarouchis, C. Kitahara, S. Denazis and J. L. Mañas. *A Policy-Based Management Architecture for Active and Programmable Networks.* IEEE Network Magazine, Special issue on Network Management of Multiservice Multimedia IP-Based Networks, May 2003, Vol. 17 No.3
7. Y. Nikolakis, Edgar Magaña, M. Solarski, A. Tan, E. Salamanca, J. Serrat, C. Brou, and A. Galis. *A Policy-Based Management Architecture for Flexible Service Deployment in Active Networks.* International Working Conference on Active Networks (IWAN03) Kyoto, Japan 2003.
8. EU IST MANTRIP Project Web Site http://www.solinet-research.com/mantrip/
9. E-toile Project Web Site http://www.ens-lyon.fr/LIP/RESO/Projects/Etoile/
10. K. Yang, A. Galis, C. Todd. *Policy-Based Active Grid Management Architecture.* 10th IEEE International Conference on Networks ICON 2002.
11. I. Foster, A. Roy, and V. Sander. *A Quality of Service Srchitecture that Combines Resource Reservation and Application Adaptation.* In 8th International Workshop on Quality of Service (IWQoS '00), June 2000.
12. R. Buyya. Nimrod-G *An Architecture for a Resource Management and Scheduling System in a Global Computational Grid.* 4th International Conference on High Performance Computing in Asia Pacific Region (HPCAsia 2000) IEEE CS Press, Los Alamitos, USA 2000.

Self-Management GRID Services – A Programmable Network Approach

L. Cheng[1], A. Galis[1], A. Savanović[2], B.J. Blažič[2], and J. Bešter[3]

[1] University College London, Electrical Engineering Dept., Torrington Pl., London, U.K.
{l.cheng,a.galis}@ee.ucl.ac.uk)
[2] Jozef Stefan Institute, Laboratory for Open Systems and Networks,
Jamova 39, 1000 Ljubljana, Slovenia
{arso,borka}@e5.ijs.si
[3] University of Ljubljana, Laboratory for Telecommunications,
Trzaska 25, 1000 Ljubljana, Slovenia
besterji@fe.uni-lj.si

Abstract. Due to the complexity and size of service oriented GRIDs, it is essential that GRID systems should be autonomous i.e. a self-management system is needed. This paper identifies the requirements of such a self-management GRID system and the required supporting services. This paper suggests that these supporting services should be deployed in the form of software modules through programmable techniques. This paper presents a communication protocol for dynamic self-configuration in programmable GRIDs as an example for supporting new network services.

1 Introduction to GRID

The concept of GRID networks [10], [11] was coined to enable transparent coupling of geographically-dispersed resources – such as computers, networks, remote data storage, scientific devices and instruments - for large-scale distributed applications in a highly distributed network environment, such as the Internet [1], [3]. There have been attempts to integrate GRID technologies with existing Web technologies by the Open Grid Service Architecture (OGSA) group [21], [24-28]. A major feature of the OGSA is that it is service-oriented, and the process of data exchange within a GRID should be transparent, efficient, and secure [2], [5], [7], [8], [9], [23]. The OGSA has defined three groups of services that occur within a wide variety of GRID systems [27]: *core services*, *data / information related services*, and *computation management related services*. This paper focuses on the last aspect. OGSA has also identified a range of supporting services, including data access agreement, resource provisioning and resource reservation, for supporting the program execution and resource management services in GRIDs [27]. Given the size of a Global GRID, manual management is not feasible. This paper suggests that there is a need for an *efficient, scalable, secured* [12] *self-managed* management system that is capable of dynamically compositing, installing, controlling and managing these supporting services; in turn these supporting services should provide services to the OGSA defined *program execution* and resource management services.

M. Bubak et al. (Eds.): ICCS 2004, LNCS 3038, pp. 141–148, 2004.
© Springer-Verlag Berlin Heidelberg 2004

2 Self-Management in GRID

Self-management is defined as using autonomic principles to provide management functionalities. An important corollary is that the management function itself is an autonomic service, and is therefore subjected to the same requirements as other autonomic services. Self-management takes many forms [4]. Some fundamental building blocks include: a) self-configuration – this involves policies that direct changes to the configuration of a component, maintain desired functionalities or provide new functionality; b) self-healing – the system can detect, diagnose, and repair hardware, software, and firmware problems; c) self-protection, self-securing – the system can automatically defend against attacks, as well as detect and stop cascading internal failures; d) self-optimisation – the system and its components can continually seek ways to improve their behaviour and becoming more efficient; e) self-learning – the system learns from past events and morphs active policies to improve behaviour. Some of these building blocks combine into more powerful functions. For example, self-maintenance can be viewed as the combination of all of the above. The following are suggested supporting self-management services that should be deployed in a self-management system in a GRID environment [6]: a) monitoring of autonomic resources and workloads, for the detection of network operation glitches, monitoring dynamic network status, collecting required network information, locating resources and ensuring their availability, real-time memory management etc; b) intelligent decision-making facility - based on real-time network status, an intelligent module should decide when to switch tasks in order to utilise resources efficiently; c) dynamic reconfiguration of network entities' behaviour: automatically re-configuration of network entities' status to ensure real-time on-demand requests of users' tasks are satisfied; d) general system administrative functions: automatic system activity logging for administrators to debug; e) security services: entities on heterogeneous networks of a GRID should create security associations between each other. An autonomous decision-making facility should also decide whether to accept a service request based on pre-defined security policy. Note that in order to provide these supporting services, additional software modules should be installed on existing network devices. We shall discuss shortly how the dynamic composition, installation and management of these software modules could be achieved through programmable techniques

3 Programmable Techniques for Self-Management

3.1 Programmable GRID

As discussed previously, GRID computing requires distributed resources to be adaptively congregated and used based on service's needs, and a GRID management system should be self-managed due to the size and complexity of GRID. We have already outlined the requirements and the required supporting services that are needed for a GRID self-management system. Supporting services should be introduced to the current network in form of software modules. One approach to self-management is the creation of a programmable infrastructure [13] of computing and storage resources that are dynamically reconfigurable during the execution lifetime of user applications; the dynamic feature of a programmable infrastructure enables the composition, installation and configuration of the service components that are required by the GRID self-management system. This is because programmable technologies [13-20] have been developed as an efficient and flexible tool for dynamically deploying new services and functionalities. With programmable technologies, the traditional store-and-forward network is transformed into a store-compute-and-forward type network: in the latter type of network, packets are no longer simple data packets but are now control packets that carry control information. Control information carried in a packet will be executed on programmable network nodes for node reconfiguration purposes and to meet service demands. New services in form of software modules and new network node functionalities may be installed and controlled on-demand. Thus programmable techniques provide the opportunity for dynamic loading and installation of software modules that provide and support the necessary self-management services in GRID. In the remainder of the paper an example self-management service is discussed, which is based on programmable techniques.

3.2   Overlay Autoconfiguration in GRIDs

A programmable GRID infrastructure should be able to operate as autonomously as possible in order to simplify management and facilitate scalability of these dynamic and potentially extremely complex systems. This requires developing a set of mechanisms and protocols for automation of different tasks in the management of programmable GRIDs. One such task is the automatic configuration of (core) overlays in programmable GRIDs, which is addressed in this paper.

Programmable GRIDs are network aware in the sense that they include active network nodes from the underlying communications infrastructure and explicitly use their processing and caching capabilities. This network awareness makes it possible to achieve performance gains for GRID applications simply by configuring the programmable GRID topology to match the underlying network infrastructure topology, i.e. directly connecting only neighbour active network nodes (ANNs) in the programmable GRID. On one hand, this enhances the performance of programmable GRID applications, because communications paths between GRID nodes are shorter and consequently communication delays are smaller. On the other hand, this also preserves the bandwidth in the underlying network, because information exchanged between GRID nodes is transfered along shorter paths and traverses each network path only once. With non-matching GRID and network topologies, data between GRID nodes can traverse the same network path more than once. Just as an example, consider a star GRID topology on top of the ring network topology. Last but not least, application of some active network facilities to programmable GRID systems, e.g. solutions for security [30] and interoperability [29] issues, requires that an ANN keep an up to date list of neighbour ANNs. Generally, neighbour ANNs in an programmable GRID overlay are generally not directly connected, i.e. they may be several hops apart. This makes neighbour discovery in an programmable GRID overlay a non-trivial problem and fundamentally different from physical neighbour discovery, which is based on a simple principle of boradcasting a query message on the physical link.

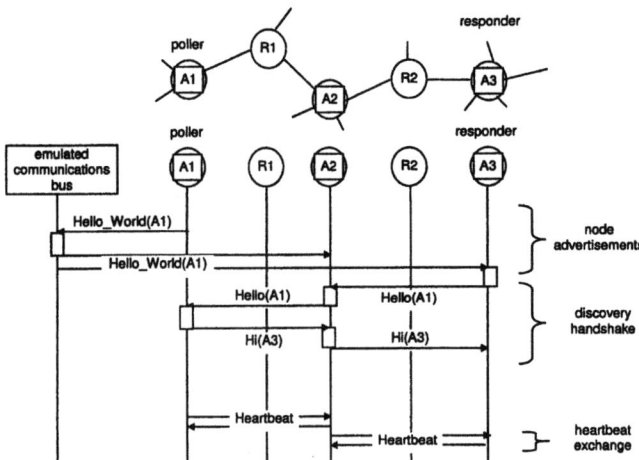

Fig.1. Protocol operation overview

3.3 Neighbour Discovery Protocol

The neighbourhood relation between a pair of ANNs in a programmable GRID is defined by the underlying network. Two nodes A1 and A2 are neighbours in a GRID overlay if network routes are defined such that there are no other ANNs on the path that packets traverse between A1 and A2. We have developed a protocol for automatic discovery of neighbour ANNs [36], which has three operation modes: node advertisements, discovery handshake, and heartbeat exchange.

The protocol uses four messages. The message header is common to all protocol messages, while messages differ in the format and semantics of the message body. Only a short description of the most important header fields is given here.

Message Type (MT, 4 bits): Specifies the type of the protocol message.
Intermediary (I, 1 bit): The main purpose of this flag is to trigger the discovery handshake, if set in the Heartbeat message.
Overlay ID (32/128 bits): Identifier of the programmable GRID overlay.
Source ID (32/128 bits): Identifier of the source node (see next paragraph).

Nodes use a single, globally unique ID to identify themselves to other nodes. For example, much like in the OSPF routing protocol [31], a node can select its ID to be the lowest or the highest among its IP-addresses. The format of all node ID fields in the message is determined by the value of the Address Type (AT) field in the header. Currently, the message can carry either a 32-bit IPv4 address or a 128-bit IPv6 address, which is defined by the value of the header field AT.

ANNs advertise themselves by periodically transmitting the Hello World protocol message to an emulated communications bus, which distributes the advertisement to all overlay members. The implementation of an emulated communications bus can be multicast-based [32, 33], with one dedicated multicast group per overlay, server based, or a combination of both. The Hello World message has an empty message body and the timeout value TO_{HW} defines the length of the period between the transmission of two consecutive messages Hello World by a node.

When the Hello World message is received, the receiving node (say, node X) only starts a discovery handshake with the sending node (say, node Y), if it does not already keep state for that node, i.e. keep it in neighbour list or keep state on a running handshake with Y. The handshake is started with probability p, which is used to limit the load on the node Y: if there are N members of a programmable GRID overlay, then on average only *(p*N)* nodes instead of all N nodes start a discovery handshake with sender Y simultaneously. Note that this does not mean that neighbour ANNs are chosen in a random fashion.

The discovery handshake is the main protocol mechanism for neighbour discovery and detection of network changes. It involves an exchange of two messages, Hello and Hi, between the node, which initiates the handshake (poller), and the node, which responds (responder). Both messages are processed and modified by every ANN en route between the poller and the responder, i.e. neighbour discovery is based on *packet interception* (see later on section).

The Hello and Hi messages contain two additional fields in the message body: the Destination ID and the Hop ID. Destination ID is the identifier (address) of the responder. Hop ID is the identifier of the current ANN en route between the initiator and responder. The transmission of the Hello message is always triggered, either by the message Hello World or by the I flag in the Heartbeat message. Similarly, the Hi message is sent only by the responder, indicated by the Destination ID in the Hello message, and only in response to the Hello message.

Upon interception of the Hello message an intermediate node first adds the previous hop ANN (Hop ID) to its neighbour list, unless it is already there. Then the node replaces the Hop ID value with its own identifier and forwards the Hello message towards the responder. The responder also adds the previous hop ANN into neighbour list and sends to the poller the Hi message, which is processed en route in the same manner.

The discovery handshake uses checksums, timeouts, retransmissions, and implicit acknowledgments to provide reliable message delivery service best suited to its needs. Use of TCP for this purpose is inappropriate for two reasons: its overhead is unacceptably high and its reliability service is not well suited for discovery handshake.

A node periodically sends the Heartbeat message to each of its known neighbours to indicate that it is still alive. The message body contains one additional field. the address of the destination node (Destination ID). This message is also processed by all ANNs en route to destination. ANNs use this message to refresh soft states in their neighbour list, detect changes in the network, and trigger a discovery handshake, when change is detected.

A node knows that it is intermediary, if its ID is not equal to the Destination ID in the Heartbeat message. If two nodes exchange Heartbeats, then they are neighbours and there should be no intermediate ANNs between them. Therefore an intermediate ANN sets the I flag in Heartbeat, thus indicating that the network change has occurred and that the receiver should start a discovery handshake.

We have seen that both discovery handshake and heartbeat exchange are based on packet interception. That is, an intermediate ANN en route between the poller X and the responder Y must intercept and process the protocol packet (message), even though it is not addressed to this intermediate ANN.

Contrary to traditional GRIDs, which are based on traditional IP networks, packet interception [34], [35] is inherently supported in programmable GRIDs, since ANNs embody the flexible *store-compute-an-forward* model, which is in contrast to the *store-and-forward* model of traditional IP routers. Thus, by definition, ANNs in programmable GRIDs can intercept packets destined for some other IP address, process them in a specific way (aside from routing), and then forward them towards their destination.

3.4 Proof-of-Concept Implementation and Initial Test Results

A prototype implementation of the protocol with the ANTS toolkit [37] is being developed. Each protocol message is implemented in the form of a distinct ANTS capsule with simple forwarding routine. The core of the protocol is implemented as a separate module, which operates on top of the ANTS system and provides features for discovery of neighbour ANNs. By varying different protocol parameters, such as timeout values, retransmit counters, and probability p, the protocol convergence delay is within the range of 30 - 40 seconds, and network losses are within the range of 2% - 5%. The observed protocol overhead corresponding to this convergence delay is within the range 0.5 - 1 kb/s per node for the test configuration. Given these results and that the above convergence delay is within our target value for initial experiments, we find that the protocol performance (i.e. the observed protocol convergence delay and the network losses) meets our requirements. The above target value for protocol convergence was chosen, because the convergence delay of current routing protocol implementations ranges from around 30 seconds for OSPF [38] to several minutes for BGP [39]. We find the comparison with the routing protocols interesting due to the following reason: as far as the authors are aware, no results on convergence delay are currently available for related work in the area of automatic discovery of neighbour ANNs (e.g. [40]). On the other hand, routing protocols provide analogous functionality to the neighbour discovery protocol: after a perturbation occurs in the network, the routing protocol has to accordingly reconfigure the physical network (i.e. routes in the network) as soon as possible in order to minimise the disruption in communication caused by the perturbation.

4   Conclusions

The highly distributed feature and flexibility of GRID services makes them subject to the limitations of distributed programming such as performance, scalability, security. Existing GRIDs tend to be homogeneous. As such, GRID services would require some sort of autonomous self-managed management components and systems. These self-management capabilities should include: self-configuration, self-optimisation, self-healing, self-securing, self-protecting, and self-learning. This paper has identified a set of supporting services that are required for a self-management system for a GRID, and has suggested that these supporting services must be installed dynamically in GRIDs. We have presented a protocol that enables ANNs in programmable GRIDs to automatically discover neighbouring ANNs. This automates the management of core overlays in programmable GRIDs, improves the performance of GRID applications by reducing communications delay, enables specific solutions from active and programmable networks to be applied to GRIDs, and preserves bandwidth by constantly adapting the overlay to match the underlying network topology in face of changes in the network infrastructure.

Acknowledgement. This paper partly describes work in progress in the context of the EU IST project CONTEXT [22]. The IST programme is partially funded by the Commission of the European Union.

References

1. http://www.hoise.com/primeur/03/articles/monthly/UH-PR-02-03-7.html
2. http://wwws.sun.com/software/cover/2003-1104/
3. http://www.datasynapse.com/solutions/diff.html
4. Waldrop, M.: Autonomic Computing – The Technology of Self-Management, http://www.thefutureofcomputing.org/Autonom2.pdf
5. http://eu-grasp.net/english/dissemination/presentations/Valles_industry.pdf
6. IST Project proposal, "Programmable GRID Network Solutions and Prospects"
7. http://www.gridcomputingplanet.com/news/article.php/2226121
8. http://setiathome.ssl.berkeley.edu/
9. http://biogrid.icm.edu.pl/
10. http://www.pallas.com/e/products/unicore-pro/index.htm
11. Galis, A., Gelas, J., Lefevre, J., Yang, K.: Active Network Approach to GRID Management & Services, in Workshop on Innovative Solutions for Grid Computing - ICCS 2003 Conference, LNCS 2658, ISBN 3-540-40195-4, Melbourne, Australia, June 02-04, (2003) 1103-1113
12. Smith, R.E.: Authentication: From Passwords to Public Keys, Addison-Wesley, ISBN: 0-201-61599-1, 110-112
13. FAIN (Future Active IP Networks) www.ist-fain.org
14. SNAP (Safe and Nimble Active Packets) http://www.cis.upenn.edu/~dsl/SNAP/
15. FAIN Internal Report R25.2 "SNAP, ANEP, and Security", Nov 2002 http://www.ist-fain.org
16. DARPA Program: www.darpa.mil/ito/research/anets/projects.html
17. Open Signalling Working Group, http://www.comet.columbia.edu/opensig/.
18. Galis, A., Denazis, S., Brou, C., Klein, C. (ed): Programmable Networks for IP Service Deployment, ISBN 1-58053-745-6; pp450, contracted for publishing in Q1 2004 by Artech House Books, 46 Gillingham Street, London SW1V 1AH, UK; www.artechhouse.com (2004)
19. Cheng, L., Galis, A., Eaves, W., Gabrijelcic, D.,: Strong Authentication for Active Networks, SoftCom 2003- 7-10 October 2003, Split (2003)
20. Suzuki, T., Kitahara, C., Denazis, S., Cheng, L., Eaves, W., Galis, A., Becker, T., Gabrijelcic, D., Lazanakis, A., Karetsos, G. -"Dynamic Deployment & Configuration of Differentiated Services Using Active Networks"- IWAN2003 Conference, 10-12 December 2003, Kyoto (2003)
21. Foster, I., Kesselman, C., Nick, J.M., Tuecke, S.: The Physiology of the Grid – www.globus.org/research/papers/ogsa.pdf
22. CONTEXT Project WWW Server: http://ontext.upc.

23. Berman. F.,, Fox, G.C., Hay, A.J.G.: Grid Computing- Wiley 2003, ISBN 0-470-85319-0,
24. http://www.globus.org/ogsa/
25. Christensen, E., Curbera, F., Meredith, G., Weerawarana, S.: Web Services Description Language (WSDL) 1.1, W3C, 2001, www.w3.org/TR/wsdl (2001)
26. Ballinger, K, Brittenham, P., Malhotra, A., Nagy, W, Pharies, S.: Web Services Inspection Language Specification (WS-Inpection) 1.0, IBM and Microsoft (2001) http://www-106.ibm.com/developerworks/webservices/library/ws-wsilspec.html
27. Curbera, F., Goland, Y., Klein, J., Leymann, F., Roller, D., Thatte, S., Weerawarana, S.: Business Process Execution Language for Web Services, Version 1.0 BEA Systems, IBM, Microsoft (2002) http://msdn.microsoft.com/library/default.asp?url=/library/enus/dnbiz2k2/html/bpel1-0.asp
28. Foster: The Open Grid Services Architecture, http://www.nesc.ac.uk/teams/OGSAreviewPeterKunszt.pdf (2003)
29. Keneth L. Calvert.: Architectural Framework for Active Networks version 1.1. Technical report, University of Kentucky (2001)
30. Savanovic, A., Gabrijelcic, D., Blazic, B.J., Karnouskos, S.: An Active Networks Security Architecture. Informatica (2002) 26(2):211.221
31. Moy, J.: OSPF Version 2. RFC 2328 (1998)
32. Kosiur, D.: IP Multicasting. John Wiley & Sons, Inc. (1998)
33. Maimour, M., Pham, C.D.: An Active Reliable Multicast Framework for the Grids. In International Conference on Computational Science (2002) 588-597
34. Lefeevre, L., et. al.: Active networking support for the Grid. Lecture Notes in Computer Science (2001) 2207:16.33
35. Mohamed, N.: Active Networks and Their Utilization in the Computational Grid. Technical Report TR03-04-01, Engineering University of Nebraska-Lincoln, Lincoln, NE 68588-0115 (2003)
36. Savanovic, A.: Automatic Discovery of Neighbour Active Network Nodes. Technical Report IJS DP-8725, Jozef Stefan Institute, Jamova 39, 1000 Ljubljana, Slovenia, January (2003)
37. ANTS. http://www.cs.utah.edu/flux/janos/ants.html.
38. Basu, A., Riecke, J.G.: Stability Issues in OSPF Routing. In Proceedings of SIGCOMM (2001) 225-236
39. Labovitz, C., Ahuja, A., Bose, A., Jahanian, F.: Delayed Internet Routing Convergence. In Proceedings of SIGCOMM (2000) 175-187
40. Martin, S., Leduc, G.: RADAR: Ring-based Adaptive Discovery of Active neighbour Routers. In Proceedings of IWAN 2002 (2002) 62-73

Application-Specific Hints in Reconfigurable Grid Scheduling Algorithms

Bruno Volckaert[1], Pieter Thysebaert[2], Filip De Turck[3],
Bart Dhoedt[1], and Piet Demeester[1]

[1] Department of Information Technology, Ghent University - IMEC
Sint-Pietersnieuwstraat 41, B-9000 Gent, Belgium
Tel.: +32 9 267 3587, Fax.: +32 9 267 35 99
{bruno.volckaert,pieter.thysebaert}@intec.ugent.be
[2] Research Assistant of the Fund of Scientific Research - Flanders (F.W.O.-V.)
[3] Postdoctoral Fellow of the Fund of Scientific Research - Flanders (F.W.O.-V.)

Abstract. In this paper, we investigate the use of application-specific hints when scheduling jobs on a Computational Grid, as these jobs can expose widely differing characteristics regarding CPU and I/O requirements. Specifically, we consider hints that specify the relative importance of network and computational resources w.r.t. their influence on the associated application's performance. Using our ns-2 based Grid Simulator (NSGrid), we compare schedules that were produced by taking application-specific hints into account to schedules produced by applying the same strategy for all jobs. The results show that better schedules can be obtained when using these scheduling hints intelligently.

1 Introduction

Computational Grids consist of a multitude of heterogeneous resources (such as Computational, Storage and Network resources) which can be co-allocated for the execution of applications or jobs. The allocation of resources to particular jobs and the order in which these jobs are processed on the Grid are determined by the Grid's management infrastructure through the application of a *scheduling algorithm*. Most jobs will need access to different resource types during their execution, meaning job execution progress depends on the quality of service delivered to that job by *every* resource involved. The exact sensitivity of a job's computational progress w.r.t. the individual resources' performance depends on the "nature" of that job: jobs that require huge amounts of CPU power, but perform (relatively) little I/O operations, will only suffer lightly from a temporary degradation of e.g. available network bandwidth, but cannot withstand a sudden loss of CPU power. Conversely, the computational progress made by an I/O-bound job is influenced dramatically by the network bandwidth available to that job, and to a lesser extent by the variation in available computing power. This leads us to the observation that:

M. Bubak et al. (Eds.): ICCS 2004, LNCS 3038, pp. 149–157, 2004.
© Springer-Verlag Berlin Heidelberg 2004

1. algorithms that schedule jobs on a Computational Grid ought to take into account the status of multiple different resource types instead of solely relying on e.g. the available computational power.
2. using the same scheduling algorithm with rigid constraints for all job types can be outperformed by applying different scheduling algorithms for different job types; each job-specific algorithm only performs rigid resource reservation with *critical* resources, but allows for relaxed resource availability constraints when dealing with non-critical resources.

This indicates that programmable architectures, where the job scheduling mechanism is provided (at least partly) by the application (and where the algorithms could even be adapted on the fly), are a promising avenue towards grids offering a wide variety of services (each having their specific service metrics and quality classes). In this approach, the grid infrastructure is coarsely managed by cross-service components, supplemented by on-the-fly configurable service specific management components. The latter components manage the resources allocated to the service on a fine grained level, optimizing job throughput and service quality simultaneously according to service specific attributes and metrics. In this paper we show how *scheduling hints* can be incorporated into job descriptions. The goal of these hints is to enable a Grid scheduler to estimate the critical level of the different resource types w.r.t. that job. Because hints are contained in the job description, they are available at each scheduler in the Grid to which the job is submitted or forwarded.

This paper continues as follows: Sect. 2 starts with a short description of the related work. In Sect. 3, we give an overview of the relevant simulation models used: the Grid, Resource, VPN, Job and Scheduling Hint models are explained in detail. In Sect. 4, we discuss the various algorithms that we compared; they differ from each other in (i) the types of resources they take into account and (ii) whether or not they treat all jobs equally. Our simulated scenario and corresponding results are presented in Sect. 5, leading to the conclusions in Sect. 6.

2   Related Work

Well-known Grid Simulation toolkits include *GridSim* [1] and *SimGrid* [2]. The key difference with *NSGrid* [3] is that NSGrid makes use of a network simulator which allows for accurate simulation down to the network packet level (ns-2 [4]).

Scheduling jobs over multiple processing units has been studied extensively in literature. Machine scheduling [5][6] is concerned with producing optimal schedules for tasks on a set of tightly-coupled processors, and provides analytical results for certain objective functions. Jobs are commonly modelled as task graphs, or as continuously divisible work entities. As these models do not deal with "network connections" or "data transfers", they do not capture all the Grid-specific ingredients described in the previous section. Grid scheduling strategies which take both computational resource load and data locality into account are extensively discussed in [7]. The use of Application-specific scheduling hints is not considered however.

The *Metacomputing Adaptive Runtime System* (MARS) [8] is a framework for utilizing a heterogeneous WAN-connected metacomputer as a distributed computing platform. When scheduling tasks, the MARS system takes into account Computational Resource and Network load, and statistical performance data gathered from previous runs of the tasks. As such, the MARS approach differs from the scheduling model simulated by NSGrid, as NSGrid allows for *user preferences* to be taken into account as well.

Application-level scheduling agents, interoperable with existing resource management systems have been implemented in the *AppLeS* [9] work. Essentially, one separate scheduler needs to be constructed per application type. Our simulation environment allows the simulation of multiple scheduling scenarios, including those using a single centralized schedule as well as those having multiple competing schedulers (not necessarily one per application type).

3 Simulation Model

3.1 Grid Model

Grids are modelled as a collection of interconnected and geographically dispersed *Grid sites*. Each Grid Site can contain multiple *resources* of different kinds such as Computational Resources (CRs) and Storage Resources (SRs) interconnected by VPN links. At each Grid Site, resource properties and status information are collected in a local *Information Service*. Jobs are submitted through a *Grid Portal* and are scheduled on some collection of resources by a *Scheduler*. To this end, the scheduler makes *reservations* with the appropriate *Resource Managers*.

3.2 Grid Resource Models

Each Grid site can offer one or more CRs and/or SRs. A CR is a monolithic entity, described by its total processing power in MIPS, the maximum number of jobs that it can handle simultaneously and the maximum slice of processing power (in MIPS) that can be reserved for a single job. An SR on the other hand, serves the purpose of providing disk space to store input and output data. In our model, their basic properties include the total available storage space, the input data sets currently stored at the resource and the speed at which the resource can read and write data. While a SR does not perform computational work, it can be attached to the same network node as some CR. Interconnections between local resources are modelled as a collection of point-to-point VPN links, each offering a guaranteed total bandwidth available to Grid jobs. Of course, these VPN links can only be set up if, in the underlying network, a route (with sufficient bandwidth capacity) exists between the nodes to which these resources are attached. Different Grid Sites can be interconnected by a VPN link. These models are covered in more detail in [3].

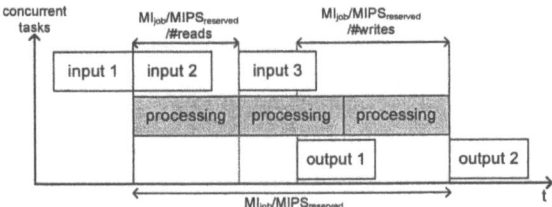

Fig. 1. Simulated job lifespan with indication of start-of-I/O events; non-blocking job

3.3 Job Model

The atomic (i.e. that which cannot be parallelized) unit of work used throughout this paper is coined with the term *job*. Each job is characterized by its length (measured in *instructions*), its required *input data sets*, its need for *storage*, and the *burstiness* with which these data streams are read or written. During a job's execution, a certain minimal computational progress is to be guaranteed at all times (i.e. a deadline relative to the starting time is to be met).

Knowing the job's total length (in million instructions, MI) and the frequency at which each input (output) stream is read (written), the total execution length of a job can be seen as a concatenation of instruction "blocks". The block of input data to be processed in such an instruction block is to be present before the start of the instruction block; that data is therefore transferred from the input source at the start of the previous instruction block. In a similar way, the output data produced by each instruction block is sent out at the beginning of the next instruction block. We assume these input and output transfers occur in parallel with the execution of an instruction block. Only when input data is not available at the beginning of an instruction block or previous output data has not been completely transferred yet, a job is suspended until the blocking operation completes. The presented model allows us to mimic both *streaming* data (high read or write frequency) and *data staging* approaches (read frequency set to 1). A typical job execution cycle (one input stream and one output stream) is shown in Fig. 1.

3.4 Scheduling Hints Model

From the job model described in the previous section, it is clear that the computational progress made by a job is determined by both the computational power and network bandwidth available to that job. As such, scheduling hints (distributed together with the job description) describe

 - the resource types that are to be taken into account when scheduling this job; any subset of {Computational / Network Resource} can be specified.
 - for each of the above resource types, the size of an acceptable (not preventing the job from being scheduled on that resource) deviation from the resource's performance delivered to that job (described in the job requirements).

It is not desirable to have critical resources deliver a less-than-minimal performance to the job, while this may not matter much for non-critical resources.

4   Algorithms

When jobs are submitted to a Grid Portal, a Scheduler needs to decide where to place the job for execution. As has been mentioned previously, we discriminate between algorithms using two criteria: the type of resources they take into account and whether or not they take into account scheduling hints. If the scheduler is unable to allocate the needed resources for a job, the job gets queued for rescheduling in the next scheduling round. The time between two scheduling rounds can be fixed, but it is also possible to set a threshold which triggers the next scheduling round. During each scheduling round, every algorithm processes submitted yet unscheduled jobs in a greedy fashion, attempting to minimize job completion time. Once scheduled, our scheduler does not pre-empt jobs.

4.1   Algorithm "NoNetwork"

As the name implies, this algorithm does not take into account the status of Network Resources when scheduling jobs. Rather, it assumes that only CRs are critical for each job. Furthermore, it will treat minimal job requirements as hard constraints; it disregards hints that might propose a softer approach. At first, "NoNetwork" will attempt to place a job on a site's local CRs, only using remote resources when strictly necessary (we believe this to be a plausible approach from an economic viewpoint). If this is impossible, and at least one remote CR is available, that job will be scheduled on the remote CR offering most processing power. It is therefore expected that this algorithm will perform badly when dealing with a significant amount of "I/O-bound" jobs: due to "blocking", these jobs will finish considerably later than predicted by the scheduling algorithm.

4.2   Algorithm "PreferLocal"

Similar to the "NoNetwork" algorithm, "PreferLocal" will a priori attempt to place a job on a site's local CRs. If this turns out to be impossible, remote CRs will be considered. While looking for the best resources for a particular job, however, "PreferLocal" not only considers the status of CRs, but also the residual bandwidth on network links connecting Computational and Storage Resources. The best resource combination is the one that maximizes the job's computational progress. For a job requiring one CR and one SR (in different Grid Sites, connected through a VPN link), the maximal computational progress (expressed in MIPS) that can be delivered to that job is given by

$$MIPS_{eff} = \min_{CR, VPN}(MIPS_{CR}, \frac{MI * BW_{VPN}}{8 * DATASIZE})$$

It is easily verified that it makes no sense to allocate a bigger portion of the best CR's power to this job, as due to network limitations, the job cannot be processed at a higher rate. In a similar way, due to CR limitations, it makes no sense to allocate more bandwidth to the job than $\frac{8*DATASIZE*MIPS_{eff}}{MI}$. Like "NoNetwork", "PreferLocal" does not adapt its strategy using job-specific hints.

4.3   Algorithm "Service"

Algorithm "Service", like "PreferLocal", considers both Computational and Network Resources when scheduling jobs. However, instead of immediately rejecting those resource combinations that would not allow some of the job's minimal requirements to be met, it can still select those resources (if none better are found) if this is declared "acceptable" by the appropriate job hint. For instance, jobs can specify that their available network bandwidth requirements are less important than their computational requirements (and/or quantify the relative importance), or that there is no gain in finishing the job before its deadline (i.e. no gain in attempting to maximize that job's $MIPS_{eff}$). Using these hints, jobs can be divided into different classes, where all jobs in one class have similar associated hints. The algorithm can then be seen as delivering the same service to each of those jobs in a single class.

5   Simulation Results

5.1   Simulated Grid

A fixed Grid topology was used for all simulations presented here. This topology is depicted in Table 1. Grid control components are interconnected by means of dedicated network links providing for out-of-band Grid control traffic (as shown by the dotted network links).

Table 1. Sketch of simulated scenario

Service type	I/O size	MI
I/O-bound	6100 MB	12500000
CPU-bound	0.4 MB	25000000

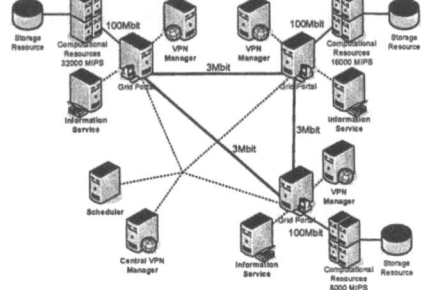

5.2   Simulated Jobs

In our simulations, two types of hints were used (i.e. two service types). The first type of hint is supplied with CPU-bound jobs, and specifies that the job should be scheduled on the fastest CR, even if this means scheduling the job on a remote resource when it could have been scheduled locally. The second type of hint is distributed with I/O-bound jobs, stating that these jobs are better off being scheduled using only local resources, as this offers better chances of allocating sufficient network bandwidth. In both cases, however, resource loads are not ignored; rather, the *preferred* execution rate for a job is no longer treated as a rigid minimum.

We have compared the "Service" algorithm (which understands and uses the hints as specified) for this job set with the "NoNetwork" and "PreferLocal" algorithms, both disregarding the hints.

5.3   Per-Class Response Times

The average job response time for each algorithm is shown in the left part of Fig. 2. The figure shows both the overall response rate and the per-class response rates when 20% I/O jobs and 80% CPU intensive jobs are submitted to the Grid. As expected, "NoNetwork" fails to produce good schedules for I/O-bound jobs, as it ignores network loads (which are, of course, of particular importance to this type of job). In addition, notice that the use of hints (algorithm "Service") improves the average response time for CPU-bound jobs (which make up most of the jobs in this simulation). Indeed, some jobs are now processed at a rate (slightly) lower than the preferred one (the goal of the hints is exactly to specify that this is allowed), but finish sooner than if they were delayed in time.

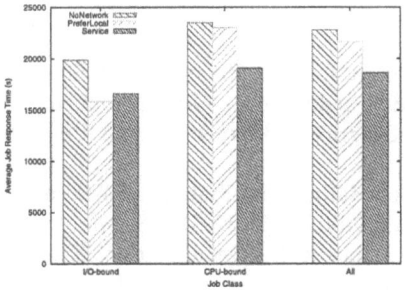   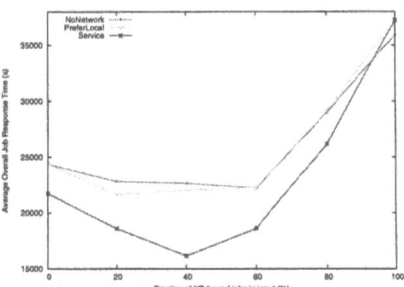

(a) average job response time   (b) average job response time as function of input job set

Fig. 2. NoNetwork, PreferLocal and Service schedule performance

5.4   Response Times with Varying Class Representation

In these simulations we stressed the three Grid sites by submitting a heavy job load (parameterized by the percentage of I/O-bound jobs in the total job load). The resulting average job response time for the three algorithms (as a function of the percentage of I/O-bound jobs) is shown in Fig. 2 (right side).

When only CPU-bound jobs are submitted, "NoNetwork" performs like "PreferLocal", as we remarked previously. Note that "Service" performs slightly better, as this algorithm does not prefer local resources over remote. When the amount of I/O-bound jobs is increased, "NoNetwork" performance degrades as network load status gains importance when scheduling jobs. Since "PreferLocal" always tries local resources first (instead of the best resources), it will schedule more I/O-bound jobs remotely (i.e. using lower-bandwidth links) as CPU-bound jobs (the majority) use up these local resources; this accounts for the difference with "Service". Once the fraction of I/O-bound jobs passes 60%, the network links of our simulated Grid saturate, and the three algorithms' performance converges.

6   Conclusions

In this paper we have shown the benefits of using active scheduling mechanisms in a service oriented Grid environment. In particular, we used NSGrid to compare the efficiency of schedules produced by algorithms that do not take into account the service specific needs of a job, to schedules produced by algorithms that use service scheduling hints. It was shown that when using the latter algorithms, average job response times in our simulated scenario improved significantly (up to 30% in some cases).

References

1. Buyya, R., Murshed, M.: GridSim: A Toolkit for the Modeling and Simulation of Distributed Resource Management and Scheduling for Grid Computing, The Journal of Concurrency and Computation: Practice and Experience (CCPE), Wiley Press, (2002)
2. Legrand, A., Marchal, L., Casanova, H.: Scheduling Distributed Applications: the SimGrid Simulation Framework, Proc. of CCGrid 2003 (2003) 138–145
3. Volckaert, B., Thysebaert, P., De Turck, F., Dhoedt, B., Demeester, P.: Evaluation of Grid Scheduling Strategies through a Network-aware Grid Simulator, Proc. of PDPTA 2003 (2003) 30–35
4. "The Network Simulator - NS2", website, http://www.isi.edu/nsnam/ns
5. Feitelson, D. G., Rudolph, L., Schwiegelshohn, U., Sevcik, K. C., Wong, P.: Theory and Practice in Parallel Job Scheduling, Job Scheduling Strategies for Parallel Processing, (1997) 1–34
6. Hall, L. A., Schulz, A. S., Shmoys, D. B., Wein, J.: Scheduling to Minimize Average Completion Time: Off-line and On-line Approximation Algorithms, Mathematics of Operations Research, **22-3** (1997) 513–544

7. Ranganathan, K., Foster, I.: Simulation Studies of Computation and Data Scheduling Algorithms for Data Grids, Journal of Grid Computing, Kluwer Academic Publishers, **1-1** (2003) 53–62
8. Gehring, J., Reinfeld, A.: Mars - a framework for minimizing the job execution time in a metacomputing environment, Proc. of Future General Computer Systems '96 (1996)
9. Berman, F., Wolski, R., Figueira, S., Schopf, J., Shao, G.: Application-Level Scheduling on Distributed Heterogeneous Networks, Proc. of SuperComputing 96 (1996)

Self-Configuration of Grid Nodes Using a Policy-Based Management Architecture

Félix J. García[1], Óscar Cánovas[2], Gregorio Martínez[1],
and Antonio F. Gómez-Skarmeta[1]

[1] Department of Information and Communications Engineering
[2] Department of Computer Engineering
University of Murcia, 30071, Murcia, Spain
{fgarcia,gregorio,skarmeta}@dif.um.es, ocanovas@ditec.um.es

Abstract. During the past years, Grid Computing has emerged as an important research field concerning to large-scale resource sharing. Several research and academic institutions have widely adopted this technology as a mechanism able to integrate services across distributed, heterogeneous, and dynamic virtual organizations. However, this rapid expansion is exposing the need to provide effective means to manage virtual organizations, especially those questions related to the insertion and maintenance of grid nodes. This paper proposes a flexible and automated method based on policies to cope with these requirements. Policies constitute a mechanism for specifying the behaviour a grid node must exhibit. In this work we present a policy-based management architecture, which makes use of COPS-PR to exchange information about specific parameters related to grid nodes based on Globus Toolkit 2.4. Moreover, we have also added new classes to the Framework PIB (Policy Information Base) in order to represent configuration data about the different components of this toolkit.

1 Introduction and Rationale

Increasingly, computing systems are becoming more decentralized and distributed. In addition, companies are realizing that they can save costs by outsourcing some elements of their computing systems. This continuous decentralization makes essential the control on how the resources are dynamically assembled. Work within this field has lead to the development of Grid Technologies [1], widely adopted by scientific and academic institutions.

These collections of resources and users from geographically distributed organizations unified by a common goal are called *virtual organizations (VO)* [2]. VOs introduce challenging management and policy issues, resulting from complex relationships between local sites and the goals of the VO with respect to resource allocation or security administration. Since VO resources controlled by grid nodes are located within multiple organizations, maintaining a consistent system-wide configuration among the grid nodes requires a consistent manner to join an existing virtual organization. A VO is defined by the set of elements it is composed of, and also by the different policies governing these elements. Therefore, on the one hand, any potential grid node joining the VO must initially conform to the overall policy, which involves an appropriate distribution of the related policy and its enforcement. On the other hand, this situation can be more complex if we consider the fact that resources may be highly dynamic, which can lead to the definition of policies that can vary during time. Changes on the VO policy must be notified to the existing grid nodes in order to adapt their behaviour to the current policy. Despite its importance, relatively little system-oriented research has addressed this topic. An exception that has encouraged our work is [3], although it follows a different approach based on active networks.

Thus, the management of the overall Grid system in an automated and flexible manner is becoming more and more important. Manual configuration of systems is an arduous task and error-prone. The concept of policy-based management (PBM) addresses some of these issues and offers possible solutions. Policies allow the administrators of a Grid node to specify the behaviour they want it to exhibit, through a set of selections made from a range of parameters related to the nodes. Moreover, policies can be dynamically changed according to the evolution of the system or management strategy (for example, with the insertion of a new trusted third party related to a new organization in the VO, or a new information server). The main goal is to adapt a dynamic VO to different configurations rather than hard-coding a particular one. This can be accomplished defining the different roles that can be played by the grid nodes, and therefore specifying the policy that will be associated to those roles. In next sections we will present several roles for grid nodes depending on the components they include.

Policies are defined using high-level tools and languages, and then are distributed, either directly or through the use of an intermediate repository, to special policy servers called policy decision points (PDPs). Finally, the grid nodes, acting as policy enforcement points (PEP), contact the PDPs in order to obtain an instance of the policy. It is possible that a particular grid node (or set of nodes) has some preconfigured internal policy statements and then a conflict can appear with the VO general policy. For solving this, policies have a level of priority. In the case that a conflict cannot be solved with this procedure, it will be properly reported to the network managers involved in the policy definition.

In this work, regarding the communication protocol between PDPs and PEPs, we have selected the COPS (Common Open Policy Service) protocol [4], and specifically its extension for provisioning policy models, COPS-PR (COPS Usage for Policy Provisioning) [5]. As we will show in next sections, Grid systems can be integrated with COPS by extending a PIB (Policy Information Base), i.e., a data structure containing policy data.

This paper is organized as follows. Section 2 provides an overview of the Globus Toolkit and specifies how the use of policies can be integrated into the management of the different components of the toolkit. Section 3 presents the elements and protocols of the policy-based management architecture for Grid Computing. Section 4 outlines our proposed Grid information model and the PIB extensions that have been designed to exchange data about configuration of grid nodes making use of the Globus Toolkit. Finally, we conclude the paper with our remarks and some future directions derived from this work.

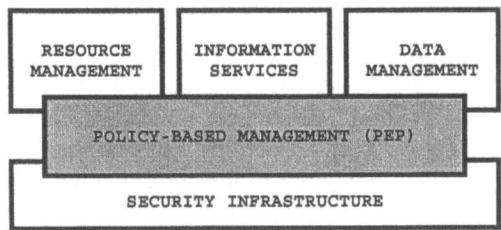

Fig. 1. View of the Globus Toolkit with PBM support

2 PBM Support for the Globus Toolkit

The software toolkit developed within the Globus project comprises a set of components that implement basic services for security, resource location, resource management, data access, communication, and so on. Globus is built as a layered architecture of services, from which developers can select to build custom tailored tools or applications on top of them. These services are distinct and have well defined interfaces, and therefore can be integrated into applications/tools in an incremental fashion. This paper is based on GT 2.4 since, nowadays, it is in use at hundreds of sites and by dozens of major Grid projects worldwide, which can benefit from the integration of policy-based management architecture for self-configuration of their grid nodes. Although OGSA (Open Grid Services Architecture) [6] represents the natural evolution of GT 2.4, and it is being the main standardization effort for Grid technologies, in this paper we focus on GT 2.4 due to its high degree of adoption by the community.

Globus Toolkit components (Figure 1) include the Grid Security Infrastructure (GSI) [7], which provides a single-sign-on, run-anywhere authentication service, with support for delegation of credentials, local control over authorization, and mapping from global to local user identities; the Grid Resource Access and Management (GRAM) [8] protocol and service, which provides remote resource allocation and process creation, monitoring, and management services; the Metacomputing Directory Service (MDS) [9], an extensible Grid information service that provides a uniform framework for discovering and accessing system configuration and status information; and data management services, that is, a high-speed data movement protocol like GridFTP and mechanisms related to dataset replicas.

As it is shown in Figure 1, the policy-based management component overlaps the three pillars of the toolkit and the security infrastructure, establishing their configuration parameters according to the VO general policy. It is layered on top the GSI since it uses its security services, while acting as policy enforcement point, in order to contact to the PBM architecture, as we will introduce in Section 3. This module modifies the configuration parameters either during the join operation to a VO, or as a consequence of a maintenance task in a VO, as for instance the insertion, deletion or modification of other grid nodes. Going into details, some of the configuration values of MDS that might be established during the join operation are network parameters, distinguished names, GIIS servers, valid GRIS reporters, information providers, etc. However, some of these parameters can change over time with the insertion or deletion of GIIS servers, GRIS reporters, job managers, access rules to the directory information tree, or bind methods. Analogously, GRAM parameters can be initially specified (such as, network parameters, list of supported job managers, nodes acting as resource coordinators) and can be further modified (new resource coordinators, information to be published in the MDS). The same procedure is applied for the data management services and the security infrastructure. Therefore, on the one hand, we need a model able to reflect, in a structured manner, these and other configuration parameters and, on the other hand, a mechanism to distribute and translate that model to a data representation format, as we will see in Sections 3 and 4.

3   A PBM Architecture Providing Self-Configuration

As stated before, policy-based management (PBM) is a term describing a new concept and architecture to control and provision dynamic configurations and behaviours for network nodes and services. In the context of this research it is used for providing self-configuration to grid nodes.

The proposed architecture is composed of five elements as depicted in Figure 2. It was designed following the recommendations defined by the IETF, mainly through the Policy Framework working group, which provides a way of representing and managing policies in a vendor-independent, interoperable, and scalable manner. These features have a key significance in our research.

The administrator interacts, using his policy console (usually a web browser) with one of the policy management tools (PMT) existing in the system. This allows him to create, modify and/or delete policies specifying how the grid nodes belonging to one particular virtual organization need to be configured. In this definition phase, the communication protocol in use is HTTPS (i.e. HTTP + SSL), which provides confidentiality, integrity, and mutual authentication between the policy console and the PMT server. Moreover, the administrator using his private key digitally signs policy documents. This signature is validated by the intermediate nodes (PMTs, PDPs, and PEPs) existing in the PBM architecture.

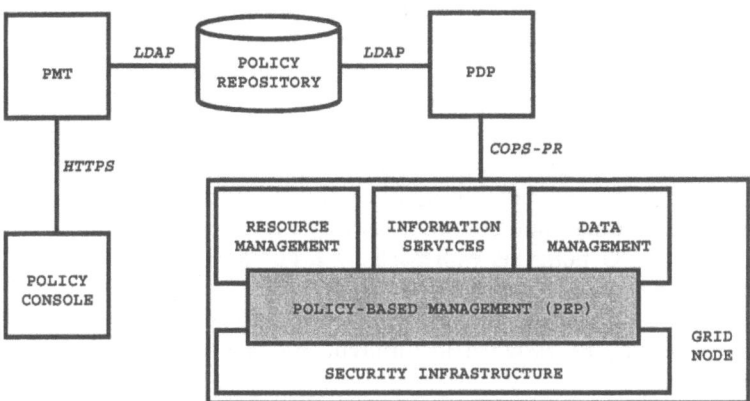

Fig. 2. General Policy-Based Management Architecture

Each PMT is composed of two main components: a policy GUI, which is an editor that can generate or edit policy documents, and a policy validator that validates every policy specification before it is stored in the policy repository. This validation process is done using a policy model, which defines the high-level syntax of every grid configuration parameter being managed using policies. Its definition, presented in next section, is based on the PCIM [10] and PCIMe [11] information models defined by the IETF.

Once the relevant digitally signed policies are stored in the policy repository using LDAP, they are conveniently retrieved by the servers in charge of taking decisions in the management system. They are called policy decision points (PDPs).

Each time one of these servers receives a new policy information (because a new document has been created, or an existing one has been modified or deleted), it has to validate its signature and decide whether this policy needs to be applied to each of the policy enforcement points (PEPs) connected to it. Moreover, when a new PEP is active in the network it needs to get or update the internal configuration it is using. As commented above, in our proposal a PEP is a component existing in every grid node.

In both cases, a communication, preferably secure, needs to be established between the PDP and the PEP. For this, we are making use of COPS-PR as a policy transfer protocol. The main reason behind this decision is that policies in COPS-PR are defined, not per individual interface (as it is in SNMP, for example), but on a per-role basis. This allows us to manage and configure grid nodes as a whole and not as a collection of individual hosts. This is also one of the major reasons of applying the PBM paradigm in this scenario.

The exchange of COPS-PR messages between a PDP server and a PEP (grid node) is secured with a TLS connection from the PEP to the PDP. This secure socket is provided by the GSI component (and its public key related information) existing in every grid node.

4   GRID-VO Policy Configuration Information Model

VO policies that are exchanged between the components of a PBM system may assume a variety of forms as they travel from a PMT to a repository, or from a PDP to a PEP. At each level, policies are represented in a way that is convenient for the current task. This means that there are several translation levels from the abstract policy to the particular PEP (grid node) configuration. Thus, it is convenient to define a configuration information model, which by definition is independent of any particular data storage mechanism and access protocol, and that can be used as a bridge in the translation process between the abstract policy and the particular configuration. The translation process is roughly expressed as follows:

1. The informal VO policy is expressed by a human policy maker (e.g., 'All job information must not be published').
2. An administrator analyzes the grid components and determines the existing roles [10]. A role may be assigned to different grid nodes, and each node may play several roles.
3. The VO administrator models the informal policy using the GRID-VO policy configuration information model (GRID-PCIM), thus creating a formal

representation of the abstract policy (e.g. *'If there is GRAM support then it does not permit to publish jobs'*).

4. The VO administrator assigns roles to the policy groups created in the previous step matching the roles of the network elements assigned in step 2.
5. A PDP translates the abstract policy created in step 3 into specific configuration parameters for all nodes affected by the new policy (i.e. grid nodes that have a role matching the policy role).
6. For each PEP (grid node) in the VO, the PDP (or an agent of the PDP) issues the appropriate node-specific instructions needed to enforce the policy.

During the second step, the administrator needs to know the roles that can be assigned to the grid nodes. Although some grid scenarios may present richer roles, we have identified four basic grid roles (which can also be combined). These are the following:

- *Individual-node*. Independent grid node not pertaining to a cluster of workstations (e.g. PCs belonging to individual users).
- *Cluster-node*. A cluster node makes reference to a computing device, which is part of a cluster and is controlled by the configuration policy related to the cluster. Software products for management of clusters (e.g. CONDOR, PBS) sometimes impose specific requirements about the behaviour of a grid node.
- *Coordinator-node*. A coordinator is a role assigned to those nodes acting as resource brokers, co-allocators, or, more generally, nodes which need to have a wider vision about the configuration and features of the VO in order to coordinate several nodes.
- *Index-node*. Nodes containing information about the VO, for monitoring or discovering services (e.g. GIIS servers).

During the third step, the administrator uses GRID-PCIM, which defines an information model for enforcing grid configuration. We have proposed GRID-PCIM based on the IETF Policy Core Information Model (PCIM) [10] an its extensions [11], and all models derive from and uses classes defined in the DMTF Common Information Model (CIM). Figure 3 shows the GRID-VO policy classes representing the set of policies contained in a grid node.

The PolicyGroup class represents the set of policies that are used in a grid node. The VO administrator, in step 4, associates this set of policies to a grid node, which is represented by the class System, depending on the roles. In the model, the class GRIDAction represents GRID actions that are policy-enforced configurations. Classes MDSAction, GRAMAction, DataServicesAction and GSIAction models the actions related to each grid component. The rest of classes are fully specified in [10] and [11].

As commented before, our architecture makes use of COPS-PR in order to exchange policy information between a PDP and a PEP. Both elements uses a general policy structure termed Policy Information Base (PIB). PIB defines a low-level policy representation and is described by a well-defined model of classes, which are referred to Provisioning Classes (PRCs). The instances of these classes are referred to Provisioning Instances (PRIs), and form the PIB. By using PIB in conjunction with the COPS-PR protocol, a PEP can inform to the PDP about its capabilities and roles. On the other hand, the PDP uses the PIB to provide configuration information to the PEP. If a change occurs in the virtual organization (i.e. a new GIIS is added or replaced), configurations are automatically updated from the PDPs to ensure operational consistency without any human intervention.

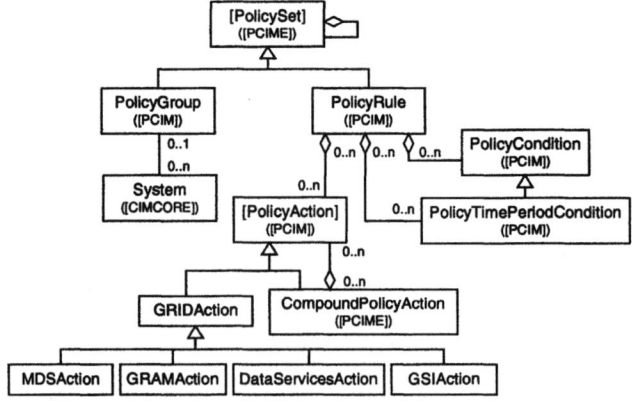

Fig. 3. UML Diagram of GRID Policy Classes

We have defined a set of the provisioning classes (PRC), based on GRID-PCIM, containing parameters for GRAM, MDS, Data Services and GSI. We have formed the following PRC groups for each component:

- *GRAM PIB classes Group*. These PRCs specify the parameters related to resource management: supported job managers, publication of job status, network parameters, etc.
- *MDS PIB classes Group*. These classes specify the configuration parameters related to the MDS: network configuration of GRIS and GIIS servers, valid schemas, access control rules, valid registrants, trusted GIIS, protection of communications between GRIS and GIIS, etc.
- *Data Services PIB classes Group*. Configuration of data management services: parameters related to the GridFTP protocol and replicas of datasets.
- *GSI PIB classes Group*. PRCs specifying the cryptographic options, trusted third parties, authorization services, delegation restrictions, etc.

5   Conclusions and Future Work

This paper contributes to the definition of a policy-based management (PBM) architecture intended to provide a flexible and automated self-configuration method for the grid nodes belonging to a virtual organization. It allows managers to partially delegate management tasks in the form of specific policies.

VO administrators interact with this architecture by defining high-level policies in a Policy Management Tool (PMT). These policies are then directly distributed from the PMTs to special policy servers, called *Policy Decision Points* (PDPs). The PDPs process these policies and take policy decisions. These policies are properly distributed using COPS-PR to the grid nodes, which act as *Policy Enforcement Points* (PEPs).

As a statement of direction, we are currently working on the integration of XML technologies with the PBM architecture and the PIB model presented in this paper. They will be used for self-configuring grid nodes belonging to a Globus 3.0-based virtual organization.

Acknowledgments. This work has been partially funded by the SENECA Piramide project (PB/32/FS/02) and the EU Euro6IX project (IST-2001-32161)

References

1. Foster, I., Kesselman, C. (eds.): The Grid: Blueprint for a New Computing Infrastructure. Morgan Kaufmann (1999)
2. Foster, I.: The Anatomy of the Grid: Enabling Scalable Virtual Organizations. Proceedings of Euro-Par (2001)
3. Galis, A., Gelas, J., Lefèvre, L., Yang, K.: Active Network Approach to Grid Management. Proceedings of International Conference on Computational Science (2003) 1103-1112
4. Durham, D., Boyle, J., Cohen, R., Herzog, S., Rajan, R., Sastry, A.: The COPS (Common Open Policy Service) Protocol. Request For Comments RFC 2748. IETF (2000)
5. Chan, K., Durham, K., Gai, S., Herzog, S., McCloghrie, K., Reichmeyer, F., Seligson, J., Smith, A., Yavatkar, R.: COPS Usage for Policy Provisioning. Request For Comments RFC 3084. IETF (2001)
6. Foster, I., Kesselman, C., Nick, J. M., Tuecke, S.: The Physiology of the Grid. Global Grid Forum (2002)
7. Welch, V., Siebenlist, F., Foster, I., Bresnahan, J., Czajkowski, K.: Security for Grid Services. Proceedings of 12th IEEE Internation Symposium on High Performance Distributed Computing (2003) 48-57
8. Czajkowski, K., Foster, I., Karonis, N., Kesselman, C., Martin, S., Smith, W., Tuecke, S.: A Resource Management Architecture for Metacomputing Systems. Proceedings of IPPS/SPDP(1998), 62-82
9. Czajkowski, K., Fitzgerald, S., Foster, I., Kesselman, C.: Grid information services for distributed resource sharing. Proceedings of 10th Internation Symposium on High Performance Distributed Computing (2001)
10. Moore, B., Ellesson, E., Strassner, J., Westerinen, A.: Policy Core Information Model - Version 1 Specification. Request For Comments RFC 3060. IETF (2001)
11. Moore, B.: Policy Core Information Model (PCIM) Extensions. Request for Comments RFC 3460. IETF (2003)

Context-Aware GRID Services: Issues and Approaches

Kerry Jean, Alex Galis, and Alvin Tan

University College London, Electrical Engineering Department,
Torrington Place, London, WC1E 7JE, U.K.; Tel: +44-20-7679573;
{kjean,agalis,atan}@ee.ucl.ac.uk

Abstract. While Grid services introduce efficiency in addressing the changing and often large resource demands of Grid applications, making the system context-aware takes the approach a step further by incorporating the important aspects of customisation and personalisation into Open Grid Services Architecture (OGSA) services. This is achieved by extending the concept of virtual organisations (VO) to Grid context, which link context information to users, policies and resources. Context is defined as any information that can be used to characterise the situation of an entity. When this information is modelled and properly managed, the OGSA services can become context-aware and context-aware services can be offered by the Grid.

1 Introduction

Large amounts of computational power and storage are required for solving complex scientific and engineering problems [1]. The advent of the Internet has caused a proliferation in the growth of computers but much of this computing power is not fully utilised most of the time. The Grid concept seeks to harness this latent computing power by providing technology that enables the widespread sharing and coordinated use of networked computing resources to solve complex computing tasks [1,2]. A Grid is a geographically distributed platform, accessible to their users via a single interface, thus providing massive computation power and storage for the solving of complex tasks [3].

Grids provide coordinated sharing of computing resources like memory, software, data, processing power and other resources that are required for collaborative problem solving. This sharing arrangement is highly controlled with clearly defined rules as to what is shared, who is allowed to share and the conditions under which this sharing takes place. The set of individuals and/or institutions that abide by these rules is termed a virtual organisation (VO) [4]. Virtual Organisations supply a "context" ((the definition of context here is one of association) for operation of the Grid that can be used to associate users, their requests, and resources. These "contexts" can be used by services as an assertion of identity, e.g. this user or resource, is a member of this VO. This "context" can be extended to take into account any information that can be used to characterise Grid entities. This rich context information when modelled and managed introduces context-awareness into the Grid and enables both the customisation of the Open Grid Services Architecture (OGSA) services [6] and the provision of context-aware services (CAS) by the Grid to end users.

M. Bubak et al. (Eds.): ICCS 2004, LNCS 3038, pp. 166–173, 2004.
© Springer-Verlag Berlin Heidelberg 2004

The development of Grid computing was found on a basic architectural premise: a computing service platform is used as the means to interconnect intelligent end-systems to support large-scale applications. Most of the current research into Grid focuses on the Grid services architecture [6,7] or the development of Grid tools and protocols [1]. Other research has focussed on the management of the Grid [10,11,12] and the use of the Grid to solve complex scientific problems [9].

The ability of the Grid infrastructure to adapt to new functional requirements has become difficult to achieve. In addition, the current Grid computing infrastructure has insufficient flexibility, support for multiple applications and rapid deployment of applications on the infrastructure. With this in mind, it has been proposed to introduce context management into the Grid to make the OGSA services and by extension the Grid context-aware.

This paper presents the authors' position on context-awareness in Grids and will introduce the idea of context-awareness in the Grid; first enabling the OGSA services to become context-aware and eventually the provision of context-aware services (CAS) by the Grid. The next section will introduce the OGSA while the third section will define context and context management. Then the context Grid service will be detailed. The paper will conclude with a summary and the proposed work needed to create a context-aware Grid.

2 Open Service Grid Architecture

The establishment, management, and exploitation of dynamic, distributed heterogeneous VOs require an enabling service architecture for Grid systems and applications. Such an architecture is the Open Grid Services Architecture (OGSA) defined by the Open Grid Services Architecture Working Group (OGSA-WG) [6] of the Global Grid Forum [8]. The OGSA defines an integrated set of web service (WS)-based service definitions designed both to simplify the creation of secure, robust Grid systems, and to enable the creation of interoperable, portable, and reusable components and systems via the standardisation of key interfaces and behaviours.

The OGSA defines, in terns of web services description language (WSDL) standard approaches to, and mechanisms for solving problems such as communicating with other services, establishing identity, negotiating authorisation, service discovery, and managing service collections. The three principal elements of the OGSA Platform are the Open Grid Services Infrastructure, the OGSA Platform Interfaces, and OGSA Platform Models [6].

Open Grid Services Infrastructure (OGSI) defines mechanisms for creating, managing, and exchanging information among entities called Grid services [7]. A Grid service is a web service that conforms to a set of conventions (interfaces and behaviours) that define how a client interacts with the Grid. These conventions, and other OGSI mechanisms, provide for the controlled, fault resilient, and secure management of the distributed and often long-lived state that is commonly required in distributed applications.

OGSA Platform Interfaces build on OGSI mechanisms to define interfaces and associated behaviours for various functions not supported directly within OGSI, such as service discovery, data access, data integration, messaging, and monitoring.

OGSA Platform Models support these interface specifications by defining *models* for common resource and service types.

The OGSA Platform components are supported by a set of OGSA Platform Profiles addressing issues such as the following [6].

Protocol bindings: These work toward attaining interoperability among different Grid services by defining common mechanisms for transport and authentication. An example of such a Grid service transport profile is "SOAP over HTTP".

Hosting environment bindings: Profiles of this sort enable portability of Grid service implementations. For example, an "OGSA J2EE Profile" might define standardised Java APIs to allow the portability of Grid services among OGSI-enabled J2EE systems.

Sets of domain-specific services: These profiles define interfaces and models to address the needs of specific application domains. For example, an "OGSA Database Profile" might define a set of interfaces and models for distributed database management.

3 Context and Context-Aware Services

Context is defined as any information that can be used to characterise the situation of an entity, where an entity can be a person, place, physical or computational object. From this point of view, almost any information that occurs in the surroundings of a specific system usage can be termed context (e.g. the current time and date, the usage history, the user profile, the temperature at the user location, the position of the user, etc.). A context element is an autonomous component that also can be used by a number of application services, but that exists independently of the application service. Typical examples are: (1) calendar information (e.g. activity type, time, actors, location, etc.); (2) user location information (e.g. outdoors/indoors, building, room, etc.); (3) weather information (e.g. outside temperature, whether it is sunny, raining, snowing, etc.); (4) social context (e.g. role –student/staff/faculty; wife; friend; boss; colleague, director, etc.); (4) personal preferences (e.g. food preferences, favourite sports, etc.); (5) permission profiles (e.g. what services can contact the user and when, what contextual information can be made available to different types of services and under which circumstances); (6) user's behaviour (e.g. task, habits); (7) device and network characteristics (e.g. network elements, bandwidth).

Context-awareness cannot be achieved without an adequate methodology and a suitable infrastructure. There has been little work on the management of context within the Grid or on efforts to make the Grid context-aware. But there has been some work on CAS and its development [13,14]. However, this work has been hampered by the need to develop a custom context infrastructure for each application. Presently, there is an initiative to create an active, programmable platform for the creation and delivery of context aware services in the Context project [15]. The introduction of context-awareness into the Grid will easily enable the personalisation and customisation of OGSA services. These services will easily adapt to a change in their context. When OGSA services become context-aware, the Grid could then offer context-aware services to customers. Hence, context-awareness will enable new types of services in Grid computing environments.

One of the most difficult aspects to cater in putting context in the Grid is that the context of people, networks, and applications can be time varying and sometimes can vary quite rapidly. This poses the challenge of tracking different types of environment changes in real time and also to react accordingly to these changes [14]. This is a key feature of context-aware applications and systems so as to fully exploit the richness of context-information. In view of this, we have proposed a flexible Grid context service that fits neatly into the Grid service architecture that was discussed in the previous section. Our proposal, which will be discussed in the next section, entails defining a new OGSA service, a Grid context service to manage context and make the Grid context aware.

4   Grid Context Service

In order to make a Grid context-aware, we introduce the context service into the OGSA. This is an extension of the virtual organisation's concept of associating users and resources to VOs by supplying a "context" (the definition of context here is one of association). This concept is extended by looking at "context" as not just an association of users or resources but as any information that characterises the situation of an entity be it associations, relationships or characteristics. Hence, the context Grid service is as an extension of this VO "context" linking contextual information, users, resources and policies. This enables the creation of services that are context-aware, and hence, easily customisable and personalisable.

The context Grid service creates a general framework for the gathering, storage and management of context information as well as an infrastructure for the creation and deployment of CAS offerings to end users. The context service is implemented through a host of services. The context management service ensures the efficient gathering, storage, transport and timely use of context information inside the Grid. Through interaction with the context information management service, the OGSA services become context-aware and hence their context can be used to make them easily customisable and personalisable. CAS management is concerned with the definition, deployment and operation of CAS by the Grid. It uses the context management service and other OGSA services to enable the Grid to offer CAS to end users, thus increasing the scope of the services that the Grid can offer.

4.1   Context Management Service

The context management service uses the context gathering and publishing service, the context retrieval service and the context storage service. Context information is obtained from a wide plethora of applications, services and sensors (collectively called context sources) spread all over the network. The job of the context gathering service is to collect this information and put it into a common model understood and accessible by all components of the context service. This highly distributed characteristic introduces a great challenge for the manageability of all this information. Hence there is a need for a common context information model and a context publishing protocol. The context is then stored in a context information base for access by context consumers through the context retrieval service. The relationship between the context sources and consumers are shown in Fig. 1. below.

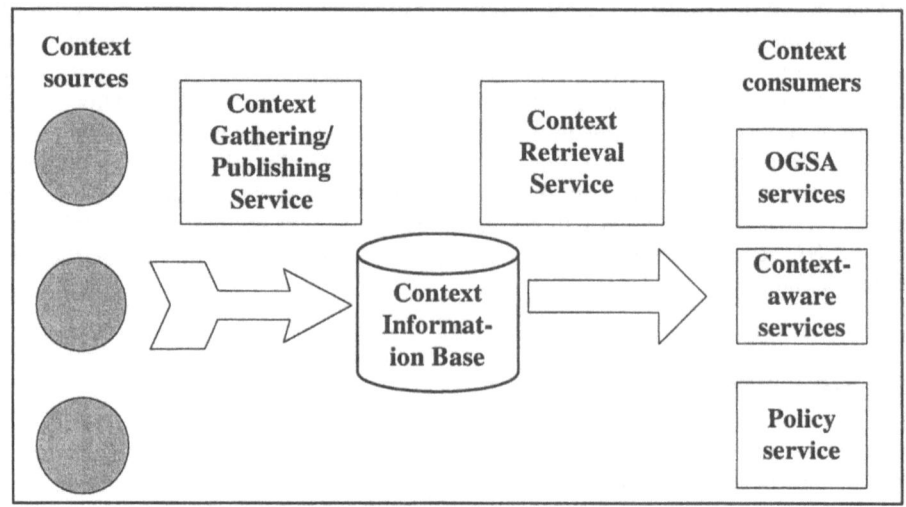

Fig. 1. Relationship between context sources and context consumers

The interaction between context consumers and context sources could be described as follows: (1) in a *push mechanism*, context sources periodically push updated context information to the context management service. This information is maintained in a context information base and client inquiries are serviced from a local store. (2) In a *pull mechanism*, the context service must explicitly request context information. It can either make this request on a periodic basis (polling) or when an application demand arises. A *push system* collects data ahead of need and thus may offer better performance. However, it may consume substantial resources transferring and storing information that is never required. A *pull system* may use fewer resources by obtaining only the data that is required. However, this exposes the context service to inevitable network delays and unavailability.

4.2 Context Information Modelling

A context information model must be defined in order to map context information, for common gathering, storage and parsing. The context information model must be rich and flexible enough to accommodate not only the current but also future aspects of context information. The model should scale well and be based on standards as much as possible. The context model proposed is based on an entity and relationship approach. This model is easily translated into an object-oriented information model and fits well into object-oriented programming (OOP). An entity is composed of a set of intrinsic characteristics or attributes that define the entity itself, plus a set of relationships with other entities. Context is classified into four main types of entities: personal context, place context, object context and task context. These main entities form the base entity classes. Subclasses of these entities are defined and are called sub-entities. The syntax of both entities and sub-entities are defined in XML.

4.3   Context-Aware Management Service

Grid CAS management facilitates the management and provision of CAS by linking providers of Grid resources to appropriate context information via predefined policies or agreements. Users are associated with the CAS as a means to exploit those Grid resources. The CAS management service provides a set of basic service components for the creation, lifetime management and termination of CAS. They include, as shown in Fig. 2 below; the CAS creation/destruction service, the CAS registry service, the subscription service and a policy-based CAS lifetime management service. For successful operation, these services rely on the context management service for their context information and other OGSA services especially the policy and agreements service.

The CAS creation/destruction service deals with the definition of the CAS, their deployment and their termination. CAS are created by the composition or customisation of predefined basic services stored in policy format. This composition mechanism is driven by a creation policy. The subscription/customisation service deals with customer subscription and service customisation. The CAS registry service keeps a directory of all the CAS that have been defined by the Grid context service. The policy-based service lifetime management manages the services as they operate. It provides the extensibility, easy management and operability that CAS require to be functional and efficient.

The Grid context service makes great use of the OGSA policy and agreement services. As stated earlier, policies are used to specify the requirements for creation and management of CAS. These policies may contain information that describes how to create and manage the CAS as well as how these services respond to changes in context and how they behave under specific conditions. A policy-based manager is used for service management while basic CAS are stored in policy format. Policy decision and enforcement points are also implemented. Hence, the context Grid service can be implemented using policy-based management to manage the gathering, publishing, storage and retrieval of context information as well as for the creation and management of CAS.

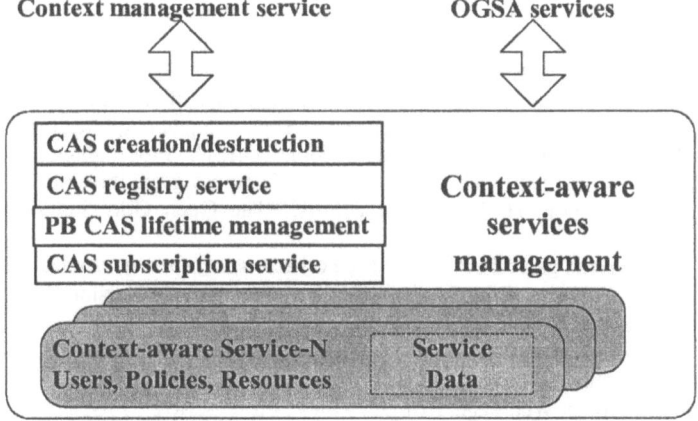

Fig. 2. GRID Context-aware Management Services

5 Conclusions and Future Work

Grid technology enables widespread sharing and coordinated use of networked computing resources. Grids are large, distributed, dynamic and heterogeneous platforms used for collaborative problem solving. By adding context-awareness to the Grid infrastructure, we expect to add the extra dimension of easily personalisable and customisable context-aware OGSA services. Such a Grid can then offer context-aware end services, which are very relevant today especially with the proliferation of mobile devices and in context rich situations. This is achieved through the introduction of the context Grid service.

As future work, protocols must be defined for interactions between the Grid, the context information management service, the context-aware services management services and the policy service. Further work must also be done on policy-based management of the context service. There is ongoing work on the creation of an active policy based system for the efficient creation of context-aware services in the Context project [15]. The results of this initiative are eagerly awaited and will be used as a guideline to implement context-awareness in the Grid. A proof of concept of the context Grid service will be provided with the future implementation of a context-aware virtual private Grid. However, there is also a need to investigate alternatives for context information management architectures and context Grid service implementations.

Acknowledgements. This paper partly describes work in progress in the EU IST project CONTEXT [15]. The IST programme is partially funded by the Commission of the European Union.

References

1. I. Foster, C. Kesselman, "Globus: A MetaComputing Infrastructure Toolkit", International Journal of Supercomputer Applications, pp. 115–128, Vol. 11 No. 2, 1997.
2. F. Berman, G.C. Fox, A. J. G. Hay, "Grid Computing", Wiley 2003.
3. I. Foster, C. Kesselman editors, "The Grid: Blueprint for a Future Computing Infrastructure", Morgan Kaufman, San Francisco, USA, 1999.
4. I. Foster, C. Kesselman, and S. Tuecke, "The Anatomy of the Grid, Enabling Scalable Virtual Organisations", International Journal of Supercomputer Applications, 2001.
5. D. B. Skillicorn, "Motivating Computational Grids", Proceedings of the Second IEEE/ACM International Symposium on Cluster Computing and the Grid, IEEE 2002.
6. I. Foster, D. Gannon, H. Kishimoto, editors, "The Open Grid Services Architecture Version 0.13a",
   https://forge.gridforum.org/projects/ogsa-wg/docman, Nov. 2003.
7. S. Tuecke et al, "Open Grid Services Infrastructure (OGSI) Version 1.0", Global Grid Forum Draft Recommendation, June 2003
   http://www.globus.org/research/papers.html.
8. Global Grid Forum Web Page: http://www.gridforum.org.
9. EU DataGrid Project Web Page: http://www.eu-datagrid.org.

10. J. Frey et al, "Condor-G: A Computational Management Agent for Multi-Institutional Grids", Proceedings of the Tenth IEEE Symposium on High Performance Distributed Computing (HPDC10), San Francisco, USA, 2001.
11. K. Yang, A. Galis, and C. Todd, "A Policy-Based Active Grid Management Architecture", Proceedings of the Tenth IEEE International Conference on Networks (ICON02), pp. 243-248, IEEE, Singapore, 2002.
12. A. Galis, J. P. Gelas, L. Lefèvre, and Y. Yang, " Programmable Network Approach to Grid Management and Services", International Conference on Computational Science 2003, LNCS 2658, pages 1103-1113, Melbourne, Australia, June 2003, www.science.uva.nl/events/ICCS2003/.
13. Gregory D. Abowd et al, "Context-Awareness in Wearable and Ubiquitous Computing", First International Symposium on Wearable Computers, pp. 179-180, 1997.
14. P. Gray and D. Salber, "Modelling and Using Sensed Context Information in the Design of Interactive Applications", Proceedings of the Eighth IFIP Working Conference on Engineering for Human-Computer Interaction (EHCI'01), Toronto, Canada, May 2001.
15. CONTEXT Project website, http://context.upc.es.

Security Issues in Virtual Grid Environments

Jose L. Muñoz, Josep Pegueroles, Jordi Forné, Oscar Esparza, and Miguel Soriano

Technical University of Catalonia (UPC)*
Telematics Engineering Department (ENTEL)
1-3 Jordi Girona, C3 08034 Barcelona (Spain)
{jose.munoz,josep,jordi.forne,oscar.esparza,soriano}@entel.upc.es

Abstract. Computational Grids (or simply Grids) enable access to a large number of resources typically including processing, memory, and storage devices. Usually, Grids are used for running very specific applications (most of them related to some kind of scientific hard problem); however, not much attention has been paid to commercial Grid applications. The massive use of such commercial services will depend on fulfilling their special security, usability and quality of service requirements. In this sense, Virtual Private Grid (VPG) provides a way of dynamically create a virtual grid environment with dedicated network resources. In this paper VPG is compared with related work such as the Grid over VPN (GoVPN), the Grid Community (GC) and the Ad-hoc Grid (AG) and the security challenges for VPGs are analyzed.

1 Introduction

A Grid is a collection of heterogeneous computing resources including processing, memory, and storage devices, all of which are geographically and organizationally dispersed. According to Foster et al. [6] "Computational Grids have emerged as an important new field, distinguished from conventional Distributed Computing by its focus on large-scale resource sharing, innovative applications, and *in some cases, high performance orientation*".

The main focus of current generation Grids is the transparent access to resources; on the contrary offering high performance services to a wide range of non-expert users has not been among the main priorities of the Grid. Current testbeds are either demo systems, with few users and little functionality, or scientific/engineering systems, with specialized users and high QoS demands. Virtual Private Grid (VPG) provides a way of dynamically create a virtual grid environment with dedicated network resources. In this paper VPG is compared with related work such as the Grid over VPN (GoVPN), the Grid Community (GC) and the Ad-hoc Grid (AG) and the security challenges for VPGs are analyzed.

The rest of the paper is organized as follows. Section 2 provides an overview of the related work. Section 3 describes the VPG security architecture requirements and gives some hints on managing privacy, a new relevant requirement for VPGs. Finally, we conclude in Section 4.

* This work has been supported by the Spanish Research Council under the projects DISQET (TIC2002-00818) and ARPA (TIC2003-08184-C02-02).

2   Related Work

Grid over VPN (GoVPN) is a possible solution for building a virtual grid environment with dedicated resources. In GoVPN, first a VPN is created and then a Grid is deployed over it. The private environment is created at the network level so that users outside this environment do not have network connectivity to the resources. There are several mechanisms that VPNs use to enable private data going through a public network, the most important among them are authentication, privacy and tunneling. IPSec, TLS and SSH are ways of performing these mechanisms. On the other hand, technologies such as RSVP and DiffServ may be used to warrant QoS to VPN tunnels. Some practical approaches of GoVPN are [9], where the authors use SSH for building a "shell-like Grid application", and [4,3], where it is proposed a "virtual private grid file system". GoVPN has several advantages: it can provide warranted QoS at the network level, it relies on existing technologies and it allows seamless integration of legacy applications. However, GoVPN has also some drawbacks. Allowing the Grid level to become aware of changes in the network level and vice versa is tedious because Grid and network level are independently deployed. This fact limits GoVPN to pretty static environments. Moreover, each level (Grid and network) performs its own security services, leading to either duplicated schemes and/or mismatching security policies.

Grid Community (GC) [11] and **Ad-hoc Grid (AG)** [10] both create the private grid environment using only grid-level mechanisms. In GC each site grants coarse-grained access of its resources to a community account requiring the existence of a Community Authentication Server (CAS). AG deploys a similar scheme but allows more spontaneous and short-lived collaborations. However, privacy and QoS at the network level are not addressed by any of them.

The **Virtual Private Grid (VPG)** provides a solution for building private grid environments comprising both network and grid levels. In this sense, VPG can be understood as a middleware interacting with these two layers. Hence, the middleware has to include two modules: one for managing network resources and another controlling security. Section 3 will further describe the requirements of the latter module, the VPG Security Architecture (VPG-SA). Figure 1 shows the different mechanisms for building the private grid environment with dedicated resources.

3   VPG Security Architecture

The VPG Security Architecture (VPG-SA) must provide the basis for a dependable grid operation. Among other general security requirements [5] as Authentication, Authorization, Assurance, Accounting, etc. the VPG-SA must support all the particular aspects of current generation Grids [7]. These security topics are the following: (1) The VPG must grant access to resources distributed over multiple administrative domains. (2) The inter-domain security solution used for VPG must be able to *interoperate with the local security solutions* encountered in individuals domains, enforcing the support for *multi-domain trust and policies*. (3) A user should be able to authenticate once and subsequently initiate computations that acquire resources without the need for further user authentication (*Single sign-on*). (4) Multi-domain support and single sign-on will need

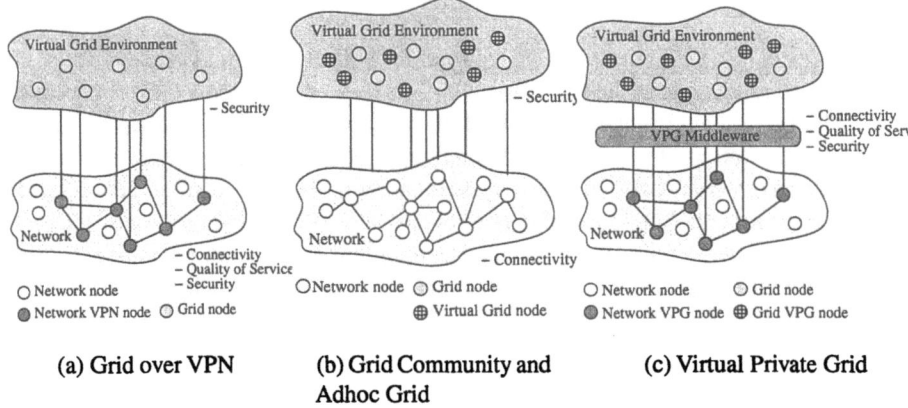

(a) Grid over VPN   (b) Grid Community and   (c) Virtual Private Grid
                              Adhoc Grid

Fig. 1. Technologies for building a virtual Grid

the usage of a common way of expressing the identity of a principal such as a user or a resource, that is a *Uniform credentials/certification infrastructure*.

Moreover, it must be taken into account that the VPG runs over an untrusted network and that *data privacy* must be assured at the Grid level. Usually, applications running inside a VPG require to send a computational result to a group of nodes for use during the next computation. Nevertheless, privacy management for large groups can be expensive and it can harmfully affect the VPG performance. Currently grid implementations and toolkits provide some specific security features, mainly for resource access control, user authentication, and hop-by-hop encryption of communication links. The issue of end to end privacy has not been supported in any of these works.

End to End Privacy. The use of multicast at the network layer and the data encryption in the grid level is the efficient way of managing privacy in VPG. The encryption will use a common key shared by all the nodes in the virtual grid environment. The key should be known by all nodes in the VPG and must be updated every time the group of nodes changes. This is the only way of achieving perfect forward and backward secrecy. The creation of the VPG multicast channel, the key stablishment, and the key update, have to be assumed by the VPG-SA.

VPG creation. In VPG each node is considered a member of a multicast group. When the virtual grid environment has to be created, a secure multicast group has also to be set up. A node will be considered member of the group if it knows the common secret shared by all the group (weak authentication). The VPG-SA has the responsibility of delivering the secret key to every initial member. Each member has to be previously authenticated and it is mandatory that all the VPG nodes support multicast connections. The only way of performing this action is by means of a unicast connection with each one of the initial nodes of the virtual environment. As long as this action only takes place once, bandwidth and computation issues are not very relevant.

VPG management. If only unicast connections were used, the VPG key update would use many bandwidth and computation resources. In this sense, multicast along with logical key tree based schemes [8,1,2] can be used to improve performance of security management in the Grid. The VPG-SA must use two types of keys: Session Encryption Keys (SEK) and Key Encryption Keys (KEK). The SEK is the key that each node in the grid has to know for weak authentication and encrypting the exchanged data. KEKs are keys that are used to decrease the bandwidth requirement for VPG management. The knowledge of these keys will allow nodes in the grid to get the updated SEK, that is, a VPG-SA can exclude or include nodes in the Grid by using these keys.

Addition of a node to the VPG environment. The VPG-SA maintains a logical binary-tree structure including all the nodes in the grid. Each node of the grid is logically placed in a leaf of the tree. Each junction in the tree structure will contain a different KEK. Every node in the grid must know its own key plus all the KEKs located in the junctions from his leaf to the tree root. When a node wishes to leave the VPG all the KEKs that it knew and the SEK must be updated, so the VPG-SA has to compute the new KEKs. After that, it has to notify the changes to the remaining nodes. In order to do that, it sends a multicast message containing the updated KEKs. This message is encrypted so that each node can only decrypt the data that concerns to it. This is achieved by encrypting each new updated KEK using the not-updated key under its junction. This mechanism can be understood seeing Figure 2.

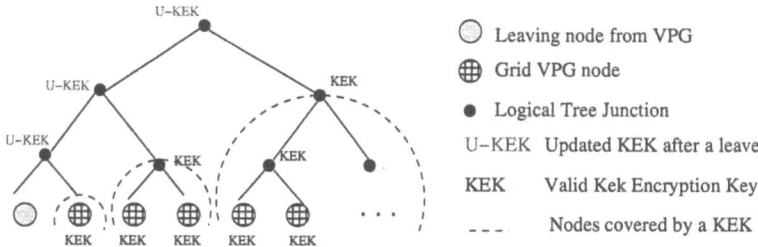

Fig. 2. Example of logical key in VPG-SA

Algorithm efficiency considerations. VPG-SA could be simpler by just updating the SEK and sending it to all the remaining nodes. This will imply as many unicast connections as remaining nodes the grid has, but it cannot be afforded neither in terms of bandwidth nor latency.

Consider a grid made of 1000 nodes. Every time a change in the composition of the grid happened, the VPG-SA would have to send 1000 messages before another node wanted to join or leave the VPG. In case this could not be achieved, a rekeying process would be aborted by the beginning of another rekeying process and the system would collapse. Using key trees, the number of messages to be sent are reduced to the order of the logarithm of the number of VPG nodes, so the VPG-SA reduces the rekeying time. A specially suitable algorithm for Grid environments was proposed by some of the authors in [12] and more detailed explanation about this topic can be found in [2].

4   Conclusions

Currently, not many attention has been paid to commercial Grid applications. The massive use of such commercial services will depend on fulfilling the special security, usability and quality of service requirements. In this sense, the VPG provides users a way of dynamically create their own Grid environment with dedicated network resources.

This paper compares VPG with GoVPN, GC, and AG. Basically, VPG operates in both grid and network levels while the others are focused in just one of them. This fact allows VPG to better solve security issues, in particular privacy which has been poorly addressed by most of the current solutions. In these sense, VPG-SA includes enhanced functionalities for efficiently creating and managing VPGs that support end to end privacy.

References

1. D. Ballenson, D. McGrew, and A. Sherman. Key Management for Large Dynamic Groups: One Way Function Trees and Amortized Initialization, 2000. Internet Draft. Work in Progress.
2. R. Canetti, J. Garay, G. Itkis, D. Micciancio, M. Naor, and B. Pinkas. Multicast security: A taxonomy and some efficient constructions. In *INFOCOMM'99*, 1999.
3. R. Figueiredo, P. Dinda, and J. Fortes. A case for grid computing on virtual machines. In *International Conference on Distributed Computing Systems (ICDCS)*, pages 550–559. IEEE Computer Society, May 2003.
4. Renato J. Figueiredo. Vp/gfs: an architecture for virtual private grid file systems. Technical Report TR-ACIS-03-001, University of Florida, May 2003.
5. I. Foster and C. Kesselman. *The Grid. Blueprint for a new computing infrastructure*. Morgan Kaufmann, 1999.
6. I. Foster, C. Kesselman, and S. Tuecke. The anatomy of the Grid. *International Journal of Supercomputer Applications*, 15(3):200–222, 2001.
7. Ian T. Foster, Carl Kesselman, Gene Tsudik, and Steven Tuecke. A security architecture for computational grids. In *ACM Conference on Computer and Communications Security*, pages 83–92, 1998.
8. H. Harney and E. Harder. Logical Key Hierarchy Protocol, 1999. Internet Draft. Work in Progress.
9. Kenji Kaneda, Kenjiro Taura, and Akinori Yonezawa. Virtual private grid: a command shell for utilizing hundreds of machines efficiently. *Future Generation Computer Systems*, 19(4):563–573, 2003.
10. Markus Lorch and Dennis Kafura. Supporting secure ad-hoc user collaboration in grid environments. In *Grid Computing*, volume 2536 of *LNCS*, pages 181–193. Springer-Verlag, November 2002.
11. Laura Pearlman, Von Welch, Ian Foster, and Carl Kesselman. A community authorization service for group collaboration. In *IEEE Third International Workshop on Policies for Distributed Systems and Networks*, pages 50–59, June 2002.
12. Josep Pegueroles, Wang Bin, Miguel Soriano, and Francisco Rico-Novella. Group rekeying algorithm using pseudo-random functions and modular reduction. In *Grid and Cooperative Computing GCC2003 LNCS Springer-Verlag*, 2003.

Implementation and Evaluation of Integrity Protection Facilities for Active Grids

Arso Savanović[1], Dušan Gabrijelčič[1], Borka Jerman Blažič[1], and Janez Bešter[2]

[1] Jožef Stefan Institute, Laboratory for Open Systems and Networks,
Jamova 39, 1000 Ljubljana, Slovenia,
[2] University of Ljubljana, Laboratory for Telecommunications,
Tržaška 25, 1000 Ljubljana, Slovenia

Abstract. Some recent studies have shown that Grid systems can benefit from active networks technologies, thus a novel networking paradigm called Active Grid has been proposed. On the other hand, the development of appropriate security components for Grid systems is still in an early stage due to complexity and mostly academic use of Grids to date. However, security issues are exacerbated in Active Grids due to the availability of processing and caching capacity in intermediate network nodes, which adds to Grid complexity. In this paper we describe a set of facilities for protection of data integrity in Active Grids, and then present and discuss some experimental results from validation and evaluation test with the prototype implementation of these facilities.

1 Introduction

Some recent studies have shown that Grid systems can benefit from active networks technologies [12,15,13], thus a novel networking paradigm called Active Grid has been proposed. The main idea is to place active network nodes (ANNs) at strategic points in the network and to make processing and caching capabilities of ANNs explicitly available to Grid applications. Examples of such Active Grid configurations include Meta-cluster computing, where an ANNs is used as a network head of each cluster or parallel computer, and Global computing, where an ANN is used to manage a single Grid node or a collection of Grid nodes [12].

On the other hand, the development of appropriate security components for Grid systems is still in an early stage, which is largely due to Grid complexity and mostly academic use of Grids to date. Existing Grid toolkits do provide some security features, e.g. [8,7,2], however many important challenges still lie ahead, including but not limited to the Grid security requirements listed in [17]. These open issues are even more challenging in Active Grids due to the availability of processing and caching capacity in intermediate network nodes, which adds to the already recognised complexity of Grids. In this paper we focus on one aspect of security, the integrity service, and present a set of facilities for integrity protection, which can be applied to Active Grids. These facilities are a part of a comprehensive security architecture for active networks based systems [19], which has been developed, implemented and tested within the FAIN project [6] from the EU VI. Framework Programme.

M. Bubak et al. (Eds.): ICCS 2004, LNCS 3038, pp. 179–186, 2004.
© Springer-Verlag Berlin Heidelberg 2004

2   On Integrity in Active Grids

An integrity service protects against unauthorised changes of data while in transit over the network, including both intentional change or destruction and accidental change or loss. Unauthorised changes include modifications, replay, and spoofing of data.

With the growing industry support an important tendency in Grid development is gradual commercialisation of Grids, which increases the incentive for unauthorised change of data, since this kind of mischief generates direct economic benefits for the adversary. For illustration consider the following example. A provider of a commercial computational Grid offers processing capacity to its customers in return for a financial compensation. Assume that the use of Grid is based on prepaid digital tokens. Obviously, there is an incentive for Grid users to tamper with these tokens in order to reduce or even eliminate the costs related to the use of commercial Grid resources. Without integrity protection facilities in place, a skillful adversary would be able to cheat in different ways, e.g. : (a) buy a single token, but use it several times by replaying it to the Grid accounting system, (b) change the nominal value of the token to allow the use of more resources than what was originally paid for, or (c) sniff the network and intercept someone else's tokens, then change the owner ID in these tokens, effectively stealing them.

Traditionally, integrity has been provided on an end-to-end basis, which is a valid approach for current Grid systems, where Grid nodes are mapped to end nodes and Grid application data is processed only at these end nodes. However, the availability of ANNs in Active Grids makes it possible to process application data at intermediate network nodes as well. Thus, Grid application data can *legally* change while in transit over the network, which requires a different approach to integrity protection. In fact, this requirement is not specific to Active Grids but is valid in general for systems based on active networks technologies.

3   Integrity Protection Facilities for Active Grids

Even though Active Grids offer Grid applications processing capabilities at intermediate network nodes, not all Grid application data require processing in these nodes, i.e. Grid application data can be divided into two classes: static (end-to-end) data and dynamic data. Consequently, our integrity protection service is based on a combination of end-to-end protection and hop-by-hop protection.

Both types of protection are based on a specific packet format, which has to provide for different treatment of static and dynamic Grid application data. We have adopted the ANEP, Active Networks Encapsulation Protocol [3], because of its extensibility and because the majority of active network approaches support it. In the context of ANEP we have defined several options to support the FAIN security architecture. These options carry a service identifier (eg. label), hop protection and credentials option related information, service variable data and resource usage information. Integrity protection is based on the hop protection, credential and variable data options.

3.1   End-to-End Protection

The end-to-end protection is facilitated through the use of the ANEP credential option. The credential option is generated at the source and is never modified. Every receiver of the ANEP packet can verify the integrity of static data in the ANEP packet by processing this option. The credential option is defined by the credential identifier and type, location field (specifying where the credentials can be fetched), target field (where the user can specify specific targets as nodes, system layer or a packet itself), optional timestamp (protects against replays), and the digital signature. The digital signature covers only static data within an ANEP packet, which includes the first 32 bits of basic ANEP header, source address, service identifier, ANEP payload, and the credential option itself except digital signature data. There can be zero, one, or more credential options in a single ANEP packet. On each visited ANN the relevant credentials are fetched, the certification path is validated, and the digital signature is verified. We note that the digital signature mechanism provides authentication of data origin and non-repudiation security services in addition to the end-to-end integrity protection.

Various credential types are supported: X.509 certificates, attribute certificates [1], SPKI certificates or Keynote credentials [5]. Unless included in the packet, the credentials can be fetched from a suitable store, e.g. DNS or LDAP. In our case, we have designed and implemented a simple protocol that enables fetching credentials from the previous hop ANN. In this way, the originating entity is responsible for supplying all the credentials, which may be required later on the nodes that a packet traverses. To be able to supply credentials on the intermediate nodes we have designed and implemented node credentials cache. After successful validation, the credentials are cached on the node for the time of their validity or according to the cache policy on cache size and maximum time period of a cache entry. Caching credential has also other benefits: if the cache entry is valid, there is no need to validate the credentials. In this way we can reduce required digital signature validation to only one per credential option in the packet, which results in a significant speed improvement, after the first principal packet has passed the node. Additionally, we cache bad credentials in a separate cache in the cases, when the credentials cannot be verified. Subsequent packets with bad credentials are discarded immediately.

3.2   Hop-by-Hop Protection

While the ANEP data option serves as a container for transporting dynamic Active Grid application data, the hop-by-hop integrity protection is facilitated with the ANEP hop protection option. Each ANN in an Active Grid maintains a (separate) shared secret key with each of its neighbour ANNs. When an ANN receives an ANEP packet, it a) uses the secret key it shares with the previous hop ANN to process the hop protection option and thus check the integrity of dynamic data in the ANEP packet, and b) uses the secret key it shares with the next hop ANN to generate a new hop protection option before forwarding the ANEP packet towards its destination.

The hop protection option is defined by a triple of Security Association (SA) identifier, which points to the respective association with a neighbour node, sequence number field, which protects against replays, and keyed hash. The keyed hash covers the entire ANEP packet except the keyed hash itself. The per-hop protection protects all ANEP packets exchanged between two neighbour ANNs. When the packet leaves the node a new hop protection option is built based on the respective SA for the next hop ANN.

Dynamic and scalable hop protection requires at least two supporting protocols. First, a SA setup protocol that supports setup and maintenance of SA (including key exchange) between two neighbour ANNs, and second, a neighbour discovery protocol. For the former we have adopted the Station-to-Station protocol [20], additionally protected with end-to-end protection, as described in section 3.1. Neighbour discovery is described in the next section.

3.3   Neighbour Discovery

As seen above, the hop-by-hop integrity protection requires each ANN in an Active Grid to keep an up-to-date list of all its neighbour ANNs. Manual configuration of ANN neighbour lists is not acceptable due to scalability issues and dynamic nature of Active Grids. We have therefore developed a protocol, which enables ANNs in an Active Grid to automatically create a list of neighbour ANNs and update it whenever necessary due to changes in the network infrastructure [18].

The protocol has three operation modes: node advertisements, discovery handshake, and heartbeat exchange. ANNs advertise themselves by periodically transmitting the advertisement message to an emulated communications bus, which distributes the advertisement to all ANNs. The implementation of an emulated communications bus can be multicast-based, with one dedicated multicast group per overlay, server-based, or a combination of both. Upon reception of an advertisement an ANN starts a discovery handshake with the sender with probability p only in order to limit the load on the sender of the advertisement. The discovery handshake is the main protocol mechanism for neighbour discovery and detection of network changes. It involves an exchange of two messages between the node, which initiates the handshake (poller), and the node, which responds (responder). Both messages carry the address of the previous hop ANN, which is used by ANNs en route for updating their neighbour tables when necessary. A node periodically sends the heartbeat message to each of its known neighbours to indicate that it is still alive. Neighbour ANNs use this message to refresh the soft state in their neighbour tables. We note that neighbour discovery protocol employs packet interception, where an intermediate ANN intercepts and processes the protocol message, even though the message is addressed to some other node. Packet interception is inherently supported by ANNs [21] and is an example of how Grid systems can benefit from active network technologies.

4 Proof-of-Concept Implementation and Experimental Results

The proof-of-concept implementation of the facilities described above has been developed in order to verify end evaluate the operation of the integrity service provided by these facilities. The facilities have been tested in a series of lab tests and demonstrated at the FAIN project Final Review and Demonstration Event.

The testing environment was set up on a commodity PC, with Intel P4 2.2 GHz processor, 512 MB RAM, Red Hat Linux 8.0, kernel 2.4.18-14, Java SDK 1.3.1_3, the Bouncy Castle crypto-library version 1.17, and the FAIN ANN code. In our tests ANEP packets carried basic ANEP header, a hop protection option, a service identifier option, a single credential option, a resource option and a zero length variable option and payload. The keyed hash algorithm used in the hop protection option was HMAC-SHA-1 [9]. The credential used was an X.509 [1] based certificate signed with a combination of RSA encryption and MD5 hash. The signature in the credential option, used for end-to-end protection, was an RSA encrypted SHA-1 hash. The RSA key length was 768 bits and the certification path length was one.

Figure 1 shows the breakdown of the measured security costs related to an ANEP packet traversing an ANN. The "hop" part represents the costs of processing the hop protection option, i.e. validating the option on the receiving side and building a new option on the sending side. The "encoding/decoding" part represents the costs of parsing a packet at the receiving side and forming a new packet at the sending side. The "other costs" part represents the costs of ANN building a security context of the packet. The "signature" part refers to the costs of validating a credential and verifying the digital signature in the packet.

The left hand side of the figure depicts a case, when the user credential (certificate) is contained in the ANEP packet (but not yet verified), while the right hand side of the figure depicts a case, when the user credential is already in the cache. As expected, the largest overhead is due to the credential option processing, since this involves the costly asymmetric cryptography (RSA signatures validation). With no caching of credentials, these costs represent as much as 85% of all security related costs, while the hop-by-hop protection related costs represent around 8%. When credentials are cached (from previous packets), the digital signature related costs are reduced to 50%. The hop-by-hop protection related costs remain the same, however their percentage raises to 25% due to a reduction of overall costs.

In absolute numbers, we were able to reach the ANN throughput of 396 packets per second without the credential cache. The use of the credential cache yields the threefold improvement, i.e. an ANN is capable of handling 1190 packets per second. When only hop-by-hop protection is activated, the ANN can handle more than 7000 packets per second. Our measurements show that security related costs are not drastically related to the RSA key size in current setup: in the case of RSA512 the node can pass 1340 and in case of RSA1024 1003 packets per second.

We have also looked into the ways of reducing the performance overhead induced by our protection facilities. A comparison of the Bouncy castle crypto-

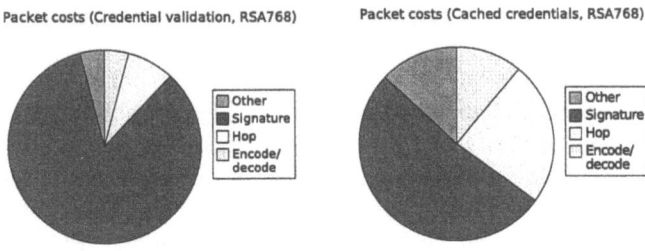

Fig. 1. Security related packet processing costs

graphic library with the OpenSSL library[1] shows that the OpenSSL outperforms the Bouncy castle by the factor of two for digital signature verification (1428 vs. 3550 verifications per sec). Cryptographic accelerators like Broadcom based BCM5821 cryptographic accelerator,[2] show additional speed improvements with very light utilisation of the main CPU (8142 signatures verification per sec, less then 1% CPU). Thus, integrating native libraries or accelerators should considerably improve performance of protection facilities presented in this paper and enable ANNs to handle few thousands ANEP packets per second even with the costly checking of digital signatures.

In comparison, the existing performance results for active networks based systems are for simpler packet structures as in our case, and range from 6000 to 11000 packets per second in a similar environment, e.g ANTS [23] and Bees [22]. As far as authors are aware, so far no security nor cryptography related costs were calculated or reported for active networks based systems.

With regards to the neighbour discovery, we were mostly interested in the speed of neighbour discovery, i.e. the delay between the network change and the corresponding updates of neighbour ANN tables. By varying different protocol parameters in our tests we were able to reach convergence delay in the range of 30–40 seconds, i.e. our target values. We chose this target value because the convergence delay of current routing protocol implementations ranges from around 30s for OSPF [4] to several minutes for BGP [11]. We find the comparison with the routing protocols interesting due to the following. As far as the authors are aware, no results on convergence delay are currently available for related efforts in the area of automatic discovery of neighbour ANNs (e.g. [14]). On the other hand, routing protocols provide analogous functionality to neighbour discovery protocol: after a perturbation occurs in the network, the routing protocol has to accordingly reconfigure the physical network (i.e. routes in the network) as soon as possible in order to minimise the disruption in communication caused by the perturbation.

---

[1] http://www.openssl.org
[2] http://www.broadcom.com/products/5821.html

5 Related Work and Conclusions

Some of the existing Grid toolkits provide support for integrity protection, see e.g. [8,7,2], however these protection facilities are based on an end-to-end approach, i.e. they completely preclude modifications of data while in transit over the network. This breaks the Active Grid model, which renders these integrity protection facilities unsuitable for Active Grids.

In [10] authors have presented a framework for providing hop-by-hop security in active network testbeds. This framework is heavily based on off the shelf security solutions. More importantly, the framework raises serious scalability concerns, since it is centralised and involves manual configuration of active network nodes. Thus, this work has demonstrated a more practical and short term solution for existing testbeds based on active networks technologies rather than a true long term security solution.

SANTS [16] provides a strong and relatively complete security solution for active networks based systems. Integrity protection in SANTS is also based on the separation of end-to-end and hop-by-hop concerns, and their solution is similar to ours. However, in SANTS there is no mechanisms for automatic discovery and configuration of neighbour active nodes, which is a basic prerequisite for (scalable) hop-by-hop protection. Neither do they report on performance of their prototype.

In this paper, we have presented a working example of a complete set of integrity protection facilities suited for communications systems based on active networks technologies, including the novel paradigm of Active Grids. The facilities are a part of a comprehensive security architecture, which has been designed, implemented and evaluated within the EU FAIN project. This work does not replace but rather complements the work in other Grid security efforts, e.g. within GGF.

In general, providing security is costly in performance terms, since strong security solutions employ cryptographic techniques. We believe that performance results presented in this paper contribute towards better understanding of performance limits imposed by security facilities. Based on the performance results from our evaluation tests with the prototype and existing reports on performance of some active networks systems, we find the performance of integrity protection facilities completely suitable for the control and management planes of Active Grids. In the transport plane, however, the integrity protection facilities may not be able to meet the requirements of those Active Grid applications demanding higher performance.

References

1. Information Technology - Open Systems Interconnection - The Directory: Public-key and Attribute Certificate Frameworks. International Standard, March 2000.
2. Uniform Interface to Computing Resources. UNICORE Plus Final Report, December 2002. http://www.unicore.org/.

3. D. S. Alexander *et. al.* Active Network Encapsulation Protocol (ANEP). Active Network Group draft, July 1997.
4. A. Basu and J. G. Riecke. Stability Issues in OSPF Routing. In *Proceedings of SIGCOMM 2001*, pages 225–236, August 2001.
5. M. Blaze, J. Feigenbaum, J. Ioannidis, and A. D. Keromytis. The KeyNote Trust-Management System, Version 2. RFC 2704, September 1999.
6. FAIN—Future Active IP Networks. http://www.ist-fain.org.
7. A. Ferrari *et. al.* A Flexible Security System for Metacomputing Environments. Technical Report CS-98-36, University of Virginia, Charlottesville, VA 22903, USA, December 1998.
8. Grid Security Infrastructure (GSI). http://www.globus.org/security/.
9. H. Krawczyk, M. Bellare, and R. Canetti. HMAC: Keyed-Hashing for Message Authentication. RFC 2104, February 1997. Informational.
10. S. Krishnaswamy, J. B. Evans, and G. J. Minden. A Prototype Framework for Providing Hop-by-Hop Security in an Experimentaly Deployed Active Network. In Werner Bob, editor, *Proceedings of DANCE 2002*, pages 216–224. IEEE Computer Society, 2002. May 29–30, 2002, San Francisco, USA.
11. C. Labovitz, A. Ahuja, A. Bose, and F. Jahanian. Delayed Internet Routing Convergence. In *Proceedings of SIGCOMM 2000*, pages 175–187, 2000.
12. L. Lefèvre *et. al.* Active Networking Support for the Grid. *Lecture Notes in Computer Science*, 2207:16–33, 2001.
13. M. Maimour and C. Pham. An Active Reliable Multicast Framework for the Grids. In *International Conference on Computational Science (2)*, pages 588–597, 2002.
14. S. Martin and G. Leduc. RADAR: Ring-based Adaptive Discovery of Active neighbour Routers. In *Proceedings of IWAN 2002*, pages 62–73, December 2002.
15. N. Mohamed. Active Networks and Their Utilization in the Computational Grid. Technical Report TR03-04-01, Engineering University of Nebraska-Lincoln, Lincoln, NE 68588-0115, April 2003.
16. S. L. Murphy, E. T. Lewis, and N. M. R. Watson. Secure Active Network Prototypes. In Werner Bob, editor, *Proceedings of DANCE 2002*, pages 166–181. IEEE Computer Society, 2002. May 29–30, 2002, San Francisco, USA.
17. N. Nagaratnam *et. al.* The Security Architecture for Open Grid Services. GGF Specification, July 2002.
18. A. Savanović and B. Jerman Blažič. A Protocol for Adaptive Autoconfiguration of Active Networks. *WSEAS Transactions on Communications*, 2:78–83, 2003.
19. A. Savanović, D. Gabrijelčič, B. Jerman Blažič, and S. Karnouskos. An Active Networks Security Architecture. *Informatica*, 26(2):211–221, 2002.
20. B. Schneier. *Applied Cryptography: Protocols, Algorithms, and Source Code in C.* John Wiley and Sons, Inc., second edition, 1996.
21. J. M. Smith *et. al.* Activating Networks: A Progress Report. *IEEE Computer Magazine*, pages 32–41, April 1999.
22. T. Stack, E. Eide, and J. Lepreau. Bees: A Secure, Resource-Controlled, Java-Based Execution Environment, December 2002. http://www.cs.utah.edu/flux/janos/bees.html.
23. D. Wetherall, J. Guttag, and D. Tennenhouse. ANTS: Network Services Without Red Tape. *IEEE Computer*, pages 42–48, Apr 1999.

A Convergence Architecture for GRID Computing and Programmable Networks

Christian Bachmeir, Peter Tabery, Dimitar Marinov,
Georgi Nachev, and Jörg Eberspächer

Munich University of Technology, Institute of Communication Networks
bachmeir@ei.tum.de, http://www.lkn.ei.tum.de

Abstract. GRID computing in the Internet faces two fundamental challenges. First the development of an appropriate middleware for provision of computing services in a scalable and secure manner. Second the distributed, scalable and fast connection of GRID processing components to the network. The latter is important to enable fast data exchange among the distributed components of virtualized super-computers.
In this work, we focus on the connection aspects of GRID architectures and advocate to enhance edge-routers with clusters of high performance computing machines.
We propose enhancing available GRID resources to a component-based programmable node. Our approach delivers three advantages: Use of the same hardware/software for GRID computing and Programmable Networks. Support of GRID computing through–then available–programmable services, like data transport and distribution in the GRID through Programmable Networks. Finally we see GRID computing as a leverage for the future deployment of Programmable Networks technology.

1 Introduction

In this paper we propose a convergence architecture for GRID computing and Programmable Networks. The primary design goal of our approach is to provide a flexible platform that can be integrated in a GRID (see Figure 1) and a Programmable Network simultaneously. The major benefit of our approach is the convergence of the two network-based, distributed computing technologies. Our approach can provide both, the execution of GRID processing jobs as well as the provision of Programmable Network services.

Basic idea of our approach is a spatial separation between *routing* of traffic, the actual *processing* of services and the *control* of the system across different machines.

Through this separation we provide a powerful, scalable programmable architecture that is integrated into a GRID. We introduce dedicated signaling hosts in our architecture, that provide access to the programmable node, based on standard web services. Using this open signaling approach our architecture can be integrated in GRID middleware.

M. Bubak et al. (Eds.): ICCS 2004, LNCS 3038, pp. 187–194, 2004.
© Springer-Verlag Berlin Heidelberg 2004

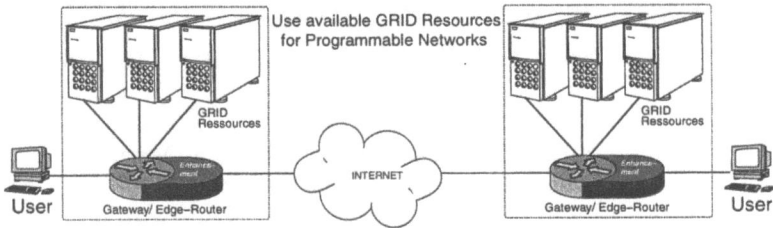

Fig. 1. Available Grid Ressources

Main contribution of this work is the integration of a new layer 4 encapsulation and proximity switching mechanism in programmable architectures, which **obsoletes explicit kernelspace-to-userspace communication** on processing machines. Based on layer 4 encapsulation and decapsulation, IP data packets are forwarded to to processing components by the router component.

Using standard UDP sockets, we introduce the concept of *Operating System as Execution Environment (OSaEE)*. We propose to execute service modules as standard userspace programs. These service modules can mangle original IP packets by receiving and sending them using **standard UDP sockets**. As we do not use kernel-userspace communication, the principle of OSaEE is inherently the same as processing GRID jobs.

Based on three mechanisms (layer 4 forwarding, OSaEE and a reliable signaling architecture) we present a robust architecture that bypasses security and fault tolerance limits which are currently faced by proposed programmable architectures, and which prohibit to use the resources of programmable architectures in a GRID.

This paper is organized as follows: In the next Section an overview of our architecture and the interworking with the GRID is presented. In Section 3 we show performance measurements of our prototype implementation.

2   Overview of the Proposed Architecture

In Figure 2, an overview of the programmable node architecture is shown. We derive our architecture from HArPooN [1], and propose to spatially separate routing functions from computing functions. The routing resp. switching is done in the **Routing Component**, an enhanced standard router. With *computing* we subsume the *signaling*, necessary for access and control, the provision of *programmable services* and the use of the resources in a GRID context. Our architecture executes these computing functions on machines in the **Computing Component**.

Key component of our architecture is a lightweight interface between the router and the computing cluster. We propose to enhance the architecture of the router with a traffic encapsulation and decapsulation module (see paragraph 2.1). When using our architecture as a programmable node this module is used to forward data packets to processing machines, which provide active and programmable services. When accessing GRID components, the module is bypassed.

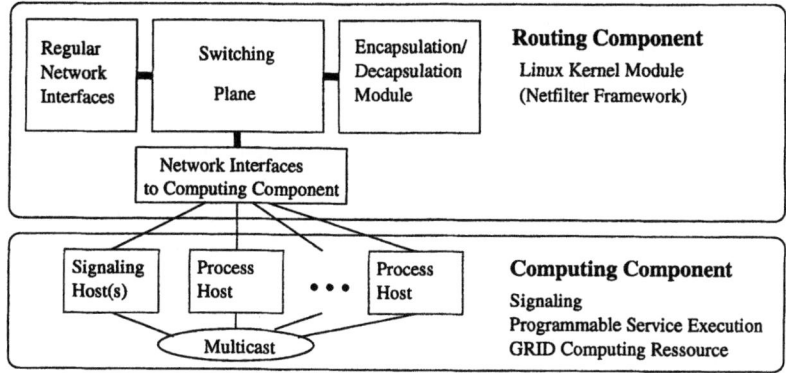

Fig. 2. Component based Programmable Router & GRID Architecture

2.1 Routing Component and OSaEE

In this section we elaborate on the encapsulation and decapsulation module, located in the routing component. Every time an IP packet arrives at the routing component, it is first checked whether the packet originates from the computing cluster of our proposed programmable architecture. If the IP packet does not originate from the computing cluster, the local database (provided by Module Control) is queried whether this IP packet belongs to a traffic flow, for which our architecture currently provides a service. If this is the case, the entire packet is encapsulated within a UDP packet[1]. The UDP/IP header is set to address the process host and the port number of the corresponding service module. Because of the encapsulation header, the packet does not continue its original way through the network, but is routed towards the processing machine and received by a service module with a corresponding UDP socket.

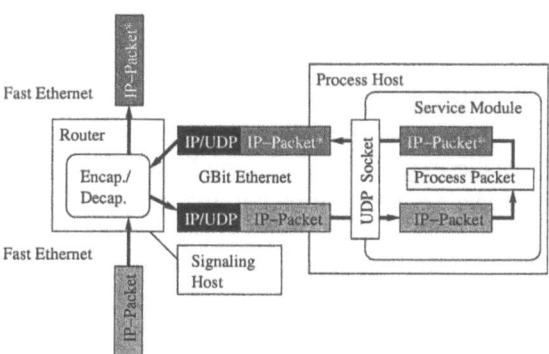

Fig. 3. Data flow in the component based Programmable Node

---

[1] Fragmentation due to encapsulation and enlargement of the packet size does not happen as we use GBit Ethernet between Routing and Computing Component, which allows "Jumbo packets" with a frame size up to 9 Kbyte

The processed IP packet is then sent by the service module as a payload using again the UDP socket. The IP destination address and the destination port number are set to the IP address of the routing component and a determined port number.

When the encapsulation/decapsulation mechanism of the routing component receives this encapsulated packet, it decapsulates the packet (deleting exterior IP and UDP headers). The remainder of the packet–the IP packet constructed by the service module–is routed by the routing component on its way through the Internet.

In some proposed architectures the loadable service modules are granted administrator privileges, due to kernel-userspace communication implementation issues. Obviously this is a back door for all kinds of attacks towards the system architecture. Therefore, security mechanisms are engineered around the service modules (e.g., the service module is prevented from deleting certain files on the system, it is not allowed to fork processes, etc.).

In case of architectures where loadable service modules are proposed as kernel modules [2], the issues concerning security is even a more severe one. Kernel modules cannot be controlled against security violations at all. Therefore architectures have been proposed which are using access policies, primarily building on trusted code sources.

When discussing security of programmable nodes against *internal* attacks, we felt like "reinventing the wheel". Therefore we propose to *make our architecture independent towards security issues like internal attacks*, through OSaEE (Operation System as Execution Environment)[2].

We consider the development of the security of standard operating systems (e.g. Linux) against internal attacks already at a mature stage. When developing OSaEE we intend to leverage that basis for our architecture.

The general goal of OSaEE is to use a standard operating systems on machines in the computing cluster, without any modifications as a platform to execute loadable programs. On that basis OSaEE enables us to use standard security features of the deployed operating system, e.g. the user management. Each service module (userspace program) is then started under a different user. Once a certain user consumes "too much" system resources or generates "too much" traffic, the control process of the system reacts, e.g. by killing offending processes of that specific user. By introducing OSaEE we see major advantages compared to state-of-the-art programmable architectures:

- Our architecture is simplified, leading to a higher robustness and usability.
- Using OSaEE we make our architecture independent towards security requirements regarding internal attacks of the system.
- Flexibility: Service modules can be written in a variety of programming languages

---

[2] External attacks, targeting or originating by the programmable node are clearly out of the scope of this work. We state that proposed mechanisms can be integrated in our architecture.

- Due to abstraction of network stack: Composition of complex services using special libraries (e.g. transcoding) in the userspace.
- Process Hosts may utilize different operating systems.
- Computing Resources can be used simultaneously in a GRID.

In Figure 3 the exemplary data flow of a single IP packet through HArPooN as described in paragraph 2.1 is shown. Due to the *Encap./ Decap. mechanism*, regular userspace programs are able to intercept entire IP packets that want to bypass the Router Component. On the processing machines in the computing cluster there is no modification necessary to enable this mechanism as the userspace programs only need to be able to use standard UDP sockets.

Besides improving issues for Programmable Networks, the adaptability of OSaEE also enables us to offer computing resources of our programmable architecture to a Computing GRID. The necessary flexible signaling mechanism is presented in the next paragraph.

2.2   Interworking: Layered Signaling

Because of the spatial separation between routing component and computing cluster, there is a need of synchronization between both. In our approach we introduce dedicated **signaling hosts** (see Figure 2) to perform this task. To enable fault tolerance our approach foresees supplemental signaling hosts, running in hot standby. The signaling hosts are basically responsible for three main tasks:

1. Control of the Router Component (entries in Encap./ Decap module)
2. Basic external signaling (load, start, and monitor a service or GRID jobs on our architecture).
3. Internal multicast-based signaling (for monitoring and controlling the Process Hosts)

These three mechanisms are general and service unspecific, therefor we subsume all three under the expression **Basic Signaling**.

It might be the case that for a specific service to work, there is a **Service or Job Specific Signaling** necessary. In case of a service (or a GRID job) that consists of modules which are distributed on numerous machines in the Internet, it is necessary to allow e.g. TCP connections between the different service modules for data exchange. We consider that kind of signaling as part of the service (or job) itself. In our approach, distributed service modules can communicate with each other, using standard sockets. The Routing Component does not affect these data streams. This separation between **general basic signaling** and (loadable) **service or job specific signaling** delivers a layered signaling architecture presented in Figure 4. In our implementation of the system, we use Java technologies[3] to realize the *general basic signaling*. Using the Java Web Services mechanism, the basic external interface can easily be provided, because

---

[3] Java 1.3.1, Netbeans IDE 3.5

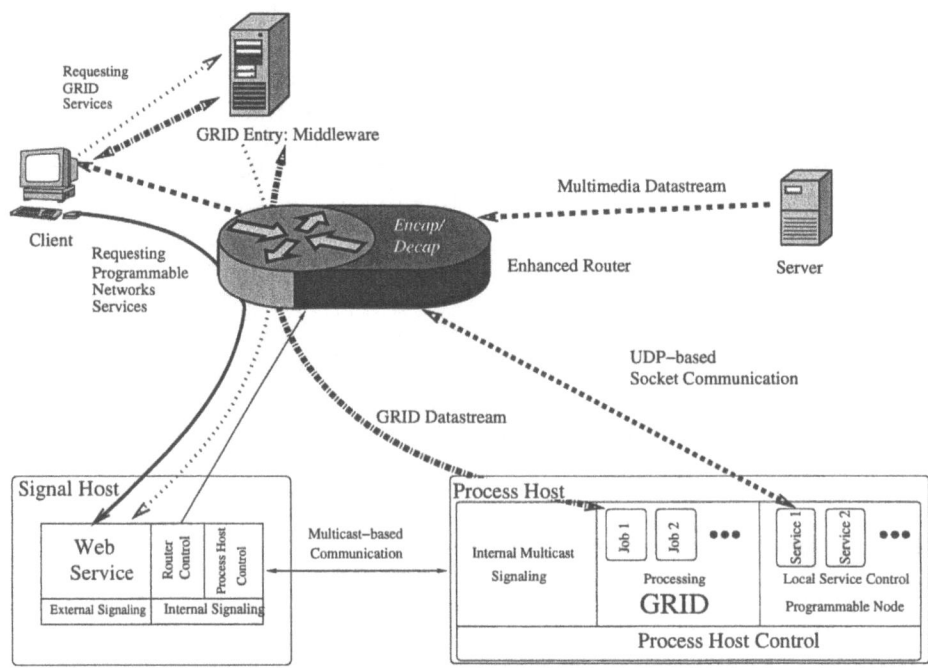

Fig. 4. Layered Signaling and embedding of signaling components in the architecture

the basic functionality is already implemented by the web service frameworks (like JAX-RPC[4], that we used).

The *Process Hosts Control Module*[5] is using the internal multicast-based signaling interface to access Process Hosts. The basic task of the multicast protocol is, to start and stop programs representing active services on the process hosts. Also other general, service unspecific information, e.g. monitoring, is exchanged via this interface. We implement that protocol based on multicast solely for reasons of fault tolerance (hot standby) and scalability of the whole system, when employing several Process Hosts. Our approach also provides a second signaling host, which can take over immediately the control of the respective services in case of a failure. Moreover the strict separation between Basic External Signaling and Internal Multicast based Signaling is also a security feature of our approach. It is not possible to start a service (or a GRID job) directly on a process host from outside of our architecture, only signaling hosts are allowed to do so.

3   Performance Evaluation of the Programmable Node

In this section, performance evaluation results of our approach are presented. In Figure 5 the perceived packet loss in our testbed is shown. As outlined in

---

[4] Java API for XML-based RPC: Simplifies Programming of Web Services
[5] Realized with Enterprise Java Beans: Performant execution of parallel beans

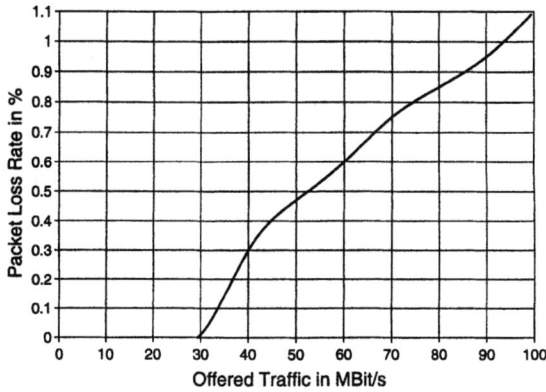

Fig. 5. Packet Loss in our prototype

figure 4, our prototype consists of five machines. One is a 700 MHz Linux-based Pentium 3 machine with four 1 GBit Ethernet cards, which represents the modified router. the others are Pentium 4 machines, acting as client, server, signaling host and one process host. Client and server are connected using 100 MBit Ethernet to the router. We send multimedia datastreams at constant bit rates from server to client. The data packets of the stream where intercepted at the router, encapsulated and forwarded to the Process Host. At the process host data packets were received, using UDP sockets and sent back[6] to the router. There they where decapsulated and forwarded to the client.

Although we used hardware with relatively low performance for the prototype implementation of the router, we perceived a considerably low packet loss. Even when offering 100 MBit/s of data only 1.1 % of the packets were lost. Compared to measurements of state-of-the-art implementations of Programmable Networks [3] our approach is clearly competitive.

We state that perceived losses will vanish entirely, when using a modified standard router hardware, instead of the PCI-Bus PC used in our prototype.

We did not measure our approach against multi machine based active and programmable node architectures like e.g. [4], as the used hardware cannot be compared. However we state if a high performance router–with proposed modifications–would be embedded in our architecture, the performance should be at least comparable to proposed architectures.

4   Conclusion

In this work we present a new architecture of a programmable node based on GRID computing resources. Foundation of our approach is, to keep changes at

---

[6] Data packets where not processed resp. modified at the Processing host, as we only measured for maximum throughput. Due to possible load balancing our architecture improves execution time of services anyway [1].

the actual router at a minimum, when transforming a regular router into an active and programmable router. Therefore we see a good chance to implement our proposal in future products and enable deployment of active and programmable services outside research environments.

In this work we present two major contributions: First, we provide a flexible platform that can be integrated in a GRID. We see a major benefit in this approach as available resources can be used concurrently for programmable network services as well as for computation of jobs in a GRID context. Besides the programmable part, our architecture delivers means to enhance the GRID, e.g. through the provision of dynamic configurable data transport mechanisms.

Second, we provide a new high performance architecture of a component based programmable router. Through the separation of routing, signaling and processing on different machines we improve local security and fault tolerance of the programmable node. The primary benefit of our architecture is in the provision of complex, resource-demanding active and programmable services.

We think that our approach is a powerful enhancement to the GRID. Based on the emerging deployment of GRID computing services, we see a potential for leveraging programmable networks technology.

References

[1] Bachmeir, C., Tabery, P., Sfeir, E., Marinov, D., Nachev, G., Eichler, S., Eberspächer, J.: HArPooN: A Scalable, High Performance, Fault Tolerant Programmable Router Architecture. In: Poster Session in IFIP-TC6 5th Annual International Working Conference on Active Networks, IWAN'2003, Kyoto, Japan (2003)

[2] Keller, R., Ruf, L., Guindehi, A., Plattner, B.: PromethOS: A dynamically extensible router architecture supporting explicit routing. In: 4th Annual International Working Conference on Active Networks, IWAN 2002, Zurich, Switzerland (2002)

[3] Conrad, M., Schöller, M., Fuhrmann, T., Bocksch, G., , Zitterbart, M.: Multiple language family support for programmable network systems. In: Proceedings of the 5th Annual International Working Conference on Active Networks, IWAN'2003, Kyoto, Japan (2003)

[4] Kuhns, F., DeHart, J., Kantawala, A., Keller, R., Lockwood, J., Pappu, P., Richards, D., Taylor, D., Parwatikar, J., Spitznagel, E., Turner, J., Wong, K.: Design of a high performance dynamically extensible router. In: DARPA Active Networks Conference and Exposition (DANCE), San Francisco (2002)

Programmable Grids Framework Enabling QoS in an OGSA Context

John Soldatos[1], Lazaros Polymenakos[1], and George Kormentzas[2]

[1] Athens Information Tchnology
19.5 Km, Markopoulou Ave., Peania GR-19002, P.O Box: 68, Greece
{jsol,lcp}@ait.gr
[2] University of the Aegean
Karlovasi, Samos, 83200, Greece
gkorm@aegean.gr

Abstract. Configuring, monitoring and guaranteeing service levels is a key prerequisite towards supporting the wave of emerging Grid computing applications, especially those relating to e-business. Managing QoS across a Grid infrastructure is particularly complex since QoS must be addressed at three different levels: network, middleware and application. Programmable Grids extend programmable networks concepts to Grid infrastructures and can alleviate this complexity. In this paper we elaborate on the importance of network, middleware and application level QoS, and accordingly provide a framework for designing and implementing programmable APIs that can handle QoS. The framework leverages the structure and properties of OGSA services, while exploiting functionality of OGSI containers. Moreover, early implementation experiences with this framework and Globus (GT3) are reported.

1 Introduction

Grid computing constitutes a distributed computing paradigm enabling virtual organizations (VO) to share resources towards achieving their goals [1]. Existing deployments have demonstrated benefits for various user groups. The emphasis is nowadays on the standardization and the commercialization of the Grid towards accelerating its development and adoption. The Global Grid Forum (GGF) has taken a significant standardization step through the initial specification of the Open Grid Services Architecture (OGSA) [2], which achieves a standardized way for virtualizing and accessing resources.

OGSA boosts the commercialization of the Grid through supporting applications that deliver improved business results [3], [7]. The commercialization of the Grid poses however new requirements and associated technical challenges. Enterprise applications demand guaranteed service levels. As a result Grid Service Level Agreements (GSLA), along with mechanisms, tools and techniques for managing them, need to be introduced [4]. Managing GSLAs can be greatly facilitated through a programmability layer on top of a Grid. This gives rise to grid architectures consisting of dynamic, reconfigurable on-demand and highly customizable computing, storage and networking environments, called programmable Grids. Programmable Grids have their roots in the programmable and active networks paradigms.

M. Bubak et al. (Eds.): ICCS 2004, LNCS 3038, pp. 195–201, 2004.
© Springer-Verlag Berlin Heidelberg 2004

The purpose of this paper is to introduce an infrastructure enabling programmable operations across a Grid. We view these operations in a programmability context (i.e. based on the IEEE P1520 initiative [5]) rather than in an active networks context. We describe a framework for offering open programmable APIs (Application Programming Interfaces) on Grid systems towards automatically provisioning and managing QoS. The notion of QoS and related metrics is familiar to the networking community, since it is an integral component of broadband networks. Nevertheless, Grid QoS is not confined to networking QoS, but extends to middleware and application level QoS [6]. Our framework for programmable operations covers all levels and focuses on resource management. Other issues (e.g., security management for pre-authorized and authenticated QoS services) are out of the scope of this paper.

Early research work has already produced results regarding mechanisms for pooling and using resources (e.g., [8]), as well as for defining and negotiating generalized GSLAs (e.g., [10]). With respect to QoS, other works have come up with mature general-purpose architectures allowing management of diverse Grid resources (e.g., [9]). The most characteristic example is the GARA architecture, which defines APIs for manipulating reservations of different resources in uniform ways. Our design focuses on an infrastructure for open, programmable OGSA based APIs that can be employed to implement GLSA models and related resource management policies.

The rest of the paper is structured as follows: Section 2 discusses the main requirements from a Grid QoS programmable framework. Section 3, presents our framework for exporting programmable OGSA based APIs. Section 4, reports on work conducted using the Globus Toolkit version 3 (GT3), towards implementing certain aspects of this framework. Finally, section 5 concludes the paper.

2 Programmable Grids for QoS

2.1 Network QoS

State of the art networking infrastructures provide immense capacity capable of supporting emerging Gird applications. Future broadband networks are expected to provide improved support for larger scale Grid infrastructures. Programmable Grids can provide the means for leveraging the capabilities of both existing and emerging networking infrastructures. The programmable infrastructure should virtualize the networking resources and provide a set of interfaces for accessing and managing these resources. Network resources and metrics of interest include bandwidth, throughput, as well as delay and loss figures. Emphasis is put on the capacity, utilization and availability statistics on particular links, interfaces and routing paths (e.g., [6]).

Exposing programmable interfaces for accessing and managing network resources, hinges on the communication with entities that can provide this information (i.e. information providers), while also allowing execution of configuration operations. Such entities are element layer management (ELM) systems, network layer management (NLM) systems, network resource managers (NRM), bandwidth brokers (BB), as well as directory systems (e.g., LDAP directories). The scope of these systems depends on the network topology, the technologies deployed, as well as on administrative policies. Usually NLM, BB and NRM systems handle a whole administrative domain.

Programmable interfaces must be exported through the Grid middleware layer (e.g., Globus). In OGSA terms, this implies that operations accessing network resources and statistics should be available as Grid Services within an OGSI [12] compliant container, (e.g., GT3). As a result, an OGSA based programmable Grids framework delivers network control and management functionality as a set of Grid Services. This delivery can leverage recent work on network management operations based on Web Services (e.g., [13]).

2.2 Middleware QoS

Middleware layer QoS deals with the lifecyle and execution of Grid services, and the related management and allocation of IT resources (i.e. servers, storage). Grid middleware such as the Globus Toolkit provides the infrastructure for managing middleware QoS. In particular, Grid middleware platforms provide the means for allocating virtualized resources to data and/or compute centric tasks.

An OGSA based programmable framework must export APIs for managing allocation of CPU and storage resources to the Grid Services. GT3 provides a set of management mechanisms enabling job submission on distributed resources, namely the Globus Resource Allocation Manager (GRAM). GRAM defines and implements APIs and protocols allowing Grid clients to securely instantiate jobs, according to the functionality of remote schedulers. GRAM constitutes a readily available resource management mechanism, which however can be enriched with more advanced resource management features (e.g., [11]). GRAM is used in conjunction with a wide range of other schedulers (e.g., the Sun 's Grid Engine (SGE), IBM's LoadLeveler, or the Portable Batch System (PBS) [14]). Instead of using GRAM, a new programmable API could be defined and implemented.

2.3 Application QoS

Application QoS deals with the performance of the operations entailed in the Grid application. In an OGSA context a Grid application consists of a set of Grid services, which interact to achieve an application specific goal. Accomplishing this goal includes: (a) configuration of a grid service, (b) data transfer to a grid service (from a calling service), (c) execution of a remote service, (d) data transfer from the grid services (back to the calling service). QoS for each of these operations depends on the middleware and network QoS. As a result an API for managing application level QoS, hinges on network and middleware QoS capabilities.

3 Programmable OGSA Grid Architecture

Taking into account the requirements outlined in the previous section, we devised an OGSA based framework for delivering programmable APIs for QoS management. Figure 1, depicts a programmable framework for managing network layer QoS across a loosely coupled heterogeneous Grid infrastructure. From a network perspective the various devices of the Grid are clustered into distinct administrative domains. Each domain consists of several devices (e.g., routers, switches, homogeneous clusters), which are controlled by a single administrative entity. In the scope of each administrative domain, we envisage a network model and resource management structure resembling current networking infrastructures. In particular, networking resources of the administrative domain are controlled through a centralized or distributed NMS. A distributed NMS will most likely rely on ELM systems to collect management information from the network elements. QoS parameters relating to inter-domain SLAs are controlled by Bandwidth Brokers, which is perfectly in line with the IETF DiffServ model.

The programmable framework accesses NMS, ELM and BB functionality based on a set of OGSA Grid Services. These services are capable of taking and combining raw network measurements, as well as of configuring network parameters. Delivering such Grid Services hinges on the following steps:

1. Selecting operations of interest, as well as the network information pertaining to these operations.
2. Defining the Grid services corresponding to the network resource management operations of interest. This step hinges the specification of the programmable APIs for network QoS. This specification entails a degree of abstraction, since the API is defined independently of particular domains and systems (e.g., NMSs, BBs).
3. Specifying the network resource parameters and metrics involved in the Grid API operations (e.g., capacity, utilization, delay). These parameters will form the major part ServiceData of the OGSA Grid Services.
4. Implementing methods (in the form of low-level drivers) for accessing the respective operations of the NMS, ELM and BB systems. The OGSA based implementation of these methods entails an OGSI middleware platform (e.g., GT3), as well as a container where such services can be deployed.

The implementation produces an API for controlling Network QoS parameters. This API will be independent of the peculiarities of the network resource management systems of the domains engaged in the Grid implementation. The virtualization layer on top of these systems allows translation between operations and parameters specific to the domains, to domain and system independent operations and parameters. Specifically, the Grid Services API (API Programmability Layer in Figure I), constitutes the domain independent implementation of an API for network QoS control. The Service Data defined in the scope of these Grid Services forms the domain independent Grid parameters.

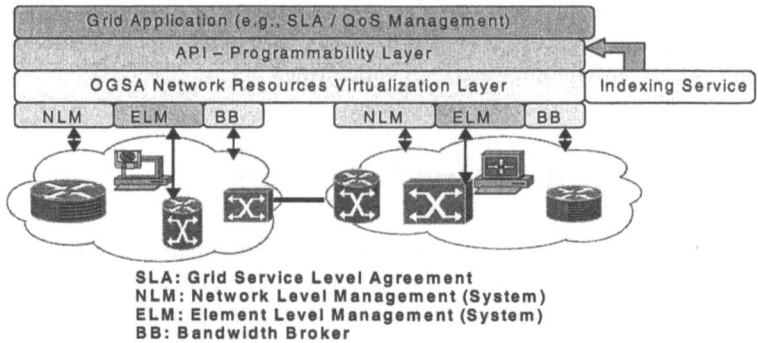

SLA: Grid Service Level Agreement
NLM: Network Level Management (System)
ELM: Element Level Management (System)
BB: Bandwidth Broker

Fig. 1. Programmable Framework for Network QoS Management

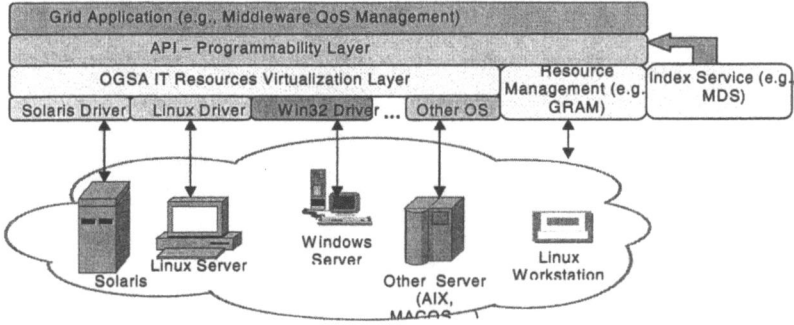

Fig. 2. Programmable Framework for Middleware QoS Management

The set of OGSA services for configuring, monitoring and managing network QoS parameters can be registered with an information indexing service (e.g., the Globus Monitoring and Discovery Service (MDS)). Accordingly, Grid applications can query the indexing service based on `Service Data` parameters.

The framework for managing Middleware QoS (Figure2) reveals many similarities to the framework for Network QoS control. In particular, it aims at delivering functionality for accessing QoS information about computing resources (e.g., storage, CPU utilization), and accordingly processing this information towards scheduling jobs. This implies that methods for accessing low level capabilities of the Grid machines (e.g., the operating systems, processors) are available. Thus, interfacing with system level services relating to the Grid donors is required. This depends on implementing drivers for interfacing with each one of the Grid constituents. As a result, the steps towards delivering OGSA Grid Services as a set of programmable APIs involve:

1. Specifying the parameters engaged in the job scheduling and infrastructure resources management operations.
2. Defining these parameters as OGSA `ServiceData`, and design Grid Services that access and configure those resources.
3. Implementing methods for accessing and altering the information specified within `Service Data`, for all of the platforms (e.g., Linux, Windows, Sun Solaris, IBM AIX, Mac OS) participating in the Grid.

Grid services relating to middleware QoS can also be registered with an Index Service (e.g., GT3 MDS) to allow scheduling applications to query for services based on `Service Data` parameters.

Using APIs for middleware QoS control and resource management, advanced scheduling mechanisms can be implemented. Commercial Grid platforms, as well as Globus provide readily available resource management and job scheduling functionality. Nevertheless, the proposed approach can still be followed towards enhancing existing functionality, or even implementing totally new schedulers (e.g., customized for particular Grid applications).

Auditing, controlling and provisioning applications' QoS is a matter of appropriately managing the underlying operation impacting network and middleware QoS. Hence, APIs for controlling application QoS can be delivered as shown in Figure 3. Application QoS APIs will be implemented as sets of other API calls concerning network and middleware QoS. These APIs will be used in the scope of

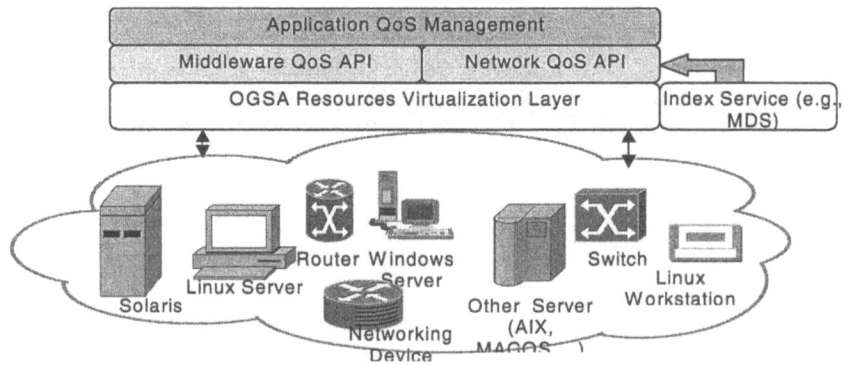

Fig. 3. Programmable Framework for Grid Application QoS Administration

resource management applications defining and setting QoS levels for Grid applications. Thus, they constitute a valuable interface for Grid applications administration.

4   OGSA Implementation Experience

As a first step, we have implemented low level drivers accessing network
management systems and system level functionality. Implementation of such drivers
has been up to date limited to a single administrative domain containing three
machines equipped with GT3.

As far as network level QoS is concerned, OGSA Grid Services accessing
networking parameters through interfacing with an ELM system have been
implemented. This ELM system executes SNMP operations on SNMP enabled
network element [13]. The ELM system exports a Web Services interface, which
facilitates the implementation of OGSA Grid Services. Based on Grid Services
exposing network management functionality we have implemented simple OGSA
clients querying and configuring element level network management information.

At the middelware layer we have implemented drivers allowing Grid services to
access metrics and statistics regarding computing resources (i.e. CPU in particular).
These metrics have been used to enrich the Service Data of OGSA services.
Towards interfacing (Java) GT3 OGSI services to system level functions pertaining to
our Solaris machines, we relied on the Java Native Interface (JNI). Accordingly, we
implemented simple job schedulers (i.e. load balancers based on processor utilization
metrics). This implementation provided a proof of concept regarding the lower layer
of our programmable middleware QoS framework. As part of ongoing and future
work, we intend to implement all aspects of the framework.

5   Conclusions

Future Grid infrastructures must guarantee QoS levels. QoS is not only concerned
with the networking infrastructure, but also with the middleware and application
structure. Programmable Grids can leverage the OGSA towards designing and

implementing programmable APIs for QoS control. OGSA imposes a structure for delivering a programmable framework, while OGSI application containers provide essential supporting features. In this paper we have presented a framework for structuring, designing and implementing OGSA compliant programmable functionality for QoS management. Early experimentation with GT3 has demonstrated several of concepts of the framework. We expect future work to provide a more complete underpinning.

References

1. I. Foster, C. Kesselman, S. Tuecke, 'The Anatomy of the Grid: Enabling Scalable Virtual Organizations', International Journal of Supercomputer Applications, 15(3), 2001.
2. I. Foster, D. Gannon, H. Kishimoto (editors), 'The Open Grid Services Architecture', Global Grid Forum Draft, draft-ggf-ogsa-ogsa-011, September 23, 2003.
3. I. Foster, D. Gannon, H. Kishimoto, Jeffrin J. Von Reich, 'Open Grid Services Architecture Use Cases', GWD-C (draft-ggf-ogsa-usecase-1.7), September 18, 2003.
4. Akhil Sahai, Sven Graupner, Vijay Machiraju, Aad van Moorsel, 'Specifying and Monitoring Guarantees in Commercial Grids through SLA', Hewlett-Packard Laboratories, White Paper, November 2002.
5. Biswas, J., Lazar, A. et al "The IEEE P1520 Standards Initiative for Programmable Network Interfaces", IEEE Communications Magazine, pp. 64-71, October 1998.
6. G-QoSM: Grid Service Discovery Using QoS Properties, Al-Ali RJ, Rana OF, Walker DW, Jha SK, Sohails S, Computing and Informatics Journal, Special Issue on Grid Computing, Institute of Informatics, Slovak Academy of Sciences, Slovakia, 2002, 21(4), pp. 363-382.
7. J.Soldatos and Lazaros Polymenakos, 'Enterprise Grids: Drivers and Implications', Athens Information Technology Report, December 2003.
8. Jacek Gomoluch and Michael Schroeder, 'Market-based Resource Allocation for Grid Computing: A Model and Simulation' in the Proc. of the 1st International Workshop on Middleware for Grid Computing, Rio de Janeiro, Brazil, 2003.
9. I. Foster, C. Kesselman, C. Lee, B Lindell, K. Nahrstedt, A. Roy, "A Distributed Resource Management Architecture that Supports Advance Reservation and Co-Allocation", in the Proceedings of the International Workshop on QoS, pp. 27-36, 1999.
10. Czajkowski, K., Foster, I., Kesselman, C., Sander, V., and Tuecke, S., "SNAP: A Protocol for Negotiating Service Level Agreements and Coordinating Resource Management in Distributed Systems", 8th Workshop on Job Scheduling Strategies for Parallel Processing, Edinburgh, Scotland, July 2002.
11. K. Czajkowski, I. Foster, and C. Kesselman, 'Co-allocation services for computational grids'. In the Proceedings of the 8th IEEE Symposium on High Performance Distributed Computing. IEEE Computer Society Press, 1999.
12. S. Tuecke, I. Foster, J. Frey, S. Graham, C. Kesselman, T. Maquire, T. Sandholm, D. Snelling, P. Vanderbilt, 'Open Grid Services Infrastructure (OGSI), Version 1.0' (draft), GWD-R (draft-ggf-ogsi- gridservice-29), April 5, 2003.
13. Dimitris Alexopoulos and John Soldatos, 'Open Programmable XML based Network Management: Architecture and Prototype Instantiation', submitted for publication in the IEEE Communications Magazine, special issue on XML based Management of Networks and Services, under review.
14. Luis Ferreira et. al. 'Introduction to Grid Computing with Globus', IBM redpaper, December 2002.

Active and Logistical Networking for Grid Computing: The E-toile Architecture

Alessandro Bassi[1], Micah Beck[3], Fabien Chanussot[1], Jean-Patrick Gelas[1], Robert Harakaly[2], Laurent Lefèvre[1], Terry Moore[3], James Plank[3], and Pascale Primet[1]

[1] INRIA / LIP (UMR CRNS, INRIA, ENS, UCB Lyon1)
Ecole Normale Supérieure de Lyon - 46, allée d'Italie - 69364 LYON Cedex 07 - France
`laurent.lefevre@inria.fr`, {`fabien.chanussot,pascale.primet`}`@ens-lyon.fr`
[2] UREC CNRS
Ecole Normale Supérieure de Lyon - 46, allée d'Italie - 69364 LYON Cedex 07 - France
`robert.harakaly@urec.cnrs.fr`
[3] LoCI Laboratory - University of Tennessee
203 Claxton Building - 37996-3450 Knoxville, TN, USA
{`abassi,mbeck,tmoore, plank`}`@cs.utk.edu`

Abstract. While active networks provide new solutions for the deployment of dynamic services in the network by exposing network processing resources, logistical networks focus on exposing storage resources inside networks by optimizing the global scheduling of data transport, and data storage. In this paper, we show how active and logistical environments working together can improve Grid middleware and provide new and innovative high-level services for Grid applications[1]. We validate and experiment this approach combining the Internet Backplane Protocol[2] suite with the Tamanoir Active Node environment. Our target architecture is the French e-Toile Grid infrastructure [1] based on high performance backbone (VTHD) [2].

1 Introduction

The emergence of Grid computing as the ultimate solution for the scientific computing community has driven attention and research in the recent past, and especially since the proliferation of high performance network capabilities. The common purpose is to aggregate geographically distant machines and to allow them to work together to solve large problems of any nature.

---

[1] This work is supported and carried out within the framework of the RNTL e-Toile project and the RNRT VTHD++ project. More information on e-Toile project are available on http://www.urec.cnrs.fr/etoile/

[2] This work is supported by the National Science Foundation Next Generation Software Program under grant # EIA-9975015, the Department of Energy Scientific Discovery through Advanced Computing Program under grant # DE-FC02-01ER25465, and the NSF Internet Technologies Program under grant # ANI-9980203.

M. Bubak et al. (Eds.): ICCS 2004, LNCS 3038, pp. 202–209, 2004.
© Springer-Verlag Berlin Heidelberg 2004

The optimization of the data transfers within a Grid, which we define as High Performance Grid Networking, is the focus of our interest in this area. As the load, the capacity and the availability of network links used during data transfers may heavily affect Grid application performances, a fully functional Grid is dependent on the nature and quality of the underlying network. Appropriate performance, together with a high degree of flexibility, are therefore key factors of a successful Grid project. One of the important characteristic of the data transfers within a Grid environment is that the spectrum of exchanged data volume can spread over 9 orders of magnitude, ranging from few bytes in interactive traffic up to terabytes scale bulk data transfer. Requirements in terms of reliability and delay can also be very heterogeneous.

To introduce our work in this area, and present our research efforts within the e-Toile [1] grid environment, we would like to step back and reason about some fundamental characteristic that current network services show. In particular, we would like to focus the attention on the *encapsulation* of current network services, while our approach allows low-level functionalities to be *exposed* to higher levels.

The architectural design of the e-Toile grid platform, based on a very high speed network and on active and logistical network technologies, allows the study of the existing communications limits regarding services and protocols and the validation of more efficient approaches aiming to carry gigabit performance to the grid user level and taking into consideration the specific needs of grid flows; these goals are achievable through services which are made possible by the exposure of certain equipments' functionalities. While in the context of the e-Toile project different innovative approaches are explored, such as network performance measurement, differentiated services evaluation, high performance and enhanced transport protocols, active networking technology and services and logistical networking, this paper will focus on the last two aspects.

The paper is organized as follows: in section 2 we will describe our driving philosophy about network services, while section 3 will focus on the concept of an active grid. in section 4 we will describe the e-Toile grid platform and active and logistical deployment on the e-Toile platform (section 5).

2   Exposing Network Services

To the extent that the scientific computing community is already using the network as a computer, the Internet provides a ubiquitous communication substrate connecting its components (with routers acting as special-purpose elements invisible in the architecture), while network servers provide all access to storage and computation. Illustrations of such servers and services are plentiful: FTP, NFS, and AFS provide access to storage; Condor, NetSolve, Ninf provide lightweight access to processing; HTTP provides access to both; GRAM provides access to heavyweight computing resources; LDAP provides access to directory services; and so on. What is notable about these instances, and is equally true of almost all the other cases we could add to the list, is that they represent relatively *encapsulated network services*, where with encapsulated network service we mean architectures implementing functionalities that do not closely model the underlying network resource, but have to be implemented by aggregating the resource and/or applying significant additional logic in its utilization.

The best effort delivery of datagrams at the IP level, on the other hand, can be taken as example of a relatively *exposed* network service. An exposed network service adds enough additional abstraction to the underlying network resource to allow it to be utilized at the next higher level, but does not aggregate it or add logic beyond what is necessary for the most common and indispensable functionality that uses it.

An important difference between the two approaches emerges when we need to extend the functionality of a given service. Encapsulated services tend to be implemented by heavyweight servers and have APIs designed at a high semantic level, interposing themselves between the client and low overhead, transparent access to the underlying resources As a result, it can be difficult, inefficient, or in some cases impossible to build new functionality on top of such APIs. Instead, encapsulated services tend to implement new functionality through plug in modules that extend the functionality of the server, introducing new code that has access to low level interfaces within the server. These plug-in modules are the server equivalent of microcode in CISC processors, raising a familiar set of questions about access control and security for the management of such code.

Extending the functionality of an exposed service makes different demands because exposed services have lighter weight servers and APIs designed at a simpler semantic level. Since these factors are conducive to low overhead and more transparent access to the underlying resources, it tends to be much easier and more efficient to build new functionality on top of exposed services. Exposed services promote the layering of higher-level functionality on top of their APIs, either in higher-level servers or in client code. This layering of services, which is analogous to the user-level scheduling of a RISC processor by a compiler, is perhaps most familiar in construction of a network services stack.

2.1 Exposing Network Storage Resources: The Internet Backplane Protocol

Despite the familiarity of the exposed approach for the network services stack, it may still not be obvious how to apply it to a resource such as storage. After all, almost every technology for the access and/or management of network storage one can think of (FTP, HTTP, NFS, AFS, HPSS, GASS, SRB, NAS, etc.) encapsulates the storage behind abstractions with relatively strong semantic properties. For that reason, our research in this area had to start by creating a protocol, the Internet Backplane Protocol (IBP) [3], that supports the management of remote storage resources while leaving them as exposed as possible. IBP is a network service that provides an abstraction of shared network storage. Each IBP server (called also depot) provides access to an underlying storage resource to any client that connects to the server. In order to enable sharing, the depot hides details such as disk addresses, and provides a very primitive capability-based mechanism to safeguard the integrity of data stored at the depot. IBP's low level, low overhead model of storage is designed to allow more complex structures, such as asynchronous networking primitives and file and database systems, to be built on top of the IBP API. This key aspect of the IBP storage model, the capacity of allocating space on a shared network resource, can be seen as doing a C-like malloc on an Internet resource, with some outstanding differences, such, for instance, time-limitation, to prevent Denial-of-Use attacks.

With IBP in place the question becomes how easy or difficult it is to layer storage services with strong semantic properties on top of the weak underlying storage resources provided by IBP depots. Our experience shows that the answer varies between different cases. In some cases (e.g. building a file abstraction) earlier models can be followed and the design is relatively straightforward; in other cases (e.g. point-to-multipoint communication) the inclination to take a more encapsulated approach remains strong, and the consequently the design choice is more difficult.

2.2   Exposing Network Processing Resources: Tamanoir

The growing interest in the Active Networking field might be seen as a natural consequence of the difficulties experienced when integrating into a shared network infrastructure the existing technologies with new ones. In "active" networking vision, network equipments can perform computations on user data in transit, therefore exposing their computation capabilities accessible by the end users by supplying programs called *services*, that modify the behavior of the network. These equipments are called *active nodes (AN)* and show a greater flexibility towards the deployment of new functionalities, more adapted to the architecture, the users and the service providers' requirements. The price to pay to have this greater flexibility is, generally, an increased attention needed towards security and performance.

The Tamanoir [4] framework is an high performance active environment based on active edge routers, able to handle different applications and various data stream at the same time. The two main transport protocol, TCP and UDP, are supported by the Tamanoir Active Node (TAN) for carrying data. One of the characteristics of Tamanoir is the capacity of making use and exposing logistical storage for optimizing end-user services requests, especially in terms of performance. As explained in section 5, each Tamanoir node can take advantage not only of IBP depot located on the same node, but also with any depot participating in backbones such as the Logistical Backbone.

3   Concepts of an Active Grid

Our attention towards an active grid paradigm was driven by the complains of Grid application designers about standard network characteristics, such as reliable packet transport between nodes using the TCP/IP protocol suite, not being suited for typical Grid applications. The active grid architecture we envision is based on a virtual topology of active network nodes spread on programmable routers of the network. Active routers are deployed on network periphery, and allow data stream processing and storage, either in an explicit way (following a request by the application) or encapsulated one. Considering that the future of WAN backbones lies on all-optical equipment, we concentrate active operations on routers and nodes mapped at network periphery.

Active nodes are connected between each other and each AN manages communications for a small subset of Grid nodes. Grid data streams cross the active layer twice, before and after passing through the passive backbone.

3.1   Active Network Benefits for Grid Applications

The communications needs of Grid applications can be improved by the use of an Active Grid architecture, in the following areas: application deployment (needs of active reliable multicast), Grid management and support, wide-area parallel processing (needs of QoS and data conversion services...).

Most of services needed by Grid environments, such as high performance transport, dynamic topology adapting, QoS, on-the-fly data compression, data encryption, data multicast, data conversion, and errors management can be easily and efficiently deployed on demand with an Active Grid architecture.

3.2   Active Grid Architecture

To support most Grid applications, an Active Grid architecture have to deal with the two main Grid configurations, Meta Cluster Computing and Global Computing. In the first case, where the configuration shows an high level of coupling, an active node is mapped on network head of each cluster or parallel machine. This node manage all data streams coming or leaving a cluster. All active nodes are linked with other AN mapped at backbone periphery. An Active node delivers data streams to each node of a cluster and can aggregate output streams to others clusters of the Grid. In the second one, characterized by a loosely coupled configuration, an AN can be associated with each Grid node or can manage a set of aggregated Grid nodes. Hierarchies of active nodes can be deployed at each network heterogeneity point. Each AN manages all operations and data streams coming to Grid Nodes, such as subscribing operations of voluntary machines, results gathering, nodes synchronization or checkpointing. For both configurations, active nodes could manage the Grid environment by deploying dedicated services adapted to Grid requirements : management of nodes mobility, dynamic topology re-configuration, fault tolerance.

4   The E-toile Grid Platform

The e-Toile project, funded by the French Ministry of Research in the realm of the RNTL (Réseau National des Technologies Logicielles) [5] initiative, focuses on three complementary objectives:

- to build an experimental high performance grid platform that scales to nation-wide needs and geographical distribution, providing an high performance network and software support for the ACI-GRID [6] initiative
- to develop original Grid services in order to go beyond the various limits of existing middleware and to exploit completely the services and capacities offered by a very high performance network. The e-Toile middleware integrates the most recent and relevant works of the French computer science laboratories (INRIA, CNRS) on the area of advanced communication services.
- to evaluate the deployment cost of chosen computing intensive and data-intensive applications and to measure the gain obtained thanks to the grid.

The partners of the project, providing various resources (clusters, storage space...) and federating the available tools in an integrated plate-form, are INRIA (Reso, Remap, Apache, Paris), CNRS (PRISM, IBCP), Communication & Systems, SUN Labs Europe, EDF (Electricite De France, French Electricity Board) and CEA (Atomic Energy Center).

The e-Toile middleware relies upon existing middleware, in particular Globus and the Sun Grid Engine, and integrates independent building blocks developed in the context of ACI-GRID. The most innovative grid components are in the areas of grid services (in terms of resource allocation), performance measurement, security, communication paradigms, distributed storage and active networking.

The figure 1 represents the architecture of the testbed with the resources currently interconnected. The platform is composed by two distinct testbed:

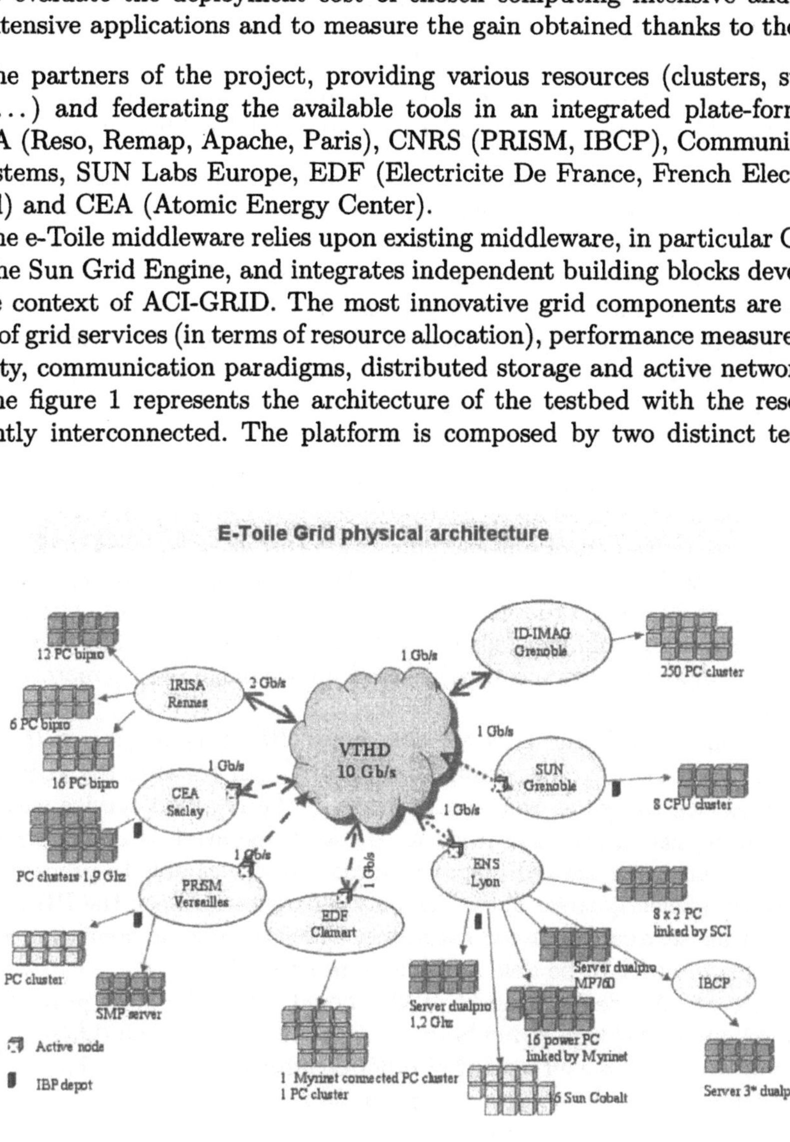

Fig. 1. Physical Architecture of the e-Toile grid

a development testbed in order to allow testing of new middleware component and a production one for grid applications. The VTHD (Vraiment Trés Haut Débit) [2] network infrastructure interconnects the grid nodes with 1 or 2 Gbps links. This existing french high performance research backbone has been extended to all participant sites.

Network monitoring and performance measurement tools, developed by IN-RIA and CNRS within the European DataGRID project are deployed in e-Toile. The MapCenter [7] Grid status visualization tool permit users and managers to visualize the available resources in real time and services.

One of the original aim of the e-Toile project is to focus on the High Performance Grid Networking dimension and to evaluate the benefits grid middleware and grid application can extract from enhanced networking technology. The performance problem of the grid network can be studied from different point of view:

- Measuring and monitoring the end-to-end performances helps to characterize the links and the network behaviour. Network cost functions and forecasts, based on such measurement information, allow the upper abstraction level to build optimization and adaptation algorithms.
- Evaluating the advantages of differentiated services (Premium or Less than Assured Services) offered by the network infrastructure for specific streams.
- Creating enhanced and programmable transport protocols to optimize heterogeneous data transfers within the grid.

5   Active and Logistical Resource Deployment in E-toile

Logistical and active networking equipments are currently deployed in the e-Toile architecture. Each service willing to use logistical storage has to instantiate its own IBP client classes in order to communicate with an IBP depot. These classes provide constructors and methods to create capabilities on any IBP depot, with whom the Tamanoir service can write, read and manage data remotely on the IBP depot.

A first and very basic TAN service, called *IBP_Service* uses a *IBP_store* operation to redirect the beginning of a data stream towards the IBP depot. The IBP service checks as well the presence of the required service each time that a new packet arrive, and if so, a *IBP_load* operation is done to redirect all the data cached in the IBP depot towards the service able to process, route and forward efficiently these data. The only difference between the IBP_Service and any other service lies in the load-time, which is done at boot time for the IBP_Service, in order to be able to cache data immediately.

The use of IBP depot on each TAN should allow the storage of data to provide reliability through redundancy. If data are replicated on different server and one of them become either out of order or unreachable, data should still be downloadable from another server transparently.

Explicit routing and optimized scheduling are also possible through logistical active nodes. Several experiments made by the IBP team show that, in order to transmit as fast as possible huge amount of data, the path chosen by "classic" routers might show well below performance than the optimal one. Adding a staging point in the middle of two fast connections, knowing the topology of the underlying network, often improves performance dramatically. Unfortunately current transport protocols encapsulate the path, offering an end-to-end service that is not well adapted to the performance needs a grid application often has. A Tamanoir service, upon reception of a data stream, could store the stream on a number of different depots, participating in directories such as the Logistical Backbone (L-Bone), optimizing the total transfer time explicitly routing the data packets towards faster connections, and staging data in middle points at the boundaries of high-speed connections. We run several experiments mixing active and logistical equipment, and the latency measured [8] for packet caching of an active stream shows a very small overhead.In case of a change in the application scheduling, data stored at those intermediate nodes could be re-routed towards the new processing node in a much more efficient way.

6   Conclusions and Future Works

A grid, empowered with active and logistical networking, can not only improve significantly its global performances, but also and foremost provide new and innovative services to grid applications and middleware, giving an easy and direct control over high level services, such as reliable multicast and active QoS.

In addition to the technical challenges outlined in this article, the direction of our research is towards the integration of our active and logistical technologies with existing middleware, such as Globus and the Sun Grid Engine, giving them the ability to encapsulate active and logistical technologies for their internal needs, and to expose active and logistical services to upper layer's applications.

Our plan is also to integrate these technologies further, especially adding active and intelligent services to the logistical layer in fields such as efficient and reliable data deployment for the grid. IBP depots, through the *Data Mover* (DM) plug-in module, by sending a request to a Tamanoir Active Node could take advantage of any active services and could improve data transport.

References

1. RNTL e-Toile French Grid Project - http://www.urec.cnrs.fr/etoile.
2. Very high broadband RNRT VTHD project: (http://www.vthd.org)
3. Beck, M., Moore, T., Plank, J.: An end-to-end approach to globally scalable network storage. In: ACM SIGCOMM 20002 Conference, Pittsburgh, PA, USA. (2002)
4. Gelas, J.P., El Hadri, S., Lefèvre, L.: Towards the design of an high performance active node. Parallel Processing Letters **13** (2003)
5. RNTL: (http://www.industrie.gouv.fr/rntl)
6. ACI Grid initiative: (Initiative of french ministry of research for globalisation of computer resources and data - http://www-sop.inria.fr/aci/grid/public/)
7. MapCenter Grid visualization tool: (http://ccwp7.in2p3.fr/mapcenter/)
8. Bassi, A., Gelas, J.P., Lefèvre, L.: Tamanoir-IBP: Adding Storage to Active Networks. In: Active Middleware Services, Edinburgh, Scotland, IEEE computer society (2002) 27–34 ISBN: 0-7695-1721-8.

Distributed Resource Discovery in Wide Area Grid Environments

T.N. Ellahi and M.T. Kechadi

Parallel Computational Research Group,
Department of Computer Science,
University College Dublin, Belfield, Dublin 4, Ireland
{tariq.ellahi,tahar.kechadi}@ucd.ie

Abstract. Multitude of resources pooled in grid environments require a scalable, efficient, low-cost and self-organizing resource discovery mechanisms. P2P networks possess overlapping characteristics like large scale resource sharing and dynamic nature of participation. P2P style resource discovery approaches have been employed in the grid environments. We propose a P2P based resource discovery approach with modified features like customized neighbourhood defined according to the user preferences and distributed search using directed flow of requests contrary to the blind propagation of P2P systems.

1 Introduction

Substantial advancement in networking and hardware technologies have resulted in using computers for modelling and simulating complex scientific and engineering problems, diagnosing medical conditions, forecasting the weather and many other purposes. Yet, although there are certain challenging problems that exceed the existing technical ability to solve them. The technological advancements have led to the possibility of using wide-area distributed computing infrastructure for solving large-scale sophisticated problems.

Grid environments may consist of millions of resources pooled together by thousands of individuals and organizations shared under locally defined sharing policies to form Virtual Organizations [1]. Resources shared in virtual organizations include computers, storage elements, software applications, data sets, custom scientific instruments etc. Such multitude of resources shared in grid environments dictate the availability of a resource discovery mechanism, which can locate resources according to the user specified criteria. The Resource discovery is made challenging by the absence of a global resource information repository, resource heterogeneity and dynamic nature of participation. We are proposing a purely decentralized resource discovery solution, which equips the users with the capability to customise their environments according to their preferences and performance requirements.

The rest of the paper is structured as follows: Section 2 talk about the related work in this field. Section 3 and 4 discuss about the customization of neighborhood. Distributed search is presented in Section 5 and results in Section 6. Section 7 concludes the paper.

M. Bubak et al. (Eds.): ICCS 2004, LNCS 3038, pp. 210–217, 2004.
© Springer-Verlag Berlin Heidelberg 2004

2   Related Work

Different systems have adopted different approaches for resource discovery. Globus uses MDS [2] which is LDAP based and hierarchical. GRIS provides information about resources and GIIS accumulate information from GRIS and GIIS of lower level. Users can query both components to obtain information about resources. Legion uses Collections [3] to store and lookup resource information.

Among P2P systems, Napster used a centralized directory of resource information. Gnutella used query flooding to discover resources. Both [4] [5] have suggested alternative request forwarding strategies. [6] uses routing indices to forward queries. Distributed Hash Tables (DHT) which include [7] [8] [9] [10] assign unique identifiers and organize nodes into a search-optimized overlay. Queries are forwarded based on identifier proximity.

[11] proposed P2P style resource discovery in grid environments with four alternative request forwarding strategies. [12] extends CAN but uses multiple indices for multiple attributes of the resources. [13] propose Chord based system with a single index.

Among approaches which have similarities with our approach, [14] adds shortcut on top of Gnutella topology based on locality of interest. Failure to find resource through shortcuts result in using underlying Gnutella topology. In [15] authors proposed the idea of using trader-based approach. Traders connect to other traders based on their preferences and forward the locally unsuccessful queries to other neighbouring traders.

3   Problem Statement and Model

Large scale resource sharing environments are characterized with the following properties:

- Existence of very large number of distributed resources shared by thousands of participants.
- Dynamic nature of participation and great deal of heterogeneity of resources and the sharing policies.
- Absence of a global naming scheme or a global repository hosting information about the resources. Every node is aware of a subset of nodes. This subset is called the neighbourhood of that node.
- Requests for the resources by specifying the desired attributes rather than the name of the resource.

We are proposing a P2P based resource discovery approach which besides coping with the above mentioned properties also provides to the user the ability to customise their neighborhood according to their preference. Following is the general overview of our approach.

Every node specifies its preferences and performance requirements. Its neighbourhood will be selected among the nodes it knows about. When requests are issued, local neighbourhood will be searched first. As neighbourhood of every node is tailored to its needs, there is a high probability that resources will be discovered within the neighbourhood. If unsuccessful, request will be forwarded to a subset of the neighbourhood and those nodes can start the discovery in a recursive manner.

The above approach works in a distributed and decentralized fashion and discovers the desired resources but raises two issues which needs to be resolved which are: selection criteria of the neighbourhood and selection of neighbours to forward the queries. Next two sections discuss both issues.

4   Neighbourhood

Contrary to the existing P2P systems in which the neighbourhood connectivity is either random or system imposed, our system adopts a different methodology for creating the neighbourhood connectivity. We believe that the participants of a P2P system in general and grid environments in particular have fixed preferences and performance requirements. So if the neighborhood of a node is selected based on user's preference, there is a high probability that user requests can be satisfied in a single lookup. We need to define some elements before we give a formal definition of neighborhood.

4.1   Distance Measure (Norm)

As stated earlier neighbourhood customisation is done by selecting nodes, which satisfy a user requirements. This leads to two questions, which are:

– what criteria should be used to measure the performance of a node?
– what procedure is used to gather user's requirements and the criteria to analyze a node performance against those requirements?

This section answers the first question and also give partial answer of the second question whereas next section will answer the second question in detail.

Numerous factors both individually or in a combined fashion can be used to evaluate the performance of a node. Sample performance factors include, network latency, response time etc. The next issue is the selection of suitable factors. This is where the second question is partially answered. Since the main goal of our approach is to allow users to customise their neighbourhoods, this decision is vested to the user to select factors according to which they want to customise their environment.

All the factors selected should be independent of each other. We can segregate the factors in two sets, factors which need to be maximized and factors which need to be minimized.

$$\text{Factors to be Maximized: } F_{Max} = \{X_1, ..., X_l\}$$
$$\text{Factors to be Minimized: } F_{Min} = \{x_1, ..., x_k\}$$

We can calculate the performance of a node using the following formula, the Norm of the node:

$$D(n) = \sqrt{\sum_{i=1}^{k}(x_{ni})^2 + \sum_{j=1}^{l}\frac{1}{(X_{nj})^2}} \tag{1}$$

To cope with the heterogeneities of scales and measurements among values of the factors, we can normalize the values to their average value.

$$D(n) = \sqrt{\sum_{i=1}^{k}\left(\frac{x_{ni}}{\bar{x}_{ni}}\right)^2 + \sum_{j=1}^{l}\left(\frac{\bar{X}_{nj}}{X_{nj}}\right)^2} \tag{2}$$

Factors can be assigned weights $\mathcal{W}$ to put emphasis on certain factors based on their importance. Above equation would become

$$D(n) = \sqrt{\sum_{i=1}^{k}\mathcal{V}_{ni}\left(\frac{x_{ni}}{\bar{x}_{ni}}\right)^2 + \sum_{j=1}^{l}\mathcal{W}_{nj}\left(\frac{\bar{X}_{nj}}{X_{nj}}\right)^2} \tag{3}$$

4.2   Node Radius

This section presents the detailed answer of the second question. We define the concept of a user requirement radius and its lower and upper bounds. This will be used to analyze other nodes likelihood of being selected in the neighbourhood. Besides choosing the factors, lower and upper bounds could also be set by the user:

$$\mathcal{Z_L} = \{X_1 = \alpha_1, X_2 = \alpha_2, ..., X_n = \alpha_n\}$$
$$\mathcal{Z_U} = \{X_1 = \beta_1, X_2 = \beta_2, ..., X_n = \beta_n\}$$

where $\{X_i, X_2, ..., X_n\}$ is a set of factors selected by the user and $\alpha_1, \alpha_2, ..., \alpha_n$ and $\beta_1, \beta_2, ..., \beta_n$ represent the lower and upper bounds for each of these factors respectively.

Radius lower bound is be defined to be the lower limit values of the performance requirement: $\delta_l = D(\mathcal{Z_L})$.

Radius upper bound is defined in the similar way by calculating the upper value of the norm: $\delta_u = D(\mathcal{Z_U})$.

4.3   Neighbourhood

Given a set of nodes N known to a node p, neighbourhood of node p denoted by Nr(P) can be defined as follows

$$\text{Nr(p)} : \forall n \in N, \delta_l \leq D(n) \leq \delta_u$$

A neighbourhood of a node will comprise all the nodes whose norm falls between the lower and upper radius of that node. The neighbourhood information will be maintained in a soft-state manner thus ensuring that stale information is erased automatically.

5   Distributed Search

We used distributed search methodology to process the requests issued by the users. It works in two phases as follows:

Local phase: The locally maintained indices of the resources in neighbourhood is searched for the availability of the resources requested by the user. Customized neighborhood will result in finding resources in this phase with a high probability.

Forwarding Phase: If the local phase is unsuccessful, the requests will be forwarded to nodes in the neighbourhood which can start the discovery process in a recursive manner. Contrary to P2P systems where request is either forwarded blindly or past knowledge is used while selecting neighbours to forward request, we can exploit the neighborhood connectivity pattern to use the informed forwarding strategies. Next section provides overview of two such strategies.

5.1   Request Forwarding

Following is the description of request forwarding strategies.

- **Request forwarding using neighbourhood:** In this request forwarding strategy, neighbours whose radius intersects the request characteristics are selected to forward the request. If no such node is found, request is forwarded to a subset of nodes. This can be a system parameter which can be set dynamically. In our experiments we set the default value to 20%.
- **Request forwarding using Routing Neighbours:** This strategy works by selecting some additional routing neighbours based on the congruence of their radius. Following is the procedure of calculating the radius congruence: Suppose a node 'n' is looking for routing neighbours, then for each node v $\in$ N, where N is the set of nodes known to n, n will calculate the radius congruence as:

Lower Bound Deviation. Lower bound deviation λ_l is the difference in the lower bound of radius of nodes n and v.

$$\lambda_l = \begin{cases} 0 & : \quad \delta_{ln} - \delta_{lv} \geq 0 \\ \delta_{lv} - \delta_{ln} & : \quad otherwise \end{cases} \tag{4}$$

Upper Bound Deviation. λ_u is the difference of node v's upper bound of radius to upper bound radius of node n.

$$\lambda_u = \begin{cases} 0 & : \quad \delta_{un} - \delta_{uv} \leq 0 \\ \delta_{uv} - \delta_{un} & : \quad otherwise \end{cases} \tag{5}$$

Radius Congruence. The radius congruence σ is the measurement of node v's radius overlap with node n using the lower and upper bound deviations.

$$\sigma = \begin{cases} 0 & : (\delta_{un} - \delta_{ln} - \lambda_l - \lambda_u) \leq 0 \\ (\delta_{un} - \delta_{ln} - \lambda_l - \lambda_u) : otherwise \end{cases} \tag{6}$$

Normalized Radius Congruence. σ can be normalized to bring the value between 0 and 1.

$$\sigma_N = \frac{\sigma}{(\delta_{un} - \delta_{ln})} \tag{7}$$

Nodes are ordered according to their normalized radius congruence values and certain number of top nodes are selected. We set this number to be 5 as default.

6   Experimental Evaluation

6.1   Simulation Methodology

Simulations were performed on networks of different scales, ranging from 300 to 1500 nodes. Resource distribution to the nodes was random. Following three factors were chosen for selecting the neighbors:

Network Latency: Nodes were spread randomly in a 2D grid, network latency was modelled as the Euclidean distance between the coordinates.

Processing Speed: Processing speed is the speed of the processor at every node.

Node Availability: This factor models the time for which the node is available for grid usage.

Besides using the above mentioned request forwarding, a third strategy was also used for comparison. This strategy works by propagating requests to 20% random nodes. A certain percent of random nodes were chosen to start the resource discovery process. Following metrics were used to analyze the system.

Success Rate: Probability of requests being satisfied out of the total number of requests issued.

Local Success Probability: Probability of discovering resources within the local neighbourhood.

Average Path Length: Average number of times a request was forwarded before finding resources.

Average # Nodes Visited: The average number of nodes which every request have visited.

Average # Messages Per Node: This is the average number of requests each node in the network processes.

6.2   Simulation Results

Average Path Length: In Fig. 1a average path length is displayed. In the largest network consisting of 1500 nodes, a resource can be found in 2 hops on average. Our request forwarding strategies exploiting the neighbourhood perform quite well in all the network sizes.

Success Rate: Fig. 1b shows the success rate of each forwarding strategy. In a small network nodes find less nodes with congruent radius so a portion of requests are not satisfied. As network grows bigger, nodes find more and more compatible nodes thus increasing the success rate.

Local Success Probability: Fig. 1c represents the probability of discovering resources within neighbourhood. Large network results in finding more suitable nodes to qualify as neighbours resulting in increased success in neighbourhood. In the largest network consisting of 1500 nodes this value is 44% which is a reasonable value but not extremely high. The reasons for not getting very high values are: 1)all the request were distinct which means a new resource was being requested. 2)Requests contained many attributes, in the case of small number of attributes especially compatible to factors already selected will result in very high rate.

Average # nodes visited: Graph in Fig. 1d depicts the average load on the network generated by every request issued. Flooding 20% has generated the highest load, whereas, our strategies generate a lot less network traffic.

Average # messages per nodes: Fig. 1e displays the other aspect of the cost of resource discovery process which is the load each network participant have to incur.

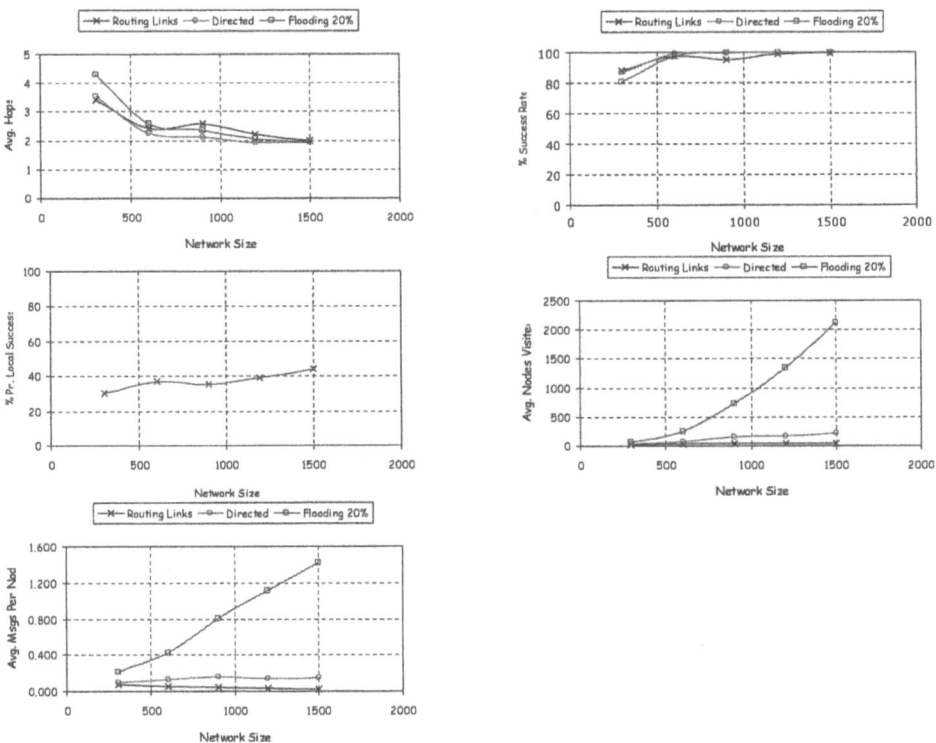

Fig. 1. a) Avg. Path Length b) Success Rate c) Local Success d) Avg. Nodes Visited e) Avg. Msgs Per Node

7   Conclusion

In this paper we have presented a new approach of discovering resources in wide area grid systems. Description of providing a customized neighborhood and distributed search was given. Simulation results were shown to prove the efficiency of our approach. Our approach works efficiently and generates less network traffic thus increasing the productivity of the system.

References

1. Ian Foster. The anatomy of the Grid: Enabling scalable virtual organizations. *LNCS*, pages 1–??, 2001.
2. K. Czajkowski, S. Fitzgerald, I. Foster, and C. Kesselman. Grid information services for distributed resource sharing. In *Proc. 10th IEEE Symp. On High Performance Distributed Computing*, 2001.
3. Steve J. Chapin, Dimitrios Katramatos, John Karpovich, and Andrew Grimshaw. Resource management in Legion. *Future Generation Computer Systems*, pages 583–594, 1999.
4. Qin Lv, Pei Cao, Edith Cohen, Kai Li, and Scott Shenker. Search and replication in unstructured peer-to-peer networks. In *Proc. of the 16th Int'l conference on Supercomputing*, 2002.
5. Beverly Yang and Hector Garcia-Molina. Improving search in peer-to-peer networks. In *Proc. of the 22nd Int'l Conference on Distributed Computing Systems*, 2002.
6. A. Crespo and H. Garcia-Molina. Routing indices for peer-to-peer networks. In *Proc. of the 22nd Int'l Conference on Distributed Computing Systems*, 2002.
7. Antony Rowstron and Peter Druschel. Pastry: Scalable, decentralized object location, and routing for large-scale peer-to-peer systems. *LNCS*, pages 329–350, 2001.
8. B. Y. Zhao, J. D. Kubiatowicz, and A. D. Joseph. Tapestry: An infrastructure for fault-tolerant wide-area location and routing. Technical report, UC Berkeley, 2001.
9. Sylvia Ratnasamy, Paul Francis, Mark Handley, Richard Karp, and Scott Shenker. A scalable content addressable network. In *Proc. of ACM SIGCOMM*, 2001.
10. Ion Stoica, Robert Morris, David Karger, M. Francs Kaashoek, and Hari Balakrishnan. Chord: A scalable peer-to-peer lookup service for internet applications. *ACM SIGCOMM*, pages 149–160, 2001.
11. Adriana Iamnitchi and Ian Foster. On fully decentralized resource discovery in grid environments. In *Int'l Workshop on Grid Computing*, 2001.
12. Artur Andrzejak and Zhichen Xu. Scalable, efficient range queries for grid information services. In *Proc. of the 2nd IEEE Int'l Conference on P2P Computing*, 2002.
13. C. Schmidt and M. Parashar. Flexible information discovery in decentralized distributed systems. In *12th IEEE Int'l Symp. on High Performance Distributed Computing*, 2003.
14. K. Sripanidkulchai, B. Maggs, and H. Zhang. Efficient content location using interest-based locality in peer-to-peer systems. In *INFOCOM*, 2003.
15. Ok-Ki Lee and Steve Benford. An explorative model for federated trading in distributed computing environments. In *Proc. of the Int'l Conference on Open Distributed Processing*, 1995.

Trusted Group Membership Service for JXTA

Lukasz Kawulok, Krzysztof Zielinski, and Michal Jaeschke

Department of Computer Science
AGH-University of Science and Technology
Krakow Poland

Abstract. This paper presents Group Membership Service for JXTA extended with single or bi-directional authentication. The proposed solution exploits certificates and PKI (Public Key Infrastructure). All information about the system's users is kept in the LDAP. An attention has been also paid to the problem of a private key secure protection.

1 Introduction

The concept of a peer group is very important to all aspects of the JXTA platform. Peer groups provide a way of segmenting the network space into distinct communities of peers organized for specific purpose providing the context for operations performed by services.

The goal of this paper is to present Group Membership Service for JXTA extended with single or bi-directional authentication. The proposed service will be called trusted Group Membership Service. The concept of the service assumes that a new member of group should be authenticated by a group in case of one-sided authentication and the group should be authenticated by a joining member in case of two-sided authentication. The proposed solution, in contrast to two already existing [1][2] implementations of a group access authentication for JXTA using passwords or null authentication, exploits certificates and PKI (Public Key Infrastructure).

The structure of the paper is as follows. First in Section 2 the concept of peer group and Membership service in JXTA has been briefly described. In this context the Trusted Group Membership Service concept has been formulated. Next in Section 3 implementation details of the proposed service are presented. The paper is ended with conclusions.

2 Trusted Groups versus Membership Service (Concept)

The first thing that the JXTA platform requires when bootstrapping is a World Peer Group, which is a peer group identified by a special Peer Group ID. The World Peer Group defines the endpoint protocol implementations supported by the peer, the World Peer Group itself can't be used to perform P2P networking. The World Peer Group is basically a template that can be used to either discover or generate a Net Peer Group instance. The Net Peer Group is a common peer group to peers in the network that allows all peers to communicate with each other.

M. Bubak et al. (Eds.): ICCS 2004, LNCS 3038, pp. 218–225, 2004.
© Springer-Verlag Berlin Heidelberg 2004

Before a peer can interact with the group, it needs to join the peer group, a process that allows the peer to establish its identity within the peer group. This process allows peer groups to permit only authorized peers to join and interact with the peer group.

Each peer group has a membership policy that governs who can join the peer group. When a peer instantiates a peer group, the peer group's membership policy establishes a temporary identity for the peer. This temporary identity exists for the sole purpose of allowing the peer to establish its identity by interacting with the membership policy. This interaction can involve the exchange of login information, exchange of public keys, or any other mechanism that a peer group's membership implementation uses to establish a peer's identity. After a peer has successfully established its identity within the peer group, the membership policy provides the user with credentials. The membership policy for a peer group is implemented as the Membership service. In addition to the MembershipService class and its implementations, the reference implementation defines a Credential interface and an implementation called AuthenticationCredential. These classes, defined in net.jxta.credential, are used in conjunction with the MembershipService class to represent an identity and the access level associated with that identity [3]. Two steps are involved in establishing an identity within a peer group using the peer group's MembershipService instance:

1. **Applying for membership**. This involves calling the peer group's MembershipService's apply method. The apply method takes an AuthenticationCredential argument, which specifies the authentication method and desired identity. The method returns an Authenticator implementation that the caller can use to authenticate with the peer group.
2. **Joining the group**. The peer must provide the Authenticator implementation with the information that it requires to authenticate. When the peer has completed authentication using the Authenticator, the Authenticator's isReadyForJoin method returns true. The peer now calls the MembershipService's join method, providing the Authenticator as an argument. The join method returns a Credential object that the peer can use to prove its identity to peer group services.

When applying for membership, the peer making the request must know the implementation of Authenticator to interact with it. This is required because the Authenticator interface has no mechanism for the peer to interact with it. Using only the Authenticator interface, a peer can only determine whether it has completed the authentication process successfully and then proceed with joining the peer group.

The current implementations of Membership Service aren't especially useful. To provide proper authentication, a developer must develop a Membership Service of his own to manage the creation and validation of authentication credentials. In addition, the developer must provide a mechanism in his service to use the credentials and validate the credentials passed by other peers in requests to the service.

Although the Protocol Specification [3] outlines the concept of an Access service whose responsibility it is to verify credentials passed with requests, no implementation is provided in the reference JXTA implementation. The Resolver service's Resolver Query and Response Messages support a Credential element, but the contents of the element are never verified. For now, it appears that it is the responsibility of a developer to define his own Access service implementation and use it to verify credentials passed to his custom peer group services. As such, a developer currently needs only to instantiate a peer group and can skip the steps of applying for membership and joining the peer group.

The proposed Trusted Membership Service concept assumes that Peer Advertisement issued by peer wanted to create trusted group contains additionally to standard information, already described, data specifying: localization of LDAP server, where certificates are stored; localization of Sign Server needed in case of bi-directional authentication; subject of group certificate.

This information is used in the process of joining trusted group. In case of the one-sided authentication the Trusted Membership Service activity follows the steps:

1. A peer invokes MembershipService's apply method of chosen peer group presenting its AuthenticationCredential in the form of Distinguished Name.
2. The peer group obtains certificate, containing its public key, corresponding to presented name from LDAP server and if the given peer has access permission to the group generates stochastic data that should be signed by the applying peer. This data are returned to peer by apply method.
3. The peer singes the data with its private key and returns back invoking the MembershipService's join method.
4. The peer group authenticate the peer using this data and its public key obtained early from certificates.

In case two-sided authentication the presented steps are extended as follows:

5. After successful peer authentication join method returns the peer group certificate.
6. The peer generates stochastic data which are signed using the peer group private key and sent back.
7. The authentication of the peer group is finally made using the peer group public key obtained early in the peer group certificate.

The proposed system architecture is depicted in Fig.1. The last problem which has to be solved is that P2P system concept doesn't not allow existing of any central point. A peer group could be created by every peer in any place. It means that in case of bi-directional authentication a private key has to be populated to each peer. It is rather risky solution from security point of view. In our service has been chosen the different solution. It exploits the concept of sign server which knows the group private key and signs data sent by a peer wanting to authenticate a peer group.

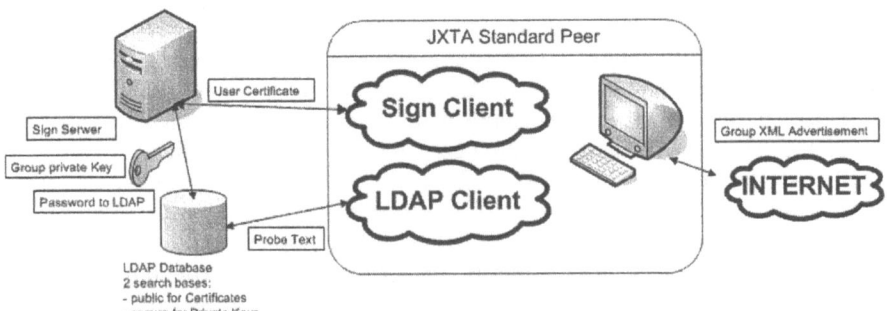

Fig. 1. Architecture of Trusted Membership Service implementation

3 JXTA Trusted Membership Service Implementation Details

In this section the details of JXTA trusted groups implementation are shortly described. We started from a new form of the trusted group advertisement specification and LDAP configuration description. Next activity of the Sign Server is specified. Finally the joining to the trusted group process is briefly described for one-sided and two-sided authentication case.

3.1 Trusted JXTA Group Advertisement

While creating a trusted group, a PeerGroupAdvertisment is created that contains additional information indispensable for proper operation of the membership service. Form of an example advertisement is:

```
<?xml version="1.0"?>
<!DOCTYPE jxta:PGA>
<jxta:PGA xmlns:jxta="http://jxta.org">
<GID>urn:jxta:uuid-8A12B3AEC66E47BAB739999684D1705302</GID>
<MSID>urn:jxta:uuid-DEADBEEFDEAFBABAFEEDBABE000000016FC5C1F3F
    7434F658A1C8C6CADFC43DB06</MSID>
<Name>VoteGroup</Name>
<Desc>Peer Group using Certificate Authentication</Desc>
<Svc>
  <MCID>urn:jxta:uuid-DEADBEEFDEAFBABAFEEDBABE0000000505</MCID>
    <Parm>
      <GroupDNName>EMAILADDRESS=CertGroup1@agh.edu.pl, zN=CertGroup
1, OU=ICS, O=AGH, L=Krakow, ST=Malopolska, C=PL
      </GroupDNName>
      <LdapServerPort>389</LdapServerPort>
      <LdapServerHost>192.168.2.4</LdapServerHost>
      <LdapServerSearchBase>dc=PP</LdapServerSearchBase>
      <SignServerPort>500</SignServerPort>
    </Parm>
  </Svc>
</jxta:PGA>
```

Fig. 2. Trusted group advertisment in XML format

Group advertisement contains information about the MembershipService. It is certified by the reserved UUID service number equal
DEADBEEFDEAFBABAFEEDBABE0000000505 in our implementation.

The similar parameters describing LDAP Server and Sign Server are specified in the properties file named mobisec.properties. They are used for the suitable servers configuration during JXTA system start up process.

3.2 LDAP Configuration Description

In order to keep the information about the system's users in the LDAP [4] database in a convenient way, we have defined two object classes which correspond to the user and group objects respectively. The classes contain the X509Certificate attribute we have created for keeping certificate in PEM or DER format [10].

Beside the X509Certificate attribute, the both classes possess attributes corresponding to all the elements of the distinguished name fields from the X.509 certificate, i.e. c, cn, mail, o, ou, st. Using these fields it is possible to search the given person. With the presented classes, we have created LDAP hierarchical database storing the group and user data. A similar structure beginning from the dc=PP root, which is used for storing private keys of individual groups is also constructed. For this purpose an additional attribute named PrivateKey and a new class for storing the elements discussed has been defined .

3.3 Sign Server

Sign Server is an independent application which performs a group authentication in case of the two-sided authentication. It functionality is to sign a text sample with a group private key. The activity of the Sign Server could described as follows – Fig. 3:

1. A dedicated thread reads the input from selected port (requests from peers);
2. After receiving a message, a dedicated thread for serving the request which contains the group certificate subject name and the text sample for signing is created.
3. The dedicated thread connects with the LDAP server to get the group private key.
4. Using this key the sample text is signed.
5. The signed text is sent back to the requesting peer.

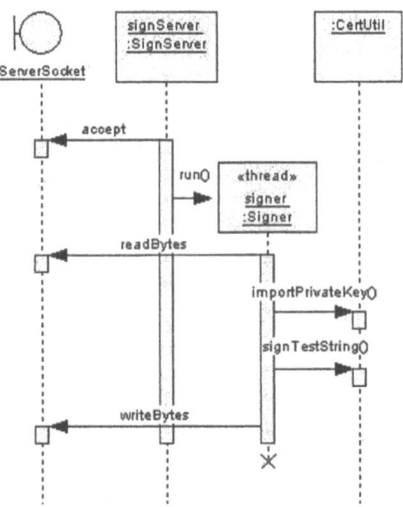

Fig. 3. Diagram of Sign Server operation sequence

There is a class named SignClient in the our utility package. After the correct initialization of the SignClient class, it may be used for communication with Sign Server. In the described project this class was used by the authenticator named CertAuthenticator in the middle of the group authenticating process.

3.4 One-Sided Authentication Sequence

This section presents how the authentication mechanism with certificates in the process of joining a group is used. Before the group starts operating, it must issue certificates to the authorized users and the group must possess itself a certificate signed by a trusted certification authority. In the simplified version, i.e. without authentication carried out by a group, the process is performed as follows [3]:

1. A group uses the CertMembershipService class to perform the authentication services. A participant who wants to join the group invokes an apply method of the CertMembershipService class, in order to establish an authentication method. This class allows only for authentication with use of the certificate defining the user's identity with his/her public key attached to it.
2. An apply method of the CertMembershipService class sends back to the user an object type CertAuthenticator (extension of the Authenticator interface), containing a group certificate and a random piece of data designed to be signed with a peer's private key. A participant fulfills a CertAuthenticator object with use of setAuthCertificate method of that object. This causes a peer's certificate (in the form of the object of the StructuredDocument class) and a piece of data sent by the group and signed with the peer's private key to be placed in the CertAuthenticator object.
3. A participant, in order to verify whether the CertAuthenticator object has been fulfilled properly, calls the method isReadyForJoin.
4. A participant gives a CertAuthenticator object containing its certificate and piece of the group data signed with a private key as a method parameter join of CertMembershipService class.
5. A group checks the certificate, verifying it by means of its public key (since the key was issued by the group) or with use of PKI and certificate issued by a trusted certification authority.
6. A group verifies the peer's identity, decoding a communicate with the peer's public key (attached to the certificate in the moment of authentication).

After the above sequence of operations is finished, a group is sure about the identity of the peer, which wants to join it.

3.5 Two-Sided Authentication Sequence

In the version with authentication carried out by a group (two side authentication) we have some additional elements in the presented above sequence. Above steps have to be repeated by the other side. It is peer who sent random piece of data designed to be signed with a group's private key. The group fulfills the CertAuthenticator object with use of setAuthCertificate method of that object. This causes a group's certificate and the piece of data sent by the peer and signed with the group's private key to be placed in the CertAuthenticator object and send back to the peer, and in the end it is the peer who decode the communicate with the group's public key and verifies the group's identity. In summary the following additional steps must be performed in the two-sided the authentication [3]:

7. While calling the CertAuthenticator object, defined in step one, the setAuth1GroupVerificationString method must be used in order to set random piece of information to be signed by a group.
8. In the very end the step verifying a group must be added. It requires the connection with the SignSerwer. The SignServer receives a sample text for signing. Then the Signserver takes a group private key from the LDAP database. Only the SignServer can make use of the additional database located on the LDAP server.

The sequence diagram for two-sided authentication is depicted in Fig.4. It presents the role of each component proposed by the trusted Group Membership Service for JXTA.

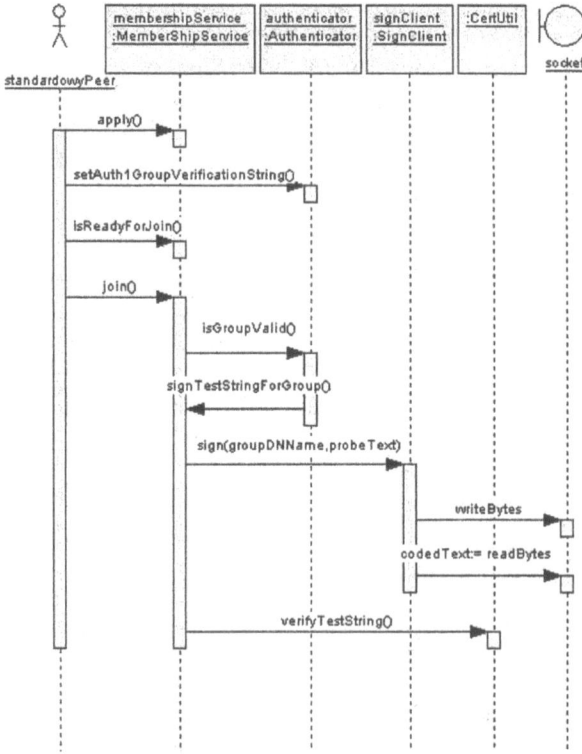

Fig. 4. Diagram of authenticator operation sequence in case of the two-sided authentication

4 Summary

The proposed implementation of Trusted Membership Service is fully compliant with JXTA reference implementation. Every effort has been put to make the system easy to use and to install. The service has been built as a separate package that can be used whit standard JXTA packages. An attention has been also paid to the problem of a private key secure protection. This problem has not been discussed in more details as it rather a general issue. It is a problem for peers and for groups how to prevent the keys before stilling. For the peers we decided to encode the keys with a password – passphrase. For the group we have implemented a proxy object which has access to the key encoded with the password typed only once. The lifetime of the proxy object is limited for example by the user which is creating the group. This solution let us to avoid storing private key decoded on the disk. Now we are working on the proposed system practical verification by developing a range of applications. Existing related work is presented in [11], [12].

References

1. JXTA Project, http://www.jxta.org/
2. JXTA v2.0 Protocols Specification,
   http://spec.jxta.org/v1.0/docbook/JXTAProtocols.html
3. JXTA Book, http://www.brendonwilson.com/projects/jxta/
4. Iplanet LDAP description,
   http://www.cc.com.pl/prods/netscape/direct.html
5. Netscape Directory SDK 4.0 for Java Programmer's Guide,
   http://developer.netscape.com/docs/manuals/dirsdk/jsdk40/contents.htm
6. Security in Java Technology,
   http://java.sun.com/j2se/1.4.1/docs/guide/security/index.html
7. The Java Developers Almanac,
   http://javaalmanac.com/egs/javax.naming.directory/Filter.html
8. Certificates structure
   http://java.sun.com/j2se/1.4.1/docs/tooldocs/windows/keytool.html -
   Certificates
9. The PKI page, http://www.pki-page.org/
10. OpenSSL Project, http://www.openssl.org/
11. Security and Project JXTA, original paper January 2002
    www.jxta.org/project/www/docs/SecurityJXTA.PDF
12. Poblano – A Distributed Trust Model for Peer-to-Peer Networks
    www.jxta.org/docs/trust.pdf

An Implementation of Budget-Based Resource Reservation for Real-Time Linux*

C.S. Liu, N.C. Perng, and T.W. Kuo

Department of Computer Science and Information Engineering
National Taiwan University, Taipei, Taiwan 106, ROC

Abstract. The purpose of this paper is to propose a budget-based RTAI (Real-Time Application Interface) implementation for real-time tasks over Linux on x86 architectures. Different from the past work, we focus on extending RTAI API's such that programmers could specify a computation budget for each task, and the backward compatibility is maintained. Modifications on RTAI are limited to few procedures without any change to Linux kernel. The feasibility of the proposed implementation is demonstrated by a system over Linux 2.4.0-test10 and RTAI 24.1.2 on PII and PIII platforms.

1 Introduction

Various levels of real-time support are now provided in many modern commercial operating systems, such as Windows XP, Windows CE .NET, and Solaris. However, most of them only focus on non-aging real-time priority levels, interrupt latency, and priority inversion mechanisms (merely at very preliminary stages). Although real-time priority scheduling is powerful, it is a pretty low-level mechanism. Application engineers might have to embed mechanisms at different levels inside their codes, such as those for frequent and intelligent adjustment of priority levels, or provide additional (indirect management) utilities to fit the quality-of-services (QoS) requirements of each individual task.

In the past decade, researchers have started exploring scheduling mechanisms that are more intuitive and better applicable to applications, such as budget-based reservation [1,2,3] and rate-based scheduling [4,5,6,7]. The concept of budget-based reservation, that is considered as an important approach for applications' QoS support, was first proposed by Mercer, et al. [8]. A microkernel-based mechanism was implemented to let users reserve CPU cycles for tasks/threads. Windows NT middlewares [1,2] were proposed to provide budget reservations and soft QoS guarantees for applications over Windows NT. REDICE-Linux implemented the idea of hierarchical budget groups to allow tasks in a group to share a specified amount of budget [3]. There were also many other research and implementation results on the QoS support for real-time applications. In particular, Adelberg, et al. [9] presented a real-time emulation

---

* This research was supported in part by the National Science Council under grant NSC91-2213-E-002-104

M. Bubak et al. (Eds.): ICCS 2004, LNCS 3038, pp. 226–233, 2004.
© Springer-Verlag Berlin Heidelberg 2004

program to build soft real-time scheduling on the top of UNIX. Childs and Ingram [10] chose to modify the Linux source code by adding a new scheduling
class called SCHED_QOS which let applications specify the amount of CPU
time per period[1]. Abeni, et al. presented an experimental study of the latency
behavior of Linux [11]. Several sources of latency was quantified with a series of
micro-benchmarks. It was shown that latency was mainly resulted from timers
and non-preemptable sections. Swaminathan, et al. explored energy consumption
issues in real-time task scheduling over RTLinux [12].

The purpose of this paper is to explore budget-based resource reservation for
real-time tasks which run over Linux and propose its implementation. We first
extend Real-time Application Interface (RTAI) API's to provide hard budget
guarantees to hard real-time tasks under RTAI. We then build up an implementation for soft budget guarantees for Linux tasks with LXRT over that for
hard real-time tasks under RTAI. Backward compatibility is maintained for the
original RTAI (and LXRT) design. The software framework and patch for the
proposed implementation is presented, and we try to minimize the modifications
on RTAI without any change to the Linux source code. Modifications on RTAI
are limited to few procedures, such as the timer interrupt handler, the RTAI
scheduler, and rt_task_wait_period(). The feasibility of the proposed implementation is demonstrated over Linux 2.4.0-test10 and RTAI 24.1.2 on PII and
PIII platforms.

The rest of the paper is organized as follows: Section 2 summarizes the layered
architecture and the functionalities of RTAI. Section 3 presents our motivation
for this implementation design. Section 4 is the conclusion.

2   RTAI

RTAI is one the most popular real-time patches to let Linux provide deterministic
and preemptive performance for hard real-time tasks [13]. LXRT is an advanced
feature for RTAI. It allows users to develop real-time tasks using RTAI's API
from the Linux user space. While in the user space, real-time tasks have the full
range of Linux system calls available. Besides, LXRT also provides the same set
of RTAI API calls available for RTAI applications in the Linux user space. When
a real-time task over Linux is initialized (by invoking rt_task_init), a *buddy
task* under RTAI is also created to execute the RTAI function invoked by the
soft real-time task running in the Linux user space. The rt_task_wait_period()
serves as an example function to illustrate the interactivity between a real-
time task and its corresponding buddy task, where rt_task_wait_period() is
to suspend the execution of a real-time task until the next period. When a
real-time tasks over Linux invokes rt_task_wait_period() via 0xFC software
trap, the corresponding buddy task is waken up and becomes ready to execute

---

[1] The approach is very different from that of this paper. We intend to propose an
RTAI-based implementation to deliver budget-based reservations to hard and soft
real-time applications.

rt_task_wait_period(). In the invocation, the buddy task delays its resumption time until the next period, and LXRT executes lxrt_suspend() to return CPU control back to Linux.

We refer real-time tasks (supported by RTAI) in the Linux user space as *real-time LXRT tasks* for future discussions, distinct from real-time tasks under RTAI (referred to as *real-time RTAI tasks*). Note that all tasks under RTAI are threads. For the rest of this paper, we will use terms tasks and threads interchangeably when there is no ambiguity.

3 Budget-Based QoS Guarantee

The concept of budget-based resource reservation is considered as a high-level resourse allocation concept, compared to priority-driven resource allocation. Each real-time task τ_i is given an *execution budget* W_i during each specified period P_i. The computation power of the system is partitioned among tasks with QoS requirements. The concept of budget-based resource reservation could provide an intuitive policy that ensures an application with resource allocation can always run up to its execution budget. Although some researchers have proposed excellent implementation work and designs for budget-based resource reservations in Linux (or other operating systems) or emulation of real-time support over Linux, little work is done on the exploring of the implementations for resource reservations over both Linux and RTAI (that is considered under Linux), and even for real-time LXRT tasks. Note that real-time LXRT tasks run over Linux, but they also invoke RTAI services.

3.1 Semantics and Syntax of Budget-Based RTAI APIs

A real-time RTAI task can be initialized by the RTAI API rt_task_init() with the entry point of the task function, a priority, etc. The invocation of rt_task_init() creates the corresponding real-time RTAI task but leaves it in a suspended state. Users must invoke rt_task_make_periodic() to set the starting time and the period of the task. A periodic real-time RTAI task is usually implemented as a loop. At the end of the loop, the real-time RTAI task invokes rt_task_wait_period() to wait for the next period.

Non-root users develop real-time tasks in the Linux user space with LXRT. All inline functions in rtai_lxrt.h do a software interrupt 0xFC to request RTAI services, where Linux system calls use the software interrupt 0x80. The interrupt vector call rtai_lxrt_handler() is to pass parameters and to transfer the execution to the corresponding buddy task in the kernel space. A real-time LXRT task can be also initialized by rt_task_init(), which resides in the header file rtai_lxrt.h and differs from the RTAI counterpart. This API generates a buddy task in the kernel space for the real-time LXRT task to access RTAI services. Users then do the same work as they program a real-time RTAI task.

In order to support budget reservation, we revise and propose new RTAI API's as follows: A real-time RTAI/LXRT task with budget-based resource reservation could be initialized by invoking rt_task_make_periodic_budget(), instead of the original rt_task_make_periodic(). The format of this new function is exactly the same as the original one, except that an additional parameter for the requested execution budget is provided. Under this model, a real-time task τ_i can request an execution budget W_i for each of its period P_i. Suppose that the maximum execution time of τ_i is C_i, and $C_i \leq W_i$. The execution of τ_i will remains the same without budget reservation because τ_i always invokes rt_task_wait_period() before it runs out of its budget. It is for the backward compatibility of the original RTAI design. When $C_i > W_i$, the execution of τ_i might be suspended (before the invocation of rt_task_wait_period()) until the next period because of the exhaustion of the execution budget. The remaining execution of the former period might be delayed to execute in the next period. If that happens (e.g., $C_{i,j} > W_i$, where $C_{i,j}$ denotes the execution time of the task in the j-th period), then the invocation of rt_task_wait_period() (that should happen in the former period) in the next period (i.e., the (j+1)-th period) will be simply ignored. The rationale behind this semantics is to let the task gradually catch up the delay, due to overrunning in some periods (that is seen very often in control-loop-based applications).

3.2   An Implementation for Budget-Based Resource Reservation

Hard Budget-Based Resource Reservation. Additional attributes are now included in the process control block of real-time RTAI tasks (rt_task_struct). We revise some minor parts of the RTAI scheduler (rt_schedule()) and the timer interrupt handler (rt_timer_handler()) to guarantee hard budget-based resource reservation.

```
struct rt_task_struct {
    ...
    // appended to provided budget-based resource reservation
    RTIME assigned_budget;
    RTIME remaining_budget;
    int rttimer_flag;
    int wakeup_flag;
    int if_wait;
    int force_sig;
};
```

Additional members assigned_budget and remaining_budget are for the reserved execution budget and the remaining execution budget of the corresponding task in a period, respectively. The remaining execution budget within a period is modified whenever some special event occurs, such as the changes of the task status. rttimer_flag serves as a flag to denote whether the corresponding real-time task is suspended by a timer expiration or a rt_task_wait_period() invocation. wakeup_flag is to denote that the corresponding real-time task is suspended because of budget exhaustion. if_wait is used to count the number of delayed invocations of rt_task_wait_period(), where an invocation is considered delayed if it is not invoked in the supposed period because of budget exhaustion. `force_sig` is set when a budget-related signal is posted to the corresponding LXRT task.

The implementation of budget-based resource reservation in RTAI must consider two important issues: (1) The correct programming of the timer chip (e.g., 8254). (2) The behavior of `rt_task_wait_period()`. Besides, the RTAI scheduler (`rt_schedule()`) and the timer interrupt handler (`rt_timer_handler()`) must also be revised to guarantee hard budget-based resource reservation: Whenever the timer expires, `rt_timer_handler()` is invoked. `rt_timer_handler()` is used to re-calculate the next resume times of the running task and the to-be-awaken task and then trigger rescheduling. We propose to revise `rt_timer_handler()` as follows to include the considerations of task budgets.

```
1  static void rt_timer_handler(void) {
2      ...
3      // calculate the remaining budget
4      if (new_task->tid != 0) {
5          temp_time = rt_times.tick_time + new_task->remaining_budget;
6          if (temp_time < rt_times.intr_time) {
7              // assigned_budget is used up
8              new_task->remaining_budget = 0;
9              rt_times.intr_time = temp_time;
10         } else {
11             // assigned_budget is not used up
12             new_task->remaining_budget -= (rt_times.intr_time-rt_times.tick_time);
13         }
14     }
15     ...
16 }
```

The value in `rt_times.intr_time` denotes the next resume time of the running real-time RTAI task. Line 5 derives the resume time based on the remaining budget of the task, where `rt_times.tick_time` is the current time. If the task uses up its budget before the next resume time (i.e., `rt_times.intr_time`), then code in Lines 8 and 9 should modify the next resume time to the time when budget is used up. Otherwise, the remaining budget time at the next resume time is recalculated in Line 12.

As explained in previous paragraphs, the semantics of `rt_task_wait_period()` is to ignore the invocation of the function that should happen in some former periods. `rt_timer_handler()` is also revised to reflect such semantics: If there is no budget remained, and the invocation is for hard budget-based resource reservation, then the remaining budget and the next ready time (i.e., the resume time) of the running real-time RTAI task must be reset.

Soft Budget-Based Resource Reservation. Let a real-time LXRT task *Task*1 be initialized by invoking `rt_task_init()`, and a corresponding buddy RTAI task *RT_Task*1 is created. When the next period of *Task*1 arrives, *Task*1 resumes through the following sequence, as shown in Fig. 1: When the timer expires, let RTAI schedule *RT_Task*1. *RT_Task*1 is suspended right away be-

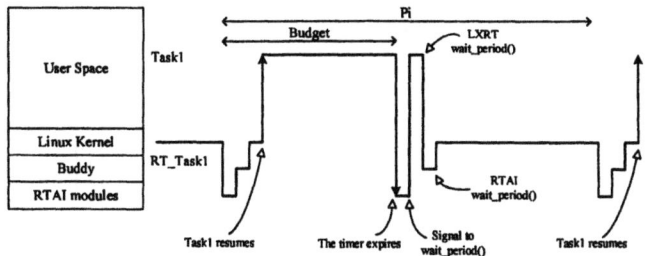

Fig. 1. Scheduling flowchart of the revised LXRT

cause it is a buddy task. The CPU execution is then transferred to the Linux kernel such that $Task1$ is scheduled. On the other hand, when $Task1$ requests any RTAI services through LXRT, such as `rt_task_wait_period()` (in the user space), $Task1$ is suspended, and RT_Task1 resumes its execution to invoke the requested RTAI service (i.e., `rt_task_wait_period()` in RTAI). The budget information of each real-time LXRT task is maintained in the `rt_task_struct` of its corresponding buddy task.

There are two major challenges in the implementation of soft budget-based resource reservation for real-time LXRT tasks: (1) How to interrupt a real-time LXRT task when its budget is exhausted and to transfer the CPU execution right to a proper task. (2) When a higher-priority real-time LXRT task arrives, how to interrupt a lower-priority real-time LXRT task and dispatch the higher-priority real-time LXRT task.

We propose to use *signals* to resolve the first challenging item. Note that we wish to restrict modifications on RTAI without any change to the Linux source code. When the budget of $Task1$ is exhausted in the current period, the timer will expire such that a signal of a specified type, e.g., SIGUSR1, is posted to $Task1$. The signal handler of the specified type is registered as a function `wait()`, that only contains the invocation of `rt_task_wait_period()`. The signal type for such a purpose is referred to as $SIGBUDGET^2$. The signal posting is done within `rt_timer_handler()`. The catching of the SIGBUDGET signal will result in the invocation of `rt_task_wait_period()` such that $Task1$ and its buddy task RT_Task1 are suspended until the next period (for budget replenishing), as shown in Fig. 2.

The implementation for second challenge item is also based on the signal posting/delivery mechanism. A different signal number, e.g., SIGUSR2, is used for the triggering purpose of rescheduling. The signal type for such a purpose is referred to as $SIGSCHED$. The signal handler of SIGSCHED is registered as a function `preempted()`, that only contains the invocation of `lxrt_preempted()`. Consider two real-time LXRT tasks $Task1$ and $Task2$, where RT_Task1 and RT_Task2 are the corresponding buddy RTAI tasks. Suppose that the priority of $Task1$ is lower than that of $Task2$. We must point out that the arrival of any real-time RTAI task will result in the expiration of the timer because

---

[2] Note that new signal numbers could be created in Linux whenever needed (under limited constraints).

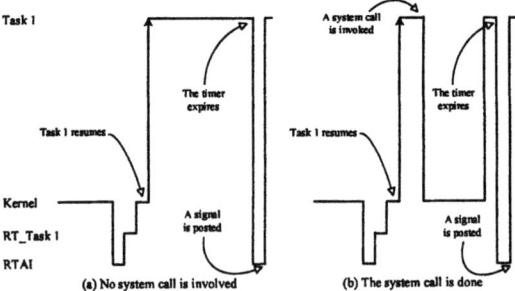

Fig. 2. The timer expires when the running real-time LXRT task is in the user space.

of the setup of the resume time of the task (initially, aperiodically, or periodically). When $Task2$ arrives (while $Task1$ is executing), the timer will expire, and rt_timer_handler() is executed to post a signal to the running task, i.e., $Task1$. The SIGSCHED signal is delivered to lxrt_preempted() (similar to the signal delivery in Fig. 2). As a result, lxrt_preempted() is executed such that $Task1$ is suspended, and the CPU execution right is transferred to RTAI. RTAI realizes the arrival of $Task2$ and dispatches RT_Task2, instead of RT_Task1, because the priority of RT_Task1 is lower than that of RT_Task2, due to the priority order of $Task1$ and $Task2$. The dispatching of RT_Task2 results in the transferring of the CPU execution right to $Task2$ through LXRT.

The following example code is for the implementation of a real-time LXRT task. The registerations of signal handlers for SIGBUDGET (i.e., SIGUSR1) and SIGSCHED (i.e., SIGUSR2) are done in Lines 4 and 5. Line 7 is for the initialization of a real-time LXRT task. Line 9 sets up the budget for the task. The loop from Line 10 to Line 13 is an example code for the implementation of a periodic task.

```
1 void wait(){ rt_task_wait_period(); }
2 void preempted(){ lxrt_preempted(); }
3 void main(void) {
4     signal(SIGUSR1,wait);
5     signal(SIGUSR2,preempted);
6     //initialization codes
7     rt_task_init();
8     //use the system call to gain the budget
9     rt_task_make_periodic_budget(srt_task, now + period, period, budget);
10    while(1) {
11        //computation codes that needs budget service.
12        rt_task_wait_period();
13    }
14 }
```

4   Conclusion

We extend RTAI API's to provide hard budget guarantees to hard real-time tasks under RTAI and soft budget guarantees for LXRT tasks over that for real-time RTAI tasks. Backward compatibility is maintained for the original RTAI (and LXRT) design. We try to minimize the modifications on RTAI without any change to the Linux source code. The feasibility of the proposed implementation is demonstrated over Linux 2.4.0-test10 and RTAI 24.1.2 on PII and PIII platforms. For the future research, we shall further extend the implementation work to multi-threading processes for the sharing of a single budget. We shall also explore the synchronization issues for cooperating processes, especially when a budget is reserved for each of them. A joint management scheme for multiple resources such CPU and devices will also be explored.

References

1. Jones, M., Rosu, D., Rosu, M.: Cpu reservation and time constraints: Efficient, predictable scheduling of independent activities. ACM Symposium on Operating Systems Principles (1997) 198–211
2. Kuo, T.W., Huang, G.H., Ni, S.K.: A user-level computing power regulator for soft real-time applications on commercial operating systems. Journal of the Chinese Institute of Electrical Engineering 6 (1999) 13–25
3. Wang, S., Lin, K.J., Wang, Y.: Hierarchical budget management in the red-linux scheduling framework. 14th Euromicro Conference on Real-Time Systems (2002)
4. Deng, Z., Liu, J.W.S.: Scheduling real-time applications in an open environment. IEEE Real-Time Systems Symposium (1997)
5. Spuri, M., Buttazzo, G., Sensini: Scheduling aperiodic tasks in dynamic scheduling environment. IEEE Real-Time Systems Symposium (1995)
6. Stoica, I., Abdel-Wahab, H., Jeffay, K., Baruah, S., Gehrke, J., Plaxton, C.: A proportional share resource allocation algorithm for real-time, time-shared systems. IEEE Real-Time Systems Symposium (1996) 288–299
7. Waldspurger, C.: ottery and stride scheduling: Flexible proportional-share resource management. Technical report, Ph.D. Thesis, Technical Report, MIT/LCS/TR-667, Laboratory for CS, MIT (1995)
8. Mercer, C.W., Savage, S., Tokuda, H.: Processor capacity reserves: An abstraction of managing processor usage. In Proceedings of the Fourth Workshop on Workstation Operating Systems (WWOS-IV) (1993)
9. Adelberg, B., Garcia-Molina, H., B.Kao: Emulating soft real-time scheduling using traditional operating systems schedulers. IEEE 15th Real-Time Systems Symposium (1994) 292–298
10. Childs, S., Ingram, D.: The linux-srt integrated multimedia operating systems: Bring qos to the desktop. IEEE Real-Time Technology and Applications Symposium, Taipei, Taiwan, ROC (2001) 135–140
11. Abeni, L., Goel, A., Krasic, C., Snow, J., Walpole, J.: A measurement-based analysis of the real-time performance of linux. Eighth IEEE Real-Time and Embedded Technology and Applications Symposium (2002)
12. Swaminathan, V., Schweizer, C.B., Chakrabarty, K., Patel, A.A.: Experiences in implementing an energy-driven task scheduler in rt-linux. Eighth IEEE Real-Time and Embedded Technology and Applications Symposium (2002)
13. Cloutier, P., Mantegazza, P., Papacharalambous, S., Soanes, I., Hughes, S., Yaghmour, K.: Diapm-rtai position paper. Real Time Operating Systems Workshop (2000)

Similarity Retrieval Based on SOM-Based R*-Tree

K.H. Choi[1], M.H. Shin[1], S.H. Bae[1], C.H. Kwon[2], and I.H. Ra[3]

[1] 375, Seosuk-Dong, Dong-Gu Kwangju , Computer Science & Statistics, Chosun University,
Korea, 501-759
ckhplc@hanmir.com, minandjih@hanmail.net, shbae@chosun.ac.kr
[2] 604-5, Dangjung-Dong,Gunpo-si,Kyunggi-do, Division of IT, Computer Engineering , Hansei
University, Korea, 435-742
kwmch@hotmail.com
[3] Depts. Electronic and Information Engineering, Kunsan National, Korea, 573-701

Abstract. Feature-based similarity retrieval has become an important research issue in multimedia database systems. The features of multimedia data are usually high-dimensional data. The performance of conventional multidimensional data structures tends to deteriorate as the number of dimensions of feature vectors increases. In this paper, we propose a SOM-based R*-tree(SBR-Tree) as a new indexing method for high-dimensional feature vectors. The SBR-Ttree combines SOM and R*-tree to achieve search performance more scalable to high dimensionalities. When we build an R*-tree, we use codebook vectors of topological feature map which eliminates the empty nodes that cause unnecessary disk access and degrade retrieval performance. We experimentally compare the retrieval time cost of a SBR - Tree with that of an SOM and an R*-tree using color feature vectors extracted from 40,000 images. The result show that the SOM-based R*-tree outperforms both the SOM and R*-tree due to the reduction of the number of nodes required to build R*-tree and retrieval time cost.

1 Introduction

With the increasing use of new database applications for dealing with highly multidimensional data sets, technology to support effective query processing with such a data set is considered an important research area. Such applications include multimedia databases, medical databases, scientific databases, time-series matching, and data analysis/data mining. For example, in the case of image searches, a typical query of content-based image retrieval [1] is "find images with similar colors, texture, or shapes in a collection of color images". The features used in this query are useful to discriminate multimedia objects (e.g., documents, images, video, music score etc.). A feature vector is a vector that contains a set of features, and usually hold high-dimensional data. Many indexing techniques [2][3][4][5] have been proposed to access such high-dimensional feature vectors effectively. These index trees wok effectively in low to medium dimensionality space (up to 20-30 dimensions). However, even a simple sequential scan performs better at higher dimensionalities[5].

This paper is organized as follows : In section 2, we provide an overview of related work. In Section 3, we present the algorithm of the SOM and R*-tree, and describe

M. Bubak et al. (Eds.): ICCS 2004, LNCS 3038, pp. 234–241, 2004.
© Springer-Verlag Berlin Heidelberg 2004

the SOM-based R*-tree proposed in this research. We experiment in order to compare the SOM-based R*-tree with the SOM and -tree alone in terms of retrieval time cost using color feature vectors extracted from 40,000 image. The experimental results are discussed in Section 4, and Section 5 presents the concluding remarks.

2 Related Works

In this Section, we describe the related work on clustering methods and high-dimensional index structures.

Clustering: There are supervised and unsupervised clustering methods for clustering similar data. During the training phase in supervised clustering, both the input data and the desired output are presented to the neural network. If the output of neural network differs from the desired output, the internal weights of the neural network are adjusted. In unsupervised clustering, the neural network is given only the input vectors and the neural network is used to create abstractions in the input space.
SOM is an unsupervised self-organizing neural network that is widely used to visualize and interpret large high-dimensional data sets. In this study, the reasons for using SOM are as follows: (i) No prior assumption is needed for distribution of data, (ii) the learning algorithm is simple, (iii) we do not need an external supervised signal for input and it learns self-organizationally, and (iv) similarity can be found automatically from multidimensional feature vector, and similar feature vectors are mapped onto neighboring regions on the topological feature map, in particular, highly similar feature vectors are mapped on the same node.

High-dimensional index structure: Many techniques, including R-tree, R^+-tree, SS-tree, SR-tree, X-tree, TV-tree, and Hybrid tree have been proposed to index feature vectors for similarity search. Despite various attempts at accessing high-dimensional feature vectors effectively, the current solutions are far from satisfactory. Although these index structures can scale to medium dimensionalities, above a certain dimensionality they are outperformed by a simple sequential scan through the database. This occurs because the data space becomes sparse at high dimensionalities, causing the bounding regions to become large[7]. We selected the R*-tree among other index techniques for the following reasons: (i) This index structure is the most successful variant of the R-tree, (ii) it can be applied to spatial data such as geography and CAD data, and (iii) it can be used as an index structure for feature space such as image retrieval currently.

3 SOM-Based R*-Tree

The construction of a SOM-based R*-tree consists of two processes; clustering similar images and construction of R*-tree, as following.

Clustering similar images: We first generate the topological feature map using the SOM. We generate the BMIL by computing the distances between the feature vectors and codebook vectors from the topological feature map. The Best-Match Nodes(BMN) and its topological neighbors are moved closer to the input feature vector. As a result of learning ; the vector, which is generated on each node of the map, is called a codebook vector, and is represent by

$$CBV_i = [cv_{i1}, cv_{i2}, ..., cv_{ij}, ..., cv_{im}]^T,$$

where i($1 \le i \le k$) is the node number of the map, m is the number of input nodes, i. e., the dimensionality of the feature vector, and k is the number of map nodes.

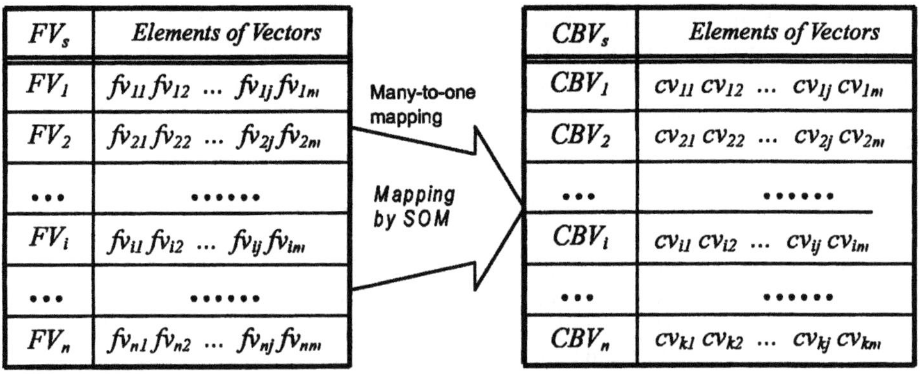

FV_s	Elements of Vectors		CBV_s	Elements of Vectors
FV_1	$fv_{11} fv_{12}$ ... $fv_{1j} fv_{1m}$	Many-to-one mapping	CBV_1	$cv_{11} cv_{12}$ ... $cv_{1j} cv_{1m}$
FV_2	$fv_{21} fv_{22}$ ... $fv_{2j} fv_{2m}$		CBV_2	$cv_{21} cv_{22}$ ... $cv_{2j} cv_{2m}$
...	......	Mapping by SOM	...	......
FV_i	$fv_{i1} fv_{i2}$ ... $fv_{ij} fv_{im}$		CBV_i	$cv_{i1} cv_{i2}$ ... $cv_{ij} cv_{im}$
...	......		...	......
FV_n	$fv_{n1} fv_{n2}$ ... $fv_{nj} fv_{nm}$		CBV_n	$cv_{k1} cv_{k2}$ ... $cv_{kj} cv_{km}$

(a) Feature vectors extracted from images   (b) Codebook vectors generated by SOM

Fig. 1. Relationship between feature vector and codebook vector

Using the topological feature map, we classify similar image to the nearest node, which has the minimum distance between a given feature vector and all codebook vectors. This classified similar image of each node is called the best-matching-image-list (BMIL). Similarity between feature vectors and codebook vectors is calculated by the Euclidean distance. Best-match-node BMN_i is

$$BMN_i = \min_i \{\| FV - CBV_i \|\}$$

where FV is a feature vector. The relationship between feature vectors and codebook vectors is shown in Figure 1. Between these two kinds of vectors, there are many-to-one relationships based on the similarity between each feature vector. This means that empty nodes occur in a topological feature map when the BMIL is generated. Empty nodes refer to the portion of the node (vector) spaces that contains no matching feature vectors. Empty mode indexing causes unnecessary disk access, thus degrading search performance. The space requirement can be reduced by indexing only live nodes (in contrast to empty nodes).

Construction of R*-tree: In the construction of a SBR-Tree, we use the R*-tree algorithm[5]. Here, let one point on the n-dimensional space correspond to each codebook vector of the topological feature map and the space covering all codebook vectors corresponds to the root node. In order to construct the R*-tree, we select a codebook vector from the topological feature map an entry. If it is an empty node, we select the next codebook vector. Otherwise, we determine the leaf node which insert codebook vector. The insertion algorithm of the R*-tree determines the most suitable subtree to accommodate the new entry(i. e., codebook vector) by choosing a subtree whose centroid is the nearest to the new entry. When a node or a leaf is full, the R*-tree reinsets or splits its entries. Otherwise, the new entry is add in the node or leaf. A leaf of the SOM-based R*-tree has the following structure:

$$L : (E_1,...,E_i,...,E_p) \quad (m \leq p \leq M)$$
$$E_i : (OID, \mu)$$

A leaf L consists of entries $E_1,...,E_i,...,E_p$ (m≤p≤M), where m and M are the minimum and the maximum number of entries in a leaf. Each entry contains an OID and its MBR μ. The node structure of the SBR-Tree is same as that of the R*-tree as shown in Fig. 2.

 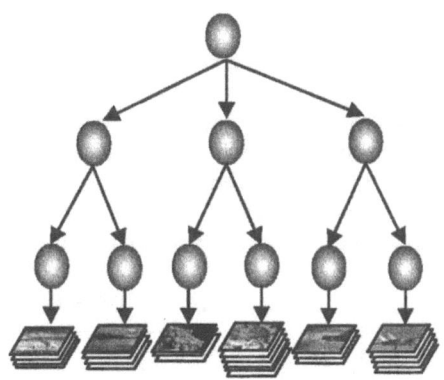

(a) example of topological feature map   (b) SOM-based R*-tree structure using topological feature map eliminated empty nodes

Fig. 2. SOM-based R*-tree structure

4 Experiments

We performed experiments to compare the SOM-based R*-tree with a normal SOM and R*-tree. Our image database contains still color images. The experimental image database currently consists of 40,000 artificial/natural images, including landscapes, animals, buildings, people, plants, CG, etc., from H^2soft and Stanford University. We fixed the image size at 128×128 pixels. All experiments were performed on a COMPAQ DESKPRO(OS:FreeBSD 3.4-STABLE) with 128 Mbytes of memory and all data was stored on its local disk.

4.1 Experimental Methodology

Feature Extraction: In this study, we extract practice[1]. One disadvantage of using Haar wavelets is that the computation tends to produce blocky image artifacts in the most important subbands. However, this drawback does not noticeably affect similarity retrieval[13]. The color space used in this paper for feature vectors is the TIQ-space (NTSC transmission primaries)[14] with luminance and color features from the image data, and use it in the experiments. To computer feature vectors, we use Haar wavelets[12], which are a kind of wavelet transform. Haar wavelets provide the fastest computations and have been found to perform well in practice[1]. One disadvantage of using Haar wavelets is that the computation tends to produce blocky image artifacts in the most important subbands. However, this drawback does not noticeably affect similarity retrieval[13]. The color space used in this paper for feature vectors is the TIQ-space (NTSC transmission primaries)[14] with luminance and chrominance information. We computed 5 level two-dimensional wavelet transforms for each of the three color spaces using Haar wavelets. Extracting the lowest submatrix for the color feature, we generated this submatrix as part of the feature vector. Each element of this feature vector represents an average of 32×32pixels of the original image. The color feature vector has 48 dimensions (=4×4×3,where 3 is the three channels of YIQ space).

Construction of SOM-based R*-tree. As shown in Table 1, we determine that the map size is almost the same as the number of images in the image databases. We generated the topological feature map using color feature vectors via the learning of the SOM, and the BMIL is generated using this feature map. The empty nodes occupied 53% to 60% of the original map size. As the map size becomes larger, the number of empty nodes increases. This means that images of high similarity are classified in the same node mutually regardless of map size.

Table 1. Tree Structure

		Data Set (×1000)					
		1	5	10	20	30	40
Total No. of nodes	R*-tree	119	499	972	2089	3042	3980
	SOM-based R*-tree	51	241	483	928	1300	1705
Heights of nodes	R*-tree	3	4	4	5	5	5
	SOM-based R*-tree	3	4	4	4	4	5
Time cost(sec)	R*-tree	7.55	50.44	111.89	233.89	350.66	476.49
	SOM-based R*-tree	3.27	19.12	40.76	81.95	115.49	153.92

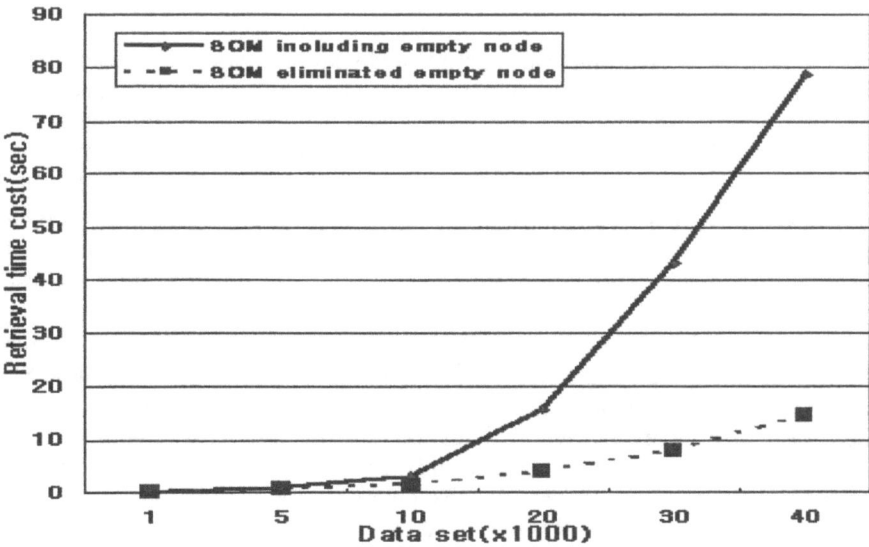

Fig. 3. Retrieval time cost

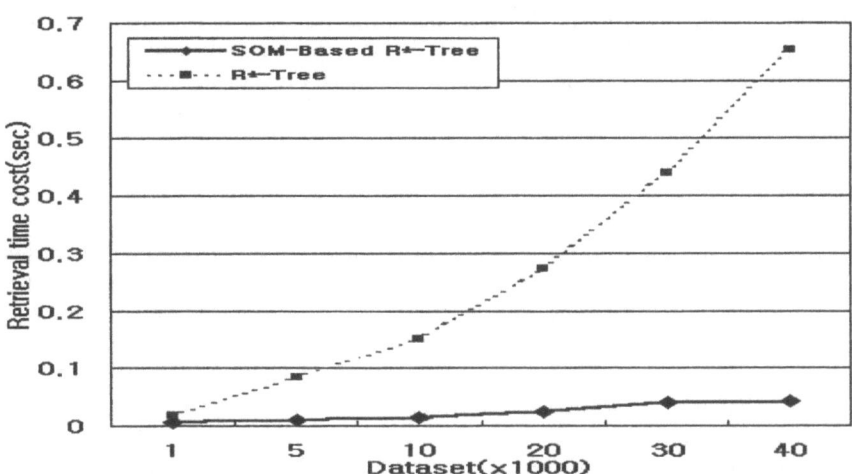

Fig. 4. Comparison of retrieval time cost between SOM-based R*-Tree and R*-Tree

Therefore, reducing the number of nodes and speeding up search time can be realized by eliminating empty nodes; an R*-tree built with this pruned set of nodes will have a smaller overall index size. The height of the tree is not that different, however both the total number nodes and the time cost of building the index decrease. These observations reduce memory usage and retrieval access time. The larger the data set, the more efficient the index.

4.2 Experimental Results

In order to measure search time, we experimented with four types of searches ; search for (i) normal SOM including empty nodes, (ii) normal SOM with eliminated empty nodes, (iii) normal R*-tree, and (iv) SOM-based R*-tree with eliminated empty nodes. The data set size was from 1,000 to 40,000 images. The search method used was the κ-Nearest Neighbor(NN)[15] method, which searches for κ ($\kappa>1$) objects nearest to the given query. In SOM, an exhaustive search of the topological feature map is performed, and finding κ ($\kappa=10$) nodes nearest to the given query. In the same manner, the normal R*-tree and SOM-based R*-tree are applied using the κ-NN ($\kappa=10$) search.

A comparison of retrieval time cost is shown in Figures 3 and 4. In both figures, the horizontal axis is the dataset size. As shown in Figure 3, the retrieval time of SOM with empty nodes, as compared to the SOM without empty nodes, grows drastically as the dataset size increases, over 5 times the retrieval time cost at 40,000 images. Thus, eliminating empty nodes significantly reduces retrieval time by removing unnecessary distance computations. We also compared the performance of the SOM-based R*-tree with that of the R*-tree based on 10-NN retrieval time cost, as shown in Figure 4. In this comparison the nearest OID was obtained for a given query. The retrieval time of the SOM-based R*-tree is far shorter compared to the R*-tree, by 3 to 15 times. The results show that building the R*-tree with overall original feature vectors improves retrieval performance.

Furthermore, the SOM-based R*-tree performs much better than SOM alone, which sequentially searches feature vectors. These experimental results clearly show that a SOM-based R*-tree is more efficient for similarity retrieval.

5 Conclusions

In this study, we proposed SOM-based R*-tree for dealing with similarity retrieval from high-dimensional data sets. Using a topological feature map and a best-matching-image-list (BMIL) obtained via the learning of a SOM, we constructed an R*-tree. The major finding of this study is that building an R*-tree in which the empty nodes in the topological feature map are removed yields an R*-tree with fewer nodes, thus enhancing performance by decreasing unnecessary access, node visits, and overall search times.

In an experiment, we performed a similarity search using real image data and compared the performance of the SOM based R*-tree with a normal SOM and R*-tree, based on retrieval time cost. The R*-tree with fewer nodes experimentally verified to shorter search tome, and search efficiency was improved due to the use of a k-NN search, compared to SOM.

References

1. A.F.C.E. Jacobs and D.H Salesin. Fast Multiresolution Image Querying. In Proc. SIGGRAPH95 , pages 6-11, New York, August 1995. ACM SIGGRAPH.

2.   C. Faloutsos, W. Equitz, M. Flickner, W. Niblack, D. Petkovic, and R. Barber. Efficient and Effective Query by Image Content. J. of Intell. inform. Syst., 3:231-262,1994.
3.   V. N. Gudivada and V. V. Raghavan. Content-based Image Retrieval system. IEEE Computer, 28(9): 18-22, September 1995.
4.   Guttman. R-tree: a dynamic index structure for spatial searching. In Proc. of ACM SIGMOD Int. Conf. on Management of Data, pages 45-57,1984.
5.   N. Beckmann, H. Kriegel, R. Schneider, and B. Seeger. R*-tree: an efficient and robust access method for points and agement of Data, pages 322-331, Atlantic City,NJ, May 1990.
6.   S. Berchtold, C. Bohm, and H.-P. Kriegal. The pyramid technique: towards breaking the curse of dimensionality. In Proc. of ACM SIGMOD int. conf. on Management of data, pages 142-153, Seattle, WA USA, June 1998.
7.   K. Chakrabarti and S. Mehrotra. High dimensional feature indexing using hybrid trees. In Proc. of ICDE1999, March 1999.
8.   T. Kohonen. Self-Organizing Maps. Springer, Berlin, 1997.
9.   T. Kohonen. Self-organizing maps. Proc.of The IEEE, 78(9):1464-1480,1990.
10. M. Flickner, H. Sawhney, W. Niblack, J. Ashley, Q. Huang, B. Dom, M. Gorkani, J. Hafner, D. Lee, D. Petkovic, D. Steele, and P. Yanker. Query by Image and Video Content: The QBIC System. IEEE Computer, 28(9):23-32, September 1995.
11. M. Koskelar. Content-Mased Images Retrieval with Self-Organizing Maps. Master's thesis, Helsinki University of Technology, Department of Engineering Physics and Mathematicd, 1999.
12. S.G. Mallat. Multifrequency Channel Decompositions of Images and Wavelet Models. IEEE. Trans., Acoust., Apeech and Signal Proc., 37(12):2091-2110, December 1989.
13. Natsev, R. Rastogi, and K. Shim. WALRUS: A Similarity Retrieval Algorithm for Image Databaese. In Proc. ACM SIGMOD International Conference on Management of Data, pages 396-406, Philadephia, PA, June 1999. ACM SIGMOD.
14. J. C. Russ. The Image Processing Handbook. CRC Press, Boca Raton, 1995.

Extending the Power of Server Based Computing

H.L. Yu, W.M. Zhen, and M.M. Shen

High-Performance-Computing Group, Computer Science Department,
Tsinghua University, 100084, Beijing, China
hlyu@tsinghua.edu.cn
http://hpc.cs.tsinghua.edu.cn

Abstract. In this paper, we review the evolution of computing, especially the development of thin client computing. Then, our study is focused on one of the most popular thin-client computing mode: server based computing. After an analysis of the problems existing in server based computing, a revised server based computing model is given, which combines the advantages of other thin-client model. As proof of our concept, our ideas are implemented in detail on our embedded platform: THUMP-NC. Some open source software, including rdesktop and VNC, are used in our implementations.

1 Introduction

With the development of computer technology, the speed of computer network is improving dramatically, network based applications have been developing very fast too. Thin-client computing is originated in this background. 'Thin client' concept, which most processing will be done by server and the client mostly take charge of displaying screen image and sending input events to server, has been changing all the ways too. Thin-client system has all the features on many aspect of a embedded system indeed.

Our study is focused on one of the most popular thin-client computing architecture: server based computing. Firstly, we will look back on the development of computing paradigms. We will see how thin-client computing originated and developed from it. Then, we will give a new paradigm for server based computing, which is more functional and easy-used. Finally, an implementation of this paradigm will be introduced.

2 Evolution of Computing

The evolution of computing has great impact on the way that server based computing does. If you want to change it, you must firstly know how it is originated.

Computing technology has evolved considerably over time. Although the process of evolution has been continuous, it can be classified as three distinct computing paradigms[1][2].

M. Bubak et al. (Eds.): ICCS 2004, LNCS 3038, pp. 242–249, 2004.
© Springer-Verlag Berlin Heidelberg 2004

The first computing paradigm, the mainframe paradigm, was popular through the 1960s and 70s. In this paradigm, computer processing power was provided by mainframe computers located in air-conditioned computer rooms. All resources were centralized, the main way to access the mainframe is through some character-based terminals; and the speed of the links between terminals to the mainframe is quite slow. This paradigm is greatly suffered from the costs of maintaining and administrating the system, and the poor price-to-performance ratio too.

The second computing paradigm is involved with powerful PCs and workstations. These standalone computers were single-user systems that executed a wide range of applications; they have their own computation and storage units, so they can work independently. With the advent of fast networking technologies, PC-LAN became more popular, which connect the PCs and workstations together. Client-server computing originated from this; a server in the LAN can provide centralized services to other PCs or workstations. Generally, client machines store and run applications locally while the server provides some file and print services. The pioneers of thin-client computing, such as InfoPad[3] system and Java NC system[4], are some kinds of client-server computing indeed.

Network-centric computing developed in early 1990s. In this paradigm, users' machines access both applications and data on networked servers. The most influential form in network-centric computing is Internet computing, which changed our application-development and content-delivery modes. The Internet has shifted the distributed computing paradigm from the traditional closely coupled form to a loosely coupled one[2]. With the development of Internet Computing, more and more applications are moved from client-server mode to web-based browser-server mode, which needs no client software installations in client machines. This is another kind of thin-client system.

However, commercial requirements and realities have revealed that professional applications are not always compatible with a web-based mode. For example, certain applications required the transmission of huge amounts of data between server and client, and cannot be redesigned for web-based mode. For this and other reasons, many businesses chose not to redevelop their applications and continue to use the traditional client-server model. Then, a new thin-client computing mode: server-based computing(SBC)[5], was populated to fill in the gap.

3   Related Works on Server Based Computing

The first version of server-based computing was provided by the X window system[6], it was originally developed for UNIX and enabled interactive applications running on large servers to accessed from low-cast workstations. But Windows system dominated the world soon. The server-based computing architecture from CITRIX[6] allows a variety of remote computers, regardless of their platform, to connect to a Windows NT terminal server to remotely access a powerful desktop and its applications. A server called MetaFrame runs under Windows NT in the desktop machine and communicates with the thin clients executing at the remote computers using the Independent Computing Architecture protocol (ICA). The ICA client and the MetaFrame server collaborate to display the virtual desktop on the remote computer screen. They also collaborate to process mouse and keyboard events, and to execute programs and view data stored at the server. All executions are remote and none takes place at the client portable computer. A research project at Motorola [7] extended CITRIX's thin client architecture so that it is optimized in the wireless environment. The work pointed out that bandwidth limitation is not as detrimental to the thin client performance as network latency. This is because the thin clients' use of bandwidth is limited. Other server-based computing implementation includes Remote Display Protocol(RDP) of Microsoft[8], Tarantella AIP of Tarantella [9], and Virtual Network Computing (VNC) of AT&T [10][11]. All these systems are quite similar, that is to say, all processing will be done on server and terminal devices works as a remote display. Each system has each proprietary protocol and data compression between server and terminal to reduce both network traffic and terminal side processing. Reduction of network traffic is usually achieved by sending only updated image.

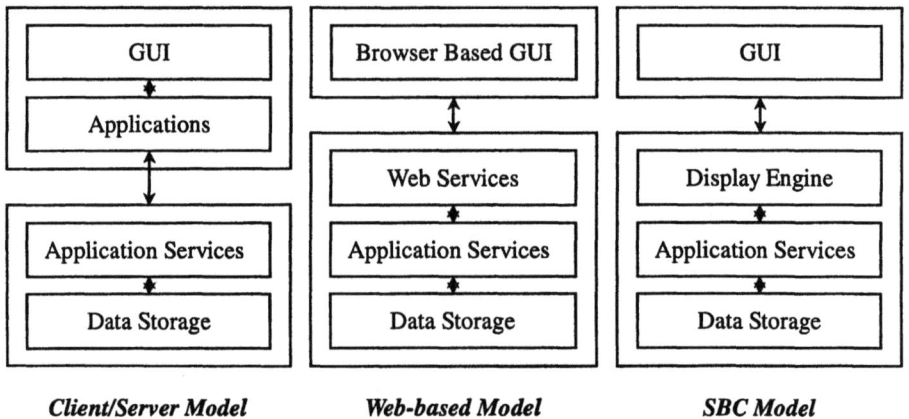

Fig. 1. Three different thin-client computing model

As we have mentioned above, three kind of thin-client computing system exists now: client-server based thin-client computing, web-based thin-client computing and server based computing(Figure 1). Server based computing(SBC) system is the thinnest system of all kinds, client device only displays the remote user interface locally; this is too big a luxury when thin-client like embedded system becomes more and more powerful.

4 The Problems in Server Based Computing

Server based computing technology is becoming mature now, it allows the deployment of traditional client-server applications over the Internet and WANs while reducing maintenance and support costs. But it's weakness is all appearance too, we will check it below.

Firstly, in most SBC system, no differentiation is made between video data and regular graphics data. This means when server decode the video data into separate frames and display it, these frames will be encoded as some static pictures by SBC scheme and transferred to client machines. This consumes lots of computing and network resources on SBC server. Some systems, like Citrix VideoFrame, recognize video data as a different type and supports it through a proprietary codec format. The stream is decoded from its original format, encoded for transmission using the proprietary codec and decoded again at the client side. In this process, the server will be burdened with some unnecessarily transcoding operations [12].

Secondly, remote display protocol has become essential to server based computing. There are many such protocols like RDP, ICA, AIP, VNC, for etc., but only windows based ICA and RDP(developed from ICA) are actually working well for the support of Microsoft. Other protocols like VNC perform well on UNIX, but quite poor on Windows environment. While some works has been done to solve the video playback in VNC system[13], it is quite meaningful to find some common ways for all main systems(especially RDP on UNIX) to share the advantages of SBC.

5 Computing Model

We have known that client-server computing based thin-client system like Java NC has no software installed locally, but they can download the operating system and application software from the server. After downloading, they can execute such applications locally. For example, if you have suitable software written in Java, you can do word processing, draw pictures, or even play videos locally. But such kind of thin client systems are generally considered as an unsuccessful initiative, because they cannot be compatible with Windows system, which is widely used and accepted. SBC system generally uses a windows system, so they become more successful. But for the lack of client local processing capability, the server is often overloaded. So, if we can combine the advantages of two computing mode, we will have a good answer.

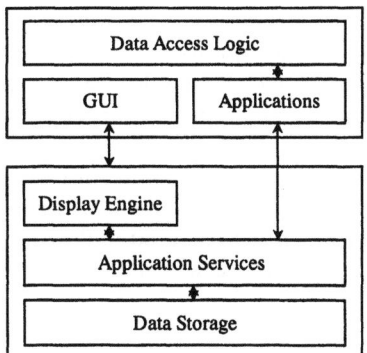

Fig. 2. Revised SBC Model

We can see from figure 2 that this computing model is developed from server based computing. Thin clients connect to a server and display the graphic-user-interface of the server. But they have two functions indeed. One is to act as a graphics terminal; the other is running as a full functional machine. Thin client functions are extended through local application execution capabilities. In this mode, thin client can even has a data access logic, which can store their private data through network.

The model is quite simple; we will show it in detail below with our implementation of a prototype system.

6   Implementation

Thin client of server-based computing runs UNIX or Windows operating system; as for Windows, Microsoft has all the solutions for it. We mainly solve the problem for client running UNIX. Then, we assume that Windows system is running on the server, so the SBC client can display a windows-like interface.

We will describe the current prototype implementation of the system below.

6.1   Hardware

We designed an embedded system for server-based computing, which is named THUMP-NC. It is powered by our embedded micro-processor: THUMP running at 400MHz. THUMP uses a MIPS instruction set, and is fully compatible with MIPS-4KC.

IT8172G is used in our mainboard. It performs high speed, high performance core logic control and data accessing between the CPU bus, system SDRAM bus, and PCI bus.

6.2   OS and Applications Installation

As a thin client system, no hard disk exists in our client machine. So, we have designed two booting mode for it: one is booting from a FLASH module, the other is booting from network.

An optional installed FLASH memory(8MB) can be used to store the OS kernel and some critical applications, which can be used as the booting source. The advantage of this mode is the speed, but the total cost of thin client hardware will be added too.

We have a BIOS like module named PMON in our hardware, it can be used to boot the system from network. In this mode, thin client will get their IP address through a DHCP request; and then, TFTP protocol is used to download the OS kernel from the server. Finally, the system is booted and some remote disks will be used locally through NFS.

6.3   Graphic-User-Interface Seamless Integration

We designed a framework named graphic-user-interface seamless integration to combine the advantages of traditional server-based computing and classical client-server computing.

Our prototype is built on top of rdesktop[14] for UNIX. Rdesktop is an open source UNIX X11 client for Windows NT Terminal Server, capable of natively speaking its Remote Desktop Protocol (RDP) in order to present the user's NT desktop. Unlike Citrix ICA, no server extensions are required.

After booted, our thin client machine will be connected to a Windows server automatically through rdesktop. So, we can log on to the server immediately. After that, a windows graphic-user-interface will be showed on thin client display, but it is partly changed by us.

Firstly, we have modified some of the system icons and shortcuts. In traditional server based computing, all icons and shortcuts on thin client displays mean some corresponding server applications. But in our system, they will be divided into two categories; one is for server applications, the other stands for some local applications on thin client machines. We design a client-server style mechanism for the execution of local applications in our thin client machine. Any request for the execution of corresponding applications of these icons and shortcuts will be passed to a process named PIQUET in SBC server; then, this process will inform the request to a DAEMON running on thin client through network; after receiving the request, the DAEMON will start the correct application on thin client machine, and commit the request at the same time. After that, any window operations like window maximize, window minimize, window close and window move will all be informed between the PIQUET and the DAEMON, in order to keep the whole graphic-user-interface seamless.

Secondly, some file associate relations are modified too. For example, if you find a video file in windows explorer, you may double click it for a playback. Traditional server based computing use a windows media player for this task, but in our system, another media player running on thin client will be in charge. Other applications like Mozilla will be used in place of IE too.

As for users, the real execution locations of applications are not so important; the important thing is: they can use them in one style.

6.4 Local Video Processing

We use VideoLAN[15] for video streaming and playback in our system. VideoLAN is free software, and is released under the GNU General Public License. The VideoLAN project targets multimedia streaming of MPEG-1, MPEG-2, MPEG-4 and DivX files, DVDs, digital satellite channels, digital terrestial television channels and live videos on a high-bandwidth IPv4 or IPv6 network in unicast or multicast under many operating systems. VideoLAN also features a cross-plaform multimedia player, VLC, which can be used to read the stream from the network or display video read locally on the computer under all main operating systems.

We design a streaming server for better using of VideoLAN in our system. In our revised SBC model, local applications in thin clients communicate with thin server in client-server mode. We can look the local applications as a streaming media client, and thin server a streaming server. The easies way to implement a streaming server is through the HTTP service of Windows system. But it is not so reliable. If the server could not send enough media data to the client, video playback will be paused. And then, we studied from traditional video on demand technologies. Interval caching[16] is used in our system. The final stage of our implementation is a file mapping

schemes, we look the whole file system in thin server as a video database, any request to the video file will be redirected through the PIQUET and the DAEMON described earlier. This is another main step to approach our seamless GUI.

7 Some Results

S.Jae Yang[12] gave some results to value the effectiveness of a number of server based computing design and implementation choices across a broad range of thin-client platforms and network environments.

Our implementations are only complementarities to server based computing. So, we only care about the differences. We use a machine equipped with PIII800, 256M RAM to act as a SBC server. Two clients are used for test too; one is a Windows PC, the other is our THUMP-NC.

We test the decoding process in three modes. Firstly, we decode the file on SBC server locally; secondly, we connect to SBC server through RDP and run the decoding; finally, we connect to SBC server from our THUMP-NC and do the same thing. That's our results below(Table 1). We can see that the CPU utilization of SBC server is greatly reduced, because neither decoding nor rendering are needed on SBC server in later mode.

Table 1. CPU Utilizations of video playback in different computing mode

Video Source	Local	Windows PC	THUMP-NC
1.2Mbps MPEG1	14%	65%	2%
5.4Mbps MPEG2	56%	90%	4%

8 Conclusions

In server based computing, a terminal server deploys, manages, supports and wholly runs the client part of application. Thin client only takes charge of the display of the graphics user interface of the running result. With the development of embedded systems, the power of traditional thin client devices is greatly boosted up. Some attempts to use such computing power more efficiently have been given in this paper. Such attempts include a seamless integrated GUI and the video support on thin client system, which lighten the load on corresponding server, and make the system easy-of-use.

References

1. Network computing: a tutorial review, Revett, M.; Boyd, I.; Stephens, C.; Electronics & Communication Engineering Journal , Volume: 13 Issue: 1 , Feb 2001, Page(s): 5 -15
2. Network Computers: The Changing face of computing, Ishfaq Ahmad, IEEE Concurrency, October-December, 2000

3. Software architecture of the InfoPad system. In Proceedings of the Mobidata Workshop on Mobile and Wireless Information Systems (Rutgers, NJ, Nov.), 1994.
4. Java NC Computing, `http://java.sun.com/features/1997/july/nc.html`
5. Server-based computing opportunities, Volchkov, A.; IT Professional , Volume: 4 Issue: 2, March-April 2002, Page(s): 18 -23
6. `http://www.citrix.com`, 1998
7. Duran, J. and Laubach, A. 1999. Virtual personal computers and the portable network. In Proceedings of the IEEE Conference on Performance, Communication, and Computing (Phoenix, AZ). IEEE Computer Society Press, Los Alamitos, CA.
8. Microsoft RDP & Citrix ICA Feature Overview, `http://www.microsoft.com/windows2000`
9. `http://www.tarantella.com`
10. Virtual network computing, Richardson, T.; Stafford-Fraser, Q.; Wood, K.R.; Hopper, A.; Internet Computing, IEEE , Volume: 2 Issue: 1 , Jan.-Feb. 1998 Page(s): 33 -38
11. VNC tight encoder-data compression for VNC, Kaplinsky, K.V.; Modern Techniques and Technology, 2001. MTT 2001. Proceedings of the 7th International Scientific and Practical Conference of Students, Post-graduates and Young Scientists , 26 Feb.-2 March 2001
12. S. Jae Yang, Jason Nieh, Matt Selsky, and Nikhil Tiwari, "The Performance of Remote Display Mechanisms for Thin-Client Computing", Proceedings of the 2002 USENIX Annual Technical Conference, Monterey, CA, June 10-15, 2002, pp. 131-146.
13. Chia-Chen Kuo,Ping Ting. Design and Implementation of A Network Application Architecture for Thin Clients, Proceeding of the 26th Annual International Computer Software and Applications Conference(COMPSAC'2002)
14. `http://sourceforge.net/projects/rdesktop/`
15. `http://www.videolan.org`
16. Dan A , Dias D M, Mukherjee R, etc., Buffering and caching in large-scale video servers. Proceedings of COMPCON. San Francisco USA: IEEE Computer Society Press, 1995. 217 224.

Specifying Policies for Service Negotiations of Response Time

T.K. Kim[1], O.H. Byeon[2], K.J. Chun[3], and T.M. Chung[1]

[1]Internet Management Technology Laboratory,
School of Information and Communication Engineering,
Sungkyunkwan University,
Chunchun-dong 300, Jangan-gu, Suwon, Kyunggi-do,
Republic of Korea
tkkim@rtlab.skku.ac.kr, tmchung@ece.skku.ac.kr
[2]Korea Institute of Science and Technology Information
ohbyeon@kisti.re.kr
[3]Sangmyung University
chunkj@smu.ac.kr

Abstract. The use of services in large-scale and cross-organizational environments requires the negotiation of agreements that define these services. This paper proposes to specify a negotiation policy for response time of distributed network. The analysis of distributed networks has become more and more important with the evolution of distributed network technologies resulting in increasing capacities. To monitor and manage service in distributed network, we must identify the relationships between network/application performance and QoS parameters. Therefore, we provide a statistical analysis on mapping user level response time to application and network level parameters by using some queueing models and suggest the negotiation of policy specification for response time of distributed network. Hence, the use of guaranteed services becomes feasible.

1 Introduction

The evolution of communication technologies results in continuously increasing capacities and higher concentration of traffic on relatively fewer network elements. Consequently, the failures of these elements can influence an increasing number of customers and can deteriorate the quality of service provided by the operator. The performance of distributed network is very important to network users for running their applications and to network manager for managing the distributed network. In general, the users do not know how to efficiently map their performance requirements to a complex QoS metric. Moreover, many of the sophisticated QoS and pricing mechanisms are complex to implement and therefore infeasible. As customers have begun to demand higher level of Quality of Service (as opposed to the best effort service) from the service providers, service level agreements (SLAs) between customers and service providers have become the norm. A service level agreement [5] is an agreement regarding the guarantees of a service. It defines mutual understandings and expectations of a service between the service provider and service consumers. These SLAs specify the quality of service and pricing information [7].

M. Bubak et al. (Eds.): ICCS 2004, LNCS 3038, pp. 250–257, 2004.
© Springer-Verlag Berlin Heidelberg 2004

Therefore, in distributed network, "best effort service" is no longer sufficient for guaranteeing QoS. Thus it is required to satisfy quality of service (QoS) in distributed network that a QoS specification at the application level permits the selection of appropriate network level parameters [3]. To allow the provision of a certain QoS, the parameters of one layer have to be mapped to those of other layers and of system resources [4]. For instance, typical QoS metrics include committed bandwidth, transit delay, packet loss rate and availability.

In this paper, we focus on network latency to represent the mapping of applications performance parameters to network performance parameters and negotiation of agreements that define this service. Negotiations are mechanisms that increase the flexibility of possible service contracts. In the context of dynamically setting up service relationships, it is important to use an efficient decision-making process that reduces cost and time of the setup. Decision-making is involved in deciding the acceptance of an offer and in the selection of an outgoing offer among multiple candidates [1].

In section2, related works are describes. Section 3 presents the response time by mapping the application level parameters to network level parameters. In section 4 we introduce the model of decision-making. Section 5 summarizes our work and identifies future research directions.

2   Related Works and Negotiation Issues

2.1   Related Works

To support the negotiation of service and QoS of application, many projects are processed in this field. Liu et al. [2] presented a formal statistical methodology for the mapping application level SLA to network level performance. They took the response time as the application level SLA and link bandwidth and router throughput/utilization at the network layer for their preliminary analysis and presented a function which directly links the response time at the application level to the network parameters and does not address the efficiency of a formal statistical methodology using simulation or real execution. Some other projects have presented QoS mapping in multimedia networks [8][9].The negotiation server by Su et el. uses rules to describe how to relax constraints defining acceptable offers in the course of the negotiation[13]. The complexity of utility functions and contract implementation plans is addressed by Boutilier et al. [6]. This approach is used for collaborative resource allocation within an organization and does not address negotiation across organizational boundaries.

2.2   Negotiation Issues

Negotiations are mechanisms that increase the flexibility of possible service contracts and negotiations are used as comprising all exchanges of messages, such as offers and acceptance messages, between two or more parties intended to reach an agreement. A common way of analyzing negotiations is differentiating the negotiation protocol, comprising the rules of the encounter, the negotiation object, which is the set of negotiated attributes, and the decision making model. A number of simple negotiation protocols are used for match making reservations such as SNAP (Service Negotiation and Acquisition Protocol) has been proposed for resource reservation and use in the context of the Grid. A common problem in negotiations is the ontology problem of electronic negotiations [10]. It deals with the common understanding of the issues among negotiating parties. One approach of solving the ontology problem is the use of templates. Templates are partially completed contracts whose attributes are filled out in the course of negotiation [1].

3   Mapping Response Time to Application and Network Parameters

We can know the response time using the information of network latency, system latency, and software component latency. Network latency is composed of propagation delay, transmission delay, and queueing delay; system latency is composed of disk I/O, and CPU processing delay; software component latency is composed of server and database transaction delays.

$$Response\_time = Network\_latency + System\_latency + Software\_component\_latency$$

We will focus on the network latency, and assume system latency and software components latency are known. We can think of latency as having three components [11]. First, there is the speed-of-light propagation delay. This delay occurs because nothing can travel faster than the speed of light. Second, there is the amount of time it takes to transmit a unit of data. This is a function of the network bandwidth and the size of the packet in which the data is carried. Third, there may be queuing delays inside the network, since packet switches need to store packets for some time before forwarding them on an outbound link. So, we define the network latency as:

$$Network\_latency = propagation\_delay + transmission\_delay + transmission\_delay + queuing\_delay$$

We can map the network latency of user level to application level and network level elements.

- user layer: response time - the elapsed time from the user requests the service to the user accepts the results of application.
- application layer: network latency, system latency, software component latency
- network layer: propagation delay, transmission delay, and queueing delay. Network layer can be modeled as end-to-end latency partitioned into a speed-of-light propagation delay, a transmission delay based on the packet volume sent and the bandwidth, and queueing delay including host and router.
- network layer elements: distance, bandwidth, utilization, throughput, packet size, arrival rate, and number of servers.

Propagation delay is related with distance and transmission delay is related with bandwidth and packet size. And queuing delay is related with bandwidth, packet size, arrival rate, utilization, throughput, and number of servers. Figure 1 shows the relations among the user level, application level, and network level parameters.

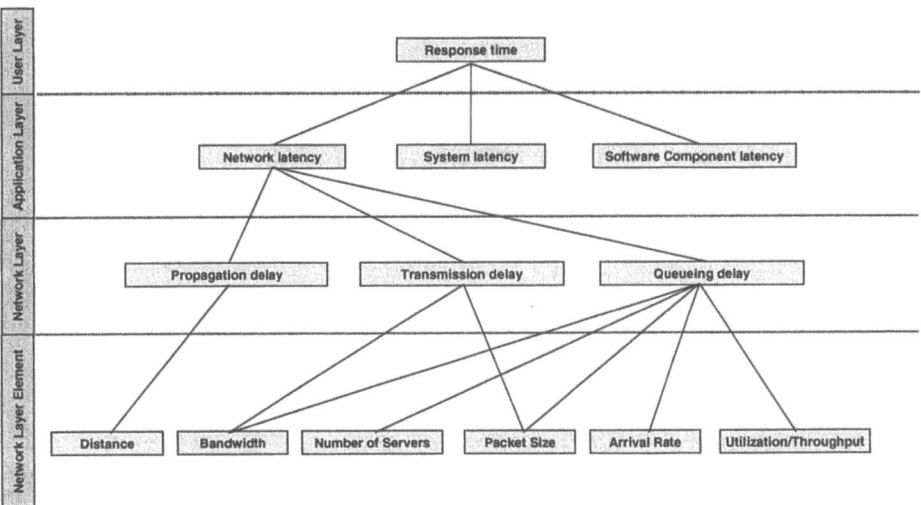

Fig. 1. Mapping of latency elements

We can calculate the response time using network elements mapped with application elements. So, we can statistically analyze the network element's contribution to the whole response time for an application running in a distributed computing environment. We assume the system to be a markovian system, which means the distribution of the interarrival times and the distribution of the service times are exponential distributions that exhibit markov property.

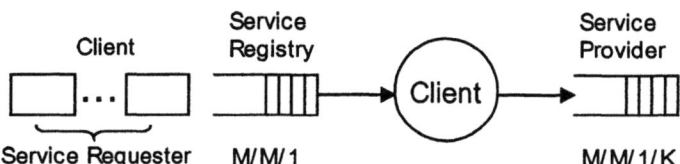

Fig. 2. The modeled distributed Network

We also modeled this system as the M/M/1 Queue for the process of getting the useful information of service; M/M/1/K for the processing of service provider.

Thus we can write the steady state probabilities as follows:

Propagation delay: $\displaystyle\sum_{j=1}^{j} \frac{D}{2.3 \times 10^{8}}$

Transmission delay: $\displaystyle\sum_{k=1}^{k} \frac{\overline{M}}{B}$

Queueing delay [2]: M/M/1, M/M/1/K - $\displaystyle\sum_{l=1}^{l} \frac{\overline{M}}{B - \lambda\overline{M}}$

$$Network\_latency = \sum_{j=1}^{j} \frac{D}{2.3 \times 10^8} + \sum_{k=1}^{k} \frac{\overline{M}}{B} + \sum_{l=1}^{l} \frac{\overline{M}}{B - \lambda \overline{M}} \qquad (1)$$

where D is distance, $\overline{M}$ is the mean size of the packet, B is the bandwidth at which the packet is transmitted, and λ is the arrival rate of the client request.

We calculated the queueing delay and network latency. The parameters used in this calculation are like this: bandwidth: 10Mbps; packet size: 1500byte; distance: 50km; arrival rate: 42; service rate: 874. The arrival rate and service rate were calculated

using $\mu = \dfrac{1}{T_S} = \dfrac{B}{M}$ and M/M/1 system queueing delay $\dfrac{\overline{M}}{B - \lambda \overline{M}}$.

The propagation and transmission delay were calculated like these:
- Propagation delay: 0.000217391
- Transmission delay: 0.0011444

Then, we calculated the value of queueing delay using the equation of (1). Also, to check the value of calculation, we used the NS-2 (Network Simulator). NS-2 is an open-source simulation tool that runs on Linux [12]. We made Tcl script of each queueing model (M/M/1, M/M/1/K) and simulated the queueing delay. The Linux server used in this simulation has dual CPU of 1G and 256MB RAM. We repeated this simulation 20 times to calculate the mean queueing delay and network latency time of simulation.

Table 1. The value of calculating, simulation, and error of the network latency

	Propagation delay	Transmission Delay	M/M/1	M/M/1/K (K=50)	Network latency
Calculation			0.001202193	0.001202	0.003766
Simulation	0.000217391	0.0011444	0.001202	0.001197	0.003761
Error			0.016%	0.42%	0.133%

To compare the network latency of calculation value and simulation value, we calculated the average error for the above type of measurements as:

$Error = average(abs(network\_latency\_function - queueing\_simulation) /$
$queueing\_simulation) \times 100$

The value of network latency calculation of (1) was relative similar to the value of queueing simulation and the error was relatively low. And distance, bandwidth, packet size, and the number of host are related to the network latency in distributed network.

4   Decision-Making Framework

The model of decision-making is designed as an object-oriented framework. The framework assigns evaluation function to check the response time, and specifies the object that can be accessed for rule-based reasoning.

Rules are expressed in a high-level language specified by an XML schema. This helps a business domain expert to specify negotiation strategies without having to deal with the programmatic implementation of the decision making system. The basic structure is a bilateral message exchange and follow-up messages are of the types accept, reject, offer, withdraw, or terminate. Accept leads to a contract based on the other's last offer, reject to the rejection of the last offer. Offer indicates that a filled template is sent as proposed contract, withdraw annuls the last offer, and terminate ends the entire negotiation process immediately [1]. Figure 3 exemplifies response time negotiation system architecture.

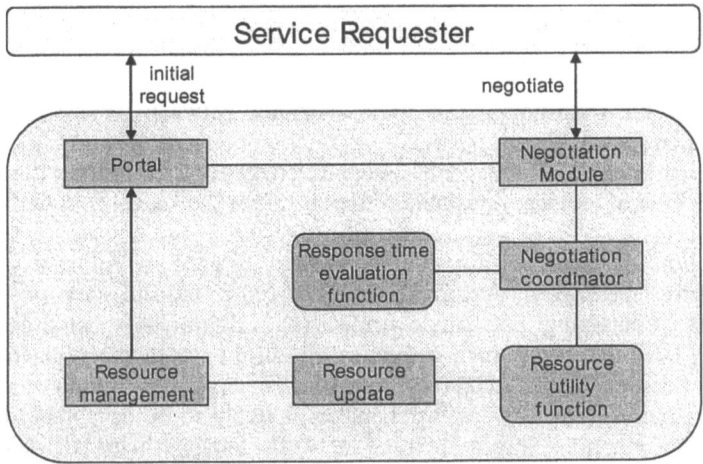

Fig. 3. Decision-maker components

A negotiation is initiated through the portal. The portal sends information about new interaction to the negotiation module and endows it with information on the service requester. After receiving new interaction, negotiation module takes control of the interaction with the service requester.

- Negotiation module: The negotiation module performs some administrative tasks such as checking the validity of the incoming message before proceeding to more sophisticated message handling. Upon reception of a termination or acceptance the procedure is straight forward: cleaning up all negotiation dependent objects and possibly canceling resource reservations, or passing the final contract on to for deployment. Otherwise, the negotiation module processes its rule corpus for producing a response message to send or for deciding to wait [1].
- Negotiation coordinator: The negotiation coordinator is designed to coordinate multiple negotiation modules and can measure the real-time response time of service using response time evaluation function. Resource utility function checks the status of usage of system and network resources. This coordinates the resource reservation to keep the contract of negotiation. If response time is more than that of contract, negotiation coordinator requests to assign more system resources to guarantee the contract.
- Resource update: resource update establishes the interface to resource management and is able to handle reservation and information requests. Resource management controls the resources of systems.

Table 2. Example of an offer sequence during service negotiation

Sender		...	Negotiation module	Service requester	Negotiation module	...
Message No.			3	4	5	
Response time			≥4	≤4	4	
Assignable resources	bandwidth		40-60%	80%	60%	
	cpu		40-50%	48%	48%	
	memory		50-55%	57%	57%	
Available resources	Bandwidth		75%	72%	73%	
	Cpu		60%	56%	55%	
	memory		68%	68%	71%	

Table 2 gives a snapshot of the offers exchange between service requester and negotiation module.

Response time is given in second, assignable resources are required to guarantee the response time of service. Available resources mean the status of system resources to be used.

The creation of negotiation module's number 5 goes as follows: negotiation module receives message 4, approves it as valid offer, and the starts processing its rule corpus containing a single rule set. Negotiation module invokes LEVEL_OF_DIFFERENT, which computes the difference of offers 3 and 4. After calculation of response time using resource utility function, negotiation module offers 4 seconds as message 5. The last offer makes it likely to be acceptable by service requester. The mapping information of network latency helps to calculate the response time and assign required system resources.

5   Conclusion and Future Works

The use of services in large-scale and cross-organizational environments requires the negotiation of agreements that define these services. Today, negotiation is complex and difficult to implement. In this paper, we provide a statistical analysis on mapping user level response time to application and network level parameters by using some queueing models and suggest the negotiation of policy specification for response time of distributed network. Using this approach, decision-making framework allows the specification of sophisticated negotiation behavior for response time in a manageable way. Also, we suggested a network latency function of (1) to calculate the network latency and showed the validity of the function by comparing the results of simulation using NS-2.

Numerous challenges still remain in this area. There are other user level parameters like availability, reliability, etc., that we haven't pursued in this paper. Future research will focus on presenting the user level parameters of SLA as numerical formula and extend negotiation framework can specify negotiation for all services in a concise, and easy way.

References

1. H. Gimpel, H. Ludwig, A. Dan and B. Kearney, "PANDA: Specifying Policies for Automated Negotiations of Service Contracts", ICSOC 2003, Trento, Italy, December 2003.
2. B. Hua Liu, P. Ray, S. Jha, "Mapping Distributed Application SLA to Network QoS Parameters," IEEE 2003.
3. J.-F. Huard, A. A. Lazar, "On QOS Mapping in Multimedia Networks", Proceedings of the 21th IEEE International Computer Software and Application Conference (COMPSAC '97), Washington, D.C., USA, August 1997.
4. S. Fischer, R. Keller. "Quality of Service Mapping in Distributed Multimedia Systems", In Proceedings of the IEEE International Conference on Multimedia Networking (MmNet95), Aizu-Wakamatsu, Japan, September 1995.
5. L. Jin, V. Machiraju, A. Sahai, "Analysis on Service Level Agreement of Web Services," Software Technology Laboratory, June 2002.
6. C. Boutilier, R. Das, J.O. Kephart, G. Tesauro, W.E. Walsh: Cooperative Negotiation in Autonomic Systems using Incremental Utility Elicitation. Proceedings of Nineteenth Conference on Uncertainty in artificial Intelligence (UAI 2003). Acapulco, 2003.
7. M. Singh Dang, R. Garg, R. S. Randhawa, H. Saran, "A SLA Framework for QoS Provisioning and Dynamic Capacity Allocation," Tenth International Workshop on Quality of Service (IWQoS 2002), Miami, May 2002.
8. Luiz A. DaSilva, "QoS Mapping along the Protocol Stack: Discussion and Preliminary Results," Proceedings of IEEE International Conference on Communications (ICC'00), June 18-22, 2000, New Orleans, LA, vol. 2, pp. 713-717.
9. T. Yamazaki, J. Matsuda, "On QoS Mapping in Adaptive QoS Management for Distributed Multimedia Applications," Proc. ITCCSCC'99, vol.2, pp.1342-1345, July, 1999.
10. M. Strobel: Engineering electronic negotiations, Kluwer Academic Publishers, New York, 2002.
11. L. Peterson, B. Davie, Computer Networks: A System Approach, Morgan Kaufmann, 2000, second edition.
12. NS network simulator. http://www-mash.cs.berkeley.edu/ns.
13. S. Y. W. Su, C. Huang, J. Hammer: A Replicable Web-based Negotiation Server for E-Commerce. Proceedings of the Thirty-Third Hawaii International Conference on System Sciences (HICSS-33). Maui, 2000.

Determination and Combination of Quantitative Weight Value from Multiple Preference Information

Ji Hyun Yoo[1], Byong Gul Lee[1], and Hyuk Soo Han[2]

[1]Department of Computer Science and Engineering,
Seoul Women's University, 126 Nowon-Gu, GongReung 2 dong, Seoul, Korea
{jhyu,byongl}@swu.ac.kr
[2]Department of Software Science, Sang Myung University,
Hongji-dong 7, Seoul, Korea
hshan@smu.ac.kr

Abstract. Recently, the concern of software quality increases rapidly. Although there have been many efforts to establish standards for software quality, such as ISO/IEC 9126, they provide only a framework for quality characteristics and evaluation process. They do not provide practical guidance for deriving reasonable weight value criteria for quality evaluation. [1] This paper presents a method to draw the quantitative weight values from evaluator's subjective data in software evaluation in compliance with ISO/IEC 9126 standard. To eliminate evaluators' subjectiveness and uncertainty, the Dempster-Shafer (D-S) theory is improvised and utilized. The D-S theory is improved with merge rule to reduce the bias of weight value when they are merged with other evaluator's weight value. The proposed merge rule has been tested for it's effectiveness with actual evaluation data.

1 Introduction

Software quality is defined as a specification of functions of which software performs. High quality software means that it not only satisfies its specification, but also achieves its quality characteristics, such as functionality, reliability, usability, efficiency, maintainability, portability. To evaluate software quality based on these quality characteristics, various standards and techniques can be applied. ISO/IEC 9126 [1], for example, provides a standard for establishing software quality characteristics and metrics. ISO/IEC 14598 [2] provides methods and procedures for performing of quality authentication. These standards, however, does not specify specific evaluation technique such as the calculation of weight value for evaluation items. Consequently, software quality evaluation tends to depend on assessor's subjective judgment and knowledge. That is, the interpretation of evaluation criteria, semantic relation between evaluation items, and measurement depends on evaluators' viewpoint [3], [4], [5]. Existing studies also overlook the problem of producing incorrect result in combining different weight values from multiple assessors [6], [7], [8].

---

[1] This research was supported by University IT Research Center Project

M. Bubak et al. (Eds.): ICCS 2004, LNCS 3038, pp. 258–265, 2004.
© Springer-Verlag Berlin Heidelberg 2004

This paper describes a quantitative method for calculating the weight value of evaluation items by combining different evaluation values which is retrieved from multiple assessors' subjective opinion. Specifically, this paper discusses how to improve the way of aggregating and combining multiple assessors' opinions considering the relations between the evaluation items.

2 Related Works

2.1 Inference Using Bayes' Theory

Bayes' theory [11], [12], [13] is a statistical inference technique to estimate the probability of certain event occurring in which different hypotheses are given for each evidence. Assuming that there exists mutually exclusive hypothesis H_k (k=1,2,...,n) for a certain evidence E, then Bayes' theory is described as following (1).

$$P(H_n \mid E) = \frac{(P(E \mid H_n) \cdot P(H_n))}{\sum_{k=1}^{n} P(E \mid H_k) \cdot P(H_k)} \tag{1}$$

From the above expression, H_n and E are related in cause and effect relation. Then, $P(H_n)$ is called as a priori probability and $P(H_n|E)$ as a posteriori probability. Bayes' theory, however, has a few problems in its application. First, if there are n numbers of evidence and m numbers of hypothesis, it is necessary to know in prior n*m numbers of probability. This means that a large number of evidence and hypothesis requires a great number of priori probability calculation. In reality, it is not possible to provide all priori probability values in advance. Second, the probability value in Bayes' theory is assumed to be mutually exclusive. That is, if the probability of a certain event occurring be 0.5, then the probability for the event not occurring is 1-0.5=0.5. However, in real situation, the probability for an event not occurring isn't always 0.5 since it is not known whether a certain event occurrs or not. Finally, there is no rule for successively accruing the probability value in Bayes' theory. That is, if a new hypothesis is introduced, the Bayes theory has to compute the probability from the scratch again.

2.2 Quantitative Translation from Qualitative Relationship

The technique of deriving quantitative value from qualitative relationship information can be achieved by using Dempster Shafer (DS) theory [9], [10], [14], [17]. It uses the qualitative preference relations such as "A is more important than B" or "A is similar to C."

DS theory compensates the drawbacks of Bayes' theory by defining the belief of hypothesis on H to an interval between Bel(H) and Pl(H). Bel(H) is a belief value of a given evidence and Pl(H) is a plausibility value based on evidence. DS theory also provides a combination rule on merging two random variables that are independent each other (2). A function m: $2^S \rightarrow [0,1]$ (S is a frame of discernment) is called a basic probability assignment [11], [12], [13], [15], [16].

$$m3 = \frac{\sum\limits_{s1 \cap s2 = s3} m1(s1) \cdot m2(s2)}{1 - \sum\limits_{s1 \cap s2 = \phi} m1(s1) \cdot m2(s2)} \qquad (2)$$

This combination rule, however, may generate an empty set after intersection of two discernments [7], [8]. This can degrade the correctness of combination results. In addition, none of intending hypothesis can have total value, 1, after normalization process. This means that the lower supporting discernments can produce the higher supporting value after the combination as shown in table 1.

Table 1. Result of normal combination rule in DS theory

m2 \ m1	m1({a}) = 0.1	m1({b}) = 0.9
m2({a}) = 0.1	m3({a}) = 0.01	m3(ø) = 0.09
m2({c}) = 0.9	m3(ø) = 0.09	m3(ø) = 0.81
Normalization	m3({a}) = 0.01 / 0.01 = 1	

3 Improved Combination Rule in DS

As described earlier, the normal combination rule in DS theory can generate empty sets when different discernments are combined. The improved combination rule can reflect assessor's different opinions by redistributing the empty set value to the discernment. Table 2 shows the result of combining the exclusive discernments using the improved combination rule. Let the basic probability assignment (bpa) assigned by assessor A be m1 and bpa by assessor B be m2. From the table 2, since m1({b})=0.9, m2({c})=0.9, the result of combination of m1 and m2 is m3({∅})=0.81. This means that assessor A assigns the wider opinion to discernment b and assessor B assigns the wider opinion to discernment c. In common, it assigns the narrower opinion to discernment a since the value of m3({∅}) is redistributed to the discernment b and c. In the same way, m3({b})=0.405 and m3({c})=0.405. The final result after the combination shows in table 2.

4 Weight Value Determination in Software Quality Evaluation

This chapter describes an example in calculating the weight value of software quality characteristics such as reliability defined in ISO/IEC 9126. For calculating the weight value, the first step is to determine the focal elements from assessors' qualitative preference evaluation. After determining the focal elements, the basic probability assignment (bpa) value is computed [9], [10], [14], [17]. If there is more than one assessor, then apply the improved combination rule in combining the bpas. Finally, compute a weight value using the belief function (Bel) and the probability function (Pl).

Table 2. Result of combination about exclusive discernment using improved combination rule in DS theory

m2 \ m1	m1({a}) = 0.1	m1({b}) = 0.9
m2({a}) = 0.1	m3({a}) = 0.01	m3(ø) = 0.09
		m3({a}) = 0.009
		m3({b}) = 0.081
m2({c}) = 0.9	m3(ø) = 0.09	m3(ø) = 0.81
	m3({a}) = 0.009	**m3({b}) = 0.405**
	m3({c}) = 0.081	**m3({c}) = 0.405**
Result of improved combination rule	m3({a}) = 0.01 + 0.009 + 0.009 = **0.028**	
	m3({b}) = 0.081 + 0.405 = **0.486**	
	m3({c}) = 0.801 + 0.405 = **0.486**	
	m3({a}) + m3({b}) + m3({c}) = 0.028 + 0.486 + 0.486 = 1	

(1) Determining the Focal Elements and bpa

Given assessor's qualitative preference relationship, the focal element can be determined using the following definition (3) and (4).

$$A \cdot > B \leftrightarrow A \text{ is higher weight value than } B. \tag{3}$$

$$A \text{ S } B \leftrightarrow (\jmath(A > B) \& \jmath(B > A)) \leftrightarrow A \text{ is similar weight value to } B \tag{4}$$

Let T={a, b, c, d} be the discernment and assume that an assessor has defined preference relationship as in (5).

$$\{a\} > \{d\}, \{b,c\} > \{b\}, \{c,d\} < \{d\}, \{a,b,c\} \text{ S } \{b,c\}, \{b\} \text{ S } \{d\}, \{b\} > 0, \{d\} > 0 \tag{5}$$

To determine the focal elements, it is necessary to eliminate illogical elements by using the elimination theorem [9], [10], [14], [17]. From the theorem, since {c, d}<{d} and {a, b, c} S {b, c}, {c, d} and {a, b, c} are not focal elements. In this way, the complete focal elements can be achieved: {a}, {b}, {b, c}, {c}, {d}.

After achieving the focal elements, the bpa of m1 can be obtained using the equality and perceptron algorithm as following.

$$bpa \text{ of } m1 = \begin{bmatrix} m1(\{a\}) \\ m1(\{b\}) \\ m1(\{b,c\}) \\ m1(\{c\}) \\ m1(\{d\}) \end{bmatrix} = \begin{bmatrix} 2 \\ 1 \\ 2 \\ 2 \\ 1 \end{bmatrix} = \begin{bmatrix} 0.25 \\ 0.125 \\ 0.25 \\ 0.25 \\ 0.125 \end{bmatrix}$$

(2) Application of the Improved Combination Rule

In case there is more than one assessor, then the improved combination rule is applied. Let's assume that other assessor's qualitative preference relationship is given as (6).

$$\{a\} \text{ S } \{b\}, \{b,c\}>\{a\}, \{b, c\}<\{b\}, \{c\} \text{ S } \{d\}, \{a\}>0, \{d\}>0 \tag{6}$$

The bpa and m2 can be obtained by repeating step (1).

$$bpa \text{ of } m2 = \begin{bmatrix} m2(\{a\}) \\ m2(\{b\}) \\ m2(\{b,c\}) \\ m2(\{c\}) \\ m2(\{d\}) \end{bmatrix} = \begin{bmatrix} 1 \\ 1 \\ 2 \\ 2 \\ 2 \end{bmatrix} = \begin{bmatrix} 0.125 \\ 0.125 \\ 0.25 \\ 0.25 \\ 0.25 \end{bmatrix}$$

Applying the improved combination rule to the bpa of m1 and m2, the results of the bpa of m3 is shown blow. To evaluate other assessor's value, repeat step (1) and (2).

$$bpa \text{ of } m3 = \begin{bmatrix} m3(\{a\}) \\ m3(\{b\}) \\ m3(\{b,c\}) \\ m3(\{c\}) \\ m3(\{d\}) \end{bmatrix} = \begin{bmatrix} 0.18 \\ 0.15 \\ 0.16 \\ 0.33 \\ 0.18 \end{bmatrix}$$

(3) Computation of a Weight Value Using Bel and Pl

The values of Bel and Pl function can be computed by using bpa of m3.

$Bel(\{a\}) = m3(\{a\}) = 0.18$, $Bel(\{b\}) = m3(\{b\}) = 0.15$,
$Bel(\{c\}) = m3(\{c\}) = 0.33$, $Bel(\{d\}) = m3(\{d\}) = 0.18$
$Pl(\{a\}) = m3(\{a\}) = 0.18$, $Pl(\{b\}) = m3(\{b\}) + m3(\{b,c\}) = 0.15 + 0.16 = 0.31$,
$Pl(\{c\}) = m3(\{c\}) + m3(\{b,c\}) = 0.33 + 0.16 = 0.49$, $Pl(\{d\}) = m3(\{d\}) = 0.18$

The interval of focal elements using the values of Bel and Pl function is then

$$\{a\}=[0.18, 0.18], \{b\}=[0.15, 0.31], \{c\}=[0.33, 0.49], \{d\}=[0.18, 0.18]$$

In this case, the focal element {b} ranges from 0.15 to 0.31. This means that the belief interval of focal element {b} is between 0.15 and 0.31. Then the weight value of focal element is determined by averaging the values of belief interval. The final weight value is given below.

Weight value of {a}=0.18, Weight value of {b}=0.23,
Weight value of {c}=0.41, Weight value of {d}=0.18

Table 3. Result after applying the combination rule

Assessor	Preference relationship	Focal element	Bpa
A	{a}>{b}	{a}, {b}	{a}=0.7, {b}=0.3
B	{c}>{d}	{c}, {d}	{c}=0.7, {d}=0.3
C	{a}>{c}	{a}, {c}	{a}=0.7, {c}=0.3
Combination of A and B (normal combination)	-	-	-
Combination of A and B (improved combination)	-	a},{b},{c},{d}	{a}=0.392, {b}=0.108, {c}=0.392, {d}=0.108
Combination of A and C (normal combination)	-	{a}	{a}=0.49 (before normalization) {a}=1 (after normalization)
Combination of A and C (improved combination)	-	{a},{b},{c}	{a}=0.784, {b}=0.108, {c}=0.108

Table 4. Combination result of multi evaluation items

Assessor	Preference relationship	Focal element	Bpa
A	{a}>{b},{b}>{c}	{a},{b},{c}	{a}=0.5, {b}=0.3, {c}=0.2
B	{a}>{b,c}	{a},{b,c}	{a}=0.7, {b,c}=0.3
Combination of A and B (not exclusive combination)	-	{a},{b},{c},{b,c}	{a}=0.701, {b}=0.152, {c}=0.09, {b,c}=0.056

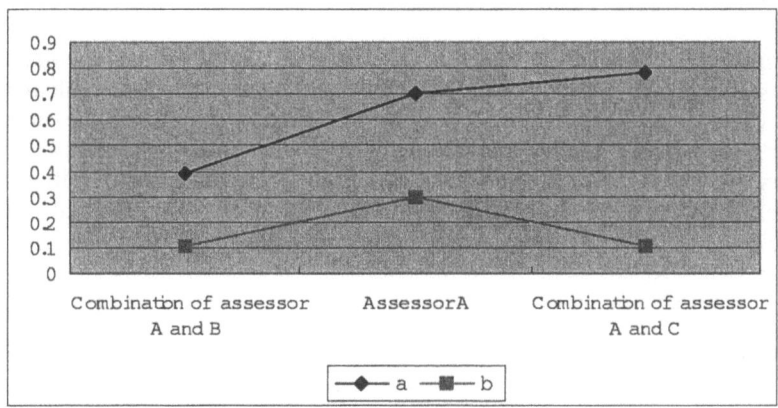

Fig. 1. Change of bpa on the focal element {a} using the improved combination

5 Evaluation Result and Analysis

5.1 Comparison of the Combination Rule

From the table 3, the results of normal combination of A and B are empty for any focal element because it is not possible to compute a weight value. However, the improved combination shows no empty set thanks to redistribution of empty set value.

5.2 Analysis of Improved Combination Rule

As shown in figure 1, the change of bpa on the focal element {a} using the improved combination is reduced from 0.7 to 0.392. This result indicates that assessor A assigns a high value to {a} but assessor B doesn't assigns any value to {a}. The change of bpa on the improved combination for A and C in table 3 is raised from 0.7 to 0.784. This result explains that both assessor A and assessor B assigns high value to {a}.

Two combination results on bpa with focal element {b} are same since the focal element {b} is not common on the two combinations. On the other hand, bpa of focal element {a} is 0.784 because focal element {a} is common on the combination of both assessor A and C. Table 4 illustrates that combination result of multi evaluation items.

The change of bpa on the combination A and B in table 4 is also raised from 0.5 to 0.701. Although assessor B had given preference relationship including multiple evaluation items such as {a} > {b,c}, we can obtain proper combination result.

6 Conclusion

The weight value calculation plays a key role in evaluation and selection of good quality software. This paper describes an quantitative method for calculating the weight value using DS theory. The proposed method eliminates the problem of assessor's subjective opinion and also improves the way of combining multiple assessors' opinion. The effectiveness of the new method has been verified with an example. The improved DS theory, however, still suffers from a great amount of numeric calculation ($O(n^2)$) since it uses every exclusive *bpa*s.

References

1. Software Quality Characteristics and Metrics - Part 1 : Quality Characteristics & Sub-Characteristics, TTAS.IS-9126.1, October, 1998.
2. Evaluation of Sotfware Product - Part 5 : Process for Evaluators, TTAS.IS-14598.5, October, 1998.
3. Hae Sool Yang, Ha Yong Lee : Design and Implementation of Quality Evaluation Toolkit in Design Phase, Korea Information Science Society Journal (C), Vol. 3, No. 3, pp. 262-274, June, 1997.

4. Hae Sool Yang, Yong Gn Lee : Design and Implementation of Quality Evaluation supporting Tool for Software Specification, Information Science Society Journal (C), Vol. 3, No. 2, pp. 152-163, April, 1997.
5. Hae Sool Yang, Kee Hyun Kwon, Ha Yong Lee, Young Sik Jo, Yong Gn Lee, Jong Ho Park, Tae Gyoung Heo : Design and Implementation of Software Quality Evaluation Toolkit, Korea Information Processing Society Journal, Vol. 2, No. 2, pp. 185-198, 1995.
6. Catherine K. Murphy : Combining belief function when evidence conflicts, Decision Support System 29, pp 1-9, 2000.
7. L.A. Zadeh : Review of Mathematical theory of evidence, by G Shafer, AI Magazine5(3), pp 81-83, 1984.
8. Joseph Giarratano, Gray Riley, "Expert Systems", PWS Publishing Company, 1994.
9. Kum Sook We : Design of an Expert System for Software Quality Evaluation with Easy Weighting of Quality Element, Graduate School of Dongguk University Computer Engineering, Doctoral Dissertation, December, 1995.
10. Kum Sook We, Keum Suk Lee : Design of an Expert System for Software Quality Evaluation, Information Science Society Journal (B), Vol. 22, No. 10, pp. 1434-1444, October, 1995.
11. Yong Tae Do, Il Gon Kim, Jong Wan Kim, Chang Hyun Park : Artificial Intelligence Concept and Application, SciTech Press, pp. 77-96, 2001.
12. Hee Seung Kim : Artificial Intelligence and Application, Saengnung Press, pp. 199-216, 1994.
13. Jin Hyung Kim, Seung Su Park, En Ok Paek, Jung Yun Seo, Il Byong Lee : Artificial Intelligence Theory and Practice, SciTech Press, pp. 373-384, 1998.
14. Wong, S.K.M. and P.Lingras, "Representation of Qualitative User Preference by Quantitative Belief Functions", IEEE Transaction on Knowledge and Data Engineering, Vol.6, No.1, 1994, pp. 72-78.
15. Gye Sung Lee : An Efficient Dempster-Shafer Evidence Combination Scheme for Uncertainty Handling, Korea Information Processing Society Journal, Vol. 3, No. 4, pp. 908-914, July, 1996.
16. Gye Sung Lee : An Approximate Evidence Combination Scheme for Increased Efficiency , Korea Information Processing Society Journal(B), Vol.9-B, No. 1, pp. 17-22, February, 2002.
17. Jong Moo Lee, Ho Won Jung : An Application of Qualitative Preference to Software Quality Evaluation, Korea Operation Research and Management Science Society Journal, Vol. 25, No. 3, pp. 109-124, September, 2000.

Forwarding Based Data Parallel Handoff for Real-Time QoS in Mobile IPv6 Networks*

Hye Yeon Jeong, Jongha Lim, Jung Deok Park, and Hyunseung Choo

Networking Lab., School of Information and Communication Engineering
Sungkyunkwan University 440-746, KOREA +82-31-290-7145
choo@ece.skku.ac.kr

Abstract. Many RSVP based handover schemes in MIPv4 were studied in the literature. However, the buffering overhead is unavoidable for the latency to optimize the route in new path establishment. Even though the data forwarding based schemes minimize the data loss and provide faster handoff, there are still some overheads when forwarding them and limitation on MIPv4. In this paper we propose a novel handoff scheme in MIPv6 based on forwarding which balances route traffic and reduces the overhead. The comprehensive performance evaluation shows that the disruption time and the signaling overhead are significantly reduced up to about 62% and 73%, respectively. Furthermore, it is able to transmit data with minimum latency and guarantee the fast and secure seamless services.

1   Introduction

RSVP based hand over schemes in Mobile IP ver. 4 (MIPv4) to provide faster services with QoS have been studied[2]. However, the buffering overhead is unavoidable during the latency to optimize the route in path establishment on the movement of the mobile node(MN). If the MN moves frequently in a certain area like cell border, *zigzag* makes the significant overhead. Besides, there are some disruption time needed in the new path establishment which is not preferable for the seamless services. There are many researches to reduce the MIP handoff latency in the literature[1-3].

We briefly discuss low latency handoff scheme with Neighbor Casting[2], the two path handoff scheme[2], and optimized handoff scheme[3] as related works. We observe the data loss caused by the disruption time and the bandwidth overhead due to RSVP channel. Even though the data forwarding based low latency handoff scheme[2] minimizes the data loss and provides the faster handoff between old foreign agent(old FA) and new one(new FA), there are some overheads when forwarding data. It also has a limitation due to based on MIPv4. The motivation of our works is based on them.

In this paper we propose a novel handoff scheme in MIPv6 for the faster and more reliable services. This scheme minimizes the possible latency occurred in

---

* This paper was supported in parts by Brain Korea 21 and University ITRC project. Dr. H. Choo is the corresponding author.

M. Bubak et al. (Eds.): ICCS 2004, LNCS 3038, pp. 266–273, 2004.
© Springer-Verlag Berlin Heidelberg 2004

the route optimization based on forwarding, load-balances in routers involved, and thus reduces the overhead. This scheme performs in parallel the data forwarding and the route optimization. The performance evaluation shows that the disruption time and the signaling overhead are significantly reduced up to about 62% and 73%, respectively. Furthermore, it is able to transmit data with minimum latency and guarantee the fast and secure seamless services.

2   Related Works

The Low Latency Scheme: To minimize the latency of handoff, the low latency scheme which uses the data forwarding between the new FA and the old FA is presented in [2]. The low latency scheme is built upon MIPv4 and data forwarding procedure. First, Neighbor FA discovery mechanism means how FAs discover their neighbor FAs dynamically and forward data in mobile networking environment. Each FA knows its neighbor FAs by MNs since the MNs move around and eventually from neighbor FAs. That is, the MN keeps the address of the old FA and transmits this information to the new FA whenever it hands over from a FA to another one. Each FA maintains a table for its neighbor agents. The MN sends the registration message which includes the information for the old FA to the new one. Then the new FA recognizes the old FA as its neighbor agent and sends a neighbor FA Notification Message to the old FA. Therefore, the old FA is informed of the new FA and updates its neighbor FA table. The handoff mechanism is based on the L2 handoff including Neighbor FA discovery. When the MN moves from the old FA to the new one, the old FA is notified from the new FA by Neighbor FA discovery mechanism. If CN sends the data, the old FA can start forwarding the data to the new FA. Therefore, the MN receives the data shortly after L2 handoff without establishing a new path between the old FA and the new one when it hands over.

The Two Path Scheme: In MIPv4, incoming packets from CN are received by the MN through the HA. As a result, it takes too much time to support real-time traffic flows with guaranteed QoS. However, the two path scheme is available to support large-scale mobility with faster handoff. This mechanism requires new procedures and signaling messages. When the MN moves into a new subnetwork, the MN's new CoA is sent to both the HA and the CN. And then, the CN which receives the MN's new CoA sends a PATH message to the MN to establish a new mobile path. Since the mobile path which is a direct RSVP path does not depend on forwarded packets from the HA, it is more efficient, however, incoming packets may be lost or arrived out of sequence.

Therefore, this scheme alternately uses the home path, i.e., from the CN through the HA to the MN. Since the MN maintains a connection with the HA, it is able to establish Home path whenever needed. When this approach uses the home path, it ensures correct and ordered forwarding of incoming packets, and reduces the packet loss.

Optimized Smooth Handoff Scheme: In [3], it proposes several methods to alleviate triangle routing and to reduce data loss. First, to reduce data loss during a handoff, it presents a buffering scheme at the FAs and smooth handoff scheme. The FA buffers any data packets forwarded to a MN. When a handoff occurs, the MN includes a handover request in its registration, and the new FA in turn requests the old FA handoff the buffered packets to the new location. To reduce duplicates, the MN buffers the identification and source address fields in the IP headers of packets. It receives and includes them in the buffer handoff request so that the old FA does not need to transmit those packets which have already been received by the MN. Also, Smooth handoff uses binding update to reduce packet loss during a handoff. Including FA buffering mechanism, it decapsulates the tunnelled packets and delivers them directly to the MN.

Second, hierarchical FA management reduces the administrative overhead of frequent local handoffs, using an extension of the Mobile IP registration process so security can be maintained. The FAs in a domain are organized into a hierarchical tree to handle local movements of the MNs within the domain. Meanwhile, when a packet for the MN arrives at its home network, the HA tunnels it to the root of the FA hierarchy. When an FA receives such a tunnelled packet, it retunnels the packet to its next lower-level FA. Finally the lowest-level FA delivers it directly to the MN. When a handoff occurs, the MN compares the new vector of CoA with the old one. It chooses the lowest-level FA that appears in both vectors, and sends a Regional Registration Request to that FA. Any higher-level agent need not be informed of this movement since the other end of it forwarding tunnel still points to the current location of the MH.

3   The Proposed Scheme

In the new forwarding based data parallel handoff technique, the CN has the option of using two different paths to reach the MN. The first path is called the R.new path, which is a direct path to any new location of the MN. Since the R.new path does not depend on the packet from HA, it is the more efficient and preferred path to use. However, while the MN is moving to a new location, incoming packet may be lost. Therefore we have to maintain the R.old path, i.e. from the CN to the R.old to the MN, to ensure uniform and fast forwarding of incoming packets, and to reduce the packet latency and the disruption.

The new technique requires a new procedure and signaling messages to coordinate the use of the paths. These are illustrated in figures 1, 2 and 3. We describe each operation along with the figure as shown below:

Step 1. Fig. 1 (1-1, 1-2): The MN located at R.old is communicating through an existing session with the CN. When the MN moves into a new subnetwork of R.new, the MN's new CoA, the handoff notification, is sent to both the R.new and the R.old. **Step 2.** Fig. 1 (2-1, 2-2, 2-3): When the R.old receives the MN's new CoA, it begins to forward the data packet to the R.new. And R.old's new path request message is sent to the CN. When R.new receives the MN's new CoA, R.new sends the CoA to the HA to register MN's new location (binding update). **Step 3.** Fig. 2 (3): When the CN receives the R.old's *new path request*, the CN sends a *new path establishment* message to the R.new to

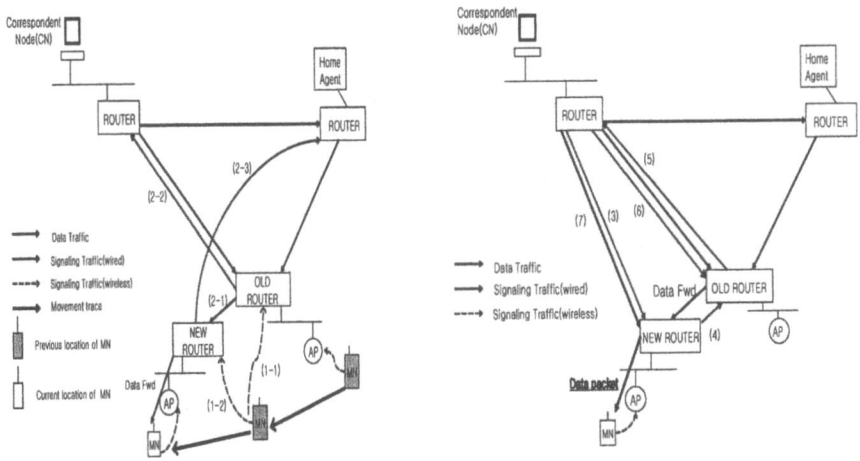

Fig. 1. Before route establishment   **Fig. 2.** After route establishment

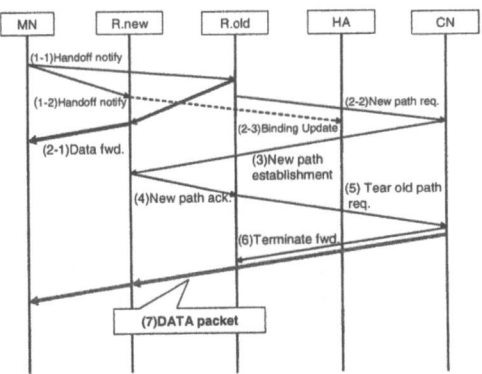

Fig. 3. Message flow of the proposed scheme

establish a new optimized path. **Step 4.** Fig. 2 (4): If the R.new path attempt is successful, the R.new sends *new path acknowledgement* message to R.old. **Step 5.** Fig. 2 (5): When the R.old receives acknowledgement message from R.new, R.old sends *the request for tear down of the old path* to CN. **Step 6.** Fig. 2 (6): When this happens, the CN stops sending data packets to the R.old and R.old explicitly tears down the forwarding path to R.new after it sends the final data packet to R.new. **Step 7.** Fig. 2 (7): When the CN switches to the R.new path from the R.old path, the new path can be used. The R.new buffers incoming packets for the MN from the CN until the forwarding path is destroyed.

In the design of this scheme, we apply the following approaches. *Mobile IPv6 compatibility:* Mobile IPv6 is the proposed by IETF, and this scheme is designed to be compatible with it. *Resource preallocation:* the handoff process requests a resource preallocation to guarantee that MNs obtain the requested resources right away after they enter the new cell/network. *Buffering:* To avoid data loss and unordered packets, the R.new is instructed to buffer packets for the incoming MN. *Forwarding to an adjacent router:* When a MN handoffs to adjacent router, its packets can be forwarded to the R.new by the R.old. *Less location update only with HA:* It is a frequent case that a MN moves to another networks or cells. This scheme does not make frequent HA updates when a handoff is performed. It contacts the HA for the update of the information for the care-of address. *Triangular message exchange:* It has to exchange messages for the smooth handoff. When a router sends a message to next router, it utilizes a next path resource to complete the message exchange.

This scheme introduces the disruption time while the R.old , R.new, and the CN exchange messages. Such problems are ignored for wired networks since wired links are much more reliable and the disruption time is short in duration. In wireless network, we assume that routers have MN's location table. They perform fast forwarding to the MN by sending *MN's location* packets each other. In the next section, we describe a framework to measure the disruption time and total overhead use for the new technique.

4   Performance Evaluation

Signaling Time: The time to send a message over one hop, M, can be calculated as $M = \alpha + \beta + \gamma$, where α is the transmission time, β is the propagation time, and γ is the processing time. The transmission time, α, is computed by b/B, where b is the size of the control message in bits, and B is the bit rate of the link on which the message is sent. B has two kinds of values which are for wired links and for wireless ones. The propagation time, β, has two kinds, too. Suppose that the processing time, γ, always has a fixed value when the message is sent. The signaling time which is the total time needed to transmit and process the messages between router i and j is computed by $T_{i,j} = T_{sig} \times D_{i,j}$, where T_{sig} is the signaling time over one hop and $D_{i,j}$ is the node distance between router i and j. T_{sig} is computed as following. Wired link($T_{RSVP} = M \times \frac{1+p}{1-p}$) and wireless link($T_{RSVP} = M \times \frac{1+q}{1-q}$) where q is the probability of wireless link failure. As mentioned, $D_{i,j}$ is the number of hops and computed by $D_{i,j} = N_{R_i,R_j} = N_{R_i,R_{join}} + N_{R_{join},R_j}$. N_{R_i,R_j} is the number of hops from R_i to R_j and R_{join} is the router placed at MAP(merged access point). The number of hops over routers on which the signaling passes are calculated by summing these two values. The RSVP signaling time is computed by $T_{i,j} = T_{RSVP} \times D_{i,j}$, where T_{RSVP} is the time to perform RSVP signaling over one hop. Wired link($T_{RSVP} = M \times \frac{1+p}{1-p}$) and wireless link($T_{RSVP} = M \times \frac{1+q}{1-q}$), here p is the probability of resource assignment failure on RSVP. When the resource failure occurs, the RSVP signaling is assumed to be repeated until either a session is successfully established, or the MN moves to a new router location.

Disruption Time and Total Signaling cost: The disruption time is to measure the time that a node cannot get packets. The shorter this time, the

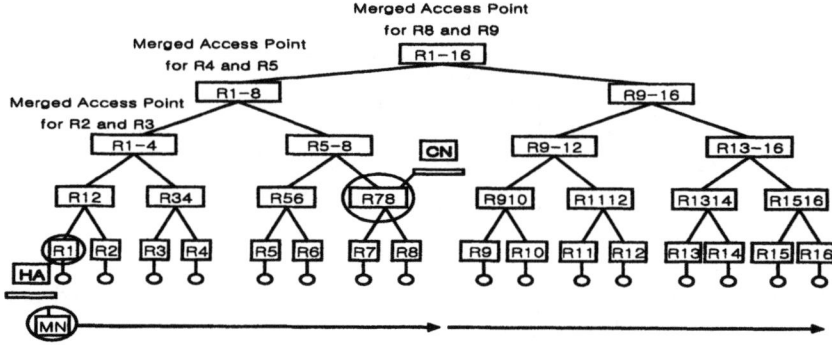

Fig. 4. Hierarchical router configuration

better the performance. The proposed scheme does not get data when the MN receives forwarded data through New.FA from Old.FA while data forwarding occurs. This time is computed by the previous two equations($T_{i,j}$). Namely, if the disruption time is short, we can say that data flow is continuously and performance is good.

Signaling overhead is due to sending and receiving signaling from neighbor FA and increases in proportion to the time sending signaling. It is proportional to generated signal in one cell. Then if signaling overhead is larger, the resource use rate becomes larger. Accordingly small signaling overhead is more efficient. Consider the case that each signal source generates UDP packets at constant signal rate c continuously. Then the amount of generated signal during a handoff of mobile node is $n \cdot \tau \cdot c$ bytes. n is the number of neighboring FAs which MN sends signal. (Suppose that each cell is hexagonal form.) In our paper signal is transferred one cell which MN moves, then n is 1. τ is the total signaling period during which all signaling and forwarding processes finish for receiving datas by the new path over one handoff process. The time which takes the parallel process is included in total signaling period. This is because we need to calculate all signaling time for the entire signaling process. The total amount of generated signal from a cell for a unit time period is $n \cdot \tau \cdot c \cdot \lambda$ where λ is the handoff rate. And the total amount of signal sent by the sources to the MNs in the cell for the same unit time period is $m \cdot c$ where m is the total number of active MNs in the cell. The handoff rate, λ, is computed by $\lambda = \frac{\rho v L}{\pi}$ where λ is the handoff rate or cell boundary crossing rate(1/sec), ρ is the active MN density in $1m^2 (1/m^2)$, v is the MN moving speed(m/sec), and L is the cell perimeter(m). The signaling overhead ratio is defined as the number of bytes generated from a cell divided by total number of bytes sent by the source, i.e. $\xi = \frac{n\tau c \lambda}{mc} = \frac{n\lambda\tau}{m} = \frac{\rho v L n \tau}{\pi m}$. Denoting r as the cell radius for a hexagonal cell scenario and $L = 6r$. ρ is computed by $\rho = \frac{m}{cellarea} = \frac{m}{\frac{3\sqrt{3}}{2}r^2}$. λ is computed based on previous two equations(λ, ρ), and we get the following. $\lambda = \frac{\rho v L}{\pi} = \frac{4mv}{\sqrt{3}r\pi}$ We can substitute this values to ξ, then get the result below. $\xi = \frac{4vm\tau}{\sqrt{3}r\pi}$ where τ is the total signaling pe-

Table 1. Performance parameters

SystemParameters	Values	Packet size, S	50Bytes
Wireless propagation time, $\beta 1$	2msec	Probability of link failure, q	0.5
Wired propagation time, $\beta 2$	0.5msec	Probability of resource denial, p	0.5
Signaling processing time, γ	0.05msec	Wireless bit rate, B1	144Kbps
RSVP processing time, γ_{RSVP}	0.5msec	Wired bit rate, B2	155Mbps

riod. This is the sum of the time taken for signaling processes included parallel signaling process based on Figure 3. $\tau 1 = time\_handoff\_notify(step1 - 1) + time\_path\_req(step2 - 2) + time\_path\_establish(step3) + time\_path\_ack(step4) + time\_stop\_transfer(step5) + time\_stop\_fwd(step6)$ where we disregard handoff latency for handoff process. $\tau 1$ is the time which takes fundamental signaling process. With mentioned above the MN notifies handoff to R.new, then R.new begins binding updates. This parallel processes are generated signaling in handoff. So it must be contained in signaling time. $\tau 2 = time\_handoff\_notify(step1 - 2) + time\_binding\_update(step2 - 3)$. Total signaling time is the sum of above two time periods. $\tau = \tau 1 + \tau 2$. This is computed based on the first two equations$(T_{i,j})$. We use 200m-radius hexagonal cell topology and assume that the mobile speed is fixed at 5m/sec.

Numerical Results: Figure 5 shows the comparison of the proposed scheme with the related ones[1, 2] with respect to the disruption time based on Table 1 and Figure 4. This results are computed based on the first two equations$(T_{i,j})$. As Figure 5, disruption time for the two path scheme is greater than other schemes applying forwarding. This is due to the character of forwarding method which it is possible to get data while the new path is established. The proposed scheme has better performance than the low latency scheme using forwarding. For each MN location, the disruption time for the proposed scheme is in average 53% and 3% smaller than [1] and [2], respectively.

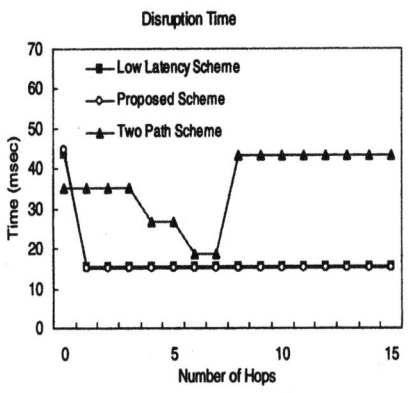

Fig. 5. The disruption time

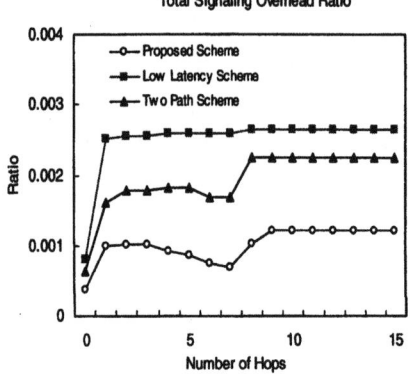

Fig. 6. Total signaling overhead

Figure 6 is the graph which shows the comparison of the proposed scheme with [1, 2] with respect to the total signaling overhead. As in Figure 6, the proposed scheme has the best performance compared with the previous schemes. The signaling overhead of the low latency scheme based on MIPv4 is computed only for the signaling related to the forwarding, hence it has a fixed value. However the proposed scheme and the two path scheme are affected by the time for path establishment from CN to R.New, so signaling traffic is changed according to the model. Therefore results of two schemes have similar trends. In Figure 6, total signaling overhead for the proposed scheme is in average 48% and 59% smaller than the two path scheme and the low latency scheme, respectively.

5   Conclusion

In this paper, we propose the new handoff scheme with guaranteed QoS which data transmission is faster and data loss is smaller than previous ones. Basically it receives data fast using data forwarding and makes the new path from the corresponding node to the new router while forwarding data. The proposed scheme is efficient in terms of the disruption time and total signaling overhead as discussed in performance evaluation and offers services of more improved quality in mobile communication environment which the real-time transmission is very important.

References

1. J. McNair, I.F. Akyldiz, and M.D. Bender, "Handoffs for Real-Time Traffic in Mobile IP Version 6 Networks," IEEE GLOBECOM, vol. 6, pp. 3463-3467, 2001.
2. E. Shin, H.-Y. Wei, Y. Chang, and R.D. Gitlin, "Low Latency Handoff for Wireless IP QOS with Neighbor Casting," IEEE ICC, vol.5, pp. 3245-3249, 2002.
3. C.E. Perkins and K.-Y. Wang, "Optimized Smooth Handoffs in Mobile IP," IEEE ISCC, pp. 340-346, 1999.
4. C.E. Perkins, "Mobile networking in the Internet," ACM/Baltzer Mobile Networks and Applications (MONET) Journal, vol. 3, pp. 319-334, 1998.
5. J. McNair, I.F. Akyldiz, and M.D. Bender, "An Inter-System Handoff Technique for the IMT-2000 System," IEEE INFOCOM, vol. 1, pp. 208-216, 2000.
6. I.F. Akyldiz, J. McNair, J.S.M. Ho, H. Uzunalioglu, and W. Wang, "Mobility management in next generation wireless systems," IEEE, vol. 87, no. 8, pp. 1347-1384, 1999.
7. I. Okajima, N. Umeda, and Y. Yamao, "Architecture and mobile IPv6 extensions supporting mobile networks in mobile communications," IEEE VTC, vol. 4, pp. 2533-2537, 2001.
8. W.-T. Chen and L.-C. Huang, "RSVP mobility support: a signaling protocol for integrated services Internet with mobile hosts," IEEE INFOCOM, vol. 3, pp. 1283-1292, 2000.

Mobile Agent-Based Load Monitoring System for the Safety Web Server Environment

H.J. Park[1], K.J. Jyung[2], and S.S. Kim[3]

[1] School of Computer Information and Communication Engineering,
Sangji University, Woosandong, Wonjusi, Kangwondo, Korea
hjpakl@sangji.ac.kr
[2] System Technology Laboratory, Dacom Corporation, Seoul, Korea
jkjin@origio.net
[3] School of Information and Communication Engineering,
Halla University, San 66, Heungup-Li, Heungup-myon, Wonjusi, Kangwondo, Korea
sskim@halla.ac.kr

Abstract. The importance of the server is growing with the growth of the internet and the load on the server has rapidly grown with the increase in the number of web server user. Load monitoring has become an important technical aspect in the performance of the server for these reasons. It has proven to be more efficient to use a java mobile agent that operates by moving between the systems for load monitoring of a web server rather than using the existing client-server paradigm. A mobile agent moves around the web servers to measure the load and sends the load information to a single node.

1 Introduction

The importance of the web server that provides web service to users all around the world has grown rapidly with the growth of the internet. Mirroring method has also been applied on the web servers for load balancing but has proven inefficient due to the difficulty in managing the performance of the web servers in remote location. Load monitoring technique on many distributed web servers is an important aspect in the management of web server performance.

A web server must always provide the user with the appropriate service and speed of the service is the most important aspect. If the load on the web server increases by the increase in the number of user, size of data, number of process or the use of memory, the server would not be able to provide the user with the appropriate service and would cause a waste of money put into developing the web server[1]. The load monitoring system must continuously check if the web server is operating properly, and must apply the optimization process to restrain the number of user or size of data for the server to operate efficiently. Accurate load information can be applied to efficiently distribute the load on a server to another mirroring web server and the web server monitoring technique can be applied to promptly detect and resolve an error that has occurred within the system. The system administration condition can be predicted based on the historical data stored in the DB and this data can also be used as the base information for analysing and expanding the system.

M. Bubak et al. (Eds.): ICCS 2004, LNCS 3038, pp. 274–280, 2004.
© Springer-Verlag Berlin Heidelberg 2004

This paper develops a load monitoring system using a java mobile agent that can efficiently monitor the load on the web server and resolve the problems that occurred on the systems based on the client-server paradigm.

This paper consists of the following parts. Chapter 2 deals with the related research, chapter 3 with the design and implementation of the monitoring system using a mobile agent and chapter 4 concludes the paper by analyzing the implemented system.

2   Related Research

2.1   Load Monitoring in the Client-Server Paradigm

The client-server paradigm has been the base for most web-server load monitoring systems. The client-server paradigm is a centralized method in the distributed computing environment and must install a program that measures the load of the web server on each server and send the load information on a demand from the client[2]. Therefore, it is impossible for a user to change the monitoring method on a single node and to manage web servers that are remotely located. Also, the monitoring program being operated on the web server may become a load on the server providing the web service and the message shared between the servers increases the network traffic. A network connection must be maintained between the client and the web server and therefore becomes hard for the web server to adapt to the change in the environment[3][4]. The mobile agent is applied to resolve such problems of the client-server program by gathering information by moving through the servers.

2.2   Java Mobile Agent

The java mobile agent is a program that can be operated by moving itself to another host on the network. The mobile agent can transfer its state information and code from its initial environment to the system it is to be operated on. The state information is the attribute value used to determine the job to be processed by the agent and the code is a class code needed for the agent's operation. The mobile agent can operate efficiently in an environment where systems connected through the network have a low bandwidth and high delay time. The agents can also share their resource through communicating with each other and returns to the system it was created or self-destructs after the finishing its job. The use of a mobile agent provides a easier and more efficient programming paradigm than the client-server paradigm in constructing an application in the distributed computing environment[5].

The agent takes an efficient approach of moving around the system through a connected network for gathering data and utilizing resource instead of requesting other systems and waiting for a reply. Network connection is only required when the mobile agent is actually moving through the system and the user does not need to take the location of the operation taking place or the communication channel into account. This enables the development , testing and locating of the distributed application to be operated more efficiently (Fig. 1).

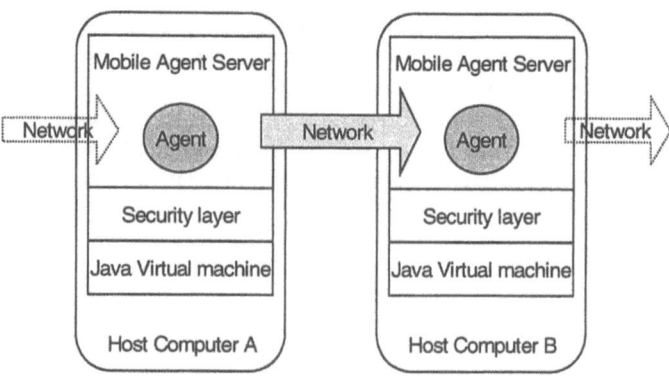

Fig. 1. Java mobile agent

2.3   Advantage of the Java Mobile Agent

Following are the advantages of using a Java mobile agent[6].
- Reduced load on the network: The network traffic increases on a distributed system environment where a lot of interaction takes place between the systems, whereas a mobile agent only requires a small amount of data to be transferred within the system.
- Resolve Network Delay: Network delay occurs on a real-time system on a process through the network. However, a mobile agent can directly process the job issued by the central manager by moving itself.
- Asynchronous and Free operation: A mobile agent only requires a network connection on a movement and operates freely and asynchronously within the system it has moved to.
- Protocol Encapsulation: An I/O related protocol must be formed to exchange data in a distributed system environment, whereas a mobile agent uses an independent protocol to move to a remote host.
- Dynamic Adaptation: Detect the operating environment and automatically respond to the change.
- Heterogeneous: Network computing through the internet contains a heterogeneous nature.
- Fault Tolerance: Able to construct a fault-tolerant distributed system even when a problem has occurred.

3   Web Server Load Monitoring System Using a Mobile Agent

This chapter discuss about the advantages of monitoring using a mobile agent and the design and implementation of a web server load monitoring system.

3.1 Mobile Agent Based Monitoring

Using a mobile agent with the above characteristics for monitoring the load on a web server is far more efficient than implementing the existing client-server paradigm.

All the load data can be gathered into a single system without installing a program on each web server to be monitored by using a mobile agent. The execution code of the mobile agent moves to a web server through the network to internally measure the load and sends the result to the monitoring server before moving onto another web server. The mobile agent only contains the code that is crucial to measuring the load to become a lightweight program and requires only a small amount of data upon a network transaction. It also puts only a little extra load on the web server by only taking up little room for execution upon a movement. A large-scale log file is required for measuring the load on the web server. However, by using a mobile agent, the agent will carry the log file itself instead of transmitting it through the network for a quicker analysis of the load data.

The mobile agent is an independent protocol that moves through the ATP(Agent Transfer Protocol). This enables the operation of the agent to proceed regardless of the physical location of the web server to be monitored and is not restricted by the number of web servers. The load of mirroring web servers that are physically distributed can be monitored from a single node and the resulting data can be utilized for the load balancing technique of a mirroring web server. Therefore, it can be concluded that the monitoring method using a mobile agent is more efficient than the existing client-server program.

3.2 Design

A distributed computing method using a mobile agent is required for an efficient load monitoring on web servers that are remotely located on the internet. The system consists of web servers to be monitored, and a monitoring system used to monitor the load information of the web servers. Each system contains a MAS (Mobile Agent Server). MAS (Mobile Agent Server) provides each host with an environment for the mobile agent to be executed and manages the creation, movement and extinction of the mobile agent. The agent is designed by combining the itinerary pattern and the master-slave pattern related to the operation. The Master agent is executed on the MAS of the monitoring system, and creates the Slave agent that will move to the web server. The Slave agent moves to the first web server to analyze the log file, measure the load on the system and sends the load information to the Master agent. The Slave agent then moves to another agent to perform the same job after transmitting the load information to the Master slave.

The master agent resides in the monitoring system to provide the GUI function, display the web server's load information gathered from the Slave agents to the users and periodically create a new Slave agent. These functions of the mobile agent are shown in Fig. 2.

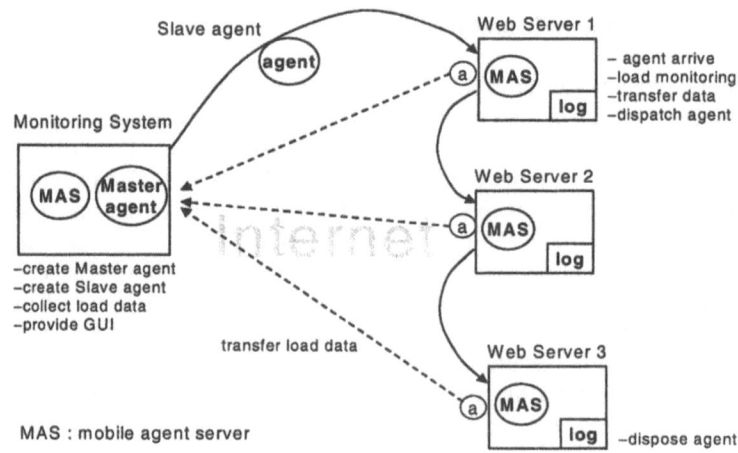

Fig. 2. Function of master and slave agent using itinerary pattern

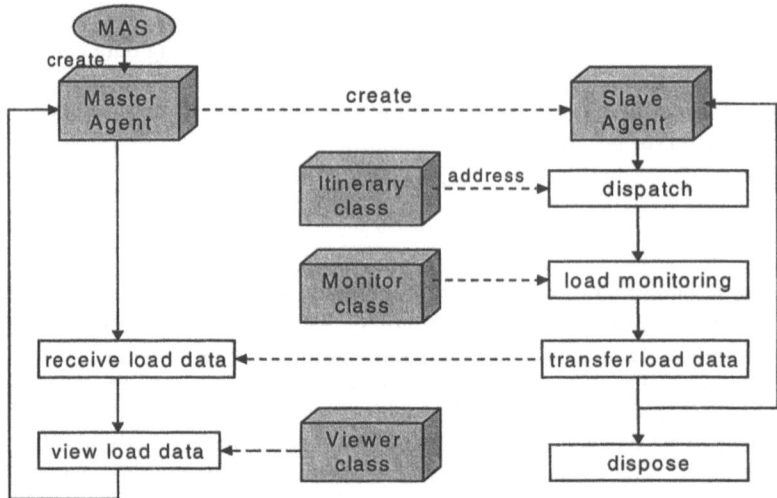

Fig. 3. Classes of load monitoring system

3.3   Implementation

The web server monitoring system using a java mobile agent consists of a master agent and a slave agent. The master agent displays the load information obtained from the slave agent through the viewer class(GUI function) and periodically creates a new slave agent for monitoring. The slave agent created by the master agent moves itself to the web server. The address of the web server the slave agent is to move to is provided by the itinerary class. The itinerary class contains the address of the web servers to be monitored and is transmitted upon the creation of the slave agent by the master agent. The itinerary class delivers the agent to the next address and provides the function for moving on to the following server if a problem occurs at the web server (Fig. 3).

The slave agent that moved to the web server analyses the log data of the web server and measures the load on the system itself. It moves to another web server after transmitting all the load data to the Master slave, and self-terminates if there is no other job left to process. The monitor class measures the number of current users, current transmission rate, number of currently occurred errors, average number of jobs processed in 1·5·15 minutes interval, size of the swap memory, size of free memory, rate of disc usage, amount of memory left in the disc and the amount of memory left to be used in the disc. The load information of the web server is displayed in the monitoring system as shown in Fig. 4.

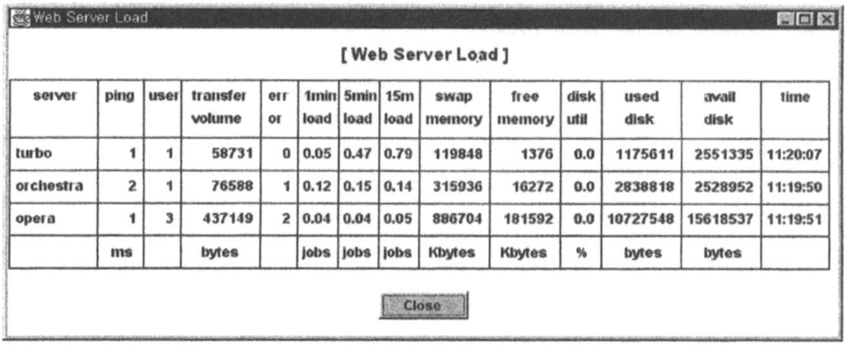

server	ping	user	transfer volume	err or	1min load	5min load	15m load	swap memory	free memory	disk util	used disk	avail disk	time
turbo	1	1	58731	0	0.05	0.47	0.79	119848	1376	0.0	1175611	2551335	11:20:07
orchestra	2	1	76588	1	0.12	0.15	0.14	315936	16272	0.0	2838818	2528952	11:19:50
opera	1	3	437149	2	0.04	0.04	0.05	886704	181592	0.0	10727548	15618537	11:19:51
	ms		bytes		jobs	jobs	jobs	Kbytes	Kbytes	%	bytes	bytes	

Fig. 4. Load information display of web servers

4 Conclusion

The size of the program operated in each web server is as follows. The program in the Slave agent takes up 3.2KB, in the Monitor class takes up 4.3KB, and the program in the Itinerary class takes up 2KB which adds up to a total size of 9.5KB. 9.5KB of class is executed during the measurement and is destroyed after the process which is the main reason this system fares better than the client-server program in which memory space is taken up the whole time. The whole class is only transmitted once upon the first agent movement and only the Slave agent is transmitted after the first transmission which results in 9.5KB of data only being transmitted once and only 3.2KB of data being transmitted after the initial transmission. This reduces the network traffic.

The purpose of this paper is on effectively monitoring the load on the web servers physically distributed on the internet using a java mobile agent. By using the java mobile agent, it becomes possible to monitor more than one web servers from a single node without installing any other program and enables the user to modify the monitoring method. It provides a new mechanism that was not provided in the existing client-server paradigm.

References

1. Web Server Monitoring, `http://www.freshtech.com`
2. M. Caprini, R. Nacach, Z. Qian, "Java Mobile Agent for monitoring task : evaluation report",
   `http://atddoc.cern.ch/Atlas/Notes/078/Note078-1.html`, 1998
3. MARS(Monitoring Application for Resources and Servers),
   `http://www.altara.org/mars.html`
4. E. Anderson, D. Patterson, "Extensible, Scalable Monitoring for Clusters of Computerts", Proceedings of 1997 LISA Conference, 1997
5. R. Gray, D. Kotz, S. Nog, D. Rus, G. Cybenko, "Mobile agents for mobile computing", Technical Report PCS-TR96-285, 1996
6. D. B. Lange, Mitsuru Oshima, Programming and Deploying Java Mobile Agents with Aglets, Addison Wesley press, 1998

A Study on TCP Buffer Management Algorithm for Improvement of Network Performance in Grid Environment

Yonghwan Jeong[1], Minki Noh[2], Hyewon K. Lee[1], and Youngsong Mun[1]

[1] School of Computing, Soongsil University
1-1, Sando 5Dong, Dongjak-Gu, Seoul, 156-743, Korea (South)
{paul7931,kerenlee}@sunny.ssu.ac.kr, mun@computing.ssu.ac.kr
[2] Korea Institute of Science and Technology Information (KISTI)
Eoeun-dong 52, Yuseong-gu, Daejeon city, Korea (South)
mknoh@kisti.re.kr

Abstract. The Grid is the environment that connects high performance computing resource, which are scattered geographically as related network. The Grid, which started in the mid of 1990, has studied and laid across in BT, NT, and ET fields. The Grid applications are developed to act as global infrastructure, especially, linked as high performance network. Nevertheless, the Grid network envionment are consists of high speed researches, it uses network management method of former old internet envionment, and it cannot take the full advantage of high performance network. This research suggests TCP buffer control mechanism that is more appropriate in high performance network for better performance in The Grid network. In addition, controlled method analyzes the network performance using the Globus Toolkit 3.0, which is the most recent Grid middleware.

1 Introduction

The Grid is the environment that connects high performance computing resource, which are scattered geographically as related network. Grid, which started in the mid of 1990, has studied and laid across in BT, NT, and ET fields. The Grid computing has concepts that it makes geographycally distributed and unused high perfomance computing resourcese into available things. Its resource sharing and cooperation is accomplished by Grid network. American Abilene and vBNS, European TEN-155, SINET of Japan and KREONet of Korea achieve Grid network's functions. These networks affect high performance network's form.

In order to guarantee the network QoS of Grid application in high performance network, Grid environment provides GARA (General Purpose Architecture for Reservation). GARA provide uniform API to various types of resources to network QoS. Also, GARA adopts Differentiated Service (DiffServ) infrastructure in IETF. DiffServ guarantees Grid application's end-to-end network QoS by establishing ToS field in IP header. Nevertheless, the Grid network envionment are consists of high speed researches, Which uses network management method of former (old) internet envionment, and it cannot take the full advantage of high performance network.

M. Bubak et al. (Eds.): ICCS 2004, LNCS 3038, pp. 281–288, 2004.
© Springer-Verlag Berlin Heidelberg 2004

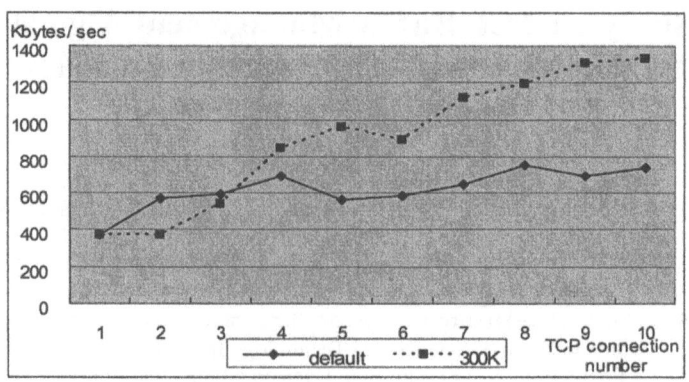

Fig. 1. TCP transmission rate variation according to number of connection

Fig. 1. shows the data transmission amount by the increasing number of TCP connection in case of configuring TCP buffer size to default size, 64 Kbytes or default size, 300Kbytes. Each case is represented as a solid line and dotted line respectively. According to Fig. 1., TCP buffer size is necessary to be modified to adopt the state of network in the high performance network such as Grid environment. As a result the data transmission rates can be improved.

This research suggests that TCP buffer control mechanism is more appropriate in high performance network for better Grid network performance. In addition, controlled method analyzes the network performance using the Globus Toolkit 3.0, which is the most recent Grid middleware.

2   Related Works

2.1   GridFTP

In Grid environment, access to distributed data is typically as important as access to distributed computational resources. Distributed scientific and engineering applications require two factors. The first factor is to transfer of large amounts of data, such as in terabytes or petabytes length data between storages systems. And the second factor is to access to large amounts of data, such as in gigabytes or terabytes data by many geographically distributed applications and users for analysis and visualizations. The lack of standard protocols for transfer and access of data in Grid has led to a fragmented Grid storage community. Users who wish to access different storage systems are forced to use multiple APIs and protocols. The performance of data transmission between these different storage systems shows a drop in efficiency; drop off in efficiency.

GridFTP, which has functionality of command data transfer and access protocol, provides secure and efficient data transmission in Grid environment. This protocol, which extended the standard FTP protocol, provides a superset of the features offered by the various Grid storage systems currently in use. The GridFTP protocol includes following features:

- Grid Security Infrastructure (GSI) and Kerberos support
- Third-party control of data transfer
- Parallel data transfer
- Support for reliable and restartable data transfer
- Integrated instrumentation

2.2 TCP Buffer Tuning

Buffer Tuning Overview. In order to decide how many packets can be sent, TCP uses "cwnd(congestion window)" parameter. The larger window size is, it is more throughputs at once. Both TCP slow start mechanism and congestion avoidance algorithm decide congestion window size. The maximum window size has relation with the buffer size of each socket. The buffer size of each socket is assigned by Kernel and has default value; this value can be modified by application before socket establishment. The application can be one of the programs using system library call. The Kernel forcese to use maximum buffer size. The buffer size can be modified by both sender and receiver.

In order to get maximum throughput, it is important to use adequate TCP sender/receiver buffer size for link. If buffer size is too small, the TCP congestion window will not open enough. On the contrary, if buffer size is too large, the TCP congestion window will close, and receiver's buffer will be overflow. The same results will be happen if sender is faster than receiver. However, whether sender's window is too large size doesn't affair in the case of enough memory. (1) means adequate buffer size.

$$\text{Buffer size} = 2 * \text{bandwidth} * \text{delay} \qquad (1)$$

$$\text{Buffer size} = \text{bandwidth} * \text{RTT} \qquad (2)$$

"ping" application is used to get delay value, and "pipechar" or "pchar" is used to get the last link bandwidth. (2) is identical to the (1) because RTT (round trip time) is obtained by "ping" application.

If ping time is 50ms on end-to-end network, which is composed of 100BT and OC3 (155Mbps), adequate TCP buffer size is 0.05sec * (100Mbits / 8bits) = 625 Kbytes for this connection.

There are two things to notice. They are default TCP sender/receiver buffer size, and maximum TCP sender/receiver buffer size. For the most UNIX OS, default maximum TCP buffer size is 256KB. Table 1 shows default maximum buffer size and default TCP socket buffer size at various OSs.

Table 1. Buffer size comparison at various to OSs

Type of Operating System	Default max socket buffer size	Default TCP socket buffer size
FreeBSD 2.1.5	256 Kbytes	16 Kbytes
Linux 2.4.00	64 Kbytes	32 Kbytes
Sun Solaris 7	256 Kbytes	8 Kbytes
MS Win2000 or Win XP	1 Gigabyte	8 Kbytes

Buffer Share Algorithm. At the first stage of TCP Auto-tuning implementation, each connection decides expected socket buffer size by increasing window size appropriately. Memory pool is used to assign resources to each connection. The connection that requires smaller buffer size than "Fair share" reserves expected buffer size. The remaining memory is assigned to connections that require larger size fairly. In order to assign buffer size at next negotiation, this fair share algorithm is configured as "current share."

3 TCP Buffer Management Algorithms

3.1 Buffer_size_negotiator

This research is focus on the Access Grid and the Data Grid, and both they require large data transmission such as gigabytes or terabytes level transmission in Grid application. Grid network affects high speed/high performance network's form, so TCP connection management algorithm is required, which is differentiated with large data transmission method of general internet environment using TCP. In this paper, If large TCP data transmission is generated by Grid application, analysis the characteristic of each TCP connection and compose module that can improve performance by automatic reconfiguration of The TCP buffer size, which appropriate for characteristic of The connection.

Fig. 2. shows buffer_size_negotiator's function on option. Traffic Analyzer achieves function that uses system commands ("netstat" or "lsof") or system information ("/proc" in system) to get information about each TCP connection. These information are entered into "buffer_size_negotiator" by "Information Provider" in Globus and then negotiate the send buffer size of each TCP connections according to setting option. Actually, GridFTP that takes charge Grid application's data transmission applies the negotiated buffer size to each TCP connections. Buffer_size_negotiator has four options as follows;

– default: none
– Max-min fair share: –m
– Weighted fair share: –w
– Minimum threshold: –t

3.2 Buffer_size_negotiator

This section explains buffer management algorithm that corresponds to buffer_size_negotiator's each option. This buffer_size_negociator accepts information form the MDS, which is resource information provider. And then, it negociate each TCP connections' receiving buffer size. The following subsections are the buffer management algorithm provided by buffer_size_negotiator.

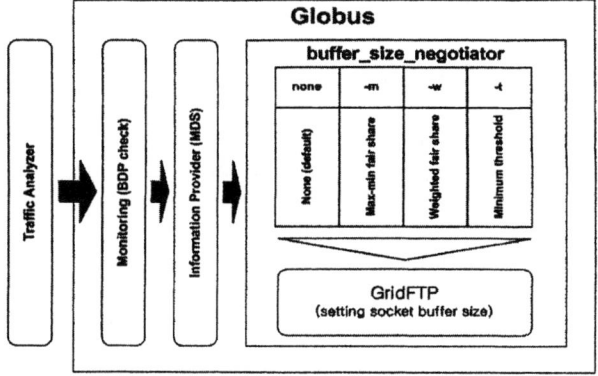

Fig. 2. Buffer_size_negotiator's options

Max-min Fair Share Algorithm. Max-min fair share algorithm is a method that equally establishes re-assignable buffer size to each TCP connection. Table 2 shows the results of the negotiated buffer size using max-min fair share algorithm.

Table 2. Buffer allocation result using –m option

Unit : Kbytes

P	N		Connection A	Connection B	Connection C
1024	3	D(i)	64	1024	768
		B(i)	64	480	480

Weighted Fair Share Algorithm. Weighted fair share algorithm is a method that establishes re-assignable buffer size to be proportional to the requested buffer size of TCP connections. Table 3 shows results of the negotiated buffer size using weighted fair share algorithm.

Table 3. Buffer allocation result using –w option

Unit : Kbytes

P	N		Connection A	Connection B	Connection C
1024	3	D(i)	64	1024	768
		B(i)	64	500	460

Minimum Threshold Algorithm. If one TCP connection's the requested buffer size is smaller than 5% of the requested buffer size's sum of all TCP connections, minimum threshold algorithm guarantees the connection's buffer size to be 5% of all available buffer size (P). Table 4 shows results of the negotiated buffer size using minimum threshold algorithm.

Table 4. Buffer allocation result using –m option

Unit : Kbytes

P	N		Connection A	Connection B	Connection C
1024	3	D(i)	64	1024	768
		B(i)	53	520	451

4 Tests and Analyses

4.1 GridFTP Data Transmission Tests

In this paragraph, the large data transmitting tests using GridFTP are experimented with executed buffer_size_negotiator on each option. We measured the amount of data transmission per second for 3 TCP connections that demand different bandwidth during.

Fig. 3. shows data transmitted amount which measured by GridFTP per second when buffer_size_negotiator configures each option (default, -m, -w and –t).

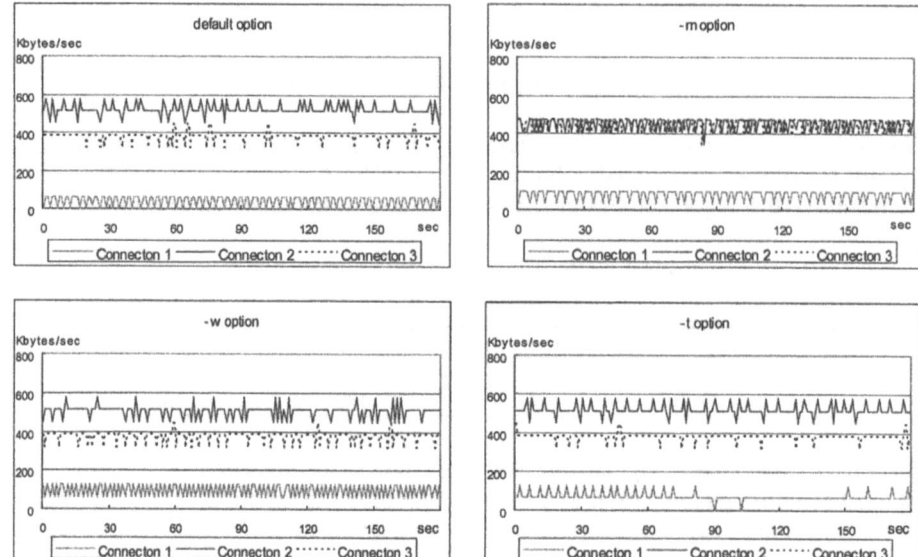

Fig. 3. Results of data transmission test using each option

4.2 Result Analysis

Negotiated Buffer size Comparative Analysis. Through experiment result of 4.1, the requested buffer size of connections and the negotiated buffer size by each buffer management's algorithm are presented at Tab. 5.

Table 5. Comparison of negotiated buffer size using each option

Unit : Kbytes

Option / Conn. name	Requested size	-m option Negotiated size	Rate of increase	-w option Negotiated size	Rate of increase	-t option Negotiated size	Rate of increase
Conn. 1	64	64	0 %	64	0 %	53	-17 %
Conn. 2	1024	480	-53 %	500	-51 %	520	-49 %
Conn. 3	768	480	-37 %	460	-40 %	451	-41 %

The following express three kinds of connection features are:

– Conn. 1: TCP connection that requires the small buffer size
– Conn. 2: TCP connection that requires overbalance of maximum buffer size
– Conn. 3: TCP connection that does not exceed maximum buffer size

The buffer size entered into an agreement is established within maximum buffer size when tested TCP connections on each option. When use "-m" and "-w" option, "Conn. 1" allocates the requested buffer size to the negotiated buffer size. However, in case of using "-t" option, it setablished by 53Kbytes, decreased 17% than the requested buffer size of original.

"Connection 2" and "Connection 3" shows that the negotiated buffer size for all options is established lower than the requested buffer size. There is advantage that "-w" and "-t" options could assign more buffer size for connections that could send large data per second than "-m" option. Also, in case of "-t" option, as sum of the requested buffer size (S) is changed. There is advantage that the negotiated buffer size of the connection which requires small buffer size, could be dynamically changed.

Data Transmission Rate Comparative Analysis. The data transmission amount by GridFTP is same with Table 6 in each option.

Table 6. Comparison of data transmission amount using each option

Unit : Kbytes

Option / Conn. name	None	-m Transmission amount	-m Rate of increase	-w Transmission amount	-w Rate of increase	-t Transmission amount	-t Rate of increase
Conn. 1	7,872	14,400	+82.9%	17,664	+124.4%	12,864	+63.4%
Conn. 2	93,760	80,673	-14.0%	90,688	-3.3%	93,440	-0.3%
Conn. 3	67,712	80,736	+19.2%	66,752	-1.4%	68,224	+0.8%
Sum	169,344	175,808	+ 3.8%	175,104	+3.4%	174,528	+3.1%

Such as table 6, if established send buffer size of each TCP connections in GridFTP to the negotiated buffer size by buffer_size_negotiator buffer size, data transfer rate increases in data transmission. Could heighten transfer efficiency of about 3.8% when used "-m" option, and case of 3.4% when used "-w" options and improved performance of 3.1% when used "-t" option

5 Conclusions

In the traditional Internet environment, there was no change in transmission amount of data even on the modified TCP buffer size in case of transmitting large data. On the other hand, the high-performance networks, such as STAR-TAP, KREONET, Grid networks make a profit on the change of TCP buffer size according to network environment. Therefore, Grid Applications in high-performance network is needs dynamic configuration of TCP send/receive buffer size.

connection controls. We implement buffer_size_negotiator for dynamic configuration of send buffer size among TCP connections in GridFTP. In case of transmitting data of The Grid applications by GridFTP, each TCP connection transmits data using buffer size which set by "buffer_size_negotiator." Improved GridFTP performs much better than standard general GridFTP, Which achieves an improvement of 3~4%.

References

1. Hasegawa, T. Terai, T. Okamoto and M. Murata, "Scalable socket buffer tuning for high performance Web servers," Proc. of IEEE ICNP 2001, Nov. 2001.
2. Hethmon, P. and Elz, R., "Feature negotiation mechanism for the File Transfer Protocol", RFC 2389, August 1998.
3. J.Semke, J. Mahdavi, and M. Mathis, "Automatic TCP Buffer Tuning", ACM Sigcomm '98/Computer communications Review, Volume 28, Oct. 1998.
4. Jeffrey Semke, "Implementation Issues of the Autotuning Fair Share Algorithm", PSC Technical Report, May 2000.
5. Qingming Ma, Petter Steenkiste and Huizhang, " Routing High bandwidth Traffic in Max-min Fair Share Networks", ACM Sigcomm '96/Computer communications Review, Volume 26, Aug. 1996.
6. T. Dunigan, M. Mathis and B. Tierney, "A TCP Tuning Daemon", Proceeding of IEEE Supercomputing 2002 Conference, Nov. 2002.
7. V. Jacobson, R. Braden and D. Borman, "TCP Extensions for High Performance", IETF RFC 1323, May.1992.
8. W.Allcock, J.Bester, J.Bresnahan, A.Chervenak, L.Limin, S.Tuecke, "GridFTP: Protocol Extensions to FTP for the Grid", GGF draft, March 2001.
9. http://www-didc.lbl.gov/TCP-tuning
10. http://www.psc.edu/networking/auto.html, "Automatic TCP Buffer Tuning Research", web page.

Evaluating the Performance of Skeleton-Based High Level Parallel Programs

Anne Benoit, Murray Cole, Stephen Gilmore, and Jane Hillston

School of Informatics, The University of Edinburgh, James Clerk Maxwell Building,
The King's Buildings, Mayfield Road, Edinburgh EH9 3JZ, UK
enhancers@inf.ed.ac.uk,
http://homepages.inf.ed.ac.uk/stg/research/ENHANCE/

Abstract. We show in this paper how to evaluate the performance of skeleton-based high level parallel programs. Since many applications follow some commonly used algorithmic skeletons, we identify such skeletons and model them with process algebra in order to get relevant information about the performance of the application, and be able to take some "good" scheduling decisions. This concept is illustrated through the case study of the Pipeline skeleton, and a tool which generates automatically a set of models and solves them is presented. Some numerical results are provided, proving the efficiency of this approach.

1 Introduction

One of the most promising technical innovations in present-day computing is the invention of Grid technologies which harness the computational power of widely distributed collections of computers [6]. Designing an application for the Grid raises difficult issues of resource allocation and scheduling (roughly speaking, how to decide which computer does what, and when, and how they interact). These issues are made all the more complex by the inherent unpredictability of resource availability and performance. For example, a supercomputer may be required for a more important task, or the Internet connections required by the application may be particularly busy.

In this context of Grid programming, skeleton based programming [3,11,5] recognizes that many real applications draw from a range of well known solution paradigms and seeks to make it easy for an application developer to tailor such a paradigm to a specific problem. In this high-level approach to parallel programming, powerful structuring concepts are presented to the application programmer as a library of pre-defined 'skeletons'. As with other high-level programming models the emphasis is on providing generic polymorphic routines which structure programs in clearly-delineated ways. Skeletal parallel programming supports reasoning about parallel programs in order to remove programming errors. It enhances modularity and configurability in order to aid modification, porting and maintenance activities. In the present work we focus on the Edinburgh Skeleton Library (eSkel) [4]. eSkel is an MPI-based library which has been designed for

M. Bubak et al. (Eds.): ICCS 2004, LNCS 3038, pp. 289–296, 2004.
© Springer-Verlag Berlin Heidelberg 2004

SMP and cluster computing and is now being considered for Grid applications using Grid-enabled versions of MPI such as MPICH-G2 [10].

The use of a particular skeleton carries with it considerable information about implied scheduling dependencies. By modelling these with stochastic process algebras such as PEPA [9], and thereby being able to include aspects of uncertainty which are inherent to Grid computing, we believe that we will be able to underpin systems which can make better scheduling decisions than less sophisticated approaches. Most significantly, since this modelling process can be automated, and since Grid technology provides facilities for dynamic monitoring of resource performance, our approach will support *adaptive* rescheduling of applications.

Some related projects obtain performance information from the Grid with benchmarking and monitoring techniques [2,12]. In the ICENI project [7], performance models are used to improve the scheduling decisions, but these are just graphs which approximate data obtained experimentally. Moreover, there is no upper-level layer based on skeletons in any of these approaches.

Other recent work considers the use of skeleton programs within grid nodes to improve the quality of cost information [1]. Each server provides a simple function capturing the cost of its implementation of each skeleton. In an application, each skeleton therefore runs only on one server, and the goal of scheduling is to select the most appropriate such servers within the wider context of the application and supporting grid. In contrast, our approach considers single skeletons which span the grid. Moreover, we use modelling techniques to estimate performance.

Our main contribution is based on the idea of using performance models to enhance the performance of grid applications. We propose to model skeletons in a generic way to obtain significant performance results which may be used to reschedule the application dynamically. To the best of our knowledge, this kind of work has never been done before. We show in this paper how we can obtain significant results on a first case study based on the pipeline skeleton.

In the next section, we present the pipeline and a model of the skeleton. Then we explain how to solve the model with the PEPA workbench in order to get relevant information (Section 3). In section 4 we present a tool which automatically determines the best mapping to use for the application, by first generating a set of models, then solving them and comparing the results. Some numerical results on the pipeline application are provided. Finally we give some conclusions.

2   The Pipeline Skeleton

Many parallel algorithms can be characterized and classified by their adherence to one or more of a number of generic algorithmic skeletons [11,3,5]. We focus in this paper on the concept of pipeline parallelism, which is of well-proven usefulness in several applications.

We recall briefly the principle of the pipeline skeleton, and then we explain how we can model it with a process algebra.

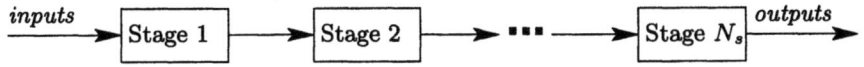

Fig. 1. The pipeline application

2.1   Principle of Pipeline

In the simplest form of pipeline parallelism [4], a sequence of N_s *stages* process a sequence of *inputs* to produce a sequence of *outputs* (Fig. 1). Each input passes through each stage in the same order, and the different inputs are processed one after another (a stage cannot process several inputs at the same time). Note that the internal activity of a stage may be parallel, but this is transparent to our model. In the remainder of the paper we use the term "processor" to denote the hardware responsible for executing such activity, irrespective of its internal design (sequential or parallel).

We consider this application in the context of computational Grids, and so we want to map this application onto our computing resources, which consist of a set of potentially heterogeneous processors interconnected by an heterogeneous network.

Considering the pipeline application in the eSkel library [4], we focus here on the function `Pipeline1for1`, which considers that each stage produces exactly one output for each input.

2.2   Pipeline Model

To model this algorithmic skeleton Grid application, we decompose the problem into the stages, the processors and the network. The model is expressed in Performance Evaluation Process Algebra (PEPA) [9].

The Stages
The first part of the model is the *application model*, which is independent of the resources on which the application will be computed. We define one PEPA component per stage. For $i = 1..N_s$, the component $Stage_i$ works sequentially. At first, it gets data (activity $move_i$), then processes it (activity $process_i$), and finally moves the data to the next stage (activity $move_{i+1}$).

$$Stage_i \stackrel{def}{=} (move_i, \top).(process_i, \top).(move_{i+1}, \top).Stage_i$$

All the rates are unspecified, denoted by the distinguished symbol $\top$, since the processing and move times depend on the resources where the application is running. These rates will be defined later, in another part of the model.

The pipeline application is then defined as a cooperation of the different stages over the $move_i$ activities, for $i = 2..N_s$.

The activities $move_1$ and $move_{N_s+1}$ represent, respectively, the arrival of an input in the application and the transfer of the final output out of the pipeline.

They do not represent any data transfer between stages, so they are not synchronizing the pipeline application. Finally, we have:

$$Pipeline \overset{def}{=} Stage_1 \underset{\{move_2\}}{\bowtie} Stage_2 \underset{\{move_3\}}{\bowtie} \cdots \underset{\{move_{N_s}\}}{\bowtie} Stage_{N_s}$$

The Processors

We consider that the application must be mapped on a set of N_p processors. Each stage is processed by a given (unique) processor, but a processor may process several stages (in the case where $N_p < N_s$). In order to keep a simple model, we decide to put information about the processor (such as the load of the processor or the number of stages being processed) directly in the rate μ_i of the activities $process_i$, $i = 1..N_s$ (these activities have been defined for the components $Stage_i$).

Each processor is then represented by a PEPA component which has a cyclic behaviour, consisting of processing sequentially inputs for a stage. Some examples follow.

- In the case when $N_p = N_s$, we map one stage per processor:

$$Processor_i \overset{def}{=} (process_i, \mu_i).Processor_i$$

- If several stages are processed by a same processor, we use a choice composition. In the following example ($N_p = 2$ and $N_s = 3$), the first processor processes the two first stages, and the second processor processes the third stage.

$$Processor_1 \overset{def}{=} (process_1, \mu_1).Processor_1 + (process_2, \mu_2).Processor_1$$
$$Processor_2 \overset{def}{=} (process_3, \mu_3).Processor_2$$

Since all processors are independent, the set of processors is defined as a parallel composition of the processor components:

$$Processors \overset{def}{=} Processor_1 || Processor_2 || \ldots || Processor_{N_p}$$

The Network

The last part of the model is the network. We do not need to model directly the architecture and the topology of the network for what we aim to do, but we want to get some information about the efficiency of the link connection between pairs of processors. This information is given by affecting the rates λ_i of the $move_i$ activities ($i = 1..N_s + 1$).

- λ_1 represents the connection between the user (providing inputs to the pipeline) and the processor hosting the first stage.

- For $i = 2..N_s$, λ_i represents the connection between the processor hosting stage $i - 1$ and the processor hosting stage i.

- λ_{N_s+1} represents the connection between the processor hosting the last stage and the user (the site where we want the output to be delivered).

When the data is "transferred" on the same computer, the rate is really high, meaning that the connection is fast (compared to a transfer between different sites).

The network is then modelled by the following component:

$$Network \stackrel{def}{=} (move_1, \lambda_1).Network + \ldots + (move_{N_s+1}, \lambda_{N_s+1}).Network$$

The Pipeline Model

Once we have defined the different components of our model, we just have to map the stages onto the processors and the network by using the cooperation combinator. For this, we define the following sets of action types:

- $L_p = \{process_i\}_{i=1..N_s}$ to synchronize the *Pipeline* and the *Processors*
- $L_m = \{move_i\}_{i=1..N_s+1}$ to synchronize the *Pipeline* and the *Network*

$$Mapping \stackrel{def}{=} Network \underset{L_m}{\bowtie} Pipeline \underset{L_p}{\bowtie} Processors$$

3   Solving the Models

Numerical results can been computed from such models with the Java Version of the PEPA Workbench [8].

The *performance result* that is pertinent for us is the throughput of the $process_i$ activities ($i = 1..N_s$). Since data passes sequentially through each stage, the throughput is identical for all i, and we need to compute only the throughput of $process_1$ to obtain significant results. This is done by adding the steady-state probabilities of each state in which $process_1$ can happen, and multiplying this by μ_1. This result can be computed by using the command line interface of the PEPA workbench, by invoking the following command:

```
java pepa.workbench.Main -run lr ./pipeline.pepa
```

The `-run lr` (or `-run lnbcg+results`) option means that we use the linear biconjugate gradient method to compute the steady state solution of the model described in the file `./pipeline.pepa`, and then we compute the performance results specified in this file, in our case the throughput of the pipeline.

4   AMoGeT: The Automatic Model Generation Tool

We investigate in this paper how to enhance the performance of Grid applications with the use of algorithmic skeletons and process algebras. To do this, we have created a tool which automatically generates performance models for the pipeline case study, and then solves the models and provides to the application significant results to improve its performance.

We describe the tool succinctly and then provide some numerical results for the pipeline application.

AMoGeT

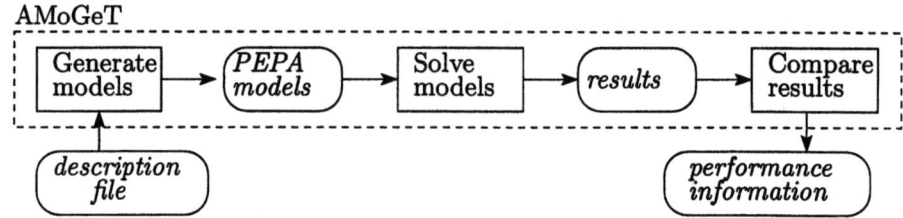

Fig. 2. Principle of AMoGeT

4.1   AMoGeT Description

Fig. 2 illustrates the principle of the tool. In its current form, the tool is a standalone prototype. Its ultimate role will be as an integrated component of a run-time scheduler and re-scheduler, adapting the mapping from application to resources in response to changes in resource availability and performance. The tool allows everything to be done in a single step through a simple Perl script: it generates and solves the models, and then compares the results. This allows us to have feedback on the application when the performance of the available resources is modified.

Information is provided to the tool via a *description file*. This information can be gathered from the Grid resources and from the application definition. In the following experiments, it is provided by the user, but we can also get it automatically from Grid services, for example from the Network Weather Service [12]. We also define a set of candidate mappings of stages to processors. Each mapping specifies where the initial data is located, where the output data must be left and (as a tuple) the processor where each stage is processed. For example, the tuple $(1, 1, 2)$ means that the two first stages are on processor 1, with the third stage on processor 2.

One model is then generated from each mapping, as described in Section 2.2. To compute the rates for a given model, we take into account the number of stages hosted on each processor, and we assume that the work sharing between the stages is equitable. We use also all the other information provided by the description file about the available resources and the characteristics of the application. The models can then be solved with the PEPA workbench, and the throughput of the pipeline is automatically computed (Section 3).

During the resolution, all the results are saved in a single file, and the last step of results comparison finds out which mapping produces the best throughput. This mapping is the one we should use to run the application.

4.2   Numerical Results

We have made some experiments and we give here a few numerical results on an example with 3 pipeline stages (and up to 3 processors). The models that we need to solve are really small (in this case, the model has 27 states and 51 transitions). The time taken to compute these results was very small, being less than one second on this and similar examples involving up to eight pipeline stages. Clearly this is an important property, since we must avoid taking longer to compute a rescheduling than we save in application time by implementing it.

We define l_{i-j} to be the latency of the communication between processors i and j, in seconds. We suppose that $l_{i-i}=0.0001$ for $i = 1..3$, and that there is no need to transfer the input or the output data. We suppose that all stages are equivalent in term of amount of work required, and so we define also the time required to complete a stage on each processor t_i ($i = 1..3$), if the stage can use all the available processing power (this time is longer when several stages are on the same processor).

We compare the mappings (1,1,1), (1,1,2), (1,2,2), (1,2,1), (1,1,3), (1,3,3), (1,3,1) and (1,2,3) (the first stage is always on processor1), and we only put the optimal mapping in the relevant line of the results table.

Set of results	Parameters						Mapping &
	11-2	12-3	11-3	t1	t2	t3	Throughput
1	0.0001	0.0001	0.0001	0.1	0.1	0.1	(1,2,3): 5.63467
	0.0001	0.0001	0.0001	0.2	0.2	0.2	(1,2,3): 2.81892
2	0.0001	0.0001	0.0001	0.1	0.1	1	(1,2,1): 3.36671
	0.1	0.1	0.1	0.1	0.1	1	(1,2,2): 2.59914
	1	1	1	0.1	0.1	1	(1,1,1): 1.87963
3	0.1	1	1	0.1	0.1	0.1	(1,2,2): 2.59914
	0.1	1	1	1	1	0.01	(1,3,3): 0.49988

In the first set of results, all the processors are identical and the network links are really fast. In these cases, the best mapping always consists of putting one stage on each processor. If we double the time required to complete a stage on each processor (busy processors), the resulting throughput is divided by 2, since only the processing power has an impact on the throughput.

The second set of results illustrates the case when one processor is becoming really busy, in this case processor3. We should not use it any more, but depending on the network links, the optimal mapping may change. If the links are not efficient, we should indeed avoid data transfer and try to put consecutive stages on the same processor.

Finally, the third set of results shows what happens if the network link to processor3 is really slow. In this case again, the use of the processor should be avoided, except if it is a really fast processor compared to the other ones (last line). In this case, we process stage2 and stage3 on the third processor.

5   Conclusions

In the context of Grid applications, the availability and performance of the resources changes dynamically. We have shown through this study that the use of skeletons and performance models of these can produce some relevant information to improve the performance of the application. This has been illustrated on the pipeline skeleton, which is a commonly used algorithmic skeleton. In this case, the models help us to choose the mapping, of the stages onto the processors, which will produce the best throughput. A tool automates all the steps to obtain the result easily.

We are currently working at getting the performance information needed by the tool from the Grid and from the application, to make it more realistic. Moreover, some detailed results on the timing of the tool will be provided. This approach will also be developed on some other skeletons so it may be useful for a larger class of applications. However this first case study has already shown that we can obtain relevant information and that we have the potential to enhance the performance of Grid applications with the use of skeletons and process algebras.

References

1. M. Alt, H. Bischof, and S. Gorlatch. Program Development for Computational Grids Using Skeletons and Performance Prediction. *Parallel Processing Letters*, 12(2):157–174, 2002.
2. R. Biswas, M. Frumkin, W. Smith, and R. Van der Wijngaart. Tools and Techniques for Measuring and Improving Grid Performance. In *Proc. of IWDC 2002*, volume 2571 of *LNCS*, pages 45–54, Calcutta, India, 2002. Springer-Verlag.
3. M. Cole. *Algorithmic Skeletons: Structured Management of Parallel Computation*. MIT Press & Pitman, 1989.
   http://homepages.inf.ed.ac.uk/mic/Pubs/skeletonbook.ps.gz.
4. M. Cole. eSkel: The edinburgh Skeleton library. Tutorial Introduction. *Internal Paper, School of Informatics, University of Edinburgh*, 2002.
   http://homepages.inf.ed.ac.uk/mic/eSkel/.
5. M. Cole. Bringing Skeletons out of the Closet: A Pragmatic Manifesto for Skeletal Parallel Programming. *To appear in Parallel Computing*, 2004.
6. I. Foster and C. Kesselman. *The Grid: Blueprint for a New Computing Infrastructure*. Morgan Kaufmann, 1998.
7. N. Furmento, A. Mayer, S. McGough, S. Newhouse, T. Field, and J. Darlington. ICENI: Optimisation of Component Applications within a Grid Environment. *Parallel Computing*, 28(12):1753–1772, 2002.
8. N.V. Haenel. User Guide for the Java Edition of the PEPA Workbench. *LFCS, University of Edinburgh*, 2003. http://homepages.inf.ed.ac.uk/s9905941/.
9. J. Hillston. *A Compositional Approach to Performance Modelling*. Cambridge University Press, 1996.
10. N. Karonis, B. Toonen, and I. Foster. MPICH-G2: A Grid-Enabled Implementation of the Message Passing Interface. *JPDC*, 63(5):551–563, May 2003.
11. F.A. Rabhi and S. Gorlatch. *Patterns and Skeletons for Parallel and Distributed Computing*. Springer Verlag, 2002.
12. R. Wolski, N.T. Spring, and J. Hayes. The network weather service: a distributed resource performance forecasting service for metacomputing. *Future Generation Computer Systems*, 15(5–6):757–768, 1999.

Towards a Generalised Runtime Environment for Parallel Haskells*

Jost Berthold

Philipps-Universität Marburg, Fachbereich Mathematik und Informatik
Hans-Meerwein-Straße, D-35032 Marburg, Germany
berthold@informatik.uni-marburg.de

Abstract. Implementations of parallel dialects (or: coordination languages) on a functional base (or: computation) language always have to extend complex runtime environments by the even more complex parallelism to maintain a high level of abstraction. Starting from two parallel dialects of the purely functional language Haskell and their implementations, we generalise the characteristics of Haskell-based parallel language implementations, abstracting over low-level details. This generalisation is the basis for a shared runtime environment which can support different coordination concepts and alleviate the implementation of new constructs by a well-defined API and a layered structure.

1 Introduction

The area of parallel functional programming exhibits a variety of approaches with the common basis of referential transparency of functional programs and the ability to evaluate subexpressions independently. Some approaches pursue the target of (half-)automatic parallelisation of operations on list-like data structures (i.e. *data parallelism*). Other dialects are more explicit in their handling of parallelism and allow what we call *general-purpose parallelism*, able to capture schemes of parallelism which are not data-oriented. Whereas machine-specific optimisation is easier with specialised structures and operations, these more general approaches present a considerable advantage in language design: It is generally accepted [1,2] that functional languages allow a clean distinction between a computation (or "base") language and independent coordination constructs for parallelism control. The more special-purpose data structures enter the language, the more vague this important distinction will become.

Implementations of general-purpose languages often resemble each other, differing in syntactic sugar or providing special domain-specific features. Thus, in the implementation, many concepts can be reduced to a common infrastructure enriched with a small amount of special features depending on the concrete, language-specific coordination concept. In this paper, we present an overview of the two main-stream general-purpose parallel extensions to the functional language Haskell, focusing on implementation aspects. Our aim is to bring together

* Work supported by ARC/DAAD Grant No. D/03/20257.

M. Bubak et al. (Eds.): ICCS 2004, LNCS 3038, pp. 297–305, 2004.
© Springer-Verlag Berlin Heidelberg 2004

related implementations in a generalised runtime environment (RTE), capable of supporting different language concepts from the same configurable base system.

The paper is organised as follows: Section 2 presents the two parallel languages based on Haskell, the starting point of our work. In Section 3, we systemise the functionality a common base system must provide to implement both languages, or even a combination of their features. A short example in Section 4 indicates the potential use of such an API. Section 5 concludes.

2   General-Purpose Parallelism in Haskell

The Haskell Community Report [3] mentions two major approaches to parallel computation based on Haskell: *Glasgow parallel Haskell* [4] and *Eden* [5]. These two languages show major differences in their coordination concept and, in consequence, in their implementation, while on the other hand, they both capture the main idea of evaluating independent subexpressions in parallel. Another common point is that, in contrast to other parallel Haskells [2], both GpH and Eden are designed as general-purpose coordination languages. Neither of them is dedicated to a certain paradigm such as task- or data-parallelism or pure skeleton-based programming, though the respective coordination schemes can be expressed by both. Both language extensions are implemented using GHC [6] as a platform. The implementation is even sharing infrastructure code, e.g. for basic communication and system setup, but diverging whenever the different languages require specific features.

2.1   Glasgow Parallel Haskell

Glasgow Parallel Haskell (GpH) [4] is a well-known parallel dialect of Haskell investigated since the 90's. The overall paradigm of GpH is semi-implicit data and task parallelism, following annotations in the source program. In every definition, subexpressions can be marked as "suitable for parallel evaluation" by a par-expression in the overall result. The coordination construct par takes 2 arguments and returns the second one after recording the first one as a "spark", to evaluate in parallel. An idle processor can fetch a spark and evaluate it. The built-in seq is the sequential analogon, which forces evaluation of the first argument before returning the second one.

```
par,seq :: a -> b -> b
```

These coordination atoms can be combined in higher-order functions to control the evaluation degree and its parallelism without mixing coordination and computation in the code. This technique of *evaluation strategies* described in [7] offers sufficient evaluation control to define constructs similar to skeleton-based programming. However, as opposed to usual skeletons, parallelism always remains semi-implicit in GpH, since the runtime environment (RTE) can either ignore any spark or eventually activate it.

The implementation of GpH, GUM [4], relies essentially on the administration of a distributed shared heap and on the described two-stage task creation mechanism where potential parallel subtasks first become local sparks before they may get activated. The only access point to the system is the spark creation primitive; parallel computations and their administrative requirements are completely left to the RTE and mainly concern spark retrieval and synchronisation of a distributed heap. Once a spark gets activated, the data which is evaluated in parallel could subsequently reside on a different processor and thus must get a global address, so it can be sent back on request.

The main advantage of the implicit GpH concept is that it dynamically adapts the parallel computation to the state and load of nodes in the parallel system. The GpH implementation would even allow to introduce certain heuristics to reconfigure the parallel machine at runtime. However, parallel evaluation on this dynamic basis is hardly predictable and accessible only by simulation and tracing tools like GranSim [8].

2.2   Eden

The parallel Haskell dialect Eden [5] allows to define *process abstractions* by a constructing function **process** and to explicitly *instantiate* (i.e. run) them on remote processors using the operator (#). Processes are distinguished from functions by their operational property to be executed remotely, while their denotational meaning remains unchanged as compared to the underlying function.

```
process :: (Trans a, Trans b) => (a -> b) -> Process a b
( # )   :: (Trans a, Trans b) => Process a b -> a -> b
```

For a given function **f**, evaluation of the expression (**process f**) **# arg** leads to the creation of a new (remote) process which evaluates the application of the function **f** to the argument **arg**. The argument is evaluated locally and sent to the new process. The implementation of Eden [9] uses implicit inter-process connections (*channels*) and data transmission in two possible ways: single-value channels and stream channels. Eden processes are encapsulated units of computation which only communicate via these channels. This concept avoids global memory management and its costs, but it can sometimes duplicate work on the evaluation of shared data structures.

Communication between processes is automatically managed by the system and hidden from the programmer, but additional language constructs allow to create and access communication channels explicitly and to create arbitrary process networks. The task of parallel programming is further simplified by a library of predefined skeletons [10]. Skeletons are higher-order functions defining parallel interaction patterns which are shared in many parallel applications. The programmer can use such known schemes from the library to achieve an instant parallelisation of a program.

The implementation of Eden extends the runtime environment of GHC by a small set of primitive operations for process creation and data transmission between processes (including synchronisation and special communication modes). While the RTE shares basic communication facilities with the GpH implementation, the implementation concept is essentially different in that the needed primitives only provide simple basic actions, while more complex operations are encoded by a superimposed functional module. This module, which encodes process creation and communication explicitly, can abstract over the administrative issues, profiting from Haskell's support in genericity and code reuse [9], moreover, it protects the basic primitives from being misused.

3   Generalising the Parallel Runtime Environment

3.1   Layered Implementation

To systemise the common parts of parallel Haskell implementations, we follow the approach of Eden's layered implementation, i.e. thick layers of functionality exploited strictly level-to-level to avoid dependencies across abstraction levels. Apart from maintenance of only one system, the concept of layered implementation is promising for the implementation of other coordination languages based on Haskell, since it facilitates maintenance and experimental development. With one flexible basic layer, different language concepts can be easily implemented by a top-layer module (making use of the underlying RTE support) where the appropriate coordination constructs are defined. As shown in Fig. 1, the *Evaluation Strategy* module for GpH [7] is just an example of these high-level parallelism libraries. As well as Eden, one could implement explicit parallel coordination e.g. for some data-parallel language by adding an appropriate module for all parallel operations to their sequential implementation.

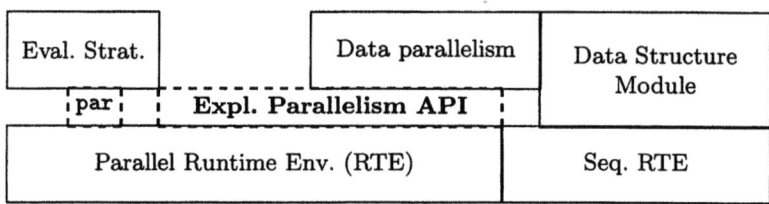

Fig. 1. Layered implementation of parallel coordination (example)

These top-layer modules use parts of a common ParallelAPI provided by the underlying runtime layer, which we will describe now in terms of the supported operations. A more basic concept is the overall coordination in Haskell-based parallel languages, which relies on parallel graph reduction with synchronisation nodes representing remote data (not described further due to space limitations, see [1]). In order to support implicit parallelism, a generalised RTE will also have to support virtual shared memory and implicit, load-dependent task creation. Concepts for implicit coordination use this basic RTE support, which is not accessible from language level and thus not part of the API.

Table 1. API Operations

Primitive :: Type	Description
rTask :: PE -> a -> ()	actively spawn a process on a remote PE
fork :: a -> b -> b	spawn new thread (same process)
par :: a -> ()	passively mark data as "potentially parallel"
createDC :: ChanMode -> Int -> (ChanName a, a)	create new channel and sync. nodes in the local heap (type a expected)
connectC :: ChanName a -> ()	connect to a channel to write data
dataSend :: SMode-> a -> ()	send subgraph from one PE to another
conSend :: a -> ChanName b -> ()	send top-level constructor and reply channel for destination of its components
noPE :: Int	number of available nodes in parallel system
myPE :: PE	node ID where caller is located (unsafe)

3.2 ParallelAPI of Primitive Operations

Functionality to coordinate a parallel computation will be implemented in a set of primitive operations (i.e. the ParallelAPI) which exposes functionality to define high-level coordination, intended for language developers. It should be complete in that it allows a variety of different coordination possibilities and ideally orthogonal in that it provides only one way to accomplish a certain task. Functionality exposed by the API falls into three categories: task management, explicit communication and system information, shown in Tab.1.

Task Management. The obvious issue in task management is to start remote computations, which requires a primitive rTask to actively transfer a subgraph to a remote processor for evaluation. The receiver of a transferred task creates a thread to evaluate the included subgraph, i.e. solve the task. The primitive in itself does not imply any communication between parent and child process, it is due to the caller and the transferred task to establish this communication. Tasks can be explicitly placed on certain nodes of the parallel system when required, otherwise, the RTE uses configurable placement patterns or random placement. The conceptual entities of computation in this description follow the style of Eden: The primitive rTask sends a subgraph (task) to a remote node, which creates a thread for its evaluation. A *task* (subgraph to be evaluated) thereby turns into a *process* containing initially one evaluation *thread*. This first thread can then fork to create other threads, which all reside conceptually in the same process.

An implicit variant of active task creation is the two-stage concept of GpH, which only records subgraphs as "sparks" suitable for remote evaluation. The API supports spark creation by an annotation primitive par, but the action of turning a spark into a thread cannot be exposed to language level, as well as all necessary data must be shared via global memory, since it cannot be explicitly transferred.

Communication. The API will expose communication functionality for (completely or partially) transferring subgraphs from one process to another via Eden-style *channels*, linking to placeholders in the evaluation graph. Channels are created using the primitive `createDC`, and configured for 1:1 or n:1-communication upon their creation. Once created, channels have a representation at the language level and can be transferred to other threads just as normal data, enabling to build arbitrary process networks. Threads in different processes can then use `connectC` to connect to the channel and send data.

Values (i.e. subgraphs) are sent through the created channels using a primitive `dataSend`, which can have several modes: Either a *single value* is sent, which implies that the channel becomes invalid (since the receiver replaces the respective placeholder by received data), or the data is an element in a *stream* and leaves the channel open for further elements, recomposed to a list in the receiver's heap. As a general rule, data transmitted by `dataSend` should be in *normal form*, i.e. sent only after complete evaluation by the sender. Fully evaluated data can be duplicated without risk, whereas data shared via global memory synchronisation inside the RTE can have any state of evaluation, and unevaluated data should be *moved* instead of copied to keep global references synchronised.

Another, more complex, variant of data transmission is to send only the top-level constructor using `conSend`, which requires the receiver to open and send back new channels for the arguments of this constructor. This rather complex implementation is the way to overrule the principle of normal form evaluation prior to channel communication, and we can imagine useful applications.

System Information. In order to profit from explicit process placement, the runtime setup must be accessible by the API to determine how many nodes are available and, for every process, on which node it is executed. While the primitive for the former, `noPE`, is at least constant during the whole runtime, the latter information, determined by `myPE`, ultimately destroys referential transparency (unsafe). On the other hand, this feature is useful to place processes on appropriate nodes; it should be used only for coordination purposes.

3.3   Functionality of the Basic System Layer

As mentioned, implicit, load-driven task creation and synchronisation must enable all threads to share data, and use a two-stage task creation in order to allow decent evaluation control by the programmer. The global memory management, as well as other system tasks such as basic communication and system management, reside in the RTE alone and are not exposed to the programmer. We will briefly outline these parts as well.

Heap (Memory) Management. Our approach inherently needs a virtual shared heap for transferred data, since implicit parallel evaluation is not within reach of the API, but managed by the RTE. However, not all data must be globalised, but only the parts which are required for remote evaluation and not addressed by explicit channel communication.

This distinction becomes manifest in the placeholder nodes in the graph heap. Nodes for globalised data, which the RTE fetches *actively* from other nodes when a thread needs it, are opposed to nodes for channel communication, which passively wait for data to arrive via the connected channel. For the latter, the RTE only needs a table for the respective channels, whereas management of the former needs more attention: besides the mapping from nodes to global addresses and vice-versa, the RTE must prevent duplication of data in one node's local heap. Furthermore, data fetching can be implemented with different strategies (bulk or incremental fetching), which yields platform-dependant results, according to [11]. Fetching strategy and data size, as well as rescheduling and spark retrieval strategy, are good examples for configuration options of the RTE.

Communication and System Management. Communication inside the RTE is needed to synchronise the global heap and to exchange data between threads. We do not need a particular standard or product; any middleware capable of exchanging raw data between nodes in a network can be used (even raw sockets). However, targeting a standard such as MPI [12] makes the system far more portable, since implementations for various platforms are freely available. We intend to build a modular communication subsystem and exchangeable adapters to common message passing middleware. Necessary startup actions and node synchronisation are another reason why existing middleware is used instead of a customised solution.

4   Prospected Coordination Languages

The variety of possible coordination languages using the described API would fill a considerable amount of pages, so we set it aside and only give a small example. The Eden implementation described in [9] is a more complex instance.

Example: We define rFork as an explicit variant of the par-construct: It spawns a process for a task on the indicated node and continues with the cont. The RTE assigns a global address to task before sending, and it is evaluated by a new thread on node. If results are needed somewhere else, data will be fetched from there. This construct can be used in Haskell code to define higher-level constructs, e.g. a parallel fold-like operation shown here:[1]

```
rFork :: PE -> a -> b -> b
rFork node task cont = (rTask# node task) 'seq' cont
foldSpread :: ( [a] -> b ) -> [a] -> [b]
foldSpread f xs = let size = ..--arbitrarily fixed, or by input length
                tasks = [ f sublist | sublist <- splitList size xs ]
            peList= drop (toInt myPE#) (cycle [ 1..noPE# ])
            in zipWith rFork peList tasks
```

The effect of foldSpread is to distribute its input list (in portions of a certain size) over several nodes, which, in turn, evaluate a sub-result for each sublist. Sublists and results will definitely reside on the nodes we indicated: we can do further processing on these same nodes explicitly.   ◁

---

[1] The symbol # indicates primitive operations and data types.

The described ParallelAPI allows to express parallel coordination in easy, declarative terms and gives much expressive power to language designers. The most ambitious and interesting area to explore with it is to combine both *explicit* coordination and *implicit* parallelism, with freedom for load-balancing by the RTE. This can be applied when computations are big enough for wide-area distribution, but in a heterogeneous setup where dynamic load-balancing is required; found e.g. in scientific Grid-Computing with high communication latency and a rapidly changing environment.

5   Conclusions

We have described a future implementation for parallel Haskells, which is based on a generalised runtime environment (RTE) and a layered implementation concept. The described API and RTE can express general-purpose parallelism declaratively and is based on experiences with two existing parallel Haskell systems. Since the approach only refers to runtime support, it can be freely combined with static analysis techniques for automatic parallelisation, and skeleton programming. Hence, the outcome of the prospected work is a standardised implementation for Haskell-based coordination concepts, for which this paper gives the guideline and points at similarities in existing parallel Haskells.

The concepts described in this paper are *work in progress*, i.e. details of the generalised RTE may change with further research. Our prospects for future work are to implement and use the described API for an integration of Eden and GpH in a common language, to express high-level coordination for large-scale parallel computations with the possibility of dynamic load-balancing and coordination control by the programmer. For this purpose, the RTE will be extended with more task placement policies, adaptive behaviour and runtime reconfiguration.

References

1. Hammond, K., Michaelson, G., eds.: Research Directions in Parallel Functional Programming. Springer (1999)
2. Trinder, P., Loidl, H.W., Pointon, R.: Parallel and Distributed Haskells. J. of Functional Programming **12** (2002)
3. Claus Reinke (ed.): Haskell Communities and Activities Report. Fifth Edition (2003) www.haskell.org/communities.
4. Trinder, P., Hammond, K., Mattson Jr., J., Partridge, A., Peyton Jones, S.: GUM: a Portable Parallel Implementation of Haskell. In: PLDI'96, ACM Press (1996)
5. Breitinger, S., Loogen, R., Ortega-Mallén, Y., Peña, R.: The Eden Coordination Model for Distributed Memory Systems. In: HIPS. LNCS 1123, IEEE Press (1997)
6. Peyton Jones, S., Hall, C., Hammond, K., Partain, W., Wadler, P.: The Glasgow Haskell Compiler: a Technical Overview. In: JFIT'93. (1993)
7. Trinder, P., Hammond, K., Loidl, H.W., Peyton Jones, S.: Algorithm + Strategy = Parallelism. J. of Functional Programming **8** (1998)
8. Loidl, H.W.: Granularity in Large-Scale Parallel Functional Programming. PhD thesis, Department of Computing Science, University of Glasgow (1998)

9. Berthold, J., Klusik, U., Loogen, R., Priebe, S., Weskamp, N.: High-level Process Control in Eden. In: EuroPar 2003 – Parallel Processing. LNCS 2790, Klagenfurt, Austria, Springer (2003)
10. Loogen, R., Ortega-Mallén, Y., Peña, R., Priebe, S., Rubio, F.: Parallelism Abstractions in Eden. In Rabhi, F.A., Gorlatch, S., eds.: Patterns and Skeletons for Parallel and Distr. Computing. LNCS 2011, Springer (2002)
11. Loidl, H.W., Hammond, K.: Making a Packet: Cost-Effective Comm. for a Parallel Graph Reducer. In: IFL'96. LNCS 1268, Springer (1996)
12. MPI Forum: MPI 2: Extensions to the Message-Passing Interface. Technical report, University of Tennessee, Knoxville (1997)

Extending Camelot with Mutable State and Concurrency

Stephen Gilmore

Laboratory for Foundations of Computer Science, The University of Edinburgh
King's Buildings, Edinburgh, EH9 3JZ, Scotland

Abstract. Camelot is a resource-bounded functional programming language which compiles to Java byte code to run on the Java Virtual Machine. We extend Camelot to include language support for Camelot-level threads which are compiled to native Java threads. We extend the existing Camelot resource-bounded type system to provide safety guarantees about the heap usage of Camelot threads.

1 Introduction

Functional programming languages allow programmers to express algorithms concisely using high-level language constructs operating over structured data, secured by strong type-systems. Together these properties support the production of high-quality software for complex application problems. Functional programs in strongly-typed languages typically have relatively few programming errors, when compared to similar applications implemented in languages without these beneficial features.

These desirable language properties mean that developers shed the burdens of explicit memory management but this has the associated cost that they typically lose all control over the allocation and deallocation of memory. The Camelot language provides an intermediate way between completely automatic memory management and unassisted allocation and deallocation in that it provides type-safe storage management by re-binding of addresses. The address of a datum is obtained in pattern matching and used in an expression (to store a different data value at that address), overwriting the currently-held value. An affine linear type system prevents addresses from being used more than once in an expression.

The Camelot compiler targets the Java Virtual Machine but the JVM does not provide an instruction to free memory, consigning this to the garbage collector, a generational collector with three generations and implementations of stop-and-copy and mark-sweep collections. The Camelot run-time disposes of unused addresses by adding them to a *free list* of unused memory. On the next allocation caused by the program the storage is retrieved from the head of the free list instead of being allocated by the JVM **new** instruction. When the free list becomes empty the necessary storage is allocated by **new**.

This storage mechanism works for Camelot, but not for Java, because Camelot uses a uniform representation for types which are generated by the

M. Bubak et al. (Eds.): ICCS 2004, LNCS 3038, pp. 306–313, 2004.
© Springer-Verlag Berlin Heidelberg 2004

compiler, allowing data types to exchange storage cells. This uniform representation is called the *diamond type* [1,2], implemented by a `Diamond` class in the Camelot run-time. The type system of the Camelot language assigns types to functions which record the number of parameters which they consume, and their types; the type of the result; and the number of diamonds consumed or freed.

One example of a situation where type-safe reuse of addresses can be used is in a *list updating* function. As with the usual non-destructive list processing, this applies a function to each element of a list in turn, building a list of the images of the elements under the function. In contrast to the usual implementation of a function such as map, the destructive version applies the function *in-place* by overwriting the contents of each cons cell with the image of the element under the function as it traverses the list.

The following simple function increments each integer in an integer list. The Camelot concrete syntax is similar to the concrete syntax of Caml. Where addresses are not manipulated, as here, a Camelot function can also be compiled by Caml.

```
let rec incList lst =
    match lst with
        []    -> []
      | h::t  -> (h + 1) :: incList t
```

This non-destructive version of list processing allocates as many cons-cells as there are elements in the list. With the destructive implementation the storage in the list is reused by overwriting the stored integers with their successors. Thus this version does not allocate any storage.

```
let rec destIncList lst =
    match lst with
        []         -> []
      | (h::t)@d   -> ((h + 1) :: destIncList t)@d
```

In a higher-order version of this function, a *destructive map*, we would have the memory conservation property that if the function parameter does not allocate storage then an application of the destructive map function would not either.

Selective use of in-place update in this way can be used to realise *deforestation*, a program transformation which eliminates unnecessary intermediate data structures which are built as a computation proceeds.

As an example of a function which is *not* typable in Camelot we can consider the following one. This function attempts to create a modified copy of a list, interleaved with the original list. The (deliberate) error in implementing this function is to attempt to store the cons cells at the front of the list and the cons cell in second place at the same location, d.

```
let rec incListCopy lst =
    match lst with
        []       -> []
      | (h::t)@d -> let tail = ((h + 1) :: t)@d
                    in (h :: tail)@d      (* Error: d used twice! *)
```

This function is faulted by the Camelot compiler with the following diagnostic error message.

```
File "incListCopy.cmlt", line 4-5, characters 18-80:
! .................let tail = ((h + 1) :: t)@d
!                  in (h :: tail)@d............
! Variable d of type <> used non-linearly
```

The destIncList function above demonstrates storage re-use in Camelot. As an example of programmed control of storage deallocation consider the destructive sum function shown below. Summing the elements of an integer list—or more generally folding a function across a list—is sometimes the last operation performed on the list, to derive an accumulated result from the individual values in the list. If that is the case then at this point the storage occupied by the list can be reclaimed and it is convenient to do this while we are traversing the list.

```
let rec destSumList lst =
    match lst with
          []          ->  0
       | (h::t)@_  ->  h + destSumList t
```

Matching the location of the object against a wildcard pattern (the _ symbol) indicates that this address is not needed (because it is not bound to a name) and thus it can be freed. The destSumList function frees the storage which is occupied by the spine of the list as it traverses the list. In a higher-order version such as *destructive fold* we would have the memory reclamation capability that the function passed in as a parameter could also free the storage occupied by the elements of the list, if these were other storage-occupying objects such as lists or trees.

2   Using Threads in Camelot

Previously the JVM had been used simply as a convenient run-time for the Camelot language but a recent extension to Camelot [3] allows the Java namespace to be accessed from a Camelot application. Thus a Camelot application can now create Java objects and invoke Java methods. Additionally, the Camelot language has now been extended with syntactic support for the definition of classes and objects. Figure 1 shows the implementation of a simple clock in Camelot.

This example shows the Camelot syntax for method invocation (obj#meth()), field access (obj#field) and mutable field update (f <- exp). The application simply shows a small window into which is written the current date and time. The parameter to the built-in sleep function for threads is given in milliseconds so the application sleeps for a tenth of a second before refreshing the display in the text area on screen with the current date and time.

In the object-oriented fragment of the language the types of parameters and results typically need to be specified by the programmer whereas in the functional part types are inferred. Polymorphic functions are monomorphised in compilation to Java byte code.

```
(* The ticker class, a thread implementing a clock *)
class ticker = java.lang.Thread
with
    field ta : javax.swing.JTextArea

    field formatter : java.text.SimpleDateFormat

    method setTextField (f : javax.swing.JTextArea) : unit =
        ta <- f

    method setFormatting (s : string) : unit =
        formatter <- new java.text.SimpleDateFormat s

    method run() : unit =
        let _ = sleep 100 in
        let now = new java.util.Date() in
        let s = this#formatter#format now in
        let _ = this#ta#setText s in this#run()
end


(* The main function of the application *)
let main args =
    let frame = new javax.swing.JFrame "Camelot clock" in
    let ta = new javax.swing.JTextArea() in
    let clock = new ticker() in
    let _ = clock#setTextField ta in
    let _ = clock#setFormatting "EEE MMM dd hh:mm:ss yyyy" in
    let _ = clock#start() in
    let f = new javax.swing.JFrame() in
    let _ = f#setContentPane ta in
    let _ = f#setSize 180 60 in
        f#setVisible true
```

Fig. 1. An implementation of a threaded clock in Camelot

3   Management of Threads

In designing a thread management system for Camelot our strongest requirement was to have a system which works harmoniously with the storage management system already in place for Camelot. One aspect of this is that the resource consumption of a single-threaded Camelot program can be computed in line with the reasoning explained in Section 1.

In moving from one to multiple threads the most important question with respect to memory usage is the following. Should the free list of storage which can be reused be a single static instance shared across all threads; or should each thread separately maintain its own local instance of the free list?

In the former case the accessor methods for the free list must be synchronised in order for data structures not to become disordered by concurrent write operations. Synchronisation incurs an overhead of locking and unlocking the parent of the field when entering and leaving a critical region. This imposes a penalty on program run-time.

In the latter case there is no requirement for access to the free list to be synchronised; each thread has its own free list. In this case, though, the free memory on each free list is private, and not shared. This means that there will be times when one thread allocates memory (with a Java **new** instruction) while another thread has unused memory on its local free list. This imposes a penalty on the program memory usage, and this form of thread management would lead to programs typically using more memory overall.

We have chosen the former scheme; we have a single static instance of a free list shared across all threads. Our programs will take longer than their optimum run-time but memory performance will be improved. Crucially, predictability of memory consumption is retained.

There are several possible variants on this second scheme which we considered. They were not right for our purposes but might be right for others. One interesting alternative is a hybrid of the two approaches is where each thread had a bounded (small) local free list and flushes this to the global free list when it becomes full. This would reduce the overhead of calls to access the synchronised global free list, while preventing threads from keeping too many unused memory cells locally. This could be a suitable compromise between the two extremes but the analysis of this approach would inevitably be more complicated than the approach which we adopted (a single static free list).

A second alternative would be to implement weak local free lists. In this construction each thread would have its own private free list implemented using *weak references* which are references that are not strong enough by themselves to keep an object alive if no genuine references to it are retained. Weak references are typically used to implement caches and secondary indexes for data structures. Other high-level garbage-collected languages such as O'Caml implement weak references also. This scheme was not usable by us because the Camelot compiler also targets small JVMs on handheld devices and the J2ME does not provide the necessary class (`java.lang.ref.WeakReference`).

The analysis of memory consumption of Camelot programs is based on the consumption of memory by heap-allocated data structures. The present analysis of Camelot programs is based on a single-threaded architecture. To assist with the development of an analysis method for multi-threaded Camelot programs we require that data structures in a multi-threaded Camelot program are not shared across threads. For example, it is not possible to hold part of a list in one thread and the remainder in another. This requirement means that the space consumption of a multi-threaded Camelot program is obtained as the sum of per-thread space allocation plus the space requirements of the threads themselves.

At present our type system takes account of heap allocations but does not take account of stack growth. Thus Camelot programs can potentially (and sometimes do in practice) fail at runtime with a `java.lang.StackOverflowError` exception because tail calls of Java methods are not optimised by the JVM.

4   A Simple Thread Model for Camelot

To retain predictability of memory behaviour in Camelot we restrict the multi-threaded programming model significantly from that offered by Java's threads.

Firstly, we disallow use of the `stop` and `suspend` methods from Java's threads API. These are deprecated methods which have been shown to have poor programming properties in any case. Use of the `stop` method allows objects to be exposed in a damaged state, part-way through an update by a thread. Use of `suspend` freezes threads but these do not release the objects which they are holding locks on, thereby often leading to deadlocks. Dispensing with pre-emptive thread interruption means that there is a correspondence between Camelot threads and lightweight threads implemented using first-class continuations, `call/cc` and `throw`, as are usually to be found in multi-threaded functional programming languages [4,5].

Secondly, we require that all threads are run, again for the purposes of supporting predictability of memory usage. In the Java language thread allocation (using **new**) is separated from thread initiation (using the **start** method in the `java.lang.Thread` class) and there is no guarantee that allocated threads will ever be run at all. In multi-threaded Camelot programs we require that all threads are started at the point where they are constructed.

Finally, we have a single constructor for classes in Camelot because our type system does not support overloading. This must be passed initial values for all the fields of the class (because the thread will initiate automatically). All Camelot threads except the main thread of control are daemon threads, which means that the Java Virtual Machine will not keep running if the main thread exits.

This simplified idiom of thread use in Camelot allows us to define *derived forms* for Camelot threads which abbreviate the use of threads in the language. These derived forms can be implemented by *class hoisting*, moving a generated class definition to the top level of the program. This translation is outlined in Figure 2.

5   Threads and (Non-)Termination

The Camelot programming language is supported not only by a strong, expressive type system but also by a program logic which supports reasoning about the time and space usage of programs in the language. However, the logic is a

```
    let rec threadname(args) =
        let locals = subexps in threadname(args)
    ...
    let threadInstance =
        new threadname(actuals) in ...

↝

    class threadnameHolder(args) = java.lang.Thread
    with
        let rec threadname() =
            let locals = subexps in threadname()
        method run() : unit =
            let _ = this#setDaemon(true)
            in threadname()
    end
    ...
    let threadInstance =
        new threadnameHolder(actuals) in
    let _ = threadInstance#start() in ...
```

Fig. 2. Derived forms for thread creation and use in Camelot

logic of partial correctness, which is to say that the correctness of the program is guaranteed only under the assumption that the program terminates. It would be possible to convert this logic into a logic of total correctness which would guarantee termination instead of assuming it but proofs in such a logic would be more difficult to produce than proofs in the partial correctness logic.

It might seem nonsensical to have a logic of partial correctness to guarantee execution times of programs ("this program either terminates in 20 seconds or it never does") but even these proofs about execution times have their use. They are used to provide a bound on the running time of a program so that if this time is exceeded the program may be terminated forcibly by the user or the operating system because after this point it seems that the program will not terminate. Such *a priori* information about execution times would be useful for scheduling purposes. In Grid-based computing environments Grid service providers schedule incoming jobs on the basis of estimated execution times supplied by Grid users. These estimates are sometimes significantly wrong, leading the scheduler either to forcibly terminate an over-running job due to an under-estimated execution time or to schedule other jobs poorly on the basis of an over-estimated execution time.

6   Conclusions and Further Work

Our programme of research on the Camelot functional programming language has been investigating resource consumption and providing static guarantees of resource consumption at the time of program compilation. Our thread management system provides a layer of abstraction over Java threads. This could allow us to modify the present implementation to multi-task several Camelot threads onto a single Java thread. The reason to do this would be to circumvent the ungenerous thread limit on some JVMs. This extension remains as future work but our present design strongly supports such an extension.

We have discussed a very simple thread package for Camelot. A more sophisticated one, perhaps based on Thimble [6], would provide a much more powerful programming model.

Acknowledgements. The author is supported by the Mobile Resource Guarantees project (MRG, project IST-2001-33149). The MRG project is funded under the Global Computing pro-active initiative of the Future and Emerging Technologies part of the Information Society Technologies programme of the European Commission's Fifth Framework Programme. The other members of the MRG project provided helpful comments on an earlier presentation of this work. The implementation of Java class support in the Camelot compiler is due to Nicholas Wolverson. Java is a trademark of SUN Microsystems.

References

1. Martin Hofmann. A type system for bounded space and functional in-place update. *Nordic Journal of Computing*, 7(4):258–289, 2000.
2. Martin Hofmann and Steffen Jost. Static prediction of heap space usage for first-order functional programs. In *Proc. 30th ACM Symp. on Principles of Programming Languages*, 2003.
3. Nicholas Wolverson. O'Camelot: adding objects to a resource-aware functional language. In *On-site proceedings of the Fourth Symposium on Trends in Functional Programming*, Edinburgh, Scotland, September 2003.
4. Edoardo Biagioni, Ken Cline, Peter Lee, Chris Okasaki, and Chris Stone. Safe-for-space threads in Standard ML. *Higher-Order and Symbolic Computation*, 11(2):209–225, 1998.
5. Peter Lee. Implementing threads in Standard ML. In John Launchbury, Erik Meijer, and Tim Sheard, editors, *Advanced Functional Programming, Second International School, Olympia, WA, USA, August 26-30, 1996, Tutorial Text*, volume 1129 of *Lecture Notes in Computer Science*, pages 115–130. Springer, 1996.
6. Ian Stark. Thimble — Threads for MLj. In *Proceedings of the First Scottish Functional Programming Workshop*, number RM/99/9 in Department of Computing and Electrical Engineering, Heriot-Watt University, Technical Report, pages 337–346, 1999.

EVE, an Object Oriented SIMD Library

Joel Falcou and Jocelyn Sérot

LASMEA, UMR 6602 CNRS/Univ. Blaise Pascal, Clermont-Ferrand, France,
{falcou,jserot}@lasmea.univ-bpclermont.fr

Abstract. This paper describes EVE (Expressive Velocity Engine), an object oriented C++ library designed to ease the process of writing efficient numerical applications using AltiVec, the SIMD extension designed by Apple, Motorola and IBM for PowerPC processors. Compared to the Altivec original C API, EVE, offers a significant improvement in terms of expressivity. By relying on template metaprogramming techniques, this is not obtained at the expense of efficiency.

1 Introduction

AltiVec [4] is an extension designed to enhance PowerPC[1] processor performance on applications handling large amounts of data. The AltiVec architecture is based on a SIMD processing unit integrated with the PowerPC architecture. It introduces a new set of 128 bit wide registers distinct from the existing general purpose or floating-point registers. These registers are accessible through 160 new "vector" instructions that can be freely mixed with other instructions (there are no restriction on how vector instructions can be intermixed with branch, integer or floating-point instructions with no context switching nor overhead for doing so). Altivec handles data as 128 bit vectors that can contain sixteen 8 bit integers, eight 16 bit integers, four 32 bit integers or four 32 bit floating points values. For example, any vector operation performed on a `vector char` is in fact performed on sixteen `char` simultaneously and is theoretically running sixteen times faster as the scalar equivalent operation. AltiVec vector functions cover a large spectrum, extending from simple arithmetic functions (additions, subtractions) to boolean evaluation or lookup table solving.

Altivec is natively programmed by means of a C API [2]. Programming at this level can offer significant speedups (from 4 to 12 for typical signal processing algorithms) but is a rather tedious and error-prone task, because this C API is really "assembly in disguise". The application-level vectors (arrays, in variable number and with variable sizes) must be explicitly mapped onto the Altivec vectors (fixed number, fixed size) and the programmer must deal with several low-level details such as vector padding and alignment. To address this programmability issue, we have investigated the possibility to provide a higher level API, in the form of a C++ class library. This class library should encapsulate all the low-level details related to the manipulation of Altivec vectors

---

[1] PPC 74xx (G4) and PPC 970 (G5).

M. Bubak et al. (Eds.): ICCS 2004, LNCS 3038, pp. 314–321, 2004.
© Springer-Verlag Berlin Heidelberg 2004

and provide a fully abstract `Array`/`Vector` object and the associated functions
as overloaded operators – allowing for example code to be written in the style
of Fig. 1.

```
Vector<float> a(1000),b(1000),c(1000),r(1000);
r = a * b + c;
```

Fig. 1. Vector processing with a high-level API

It is well known, however, that the code generated by such "naive" class
libraries is often very inefficient [7], due to the unwanted copies caused by tem-
poraries[2]. This has led, at least in the domain of C++ scientific computing,
to the development of *Active Libraries* [13,12,8,9], which both provide domain-
specific abstractions and dedicated code optimization mechanisms. This paper
describes how this approach can be applied to the specific problem of efficient
Altivec code generation from a high-level C++ API. It is organized as follows.
Sect. 2 explains why generating efficient code for vector expressions is not trivial and introduces the concept of template-based meta-programming. Sect. 3
explains how this can used to generate optimized Altivec code. Sect. 4 rapidly
presents the API of the library we built upon these principles. Performance results are presented in Sect. 5. Sect. 6 is a brief survey of related work and Sect. 7
concludes.

2   Template Based Meta-programming

The problem of generating efficient code for vector expressions can be understood
– in essence – starting with the code sample given in Fig. 1. Ideally, this code
should be inlined as:

```
for ( i=0; i<1000; i++ ) r[i] = a[i] * b[i] + c[i];
```

In practice, due to the way overloaded operators are handled in C++, it is
developed as:

```
Vector<float> __t1(1000), __t2(1000);
for (int i=0; i < 1000; ++i)  __t1[i] = a[i] * b[i];
for (i=0; i < 1000; ++i) __t2[i] = __t1[i] * c[i];
for (i=0; i < 1000; ++i) r[i] = __t2[i];
```

The allocation of temporary vectors and the redundant loops result in poor efficiency. For more complex expressions, the performance penalty can easily reach
one order of magnitude. In this case, it is clear that expressiveness is obtained
at a prohibitive cost. This problem can be overcome by using an advanced C++
technique known as *expression templates*. The basic idea of *expression templates*

---

[2] In our case, the observed speedup for code like the one given in Fig. 1 went down
from 3.9 – when coded with the native C API – to less than 0.8 – with a "naive"
`Vector` class.

is to create parse trees of vector expressions at compile time and to use these parse trees to generate customized loops. This is actually done in two steps[3]. In the first step, the parse trees – represented as C++ types – are constructed using recursive template instantiations and overloaded versions of the vector operators (+, *, etc.). In the second step, these parse trees are "evaluated" using overloaded versions of the assignment (=) and indexing ([]) operators to compute the left-hand side vector in a single pass, with no temporary. Both steps rely on the definition of two classes:

- an Array class, for representing application-level vectors (arrays); this class has a member data_, where the array elements are stored and an operator [] for accessing these elements:

  ```
  float Array::operator[](int index) { return data_[index]; }
  ```

- an Xpr class, for representing (encoding) vector *expressions* :

  ```
  template<class LEFT,class OP,class RIGHT> class Xpr {};
  ```

Consider for example, the statement r=a*b+c given in the code sample above. Its right-hand side expression (a*b+c, where a, b and c have type Array) will be encoded with the following C++ type:

```
Xpr<Xpr<Array,mul,Array>,add,Array>
```

This type will be automatically built from the expression syntax a*b+c using overloaded versions of the + and * operators:

```
template<class T> Xpr<T,add,Array> operator+(T, Array)
  { return Xpr<T,add,Array>(); }
template<class T> Xpr<T,mul,Array> operator*(T, Array)
  { return Xpr<T,mul,Array>(); }
```

The "evaluation" (at compile time) of the encoded vector expression is carried out by an overloaded version of the assignment operator (=):

```
template<class T> Array& Array::operator=( const T& xpr ) {
  for(int i=0;i<size;i++) data_[i] = xpr[i];
  return *this;
}
```

For this, the Xpr class provides an operator[] method, so that each element of the result vector (data_[i]) gets the value xpr[i]:

```
template<class LEFT,class OP,class RIGHT>
float X<LEFT,OP,RIGHT>::operator[](int index)
  { return OP::eval(left_[index],right_[index]); }
```

---

[3] The presentation given here is deliberately simplified, due to space limitations. More details can be found, for example, in Veldhuizen's papers [5,6,7].

where `left_` and `_right` are the members storing the left and right sub-expressions of an `Xpr` object and `eval` the static method of the C++ functor associated with the vector operation `OP`. Such a functor will be defined for each possible operation. For example, the `add` and `mul` functors associated with the `+` and `*` vector operators are defined as:

```
class add { static float eval(float x,float y) { return x+y; } };
class mul { static float eval(float x,float y) { return x*y; } };
```

Using the mechanism described above, a standard C++ compiler can reduce the statement `r=a*b+c` to the following "optimal" code:

```
for ( i=0; i<1000; ++i) r[i] = a[i]*b[i]+c[i];
```

3 Application to the Generation of Efficient AltiVec Code

The template-based meta-programming technique described in the previous section can readily be adapted to support the generation of efficient Altivec code. For this:

- the `Array` class must provide a `load` method returning an Altivec vector instead of a scalar,
- the `add` (resp. `mul`, etc.) functor must call the native `vec_add` (resp. `vec_mul`, etc.) instruction,
- the assignment operator must use the native `vec_st` instruction to store the result:

```
vector float Array::load(int index) { return vec_ld(data_,index*16); }

class add {eval(vector float x,vector float y) { return vec_add(x,y); }};

template<class T>
Array& Array::operator=(T xpr) {
  for(int i=0;i<size/4;i++) vec_st(xpr.load(i),0,data);
  return *this;
}
```

With this approach, the assembly code resulting from the compilation of the previous example (`r=a*b+c`) contains three vector load operations, one vector addition, one vector multiplication and one vector store, which is clearly optimal.

4 The EVE Library

Using the code generation technique described in the previous section, we have produced a high-level array manipulation library aimed at scientific computing and taking advantage of the SIMD acceleration offered by the Altivec extension on PowerPC processors. This library, called EVE (for *Expressive Velocity Engine*) basically provides two classes, `array` and `matrix` – for 1D and 2D arrays –, and a rich set of operators and functions to manipulate them. This set can be roughly divided in four families:

1. **Arithmetic and boolean operators**, which are the direct vector extension of their C++ counterparts. For example:

```
array<char>  a(64),b(64),c(64),d(64);
d = (a+b)/c;  // d[i] = (a[i]+b[i])/c[i], for i=0...63
```

2. **Boolean predicates.** These functions can be used to manipulate boolean vectors and use them as selection masks. For example:

```
array<char>  a(64),b(64),c(64);
c = where(a<b, a, b); // c[i] = a[i]<b[i] ? a[i] : b[i], for i=0...63
```

3. **Mathematical and STL functions.** These functions work like their STL or math.h counterparts. The only difference is that they take an array (or matrix) as a whole argument instead of a couple of iterators. Apart from this difference, EVE functions and operators are very similar to their STL counterparts (the interface to the EVE array class is actually very similar to the one offered by the STL valarray class. This allows algorithms developed with the STL to be ported (and accelerated) with a minimum effort on a PowerPC platform with EVE. Example:

```
array<float>  a(64),b(64);
b = tan(a);  // b[i] = tan(a[i]) for each i=0..63
float r = inner_product(a, b); // r = a[0]*b[0]+...+a[63]*b[63]
```

4. **Signal processing functions.** These functions allow the direct expression (without explicit decomposition into sums and products) of 1D and 2D FIR filters. For example:

```
array<float>  a(64),b(64);
filter< 3,1,mask<1,2,1> >  gaussian;
res = gaussian(image);
```

5   Performance

Two kinds of performance tests have been performed: basic tests, involving only one vector operation and more complex tests, in which several vector operations are composed into more complex expressions. All tests involved vectors of different types (8 bit integers, 16 bit integers, 32 bit integers and 32 bit floats) but of the same total length (16 Kbytes) in order to reduce the impact of cache effects on the observed performances[4]. They have been conducted on a 1.2GHz PowerPC G4 with gcc 3.3.1 and the following compilation switches: -faltivec -ftemplate-deph-128 -O3. A selection of performance results is given in Table 1. For each test, four numbers are given: the maximum theoretical speedup[5] (TM), the measured speedup for a hand-coded version of the test using the native C API (NC), the measured speedup with a "naive" vector library – which does not use the expression template mechanism described in Sect. 2 (NV), and the measured speedup with the EVE library.

---

[4] I.e. the vector size (in elements) was 16K for 8 bit integers, 8K for 16 bit integers and 4K for 32 bits integers or floats.

[5] This depends on the type of the vector elements : 16 for 8 bit integers, 8 for 16 bit integers and 4 for 32 bit integers and floats.

Table 1. Selected performance results

Test	Vector type	TM	NC	NV	EVE
1. v3=v1+v2	8 bit integer	16	15.7	8.0	15.4
2. v2=tan(v1)	32 bit float	4	3.6	2.0	3.5
3. v3=v1/v2	32 bit float	4	4.8	2.1	4.6
4. v3=v1/v2	16 bit integer	8(4)	3.0	1.0	3.0
5. v3=inner_prod(v1,v2)	8 bit integer	8	7.8	4.5	7.2
6. v3=inner_prod(v1,v2)	32 bit float	4	14.1	4.8	13.8
7. 3x1 FIR	8 bit integer	8	7.9	0.1	7.8
8. 3x1 FIR	32 bit float	4	3.7	0.1	3.7
9. v5=sqrt(tan(v1+v2)/cos(v3*v4))	32 bit float	4	3.9	0.04	3.9
10. Image processing algorithm	16 bit integer	8	6.9	0.1	2.7

It can be observed that, for most of the tests, the speedup obtained with EVE is close to the one obtained with a hand-coded version of the algorithm using the native C API. By contrast, the performances of the "naive" class library are very disappointing (especially for tests 7-10). This clearly demonstrates the effectiveness of the metaprogramming-based optimization.

Tests 1-3 correspond to basic operations, which are mapped directly to a single AltiVec instruction. In this case, the measured speedup is very close to the theoretical maximum. For test 3, it is even greater. This effect can be explained by the fact that on G4 processors, and even for non-SIMD operations, the Altivec FPU is already faster than the scalar FPU[6]. When added to the speedup offered by the SIMD parallelism, this leads to super-linear speedups. The same effect explains the result obtained for test 6. By contrast, test 4 exhibits a situation in which the observed performances are significantly lower than expected. In this case, this is due to the asymmetry of the Altivec instruction set, which does not provide the basic operations for all types of vectors. In particular, it does not include division on 16 bit integers. This operation must therefore be emulated using vector float division. This involves several type casting operations and practically reduces the maximum theoretical speedup from 8 to 4.

Tests 5-9 correspond to more complex operations, involving *several* AltiVec instructions. Note that for tests 5 and 7, despite the fact that the operands are vectors of 8 bit integers, the computations are actually carried out on vectors of 16 bit integers, in order to keep a reasonable precision. The theoretical maximum speedup is therefore 8 instead of 16.

In order to show that EVE can be used to solve realistic problems, while still delivering significant speedups, we have used it to vectorize several complete image processing algorithms. Test 10, for example, give the performances obtained with an algorithm performing the detection of *points of interest* in grey scale images using the Harris filter [1]. This algorithm involves several filtering steps (on both directions of the image) and matrix computations. The measured speedup, while being lower than the one obtained with the hand-coded version,

---

[6] It has more pipeline stages and a shortest cycle time.

is *still* satisfactory if we take into account the large difference in code size and complexity between the two versions (15 lines of C++ with EVE, 80 lines of C with the Altivec native API).

6   Related Work

Since its introduction, most of the development for the Altivec has been conducted using the native C API and very few projects have proposed higher-level alternatives to this design flow.

Apple proposes the **VecLib** [3] library, as a complement to the native C API. This framework provides software equivalent of some missing functions and more complex operations such as FFT or convolution. Compared to EVE, the level of expressiveness is lower (VecLib does not support the construction of complex vector expressions by means of overloaded operators, in particular). Correlatively, no syntax directed optimization can be performed and performances drop when a lot of function calls are sequenced in the same program.

The **VAST** code optimizer [10] – which offers automatic vectorization and parallelization from source – has a specific back-end for generating Altivec code. This tool automatically replaces loops in C/C++ programs with inline vector extensions, and loops in Fortran programs with calls to newly-generated C functions with inline vector extensions. These vector extensions allow VAST to access the AltiVec unit at close to instruction level efficiency while remaining at the source code level. VAST's speedups are generally very close to those obtained with hand-vectorized code. VAST is a commercial product and costs 3000$.

The **Mac STL** [11] C++ library is very similar, in goals and design principles, to EVE. It provides a fast `valarray` class optimized for Altivec and relies on template-based metaprogramming techniques for code optimization. The only difference is that MacSTL only provides STL-compliant functions and operators (it is viewed as a specific implementation of the STL for G4/G5 computers) whereas EVE offers additional domain-specific functions (for signal processing, for example). Mac STL is available for a low-cost license.

7   Conclusion

We have shown how a classical technique – template-based metaprogramming – can be applied to the design and implementation of an efficient high-level array manipulation library aimed at scientific computing on PowerPC platforms. This library offers a significant improvement in terms of expressivity over the native C API traditionnaly used for taking advantage of the SIMD capabilities of this processor. It allows developers to obtain significant speedups without having to deal with low level implementation details. The EVE API is largely compliant with the STL standard and therefore provides a smooth transition path for applications written with other scientific computing libraries. A prototype version of the library can be downloaded from the following URL: http://wwwlasmea.univ-bpclermont.fr/Personnel/falcou/EVE/download.html. We are currently working on improving the performances obtained with this prototype for complex, realistic applications. This involves, for instance, globally minimizing the number of vector load and store operations, using more judiciously Altivec-specific cache manipulation instructions or taking advantage of fused operations (e.g. multiply/add). Finally, it can be noted that, although the current version of EVE has been designed for PowerPC processors with Altivec, it could be retargeted, with a moderate effort, to Pentium 4 processors with MMX/SSE2 because the code generator itself (using the expression template mechanism) can be made largely independent of the SIMD instruction set.

References

1. C. Harris and M. Stephens. A combined corner and edge detector. In 4th Alvey Vision Conference, 1988.
2. Apple. AltiVec Instructions References, Tutorials and Presentation. http://developer.apple.com/hardware/ve.
3. Apple. The VecLib framework. http://developer.apple.com/hardware/ve/vector_libraries.html
4. I. Ollman. AltiVec Velocity Engine Tutorial. http://www.simdtech.org/altivec. March 2001.
5. T. Veldhuizen. Using C++ Template Meta-Programs. In C++ Report, vol. 7, p. 36-43,1995.
6. T. Veldhuizen. Expression Templates. In C++ Report, vol. 7, p. 26-31, 1995.
7. T. Veldhuizen. Techniques for Scientific C++. http://osl.iu.edu/~tveldhui/papers/techniques/
8. T. Veldhuizen. Arrays in Blitz++. In Dr Dobb's Journal of Software Tools, p. 238-44, 1996.
9. The BOOST Library. http://www.boost.org/.
10. VAST. http://www.psrv.com/vast_altivec.html/.
11. G. Low. Mac STL. http://www.pixelglow.com/macstl/.
12. The POOMA Library. http://www.codesourcery.com/pooma/.
13. T. Veldhuizen and D. Gannon. Active Libraries: Rethinking the roles of compilers and libraries Proc. of the SIAM Workshop on Object Oriented Methods for Interoperable Scientific and Engineering Computing SIAM Press, 1998

Petri Nets as Executable Specifications of High-Level Timed Parallel Systems

Franck Pommereau

LACL, Université Paris 12 — 61, avenue du général de Gaulle
94010 Créteil, France
pommereau@univ-paris12.fr

Abstract. We propose to use *high-level Petri nets* for the semantics of *high-level parallel systems*. This model is known to be useful for the of verification and we show that it is also executable in a parallel way. Executing a Petri net is easy in general but more complicated in a *timed context*, which makes necessary to *synchronise* the *internal time* of the Petri net with the *real time* of its environment. Another problem is to relate the execution of a Petri net, which has its own semantics, to that of its environment; *i.e.*, to properly handle *input/output*.

This paper presents a parallel algorithm to execute Petri nets with time enforcing the even progression of the internal time with respect to the real time and allowing the exchange of information with the environment. We define a class of Petri nets suitable for a *parallel execution machine* which preserves the *step sequence semantics* of the nets and ensures time consistent executions while taking into account the solicitation of its environment. The question of the efficient verification of such nets has been addressed in a separate paper [4], the present one is focused on the practical aspects involved in the execution of so modelled systems.

Keywords: Petri nets, parallelism, real-time, execution machines.

1 Introduction

Petri nets are widely used as a model of *concurrency*, which allows to represent the occurrence of *independent* events. They can be as well a model of *parallelism*, where the *simultaneity* of the events is more important, when we consider their *step sequence semantics* in which an execution is represented by a sequence of *steps*, *i.e.*, simultaneous occurrences of transitions. In this paper, we consider high-level Petri nets for modelling high-level parallel systems, with the aim to allow both *verification and execution of the specification*. Petri nets like those used in this paper are already used has a semantical domain for parallel programming languages or process algebra, *e.g.*, [3,4]. These approaches could be directly applied to massively parallel languages or formalisms.

Executing a Petri net is not difficult when we consider it alone, *i.e.*, in a closed world. But as soon as the net is *embedded in an environment*, the question becomes more complicated. The first problem comes when the net is timed: we

M. Bubak et al. (Eds.): ICCS 2004, LNCS 3038, pp. 322–330, 2004.
© Springer-Verlag Berlin Heidelberg 2004

have to ensure that its time reference matches that of the environment. The second problem is to allow an exchange of information between the net and its environment. Both these questions are addressed in this paper.

This work is set in the context of *causal time* which allows to use "ordinary" (untimed) Petri nets by explicitly introducing a *tick transition* which increments *counters* used as clock watches by the rest of the system [2,5]. It was shown in [1, 4] that the causal time approach is highly relevant since it is simple to put into practice and allows for *efficient verification* through model checking. In this context, the tick transition of a Petri net may causally depend on the other transitions in the net, which results in the so called *deadline paradox* [2]: tick is disabled until the system progresses. In the closed world of verification, this statement is logically equivalent to "the system has to progress before the next tick", which solves the paradox. But, this is not the case in the open world of execution.

In this paper, we define a *parallel execution machine* whose role is to run a Petri net with a tick transition in such a way that the ticks occur evenly with respect to the real time. We show that this can be ensured under reasonable assumptions about the Petri net. The other role of the machine is to allow the communication between the Petri net and the environment. Producing output is rather simple since the net is not disturbed; but reading input (*i.e.*, changing the behaviour of the net in reaction to the changes in the environment) is more difficult and may not be always possible. We will identify favourable situations, very easy to obtain, in which the reaction to a message is ensured within a short delay. An important property of our execution machine is that it preserves the step sequence semantics of the Petri net: this machine can be seen as an implementation of the Petri net execution rule including constraints related to the environment (real time and communication).

This paper is an extended abstract of a technical report which can be found at http://www.univ-paris12.fr/lacl where more details (and proofs) are given.

2 Basic Definitions about Petri Nets

This section briefly introduces the class of Petri nets that will be used in the following. We assume that the reader is familiar with the notion of multisets and we denote by $\text{mult}(X)$ the set of all finite multisets over a set X.

Let $\mathbb{S}$ be a set of *actions symbols*, $\mathbb{D}$ a finite set of *data values* (or just *values*) and $\mathbb{V}$ a set of *variables*. For $F \subseteq \mathbb{S}$ and $X \subseteq \mathbb{D} \cup \mathbb{V}$, we denote by $F \otimes X$ the set $\{a(x) \mid a \in F, x \in X\}$. Then, we define $\mathbb{A} \stackrel{\text{df}}{=} \mathbb{S} \otimes (\mathbb{D} \cup \mathbb{V})$ as the set of *actions* (with parameters). These four sets are assumed pairwise disjoint.

A *labelled marked Petri net* is a tuple $N = (S, T, \ell, M)$ where:

- S is a non-empty finite set of *places*;
- T is a non-empty finite set of *transitions*, disjoint from S;
- ℓ defines the *labelling* of places, transitions and *arcs* (elements of $(S \times T) \cup (T \times S)$) as follows:

- for $s \in S$, the labelling is $\ell(s) \subseteq \mathbb{D}$ which defines the tokens that the place is allowed to carry (often called the *type* of s),
- for $t \in T$, the labelling is $\ell(t) \stackrel{\text{df}}{=} \alpha(t)\gamma(t)$ where $\alpha(t) \in \mathbb{A}$ and $\gamma(t)$ is a boolean expression called the *guard of t*,
- for $(x,y) \in (S \times T) \cup (T \times S)$, the labelling is $\ell(x,y) \in \text{mult}(\mathbb{D} \cup \mathbb{V})$ which denotes the tokens flowing on the arc during the execution of the attached transition. The empty multiset $\varnothing$ denotes the absence of arc;

- M is a *marking* function which associates to each place $s \in S$ a multiset in $\text{mult}(\ell(s))$ representing the tokens held by s.

Petri nets are depicted as usual with several simplifications: the two components of transition labels are depicted separately; true guards and brackets around sets are omitted; arcs may be labelled by expressions as a shorthand, like $n + 1$ in the figure 1 page 325 which could be replaced by $y \in \mathbb{V}$ by adding a guard $y = n + 1$ to the transition t_r.

A *binding* is a function $\sigma : \mathbb{V} \to \mathbb{D}$ which associates concrete values to the variables appearing in a transition and its arcs. We denote by $\sigma(E)$ the evaluation of the expression E bound by σ. Let (S, T, ℓ, M) be a Petri net, and $t \in T$ one of its transitions. A binding σ is *enabling* for t at M if the guard evaluates to true, *i.e.*, $\sigma(\gamma(t)) = \top$, and if the evaluation of the annotations on the adjacent arcs respects the types of the places, *i.e.*, for all $s \in S$, $\sigma(\ell(s,t)) \in \text{mult}(\ell(s))$ and $\sigma(\ell(t,s)) \in \text{mult}(\ell(s))$.

A *step* corresponds to the simultaneous execution of some transitions, it is a multiset $U = \{(t_1, \sigma_1), \ldots, (t_k, \sigma_k)\}$ such that $t_i \in T$ and σ_i is an enabling binding of t_i, for $1 \leq i \leq k$. U is *enabled* if the marking is sufficient to allow the flow of tokens required by the execution of the step. It is worth noting that if a step U is enabled at a marking, then so is any sub-step $U' \leq U$. A step U enabled by M may be *executed*, leading to the new marking M' defined for all $s \in S$ by $M'(s) \stackrel{\text{df}}{=} M(s) - \sum_{(t,\sigma) \in U} U((t,\sigma)) * \sigma(\ell(s,t)) + \sum_{(t,\sigma) \in U} U((t,\sigma)) * \sigma(\ell(t,s))$. This is denoted by $M[U\rangle M'$ which naturally extends to sequences of steps. A marking M' is *reachable* from a marking M if their exists a sequence of steps ω such that $M[\omega\rangle M'$; we will say in this case that M enables ω.

The *labelled step* associated to a step U is defined as $\sum_{(t,\sigma) \in U} U((t,\sigma)) * \sigma(\alpha(t))$, which allows to define the *(labelled) step sequence semantics* of a Petri net as the set containing all the sequences of (labelled) steps enabled by a net.

A Petri net (S, T, ℓ, M) is *safe* if any marking M' reachable from M is such that, for all $s \in S$ and all $d \in \ell(s)$, $M'(s)(d) \leq 1$, *i.e.*, any place holds at most one token of each value. The class of safe Petri nets is very interesting for both theoretical and practical reasons. In particular, they have finitely many reachable markings, each of which enabling finitely many steps whose sizes are bounded by the number of transitions in the net. As many previous works [5,1,4,3], this paper only considers safe Petri nets.

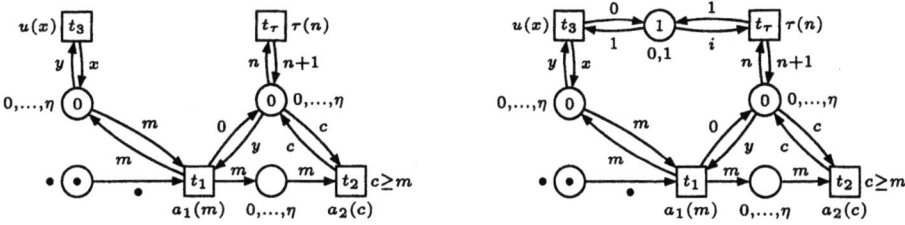

Fig. 1. On the left: an example of a CT-net, where $\eta > 0$, $\{a_1, a_2, u, \tau\} \subseteq \mathbb{S}$, $\{c, n, m, x, y\} \subseteq \mathbb{V}$ and $\{0, \ldots, \eta\} \cup \{\bullet\} \subseteq \mathbb{D}$. On the right: the tick-reactive version of this net, where $i \in \mathbb{V}$

3   Petri Nets with Causal Time: CT-Nets

The class of Petri nets we are actually interested in consists in safe labelled Petri nets, with several restrictions, for which we will define some specific vocabulary related to the occurrence of ticks. We assume that there exists $\tau \in \mathbb{S}$.

A *Petri net with causal time* (*CT-net*) is a safe Petri net (S, T, ℓ, M) in which there exists a unique $t_\tau \in T$ such that $\alpha(t_\tau) \in \{\tau\} \otimes (\mathbb{D} \cup \mathbb{V})$ and, for all $t \in T \setminus \{t_\tau\}$, we have $\alpha(t) \notin \{\tau\} \otimes (\mathbb{D} \cup \mathbb{V})$. Moreover, we impose that t_τ has at least one incoming arc labelled by a singleton. This transition t_τ is called the *tick transition* of the net. A *tick-step* is a step U which involves the tick transition, i.e., such that $\tau(d) \in U$ for a $d \in \mathbb{D}$. Thanks to the safety and the last restriction on t_τ, any tick-step contains exactly one occurrence of the tick transition.

The left of the figure 1 gives a toy CT-net in which the role of the tick transition is to increment a counter located in the top-right place. When the transition t_1 is executed, it resets this counter and picks in the top-left place a value which is bound to the variable m. This value is transmitted to the transition t_2 which will be executable when at least m ticks will have occurred. Thus, m specifies the minimum number of ticks between the execution of t_1 and that of t_2. At any time, the transition t_3 may randomly change the value of this minimum while emitting a visible action $u(x)$ where x is the new value. Notice that the maximum number of ticks between the execution of t_1 and that of t_2 is enforced by the type of the place connected to t_τ which specifies that only tokens in $\{0, \ldots, \eta\}$ are allowed (given $\eta > 0$).

Assuming $\eta \geq 5$, a possible execution of this CT-net is $\{\tau(0)\}\{u(2)\}\{a_1(2)\}$ $\{\tau(0), u(1)\}\{\tau(1)\}\{u(5)\}\{\tau(2)\}\{\tau(3)\}\{a_2(3), u(0)\}\{\tau(4)\}$.

A CT-net (S, T, ℓ, M) is *tractable* if there exists an integer $\delta \geq 2$ such that, for all marking M' reachable from M, any sequence of at least δ non-empty steps enabled by M' contains at least two tick-steps. In other words, the length of an execution between two consecutive ticks is bounded by δ whose smallest possible value is called the *maximal distance between ticks*. This notion of tractable nets is important because it allows to distinguish those nets which can be executed on a realistic machine: indeed, a non-tractable net may have potentially infinite runs between two ticks (so called *Zeno runs*), which cannot be executed on a finitely fast computer without breaking the evenness of ticks occurrences. For example, the CT-net of our running example is not tractable because the transition t_3 can be executed infinitely often between two ticks: in the execution given above, the step $\{u(5)\}$ could be arbitrarily repeated.

The communication between a CT-net and its environment is modelled using some of the actions in transitions labels. We distinguish for this purpose two finite disjoint subsets of $\mathbb{S}$: $\mathbb{S}_i$ is the set of *input action symbols* and $\mathbb{S}_o$ is that of *output actions symbols*. We assume that $\tau \notin \mathbb{S}_i \cup \mathbb{S}_o$. We also distinguish a set $\mathbb{D}_{io} \subseteq \mathbb{D}$ representing the values allowed for input and output. Intuitively, the distinguished symbols correspond to communication ports on which values from $\mathbb{D}_{io}$ may be exchanged. Thus the execution of a transition labelled by $a_o(d_o) \in \mathbb{S}_o \otimes \mathbb{D}_{io}$ is seen as the sending of the value d_o on the output port a_o. Conversely, if the environment sends a value $d_i \in \mathbb{D}_{io}$ on the input port $a_i \in \mathbb{S}_i$, the net is expected to execute a step containing the action $a_i(d_i)$. In general, we cannot ensure that such a step is enabled, in the worst case, it may happen that no transition has a_i in its label.

A CT-net is *reactive* to a set of action symbols $R \subseteq \mathbb{S}_i$ if: (1) either, for all $a \in R$ and all $d \in \mathbb{D}_{io}$, it always allows, but never forces, the execution of a step containing $a(d)$; (2) or the net is in a marking M from which only actions in $\mathbb{S}_i \cup \mathbb{S}_o$ may ever be executed (M is called *final*). Thus, a net which is reactive to some actions will always allow a good responsiveness to the solicitation of the environment using these actions. It turns out that building such a net is very easy in general: for instance, the net given on the left of the figure 1 is reactive to $\{u\}$ (assuming $\mathbb{D}_{io} \subseteq \{0, \ldots, \eta\}$) thanks to the self loop on t_3.

Unfortunately, it can be shown that any CT-net N which is reactive $R \subseteq \mathbb{S}_i$ is not tractable. This negative result shows that the intuitive notion of reactiveness is too strong: the non-tractability actually indicates that a reactive CT-net is expected to be able to respond instantaneously to all the messages that the environment would send on a port in R. But if the number of such messages sent in a given amount of real time is not bounded then a finitely fast computer cannot avoid to miss some of them. We thus assume that the environment may not produce more than one message on each port between two ticks, which leads to the new notion of tick-reactiveness.

We denote by $U[a]$ the number of occurrences of the action symbol a in a step U, i.e., $U[a] \stackrel{\text{df}}{=} \sum_{a(x) \in U} U(a(x))$. Let N be a CT-net whose marking is M_0 and $R \subseteq \mathbb{S}_i$, consider an execution $M_0[U_1\rangle M_1 \cdots [U_k\rangle M_k$ of N such that only U_k may be a tick-step (if it is not, M_k should enable only the empty step), and define $U_0 \stackrel{\text{df}}{=} \varnothing$. Then, N is *tick-reactive* to R if: (1) for $0 \leq i < k$, the marking M_i is reactive to $R \setminus \cup_{0 \leq j \leq i} \{a \in R \mid U_j[a] > 0\}$; (2) N with the marking M_k is tick-reactive to R. This definition is inductive and holds over the executions of a CT-net. Intuitively, it states that N can react to any message sent on $a \in R$ after what it may miss them until the next tick, being then able to react again. This guarantees that one message on a may always be handled between two ticks, which exactly matches our assumption. It turns out that it is generally easy to transform a reactive CT-net into a tick-reactive one. For instance, the right of the figure 1 shows a modified version of our running example which is tick-reactive to $\{u\}$ and tractable (the step $\{u(5)\}$ could not be duplicated now).

A step U is *consistent* if $U[a] \leq 1$ for all $a \in \mathbb{S}_i \cup \mathbb{S}_o$. A CT-net is *consistent* if none of its reachable markings enables a non-consistent step. Non-consistent steps are those during the execution of which several communications can take place on the same port. Since the transitions executed by a single step occur simultaneously, this means that several distinct values may be sent or received on the same port at the same time. This is certainly something which is not realistic and should be rejected. The nets given in the figure 1 are both consistent.

4   Compilation and Execution

We now show how to transform a tractable and consistent CT-net into a form more suitable to the execution machine. This corresponds to a compilation whose result will be an automaton (non-deterministic in general), called a *CT-automaton*, whose states will be the reachable markings of the net and whose transitions will correspond to the steps allowing to reach one marking from another. It should be remarked that this compilation is not required but allows to simplify things a lot, in particular in an implementation of the machine: with respect to its corresponding CT-net, a CT-automaton has no notion of markings, bindings, enabling, etc., which results in a much simpler model.

In order to record only the input and output actions in a step U of a CT-net, we define the set of the *visible actions in U* by $\lfloor U \rfloor \stackrel{\text{df}}{=} U \cap (((\mathbb{S}_i \cup \mathbb{S}_o) \otimes \mathbb{D}_{io}) \cup (\{\tau\} \otimes \mathbb{D}))$. Because of the consistency, $\lfloor U \rfloor$ could not be a multiset.

Let $N = (S, T, \ell, M)$ be a tractable and consistent CT-net, the *CT-automaton of N* is the finite automaton $\mathcal{A}(N) \stackrel{\text{df}}{=} (S_\mathcal{A}, T_\mathcal{A}, s_\mathcal{A})$ where:

- $S_\mathcal{A}$ is the set of states defined as the set of all the reachable markings of N;
- the set of transitions is $T_\mathcal{A} \subseteq S_\mathcal{A} \times L_\mathcal{A} \times S_\mathcal{A}$, where $L_\mathcal{A} \stackrel{\text{df}}{=} \{A \subseteq ((\mathbb{S}_i \cup \mathbb{S}_o) \otimes \mathbb{D}_{io}) \cup (\{\tau\} \otimes \mathbb{D})\}$, and is defined as the set of all the triples (M', A, M'') such that $M', M'' \in S_\mathcal{A}$ and there exists a non-empty step U of N such that $M[U\rangle M'$ and $A = \lfloor U \rfloor$;
- $s_\mathcal{A} \stackrel{\text{df}}{=} M \in S_\mathcal{A}$ is the initial state of $\mathcal{A}(N)$, *i.e.*, the initial marking of N.

We now describe the execution machine. In order to communicate with the environment, a symbol $a_o \in \mathbb{S}_o$ is considered as a port on which a value $d \in \mathbb{D}_{io}$ may be written, which is denoted by $a \leftarrow d$ (more generally, this is used for any assignment). Similarly, a symbol $a_i \in \mathbb{S}_i$ is considered as a port on which such a value, denoted by $a_i?$, may be read; we assume that $a_i? = \circ \notin \mathbb{D}$ when no communication is requested on a_i. Moreover, in order to indicate to the environment if a communication have been properly handled, we also assume that each $a \in \mathbb{S}_i$ may be marked "accepted" (the communication has been correctly handled), "refused" (the communication could not been handled), "erroneous" (a communication on this port was possible but with another value, or that a communication was expected but not requested) or not marked, which

```
 1: s ← s_A                          procedure execute(A, I) :
 2: Θ ← now                         17: for all a(d) ∈ A (a ≠ τ) do
 3: while s has successors do       18:    if a ∈ S_o then
 4:    for all a ∈ S_i do           19:       a ← d
 5:       a ← "no mark"             20:    else if a ∈ S_i and a? = d then
 6:    end for                      21:       a ← "accepted"
 7:    I ← {a ∈ S_i | a? ≠ ∘}       22:    else
 8:    choose a transition (s, A, s') 23:       a ← "erroneous"
 9:    if A is a tick step then     24:    end if
10:       wait until now = Θ + Δ    25:    I ← I \ {a}
11:       Θ ← now                   26: end for
12:    end if                       27: for all a ∈ I do
13:    execute(A, I)                28:    a ← "refused"
14:    s ← s'                       29: end for
15: end while
```

Fig. 2. The main loop of the execution machine (on the left) and the execution of a step A with respect to requested inputs given by I (on the right).

is represented by "no mark". We also use the notation $a_i \leftarrow$ "mark" when an input port is being marked.

Let (S_A, T_A, s_A) a CT-automaton and let Δ be an amount of time (defined below). We will use three variables: Θ is a time corresponding to the occurrences of ticks; $s \in S_A$ is the current state; $I \subseteq S_i$ is a set of input ports. The statement "now" evaluates to the current time when it is executed.

The behaviour of the machine is given in the figure 2. Several aspects of this algorithm should be commented:

- the "**for all**" loops are parallel loops;
- each execution of the "**while**" loop performs a bounded amount of work, in particular the following numbers are bounded: the number of ports; the number of transitions outgoing from a state; the number of actions in each step. Assuming that choosing a transition requires a fixed amount of time (see below), Δ is the maximum amount of time required to execute the "**while**" loop $\delta - 1$ times;
- no tick is explicitly executed but its occurrence actually corresponds to the execution of the line 11;
- one can show that the even occurrence of the ticks is ensured.

We still have to define how a transition may be chosen, in a fixed amount of time, in order to mark "accepted" as much as possible input ports in I. To start with, we assume a total order on S_i. This corresponds to a priority between the ports: when several communications are requested but not all are possible, we first serve those on the ports with the highest priorities. Then, given I, we define a partial order $\prec$ on the transitions outgoing from a state and the machine chooses one of the smallest transitions according to $\prec$. This choice may be random or driven by a scheduler. For instance, we may choose to execute steps as large as possible, or steps no larger than the number of processors, etc. But this discussion is out of the scope of this paper.

The partial order $\prec$ is based on the lexicographic order on the vectors $V_A \in$ {"accepted", "no mark", "refused", "erroneous"}^{S_i} obtained by simulating the execution of each step A and assuming that the marks are ordered as given above. Again, it is clear that building these vectors and choosing the smallest is feasible in a fixed amount of time since the number of transitions outgoing from a given state is bounded. This is also feasible in parallel: all the V_A's can be computed in parallel and the selection of the smallest one is a reduction. Notice that if $\prec$ allows to define a total order on steps, it is not the case for the transitions since several transitions may be labelled by the same step.

Proposition 1. *Let $a \in S_i$ be an input action symbol and N be a CT-net which is tick-reactive to $R \ni a$. Then, the execution of $\mathcal{A}(N)$ will never mark a as "erroneous" nor "refused" except from a state which is a final marking of N. Moreover, if $a? = d \neq \circ$ before the execution of the line 7 in the figure 2, then a is marked "accepted" after the line 13 has executed.*

5   Conclusion

We defined a parallel execution machine which shows the adequacy of causal and real time by allowing time-consistent executions of causally timed Petri nets (CT-nets) in a real-time environment. We also shown that it was possible to ensure that the machine efficiently reacts to the solicitation of its environment by designing CT-nets having the property of tick-reactiveness, which is easy to ensure. In order to obtain these results, several restrictions have been adopted: (1) only safe Petri nets are considered; (2) the nets must be *tractable*, *i.e.*, they cannot have unbounded runs between two ticks; (3) the nets must be *consistent*, *i.e.*, they cannot perform several simultaneous communications on the same port; (4) the machine must be run on a computer fast enough to ensure that the environment cannot attempt more than one communication on a given port between two ticks. We do not consider the tractability and consistency requirements as true restrictions since they actually correspond to what can be performed on a realistic machine. The last restriction is actually a prescription which can be ensured after measuring physical properties of the environment and choosing an appropriate computer. Future works may consider using non-safe Petri nets, but this class happens to be expressive enough for many interesting problems and there may be no real need to remove the first restriction.

Petri nets like CT-nets have been used for a long time as a semantical domain for high-level parallel programming languages and process algebras (see, *e.g.*, [3, 4]) and these techniques could be directly applied to massively parallel languages or formalisms. Considering the features of the execution machine combined with the fact that CT-nets allow for efficient verification, we obtain a framework in which the analysed model and the executed code are the same object, which saves from the risk of implementation errors.

References

1. C. Bui Thanh, H. Klaudel and F. Pommereau. *Petri nets with causal time for system verification.* MTCS 2002. ENTCS 68(5), Elsevier, 2003.
2. R. Durchholz. *Causality, time, and deadlines.* Data & Knowledge Engineering, 6. North-Holland, 1991.
3. H. Klaudel. *Compositional High-Level Petri nets Semantics of a Parallel Programming Language with Procedures.* SCP 41, Elsevier, 2001.
4. F. Pommereau. *Causal Time Calculus.* FORMATS'03. LNCS, Springer, to appear.
5. G. Richter. *Counting interfaces for discrete time modelling.* Technical report 26, GMD. September 1998.

Parallel I/O in Bulk-Synchronous Parallel ML

Frédéric Gava

LACL, University Paris XII, Créteil, France
gava@univ-paris12.fr

Abstract. Bulk Synchronous Parallel ML or BSML is a functional data-parallel language for programming bulk synchronous parallel (BSP) algorithms. The execution time can be estimated and dead-locks and indeterminism are avoided. For large scale applications where parallel processing is helpful and where the total amount of data often exceeds the total main memory available, parallel disk I/O becomes a necessity. We present here a library of I/O features for BSML and its cost model.

1 Introduction

External memory (EM) algorithms are designed for *large computational* problems in which the size of the internal memory of the computer is only a small fraction of the size of the problem. Important applications fall into this category ([12] for a survey). Parallel processing is an important issue for EM algorithms for the same reasons that parallel processing is of practical interest in non-EM algorithm design. The combination of I/O computing, on multiple disks, with multiprocessor parallelism is a challenge for the "Large-Scale" computing community. Bulk-Synchronous Parallel ML or *BSML* is an extension of ML for programming *Bulk-Synchronous Parallel* (BSP) algorithms as functional programs associated with a compositional cost model. BSP computing is a parallel programming model introduced by Valiant [11] to offer a high degree of abstraction like PRAM models. Such algorithms offer portable, predictable and scalable performances on a wide variety of architectures ([8] for a survey). BSML expresses them with a small set of primitives taken from the *confluent* BSλ-calculus. Those operations are implemented as a parallel library (http://bsmllib.free.fr) for the functional programming language Objective Caml (http://www.ocaml.org).

Parallel disk I/O has been identified as a *critical component* of a suitable *high performance* computer. [2] showed how an EM machine can take full advantage of parallel disk I/O and multiple processors. This model is based on an extension of the BSP model for I/O accesses. To take advantage of these new results, we have to extend the BSML language with parallel I/O features for programming this new kind of algorithms. This paper describes our first work in this direction. The remainder of this paper is organized as follows. In Section 2 we briefly present the BSML language. In Section 3 we introduce the EM-BSP model and the problems that appear in BSML. We then give in Section 4 the new primitives for BSML, the associated cost model and an example. We discuss related work and conclude (section 5). This paper is an extended abstract of a technical report which can be found at http://www.univ-paris12.fr/lacl where more details are given.

M. Bubak et al. (Eds.): ICCS 2004, LNCS 3038, pp. 331–338, 2004.
© Springer-Verlag Berlin Heidelberg 2004

2   Functional Bulk-Synchronous Parallel ML

A BSP computer contains a set of *processor-memory* pairs, a *communication network* allowing inter-processor delivery of messages and a *global synchronization unit* which executes collective requests for a *synchronization barrier*. In this model, a parallel computation is subdivided in *super-steps*, at the end of which a barrier synchronization and a routing is performed. Hereafter all requests for data were posted during a preceding super-step are fullfilled.

There is currently no implementation of a full BSML language but rather a partial implementation as a library (the BSMLlib library) for Objective Caml. In particular, it offers the function bsp_p:unit->int such as the value of bsp_p() is p, the *static number* of processes of the parallel machine. There is also an *abstract polymorphic* type 'a par which represents the type of p-wide parallel vectors of objects of type 'a, one per process. The nesting of par types is prohibited. Our static analysis enforces this restriction [3]. The BSML parallel constructs operate on parallel vectors. Parallel vectors are created by:

$$\text{mkpar: (int -> 'a) -> 'a par}$$

so that (mkpar f) stores (f i) on process i for i between 0 and $(p-1)$ ((mkpar f)=$\langle (f\ 0),\ldots,(f\ (p-1))\rangle$). These values are said to be *local*. The expression (mkpar f) is a parallel object and it is said to be *global* (which is similar on each processor). A BSP algorithm is expressed as a combination of asynchronous local computations (first phase of a superstep) and phases of global communication (second phase of a superstep) with global synchronization (third phase of a superstep). Asynchronous phases are programmed with mkpar and with:

$$\text{apply: ('a -> 'b) par -> 'a par -> 'b par}$$

so that apply (mkpar f) (mkpar e) stores (f i) (e i) on process i. The communication and synchronization phases are expressed by:

$$\text{put:(int->'a option) par -> (int->'a option) par}$$

where 'a option is defined by: type 'a option = None | Some of 'a. Consider the expression: put(mkpar(fun i->fs$_i$))(∗). To send a value v from process j to process i, the function fs$_j$ at process j must be such that (fs$_j$ i) evaluates to Some v. To send no value from process j to process i, (fs$_j$ i) must evaluate to None. Expression (∗) evaluates to a parallel vector containing a function fd$_i$ of delivered messages on every process. At process i, (fd$_i$ j) evaluates to None if process j sent no message to process i or evaluates to Some v if process j sent the value v to the process i. The full language would also contain a synchronous global conditional, omitted here for the sake of conciseness.

3   External Memories in BSML

3.1   The EM-BSP Model

In the BSP model, the performance of a parallel computer is characterized by only three parameters (expressed as multiples of the local processing speed): p the number of processors, l the time required for a global synchronization and g the time for collectively delivering a 1-relation (communication phase where

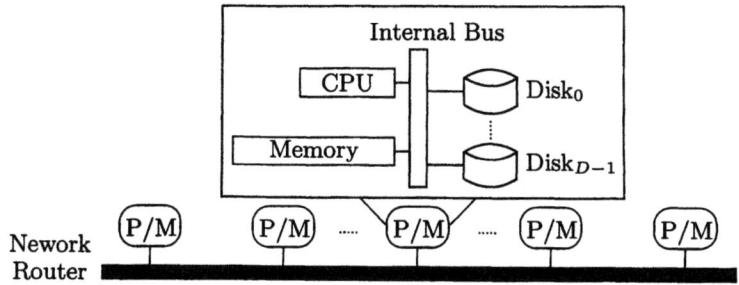

Fig. 1. A BSP Computer with External Memories

every processor receives/sends at most one word). The network can deliver an h-relation in time gh for any arity h.

[2] extended the BSP model to include secondary local memories. The basic idea is very simple and it is illustrated in Figure 1. Each processor has, in addition to its local memory, an EM in the form of a set of *disks*. Modern computers typically have several layers of memory which include main memory and caches as well as disks. We restrict ourselves to the two-level model because the speed difference between disk and main memory is much more signifiant than between the other layers of memory. This idea is applied to extend the BSP model to its EM version **EM-BSP** by adding the following parameters to the standard BSP parameters: M is the local memory *size* of each processor, D the number of *disk drives* of each processor, B the *transfer block size* of a disk drive and G is the *ratio* of local computational capacity (number of local computation operations) divided by local I/O capacity (number of blocks of size B that can be transferred between the local disks and memory) per unit time. In many practical cases, all processors have the *same number* of disks and, thus, the model is restricted to that case (although the model forbids different numbers of drives and memory sizes for each processor). The disk drives of each processor are denoted by $\mathcal{D}_0, \mathcal{D}_1, \ldots, \mathcal{D}_{D-1}$. Each processor can use all its D disk drives concurrently, and transfer $D \times B$ items from the local disks to its local memory in a single I/O operation being at cost G (thus, we do not deal with the intricacies of blocked secondary memory accesses). Each processor is assumed to be able to store in its local memory at least one block from each disk at the same time, i.e., $DB << M$. Like computation on the BSP model (for the sake of conciseness, we refer to [8] for more details), the computation on the EM-BSP model proceeds in a succession of super-steps and it allows multiple I/O operations during the computation phase of the super-step.

3.2 Problems by Adding I/O in BSML

The main problem in adding external memory (and so I/O operators) to BSML is to keep safe the fact that in the global context, the values on each processor are the same. For example, take the following expression:

```
let our_channel= open_in "file.txt" in
 let our_value=(input_value our_channel)) in ...
```

It is not true that the files (or the associate *channel*) on each processor contain the same value and in this case, each processor reads on its secondary memory a different value. If this expression had been evaluated with the BSMLlib library, and we would have obtained an incoherent result (and a crash of the BSP machine). Another problem come from *side-effects* that can occur on each processor:

```
let a=mkpar(fun i->if(i=0)then(open_in "file.txt");skip else skip)
 in (open_out "file.txt")
```

where only the first processor has opened a file in read mode and after, each processor opened the same file in write mode except the first processor: the file has already been open and we have an incoherent result. Our solution is to add two kinds of files: global and local ones. In this way, we also add two kinds of I/O operators. Local I/O operators do not have to occur in the global context (also global I/O operators do not have to occur locally) and the global files need to be the same on each node (in a *shared disks* or as a copy of the files on each process). An advantage having shared disks is in the case of some algorithms (as those which sort) when we have only one file (the list of data to sort) at the beginning of the program and one file (the sorted data) at the end. On the other hand, in the case of a distributed global file system, the global data are also distributed and programs are less sensitive to the problem of *faults*.

Thus, we have two important cases for the global file system which could be seen as new parameters of the EM-BSP machine: we have shared disks or not. In the first case, the condition that the global files are the same for each processor point of view requires some synchronizations for some global I/O operators. For example, when the program either created or deleted a file because it seems to be impossible (un-deterministic) for a process to create a file in the global file system if at the same time another process deleted it. On the other hand, reading (resp. writing) values from files do not need any synchronization (only one of the processors need to really write the value on the shared disks). In the second case, all the files are distributed and no synchronization is needed (each processor read/write/delete etc. in its own file system) but at the beginning, the global files systems need to be empty. By this way, a new *ratio* (G^g) is needed which could be G is their is no shared disks. We supposed to have the same D and B for the shared disks (if not we can simulating these by cut the shared disks and change the G^g constant).

4   New Primitives and Their Costs

4.1   New Primitives

In this section we describe the core of our library, i.e, the minimal set of functions for programming EM-BSP algorithms. This library will be incorporated in the next release of the BSMLlib. As in the BSMLlib library, we used MPI and we have functions to access to the EM-BSP parameters of the underlining architecture. In particular, if offers the functions embsp_D:unit->int which gives the number of disks and is_global_shared: unit->bool which gives if the global file system is shared or not. Since having two file systems, we need two abstract types of input channels and output channels: glo_in_channel (resp. loc_in_channel) and glo_out_channel (resp. loc_out_channel) to read or write in a global (resp. local) file. Therefore, we can open the named files for writing, and return a new output channel on that file, positioned at the beginning of the file. For this, we have two kinds of functions for global and local files:

```
glo_open_out : string -> glo_out_channel
loc_open_out : string -> loc_out_channel
```

The file is truncated to zero length if it already exists. It is created if it does not already exist. Raise Sys_error if the file could not be opened. In the same manner, we have two functions for opening a named file in read mode which returns a new input channel positioned at the beginning of the file. Now, with our channel, we can read and write values to the files. To do this, we need to "serialize" our values, i.e. transform our values to be written on a file: the module Marshal of the Objective Caml language provides functions to encode arbitrary data structures as sequences of bytes, which can then be written on a file and can then be read back, possibly by another process. To Write the representation of a structured value of any type to a channel (global or local), we used the following functions:

```
glo_output_value : glo_out_channel -> 'a -> int
loc_output_value : loc_out_channel -> 'a -> int
```

which return the number of I/O operations used to write the values. The object can be read back, by the read functions:

```
glo_input_value : glo_in_channel -> int * 'a
loc_input_value : loc_in_channel -> int * 'a
```

which read the representation of a structured value and return the corresponding value with the number of I/O operations that have been done (we refer to [7] in order to have type safe values in channel and read it). To write (or read) on the D-disks of the machines, we used the thread facilities of Objective Caml: we create D-threads which write (or read) on the D-disks. The last primitive copies a local file from a processor to the global files system:

```
glo_copy : int -> string -> string -> unit par
```

and could be used at the end of the BSML program to copy the result to the global file system. It is not a communication primitive because this method has a more expensive cost than any communication primitive. As in any programming language, we also have some functions close channels, to set the current writing/reading position for channel, to return the total length (number of characters) of the given channel, or to return the total size, in characters of a value if it would be serialized to be written on a disk (and, thus, have the number of I/O operations needed). But for the sake of conciseness, we did not present them.

operator	cost
`loc_open_in`	constant time t^l_{or}
(`loc_output_value` v)	$G \times \lceil \frac{size(v)}{DB} \rceil$
`glo_open_in`	$\begin{cases} p \times t^g_{or} + l \text{ If shared global file system} \\ t^l_{or} \qquad\qquad \text{Else} \end{cases}$
(`glo_output_value` v)	$\begin{cases} G^g \times \lceil \frac{size(v)}{DB} \rceil \quad \text{ If shared global file system} \\ p \times G \times \lceil \frac{size(v)}{DB} \rceil \text{ Else} \end{cases}$
(`glo_copy` $file$)	$\begin{cases} \lceil \frac{size(file)}{DB} \rceil \times (G + G^g) + l \text{ If shared global file system} \\ \lceil \frac{size(file)}{DB} \rceil \times 2 \times G + size(file) \times g + l \text{ Else} \end{cases}$

Fig. 2. Cost of some operators

4.2 Formal Cost Model

Given the *weak call-by-value strategy*, a program is always reduced in the same way. In this case, costs can be associated to the parallel and I/O operators. The cost model associated to our programs follows our extention of the EM-BSP cost model. If the sequential evaluation time of each component of the parallel vector is $w_i + m_i$ (computional time and local I/O time), the parallel evaluation time of the parallel vector is $\max_{0 \le i < p} w_i + \max_{0 \le i < p} m_i$. Provided the two arguments of the parallel application are vectors of values, the parallel evaluation time of (**apply** $\langle f_0, \ldots, f_{p-1} \rangle \langle v_0, \ldots, v_{p-1} \rangle$) is $\max_{0 \le i < p} w_i + \max_{0 \le i < p} m_i$ where w_i (resp. m_i) is the computational time (resp. I/O time) of $(f_i\ v_i)$ at processor i. To evaluate **put** $\langle f_0, \ldots, f_{p-1} \rangle$, first each processor evaluates the p local terms $(f_i\ j)$, $0 \le j < p$ leading to p^2 sending values v^j_i. Once all values have been exchanged, a synchronization barrier occurs. At the beginning of this second super-step, each processor i constructs the function from the v^i_j received values. So, the parallel evaluation time of **put** $\langle f_0, \ldots, f_{p-1} \rangle$ is:

$$\max_{0 \le i < p} w^1_i + \max_{0 \le i < p} h_i \times g + \max_{0 \le i < p} m_i + \max_{0 \le i < p} w^2_i + l$$

where w^1_i (resp. m_i) is the computation time (resp. I/O time) of $(f_i\ j)$, h_i is the number of words transmitted (or received) by processor i and w^2_i is the computation time at processor i to constructing the result function from the v^i_j values. Our I/O operators have naturally some I/O costs and some computational time. We also provided that the arguments of the I/O operators have been evaluated first (italweak call-by-value strategy). As explained in the EM-BSP model, each transfer from (resp. to) the local files to (resp. from) its local memory has the cost G for DB items and depending if the global files system is shared or not, the global I/O operators have different costs and some barrier synchronisations is needed. The Figure 2 gives the costs of some selected operators (for more details, see the technical report). The cost (parallel evaluation time) above are context independent. This is why our cost model is compositional. The compositional nature of this cost model relies on the absence of nesting of parallel vectors (our static analysis enforces this condition [3]) and the fact of having two kinds of file systems: a global I/O operator which access to a global file (and could make some communications or synchronization) never occurs locally.

4.3   Example

Our example is the classical *reduction* of lists. Each processor performs a local reduction, then sends its partial results to the following processors and finally localy reduces the partial results with the sended values:

em_scan_list (+) <[1;2], [3;4]> = <[1;1+2], [1+2+3, 1+2+3+4]>

for a reduction of two processors. But to take advantage of the disks and I/O operators, we suppose having large lists on each processor (6 billions of elements). These lists are supposed to be in a file on each processor and they are cutted out on sub-lists of sizes DB. The final result would be a file on each processor which contain sub-lists of size DB. Preliminary experiments on a cluster of PCs (7 bi-Pentium III with 512MB of RAM and ethernet network, see technical report) has been done to show a performance comparison between a BSP algorithms using only the BSMLlib and the corresponding EM-BSP code using our library to have some parallel virtual memories. For small lists, the overhead for the external memory mapping makes the BSML program outperform the EM-BSML one. However, once the main memory is all utilized, the performance of the BSML program degenerates (cost of the paging mechanism). The EM-BSML program continues "smoothly" and clearly outperforms the BSML code.

5   Conclusions and Future Works

With few exceptions (e.g. [12]), previous authors focused on a uniprocessor EM model. The *Parallel Disk Model* (PDM) introduced by Vitter and Shriver [12] is used to model a two-level memory hierarchy consisting of D parallel disks connected to $v \geq 1$ processors via a shared memory or a network. The PDM cost measure is the number of I/O operations required by an algorithm, where items can be transferred between internal memory and disks in a single I/O operation. While the PDM captures computation and I/O costs; it is designed for a specific type of communication network, where a communication operation is expected to take a single unit of time, comparable with a single CPU instruction. BSP and similar parallel models capture communication and computational costs for a more general class of interconnection networks, but do not capture I/O costs. Some other parallel functional languages like SAC [4], Eden [6] or GpH [5] offer some I/O features but without any cost model, and parallel EM algorithms need to be carefully hand-crafted to work optimally and correctly in EM environments. In [1], the authors have implemented some I/O operations to test their models but in a low level language and low level data. To our knowledge, our library is the first for an extension of the BSP model with I/O features (called EM-BSP) and for a parallel functional language with a formal cost model.

The Bulk Synchronous Parallel ML allows direct mode Bulk Synchronous Parallel programming and the current implementation of BSML is the BSMLlib library. But for some applications where the size of the problem is very significant, external memory is needed. We have presented in this paper an extention of BSP model named the EM-BSP for external memory and how to extend the BSMLlib for I/O access in this external memory. The cost model of these new operators and a formal semantics (see technical report) have been investigated. This library is the continuity of our work about imperative and persistant features on our functional data-parallel language. To ensure safety and the compositional cost model, two kinds of I/O operators are needed (global and local ones) and those operators need not occur in another context (local or global). We are currently working on a flow analysis [9] for BSML to avoid this problem statically and to forbid nesting of parallel vectors. We are also working on the implementation of BSP algorithms [8] [10] and their transformations into EM-BSP algorithms as described in [2] to have a new library of classical programs as in the BSMLlib library to be used with large computational problems.

Acknowledgments. The authors wish to thank the anonymous referees for their comments. This work is supported by a grant from the French Ministry of Research and the ACI Grid program, under the project CARAML (www.caraml.org).

References

1. F. Dehne, W. Dittrich, D. Hutchinson, and A. Maheshwari. Parallel virtual memory. In *10th Annual ACM-SIAM Symposium on Discrete Algorithms*, pages 889–890, Baltimore, MD, 1999.
2. F. Dehne, W. Dittrich, D. Hutchinson, and A. Maheshwari. Bulk synchronous parallel algorithms for the external memory model. *Theory of Computing Systems*, 35:567–598, 2003.
3. F. Gava and F. Loulergue. A Polymorphic Type System for Bulk Synchronous Parallel ML. In *PaCT 2003*, LNCS, pages 215–229. Springer Verlag, 2003.
4. C. Grelck and Sven-Bodo Scholz. Classes and objects as basis for I/O in SAC. In *Proceedings of IFL'95*, pages 30–44, Gothenburg, Sweden, 1995.
5. P.W. Trinder K. Hammond and all. Comparing parallel functional languages: Programming and performance. *Higher-order and Symbolic Computation*, 15(3), 2003.
6. U. Klusik, Y. Ortega, and R. Pena. Implementing EDEN: Dreams becomes reality. In K. Hammond, T. Davie, and C. Clack, editors, *Proceedings of IFL'98*, volume 1595 of *LNCS*, pages 103–119. Springer-Verlag, 1999.
7. X. Leroy and M. Mauny. Dynamics in ML. *Journal of Functional Programming*, 3(4):431–463, 1994.
8. W. F. McColl. Scalability, portability and predictability: The BSP approach to parallel programming. *Future Generation Computer Systems*, 12:265–272, 1996.
9. F. Pottier and V. Simonet. Information flow inference of ML. *ACM Transactions on Programming Languages and Systems*, 25(1):117–158, 2003.
10. J. F. Sibeyn and M. Kaufmann. BSP-Like External-Memory Computation. In *Proc. 3rd Italian Conference on Algorithms and Complexity*, volume 1203 of *LNCS*, pages 229–240. Springer-Verlag, 1997.
11. L. G. Valiant. A bridging model for parallel computation. *Communications of the ACM*, 33(8):103, August 1990.
12. J.S. Vitter and E.A.M. Shriver. Algorithms for parallel memory, two -level memories. *Algorithmica*, 12(2):110–147, 1994.

File Replacement Algorithm for Storage Resource Managers in Data Grids

J.H. Abawajy

Deakin University
School of Information Technology
Geelong, Victoria, Australia.

Abstract. The main problem in data grids is how to provide good and timely access to huge data given the limited number and size of storage devices and high latency of the interconnection network. One approach to address this problem is to cache the files locally such that remote access overheads are avoided. Caching requires a cache-replacement algorithm, which is the focus of this paper. Specifically, we propose a new replacement policy and compare it with an existing policy using simulations. The results of the simulation show that the proposed policy performs better than the baseline policy.

1 Introduction

Grid computing enables sharing, selection, and aggregation of geographically distrusted autonomous resources (e.g., super-computers, cluster-computing farms and workstations). One class of grid computing and the focus of this paper is the data grid systems [1] [7] that facilitate the sharing of data and storage resources (e.g., direct attached storage, a tape storage system, mass storage sys-tem, etc.) for large-scale data-intensive applications. Research in data grid computing is being explored worldwide and data grid technologies are being applied to the management and analysis of large-scale data intensive applications. The main problem in data grids is how to provide good and timely access to such huge data given the limited number and size of storage devices and high latency of the interconnection network. One approach to address this problem is to cache the files locally such that remote access overheads are avoided. The motivation for disk caching in data grid is that it tasks several minutes to load data on to the mass storage system. Also, it takes a very long time (up to a few hours) to complete file transfers for a request over WAN. In addition, the user's personal workstation and in some cases the local computer centers may not be able to store all the dataset for a long time for his needs.

In this paper, we address the problem of disk cache replacement policy in data grids environments. The motivation for studying this problem is that the performance of the caching techniques depends heavily on the cache replacement policy that determines which file(s) are evicted to create space for incoming files. Caching techniques have been used to improve the performance gap of storage hierarchies in computing systems, databases and web-caching. Note that, unlike cache replacement policies in virtual memory paging or database buffering or web-caching, developing an optimal replacement policy for data grids is complicated by the fact that the file

M. Bubak et al. (Eds.): ICCS 2004, LNCS 3038, pp. 339–346, 2004.
© Springer-Verlag Berlin Heidelberg 2004

objects being cached have varying sizes and varying transfer and processing costs that vary with time. For example, files or objects in web-caching can be of any media type and of varying sizes. However web caching in proxy servers are realistic only for documents, images, video clips and objects of moderate size in the order of a few megabytes. On the other hand, the files in data grids have sizes of the order of hundreds of megabytes to a few giga-bytes. Some of the differences between caching in data grids and web-caching are given in [2].

The rest of the paper is organized as follows. Section 2 describes related work. Section 3 presents the proposed file replacement algorithm. Section 4 discusses the performance evaluation framework and the results of the proposed policy. Conclusion and future directions are presented in Section 5.

2 Related Work and System Model

The problem of disk cache replacement policy is one of the most important problems in data grids. This is because effectively utilizing disk caches is critical for delivering and sharing data in data-grids considering the large sizes of re-quested files and excessively prolonged file transmission time. Ideally, one strives for an optimal replacement policy for a particular metric measure. An optimal replacement policy evicts a file that will not be used for the longest period of time. This requires future knowledge of the file references, which is quite difficult, if not impossible, to determine.

2.1 System Model

The data grid model used in this paper has n sites interconnected by a wide-area network (WAN). Each site has one or more storage subsystems (e.g., a tape drive and disk) and a set of processors. The resources within a site are interconnected through a local area network (LAN). We assume that the disk subsystem is used as data cache while all other storage subsystems are used as archives. Each site maintains a set of files, which have varying sizes. A given file, fi, can be in one or more sites in the grid at the same time. We use a nomenclature hard replica (HR) to designate a copy of fi file that has not been modified and kept in one or more well-known sites in the system while those file replicas that may have been modified or subject to be purged from a site (i.e., a file copy is not guaranteed to be available all the time) as soft replica (SR). For example, in European Data Grid system [3], the original copy of the files (i.e., FH replica) is always kept in at least the European Organization for Nuclear Research (CERN) site while other sites can keep a copy of the files if need be.

Each site has a storage resource manager (SRM) that is responsible for the management of the local storage subsystem usage as well as maintaining HR (may be SR as well) file replica meta-data information (e.g., location, cost, size, etc.) in replica table. The storage resources in data grids are accessible to a user, remotely or locally, through a middle-ware service called storage resource manager (SRM). The information on HR and SR file replicas is collected in two ways. First, the sites that are designated as keepers of HR replicas can inform all other sites by sending a notification message when a new file is created. Also, every time these sites send a copy of a given file to a remote SRM upon a request from the latter, a notification message along the identity of the SRM is sent to all sites within certain distance (e.g., region) from SRM. Alternatively, each SRM will send its own notification message to certain sites only when it receives a copy of the file. The notification message includes the cost of preparing the file for transferring (e.g., the time taken to stage a file from a tape to a disk). When a given SRM receives the notification message, it will estimate the cost of a file acquisition based on the time taken for the notification message using the following function:

$$Cost\ (f_i, v, u) = T_{process}(v) + \left(Latency + \frac{size(f_i)}{bandwidth(u,v)}\right) \qquad (1)$$

Data requests from the local system users are sent to the local SRM for processing. We assume that when a user submits a request for data, the size of storage locally available (i.e., reserved) for the data is also declared. The file request scheduler (FRS) is responsible for the admission of data requests, R_i, from users. All requests that can be satisfied from the local disk are called local request while those that are fetched from remote sites are called remote requests.

Each admitted request, R_i, is stored in the ready request queue and serviced, based on the scheduling approach used (e.g., first come first service), by the file request placement (FRP) component of the SRM. When the file has to be brought in from another site, the SRM searches the replica table for the best site to get the file from. In our system, local requests have priority over remote requests. We refer to files cached locally that are in use as active files while those that are not in use as passive files.

The key function of the SRM is the management of a large capacity disk cache that it maintains for staging files and objects of varying sizes that are read from or written to storage resources that are either at the same local site or some remote site. Two significant decisions govern the operation of an SRM: file request scheduling and file re-placement. Each of file requests that arrive at an SRM can be for hundreds or thousands of objects at the same time. As a result, an SRM generally queues these requests and subsequently makes decisions as to which files are to be retrieved into its disk cache. Such decisions are governed by a policy termed the file admission policy. When a decision is made to cache a file it determines which of the files currently in the cache may have to be removed to create space for the incoming file. The latter decision is generally referred to as a cache replacement policy, which is the subject of this paper.

2.2 Related Work

The Least Recently Used (LRU) and Least Frequently Used (LFU) replacement policies are two extreme replacement policies. The LRU policy gives weight to only one reference for each file, that is, the most recent reference to the file while giving no weight to older ones representing one extreme. In contrast, the LFU gives equal weight to all references representing the other extreme. These extremes imply the existence of a spectrum between them. A number of replacement policies that fall within such a spectrum have been proposed in the literature.

A replacement policy referred to as Least Cost Beneficial based on K backward references (LCB-K) is proposed in [2]. LCB-K is an always cache policy, which means that all files retrieved from remote sites will be cached. It uses a utility function that probabilistically estimates which files would be accessed relatively less soon for ranking the candidate files for eviction. Whenever a file in the cache needs to be evicted, the algorithm orders all the files in non-decreasing order of their utility functions and evict the first files with the lowest values of utility function and whose sizes sum up to or just exceed that of the incoming file. The main problem with this policy is that the utility function used to predict the future references of the file is based on global information such as the file reference counts.

An economic-based cache replacement (EBR) policy is proposed in [5]. EBR uses probability-based utility function technique to measure relative file access locality strength for the files and makes a decision as to cache a file or not. If there enough space to store a file, a newly arrived file is automatically stored on the disk. However, if there is no space left on the disk, EBR evicts the least valuable file from the disk. In order to make this decision, the Replica Optimiser keeps track of the file requests it receives and uses this history as input to a future revenue prediction functions. The prediction function returns the most probable number of times a file will be requested within a time window W in the future based on the requests (for that or similar files) within a time window W' in the past. The algorithm keeps track of the access history and uses a prediction function for estimating the number of future access for a given file in the next n requests based on the past r request in the history.

In [6], a replacement policy called the Least Value-based on Caching Time (LVCT) that exploits access regularities (i.e., one-timers versus multiple-timers) in references is proposed. A single stack with two fields (i.e., caching time and file size) is used for storing every file request in the site. The size of the stack is adjusted by pruning it when the total size of files represented in the stack is greater than twice the cache size or the number of files represented in the stack is greater than two times of the number of files in the cache. LVCT is shown, through simulation, that it outperforms LRU, GDS and LCB-K policies. However, this performance comes at additional costs for taking replacement decisions as updating the cache state at each reference has overhead. Moreover, there are several shortcomings of this policy. First, it assumes that clients have enough storage to store the file, which is not the case most of the time. Second, a file with multiple accesses must be brought from remote size at least twice. Third, maintaining the stack takes time. Even if there is enough disk space, first time access file is not stored. Decision to evict or not is made after a remote file has been completely received.

We propose a new replacement policy that takes into account factors such as the latency delays in retrieving, transferring and processing of the files. We compare the proposed replacement policy with several existing policies through simulation under varying system and workload parameters (e.g., account for delays in cache space reservation, data transfer and processing). The results of the experiment shows that the proposed policy performs substantially better than the baseline policies used in the experiments.

3 New Disk Cache Replacement Policy

The proposed policy is called New Cache Replacement (NRP) policy. It combines locality, size and cost of files when making a replacement decision. Also, only events that matter for replacement decisions are taken into account. Note that this is not the case in the other policies discussed in the previous section. For each file f, we keep its size S and the cost of fetching it from remote site C. Note that C can be the actual cost incurred of a file or estimated as per Equation (1). This information is used in determining the set of files to be evicted if this need be. In the proposed algorithm, only passive files in the cache are potential candidates for eviction. Also, we assume that SRM serves user requests based on FCFS with local requests having priority over remote requests.

The proposed policy has three phases: request admission phase (RAP), candidate selection phase (CSP) and file eviction phase (FEP). The RAP decides if the request would be a remote-fetch-only (i.e., no caching of the file occurs) or remote-fetch-store (i.e., fetch the file from remote site and then store it locally). If there is enough disk space, the request automatically admitted. When there is a miss, the algorithm invokes the CSP to select possible files to be replaced. That is, for each remote-fetch-store request i, CSP selects a set of candidate files that are passive with their size greater than or equal to $S_i / 2^K$ where K >= 0. Initially, we set k=0 and if no file is found, we increment k by one and this process is continued until a set of files that meet the criterion is located. The FEB uses the list of candidate files returned by the CSP to replace appropriate files. The algorithm evicts a set of files from the list (i.e., L) whose sizes sum up to or just exceed that of the incoming file.

The advantage of the NRP is that it uses local information as opposed to global information when making the replacement decisions. It also incorporates locality, size and cost considerations effectively to achieve the best performance possible. As it selects the file that have size equal to or greater than the documents to be replaced; this technique of evicting the files saves many misses by not evicting the small files that are least recently used. In addition, it tries to minimize the number of files replaced by evicting files that have not been used recently. The main and more important thing is that the above algorithm also takes in to account the cost of fetching the file to cache from its original severs. By doing so, this algorithm also considers the cost of the file before it replaces a particular page. The above algorithm has some drawbacks when it takes into the account the cost of fetching the document. Because the above algorithm considers the cost of the document before it replaces the page, it sometimes replaces a page that has been accessed recently but doing so it also saves the cost for the new document.

4 Performance Analysis

As in [2] [6] [5], we used the discrete event simulation to compare the performance of the proposed replacement policy against the economic-based policy [5]. As in [2][6], we used the hit ratio and the byte hit ratio to measure the effectiveness of the cache replacement policies.

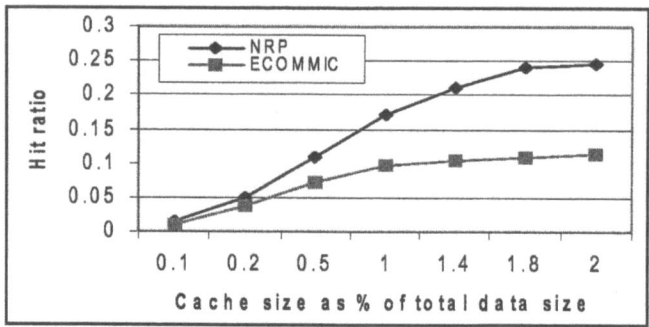

Fig. 1. An overview of the data grid model used in the paper.

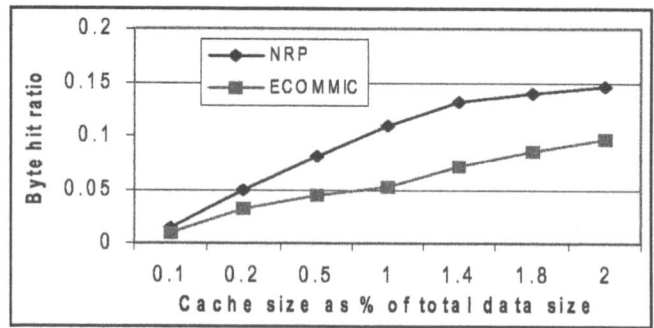

Fig. 2. An overview of the data grid model used in the paper.

4.1 Experimental Setup

The simulator is a modified version of [8] that also accounts for the latency incurred at the source of the file; the transfer delay in reading the file into the disk cache; and the holding or pinning delay incurred while a user processes the file after it has been cached. The simulator takes a stream of requests as an input, calculates the hit ratio and byte hit ratio for each algorithm, under cache sizes being various percentages of the total data set size. The data for the simulation is based on [2]. The sizes of files ranged from about 1.0 to 2.1 gigabytes and the time scales are in the order of minutes. We use a Poisson inter-arrival time with mean 90 seconds for the requests. The file sizes are uniformly distributed between 500,000 bytes and 2,147,000,000 bytes. The entire period of request generation is broken into random intervals and we inject locality of reference using the 80-20 rule, which within each interval of the request stream, 80% of the requests are directed to 20% of the files. The length of an interval is uniformly distributed between 1% and 5% of the generated workload.

As in [5], we used the European DataGrid testbed sites and the associated network performance. In the simulation 20 sites were used with the total disk capacity of 1.1 terabits distributed in such a way two sites have a disk of 100 GB each while the remaining 18 sites have 50 GB each. The two sites with the 100 terabit disk are used to store the original copy of the files in the system. A single copy of each data file was initially placed at CERN. The storage facilities of each site were set at a level to prevent any site (except CERN) holding all of the files, but allowing the site room for the preferred job files.

4.2 Results and Discussions

Figure 1 show the results of the experiments. The graphs are generated from runs using variance reduction technique. Each workload has just over 500,000 entries. For each specified value of the cache size and for each workload, a run generates 5 values of the required performance metric, at intervals of about 100,000 requests. Each point in the graphs, for any particular measure, is the average of the 5 recorded measures in a run. Figure 1 shows the performance of the two policies as a function of the hit ratios.

Figure 2 shows the performance of the two policies as a function of the byte hit ratios. From the data on these figures, it can be observed that the proposed policy substantially outperforms than the economic-based policy. This can be explained by the fact that eviction in the economic-based policy is driven by probability that a file can be used many times. Also, there is a tendency in the economic-policy for bringing in a file even if there is no space to accommodate it while our policy do not. Moreover, our policy was able to simply transfer one-timer requests without storing them in the cache while the economic-based policy did not.

5 Conclusions and Future Direction

A replacement policy is applied to determine which object in the cache needs to be evicted when space is needed. As a replacement policy is an essential component of the disk cache management system, we have presented a new policy and shown through simulation that it performs substantially better than an existing one. Future work in this area would involve more extensive testing with real workloads from other mass storage systems. Also, we will include the policies proposed in [2][66]. In this paper, we assumed that SRM serves user requests based on FCFS with local requests having priority over remote requests. We are currently studying the impact of different request scheduling on the performance on the cache replication policies.

References

1. Chervenak, A., Foster, I., Kesselman, C., Salisbury, C., Tuecke, S.: The Data Grid: Towards an architecture for the distributed management and analysis of large scientific data-sets. Journal of Network and Computer Applications (2000) 187-200
2. Otoo, E.J., Olken, F., Shoshani, A.: Disk Cache Replacement Algorithm for Storage Resource Managers in Data Grids. In Proceedings of the SC (2002) 1-15
3. EU Data Grid Project. http://www.eu-datagrid.org.
4. Abawajy, J. H.: Placement of File Replicas in Data Grid Environments. In Proceedings of PParGMS Workshop (2004)

5. Carman, M., Zini, F., Serafini, L., Stockinger, K.: Towards an Economy-Based Optimisation of File Access and Replication on a Data Grid. In Proceedings of 2nd CCGRID (2002) 120-126
6. Jiang, S., Zhang, X.: Efficient Distributed Disk Caching in Data Grid Management. In Proceedings of Cluster Computing (2003) , Hong Kong, China.
7. Hoschek, W., Jaén-Martínez, F. J., Samar, A., Stockinger H., Stockinger, K.: Data Management in an International Data Grid Project. In Proceedings of the 1st Workshop on Grid Computing (2000) 77-90
8. P. Cao and S. Irani, "Cost-aware WWW proxy caching algorithms," In USENIX Symposium on Internet Technologies and Systems, 1997.

Optimizations Based on Hints in a Parallel File System

María S. Pérez, Alberto Sánchez, Víctor Robles, José M. Peña, and Fernando Pérez

DATSI. FI. Universidad Politécnica de Madrid. Spain
{mperez,ascampos,vrobles,jmpena,fperez}@fi.upm.es

Abstract. Existing parallel file systems provide applications a little control for optimizing I/O accesses. Most of these systems use optimization techniques transparent to the applications, limiting the performance achieved by these solutions. Furthermore, there is a big gap between the interface provided by parallel file systems and the needs of applications. In fact, most of the parallel file systems do not use intuitive I/O hints or other optimizations approaches. In this sense, applications programmers cannot take advantage of optimization techniques suitable for the application domain. This paper describes I/O optimizations techniques used in MAPFS, a multiagent I/O architecture. These techniques are configured by means of a double interface for specifying access patterns or hints that increase the performance of I/O operations. An example of this interface is shown.

Keywords: Parallel I/O, optimizations, caching, prefetching, hints.

1 Introduction

I/O systems have traditionally been the bottleneck of computing systems because of the difference between I/O devices access and computing time, avoiding data-intensive applications achieve enhanced improvements. The usage of parallel I/O architectures and, in this particular case, parallel file systems has involved a widely used approach that mitigates in some sense this problem, known as *I/O crisis*.

Parallel file systems achieve good performance in some scientific applications, by means of *ad hoc* solutions, tailored to the particular problem by an I/O expert. This is not the usual scenario, since applications are developed by software teams, which do not usually know the underlying I/O architecture deeply. On the other hand, data-intensive applications have very different data access patterns. Therefore, it is not possible to improve these applications performance with a non-flexible architecture, which applies the same techniques to applications, without taking into account their requirements.

I/O models arise within MAPFS (MultiAgent Parallel File System) [10] with the aim of fixing the following problems:

- Applications have different data access patterns, depending on their domain.
- Additional information for increasing data access performance is also based on the particular application.
- Most of the studied access patterns correspond to scientific applications. In fact, a great amount of studies has been performed on scientific I/O workloads, as much in parallel as sequential frameworks.

M. Bubak et al. (Eds.): ICCS 2004, LNCS 3038, pp. 347–354, 2004.
© Springer-Verlag Berlin Heidelberg 2004

Although MAPFS is a general purpose file system, it is very flexible and it can be configured for adapting to different I/O access patterns. In MAPFS, data items from secondary storage are "cached" in memory for faster access time. This approach allows MAPFS to increase the performance of the I/O operations. This operation is made in MAPFS by means of an independent subsystem, constituted by a multi-agent system. This task can be made in a more efficient way using hints on future access patterns.

The outline of this paper is as follows. Section 2 describes the I/O optimization techniques implemented by MAPFS. Section 3 presents MAPFS configuration for using these optimization techniques. Section 4 shows the evaluation of a scientific application using MAPFS optimization techniques. Section 5 describes the related work. Finally, Section 6 summarizes our conclusions and suggests further future work.

2   I/O Optimizations Techniques

Although I/O systems have traditionally tackled the I/O phase in the same way, independent of the applications domain and their access pattern, some studies [8], [5] have demonstrated that a higher control of the user applications over the I/O system can increase their performance. MAPFS provides this control as part of its functionality. Furthermore, MAPFS increases its interface in such way that user applications can make use of this control. Next we are going to analyze these MAPFS capacities.

2.1   I/O Caching and Prefetching

If the behaviour of an algorithm is known, I/O requirements could be achieved in advance before the algorithm actually needs the data. An optimal usage of computational resources, like I/O operations or disk caches is also a key factor for high-performance algorithms. With the usage of data access knowledge both accurate prefetching and resource management could be achieved.

For using these features, it is necessary to have a cache structure, where the I/O data items are stored in memory. The effectiveness of the cache is determined largely by the replacement policy. A typical cache replacement policy is *Least Recently Used (LRU)*. This policy assumes that data items used recently are likely to be used again. Other alternatives widely used are the FIFO policy, which replaces the oldest item and MRU policy, which replaces the most recently used item. Historical information is often used for both file caching and prefetching. In the case of the prefetching, the most used approach is sequential readahead. All these policies are used in conventional systems. Nevertheless, these policies are not suitable for all the different access patterns of applications. A more general scheme consists in using hints about applications with the aim of increasing of the caching and prefetching phases.

2.2   Usage of Hints

As we have mentioned previously, MAPFS uses hints in order to access in an efficient way to files. Hints are structures known and built by the file system, which are used for improving the read and write routines performance. In MAPFS, hints can be determined in two ways: (i) they can be given by the user, that is, the user application provides the necessary specifications to the file system for increasing the performance of the I/O routines, and (ii) they can be built by the MAPFS multiagent subsystem. If this option is selected, the multi-agent system must study the access patterns of the applications in order to build hints that improve both the caching and prefetching stages.

The semantic of the applications that use the I/O system must configure the hints of the underlying I/O system. Therefore, this I/O system must be flexible enough to allow the application to modify the hints. In a generic way, hints can be based on different attributes values related to I/O configuration parameters.

3   MAPFS Configuration

As we have mentioned previously, most of the data accesses can be settled by means of *attributes*. For example, in Data Mining applications, the queries are based on values of certain database attributes. In other kind of applications, metadata, which stores additional information about data, can be used to access them in a more efficient fashion. This metadata also can be structured by means of attributes. Therefore, hints in MAPFS must be able to store simple regular expressions, composed of boolean operators applied to these attributes. These expressions allow stored data to be classified.

MAPFS provides a double interface for specifying access patterns or hints. The first one (*user control operations*) is a high level interface, adapted to applications, which may be used in terms of attributes. Applications can manage the performance of the system only by means of this API. The second one (*hints setting operations*) is nearer to the file system and may be used for subsystems that belong to the parallel file system or expert I/O programmers. In [11], a complete syntax of the MAPFS configurations operations is shown. This paper is going to describe a sample of scientific application using the MAPFS interface.

3.1   Matrix Multiplication as Sample Application

Operations about matrices are significative examples of scientific applications. From the point of view of hints usage, the addition of matrices offers the following alternatives. When the first element of the row i of the first matrix is accessed, the I/O operation may be optimized if complete rows i of both matrices were prefetched. The problem appears when a row does not fit in the cache structure. In this case, this rule is not so trivial. This example will be analyzed in Section 3.2. In the case of the multiplication of matrices, rules for achieving optimal prefetching are not obvious either.

Let be a multiplication operation between matrices: $A[M, N] * B[N, P] = C[M, P]$. in such way that the matrices are stored by rows in their respective files and they have a considerable size, so that a row or column does not fit completely in the cache structure.

On the other hand, let be a traditional multiplication algorithm:

```
for (i=0; i<M; i++)
  for (j=0; j<P; j++)
    for (k=0; k<N; k++)
      C[i][j]+=A[i][k]*B[k][j];
```

Next, possible prefetching strategies are analyzed.

3.2 Prefetching Rules on the Matrix A

Regarding to the matrix A, instead of reading in advance the row corresponding to the requested element, other strategies are more suitable because of the following facts:

- Whenever the current row has been used, that is, the index i advances one position after using the row P times, a hole has been left in the prefetching. In this case, it would be advisable to request in advance elements of the row $i+1$.
- It is possible that the cache has not enough space to store the complete row in the cache. Then, it would be better reading in advance only one window of K values, which would correspond to the K following elements of the request, but in a circular fashion (N module), where K can be determined by different factors, such as the cache size or the system load.

According to this, the following rules can be defined:

- **RULE 1:**
  Row prefetching with circular window K (N module), a P-times cycle by row, sequential advance by rows (and without matrix repetition cycle). This rule is equivalent to:

  $read(A[i,j]) \implies$
  If the current cycle is not the last one (number of accesses $A[i,j] < P$) :
  • Prefetching from $A[i, j + 1]$ until $A[i, (j + K)\%N]$
  If the current cycle is the last one (=P) $\implies$ advance to the following row:
  • Prefetching of the K following elements both elements of the present row,
  if they are left, and first elements of the "$i + 1$" row, if it is necessary and $i < M$

3.3 Prefetching Rules on the Matrix B

Prefetching on the matrix B is more complex, because it is necessary to access matrix data by columns, which decreases the performance, since small accesses are made. In this case, it is possible to raise different strategies.

- **Horizontal Prefetching**
  When the element $B[i, j]$ is accessed, it is advisable to prefetch a small data window of size K' of that same row, corresponding to later elements than required element $B[i, j]$, in a circular fashion (module P). These values will not be used inmediately, but they are used when the index "j" of the loop is increased, that is, when an element of C is calculated. Thus, K' must not be a large value because, otherwise, this window is bigger than the cache size.
- **RULE 2**
  On the matrix B, it can be made a row prefetching with circular window of size K'(module P), a cycle of M repetitions by row and without advance by rows (nor matrix repetition cycle), because after the M cycles, the algorithm will have finished.

- **Vertical or Column Prefetching**

  It seems reasonable that the read request of $B[i, j]$ causes a prefetching in that column ($B[i + 1, j]$, $B[i + 2, j]$, ...). Although the vertical prefetching is theoretically feasible, it can involve small reads, achieving a poor performance. On the other hand, if the prefetching is transitive (see Section 3.4), when the rule that activates the prefetching of $B[i + 1, j]$ is triggered, the previous rule (RULE 2) will be activated by transitivity, requesting the horizontal K'-size window. This will improve the I/O performance. Considering that the column can be bigger than the cache and it is not possible to prefetch it completely, the following two rules can be used:

  1. Row prefetching with K'-size window according to RULE 2.
  2. The new rule corresponding to vertical prefetching (RULE 3).

- **RULE 3**

  On the matrix B, it would be made a column prefetching with K"-size window, without repetition cycle by column, with sequential advance by columns and a cycle of M repetitions of matrix. In the case that the complete column does not fit in the cache, the positive effect of RULE 2 disappears, because prefetched data is discarded. Therefore, K' would have to be equal to 0.

3.4 Transitive or Induced Prefetching

A prefetching rule provides prefetching of probably referenced data in next read operations. This prefetching rule is associated to a concrete data block. In the case of the matrix multiplication, it is associated to an element or a set of elements. If transitives rules are allowed, the access to a concrete data block, which is associated to a certain prefetching rule, involves another prefetching rule, associated to the data block that has been prefetched by the first rule. Nevertheless, it is necessary to control this transitivity, because in certain cases, it is possible that the complete file may be read.

3.5 Induced Prefetching between Files

This kind of prefetching implies prefetching data of a file, induced by accesses in another file.

- **RULE 4**

  In the case of the multiplication of matrices, it is possible to do an induced P-cycle prefetching, so that the first access to $A[i, j]$ causes the prefetching of $B[j, 0]$, $B[j, 1]$ and so on.

3.6 Combination of Prefetching Rules

Joining all the previous rules, an access to $A[i, j]$ causes the following accesses:

1. Using RULE 1, a prefetching of the K next elements in the row "i" is made in a circular fashion. In the last cycle, we advance to the row "i+1".
2. Using RULE 4, a prefetching of $B[j, x]$ is made (if it is the x-th time that the element is accessed)
3. By transivity, the following rules can be applied:
   a) RULE 2, that is, the K' elements of the row i of B are accessed.
   b) RULE 3, that is, the K" elements of the column j of B are accessed.

4 Hints Evaluation

As is described in the previous section, in the case of the matrix multiplication, it is possible to apply different rules and even a combination of such rules. These rules use different configuration parameters, which are the basic expressions or component of such syntactic rules. In the case of the prefetching, these parameters or attributes are:

- Type: It may be per row (horizontal) or per column (vertical).
- Window type: It may be circular or not.
- Window size: Number of elements to read in every cycle of prefetching.
- Cycle per row or per column: It specifies if there exists cycle per row or per column or not. This last case is indicated with a value 0.
- Cycle per matrix: It specifies if there exists cycle per matrix or not. This last case is indicated with a value 0.

The user control structure provided by the application follows this boolean expression:

```
Type=Row AND Window Type=Circular AND Window Size=K
        AND Cycle Per Row=P AND Cycle Per Matrix=0
```

This user control structure is translated to hints by the multiagent subsystem. Hints are used as data tags for processing the elements in an efficient manner, prefetching data provided by the corresponding rule. In this case, hints have been used in order to increase the performance of the prefetching phase. However, the hints can be used together with other techniques or algorithms.

In order to evaluate the performance of the MAPFS hints, we have implemented the matrix multiplication in MAPFS and in the native interface of the *Parallel Virtual File System* PVFS [1]. This section shows the comparison between the time of the matrix multiplication in MAPFS and in PVFS.

Figure 1 represents this comparison for the multiplication of a single row of two 100 MB-size matrices. As this figure represents, we can see that MAPFS solution without hints is lightly slower than PVFS solution. However, MAPFS solution with hints is faster than PVFS solution. Therefore, the usage of hints increases the performance in a flexible way.

The more rows are processed, the more differences are in the execution times. Figure 2 shows the evaluation of the multiplication of 100 MB-size matrices. As can be seen, in this case the differences are more significative.

5 Related Work

A suitable data placement is extremely important in the increase of the performance of I/O operations. Most of the file systems provide transparency to user applications, hiding details about the data distribution or data layout. Some file system such as nCUBE [4] or Vesta [3] provide programmer a higher control over the layout.

Panda [2] hides physic details of the I/O systems to the applications, defining transparent schemas known as *Panda schemas*.

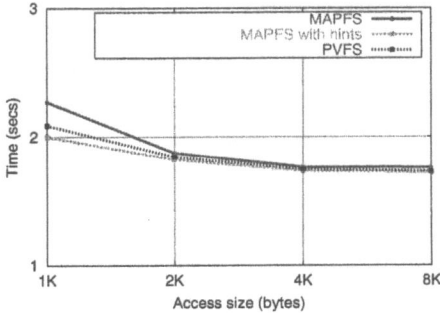

Fig. 1. Comparison of the matrix multiplication in PVFS and MAPFS (a single row)

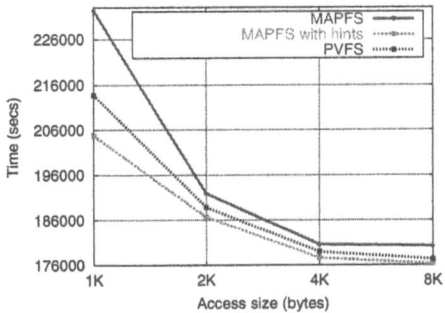

Fig. 2. Comparison of the matrix multiplication in PVFS and MAPFS (100 MB-size file)

PPFS defines a set of access pattern specified when a parallel file is opened [6].

MPI-IO uses *MPI datatypes* to describe data layout both in memory and in file, in such way that it is possible to specify non-contiguous access [12]. Data layout in memory is specified in every I/O call in MPI-IO. Data layout in file is defined by a file view. The function MPI_File_set_view() is used for setting this view.

On the other hand, hints are widely used for increasing the system performance, in general topics in computing systems. In the context of the file systems, hints are usually used as historical information for optimizing caching and prefetching techniques. For example, hints can be used in the prefetching phase for deciding how many blocks are read in advance. OSF/1 system prefetchs up to 64 data blocks when large sequential request are detected [9]. Other works detect more complex access patterns for non-sequential prefetching. There exists a great number of works which infer future accesses based on past accesses. In [7], historical information is used in order to predict the I/O workload and, thus, increase the prefetching stage.

6 Conclusions and Future Work

This paper presents the usage of hints in MAPFS, showing a sample application, a multiplication of matrices. We use the MAPFS configuration interface in order to provide hints to the I/O system. As the experimentations show, the usage of hints is a flexible way of increasing the performance of the I/O phase. MAPFS provides a double interface, a high level interface, adapted to applications and a second one, used for subsystems belong to the parallel file system or expert I/O programmers.

Nevertheless, the hints configuration is not an obvious task. As future work, we are applying autonomic computing in order to allow the I/O system to be configured itself.

References

1. P. H. Carns, W. B. Ligon III, R. B. Ross, and R. Thakur. PVFS: A parallel file system for linux clusters. In *Proceedings of the 4th Annual Linux Showcase and Conference*, pages 317–327, October 2000.
2. Yong E. Cho. *Efficient resource utilization for parallel I/O in cluster environments*. PhD thesis, University of Illinois at Urbana-Champaign, 1999.
3. P. Corbett, D. Feitelson, J. Prost, and Johnson S. Parallel access to files in the Vesta file system. In *Proc. of the 15th. Int. Symp. on Operating Systems Principles*, pages 472–481. ACM, 1993.
4. E. DeBenedictis and J. M. del Rosario. nCUBE parallel I/O software. In *Eleventh Annual IEEE International Phoenix Conference on Computers and Communications (IPCCC)*, pages 117–124, Apr 1992.
5. J.M. del Rosario and A.N. Choudhary. High-performance I/O for massively parallel computers: Problems and prospects. *IEEE Computer*, 27(3):59–68, 1994.
6. C. Elford, C. Kuszmaul, J. Huber, and T. Madhyastha. Scenarios for the Portable Parallel File System, 1993.
7. David Kotz and Carla Schlatter Ellis. Practical prefetching techniques for parallel file systems. In *Proceedings of the First International Conference on Parallel and Distributed Information Systems*, pages 182–189. IEEE Computer Society Press, 1991.
8. Barbara K. Pasquale and George C. Polyzos. A static analysis of I/O characteristics of scientific applications in a production workload. In *Proceedings of Supercomputing '93*, pages 388–397, 1993.
9. R. Hugo Patterson, Garth A. Gibson, Eka Ginting, Daniel Stodolsky, and Jim Zelenka. Informed prefetching and caching. In Hai Jin, Toni Cortes, and Rajkumar Buyya, editors, *High Performance Mass Storage and Parallel I/O: Technologies and Applications*, pages 224–244. IEEE Computer Society Press and Wiley, New York, NY, 2001.
10. María S. Pérez, Jesús Carretero, Félix García, José M. Peña, and Víctor Robles. A flexible multiagent parallel file system for clusters. *International Workshop on Parallel I/O Management Techniques (PIOMT'2003) (Lecture Notes in Computer Science)*, June 2003.
11. María S. Pérez, Ramón A. Pons, Félix García, Jesús Carretero, and Víctor Robles. A proposal for I/O access profiles in parallel data mining algorithms. In *3rd ACIS International Conference on SNPD*, June 2002.
12. Rajeev Thakur, William Gropp, and Ewing Lusk. Optimizing noncontiguous accesses in MPI-IO. *Parallel Computing*, 28(1):83–105, 2002.

Using DMA Aligned Buffer to Improve Software RAID Performance

Zhan Shi[1], Jiangling Zhang[1], and Xinrong Zhou[2]

[1] National Storage Lab, Computer Science Department
Huazhong University of Science and Technology, 430074 Wuhan, P.R.China
jlzhang@hust.edu.cn
[2] Turku Centre for Computer Science, Åbo Akademi,
20520 Turku, Finland
xzhou@abo.fi
http://www.abo.fi/~xzhou

Abstract. While the storage market grows rapidly, software RAID, as a low-cost solution, becomes more and more important nowadays. However the performance of software RAID is greatly constrained by its implementation. Varies methods have been taken to improve its performance. By integrating a novel buffer mechanism – DMA aligned buffer (DAB) into software RAID kernel driver, we achieved a significant performance improvement, especially on small I/O requests.

1 Introduction

Most business computers running critical applications use SCSI disks with hardware RAID controllers. With the power of processors keeping on growing, more and more developers place greater loads on system CPUs. Nowadays, there are companies that provide software ATA RAID using cheaper IDE disks instead of the expensive SCSI disks. Lots of motherboard manufacturers are also including ATA RAID chips on board. This means that even a home user can afford a RAID solution today. Provided with benefits of any RAID system [1] such as increased write speeds while writing to a striped partition, software ATA RAID is now been widely accepted. [2].

However, there are some drawbacks to implement RAID in software. First, the issue of portability. Since a software implementation will have some OS-specific components, those components have to be re-written for each different OS. Second, the kernel-mode program. This is the main problem that troubles kernel-mode software developers. Unlike applications, kernel-mode program can execute privileged instructions and manipulate the contents of any virtual address, the whole system is exposed to these programs without any safeguards against programming errors. A simple mistake may even lead to a system crash. The third, performance. As we have already stated, software RAID solutions are typically implemented as kernel mode components. In fact it is embedded into the kernel itself in Linux. [3]. Many OS already have build-in buffers and caches to improve its I/O performance [4][5], but they may not cooperate with software RAID well.

M. Bubak et al. (Eds.): ICCS 2004, LNCS 3038, pp. 355–362, 2004.
© Springer-Verlag Berlin Heidelberg 2004

The remainder of this paper is organized as follows. Section 2 introduces the related works. Section 3 presents how we can improve the performance using DMA-aligned buffer in detail. Section 4 describes the implementation of DAB in Windows 2000 based software RAID. Section 5 evaluates DAB under a variety of circumstances. Section 6 ends this paper with a conclusion .

2   Related Work

Caches and buffers are widely used to improve I/O performance in modern operating systems. Data prefetching is also a popular method to process data inside the cache and buffer. Lots of work have been done by many researchers. Sequential prefetching [9] simply look ahead to prefetch a variable number of sequential blocks past the current reference. Track-aligned prefetching utilizes disk-specific knowledge to match access patterns to the strengths of modern disks[10]. By allocating and accessing related data on disk track boundaries, a system can avoid most rotational latency and track crossing overheads. Avoiding these overheads can increase disk access efficiency by up to 50% for mid-sized requests (100–500KB). To be the most successful, prefetching should be based on knowledge of future I/O accesses, not inferences. Such knowledge is often available at high levels of the system. Programmers could give hints about their programs' accesses to the file system. Thus informed, the file system could transparently prefetch the needed data and optimize resource utilization. This is what Transparent Informed Prefetching (TIP) has done[11].

Currently, our problem is that I/O requests generated by most applications such as database, daily transaction, etc, are fairly small, always lesser than 64KB. The prefetching methods mentioned above do not work well in our situation. How to improve the small I/O performance while keeping larger I/O least affected in our software RAID implementation is the main topic of this paper.

3 DMA Aligned Buffer

When surveying the characteristic of access patterns of standard software RAIDs, we found that there is a significant change in transfer rate while we adjust the I/O packet size between its minimum value and maximum IDE DMA transfer length. If we take a close look at many Operating Systems, we will find that every data request is sent to the low-level drivers such as port/miniport driver. In modern IDE disks, every device I/O requires a DMA operation to complete. [12] Some applications are likely to generate a lot of small I/O requests, for example, database application, web transaction, etc. Operating Systems have to handle these requests by sending them to their driver stacks, from the highest file system drivers to the lowest HBA drivers [6]. File cache in most operating systems can mitigate the problem when processing most application-level I/O requests, but it cares little about low-level I/O efficiency. Low-level fragmental requests can hardly benefit from that.

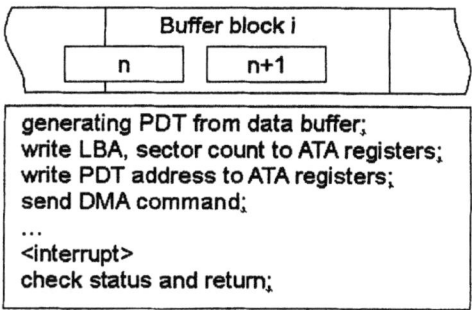

Fig. 1. DMA aligned buffer and DMA working process from the programmers point of view, PDT: Physical Descriptor Table. LBA: Logical Block Address

A typical DMA operation in OS kernel requires several steps to complete, which is shown in Figure 1. Although our program is running on Windows 2000, the steps are the same. PDT is an array of descriptor data structure which stands for physical memory blocks used to perform DMA. According to ATA specification [13], the sector count register has 8 bits currently, thus the DMA data transfer of IDE interface is restricted by a maximum length of 256 sectors by now (set the SECTOR COUNT register to 0).

As we've mentioned above, a large number of fragmental requests sent to low-level drivers may cause too much device I/O, and every device I/O consists of generating PDT, writing ATA registers, starting DMA, seeking and rotational latency and waiting for interrupt, therefore they result in a performance decline. We proposed a buffer method to solve this problem. The buffer format is demonstrated in Figure 1. Buffer block i acts as a generic LRU buffer together with other buffer blocks. n and n+1 represent two sequential small requests sent to HBA miniport. If request n isn't in the buffer (missed), it'll then be buffered in one of these blocks and hoping it'll be visited again, request n+1 is the same. A prefetching method can be applied here so that request n can bring on a prefetch to benefit following requests, such as request n+1. For the small I/O we observed, larger prefetching may service more future requests. But this differs to typical prefetching, because larger prefetching can also cause multiple device I/O to complete, not only the longer data transferring. So the most economical way is to limit prefetching length to just one time larger than the maximum DMA transfer length (maybe multiple times according to application, and this needs further working). In order to implement the buffer in a more efficient way, the starting address we used to prefetch is aligned to the 256 sectors boundary, thus avoiding buffer block overlapping (In Figure 1, this means all the buffer blocks, including buffer i are 256 sectors in size and all their mapped starting address are 256 sectors aligned.). Since the buffer is composed of buffer blocks aligned with maximum DMA transfer length, we call this approach DMA-Aligned Buffer.

However the time spent on data copy in memory and excess device I/O may reduce the benefit. Additionally, new ATA standard may break DMA transfer barrier by extending sector count register from 8 bits to 16 bits, hence the maximum data transfer length in one DMA operation can be increased to 65536 sectors in the future [13][14]. Thereby more work is needed to improve buffer method in the future.

Besides, we may also use other access characteristics to adjust our buffer mechanism. Our experimental software RAID is a part of network storage platform, we can also use special designed aligned-buffer to cut down possible fragmentary requests which may cause noticeable transfer overhead.

4 Implementation

4.1 Buffer Data Structure

The main buffer of software RAID consists of several fixed-size memory blocks which act as R/W buffers. The number and size of these memory blocks are described as BufferNum and BufferLen respectively. These memory blocks form up a buffer queue, and every buffer block is managed by a data structure called _BUFFERBLOCK, which is shown in the following structure.

```
typedef struct_BUFFERBLOCK
{ PVOID      PreReadBuffer;      //Buffer virtual address pointer
  PVOID      Phy_PreReadBuffer;//Physical address  pointer
  UCHAR      Adapter;            //HBA ID
  UCHAR      Device;             //Device ID
  PREGISTERS_ADDR BaseAddress; // Base address of PCI
configuration space
  ULONG      StartLBA;           //Start LBA of data stored in
this buffer block
  ULONG      SectorCount;        //Sector number contained in
this buffer block
  USHORT     HitCount;           //Buffer hit counter
  USHORT     WriteCounter;       //Dirty counter (how many
writes since last refresh)
}BUFFERBLOCK, *PBUFFERBLOCK
```

According to Microsoft Windows Device Driver Development Kit (DDK) documentation, we may call ScsiPortGetUncachedExtension to allocate BufferNum * BufferLen bytes memory while SCSI port/miniport driver loading for buffer use. Maximum allocable memory depends on available system resources and can influence buffer hit rate, in most cases, the larger the better. BufferLen is decided by device interface. For example, IDE uses an 8-bit register as sector count, hence the maximum read/write length is 256 sectors * 512 bytes per sector = 128KB.

4.2 The DAB Scheme

The operation of the DAB scheme can first be divided into four cases depending on whether the sector of requested block (whether a read or write) is found in the buffer.

(1) Both the first and the last sector of requested block are found in the buffer:
The requested block is located inside of one buffer block since it's smaller than 128KB. So the request is hit and no actual device I/O is performed, we just return the buffered data to complete the request.

(2) When the first sector of requested block is found in the buffer and the last sector isn't: The former part of requested block is located in one of the buffer blocks. So the request is partially hit, device I/O is needed to fetch its latter part. This will cause a DMA operation in order to prefetch 128K data that begin with this missing part. After necessary buffer block replacement and prefetching, this request is completed.

(3) When the first sector of requested block isn't found in the buffer but the last sector is: The latter part of requested block is located in one of the buffer blocks. So the request is also treated as partial-hit, device I/O is needed to fetch its former part. This will cause a DMA operation in order to prefetch 128K data that end with this missing part. After necessary buffer block replacement and prefetching, this request is completed.

(4) Neither the first sector nor the last sector of requested block is found in the buffer: The requested block isn't in any buffer block. Act as a traditional buffer and use 128K-aligned address to prefetch.

The action of the DAB scheme is presented below in algorithmic form.

```
if (request beginning sector in buffer) then
if (request ending sector in buffer) then{
request is satisfied and LRU stack updated;
}else{
fetch 128K-aligned data that contains the latter part
of request to buffer;
}
elseif (request ending sector in buffer) then{
fetch 128K-aligned data that contains the former part
of request to buffer;
}else{
use 128K-aligned address to prefetch;}
```

All prefetching can be completed in one DMA operation. We may use either write through (WT) or write back (WB) policy to handle write requests. [15] Missed write request will cause its referenced 128K block be prefetched in WB buffer, which is then updated in place in the buffer, so that later read/write requests may hit, saving its DMA. For WT buffer, write requests will be sent directly to HBA, thus write performance will not be affected.

5 Performance Evaluations

We realize our DAB method under Windows 2000. Low-level raw I/O accesses with different request size are tested. Block-level I/O access is achieved through SRBs (SCSI Request Block) It is sent from disk class driver to SCSI port/miniport driver, which is fully complied with Windows internal standards [16].

The benchmark we used to measure system throughput is Iometer [17] which is a popular file system benchmark developed by Intel. Iometer is an I/O subsystem measurement and characterization tool for single and clustered systems. It measures performance under a controlled load. Iometer is both a workload generator and a measurement tool. It can be configured to emulate the disk or network I/O load, or can be used to generate synthetic I/O loads. It can generate and measure loads on single or multiple (networked) systems.

Table 1. Iometer test options

Sequential Distribution	100% Sequential
Read/Write Distribution	100% Read
Transfer Delay	0ms
Burst Length	1 I/Os
Transfer Request Size	from 1KB to 16KB

Our experimental settings for the purpose of evaluating the performance of DAB are shown in Figure 2. We load our driver to the test machine for DAB testing with the Iometer options shown in Table 1. The test machine runs windows 2000 with two Tekram DC200 dual-channel UltraDMA IDE adaptors. Its configuration is described in Table 2 and the parameters of individual disks are summarized in Table 3.

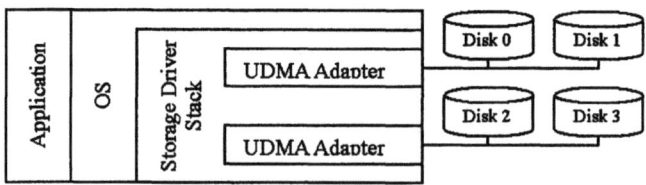

Fig. 2. Experimental platform –Software RAID based on Windows 2000 workstation. We have four hard disks connected via two dual-channel UDMA66 adapters. This software RAID can be configured as RAID0 with 2 or 4 disks, RAID1 or RAID1+0

Table 2. Test machine configuration

CPU	RAM	OS	FS
Intel Celeron 700	128M	WIN2000	NTFS

Table 3. Disk parameters

Disk Model	DJNA371350	RPM	7200
Interface	IDE ATA/66	Latency (ms)	4
Capacity	13.5G	Transfer rate (MB/s)	17
Data Buffer	2MB	Seek Time (ms)	9
Manufacturer	IBM		

We compared DAB software RAID with original software RAID and single disk in four aspects: IRP number, Read throughput, IRP response time and CPU utilization, which stand for total I/Os per second, total MBs data per second, request delay, host CPU load, respectively. As can be seen in Figure 3, compared to single disk (roughly the same as RAID1 configuration), both software RAIDs work better. If DAB isn't used, software RAID will suffer a significant performance decline on both the transfer rate and the number of request when the data block size is small. However, by using DAB, software RAID driver will consume more CPU cycles and the CPU utilization increases obviously on large data blocks.

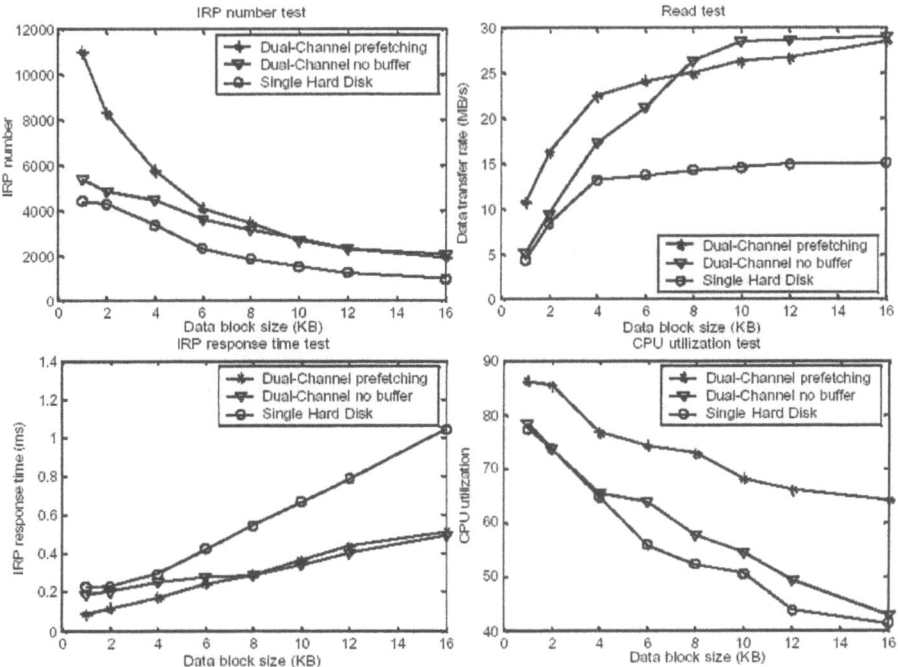

Fig. 3. Software RAID performance test results under windows 2000, using two-disk RAID0 configuration, based on dual-channel UDMA adapter. IRP represents an I/O operation

6   Conclusion and Future Work

According to our previous research on storage system, we noticed that I/O request's characteristic has a notable impact on I/O performance. As for software ATA RAID, adjusting its accessing block size around maximum DMA transfer length may result in transfer rate change. We proposed DMA-aligned Buffer to make use of this issue, by aligning small requests with DMA transfer boundary.

We have also shown that DAB can improve small block transferring, both the number of serviced request per second and the data transfer rate are improved when accessing block size is small. For larger accessing block size, the number of serviced request per second and data transfer rate are least affected by DAB, the downside happens only when CPU utilization is increased. The gap between the implementation with and without DAB grows wider as accessing block size increases.

One of our future work is to implement X-aligned buffer in our network software RAID, where X refers to various access pattern, such as network transfer block size. Especially for iSCSI protocol [18], its performance is greatly affected by different block sizes because a large number of fragmentary requests may result in considerable high overhead.

Although we have proved that our approach is able to enhance software RAID performance, the work presented in this paper is only the tentative step to the solution for storage system software implementation with better Cost/Performance Ratio.

References

1. P. M. Chen, E. K. Lee, G. A. Gibson, R. H. Katz, and D. A. Patterson.: RAID: High-Performance, Reliable Secondary Storage. ACM Computing Surveys, (1994) 145–185
2. Software RAID vs. Hardware RAID, http://www.adaptec.com/ (2002)
3. Jakob Østergaard, "The Software-RAID HOWTO", v0.90.8, http://www.tldp.org/HOWTO/Software-RAID-HOWTO.html, August 5, (2002)
4. David A. Solomon and Mark E. Russinovich.: Inside Microsoft Windows 2000, Third Edition, Microsoft Press (2000)
5. Alessandro Rubini, Jonathan Corbet, "Linux Device Drivers", 2nd Edition, O'Reilly (2001)
6. Microsoft Windows2000 DDK documents, Microsoft Press (2000)
7. Kotz, D., Ellis, C.S., "Practical Prefetching Techniques for Parallel File Systems" Proc. First Int'l Conf. on Parallel and Distributed Information Sys, Florida, 1991, pp. 182-189.
8. Smith, A.J., "Disk Cache--Miss Ratio Analysis and Design Consideration". ACM Trans. on Computer Systems (1985) 161-203.
9. Knut Stener Grimsrud, James K. Archibald, and Brent E.Nelson.: Multiple Prefetch Adaptive Disk Caching, IEEE Transactions of knowledge and data engineering, Vol 5. NO.1 (1993)
10. Jiri Schindler, John L. Griffin, Christopher R. Lumb, and Gregory R. Ganger.: Track-aligned extents: matching access patterns to disk drive characteristics. Conference on File and Storage Technologies, Monterey, California, USA (2002)
11. Russel Hugo Patterson III.: Informed Prefetching and Caching. PhD Thesis (1997)
12. Leonard Chung, Jim Gray, Bruce Worthington, Robert Horst.: Windows 2000 Disk IO Performance, Microsoft Research Advanced Technology Division (2000)
13. Information Technology - AT Attachment with Packet Interface – 7. Revision 3e, http://www.t10.org/ (2003)
14. Serial ATA: High Speed Serialized AT Attachment, http://www.t10.org/ (2003)
15. Stallings, Wm.: Computer Organization and Architecture: Principles of Structure and Function, Macmillan, New York (1993)
16. Walter Oney.: Programming the Microsoft Windows Driver Model, Microsoft Press (2002)
17. Iometer project, http://iometer.sourceforge.net/
18. Kalman Z. Meth, Julian Satran.: Design of the iSCSI Protocol, MSS'03 (2003)

mNFS: Multicast-Based NFS Cluster*

Woon-Gwun Lee, Chan-Ik Park, and Dae-Woong Kim

Department of Computer Science and Engineering/PIRL
Pohang University of Science and Technology
Pohang, Kyungbuk 790-784, Republic of Korea
{gamma,cipark,woong}@postech.ac.kr

Abstract. NFS is a distributed file system which is widely used in UNIX environments. Many studies have been worked to improve its I/O performance and scalability. However, existing architectures, which have either single load balancer or meta server, could suffer from single-point-of-failure and unnecessary network delay. We propose a new NFS architecture called mNFS cluster to improve its scalability. Experimental results show that mNFS outperforms both Linux NFS and NFSp by 64% and 99%, respectively.

1 Introduction

NFS is a distributed file system, which is widely used in UNIX operating system. NFS performance depends on network bandwidth, disk I/O throughput and CPU power for processing NFS protocol. Unlike AFS and CODA distributed file system, NFS server was designed to work on a single host resulting in relatively poor scalability. Several research such as NFSp[1] and DirectNFS[2] has been conducted to improve its performance and scalability. Bigfoot-NFS[3] allows multiple NFS file servers to use transparently their aggregate space as a single file system. It applies RAID techniques to network file systems. Bigfoot-NFS uses client-driven approach and runs without any central meta server. The server side is composed of normal NFS servers. Each NFS server has distinct files, that is, files are the unit of interleaving. To provide a single name space, all of the NFS servers have the same directory hierarchy. When multiple files are accessed, I/O loads are distributed into the NFS servers, and I/O performance can be improved. However, according to the result of initial implementation, 24 nodes Bigfoot-NFS performs worse than SunOS kernel-level NFS implementation and single node Bigfoot-NFS.

NFSp is an extension of NFS, which enables the use of the disk space of the nodes of a PC-based beowulf cluster. Early developed distributed file systems, such as AFS, CODA and xFS, have too many features to apply for cluster system

---

* The authors would like to thank the Ministry of Education of Korea for its support toward the Electrical and Computer Engineering Division at POSTECH through its BK21 program. This research was also supported in part by HY-SDR IT Research Center and in part by grant No. R01-2003-000-10739-0 from the Basic Research Program of the Korea Science and Engineering Foundation.

M. Bubak et al. (Eds.): ICCS 2004, LNCS 3038, pp. 363–370, 2004.
© Springer-Verlag Berlin Heidelberg 2004

in the private network. NFSp is divided into several kinds of dedicated servers. A meta server holds file attributes, whereas an I/O node stores the content of files physically. NFSp server acts as a standard NFS server. All I/O requests are sent to the meta server and then forwarded to I/O nodes. Hence, the meta server can be a bottleneck and forwarding I/O requests to I/O nodes causes additional network delay. Moreover, NFSp is based on the NFS protocol version 2 and user-level NFS servers, which limits the I/O performance. An experiment shows that a NFSp system, which have 32 I/O nodes, performs worse than a Linux kernel-level NFS server.

DirectNFS[2] is a hybrid architecture of both NAS and SAN storage system. It extends NFS protocols. DirectNFS is composed of a meta server and SAN disks. Legacy NFS clients are able to access the data in the SAN disks via the meta server which supports the standard NFS protocol. DirectNFS clients, however, have direct connection to SAN and they are able to bypass the meta server, once they obtain meta data from the meta server. Results with the initial implementation shows that DirectNFS performs as good as local file systems, EXT2 and ReiserFS. But DirectNFS system costs high because DirectNFS clients require SAN connectivity.

Bigfoot-NFS, NFSp and DirectNFS are all server driven architectures which require either load balancer or meta server. Therefore, they suffer from single-point-of-failure or additional network delay. To overcome these disadvantages, this paper proposes a multicast-based NFS cluster called *mNFS*. The mNFS is a client driven approach. A mNFS client sends each NFS request to multiple servers via multicasting. Thus, it can perform more efficiently transmission to multiple cluster nodes. The rest of this paper is organized as follows. Sections 2 describes the design of multicast-based NFS cluster and its prototype implementation. Section 3 gives experimental results. Finally, concluding remarks are given in Section 4.

2   mNFS: A Multicast-Based NFS Cluster System

To begin with, mRPC is designed for mNFS and it is a one-to-multiple variant of the normal one-to-one RPC. To support *portmapper* lookup, RPC protocol[4] has one-to-multiple RPC, called broadcast RPC. The broadcast RPC does not decide the set of receiving host and its transmission is limited in local subnet due to the characteristics inherited from broadcasting. mRPC is based on multicasting with IGMP[5] and is able to control the set of receiving hosts owing to the multicast group management facility of IGMP. If all the nodes of a mNFS cluster are joined to a multicast group, than they are capable to receive a NFS request simultaneously. When receiving a NFS request, each cluster node processes it selectively according to request type and load balancing policy of the cluster system.

To provide a single file system image, each cluster node maintains identical metadata of virtual file system for the mNFS cluster where each node has its own local disk and its local file system.

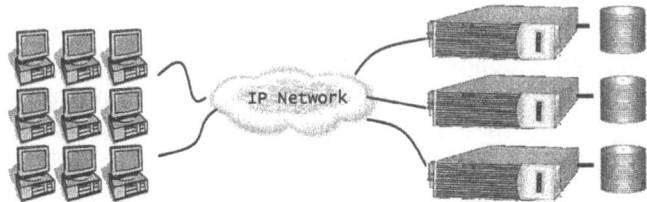

Fig. 1. Dedicated Disk Architecture

The metadata write operations like MKDIR, RMDIR and CREATE, must be sent to and processed by all cluster nodes. In the case of metadata read operations such as READDIR and GETATTR, it is enough for only one cluster node to handle the request and return the result to a client. A cluster node which is responsible for each file block I/O is determined by the offset of the block. The mNFS cluster system stripes data among cluster nodes. When a block read or write request is issued, a given file offset determines which cluster node should handle the block request.

2.1   Considerations of mNFS

Single Filesystem Image. A NFS client gets NFS server-side file system attribute information like Table 1 through a *STATFS* request. When a *STATFS* request is sent by a client, a corresponding cluster node returns virtual file system's attribute information configured by the mNFS cluster.

Metadata Operations. There is no control node in a mNFS cluster system, and each cluster node runs autonomously in co-operation with each others. Decision for I/O load balancing is made by each cluster node independently. Hence, the mNFS cluster must have a predefined load balancing policy to resolve conflict between which node is going to handle an I/O request. On the mNFS prototype, the RPC transaction ID is used as a unique identifier for round-robin load balancing policy.

Table 1. Information of File system attributes

Attribute	Description
bsize	block size of file system
blocks	the total number of blocks
bfree	the number of free blocks
bavail	the number of blocks available to non-privileged users
files	the number of used *inode*
ffree	the number of free *inode*

File I/Os. Performance improvement of the mNFS cluster system depends mainly on file I/O operations. The load balancing of file related operations can't be achieved by simple round-robin policy with RPC transaction ID because cluster node does not handle the whole of a file. A file is divided into several chunks of a fixed length, and they are distributed and stored into the cluster nodes. Therefore, the mNFS cluster can determine which cluster node are responsible for a (read or write) operation of a chunk of data as follows:

$$chunk = \lfloor offset \ / \ su \rfloor$$
$$i \equiv (chunk + ino) \bmod n$$

In the above equation, *offset* refers to a location of data block in a file, *chunk* refers to a processing unit of contiguous data blocks, *su* is the size of a chunk, *ino* is an inode number, and *i* refers to cluster node ID.

Inode Number Manipulation. The prototype mNFS cluster is based on dedicated disk architecture. So, each cluster node has its own disks and local file systems. In this environment, each file should be assigned a unique identifier which is common over the multiple cluster nodes. In the NFS protocol, an inode number, *ino* is used as an unique identifier of a NFS I/O operation. Even though each cluster node makes use of its own local file system, if it uses the same kind of disk and file system, then the inode number of newly created file becomes common over all cluster nodes. On the Linux, EXT2[6] is one of the most widely used local file systems. In EXT2 file system, the inode number of a file is allocated sequentially and can be synchronized easily among cluster nodes. In the case of directory, inode numbers are allocated on a less occupied block group. And each cluster node may have different block group usages due to the policy of file interleaving. Consequently, different inode numbers can be allocated for the same directory. Thus, in the prototype, we removed dynamic factors of inode number allocation for directories.

File Write and Inode Object. The metadata update operation such as *SE-TATTR, CREATE, REMOVE, MKDIR, RMDIR,* is achieved by the mRPC. The mRPC is performed at the same time and maintains the identity of a node metadata in all cluster node. But the mNFS cluster needs to update additional metadata changes. If file write operation occurs on a chunk of file data, then it implicitly updates the corresponding inode's fields such as file length and modification time. In a mNFS system, a file write request is dealt with only in a cluster node which is selected by load balancing policy. However, the change of inode object should be updated by all cluster nodes to keep identical metadata over the system.

2.2 The Pros and Cons of mNFS

The mNFS has the following advantages owing to multicast-based clustering technique.

- Standard NFS Protocol.
  Both server-side and client-side of the mNFS cluster requires some modification. But it does not modify any NFS protocol and not define any additional protocol. Thus the mNFS can inherit the stability from the standard NFS protocol.
- Improvement of File I/O Throughput.
  Generally, when processing a request in network file systems such as NFS and CIFS, the disk time is longer than the request transmission time. For a file read/write request, NFS does not consider an order of I/O requests. So the mNFS is able to improve I/O performance because file I/O requests are distributed into multiple cluster nodes.
- Scalability.
  System scalability has been restricted by its physical scalability. If a system is composed of multiple hosts, it could be more scalable than a single host system. Besides the mNFS cluster is based on IP multicast. Therefore the number of cluster nodes does not affect the time of NFS request transmission.
- Elimination of Additional Network Delay.
  If it has any control node for storage cluster such as load balancer or meta server in NFSp, at first, an I/O request is sent to a control node, and then the real I/O node, which is responsible for the request, is determined. At the end, the I/O request is handled at the I/O node. So, it requires two network transmission to deliver an I/O request to the corresponding I/O node. Whereas, mNFS cluster sends an I/O request through a multicasting, this mechanism delivers I/O request to all cluster nodes with just one transmission. And then each cluster node determines whether it should handle the request or not. In this way, there is no additional network delay.

Transmitting NFS requests via mRPC forces a mNFS client to be modified, increasing the maintenance cost of system. When a block write request is issued, the request is sent to all cluster nodes through multicasting and valuable network bandwidth is wasted by the multiple copies of the data block. Also, because metadata write operations must be performed synchronously by all of the cluster nodes, its response time are affected by the synchronization overhead.

3   Evaluation

A prototype of the mNFS cluster is based on the NFS implementation included in Linux kernel 2.4.20. Ext2, which is widely used in the Linux environments, is used as the local file system of mNFS cluster nodes. The mNFS cluster server consists of up to three cluster nodes. The cluster nodes and a single client are connected by gigabit ethernet. The multicast group address for the mNFS cluster is configured with *224.100.1.1*. The cluster nodes and client are used with PC-based host as shown in Table 2. Each host has two disks one for operating system and the other for storage space.

The Intel NetStructure 470T gigabit ethernet switch has been used, providing sufficient network bandwidth. The NetStructure has six 1000base-T ethernet

Table 2. Testbed specification

Processor	Intel Pentium 4 2.0GHz
Main Memory	256MB DDR SDRAM
HDD	IBM Deskstar 80GB ATA HDD
RAID Controller	Promise 20276 ATA RAID Controller
NIC	Intel 82540EM Gigabit ethernet adapter

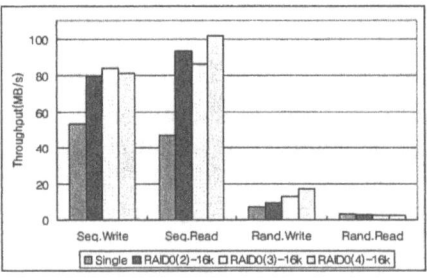

(a) Local disk

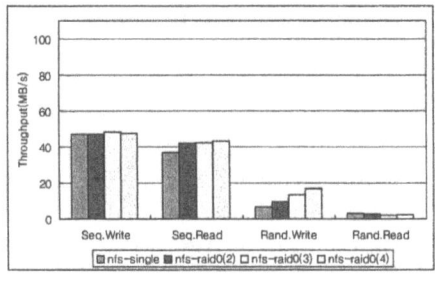

(b) Linux NFS

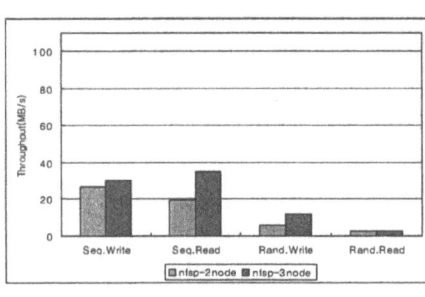

(c) NFSp

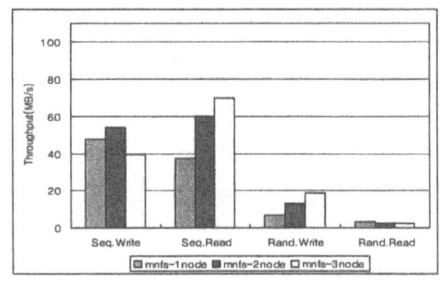

(d) mNFS

Fig. 2. I/O throughput

ports. Theoretically, it can support 21.2Gbps of internal bandwidth and forward about 1.4 million packets per second. Also, to support more efficient multicasting, it has the packet filter capability through *IGMP Snooping*[7].

In this evaluation, the throughput of the mNFS prototype is measured and compared with those of Linux NFS and NFSp. Comparison measures are file I/O throughput via NFS protocol and scalability when the number of cluster nodes or disks increases The mNFS and NFSp allocates a single disk to each cluster node. On the other hands, Linux NFS has a RAID level 0 disk array which has the same number of disks as cluster systems. IOzone[8] was used as a performance benchmark program. To exclude the effects of the difference in the total memory size between the experiments, size of experiment data sets are adjusted, that is, the size of data sets are three times of the total memory size.

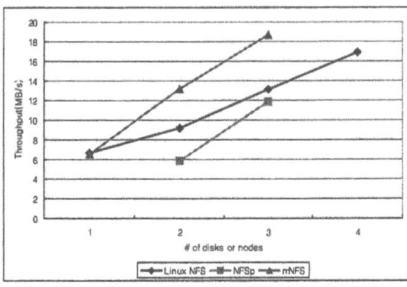

Fig. 3. Scalability of Random write

Table 3. Throughput of Linux NFS and mNFS when they contains 3 disks in total

	Linux NFS		mNFS	
Sequential Write	48.264	1	39.478	0.82
Sequential Read	42.567	1	69.648	1.64
Random Write	13.132	1	18.669	1.42
Random Read	2.11	1	2.557	1.21

In Figure 2(a) and 2(b), nevertheless the number of disks increases, a sequential I/O performance of Linux NFS is almost not changed. But the performance of random I/O is similar to that of local disk. Thus, a sequential I/O performance is limited by both gigabit ethernet and NFS protocol overheads.

For the convenience of implementation, NFSp is based on a user-level NFS server implementation. Due to frequent memory copy between user memory space and kernel memory space, it is difficult to expect good I/O performance. And it supports only NFS protocol version 2 where all file writes are handled synchronously to provide stability of data. Figure 2(c) shows that I/O performance improves when the number of server nodes or disks increases, But the performance of 3 nodes NFSp is lower than that of Linux NFS.

Figure 2(d) shows the result of mNFS experiments. In the case of sequential write, 2 nodes mNFS performed better than a single node. But 3 nodes mNFS performed worse than 2 nodes system and even a single node system. As the number of node increases, file write requests are prone not to be delivered in order. Thus sequential writes can not be processed sequentially at the I/O nodes and performance degradation can occur. In the case of random I/O, mNFS outperforms both Linux NFS and NFSp as shown in Figure 3. mNFS and NFSp seems to have similar scalability but the throughput of mNFS is much better than that of NFSp. Table 3 shows the throughput of Linux NFS and mNFS when they contains 3 disks in total. It is shown that mNFS provides better sequential write and read performance than Linux NFS by 18% and 64%, respectively. In random read and write case, mNFS performs better than Linux NFS by 21% and 42%, respectively.

As a whole, A mNFS cluster provides high scalability and good I/O performance.

4   Conclusions

The mNFS cluster system is a storage system which enhances the NFS that is an internet standard distributed file system. Through multicast RPC of client side and clustering of server side, the mNFS improves I/O performance and scalability. The mNFS cluster does not need any additional hardware. On the existing TCP/IP network, it can achieve good price/performance ratio. We show a prototype implementation of the mNFS. The mNFS does not have any dedicated control node, but it is controlled by autonomous co-operation between cluster nodes. Because the mNFS does not have any single-point-of-failure node, it can provide much higher availability. Although there are overheads of meta data update, the mNFS can improve file I/O performance and scalability. We expect that a mNFS system can be used successfully as a virtualized high-end storage system.

References

1. Lombard, P., Denneulin, Y.: nfsp: A distributed nfs server for clusters of workstations. In: Proc. of International Parallel and Distributed Processing Symposium. (2002)
2. Bhide, A., Engineer, A., Kanetkar, A., Kini, A.: File virtualization with directnfs. In: Proc. of the 19th IEEE Symposium on Mass Storage Systems and Technologies. (2002)
3. Kim, G.H., Minnich, R.G., McVoy, L.: Bigfoot-nfs: A parallel file-striping nfs server (1994)
4. Sun Microsystems, Inc.: RFC 1057: RPC: Remote procedure call protocol specification, version 2 (1988)
5. Fenner, W.: RFC 2236: Internet Group Management Protocol, version 2 (1997)
6. Bovet, D.P., Cesati, M.: Understanding the LINUX KERNEL. O'Reilly (2001)
7. Intel Corp.: Intel netstructure 470 switch user guide, 2nd edition (2001)
8. Norcott, W.: Iozone filesystem benchmark. URL: *http://www.iozone.org/* (1998)

Balanced RM2: An Improved Data Placement Scheme for Tolerating Double Disk Failures in Disk Arrays*

Dae-Woong Kim, Soon-Ho Lee, and Chan-Ik Park

Department of Computer Science and Engineering/PIRL
Pohang University of Science and Technology
Pohang, Kyungbuk 790-784, Republic of Korea
cipark@postech.ac.kr

Abstract. There is a growing demand for high data reliability beyond what current RAIDs can provide and users need various levels of data reliability. In order to meet these requirements, an efficient data placement scheme called RM2 has been proposed in [1], which enables a disk array system to tolerate double disk failures. However, RM2 has high data reconstruction overhead due to data dependency between parity groups in the case of double disk failures, suffering from large performance fluctuation according to disk failure patterns. To minimize the reconstruction overhead, this paper presents an improved data placement scheme called *Balanced RM2*. Experimental results show that the performance fluctuation of the Balanced RM2 becomes much less than that of the RM2 regardless of read or write. It is also shown that Balanced RM2 provides better performance than RM2 by 5 ~ 20% in the case of double disk failures.

1   Introduction

There has been much research on improving I/O performance, known as data declustering and disk striping, in disk array systems [2,3,4]. However, incorporating a large number of disks into a disk array system makes the system more prone to failure than a single disk would [3].

In general, high reliability can be achieved by redundant information. For example, RAID (Redundant Arrays of Independent Disks) enables a disk array system to tolerate a single disk failure by maintaining parity information. In order to meet reliability requirement beyond what RAID can provide, that is, tolerate more than one disk failures, there has been much research on high reliability of disk arrays [1,5,6]. Blaum *et al.* have proposed EVENODD in [6]. This

* The authors would like to thank the Ministry of Education of Korea for its support toward the Electrical and Computer Engineering Division at POSTECH through its BK21 program. This research was also supported in part by HY-SDR IT Research Center and in part by grant No. R01-2003-000-10739-0 from the Basic Research Program of the Korea Science and Engineering Foundation.

M. Bubak et al. (Eds.): ICCS 2004, LNCS 3038, pp. 371–378, 2004.
© Springer-Verlag Berlin Heidelberg 2004

method maintains two types of redundant information: horizontal redundancy and diagonal redundancy. Note that EVENODD requires only two additional disks (which is optimal). However, EVENODD has an apparent restriction; the number of information disks has to be a prime number. Park has proposed RM2 in [1] which does not exhibit any explicit I/O bottleneck by enforcing the even distribution of parity information across all disks. However, RM2 has high data reconstruction overhead due to data dependency between parity groups in the case of double disk failures, suffering from large performance fluctuation according to disk failure patterns.

This paper presents an improved data placement scheme called *Balanced RM2* which distributes the load of data reconstruction as evenly as possible over all the remaining disks under disk failures.

The rest of the paper is organized as follows. Section 2 describes the RM2 architecture. Section 3 presents the Balanced RM2 architecture and its data reconstruction method. Section 4 gives experimental results for various cases. Finally, concluding remarks and future works are given in Section 5.

2   Backgrounds

A disk array is assumed to consist of N disks each of which handles independent requests in parallel. The address space of a disk array is logically structured with a set of stripe units. A *stripe unit* is a basic unit of data striping consisting of a group of logically contiguous blocks. It is placed consecutively on a single disk before placing another stripe unit on a different disk. A *stripe* consists of N stripe units each of which has the same physical address in each disk. If a stripe contains only data units, then it is called a *data stripe*. If a stripe consists of parity units only, then it is called a *parity stripe*. A *parity group* is defined as a set of data units and a parity unit, where the parity unit is computed from the set of data units. We define a *stripe group* as a set of data stripes and a parity stripe, which covers all stripe units in a given parity group. In RM2, data and parity mapping of a stripe group are defined by an $N \times N$ redundancy matrix. In the redundancy matrix, each column corresponds to a disk and each row corresponds to a parity group. Each column has one entry with a value of -1 and two entries with a value of k for $1 \leq k \leq M - 1$ where M is the size of a stripe group. Let $RM_{i,j}$ be an element of the redundancy matrix with a value ranging from -1 to $M - 1$. According to the value of $RM_{i,j}$, the interpretation differs as follows:

- $RM_{i,j} = -1$: A parity unit of the disk j belongs to the parity group i.
- $RM_{i,j} = 0$: No meaning.
- $RM_{i,j} = k$ for $1 \leq k \leq M - 1$: The k-th data unit of the disk j belongs to the parity group i.

Figure 1 represents the data layout of the RM2 architecture and its corresponding 7×7 redundancy matrix when the number of disks is seven. Note that each disk in RM2 has equal amount of parity information. There are two data

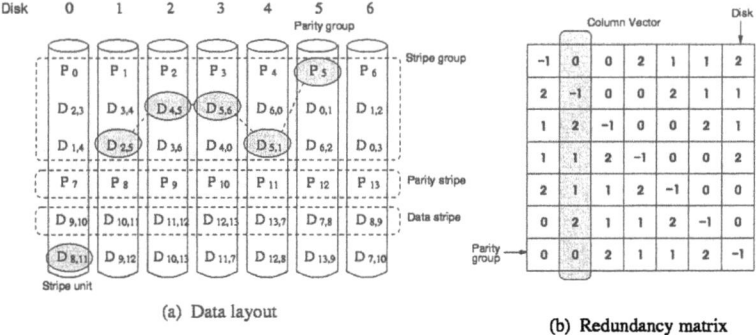

(a) Data layout

(b) Redundancy matrix

Fig. 1. Data layouts of RM2 and its redundancy matrix for $N = 7$ and $M = 3$ where N is the number of disks and M is the size of a stripe group

stripes and one parity stripe in a stripe group, i.e., $M = 3$. The amount of disk space used to store parity information is represented by the reciprocal of the size of a stripe group (i.e., $1/M$), called *redundancy rate*. For example in Figure 1, the redundancy rate is $1/3$ since a stripe group consists of two data stripes and one parity stripe, i.e., the size of a stripe group is 3. P_i represents the parity unit in the i-th parity group and $D_{i,j}$ means that the data unit is involved in computing parity information of both the i-th and j-th parity groups. For example, in order to compute the parity unit P_5, we have to consider the four data units $D_{2,5}$, $D_{4,5}$, $D_{5,6}$, and $D_{5,1}$. And, the data unit $D_{8,11}$ belongs to both parity groups P_8 and P_{11}.

3   Balanced RM2

The main problem of RM2 is that it has high data reconstruction overhead due to data dependency between parity groups in the presence of double disk failures. To minimize the reconstruction overhead, this section describes how to place data and parity to address the data dependency problem.

3.1   Dual-Symmetry

In RM2, *c-independence* was introduced to formulate how to place In order to deal with the efficiency of data reconstructing operation, we define *dual-symmetry* of RM2.

Definition 1 (Dual-Symmetry). *The column vector of a redundant matrix is dual-symmetric if every positive entries are symmetrically positioned both around the entry with a value of −1 (parity unit) and around a sequence of zero entries simultaneously. And the redundant matrix is said to be dual-symmetric if its column vector is dual-symmetric.*

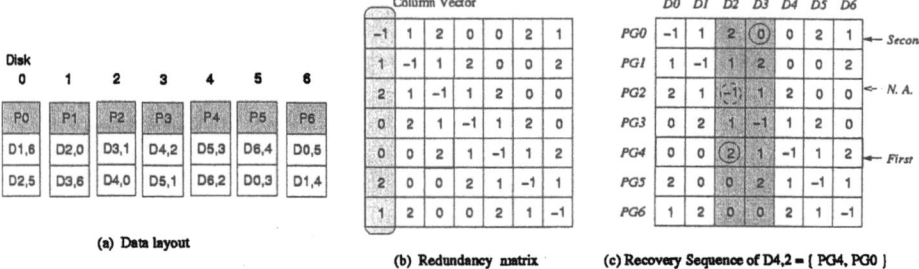

(a) Data layout

(b) Redundancy matrix

(c) Recovery Sequence of D4,2 = { PG4, PG0 }

Fig. 2. Data layouts of the Balanced RM2 and its redundancy matrix for $N = 7$ and $M = 3$

For example, Figure 1(b) shows a redundancy matrix that is c-independent, but whose column vector is not dual-symmetric. Figure 2(b) represents the redundancy matrix that is both c-independent and dual-symmetric.

For the Balanced RM2 architecture, it is necessary for a redundancy matrix to meet the following two conditions:

- any pair of columns is *c-independent*.
- the column vector is *dual-symmetric*.

The first condition guarantees that all failed data blocks can be reconstructed in the case of one or two disk failures [1]. And, the second condition enforces that the load of data reconstruction are evenly distributed, which will be described in detail later.

Figure 2 represents the data layout of the Balanced RM2 architecture and its redundancy matrix when $N = 7$ and $M = 3$. It looks similar to that of RM2 (see Figure 1), but note that all stripe units of a parity group are symmetrically positioned around the parity unit.

3.2   Recovery Points and Sequences

A *recovery point* is defined as a parity group that has only one inaccessible data unit owing to disk failures. If the inaccessible data unit is a parity unit, then the recovery point will not be used as a starting sequence for data reconstruction. Otherwise, it can be regarded as a starting sequence for data reconstruction. A *recovery path* is defined as the sequence of all parity groups reachable from a recovery point. For example shown in Figure 2(c), parity groups $PG0$ and $PG5$ become recovery points when disks $D2$ and $D3$ fail. Thus, two recovery paths $\{PG0, PG4, PG2\}$ and $\{PG5, PG1, PG3\}$ are found. If a recovery path ends up with another recovery point, then it is called a *bi-directional recovery path*. Otherwise, it is called a *uni-directional recovery path*.

A *recovery sequence* is defined as the shortest recovery path which starts from a recovery point and ends up with a parity group including the failed data unit. Following the recovery sequence, we can reconstruct the failed data unit. The *length of a recovery sequence* represents the number of parity groups involved in

Table 1. The length of minimum and maximum recovery sequence depending on the distance between two failed disks, where $N = 7$ and $M = 3$

Distance between two failed disks	RM2		Balanced RM2	
	minimum	maximum	minimum	maximum
1	1	4	1	2
2	1	2	1	1
3	1	1	1	1
4	1	1	1	1

the recovery sequence. For example, the recovery sequence length in RAID5 is one, that is, accessing only one parity group is enough for data reconstruction.

For example shown in Figure 2, consider the case where we want to reconstruct the first data unit, $D_{4,2}$, in the disk $D3$ when the disks $D2$ and $D3$ fail (denoted as shaded columns). Both the parity group $PG2$ and $PG4$ are the parity groups that include the failed data unit $D_{4,2}$ since $RM_{2,3}$ and $RM_{4,3}$ have a value of one in the redundancy matrix. From the parity group $PG4$, we can find the second data unit of $D2$, $D_{4,0}$, as a pivot because $PG4$ has another failed data unit. Then, the recovery sequence includes the parity group $PG0$ because $PG0$ also depends on the data unit selected as a pivot. Since $PG0$ has only one broken data unit, $PG0$ is then a recovery point that can be a starting sequence of data reconstruction. Thus, we get the recovery sequence that includes the parity group $PG4$ and $PG0$. However, from the parity group $PG2$, we cannot further construct a recovery sequence because the pivot for the parity group $PG2$ is a parity unit.

We know that the length of a recovery sequence has effects on the data reconstruction performance in the case of disk failures.

Lemma 1. *The Balanced RM2 has at least two recovery points in the case of double disk failures.*

Theorem 1. *The length of any recovery sequence in the Balanced RM2 is at most $M - 1$, where M is the size of a stripe group.*

For the proof of Lemma 1 and Theorem 1, please refer to [7].

Consequently, the Balanced RM2 architecture is expected to provide better performance than RM2 in the case of double disk failures since the worst case of recovery sequence lengths in the RM2 is $2(M - 1)$ while only $M - 1$ in Balanced RM2, where M is the size of a stripe group. Table 1 shows the minimum and maximum recovery sequence depending on the distance between two failed disks.

4   Performance Evaluation

This section evaluates the performance, in terms of throughput and response time, of RM2, EVENODD, and Balanced RM2 under a highly concurrent workload of small reads and writes.

Table 2. Disk operation overheads in each disk array architecture according to the number of disk failures. C_R and C_W represent the cost of read operations and write operations, respectively. N denotes the number of disks and G represents the size of a parity group. L and L^* denote the number of parity groups involved in reconstructing the failed data unit in RM2 and Balanced RM2 each other

Disk arrays	Single failure		Double failures	
	Read cases	Write case	Read case	Write case
EVENODD	$(N-1)\,C_R$	$(N-1)\,C_R+2C_W$	$2\,(N-1)\,C_R$	$2\,(N-1)\,C_R+2C_W$
RM2	$(G-1)\,C_R$	$(G-1)\,C_R+2C_W$	$L\,(G-1)\,C_R$	$L\,(G-1)\,C_R+2C_W$
Balanced RM2	$(G-1)\,C_R$	$(G-1)\,C_R+2C_W$	$L^*\,(G-1)\,C_R$	$L^*\,(G-1)\,C_R+2C_W$

Simulation Environments. Our experiments were run on *RAIDframe* [8], a simulator in which various RAID architectures can be modeled and evaluated. RAIDframe's synthetic workload generator makes highly concurrent sequences of user I/O requests for disk accesses from multiple client threads. Each individual user request is converted into a set of read and write accesses on the specific disk in the disk array.

The individual disk of the array was modeled with *HP 2247* whose *revolution time, single cylinder seek time* and *average seek time* are 11.12 ms, 2.5 ms and 10 ms, respectively. The stripe unit was configured to be of 8 sectors, i.e., 4 *Kbytes*. For disk queue management, the shortest-seek-time-first (SSTF) queueing policy is adopted in a disk queue that can hold 10 outstanding requests. The synthetic workload generator incessantly makes random and independent 4 *Kbyte* accesses aligned to the stripe unit boundaries. These workloads are chosen because small requests are typically the most demanding for many RAID architectures [9,10,11].

Each disk array is configured as $N = 15$ and $M = 3$. The experiments are performed according to the increase of client threads, which shows us the saturation point of throughput in each disk array architecture.

Analytic Comparison. Table 2 shows how many disk operations are needed to serve small-sized user requests in each architecture depending on the number of disk failures. Note that each array has different values in the size of a parity group as well as the length of a recovery sequence. For example, the size of a parity group in the RM2 architecture is the same as that in the Balanced RM2 architecture whereas the RM2 has the longer recovery sequence than the Balanced RM2.

Comparison under Double Disk Failure. In the case of read operations, it is expected that the RM2 and the Balanced RM2 have much better performance than the EVENODD due to the declustering effect [10]. In detail, the Balanced RM2 architecture provides better performance than the RM2 architecture since it needs smaller number of parity groups in reconstructing the failed data unit. Figure 3(a) shows that the Balanced RM2 architecture reaches

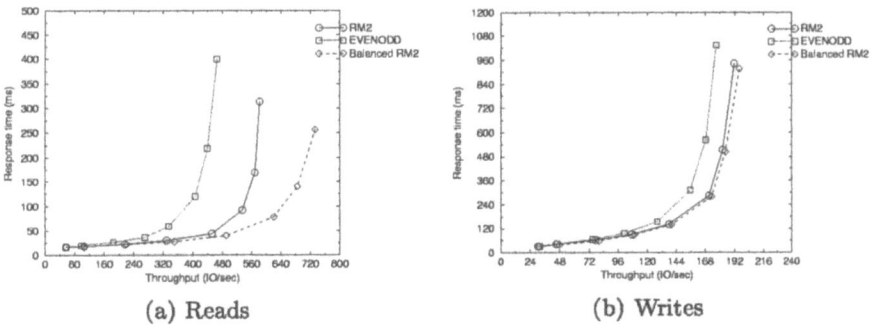

(a) Reads   (b) Writes

Fig. 3. Performance comparison for the case of double disk failures

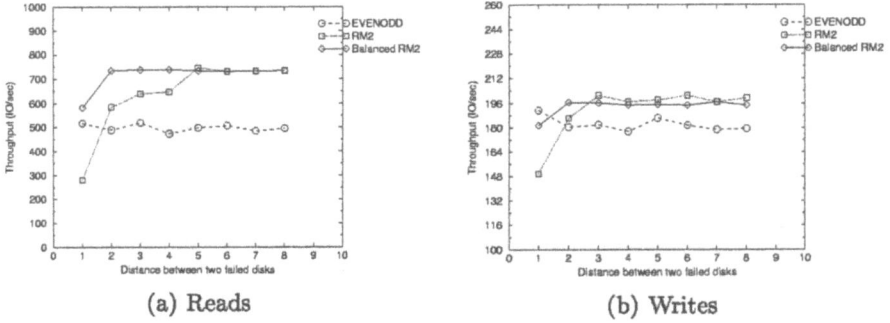

(a) Reads   (b) Writes

Fig. 4. Throughput according to the distance between two failed disks

its point of saturation at about 740 *reqs/sec* whereas the RM2 architecture at about 580 *reqs/sec* and the EVENODD architecture at about 460 *reqs/sec*.

In the case of the write operations, RM2 and Balanced RM2 have non trivial performance advantage over EVENODD as shown in Figure 3(b). Furthermore, the performance of Balanced RM2 is slightly better than RM2 due to the smaller number of parity groups involved in reconstructing the failed data unit. In the case of write operations, the Balanced RM2 architecture reaches its point of saturation at about 194 *reqs/sec*, compared to about 190 *reqs/sec* for the RM2 architecture and 175 *reqs/sec* for the EVENODD architecture.

Comparison According to the Distance between Two Failed Disks.
Figure 4 demonstrates how the distance between two failed disks affects on the performance of read and write operation in the case of double disk failures. It is shown that Balanced RM2 provides a better throughput than RM2. Remember that the maximum length of a recovery sequence in the Balanced RM2 is $(M - 1)$ whereas $2(M - 1)$ in RM2. The Balanced RM2 architecture has better performance than the EVENODD architecture because of the declustering effect. However, the declustering effect becomes negligible due to long recovery sequences in the case of adjacent disk failures.

5   Conclusions

This paper has presented an improved data and parity placement scheme called *Balanced RM2* to minimize high data reconstruction overhead in RM2. Minimizing data reconstruction overhead is achieved by reducing the number of parity groups involved in reconstructing the failed data unit.

The experiments have been conducted by the RAIDframe under a highly concurrent workload of small-sized reads and writes. It has been shown that the Balanced RM2 provides better performance than the RM2 and the EVENODD in data reconstruction operations.

The data and parity placement schemes using the redundancy matrix are applicable to many research fields for disk arrays. Currently, we are considering the data and parity placement schemes to tolerate n disk failures in disk arrays.

References

1. Park, C.I.: Efficient placement of parity and data to tolerate two disk failures in disk array systems. IEEE Transactions on Parallel and Distributed Systems **6** (1995) 1177–1184
2. Kim, M.Y., Tantawi, A.N.: Asynchronous disk interleaving. IEEE Transactions on Computers **40** (1991)
3. Patterson, D.A., Gibson, G.A., Katz, R.H.: A case for redundant array of inexpensive disks (raid). In: ACM SIGMOD Conference Proceedings, Chicago, Illinois (1988)
4. Narasimha, R.A.L., Banerjee, P.: An evaluation of multiple-disk i/o systems. IEEE Transactions on Computers **38** (1989) 1680–1690
5. Alvarez, G.A., Burkhard, W.A., Cristian, F.: Tolerating multiple failures in raid architectures with optimal storage and uniform declustering. In: Proceedings of the 24th Annual ACM/IEEE International Symposium on Computer Architecture(ISCA '97), Denver, Colorado (1997)
6. Blaum, M., Brady, J., Bruck, J., Menon, J.: Evenodd: An efficient scheme for tolerating double disk failures in raid architectures. IEEE Transactions on Computers **44** (1995)
7. Kim, D.W., Lee, S.H., Park, C.I.: Balanced rm2: An improved data placement scheme for tolerating double disk failures in disk arrays. Technical Report CSE-SSL-2003-08, System Software Lab., Department of Computer Science and Engineering, Pohang University of Science and Technology (2003)
8. Courtright, W.A., Gibson, G., Zelenka, J.: Raidframe: Rapid prototyping for disk arrays. In: Proceedings of the 1996 Conference on Measurement and Modeling of Computer Systems (SIGMETRICS). Volume 24. (1996) 268–269
9. Stodolsky, D., Gibson, G., Holland, M.: Parity logging: Overcoming the small write problem in redundant disk arrays. In: Proceedings of International Symposium on Computer Architecture, San Diego CA (1993)
10. Holland, M., Gibson, G.A.: Parity declustering for continuous operation in redundant disk arrays. Journal of Distributed and Parallel Databases **2** (1994)
11. Courtright, W.A., Gibson, G., Zelenka, J.: A structured approach to redundant disk array implementation. In: Proceedings of the International Symposium on Performance and Dependability. (1996)

Diagonal Replication on Grid for Efficient Access of Data in Distributed Database Systems

M. Mat Deris, N. Bakar, M. Rabiei, and H.M. Suzuri

University College of Science and Technology,
Department of Computer Science, 21030, MengabangTelipot, Kuala Terengganu, Malaysia
{mustafa,suzuri}@kustem.edu.my

Abstract. Data Replication can be used to improve data availability in distributed database environments. In such a system, a mechanism is required to maintain the consistency of the replicated data. Grid structure technique based on quorum is one of the solutions for performing this while providing a high data availability of the system. It was shown in the previous study that, it still requires a bigger number of copies be made available to construct a quorum. So it is not suitable for the large systems. In this paper, we propose a technique called diagonal replication on grid (DRG) technique where a data will be replicated in a diagonal manner in a logical grid structure. In comparison to the Grid structure technique, DRG requires lower communication cost for an operation, while providing higher read/write availability, which is preferred for large distributed database systems.

1 Introduction

With the proliferation of computer networks, PCs and workstations, new models for workplaces are emerging [1]. In particular, organizations need to provide current data to users who may be geographically remote and to handle a volume of requests that might be high for a single server. Consequently, the availability and the consistency of data in the systems are extremely important. To maintain the consistency and integrity of data, expensive synchronization mechanisms are needed [2,4]. One way to provide access to such data is through replication.

One of the simplest techniques for managing replicated data is read-one write-all (ROWA) technique. Read operations on a data object are allowed to read any copy, and write operations are required to write all copies of the data object. The ROWA protocol is good for environments where the data is mostly read-only. This is because it provides read operation with high degree of availability at low communication overhead but vulnerable to the write availability. It is very popular and has been used in mobile and peer-to-peer environments [3] and database systems [6]. Dynamic quorum techniques have also been proposed to further increase availability in replicated databases [5,7]. Nevertheless, these approaches do not address the issue of low-cost read operations.

Recently, Computational Grids are becoming ubiquitous computing infrastructure for the coordinated and secure use of shared resources in a distributed environment. Based on this new computing platform, data replication technique proposed in [4]

M. Bubak et al. (Eds.): ICCS 2004, LNCS 3038, pp. 379–387, 2004.
© Springer-Verlag Berlin Heidelberg 2004

provides high data availability of the system. However, this technique still requires that a bigger number of copies be made available to construct a read or write quorum, which is not suitable for the large systems.

Without loss of generality, the terms node and site will be used interchangeably. If data is replicated to all sites, the storage capacity becomes an issue, thus an optimum number of sites to replicate the data is required with non-tolerated read or write availability of the system. The focus of this paper is on modeling a technique to optimize the read or write availability of the replicated data in the distributed database systems. We describe the Diagonal Replication on Grid (DRG) technique considering only binary vote assigned with vote one to the primary site and its diagonal site, and zero otherwise. This assignment provides a higher availability of executing read or write operations in replicated database due to the minimum number of quorum size required. Moreover, it can be viewed as an allocation of replicated copies such that a copy is allocated to a site if an only if the vote assigned to the site is one.

This paper is organized as follows: In Section 2, we review the grid structure technique. In Section 3, the model and the technique of the DRG is presented. In Section 4, the performance of the proposed technique is analyzed in terms of communication cost and availability, and a comparison with other technique is given.

2 Review of Grid Structure (GS) Technique

In this section, we review the GS technique, which is then compared to the proposed DRG technique.

Maekawa [8] has proposed a technique by using the notion of finite projective planes to obtain a distributed mutual exclusion algorithm where all quorums are of equal size. Cheung et al. [9], extended Maekawa's grid technique for replicated data that supports read and write operations. In this technique, all sites are logically organized in the form of two-dimensional n x n grid structure as shown in Fig. 1 if there are N = n x n sites in the system. Read operations on the data item are executed by acquiring a read quorum that consists of a copy from each column in the grid. Write operations, on the other hand, are executed by acquiring a write quorum that consists of all copies in one column and a copy from each of the remaining columns. The read and write operations of this technique is of the size $O(n)$; this technique is normally referred to as the *sqrt(R/W)* technique. In Fig. 1, copies {1,2,3,4,5} are

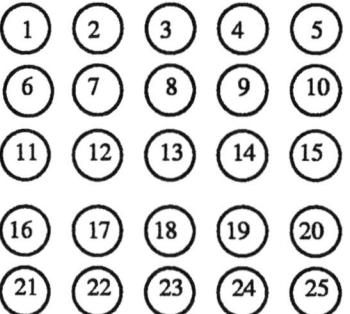

Fig. 1. A grid organization with 25 copies of a data object

sufficient to execute a read operation whereas copies {1,6,11,16,21,7,13,19,25} will be required to execute a write operation. However, it still has a bigger number of copies for read and write quorums, thereby degrading the communication cost and data availability. It is also vulnerable to the failure of entire column or row in the grid [4].

3 Diagonal Replication on Grid (DRG) Technique

3.1 Model

A distributed database consists of a set of data objects stored at different sites in a computer network. Users interact with the database by invoking transactions, which are partially ordered sequences of atomic read and write operations. The execution of a transaction must appear atomic: a transaction either commits or aborts [10,11].

The correctness criteria for replicated database is one-copy serializability [11], which ensures the serializable execution of transactions. In order to ensure one-copy serializability, a replicated data object may be read by reading a quorum of copies, and it may be written by writing a quorum of copies. The selection of a quorum is restricted by the quorum intersection property to ensure one-copy equivalence: For any two operations o[x] and o'[x] on a data object x, where at least one of them is a write, the quorum must have a non-empty intersection. The quorum for an operation is defined as a set of copies whose number is sufficient to execute that operation.

Briefly, a site s initiates a DRG transaction to update its data object. For all accessible data objects, a DRG transaction attempts to access a DRG quorum. If a DRG transaction gets a DRG write quorum without non-empty intersection, it is accepted for execution and completion, otherwise it is rejected. We assume for the read quorum, if two transactions attempt to read a common data object, read operations do not change the values of the data object. Since read and write quorums must intersect and any two DRG quorums must also intersect, then all transaction executions are one-copy serializable.

3.2 The DRG Technique

All sites are logically organized in the form of two-dimensional grid structure. For example, if a DRG consists of twenty-five sites, it will logically organized in the form of 5 x 5 grid as shown in Fig. 2. Each site has a master data file. In the remainder of this paper, we assume that replica copies are data files. A site is either operational or failed and the state (operational or failed) of each site is statistically independent to the others. When a site is operational, the copy at the site is available; otherwise it is unavailable. The data file will replicate to *diagonal sites*. The circles in the grid represent the sites under the distributed database environment and $a,b,...,$and y represent the master data files located at site s(1,1),s(1,2),...,and s(5,5) respectively.

Definition 3.2.1: Assume that a database system consists of n x n sites that are logically organized in the form of two dimensional grid structure. All sites are labeled

$s(i,j)$, $1 \le i \le n$, $1 \le j \le n$. The *diagonal site* to $s(i,j)$ is $\{s(k,l) \mid k=i+1,\ l=j+1,\ \text{and}\ k,l \le n,\ \text{if}$ $i=n$, initialized $i=0$, if $j=n$, initialized $j=0\}$. A *diagonal set*, $D(s)$, is a set of diagonal sites.

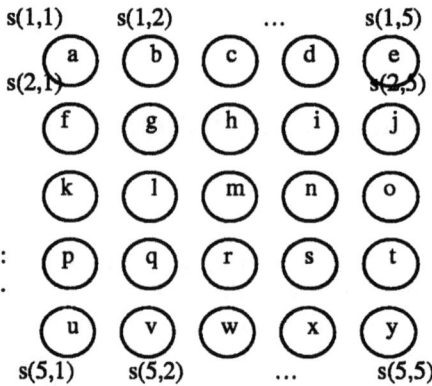

Fig. 2. A grid organization with 25 sites, each of which has a master data file $a,b,...,$and y respectively.

For example, from Fig. 2, the diagonal site to $s(1,1)$ is $s(2,2)$, the diagonal site to $s(2,1)$ is $s(3,2)$, etc. Thus, based on this technique, sites in the diagonal set will have the replica copies in common. One such diagonal set is $\{s(1,1), s(2,2), s(3,3), s(4,4), s(5,5)\}$, and each site will have the same replica copies i.e., $\{a,g,m,s,y\}$. Also the number of diagonal sets in the system equals to n. Let $D^m(s)$ be the mth diagonal set, for m=1,2,...n, then from Fig. 2, the diagonal sets are;

$$D^1(s) = \{s(1,1), s(2,2), s(3,3), s(4,4),s(5,5)\},$$
$$D^2(s) = \{s(2,1), s(3,2), s(4,3), s(5,4),s(1,5)\},$$
$$D^3(s) = \{s(3,1), s(4,2), s(5,3), s(1,4),s(2,5)\},$$
$$D^4(s) = \{s(4,1), s(5,2), s(1,3), s(2,4),s(3,5)\},\ \text{and}$$
$$D^5(s) = \{s(5,1), s(1,2), s(2,3), s(3,4),s(4,5)\}.$$

Since the data file is replicated to only the diagonal sites, then it minimize the number of database update operations, misrouted and dropped out calls. Also, sites are autonomous for processing different query or update operation, which consequently reduces the query response time. The primary site of any data file and for simplicity, its diagonal sites are assigned with vote one and vote zero otherwise. A vote assignment on grid, B, is a function such that

$$B(s(i,j)) \in \{0,1\},\ 1 \le i \le n, 1 \le j \le n$$

where $B(s(i,j))$ is the vote assigned to site $s(i,j)$. This assignment is treated as an allocation of replicated copies and a vote assigned to the site results in a copy allocated at the diagonal site. That is,

$$1\ \text{vote} \equiv 1\ \text{copy}.$$

Let

$$L_B\ =\ \sum_{s(i,\,j)\,\in\,D(s)} B(s(i,j))$$

where, L_B is the total number of votes assigned to the primary site and its diagonal sites. Thus, $L_B = d$.

Let r and w denote the read quorum and write quorum, respectively. To ensure that the read operation always gets up-to-date values, r + w must be greater than the total number of copies (votes) assigned to all sites. The following conditions are used to ensure consistency:

  1.) $1 \leq r \leq L_B, 1 \leq w \leq L_B,$
  2.) $r + w = L_B + 1.$

Conditions (1) and (2) ensure that there is a nonempty intersection of copies between every pair of read and write operations. Thus, the conditions ensure that a read operation can access the most recently updated copy of the replicated data. Timestamps can be used to determine which copies are most recently updated.

Let S(B) be the set of sites at which replicated copies are stored corresponding to the assignment B. Then

$$S(B) = \{s(i,j)| \ B(s(i,j)) = 1, \ 1 \leq i \leq n, \ 1 \leq j \leq n \ \}.$$

Definition 3.2.2: For a quorum q, a *quorum group* is any subset of S(B) whose size is greater than or equal to q. The collection of quorum group is defined as the *quorum set*.

Let Q(B,q) be the quorum set with respect to the assignment B and quorum q, then

$$Q(B,q) = \{ \ G| \ G \subseteq S(B) \ and \ |G| \geq q\}$$

For example, from Fig. 1, let site s(1,1) be the primary site of the master data file *a*. Its diagonal sites are s(2,2),s(3,3)),s(4,4), and s(5,5). Consider an assignment B for the data file *a*, such that

$$B_a(s(1,1))=B_a(s(2,2))=B_a(s(3,3))=B_a(s(4,4))=B_a(s(5,5)) = 1$$

and $L_{B_a} = B_a(s(1,1))+B_a(s(2,2))+B_a(s(3,3))+ B_a(s(4,4)) + B_a(s(5,5)) = 5.$

Therefore, $S(B_a) = \{ \ s(1,1),s(2,2),s(3,3), \ s(4,4),s(5,5) \ \}.$

If a read quorum for data file *a*, r =2 and a write quorum w = L_{B_a}-r+1 = 4, then the quorum sets for read and write operations are Q(B_a,2) and Q(B_a,4), respectively, where

Q(B_a,2)={{s(1,1),s(2,2)},{s(1,1),s(3,3)},{s(1,1),s(4,4)},{s(1,1),s(5,5)},{s(2,2),s(3,3)}, {s(2,2),s(4,4)},{s(2,2),s(5,5)},{{s(3,3),s(4,4)},{s(4,4),s(5,5)},{s(4,4),s(5,5)}, {s(1,1),s(2,2),s(3,3)},{s(1,1),s(2,2),s(4,4)},{s(1,1,),s(2,2),s(5,5)},{s(1,1), s(3,3),s(4,4)},{s(1,1,),s(3,3),s(5,5)},{s(1,1),s(4,4),s(5,5)},{s(2,2),s(3,3), s(4,4)}, {s(2,2), s(3,3),s(5,5)}, {s(2,2),s(4,4),s(5,5)},{s(3,3),s(4,4), s(5,5)}, s(1,1),s(2,2),s(3,3),s(4,4)},{s(1,1),s(2,2),s(3,3),s(5,5)},{s(1,1),s(2,2),s(4,4), s(5,5)},{s(1,1),s(3,3),s(4,4),s(5,5)},{s(2,2),s(3,3),s(4,4),s(5,5)},{s(1,1), s(2,2),s(3,3),s(4,4),s(5,5)}

and
Q(B_x,4)={{s(1,1),s(2,2),s(3,3),s(4,4)},{s(1,1),s(2,2),s(3,3),s(5,5)},{s(1,1),s(2,2), s(4,4),s(5,5)},{s(1,1),s(3,3),s(4,4),s(5,5)},{s(2,2),s(3,3),s(4,4),s(5,5)}, {s(1,1),s(2,2),s(3,3),s(4,4),s(5,5)}}

4 Performance Analysis and Comparison

In this section, we analyze and compare the performance of the DRG technique with grid structure technique on the communication cost and the data availability.

4.1 Availability Analysis

In estimating the availability, all copies are assumed to have the same availability p. Let SA(t) be the read/write availability of technique t. If the probability that an arriving operation of read and write for data file x are f and $(1-f)$, respectively, then the read/write availability can be defined as,

$$SA(t) = f SA(t_{read}) + (1-f) SA(t_{write})$$

4.1.1 The GS Technique

Let N be the number of copies which are organized as a grid of dimension n x n. In the case of the quorum technique, read quorums can be constructed as long as a copy from each column is available. Then, the read availability in the GS technique, $A_{GS,R}$, as given in [4] equals to $[1-(1-p)^n]^n$. On the other hand, write quorums can be constructed as all copies from a column and one copy from each of the remaining columns are available. Let $A_{X,Y}$ be the availability with X technique for Y operation, then the write availability in the GS technique, $A_{GS,W}$, as given in [4] equals to
$$[1-(1-p)^n]^n - [(1-(1-p)^n-p^n]^n.$$
Then,

$$SA(GS) = f A_{GS,R} + (1-f) A_{GS,W} \tag{1}$$

4.1.2 The DRG Technique

Let p_i denote the availability of site i. Read operations on the replicated data are executed by acquiring a read quorum and write operations are executed by acquiring a write quorum. For simplicity, we choose the read quorum less than or equals to the write quorum. For example, from Fig. 2 the $D^1(s)$ has five diagonal sites, each of which has vote one, then $C_{DRG,R} = C_{DRG,W} = \lfloor 5/2 \rfloor = 3$.
For any assignment B and quorum q for the data file x, define $\varphi(B_x,q)$ to be the probability that at least q sites in $\Omega(B_x)$ are available, then

$$\varphi(B_x,q) = \Pr \{\text{at least q sites in } \Omega(B_x) \text{ are available}\}$$

$$= \sum_{G \in Q(B_x,q)} \left(\prod_{j \in G} p_j \prod_{j \in S(B_x)-G} (1-p_j) \right). \tag{2}$$

Thus, the availability of read and write operations for the data file x, are $\varphi(B_x,r)$ and $\varphi(B_x,w)$, respectively. Then,

$$SA(DRG) = f \varphi(B_x,r) + (1-f) \varphi(B_x,w). \tag{3}$$

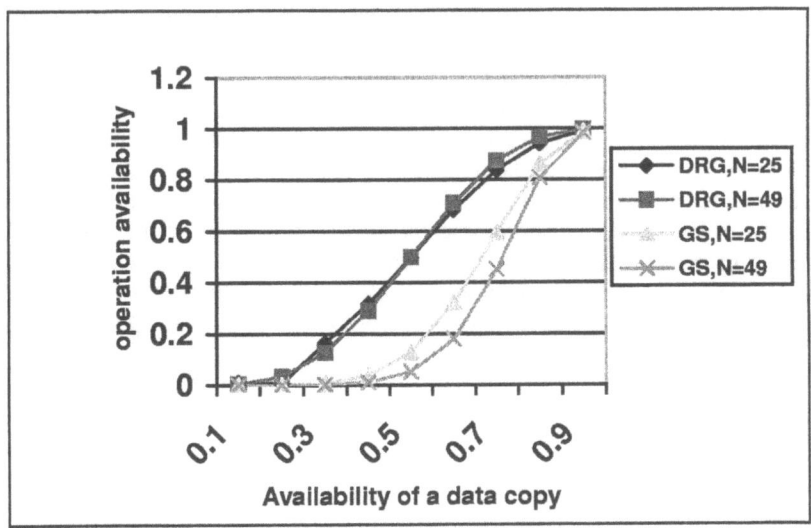

Fig. 3. Comparison of the write availability between GS and DRG, for N=25

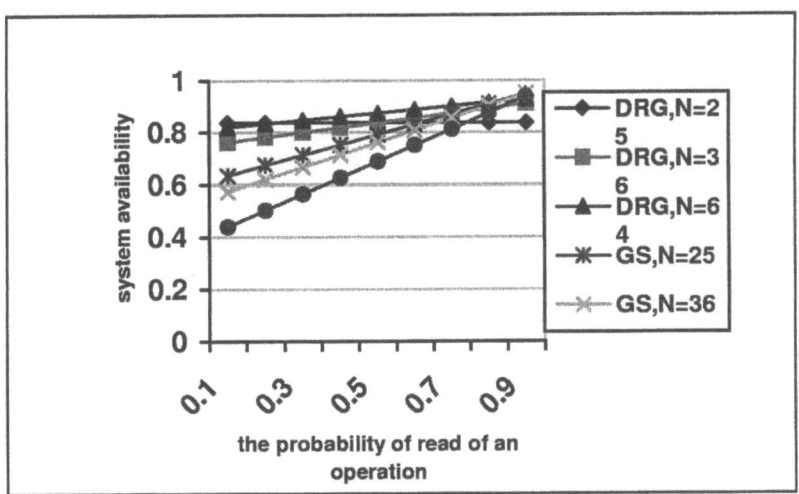

Fig. 4. Comparison of the read/write availability between GS and DRG for $p = 0.7$

4.2 Performance Comparisons

In this section, we will compare the performance on the read/write availability of the GS technique based on equations (1), and our DRG technique based on equations (2) and (3) for the case of N = 25, 36, 49, and 64. In estimating the availability of operations, all copies are assumed to have the same availability.

Fig. 3 shows that the DRG technique outperform the GS technique. When an individual copy has availability 70%, write availability in the NRG is approximately 84% whereas write availability in the GS is approximately 60% for N=25. Moreover, write availability in the GS decreases as N increases. For example, when an individual copy has availability 80%, write availability is approximately 86% for N=25 whereas write availability is approximately 80% for N = 49.

Fig. 4 shows that the DRG technique outperform the GS technique when the probability of an arriving operation is read, $f \leq 0.7$. This shows that the read/write availability is very sensitive to the read and write probability. For example, when $f = 0.5$, read/write availability in GS is approximately 69% whereas read/write availability in the DRG is approximately 87% for N = 64. Moreover, read/write availability in the GS decreases as N increases. This is due to the decreases of write availability as N increases. For example, when $f = 0.6$, read/write availability in GS is approximately 83% for N=25 whereas read/write availability is approximately 75% for N = 64.

5 Conclusions

In this paper, a new technique, called diagonal replication on grid (DRG) has been proposed to manage the data replication in the distributed database systems. The analysis of the DRG technique was presented in terms of read/write availability and communication costs. It showed that, the DRG technique provides a convenient approach to high availability for write-frequent operations by imposing a diagonal binary vote assignment to the logical grid structure on data copies. This is due to the minimum number of quorum size required. This technique also provides an optimal diagonal vote assignment which is less computational time required. In comparison to the Grid structure, DRG requires significantly lower communication cost for an operation, while providing higher read/write availability which is preferred for large distributed database systems.

References

1. J. Holliday, R.Steinke, D.Agrawal, A.El Abbadi, "Epidemic Algorithms for Replicated Databases", *IEEE Trans. On Knowledge and Data Engineering*, vol.15, no.5, (2003), 1218-1238.
2. O. Wolfson, S. Jajodia, and Y. Huang, "An Adaptive Data Replication Algorithm," *ACM Transactions on Database Systems*, vol. 22, no 2 (1997), 255-314.
3. Budiarto, S. Noshio, M. Tsukamoto, "Data Management Issues in Mobile and Peer-to-Peer Environment," *Data and Knowledge Engineering, Elsevier*, 41 (2002),183-204.
4. D Agrawal and A.El Abbadi, "Using Reconfiguration For Efficient Management of Replicated Data," *IEEE Trans. On Knowledge and Data Engineering*, vol.8, no. 5 (1996), 786-801.
5. J.F. Paris and D.E. Long, "Efficient Dynamic Voting Algorithms," *Proc. Fourth IEEE Int'l Conf. Data Eng*, (1988), 268-275.
6. D Agrawal and A.El Abbadi, "The Tree Quorum technique: An Efficient Approach for Managing Replicated Data," *Proc.16ʰ Int'l Conf. On Very Large Data Bases*, (1990), 243-254.
7. S.Jajodia and D. Mutchles, "Dynamic Voting Algorithms for Maintaining the Consistency of a Replicated Database," *ACM Trans. Database Systems*, vol 15, no. 2 (1990), 230-280.

8. M. Maekawa, "A √n Algorithm for Mutual Exclusion in Decentralized Systems," *ACM Trans. Computer Systems*, vol. 3, no. 2, (1992), 145-159.
9. S.Y. Cheung, M.H. Ammar, and M. Ahmad, "The Grid Technique: A High Performance Schema for Maintaining Replicated Data," *IEEE Trans. Knowledge and Data Engineering*, vol. 4, no. 6, (1992), 582-592.
10. B. Bhargava, "Concurrency Control in Database Systems," *IEEE Trans. Knowledge and Data Engineering*, vol 11, no.1, (1999), 3-16.
11. P.A. Bernstein, V. Hadzilacos, and N.Goodman, Concurrency *Control and Recovery in Database Syatems*, Addison-Wesley, (1987).

Performance Comparison between OpenMP and MPI on IA64 Architecture

Lin Qi, Meiming Shen, Yongjian Chen, and Jianjiang Li

Institute of High Performance Computing, Tsinghua University
Qilin97@mails.tsinghua.edu.cn

Abstract. In order to evaluate performances of OpenMP and MPI on 64-bits shared memory architecture, we made a comparison between OpenMP and MPI on IA64 SMP with NPB2.3 benchmark suite (excluding IS kernel). Assisted by Vtune(TM) Performance Analyzer 7.0, we raised comparisons in execution time, detailed performance data and overheads between these two paradigms, and managed to optimize some of these kernels.

1 Introduction

The opportunities of shared-memory programming paradigms such as OpenMP for parallel programming on distributed shared memory architectures like SMP clusters are still not clear. Although experiments of mixed mode programming for SMP clusters, often using MPI and OpenMP, have been done on several benchmarks and applications, the results don't directly deny or encourage such combinations. To achieve the final goal, our current works focus on investigating this problem by studying the overhead characteristics of OpenMP and MPI.

Recently, detailed overhead analysis technologies have shown that the overhead ratios and characteristics of OpenMP and MPI programs are quite different, which means that the program models used by MPI and OpenMP may affect program behaviors in different ways, and the combination of these two models may compensate each other thus to reduces overheads. So our works concentrated on how to use overhead analysis technologies to identify overheads in both models in order to find out opportunities for mixed mode paradigms on SMP clusters. After trying several software tools we finally used Vtune from Intel to assist our performance analysis work.

2 Experimental Conditions

2.1 Programming Model

OpenMP
Along with the appearance of scalable shared-memory multiprocessor, people realized that there is more direct and simple way to program on SMP. OpenMP was

M. Bubak et al. (Eds.): ICCS 2004, LNCS 3038, pp. 388–397, 2004.
© Springer-Verlag Berlin Heidelberg 2004

proposed for a transplantable standard to substitute for MPI. The original objective of OpenMP was to provide a transplantable and scalable programming mode on shared memory platform and to offer a simple and flexible programming interface for development of parallel applications. It is composed of directives, environmental variables and runtime library. It supports incremental parallelization so it is very easy to transform a serial application to a parallel one and gain considerable performance enhancement on shared memory architecture.

MPI

In a long time people considered the only way to develop parallel programs is utilizing message passing model. Message passing programming model is a widely used programming method explicitly controlling parallelism on distributed memory system. All parallelism is explicit: the programmer is responsible for correctly identifying parallelism and implementing parallel algorithms using MPI constructs. The most familiar programming style is SPMD. MPI (Message Passing Interface) defines a whole set of functions and procedures to implement message passing model.

2.2 Methodology

We mainly concentrated on program performances in this paper. Each of NPB2.3 kernels was tested on four threads or processes when it monopolized the IA64 SMP. At first we measured the execution time of each kernel with OpenMP and MPI versions respectively. We call the execution time measured by function w_time wall clock. They are listed in Fig. 1. Then we used Vtune(TM) Performance Analyzer 7.0 to obtain more detailed performance data in each thread or process of the kernel. In this paper we focused on CPU_CYCLE events, CPU_CYCLE %, IA64_INST_RETIRED-THIS events, IA64_INST_RETIRED-THIS % and IPC (IA64_INST_RETIRED-THIS per CPU_CYCLE). These performance data obtained by Vtune were on function level, that is, they gave amount of CPU cycles and instructions in every function. We classified these functions into two categories, one is explicitly existent functions programmed by developers and the other is runtime library calls which are implicitly called by the system, we call them runtime library overheads. Theoretically, with the same algorithm the execution times of explicitly existent functions should be similar between OpenMP and MPI versions, and the different parts should come from the runtime library overheads.

2.3 Benchmark

We chose NPB2.3 suite excluding kernel IS to launch our experiments. We parallelized NPB2.3 serial version with OpenMP directives in Fortran referring to the OpenMP version in C by NASA. Noticing that the corresponding MPI version is highly tuned in Fortran, comparisons between them should be challenging. We tested all of these kernels in their default class. BT, LU and SP are in class W, the others are in class A.

2.4 Platform

Our experiment environment was established on IA64 platform with a single SMP. With total 4 GB memory, it had four Itanirum2 processors with 900MHz which were connected by the system bus with bandwidth up to 6.4 GB/s. For each processor, the caches were 16 KB of L1I, 16KB of L1D, 256KB of L2 and 3MB of L3 on die with 32 GB/s bandwidth. The Itanium processor's EPIC features made higher instruction level parallelism up to 6 instructions per cycle.

We used mpich-1.2.5 for the MPI versions. This shared memory version of the MPI library made performance of MPI versions on SMP scale as high as it could.

In our experiments all of the kernels were compiled with intel's efc and the compiler options are configured as:

F77 = efc –openmpP –O3 –g –w for OpenMP version; MPIF77 = mpif90 –O3 for MPI version. The –g option was to obtain debug information for linux binary during Vtune's source drill down.

3 Understanding the Performance Results

3.1 Execution Time Comparison

Fig. 1 demonstrates the seven NPB2.3 kernels's performance differences between MPI and OpenMP on the IA64 SMP.

Among the seven kernels, BT, CG, EP and FT exhibited a slightly superior performance with OpenMP version compared with their MPI version; two versions of SP showed almost the same performance, while LU and MG obviously did better with MPI. Being highly tuned with shared memory version of MPI library, these seven kernels all showed fairly high performance with MPI on SMP. However, except LU and MG the other five kernels also exhibited competitive performances with OpenMP. Optimizing LU and MG will be a part of our works in the near future.

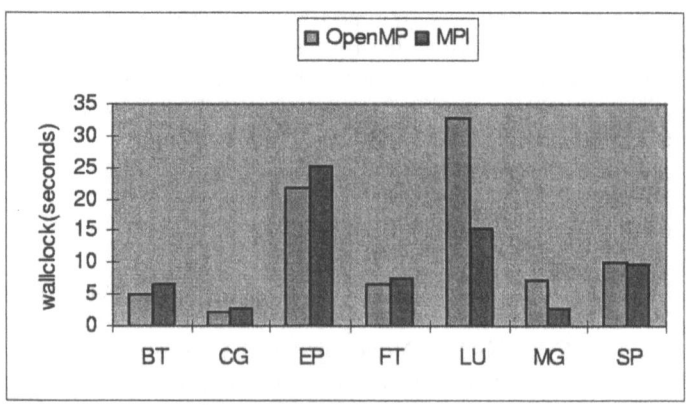

Fig. 1. Execution time of NPB2.3 suite on IA64 SMP

Table 1. Execution time table

	OpenMP(s)	MPI(s)
BT	5.06	6.63
CG	2.28	2.80
EP	21.61	25.07
FT	6.53	7.66
LU	32.81	15.44
MG	7.14	2.97
SP	9.94	9.67

3.2 Comparison of Detailed Performance Data between OpenMP and MPI

To find the causes of such performance distinction, we used Vtune(TM) Performance Analyzer7.0 to measure more detailed data which could indicate program and system behaviors in both programming paradigms.

In figures below, we showed two basic events and their proportion of BT and MG. They represented different performance comparison results. The basic events were CPU_CYCLE events, IA64_INST_RETIRED-THIS events and IPC respectively. Because we tested every kernel on four threads or processes only, the 0#, 1#, 2# and 3# mean to four threads or processes of OpenMP or MPI. The precise data were listed below in the tables following the corresponding figures.

Upon further analysis of Vtune results, we found although the workloads were not distributed averagely OpenMP performs better than MPI. According to the first three threads, retired instructions of OpenMP version were less than those of MPI version. Therefore with the similar IPC, the number of CPU cycles was also less than that of MPI version. However the fourth thread has much more retired instructions than the first three. Vtune Analyzer collected 9.150E+9 retired instruction events, outweighing 8.003E+9, retired instruction events of the 3# thread of MPI version. In this thread __kmp_wait costed 4.088E+9 instructions, 2.206E+9 CPU cycles, and it accounted for 44.69% of the total retired instructions, 42.60% of the total CPU cycles. Because the IPC of __kmp_wait in this thread was 1.85, it pulled the average IPC of all functions in 3# thread up to 1.767. Finally the higher IPC makes the 3# thread's CPU_CYCLES events of OpenMP version less than that of MPI version.

For MG, IPCs of OpenMP code were higher than IPCs of MPI code which showed better instruction level parallelism with OpenMP. However we also observed that its workloads were not very balanced among the four threads as the MPI version did as it was shown in Table 3. More instructions and lower IPCs in function reside(), vranlc(), zran3() of 3# thread induced more CPU cycle events in this thread and longer execution time of the whole program.

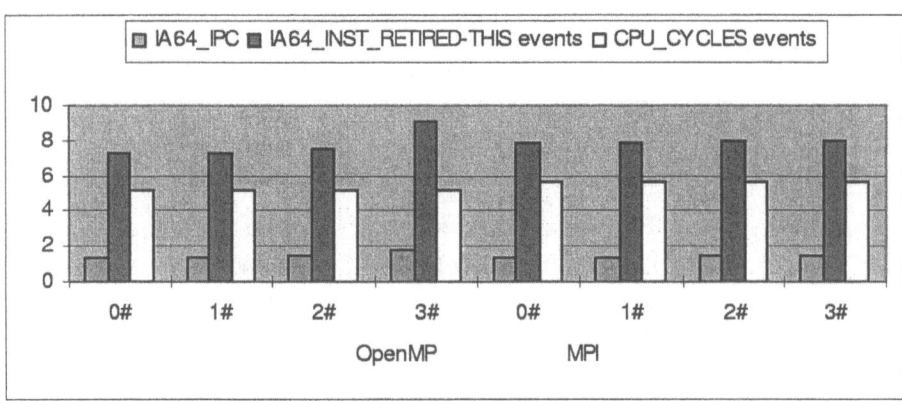

Fig. 2. BT – class W

Table 2. BT – class W

version	proc/thr No.	IPC	INST_RETIRED	CPU_CYCLES
OpenMP	0#	1.395	7.255E+9	5.202E+9
	1#	1.396	7.273E+9	5.209E+9
	2#	1.450	7.531E+9	5.193E+9
	3#	1.767	9.150E+9	5.179E+9
MPI	0#	1.395	7.854E+9	5.629E+9
	1#	1.403	7.891E+9	5.624E+9
	2#	1.421	7.958E+9	5.602E+9
	3#	1.432	8.003E+9	5.588E+9

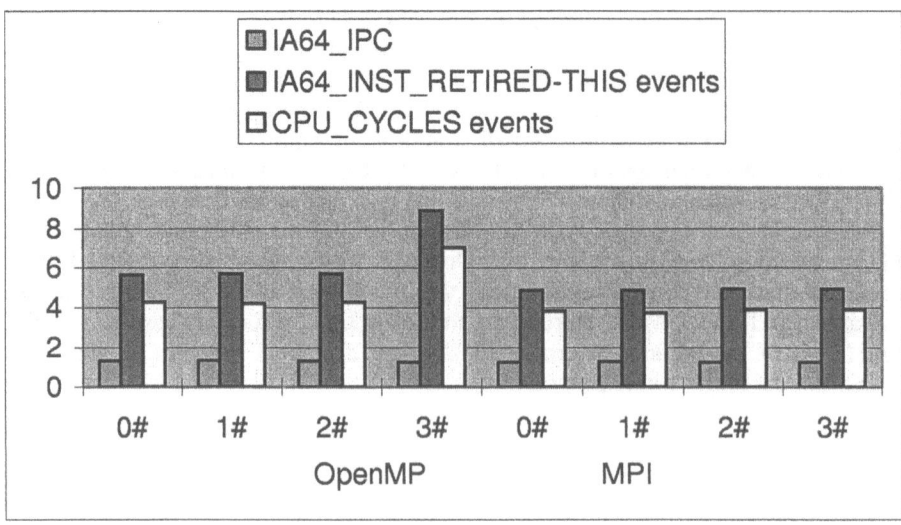

Fig. 3. MG - class A

Table 3. MG – class A

version	proc/thr No.	IPC	INST_RETIRED	CPU_CYCLES
	0#	1.319	5.662E+9	4.292E+9
OpenMP	1#	1.359	5.729E+9	4.216E+9
	2#	1.331	5.730E+9	4.306E+9
	3#	1.267	8.891E+9	7.016E+9
	0#	1.275	4.896E+9	3.841E+9
MPI	1#	1.303	4.908E+9	3.765E+9
	2#	1.272	4.955E+9	3.896E+9
	3#	1.267	4.971E+9	3.922E+9

3.3 Breakdowns of Execution Time

When testing with a benchmark program, Vtune provides performance events on function level. These functions are not only composed of those which are programmed by developers but also some runtime library functions which are implicitly called. We call the latter ones overheads.

For example, when we tested with kernel BT, we obtained performance data of many functions, lhsx, lhsy, lhsz, compute_rhs, x_backsubstitute and so on, all of which could be found in BT's source code. But there were still some other functions such as MPI_SHMEM_ReadControl, MPI_Waitall, kmp_wait, kmp_yield which could not be found in source code. So we classified all functions into algorithm functions and overheads. Through quantitative analysis it was clear that MPI_SHMEM_ReadControl and kmp_wait were the main overheads of MPI and OpenMP respectively. According to measurement results of all NPB2.3 kernels, CPU cycles costed by MPI_SHMEM_ReadControl were all less than, or at most similar with those costed by __kmp_wait. This was because most of overheads in MPI were resulted of communication and synchronization among processes, for example mpi_send, mpi_receive, mpi_wait, but all of these operations were explicitly controlled, so they were not included in MPI_SHMEM_ReadControl. On the other hand many causes generated OpenMP program's overheads, for example synchronization, load imbalance and improper memory access strategies. Obviously most of overheads in OpenMP rose from synchronization and load imbalance, for example omp_barrier and single, critical operations which resulted in the other threads were idle while only one thread was busy. Most of these overheads were included into __kmp_wait. Anyway, generally when overhead's quantity is large, it means the program's performance is low and should be optimized.

Fig. 4 shows execution time breakdowns of LU, the overheads of its OpenMP version account for over 50% of its execution time. We are trying to find the sources and optimize it. It will be discussed in the future papers.

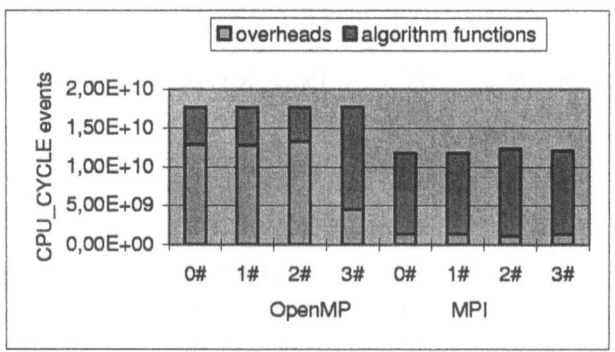

Fig. 4. Execution time breakdowns of LU with OpenMP and MPI

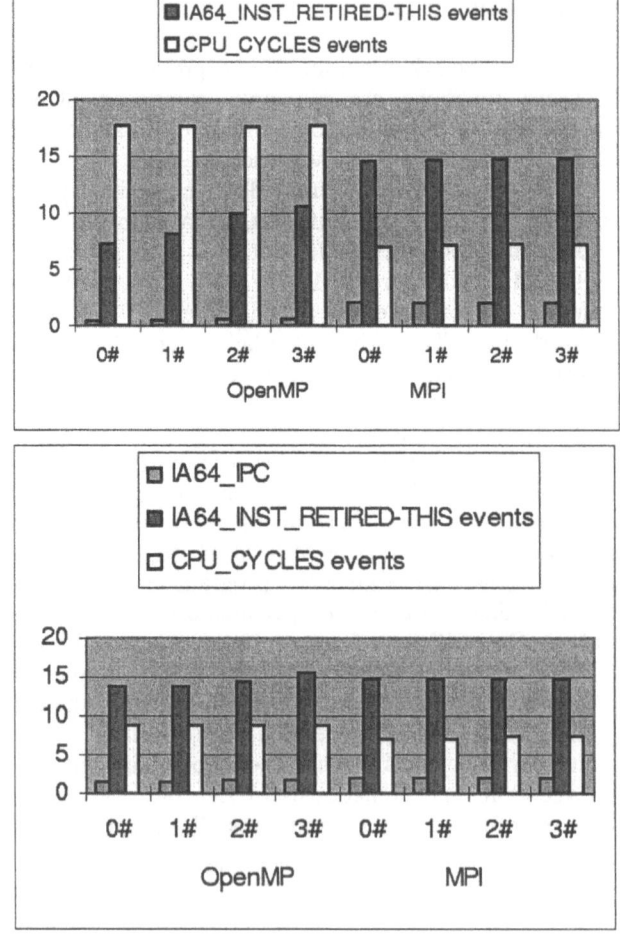

Fig. 5. Performance before and after optimization

3.4 Performance Enhancement from Optimization

The upper one of Fig. 5 is performance exhibition of SP OpenMP version before optimization. For OpenMP version, the IPC were very low and there were obviously much more CPU cycle events measured. We observed that the CPU cycle events of __kmp_wait were 7.53E+9, 7.31E+9, 8.66E+9 and 4.48E+9 in every thread respectively, being the most time-consuming function. And functions such as x_solve, y_solve, z_solve, lhsx, lhsy and lhsz accounted for the main part of execution time. The optimization focus on the six functions mentioned above and the optimizing method was simply called loop inversing and combination. We took the x-series functions for example:

x_solve.f:
The original code segment is:

```
do i = 0, grid_points(1)-3
    i1 = i + 1
    i2 = i + 2
!$omp do
    do k = 1, grid_points(3)-2
      do j = 1, grid_points(2)-2
        fac1 = 1.d0/lhs(i,j,k,n+3)
        lhs(i,j,k,n+4) = fac1*lhs(i,j,k,n+4)
        lhs(i,j,k,n+5) = fac1*lhs(i,j,k,n+5)
        ...
      end do
    end do
!$omp end do
    end do
```

The optimized code segment is:

```
!$omp do
  do k = 1, grid_points(3)-2
    do j = 1, grid_points(2)-2
      do i = 0, grid_points(1)-3
        i1 = i + 1
        i2 = i + 2
        fac1 = 1.d0/lhs(i,j,k,n+3)
        lhs(i,j,k,n+4) = fac1*lhs(i,j,k,n+4)
        lhs(i,j,k,n+5) = fac1*lhs(i,j,k,n+5)
        ...
!$omp end do
```

lhsx.f:
The original code segment as below:

```
do k = 1, grid_points(3)-2
   do j=1, grid_points(2)-2
!$omp do
     do i = 0, grid_points(1)-1
```

```
        ru1 = c3c4*rho_i(i,j,k)
        cv(i) = us(i,j,k)
        rhon(i) =
            dmax1(dx2+con43*ru1,dx5+c1c5*ru1,dxmax+ru1,dx1)
      end do
!$omp end do
!$omp do
      do i = 1, grid_points(1)-2
      ...
      end do
!$omp end do
end do
```

This code segment was optimized as follows:

```
!$omp do
  do k = 1, grid_points(3)-2
    do j = 1, grid_points(2)-2
      do i = 0, grid_points(1)-1
      rhon(i,j,k) = dmax1(dx2+con43*c3c4*rho_i(i,j,k),
dx5+c1c5*c3c4*rho_i(i,j,k),dxmax+c3c4*rho_i(i,j,k),dx1)
      end do
      do i = 1, grid_points(1)-2
      ...
      end do
!$omp end do
```

The nether one of Fig. 5 is the performance exhibition after optimization. The IPC of each thread was lifted, from 0.41, 0.46, 0.57, 0.60 to 1.57, 1.58, 1.66, 1.78 respectivly. The CPU cycle events of __kmp_wait were reduced to 1.7E+9, 1.7E+9, 2.7E+9 and 7.1E+8 respectively. And the execution times of those time-consuming functions were also cut down to some extent. Because of the enhancement of IPCs, although our optimization made retired instructions increased CPU_cycle events of the whole program were decreased.

4 Conclusion and Future Work

In this paper we made a comparison between OpenMP and MPI on IA64 SMP with NPB2.3 benchmark suite excluding IS. Our study shows that with simple OpenMP directives people can gain as good performance as shared-memory MPI on IA64 SMP. The merits of easier programming and satisfying performance make OpenMP a profitable paradigm for 64-bits shared memory architectures.

Furthermore there are some noteworthy questions in OpenMP. Although the easy programming is the obvious merit of OpenMP, improper programming is more likely to happen and that often leads to inferior performance. Another common problem is the imbalanced load distribution in OpenMP program. Furthermore, whether OpenMP can gain superior performance than shared-memory MPI on SMP is still an unsolved topic since OpenMP was proposed especially for shared memory architecture.

Our future work will continue to focus on performance evaluation and analysis of OpenMP and MPI in order to explore the possibilities in mixed mode programming on SMP cluster.

References

1. Geraud Krawezik and Franck Cappello, Performance comparison of MPI and three OpenMP Programming Styles on Shared Memory Multiprocessors, SPAA'03, June 7-9, 2003.
2. F. Cappello and D. Etiemble, MPI versus MPI+OpenMP on IBM SP for the NAS Benchmarks. Procs. of the international Conference on Supercomputing 2000: High-Performance Networking and Computing (SC2000), 2000.

Defining Synthesizable OpenMP Directives and Clauses

P. Dziurzanski and V. Beletskyy

Faculty of Computer Science and Information Systems,
Technical University of Szczecin, Zolnierska 49, 71-210 Szczecin, POLAND
{pdziurzanski,vbeletskyy}@wi.ps.pl

Abstract. Possibilities of synthesizing parallel C/C++ codes into hardware are presented provided that the code parallelism is represented by means of the directives of OpenMP, a de-facto standard that specifies portable implementation of shared memory parallel programs. The limitations of the hardware realizations of OpenMP directives are described and implementation details are stressed.

1 Introduction

Sources in different hardware description languages (HDLs) are used as input to behavioral synthesis. The most commonly used languages are VHDL and Verilog, but since designers often write system level models using programming languages, using software languages are of mounting importance. Applying software languages makes easier, performing SW/HW cosynthesis, which accelerates the design process and improves the flexibility of the software/hardware migration. Moreover, the system performance estimation and verification of the functional correctness is easier and software languages offer fast simulation and a sufficient amount of legacy code and libraries which facilitate the task of system modelling.

To implement parts of the design modelled in C/C++ in hardware using synthesis tools, designers must present these parts into a synthesizable subset of HDL, which then is synthesized into a logic netlist.

C/C++ defines sequential processes (procedures), whereas HDLs processes are run in parallel. Consequently, in order to synthesize hardware of high performance, there is the need of establishing groups of C/C++ functions which can be executed in parallel. To achieve the expected efficiency, each C/C++ function has to be treated as a separate process. This allows all functions to be executed concurrently taking into account that there exist no data dependences between functions. Otherwise, data synchronization techniques have to be utilized. Blocking actions before accessing to a shared variable can be implemented as a *wait* statement, which can be left when a synchronization signal from other module is set.

In order to control the calculations executed by the entities synthesized from C/C++ functions, distributed or centralized control schemes can be applied. They require that each entity includes both functional logic and a local controller. In the distributed scheme, this controller is connected with other blocks in a form of additional input and output ports. The input port determines whether all the preceding entities in a control flow have finished their tasks. After execution, the controller informs all the following entities

M. Bubak et al. (Eds.): ICCS 2004, LNCS 3038, pp. 398–407, 2004.
© Springer-Verlag Berlin Heidelberg 2004

in the control flow about finishing its task. In the centralized scheme, one controller determines which entities are executed at any given time.

More details on communication schemes are given in [7].

In this paper, we consider the following two-stage approach: (i) parallelizing an C/C++ code and presenting parallelism by means of OpenMP directives, (ii) transforming the C/C++ code with OpenMP pragmas into an HDL code.

We focus on the second stage, i.e., on transforming a parallelized C/C++ code into an HDL code and assume that an input code includes OpenMP directives representing C/C++ code parallelism.

2   Related Works

Recently, there has been a lot of work in the use of the C programming language to generate hardware implementations. However, as the C and C++ languages do not allow us to describe parallelism directly, the aspect of treating concurrency is of great importance in high-level synthesis approaches.

For example, in Transmogrifier C compiler [3], there exists only one thread of control. In order to compile multiple threads, one have to compile each thread separately into a netlist and then use the I/O ports for communication.

The PACT HDL compiler does not perform data dependency analysis and thus does not support function replication and parallel execution [4].

Despite the systems, described in [5,6], perform their own analysis of data dependency, they do not allow the designer to indicate parallel regions. Consequently, the information about parallelism known before compilation is ignored.

To our knowledge, there is no publications which describe hardware implementations of C/C++ code with standard OpenMP directives permitting us to present algorithm parallelism.

3   Principles of OpenMP

OpenMP [1] is a de-facto standard that specifies portable implementation of shared memory parallel programs. It includes a set of compiler directives, runtime library functions supporting shared memory parallelism in the C/C++ and Fortran languages.

OpenMP is based on the fork-and-join execution model in which a program is initialized as a single process (master thread). This thread is executed sequentially until the first parallel construct is encountered. Then, the master thread creates a team of threads that executes the statements concurrently. There is an implicit synchronization at the end of the parallel region, after which only the main thread continues its execution.

4   Influence of OpenMP Pragmas on Behavioral Synthesis

Similarly to C/C++ clauses, directives from the OpenMP standard [1] can be split into synthesizable and nonsynthesizable subsets. Certain OpenMP constructs have no equivalence in hardware realizations, whereas others can lead to hardware utilizing the enormous amount of resources.

if omp_get_dynamic() is equal to false {
 if num_thread clause is present
  return the parameter of the clause
 else if omp_set_num_threads function has been called
  return omp_get_num_threads()
 else if OMP_NUM_THREADS is defined
  return OMP_NUM_THREADS}
return omp_get_max_threads()

Fig. 1. Algorithm for determining the number of instances, NoOfInstances

In order to follow our results, the reader should be familiar with OpenMP attributes and environment variables (this knowledge is comprised in [1].)

4.1 Parallel Constructs

A parallel region is a region which can be executed by multiple threads concurrently.

The thread number, which corresponds to the number of multiple instances of elements in hardware implementations, can be determined statically or dynamically.

In the case of the statical thread number, to determine the number of requested threads, the **num_threads** clause, or the **omp_set_num_threads** OpenMP run-time function, or the **OMP_NUM_THREADS** environment variable can be used. In order to achieve an expected efficiency in hardware realizations, multiple threads should be implemented as multiple instances of the hardware realizing the same functionality. Consequently, it is requested that the number of threads is evaluated to a constant value during synthesis.

If a dynamic adjustment of the thread number is chosen, the number of threads is set to the maximum possible value, returned by the **omp_get_max_threads** OpenMP run-time function (described in section 4.5), unless the synthesizer determines that the maximum number of iterations is less than this value or a synthesized circuit does not fit into the destined chip.

The algorithm for determining the number of instances of parallel constructs is given in Fig. 1.

Although synthesis from nested parallel regions can lead to hardware with a very large area, an implementation of nested parallelism may improve the performance greatly [2]. Thus, nested parallelism should not be prohibited in hardware realizations.

The parallel clause can be modified with the following clauses **if, private, firstprivate, default, shared, copyin, reduction,** and **num_threads**. The majority of them are considered in corresponding sections below. In this subsection, only the clauses that are not used with work-sharing constructs are described.

In OpenMP, parallel execution can be conditional when an **if** clause is present. This clause usually defines a *breaking condition*. If this condition is not satisfied, a data dependency appears and the section cannot be run concurrently. If the condition can be evaluated during synthesis, then a synthesizer creates either single or multiple instances of hardware realizing the corresponding section body, according to the value of the condition. Otherwise, the decision on sequential or parallel execution is postponed up to runtime by the transformation which moves the condition from an OpenMP directive

if(condition) {

#pragma omp parallel　　　　　*#pragma omp parallel*

if(condition)　　　　　　　　*// parallel section body*}

// parallel section body　　　　*else*

　　　　　　　　　　　　　　　// sequential section body

　　　　　(a)　　　　　　　　　　　　　(b)

Fig. 2. Replacing the OpenMP if clause (a) with the C code condition (b)

if **if(condition)** *clause is present in an OpenMP* **parallel** *directive* {
  if the condition is evaluated to the zero value {
  remove the corresponding OpenMP **parallel** *directive*}
  else if the condition is evaluated to a nonzero value {
  remove the **if(condition)** *clause from the corresponding OpenMP* **parallel** *directive* }
  else if the condition cannot be evaluated during synthesis {
  move the condition from an OpenMP directive to a regular C/C++ code
  provide parallel and serial versions of the parallel section body (Fig. 2) } }

Fig. 3. Algorithm for the transformation of a conditional parallel construct

to a regular C/C++ code, as shown in Fig. 2. The algorithm for the conditional parallel construct transformation is given in Fig. 3.

The data sharing of variables is specified at the beginning of a parallel block using the **shared** or **private** clauses.

The **shared** clause indicates that the following variables which this clause defines are shared by each thread in a team. In the hardware realization, this may cause problems with global routing, as a number of hardware elements are to be connected to registers keeping one variable. On the other hand, this can decrease the number of cells in the hardware.

The **copyin** clause provides a mechanism for assigning the same value to multiple instances of one variable, each in one thread executing a parallel region. Each instance of the private variables are to be connected (by routing channels) with the hardware implementing the value of the variable instance from the instance which corresponds to the main thread.

The **default** clause informs whether the default data-sharing attributes of variables are **shared**. It is treated as a preprocessor command, which determines either shared or private variables and hence can be used in synthesis.

4.2 Work-Sharing Construct Clauses

OpenMP defines three work-sharing directives: **for, sections,** and **single**.

The **for** directive indicates that the iterations of the associated loop should be executed in parallel. In order to execute the loop iterations in parallel, the number of iterations has to be evaluated at the synthesis stage. The algorithm for the synthesizing of the **for** loops is presented in Fig. 4.

If the number of the loop iterations is not greater than the value $NumberOf$-$Instances$, returned by the NoOfInstances algorithm (Fig. 1), then all the loop iterations are executed in parallel. Otherwise, the loop either can be unrolled so as the

if a number of iterations can be evaluated during synthesis {
 NumberOfInstances = return value of the NoOfInstances algorithm
 if NumberOfInstances is greater than the number of iterations
  NumberOfInstances=number of iterations;
 if **private** *clause is present*
  create one instance of each private variable for each occurrence
  of parallel realization
 if **firstprivate** *clause is present*
  create one instance of each private variable for each occurrence
  of parallel realization which initialize their values
  with the value of the variable original object
 if master and synchronization directive is present
  create appropriate synchronization links between hardware
  modules (algorithm MasterSynchr)
 Compute IterPerInst, the number of iterations which should be executed
 by one instance by rounding up the number of iteration
 divided by NumberOfInstances and Synthesize the loop body
 with NumberOfInstances instances each executing IterPerInst iterations;
 if **ordered** *clause is present*
  create synchronization links between hardware modules
  to execute bodies in the appropriate order
 if **lastprivate** *clause is present*
  create hardware elements for coping the value of private variable in the last
  instance to the value of the variable original object
 if **nowait** *clause is not present*
  create hardware elements for waiting on finishing execution
  of each instance }
else synthesize the loop body with a single instance;

Fig. 4. Algorithm for synthesis of the **for** clause, MasterSynchr

number of the loop iterations is equal to $NumberOfInstances$ or the loop can be split into $NumberOfInstances$ independent loops.

According to the OpenMP specification, the execution of each iteration must not be terminated by a **break** statement and the value of the loop control expressions must be the same for all iterations.

With the **for** directive, the following clauses can be used:

- **private** indicates that the variables which this clause defines are private to each thread in a team. In the hardware realization, it usually results in multiple instances of hardware holding the variable.
- **firstprivate (lastprivate)** provides a superset of the **private** clause functionality, which causes that the initial (final) value of the new private variables is copied from (to) the value of the object existing prior to the section. Their hardware synthesis conditions are similar to those of the **private** clause.
- **reduction** which calculates one scalar value by performing a given operation on all elements of the vector. It is synthesizeable as the loop with a reduction variable can be split into $NumberOfVariables$ independent loops and an additional loop forming the final value of the reduction variable.

if **private** *clause is present*
  create one instance of each private variable for each section
if **firstprivate** *clause is present*
  create one instance of each private variable for each section
  of parallel realization which initialize their values
  with the value of the variable original object
Synthesize each section as a separate entity;
if **lastprivate** *clause is present*
  create hardware elements for coping the value of private variable in the last
  instance to the value of the variable original object
if **reduction** *clause is present*
  create hardware elements for performing the reduction operation and copy its result
  to the value of the variable original object
if **nowait** *clause is not present*
  create hardware elements for waiting on finishing execution
  of each instance

Fig. 5. Algorithm for synthesis of the **sections** clause

if **private** *clause is present*
  create an instance of each private variable for the single region
if **firstprivate** *clause is present*
  create an instance of each private variable for the single region
  which initialize their values with the value of the variable original object
Synthesize a body as a single instance;
if **nowait** *clause is not present*
  create hardware elements for waiting on finishing execution of each occurence
  of parallel realization
if **copyprivate** *clause is present*
  copy the value of the private variables to corresponding private
  variables of other instances of the parallel region

Fig. 6. Algorithm for synthesis of the **single** clause

- **ordered** which causes the following region that it defines to be executed in the order in which iterations are executed in a sequential loop. It is considered in Section 4.4.
- **schedule** specifies dividing threads into teams. According to the OpenMP specification, the correctness of a program must not depend on scheduling, so it can be ignored in hardware realizations because it defines schedules for executing the loop iterations on a numbers of sequential processors and is not useful for hardware synthesis.
- **nowait** which causes that there is no barrier at the end of the parallel section. It is synthesizable by eliminating the need of adding the hardware for synchronization at the end of the parallel region.

The **sections** directive indicates that the corresponding region includes separate sections (declared with the **section** pragma) which can be executed in parallel. In hardware realization, for each section, separate instances should be synthesized. With this directive, the clauses **private**, **firstprivate**, **lastprivate**, **reduction**, and **nowait** can be used with the same limitations as in the case of the **for** directive. The algorithm of the **sections** synthesis is presented in Fig. 5.

The **single** directive indicates that the following structured block should be executed by only one thread in the team. Consequently, it is realized by the single instantiation of hardware realizing the functionality of the block. The clauses **private, firstprivate**, and **nowait** can be used with the same limitations as in the case of the **for** directive. The **copyprivate** clause provides a mechanism for broadcasting values between threads of one team with their private variables. The **single** directive can be synthesized by implementing the hardware realizing the corresponding structured block in only one (arbitrary) instance of the hardware realizing the corresponding parallel region and synchronizing its execution with finishing the execution of functionality given prior to the single directive.

4.3   Data Environment Directives

In OpenMP, there are the following clauses for controlling the data environment in parallel regions

- the **threadprivate** directive for making file-scope, namespace-scope, or static-block scope enumerated variables local to a thread, so that each thread obtains its own copy of the common block. In the hardware realization, the common region is synthesized in multiple units.
- the **private, firstprivate, lastprivate, shared, default, reduction, copyin**, and **copyprivate**; their synthesis is described in the previous section.

4.4   Master and Synchronization Directives

Within the parallel section, the master and synchronization directives can be used. They change the standard execution flow and are as below:

- **master** specifies a block that is executed by the master thread of the team. In hardware, the functionality of this block is implemented in the only instance of the hardware, which corresponds to the master thread. Its implementation resembles the **single** construct one, synchronization is required only in the instance of the hardware which corresponds to the master thread.
- **critical** specifies a block that can be executed only by one process at a time. It is synthesizable by means of the hardware which allows only one entity to execute a given code at once.
- **barrier** synchronizes all the threads in a team. It is synthesizable by means of hardware which allows execution only when each instantiation of the parallel region terminates the functionality given prior to this clause.
- **atomic** specifies that a memory location is updated atomically. It can be synthesized by being treated as the **critical** construct.
- **flush** specifies a sequence point at which all the threads in a time are required to have a consistent view of specified objects in memory. In software realization, it results in copying the values from registers into memory or flushing write buffers. In hardware, this directive should be ignored, as such an incoherence is not permitted.

if **master** *clause is present*
  create hardware implementation only in the first
  occurrence of parallel realization
if **critical** *or* **atomic** *clause is present*
  create hardware implementation of the mutual exclusion
  and connect it with all the occurrences of parallel realization
if **barrier** *clause is present*
  create hardware implementation of the barrier
  and connect it with all the occurrences of parallel realization
if **ordered** *clause is present*
  create synchronization links between hardware modules
  to execute bodies in the appropriate order

Fig. 7. Algorithm for synthesis of the master and synchronization directives

- **ordered** causes that iterations in the parallel block are executed in the same order as in a sequential loop. The functionality is synthesized in each instance implementing the parallel region, but only execution in the first instance is not blocked. In each other instance, the execution is blocked until the previous instance finishes the execution of the corresponding block. This construct adds one synchronization wire between adjacent implementation and the synchronizing hardware.

The algorithm for parallelizing the mentioned above directives is presented in Fig. 7.

4.5   Run-Time Library Functions

In OpenMP, there exist the following execution environment functions:

- **omp_set_num_threads** sets the number of threads used during execution. In hardware realization, its value determines the maximal number of the hardware instances realizing parallel regions. This function is synthesizable as long as its parameter can be computed during the compilation process.
- **omp_get_num_threads** sets the number of threads used during execution. The value set by this function is used in the *NoOfInstances* algorithm.
- **omp_get_max_threads** returns an integer that is at least as large as the number of threads executing the parallel region. In hardware realization, it is equal to the maximal possible number of the instances realizing the parallel region. The return value of this function is substituted during the synthesis stage.
- **omp_get_thread_num**. This function is substituted during the synthesis stage with the value equal to the index of the instance realizing the parallel region.
- **omp_get_num_procs**. This function is substituted during the synthesis stage with the value equal to **omp_get_max_threads**.
- **omp_in_parallel** returns a nonzero value if it is called within the parallel region and 0 otherwise. This value is substituted during the synthesis stage.
- **omp_set_dynamic** enables or disables a dynamic adjustment of the number of threads executing a parallel region. In hardware realizations, this function should be treated as a preprocessor command which switches between the permission and the prohibition of the dynamic adjustment of the thread number.

- **omp_get_dynamic** returns a nonzero value if dynamic adjustment of threads is enabled and 0 otherwise. The return value of this function is substituted during the synthesis stage.

- **omp_set_nested** enables or disables the nested parallelism. In hardware realizations, this function should be treated a a preprocessor command, which switches between the permission and the prohibition of the nested parallelism during the synthesis state.

- **omp_get_nested** returns the value according to the actual state of nested parallelism enabled. The return value of this function is substituted during the synthesis stage.

In OpenMP, there exist the following lock functions: **omp_init_lock** initializes a lock, **omp_destroy_lock** uninitializes a lock, **omp_set_lock** blocks the thread as long as the given lock is available and then sets the lock, **omp_unset_lock** releases a lock, and **omp_test_lock** attempts to set a lock without blocking the thread. All of them can be synthesized in hardware.

Nested versions of the lock functions (**omp_init_nest_lock**, **omp_destroy_nest_lock**, **omp_unset_nest_lock**, **omp_test_nest_lock**, and **omp_set_nest_lock**) are similar to their plain versions, except that they are used for nested locks.

The usage of the timing routines **omp_get_wtime** and **omp_get_wtick** in hardware realizations are pointless and thus they should not be permitted.

Table 1. Synthesizable (a), ignored (b) and nonsynthesizable (c) OpenMP constructs

atomic construct	barrier directive	copyin attribute clause
copyprivate attribute clause	critical construct	default attribute clause
firstprivate attribute clause	for construct	lastprivate attribute clause
if clause which cannot be evaluated during synthesis	if clause evaluating to a zero value during synthesis	master construct
		nowait clause
omp_destroy_lock function	omp_destroy_nested_lock function	omp_get_dynamic function
omp_get_nested function	omp_get_num_procs function	omp_get_num_threads function
omp_get_thread_num function	omp_in_parallel function	omp_init_lock function
omp_init_nested_lock function	omp_set_dynamic function	omp_set_lock function
omp_set_max_threads function	omp_set_nested function	omp_set_nested_lock function
omp_set_num_threads function	omp_test_lock function	omp_test_nested_lock function
omp_unset_lock function	omp_unset_nested_lock function	ordered construct
parallel construct	parallel for construct	parallel sections construct
private attribute clause	reduction attribute clause	sections construct
shared attribute clause	single construct	threadprivate directive

(a)

if clause evaluating to a nonzero value during synthesis	flush directive	schedule clause

(b)

omp_get_wtick function	omp_get_wtime function

(c)

5   Conclusion

In this paper, we have presented the possibilities and limitations of the hardware realization of the OpenMP directives.

In Table 1a-c, we summarize the synthesizabilty of OpenMP directives. The constructs which can be synthesized into hardware, are given in Table 1a. The constructs enumerated in Table 1b are not relevant to synthesis and identified as **ignored**. These constructs result in displaying warnings during synthesis.

Finally, not synthesizable constructs are given in Table 1c. If a not supported construct is encountered in a program source, its synthesis is not possible.

We plan to build a compiler transforming C/C++ sources with OpenMP pragmas into synthesizable SystemC sources.

References

1. OpenMP C and C++ Application Program Interface, ver 2.0, OpenMP Architecture Review Board, 2002, *www.openmp.org*
2. R. Blikberg, T. Sorevik, Nested Parallelism: Allocation of Processors to Tasks and OpenMP Implementations, *Second European Workshop on OpenMP*, Edinburgh, Scotland, UK, 2000
3. D. Galloway, The Transmogrifier C hardware description language and compiler for FPGAs, *Proceedings of IEEE Symposium on FPGAs for Custom Computing Machines*, 1995
4. A. Jones, D. Bagchi, S. Pal, X. Tang, A. Choudhary, P. Banerjee, PACT HDL: A C Compiler with Power and Performance Optimizations, *International Conference on Compilers, Architecture, and Synthesis for Embedded Systems (CASES)*, Grenoble, France, 2002
5. J.B. Peterson, R. Brendan O'Connor, P. M. Athanas, Scheduling and Partitioning ANSI-C Programs onto Multi-FPGA CCM Architectures, *IEEE Symposium on FPGAs for Custom Configurable Computing Machines*, Napa, California, 1996
6. J. Babb, M. Rinard, C.A. Moritz, W. Lee, M. Frank, R. Barua, S. Amarasinghe, Parallelizing Applications into Silicon, *IEEE Symposium on FPGAs for Custom Computing Machines*, Los Alamitos, CA, USA, 1999, pp. 70-80
7. R. Kastner, M. Sarrafzadeh, Incorporating Hardware Synthesis into a System Compiler, Technical Report, *Department of Electrical and Computer Engineering University of California*, Santa Barbara, CA, USA

Efficient Translation of OpenMP to Distributed Memory

L. Huang[1], B. Chapman[1], Z. Liu[1], and R. Kendall[2]

[1] Computer Science Dept., University of Houston, Texas
{leihuang,chapman,zliu}@cs.uh.edu
[2] Scalable Computing Lab, Ames Laboratory, Iowa
rickyk@ameslab.gov

Abstract. The shared memory paradigm provides many benefits to the parallel programmer, particular with respect to applications that are hard to parallelize. Unfortunately, there are currently no efficient implementations of OpenMP for distributed memory platforms and this greatly diminishes its usefulness for real world parallel application development. In this paper we introduce a basic strategy for implementing OpenMP on distributed memory systems via a translation to Global Arrays. Global Arrays is a library of routines that provides many of the same features as OpenMP yet targets distributed memory platforms. Since it enables a reasonable translation strategy and also allows precise control over the movement of data within the resulting code, we believe it has the potential to provide higher levels of performance than the traditional translation of OpenMP to distributed memory via software distributed shared memory.

1 Introduction

Recently, programmer productivity has been emphasized for parallel and distributed computing. However, few robust high-level language models have been developed for parallel programming. The difficulty of designing a language that offers expressivity, portability, ease of use and high performance has inevitably led to many failures. The distributed memory paradigm with explicit message passing remains the de facto programming standard, mainly because it provides good scalability for regular applications and it addresses the price/performance driven evolution of the HPC market toward clustered architectures. This scalability comes at a high programming cost. The shared memory paradigm has a low entry cost and can be much more flexible in its ability to respond to dynamically changing characteristics of an application. OpenMP [1] is a popular parallel programming interface for medium scale high performance applications on shared memory platforms. Strong points are its APIs for Fortran, C and C++, the ease with which its directives can be inserted into a code, its ability to support incremental parallelization, features for dynamically setting the numbers of threads and scheduling strategies, and strong vendor support. This is offset by its lack of support for distributed memory.

There have been a variety of attempts to implement OpenMP on clusters, most of which are based upon a translation of OpenMP to a software DSM (Distributed Shared Memory) system which is then responsible for managing data declared to be shared [2, 8]. Such solutions tend to be inefficient, as the software DSM will perform

M. Bubak et al. (Eds.): ICCS 2004, LNCS 3038, pp. 408–413, 2004.
© Springer-Verlag Berlin Heidelberg 2004

expensive amounts of data transfer at each (explicit or implicit) barrier in the program, and are particularly problematic for codes that are hard to parallelize, such as unstructured computations. In this paper, we propose an alternative strategy for translating OpenMP code to execute on clusters. We believe that our strategy will be more efficient and of specific benefit for irregular (unstructured) computations.

The paper is organized as follows. In the next section, we introduce Global Arrays, a library for parallel programming upon which our translation strategy is based, and explain why we believe that this can help us implement OpenMP efficiently on clusters. The remainder of the paper discusses our translation strategy, summarizes related work and our future plans.

2   Global Arrays as a Basis for Translating OpenMP

Global Arrays (GA) [7] is a collection of library routines that was designed to simplify the programming methodology on distributed memory systems. GA has been available in the public domain since 1994 and has since been utilized to create parallel versions of many major scientific codes for distributed memory machines. It realizes a portable interface via which processes in an SPMD-style parallel program do not need the explicit cooperation of other processes. In contrast to other popular approaches, it does so by providing a library of routines that enable the user to specify and manage access to *shared* data structures in a program. Compared with MPI programming, GA thus simplifies parallel programming by providing users with a conceptual layer of virtual shared memory for distributed memory systems. Programmers can write their parallel program on clusters as if they have shared memory access, specifying the layout of shared data at a higher level. However, it does not change the parallel programming model dramatically since programmers still need to write SPMD style parallel code and deal with the complexity of distributed arrays by identifying the specific data movement required for the parallel algorithm. The GA programming model forces the programmer to determine the needed locality for each phase of the computation. By tuning the algorithm such that the locality is maximized, portable high performance is easily obtained. Furthermore, since GA is a library-based approach, the programming model works with most popular language environments. Bindings are currently available for Fortran, C, C++ and python, and hence for those languages that are of interest when handling OpenMP.

GA programs distribute data in blocks to the specified number of processes. The current GA is not able to redistribute data. Before a region of code is executed, the required data must be gathered from the participating processes; results are scattered back to their physical locations upon completion. Since data distributions are simple, it is easy to compute the location of any data element. The implementation focuses on achieving very high efficiency in the data gathering and scattering phases. This approach is efficient if the regions of code are sufficiently large and the code is able to compute the gather and scatter sets only when really necessary. GA relies upon MPI to provide it with the execution context.

The most innovative idea of GA is that it provides an asynchronous one-sided, shared-memory programming environment for distributed memory systems. Both this shared memory abstraction, and the specific set of features GA offers, make it quite reasonable to translate OpenMP to GA. The traditional approach to implementing

OpenMP on distributed systems is based upon software DSM, which will transfer pages of data between memories when just a single element on that page has been modified, thereby potentially unnecessarily moving a considerable amount of data at synchronization points. GA provides a higher level of control, since the routines for gathering and scattering data can be used to specify precisely which data elements are to be transferred to which processor, and they also state when to do so. There are no "hidden" data transfers and there is no need to compare sets of changes to a page in memory. OpenMP maps computation to threads (just as GA maps computation to processes) and thereby indirectly specifies which data is needed by a given thread. This attribute makes it possible to translate OpenMP to GA. If the user has taken data locality into account when creating the OpenMP code, the benefits will be realized in the corresponding GA code.

3 The Translation Process

A careful study of OpenMP and GA routines showed that almost all of the OpenMP directives, library routines and environment variables can be translated into GA or MPI library calls at source level. Using GA and MPI together is not problematic since GA was designed to work in concert with the message passing environment. GA has the concept of shared data without explicit cooperative communication between processes. Coding for GA programs are similar to NUMA (non-uniform memory architecture) shared memory systems.

OpenMP parallel regions are transformed into GA program by invoking MPI_INIT and GA_INITIALIZE routines to initialize processes and the memory needed for storing distributed array data. Note too that the program only needs to call MPI_INIT and GA_INITIALIZE once in GA program for efficiency. Similarly, GA_TERMINATE and MPI_FINALIZE routines are called once to terminate the parallel regions.

The general approach to translating OpenMP into GA is to declare all shared variables in the OpenMP program to be global arrays in GA. Private variables can be declared as local variables that are naturally private to each process in a GA. If the parallel region contains shared variables, the translation will turn them into distributed global arrays in the GA program by inserting a call to the GA_CREATE routine. GA enables us to create regular and irregular distributed global arrays, and ghost cells (or halos) if needed. OpenMP FIRSTPRIVATE and COPYIN clauses are implemented by calling the GA broadcast routine GA_BRDCST. The reduction clause is translated by calling GA's reduction routine GA_DGOP. GA library calls GA_NODEID and GA_NNODES are used to get process ID and number of computing processes respectively at run time. OpenMP provides routines to dynamically change the number of executing threads at runtime. We do not attempt to translate these currently since this would amount to performing data redistribution and GA is based upon the premise that this is not necessary.

In order to implement OpenMP parallel loops in GA, the generated GA program reduces the loop bounds according to specified schedule so as to assign work. Based on the calculated lower and upper bounds, and the array region accessed in the local code, each process in the GA program fetches a partial copy of global arrays via GA_GET, performs its work and puts back the modified local copy into global locations by calling GA_PUT or GA_ACCUMULATE. The iteration set and therefore also the shared data must be computed dynamically For DYNAMIC and GUIDED loop schedules. We use GA locking routines to ensure that a process has exclusive access to code where it gets a piece of work and updates the lower bound of the remaining iteration set; the latter must be shared and visible for every process. However, due to the expense of data transfer in distributed memory systems, DYNAMIC and GUIDED schedules may not be as efficient as a static schedule, and it may not provide the intended benefits.

GA synchronization routines will replace OpenMP synchronizations. As OpenMP synchronization ensures that all computation in the parallel construct has completed, GA synchronization will do the same but will also guarantee that the requisite data movement has completed to properly update the GA data structures. GA locks and Mutex library calls are used to protect a critical section; we use them to translate the OpenMP CRITICAL and ATOMIC directives. The OpenMP FLUSH directive is implemented by using GA put and get routines to update shared variables. This could be implemented with the GA_FENCE operations if more explicit control is necessary. GA provides the GA_SYNC library call for synchronization; it is used to replace OpenMP BARRIER as well as implicit barriers at the end of OpenMP constructs. The only directive that cannot be efficiently translated into equivalent GA routines is OpenMP's ORDERED. We use MPI library calls, MPI_Send and MPI_Recv, to guarantee the execution order of processes if necessary. Since GA works as a complement of MPI, and must be installed on a platform with GA, there is no problem invoking MPI routines in a GA program.

The translation of sequential program sections (serial regions outside parallel regions, OpenMP SINGLE, MASTER, and CRITICAL constructs) becomes non-trivial besides that of parallel regions. The program control flow must be maintained correctly in all processes so that some parts of the sequential section have to be executed redundantly by all processes. Subroutine/function calls in serial regions need to be executed redundantly if these subroutines/functions have parallel regions inside. We have identified three different strategies to implement the sequential parts: *master execution*, *replicated execution* and *distributed execution*.

In *master execution*, only the master process performs the computation, and gets/puts the global arrays before and after the computation. Exclusive master process execution of the sequential portion of programs invokes coherence issue of private data between master process and other processes; a broadcast operation is necessary after master process execution in order to achieve a consistent view of data.

In *replicated execution*, each process redundantly executes the same computation. At the end of computation, only one processor needs to update the global arrays using its own local copies, although all the processes need to fetch into local copies before the computation. The replicated execution approach has advantages of easy maintenance of the coherence of private data, and less data communication if a small number of shared variables are modified. But it has overhead of redundant computation and may cause more global synchronizations for shared memory updates and potentially too much data gathering. The approach could work when a sequential part computes mostly scalar data.

In *distributed execution*, the process that owns data performs the corresponding computation and keeps the computation in a certain order according to data dependency information. Each processor executes a portion of work of the sequential part according to constraints of sequential execution order. This may introduce considerable synchronization. The distributed computation maximizes the data locality and minimizes the shared data communication, but may also require broadcasting of some data.

4   Related Work

OpenMP is not immediately implementable on distributed memory systems. Given its potential as a high level programming model for large applications, the need for a corresponding translation has been recognized. In our previous work, we have considered various strategies for helping the user improve the code prior to any strategy translating it for distributed execution, primarily by minimizing the amount of data that is shared [6].

A number of efforts have attempted to provide OpenMP on clusters by using it together with a software distributed shared memory (software DSM) environment [2, 3,8]. Although this is a promising approach, and work will continue to improve results, it does come with high overheads. In particular, such environments generally move data at the page level and may not be able to restrict data transfers to those objects that truly require it. There are many ways in which this might be improved, including prefetching and forwarding of data, general OpenMP optimizations such as eliminating barriers, and using techniques of automatic data distribution to help carefully place pages of data. The OMNI compiler has included additional data layout directives that help it decide where to place the pages of data in the various memories[8]. An additional approach is to perform an aggressive, possibly global, privatization of data. These issues are discussed in a number of papers, some of which explicitly consider software DSM needs [3, 4, 6, 9].

The approach that is closest to our own is an attempt to translate OpenMP directly to a combination of software DSM and MPI [5]. This work attempts to translate to MPI where this is straightforward, and to a software DSM API elsewhere. The purpose of this hybrid approach is that it tries to avoid the software DSM overheads as far as possible. While this has similar potential to our own work, GA is a simpler interface and enables a more convenient implementation strategy. Because it has a straightforward strategy for allocating data, it can also handle irregular array accesses, which is the main reason for retaining a software DSM in the above work. GA data has a global "home" but it is copied to and from it to perform the computation in regions of code; this is not unlike the OpenMP strategy of focusing on the allocation of work. For both models, this works best if the regions are suitably large. If the user is potentially exposed to the end result of the translation, we feel that they should be shielded as far as possible from the difficulties of distributed memory programming via MPI. GA is ideal in this respect as it retains the concept of shared data.

5   Conclusions and Future Work

This paper presents a basic compile-time strategy for translating OpenMP programs into GA programs. Our experiments have shown good scalability of the translated GA program in distributed memory systems, even with relatively slow interconnects. This shared memory parallel programming approach introduces new overheads as it then must efficiently gather and scatter (potentially) large amounts of data before and after parallel loops. Our on-going work investigates the ability of the compiler to support the need for efficiency in these gather and scatter operations. We believe that recent advances in the MPI standard might enable GA to provide additional functionality that could increase the viability of this approach to parallel programming. We intend to explore this issue with our GA colleagues.

References

1. OpenMP Architecture Review Board, Fortran 2.0 and C/C++ 1.0 Specifications. At www.openmp.org.
2. C. Amza, A. Cox et al.: Treadmarks: Shared memory computing on networks of workstations. IEEE Computer, 29(2):18-28, 1996
3. A. Basumallik, S-J. Min and R. Eigenmann: Towards OpenMP execution on software distributed shared memory systems. Proc. WOMPEI'02, LNCS 2327, Springer Verlag, 2002
4. Chapman, B., Bregier, F., Patil, A. and Prabhakar, A.: Achieving High Performance under OpenMP on ccNUMA and Software Distributed Share Memory Systems. Currency and Computation Practice and Experience. Vol. 14, (2002) 1-17
5. R. Eigenmann, J. Hoeflinger, R.H. Kuhn, D. Padua et al.: Is OpenMP for grids? Proc. Workshop on Next-Generation Systems, IPDPS'02, 2002
6. Z. Liu, B. Chapman, Y. Wen, L. Huang and O. Hernandez: Analyses and Optimizations for the Translation of OpenMP Codes into SPMD Style. Proc. WOMPAT 03, LNCS 2716, 26-41, Springer Verlag, 2003
7. J. Nieplocha, RJ Harrison, and RJ Littlefield: Global Arrays: A non-uniform memory access programming model for high-performance computers. The Journal of Supercomputing, 10:197-220, 1996
8. M. Sato, H. Harada and Y. Ishikawa: OpenMP compiler for a software distributed shared memory system SCASH. Proc. WOMPAT 2000, San Diego, 2000
9. T.H. Weng and B. Chapman Asynchronous Execution of OpenMP Code. Proc. ICCS 03, LNCS 2660, 667-676, Springer Verlag, 2003

ORC-OpenMP: An OpenMP Compiler Based on ORC

Yongjian Chen, Jianjiang Li, Shengyuan Wang, and Dingxing Wang

Tsinghua University, Beijing 100084, P.R. China,
chenyj99@mails.tsinghua.edu.cn

Abstract. This paper introduces a translation and optimization framework for OpenMP, based on the classification of OpenMP translation types. And an open source OpenMP compiler, which implements this framework is also introduced as a high performance research platform for Linux/IA-64. Different from other open source OpenMP compilation system, this compiler has the following characteristics: First, it's integrated into the backend optimization compiler that mainly exploits Instruction Level Parallelism. This integral solution makes analyses and optimizations that require interactions between the instruction level and the thread level possible. Second, it's based on a unified model called translation type classification. This approach improves the code quality by reducing runtime overhead and code size.

1 Introduction

There is still not enough high performance OpenMP implementation with source code opened to public for further research usage. As far as we know, there are several other OpenMP compilers opened to public domain for research usage: OdinMP/CCp (translate C OpenMP programs to C programs with pThreads([2]), Omni OpenMP Compiler ([1]) and PCOMP (Portable Compiler for OpenMP, [3]). All these OpenMP compilers are implemented as source-to-source translators, and thus are loosely coupled with backend compilers. Although these source-to-source translation approaches gain the merit of portability, we find that they are not suitable as a research platform for OpenMP optimization, especially when we try to use such a platform to study thread level parallelism upon the traditional instruction level parallelism, which requires the ability to model the interactions between these two levels, since to the three compilers, the backend optimization compilers just appear as black boxes.

This paper introduces a new framework to translate and optimize OpenMP programs for shared memory systems, and based on this framework, an OpenMP implementation based on ORC (Open Research Compiler, [4]) for Linux/IA-64 (or simply called ORC-OpenMP), which is designed and implemented as a research platform on IA-64 based systems. Different from source-to-source translation strategies found in the three implementations mentioned above, our OpenMP compiler is implemented as a module inside ORC, integrated with other optimization modules of ORC. While the source-to-source approaches are more portable, implementing OpenMP as an integrated part of compilers has more opportunities to share information with other optimization modules in the compiler, and are more flexible to translate code in more complex way. Results of OpenMP

M. Bubak et al. (Eds.): ICCS 2004, LNCS 3038, pp. 414–423, 2004.
© Springer-Verlag Berlin Heidelberg 2004

benchmark show that the implementation can fulfill the performance requirement as a research platform.

The rest of the paper is organized in the following way: the framework to translate and optimize OpenMP programs is discussed in section 2. The design and implementation issues of ORC-OpenMP are presented as section 3, and section 4 presents performance results about this implementation. Section 5 summaries the work and also gives plans about future work. Bibliography comes as section 6.

2   The Translation and Optimization Framework

2.1   Dynamic Directive Nesting, Binding, and Directive Classification

To be more versatile, dynamic directive nesting and binding are introduced into OpenMP standards. It means that the binding of execution context of a specific OpenMP construct can be delayed to runtime. According to this, a program constructs may be executed sequentially, by multithreads, or by nested multithreads. Code 1 illustrates the case of dynamic nesting and binding.

Code 1. dynamic directive binding, orphaned directives and nested parallelism

```
#pragma omp parallel /*1. normal parallel */
{
#pragma omp parallel /*2. nested parallel*/
  {
  }
  foo( ); /* call site 1 */
}

foo( )
{
#pragma omp for   /*3. orphaned for */
  {
  }
}
```

In Code 1, unique numbers identifies three OpenMP directives and two call sites of function foo are also numbered. Since foo is compiled in a separate compiling unit, directive 3 is not in the lexical extent of its call sites, so its binding to the parallel region must be delayed until the call occurs in execution. At call site 1, the parallel for declared in directive 3 is bound to parallel region declared in directive 1, and at call site 2, the parallel for should be bound to master thread according to the binding rule. While these kinds of semantics tend to be more flexible, they bring more complexity into the implementation.

According to the lexical containing type relative to parallel region constructs, directives can be divided into three categories: normal directives, directives in nested parallel region, and orphaned directives. In Figure 1, if we assume the parallel region declared in directive 1 is the outmost parallel region during execution, then all the directives directly nested in its lexical extent (not include those in nested parallel regions) are called normal directives in this paper. The directive 2 declares a parallel region inside another parallel region, and all it's containing directives are shortly called nested directives. The directive 3 is not in the lexical extent of any parallel region constructs, and it's binding to parallel regions is dynamic. Such kind of directives is called orphaned directives in the OpenMP specifications.

In order to produce better code, we extend the classification according to runtime binding relationship instead of the lexical containing relationship of constructs. The directives of only one level of parallelism are called normal directives, just as stated above, and nested directives are those inside multi-level parallel regions. Orphaned directives can be normal directives, nested directives, or serialized parallel directives according to the binding situation at runtime.

While such classification is not necessary for all OpenMP implementations, it helps to generate better code by eliminating redundant code and further enable other optimizations. In this paper, these attributes of OpenMP constructs are called translation types.

2.2   Translation Type Classification and the Framework

Since at compile time, the static analysis may not be able to determine all the dynamic bindings, the translation type of OpenMP directives can not always be determined as these three types. In the compiler, one more type is added called default directives type, to represent those whose translation types are not explicitly determined. The default directives type is the default setting for all directives. So the translation type can be one the following four:

1. default directives
2. normal directives
3. nested directives
4. serialized directives

We call the process of determining the translation type of directives "translation type disambiguation". Most of the case, such a process must cross the boundary of procedures or subroutines, thus must be done with the help of InterProcedural Analysis.

The framework for OpenMP translation and optimization is presented in Figure 1.

In this framework, frontends is used to translate OpenMP constructs in different languages into a common IR form, and later phases all work on this common IR. Two phases are necessary: pre-translation and optimization and OpenMP translation phase, while two others are optional: translation type disambiguation and post optimization phase. Translation type disambiguation has been introduced above. In this phase, translation types will be derived across procedure boundaries, and necessary data structures will be set up for later usage. Pre-translation and optimization works at the OpenMP directives level (but in the form of IR), and it may transform one OpenMP constructs into equivalent constructs, or do some optimizations such as redundant directives elimination, or merging immediately adjacent parallel regions. OpenMP translation phase is the main translation phase. It translates OpenMP constructs into low level constructs. Essentially, this phase translates OpenMP programs into normal multithreaded programs. An optional post optimization phase works on the multithreaded program level, doing optimizations such as deadlock and race condition detection.

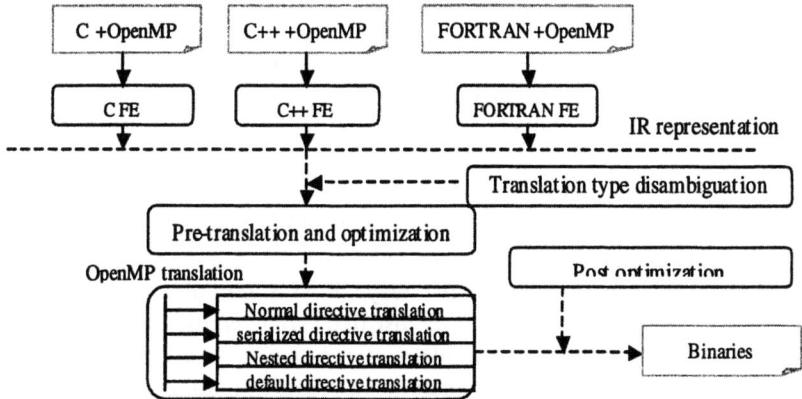

Fig. 1. The framework for OpenMP translation and optimization.

Normal directives, nested directives, serialized directives and default directives are treated differently for better performance.

2.3   Default Directives Translation

Default directives can be any of the other three translation types. So stuff code must be inserted to dynamically determine the region type and choose right code segment. An example translation is depicted in code 2. This translation implements the nested parallel regions as single-threaded groups.

Code 2. Default directives translation

```
#pragma omp for
{
/* for body */
}

(a) OpenMP constructs

if( _is_parallel()) {
  /* normal parallel do translation*/
}
else {
  /* serialized parallel do translation*/
}

(b) translated code
```

Since the OpenMP constructs are structural program constructs, the processing process of OpenMP constructs is also in a top-down manner. That means, when the compiler translates an OpenMP construct, all the OpenMP constructs inside this construct will also be translated.

2.4   Normal Directives, Serialized Directives, and Nested Directives Translation

Normal directives need not any stuff code to check the nesting type at runtime, and the directives are translated in the normal way. Further more, some common variables, such as thread id and number of threads, can be set up in a direct way at the beginning of parallel region.

Serialized directives can be stripped off from the programs, and the result code executes in sequential mode. For those directives with multiple translation types, two version of the procedure can also be generated.

Nested directives can be treated in two ways. First is to implement a pseudo-nested region, i.e., we treat the nested directives just like serialized directives and strip them away, and the nested parallelism is implemented as a one-thread group. While this is a simplified implementation, it may satisfy most practical requirements raised in SMP case. In the second case, nested parallelism must be supported. In this case, special stack must be maintained, to provide correct thread environment, such as thread group number and thread id. And further more, special thread task structures may be helpful to reduce the call stack and thus reduce the overhead of implementing multi-level thread creation.

There is still optimization space for this translation process and better translation may exist for some special cases. For example, when both the thread number of a parallel region (by using NUM_THREADS clause of PARALLEL directive) and the trip count of the do loop are known at runtime, for the static schedule type, since the loop iterations assignment can be determined totally at compile time, the prologue schedule code can be eliminated.

3   ORC-OpenMP

3.1   Introduction to ORC

ORC has an five-level IR (Intermediate Representation) named WHIRL ([5]) for its BE (BackEnd) processing, and various optimizations are carried out on different levels of WHIRL. In general, WHIRL is a tree like structure, and each WHIRL tree represents one function/subroutine in the programs. Generally, OpenMP regions are also represented as region type nodes in the WHIRL tree, and OpenMP directives are represented as pragma nodes attached to the region node. OpenMP directives without region scope are represented as single pragma nodes, just like normal statement nodes.

3.2   Framework of the OpenMP Module

The logical processing flow of OpenMP processing module in ORC is given in Fig. 2.

Four main components for OpenMP processing are indicated in the framework.

Two frontends process the OpenMP syntax in FORTRAN and C programs and translate OpenMP directives into proper WHIRL structures. The FORTRAN frontend accept FORTRAN90 syntax, and after a FORTRAN90 expansion module, the FORTRAN90 vector operations are translated into proper loops. After this point, except for special language divergences, the later processing modules don't bother with the language details anymore.

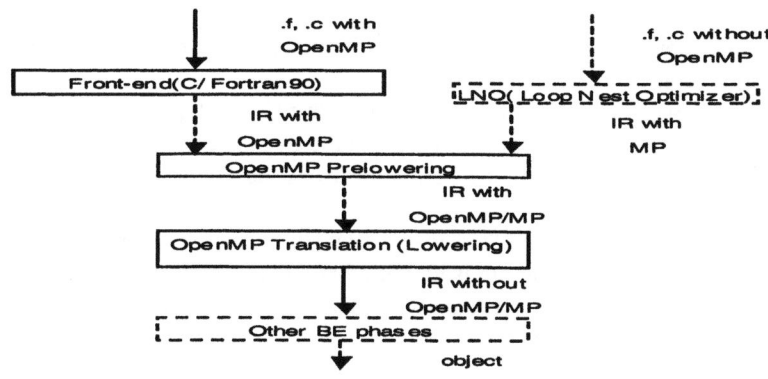

Fig. 2. OpenMP logical processing flow inside ORC

Two translation modules, OpenMP prelowering and OpenMP lowering, translate program constructs marked with OpenMP/MP pragmas to ordinary constructs using RTL API calls on IR level.

3.3 OpenMP Prelowering

OpenMP prelowering module is a pre-processing pass for OpenMP. In this module, OpenMP related program constructs are normalized and some constructs, such as SECTIONS, are translated into others for the convenience of the later OpenMP lowering module. The most important normalization operation is for loops declared to be parallel loops. The form of these loops is carefully adjusted to the form assumed by later lowering process.

In the pre-pass, OpenMP SECTIONS construct is translated into equivalent OpenMP DO construct. But in practice, more efficient program constructs such as branch-table will be used for translation.

3.4 OpenMP Lowering

The final translation of OpenMP constructs into RTL(RunTime Library) constructs is done in OpenMP lowering module. Translation (or lowering) of OpenMP constructs falls into two ways: direct one-to-one translation and WHIRL tree restructuring.

Only several OpenMP directives can be directly translated into corresponding RTL calls. An example is BARRIER. A BARRIER WHIRL node (represent a BARRIER directives in the program) lexically inside parallel regions can be replaced with one single WHIRL node with the type function call represents a call to RTL function _omp_barrier. ATOMIC and FLUSH directives with explicit arguments can also be translated in this way.

Other OpenMP region based constructs often need to restructure the corresponding WHIRL tree rather than simple replacement. This kind of constructs include PARALLEL region, DO loops, SINGLE and MASTER section, and CRITICAL section.

Nested directives are implemented in a simple way. Whenever the compiler ensure it encounter a nested parallel region, it simply strip off all the OpenMP directives contained in the region except for a few synchronization directives. Thus the nested parallelism is always expressed as a single thread group.

The lowering pass is implemented based on the translation type classification framework, to eliminate the stuff codes required to handle dynamic nesting semantics. Compiler analysis is designed to gather information from the OpenMP programs or even profiling data.

A simple analysis is implemented in the IPA (InterProcedural Analysis) module. Every parallel region in the program is identified and the information is attached to the call graph generated by the main IPA analysis phase. A typical form of these annotated call graphs may look like Figure 3:

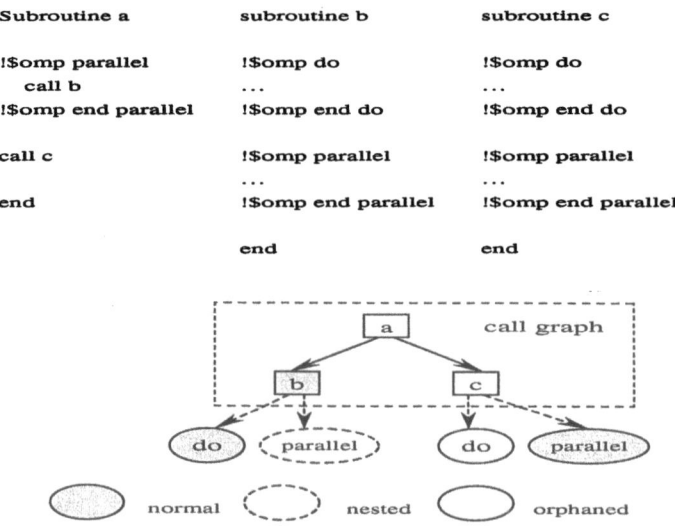

Fig. 3. Annotated call graph

For a given specific subroutine, if we know that it's a local call just like those decorated with qualifier static in C, or it will be called only in current module, then the annotated call graph can be used to extract information for further code simplification.

For example in Code 1, if we can determine from the call graph that all the calls to subroutine _foo_ are inside some parallel regions, then we can use this additional information to infer the directive attributes in subroutine _foo_. In this way, original Orphaned directives may be treated as normal directives and original normal directives will be treated as nested directives and some stuff code can be strip off because now the compiler get enough information about the context.

3.5   Data Environment Handling

Data environment constructs are not always translated into real code. More often, they specify where the specified variables are allocated and thus affect the way they are accessed. But for directives like REDUCTION, FIRSTPRIVATE, LASTPRIVATE and

THREADPRIVATE related constructs, stuff codes for initialization and finalization are also necessary.

For THREADPRIVATE variables, the official suggested implementation allocate such variables in special thread local storage sections .tdata or .tbss, and the real storage allocation and binding is done by the loader rather by compiler. In out implementation, we choose to let the program manage the allocation, binding and access, and the complexity is left to the compiler alone. The THREADPRIVATE variables are thus dynamically allocated in the heap storage.

4   Performance

4.1   Experiment Platform and Benchmarks

As a research platform, the implementation should also have satisfactory performance. NPB3.0 OpenMP FORTRAN benchmark ([6]) is used to test the FORTRAN performance. Another benchmark NPB2.3 OpenMP C version ([7]) developed as part of the Omni project is used to test the C OpenMP compilation functionality and performance. These benchmarks represent typical kinds of floating point workloads in science and engineering computation. The underlying system is a Linux SMP box with 4 itanium2 900MHz CPUs, 4GB main memory ("tiger 4" or "Bandera"). Benchmarks are compiled using Omni OpenMP compiler (1.4a) and our OpenMP compiler separately. Beyond the option to turn on OpenMP processing, for both compilers, only -O3 is specified as compiler option, and more aggressive optimizations, such as profiling and IPA are not turned on.

4.2   Performance Data

The experiment result of the NPB3.0 FORTRAN OpenMP benchmark suite is presented in figure 4. In this figure, benchmarks are presented in a form such BT.w. BT is the benchmark name, and w represents the problem size. The benchmark score is given in a metric defined as Mega-Operations per second, and the higher, the better. orf90 is our FORTRAN OpenMP compiler, and omf77 is the FORTRAN OpenMP compiler of Omni OpenMP compiler. Among the seven benchmarks, BT, SP and LU are application benchmarks, while CG, FT, MG and EP are kernel benchmarks.

The experiment result of the NPB2.3 C OpenMP benchmark suite is given in figure 5. 6 benchmarks are presented. The name convention is like Figure 4. orcc is our C OpenMP compiler, and omcc is Omni's C OpenMP compiler.

The benchmark scores simply demonstrate that our compiler's performance is much better than Omni's. And such a performance is what we need when we intend to use it as a research platform.

5   Summary

Indeed, this compiler is developed in hope to bridge the gap of traditional parallel compiler that exploits loop level parallelism and task level parallelism and traditional optimization compiler that mainly exploits instruction level parallelism. Possible research

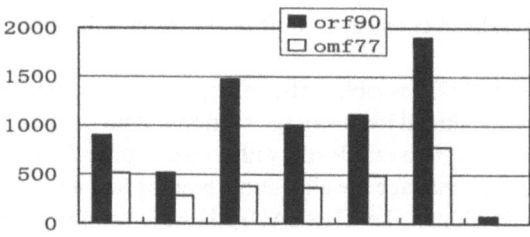

Fig. 4. OpenMP FORTRAN performance

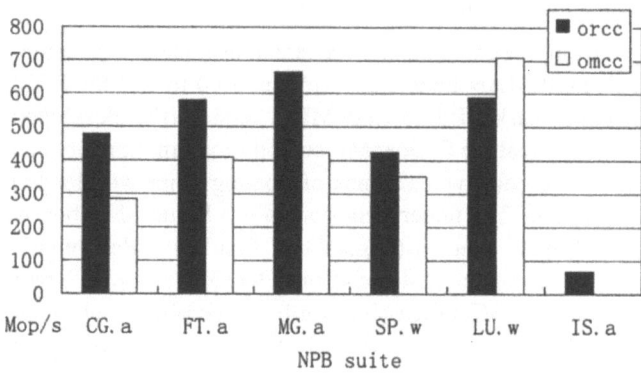

Fig. 5. OpenMP C performance. For LU.w, omcc gets better performance, but the compiled code produces wrong results and cannot pass the validity test

topics includes the interaction between thread level parallelism and instruction level parallelism, and auto-multithreading. Indeed, we are trying to use OpenMP as a tool to express thread level parallelism, and use other stuff multithreading techniques, such as helper thread to exploit the parallelism of applications at thread level, besides the efforts of exploiting instruction level parallelism. The requirement of exploiting parallelism at multiple levels demands for a unified cost model, and more direct communication between different modules. This is the motivation why we implement such an OpenMP compiler inside backend compiler, other than a standby one.

One important design decision in the design and implementation of this OpenMP compiler is the idea of translation type classification. This is in fact a static analysis for runtime context. By determine some of the contexts at compile time, the compiler can produce better code. Although the related optimizations are still not well explored, experiments show that the improvements in run time and code size is impressive.

In this paper, important design issues and tradeoffs for a special OpenMP implementation are presented, in hoping that it may help more OpenMP compilers for research use to be designed. As far as we know, this is the first open source OpenMP compiler on IA64 base platforms, and we also use this compiler as a research vehicle to study transitions from IA32 to IA64 in the HPC realm.

Acknowledgement. The work described in this paper is supported by Intel's university research funding, and partly supported by the Gelato project set up jointly by HP and Intel. We also want to thank the ORC group members for their work on ORC.

References

1. M. Sato, S. Satoh, K. Kusano and Y. Tanaka: Design of OpenMP Compiler for an SMP Cluster. In the 1st European Workshop on OpenMP (1999) 32–39
2. C. Brunschen, M. Brorsson: OdinMP/CCp-a portable implementation of OpenMP for C. Concurrency: Practice and Experience, Vol 12. (2000) 1193–1203
3. Seung Jai Min, Seon Wook Kim, M. Voss, Sang Ik Lee and R. Eigenmann.: Portable Compilers for OpenMP. In the Workshop on OpenMP Applications and Tools (2001) 11–19
4. Open Research Compiler: `http://ipf-orc.sourceforge.net`.
5. SGI Inc.: WHIRL Intermediate Language Specification. WHIRL Symbol Table Specification. (2000)
6. H. Jin, M. Frumkin, and J. Yan: The OpenMP implementation of NAS parallel benchmarks and its performance. NASA Ames Research Center Technical report, Report NAS-99-011. (1999)
7. RWCP: OpenMP C version of NPB2.3.
   `http://phase.etl.go.jp/Omni/benchmarks/NPB/index.html`.

Performance Analysis, Data Sharing, and Tools Integration in Grids: New Approach Based on Ontology⋆

Hong-Linh Truong[1] and Thomas Fahringer[2]

[1] Institute for Software Science, University of Vienna
truong@par.univie.ac.at
[2] Institute for Computer Science, University of Innsbruck
Thomas.Fahringer@uibk.ac.at

Abstract. In this paper, we propose a new approach to performance analysis, data sharing and tools integration in Grids that is based on ontology. We devise a novel ontology for describing the semantics of monitoring and performance data that can be used by performance monitoring and measurement tools. We introduce an architecture for an ontology-based model for performance analysis, data sharing and tools integration. At the core of this architecture is a Grid service which offers facilities for other services to archive and access ontology models along with collected performance data, and to conduct searches and perform reasoning on that data. Using an approach based on ontology, performance data will be easily shared and processed by automated tools, services and human users, thus helping to leverage the data sharing and tools integration, and increasing the degree of automation of performance analysis.

Keywords: Performance analysis, performance data model, Grid, ontologies.

1 Introduction

The recent emerging Grid computing raises many challenges in the domain of performance analysis. One of these challenges is how to understand and utilize performance data where the data is diversely collected and no central component manages and provides semantics of the data. Performance monitoring and analysis in Grids differ from that in conventional parallel systems in terms of no single tool providing performance data for all Grid sites and the need of conducting monitoring, measurement and analysis across multiple Grid sites at the same time. Normally users run their applications in multiple Grid sites, each is equipped with different computing capabilities, platforms, libraries that require various tools to conduct performance monitoring and measurement. Without the central component, performance monitoring and measurement tools have to provide a means for seamlessly utilizing the data they collect and provide, because many tools and services atop them need the data for specific purposes such as performance analysis, scheduling and resource matching. Current Grid performance tools focus on monitoring and measurement, but neglect data sharing and tools integration.

⋆ This research is supported by the Austrian Science Fund as part of the Aurora Project under contract SFBF1104.

M. Bubak et al. (Eds.): ICCS 2004, LNCS 3038, pp. 424–431, 2004.
© Springer-Verlag Berlin Heidelberg 2004

We take a new direction on describing the semantics of performance data and establishing performance data sharing and tools integration by investigating the use of *ontology* in performance analysis domain. Basically, ontologies provide a *shared and common* understanding of a domain that can be communicated between people and heterogeneous and widely spread application systems; ontology is developed to facilitate *knowledge sharing and reuse*. Based on sharable and extensible ontologies in the domain of performance analysis, an analysis tool, service or user is able to access multiple sources of performance and monitoring data provided by a variety of performance monitoring and measurement tools, understanding the data and making use of that data. With the expressive power provided by ontology which can describe concepts, resources in sufficient detail, supporting automatic performance analysis will also be enhanced.

The rest of this paper is organized as follows: Section 2 discusses the motivation. In Section 3 we present our proposed ontology for performance data. We describe an architecture for an ontology-based model for performance analysis, data sharing and tools integration in Section 4. Section 5 overviews our prototype implementation. The related work is discussed in Section 6 followed by the Conclusion and Future work in Section 7.

2   Motivation

Currently, several data representations with different capabilities and expressiveness, e.g. XML, XML and relational database schema, are employed by Grid performance monitoring and measurement tools. However, little effort has been done to standardize the semantics of performance data as well as the way performance tools collaborate. In Grids, data is diversely collected and no central component manages and provides its semantics. Each Grid site may be equipped with its own performance monitoring and measurement tool. Thus, the end user or the high-level tool in Grids has to interact with a variety of tools offering monitoring and measurement service. Performance monitoring and measurement tools should not simply offer well-defined operations for other services calling them, e.g. based on Open Grid Services Architecture (OGSA) [3], but they have to provide means for adding semantics to the data as Grid users and services require seamless integration and utilization of the data provided by different tools.

Existing approaches on performance data sharing and tools integration which mostly focus on building wrapper libraries for directly converting data between different formats, making data available in relational database with specific data schema, or exporting data into XML, have several limitations. For example, building a wrapper requires high cost of implementation and maintenance; wrappers convert data between representations but not always between semantics. Although XML and XML schemas are sufficient for exchanging data between parties that have agreed in advance on definitions, its use and meaning, they mostly are suitable for one-to-one communication and impose no semantic constraints on the meaning of the data. Everyone can create his own XML vocabularies with his own definitions for describing his data. However, such vocabularies and definitions are not sharable and do not establish a common understanding about the data, thus preventing semantic interoperability between various parties which is an important issue that Grid monitoring and measurement tools have to support. Utilizing relational databases to store performance data [13,11] simplifies sharing of data. However, data models represented in relational database are still very tool-specific and inextensible. Notably, XML and relational database schemas do not explicitly express meanings of data they encode. Since all above-mentioned techniques do not provide enough capability to express the semantics of performance data and to support tools integration, they might not be applicable in Grids due to the autonomy and diversity of performance monitoring and measurement tools.

We investigate whether the use of ontology can help to solve the above-mentioned issues. Ontologies are a popular research topic in various communities such as knowledge engineering, cooperative information systems. An ontology is a formal, explicit specification of a shared conceptualization [5]. An ontology typically consists of a hierarchical description of important concepts in a domain, along with descriptions of the properties of each concept. One of key features of ontology is that it provides a shared and common understanding of some domains that can be communicated between people and application systems. Another feature is that a set of ontology statements by itself can allow to conclude another facts, e.g. via description logics, while that can not be achieved with XML or database schema.

Ontology can help addressing the above-mentioned issues in many ways. Firstly, ontology can be used to directly *describe and model the data* collected, thus allowing to share a common understanding of performance data and easily correlating the data with the knowledge domain. Secondly, ontology can be used to *define mappings between different representations* employed by different Grid monitoring and measurement tools. This would allow a high level service to transparently access different types of data in a homogeneous way. This paper works on the first direction. Due to space limit, this paper discusses only main concepts and results of our approach. For more details, readers should refer to [14].

3   PERFONTO: Ontology for Describing Performance Data

While initial work on using ontology for system and network management has been introduced, e.g. in [1], to date we are not aware any ontology for describing performance data of applications in the field of performance analysis. Our starting point is that we try to propose an ontology for describing monitoring and performance data of both applications and systems. In this section, we describe PERFONTO (**ONTO**logy for **PERF**ormance data), an ontology based on OWL (Web Ontology Language) [9]. PERFONTO comprises two parts that describe *experiment-related concept* and *resource-related concept*. Here we briefly discuss main classes and properties of PERFONTO.

Experiment-related concept describes experiments and their associated performance data of applications. The structure of the concept is described as a set of definitions of *classes* and *properties*. Figure 1 demonstrates a part of classes and properties of experiment-related concept in PERFONTO. **Application** describes information about the application. **Version** describes information about versions of an application. **Source-File** describes source file of a version. **CodeRegion** describes a static (instrumented) code region. Code regions are classified into subclasses that are programming paradigm-dependent and paradigm-independent. **Experiment** describes an experiment which refers to a sequential or parallel execution of a program. **RegionInstance** describes a

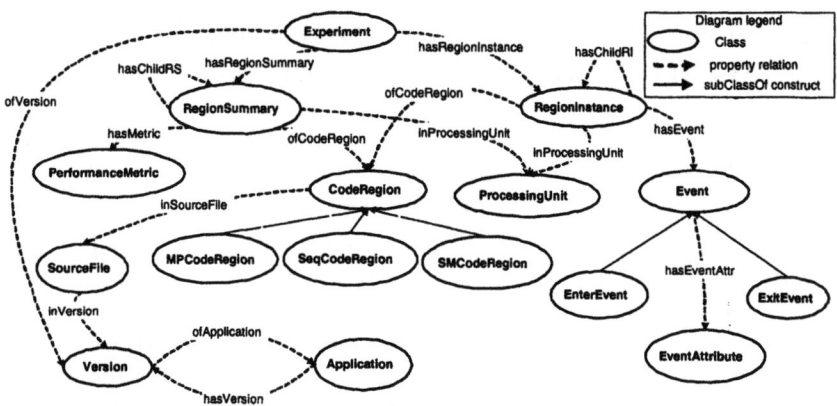

Fig. 1. Illustrative classes and properties of experiment-related concept

region instance which is an execution of a static (instrumented) code region at runtime. A code region instance is associated with a processing unit (this relationship is described by property *inProcessingUnit*) and has events (property *hasEvent*) and subregion instances (property *hasChildRI*). A processing unit, represented by class **ProcessingUnit**, describes the context in which the code region is executed; the context includes information about grid site, compute node, process, thread. **RegionSummary** describes the summary of code region instances of a static (instrumented) code region in a processing unit. A region summary has performance metrics (property *hasMetric*) and subregion summaries (property *hasChildRS*). **PerformanceMetric** describes a performance metric, each metric has a name and value. **Event** describes an event record. An event happens at a time and has event attributes (property *hasEventAttr*). **EventAttribute** describes an attribute of an event which has an attribute name and value.

Resource-related concept describes static, benchmarked, and dynamic (performance) information of computing and network systems. In the current version, resource-related concept provides classes to describe static and benchmarked data of computing and network resources. For example, **Site** describes information of (grid) computing site. **Cluster** describes a set of physical machines (compute nodes). Cluster has a subclass namely *SMPCluster* represented a cluster of SMP. **ComputeNode** describes information about physical machine. ComputeNode also has subclasses, e.g. *SMPComputeNode* represented an SMP machine. **Network** describes an available network. Subclasses of Network can be *EthernetNetwork, MyrinetNetwork*, etc. **NodeSharedMemoryPerf** describes performance characteristics of shared memory operations of a compute node. **NetworkMPColPef** and **NetworkMPP2PPerf** describe performance characteristics of collective and point-to-point message passing operations of a network, respectively.

The proposed ontology is largely based on our previous work on developing data schema for expressing experiment data in relational database [13] and on APART experiment model [2]. The development of PERFONTO should be considered as the investigation of using ontology for describing performance data, not establishing a standard for all tools. However, one of the main key advantages of ontology is that different ontologies can be reconciled [8]. Therefore, one can employ or extend PERFONTO, others may develop their own ontologies. Finally, proposed ontologies can be merged.

4   An Architecture for an Ontology-Based Model for Performance Analysis, Data Sharing, and Tools Integration

Figure 2 presents a three layers architecture for an ontology-based model for performance analysis, data sharing and tools integration. At the core of this architecture is a performance data repository service which includes:

- *PERFONTO* is ontology for representing performance data discussed in Section 3.
- *Ontological database* is a relational database which is used to hold ontologies (e.g PERFONTO) and performance data (instance data).
- *ONTO APIs* are interfaces used to store and access data in ontological database.
- *Query Engine* provides searching and querying functions on ontological data.
- *Inference Engine* provides reasoning facilities to infer, discover and extract knowledge from ontological performance data.

The performance data repository service is designed as an OGSA Grid service [3]. The *Performance Data Collector* of performance monitoring and measurement tools can store collected data (instance data) along with corresponding ontology model (e.g. PERFONTO) into the ontological database. Via service operations, any clients needed performance data such as performance analysis tools, schedulers or users can easily request the ontology model and then retrieve instance data from ontological database. The key difference from approaches of

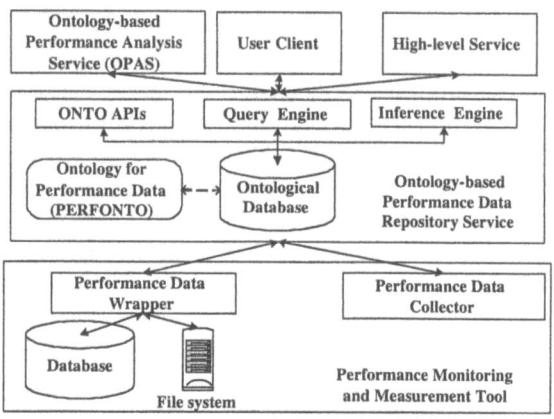

Fig. 2. Three layers architecture for an ontology-based model for performance analysis, data sharing and tools integration

using XML or relational database is that performance data is either described by a common ontology (e.g. PERFONTO) or by a tool-defined ontology, thus, with the presence of ontology model, these clients can easily understand and automatically process retrieved data. Via *Performance Data Wrapper*, data in tool-defined non-ontology format also can be extracted and transformed into ontological representation and vice versa. To implement this architecture, we select Jena [6] for processing ontology-related tasks.

4.1   Search on Ontological Data

A search engine is developed to support clients finding interesting data in the ontological database. At the initial step, we use a search engine provided by Jena that supports RDQL query language [7]. The use of RDQL in combining with ontology can simplify and provide a high-level model for searches in performance analysis in which searching query is easily understood and defined by end-users, not only by tool developers.

```
l₁:   SELECT ?regionsummary
      WHERE
l₂:       (?regionsummary perfonto:inProcessingUnit ?processingunit)
l₃:       (?processingunit perfonto:inNode "gsr410")
l₄:       (?regionsummary perfonto:hasMetric ?metric)
l₅:       (?metric perfonto:hasMetricName "wtime")
l₆:       (?metric perfonto:hasMetricValue ?value)
l₇:   AND (?value >=3E8)
l₈:   USING perfonto FOR <http://www.par.univie.ac.at/project/scalea/perfonto#>
```

Fig. 3. An example of RDQL query based on PERFONTO

```
l₁:   [rule_detect_bigmessages:
l₂:       (?regionsummary perfonto:ofCodeRegion ?codeRegion),
l₃:       (?codeRegion rdf:type perfonto:MPCodeRegion),
l₄:       (?codeRegion perfonto:hasCrType "CR_MPIP2P"),
l₅:       (?regionsummary perfonto:hasMetric ?metric),
l₆:       (?metric perfonto:hasMetricName "AvgMessageLength"),
l₇:       (?metric perfonto:hasMetricValue ?length),
l₈:       greaterThan(?length, BIG_MESSAGES_THREADHOLD)
l₉:   ->    print(?regionsummary,"Big message hold!")]
```

Fig. 4. An example of rule-based reasoning based on PERFONTO

Figure 3 presents an RDQL query, based on PERFONTO, which finds any region summary executed in compute node $gsr410$ that its wallclock time (denoted by metric name $wtime$) is greater than or equal to 3E8 microsecond. Line l_1 selects variable $regionsummary$ via $SELECT$ clause. In line l_2 information about processing unit of $regionsummary$, determined by property $perfonto{:}inProcessingUnit$, is stored in variable $processingunit$. The compute node of $processingunit$ must be *"gsr410"* as stated in line l_3. In line l_4, performance metric of $regionsummary$ is stored in variable $metric$ and line l_5 states that the name of $metric$ must be *"wtime"*. In line l_6, the value of $metric$ is stored in variable $value$ which must be greater than or equal to 3E8 as specified in line l_7. Line l_8 specifies the URI for the shortened name *perfonto*.

4.2   Reasoning on Ontological Data

The use of ontology for representing performance data allows additional facts to be inferred from instance data and ontology model by using axioms or rules. Based on ontology, we can employ inference engine to capture knowledge via rules.

Let us analyze a simple rule for detecting all MPI point-to-point communication code regions of which the average message length is greater than a predefined threshold [2]. As presented in Figure 4, line l_1 defines the name of the rule. In line l_2, a term of triple pattern specifies link between a region summary and its associated code region. Line l_3 states the code region is an instance of $MPCodeRegion$ (message passing code region) and is an MPI point-to-point communication region (denoted by mnemonic CR_MPIP2P) as specified in line l_4. Line l_5, l_6 and l_7 are used to access the average message length of the region summary. Line l_8 checks whether the average message length is greater than a predefined threshold (BIG_MESSAGES_THRESHOLD) by using a built-in function. In line l_9, the action of this rule concludes and prints the region summary having big message. This example shows how using ontology helps simplifying the reasoning on performance data.

5 Prototype Implementation

We are currently implementing the proposed ontology and ontology-based service. The ontology-based performance data repository is an OGSA-based service of which the ontological database is based on PostgreSQL. However, in current prototype this service supports only operations for retrieving and storing ontology descriptions and instance data; searching and reasoning have to be done at client side. We are working on providing searching and reasoning operations.

We develop an Ontology-based Performance Analysis Service (OPAS) which supports ontology-based searching and reasoning. Figure 5 presents an user interface for performing searches in OPAS. In the top window the user can specify queries whereas the result will be shown in the bottom window. For example, we conducted a search with the query presented in Section 4.1 with a 3D Particle-In-Cell application. In the bottom window, under the subtree of variable *regionsummary*, list of region summaries met the condition will be shown. The user can examine performance metrics in details. Also other information such as source code and machine can be visualized as needed.

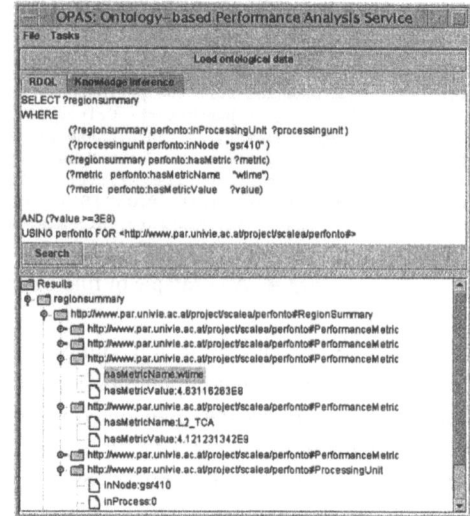

Fig. 5. GUI for conducting searches

6 Related Work

Database schemas are proposed for representing performance data, e.g. in SCALEA [13] and Prophesy [11]. However, these approaches are tool-specific rather than widely-accepted data representations. It is difficult to extend database schema structure to describe new resources. The relational database schema does not explicitly express semantics of data whereas the ontology does. As a result, building knowledge discovery via inference on ontological data is less intensive work, hard and costly.

The Global Grid Forum (GGF) Network Measurements working group has created an XML schema which provides a model for network measurement data [4]. Similarly, GLUE schema [12] defines a conceptual data model to describe computing and storage elements and networks. In [1], ontology has been applied for improving the semantic expressiveness of network management information and the integration of information definitions specified by different network managements. None of these schemas models concepts of application experiments. However, the objects modeled in GGF and GLUE schema are similar to that in our resource-related ontology. Thus vocabularies and terminologies of these schemas can be incorporated into our resource-related ontology.

Recent work in [10] describes how ontology can be used for resource matching in the Grid. Our framework can provide data for matching resources in Grids.

7 Conclusion and Future Work

In this paper, we have investigated how ontology can help overcome the lack of semantics description possessed by current techniques that are used in existing performance monitoring and measurement tools to describe performance data. Initial results show that ontology is a promising solution in the domain of performance analysis because it not only provides a means for seamless utilization and integration of monitoring and performance data but also increases the degree of automation of performance analysis.

Besides working toward the full prototype, we are currently enhancing and reevaluating our proposed ontology. We are extending resource-related concept to cover dynamic data of compute and network systems at runtime, and advancing the experiment-related ontology to describe performance properties, performance data of workflow applications, etc. In addition, we plan to study the use of ontology for mapping between different representations of performance data.

References

1. J.E. Lopez de Vergara, V.A. Villagra, J.I. Asensio, and J. Berrocal. Ontologies: Giving semantics to network management models. *IEEE Network*, 17(3):15–21, May-June 2003.
2. T. Fahringer, M. Gerndt, Bernd Mohr, Felix Wolf, G. Riley, and J. Träff. Knowledge Specification for Automatic Performance Analysis, Revised Version. APART Technical Report, http://www.kfa-juelich.de/apart/result.html, August 2001.
3. I. Foster, C. Kesselman, J. Nick, and S. Tuecke. Grid Services for Distributed System Integration. *IEEE Computer*, pages 37–46, June 2002.
4. GGF Network Measurements Working Group. http://forge.gridforum.org/projects/nm-wg/.
5. T. R. Gruber. A translation approach to portable ontology specifications. *Knowledge Acquisition*, 5(2):199–220, 1993.
6. Jena - A Semantic Web Framework for Java. http://jena.sourceforge.net.
7. RDQL: RDF Data Query Language. http://www.hpl.hp.com/semweb/rdql.htm.
8. Deborah L. McGuinness, Richard Fikes, James Rice, and Steve Wilder. An Environment for Merging and Testing Large Ontologies. In *Proceedings of the 7th International Conference on Principles of Knowledge Representation and Reasoning (KR2000)*, April 2000.
9. OWL Web Ontology Language Reference. http://www.w3.org/tr/owl-ref/.
10. H. Tangmunarunkit, S. Decker, and C. Kesselman. Ontology-based Resource Matching in the Grid—The Grid meets the Semantic Web. In *Proceedings of the Second International Semantic Web Conference*, Sanibel-Captiva Islands, Florida, USA, October 2003.
11. V. Taylor, X. Wu, J. Geisler, X. Li, Z. Lan, R. Stevens, M. Hereld, and Ivan R.Judson. Prophesy:An Infrastructure for Analyzing and Modeling the Performance of Parallel and Distributed Applications. In *Proc. of HPDC's 2000*. IEEE Computer Society Press, 2000.
12. The Grid Laboratory Uniform Environment (GLUE). http://www.cnaf.infn.it/˜sergio/datatag/glue/index.htm.
13. Hong-Linh Truong and Thomas Fahringer. On Utilizing Experiment Data Repository for Performance Analysis of Parallel Applications. In *9th International Europar Conference(EuroPar 2003)*, LNCS, Klagenfurt, Austria, August 2003. Springer-Verlag.
14. Hong-Linh Truong and Thomas Fahringer. An Ontology-based Approach To Performance Analysis, Data Sharing and Tools Integration in Grids. Technical Report AURORA TR2004-01, Institute for Software Science, University of Vienna, January 2004.

Accurate Cache and TLB Characterization Using Hardware Counters

Jack Dongarra, Shirley Moore, Philip Mucci, Keith Seymour, and Haihang You

Innovative Computing Laboratory, University of Tennessee
Knoxville, TN 37996-3450 USA
{dongarra,shirley,mucci,seymour,you}@cs.utk.edu

Abstract. We have developed a set of microbenchmarks for accurately determining the structural characteristics of data cache memories and TLBs. These characteristics include cache size, cache line size, cache associativity, memory page size, number of data TLB entries, and data TLB associativity. Unlike previous microbenchmarks that used time-based measurements, our microbenchmarks use hardware event counts to more accurately and quickly determine these characteristics while requiring fewer limiting assumptions.

1 Introduction

Knowledge of data cache memory and data TLB characteristics is becoming increasingly important in performance optimization and modeling. Cache-conscious algorithmic design is the basis of tuned numerical libraries, such as the BLAS and LAPACK and has been shown to improve full application performance significantly [4,5]. Compilers and automatically tuned software systems such as ATLAS [11] and PHiPAC [2] need accurate information about cache and TLB characteristics to generate optimal code. This information is also important for performance modeling techniques such as those described in [8,9]. In [9], it is hypothesized that much of the error in the miss surfaces used for the performance modeling lies in the creation of the cache characterization surfaces. This and other performance modeling work would benefit from more accurate cache characterization for new architectures.

Published information about detailed cache and TLB characteristics can be difficult to find or may be inaccurate or out-of-date. Thus, it will often be necessary to determine or verify this information empirically. This paper describes a methodology of instrumenting some microbenchmark codes that exercise the cache and memory subsystem to collect hardware counter data for cache and memory events. These data can then be used to give accurate numbers for cache and TLB characteristics. We describe how our methodology builds on and extends previous work based on timings of microbenchmarks, give results for the Itanium2 processor, and discuss ideas for future work and applications of the results.

M. Bubak et al. (Eds.): ICCS 2004, LNCS 3038, pp. 432–439, 2004.
© Springer-Verlag Berlin Heidelberg 2004

2   Methodology

Previous work measured the time to execute simple memory-bound loops on arrays of different sizes and with different strides, and estimated the cache and TLB characteristics from the results [7,10]. The data cache and TLB characteristics can be calculated once the number of memory references and the number of cache and TLB misses are known for the different array sizes and strides. Previous work inferred the numbers of misses from the timing results. In [7], an analytical model is developed for a single cache that identifies four cases that show up as four distinct regions on the timing curves. The model assumes a single blocking cache with an LRU replacement policy, and it can determine cache size, cache line size, and cache associativity from the timing results. The same model is used for TLB characterization, and the values for the array size and stride at which the TLB phenomena occur are different enough from those for the cache phenomena that the effects can be isolated from one another.

In [10], the analytical model from [7] is first extended from four to six cases to include transitional portions of the timing curves. The model is further extended to a system with two caches. The two-cache model assumes that the second-level cache includes the first-level, that both caches have the same block size, that associativity is non-decreasing, that there is not prefetching, and that the replacement policy for both caches is LRU. Six cases are identified and some fairly complicated timing formulas are given that allow the cache characteristics of the two caches to be calculated from timing results for the microbenchmarks. Next the model is extended to a two-cache system with a TLB. Some assumptions on the TLB operation are made, seven cases are identified, and another set of formulas is given.

The Calibrator microbenchmark described in [6] varies stride and array size in a small loop that executes a million memory reads in order to determine cache characteristics including cache size, cache line size, TLB size, page size, and miss latencies. The microbenchmark uses a pointer chasing scheme to try to prevent latency hiding. Results are given for SGI Origin 2000, Sun Ultra, Intel Pentium III, and AMD Athlon.

The above approaches break down when the simplifying assumptions do not hold – e.g., in the presence of non-blocking caches or prefetching. In some newer processors, the assumption of cache inclusivity does not always hold. These factors, combined with variability in timing results, make it difficult to obtain cache and TLB characteristics accurately on new architectures using the above approaches.

Our approach uses hardware counters to obtain precise values for the number of misses at different levels of data cache and for the data TLB. The cache characteristics can then be directly inferred from the miss counts, rather than indirectly from timing data. To collect these data, we use the PAPI portable interface to hardware performance counters [3]. PAPI defines both a standard set of routines for accessing the counters and a standard set of events. As many as possible of the PAPI standard events are mapped to native events on a given platform. On processors where they are available, we measure PAPI standard

Table 1. PAPI standard cache and TLB events

PAPI_L1_DCM	Level 1 data cache misses
PAPI_L2_DCM	Level 2 data cache misses
PAPI_L3_DCM	Level 3 data cache misses
PAPI_TLB_DM	Data TLB misses

events shown in Table 1 for our microbenchmarks. We then plot separate curves for each of these events. The cache and TLB characteristics can be read directly off each curve as well as being determined exactly by the numerical results.

For processors on which one or more of the above events are not available, we can use those events that are available to accurately determine the cache characteristics for those levels and then deduce the remaining characteristics using the formulas from [10]. On some platforms, another event may be available that is closely correlated to the missing event and thus can be used as a substitute.

Due to the fact that cache and TLB miss events exhibit little or no variability from run to run, whereas timing measurements are more variable, our hardware counter approach requires fewer trials to achieve accurate results. This savings in time is important for applications such as automatically tuned software that need to generate optimal code in a reasonable amount of time.

We have developed two microbenchmarks:

- `papi_cacheSize` which increases the size of the array being referenced with stride one until it exceeds the capacities of the various levels of cache memory. Both integer and floating point type data are tested, since on some architectures, such as the Itanium 2, these types are treated differently by the cache and memory system. The loop is timed and the number of cache misses at each level measured for each array size. A sharp increase in the number of cache misses occurs when the cache size is exceeded.
- `papi_cacheBlock` which increases the stride with which an array of fixed size is being referenced (with the array size chosen depending on the cache size) by powers of two from stride one to stride equal to the array size. Again, both integer and floating point data are tested. The total number of iterations is held constant. A sharp increase in the number of misses occurs at the cache line size, and a drop to zero occurs when the total number of array items referenced fits within one cache set.

For analysis of the `papi_cacheBlock` results, we consider the following cases, where C is the cache size, N is the array size, s is the stride, b is the cache block or line size, and a is the cache associativity:

1. $N > C$ and $1 \leq s \leq b$
   There is one miss every b/s accesses. Since we double the number of iterations as we double the stride, the total number of misses also doubles.
2. $N > C$ and $b \leq s < N/a$
   There is one miss every access. We reduce the number of different cache lines accessed by a factor of two every time we double the stride in this region,

but since we also reduce the number of cache congruence classes in use by the same factor, the number of misses stays constant in this region.

3. $N > C$ and $N/a \leq s \leq N$

Only $N/s \leq a$ array elements are accessed. Since all of them fit in a single cache set, there are no more misses.

3   Results

In this section, we show representative results for the Itanium 2 platform. Similar results can be obtained on any of the platforms on which PAPI is implemented, which include IBM POWER3/4, MIPS R10K/R12K/R14K/R16K, Sun Ultra-Sparc, HP Alpha, Pentium, Opteron, and Cray X1, although on some of these platforms we are limited by the availability of hardware counter events for different levels of cache and TLB. We show the results in graphical form for illustrative purposes. Our benchmark codes produce exact numerical results by initially increasing the array sizes for the tests by powers of two and then using a more refined search to pinpoint the exact array size where the change in cache or TLB misses occurs.

Our results for the Itanium 2 are shown in Figures 1-6. The average memory access time for each test is calculated by dividing the CPU time for the loop in the test by the total number of memory references. Figure 1 shows the results of running the papi_cacheSize microbenchmark instrumented to count the PAPI_L1_DCM, PAPI_L2_DCM, and PAPI_L3_DCM events in addition to timing the execution of the loop for different array sizes up to 8 megabytes using integer data type. The curves show the L1 data cache size of 16KB and the L2 cache size of 256KB. For the L3 cache size, we refine the search (not shown due to space limitations) to determine the L3 cache size of 1.5MB. Figure 2 shows the results for the events PAPI_L2_DCM and PAPI_L3_DCM for double precision real data type. L1 data cache misses are not shown for double precision real data type, because floating point loads bypass the L1 data cache and L1 DTLB. The faster time for real data is most likely due to the high bandwidth data paths to and from the L2 cache and the floating point register file and the L1 cache bypass for floating point data. Figure 3 shows the papi_cacheSize microbenchmark instrumented to count L1 DTLB misses and PAPI_TLB_DM (which is mapped to native L2DTLB_MISSES) using integer data type. The L1 DTLB misses are generated by adding the counts for the native events L1DTLB_TRANSFER (L1DTLB misses hit in L2DTLB) and L2DTLB_MISSES. Because the L1 DTLB is a strict subset of the L2 DTLB, a L2 DTLB miss is also a L1 DTLB miss. We see that the L1 DTLB has 32 entries since L1 DTLB misses begin to occur at an array size of $32 \times 4K = 2^{17}$ bytes (see papi_cacheBlock results below for the page size). Similarly, we see that the L2 DTLB has 128 entries since the page size is 16K and misses begin to occur at an array size of $128 \times 16K = 2^{21}$ bytes.

The remaining figures show the results of running the papi_cacheBlock microbenchmark to determine the cache line sizes, TLB page sizes, and cache

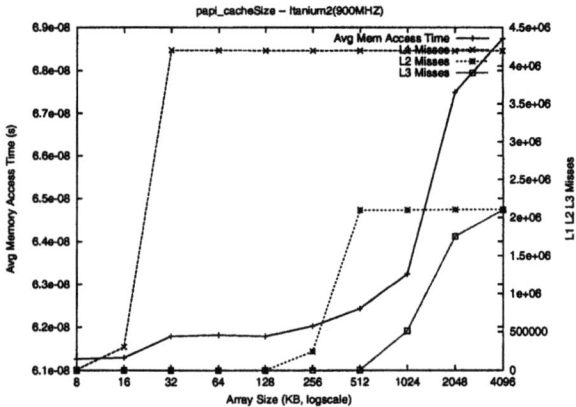

Fig. 1. Itanium2 Cache Sizes with Integer Data Type

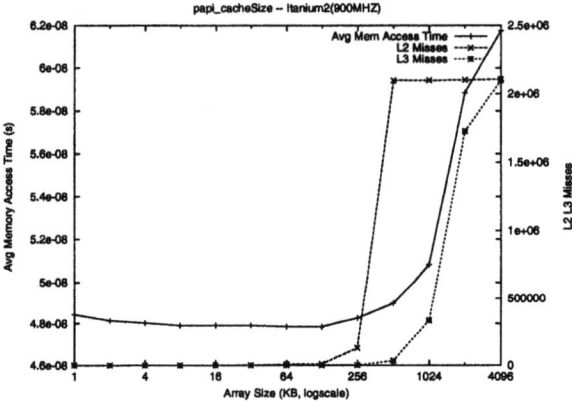

Fig. 2. Itanium2 Level 2 and 3 Cache Sizes with Real Data Type

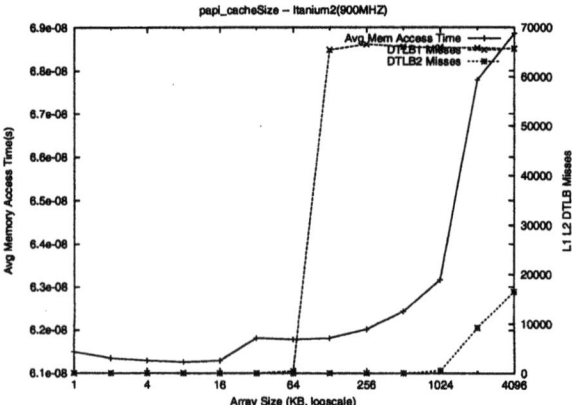

Fig. 3. Itanium2 Level 1 and 2 DTLB Sizes with Integer Data Type

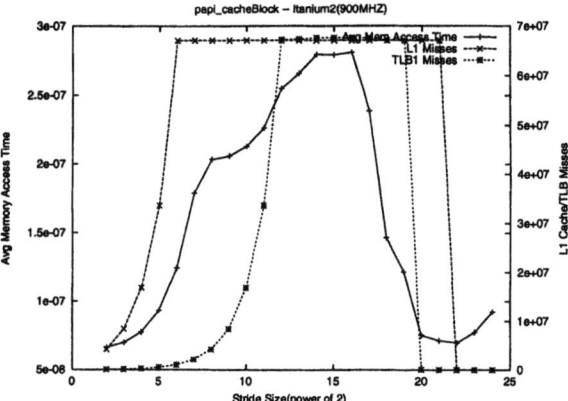

Fig. 4. Itanium2 Level 1 Cache and TLB Characteristics with Integer Data Type

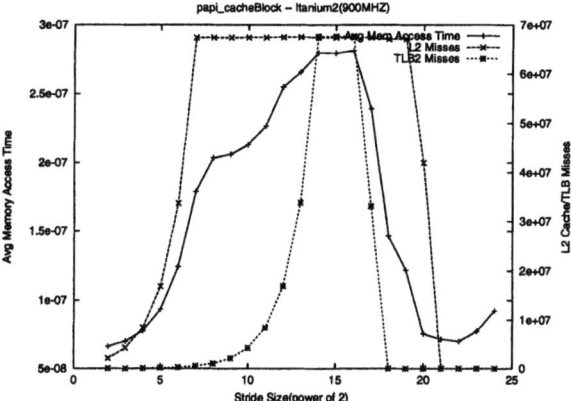

Fig. 5. Itanium2 Level 2 Cache and TLB Characteristics with Integer Data Type

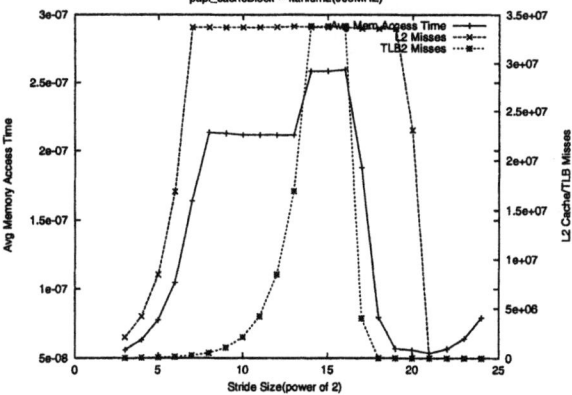

Fig. 6. Itanium2 Level 2 Cache and TLB Characteristics with Real Data Type

and TLB associativities. These quantities are calculated using the three cases described at the end of section 2. Figure 4 shows the results of running the `papi_cacheBlock` microbenchmark instrumented to count the PAPI_L1_DCM and native L1 DTLB miss events for the integer data type. We can see that the L1 cache line size is $2^6 = 64$ bytes and that the L1 cache has 4-way associativity. We can also see that the L1 DTLB page size is $2^{12} = 4K$ bytes and that the L1 DTLB is fully associative. Figure 5 shows the results of running the `papi_cacheBlock` microbenchmark instrumented to count the PAPI_L2_DCM and PAPI_TLB_DM (which is mapped to native L2DTBL_MISSES) events for integer data type, and Figure 6 shows the same results for double precision real data type. We can see that the L2 cache line size is $2^7 = 128$ bytes and that the L2 cache has 8-way associativity. We can also see that the page size is 16K bytes and that the L2 DTLB is fully associative. Note that the curves for Figures 5 and 6 have the same shape but that the execution time in Figure 6 is less. Again, this difference is most likely due to the high bandwidth data paths to and from the L2 cache and the floating point register file and the L1 cache bypass for floating point data. The reason for the difference in pages sizes for the L1 DTLB and L2 DTLB is that the L1 DTLB has a fixed page size of 4KB. Larger page sizes are supported by allocating additional L1 DTLB entries as a 4KB portion of the larger page [1].

4   Conclusions and Future Work

We have shown how hardware counter measurements can be used to generate accurate information about data cache and TLB characteristics on modern microprocessors. Even for processors where non-blocking caches, prefetching, and other latency-hiding techniques can make determining cache and TLB characteristics from timing results difficult, hardware counter measurements can produce accurate results.

We plan to release the `papi_cacheSize` and `papi_cacheBlock` microbenchmarks as part of PAPI 3.0. Scripts will be included for analyzing and plotting the data. The most portable and efficient means available will be used for determining the cache and TLB characteristics, starting with PAPI standard cache and TLB events. Next native cache and TLB events will be used, and finally other hardware events and timing results from which cache and TLB characteristics can be inferred.

Caches of some modern microprocessors are split into banks in order to allow interleaved access to data in different banks. For example, the Itanium 2 Level 2 data cache is organized into sixteen banks, with an interleaving of 16 bytes. We are investigating how to extend our microbenchmark approach to determine the number of cache banks, which can be an important factor affecting performance of the cache.

We plan to use automated determination of cache and TLB information in our research to extend semi-automatic tuning, already demonstrated for the BLAS [11] to higher level libraries, including sparse methods. The information will also be used to generate more accurate machine signatures for performance modeling efforts in which we are collaborating.

For more information about PAPI, including software and documentation, see the PAPI web site at http://icl.cs.utk.edu/papi/.

References

1. *Intel Itanium 2 Processor Reference Manual.* http://developer.intel.com/, Apr. 2003.
2. J. Bilmes, K. Asanovic, C.-W. Chin, and J. Demmel. Optimizing matrix multiply using PhiPAC: a portable high-performance ANSI C coding methodology. In *Proc. International Conference on Supercomputing*, Vienna, Austria, 1997.
3. S. Browne, J. Dongarra, N. Garner, G. Ho, and P. Mucci. A portable programming interface for performance evaluation on modern processors. *International Journal of High Performance Computing Applications*, 14(3):189–204, 2000.
4. T. M. Chilimbi, M. D. Hill, and J. D. Larus. Cache-conscious structure layout. In *Proc. 1999 ACM SIGPLAN Conference on Programming Languages and Implementation (PLDI)*, pages 1–12, 1999.
5. W. Jalby and C. Lemuet. Exploring and optimizing Itanium2 cache performance for scientific computing. In *Proc. 2nd Workshop on EPIC Architectures and Compiler Technology*, Istanbul, Turkey, Nov. 2002.
6. S. Manegold and P. Boncz. *Cache-Memory and TLB Calibration Tool.* http://homepages.cwi.nl/ manegold/Calibrator/calibrator.shtml, 2001.
7. R. H. Saavedra and A. J. Smith. Measuring cache and TLB performance and their effect on benchmark runtimes. *IEEE Transactions on Computers*, 44(10):1223–1235, 1995.
8. A. Snavely, L. Carrington, N. Wolter, J. Labarta, R. Badia, and A. Purkayastha. A framework for performance modeling and prediction. In *Proc. SC2002*, Baltimore, MD, Nov. 2002.
9. E. S. Sorenson and J. K. Flanagan. Cache characterization surfaces and predicting workload miss rates. In *Proc. 4th IEEE Workshop on Workload Characterization*, pages 129–139, Austin, Texas, Dec. 2001.
10. C. Thomborson and Y. Yu. Measuring data cache and TLB parameters under Linux. In *Proc. 2000 Symposium on Performance Evaluation of Computer and Telecommunication Systems*, pages 383–390. Society for Computer Simulation International, July 2000.
11. R. C. Whaley, A. Petitet, and J. Dongarra. Automated empirical optimizations of software and the ATLAS project. *Parallel Computing*, 27(1-2):3–25, 2001.

A Tool Suite for Simulation Based Analysis of Memory Access Behavior

Josef Weidendorfer[1], Markus Kowarschik[2], and Carsten Trinitis[1]

[1] Technische Universität München, Germany
[2] Universität Erlangen-Nürnberg, Germany
kowarschik@cs.fau.de

Abstract. In this paper, two tools are presented: an execution driven cache simulator which relates event metrics to a dynamically built-up call-graph, and a graphical front end able to visualize the generated data in various ways. To get a general purpose, easy-to-use tool suite, the simulation approach allows us to take advantage of runtime instrumentation, i.e. no preparation of application code is needed, and enables for sophisticated preprocessing of the data already in the simulation phase. In an ongoing project, research on advanced cache analysis is based on these tools. Taking a multigrid solver as an example, we present the results obtained from the cache simulation together with real data measured by hardware performance counters.

Keywords: Cache Simulation, Runtime Instrumentation, Visualization.

1 Introduction

One of the limiting factors for employing the computational resources of modern processors is excessive memory access demands, i.e. not taking advantage of cache hierarchies by high temporal and spatial locality of sequential memory accesses. Manual optimization activities trying to overcome this problem typically suffer from difficult-to-use and -to-setup tools for detailed bottleneck analysis. We believe that profiling tools based on simulation of simple cache models can significantly help in this area. Without any need for hardware access to observation facilities, the simulator produces data easy to understand and does not influence simulated results. Therefore, runtime instrumentation and sophisticated preprocessing is possible directly in the simulation phase. Optimizations based on cache simulation usually will improve locality of an application in a general way, therefore enabling better cache usage not only on a specific hardware platform, but on cache architectures in general. This gives the user more flexibility in terms of target architectures. Furthermore, it is essential to present the obtained measurement data in a way easy to understand.

In this paper, we discuss the tool suite Calltree/KCachegrind[1]. The profiling tool uses the instrumentation framework Valgrind [18], running on Linux/IA-32.

---

[1] http://kcachegrind.sf.net

M. Bubak et al. (Eds.): ICCS 2004, LNCS 3038, pp. 440–447, 2004.
© Springer-Verlag Berlin Heidelberg 2004

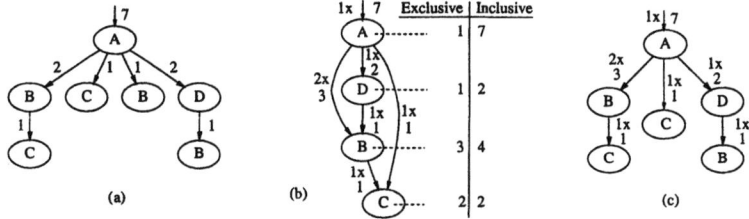

Fig. 1. A dynamic call tree, its call graph, and its context call tree

The visualization front end, which is based on QT/KDE libraries, runs on most UNIX platforms.

The paper is organized as follows: in the next section, we give a short overview on profiling methods and problems. In section 3, simulation based profiling and the implementation we use is covered in detail. Visualization concepts and implementation are presented in section 4. In our ongoing project DiME[2], memory intensive multigrid algorithms are optimized with regard to their memory access behavior. Section 5 discusses this application and presents measurements from both simulation and hardware counters. We close by presenting related work and future research directions.

2   Profiling

It is important to concentrate on code regions where most time is spent in typical executions. For this, runtime distribution is detected by *profiling*. Also, it can approve assumptions regarding runtime behavior, or show the right choice of multiple possible algorithms for a problem. A dynamic call-tree (DCT), see fig. 1 (a), represents a full trace from left to right, and has a size linear to the number of calls occurring. For profiling, the much smaller dynamic call graph (DCG), see fig. 1 (b), typically is enough. However, a DCG contains less information: in fig. 1 (b), one cannot see that D → B → C actually never happened. Therefore, [1] proposed the context call tree (CCT), see fig. 1 (c), where each node represents the occurrence of a function in a call chain, a context, starting at the root node[3]. Our tool is able to produce a "reduced" $CCT_{n_{max}}$: as a CCT can still get huge, we collapse two contexts if the trailing n_{max} contexts of the corresponding call chains are identical. Table 1 gives some numbers on the size of these reduced CCTs for various UNIX applications.

During the profiling processing, event attribution of nodes and arcs is done. Events can be memory accesses, cache misses, or elapsed clock ticks, for example. Attributes of interest for nodes are the number of events occurring inside the function (*exclusive* or *self* cost), and additionally in functions which are called

---

[2] http://www10.informatik.uni-erlangen.de/Research/Projects/DiME/
[3] In the CCT, recursive calls or mutually recursive functions are handled specially to avoid a size linear to the number of calls occurring at runtime.

Table 1. Number of nodes and arcs in reduced CCTs

Command		Call chain length limit n_{max}					
		0	1	2	5	10	20
bzip2 libm.so.6	Nodes	408	850	1 004	1 329	1 332	1 132
(Compressor)	Arcs	538	688	861	1 113	1 113	1 113
ccl ct_main-i.c	Nodes	1 519	5 157	8 352	22 060	41 164	44 899
(C compiler)	Arcs	6 741	10 881	15 905	34 282	52 191	54 722
konqueror	Nodes	21 500	55 829	91 713	251 449	420 871	507 507
(KDE Browser)	Arcs	51 052	90 629	147 958	315 838	470 032	544 487

from the given function (*inclusive* cost). For arcs, events occurring while the program is running in code called via this arc are interesting, as well as the number of times the call is executed. In fig. 1, event attribution is shown, assuming one event during the execution of each function.

Profiling should not perturb the performance characteristics of the original program, i.e. its influence on runtime behavior should be minimal. For this reason, modern CPU architectures include performance counters for a large range of event types, enabling *statistical sampling*: After a hardware counter is set to an initial value, the counter is decremented whenever a given event takes place, and when reaching zero, an interrupt is raised. The interrupt handler has access to the program counter of the interrupted instruction, updates statistical counters related to this address, resets the hardware counter and resumes the original program[4]. The result of statistical sampling is self cost of code ranges.

To overcome the limits of pure statistical sampling, one has to instrument the program code to be profiled. Several instrumentation methods exist: source modification, compiler injected instrumentation, binary rewriting to get an instrumented version of an executable, and binary translation at runtime. Instrumentation adds code to increment counters at function entry/exit, reading hardware performance counters, or even simulate hardware to get synthetic event counts. To minimize measurement overhead, only small action is performed[5]. When synthetic event counters are enough for the profiling result quality to be achieved, hardware simulation allows for runtime instrumentation with its ease of use.

3   Simulation Based Profiling

Runtime instrumentation can dramatically increase execution time such that time measurements become useless. However, it is adequate for driving hardware simulations. In our profiling tool, we use the CPU emulator *Valgrind* [18] as a runtime instrumentation framework. The instrumentation drives the cache

---

[4] For the results of statistical sampling to be meaningful, the distribution of every n-th event occurring over the code range of a program should be the same as the distribution of *every* event of this type.

[5] GProf [8] instrumentation still can have an overhead of 100%. Ephemeral Instrumentation [21], can keep instrumentation overhead smaller.

simulation engine, which is largely based on the cache simulator found in Valgrind itself: calls to the simulator are inserted on every instruction fetch, data read, and data write. Separate counters for each original instruction are incremented when an event occurs in the simulation of the simple, two-level cache hierarchy. The cache model enables the user to understand the numbers easily, but it has drawbacks compared to reality: it looks at memory accesses from user-level only, simulates a single cache, and assumes a processor without an out-of-order engine and without speculation. Despite of this, the synthetic event counters often come close to numbers of actual hardware performance counters [17]. We note that the simulation is not able to predict consumed wall clock time, as this would need a detailed simulation of the microarchitecture.

Our addition to the simulation is two-folded: First, multiple counters are maintained even for the same instruction, depending on the current thread ID or the call chain leading to this instruction. Thus, we support profiles per threads in multi-threaded code, and more fine-granular analysis by event attribution to CCTs. Second, we introduce the ability for construction of the dynamic call graph of the executed program. For every call site in the program, the list of target addresses called from that site is noted, and for each of these call arcs, the cost spent inside the called function. Optionally, this is even done for every context of a call site, where the context includes the current thread ID and call chain leading to this call site. To be able to provide the recent portion of a call chain with a given length, we maintain our own call stack for every thread and signal handler. Fortunately, thread switches can be trapped with Valgrind (using its own implementation of the POSIX Thread API). To be able to reliably trace a call stack, we always have to watch the stack pointer not only on detection of a CALL or RET instruction: most implementations of exception support in different languages (such as C++), and the C runtime functions setjmp/longjmp write the stack pointer directly. Function calls to shared libraries usually are resolved by the dynamic linker the first time they take place. After resolution, the linker directly jumps to the real function. As the notion of a jump between functions is not allowed in a call graph, we pretend two independent calls from the original call site.

4   Visualization of Profiling Data

Most profiling tools provide a post-processing command to output a text file with a list of functions sorted by cost and annotated source. While this is useful, it makes it difficult to get an overview of the performance characteristics of an application without being overwhelmed by the data measured. A GUI should enable convenient browsing in the code. Requirements are

- Zooming from coarse to fine granular presentation of profiling data, starting from event relation to all functions in a shared library, a source file or a C++ class, and going down to relation to loop structure, source code lines and machine instructions.

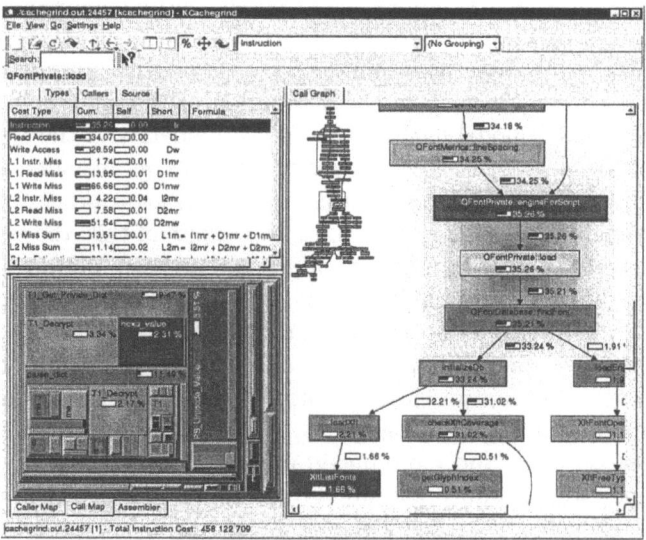

Fig. 2. Cost Types (Top), Call Graph (Right) and Call TreeMap (Bottom)

- Support for arbitrary and user specified, derived event types, including support for profile data produced by different profiling tools. Additionally, it is important to be able to specify and browse according to derived event types, which have to be calculated on the fly from the given ones (e.g. MFLOPS).
- Visualization of metrics that makes it easy to spot performance problems.
- Possibility to combine and compare multiple profile measurements produced by different tools.

Even for the simplest program, the runtime linker, the setup of the environment for the C runtime library, and implementation of high-level print functions via low-level buffering are involved. Thus, only a fraction of a call graph should be presented at once. Still, it is desirable to show multiple levels of a call chain. As a function can be part of multiple call chains, we eventually have to estimate the part of cost which happens to be spent in the given chain. This is only needed when context information, i.e. an attributed CCT, is not available. Our call graph view starts from a selected function, and adds callers and callees until costs of these functions are very small compared to the cost of the selected function. Because the functions in the call graph view can be selected by a mouse click, the user can quickly browse up or down a call chain, only concentrating on the part which exposes most of the event costs. Another view uses a TreeMap visualization [20]. Functions are represented by rectangles whose area is proportional to the inclusive cost. This allows for functions to be fully drawn inside their caller. Therefore, only functions which are interesting for performance analysis are visible. In a TreeMap, one function can appear multiple times. Again, costs of functions have to be split up based on estimation. Figure 2 shows a screenshot with both the call graph and TreeMap visualization. Additionally, on the top

Table 2. Simulation and profiling results for the 3D multigrid code

	Simulation			Real Measurement		
	Instr. exec.	L2 Misses	Runtime	Instr. retired	L2 Lines In	Runtime
Standard	11 879 M	751 421 K	1 865 s	11 879 M	777 131 K	40.4 s
Optimized	11 666 M	361 336 K	1 798 s	11 666 M	383 609 K	27.1 s

left the list of available cost types is shown, produced by the cache simulator. As "Ir" (instructions executed) is selected as cost type, these costs are the base for the graphical drawings.

To be able to show both annotated source and machine code[6], we rely on debug information generated by the compiler. Our tool parses this information and includes it in the profile data file. Thus, the tool suite is independent from the programming language. Additionally, our profiling tool can collect statistics regarding (conditional) jumps, enabling jumps visualized as arrows in the annotation view.

5   Application Example and Results

In our DiME project, research on cache optimization strategies for memory-intensive, iterative numerical code is carried out. The following experiment refers to a standard and to a cache-optimized implementation of multigrid V(2,2) cycles, involving variable 7-point stencils on a regular 3D grid with 129^3 nodes. The cache-efficient version is based on optimizations such as array padding and loop blocking [11]. Table 2 shows simulated and real events obtained on a Pentium-M with 1.4 GHz with corresponding wall clock runtimes. In both cases, the advantage of cache optimization is obvious by effectively reducing the number of cache misses/cache line loads by 50%. Reduction of runtime gives similar figures.

Simulation runtimes show that the slowdown of runtime instrumentation and cache simulation is quite acceptable.

6   Related Work

The most popular profiling tool on UNIX are prof/gprof [8]: a profile run needs the executable to be instrumented by the GCC compiler, which inserts code to get call arcs and call counts, and enables time based statistical sampling. Measurements inside of shared library code is not supported. GProf has to approximate the inclusive time by assuming that time spent inside a function grows linearly with the number of function calls. Tools taking advantage of hardware performance counters for event based sampling are available by most processor vendors (e.g. DCPI for Alpha processors [2] or VTune for Intel CPUs). Most of them relate event counters to code, but Sun's developer tools also allow data

---

[6] For machine code annotation, the standard GNU disassembler utility 'objdump' is used.

structure related measurements since recently [10]. OProfile is a free alternative for Linux [12]. PAPI [4] is a library for reading hardware performance counters. TAU [16] is a general performance analysis tool framework for parallel code, using automated source instrumentation.

For instrumentation purpose, binary rewriting of executables is possible with ATOM [7]. DynInst [15] allows for insertion of custom code snippets even into running processes. Regarding runtime instrumentation, the Shade tool [6] for SPARC inspired development of Valgrind. Sophisticated hardware simulators are MICA [9] and RSIM [19]. Using simulation, MemSpy [13] analyses memory access behavior related to data structures, and SIP [3] gives special metrics for spatial locality. A profile visualization with a web front end is HPCView [14].

7   Conclusions and Future Work

In this paper, we presented a tool suite for profiling of sequential code based on cache simulation and runtime instrumentation, and a graphical visualization front-end allowing fast browsing and recognition of bottlenecks. These tools have already been used successfully, e.g. in projects at our lab as well as in other labs in Germany, and in the open-source community.

Improvement of the simulation includes data structure based relation, even for dynamic allocated memory and stack accesses. This is feasible with the run-time instrumentation approach. In order to avoid the measured data to become huge, relation to data types seems necessary. For large arrays, relation to address differences of accesses promises to be interesting. Suited visualizations are needed. Other directions for research are more sophisticated metrics to easily detect problems regarding spatial or temporal locality. LRU stack distances of memory accesses [5] are especially useful as they are independent on cache sizes. For the visualization tool, we want to support combination of simulated and real measurement: e.g. flat profiles from event based sampling with hardware performance counters can be enriched with call graphs got from simulation, and thus, inclusive cost can be estimated. In addition, we recognize the need to further look into differences between simulation and reality.

Acknowledgement. We would like to thank Julian Seward for his excellent runtime instrumentation framework, and Nick Nethercote for the cache simulator we based our profiling tool on.

References

1. G. Ammons, T. Ball, and J. R. Larus. Exploiting Hardware Performance Counters with Flow and Context Sensitive Profiling. In *Proceedings of PLDI '97*, June 1997.
2. J. M. Anderson, L. M. Berc, J. Dean, et al. Continuous Profiling: Where Have All the Cycles Gone? *ACM Transactions on Computer Systems*, 15(4):357–390, November 1997.

3. E. Berg and E. Hagersten. SIP: Performance Tuning through Source Code Interdependence. In *Proceedings of the 8th International Euro-Par Conference (Euro-Par 2002)*, pages 177–186, Paderborn, Germany, August 2002.
4. S. Browne, J. Dongarra, N. Garner, G. Ho, and P. Mucci. A portable programming interface for performance evaluation on modern processors. *The International Journal of High Performance Computing Applications*, 14(3):189–204, Fall 2000.
5. G. C. Cascaval. *Compile-time Performance Prediction of Scientific Programs*. PhD thesis, University of Illinois at Urbana-Champaign, August 2000.
6. B. Cmelik and D. Keppel. Shade: A Fast Instruction Set Simulator for Execution Profiling. In *SIGMETRICS*, Nashville, TN, US, 1994.
7. A. Eustace and A. Srivastava. ATOM: A Flexible Interface for Building High Performance Program Analysis Tools, 1994.
8. S. Graham, P. Kessler, and M. McKusick. GProf: A Call Graph Execution Profiler. In *SIGPLAN: Symposium on Compiler Construction*, pages 120–126, 1982.
9. H. C. Hsiao and C. T. King. MICA: A Memory and Interconnect Simulation Environment for Cache-based Architectures. In *Proceedings of the 33rd IEEE Annual Simulation Symposium (SS 2000)*, pages 317–325, April 2000.
10. M. Itzkowitz, B. J. N. Wylie, Ch. Aoki, and N. Kosche. Memory profiling using hardware counters. In *Proceedings of Supercomputing 2003*, November 2003.
11. M. Kowarschik, U. Rüde, N. Thürey, and C. Weiß. Performance Optimization of 3D Multigrid on Hierarchical Memory Architectures. In *Proc. of the 6th Int. Conf. on Applied Parallel Computing (PARA 2002)*, volume 2367 of *Lecture Notes in Computer Science*, pages 307–316, Espoo, Finland, June 2002. Springer.
12. J. Levon. OProfile, a system-wide profiler for Linux systems.
13. M. Martonosi, A. Gupta, and T. E. Anderson. Memspy: Analyzing memory system bottlenecks in programs. In *Measurement and Modeling of Computer Systems*, pages 1–12, 1992.
14. J. Mellor-Crummey, R. Fowler, and D. Whalley. Tools for Application-Oriented Performance Tuning. In *Proceedings of 15th ACM International Conference on Supercomputing*, Italy, June 2001.
15. B. P. Miller, M. D. Callaghan, J. M. Cargille, et al. The Paradyn Parallel Performance Measurement Tool. *IEEE Computer*, 28(11):37–46, November 1995.
16. B. Mohr, A. Malony, and J. Cuny. *Parallel Programming using C++*, chapter TAU. G. Wilson, editor, M.I.T. Press, 1996.
17. N. Nethercote and A. Mycroft. The Cache Behaviour of Large Lazy Functional Programs on Stock Hardware. In *Proceedings of the ACM SIGPLAN Workshop on Memory System Performance (MSP 2002)*, Berlin, Germany, July 2002.
18. N. Nethercote and J. Seward. Valgrind: A Program Supervision Framework. In *Proceedings of the Third Workshop on Runtime Verification (RV'03)*, Boulder, Colorado, USA, July 2003. Available at http://valgrind.kde.org/.
19. V. S. Pai, P. Ranganathan, S. V. Adve, and T. Harton. An Evaluation of Memory Consistency Models for Shared-Memory Systems with ILP Processors. In *Proceedings of the Seventh International Conference on Architectural Support for Programming Languages and Operating Systems*, pages 12–23, October 1996.
20. B. Shneiderman. Treemaps for space-constrained visualization of hierarchies. http://www.cs.umd.edu/hcil/treemap-history/index.shtml.
21. O. Traub, S. Schechter, and M. D. Smith. Ephemeral instrumentation for lightweight program profiling. In *Proceedings of PLDI '00*, 2000.

Platform-Independent Cache Optimization by Pinpointing Low-Locality Reuse

Kristof Beyls and Erik H. D'Hollander

Departement of Electronics and Information Systems
Ghent University
Sint-Pietersnieuwstraat 41, 9000 Ghent, Belgium
{kristof.beyls,erik.dhollander}@elis.ugent.be

Abstract. For many applications, cache misses are the primary performance bottleneck. Even though much research has been performed on automatically optimizing cache behavior at the hardware and the compiler level, many program executions remain dominated by cache misses. Therefore, we propose to let the programmer optimize, who has a better high-level program overview, needed to resolve many cache problems. In order to assist the programmer, a visualization of memory accesses with poor locality is developed. The aim is to indicate causes of cache misses independent of actual cache parameters such as associativity or size. In that way, the programmer is steered towards platform-independent locality optimizations. The visualization was applied to three programs from the SPEC2000 benchmarks. After optimizing the source code based on the visualization, an average speedup of 3.06 was obtained on different platforms with Athlon, Itanium and Alpha processors; indicating the feasibility of platform-independent cache optimizations.

1 Introduction

On current processors, the execution time of many programs is dominated by processor stall time during cache misses. In figure 1, the execution time of the SPEC2000 programs is categorized into data cache miss, instruction cache miss, branch misprediction and useful execution. On average, almost 50% of the execution time, the processor is stalled due to data cache misses.

Several studies on different benchmark suites have shown that *capacity misses* are the dominant category of misses[7,2,8,4]. Therefore, we focus on reducing capacity misses.

Cache misses can generally be resolved at three different levels: the hardware level, the compiler level or the algorithm level. At the hardware level, capacity misses can only be eliminated by increasing the cache size[7], which makes it slower. Previous research indicates that state-of-the-art compiler optimizations can only reduce about 1% of all capacity misses[2]. This shows that capacity miss optimization is hard to automatize, and should be performed by the programmer. However, cache behavior is not obvious from the source code, so a tool is needed to help the programmer pin-point the causes of cache bottlenecks. In contrast

M. Bubak et al. (Eds.): ICCS 2004, LNCS 3038, pp. 448–455, 2004.
© Springer-Verlag Berlin Heidelberg 2004

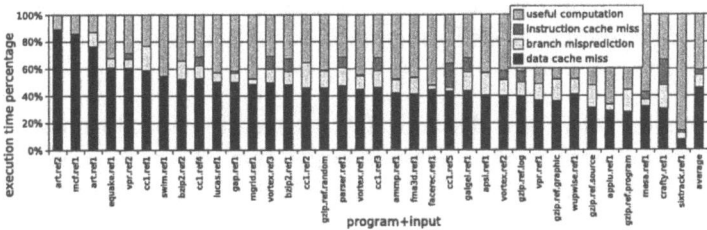

Fig. 1. Breakdown of execution time for the SPEC2000 programs, on a 733Mhz Itanium1 processor. The programs are compiled with Intel's compiler, using the highest level of feedback-driven optimization.

to earlier tools[1,3,6,10,11,12,14,15], we aim at steering the programmer towards platform-independent cache optimizations.

The key observation is that capacity misses are caused by low-locality accesses to data that has been used before. Since the data has already been accessed before, it has been in the cache before. However, due to low locality, it's highly unlikely that the cache retains the data until the reuse. By indicating those reuses, the programmer is pointed to the accesses where cache misses can be eliminated by improving locality.

Both temporal and spatial locality are quantified, by locality metrics, as discussed in section 2. Section 3 presents how the metrics are graphically presented to the programmer. The visualization has been applied to the three programs in SPEC2000 with the largest cache miss bottleneck: **art**, **mcf** and **equake** (see fig. 1). A number of rather small source code optimizations were performed, which resulted in platform-independent improvements. The average speedup of these programs on Athlon, Alpha and Itanium processors was 3.06, with a maximum of 11.23. This is discussed in more detail in section 4. In section 5, concluding remarks follow.

2   Locality Metrics

Capacity misses are generated by reuses with low locality. Temporal locality is measured by *reuse distance*, and spatial locality is measured by *cache line utilization*.

Definition 1. *A* **memory reference** *corresponds to a read or a write in the source code, while a particular execution of that read or write at runtime is a* **memory access***. A* **reuse pair** $\langle a_1, a_2 \rangle$ *is a pair of accesses in a memory access stream, which touch the same location, without intermediate accesses to that location. The* **reuse distance** *of a reuse pair* $\langle a_1, a_2 \rangle$ *is the number of unique memory locations accessed between* a_1 *and* a_2. *A* **memory line** *is a cache-line-sized block of contiguous memory, containing the bytes that are mapped into a single cache line. The* **memory line utilization** *of an access a to memory line l is the fraction of l which is used before l is evicted from the cache.* □

Lemma 1. *In a fully associative LRU cache with n lines, a reuse pair $\langle a_1, a_2 \rangle$ with reuse distance $d < n$ will generate a hit at access a_2. If $d \geq n$, a_2 misses the cache.[2]*   □

The accesses with reuse distance $d \geq n$ generate capacity misses. When the backward reuse distance of access a is larger than the cache size, a cache miss results and memory line l is fetched into the cache. If the memory line utilization of a is less than 100%, not all the bytes in l are used, during that stay in the cache. Therefore, at access a, some useless bytes in l were fetched, and the potential benefit of fetching a complete memory line was not fully exploited. The memory line utilization shows how much spatial locality can be improved.

3   Measurement and Visualization

3.1   Instrumentation and Locality Measurement

In order to measure the reuse distance and the memory line utilization, the program is instrumented to obtain the memory access trace. The ORC-compiler[13] was extended, so that for every memory access instruction, a function call is inserted. The accessed memory address, the size of the accessed data and the instruction generating the memory access are given to the function as parameters. The instrumented program is linked with a library that implements the function which calculates the reuse distance and the memory line utilization. The reuse distance is calculated in constant time, in a similar way as described in [9]. For every pair of instructions (r_1, r_2), the distribution of the reuse distance of reuse pairs $\langle a_1, a_2 \rangle$ where a_1 is generated by r_1 and a_2 is generated by r_2 is recorded.

The memory line utilization of a reuse pair is measured by choosing a fixed size CS_{min} as the minimal cache size of interest. On every access l, it is recorded which bytes were used. When line l is evicted from the cache of size CS_{min}, the fraction of bytes accessed during that stay of l in cache of size CS_{min} is recorded. In the experiments in section 4, $CS_{min} = 2^8$ cache lines of 64 bytes each = 16 KB. At the end of the program execution, the recorded reuse distance distributions, together with the average memory line utilizations are written to disk.

A large overhead can result from instrumenting every memory access. The overhead is reduced by sampling, so that reuses are measured in bursts of 20 million consecutive accesses, while the next 180 million accesses are skipped. In our experiments, a slowdown between 15 and 25 was measured The instrumented binary consumes about twice as much memory as the original, due to bookkeeping needed for reuse distance calculation.

3.2   Visualization

Lemma 1 indicates that only the reuses whose distance is larger than the cache size generate capacity misses. Therefore, only the reuse pairs with a long reuse distance (=low locality) are shown to the programmer. Furthermore, in order to guide the programmer to the most important low-locality reuse places in the

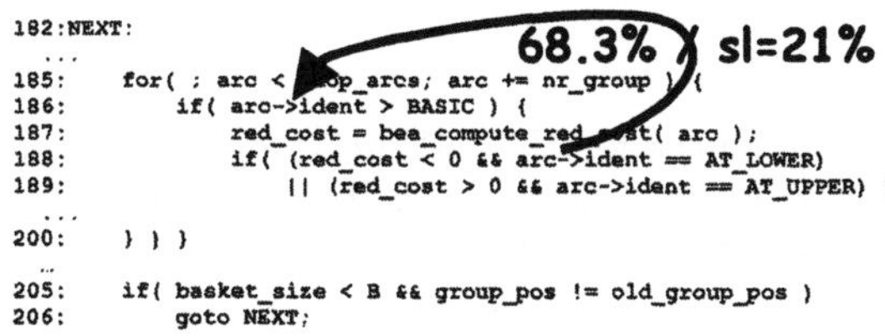

```
182:NEXT:
    ...
185:    for( ; arc <    p_arcs; arc += nr_group )(      68.3%    sl=21%
186:        if( arc->ident > BASIC ) {
187:            red_cost = bea_compute_red     t( arc );
188:            if( (red_cost < 0 && arc->ident == AT_LOWER)
189:                || (red_cost > 0 && arc->ident == AT_UPPER) ) (

200:    } } }
    ...
205:    if( basket_size < B && group_pos != old_group_pos )
206:        goto NEXT;
```

Fig. 2. A zoom in on the major long reuse distance in 181.mcf. A single pair of instructions produce 68.3% of the long reuse pairs. The second number (sl=21%) indicates that the cache-missing instruction (on line 186) has a memory line utilization of 21%.

program, only the instruction pairs generating at least 1% of the long distances are visualized. The visualization shows these pairs as arrows drawn on top of the source code. A label next to the arrow shows how many percent of the long reuse distances were generated by that reuse pair. Furthermore, the label also indicates the memory line utilization of the cache missing accesses generated by that reuse pair. A simple example is shown in fig. 2.

3.3   Program Optimization

The locality measurement and visualization is performed automatically. Based on the visualization, the capacity misses can be reduced by the programmer. There are four basic ways in which the capacity misses, or their slowdown effect, can be reduced.

1. *Eliminate the memory accesses with poor locality* altogether.
2. Reduce the distance between use and reuse for long reuse distances, so that it becomes smaller than the cache size. This can be done by reordering computations (and memory accesses), so that *higher temporal locality* is achieved.
3. *Increase the spatial locality* of accesses with poor spatial locality. This is most easily done by rearranging the data layout.
4. If neither of the three previous methods are applicable, it might still be possible to improve the program execution speed by *hiding their latency with independent parallel computations*. The most well-known technique in this class is prefetching.

4   Experiments

In order to evaluate the benefit of visualizing low-locality reuses, the three programs from SPEC2000 with the highest cache bottlenecks were considered: 181.mcf, 179.art and 183.equake (see fig. 1). Below, for each of the programs, the visualization of the main cache bottlenecks is shown. Also, it is discussed how the programs were optimized.

4.1   Mcf

The main long reuse distances for the 181.mcf program are shown in figure 2. The figure shows that about 68% of the capacity misses are generated by a single load instruction on line 187. The best way to solve those capacity misses would be to shorten the distance between the use and the reuse. However, the reuses of arc-objects occur between different invocations of the displayed function. So, bringing use and reuse together needs a thorough understanding of the complete program, which we do not have, since we didn't write the program ourselves. A second way would be to increase the spatial locality from 21% to a higher percentage. In order to optimize the spatial locality, the order of the fields of the arc-objects could be rearranged. However, this change leads to poorer spatial locality in other parts of the program, and overall, this restructuring does not lead to speedup. Therefore, we tried the fourth way to improve cache performance: inserting prefetch instructions to hide the miss penalty.

4.2   Art

The 179.art program performs image recognition by using a neural network. A bird's-eye view of the main long reuse distances in this program is shown in figure 3(a). Each node in the neural network is represented by a struct containing 9 double-fields, which are layed out in consecutive locations. The visualization shows low spatial locality. The code consists of 8 loops, each iterating over all neurons, but only accessing a small part of the 9 fields in each neuron. A simple data layout optimization resolves the spatial locality problems. Instead of storing complete neurons in a large array, i.e. an array of structures, the same field for all the neurons are stored consecutively in arrays, i.e. a structure of arrays. Besides this data layout optimization, also some of the 8 loops were fused, when the data dependencies allowed it and reuse distances were shortened by it.

4.3   Equake

The equake program simulates the propagation of elastic waves during an earthquake. A bird's-eye view on the major long reuse distance in this program is shown in figure 3(b). All the long reuse distances occur in the main simulation loop of the program which has the following structure:

- Loop for every time step (line 447–512):
  - Loop to perform a sparse matrix-vector multiplication. (line 455–491)
  - A number of loops to rescale a number of vectors. (line 493–507)

Most of the long reuse distances occur in the sparse matrix-vector multiplication, for the accesses to the sparse matrix K, declared as double*** K. The matrix is symmetric, and only the upper triangle is stored. An access to an element has the form K[Anext][i][j], leading to three loads. The number of memory accesses are reduced by redefining K as a single dimensional array and replacing

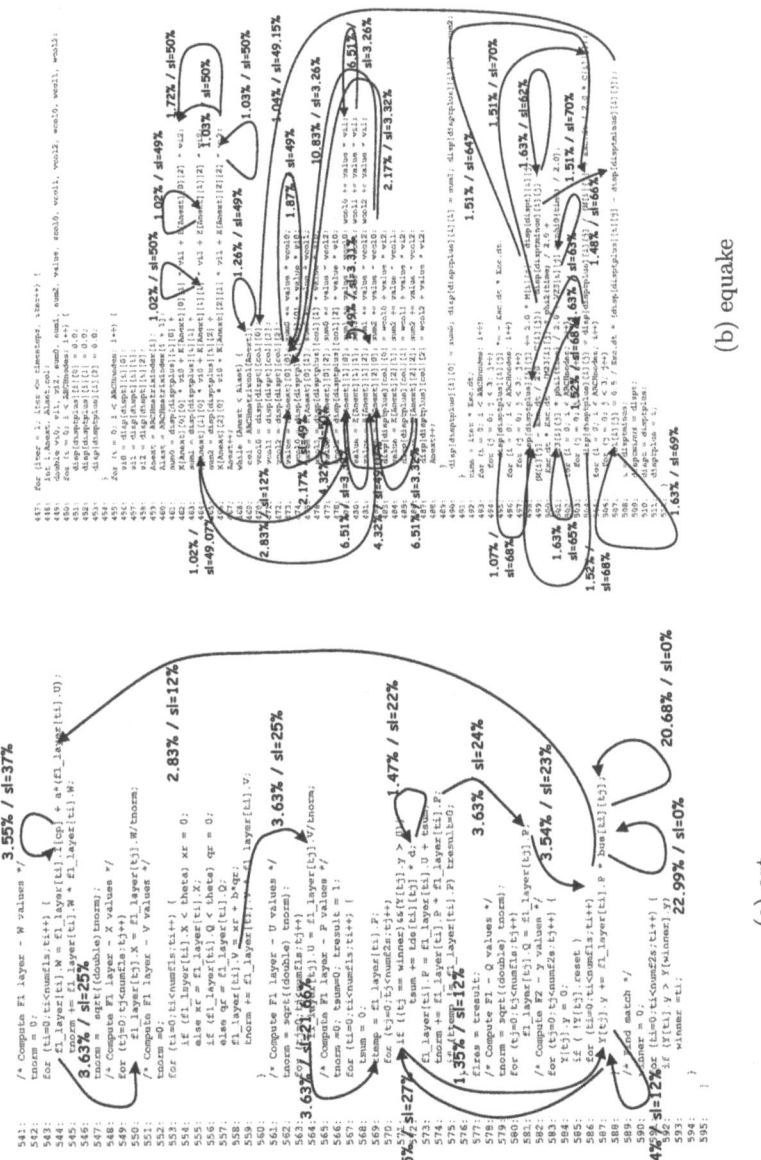

Fig. 3. A zoom-out view of the major long reuse distances in 179.art and 183.equake.

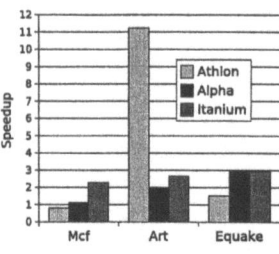
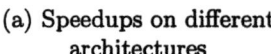
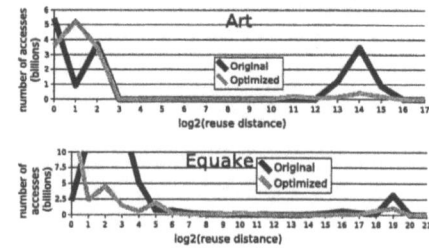

(a) Speedups on different
architectures

(b) Reuse distance distribution

Fig. 4. (a) Speedups and (b) Reuse distance distributions before and after optimization.

the accesses into K[Anext*N*9+i*3+j], leading to a single load instruction. Furthermore, after analyzing the code a little further, it shows that for most of the long reuse pairs, the use is in a given iteration of the time step loop and the reuse is in the next iteration. In order to bring the use and the reuse closer together, some kind of tiling transformation should be performed on the time-step loop (i.e. try to do computations for a number of consecutive time-steps on the set of array elements that are currently in the cache). All the sparse matrix elements were stored in memory, and not only the upper triangle, which allows to simplify the sparse code and remove some loop dependences. After this, it was possible to fuse the matrix-vector multiply loop with the vector rescaling loops, resulting in a single perfectly-nested loop. In order to tile this perfectly-nested loop, the structure of the sparse matrix needs to be taken into account, to figure out the real dependencies, which are only known at run-time. The technique described in [5] was used to perform a run-time calculation of a legal tiling.

4.4   Discussion

The original and optimized programs were compiled and executed on different platforms: an Athlon PC, an Alpha workstation and an Itanium server. For the Athlon and the Itanium, the Intel compiler was used. For the Alpha 21264, the Alpha compiler was used. All programs were compiled with the highest level of feedback-driven optimization. In figure 4(a), the speedups are presented, showing that most programs have a good speedup on most processors. Figure 3(b) shows that the long reuse distances have been effectively diminished in both art and equake. In mcf (not displayed), the reuse distances were not diminished, since only prefetching was applied. Only for mcf on the Athlon, a slow-down is measured, probably because the hardware-prefetcher in the Athlon interferes with the software prefetching.

5   Conclusion

On current processors, on average about half of the execution time is caused by cache misses, after applying hardware and compiler optimizations, making it the number one performance bottleneck. Therefore, the programmer needs to improve the cache behavior. A method is proposed which measures and visualizes the reuses with poor locality, resulting in cache misses. The visualization hints the programmer into portable cache optimizations.

The effectiveness of the visualization was tested by applying it to the three programs with the highest cache bottleneck in SPEC2000. Based on the visualization, a number of rather simple source code optimizations were applied. The speedups of the optimized programs were measured on CISC, RISC and EPIC processors, with different underlying cache architectures. Even after full compiler optimization, the visualization enabled the programmer to attain speedups of up to 11.23 on a single processor (see fig. 4). Furthermore, the applied optimizations lead to a consistent speedup in all but one case.

References

1. E. Berg and E. Hagersten. Sip: Performance tuning through source code interdependence. In *Euro-Par'02*, pages 177–186, 2002.
2. K. Beyls and E. H. D'Hollander. Reuse distance as a metric for cache behavior. In *Proceedings of PDCS'01*, pages 617–662, Aug 2001.
3. R. Bosch and C. S. et al. Rivet: A flexible environment for computer systems visualization. *Computer Graphics-US*, 34(1):68–73, Feb. 2000.
4. J. F. Cantin and M. D. Hill. Cache performance for selected SPEC CPU2000 benchmarks. *Computer Architecture News (CAN)*, September 2001.
5. C. C. Douglas and J. H. et al. Cache optimization for structured and unstructured grid multigrid. *Elect. Trans. Numer. Anal.*, 10:25–40, 2000.
6. A. Goldberg and J. Hennessy. Mtool: An integrated system for performance debugging shared memory multiprocessor applications. *IEEE Transactions on Parallel and Distributed Systems*, 4(1):28–40, 1993.
7. J. L. Hennessy and D. A. Patterson. *Computer Architecture – A Quantitative Approach*. Morgan Kaufmann Publishers, third edition, 2002.
8. M. D. Hill and A. J. Smith. Evaluating associativity in CPU caches. *IEEE Transactions on Computers*, 38(12):1612–1630, Dec. 1989.
9. Y. H. Kim, M. D. Hill, and D. A. Wood. Implementing stack simulation for highly-associative memories. In *ACM SIGMETRICS conference*, pages 212–213, 1991.
10. A. R. Lebeck and D. A. Wood. Cache profiling and the SPEC benchmarks: A case study. *IEEE Computer*, 27(10):15–26, 1994.
11. M. Martonosi, A. Gupta, and T. Anderson. Tuning memory performance in sequential and parallel programs. *IEEE Computer*, April 1995.
12. J. Mellor-Crummey and R. F. et al. HPCView: a tool for top-down analysis of node performance. *The Journal of Supercomputing*, 23:81–104, 2002.
13. Open research compiler. http://sourceforge.net/projects/ipf-orc.
14. E. Vanderdeijl, O. Temam, E. Granston, and G. Kanbier. The cache visualization tool. *IEEE Computer*, 30(7):71, 1997.
15. Y. Yu, K. Beyls, and E. D'Hollander. Visualizing the impact of cache on the program execution. In *Information Visualization 2001*, pages 336–341, 2001.

Teuta: Tool Support for Performance Modeling of Distributed and Parallel Applications*

Thomas Fahringer[1], Sabri Pllana[2], and Johannes Testori[2]

[1] Institute for Computer Science, University of Innsbruck
Technikerstraße 25/7, 6020 Innsbruck, Austria
Thomas.Fahringer@uibk.ac.at
[2] Institute for Software Science, University of Vienna
Liechtensteinstraße 22, 1090 Vienna, Austria
{pllana,testori}@par.univie.ac.at

Abstract. In this paper we describe Teuta, which we have developed to provide tool support for the UML-based performance modeling of distributed and parallel applications. Teuta features include model checking and model traversing. Model checking is used to verify whether the model conforms to the UML specification. In addition, Teuta supports semantic model checking for the domain of high performance computing. For the generation of different model representations the model traversing is used. In addition, we present our methodology for automatic generation of the simulation model from the UML model of an application. This simulation model is used to evaluate the performance of the application. We demonstrate the usefulness of Teuta by modeling LAPW0, a distributed material science application.

Keywords: Distributed and Parallel Applications, Modeling and Simulation, Performance, UML

1 Introduction

The high performance computing is usually used for solving complex problems in science and engineering. However, effective performance-oriented program development requires the programmer to understand the intricate details of the programming model, the parallel and distributed architecture, and the mapping of applications onto architectures. On the other hand, most performance tools [3, 4] provide little support at early application development stages when several designs and strategies are examined by an application programmer. We anticipate that a tool which supports the application programmer to graphically develop the performance model of application at an early development stage would help to influence design decisions without time-consuming modifications of the code of an already implemented application.

---

* The work described in this paper is supported in part by the Austrian Science Fund as part of Aurora Project under contract SFBF1104.

M. Bubak et al. (Eds.): ICCS 2004, LNCS 3038, pp. 456–463, 2004.
© Springer-Verlag Berlin Heidelberg 2004

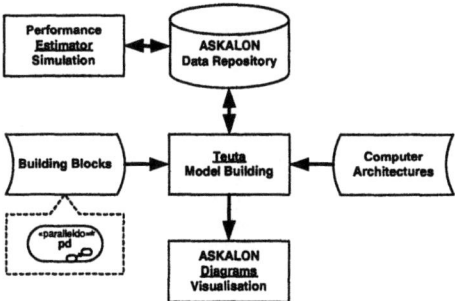

Fig. 1. Performance Prophet architecture

In this paper we give an overview of our tool *Teuta*, which supports the development of performance models for parallel and distributed applications based on the Unified Modeling Language (UML) [5]. The UML is de facto standard visual modeling language which is a general purpose, broadly applicable, tool supported, industry standardized modeling language. However, by using only the core UML, some shared memory and message passing concepts can not be modeled in an adequate manner. In order to overcome this issue we have developed an extension of UML for the domain of high performance computing [6].

Teuta is an integral part of our performance prediction system *Performance Prophet* [1]. Figure 1 depicts the architecture of Performance Prophet. The role of Teuta is to assist the user to develop the model for an application by composing existing *building blocks*. The application model is enriched with cost functions. Thereafter, Teuta transforms the annotated model to an intermediate form based on which the *Performance Estimator* of Performance Prophet evaluates the performance of the application on the computer *architecture* selected by user.

Teuta features include model checking and model traversing. Model checking is used to verify whether the model conforms to the UML specification. In addition, Teuta supports semantic model checking for the domain of High Performance Computing. Teuta makes use of model traversing for the generation of different model representations (such as XML, C++).

Furthermore, in this paper we present our methodology for automatic generation of the simulation model from the UML model of an application. This simulation model is used to evaluate the performance of the application. We demonstrate the usefulness of Teuta and our methodology by modeling LAPW0, a distributed material science application.

The rest of this paper is organized as follows. An overview of Teuta is described in Section 2. Section 3 presents our methodology for the automatic generation of the simulation model based on the UML model of an application. A case study is described in Section 4. Finally, some concluding remarks are made and future work is outlined in Section 5.

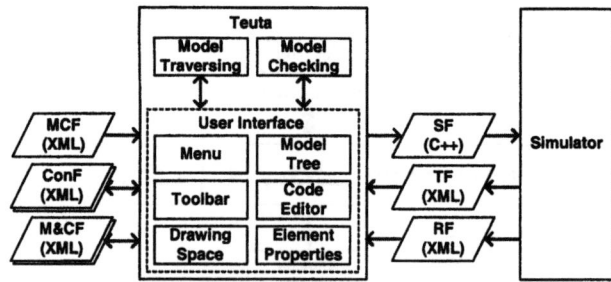

Fig. 2. The architecture of Teuta. Description of abbreviations: MCF - Model Checking File, ConF - Configuration Files, M&CF - Model and Construct Files, SF - Simulation File, TF - Trace File, RF - Result File.

2   An Overview of Teuta

Teuta[1] is a platform independent tool for UML-based modeling of parallel and distributed applications.

Because the application developers may work on various platforms, the tool should be able to run on various platforms as well. Therefore, we have used the Java language for the implementation of Teuta, based on the Model-View-Controller (MVC) paradigm [2]. MVC is a paradigm that enforces the separation between the user interface and the rest of the application. In this paradigm, the *model* represents data and the core functionality of the application. The *view* is the user interface. The *controller* interprets and delegates the interactions with the *view* (for instance button clicks) to the *model*, which performs the corresponding actions. The advantage of MVC is that it allows creation of independent *views* (user interfaces) that correspond to a single *model* of an application.

It is difficult to foresee all the types of the building blocks that the user might want to use to model his application. Therefore, we have included a basic set of building blocks, which are described in [6,7], and made it easy to extend the tool with the new building blocks. Teuta may be extended with new building blocks by modifying a set of XML-based configuration files.

Figure 2 shows the architecture of Teuta, which consists of three main parts: (i) Model checking; (ii) Model traversing; (iii) Graphical user interface (GUI). We will describe each of these parts in the subsequent sections.

Element *M&CF* in Figure 2 indicates XML files which contain application model and constructs. The application model is stored in a file, which contains all the information which is required to display the model in the editor. All the diagrams and modeling elements with their properties and geometric information are stored in this file. A construct is a set of modeling elements. For instance a set of modeling elements that represents a loop. Constructs may be used to simplify the development of the application model.

Element *ConF* in Figure 2 indicates XML files which are used for configuration of Teuta.

---

[1] Teuta (Tefta): Queen of the Illyrians (231-228 BC)

The communication between Teuta and the simulator is done via files *SF*, *TF*, and *RF* (see Fig. 2). Element *SF* indicates the C++ representation of the application performance model, which is generated by Teuta. Elements *TF* and *RF* represents the trace file and result file respectively, which are generated by the simulator.

Element *MCF* in Figure 2 indicates the XML file, which is used for model checking. The following Section describes the model checking in more detail.

2.1   Model Checking

This part of Teuta is responsible for the correctness of the model. The rules for model checking are specified by using our XML based Model Checking Language (MCL). The model checker gets the model description from an MCL file. This MCL file contains a list of available diagrams, modeling elements and the set of rules that defines how the elements may be interconnected.

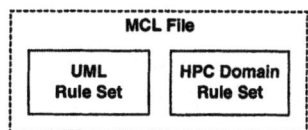

Fig. 3. Model checking rule sets

Based on the *UML Rule Set* the model checker verifies whether the application model conforms to the UML specification (see Fig 3). In addition, Teuta supports *semantic model checking* for the domain of High Performance Computing (HPC). The *HPC Domain Rule Set* specifies whether two modeling elements can be interconnected with each other, or nested one within another, based on their semantics. For instance, it is not allowed to place the modeling element *BARRIER* within the modeling element *CRITICAL*, because this would lead to the deadlock.

2.2   Model Traversing

This component of Teuta provides the possibility to walk through the model, visit each modeling element, and access its properties (for instance element name). We use the model traversing for the generation of various model representations (such as XML and C++).

The *Model Traversing* component of Teuta consists of three parts (i) the *Navigator*, (ii) the *Traverser*, (iii) and the *ContentHandler*.

The *Navigator* is responsible for interpreting the input, creating the corresponding object model and for the navigation in this object model.

The *Traverser* is responsible for traversing the object model created by the *Navigator*. It walks through the diagrams of the model by calling the *Navigator's* methods. Various types of traversing strategies can be implemented, such as depth-first or breadth-first.

The *ContentHandler* methods are called by the *Traverser* whenever the *Traverser* begins or finishes a model, or visits an element. It is responsible for the output of the traversing, for instance it can create another object model or write to a file in a specific format.

2.3   Graphical User Interface (GUI)

Figure 6(a) in Section 4 depicts Teuta GUI, which consists of: *menu, toolbar, drawing space, model tree, code editor,* and *element properties.* Because of the limited space we are not able to describe the graphical user interface of Teuta in detail. Therefore, in this section we describe only the support of Teuta for constructs. A construct is a set of modeling elements. The idea is to relinquish the user from trivial time consuming tasks, by automatic generation of constructs. We have identified two types of constructs: (i) simple constructs that are used frequently, such as loops; (ii) large constructs with regular structure, such as topologies of processes.

Figure 4 shows an example of the automatic generation of topology of processes. The user specifies the type (for instance 2D mesh) and parameters (for instance 4 rows, 4 columns) of the process topology (see Figure 4(a)). Based on this information Teuta generates the corresponding process topology (see Figure 4(b)). The process topology is represented by the UML CollaborationInstanceSet (see [7]). Without this support of Teuta, the user would spend a significant amount of time to create all the modeling elements and arrange them to represent the process topology.

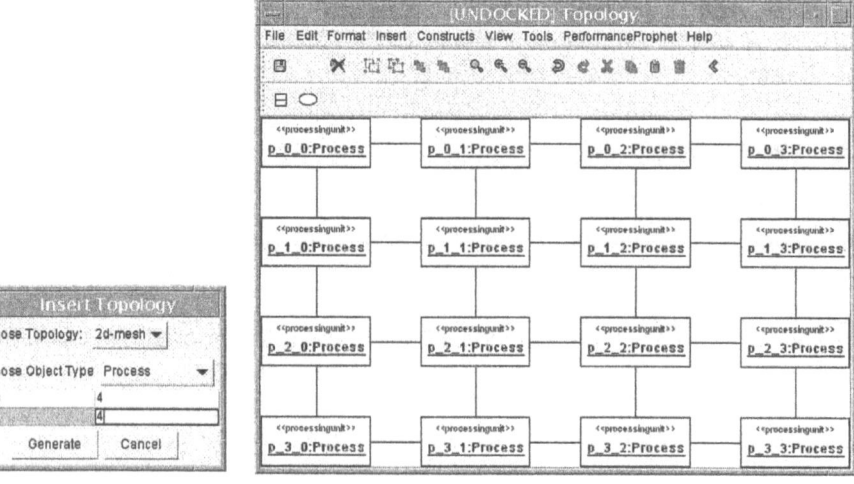

(a) Specification   (b) Topology of processes

Fig. 4. An example of the automatic generation of topology of processes

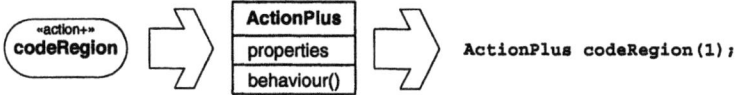

Fig. 5. From the UML model to the simulation model

3 The Development of the Simulation Model Based on the UML Model

Figure 5 depicts the process of transition from the model of the application represented by UML activity diagram, to the simulation model. The simulation model is represented in C++.

The UML element *activity* is stereotyped as *action+*. The UML stereotyping extension mechanism makes possible to make the semantics of a core modeling element more specific (see [6]). The element *action+* is used to represent a code region of an application. For this modeling element we have defined the class *ActionPlus*. The properties of the UML modeling element *action+* (for instance the element ID) are mapped to the *properties* of the class *ActionPlus*. The performance behavior of the element is defined in the method *behaviour()* of the class. This method is responsible for advancing the time in the simulation clock. This time is estimated either by a parameterized cost function or by simulating the behavior of the element. The name of the UML modeling element, in our example *codeRegion*, is mapped to the name of the instance of the class.

In the next section we present the performance modeling and evaluation of a real-world application.

4 Case Study: LAPW0

The objective of our case study is to examine whether the tool support described in this paper is sufficient to build a performance model for a real-world application. The application for our study LAPW0, which is a part of the WIEN2k package [9], was developed at Vienna University of Technology. The Linearized Augmented Plane Wave (LAPW) method is among the most accurate methods for performing electronic structure calculations for crystals. The code of LAPW0 Application is written by using FORTRAN90 and MPI.

The LAPW0 application consists of 100 file modules (a module is a file containing source code). The modeling procedure aims to identify the more relevant (from performance point of view) code regions. We call these code regions building blocks. A building block can be a sequence of computation steps, communication operations or input/output operations. In order to evaluate the execution time of the relevant code regions of LAPW0 application, we have instrumented these code regions and measured their corresponding execution times by using SCALEA [11], which is a performance analysis tool for distributed and parallel applications.

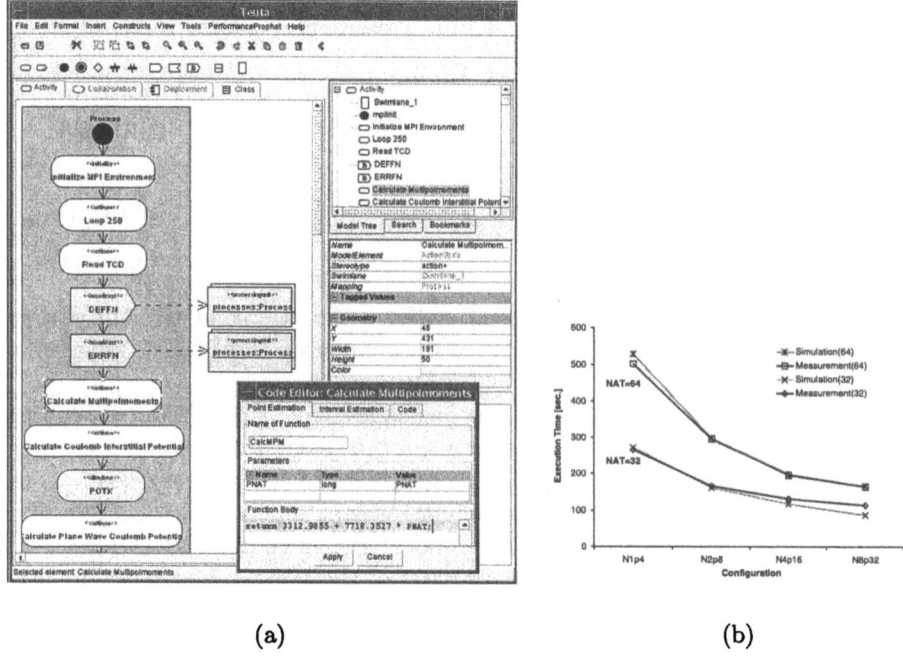

(a)   (b)

Fig. 6. Performance modeling and evaluation of the LAPW0 application

Figure 6(a) depicts the model of LAPW0, which is developed with Teuta. Due to the size of the LAPW0 model, we can see only a fragment of the UML activity diagram within the drawing space of Teuta. On the right hand side of Figure 6(a) is shown how the model of LAPW0 is enriched with cost functions by using Teuta *code editor*. A cost function models the execution time of a code region.

In order to evaluate the model of LAPW0 we have transformed the high level UML model of LAPW0 into a simulation model. Teuta automatically generates the corresponding C++ representation, which is used as input for the *Performance Estimator* (see Figure 1). The Performance Estimator includes a simulator, which models the behavior of cluster architectures. CSIM [10] serves as a simulation engine for the Performance Estimator.

Figure 6(b) shows the simulation and measurement results for two problem sizes and four machine sizes. The problem size is determined by the parameter NAT, which represents the number of atoms in a unit of the material. The machine size is determined by the number of nodes of the cluster architecture. Each node of the cluster has four CPU's. One process of the LAPW0 application is mapped to one CPU of the cluster architecture. The simulation model is validated by comparing the simulation results with the measurement results. We consider that this simulation model provides the performance prediction results with the accuracy which would be sufficient to compare various designs of the application.

5   Conclusions and Future Work

In this paper we have described Teuta, which provides the tool support for the UML-based performance modeling of parallel and distributed applications. We have demonstrated the usefulness of Teuta by modeling LAPW0, which is a distributed material science application. Based on the high level UML model of LAPW0 application the simulation model is automatically generated by Teuta. The simulation model is validated by comparing the simulation results with the measurement results. We consider that this simulation model provides the performance prediction results with the accuracy which would be sufficient to compare various designs of the application.

Currently, we are extending Teuta for modeling Grid [8] applications.

References

1. T. Fahringer and S. Pllana. Performance Prophet. University of Vienna, Institute for Software Science. Available online: http://www.par.univie.ac.at/project/prophet.
2. G. Krasner and S. Pope. A cookbook for using the Model-View-Controller interface paradigm. *Journal of Object-Oriented Programming*, 1(3):26–49, 1988.
3. D. Kvasnicka, H. Hlavacs, and C. Ueberhuber. Simulating Parallel Program Performance with CLUE. In *International Symposium on Performance Evaluation of Computer and Telecommunication Systems (SPECTS)*, pages 140–149, Orlando, Florida, USA, July 2001. The Society for Modeling and Simulation International.
4. N. Mazzocca, M. Rak, and U. Villano. The Transition from a PVM Program Simulator to a Heterogeneous System Simulator: The HeSSE Project. In *7th European PVM/MPI*, volume 1908 of *Lecture Notes in Computer Science*, Balatonfüred, Hungary, September 2000. Springer-Verlag.
5. OMG. Unified Modeling Language Specification. http://www.omg.org, March 2003.
6. S. Pllana and T. Fahringer. On Customizing the UML for Modeling Performance-Oriented Applications. In *<<UML>> 2002, "Model Engineering, Concepts and Tools", LNCS 2460, Dresden, Germany*. Springer-Verlag, October 2002.
7. S. Pllana and T. Fahringer. UML Based Modeling of Performance Oriented Parallel and Distributed Applications. In *Proceedings of the 2002 Winter Simulation Conference*, San Diego, California, USA, December 2002. IEEE.
8. S. Pllana, T. Fahringer, J. Testori, S. Benkner, and I. Brandic. Towards an UML Based Graphical Representation of Grid Workflow Applications. In *The 2nd European Across Grids Conference*, Nicosia, Cyprus, January 2004. Springer-Verlag.
9. K. Schwarz, P. Blaha, and G. Madsen. Electronic structure calculations of solids using the WIEN2k package for material sciences. *Computer Physics Communications*, 147:71–76, 2002.
10. Herb Schwetman. Model-based systems analysis using CSIM18. In *Winter Simulation Conference*, pages 309–314. IEEE Computer Society Press, 1998.
11. Hong-Linh Truong and Thomas Fahringer. SCALEA: A Performance Analysis Tool for Distributed and Parallel Program. In *8th International Europar Conference(EuroPar 2002)*, Lecture Notes in Computer Science, Paderborn, Germany, August 2002. Springer-Verlag.

MPI Application Development Using the Analysis Tool MARMOT

Bettina Krammer, Matthias S. Müller, and Michael M. Resch

High Performance Computing Center Stuttgart
Allmandring 30, D-70550 Stuttgart, Germany
{krammer,mueller,resch}@hlrs.de

Abstract. The Message Passing Interface (MPI) is widely used to write parallel programs using message passing. Due to the complexity of parallel programming there is a need for tools supporting the development process. There are many situations where incorrect usage of MPI by the application programmer can automatically be detected. Examples are the introduction of irreproducibility, deadlocks and incorrect management of resources like communicators, groups, datatypes and operators. We also describe the tool MARMOT that implements some of these tests. Finally we describe our experiences with three applications of the CrossGrid project regarding the usability and performance of this tool.

1 Introduction

The Message Passing Interface (MPI) is a widely used standard [7] to write parallel programs. The main reason for its popularity is probably the availability of an MPI implementation on basically all parallel platforms. Another reason is that the standard contains a large number of calls to solve standard parallel problems in a convenient and efficient manner. The drawback is that the MPI 1.2 standard is with 129 calls large enough to introduce a complexity that also offers the possibility to use the MPI API in an incorrect way. According to our experience there are several reasons for this: first, the MPI standard leaves many decisions to the implementation, e.g. whether or not a standard communication is blocking. Second, parallel applications get more and more complex and especially with the introduction of optimisations like the use of non-blocking communication also more error prone.

2 Related Work

Debugging MPI programs has been addressed in various ways. The different solutions can roughly be grouped in three different approaches: classical debuggers, special MPI libraries and tools.

1. Classical debuggers have been extended to address MPI programs. This is done by attaching the debugger to all processes of the MPI program. There

M. Bubak et al. (Eds.): ICCS 2004, LNCS 3038, pp. 464–471, 2004.
© Springer-Verlag Berlin Heidelberg 2004

are many parallel debuggers, among them the very well-known commercial debugger Totalview [1]. The freely available debugger gdb has currently no support for MPI, however, it may be used as a back-end debugger in conjunction with a front-end that supports MPI, e.g. mpigdb. Another example of such an approach is the commercial debugger DDT by streamline computing, or the non-freely available p2d2 [3,9].

2. The second approach is to provide a debug version of the MPI library (e.g. mpich). This version is not only used to catch internal errors in the MPI library, but it also detects some incorrect usage of MPI by the user, e.g. a type mismatch of sending and receiving messages [2].

3. Another possibility is to develop tools dedicated to look for problems within MPI applications. Currently three different tools are under active development: MPI-CHECK [6], Umpire [12] and MARMOT [4]. MPI-CHECK is currently restricted to Fortran code and performs argument type checking or finds problems like deadlocks [6]. Like MARMOT, Umpire [12] uses the profiling interface. But in contrast to our tool, Umpire is limited to shared memory platforms.

3   Description of MARMOT

Among the design goals of MARMOT are portability, scalability and reproducibility. The portability of an application should be improved by verifying that the program adheres to the MPI standard [7]. The tool issues warnings and error messages if the application relies on non-portable MPI construct. Scalability is addressed with the use of automatic techniques that do not need user intervention. The tool contains a mechanism to detect possible race conditions to improve the reproducibility. It also automatically detects deadlocks and notifies the user where and why these have occurred. MARMOT uses the MPI profiling interface to intercept the MPI calls and analyse them. MARMOT can be used with any MPI implementation that provides this interface. It adds an additional MPI process for all tasks that cannot be handled within the context of a single MPI process, like deadlock detection. Information between the MPI processes and this additional debug process are transferred using MPI. Another possible approach is to use a thread instead of an MPI process and use shared memory communication instead of MPI [12]. The advantage of the approach taken here is that the MPI library does not need to be thread safe. Without the limitation to shared memory systems the tool can also be used on a wider range of platforms. Since the improvement of portability was one of the design goals we did not want to limit the portability of MARMOT. This allows to use the tool on any development platform used by the programmer.

4   Description of the Applications

To measure its overhead when using MARMOT with a real application, we chose different applications from the CrossGrid project.

4.1   Weather Forecast and Air Pollution Modeling

The MPI parallel application [8] of Task 1.4.3 of the CrossGrid project calculates the acid deposition caused by a power plant. The used STEM-II model is one of the most complex air quality models. The transport equations are solved through the Petrov-Crank-Nicolson-Galerkin method (FEM). The chemistry and mass transfer terms are integrated using a semi-implicit Euler and a pseudo-analytic method.

The STEM application is written in FORTRAN and consists of 15500 lines of code. 12 different MPI-calls are used within the application: MPI_Init, MPI_Comm_size, MPI_Comm_rank, MPI_Type_extent, MPI_Type_struct, MPI_Type_commit, MPI_Type_hvector, MPI_Bcast, MPI_Scatterv, MPI_Barrier, MPI_Gatherv, MPI_Finalize.

4.2   High Energy Physics Application

The HEP application [10] from CrossGrid Task 1.3 performs an analysis of the physical data produced by the Large Hadron Collider (LHC) at CERN. All collisions will be recorded by detectors, the corresponding information being stored in distributed databases with a volume of millions of gigabytes. On-line filtering techniques as well as mathematical algorithms, such as neural networks, will be used to select those events and analyse them by physicists working in research centers across the world.

The MPI parallel application ANN (Artificial Neural Network) is a neural network application that is part of the data analysis described above. It currently consists of 11500 lines of C code and uses 11 different MPI calls: MPI_Init, MPI_Comm_size, MPI_Comm_rank, MPI_Get_processor_name, MPI_Barrier, MPI_Gather, MPI_Recv, MPI_Send, MPI_Bcast, MPI_Reduce, MPI_Finalize.

4.3   Medical Application

The application [11] from Task 1.1 is a system used for pre-treatment planning in vascular interventional and surgical procedures through real-time interactive simulation of vascular structure and flow. A 3D model of the arteries serves as input to a real-time simulation environment for blood flow calculations. The user will be allowed to change the structure of the arteries, thus mimicking a surgical procedure. The effects of any modification is analysed in real time while the results are presented to the user in a virtual environment. A stripped down version of the MPI parallel application calculating the blood flow consists of 7500 lines of C code. The code makes use of the following MPI calls: MPI_Init, MPI_Comm_rank, MPI_Comm_size, MPI_Pack, MPI_Bcast, MPI_Unpack, MPI_Cart_create, MPI_Cart_shift, MPI_Send, MPI_Recv, MPI_Barrier, MPI_Reduce, MPI_Sendrecv, MPI_Finalize.

5   Possible Checks for the Used MPI-Calls

This section is to present in more detail what kind of checks MARMOT performs on the calls made by an MPI application. For example, the ANN-application and the STEM-application described above employ the following calls.

- Environmental calls (MPI_Init, MPI_Get_processor_name, MPI_Finalize): A possible test is to check for illegal MPI-calls before MPI_Init and after MPI_Finalize. It is also illegal to call MPI_Init more than once in an application. Currently MARMOT does not perform any checks for MPI_Get_processor_name, because it is a local call without much room for errors. Concerning MPI_Finalize, MARMOT checks if active requests and pending messages are left over in any communicator.
- MPI_Comm_size, MPI_Comm_rank, MPI_Barrier: MARMOT checks if the communicator of the call is valid, i.e. if it is MPI_COMM_NULL or if it is a user-defined communicator, which has been created and committed properly or which might have been freed again before using it.
- Construction of datatypes (MPI_Type_extent, MPI_Type_struct, MPI_Type_hvector, MPI_Type_commit): MARMOT inspects the validity of the datatype argument, and for MPI_Type_struct and MPI_Type_hvector it also inspects if the count and the block length are greater than zero. The tool also verifies if MPI-Type_commit is used to commit a type again that has already been committed.
- Point-to-point communication (MPI_Send, MPI_Recv, MPI_Sendrecv): MARMOT inspects the correctness of the communicator, count, datatype, rank and tag arguments. Similar to the way communicators are checked, it is verified if the datatype is MPI_DATATYPE_NULL or if it has been created and registered correctly by the user. If the count is zero or negative, a warning will be issued, also if ranks and tags are beyond valid ranges. The MPI standard also requires that the program does not rely on any buffering made by the standard send and receive operations. Since the amount and type of buffering is different between the various MPI implementations this is one of the problems that would limit the portability of the MPI program.
- Other collective operations (MPI_Bcast, MPI_Reduce, MPI_Gather, MPI_Gatherv, MPI_Scatterv): The tool checks if the communicator, count, datatype and the rank of the root are valid. Additionally for MPI_Reduce, it is also checked if the operator is valid, e.g. if it has been created properly. For MPI_Gatherv and MPI_Scatterv, also the displacements are examined.

Besides these calls, MARMOT supports the complete MPI-1.2 standard, although not all possible tests have been implemented so far. It also issues warnings when a deadlock occurs and allows the user to trace back the last few calls on each node. Currently the deadlock detection is based on a timeout mechanism. MARMOT's debug server surveys the time each process is waiting in an MPI call. If this time exceeds a certain user-defined limit on all processes at the same time, the debug process issues a deadlock warning.

Possible race conditions can be identified by locating the calls that may be sources of race conditions. One example is the use of a receive call with MPI_ANY_SOURCE as source argument. MARMOT does not use methods like record and replay to identify and track down bugs in parallel programs [5].

6 Performance with the Applications

Our main interest is whether the performance penalty induced by the usage of MARMOT is small enough to use MARMOT in the daily development work of the application. A typical development run is between 10 minutes and one hour. In our experience a run like this should not exceed a time limit of two hours. In order to be useful MARMOT's overhead should therefore be less than 100%. In rare cases where a bug is only present on high processor count an overnight run (12 hours) should be feasible allowing an overhead of up to 1000%. The applications were run with and without MARMOT, on an IA32-cluster using mpich and Myrinet interconnect. For these runs, the execution times were kept in the range of several minutes by reducing the number of iterations. This does not limit the validity of the analysis because evey single iteration shows the same communication behaviour. For example, the STEM application executes 300 iterations to simulate 5 hours of real time, each iteration corresponding to a minute of real time.

6.1 Weather Forecast and Air Pollution Modelling

The scalability of the weather forecast application is limited. It has a parallel efficiency on 16 processes of only 14% (19s of 26s execution time is spent in MPI calls). The scalability with and without MARMOT is shown in Fig. 1. With MARMOT the total time increases from 69.4 s to 70.6 s on 1 process, and from 29 s to 50 s on 16 processes. MARMOT's overhead is approximately between 2 and 72% on up to 16 processes.

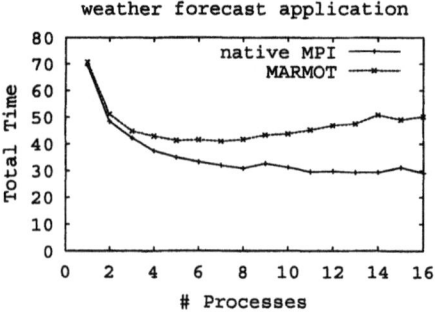

Fig. 1. Comparison of total execution times for the STEM-application between the native MPI approach and the approach with MARMOT.

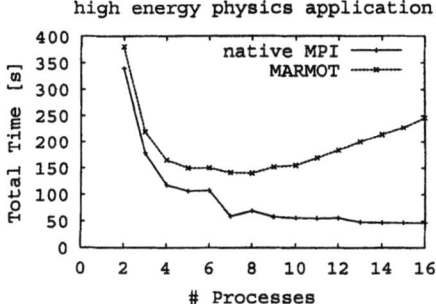

Fig. 2. Comparison of total execution times for the ANN-application between the native MPI approach and the approach with MARMOT.

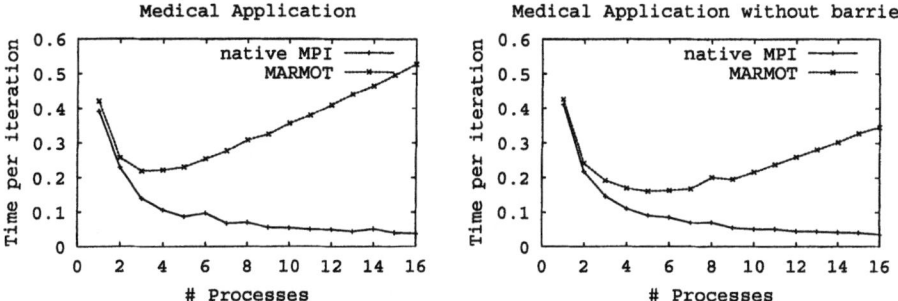

Fig. 3. Comparison of total execution times for the medical application between the native MPI approach and the approach with MARMOT. The left side shows the original result, on the right side one barrier in the application was removed.

6.2 High Energy Physics Application

The results of the HEP application can be seen in Fig. 2. The total time increases from 339 s to 380 s on 2 processes when MARMOT is linked to the application, and from 47 s to 246 s on 16 processes.

MARMOT's overhead is approximately between 20 and 420% on up to 16 processes. On 16 processes the applications runs with 50% parallel efficiency, 26s of 48s execution time are spent in MPI calls. Most of the communication time is spent for broadcast and reduce operations.

6.3 Medical Application

The results of the medical application can be seen in Fig. 3. The time per iteration increases from 0.39s to 0.42 s on 1 process when MARMOT is linked to the application, and from 0.037 s to 0.53 s on 16 processes.

MARMOT's overhead is approximately between 8 and 1312% on up to 16 processes. A detailed analysis with Vampir showed several reasons for the large

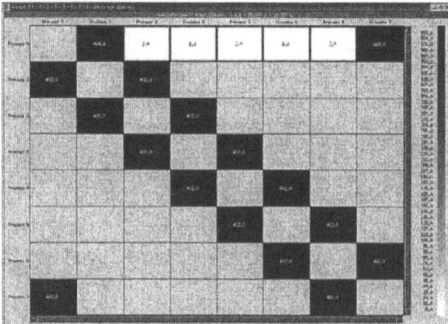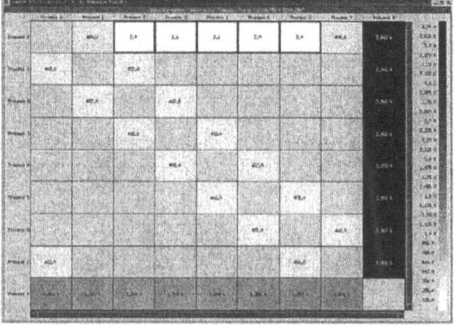

Fig. 4. Message statistics for the medical application without (left) and with MAR-MOT (right).

of the application is modified, because all application processes have to notify MARMOT's debug server about every MPI call. Figure 4 shows a typical communication pattern that is dominated by message exchanges with two neigbours. As soon as MARMOT is used the communication is dominated by the messages between all application processes and the debug server. Second, the application performs a large number of MPI_Barriers. This is a fast operation for most MPI implementations. However, inside MARMOT each client will register with the debug server to notify it about its participation in this operation. This results in a linear scaling with the number of processes. If the barrier that is called after each iteration is removed the execution time with MARMOT is reduced from 0.53s to 0.35s (see Fig. 3).

7 Conclusions and Future Work

In this paper we have presented the tool MARMOT. It analyses the behaviour of an MPI application and checks for errors frequently made in the use of the MPI API. We demonstrated the functionality of the tool with three real world applications from the CrossGrid IST project. The applications cover the C and Fortran binding of the MPI standard. The inevitable performance penalty induced by the performed analysis and checks depends strongly on the application. For the applications of the CrossGrid project used in our tests, the runtimes with MARMOT are in the range of several minutes, which allows a regular usage during the development and verification process of an application. However, especially for the communication intensive applications with a high number of collective communications the overhead caused by MARMOT was above 1000%. Future work will include improvements for this type of applications.

Acknowledgements. The development of MARMOT is supported by the European Union through the IST-2001-32243 project "CrossGrid".

References

1. WWW. http://www.etnus.com/Products/TotalView.
2. William D. Gropp. Runtime checking of datatype signatures in MPI. In Jack Dongarra, Peter Kacsuk, and Norbert Podhorszki, editors, *Recent Advances in Parallel Virtual Machine and Message Passing Interface*, volume 1908 of *Lecture Notes In Computer Science*, pages 160–167. Springer, Balatonfüred, Lake Balaton, Hungary, Sept. 2000. 7th European PVM/MPI Users' Group Meeting.
3. Robert Hood. Debugging computational grid programs with the portable parallel/distributed debugger (p2d2). In *The NASA HPCC Annual Report for 1999*. NASA, 1999. http://hpcc.arc.nasa.gov:80/reports/report99/99index.htm.
4. Bettina Krammer, Katrin Bidmon, Matthias S. Müller, and Michael M. Resch. MARMOT: An MPI analysis and checking tool. In *Proceedings of PARCO 2003*, Dresden, Germany, September 2003.
5. Dieter Kranzlmüller. *Event Graph Analysis For Debugging Massively Parallel Programs*. PhD thesis, Joh. Kepler University Linz, Austria, 2000.
6. Glenn Luecke, Yan Zou, James Coyle, Jim Hoekstra, and Marina Kraeva. Deadlock detection in MPI programs. *Concurrency and Computation: Practice and Experience*, 14:911–932, 2002.
7. Message Passing Interface Forum. *MPI: A Message Passing Interface Standard*, June 1995. http://www.mpi-forum.org.
8. J.C. Mourino, M.J. Martin, R. Doallo, D.E. Singh, F.F. Rivera, and J.D. Bruguera. The stem-ii air quality model on a distributed memory system, 2004.
9. Sue Reynolds. System software makes it easy. *Insights Magazine*, 2000. NASA, http://hpcc.arc.nasa.gov:80/insights/vol12.
10. D. Rodriguez, J. Gomes, J. Marco, R. Marco, and C. Martinez-Rivero. MPICH-G2 implementation of an interactive artificial neural network training. In *2nd European Across Grids Conference, Nicosia, Cyprus*, January 28-30 2004.
11. A. Tirado-Ramos, H. Ragas, D. Shamonin, H. Rosmanith, and D. Kranzlmueller. Integration of blood flow visualization on the grid: the flowfish/gvk approach. In *2nd European Across Grids Conference, Nicosia, Cyprus*, January 28-30 2004.
12. J.S. Vetter and B.R. de Supinski. Dynamic software testing of mpi applications with umpire. In *Proceedings of the 2000 ACM/IEEE Supercomputing Conference (SC 2000)*, Dallas, Texas, 2000. ACM/IEEE. CD-ROM.

Monitoring System for Distributed Java Applications

Włodzimierz Funika[1], Marian Bubak[1,2], and Marcin Smętek[1]

[1] Institute of Computer Science, AGH, al. Mickiewicza 30, 30-059 Kraków, Poland
[2] Academic Computer Centre – CYFRONET, Nawojki 11, 30-950 Kraków, Poland
phone: (+48 12) 617 39 64, fax: (+48 12) 633 80 54
{bubak,funika}@uci.agh.edu.pl, smentos@icslab.agh.edu.pl

Abstract. Recently, the demand for tool support (performance analyzers, debuggers etc.) for efficient Java programming is increasing. A universal, open interface between tools and a monitoring system, the On-line Monitoring Interface Specification (OMIS), and the OMIS compliant monitoring system (OCM) enable to specify such a Java oriented monitoring infrastructure which allows for an extensible range of functionality intended for supporting various kinds of tools. The paper presents an approach to building a monitoring system underlying this infrastructure.

Keywords: Java, monitoring system, monitoring interface, distributed object system, tools, OMIS

1 Introduction

Java technology has grown in popularity and usage because of its portability. This simple, object oriented, secure language supports multithreading and distributed programming including remote method invocation, garbage collection and dynamic class loading. The Java environment simplifies the development and deployment of distributed systems, but such systems are still very complex and more error prone than single process applications. The best monitoring of distributed Java programs could be achieved by using diverse observation techniques and mechanisms, therefore it is often desirable to have a suite of specialized tools such as debuggers, performance analyzers each of them addressing a different task and allowing developers to explore the program's behaviour from various viewpoints. Our *goal* is to build a comprehensive tool support for building distributed Java applications by providing uniform, extendible monitoring facilities for communication between components, for analyzing an application's execution, and for detecting bugs.

The paper is organised as follows. General considerations of Java tools functionality and implications for monitoring are presented in Section 2. Then follows an overview of OMIS, the OMIS-compliant monitoring system OCM, and the characteristics of a Java-oriented extension to OMIS. Next, an overview of the Java-oriented monitoring system architecture is presented (Section 3).

M. Bubak et al. (Eds.): ICCS 2004, LNCS 3038, pp. 472–479, 2004.
© Springer-Verlag Berlin Heidelberg 2004

Sections 4 and 5 describe how the proposed monitoring system processes requests from monitoring based tools and events generated by the target Java system. Conclusions and future work are summarised in Section 6.

2   Design Concepts of a Monitoring System

Based on the analysis of *Related work* presented in [2] we have accepted requirements which must be met by the part of the infrastructure which provides monitoring data. In order to provide comprehensive support for distributed programming, the tools need to be able to *manipulate* and *observe* the whole application distributed over different machines at runtime. To provide a flexible functionality of tools, the monitoring activities underlying the access to and manipulation of the observed application should be concentrated in a separate module which is usually called *monitoring system*. Such a system should ideally provide a uniform on-line interface for different kinds of tools, which allows to easily build tools without a need to understand the implementation of monitoring, while providing scalability and a wide range of analysis methods.

To build a versatile monitoring system we used a monitor/tool interface specification, OMIS [3], and a monitoring system, the OCM [4], which implements this specification. OMIS follows the idea of separation of a tool from a monitoring system through a standardized interface. The cooperation between the tool and the monitoring system is based on the *service request/reply* mechanism. A tool sends a service request to the monitoring system, e.g. as a coded string which describes a condition (event) (if any) and activities (action list) which have to be invoked (when the condition gets true). In this way the tool programs the monitoring system to listen for event occurrences, perform needed actions, and transfer results to the tool. OMIS relies on a hierarchy of the abstract objects: *nodes, processes, threads, messages queues* and *messages*. Every object is represented by an abstract identifier (*token*) which can be converted into other token types by the conversion functions *localization* and *expansion* which are automatically applied to every service definition that has tokens as a parameter. Each tool at each moment has a well defined scope, i.e. it can observe and manipulate a specific set of objects *attached* on a request from this tool. Due to the distributed nature of parallel application, the monitoring system must itself be distributed and needs one monitoring component per node, which in case of the OCM is called Local Monitor (LM). The OCM also comprises a component called the Node Distribution Unit (NDU) that has to analyze each request issued by a tool and split it into separate requests that can be processed locally by LMs on proper nodes. OMIS allows the monitoring system to be expanded with a *tool extension* or *monitor extension*, which adds new services and new types of objects to the basic monitoring system, for specific programming environments.

Java-bound On-line Monitoring Interface Specification (J-OMIS) is a monitor extension to OMIS for Java distributed applications, which introduces a new Java-bound object hierarchy that is divided into two kinds of objects: *execution objects*, like nodes, JVMs, threads and *application objects*, e.g. interfaces, classes, objects, methods. As in the original OMIS, J-OMIS specifies tree types of services: *information services, manipulation services,* and *event services.* J-OMIS defines relations between the objects of a running application, which are expressed by conversion functions, enlarged w.r.t. OMIS by additional operations that result from the object-oriented nature of Java.

3   Architecture of the Java-Oriented Monitoring System

Based on J-OMIS, we have designed a Java-oriented extension to the OCM, the J-OCM, by extending the functionality of the OCM, via adding new software components and adapting existing ones (see Fig.1). This approach allows to combine the existing functionality of the OCM with the Java platform.

The *Node Distribution Unit*(NDU), an unchanged part of the whole monitoring infrastructure, is still responsible for distributing requests and assembling replies. E.g. a tool may issue a request in order to run the garbage collector on specified JVMs, therefore the NDU must determine the nodes executing the JVMs and, if needed, split the request into separate sub-requests to be sent to the proper nodes. The NDU must also assemble the partial replies from local monitors into a global reply. The NDU is aimed to make the whole system manageable, thus it has to program the local monitors of all the currently observed nodes. As the set of monitored nodes may change over time, the NDU must properly react to these changes: to create local monitors on the newly added nodes or to re-arrange a list of the objects involved in the execution of requests that have been issued by tools, when some nodes are deleted.

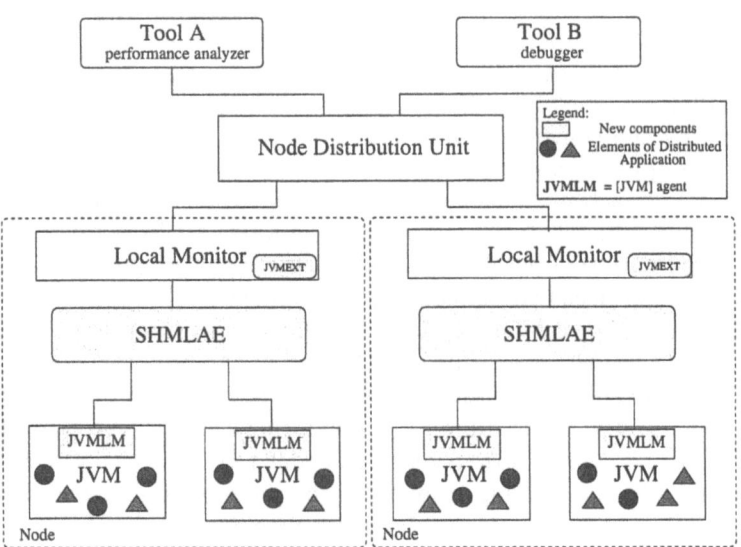

Fig. 1. J-OCM architecture

The *Local Monitor* is a monitor process, independent from the whole global monitoring infrastructure. Each monitor process provides an interface similar to that of the NDU, with the exception that it only accepts requests to operate on local objects. To support the monitoring of Java applications, the LM's extension, JVMEXT, provides new services defined by J-OMIS, which control JVM via agents. JVMEXT is linked to LMs as a dynamically linked library at runtime using the *dlopen* interface, whenever the tool issues a service request to JVMEXT.

LM stores information about the target Java application's objects such as *JVMs, threads, classes, interfaces, objects, methods* etc. referred to by tokens. The *Java Virtual Machine Local Monitor* is an agent embedded into a JVM process, which is responsible for execution of the requests received from the LM. It uses Java Virtual Machine native interfaces such as JVMPI [5], JVMDI [6], and JNI [7] that provide low level mechanisms for interactive monitoring, independent of a JVM implementation. The *Shared Memory based Local Agents Environment* (SHMLAE) is a communication layer to support cooperation between agents and the LM. This allows the components involved in communication to find each other during start-up and notify about their existence.

4 Request Processing

The proposed monitoring infrastructure allows the tool to see the whole monitored application as a set of distributed objects and the monitoring system as a higher-level software layer (middleware) that provides a standardized interface to access those objects, regardless of implementation details, like hardware platform or software language.

To deal with the distributed target system we consider the functioning of the J-OCM as a distributed system, which has to usually comprise additional architectural elements: an object interface specification, object stub and skeleton, object manager, registration/naming service and communication protocol.

Interface definition. The first stage of the process of developing a distributed application is to define the interface of a remote object (e.g. methods, data types), written in an Interface Definition Language. Similarly, a leading idea of OMIS is to provide support for building monitoring tools and systems for new parallel or distributed environments by extending its functionality. For each extension, an IDL file, called *registry*, has to be provided, which specifies new objects, new services and their relevant attributes.

Stub and Skeleton provide transparent communication between the client and the remote object. They are based on the `Proxy Design Pattern`[1], where the object is represented by another object (the proxy), in order to control access to the object. In distributed computing the stub plays the role of the proxy and allows to make a local call on the remote object. The skeleton, residing on the server side, receives an incoming call, and invokes it on the real object. The Proxy pattern is used by the J-OCM to provide the tool with transparent access

---

[1] Proxy design pattern as described in "Design Patterns" by Gamma, Helm, Johnson and Vlissides [8]

to monitored objects [2]. The monitored objects are identified by tokens which refer to the proxy object. The proxy is a representation of the real object in the monitoring system. The object proxy contains all information that is needed to deliver the tool's requests to the JVM agent (JVMLM) which directly accesses the JVM. The JVM agent acts as a skeleton, while the remote proxy which is embedded into the JVM is a platform dependent native library. The agent transforms a call and parameters received from the LM into the format required by one of interfaces used to interact with JVM.

Object manager and registration/naming service. Remote method calls issued by the client are routed through the object manager to the proper object on the server. The object manager also routes results back to the client. The *registration/naming service* acts as an intermediary layer between the object client and the object manager. Once an interface to the object has been defined, an implementation of the interface needs to be registered with the naming service so that the object can be accessed by clients using the object's name. The main components of the J-OCM — the NDU and LMs — can be classified as an object manager and provide operations similar to the naming service that is present in distributed systems. Any object that can be observed or manipulated by tools is represented by a *token*, which is a globally unique name of monitored object.

The tokens and objects' proxies of proper monitored objects are created when: (1) the JVMLM is started up and notifies the node's LM of its existence (e.g. jvm_j_1), (2) events referring to the creation of *threads, classes, objects, interfaces,* etc. are raised and the tool is interested in them, or (3) the tool issues information requests of the following syntax: {jvm, thread, class, etc.}_get_tokens() to obtain a list of tokens of all monitored objects of a given class.

5   Event Handling

An *event* is a characteristic atomic program behavior. In an event-based monitoring system, basic events are captured by sensors, which are inserted in to a target systems, and notified to the monitoring system. The monitoring system takes some action(s) - a sequence of commands associated with the event. These actions can either carry out data collection, or also manipulate the running program. In order to follow the idea of event-based monitoring both the LM and the JVMLM must support the event notification.

Java-oriented Techniques for Event-Based Monitoring. JVM notifies several internal events to the JVMLM, using JVMPI and JVMDI. These events are fired by changes in the state of Java threads, like (*started, ended, blocked on a locked monitor*), the beginning/ending of an invoked method, class loading operations, object allocation/deallocation, and the beginning/ending of JVM garbage collection, exception throwing, etc.

To support the interactive observation of the target system, all events must be processed by the JVM agent, while the agent sends the events to the LM

---

[2] The word *object* is used in terms of the J-OMIS object hierarchy.

selectively, to avoid too much overhead on the LM. This is based on a filtering mechanism introduced into the JVM agent, which selects which events should be sent to the LM.

To control the forwarding of events, the agent uses a filter in form of a table, where it stores information about what events the LM is interested in. The LM manipulating this event filter table can *stop* or *resume* the notification of specific events sent from its agents.

The OMIS Event Model, along with the *event* term, additionally defines the *event class* predicate, specifying a set of event occurrences, a pattern defining the events of interest. In the OCM, event classes are represented by an event service together with its parameters. The event processing in the OCM is based on the idea that event classes form a tree hierarchy, where the root is the universal class containing all detectable events. New event classes are derived from the existing ones via the use of filters that allow only certain events to pass. Each class in the event class tree may be associated with a set of filters, each of which is used to derive a more specific event class from a more general one. In addition, some event classes may have their associated action lists. When an event is detected by the OMIS compliant monitoring system, it is matched against all the event classes in the tree, by doing a tree traversal starting with the tree's root, which matches all events. At each traversed node, the filters associated with that node are evaluated. If a filter evaluates to true, the event matches the derived event class, which is therefore traversed, too. During this tree traversal, the action lists associated with the event classes matched by the event are scheduled for execution.

Event Processing in the J-OCM. The J-OCM as an adaptation of the OCM for Java applications extends the event tree of the OCM by its own subtree. The new event hierarchy, shown in Fig.2, consists of three types of event classes.

The jvm_any is the 'root' of Java related event classes and is triggered whenever any of JVMs registered in the monitoring system generates an event. The jvm_any is a Singleton [3], the event class that has only one instance to provide the global point of access to it in the Local Monitor.

All other event classes, which relate to other object types specified in J-OMIS, e.g. *jvms, threads, classes,* etc. are derived from this one using filters. The jvm_control, jvm_{thread, class, object, method}_control classes represent abstract object control event classes [4]. Leaves, those elements of the tree which do not have children, represent the events which are defined by the interface specification, i.e. J-OMIS. J-OMIS classifies event services based on the categories of elements of the Java program architecture the services operate on. The J-OCM event tree follows this classification, similarly the *control* services group the event services operating on a type of token, e.g. *thread, JVM.* But some

---

[3] Singleton Design Pattern

[4] The term 'abstract' is used in the sense that these event classes cannot be involved in a conditional request via the J-OMIS tool/monitor interface, whereas the term 'control' refers to all events occurrences, characteristic for the particular type of object, e.g. jvm_ thread_control comprises all events connected with JVM threads

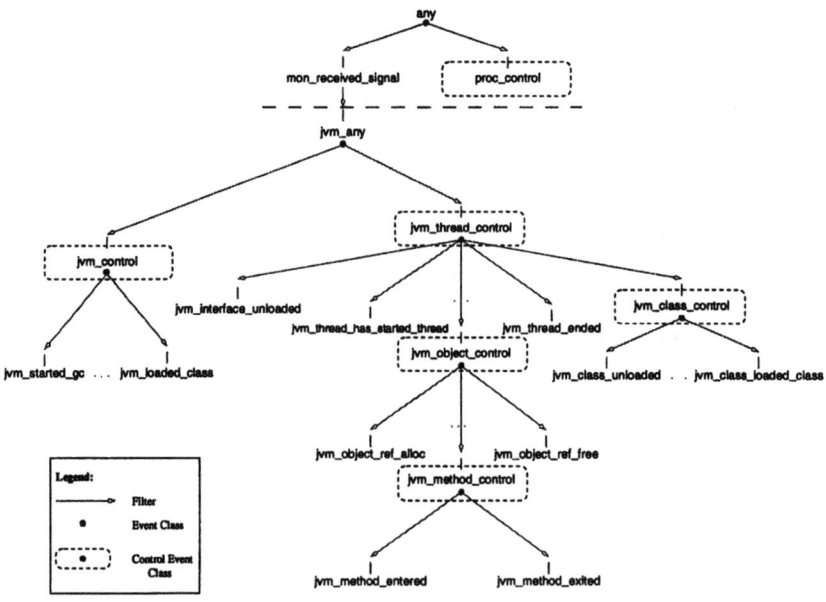

Fig. 2. Hierarchy of J-OCM events

event services have been moved over in the event hierarchy in order to recognize the situation where the event took place. The information which is needed to determine whether an event matches a given event class is the location (context) of the event, e.g. the event service jvm_method_entered indicates when a given method is called, i.e. we need to specify the most dynamic elements of a Java application execution, e.g. *JVM, thread, object* because they determine the context of the event occurrence, i.e. an event may occur on a particular *JVM*, in a *thread*, and refer to an *object*. The presented shape of the J-OCM events tree allows to narrow the event detection and to simplify extending the tree by new events.

Event-Based Interaction. The most critical part of the J-OCM event system are interactions between the Local Monitor and its Java agents (JVMLMs). The Java agents use the SHMLAE to notify the LM of the event occurrence and to transfer event-specific parameters. A specific event, before being handled by the LM, must be enabled in the specific JVMLM. This is done when a tool enables the previously defined event service request, by issuing a csr_enable service. Once the JVM agent has received the request from the LM, it starts passing events to the LM which must take care of handling them.

The OCM uses signals as asynchronous mechanism to notify the message arrival and defines a special event class, called mon_received_signal(integer sig_no) that is triggered when the monitor process receives the specified signal. The use of this event class is the solution to handle events from the JVM agents, and next to process them, according to the OMIS event model. The universal event class jvm_any relating to Java specific events derives from mon_receive_signal event class. The filter that is located between these two event classes (Fig. 2) is invoked each time the monitor gets a signal sent by an agent. It uses a non-blocking receive call to get messages containing information about an event that has occurred in a monitored JVM and executes jvm_any, i.e. the 'root' of the Java event classes hierarchy, where further event processing takes place.

6 Conclusions

The work on building a Java-oriented tools followed the idea of separating the layer of tools from a monitoring system's functionality. We extended the On-line Monitoring Interface Specification by a Java specific hierarchy of objects and a set of relevant services. The work on a Java-oriented monitoring system, the J-OCM, concentrated on extending the functionality of Local Monitors which are the distributed part of the system and introducing new software levels interfacing the J-OCM to JVM and providing a communication mechanism for the low-level layers. Our on-going work focuses on completing the implementation of the J-OCM and designing a suite of Java-oriented monitoring-based tools.

Acknowledgements. We are very grateful to Prof. Roland Wismüller for valuable comments. This research was carried out within the Polish-German collaboration and it was partially supported by the KBN grant 4 T11C 032 23.

References

1. M. Bubak, W. Funika, P. Mętel, R. Orłowski, and R. Wismüller: Towards a Monitoring Interface Specification for Distributed Java Applications. In Proc. 4th Int. Conf. PPAM 2001, Nałęczów, Poland, September 2001, LNCS 2328, pp. 315-322, Springer, 2002.
2. M. Bubak, W. Funika, M.Smętek, Z. Kiliański, and R. Wismüller: Architecture of Monitoring System for Distributed Java Applications. In: Dongarra, J., Laforenza, D., Orlando, S. (Eds.), Proceedings of 10th European PVM/MPI Users' Group Meeting, Venice, Italy, September 29 - October 2, 2003, LNCS 2840, Springer, 2003
3. T. Ludwig, R. Wismüller, V. Sunderam, and A. Bode: OMIS – On-line Monitoring Interface Specification (Version 2.0). Shaker Verlag, Aachen, vol. 9, LRR-TUM Research Report Series, (1997)
   http://wwwbode.in.tum.de/~omis/OMIS/Version-2.0/version-2.0.ps.gz
4. R. Wismüller, J. Trinitis and T. Ludwig: A Universal Infrastructure for the Runtime Monitoring of Parallel and Distributed Applications. In Euro-Par'98, Parallel Processing, volume 1470 of Lecture Notes in Computer Science, pages 173-180, Southampton, UK, September 1998. Springer-Verlag.
5. Sun Microsystems: Java Virtual Machine Profiler Interface (JVMPI)
   http://java.sun.com/j2se/1.4.2/docs/guide/jvmpi/jvmpi.html
6. Sun Microsystems: Java Virtual Machine Debugging Interface (JVMDI)
   http://java.sun.com/products/jdk/1.2/docs/guide/jvmdi/jvmdi.html
7. Sun Microsystems: Java Native Interface (JNI)
   http://java.sun.com/products/jdk/1.2/docs/guide/jni/
8. Erich Gamma, Richard Helm, Ralph Johnson, and John Vlissides. Design Patterns. Addison-Wesley, 1995.

Automatic Parallel-Discrete Event Simulation

Mauricio Marín*

Centro de Estudios del Cuaternario CEQUA
Universidad de Magallanes
Punta Arenas, Chile
Mauricio.Marin@umag.cl

Abstract. This paper describes a software enviroment devised to support parallel and sequential discrete-event simulation. It provides assistance to the user in issues such as selection of the synchronization protocol to be used in the execution of the simulation model. The software framework has been built upon the bulk-synchronous model of parallel computing. The well-defined structure of this model allowed us to predict the running time cost of synchronization protocols in accordance with the particular work-load generated by the simulation model. We exploit this feature to automatically generate the simulation program.

1 Introduction

Discrete-event simulation is a widely used technique for the study of dynamic systems which are too complex to be modelled realistically with analytical and numerical methods. Amenable systems are those that can be represented as a collection of state variables and whose values may change instantaneously upon the occurrence of events in the simulation time. It is not difficult to find real-life systems with associated simulation programs which are computationally intensive enough to consider parallel computing, namely *parallel discrete-event simulation* (PDES) [1], as the only feasible form of computation.

Over the last decade or so, parallel discrete event simulation (PDES) has been intensively performed on traditionals models of parallel computation [1,7]. Paradoxically, these models have failed to make parallel computation a paradigm of wide-spread use. Among other drawbacks, they lack realistic cost models for predicting the performance of programs. As a result, existing simulation software products have not been built upon frameworks which are able to predict performance and select accordingly the most efficient algorithms for implementing the parallel simulation program. Usually this kind of software either encapsulate just one fixed algorithm or leaves to the user the responsability of selecting one from a set of alternative ones. Neither is convenient since it is known that a particular algorithm is not efficient for all work-load cases and the target user not necessarily is an expert on efficient algorithms for parallel simulation. This has restrained potential users from using existing parallel simulation software.

---

* Partially supported by Fondecyt project 1030454.

M. Bubak et al. (Eds.): ICCS 2004, LNCS 3038, pp. 480–487, 2004.
© Springer-Verlag Berlin Heidelberg 2004

In this paper we describe a simulation environment whose design is based on the use of the bulk-synchronous model of parallel computing (BSP model) [8, 11]. The key point is that the model of computing provides a simple way to cost parallel algorithms in their computation, communication and synchronization components. This allows us to predict the performance of different synchronization protocols in accordance with the particular features of the user defined simulation model. We use this facility to automatically select the synchronization protocol that is best suited for the particular simulation.

2   Object-Oriented Approach to Modeling

We describe a simple general purpose modeling methodology (the *world-view* in simulation parlance) that we have devised to facilitate the use of our system.

The world is seen as a collection of simulation objects that communicate with each other via timestamped event-messages. Associated with each object there is a global instance identifier, called object-id, and a class identifier, called entity-id. There exists a simulation kernel that is responsible for efficiently delivering the messages in strict message timestamps chronological order. Each simulation object inherits from a base class called *Device* that provides methods for the interface with the kernel. In particular, the kernel delivers messages to the simulation objects by executing the Device's method *cause* with parameters such as event-type, object-id, entity-id, and the simulation time at which the event takes place in the target object.

For each simulation object, the user must provide an implementation of the cause method so that events are handled in accordance with the behaviour defined for the particular object. Nevertheless, we enable users to maintain a class library for pre-defined/common use simulation objects together with a graphical representation for the user interface. Groups of classes can be tailored to specific application domains. The entity-id parammeter allows the user to split the event processing task into a set of private methods, each handling different types of events for a given type of entity.

In addition, simulation objects contain output channels that they use for sending messages to other objects connected to those channels. A message is sent out by executing the Device's method *schedule* which takes parammeters such as the channel id, time at which the event must take place in the target object, and sender's object-id and entity-id. At initialization time, all output channels are connected to their respective target objects. Note that the cause method could work on input channels as well. However, we have not seen a need for including in our world-view the notion of input channels yet. In fact, the combination object-id/entity-id appears to be very flexible as it allows objects to receive messages from multiple sources without complicating too much the initialization process.

On the other hand, the notion of output channels makes it easier to the programmer to work on generic implementations of the object without worrying about the specific object connected to the output channel. Actually, all this is

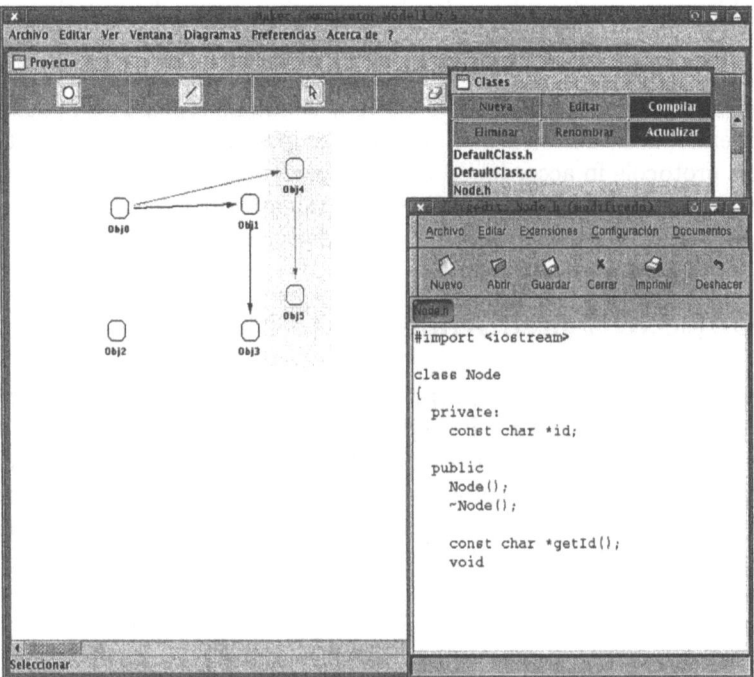

Fig. 1.

a tradeoff between generality and simplification of the initialization process and its implicancies in code debuging and maintenance. As an alternative to output channels, we also support the concept of associative arrays (e.g., C++ maps) that are used by *schedule* to go from global object ids to pointers to the actual simulation objects for direct message delivering.

Currently, we have C++ and Java implementations of this approach.

The graphical user interface is designed to ease the burden of deploying thousands of simulation objects and defining their communication relations. The model definition process starts up with the creation of a project and drawing of a few objects. Each object is an instance of a given class which can be edited and/or viewed by using a class editor as shown in figure 1. More objects of a given class can be deployed by duplicating existing objects. The user interface allows the definition and/or edition of different properties associated with the objects as well as the automatic generation of the simulation program.

The architecture of the simulation framework is as follows. On the upper layer we have the user interface which allows the simulation model to be defined in a graphical manner as well as the codification of the classes that define the behavior of the respective simulation objects. The second layer takes the user definitions and generates a specification of the model written in a mid-level language. The third layer is in charge of selecting the synchronization protocol that happears to be most suitable for the simulation model. This is effected by directly

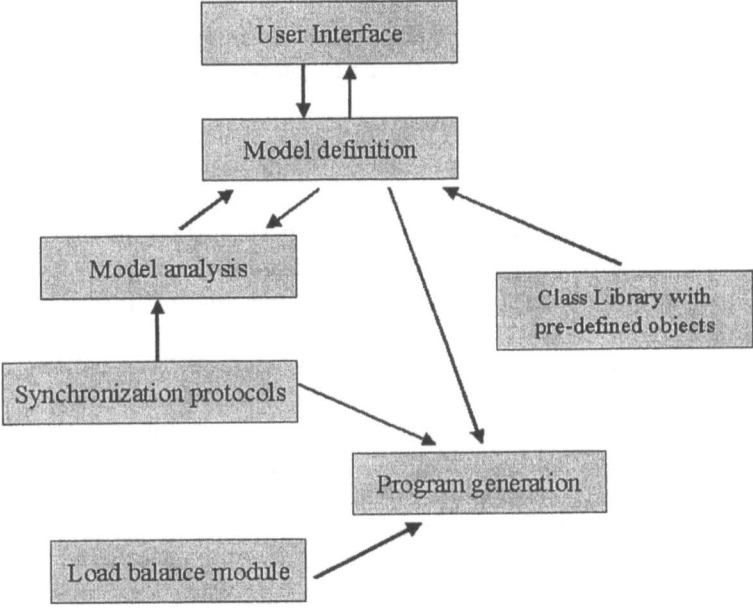

Fig. 2. Software Architecture

analyzing the mid-level definitions. The first step is to perform a pre-simulation of the model. The result is a set of parameters that predicts the model behavior. These parameters are then input to a formula that predicts whether it is better to simulate the model in parallel or just sequentially in the target parallel computer. The last step is to generate the simulation program by linking the selected synchronization protocol with the simulation objects defined by the user. Finally a C++ or java compiler is used to produce the executable simulation program.

Figure 2 shows the relationships among the main components of the simulation environment. Note that users who are not familiar with C++/Java programming can profit from the class library that contains definitions for most typical objects in, for example, queuing systems. The library can be increased by directly including new user-defined class definitions.

Usually a parallel simulation will have to deal with thousands of simulation objects. Thus the mid-level languaje only specifies general information about the objects such as the class they belong to and their communication relations. The instances themselves are stored in a symbolic manner into a file to be actually created later at simulation running time.

The methodology used to automatically generate the simulation program can be divided into the following major steps (see figure 2):

(i) Pre-simulation of the simulation model defined by the user. This uses information of the communication topology among objects defined by the user and the details about what random number generators are used to send the messages among them (simulation time of those event messages). The results of the pre-simulation are used to determine a tuple which describes the overall behaviour of the model in terms of the amount of computation, communication and synchronization it demands to the BSP computer per unit simulation time (see next section).

(ii) The tuple obtained in the previous step is plugged into a formula that predicts the feasible speedup to be achieved by the simulation on the target BSP computer (next section). The outcome can be a recomendation to simulate the model just sequentially because the predicted speedup is too modest or even less than one. The effect of the particular parallel computer hardware is included in the BSP parameters G and L obtained for the machine via proper benchmarks (next section).

(iii) In the case that the recomendation is a parallel simulation, the tuple is now plugged into a set of formulas that predict the running times of a set of synchronization protocols available in our system for parallel simulation. These protocols are optimistic and conservative ones and for each of them it is necessary to include new definitions into the simulation model. Conservative protocols need the so-called lookhaead information whereas the optimistic one requires the specification of what are the states variables that need to be handled by rollbacks.

(iv) The simulation program is generated by putting together objects and synchronization protocol. During simulation the synchronization protocols have the avility of adapting themselves to changes in the work-load evolution. Those protocols also implement a dynamic load balancing strategy we devised to re-distribute objects onto processors in order to reduce running times (figure 2). The initial mapping of objects onto the processors is uniformly at random. During running time a dynamic load balancing algorithm is executed to correct observed imbalance.

3   Technical Details

In the BSP model of computing both computation and communication take place in bulk before the next point of global synchronization of processors. Any parallel computer is seen as composed of a set of P processor-local-memory components which communicate with each other through messages. The computation is organised as a sequence of *supersteps*. In each superstep, the processors may perform sequential computations on local data and/or send messages to other processors. The messages are available for processing at their destinations by the next superstep, and each superstep is ended with the barrier synchronisation of the processors.

The total running time cost of a BSP program is the cumulative sum of the costs of its supersteps, and the cost of each superstep is the sum of three quantities: w, $h\,g$ and l, where (i) w is the maximum of the computations performed by each processor, (ii) h is the maximum number of words transmitted in messages sent/received by each processor with each one-word-transmission costing g units of running time, and (iii) l is the cost of barrier synchronising the processors.

The effect of the computer architecture is costed by the parameters g and l, which are increasing functions of P. These values can be empirically determined by executing benchmark programs on the target machine [8].

We use the above method to cost BSP computations to compute the speedup S_{up} under a demanding case for the underlying parallel algorithm. In [4] we derived the following speed-up expression,

$$S_{up} = \frac{1}{P_B\,(1 + P_M/r) + z\,P_B\,P_M\,g_e + P_S\,l_e}$$

with $\frac{1}{P} \leq P_B \leq 1$, $0 \leq P_M \leq 1$, $0 \leq P_S \leq 1$, $r \geq 1$, and $z \geq 1$. The parameter P_S is a measure of slackness since $1/P_S =$ number of simulated events per superstep. The parameter P_M accounts for locality as it is the average fraction of simulated events that results in message transmitions. The parameter P_B accounts for load balance of the event processing task. The size of messages is represented by z. In addition, $r \geq 1$ is the event granularity defined with respect to C_e which is the lowest (feasible) cost of processing an event in the target machine. Finally, g_e and l_e are defined as $g_e = g/(r\,C_e)$ and $l_e = l/(r\,C_e)$ for the BSP parameters g and l respectively. In this way simulation models can be represented by an instance of the tuple (P_B, P_S, P_M, r, z).

The most popular synchronisation protocols [2,5,6,10] base their operation on one of two strategies of simulation time advance. Synchronous time advance (SYNC) protocols define a global time window to determine the events allowed to take place in each iteration (superstep). The SYNC protocol advances its time window forward in each iteration to let more events be processed. On the other hand, asynchronous time advance (ASYNC) protocols implicitly define windows which are local to the simulation objects. Figure 3 describes sequential simulation algoritms which predict the supersteps executed by each protocol (SYNC and ASYNC).

The comparative cost of synchronization protocols is calculated as follows. Let S_p^s and S_p^a be the number of supersteps per unit simulation time required by synchronous time advance (SYNC) or asynchronous time advance (ASYNC) respectively. We measure load balance (at superstep level) in terms of the *event efficiency* E_f as follows. If a total of $M_e \gg N$ events are processed during the complete simulation with P processors, then $E_f = \frac{M_e}{SumMaxEv}\,\frac{1}{P}$ where *SumMaxEv* is the sum over all supersteps of the maximum number of events processed by any processor during each superstep (i.e., the cumulative sum of the maximum in each superstep). Both S_p and E_f can be determined empirically for the simulation model at hand by employing the algorithms shown in figure 3.

The protocols in our system are optimized BSP realizations of YAWNS [6], BTB [9], CMB-NM (null messages) [5] and TW [2]. YAWNS and BTB are SYNC protocols whereas CMB-NM and TW are ASYNC protocols.

In the optimistic protocols we increase the cost of processing each event in $\varphi \geq 1$ units in order to include the effect of state saving. Roll-backs cause re-simulation of events thus we consider that this operation increases the total

SYNC	ASYNC
Generate N initial pending events;	Generate N initial pending events;
[e is an event with time $e.t$]	[$e.s$ indicates the minimal superstep at
$T_Z := \infty$; [event horizon time]	which the event e may take place in
$S_Z \leftarrow \Phi$; [buffer]	processor $e.p$.]
loop	**loop**
if TimeNextEvent() $> T_Z$ **then**	$e :=$ NextEvent();
SStep := SStep + 1;	$p := e.p$; [e occurs in processor p]
Schedule(S_Z);	**if** $e.s >$ SStep[p] **then**
$T_Z := \infty$;	SStep[p] := $e.s$;
$S_Z \leftarrow \Phi$;	**endif**
endif	$e.t := e.t +$ TimeIncrement();
$e :=$ NextEvent();	$e.p :=$ SelectProcessor();
$e.t := e.t +$ TimeIncrement();	**if** $p = e.p$ **then**
$p := e.p$; [e occurs in processor p]	$e.s :=$ SStep[p];
$e.p :=$ SelectProcessor();	**else**
if $e.p \neq p$ **then**	$e.s :=$ SStep[p] + 1;
$S_Z \leftarrow S_Z \cup \{e\}$;	**endif**
$T_Z :=$ MinTime(S_Z);	Schedule(e);
else	**endloop**
Schedule(e);	
endif	The total number of supersteps is the
endloop	maximum of the P values in array $SStep$.

Fig. 3.

number of simulated events by a factor of ϕ events with $\phi \geq 1$. In the asynchronous protocol roll-backs also increase the message traffic. The conservative protocols do not have these overheads thus we set $\varphi = 1$ and $\phi = 1$. Synchronous time advance protocols (SYNC) require a min-reduction operation with cost R_D for each event processing superstep.

Defining N and N_m as the number of simulated events and sent messages per unit simulation respectively on a P-processors BSP computer, the total cost of a SYNC protocol (YAWNS and BTB) is given by SYNC $= \frac{\varphi^s \phi^s r N}{E_f^s P} + \frac{N_m}{E_f^s P} g + S_p^s l + S_p^s R_D$, whereas the total BSP cost of the ASYNC protocol (TW and CMB-NM) is given by ASYNC $= \frac{\varphi^a \phi^a r N}{E_f^a P} + \frac{(2\phi^a - 1) N_m}{E_f^a P} g + c S_p^a l$. where $c \geq 1$ is a factor that signal the average increase of supersteps in CMB-NM. The determination of the synchronization protocol to be suggested to the user takes into account the following cases. Conservative protocols (YAWNS, CMB-NM) have higher priority than the optimistic ones (BTB, TW). For the case in which the observed fan-in/fan-out of the communication topology among the simulation objects is large enough, the CMB-NM is discarded since this protocol loses efficiency dramatically. Also for the cases in which the pre-simulation did not find a sufficient amount of "lookahead" in the built-in random number generators for timestamps increments, the YAWNS and CMB-NM protocols are discarded. On the other hand, the cost of state saving for the optimistic protocols depends on the size of data to be periodically saved. Also roll-backs in BTB are on everage 20% lower than in TW. Thus the best choise is a tradeoff.

4   Conclusions

We have described the overall design of a simulation framework we have developed to support parallel discrete event simulation. The main objective was to assist users on the complexity associated with the selection of a proper synchronization protocol to conduct the simulation of the user-defined model.

A pre-simulation of the user model produces information about what synchronization protocol is best suited for the execution of the model. Those protocols are devised upon a model of computing that provides both independence of the architecture of the parallel computer and a method for determining the cost of parallel algorithms.

We have tested the suitability of our system using several simulations models. Those include the synthetic work-load PHold, Wind energy electricity generation systems, hard-particles models, Web crawlers, and a large toriodal queuing network. In all cases, specially in regular systems, we have observed that our prediction methodology is very effective in practice (this claim is supported by the results in [3,4]).

References

1. R.M. Fujimoto. Parallel discrete event simulation. *Comm. ACM*, 33(10):30–53, Oct. 1990.
2. D.R. Jefferson. Virtual time. *ACM Trans. Prog. Lang. and Syst.*, 7(3):404–425, July 1985.
3. M. Marín. Asynchronous (time-warp) versus synchronous (event-horizon) simulation time advance in bsp. In *Euro-Par'98 (Workshop on Theory and Algorithms for Parallel Computation*, pages 897–905, Sept. 1998. LNCS 1470.
4. M. Marín. Towards automated performance prediction in bulk-synchronous parallel discrete-event simulation. In *XIX International Conference of the Chilean Computer Science Society*, pages 112–118. (IEEE-CS Press), Nov. 1999.
5. J. Misra. Distributed discrete-event simulation. *Computing Surveys*, 18(1):39–65, March 1986.
6. D.M. Nicol. The cost of conservative synchronization in parallel discrete event simulations. *Journal of the ACM*, 40(2):304–333, April 1993.
7. D.M. Nicol and R. Fujimoto. Parallel simulation today. *Annals of Operations Research*, 53:249–285, 1994.
8. D.B. Skillicorn, J.M.D. Hill, and W.F. McColl. Questions and answers about BSP. *Journal of Scientific Programming*, V.6 N.3, 1997.
9. J.S. Steinman. Speedes: A multiple-synchronization environment for parallel discrete event simulation. *International Journal in Computer Simulation*, 2(3):251–286, 1992.
10. J.S. Steinman. Discrete-event simulation and the event-horizon. In *8th Workshop on Parallel and Distributed Simulation (PADS'94)*, pages 39–49, 1994.
11. L.G. Valiant. A bridging model for parallel computation. *Comm. ACM*, 33:103–111, Aug. 1990.

Creation of Information Profiles in Distributed Databases as a *n*-Person Game

Juliusz L. Kulikowski

Institute of Biocybernetics and Biomedical Engineering PAS, Poland.
jlkulik@ibib.waw.pl

Abstract. It is considered a problem of forming information profiles of distributed data- and/or knowledge bases as a result of satisfying the requirements representing the information profiles of the customers. It is shown that the interests of data- and knowledge bases managers are not fully convergent and that they participate in a composite, partially co-operative, partially non-co-operative game. There is given a formal description of the strategies used in this game, as well as the methods of decision making of the players.

1 Introduction

A spontaneous and dramatic development of open-access data- and knowledge bases in the last decades has led to a new situation in many areas of human activity depending on the access to information resources. This remark concerns various branches of scientific research, technology, education, administration, trade, health services, national security, natural environment protection, etc. For the existence and development of all the above mentioned areas access to information resources satisfying specific requirements of the users is necessary. Distributed databases accessible through Internet (or through any other computer networks) make possible reaching higher quality of human activities and, as a result, they open a new era in modern civilisation development. But at the same time their impact on our life has caused new problems that existed never before or existed only in a germinal state. Till databases were created by the same organisations they were dedicated to the interests of database managers were close to those of database users and the area of conflicts between them was strongly limited. New situation arose when government or other higher administrative authorities tried to initialise design and construction of computer-based information systems in order to force higher effectiveness in sub-ordered organisations. The goals of information systems' sponsors, designers and users were divergent and, as a result, many so-designed information systems failed as being not accepted by their potential users. Free information market gives another possibility of information systems creation according to the requirements of various subjects. In this case existence of no general "optimum" is assumed; information creators, systems managers, and information users can express their proper goals and they participate in a multi-person game trying to reach their individual optima. Simultaneous reaching of all so-defined optima is impossible; however, the market mechanisms make possible reaching a common balance point being a compromise between the partners' expectations.

M. Bubak et al. (Eds.): ICCS 2004, LNCS 3038, pp. 488–496, 2004.
© Springer-Verlag Berlin Heidelberg 2004

General concepts of the theory of games we owe to J. von Neumann and O. Morgenstern [1]. For more than fifty years various types of games: two- and multi-person, discrete- and continuous-strategy, differential, antagonistic and non-antagonistic, co-operative and non-co-operative, one- and multi-level, etc. were investigated [2,3,7]. Various areas of game theory applications: in business, technology, military service, etc., were also considered. The work [4] by E.A. Berzin where game theory application to distribution of resources has been investigated is close to our interests. Some aspects of games played in information systems design and maintaining were also mentioned in [5] while in [6] the role of self-organisation in distributed databases development was considered. The aim of this paper is presentation of a more detailed model of distributed information resources gathering and their profiles forming based on the theory of *n*-person games.

2 Basic Model Assumptions

It will be considered a system of primary information supply (*IS*) and a one of information distribution (*ID*) as two complementary components of information market. The *IS* system is the one where information in the form of various types of documents (electronic, photo, multi-media, hard copy manuscripts, publications, etc.) is offered to the customers. For many years this type of information distribution and exchange was prevailing.

This situation has been changed with computer networks development; the action of *ID* system is based on electronic documents exchange through computer networks. *ID* system thus consists of a set of open-access data banks (*OADB*) and of a number of dedicated local data banks (*LDB*), mutually connected by computer network, as shown in Fig. 1.

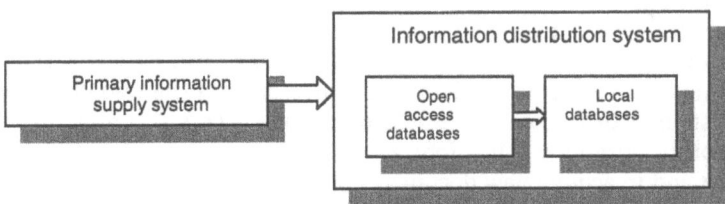

Fig. 1. A model of information storage and distribution

Information resources of *OADB*s are permanently extended and supplied with documents from the *IS* system. The role of *OADB*s consists in brokerage: they buy information documents, select information from them, transform, store and sell or distribute it among the *LDB*s, the last being explored by organisations or by individual users. Information resources of the *LDB*s are thus completed according to the needs of the users. The *LDB*s can be supplied by information not only from *OADB*s, but also from their proper information sources, as well as directly from the *IS* system.

It is assumed that each document offered at the information market can be described by the following characteristics: 1) *formal properties*: type of the document, its author(s), title, volume, editor, date of edition, etc.; 2) *contents* (described by keywords, descriptors, etc.); 3) *acquirement conditions* (name of the seller, price, forms of payment, delivery time, number of available copies of the document, etc.); 4) *supply indicator* (number of available copies of the document, if limited, infinite otherwise). The document characteristics can be formally represented as elements of a *document characteristics space DCS* being a Cartesian product:

$$DCS = C_f \times C_c \times C_a \times C_s \tag{1}$$

where C_f, C_c, C_a and C_s stand, correspondingly, for the sets of formal, contents, acquirement and supply characteristics' possible values. We shall denote by h_i, $h_i \in DCS$, the characteristic of a document x_i, $i \in [1,2,3,...]$.

The state of information market is a time-varying process: at any fixed time-instant t_0, $t_0 \in T$ (T being a real discrete time-axis), the instant-value of the process is given by a finite subset $X(t_0) \subset DCS$ of the documents that actually are offered on the market. The members x_i of $X(t)$ thus arise at certain time-instants and next, after a certain time-delay they may disappear. However, formal description of a time-process' values in the set theory terms is not very convenient. Instead of this it will be taken into account a set of time-functions: ξ_i :

$$T \rightarrow DCS \tag{2}$$

assigning to each time-instant $t \in T$ a document characteristic (an element of *DCS*) so that the components of C_f and C_c remain constant while those of C_a and C_s may be varying in time. The subsets $X(t)$ contain only the elements (documents) for which the indices δ_i of the components of C_a take positive, non-zero values. For those elements the components of C_a are also defined, otherwise the values $\varnothing$ (= *undefined*) to them are assigned. However, the vector $\xi(t)$ consisting of linearly ordered components $\xi_i(t)$ is not known but for the past and present time-instants t only. It is reasonable to consider it as a realisation of a stochastic vector process $\Xi(t)$ describing the states of the information market changing in time. The subsets $X(t)$ determine, at the same time, the areas of possible *instant decisions* of the *OADB* managers. Their aim consists in actualisation and extension of the *OADBs* according to the expected demands of the customers. Therefore, at any t they can chose the following decisions: 1^0 to select and to acquire new documents in order to include and to keep them in the data banks, 2^0 to select some documents and to delay a final decision about their acquirement, and 3^0 to reject all actual proposals concerning documents selling. In the cases 1^0 an d 2^0 the decision made by the ν-th *OADB* manager ($\nu = 1,2,...,N$, where N denotes the total number of *OADB* managers) takes the form of a subset:

$$\psi_\nu(t) \subseteq DCS \tag{3}$$

whose members correspond to the selected elements of $X(t)$ and are such that:

1^0 the projections of the members of $\psi_\nu(t)$ on $C_f \times C_c \times C_a$ are equal to the corresponding elements of $X(t)$ projected on $C_f \times C_c \times C_a$;

2^0 the values of the components β_i of C_s in the members of $\psi_\nu(t)$ satisfy the inequalities

$$0 \leq \beta_i(t) \leq \delta_i(t), \tag{4}$$

which means that the documents acquisition requirement can not exceed the corresponding supply indicators. The document suppliers then collect the requirements and try to realise them. The number of sold document's copies can not exceed the declared supply indicator. Therefore, it may happen that some document acquisition requirements are not satisfied. However, the strategy of clients' selection on the information market here will not be considered. In any case, when coming to the next $(t+1)$ time-instant the supply indicators should be actualised: reduced by the numbers of sold document copies and increased by the numbers of the new-supplied ones.

The documents acquired by the OADBs can be included into the data banks directly or after a transformation process changing their form or generating some secondary documents on the basis of information selected from the primary ones. In similar way as shown before a modified *document characteristics space* can be defined:

$$DCS^{(v)} = C^{(v)}_f \times C^{(v)}_c \times C^{(v)}_a \qquad (5)$$

Here the notions $C^{(v)}_f$, $C^{(v)}_c$ and $C^{(v)}_a$ are, in general, similar to the C_f, C_c and C_a used in (1); however, a dependence on v indicates that each (v-th) OADB may use its proper language and standards for data files characterisation and define data acquirement conditions for the customers. So, a subset $K^{(v)} \subseteq DCS^{(v)}$ plays the role of a catalogue of data offered to the managers of LDBs or to individual users. A projection of $K^{(v)}$ on the document characteristic subspace $C^{(v)}_f \times C^{(v)}_c$, $L^{(v)} \subseteq C^{(v)}_f \times C^{(v)}_c$, will be called a *profile of the v-th OADB* ($OADB^{(v)}$), and its elements will be denoted by $\lambda^{(v)}$. A substantial difference between the formerly defined characteristics h_i and the $\lambda^{(v)}$ ones consists in the fact that h_i describes the document in its original form while $\lambda^{(v)}$ describes secondary documents in electronic form. Forming the profiles $L^{(v)}$ of the OADBs is the main element of long-term strategies of OADB's managers.

Then, let us take into consideration the managers' of LDBs point of view. They represent the interests of some groups of information users (or are information users themselves). The users need to have easy access to information resources suitable for satisfying their intellectual (educational, cultural, etc.) interests or for solving some professional (technological, administrative, etc.) tasks. Let us assign index μ to a certain group of information users. Then their information needs can be formally represented by subsets of a Cartesian product describing an *information requirements space*:

$$IRS^{(\mu)} = C*^{(\mu)}_f \times C*^{(\mu)}_c \times C*^{(\mu)}_a \qquad (6)$$

Once more, the notions $C*^{(\mu)}_f$, $C*^{(\mu)}_c$ and $C*^{(\mu)}_a$ are, in general, similar to the C_f, C_c and C_a ones; however, a dependence on m shows that each (m-th) LDB may use its proper language and standards for data files characterisation and define additional conditions for data acquisition (like admissible cost, time-delay, etc.). In particular, the sets $C*^{(\mu)}_f$, $C*^{(\mu)}_c$ and $C*^{(\mu)}_a$ may contain an element * (*any possible*) to be used if some characteristics of data files or records are not fixed by the users. The subset $R^{(\mu)} \subseteq IRS^{(\mu)}$ characterising information needs of the user(s) can be used in two ways: 1) for information retrieval in $LDB^{(\mu)}$, and 2) for actualisation of *local information resources* ($LIR^{(\mu)}$) in $LDB^{(\mu)}$.

$LIR^{(\mu)}$ is also a subset of $IRS^{(\mu)}$. An information retrieval order can be realised if a certain *consistency level* between $R^{(\mu)}$ and $LIR^{(\mu)}$ is reached, otherwise it is necessary to import the necessary data from the $OADB$'s. However, the managers of $LDBs$ may conduct a more active policy of users' requirements realisation. This can be reached by a permanent monitoring of the information requirements flow: ... $R^{(\mu)}(t-3)$, $R^{(\mu)}(t-2)$, $R^{(\mu)}(t-1)$, $R^{(\mu)}(t)$ (t being the current time) in order to define a preferable *local resources profile*. The last for the given $LDB^{(\mu)}$ can be defined as a subset

$$\Lambda^{(\mu)} \subseteq C^{*}{}^{(\mu)}_{t} \times C^{*}{}^{(\mu)}_{c} \tag{7}$$

such that if $LIR^{(\mu)} = \Lambda^{(\mu)}$ then a considerable part of expected information requirements can be directly realised. The managers of $LDBs$ try to achieve this situation within their possibilities as it will be shown below.

3 Strategies for *OADB* Managers

The $OADBs$' and $LDBs$' managers are interested in realisation of information requirements of their customers. The managers of $OADBs$ are customers on the IS market and, at the same time, they are data sellers with respect to the managers of $LDBs$. On the other hand, the last ones are data suppliers for the information users. Buying a document x_i available on IS market needs covering a cost κ_i being indicated as a component of the corresponding document characteristic. The same document (or data drawn from it) included into the information resources of $OADB^{(v)}$ is expected to be distributed among a number of $LDBs$ to the information profiles of which it suits. As a consequence, the manager of $OADB^{(v)}$ expects to reach proceeds of $r_i^{(v)}$. Finally, its expected profit from buying and distributing x_i is

$$c_i^{(v)} = r_i^{(v)} - \kappa_i \tag{8}$$

It might seem that there is a simple decision rule for the $OADB^{(v)}$ manager:

$$\left.\begin{array}{l} \text{buy } x_i \text{ if } c_i^{(v)} \geq e_i^{(v)} > 0, \\ \text{delay the decision if } 0 < c_i^{(v)} < e_i^{(v)} \\ \text{do not buy it otherwise.} \end{array}\right\} \tag{9}$$

where $e_i^{(v)}$ is a threshold. However, there arise the following problems: $1°$ how to evaluate $c_i^{(v)}$ (or $r_i^{(v)}$, see (8)), and $2°$ how to establish $e_i^{(v)}$?

For answering the first question there can be made the following assumptions:

a/ expected proceeds $r_i^{(v)}$ can be described by an increasing function of the number of customers whose information profiles $\Lambda^{(\mu)}$ are *consistent* with the information characteristics h_i of x_i and of a total measure of this *consistency*;

b/ the information profiles $\Lambda^{(\mu)}$ of the $LDBs$ are not known exactly to the manager of $OADB^{(v)}$, he can only approximate them in the form of information profile $L^{(v)}$ constructed on the basis of all past and actual information requirements from the $LDBs$;

c/ $r_i^{(v)}$ is a decreasing function of other $OADBs$ in the ID system that will offer x_i and/or some secondary information drawn from it.

Then, it arises the next problem: how to evaluate the *measure of consistency* between a document characteristic h_i and the profile of $L^{(v)}$, assuming that (3) holds and $h_i \in C^{(v)}_f \times C^{(v)}_c$. Let us remark that in general the elements of $C^{(v)}_f \times C^{(v)}_c$ are not vectors in algebraic sense, but rather some strings of elementary data of various formal nature. Therefore, it is not possible to take an ordinary *distance measure* concept as a basis of a *consistency measure* definition. However, the last can be based on a generalised, *multi-aspect similarity measure* concept proposed in [8]. In this case, if A is a non-empty set, then a similarity measure between its elements can be defined as a function:

$$\sigma: A \times A \rightarrow [0,1]_c \qquad (10)$$

where $[p,q]_c$ denotes a continuous interval between p and q, such that: a/ for each $a \in A$ there is $\sigma(a,a) \equiv 1$; b/ for any $a, b \in A$ there is $\sigma(a,b) \equiv \sigma(b,a)$; c/ for any $a, b, c \in A$ there is $\sigma(a,c) \geq \sigma(a,b) \cdot \sigma(b,c)$.

If $f^{(r)} = [f^{(r)}_1, f^{(r)}_2, ..., f^{(r)}_p, ..., f^{(r)}_g]$ and $f^{(s)} = [f^{(s)}_1, f^{(s)}_2, ..., f^{(s)}_p, ..., f^{(s)}_g]$ are two strings of characteristics whose components are of various formal nature then a measure of multi-aspect similarity can be defined as a product:

$$\sigma(f^{(r)}, f^{(s)}) = \sigma_1(f^{(r)}_1, f^{(s)}_1) \cdot \sigma_2(f^{(r)}_2, f^{(s)}_2) \cdot ... \cdot \sigma_g(f^{(r)}_g, f^{(s)}_g) \qquad (11)$$

This definition can be used directly to the similarity evaluation of documents characteristics. Let $\sigma(h_i, h_j)$ be a similarity measure described on the Cartesian product $A = C^{(v)}_f \times C^{(v)}_c$. Then it will be said that a member $h_p, h_i \in A$ is *adherent* to a subset $L^{(v)} \subseteq A$ on a level ε, $0 < \varepsilon \leq 1$, if there is at least one element $h_j \in A$ such that $\varepsilon \leq \sigma(h_i, h_j) \leq 1$. Here ε is a threshold   chosen according to the application requirements.

Adherence of h_i to $L^{(v)}$ on a fixed level is necessary, but not a sufficient condition for making a positive decision according to the rule (9). For this purpose the $OADB^{(v)}$ manager should also take into account: 1° how many characteristics h_j in $L^{(v)}$, for all possible h_p, satisfy the adherence condition, 2° how many customers have declared their interests in acquiring the data characterised by h_p, and 3° how long time has been passed since the last call for h_r The corresponding, additional data can be stored and included into the information requirement characteristics (see (6)). Taking them into account a *measure of consistency* $\Gamma(h_i, L^{(v)})$ between h_i and $L^{(v)}$ can be defined as it will be illustrated below.

Let us assume that the contents of documents are characterised by keywords that are presented in a linear order: $w_1, w_2, w_3, ...$, etc. Let us take into account two documents, whose characteristics h_i, and h_j contain, correspondingly, the subsets of keywords W_i and W_p. There will be considered the sets: $W_i \cup W_p$, and $W_i \cap W_p$. If the cardinal number of a set W is denoted by $|W|$ then the similarity measure of the above-mentioned sets can be defined as

$$\sigma(W_i, W_j) = \frac{|W_i \cap W_j|}{|W_i \cup W_j|} \qquad (12)$$

This similarity measure can be taken as a basis of the $\Gamma(h_i, L^{(v)})$ definition. First, let us remark that $L^{(v)}$ can be formally interpreted as a virtual document consisting of all documents whose characteristics were in the past required by a considerable part of customers. Therefore, the formula (12) can be used for evaluation of similarity between h_i and $L^{(v)}$, as well. Let us suppose that in the given $OADB^{(v)}$ the requirements are registered in such a way that to any keyword w_α there are assigned the numbers $e_{\alpha,\tau}$, $\tau = 0,1,2,...,T$ indicating how many times w_α was mentioned in the information calls in the present ($\tau = 0$), as well as in the former time-periods (years). Then the following weight coefficient:

$$\lambda_\alpha = \sum_{\tau=0}^{T} \frac{e_{\alpha,\tau}}{\tau+1} \tag{13}$$

and, at last we can put

$$\Gamma(h_i, L^{(v)}) = \sigma(h_i, L^{(v)}) \cdot \lambda_\alpha \tag{14}$$

where the sum λ_α on the right side is taken over all keywords occurring both in h_i and $L^{(v)}$. The manager of $OADB^{(v)}$ thus can assume a proportionality:

$$c_i^{(v)} = k \cdot \Gamma(h_i, L^{(v)}) \tag{15}$$

(k being a positive coefficient of proportionality) meaning that the greater is the consistency measure between h_i and the profile $L^{(v)}$, the higher are the expected proceeds from selling information drawn from x_i. However, in the situation when the supply of x_i on the market is high (i.e. > 1) the same document can be bought and distributed by other $OADBs$. In such case, assuming that the number of $LDBs$ interested in acquiring x_i is unchanged, the incomes of the $OADBs$ will be inversely proportional to the number of $OADBs$ acquiring x_i. Therefore, the managers of $OADBs$ in this case play a competitive game with fixed positive total gain. They may all acquire the given document; in such case, if they do it at the same time, they will share the gain proportionally to the number of customers that will require access to data from the document. However, the number of $OADBs$ buying the document a priori is not known. In such case the manager of $OADB^{(v)}$ may, first, delay his decision in order to recognise the actions of other players. However, if the delay is large then the expected proceeds will be reduced as well.

The risk of acquiring documents which will not be needed by the customers may be not accessible for the $OADBs$' managers. In such case they may change the game rules by establishing a coalition among them. The coalition consists in specialization of their information resources so that the information profiles $L^{(v)}$ are (at least, approximately) disjoint. As a consequence, the $OADBs$ will share the information requirements of the customers according to their contents. In such case the expected incomes of the $OADBs$ lose the factor of uncertainty connected with the number of competitive $OADBs$ offering access to the same documents.

4 Strategies for *LDB* Managers

The relationships between the LDB managers and the information users is not the same as this one between the $OADBs$' and $LDBs$' managers. The difference consists in the fact that the $LDBs$' managers and their information users usually act within the same organisations Therefore, up to a certain degree their interests are convergent. However, a source of conflicts may be connected with the fact that the $LDBs$'

manager has at his disposal a limited amount of financial resources that can be used to a realisation of information requirements of his clients. In such case he must establish a system of priorities in acquiring new data from the *OADBs*. As a consequence, the users attached to a certain $LDB^{(\mu)}$ may have partially common and partially competitive interests. Assuming that the *LDBs'* manager has no reason for distinguishing some information users against some other ones he will try to make his decisions on basis of the information profile $A^{(\mu)}$. His problem is thus as follows: 1) He has at his disposal an amount ζ of financial means that he can spent for extending the $LIR^{(\mu)}$ for a certain time T^*; 2) He knows the actual information requirements $R^{(\mu)}$ of his clients (data users) that can not be realised on the basis of $LIR^{(\mu)}$; 3) He knows the actual information resources of the *OADBs* as well as the corresponding data selling and/or data access conditions.

How to realise the information needs of the clients within the financial and/or formal limits?

It is assumed that the prices of an incidental access to some data and of data buying for their permanent using within the organisation are different (the last being usually much higher). Therefore, if an analysis of the *OADBs'* resources shows that a certain requirement can be realised by importing data then it is necessary to determine the expected costs of: a/ data buying and b/ data access during the time-interval T^* for various *OADBs* and to find out the minimum ones. At the next step the conflict of interests of various clients should be solved: it has been determined a set of data acquisition offers: $O_1, O_2, O_3...$ etc., each one being characterised by its expected cost χ_1 and by the subset of clients that might be interested in using the given data in the time-interval T^*. The problem is: how to select a subset of the offers for their final acceptation?

The problem can be solved if a system of ordering the offers is defined. Each offer can be characterised by a vector whose components indicate: 1° the expected cost χ_1, 2° the number b_1 of clients that are interested in using the given data, and 3° the mean measure of similarity s_1 of the offered data to the actual (and expected, if possible) data requirements. Therefore, it arises the problem of semi-ordering of the vectors $u_1 = [\chi_1, b_1, s_1]$ in a three-dimensional space. The problem can be solved easily on the basis of K-space (Kantorovich space) concept [9]. Then the strategy of the *LDB* manager consists in rearranging the offers: $O_{\rho_1}, O_{\rho_2}, O_{\rho_3}$ etc. (ρ_1, ρ_1, ρ_1 etc. being some integers) and in accepting for realisation as much offers as possible within the financial limits.

5   Conclusions

We tried to show that the relationships between the managers of *OADBs* and of *LDBs* have the form of a *n*-person game of partially co-operative, partially non-co-operative type with incomplete information. It is possible to define the strategies of the players. The strategies are realised by creation of database profiles that are used in making decisions concerning documents or data acquisition. The decision rules of the players can be strongly optimised only in particular, rather simple cases. In general, it has been shown that a co-operation between the database managers may them lead to less risk in reaching their goals.

References

1. Neumann, von J., Morgenstern O.: Theory of Games and Economic Behavior. Princeton (1944)
2. Dubin, G.N., Suzdal, V.G.: Introduction to Applied Theory of Games. Nauka, Moskva (1981) (in Russian)
3. Owen, G.: Game Theory. W.B. Saunders Company, Philadelphia (1968)
4. Berzin, E.A.: Optimum Distribution of Resources and Theory of Games. Radio i Svjaz', Moskva (1983) (in Russian).
5. Kulikowski, J.L.: Problems of Information Systems Design in Social Environment (in Polish). Techniki i metody rozproszonego przetwarzania danych, cz. V (M. Bazewicz ,ed.). Wroclaw Univ. of Tech. Press (1990).
6. Kulikowski, J.L.: Self-organization in Distributed Data Bases. Information Systems' Architecture and Technology ISAT'93 (M. Bazewicz ed.). Wroclaw Univ. of Tech. Press (1993), 94-104
7. Vorob'ev, N.N.: Principles of game theory. Non-co-operative games. Nauka, Moskva (1994)
8. Kulikowski, J.L.: From Pattern Recognition to Image Interpretation. Biocybernetics and Biomedical Engineering, vol. 22, No 2-3 (2002) 177-197.
9. Kantorovich, L.V., Vulich B.Z., Pinsker A.G.: Funkcjonalnyj analiz v poluuporiadocennych prostranstvach. GITTL, Moskva (1959).

Domain Knowledge Modelling for Intelligent Instructional Systems

Emilia Pecheanu, Luminita Dumitriu, and Cristina Segal

Department of Computer Science, "Dunarea de Jos" University of Galati
Domneasca 47, GALATI 800008 – ROMANIA
{emilia.pecheanu,luminita.dumitriu,cristina.segal}@ugal.ro

Abstract. This paper presents a solution for conceptually modeling the training domain knowledge of computer-assisted instructional environments. A stringent requirement for an effective computer-assisted instructional environment is to elaborate a proper answer for the differentiated cognitive demands of users. The different cognitive style of learners imposes different modalities of presenting and structuring the information (the knowledge) to be taught. Conceptual organization of the training domain knowledge, with learning stages phasing, can constitute a better solution to the problem of adapting the instructional system interaction to users with different cognitive style and needs.

1 Learning and Individual Cognitive Styles

Unanimously, psychologists and teachers consider that learning is a process that engages all aspects of the human personality. Whatever the learning content, from the simplest learning tasks to the most complex ones, a whole chain of psychical mechanisms is triggered, in what concerns receiving, processing, interpreting and valorizing the instructional entities [5].

Learners' cognitive style leads to individual variations within the learning process: each person has his own way of thinking and, thereby, the acquisition of new knowledge is person-specific. The concept of cognitive style ignores the knowledge content, emphasizing on the manner in which a person succeeds in the cognitive process [7]. These differences can be expressed through the relationships between heredity, environment and educational factors, in which these developments are often contributing with different intensities from person to person [7] , [5].

The different cognitive style of learners imposes different modalities of presenting and structuring the knowledge to be taught. By individual cognitive style analysis, two ways of approaching the learning process by human subjects are emphasized:

1. *a closed cognitive style approach*, in which the learner follows algorithmic mental schemes and a convergent way of thinking; this style has as origin extrinsic motivations and prevalently uses memorization;
2. *an open cognitive style approach*, in which the learner mostly follows heuristic mental schemes and uses a divergent way of thinking; this second style is generated by an intrinsic motivation and it is based mostly on thinking flexibly.

M. Bubak et al. (Eds.): ICCS 2004, LNCS 3038, pp. 497–504, 2004.
© Springer-Verlag Berlin Heidelberg 2004

Differences between the users' individual cognitive styles should be reflected in the manner of structuring the domain-knowledge and in the manner of designing the user interface. Correlating the assistance capability of instructional systems with different cognitive styles of potential users has lately become an important study and research subject in the area of computer–assisted instruction.

The possibility of personalizing the educational process is the main demand in building computer-assisted learning systems for any form of education, continuous education and professional re-conversion included.

In the case of these education forms, the subject of study frequently consists of groups of users; between these users there can be significant differences concerning the training level, age or internal goals pursued through study.

Consequently, the way the educational process is approached – the cognitive style – can be extremely different among the users who form the study groups.

When computer-assisted learning systems are used for any of these forms of education, they have to be able to dynamically adapt to the various cognitive necessities and demands of the users, in order to ensure the efficiency of the educational act.

2 Implementing Adaptive Training Systems

Considerable research effort has gone into implementing adaptive (depending on the cognitive style) training systems. A direction pursued by many systems designers is to dynamically build a behavioral – cognitive profile of each user (the "Student Model"). Based on an individual profile (corroborated with training objectives), an individualized model of the system's interaction with each user is generated [10].

Another trend in implementing adaptive web-based training systems consists of a specific structuring of the content elements included in the Domain Model (the Domain Model represents the knowledge belonging to the training area).

This structuring is achieved either from the training phases of the instructional process, or from a hierarchy of clearly stated instructional goals and sub-goals. In both cases, knowledge space browsing is monitored in order to dynamically adjust it whenever the user correctly performs certain acts, or reaches certain stages [4].

The aspects mentioned above show that implementing training systems able to adapt (in certain circumstances) to various training requests, coming from users with different cognitive profiles, is possible. Anyway, there is a second basic problem, with no exhaustive solution yet: which method can lead to optimal structuring and proper sequencing of the instructional material included in the system, for each kind of user? In other words, what kind of general structure of knowledge is needed and how must it be delivered to users with such different cognitive profiles? [6].

An answer to this question cannot be given unless taking into consideration the problem of conceptual structuring the domain's knowledge, before considering its presentation through pedagogical sequences. The organization model of the training material, which is tightly connected to the representation of the training domain knowledge, has to allow multiple views of the concepts in the training domain.

Studies of computer-based training systems have shown that if they contain a structural model, they facilitate learning. The network-like model with interconnected knowledge nodes would best reflect the schemata (structure) -based learning paradigm. In Bartletts's Schemata Theory [1] it is considered that learning occurs as a result of accumulation and reorganization at the human cognitive–mental structures level. These structures would be the way knowledge objects (ideas, events, scripts etc) are represented and organized in the semantic memory of each human being.

3   A Theoretical Model for Representing the Domain Knowledge of an Instructional System

The model described herein is destined to represent the domain knowledge of an instructional system. The model was called COUL-M (COnceptual Units' Lattice Model). The mathematical formalization of the COUL-M model has been developed in [8] and has been derived mainly from two mathematical theories: the Formal Concepts Analysis, [3] and the Logical Concept Analysis,[2].

In the COUL-M approach, the key elements in modeling the domain knowledge of a CAI system are the followings: the Conceptual Unit, the Conceptual Structure, and the Conceptual Transition Path. These three elements will be defined below.

Definition 1: A **Conceptual Unit** C_i is a group of related concepts belonging to the domain knowledge of an instructional system. Conceptual Units are obtained by applying a relation (or a set of relations) over a set of concepts belonging to that domain knowledge.

Definition 2: A **Conceptual Structure** S is a tuple (N, R_N, C_S, L_S), where N is a set of concepts belonging to the domain knowledge of a course or discipline, R_N is an order relation over N, C_S is the set of Conceptual Units obtained by applying the relation R_N over the set N, and L_S is the set of traversal paths of the structure S.

Definition 3: A **Conceptual Transition Path** $T_s \in L_S$ is a sequence of conceptual units $(C_{1s}, C_{2s}, ..., C_{Ns})$ connected one after another within a Conceptual Structure S. The conceptual transition path T_s has as origin the Conceptual Unit C_{1s}, considered initial, and as destination a Conceptual Unit C_{Ns}, which encapsulates a set of concepts comprising a study goal within an interactive course.

A Conceptual Structure is a model meant to represent the domain knowledge of an interactive course. A Conceptual Structure should map the cognitive structure of the domain knowledge, and should also reflect the pedagogical vision of the teacher/author of that course. The model (the Conceptual Structure) has to allow for flexibility, i.e., to provide as many as possible transition paths for learning (assimilating) the domain's main concepts. The COUL-M model [9] is a solution for building the Conceptual Structure covering the domain knowledge of a CAI system.

The starting pointing in developing the COUL-M model (and thus, the Conceptual Structure) for a certain domain is the analysis of the pedagogical relations between that domain's concepts. From a pedagogical point of view, several types of relations can exist among the concepts within a domain knowledge: precedence, contribution, hierarchical order, implication, equivalence, influence, etc.

From an instructional point of view, two such relations might be of interest:

1. The Precedence Relation between concepts, which refers to the order of presenting the concepts within an interactive course;
2. The Contribution Relation between concepts, which refers to the explicit contribution (participation) of a concept in the presentation of another concept.

The mathematical formalization of the above relations (the Precedence and the Contribution Relation), combined with their transformation into incidence relations [8], allows for building a formal context and then, the concept lattice for this context. The terms of formal context, formal concept and concept lattice are considered in the sense of the Formal Concept Analysis approach [3]. The resulting model, a complete lattice of formal concepts, expresses the relations among the course concepts, as described by the teacher/author. The lattice-like model implicitly includes the generalization-specialization relation. The formal concept lattice is isomorphic to a complete lattice of Conceptual Units [8], which is the structure (the model) aggregating the related concepts within the domain knowledge. This aggregation model, lattice-like, can constitute a Conceptual Structure for the domain knowledge [8].

As an example we will consider the domain knowledge of a training system which is meant for assisted instruction in an IT discipline: Operating Systems. Furthermore, we will consider a set of concepts belonging to a specific chapter of this discipline (domain), the "Mutual Exclusion" chapter. Starting from the pedagogical relations which are connecting this chapter's concepts, a COUL-M model of knowledge representation will be built. The most important concepts of the "Mutual Exclusion" chapter are considered to be (in the author's pedagogical vision) the following ones: MEP, CR, CS, WA, Sem, SA, PC, where:

A = MEP = Mutual Exclusion; B = CR = Critical Resource;
C = CS = Critical Section; D = WA = Wait-Activation; E = Sem = Semaphore;
F = SA = Strict-Alternation; G = PC = Producer-Consumer.

It is assumed that the professor who authored the interactive course has specified the pedagogical precedence between these concepts (according to their signification within the subject of study and according to the professor's pedagogical vision) as an ordered list: CR, MEP, CS, Sem, SA, WA, PC.

The representation of these Mutual Exclusion chapter's concepts using the COUL-M model is shown in Fig. 1. As Fig. 1 shows, the COUL-M model (the lattice of Conceptual Units), preserves the significations of the Precedence Relation that has been established initially:

– the formal concepts C2, C3, C4, C5, C6, C7, C8 correspond, respectively, with the course concepts identified as: B, A, C, D, E, F, G (CR, MEP, CS, Sem, SA, WA, PC); the C2 formal concept is a formal representation for the course concept CR, the formal concept C3 is a formal representation for the concept MEP, and so on.
– the complete lattice making up the COUL-M model has the shape of a simple chain, which corresponds to the precedence in any training activity: learning content's pieces in succession, similarly to the linear reading of printed text.

The Contribution Relation between a domain's concepts might be modeled in a very similar way: by applying a sequence of mathematical operations (decomposition and aggregation) this relation can be transformed in an incidence relation [8]. Then,

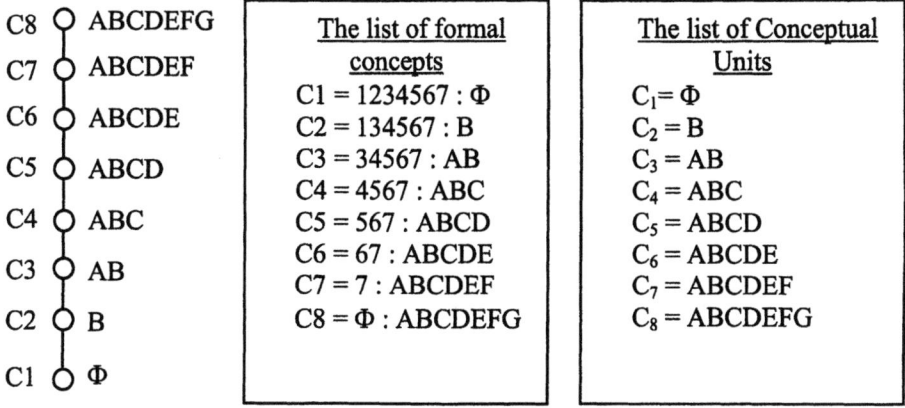

Fig. 1. The COUL-M model, based upon the Precedence Relation within a domain knowledge

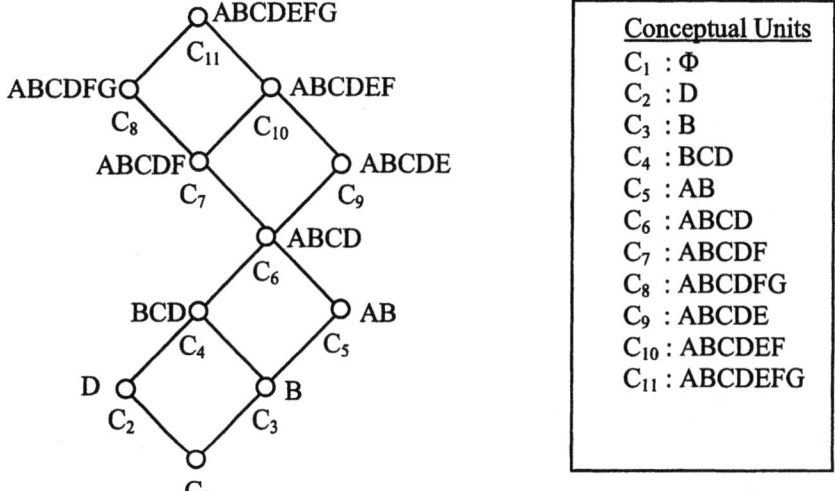

Fig. 2. The COUL-M model, based upon the Contribution Relation within the domain knowledge of an instructional environment

the formal concept and concept lattice can be built and the lattice of Conceptual Units, standing as the Conceptual Structure of the domain, can be finally derived, as shown in Fig.2.

Finally, the Fig. 3. is showing the COUL-M model of representation in case of combining the Precedence Relation and the Contribution Relation in the same structure (formal context subpozition[3].

A further development can be realized if the course concepts are considered to be teach by using several pedagogical methods, and if we consider a pedagogical method as a set of related pedagogical resources. As a result, a new, more detailed model of knowledge representation can be built: the COUL-FM, Conceptual Units Lattice – Formal Model. In the new model, COUL-FM, the Conceptual Units become logical formulae expressing the aggregation between the conceptual model of the domain knowledge, instructional methods and instructional resources.

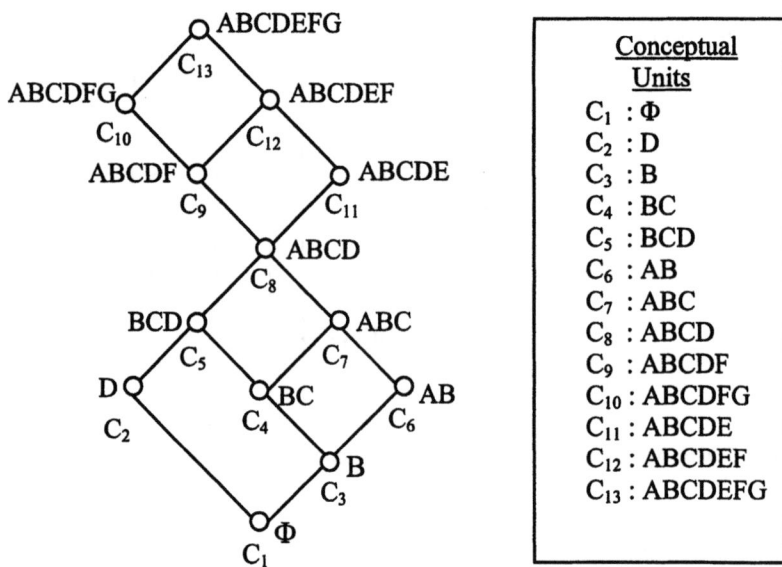

	Conceptual Units
C_1	: Φ
C_2	: D
C_3	: B
C_4	: BC
C_5	: BCD
C_6	: AB
C_7	: ABC
C_8	: ABCD
C_9	: ABCDF
C_{10}	: ABCDFG
C_{11}	: ABCDE
C_{12}	: ABCDEF
C_{13}	: ABCDEFG

Fig. 3. The COUL-M model, based upon the Precedence Relation and the Contribution Relation within the domain knowledge of an instructional environment

Table 1. Instructional methods and Pedagogical Resources for a domain chapter.

Concept	Resources-Roles	Instructional Methods	Instructional Objectives
MEP (A)	R_1^A: Introduction R_2^A: Presentation	$M_1^A = \{ R_1^A, R_2^A \}$	$O_1 \models M_1^A$
CR (B)	R_1^B: Introduction R_2^B: Presentation	$M_1^B = \{ R_1^B, R_2^B \}$	$O_2 \models M_1^B$
CS (C)	R_1^C: Presentation R_2^C: Assessment	$M_1^C = \{ R_1^C, R_2^C \}$	$O_3 \models M_1^C$
Sem (D)	R_1^D: Introduction R_2^D: Presentation R_3^D: Assessment	$M_1^D = \{ R_1^D, R_2^D \}$ $M_2^D = \{ R_1^D, R_2^D, R_3^D \}$	$O_4 \models M_1^D \vee M_2^D$
SA (E)	R_1^E: Introduction R_2^E: Presentation R_3^E: Assessment	$M_1^E = \{ R_1^E, R_2^E \}$ $M_2^E = \{ R_1^E, R_2^E, R_3^E \}$	$O_5 \models M_1^E \vee M_2^E$
WA (F)	R_1^F: Presentation R_2^F: Assessment	$M_1^F = \{ R_1^F, R_2^F \}$	$O_6 \models M_1^F$
PC (G)	R_1^G: Introduction R_2^G: Presentation R_3^G: Synthesis R_4^G: Assessment	$M_1^G = \{ R_1^G, R_2^G \}$ $M_2^G = \{ R_1^G, R_2^G, R_4^G \}$ $M_3^G = \{ R_1^G, R_2^G, R_3^G, R_4^G \}$	$O_7 \models M_1^G \vee M_2^G \vee M_3^G$

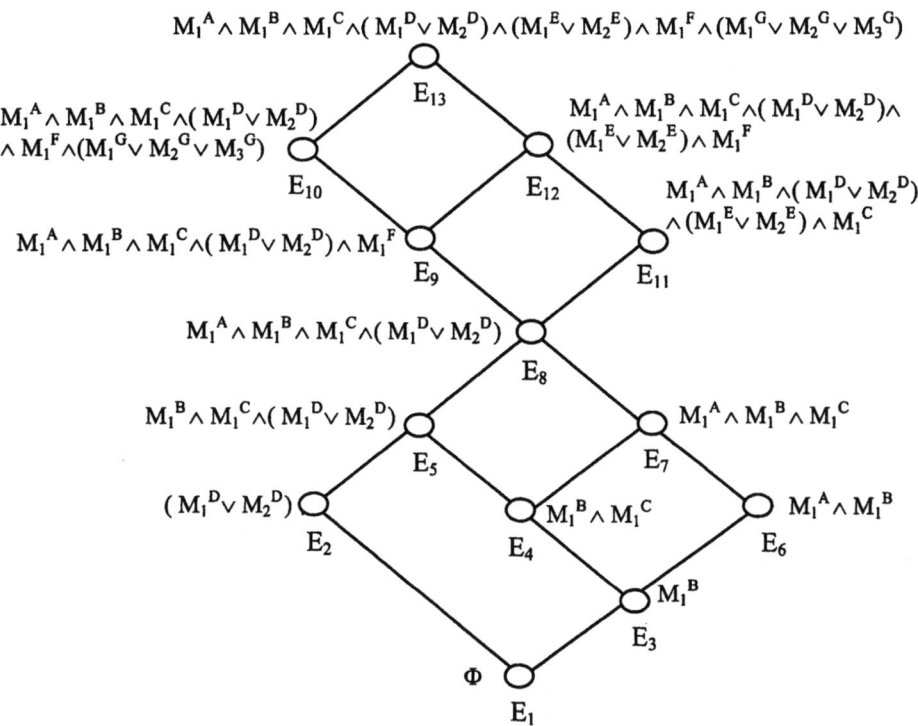

Fig. 4. The COUL-FM model of representation for the domain knowledge, instructional methods and instructional resources for a chapter of the Operating Systems discipline

Table 1 is showing a set of pedagogical methods (viewed as collection of resources with a well stated pedagogical role in a lesson's scenario, for teaching a concept) and Fig. 4 is showing the mapping of this instructional elements over the lattice-like model of the domain knowledge. In this manner, a knowledge base for an intelligent instructional system can be built, and consequently, an inference engine able to perform complex information processing in the frame of that system.

4 Authoring the Domain Knowledge and Instructional Knowledge

The COUL-FM model has been implemented by means of a software tool: a knowledge compiler named COUL-COMP (COnceptual Units' Lattice – knowledge COMPiler). COUL-COMP stands for an authoring system, able to realize the representation of the domain knowledge for a computer assisted instructional system.

A easy to use specification language, COUL-SL (Conceptual Units' Lattice-Specification Language) has been also developed, in order to allow user to perform:

– the description of the domain's knowledge structure, i.e., the specification of the relations between the domain's concepts(Precedence and Contribution Relation;
– the specifications of the instructional methods the training system might use and of the Dependence Relations between pedagogical resources and domain concepts;

- the specification of pedagogical resources and the role they can play in a an instructional method deployment.

The compiler COUL-COMP is able to lexically and syntactically check the correctness of a "program", i.e., a set of specifications written in COUL-SL language If the program is syntactically correct, the compiler will produce o set of pedagogical prescriptions, in a form of Conceptual Transition Paths. Each Transition Path is specifying the Conceptual Units- stages in the learning process, or "nodes" in the Conceptual Transition Path and the "new" concept learner has to understand and assimilate at each stage of the learning process. The methods and related resources necessary to help user to learn each "new" concept of the training domain.

5   Conclusions and Further Developments

The lattice-like COUL-M model is able to represent, in a comprehensible way, the relations between the concepts of the training domain of a CAI system. Furthermore, the COUL-FM model is including knowledge about the targeted instructional methods and the existing instructional resources.

The compiler COUL-COMP is a computational representation of the COUL-FM model, able to extract pedagogical prescriptions from the model. The COUL model can constitute o low level layer for various types of training systems. The model is providing some essential elements for developing adaptive training systems: a mapping of instructional methods and pedagogical resources over a conceptual structure of the domain knowledge. The COUL-M model can be integrated in any instructional environment including a Pool of Pedagogical resources (like the Ariadne system).

References

1. Brewer, W. F., Nakamura, G.V.: The nature and functions of schemas. In Wyer and Srull, Eds., Handbook of Social Cognition, vol. 1. Hillsdale, NJ: Erlbaum, (1984) 119-160
2. Ferre, S.: Systemes d'information logiques, IRISA, Rennes, (1999).
3. Ganter, B., Wille, R.: Formal Concept Analysis –Mathematical Foundations, Springer Verlag, Berlin (1999).
4. La Passardiere, B., Dufresne, A.: Adaptive Navigational Tools for Educational Hypermedia, Proceedings of ICCAL '92, Nova Scotia, Canada (1992)
5. Larmat, J.: Intelligence's Genetics, Scientific Printing House, Bucarest (1997).
6. Messing, J.: The Use of Teaching Strategy Control Features in Computer Assisted Learning Courseware, Charles Sturt University (1990).
7. Nicola, I.: Pedagogy, Didactical and Pedagogical Printing House, Bucarest (1992).
8. Pecheanu, E.: Conceptually Modeling the Domain Knowledge of an Computer assisted Instructional System, PhD Thesis, Galati, Romania (2003).
9. Pecheanu, E.: Content Modeling in Intelligent Instructional Environments, Proceedings of, KES 2003, Part II pp 1229-1234, Oxford ,UK, 2003, ISBN 3-540-40804-5.
10. Santiago, R.: The Effect of Advisement and Locus of Control on Achievement in Learner-Controlled Instruction, Journal of Computer-Based Instruction, Vol19, No2 (1992).

Hybrid Adaptation of Web-Based Systems User Interfaces

J. Sobecki

Department of Information Systems, Wroclaw University of Technology
Wyb.Wyspianskiego 27, 50-370 Wroclaw, POLAND,
sobecki@pwr.wroc.pl

Abstract: The interface adaptation is one of the methods for increasing the web-based system usability, especially when we consider differences among the population of users. We can distinguish demographic, content-based, and collaborative recommendations. The combination of these approaches that is called hybrid adaptation enables to overcome their disadvantages.

Keywords: Adaptive user interfaces, hybrid recommendation, web-based systems

1 Introduction

The success of the today's web-based information systems relies on the delivery of customized information for their users. The systems with this functionality are often called recommender systems [7]. We can distinguish three types of recommendations: demographic, content-based and collaborative. All of these three types use the user profiles to recommend particular items to the particular user. Recommender systems concerning user interfaces are usually called adaptive or personalized user interfaces [3] and partially belong to the more general class of intelligent user interfaces that contains also different interface agents [5].

One of the most important elements in the recommendation systems is the user model that is usually built of the two elements: the user data and the usage data [3]. The user data contains different information and it's the most common part is the demographic data containing: record data (name, address, e-mail, etc.), geographic data (zip-code, city, state, country), user's characteristics (sex, education, occupation), and some other customer qualifying data. The user data may contain also information about the users' knowledge, their skills and capabilities, their interests and preferences and also their plans and goals.

The second element of the user model, the usage data, is observed and recorded during the user's interactions with web-based systems. The usage data may concern selective operations that express users' interests, unfamiliarity or preferences, temporal viewing behavior, as well as ratings concerning the relevance of these elements.

The methods of building the user model are very differentiated. The user data is usually entered by the users themselves, but sometimes other methods from are used to associate these data with ones delivered for the other web-based system. The usage

M. Bubak et al. (Eds.): ICCS 2004, LNCS 3038, pp. 505–512, 2004.
© Springer-Verlag Berlin Heidelberg 2004

data are usually collected automatically by the client and/or system sides by means of very different technologies ranging from CGI, PHP, ASP, Flash and Cookies to DoubleClick [14]. These entries can contain a lot of different data and settings, such as URL's of visited pages, links followed by the users, data entered by them into the forms on the pages and interface settings made by the users themselves as well as the goods purchased in the e-shops.

The user model is used by the recommendation process according to the implemented approach. Demographic filtering [7] takes descriptions of people from the user profile to learn the relationship between a particular item and the type of people who like it. Content-based filtering uses descriptions of the content of the items to learn the relationship between a single user and the description of the items according to the user usage description. Finally, the collaborative filtering uses the feedback from a set of some how similar people concerning a set of items in order to make recommendations.

In this paper, first the collaborative user interface recommendation with the demographic elements that is using consensus methods is presented. Then the hybrid architecture that is some combination of the former solution with some elements of content-based approach will be presented.

2 User Interface Recommendation Approaches

There are three basic recommendation approaches: demographic, content-based and collaborative. The demographic approach is using stereotype reasoning [3] in its recommendations and is based on the information stored in the user profile that contains different demographic features [7].

According to [3] stereotype reasoning is a classification problem that is aimed at generating initial predictions about the user. Usually the data used in the classification process is of the demographic type. This type of recommendation is also used in the initial steps of the collaborative user interface recommendations [10].

The demographic recommendations have however two basic disadvantages [7]:

- for many users recommendations may be too general;
- they do not provide any adaptation to user interests changing over time.

Content-based filtering takes descriptions of the content of the previously evaluated items to learn the relationship between a single user and the description of the new items [7]. The application of content-based approach enables personalized and effective recommendations for particular users, but has also some disadvantages:

- content-based approaches depends on so called objective description of the recommended items;
- it tends to overspecialize its recommendations;
- content-based approach is based only on the particular user relevance evaluations, but users usually are very reluctant to give them explicit, so usually other implicit, possibly less adequate, methods must be used.

The collaborative recommendations are able to deliver recommendations based on the relevance feedback from other similar users. Its main advantages over the content-based architecture are the following [7]:

- the community of users can deliver subjective data about items;
- collaborative filtering is able to offer novel items, even such that user have never seen before;
- collaborative recommendation utilizes item ratings of other users to find the best fitting one.

Collaborative recommended agents have also some disadvantages:

- when the number of other similar users is small then the prediction is rather poor;
- the quality of service for users of peculiar tests is also bad; this is rather difficult to get sufficient number of similar users to be able to make proper predictions;
- lack of transparency in the process of prediction and finally the user's personal dislike may be overcome by the number of other similar users opinions.

The disadvantages of content-based and collaborative recommendation could be overcome by applying the hybrid solution. In the works [10,13], where the concept and implementation of the collaborative user interface adaptation using consensus method is presented, the disadvantage of the insufficient number of the similar users at the early stages of the system operation was overcome by application of the demographic stereotype reasoning. However in this architecture the content-based recommendation is not implemented, but individual preferences of the interface settings that are selected manually by the user and stored in the interface profile and used in every system session.

3   User Profile

In works [10,12,13] we distinguish the user profile that contains user data (mainly demographic data) delivered by the user and the interface profile that is designated by the system and may be changed by the user during the personalization proces. In this paper, however, some changes to the previous model will be introduced. Like in the work [3] the user profile will contain the whole user model and the interface profile will be used only for the implementation of the current interface for the particular user.

The user data may contain different user characteristics: demographic data (name, sex, address, occupation, education, customer and psychographic data), user knowledge, user skills and user interests. The user data may also contain direct user preferences concerning the following elements of the user interface: the interface layout, the information content and its structure. For identification and authorization purposes data such as user login and password are also included in the user profile.

The user profile could also contain the usage data that may be observed directly from the user's interaction with a web-based system. This part should contain all versions of interface settings, visited pages and if appropriate also other information on other actions concerning the system purposes, such as purchase of goods, order of catalogs, request for some attendance, etc. In order to know whether the interface settings are appropriate for the particular user we should also store the usability value associated with these settings [8].

In order to find similar users by means of clustering methods and find consensus among interface settings [10] it is necessary to define the distance function in the space of the interface profile.

3.1 Formal Description of the User Profile

In the recommender systems the user profile could be represented by many different forms: binary vectors, feature vectors, trees, decision trees, semantic networks. In our paper we propose to describe the user profile as a tuple defined in the following way: the finite set A^u of attributes of the user profile and the set V^u contains attribute values, where: $V^u = \bigcup_{a \in A^u} V_a^u$ (V_a^u is the domain of attribute a). The tuple p is a function $p : A^u \rightarrow V^u$ where $(\forall a \in A^u)(p(a) \in V_a^u)$. This type of representation is also called as a single valued information system. It is also possible to consider so-called multivalued information that enables to be the attribute values not only atomic but also sets of values. system. In this case we introduce $\Pi(V^u_a)$ that denote the set of subsets of set V_a^u and $\Pi(V^u) = \bigcup_{a \in A^i} \Pi(V_a^u)$. The interface profile is represented by a tuple that is a function $r : A^i \rightarrow \Pi(V)$ where $(\forall a \in A^i)(p(a) \in \Pi(V_a))$. We can also consider more general representation that means the ordered set or even ordered set with repetitions.

The user profile may be represented by other means. An example of application of the tree structure for the user interface description can be found in work [12].

In the user profile we can distinguish several parts represented by the attribute sets. We can distinguish the demographic attributes set D, the set of the interface setting attributes I, the set C of attributes associated with the content (for example visited pages, purchased or ordered items, retrieved elements) and the set U of the usability or relevance of the interface or contents. Finally, we can distinguish some attributes used for identification and authorization purposes T.

3.2 Distance Function Definition

The distance function between values of each attribute of the user profiles is defined as a function $\delta^{at} : V_a^u \times V_a^u \rightarrow [0,1]$ for all $a \in B \subseteq A^u$. This function should be given by the system designer and fulfill all the distance function conditions but not especially all the metrics conditions. The distance function values could be enumerated or given in any procedural form.

The distance between elements of the user profiles could be defined in many different ways. First, the distance between tuples i and j could be defined as a simple sum of distances between values of each attribute:

$$\delta(p_i, p_j) = \sum_{a \in B} \delta^{at}(p_i(a), p_j(a))$$

or we can consider the cosine distance or define the distance as a square root of sum of squares of distances (Euclidean distance). We can also indicate the importance of each attribute a by multiplying the distance by appropriate factor defined as a function $c : A \rightarrow [0,1]$:

$$\delta(p_i, p_j) = \sum_{a \in B} [c(a) * \delta^{at}(p_i(a), p_j(a))]$$

The functions shown above are devoted for the user profile representation in form of a tuple, but in the case of the tree structures we must consider other distance functions [12].

3.3 User Clustering Based on the User Profile

Clustering problem could be defined as a partition of the given set of user profiles into subsets such that a specific criterion is optimized. The criterion is often defined as the average squared Euclidean distance between a profile and the corresponding cluster center. To minimize this criterion we can use k-means clustering that partitions the set of the profiles into k non-overlapping clusters that are identified by their centers. This problem is known to be NP-hard, but it is still attractive because of its simplicity and flexibility [2]. It has however some disadvantages that reveals especially in case of a large datasets, these are: its low speed and lack of scalability; it is possible to obtain local minima instead of global ones.

In the interface recommendation implementation described in [13] Dattola clustering algorithm [1] that is known from the field of Information Retrieval was used. This algorithm is not NP-hard but produces the sub-optimal solution to the clustering problem.

In the Datolla algorithm first the initial centroids must be selected, in our case they are selected by experts. Then for each user profile the distance function (see previous section) between the profile and each centroid is determined. The profile is joined to the group with the closest the centroid and also lower that assumed threshold, those above are assigned to the class of so-called isolated elements. Then for each group the centroides are recalculated and the process is repeated until no one profile changes its class assignment. Finally all profiles from the class of isolated elements are assigned to the group with the lowest distance function values or left as a separate group.

In the field of data mining however other algorithm is used to solve k-means problem. Its name is Lloyd's algorithm [2] and its steps are following. First, select randomly k elements as the starting centers of the clusters (centroides). Second, assign each element of the set to a cluster according to the smallest distance to its centroid. Third, recompute the centroid of each cluster, for example the average of the cluster's elements. Fourth, repeat steps 2 and 3 until some convergence conditions have not been met (for example centroides do not change).

The attractiveness of this algorithm lies in its simplicity and its ability to terminate when using the above mentioned convergence condition and for configurations without equidistant elements to more than one centroid. There is, however, one important problem with k-means algorithm, namely the algorithm takes a long time to run. First, the step 2 that has to be performed in each iteration costs $O(kdN)$, where d is the dimension of each element and N is the number of elements. Second, algorithm usually needs many iterations to terminate. There are however quite many modification of this algorithm that run faster, for example bisecting k-means [2].

3.4 Interface Profile Determination by Means of Consensus Methods

The consensus methods are used to find collaborative interface recommendation for particular user. In order to do this we must have the group of similar users G. Their interface settings with assigned usability values are used for the consensus determination. Let j be the index of the a group G member, r_j be the tuple that is sub-tuple of the p_j such that $r_j(a)=p_j(a)$ for all $a \in C$ where $C \subset A$ is a set of the interface setting attribute and $u(j)$ usability measure assigned to the each of the interface users. Let assume that the usability measure falls in the [0,1] interval and the value 0 denotes completely useless interfaces and 1 denotes ideally useful interface. Then to find the consensus (recommended interface for particular group of users G) we must find the interface settings r that conform the following formula [12]:

$$\min(\sum\nolimits_{j \in G} u(j) * \delta(r, r_j))$$

This problem could be computationally difficult, but according to [9] we can reduce the computation by finding the minimal value for all attributes a of a tuple separately:

$$\min(\sum\nolimits_{j \in G} u(j) * \delta(r(a), r_j(a)))$$

4 Hybrid Recommendation System Architecture

In works [10,13] the user interface recommendation for the web-based system is based on demographic recommendation and collaborative recommendation is presented. The system designers prepare centroides that represent stereotype classes of users. Each centroid has the associated interface profile that for the designers represents the most efficient interface settings for these classes. At start each new user registers to the system. The user delivers mainly demographic data (value of the attributes from the set D) and identification data (set T) that is stored in user data part of the user profile. Then the class centroid with the closest distance to the user profile is found and associated interface profile offered to the user.

According to the interface profile the actual user interface content, layout and structure is generated. The user may start to work with the system and if he or she wishes also modify the interface settings (values of the attributes from the set I). Finally, these settings together with usability evaluation (set U) given by user are also stored in the user profile.

When the system registers required number of users, first users are clustered (using attribute values from the set D) and then according to these clusters using consensus methods on values stored in the set I, new centroides and associated interface profiles for recommendation are distinguished. These procedures may be repeated from time to time in the following occasions: many new users registering to the system or interface recommendations become poor usability ratings.

In the hybrid recommendation architecture we can introduce some modifications concerning individual relevance rating of the interface elements content recommendation. First, using the stereotype reasoning we can ascribe for specified values of the attributes D appropriate values of the attributes from the set I (i.e. when the user specifies the country in the demographic data we can recommend the language of the system or we can recommend the weather forecast concerning the user's place of living). Second, the personalization settings from the set I made by the user and those recommended by the system and presented to the user are not the subject of any further automatic adaptation without the direct user permission.

Finally, we can implement the content-based recommendation. We must remember however that the rules for efficient content-based recommendations strongly depend on the goals of the web based system. For example for web-based information retrieval systems we can consider the previous relevant items as a basis for recommendation of further retrievals. In this case many different methods can be used: fuzzy retrieval, Bayesian networks or other intelligent information retrieval method. The same methods we can use for other similar tasks such as news and e-mails filtering or spam detection. Those methods were especially applied in the field of the interface agents, for example in Letizia [5] or Apt Decisions [11].

For quite many systems however, the logic used for the retrieval systems does not hold, for example when a user buys a thing that is needed in only one item (e.g. car, TV set, DVD player, etc.). In that cases we should rather recommend items that other users who bought the same item were also interested in, for example insurance when a user buys a car. In many cases the logic of the system may be even more complicated, so for each recommended item, we shall define precise relationship between the user profile (or also other users profiles) and this item.

In the content item recommendation we can also consider application of the design heuristics concerning the content and presentation of the information:

- meaning: the information has meaning when it concerns many people;
- time: the information is delivered on time;
- publicity: the information concerns a very well known person or organization;
- adjacency: the information is adjacent when it concerns the user residence place;
- conflict: the information concerns the conflict among people;
- peculiarity: the information is interesting when it is quite unusual;
- timeliness: the information is interesting when it concerns current event.

5 Summary

Recent works on several implementations of consensus based collaborative user interface recommendations [10] have proven that it is possible to implement this method for a range of different web-based information systems. All of the implemented systems were used in the field of the product or services promotion. The system concerning a particular car model ware analyzed in more details [13], the results prove that collaborative user interface adaptation leads to usability improvements. The other systems, however, are still investigated and we have no the final result yet.

The hybrid approach was up till now only partially implemented but comparing to the pure collaborative approach, the experiments requires longer times of users work with the system (also in several separate sessions) to gather sufficient data for content-based recommendation.

References

1. Dattola RT (1968) A fast algorithm for automatic classification. In: Report ISR-14 to the National Science Foundation, Section V, Cornell University, Dep. of Computer Science.
2. Kanungo T, Mount DM, Netanyahu, NS, Piatko C, Silverman R, Wu AY (2002) An Efficient k-means clustering algorithm: analysis and implementation. IEEE Tran. On Pattern Analysis And Machine Intelligence 24(7): 881-892
3. Kobsa A, Koenemann J, Pohl W (2001) Personalized Hypermedia Presentation Techniques for Improving Online Customer Relationships. Knowledge Eng. Rev. 16(2): 111-155.
4. Langley P (1999) User Modelling in Adaptive Interfaces. Proceedings of the Seventh International Conference on User Modeling: 357-371
5. Lieberman H (1997) Autonomous Interface Agents. Proc. CHI 97, ACM: 67-74.
6. Mobasher B, Cooley R, Srivastave J (1999) Automatic personalization based on Web usage mining. Technical Report TR99010, Dep. of Computer Science, DePaul University
7. Montaner M, Lopez B, de la Rosa JP (2003) A Taxonomy of Recommender Agents on the Internet. Artificial Intelligence Review 19: 285-330.
8. Newman WM, Lamming MG (1996) Interactive System Design. Addison-Wesley, Harlow.
9. Nguyen NT (2001) Conflict Profiles' Susceptibility to Consensus in Consensus Systems. Bulletin of International Rough Sets Society 5(1/2): 217-224
10. Nguyen NT, Sobecki J (2003) Using Consensus Methods to Construct Adaptive Interfaces in Multimodal Web-based Systems. Universal Access in Inf. Society 2(4): 342-358
11. Shearin S, Lieberman H (2001) Intelligent Profiling by Example. In: Proc of the Conf on Intelligent User Interfaces, ACM Press, 2001.
12. Sobecki J (2003) XML-based Interface Model for Socially Adaptive Web-Based Systems User Interfaces. Lecture Notes in Computer Science 2660: 592-598
13. Sobecki J, Weihberg M (2004) Consensus-based Adaptive User Interface Implementation in the Product Promotion. To be published as a chapter in book "Design for a more inclusive world", by Springer-Verlag (London).
14. Whalen D (2002) The Unofficial Cookie FAQ, Version 2.54 Contributed to Cookie Central by David Whalen. http://www.cookiecentral.com/faq/#2.7.

Collaborative Web Browsing Based on Ontology Learning from Bookmarks

Jason J. Jung, Young-Hoon Yu, and Geun-Sik Jo

Intelligent E-Commerce Systems Laboratory,
School of Computer Engineering, Inha University,
253 Yonghyun-dong, Incheon, Korea 402-751
j2jung@intelligent.pe.kr, yhyu@eslab.inha.ac.kr, gsjo@inha.ac.kr

Abstract. This paper proposes the collaborative web browsing system sharing knowledge with other users. We have specifically focused on user interests extracted from bookmarks. A simple URL based-bookmark is provided with structural information by the conceptualization of the ontology. Furthermore, ontology learning based on a hierarchical clustering method can be applied to handle dynamic changes in bookmarks. As a result of our experiments, with respect to *recall*, about 53.1% of the total time was saved during collaborative browsing for seeking the equivalent set of information, as compared with single web browsing.

1 Introduction

Recently, in order to search relevant information, navigating in this overwhelming web environment is a lonely and time-consuming task [1]. There have been many kinds of studies to handle this problem such as the personal assistant agent [2]. Collaborative web browsing is an approach whereby users share knowledge with other like-minded neighbors while searching information on the web. When communicating with the others, we can have many kinds of experiences and gain knowledge such as which searching method is more useful and which steps they needed in order to search a certain piece of information. The representative collaborative browsing systems are *Let's Browse* [5], ARIADNE [6], WebWatcher [3], and BISAgent [4].

Recognizing what a user is interested in is very important in collaborative web browsing when querying relevant information from other users and helping with searching tasks. This paper proposes the extended application of a BISAgent, which is a bookmark-sharing agent system based on a modified *TF-IDF* scheme without considering user preference. According to the GVU's survey, nowadays there is no doubt that the number of bookmarks has increased more than ever. This means that the set of bookmarks in a user's folder can be considered to be enough to infer user interests [10]. Due to the lack of semantic information from simple URL-based bookmarks, we are focusing on a way of conceptualizing them by referring to ontology. When the structural information for users' bookmarks is provided, not only the precision but also the reliability of the extraction of user preferences can be improved.

M. Bubak et al. (Eds.): ICCS 2004, LNCS 3038, pp. 513–520, 2004.
© Springer-Verlag Berlin Heidelberg 2004

2   Ontology Learning from Bookmarks

An ontology is a specification of a conceptualization, which plays a role in enriching semantic or structural information [7]. In addition, a bookmark means URL information about a web site that a user wants to remember and visit again during web browsing. We try to analyze not only his bookmarks but also semantic information that each bookmark implies. Thereby, ontology is applied to conceptualize the simple URL-based bookmarks, and more importantly, hierarchical clustering is exploited to learn these conceptualized bookmarks, as shown in Fig. 1.

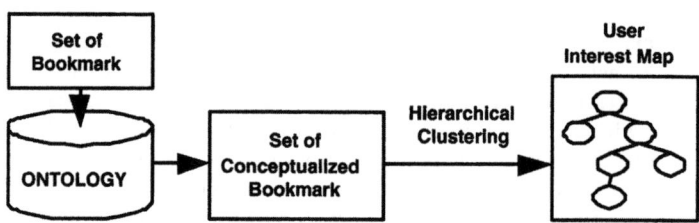

Fig. 1. Hierarchical Clustering of bookmarks

Ontology learning has four main phases which are import, extract, prune, and refine [9]. We are focusing on extracting semantic information from bookmarks based on hierarchical clustering, which is the process of organizing tree structures of objects into groups whose members are similar in some way [8]. The tree of hierarchical clusters can be produced either bottom-up, by starting with individual objects and grouping the most similar ones or top-down, whereby one starts with all the objects and divides them into groups [9]. When clustering conceptualized bookmarks, the top-down algorithm is more suitable than the bottom-up, because directory path information is already assigned to the bookmarks during conceptualization step.

Instead of ontology, the well-organized web directory services such as Yahoo and Cora can be utilized. In practice, however, these directory services have some drawbacks we have to consider as follows.

- **The multi-attributes of a bookmark.** A bookmark can be involved in more than one concept. As shown in Fig. 2 (1), a bookmark can be included in not only a concept named as A but also a concept B.
- **The complicated relationships between concepts.** The semantic relationships between concepts can be classified to
  - Redundancy between semantically identical concepts
  - Subordination between semantically dependent concepts.
  In Fig. 2 (1), the concept A is a subconcept of the root, but the concept A can be redundantly linked as subconcept of the concept P. Moreover, the concept C can be a subconcept of more than a concept like P, as shown in Fig. 2 (2).

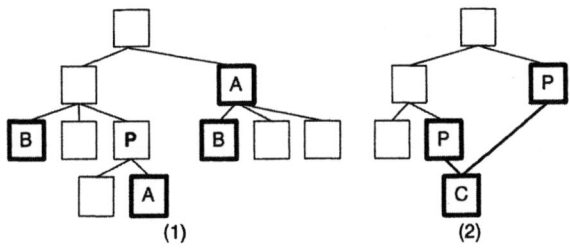

Fig. 2. (1) The multi-attribute of bookmarks; (2) The subordinate relationship between two concepts

When considering influence propagation between concepts, we define notations for semantic analysis dealing with problems caused by web directories. Let the user U_i have the set of bookmarks B_i as follows:

$$B_i = \left\{ b_1^i, b_2^i, ..., b_m^i \right\} \tag{1}$$

where m is the total number of bookmarks. To conceptualize B_i, each bookmark in this set is categorized with the corresponding concepts represented as the directory path. Therefore, the set of conceptualized bookmarks C_i is

$$CB_i = \left\{ cb_1^i, cb_2^i, ..., cb_n^i \right\}$$
$$CRB_i = \left\{ crb_1^i, crb_2^i, ..., crb_a^i \right\}$$
$$C_i = CB_i + CRB_i$$

where n is the total number of concepts including the bookmarks in B_i and a is the number of additional concepts subordinately related with CB_i. Generally, due to the drawbacks of web directories, n becomes larger than m. Here we mention the step for conceptualizing bookmarks by referring to web directories as follows:

```
Function Conceptualization (User)
  var
    counter1, counter2: integer; b: set_bookmark[];
    cb, crb: set_conceptualized_bookmark[];
  begin
    b := Bookmark(User);   counter1 := 1;
    repeat
      cb := cb + Concept(b[counter1]);
      repeat
        counter2 := 1;
        if ((isLinked(Concept(b[counter1]))) = TRUE) then
          crb := crb + Linked(Concept(b[counter1]));
      until counter2 = size(b[counter1])
      counter1 := counter1 + 1;
```

```
    until counter1 = size(b)
    return (cb, crb);
end.
```

The functions Bookmark and Concept return the set of bookmarks of an input user and the set of concepts matched with an input bookmark by looking up the ontology, respectively. The function Linked retrieves the additional concepts related with the input concept. After the function isLinked checks if the input parameter is connected from more than one parent concept on the ontology.

3 Extracting User Interests from Conceptualized Bookmarks

In order to extract user interests, the interest map (i-Map) of each user is established and DOIs (Degree Of Interest) of the corresponding concepts on the i-Map are measured, according to the following axioms:

Axiom 1. The initial DOI of a concept is the number of times that this concept is matched with the set of bookmarks through the function Conceptualization. The larger DOI of a concept means that the corresponding user is more interested in this concept. This means that this number of times is in linear proportion to user preference for that concept.

$$\text{The Number of Matched Times of Concepts} \propto DOI(C_i)$$

Axiom 2. The DOI of a concept is propagated from its subconcepts using this influence propagation:

$$Propagate[DOI(C_i)] = (\log_k(DOI(C_i) + 1))/N \tag{2}$$

where N is the number of total subconcepts of a concept and k is given by

$$k = Variance(DOI(subc(C_i))) + bias = \sigma^2 + bias \tag{3}$$

where $subc(C_i)$ is the set of subconcepts of C_i.

- The dispersion of DOI. As the number of subconcepts of a parent is increased, each of them has less influence on its parent concepts.
- The distance between concepts. The closer concepts are more tightly related with each other. In other words, the influence propagation is exponentially increasing, as the distance between concepts becomes closer.

Axiom 3. The DOI of a concept is measured from the propagations of all subconcepts and all concepts have influence on the root node.

$$DOI(C_i) = \sum_j [Propagate(DOI(subc(C_i)_j)) \times DOI(subc(C_i)_j)] \tag{4}$$

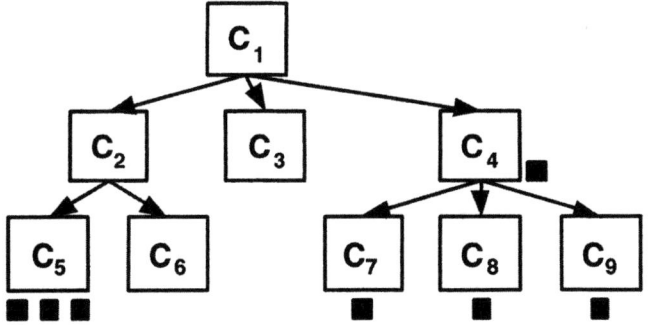

Fig. 3. An example of the conceptualized bookmarks of a user

Axiom 4. Concepts whose *DOI*s are over the predefined threshold value after normalization finally represent user interests.

In Fig. 3, as an example, the black squares indicate the bookmarks of a user, U_i, and assign the initial states, as shown in the following equations:

$$DOI(c_4) = 1, DOI(c_5) = 3, DOI(c_6) = 0,$$
$$DOI(c_7) = 1, DOI(c_8) = 1, DOI(c_9) = 1$$

According to the influence propagation equations, all *DOI*s of other concepts can be computed. The *DOI*s of c_2 and c_4 are as follows:

$$DOI(c_2) = \sum_{k-1}^{2} propagate[DOI(c_k)] \times DOI(c_k) = 1.11$$
$$DOI(c_4) = 1 + (\log_2 2/3 \times 1) \times 3 = 2.0$$

The mean of all *DOI*s is 1.44 and the *DOI* of every concept is assigned after normalization. If the threshold value is 0.2, only c_4 and c_5 are extracted as the most interested concepts for the user. In Fig. 4, the tree represents the user's *i*-Map. Each user is given an *i*-Map, and every time he inserts a bookmark, this *i*-Map is updated.

4   Collaborative Web Browsing with Recommendation

Generally, in computer supported cooperative work (CSCW), a common distinction is made between the temporal and spatial nature of activities. Activities are either co-located or remote and either synchronous or asynchronous [11]. The collaborative web browsing system proposed in this paper is remote and asynchronous because this system is based on a web environment and information about what a participant is interested in extracted from his set of bookmarks and ontology. While browsing to search information, users can be recommended from the facilitator in the following two ways:

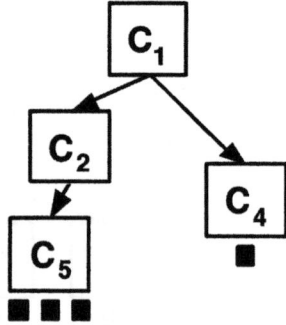

Fig. 4. An *i*-Map representing the high ranked concepts

- By *querying* specific information for the facilitator. After the information about a particular concept is requested, the facilitator can determine who has the maximum *DOI* for that concept by scanning his yellow pages.
- By *broadcasting* new bookmarks of like-minded users from the facilitator. Every time a user inserts a new bookmark, this fact, after conceptualization, is sent to the facilitator. Thereby, users can obtain information related to the common concepts in their own *i*-Map from neighbors.

As shown in Fig. 5, the whole system consists of a facilitator located between the users and the client-side web browser which communicates with the facilitator.

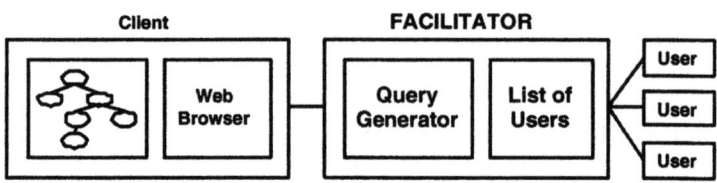

Fig. 5. The whole system architecture

Most importantly, the facilitator must create yellow pages where all users can register themselves. Then, every bookmarking activity can be automatically transmitted to the facilitator.

5   Experimentation

We made up a hierarchical tree structure as a test bed for "Home > Science > Computer Science >" from Yahoo. This tree consisted of about 1300 categories and the maximum depth was eight. For gathering bookmarks, 30 users explored

Yahoo directory pages during for 28 days. Every time users visit a web site related with their own interests, they stored URL information in their bookmark repositories. Finally 2718 bookmarks were collected. In order to evaluate this collaborative web browsing based on extracting user interests, we adopted the measurements *recall* and *precision*. After all of the bookmark sets of the users were reset, these users began to gather bookmarks again after getting the system's recommendations according to their own preferences. During this time, users were being recommended information retrieved from Yahoo based on their interests as extracted up to that moment. As a result, with information recommendations, 80% of the total bookmarks were collected in only 3.8 days, thereby 53.1% of the total time spent previously was saved.

The *precision* was measured by the rate of the inserted bookmarks among the recommended information set. In other words, this was the measurement for the accuracy of predictability. As time passed, the user preferences were changed according to the inserted bookmarks. At the beginning, the precision was especially low because the user preferences were not yet set up. While user interests were being extracted during the first 6 days, the *precision* of recommended information quickly tracked to that of the testing data.

For the rest of the experiment time, the *precision* maintained the same level as that of the testing data because the user interests had already been extracted.

6   Conclusion and Future Work

This paper proposes that bookmarks are the most important evidence to support the extraction of user interests. In order to make up for the structural information, simple URL-based bookmarks were conceptualized by ontology. Then, by establishing an *i*-Map of each user and *DOI* of the concepts on that map, we made it much easier to generate queries for relevant information and to share bookmarks among users. We have implemented a collaborative web browsing system sharing conceptualized bookmarks. Based on the information recommendation of this system, we saved about 53% of the searching time as compared with single web browsing. Moreover, a beginner in a certain field can be helped by finding out valuable hidden information from experts.

As future work, we are considering the privacy problem associated with sharing personal information such as user interests. The visualizing of an *i*-Map is also the next target of this research in order to increase users' intuition recognizing their own preferences quantitatively regarding locations.

Acknowledgement. This work was supported by the Korea Science and Engineering Foundation(KOSEF) through the Northeast Asia e-Logistics Research Center at University of Incheon.

References

1. Maes, P.: Agents that Reduce Work and Information Overload. Comm. of ACM **37**(7) (1994) 31–40
2. Lieberman, H.: Letizia: An Agent That Assists Web Browsing. Proc. of the 4th Int. J. Conf. on Artificial Intelligence (1995) 924–929
3. Armstrong, R., Freitag, T., Mitchell, T.: WebWatcher: A Learning Apprentice for the World Wide Web. AAAI Spring Sym. on Information Gathering from Heterogeneous, Distributed Environments (1997) 6–12
4. Jung, J.J., Yoon, J.-S., Jo, G.-S.: BISAgent: Collaborative Web Browsing through Sharing of Bookmark Information. Proc. of IIP 2000, 16th IFIP World Computer Congress (2000)
5. Lieberman, H., van Dyke, N., Vivacqua, A.: *Let's Browse*: A Collaborative Web Browsing Agent. Proc. of Int. Conf. on Intelligent User Interfaces (1999) 65–68
6. Twidale, M., Nichols, D.: Collaborative Browsing and Visualization of the Search Process. Electronic library and visual information research (1996) 51–60
7. Gruber, T.R.: A translation approach to portable ontologies. Knowledge Acquisition **5**(2)(1993) 199–220
8. Kaufman, L., Rousseeuw, P.: Finding Groups in Data: An Introduction to Cluster Analysis. John Wiley (1990)
9. Maedche, A.: Ontology Learning for the Semantic Web. Kluwer Academic Publishers (2002)
10. Jung, J.J., Jo, G.-S.: Extracting User Interests from Bookmarks on the Web. Proc. of the 7th Pacific-Asia Conf. on Knowledge Discovery and Data Mining (2003) 203–208
11. Rodden, T.: A survey of CSCW systems. Interacting with Computers, **3**(3) (1991) 319–354

Information Retrieval Using Bayesian Networks

Lukasz Neuman, Jakub Kozlowski, and Aleksander Zgrzywa

Department of Information Systems, Wroclaw University of Technology, Poland
{neuman,zgrzywa}@pwr.wroc.pl, topic@oporow.net

Abstract. Information retrieval (IR) systems are used for finding those documents, which satisfy user information need. By such a great increase of documents in the Internet, income of information in databases, precise and quick retrieval of relevant documents is of great significance. Artificial intelligence methods can be essential for achieving this goal. The article describes one of such methods – a model of IR based on Bayesian networks. Usage of the network and an experiment aiming in showing that using this method improves information retrieval is presented. An emphasis was made on the benefits of using the Bayesian networks and the way of adapting such a network to information retrieval system is presented.

1 Introduction

In the late 90s, World Wide Web has caused an explosion of the amount of information available for user. The number of Internet web sites has increased and is still rapidly increasing. This fact causes many problems connected with locating relevant information. Among other, implementation of ordering, classification and filtering of accessible information is needed. To do that one uses many methods helping information retrieval process such as: indexing, classification, query formulation, comparison of documents, feedback.

The main aim of web-based adaptive systems is to determine a set of documents which are relevant to given information need. The problems are well-known, but using information retrieval systems may be insufficient that's why using the mechanisms of artificial intelligence in browsers, which support the users with technology of processing the text of natural queries, classify the documents, group them and estimate their relevance, is needed.

Information retrieval gives many Internet users many benefits, so testing on improving mechanism, models and tools are still lasting. An example of such model is the Bayesian network, described in detail in the following part of paper.

2 Bayesian Networks

A Bayesian network [9] is a representation of a joint probability distribution. It consists of two components. The first component is a directed acyclic graph G (DAG), whose vertices correspond to the random variables $X_1,...,X_n$. The second

M. Bubak et al. (Eds.): ICCS 2004, LNCS 3038, pp. 521–528, 2004.
© Springer-Verlag Berlin Heidelberg 2004

component describes a conditional distribution for each variable, given its parents in G. Together, these two components specify a unique distribution on $X_1,...,X_n$.

The graph G represents conditional independence assumptions that allow the joint distribution to be decomposed on the number of parameters. The graph G encodes Markov assumption: **each variable X_i is independent of its non descendants, given its parents in G.**

By applying the chain rule of probabilities and properties of conditional independencies, any joint distribution that satisfies Markov's assumption can be decomposed into product form:

$$P(X_1,...,X_n) = \prod_{i=1}^{n} P(X_i | Pa^G(X_i))$$

(1)

where $Pa^G(X_i)$ is the set of parents of X_i in G. Below (Fig. 1) there is an example of Bayesian network.

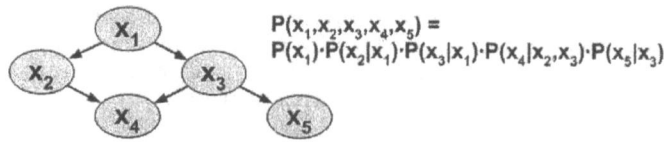

Fig. 1. A simple Bayesian network consisting of 5 nodes

Bayesian structure can be used to learn causal relationships, and hence can be used to gain understanding about a problem domain and to predict the consequences of intervention. The Bayesian network has both a causal and probabilistic semantics; it is an ideal representation for combining prior knowledge and data. In conjunction with statistical methods they offer an efficient and principled approach for avoiding the overfitting of data. All these advantages of this structure make the Bayesian network one of the most important research area in Artificial Intelligence especially in information systems [8].

2.1 Application of Bayesian Networks to Information Systems and Related Works

Probabilistic networks [5], [9] have become an established framework for representing and reasoning with uncertain knowledge. They consist of a dependency structure coupled with a corresponding set of probability tables. Distinguish oneself two types of probabilistic networks, namely Bayesian and Markov.

In this chapter we focus on Bayesian networks and their application to web-based systems, which are becoming an increasingly important area for research and application in the entire field of Artificial Intelligence. They model stochastic processes such as medical systems, military scenarios, academic advising, information retrieval [1], [2] system troubleshooting [3], [4], language understanding, business and in many more [8].

Nowadays, many expert systems taking the advantage of Bayesian approach, work very effectively. It is caused, because Bayesian nets readily permit qualitative inferences without the computational inefficiencies of traditional decision making systems, value of information and sensitivity analysis. Despite offering assistance in the searching process, they support a form of automated learning.

The most important aspect of Bayesian networks is that they are not reasoning processes but they are the direct representations of the world. The arrows in the diagram represent real causal connections and not the flow of information during reasoning [10]. Reasoning processes can operate on Bayesian networks by propagating information in any direction. This kind of graphical representation is easy to construct and interpret. It has formal probabilistic semantics making it suitable for statistical manipulation.

Also the use of Bayesian networks in Internet search browsers [7] is very important because of making connections between documents exceeding out of presence the keywords. The Bayesian networks are used documents classification, thesauri construction [6] and keywords extraction. The vocabulary problem, especially the discrepancies between terms used for describing documents and the terms used by the users to describe their information need, is the one of the key problems of modern Information Retrieval (IR) systems. We can deal with vocabulary problems by using thesaurus, because it shows us the relationships between terms, it is mostly a semantic relationship. We can see three ways of creating thesaurus: first, statistical co-occurrence analyses or the concept space approach, and finally, Bayesian networks.

Example 1. Center for Intelligent Information Retrieval in the University of Massachusetts, has developed INQUERY [1], the most interesting application of Bayesian networks to information systems, which is based on some kind of probabilistic retrieval model, later called document retrieval inference network [14], [15]. It can represent many approaches to information retrieval and combine them into a single framework [6]. It contains two kinds of networks. First net is a set of different representation techniques presenting varying levels of abstraction, and the second represents a need for information (Fig. 2). Nodes in the both nets are either true or false and are representing by values in binary system and interpreted as belief.

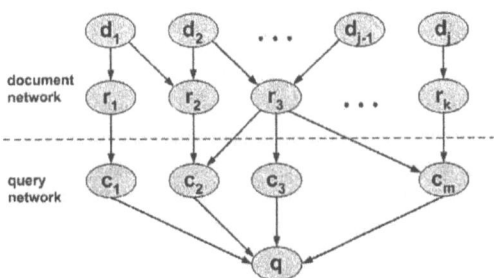

Fig. 2. Simple document retrieval inference network [1]

INQUERY is based on a probabilistic retrieval model and provides support for sophisticated indexing and complex query formulation. It has been used with database containing about 400.000 documents. Recent experiments showed that this system improves information retrieval using Bayesian approach.

3 Experiment

The aim of our experiment was to examine the use of Bayes' methods in the process of information retrieval. What seemed interesting was checking the possibilities of this structure to learn relevance feedback by the user, and on the base of that – increasing the search effectiveness.

This experiment was conducted using a collection of 3204 records from the journal "Communications of the Association for Computing Machinery" (CACM). This collection contains also 64 questions, of which 52 have an additional information about relevant documents. We followed with using the Porter algorithm [11] to remove the commoner morphological and inflexional endings from words. Below we present a table containing 8 most frequently stems in the CACM collection and a diagram fragment of keywords occurring, sorted according to occurring frequency.

Table 1. Ten most frequently stems in the CACM collection with the number of occurrence

Order	Stem	Number of occurrence
1	program	2111
2	algorithm	2015
3	system	1946
4	comput	1945
5	language	1071
6	method	1013
7	data	950
8	time	860

On the basis of common relevant documents for every pair of questions, we have defined a factor of similarity S by means of the formula:

$$S(a,b)=\frac{|A \cap B|}{\frac{1}{2}(|A|+|B|)} \qquad (2)$$

where A and B mean the set of documents relevant adequately to question a and b. For $S >= 0.25$, we have created a graph of question similarity (Fig. 3); every node of the graph has an annotated average arithmetical value of similarity to the questions represented by nodes joined with it by the edge.

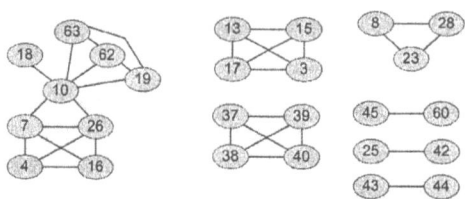

Fig. 3. Graph of question similarity; numbers indicates question number in CACM collection.

Examining the influence of learning feedback on the effectiveness of the search consists in learning the model of answering to the questions with highest similarity factor, and than on checking how this process has reflected on effectiveness of searching questions directly connected with it. Following, the desired behavior of the the intelligent IR system would consist of increasing the effectiveness of answering to particular questions.

3.1   The Implementation of the Bayesian Model

Let I be the information need and D_j denotes the nodes configuration (d) representing the documents in Bayes network, so that node j was observed $(d_j=1)$, and the others were not, meaning.

$$D_j = \langle d_1 = 0, ..., d_{j-1} = 0, d_j = 1, d_{j+1} = 0, ..., d_N = 0 \rangle \tag{3}$$

Than the ranking of documents $rank(I,d_j)$ is created counting $P(I=1|D_j)$ for $j=1....N$ as follows:

$$rank(I,d_j) = P(I=1|D_j) = \frac{P(I=1,D_j)}{P(I=1,D_j)+P(I=0,D_j)} = \frac{P(I=1,D_j)}{P(I=1,D_j)+1-P(I=1,D_j)} = P(I=1,D_j) \tag{4}$$

The information need I is expressed by the sequence of different keywords, which probability of occurring in document j is q_{ij}, $i=1...n$. Above expression can be presented in a simplified form:

$$P(I=1,D_j) = \frac{1}{n}\sum_{i=1}^{n} q_{ij} \tag{5}$$

Initially q_{ij} was initialized in accordance with TF-IDF (term-frequency, inverse document frequency) method [12]:

$$q_{ij} = d_{ij} = (1 + \log df_{ij})\log\frac{idf_i}{N} \tag{6}$$

This method was presented as one of the possibilities defined by Turtle [13]. In this case, working of the model with Bayesian's network, before the process of learning, is identical as for the vector model.

Learning of the feedback was realized by means of the MAP (maximum a posteriori) algorithm. It was based on changing values q_{ij}, which were changeable for those documents which were considered. The values q_{ij} were not changed for the documents regarded as irrelevant. The modification was given by the formula:

$$q'_{ij} = \frac{\beta q_{ij} + 1}{\beta + 1} \tag{7}$$

where β is an arbitrary chosen factor of belief, $\beta=10$.

From the test questions of the CACM collection, we have chosen three questions of the highest level of probability. They have performed the following condition: there is no connection between those questions. That means, we have chosen questions with highly similar adequates, but not conjoined between each other by the similarity relation. Above conditions were fulfilled by questions no 17, 39 and 63.

Table 2. Question chosen to the experiment

Question number	Similarity factor	The numbers of neighbor's question in the graph of the probability
17	0,50	3,13,15
39	0,53	37,38,40
63	0,62	10,19,62

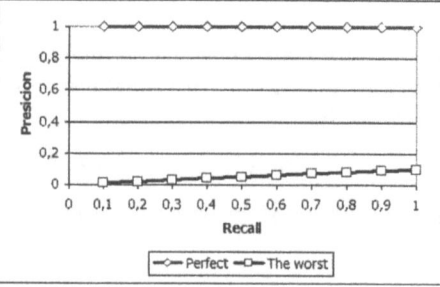

Recall	Model's precision	
	Perfect	The worst
0,1	1,0000	0,0109
0,2	1,0000	0,0217
0,3	1,0000	0,0322
0,4	1,0000	0,0425
0,5	1,0000	0,0526
0,6	1,0000	0,0625
0,7	1,0000	0,0721
0,8	1,0000	0,0816
0,9	1,0000	0,0909
1,0	1,0000	0,1000

Fig. 4. Hypothetic values of recall and precision of answers of perfect and the worst model

3.2 Evaluation of Results

We have used the basic criteria of valuation in the information retrieval systems, precisely – recall and precision. The properties of the chart are important because the foundation was using of the models of searching not only to separate the documents for relevant and irrelevant sets, but to create an order according to level of relevance list of documents. Because of that for the given question the recall will always be 1, $(r=1)$, while the precision will always equal $p=Rel/N$, where Rel is the number of relevant documents and N – the size of the collection.

A good model places the relevant documents at the beginning of the list. A wrong model will place at the beginning documents which are irrelevant. A question arises – how do the best and the worst diagrams look like $p(r)$ (Fig. 4). One can follow it by an example of two system's answers: both give 100 documents from which 10 are relevant. One of the systems places all ten at the very beginning, other – at the end.

The curve of an ideal model is constant, while the curve of the worst is increasing. It results from the dependence:

$$\underset{\substack{a>b \\ a,b \in N}}{\forall} \frac{a+1}{b+1} > \frac{a}{b} \tag{8}$$

This example presents that the better the model, the higher the curve $p(r)$ lays on the diagram. It gives an idea of using additional criteria of effectiveness, meaning the area under the curve $p(r)$. The higher is the curve, the higher the value of the area, which implies the better effectiveness of the model. In counting for the data achieved during the experiment, the value of integral $\int_0^1 p(r)dr$ was estimated by the sum:

$$\sum_{i=1}^{|R|}(r_i - r_{i-1})p(r_i) \tag{9}$$

where R means the set of the points of recall, extended with an extra point $r_0=0$. The sum takes values of the range $(0,1]$. Because usually the relevant documents number for the question is much smaller than the total collection size, one can say that for the worst model the value of above sum is close to zero. The perfect model gives value 1:

- the worst

$$\sum_{i=1}^{|R|}(r_i - r_{i-1})p(r_i) \approx 0$$
$$\tag{10}$$

- perfect

$$\sum_{i=1}^{|R|}(r_i - r_{i-1})p(r_i) = 1$$

The diagrams presented below show the results of learning done the Bayesian network and the results of non–thought network. 5 possibilities of teaching were estimated, they differ by the number of iterations of teaching steps: 1 step, 5, 10, 20, 50 steps. Because the tables of probability of terms in document occurrences were initialized by the same values as weights in the vector model, effectiveness of the non–thought network are exactly the same as the effectiveness of the vector model. Because of that, the curves $p(r)$ of the vector model were not placed in the diagram.

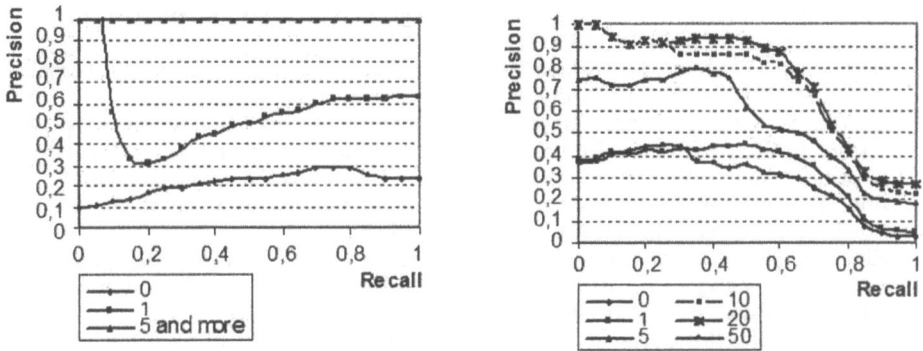

Fig. 5. The effectiveness of Bayesian networks for question 63 of the CACM collection and joined effectiveness of the web for questions 10, 19, 62 for curtain amount of iteration number of learning question 63.

The diagrams at above figure (Fig. 5) present that teaching Bayesian network of answers for specific questions gives very good results. Following iterations significantly improve the results making the net answers for thought questions almost without mistakes. What is important is the fact that the correction does not only concern specific question but also the questions relative to it. The experiment proved that the biggest improvement exists in from 0 to 10 iterations of thought algorithm.

4 Summary

To fulfill the growing needs of the information retrieval system users a lot of methods and artificial intelligence mechanisms are used. Those methods aim is helping the users and fulfilling their information requirements. This paper describes one of the information retrieval methods based on the Bayesian networks. Some of the usage of Bayesian structures applied in information retrieval were presented.

Our experiment presents an example of using the structures of the network in process of information retrieval. Adequately used structures significantly improve the effectiveness and quality. It is very important in the information retrieval.

Summarizing we claim that the Bayesian network is a simple structure, as for initiating apriori values of the distribution of probability tables, allows to be thought and propagate in time $O(n)$. Although it is happening by significant simplification.

References

1. Callan, J.P., Croft, W.B., Harding, S.M.: The INQUERY Retrieval System. Proceedings of the 3rd International Conference on Database and Expert Systems Applications (1992) 78-83
2. Fung, R., Favero, B.D.: Applying Bayesian networks to information retrieval. Communication of the ACM, 38(3) (1995).
3. Heckerman, D., Breese, J.S., Rommelse, K.: Troubleshooting under uncertainty. Technical Report MSR-TR-94-07, Microsoft Research, Redmond, WA (1994).
4. Horvitz, E.: Lumiere Project: Bayesian Reasoning for Automated Assistance. Decision Theory & Adaptive Systems Group, Microsoft Research, MS Corp. Redmond, WA (1998).
5. Jensen, F.: An Introduction to Bayesian Networks. UCL Press Ltd, London (1996).
6. Jing, Y., Croft, W.B.: An association thesaurus for information retrieval. In RIAO'94 Conference Proceedings, New York (1994) 146-160.
7. Klopotek, M.: Inteligentne wyszukiwarki internetowe. Exit, Warszawa (2001), (in Polish).
8. Neuman, L., Zgrzywa A.: Application the Bayesian Networks to Information Systems. IIAS-Transactions on Systems Research and Cybernetics II(1) (2002) 19-23.
9. Pearl, J.: Probabilistic reasoning in intelligent systems: networks of plausible inference. Morgan Kaufmann Publishers (1988).
10. Pearl, J., Russell, S.: Bayesian Networks. TR R-277, University of California (2000).
11. Porter, M.F.: An algorithm for suffix stripping. Program, 14(3) (1980) 130-137.
12. Salton, G., McGill, M.J.: Introduction to Modern Information Retrieval. New York, NY: McGraw-Hill (1983).
13. Turtle, H. R.: Inference networks for document retrieval. Ph.D. dissertation. Computer and Information Science Department, University of Massachusetts. COINS TR 90–92 (1990).
14. Turtle, H., Croft, W.B.: Efficient probabilistic inference for text retrieval. In RIAO'91 Conference Proceedings, Barcelona, Spain (1991) 644-661.
15. Turtle, H., Croft, W.B., Evaluation of an Inference Network-Based Retrieval Model. ACM Transactions on Information Systems 9(3) (1991) 187-222.

An Application of the DEDS Control Synthesis Method*

František Čapkovič

Institute of Informatics, Slovak Academy of Sciences
Dúbravská cesta 9, 845 07 Bratislava, Slovak Republic
{Frantisek.Capkovic,utrrcapk}@savba.sk
http://www.ui.sav.sk/home/capkovic/capkhome.htm

Abstract. An application of the method suitable for modelling and control of general discrete event dynamic systems (DEDS) to special kinds of communication systems is presented in this paper. The approach is based on Petri nets (PN) defined in [12] and directed graphs (DG) described in [11]. It is supported by the previous author's works [1]-[10], [13].

1 Introduction

DEDS are the systems driven by discrete events. A sequence of discrete events can modify the DEDS behaviour. There are two kinds of discrete events - spontaneous events (peculiar to the system) and controllable ones (forced from without). Typical DEDS are flexible manufacturing systems, communication systems, transport systems. Processes in Web and/or multiagent systems are special kinds of communication systems. Thus, the modelling and control methods suitable for DEDS in general can be applied to modelling and control of them. We will use the analytical PN-based model of the DEDS dynamics development as follows

$$\mathbf{x}_{k+1} = \mathbf{x}_k + \mathbf{B}.\mathbf{u}_k \quad , \quad k = 0, N \tag{1}$$

$$\mathbf{B} = \mathbf{G}^T - \mathbf{F} \tag{2}$$

$$\mathbf{F}.\mathbf{u}_k \le \mathbf{x}_k \tag{3}$$

where k is the discrete step; $\mathbf{x}_k = (\sigma_{p_1}^k, ..., \sigma_{p_n}^k)^T$ is the n-dimensional state vector of DEDS in the step k; $\sigma_{p_i}^k \in \{0, c_{p_i}\}$, $i = 1, ..., n$ express the states of the DEDS elementary subprocesses or operations - 0 (passivity) or $0 < \sigma_{p_i} \le c_{p_i}$ (activity); c_{p_i} is the capacity of the DEDS subprocess p_i as to its activities; $\mathbf{u}_k = (\gamma_{t_1}^k, ..., \gamma_{t_m}^k)^T$ is the m-dimensional control vector of the system in the step k; its components $\gamma_{t_j}^k \in \{0, 1\}$, $j = 1, ..., m$ represent occurring of the DEDS elementary discrete events (e.g. starting or ending the elementary subprocesses or their activities, failures, etc.) - i.e. the presence (1) or the absence (0) of the discrete event; $\mathbf{B}, \mathbf{F}, \mathbf{G}$ are constant matrices; $\mathbf{F} = \{f_{ij}\}_{n \times m}$, $f_{ij} \in \{0, M_{f_{ij}}\}$, $i = 1, ..., n$, $j = 1, ..., m$ expresses

* Partially supported by the Slovak Grant Agency for Science (VEGA) under grant
# 2/3130/23.

M. Bubak et al. (Eds.): ICCS 2004, LNCS 3038, pp. 529–536, 2004.
© Springer-Verlag Berlin Heidelberg 2004

the causal relations among the states of the DEDS (as causes) and the discrete events occuring during the DEDS operation (as consequences) - i.e. the nonexistence (0) or the existence and multiplicity ($M_{f_{ij}} > 0$) of the causal relations; $\mathbf{G} = \{g_{ij}\}_{m \times n}$, $g_{ij} \in \{0, M_{g_{ij}}\}$, $i = 1, ..., m$, $j = 1, ..., n$ expresses very analogically the causal relations among the discrete events (causes) and the DEDS states (consequences); $\mathbf{F}$ and $\mathbf{G}$ are the arcs incidence matrices and $\mathbf{B}$ is given by means of them according to (2); $(.)^T$ symbolizes the matrix or vector transposition.

Simultaneously, we will utilize the DG-based model in the form

$$\mathbf{X}(k+1) = \mathbf{\Delta}_k.\mathbf{X}(k) \quad , \quad k = 0, N \tag{4}$$

where k is the discrete step; $\mathbf{X}(k) = (\sigma_{\pi_1}^{(k)}(\gamma), ..., \sigma_{\pi_{n_{RT}}}^{(k)}(\gamma))^T$, $k = 0, N$ is the n_{RT}-dimensional state vector of the DG in the step k; $\sigma_{\pi_i}^{(k)}(\gamma), \in \{0, 1\}$, $i = 1, n_{RT}$ is the state of the elementary DG node π_i in the step k. Its value depends on actual enabling its input transitions. γ symbolizes this dependency; $\mathbf{\Delta}_k = \mathbf{A}_{DG_f}^T = \{\delta_{ij}^{(k)}\}_{n_{RT} \times n_{RT}}$, $\delta_{ij}^{(k)} = \gamma_{t_{\pi_i | \pi_j}}^{(k)}$, $i = 1, n_{RT}$, $j = 1, n_{RT}$ is the functional matrix; $\gamma_{t_{\pi_i | \pi_j}}^{(k)} \in \{0, 1\}$ is the transition function of the PN transition fixed on the edge oriented from the DG node π_j to the DG node π_i. It is necessary to say that the PN places p_i are completely different form the DG nodes π_i. While p_i represent the states of elementary activities inside PN, π_i represent the complete state vectors of the PN. It corresponds with the fact that the DG (with the nodes π_i, $i = 1, ... n_{RT}$) is the RG of the PN (with the places p_i, $i = 1, ... n$) and represents the alternative artificial fictive state machine (SM) with the nodes π_i, $i = 1, ... n_{RT}$ representing the reachable state vectors of the PN.

In [1], [2] the procedure enumerating the *quasi-functional* adjacency matrix $\mathbf{A}$ of the RG and the space of the PN reachable states in the form of the matrix $\mathbf{X}_{reach}$ was presented in a different depth. The columns of the matrix $\mathbf{X}_{reach}$ are the PN state vectors $\mathbf{x}_0$, $\mathbf{x}_1$, $\mathbf{x}_2$, ... reachable from the initial state $\mathbf{x}_0$. The inputs of the procedure are the PN structural matrices $\mathbf{F}$, $\mathbf{G}^T$ and the PN initial state vector $\mathbf{x}_0$. While the PN-based model in general (where any transition can have more than one input places as well as more than one output places) cannot be understood to be the classical SM (because of synchronization problems), the DG-based model (where DG is RG of the PN in question) is the classical SM.

To illustrate the operation of the above mentioned procedure let us model the client-server cooperation as simply as possible. There can be distinguished the following principal partial activities expressed by PN places: $p_1 =$ the client requests for the connection, $p_2 =$ the server is listening , $p_3 =$ the connection of the client with the server, $p_4 =$ data sent by the client to the server, $p_5 =$ the disconnection of the client by the client himself. The PN representing the problem is given on the left side in Fig. 1. The PN transitions t_1 - t_3 represent discrete events that realize the system dynamics. The order of their occurrence influences the actual development of the system dynamics. The inputs $\mathbf{F}$, $\mathbf{G}$ and $\mathbf{x}_0$ are the following as well as the outputs - i.e. the *quasi-functional* adjacency matrix $\mathbf{A}$ of the RG (given on the right in Fig. 1), the corresponding transpose of the functional matrix $\mathbf{\Delta}_k$, and the state space of its reachable states $\mathbf{X}_{reach}$.

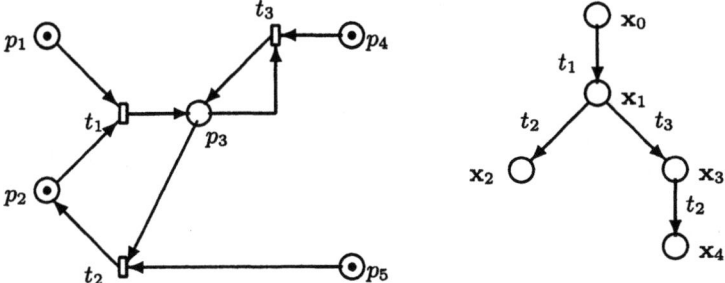

Fig. 1. The PN-based model of the client-server cooperation (on the left) and the corresponding reachability graph (on the right)

$$\mathbf{F} = \begin{pmatrix} 1 & 0 & 0 \\ 1 & 0 & 0 \\ 0 & 1 & 1 \\ 0 & 0 & 1 \\ 0 & 1 & 0 \end{pmatrix} \quad \mathbf{G} = \begin{pmatrix} 0 & 0 & 1 & 0 & 0 \\ 0 & 1 & 0 & 0 & 0 \\ 0 & 0 & 1 & 0 & 0 \end{pmatrix} \quad \mathbf{x}_0 = (1,\ 1,\ 0,\ 1,\ 1)^T$$

$$\mathbf{A} = \begin{pmatrix} 0 & 1 & 0 & 0 & 0 \\ 0 & 0 & 2 & 3 & 0 \\ 0 & 0 & 0 & 0 & 0 \\ 0 & 0 & 0 & 0 & 2 \\ 0 & 0 & 0 & 0 & 0 \end{pmatrix} \quad \mathbf{\Delta}_k^T = \begin{pmatrix} 0 & t_1 & 0 & 0 & 0 \\ 0 & 0 & t_2 & t_3 & 0 \\ 0 & 0 & 0 & 0 & 0 \\ 0 & 0 & 0 & 0 & t_2 \\ 0 & 0 & 0 & 0 & 0 \end{pmatrix} \quad \mathbf{X}_{reach} = \begin{pmatrix} 1 & 0 & 0 & 0 & 0 \\ 1 & 0 & 1 & 0 & 1 \\ 0 & 1 & 0 & 1 & 0 \\ 1 & 1 & 1 & 0 & 0 \\ 1 & 1 & 0 & 1 & 0 \end{pmatrix}$$

As we can see, the nonzero elements of $\mathbf{A}$ represent the indices of the PN transitions. In this very simple example the control synthesis is very simple. After occurrence of the discrete event represented by the transition t_1 the client is connected with the server (the state $\mathbf{x}_1$). Now the client has two possibilities - to disconnect once more (by the event t_2 to the state $\mathbf{x}_2$) or to sent data to the server (by t_3 to $\mathbf{x}_3$). After sending data the client can work on the server and after finishing the work he can disconnect (by t_2 to $\mathbf{x}_4$). In more complicated cases (with more places and transitions and/or with more complicated structure of PN) it is necessary to perform the automatic control synthesis.

2   The Procedure of the Control Synthesis

The problem of control in general is the following: to transform the system to be controlled from a given initial state $\mathbf{x}_0$ to a prescribed terminal state $\mathbf{x}_t$ at simultaneous fulfilling the prescribed control task specifications (like criteria, constraints, etc.). For DEDS control synthesis the very simple idea can be utilized. Consider a system being in the initial state $\mathbf{x}_0$. Consider the desirable terminal state $\mathbf{x}_t$. Develop the straight-lined reachability tree (SLRT) from the state $\mathbf{x}_0$ towards $\mathbf{x}_t$ directed to $\mathbf{x}_t$. Develop the backtracking reachability tree (BTRT) from the state $\mathbf{x}_t$ towards $\mathbf{x}_0$ however, directed to $\mathbf{x}_t$. Intersect both the SLRT and the BTRT. The trajectory (or several ones) starting from $\mathbf{x}_0$ and

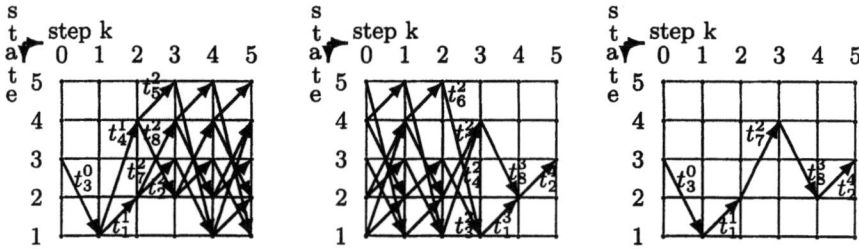

Fig. 2. The straight-lined system development from the initial state (on the left), the backtracking system development from the terminal state (in the centre), and the intersection (on the right)

finishing in $\mathbf{x}_t$ is (are) obtained. To illustrate such an approach see Fig. 2 (where the SLRT is given on the left, the BTRT in the centre, and their intersection on the right). Although in general there can be more than one trajectory after intersection, in any case we obtain the feasible possibilities of the system behaviour between the $\mathbf{x}_0$ and $\mathbf{x}_t$. When a criterion for optimality is given we can find even the optimal trajectory.

To avoid problems with symbolic operations at computer handling $\mathbf{\Delta}_k$ in (4) we will understand all of the transitions to be enabled (i.e. their transition functions having the values 1). In such a way we can replace the functional matrix $\mathbf{\Delta}_k$ by the constant matrix $\mathbf{\Delta}$. Thus, the DG-based approach operates with the constant matrix $\mathbf{\Delta}$ being the transpose of the constant adjacency matrix of the RG (representing the SM corresponding to the original PN). The constant RG adjacency matrix can be obtained from the *quasi-functional* adjacency matrix $\mathbf{A}$ by means of the replacement all of the nonzero elements by the integer 1. Hence, the SLBT can be constructed in analytical terms as follows

$$\{\mathbf{X}_1\} = \mathbf{\Delta}.\mathbf{X}_0; \ \{\mathbf{X}_2\} = \mathbf{\Delta}.\{\mathbf{X}_1\} = \mathbf{\Delta}^2.\mathbf{X}_0; \ \ldots; \ \{\mathbf{X}_N\} = \mathbf{\Delta}.\{\mathbf{X}_{N-1}\} = \mathbf{\Delta}^N.\mathbf{X}_0$$

where $\mathbf{X}_N = \mathbf{X}_t$. In general, $\{\mathbf{X}_j\}$ is an aggregate all of the states that are reachable from the previous states. According to graph theory $N \leq (n_{RT} - 1)$. The BTRT is developed from the $\mathbf{X}_t$ towards $\mathbf{X}_0$, however, it contains the paths oriented towards the terminal state. It is the following

$$\{\mathbf{X}_{N-1}\} = \mathbf{\Delta}^T.\mathbf{X}_N; \ \{\mathbf{X}_{N-2}\} = (\mathbf{\Delta}^T)^2.\mathbf{X}_N; \ \ldots; \ \{\mathbf{X}_0\} = (\mathbf{\Delta}^T)^N.\mathbf{X}_N$$

Here, $\{\mathbf{X}_j\}$ is an aggregate all of the states from which the next states are reachable. It is clear that $\mathbf{X}_0 \neq \{\mathbf{X}_0\}$ and $\mathbf{X}_N \neq \{\mathbf{X}_N\}$. It is the consequence of the fact that in general $\mathbf{\Delta}.\mathbf{\Delta}^T \neq \mathbf{I}_n$ as well as $\mathbf{\Delta}^T.\mathbf{\Delta} \neq \mathbf{I}_n$ (where $\mathbf{I}_n$ is $(n \times n)$ identity matrix). The intersection of the trees is made as follows

$$\mathbf{M}_1 = (\mathbf{X}_0, {}^1\{\mathbf{X}_1\}, \ldots, {}^1\{\mathbf{X}_{N-1}\}, {}^1\{\mathbf{X}_N\}); \ \mathbf{M} = \mathbf{M}_1 \cap \mathbf{M}_2$$
$$\mathbf{M}_2 = ({}^2\{\mathbf{X}_0\}, {}^2\{\mathbf{X}_1\}, \ldots, {}^2\{\mathbf{X}_{N-1}\}, \mathbf{X}_N); \ \mathbf{M} = (\mathbf{X}_0, \{\mathbf{X}_1\}, \ldots, \{\mathbf{X}_{N-1}\}, \mathbf{X}_N)$$

where the matrices $\mathbf{M}_1$, $\mathbf{M}_2$ represent, respectively, the SLRT and the BTRT. The special intersection both of the trees is performed by means of the column-to-

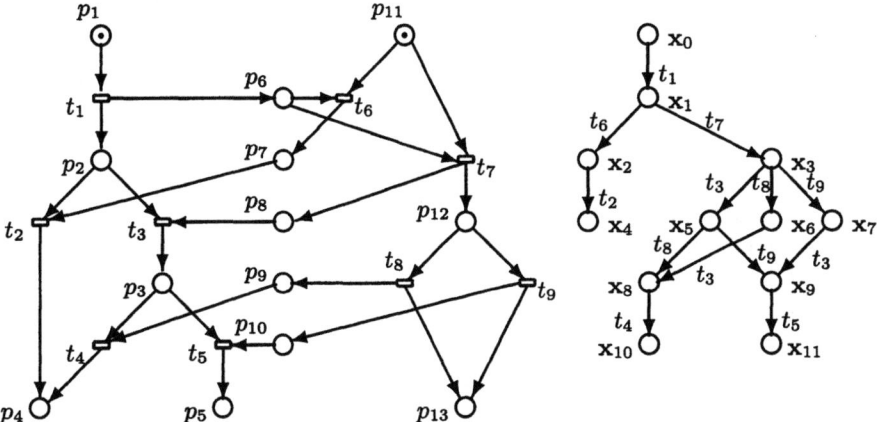

Fig. 3. The PN-based model of two agents cooperation (on the left) and the corresponding reachability graph (on the right)

column intersection both of the matrices. Thus, $\{\mathbf{X}_i\} = \min\left(^1\{\mathbf{X}_i\}, \,^2\{\mathbf{X}_i\}\right)$, $i = 0, ..., N$ with $^1\{\mathbf{X}_0\} = \mathbf{X}_0$, $^2\{\mathbf{X}_N\} = \mathbf{X}_N$.

To illustrate the approach let us model a simple cooperation of two agents A and B by PN. A needs to do an activity P, however, it is not able to do this. Therefore, A requests B to do P for him. On the base of the conversation between the agents P is either done (if B is willing to do P and it is able to do P) or not (when B refuses to do P or it is willing to do P, however, it is not able to do P). The PN places express the activities of the agents and the messages being routed between them: p_1 = the agent A wants to do P, however, he is not able to do P, $p_2 = A$ waits for an answer from B, $p_3 = A$ waits for a help from B, p_4 = the failure of the cooperation, p_5 = the satisfaction of the cooperation, $p_6 = A$ requests B to do P, $p_7 = B$ refuses to do P, $p_8 = B$ accepts the request of A to do P, $p_9 = B$ is not able to do P, p_{10} = doing P by B, $p_{11} = B$ receives the request of A, $p_{12} = B$ is willing to do P, p_{13} = the end of the work of B. The transitions correspond either to synchronization due to the receipt of a message or to conditions of the application of actions. The PN-based model is given on the left in Fig. 3 while the corresponding RG is given on the right. When the initial state vector $\mathbf{x}_0 = (1, 0, 0, 0, 0, 0, 0, 0, 0, 0, 1, 0, 0)^T$ and the structural matrices are

$$\mathbf{F}^T = \begin{pmatrix} 1 & 0 & 0 & 0 & 0 & 0 & 0 & 0 & 0 & 0 & 0 & 0 & 0 \\ 0 & 1 & 0 & 0 & 0 & 0 & 1 & 0 & 0 & 0 & 0 & 0 & 0 \\ 0 & 1 & 0 & 0 & 0 & 0 & 0 & 1 & 0 & 0 & 0 & 0 & 0 \\ 0 & 0 & 1 & 0 & 0 & 0 & 0 & 0 & 1 & 0 & 0 & 0 & 0 \\ 0 & 0 & 1 & 0 & 0 & 0 & 0 & 0 & 1 & 0 & 0 & 0 & 0 \\ 0 & 0 & 0 & 0 & 0 & 1 & 0 & 0 & 0 & 0 & 1 & 0 & 0 \\ 0 & 0 & 0 & 0 & 0 & 1 & 0 & 0 & 0 & 0 & 1 & 0 & 0 \\ 0 & 0 & 0 & 0 & 0 & 0 & 0 & 0 & 0 & 0 & 0 & 1 & 0 \\ 0 & 0 & 0 & 0 & 0 & 0 & 0 & 0 & 0 & 0 & 0 & 1 & 0 \end{pmatrix} \quad \mathbf{G} = \begin{pmatrix} 0 & 1 & 0 & 0 & 0 & 1 & 0 & 0 & 0 & 0 & 0 & 0 & 0 \\ 0 & 0 & 0 & 1 & 0 & 0 & 0 & 0 & 0 & 0 & 0 & 0 & 0 \\ 0 & 0 & 1 & 0 & 0 & 0 & 0 & 0 & 0 & 0 & 0 & 0 & 0 \\ 0 & 0 & 0 & 1 & 0 & 0 & 0 & 0 & 0 & 0 & 0 & 0 & 0 \\ 0 & 0 & 0 & 0 & 1 & 0 & 0 & 0 & 0 & 0 & 0 & 0 & 0 \\ 0 & 0 & 0 & 0 & 0 & 0 & 1 & 0 & 0 & 0 & 0 & 0 & 0 \\ 0 & 0 & 0 & 0 & 0 & 0 & 0 & 1 & 0 & 0 & 0 & 1 & 0 \\ 0 & 0 & 0 & 0 & 0 & 0 & 0 & 0 & 1 & 0 & 0 & 0 & 1 \\ 0 & 0 & 0 & 0 & 0 & 0 & 0 & 0 & 1 & 0 & 0 & 1 \end{pmatrix}$$

the output matrices $\mathbf{A}$, $\mathbf{X}_{reach}$, and $\mathbf{M}$ are, respectively, the following

$$
\begin{pmatrix}
0\,1\,0\,0\,0\,0\,0\,0\,0\,0\,0 \\
0\,0\,6\,7\,0\,0\,0\,0\,0\,0\,0 \\
0\,0\,0\,0\,2\,0\,0\,0\,0\,0\,0 \\
0\,0\,0\,0\,0\,3\,8\,9\,0\,0\,0\,0 \\
0\,0\,0\,0\,0\,0\,0\,0\,0\,0\,0 \\
0\,0\,0\,0\,0\,0\,0\,8\,9\,0\,0 \\
0\,0\,0\,0\,0\,0\,0\,3\,0\,0\,0 \\
0\,0\,0\,0\,0\,0\,0\,0\,3\,0\,0 \\
0\,0\,0\,0\,0\,0\,0\,0\,0\,4\,0 \\
0\,0\,0\,0\,0\,0\,0\,0\,0\,0\,5 \\
0\,0\,0\,0\,0\,0\,0\,0\,0\,0\,0 \\
0\,0\,0\,0\,0\,0\,0\,0\,0\,0\,0
\end{pmatrix}
;
\begin{pmatrix}
1\,0\,0\,0\,0\,0\,0\,0\,0\,0\,0\,0 \\
0\,1\,1\,1\,0\,0\,1\,1\,0\,0\,0\,0 \\
0\,0\,0\,0\,0\,1\,0\,0\,1\,1\,0\,0 \\
0\,0\,0\,0\,1\,0\,0\,0\,0\,0\,1\,0 \\
0\,0\,0\,0\,0\,0\,0\,0\,0\,0\,0\,1 \\
0\,1\,0\,0\,0\,0\,0\,0\,0\,0\,0\,0 \\
0\,0\,1\,0\,0\,0\,0\,0\,0\,0\,0\,0 \\
0\,0\,0\,1\,0\,0\,1\,1\,0\,0\,0\,0 \\
0\,0\,0\,0\,0\,0\,1\,0\,1\,0\,0\,0 \\
0\,0\,0\,0\,0\,0\,0\,1\,0\,1\,0\,0 \\
1\,1\,0\,0\,0\,0\,0\,0\,0\,0\,0\,0 \\
0\,0\,0\,1\,0\,1\,0\,0\,0\,0\,0\,0 \\
0\,0\,0\,0\,0\,0\,1\,1\,1\,1\,1\,1
\end{pmatrix}
;
\begin{pmatrix}
1\,0\,0\,0\,0\,0 \\
0\,1\,0\,0\,0\,0 \\
0\,0\,0\,0\,0\,0 \\
0\,0\,1\,0\,0\,0 \\
0\,0\,0\,0\,0\,0 \\
0\,0\,0\,1\,0\,0 \\
0\,0\,0\,0\,0\,0 \\
0\,0\,0\,1\,0\,0 \\
0\,0\,0\,0\,0\,0 \\
0\,0\,0\,0\,1\,0 \\
0\,0\,0\,0\,0\,0 \\
0\,0\,0\,0\,0\,1
\end{pmatrix}
$$

The approach can be tested very quickly in Matlab. The trajectory given in the matrix $\mathbf{M}$ represents the situation when the desired terminal state is the successful cooperation $\mathbf{x}_{11} = (0, 0, 0, 0, 1, 0, 0, 0, 0, 0, 0, 0, 1)^T$. It is graphically demonstrated on the left in Fig. 4. When the terminal state represents the failure of the cooperation $\mathbf{x}_{10} = (0, 0, 0, 1, 0, 0, 0, 0, 0, 0, 0, 0, 1)^T$ (when B is not able to do P) the result is different - see the trajectory on the right in Fig. 4.

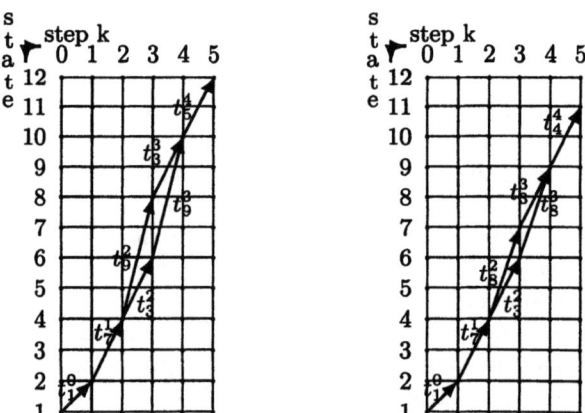

Fig. 4. The resulting trajectories - in case of the successful cooperation (on the left) and in case of the failure of the cooperation when B is not able to do P (on the right)

3   The Adaptivity

As to the applicability of the above described approach we can distinguish two kinds of adaptivity. On the one hand it is the adaptivity concerning the modifying of the system dynamics development by means of choosing a suitable trajectory from the set of feasible state trajectories obtained in the control synthesis process. Such a kind of the adaptivity can be clear e.g. from the left picture in Fig. 4 where two different feasible trajectories (expressing possibilities of the system behaviour) are presented. Because no other trajectory exists, either the trajectory corresponding to the sequence of enabled transitions $\{t_1^0, t_7^1, t_9^2, t_3^3, t_5^4\}$ or the trajectory corresponding to the sequence $\{t_1^0, t_7^1, t_3^2, t_9^3, t_5^4\}$ can be chosen in order to adapt the system behaviour to the actual demand. On the other hand it is the adaptivity of the system behaviour by means of a structural fragment added to the original system structure. Such a fragment is able to accept demands (given from without) on the system behaviour and realize them. The adaptivity is illustrated in Fig. 5. On the left the system model consisting of two processes and the structural fragment containing the place p_4 is given. The fragment is able to influence the performance of the processes (the mutual exclusion, sequencing, re-run). In the centre of Fig. 5 the RG is presented while the result of the control synthesis process transforming the system from $x_0 = (1, 0, 0, 1, 1, 0, 0)^T$ to $x_6 = (0, 0, 1, 0, 0, 1, 0)^T$ (when p_3 has priority ahead of p_7) is put on the right.

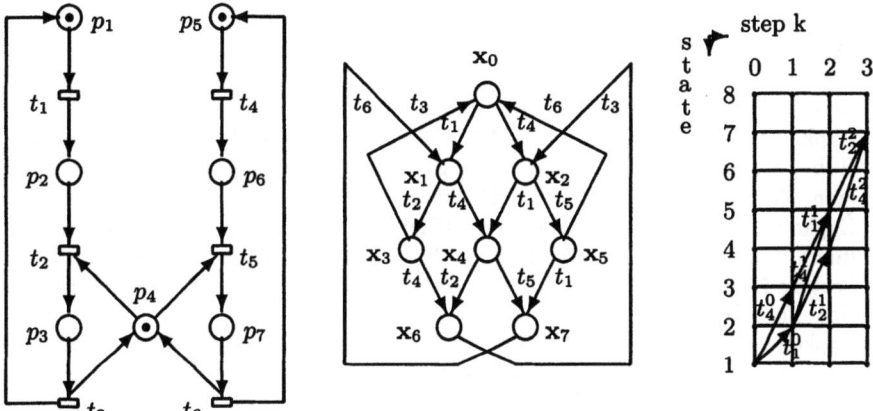

Fig. 5. The PN-based model of the system behaviour (on the left), its reachability graph (in the centre), and the feasible trajectories from x_0 to x_6 (on the right)

4   Conclusions

The approach to the modelling and control synthesis of DEDS was presented in this paper. Its applicability to the special communication systems was demonstrated. The approach is suitable for DEDS described by PN with the finite state space. In order to automate the control synthesis process the graphical tool GraSim was developed. The input of the tool is represented by the RG created by means of icons. On its output the tool yields (in the graphical form) the system trajectories from a given initial state to a prescribed terminal one.

References

1. Čapkovič, F.: The Generalised Method for Solving Problems of DEDS Control Synthesis. In: Chung, P.W.H., Hinde, C., Ali, M. (eds.): Developments in Applied Artificial Intelligence. Lecture Notes in Artificial Intelligence, Vol. 2718. Springer-Verlag, Berlin Heidelberg New York (2003) 702–711
2. Čapkovič, F., Čapkovič, P.: Petri Net-based Automated Control Synthesis for a Class of DEDS. In: ETFA 2003. Proceedings of the 2003 IEEE Conference on Emerging Technologies and Factory Automation. IEEE Press, Piscataway, NJ, USA (2003) 297–304
3. Čapkovič, F.: Control Synthesis of a Class of DEDS. Kybernetes. The International Journal of Systems and Cybernetics **31** No. 9/10 (2002) 1274–1281
4. Čapkovič, F.: A Solution of DEDS Control Synthesis Problems. Systems Analysis Modelling Simulation **42** No. 3 (2002) 405–414
5. Čapkovič, F., Čapkovič, P.: Intelligent Control Synthesis of Manufacturing Systems. In: Monostori, L., Vancza, J., Ali, M. (eds.): Engineering of Intelligent Systems. Lecture Notes in Computer Sciences, Vol. 2070. Springer-Verlag, Berlin Heidelberg New York (2001) 767–776
6. Čapkovič, F.: Intelligent Control of Discrete Event Dynamic Systems. In: Koussoulas, N.T., Groumpos, P.P., Polycarpou, M. (eds.): Proc. of IEEE International Symposium on Intelligent Control. IEEE Press, Patras, Greece (2000) 109–114
7. Čapkovič, F.: Modelling and Control of Discrete Event Dynamic Systems. BRICS Report Series, RS-00-26. University of Aarhus, Denmark (2000) 58 p.
8. Čapkovič, F.: A Solution of DEDS Control Synthesis Problems. In: Kozák, Š., Huba, M. (eds.): Control System Design. Proceedings of IFAC Conference. Pergamon, Elsevier Science. Oxford, UK (2000) 343–348
9. Čapkovič, F.: Automated Solving of DEDS Control Problems. In: El-Dessouki, A., Imam, I., Kodratoff, Y., Ali, M. (eds.): Multiple Approaches to Intelligent Systems. Lecture Notes in Computer Science, Vol. 1611. Springer-Verlag, Berlin Heidelberg New York (1999) 735–746
10. Čapkovič, F.: Knowledge-Based Control Synthesis of Discrete Event Dynamic Systems. In: Tzafestas, S.G. (ed.): Advances in Manugacturing. Decision, Control and Information Technology. Chapter 19. Springer-Verlag, London (1998) 195–206
11. Diestel, R.: Graph Theory. Springer-Verlag, New York (1997)
12. Peterson, J.L.: Petri Net Theory and Modeling the Systems. Prentice Hall, New York (1981)
13. Tzafestas, S.G., Čapkovič, F.: Petri Net-Based Approach to Synthesis of Intelligent Control for DEDS. In: Tzafestas, S.G. (ed.): Computer Assisted Management and Control of Manufacturing Systems, Chapter 12. Springer-Verlag, Berlin Heidelberg New York (1997) 325–351

Using Consistency Measures and Attribute Dependencies for Solving Conflicts in Adaptive Systems

M. Malowiecki, N.T. Nguyen, and M. Zgrzywa

Department of Information Systems, Wroclaw University of Technology, Poland
{malowiecki,thanh,mzgrzywa}@pwr.wroc.pl

Abstract. In this paper we consider two problems related to consensus determining as a tool for conflict resolution. The first problem concerns the consistency measures for conflict profiles which could be useful for taking decision if the consensus should be determined or not. The second problem refers to dependencies of attributes representing the content of conflicts, which cause that one may not treat the attributes independently in consensus determining. We show that using some kind of distance functions may enable to determine consensus in the same way when the attributes are independent on each other.

1 Introduction

Conflict resolution is one of the most important aspects in distributed systems in general, and in adaptive systems in particular. The resources of conflicts in these kinds of systems come from the autonomy feature of their sites. This feature causes that each site of a distributed or multiagent system processes a task in independent way and there may arise such situation that the for same task different sites may generate different solutions. Thus, one deals with a conflict.

For a conflict one can distinguish the following three its components: *conflict body*, *conflict subject* and *conflict content*. Consensus models seem to be useful in conflict solving. The oldest consensus model was worked by such authors as Condorcet, Arrow and Kemeny [1]. This model serves to solving such conflicts in which the content may be represented by orders or rankings. Models of Barthelemy [2] and Day [4] enable to solve such conflicts for which the structures of the conflict contents are n-trees, semilattices, partitions etc. The common characteristic of these models is that they are one-attribute, it means that conflicts are considered only referring to one feature. Multi-feature conflicts are not investigated.

In works [7],[8] the author presents a consensus model, in which multi-attribute conflicts may be represented. Furthermore, in this model attributes are multi-valued, what means that for representing an opinion of an agent on some issue one may use not only one elementary value (such as +, −, or 0) [9] but a set of elementary values. Consensus determining is often required in adaptive systems. The reason is that to be adaptive a system needs to use information from different resources for determining a better way to serve users. These resources can contain inconsistent information referring to the same class of users. Thus, for generating a more adaptive model in user serving the system has to reconcile the inconsistency. One of the most used ways for this process is based on consensus calculating.

M. Bubak et al. (Eds.): ICCS 2004, LNCS 3038, pp. 537–544, 2004.
© Springer-Verlag Berlin Heidelberg 2004

In mentioned models a consensus always is able to be determined, it means that for each profile representing a conflict one can calculate the consensus referring to some criterion. However, a question may arise: Is the chosen consensus good enough and can it be acceptable as the solution of given conflict situation? In other words, is the conflict situation susceptible to (good) consensus? We will consider the susceptibility to consensus for conflict profiles. Before defining the notion of susceptibility to consensus below we present an example. For this problem a notion of so called *susceptibility to consensus* has been defined [8], which is relied on defining a criterion allowing to assess if a conflict can have a solution. In this paper we present another approach [3],[6]to this problem. It is based on measuring the consistency of the conflict. Thus a conflict profile should have a number representing its consistency degree. A set of postulates for consistency functions and some concrete functions are presented.

Another problem is related to attribute dependencies. Consensus model presented in [7],[8] enables to process multi-feature conflicts, but attributes are mainly treated as independent. However, in many practical conflict situations some attributes are dependent on others. For example, in a meteorological system attribute *Wind_power* (with values: *weak, medium, strong*) is dependent on attribute *Wind_speed*, the values of which are measured in unit *m/s*. This dependency follows that if the value of the first attribute is known, then the value of the second attribute is also known. It is natural that if a conflict includes these attributes then in the consensus the dependency should also take place. The question is: Is it enough to determine the consensus for the conflict referring to attribute *Wind_speed*? In other words, when is it true that if some value is a consensus for the conflict referring to attribute *Wind_speed*, then its corresponding value of attribute *Wind_power* is also a consensus for the conflict referring to this attribute? In this paper we consider the answer for mentioned above question. For this aim we assume some dependencies between attributes and show their influence on consensus determining.

2 Postulates for Consistency Measures

Formally, let U denote a finite universe of objects (alternatives), and $\Pi(U)$ denote the set of subsets of U. By $\hat{\Pi}_k(U)$ we denote the set of k-element subsets (with repetitions) of the set U for $k \in N$, and let $\hat{\Pi}(U) = \bigcup_{k>0} \hat{\Pi}_k(U)$. Each element of set $\hat{\Pi}(U)$ is called a *profile*. In this paper we only assume that the macrostructure of the set U is known and a distance is a function $d: U \times U \to \Re^+$, which is: a) *Nonnegative*: $(\forall x,y \in U)[d(x,y) \geq 0]$; b) *Reflexive*: $(\forall x,y \in U)[d(x,y)=0$ iff $x=y]$ and c) *Symmetrical*: $(\forall x,y \in U)[d(x,y)=\delta(y,x)]$. For normalization process we can assume that values of function d belong to interval [0,1].

Let us notice, that the above conditions are only a part of metric conditions. Metric is a good measure of distance, but its conditions are too strong. A space (U,d) defined in this way does not need to be a metric space. Therefore we will call it a *distance space* [7]. A profile X is called *homogeneous* if all its elements are identical, that is $X=\{n*x\}$ for some $x \in U$. A profile is *heterogeneous* if it is not *homogeneous*. A profile is called *distinguishable* if all its elements are different from each other. A profile X is *multiple* referring to a profile Y (or X is a *multiple* of Y), if $Y = \{x_1,...,x_k\}$ and $X=\{n*x_1,..., n*x_k\}$ for some $n \in N$. A profile X is *regular* if it is a multiple of some distinguishable profile.

By symbol C we denote the consistency function of profiles. This function has the following signature:

$$C: \hat{\Pi}(U) \rightarrow [0,1].$$

where $[0,1]$ is the closed interval of real numbers between 0 and 1.

The idea of this function relies on the measuring the consistency degree of profiles' elements. The requirements for consistency are expressed in the following postulates. Those postulates are result of the intuition, because this is the only one way to translate of the empirical knowledge into the rule-based knowledge. Of course, founded function has not to grant all of those postulates. Mostly it depends on the desired appliance. The following postulates for consistency functions have been defined: **P1a**. *Postulate for maximal consistency*; **P1b**. *Extended postulate for maximal consistency*; **P2a**. *Postulate for minimal consistency*; **P2b**. *Extended postulate for minimal consistency*; **P2c**. *Alternative postulate for minimal consistency*; **P3**. *Postulate for non-zero consistency*; **P4**. *Postulate for heterogeneous profiles*; **P5**. *Postulate for multiple profiles*; **P6**. *Postulate for greater consistency*; **P7a**. *Postulate for consistency improvement*; **P7b**. *Second postulate for consistency improvement*. The content of these postulates can be found in work [6].

3   Consistency Functions and Their Analysis

Let $X=\{x_1, ..., x_M\}$ be a profile. We define the following parameters:

- The matrix of distances between the elements of profile X:

$$D^X = [\, d_{ij}^X \,] \text{ for } i,j = 1,...,M.$$

- The vector of average distances of knowledge states of following objects to the rest:

$$W^X = \left[w_i^X\right] = \left(\frac{1}{M-1}\sum_{j=1}^{M}d_{j1}^X, \frac{1}{M-1}\sum_{j=1}^{M}d_{j2}^X, ..., \frac{1}{M-1}\sum_{j=1}^{M}d_{jM}^X \right),$$

- Diameters of sets X and U $Diam(X) = \max\limits_{x,y \in X} d(x,y)$, and the maximal element of vector W^x: $Diam(W^X) = \max\limits_{1 \le i \le M} w_i^X$,

- The average distance of objects' knowledge states:

$$\bar{d}(X) = \frac{1}{M(M-1)}\sum_{i=1}^{M}\sum_{j=1}^{M}d_{ij}^X = \frac{1}{M}\sum_{i=1}^{M}w_i^X,$$

- The sum of distances between an element x of universe U and the elements of set X: $d(x,X) = \sum_{y \in x} d(x,y)$,

- The minimal sum of distances from an object to the elements of profile X:

$$d_{min}(X) = \min\ \{d(x,X): x \in U\},$$

- The maximal sum of distances from an object from X to its elements:

$$d_{max}(X) = \max\ \{d(x,X): x \in X\}.$$

These parameters are applied for the defining the following consistency functions:

$$c_1(X) = 1 - Diam(X),$$

$$c_2(X) = 1 - Diam(W^X),$$

$$c_3(X) = 1 - \overline{d}(X),$$

$$c_4(X) = 1 - \frac{1}{M} d_{min}(X).$$

$$c_5(X) = 1 - \frac{1}{M} d_{max}(X).$$

Values of functions c_1, c_2, c_3 and c_4 reflect accordingly:

- $c_1(X)$ – the maximal distance between two elements of profile.
- $c_2(X)$ – the maximal average distance between an element of profile X and other elements of this profile.
- $c_3(X)$ – the average distance between elements of X.
- $c_4(X)$ – the minimal average distance between an element of universe U and elements of X.
- $c_5(X)$ – the maximal average distance between an element of profile X and elements of this profile.

The table presented below shows result of functions analysis. The columns denote postulates and rows denote functions. Plus means that presented function complies the postulate and minus means that presented function doesn't comply the postulate.

Table 1. Results of functions analysis

	P1a	P1b	P2a	P2b	P2c	P3	P4	P5	P6	P7a	P7b
c_1	+	−	+	−	+	−	+	+	−	−	−
c_2	+	−	+	−	−	−	+	+	−	+	+
c_3	+	+	+	−	−	+	+	−	−	+	+
c_4	+	+	−	−	−	+	+	+	+	−	−
c_5	+	+	−	−	−	+	+	+	+	−	−

4 The Outline of Consensus Model

The consensus model which enables processing multi-attribute and multi-valued conflicts has been discussed in detail in works [7],[8]. In this section we present only some of its elements needed for the consideration of attribute dependencies. We assume that some real world is commonly considered by a set of agents that are placed in different sites of a distributed system. The interest of the agents consists of events which occur (or are to occur) in this world. The task of the agents is based on determining the values of attributes describing these events. If several agents consider the same event then they may generate different descriptions (which consist of, for example, scenarios, timestamps etc.) for this event. Thus we say that a conflict takes place. For representing potential conflicts we use a finite set A of attributes and a set V of attribute elementary values, where $V = \bigcup_{a \in A} V_a$ (V_a is the domain of attribute a). Let $\Pi(V_a)$ denote the set of subsets of set V_a and $\Pi(V_B) = \bigcup_{b \in B} \Pi(V_b)$. Let $B \subseteq A$, a tuple r_B of type B is a function $r_B: B \to \Pi(V_B)$ where $r_B(b) \subseteq V_b$ for each $b \in B$. Empty tuple is denoted by symbol ϕ. The set of all tuples of type B is denoted by $TYPE(B)$. The consensus system is defined as a quadruple

$$Consensus\_Sys = (A, X, P, Z),$$

where:

- A is a finite set of attributes, which includes a special attribute $Agent$; a value of attribute a where $a \neq Agent$ is a subset of V_a; values of attribute $Agent$ are singletons which identify the agents;
- $X = \{\Pi(V_a): a \in A\}$ is a finite set of consensus carriers;
- P is a finite set of relations on carriers from X, each relation $P \in P$ is of some type T_P (for $T_P \subseteq A$ and $Agent \in T_P$). Relations belonging to set P are classified into 2 groups identified by symbols "$+$" and "$-$" as the upper index to the relation names.
- At least Z is a set of logical formulas for which the relation system (X,P) is a model. These formulas are the conditions which should be satisfied by consensus as the solution of the conflict.

The structures of the consensus carriers are defined by means of a distance function between tuples of the same type. This function can be defined on the basis of one of distance functions δ and ρ between sets of elementary values [8]. These functions are defined as follows: Function δ measures the distance between 2 sets X and Y $(X,Y \subseteq V)$ as the minimal costs of the operation which transforms set X into set Y; the distance measured by function ρ is equal to the sum of shares of elements from V in this distance. Functions δ and ρ are called propositional (denoted by δ^P and ρ^P respectively) if the condition $(X+Y \supseteq X'+Y') \Rightarrow (\kappa(X,Y) \geq \kappa(X',Y'))$ is satisfied for any $X,X',Y,Y' \subseteq V_a$ and $\kappa \in \{\rho, \delta\}$. The distance $\partial(r,r')$ between 2 tuples r and r' of type A is equal to the number $\dfrac{1}{card(A)} \sum_{a \in A} \kappa(r_a, r'_a)$ where $\kappa \in \{\rho^P, \delta^P\}$.

A consensus is considered within a conflict situation, which is defined as a pair $s = <\{P^+,P^-\}, A \to B>$ where $A,B \subseteq A$, $A \cap B = \varnothing$ and $r_A \neq \phi$ holds for any tuple $r \in P^+ \cup P^-$.

The first element of a conflict situation (i.e. set of relations $\{P^+,P^-\}$) includes the domain from which consensus should be chosen, and the second element (i.e. relationship $A \to B$) presents the schemas of consensus subjects of the consensus content, such that for a subject e (as a tuple of type A, included in P^+ or P^-) there should be assigned only one tuple of type B. A conflict situation yields a set $Subject(s)$ of conflict subjects which are represented by tuples of type A. For each subject e two conflict profiles, i.e. $profile(e)^+$ and $profile(e)^-$, as relations of $TYPE(\{Agent\} \cup B)$ should be determined. Profile $profile(e)^+$ contains the positive opinions of the agents on the subject e, while profile $profile(e)^-$ contains agents' negative opinions on this subject.

Definition 1. Consensus on a subject $e \in Subject(s)$ is a pair $(C(s,e)^+, C(s,e)^-)$ of 2 tuples of type $A \cup B$ which fulfill the following conditions:

a) $C(s,e)_A{}^+ = C(s,e)_A{}^- = e$ and $C(s,e)^+{}_B \cap C(s,e)^-{}_B = \phi$,

b) The sums $\displaystyle\sum_{r \in profile(e)^+} \partial(r_B, C(s,e)_B^+)$ and $\displaystyle\sum_{r \in profile(e)^-} \partial(r_B, C(s,e)_B^-)$ are minimal.

Any tuples $C(s,e)^+$ and $C(s,e)^-$ satisfying the conditions of Definition 1 are called consensuses of profiles $profile(e)^+$ and $profile(e)^-$ respectively.

5 Attribute Dependencies and Consensus Determining

In Definition 1 the most important is condition b) which requires the tuples $C(s,e)^+{}_B$ and $C(s,e)^-{}_B$ to be determined in such way thus the sums $\displaystyle\sum_{r \in profile(e)^+} \partial(r_B, C(s,e)_B^+)$ and

$\displaystyle\sum_{r \in profile(e)^-} \partial(r_B, C(s,e)_B^-)$ are minimal, respectively. These tuples could be calculated in

the following way: For each attribute $b \in B$ one can determine sets $C(s,e)^+{}_b$ and $C(s,e)^-{}_b$,

which minimize sums $\displaystyle\sum_{r \in profile(e)^+} \partial(r_b, C(s,e)_b^+)$ and $\displaystyle\sum_{r \in profile(e)^-} \partial(r_b, C(s,e)_b^-)$ respectively.

Such way is an effective one, but it is correct only if the attributes from set B are independent. In this section we consider consensus choice assuming that some attributes from set B are dependent. The definition of attribute dependency given below is consistent with those given in the information system model [10]:

Definition 2. Attribute b is dependent on attribute a if and only if there exists a function $f_b^a : V_a \to V_b$ such that in any consensus system (A,X,P,Z) for each relation $P \in \mathbf{P}$ of type T_p and $a,b \in T_p$ the formula $(\forall r \in P)(r_b = \bigcup_{x \in r_a} \{f_b^a(x)\})$ is true.

The dependency of attribute b on attribute a means that in the real world if for some object the value of a is known then the value of b is also known. In practice, owing to this property for determining the values of attribute b it is enough to know the value of attribute a. Instead of $\bigcup_{x \in Y} \{f_b^a(x)\}$ we will write shortly $f_b^a(Y)$.

Consider now a conflict situation $s = <\{P^+, P^-\}, A \to B>$, in which attribute b is dependent on attribute a where $a, b \in B$. Let $profile(e)^+$ be the positive profile for given conflict subject $e \in Subject(s)$. The problem relies on determining consensus for this profile. We can solve this problem using two approaches:

1. Notice that $profile(e)^*$ is a relation of type $B \cup \{Agent\}$. The dependency of attribute b on attribute a implies that there exists a function from set $TYPE(B \cup \{Agent\})$ to set $TYPE(B \cup \{Agent\} \setminus \{b\})$ such that for each profile $profile(e)^*$ one can assign exactly one set $profile'(e)^* = \{r_{B \cup \{Agent\} \setminus \{b\}}: r \in profile(e)^*\}$.

Set $profile'(e)^*$ can be treated as a profile for subject e in the following conflict situation $s' = <\{P^*, P^-\}, A \rightarrow B \setminus \{b\}>$.

Notice that the difference between profiles $profile(e)^*$ and $profile'(e)^*$ relies only on the lack of attribute b and its values in profile $profile(e)^*$. Thus one can expect that the consensus $C(s,e)^*$ for profile $profile(e)^*$ can be determined from the consensus $C(s,e)'^*$ for profile $profile(e)'^*$ after adding to tuple $C(s,e)'^*$ attribute b and its value which is equal to $f_b^a(C(s,e)'^*_a)$. In the similar way one can determine the consensus for profile $profile(e)^-$.

2. In the second approach attributes a and b are treated equivalently. That means they play the same role in consensus determining for profiles $profile(e)^*$ and $profile(e)^-$. The consensus for profiles $profile(e)^*$ and $profile(e)^-$ are defined as follows:

Definition 3. The consensus for subject $e \in Subject(s)$ in situation $s=<\{P^*,P^-\},A \rightarrow B>$ is a pair of tuples $(C(s,e)^*,C(s,e)^-)$ of type $A \cup B$, which satisfy the following conditions:

a) $C(s,e)^*_A = C(s,e)^-_A = e$ and $C(s,e)^*_B \cap C(s,e)^-_B = \phi$,

b) $C(s,e)^*_b = f_b^a(C(s,e)^*_a)$ and $C(s,e)^-_b = f_b^a(C(s,e)^-_a)$,

c) *The sums* $\displaystyle\sum_{r \in profile(e)^+} \partial(r_B, C(s,e)^+_B)$ *and* $\displaystyle\sum_{r \in profile(e)^-} \partial(r_B, C(s,e)^-_B)$ *are minimal.*

We are interested in the cases when conditions b) and c) of Definition 3 can be satisfied simultaneously. The question is: Is it true that if set $C(s,e)^*_a$ is a consensus for profile $profile(e)^*_a$ (as the projection of profile $profile(e)^*$ on attribute a) then set $f_b^a(C(s,e)^*_a)$ will be a consensus for profile $profile(e)^*_b$ (as the projection of profile $profile(e)^*$ on attribute)? The following theorem was proven [8]:

Theorem 1. With using distance function ρ^P or δ^P if a set $C(s,e)^*_a$ is a consensus for profile $profile(e)^*_a$ then there exists a set $C(s,e)^*_b$ being a consensus for profile $profile(e)^*_b$ such that $f_b^a(C(s,e)^*_a) \subseteq C(s,e)^*_b$.

A similar theorem may be formulated and proved for consensuses of profiles $profile(e)^-_a$ and $profile(e)^-_b$.

Example 1. [5],[8] Let profiles $profile(e)^*_a$ and $profile(e)^*_b$ and function f be given:

$profile(e)^*_a$	$profile(e)^*_b$
$\{a_1,a_3\}$	$\{b_2\}$
$\{a_2,a_3\}$	$\{b_1,b_2\}$
$\{a_1,a_4\}$	$\{b_1,b_2\}$
$\{a_1,a_4,a_5\}$	$\{b_1,b_2\}$

Function $f_b^a : V_a \rightarrow V_b$	
a_1	b_2
a_2	b_1
a_3	b_2
a_4	b_1
a_5	b_1

the consensus for profile $profile(e)^{+}_{a}$ is the set $\{a_1,a_3\}$ and the consensus for profile $profile(e)^{+}_{b}$ is the set $\{b_1,b_2\}$. Notice that $f^{a}_{b}(\{a_1,a_3\}) = \{b_2\} \subseteq \{b_1,b_2\}$.

6 Conclusions

In this paper we consider 2 problems: The first refers to measuring consistency of conflict profiles. We present its definition, conditions for its determining and some concrete consistency functions. The future works should concern the deep analysis of proposed postulates and considering if a function satisfying all of them exists. The second problem is related to how dependencies of attributes influence the possibilities of consensus determining. The limitations of distance functions between attribute values were shown, guarantying determining a correct consensus despite of treating attributes independently. Using such functions provides the following profits. First of all, they enable determining a consensus for only a part of the attributes (the rest may be calculated using dependency functions). Secondly, they prevent from determining an incorrect consensus, which does not fulfill some of the dependencies of attributes. The results of these works should be useful in adaptive systems where reconciling processes are often required, for example, in user interface design [11].

References

1. Arrow, K.J.: Social Choice and Individual Values. Wiley New York (1963).
2. Barthelemy, J.P., Janowitz, M.F.: A Formal Theory of Consensus. SIAM J. Discrete Math. 4 (1991) 305-322.
3. Danilowicz, C., Nguyen, N.T., Jankowski, Ł.: Methods for selection of representation of agent knowledge states in multiagent systems. Wroclaw University of Technology Press (2002) (in Polish).
4. Day, W.H.E.: Consensus Methods as Tools for Data Analysis. In: Bock, H.H. (ed.): Classification and Related Methods for Data Analysis. North-Holland (1988) 312-324.
5. Katarzyniak, R., Nguyen, N.T.: Modification of Weights of Conflict Profile's Elements and Dependencies of Attributes in Consensus Model. Lecture Notes in Artificial Intelligence 2475 (2002) 131-138.
6. Nguyen, N.T., Malowiecki, M.: Consistency Measures for Conflict Profiles. Manuscript submitted for publication in: Advanced Rough Sets (Springer-Verlag).
7. Nguyen, N.T.: Consensus System for Solving Conflicts in Distributed Systems. Information Sciences – An International Journal 147 (2002) 91-122.
8. Nguyen, N.T.: Methods for Consensus Choice and their Applications in Conflict Resolving in Distributed Systems. Wroclaw University of Technology Press (2002), (in Polish).
9. Pawlak, Z.: An Inquiry into Anatomy of Conflicts. Information Sciences 108 (1998) 65-78.
10. Pawlak, Z., Skowron, A.: A Rough Set Approach to Decision Rules Generation. Reports of Institute of Computer Science, Warsaw University of Technology (1993).
11. Sobecki J, Weihberg M (2004) Consensus-based Adaptive User Interface Implementation in the Product Promotion. To be published as a chapter in book "Design for a more inclusive world" (Springer-Verlag).

Logical Methods for Representing Meaning of Natural Language Texts

Tatyana Batura and Fedor Murzin

A.P. Ershov Institute of Informatics Systems
Novosibirsk, Russia
www.iis.nsk.su
tbatura@ngs.ru, murzin@iis.nsk.su

Abstract. The purpose of the work is development of various algorithms of mapping predicates and formulas of the first-order predicate calculus to the texts in a natural language. The results of the work can be used in the computer-aided systems for processing the texts in a natural language, and for constructing the theory of the text sense, that is a subject of research, first of all in linguistics, and also in mathematical logic.

1   Introduction

Within frameworks of the given project it is supposed to develop methods which will help to analyze texts and separate sentences in natural language from different aspects. It is planned to use such methods as text meaning representation in the context of Melchuk's approach and lexical functions proposed by him [1], works of Apresyan [2], Markus's set-theoretical models [3], methods of classical linguistics ([4], [5]), methods used in computer translation systems [6] and to adapt some methods and constructions from mathematical logic for analyzing texts in natural language, e.g. Henkin's construction used in the Model existence theorem and in the omitting types theorem[7], finite forcing etc.

The purpose of this work is to develop different algorithms for matching predicates and formulas of the restricted predicate calculus with natural language texts. The authors also made an attempt to match finite models with text sentences and even the whole text.

In a future, the obtained results may be studied and transformed by means of methods of mathematical logic, which gives a possibility to realize a transferring from a syntactical to semantic level and in some sense teach a machine to understand a meaning of a natural language text.

The results of this work may be applied in automation systems of extracting information from natural language texts, in intellectual systems of searching information in the Internet, in constructing automated summarizing systems, electronic translators and dictionaries.

M. Bubak et al. (Eds.): ICCS 2004, LNCS 3038, pp. 545–551, 2004.
© Springer-Verlag Berlin Heidelberg 2004

The present work may also help to develop various search systems, in cases when it is needed to extract necessary information from a document by query or to select required documents from large amount of documents by a given query. On basis of this work it will be possible to develop systems that will be able to reconstruct text sense and extract knowledge from the text that may be presented to user in form of compact reports (schemes, abstracts) or referred to the knowledge base.

2   A Review of Methods for Representing Natural Language Text Meaning

One of the algorithms for predicates matching is based on lexical functions proposed by Melchuk. On syntactic level, these functions may be represented as predicates in the following form. Consider the whole set of word forms in a language which appear when nouns are declined, verbs conjugated etc. (i.e. the whole vocabulary) and suppose that x and y are words or word combinations from this set Than we have predicates of the following form:

$Syn(x, y)$, x, y are synonyms;

$Anti(x, y)$, x, y are antonyms;

$Destr(x, y)$, y is a standard name for an "aggressive" action

(x = "оса", y = "жалит").

The Markus set-theoretical models are constructed as follows. Consider some class decomposition of a natural language vocabulary (which is supposed to be a finite set). For example this decomposition may consist of classes corresponding to inflectional wordform sets. With the help of such decomposition it is possible to give a formal definition of Gender and Case. Also Markus defines the so-called "syntactic types" which correspond practically with the traditional parts of speech. On the basis of syntactic types operations there is appearing a possibility to establish grammatical correctness of a natural language sentence.

3   Structures Corresponding to Natural Language Sentences

A part of the carried out work may be described as follows. Each sentence is corresponded by several structures $structure_1, ..., structure_q$ and each structure $structure_i$ is corresponded by predicates $predicate_{i1}, ..., predicate_{ij(i)}$.

On the other hand, it is also possible to consider elements of natural language vocabulary as constants, then introduce predicates and get formulas on their basis. The predicates in their turn are at first considered on syntactic level. After that they are regarded as subsets of basic model sets in corresponding Cartesian powers. This approach gives an opportunity to construct models, i.e. to perform transition from syntactic to semantic level.

As an example, let's consider structures that correspond with verbs. They may be obtained in the following way. Suppose that there is only one verb and several nouns in different cases (which are related to this verb) in the sentence. Every case is considered to have no more than one noun. Such sentence may be matched by the following structure

| V | NNom | NGen | NDat | NAcc | NInstr | NPrep |

where *NPrep* is a noun in Prepositional Case (if there is any), etc. When there is no a noun in this case in the sentence the corresponding position of the structure may be filled up by some auxiliary information about the fact that there is no a noun in this case in the sentence but in principal it can be placed there, or there is no a noun in the given case and it cannot exist there at all.

The predicate $P(v, n_1, ..., n_6)$, corresponds to this structure where v is a verb, $n_1, ..., n_6$ are nouns. The predicate is sixtiary since there are six cases in Russian.

4 Grammatical Predicates

There is one more way of introducing predicates - matching with parts of speech. We call such predicates grammatical predicates.

For example, $N(x, y_1, ..., y_n)$, x is noun, y_i are characteristics used for dividing nouns into several groups.

Record $N(x, y_1, ..., \underset{i}{0}, ..., y_n)$ means absence of i-characteristic.

If characteristics $y_1, ..., y_n$ are alternative, we will denote this as $N(x, y)$, where $y = y_1$, if x has characteristic y_1;...; $y = y_n$, if x has characteristic y_n.

Let's take a look at noun number (singular or plural forms) as an example: it is an alternative characteristic since nouns cannot be in singular and plural form at the same time. However the noun can exist in different cases simultaneously (метро), have masculine and feminine gender (плакса), be animate and inanimate (пень) etc. We don't regard these characteristics as alternative.

Because of this, the XOR operation is defined in a different way. For predicates of the form $P(x, y_1, ..., y_n)$ the XOR is defined as conjunction of disjunctions, for example:

$Prep_1(x, y)$ means that prepositions are divided by their origin into $y =$ "непр", i.e. x is an underivative (prototypal) preposition and $y =$ "пр", i.e. x is a derivative preposition. Derivative prepositions are divided into

a) $Prep_1^1(x)$ – derived from an adverb (adverbial) (близ, около, сквозь etc.);

b) $Prep_1^2(x)$ – derived from a noun (nounal) (вследствие, по пути, по причине etc.);

c) $Prep_1^3(x)$ – derived from a verb (verbal) (благодаря, спустя etc.).

In particular we obtain

$$(\forall x) \left(Prep_1(x, np) \leftrightarrow \right.$$

$$\leftrightarrow \underset{\substack{1 \leq i, j \leq 3 \\ i \neq j}}{\&} ((Prep_1^i(x) \& \neg Prep_1^j(x)) \vee (Prep_1^j(x) \& \neg Prep_1^i(x))) \left. \right).$$

For the predicates of the type $P(x, y)$ this operation coincides with the usual "or". For example:

$N_5(x, y)$, y = "отвл", if the noun is abstract, y = "конкр", if the noun is concrete (they represent individual objects, living creatures and some phenomena of environment).

$(\forall x)(N_1(x, собст) \rightarrow \neg(N_5(x, отвл) \vee N_5(x, конкр)))$ or

$(\forall x)((N_5(x, отвл) \vee N_5(x, конкр)) \rightarrow N_1(x, нар))$ — these formulas mean that abstract and concrete nouns are nominal ones.

5   Predicates Associated with Sentence Parts

Furthermore, one can introduce **predicates associated with sentence parts.** Unary predicates of the sentence parts: $P_{sub}(x)$, where x is subject; $P_{pred}(x)$, where x is predicate; $P_{adv}(x)$, where x is adverbial modifier.

Note that a notion "predicate"is used in two senses: as usual and as a grammatical notion. In the second case, it is a word or a sequence of words, i.e. a sentence part. Here we do not consider the second order predicates.

Binary predicates of the sentence parts: $P_{sub}(x, y)$, x – subject; $P_{pred}(x, y)$, x – predicate; $P_{adv}(x, y)$, x – adverbial modifier; where y is a word or word-combination determined (explained).

It is possible to achieve formula representation of these predicates considering x, y as words or word-combinations. Upper index of Q in brackets – predicate arity (quantity of predicate places), lower index of Q shows to what part of the sentence we ask a question.

1. The determined word is a subject

$(\forall x, y)\left(Q_1^{(2)}(x, y) \leftrightarrow (P_{sub}(x, y)\&P_{sub}(x)\&P_{pred}(y))\right)$ – it is possible to raise a question from a subject to a predicate.

2. The determined word is a predicate

$(\forall x, y)\left(Q_2^{(2)}(x, y) \leftrightarrow (P_{adv}(y, x)\&P_{pred}(x)\&P_{adv}(y))\right)$ – it is possible to raise a question from a predicate to an adverbial modifier.

In a general case formulas for n heterogeneous sentence parts may be written in the following form:

$(\forall x, y_1, ..., y_n)\left(Q_1^{(n+1)}(x, y_1, ..., y_n) \leftrightarrow \left(\overset{n}{\underset{i=1}{\&}} P_{attr}(y_i, x)\&P_{sub}(x)\& \overset{n}{\underset{i=1}{\&}} P_{attr}(y_i)\right)\right)$

describes a case with heterogeneous attributes of the subject.

$(\forall x, y_1, ..., y_n)\left(Q_2^{(n+1)}(x, y_1, ..., y_n) \leftrightarrow \left(\overset{n}{\underset{i=1}{\&}} P_{adv}(y_i, x)\&P_{pred}(x)\& \overset{n}{\underset{i=1}{\&}} P_{adv}(y_i)\right)\right)$

describes a case with heterogeneous adverbial modifiers of the predicate.

Below several examples of sentences are presented in the form of predicates

I. Купить машину нам не по средствам.

$N(машину)$, $ProN(нам)$, $N(средствам)$, $V(купить)$, $Prep(по)$, $PartL(не)$; $P_{pred}(купить)$, $P_{obj}(машину)$, $P_{obj}(нам)$, $P_{adv}(не\ по\ средствам)$, $P_{obj}(нам,\ купить)$, $P_{adv}(не\ по\ средствам,\ купить)$, $P_{obj}(машину,\ купить)$;

1. $(\forall x, y) \left(Q_2^{(2)}(x, y) \leftrightarrow (P_{adv}(y, x) \& P_{pred}(x) \& P_{adv}(y)) \right)$ – if $x = $ "купить", $y = $ "не по средствам";

2. $(\forall x, y_1, y_2) \left(Q_2^{(3)}(x, y_1, y_2) \leftrightarrow \right.$

$\leftrightarrow (P_{obj}(y_1, x) \& P_{obj}(y_2, x) \& P_{pred}(x) \& P_{obj}(y_1) \& P_{obj}(y_2)) \left. \right)$ – if $x = $ "купить", $y_1 = $ "машину", $y_2 = $ "нам".

II. Она шла нетвердой походкой.

$ProN(она)$, $V(шла)$, $N(походкой)$, $Adj(нетвердой)$;

$P_{sub}(она)$, $P_{pred}(шла)$, $P_{attr}(нетвердой)$, $P_{adv}(походкой)$, $P_{sub}(она, шла)$, $P_{adv}(походкой, шла)$, $P_{attr}(нетвердой, походкой)$, $P_{pred}(шла, она)$;

1. $(\forall x, y) \left(Q_1^{(2)}(x, y) \leftrightarrow (P_{sub}(x, y) \& P_{sub}(x) \& P_{pred}(y)) \right)$ – if $x = $ "она", $y = $ "шла";

2. $(\forall x, y) \left(Q_2^{(2)}(x, y) \leftrightarrow (P_{pred}(x, y) \& P_{pred}(x) \& P_{sub}(y)) \right)$ – if $x = $ "шла", $y = $ "она";

3. $(\forall x, y) \left(Q_2^{(2)}(x, y) \leftrightarrow (P_{adv}(y, x) \& P_{pred}(x) \& P_{adv}(y)) \right)$ – if $x = $ "шла", $y = $ "походкой";

4. $(\forall x, y) \left(Q_4^{(2)}(x, y) \leftrightarrow (P_{attr}(y, x) \& P_{adv}(x) \& P_{attr}(y)) \right)$ – if $x = $ "походкой", $y = $ "нетвердой".

III. Самолет, пролетающий над нами, скрылся в облаках.

$N(самолет)$, $N(облаках)$, $ProN(нами)$, $V(скрылся)$, $PartP(пролетающий)$, $Prep(над)$, $Prep(в)$;

$P_{sub}(самолет)$, $P_{pred}(скрылся)$, $P_{attr}(пролетающий над нами)$, $P_{adv}(в облаках)$, $P_{adv}(в облаках, скрылся)$, $P_{sub}(самолет, скрылся)$, $P_{pred}(скрылся, самолет)$, $P_{attr}(пролетающий над нами, самолет)$;

see II.1. – if $x = $ "самолет", $y = $ "скрылся";

$(\forall x, y) \left(Q_1^{(2)}(x, y) \leftrightarrow (P_{attr}(y, x) \& P_{sub}(x) \& P_{attr}(y)) \right)$ – if $x = $ "самолет", $y = $ "пролетающий над нами";

see II.2. – if $x = $ "скрылся", $y = $ "самолет";

see II.3. – if $x = $ "скрылся", $y = $ "в облаках".

As an intermediate result we get that it is possible to determine syntactic valencies of a word by means of predicates introduced above.

6   Matching Text with Streams

Let us consider now not a separate sentence but text.

There is a text, i.e. final set of sentences, $p_1 p_2 ... p_N$, at the input. Some streams are formed at the output:

$S_1 = < s_{11}, s_{12}, ..., s_{1m_1}, ... >$

. . . .

$S_k = < s_{k1}, s_{k2}, ..., s_{km_k}, ... >$

An elementary auxiliary stream consists of well-ordered pairs

$< 1, p_1, 2, p_2, ..., N, p_N >$, where the first multiplier is the sentence number the second one is the sentence itself.

Information about word-formation may be placed in streams like
$< h, k_1, L_1, k_2, L_2, ... >$, where h is the stream heading, for instance a selected suffix; k_i is the sentence number, where the word with this suffix appears (i.e. k_i are numbers not for all sentences but only for those, where these words appear); L_i is the list of words with the given suffix appearing in the sentence.

Streams may be associated with lexical functions, too. We will also construct finite models matching with source text in the form of streams.

For instance, let's pick all nouns from sentences and write them in stream $< 1, n_1^1, ..., n_{l_1}^1; 2, n_1^2, ..., n_{l_2}^2; ... >$, where sentence numbers and lists of nouns present in this sentence are written in series (l_i - list size). Let's write this stream in other way $<< 1, n_1^1 >, ..., < 1, n_{l_1}^1 >, < 2, n_1^2 >, ..., < 2, n_{l_2}^2 >, ... >$.

Denote $C = \{< t, n_j^t > | t = \overline{1, N}, j = \overline{1, l_t}\}$ as set of all pairs that appear in the stream. The underlying sets of models will be ones of the following kind $C_0 / \sim$, where $C_0 \subseteq C$, $\sim$ is some kind of equivalence relation.

Equivalence relations will appear almost in the same way as they appear in Henkin's construction when proofing model existence theorem [7], i.e. pairs of the type $< t, c_j^t > (t = 1, ..., N)$ may be considered as constants and depending on different statements about these constants we regard some of them as equivalent.

In a similar manner using the stream obtained it will be possible to apply the types omitting theorem [6] and, besides, to get some models as a result.

Let us note that while using Henkin's constructions it is essential to check consistency of corresponding theories at every stage. However, only partial testing for noncontradictory can be used while running computer processing of a natural language text.

For example we check that relations like "над" or "под" are really transitive; if it is said "white" about an object, then there isn't "black" statement anywhere in the sentence and so on.

7   Conclusion

Different approaches to representing semantics of natural language texts are of great interest now. That is why we have made efforts to analyze the sense of the text on a base on a structural analysis of sentences and a text as a whole ([8], [9]).

Large amount of predicates and logic formulas of the first order there were proposed for such analysis. However we note that in the main the given predicates and formulas are concerned with a grammatical and syntactic structure of sentences.

In future the results achieved may be studied and transformed by mathematical logic means. It gives us an opportunity to make a transition from syntactic to semantic level.

This work can be used for a creation of a text sense theory, and it is possible to apply the results of this work in a mathematical logic area and in linguistic investigations.

Thus, in spite of the fact that this work stage is absolutely necessary, it is important to note that semantic text structure has not been adequately reflected in achieved formulas up till now, and the following investigations are necessary.

We also note that the large volume of factual information from classical and mathematical linguistics, and mathematical logic was used at this simplest (in our opinion) stage. It tells about difficulty of this problem in the whole. Also in this article, we omitted questions connected with computer realizations.

References

1. Melchuk, I.A.: Experience of Theory of Linguistic Models like "Sence $<->$ Text". Moscow (1974) (in Russian)
2. Apresyan, U.D.: Experimental Semantic Invesigation of a Russian Verb. Nauka, Moscow (1967) (in Russian)
3. Markus, S.: Set-theoretical Models of Languages. Nauka, Moscow (1970) (in Russian)
4. Beloshapkova, V.A.: Modern Russian Language: Manual for Philology Students of Institutes of Higher Education. Azbukovnik, Moscow (1997) (in Russian)
5. Rosental, D.E.: Modern Russian Language: Manual for Philology Students of Institutes of Higher Education. MSU Edition, Moscow (1971) (in Russian)
6. Sokirko, A.V.: The semantic dictionaries in automatic text processing. Thesis. MG-PIIA, Moscow (2000) (in Russian)
7. Sacks, G.E.: Saturated Model Theory. W. A. Benjamin Inc. (1972)
8. Batura, T.V., Erkayeva, O.N., Murzin, F.A.: To the problem of analysis of texts in a natural language. "New information technologies in science and Education", Institute of Informatics Systems, Novosibirsk (2003) (in Russian)
9. Batura, T.V., Murzin, F.A.: Logical methods of the sense representation for a text in a natural language. "New information technologies in science and Education", Institute of Informatics Systems, Novosibirsk (2003) (in Russian)

Software Self-Adaptability
by Means of Artificial Evolution

Mariusz Nowostawski[1], Martin Purvis[1], and Andrzej Gecow[2]

[1] Information Science Department, University of Otago
PO Box 56, Dunedin, New Zealand
{MNowostawski,MPurvis}@infoscience.otago.ac.nz
[2] *temporary* Institute of Paleobiology Polish Academy of Science
ul. Twarda 51/55, 00-818 Warszawa, Poland
gecow@polbox.com

Abstract. Contemporary distributed software systems are reaching extremely high complexity levels which exceeds complexities of known engineering problems to date. Especially open heterogeneous multi-agent
systems which may potentially be spread all around the globe, interacting
with different changing web-services and web-technologies are exposed to
demanding, dynamic and highly unpredictable environments. Traditional
control-based handling of adaptability may not be suitable anymore,
therefore there is a tendency for exploring different adaptability models
inspired by natural/biological phenomena. In this article we review overall design of an adaptive software system based on a simple model of artificial evolution. We propose a new paradigm for handling complexity in
dynamic environments based on a theory of self-producing self-adaptive
software systems. We have substantial evidence to believe that a bottom-
up approach based on self-production and self-maintenance may help to
build more robust and more flexible self-adapting software systems. This
paper introduces the new framework, provides analysis of some results,
implications and future research directions toward a complete and self-
contained theory of evolvable and self-adaptable software systems.

1 Introduction

In plain English *adaptation* is the act of changing something to make it suitable
for a new purpose or situation. In software systems, the term *adaptation* is
being used mostly, if not exclusively, with the second semantic meaning. What
is usually meant by software adaptation, is that the system will continue to
fulfil its original and the same purpose in a different circumstances, situation or
environment. The adaptability in such software systems may be achieved by a
set of feedback loops between the system, the controller monitoring and changing
and *adapting* the system, and the environment itself. The system purpose is pre-
defined in advance as a set of specifications, which are kept within the controller.
The behaviour of the system is automatically altered if the expected outputs are
outside of these pre-defined specifications. Such models are built analogously to

M. Bubak et al. (Eds.): ICCS 2004, LNCS 3038, pp. 552–559, 2004.
© Springer-Verlag Berlin Heidelberg 2004

a way automatic control systems work[1]. Most of them are based on top-down design and work well in limited environments, where changes in environment can be predicted and constrained in advance[2]. Such adaptive systems are tuned to particular kinds and specific levels of change in the environment.

Most of the adaptability in software systems is achieved via control mechanism like in automatics. There is a central system, with set of sensors and actuators, a controller, and an environment. Sensors sense an environment, system and controller are tied via a set of feedback loops and the controller tries to keep the system within a pre-defined boundaries. This model is very easily implementable, however it is extremely static and can be applied in situations where we can predict in advance all the changes and variations in the environment.

To make things more robust and flexible, we could implement into the controller an ability to learn, so the rules of changing the system become more dynamic, therefore the whole ensemble can follow changes in more dynamic environments. Yet, it still suffers some of the drawbacks of the simple model. Although in a different scale, there is a limit of environmental change the system can cope with, which is predefined within the learning mechanism itself.

2 Adaptability and Biological Inspirations

To fully benefit from life-like adaptability in artificial software systems, which (at least in theory) might match the levels of complexity of biological organism, we need a formal mathematical model of all the fundamental concepts like: life, organism, evolvability and adaptation. In this work we will use a formal deductive model of process of life described in details in [3,4]. Due to limited scope of this article we will only briefly highlight the main aspects of the theory.

The major step in understanding the process of evolution in natural life was done by Darwin[5], who proposed mechanisms by which purposeful adaptive changes take place via processes of random mutation and natural selection. Darwinian mechanisms postulate reproduction, statistical character of change processes, and the process of elimination. After elimination the object ceases to exist (is not alive anymore). The formal deductive model we are going to use is just based on these rudimentary darwinian mechanisms, and adaptability in software is inspired by the mechanisms which handle purposefulness in natural life.

There has been earlier attempts to formally or mathematically define *life*, *complexity*, *organism*, *organism boundary* and information content. In this work we use a theory of evolvable ensembles. Some of these ideas have been developed over the last three decades [6,4,3] with the roots of the proposed model can be traced back to the work of John von Neumann [7,8]. Von Neumann submitted that a precise mathematical definition must be given to a basic biological theories. The work of von Neumann has been, most noticeably, pursued and extended by Gregory Chaitin [9,10]. Slightly different approach in formalising process of life has been pursued by others (e.g. [11]).

Similarly to von Neumann and Chaitin, our model is based on the discrete model universe, an automata space, with a finite number of states. Note however, that the formal definition of information, which is being used throughout this article, is defined in a context of static collection of bits (as it was originally proposed in [12]) rather than an algorithmic settings (as in [10]).

3   Theoretical Foundations

The model we discuss here can be applied to different software architectures. It is suited for object-oriented web technologies or multi-agent systems. It is not constrained however to these paradigms, and it can be easily implemented in any computing paradigm, for example the presented results were obtained on a simple computing model based on finite-state automata without memory.

For sake of uniformity we will use the term *object* to denote a coarse-grained unit of processing within a given computing framework. It can be the actual object like in the object-oriented paradigm, it can be an individual agent from agent-oriented paradigm, etc. The important part is that the individual *object* is an ensemble of lower-level structures, which can be manipulated at runtime. That is, the *object* can be disassembled into its individual components, and re-assembled again within the system during the actual operation of the system. In other words a certain level of reflection is needed within a computing paradigm for the proposed model to be implementable.

We will use here the basic notion of information as introduced in [12]. *Information* is a selection from a set of available choices. Each *object* contains in its structure the information of how to react with the external stimuli. Any given relation between two sets of events or even conditional probability which maps two sets of events, so that selection in one set causes the selection in another, represents the casual relationship between respective events. Such a relationship can be represented as *code*, and we will use the term *code* to refer to this kind of relationships. The process of mapping one set into another we will call: an *encoding*.

The mapping between the conditions and the expected reactions to these conditions of our software system can be understood as if it was a specification of the system. The specification traditionally describes the general behaviour and the relationships of all objects in a given system, therefore describes the overall behaviour, or expected behaviour of the system. Specification is based on some pre-defined properties and characteristics of the interactions of the software system with its environment. In our case however, the specification (or as we said earlier, the mapping between the condition and the expected reactions) is constructed dynamically together with the process of adaptation to the changing environment.

A *purpose* is a pre-defined effect, which can be obtained by a particular set of stimuli. We use here the term *stimuli* as if they were events observed and anticipated by the object. Therefore, to keep the software system within the expected trajectory the object tries to find the mapping between specific events (causes) which lead to particular effects. In other words the system tries to find such a configurations which lead to an expected results. The object responsible for finding such configurations is called *controlling unit*, or *controller* for short.

The actual recording of the selection from a set of causes to the given effect is called *purposeful information*, and this *purposeful information* is been stored in the structure of the objects themselves.

The relations between events in the environment and effects is, in most cases, not known in advance. This is why the basic (but not the only) activity of a controller is to test many hypotheses. The testing requires controller to setup hypothetical causes which are encoded via the interactions of the object with its environment into effects, and then record the obtained results within the structure of the object (for future reference).

4 Experimental Model

The aim of this work is to propose such a mechanism, which would follow in open-ended fashion the changes in the environment. We require the process of adaptation to be uniform and long. For this we use the theory of living systems which exhibits similar (if not the same) characteristics as those required for self-adaptive software systems.

Given an ensemble of objects we want it to automatically maintain itself within a presupposed limits, and to compensate any changes in the environment which may interfere with the operation of the ensemble. We require the process to be possibly most effective, i.e. such that will lead fastest to higher values of aptness. This leads to increased amount of redundancy which allows the controller to test as many hypotheses at the same period of time as possible. (However, this may be limited to one hypothesis at a time in some application.)

We expect an ensemble to continuously improve its aptness, that is to increase its effective information content within a given environment, and to follow any changes in the environment. This can be achieved by different alterations to the ensemble itself. There are some very characteristics structural tendencies which one can observe during the development (growth, adaptation and evolution) of the ensemble, and we discuss them in the next section in more detail.

We do not want objects to exist and maintain themselves just for the sake of exiting. We want all the software components which occupy our resources to be in some way *useful*, therefore there is we require all the objects not only to maintain their existence, but also to maintain their *usefulness*. All objects not *useful* anymore should be automatically removed from the system (in a similar way to garbage collection).

Usefulness is very subjective term, and it must be specified in advance. It is not similar to specification of the system, because it has to be expressed on different, higher level. It is a meta-purpose. It can be specified by means of a CPU cycles, e.g. processes not utilising CPU should be garbage collected, by user interactions, etc. Of course, all objects referenced by *useful* objects are useful and shall not be removed.

To facilitate experiments we have used a simple model of finite-state automata. Such a model has many advantages: it is simple, easy for analysis and implementation, provides easy means to aggregate objects (automata) into higher-level structures (ensembles), re-arrange connectivity and relations, etc.

Consider a complex system, an object, transforming one set of signal into another one. The former set of signals represents the environment x of the object, the latter set y is the object's answer to the environment.

The system consists of deterministic finite-state automata without memory. Each automaton receives two or more input signals and it transforms them into two or more output signals. Signals are several bits long, in the simulation from 1 to 4. Each automaton represents a simple function operating on input signals. The number of input signals received and output signals sent may vary among automata, however during simulations we used 2-input and 2-output automata. Each output may provide input signals for only a single input, and the same for output signals. The input and outputs of automata are somehow interconnected to form an aggregate of automata. Free inputs of the whole aggregate receive signals form the environment x, while free outputs represent the answer y of the aggregate. In our model, due to the symmetry, both: an object or the environment could represent an active process of encoding of a given cause into the effect, or the input signal into the output one. Note however that this time the active part is of the situation is the object, thus the environment plays the passive role. In the stable (passive) environment x the signals y are changing because object is changing. The outcome y, is compared with the assumed ideal goal y^* and the similarity is the basis for calculated aptness of a given object.

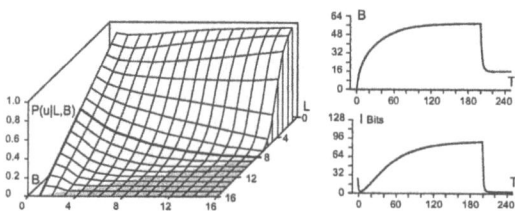

Fig. 1. Probability of accepting signal change

Figure 1: The probability $P(u|L, B)$ of accepting u of the change of L signals (2-bit each) from 16 signals of the result y, where B signals match the ideal $y*$ (on the left). We are interested in the area of higher aptness B, e.g. $B > 8$. Area shown in grey represents all the non-acceptable changes, with $P(u) = 0$. One can see the clear tendency of accepting the small changes. On the right hand side there is plotted the example of the history of growing aptness and amount of purposeful information for y with 64 2-bit signals. At the time $T = 200$ the condition of non-decreasing aptness has been cancelled and both characteristics rapidly dropped down to the point of the highest entropy.

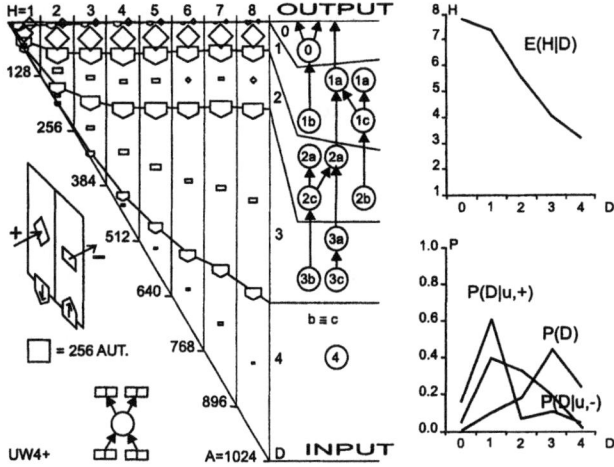

Fig. 2. Basic results of the structural tendencies simulations

5   Structural Tendencies

To investigate the structural tendencies for an evolving object we will use the
model of automata aggregate is being proposed above.

Figure 2 shows, on the left diagram, an average growth of an aggregate with
2-bit signals. This is with strict non-decreasing aptness condition when adding
and no constraints when removing automata. Aggregates were allow to grow
up to 1024 automata. The growth is divided into 8 stages 128 automata each.
Addition and removal process have had uniform probability distributions. Nevertheless, one can observe that addition was dominating, leading to increased
size of the aggregate. In the middle we have shown the *depth* of the automata,
in other words the measure of the functional order D. The depth of automaton depends on the receivers of its output signals. On the left hand side the
aggregate is divided on the layers with the stated balance of addition and removal of automata. In between, we have shown the flow balance. It is clear that
automata are pushed downward by the strong terminal addition, i.e. frequent
addition of new automata at the terminal positions. The acceptance of changes
(u) is contained only within very shallow regions, deeper it ceases (right bottom
corner). It provides consistency of the historic order and functional order (the
right upper diagram). The right diagrams cover all the higher stages from stage
4, where aggregate has already 384 automata. These diagrams are averaged over
different runs and different setups, e.g. with 2-,3-, and 4-bits signals.

The changes of the object are constrained by the improvement condition.
Before we investigate the influence of the object's changes on the changes of the
signals y, we calculated the acceptance probability distribution of change of y in
function of the change's size and the object's aptness. We have noted the "small
change tendency" [13,3] (see also Figure 1 on the left). This in itself is a base
for all further investigated structural tendencies.

The object consists of many randomly (but subject to improvement condition) aggregated automata. Signal vectors x, y and y^* are up to a couple of dozens dimensions big: 64 in our simulations. Such an aggregate is subjected to random changes of the structure, i.e. the individual automata are added and removed from the aggregate at random (see Figure 2). We have investigated also the influence of changing environment and the ideal signal y^*. The obtained tendencies are in close correlation to these observed in common day life, in computer science or in biology. In particular, in software engineering one can observe this phenomenon at work. Some software systems reached the complexity levels that it became very difficult to make any changes. The only alternative is to add new features rather than modify the existing system. In biology such phenomena are observed and referred as de Beer's *terminal changes* or Wiessman's *terminal additions*. It is somehow similar to controversial Haeckel formulation of ontogeny recapitulates phylogeny [3]. The intuitive model proposed by Haeckel seems to be very adequate to the observed phenomena [6,14]. The aggregate was growing, the addition and removal of individual automata was concentrated at the area near output terminals, which are the youngest evolutionary regions. This process is also referred as *terminal changes*. In case there is positive balance of additions to removals in this region, it is called *terminal addition*.

When the environment changes, the aggregate has only one way to continue operating properly: it must reproduce within itself the disappearing signals from the environment. In our model it is achieved by addition of new automata. Some of the automata added as terminals reproduce the important changed signal from the environment and thus maintain the aptness of the aggregate. This is a very strong tendency which we call *covering tendency* [4,3].

There is another characteristics of our proposed self-adaptable software model. The controller will inherently pick these hypothesis, which are re-enforcing discussed here structural tendencies. In other words these characterised by the highest probability of increasing aptness. This happens because the actual probability distributions are part of the information recorded in the object structure (parameters of the controller). This is in a way self-regulating mechanism of adjusting the mechanisms of selecting the hypothesis to be tested - a mechanism (tendency) of re-enforcing structural tendencies.

6   Conclusions

Interestingly, coming independently from two different set of basic definitions and assumptions, both models (Chaitin [10] and Gecow [3]) achieved same conclusions. The process of improvement and the object growth are being accomplished by carrying along all the previously developed structure, as a new pieces of the structure is being added [15]. The simulations and statistical analyses together replicate similar conclusions of Chaitin. The experimental proof of this is that ontogeny recapitulates phylogeny, i.e. each embryo to a certain extend recapitulates in the course of its development the evolutionary sequence that led to it [10]. The preliminary results based on the finite-state automata model discussed in the previous sections present very promising tendencies and robustness. In the discussed scenario the model exhibited self-adaptability and could be successfully used in some applications with binary input-output signals.

Future work will include a) formal definitions together with analysis of aggregation of aggregations, and tendencies to improve tendencies needs to be further explored and investigated; b) more experimental data needs to be collected, and bigger real-life problems must be tested and evaluated. Better understanding of the necessary reflective capabilities is also required.

References

1. Kokar, M., Baclawski, K., Eracar, A.: Control theory-based foundations of self-controlling software. IEEE Intelligent Systems (1999) 37–45
2. Meng, A.C.: On evaluating self-adaptive software. In Robertson, P., Shrobe, H., Laddaga, R., eds.: Self-Adaptive Software. Number 1936 in LNCS. Springer-Verlag, Oxford, UK (2000) 65–74 IWSAS 2000, Revised Papers.
3. Gecow, A.: Statistical analysis of structural tendencies in complex systems vs. ontogeny. PhD thesis, Instytut badań Systemowych PAN, Warsaw, Poland (1986)
4. Gecow, A., Hoffman, A.: Self-improvement in a complex cybernetic system and its implication for biology. Acta Biotheoretica **32** (1983) 61–71
5. Darwin, C.: On the Origin of Species by Means of Natural Selection. John Murray (1859)
6. Gecow, A.: A cybernetical model of improving and its application to the evolution and ontogenesis description. In: Proceedings of Fifth International Congress of Biomathematics. Paris (1975)
7. von Neumann, J.L.: The general and logical theory of automata. In Taub, A.H., ed.: John von Neumann – Collected Works. Volume V. Macmillan, New York (1963) 288–328
8. von Neumann, J.L., Burks, A.W.: Theory of self-reproducing automata (1966)
9. Chaitin, G.J.: To a mathematical definition of 'life'. ACM SICACT News 4 (1970) 12–18
10. Chaitin, G.J.: Toward a mathematical definition of "life". In Levine, R.D., Tribus, M., eds.: The Maximum Entropy Formalism. MIT Press (1979) 477–498
11. Eigen, M., Schuster, P.: The Hypercycle: A Principle of Natural Self-Organization. Springer-Verlag (1979)
12. Shannon, C.E., Weaver, W.: The Mathematical Theory of Communication. University of Illinois Press (1949)
13. Gecow, A.: Obiekt żywy jako stan odchylony od statystycznego stanu równowagi trwałej [living object as a state displaced from the stable statistical equilibrium]. In: Proceedings of I Sympozjum Krajowe CYBERNETYKA-83. PTC, Warsaw, Poland (1983)
14. Gecow, A.: Strukturalne tendencje w procesie udoskonalania [structural tendencies in a process of improvement]. In: Proceedings of I Sympozjum Krajowe CYBERNETYKA-83. PTC, Warsaw, Poland (1983)
15. Simon, H.A.: The sciences of the artificial. MIT Press (1968)

Professor:e – An IMS Standard Based Adaptive E-learning Platform

Cristina Segal and Luminita Dumitriu

COMSYS ltd., Al. Marinarilor nr. 2, Mazepa II, Galati, Romania
{Cristina.Segal,Dumitriu.Luminita}@comsys.ro

Abstract. Recently, the IMS Global Learning Consortium has provided specifications on modeling several aspects related to e-learning. Some of these aspects are separating the actual educational content from the structure of a learning unit, the learner design, the quiz testing, the learning scenario. Learning Simple Sequencing (LSS) is a scenario modeling principle, published in March 2003, meant to help e-learning platform developers to implement educational content delivery modules. Also, an extension of the Content Packaging XML manifest was provided in order to include the new standard.

1 Introduction

First, we shall review the main contributions of IMS standards to the components of an e-learning platform. Thus, we have to consider several aspects that have been specified by the IMS Global Learning Consortium. The main entities involved in an e-learning platform are the trainers, the trainees and the educational resources, as well as the administration of the entire learning process.

On the management of educational resources, IMS has issued the Content Packaging specification, facilitating the separation between the structure of the learning material and the included information.

The learning process administration is specified by the Enterprise standard that intends to ensure the interoperability between the system and the following components: Human Resources, Student Administration, Training Administration and Library Management.

The student evaluation process is specified at the quiz level, by the Question&Test Interoperability specification, providing rules for questions and tests representation.

The learning scenario delivery module should manifest the adaptive behavior of the learning system. In order to model the learning scenario (firing rules, constraints, randomization, etc.) IMS issued the Simple Sequencing standard, taking into account the most frequent and simple models leading to a student-adaptive course delivery.

In the following sections, we are presenting the basics of these specifications.

1.1 Content Packaging Standard

The fundamental notions of this specification [1] are the **package**, the **manifest** and the **resource**.

M. Bubak et al. (Eds.): ICCS 2004, LNCS 3038, pp. 560–566, 2004.
© Springer-Verlag Berlin Heidelberg 2004

A **package** designates an usable autonomous content unit. It can be a course, a part of a course or a course collection. As soon as a package is delivered to a run time service, it must allow aggregation/disaggregation into other packages. A package is a stand-alone unit, thus it has to contain all the information required in order to use its contents for learning when unpacked.

A **manifest** is a XML description of the resources. It contains a variable number of static organizing the instructional resources for presentation.

A **resource** described in a manifest is a physical content file or a collection of resources described through their own manifests. There can be web pages, documents, text files, media files, data files, as well as external URL-available resources. Each resource is described as a <resource> element in the XML manifest and each associated file as a <file> element within its resource.

From the implementation point of view, a package is defined by:
– a specially named XML file, the top-level manifest;
– all XML control documents it refers along with
– the actual physical files, all grouped together in a logical directory.

The top-level manifest describes the package itself. It contains the following sections:
– meta-data section - an XML element describing the whole manifest;
– organizations section - an XML element describing several organizations of the content within a manifest;
– resources section - an XML element referencing all of the actual resources and media elements needed for a manifest, including meta-data describing the resources, and references to any external files;
– (sub)manifest - one or more optional, logically nested manifests;
– Physical Files - these are the actual media elements, text files, graphics, and other resources as described by the manifest(s). The physical resources may be organized in sub-directories.

Characterizing an instructional content is up to the content developer. If a package is meaningful only as a whole, then it should be described by the top-level manifest only. If a package contains elements that can be used on their own, each one of these elements should be described by a nested (sub)manifest in the top-level one. Thus, disaggregation of content is allowed.

If we consider a package at the curriculum level, the top-level manifest has to comprise all course-level manifests as well as any instructional object of each course.

The benefit of complying to the Content Package standard stands in the inter-platform operability of the content as well as in its potential reusability.

1.2 Enterprise Standard

The reason behind the Enterprise specification [3] is ensuring the interoperability between the Learning Management System and other systems within an enterprise. The later ones are:
– Human Resource system;
– Student Administration system;
– Training Administration system;
– Library Management system.

The human resource system is in charge with the skill and competency management, as well as with the selection for learning.

The student administration performs the course catalog management, the class scheduling, class enrollment, attendance tracking, grading, and many other education functions;

The training administration supports course administration, course enrollment, and course completion functions for work force training;

The library management system manages collections of physical and electronic learning objects and manages the access to these materials.

This specification considers that the people-related data from the enterprise system have to be passed to the learning environment, thus creating student profiles. Class/group creation as well as enrolling students and assigning teachers are ensured. Group management can include class scheduling and membership maintenance. Finally, the evaluation and final results recording can occur.

The specification revolves around the following basic concepts: **person, group** and **group membership**. A **person**, in this context, is an individual involved in a learning activity. The required information does not aim to creating a person repository; it only considers learning-relevant information. A **group** is a collection (generic container) of objects relating to a learning activity; it may contain a class, a tutor, a curriculum, a sub-group, but there are no restrictions on the group structure. The **membership** is used to include persons and groups to a group. A group may contain any number of persons and groups, while a person or group may be members of any number of groups. The recurrent membership between groups is not allowed.

The XML instance for the Enterprise standard contains any number of persons, groups and membership structures in this order.

1.3   Learning Simple Sequencing

The simple sequencing standard defines means of representation for the behavior of a learning experience in such way that a computer-assisted learning system (CAL/CAI) has the ability to decide the sequence of learning activities in a consistent manner. A learning scenario designer can specify the relative order in which learning items will be presented to the learner, as well as the conditions governing selection, delivery and ignoring items during a presentation

The standard [4] comprises rules describing the learning activities branching and browsing through educational content with respect to the outcome of learner-content interaction. While educational content developers have to be able to create and describe the content sequencing, the CAI/CAL systems may hide the details related to the models in the standard to the user. The advantage in committing to this standard consists in interchanging learning activities sequencing between systems.

The ensemble of a system's modules used to fire rules and perform behaviors related to presenting educational content to the learner is called a **sequencing engine**.

The **simple** sequencing refers to the limited number of techniques included in the specifications, namely the most frequently used ones. Techniques based on artificial intelligence or schedule-based, or collaborative are neither included nor forbidden by this standard.

The presentation context is assumed to be a web-based one, but there is no actual restriction. Also, there is no assumption on communication and control interfaces, on the graphical aspect or the functionality of the presentation interface.

Even though the representation extends Content Packaging manifest, implementing the later is not mandatory.

1.4 Question and Test Interoperability Standard

This specification [2] aims to enable the exchange of this test and assessment data between Learning management Systems by defining a basic structure for the representation of questions and tests. Thus, the following are provided:

- the definition of questions, choices, feedback/branch and scoring, through standardized attributes (question meta-data along with identification of required and optional elements);
- the interoperability of question collections – definition for packaging and distribution;
- the representation for results reporting;
- the schema for assessment, tracking and presentation;
- APIs for dynamic interface on question retrieval and scoring/assessment engines.

The main concepts of the QTI specification are the **item**, the **section** and the **assessment**. The item contains beside the question itself other information related to rendering, scoring and providing feed-back. Different types of questions are considered like multiple choice, fill in the blanks, hot-spot, etc. A **section** is an assembly of several items or sections, while an **assessment** is a collection of sections. A participant will only interact with an assessment.

1.5 Other Related Standards

Since we have chosen that the learning environment will be web-based deployable, we have included an editing tool allowing the user (author/teacher/student) to compose text, graphics and symbolic expressions. In order to implement these features we have also considered the MathML and SVG standards of representation for graphics and expressions. These standards have an XML-based representation leading to a performance increase due to the dimension reduction of physical content (e.g.: a raster format for a drawing is considerably larger than the vectorial format).

2 Our Approach

The e-learning platform we have designed and implemented is based of the mentioned specification. There is no doubt that complying to standards ensures, on one hand, various content integration and on the other grants the soundness of the approach.

Implementing the standards meant instantiating them in a concrete manner. The interpretation we gave to the specifications above is described in this section.

2.1 Content Related Aspects

We have integrated both Content Packaging and QTI standards in order to produce learning content. We have implemented the resource concept enhancing it with an integrated editor of instructional resources. From the QTI standard we have implemented for the moment only multiple choice questions, in order to extend it later with the other possible types.

The author-user can create/manage/modify/delete packages, collection of questions and tests. We have also considered download/upload of packages and question collection to comply with the spirit of the standards.

We have introduced a concept over both standards, the **course**, which can comprise packages or package sections as well as tests, allowing the teacher/content manager to realize the ensemble of the instructional material required in an e-learning environment.

2.2 Role Related Aspects

The roles we have implemented in platform are the following: the student, the teacher, the content manager and the training administrator.

A teacher can assume the content creator/manager role, but it is not mandatory. The student can be a content creator as well, in the context of homework or projects that require editing documents.

The training administration component implements in part the Enterprise standard, meaning enrolling, group management, class creation, etc.

As we can see in Fig. 1, in the left side of the screen, we have a general group of teachers and a general group of students that can register themselves into the platform, for the moment. If a Human Resources System exists, we can consider importing the personal data into the platform, but this is not yet the case.

The training administrator can create a group around an existing course in the platform, attaching students from the general group, as well as trainers from the general teacher group. He can also schedule the classes of course study. From this moment, the student in this group can visualize the course contents, perform the assessments, while the tutor can visualize the grades/scores obtained by the students to whom he is teaching the course.

In what concerns the Library System in the Enterprise standard, we have not considered it since the platform already has Content Packaging and QTI implementation, thus providing its own contents.

2.3 Student Tracking

Only student activity is logged for the moment. The information available on student activity is:
- the grades/scores of a student, for all courses he is enrolled to;
- the grades/scores of the students taking on of the courses are visible to the course teacher.

The student access is granted to the course contents to the essay writing section, as well as to his section of the catalogue.

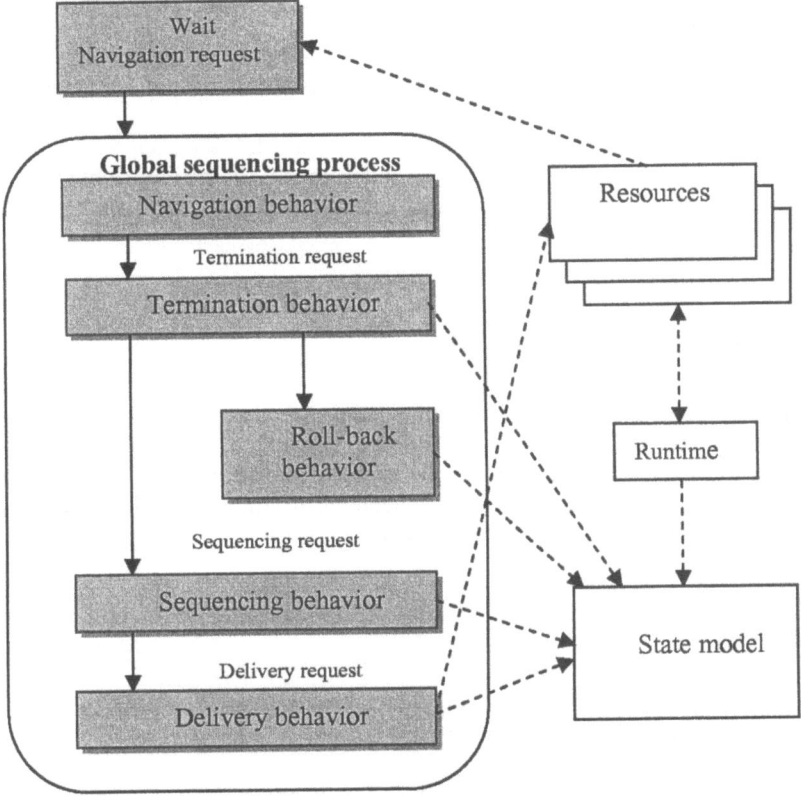

Fig. 1. The sequencing loop from IMS's LSS

2.4 Sequencing

For the moment, we have a linear, random-access scenario to the course. We are in the process of implementing the LSS standard, allowing the teacher to adapt the course presentation to the student profile. We are considering the scenario creation process, involving the editing of pre- and post-condition rules, the limit for tentative access to the contents, etc.

The Simple Sequencing approach, proposed by IMS, even called Simple, has a considerable number of modules and behaviors to be harmonized when implementing.

As we can see in Fig. 1, the main interest in the LSS standard, is modeling the navigation behavior, by firing rules that control the number of accesses to the contents, by delivering or skipping sections depending on the student profile.

The student current profile is decided upon completion of a pre-test for each chapter. This way, the teacher can specify thresholds on the mark, or partial mark, a student receives on his pre-test; if student's score is higher than the threshold then a certain section of the course is skipped for presentation.

Another dependence on student profile consists in analyzing the previous instructional path of the student in order to avoid redundancy. This instructional path is expressed either through the previous learning activity within the platform or through the instructional past of the student.

An assessment test will consist of questions according to the sections previously presented to the student, meaning that skipped sections will not be materialized as questions in the test.

Whenever the results of an assessment test or a pre-test are poor within a concept-related section the student will not be able to advance with the course but will have to consider studying recommended readings relating to that concept and retake the test later.

3 Conclusion and Future Work

We have gained a lot of insight knowledge of the e-learning process while implementing the IMS's standards. We are considering for the future to fully implement the QTI standard with all the types of questions. It is also important to build or import the student profiles, as well as to found the Simple Sequencing standard implementation on the profile structure.

References

1. C. Smythe, T. Anderson, M. McKell, A. Cooper, W. Young and C. Moffatt "IMS Content Packaging Information Model", IMS Global Learning Consortium Inc., http://www.imsglobal.org.
2. C. Smythe, E. Shepherd, "*IMS Question & Test Interoperability:ASI Information Model Specification*", IMS Global Learning Consortium Inc., http://www.imsglobal.org.
3. C. Smythe, G. Collier and C. Etesse, "*IMS Enterprise Best Practice & Implementation Guide V1.1*", Final Specification, IMS Specification, July 2002, http://www.imsglobal.org.
4. IMS Global Learning Consortium Inc. "IMS Simple Sequencing Information and Behavior Model", http://www.imsglobal.org.

Towards Measure of Semantic Correlation between Messages in Multiagent System

Agnieszka Pieczyńska-Kuchtiak and Radoslaw Katarzyniak

Institute of Control and Systems Engineering, Wrocław University of Technology,
Wybrzeże Wyspianskiego 27, 50-370 Wrocław, Poland
{agnieszka.pieczynska-kuchtiak,radoslaw.katarzyniak}@pwr.wroc.pl

Abstract. In this work an original method for the measure of semantic correlation between messages generated by communicative agents in multiagent system is presented. This method is applied to the algorithm for finding the set of agents with the maximal cognitive capacity to observe the states of the objects from external world. It is assumed that the generated messages are the result of the process of language generation, represented by the algorithm for the choice of relevant semantic messages. This algorithm relates belief formulas to the internal agents' knowledge states.

1 Introduction

In this paper a method for the measure of semantic correlation between messages in multiagent system is presented. It is assumed that each agent $a \in A = \{a_1, a_2, ..., a_z\}$ is situated in a real, ontologically independent world. In this world some objects $O = \{o_1, o_2, ..., o_s\}$ exist. States of these objects are a target of the agents' cognitive processes. Each object $o \in O$ is described by means of properties from the set $\Delta = \{P_1, P_2, ..., P_K\}$. In particular, the cognitive agent $a \in A = \{a_1, a_2, ..., a_z\}$ can think of an object $o \in O$ as having or not having a particular property $P \in \Delta$. Perceptions collected by the agents are ordered in relation to a line of time points $T = \{t_0, t_1, t_2, ..\}$ [9]. Each agent is equipped with the communication language, which makes it possible for an agent to generate the logic formulas. Each formula is interpreted from the agent's point of view as external, logical representation of beliefs on current state of the object. Each formula is built of the modal operator of belief, two names of the properties and logic operator of belief. The scope of the agents' language is given in a Table 1.

The basic assumption is that, if an agent can not observe the current state of the particular object, then he refers to the overall private knowledge and applies dedicated algorithm for the choice of relevant semantic messages. This algorithm, which reflects the process of symbol grounding, applies a new and alternative approach to define the epistemic concept of belief. The symbol grounding is understood as a mental phenomenon that assigns meaning to language and relates language symbols to the external objects in a very certain way [5], [7], [8]. The algorithm for the choice of relevant messages relates belief formulas to internal representations of an object rather than to ontologically existing entity. In consequence each formula is treated by

M. Bubak et al. (Eds.): ICCS 2004, LNCS 3038, pp. 567–574, 2004.
© Springer-Verlag Berlin Heidelberg 2004

Table 1. Language messages

	Formal Abbreviations	Spoken Language Interpretation
(1)	$B^a(P_i(o) \wedge P_j(o))$	I (agent a) believe that object o has the property P_i and the property P_j.
(2)	$B^a(P_i(o) \wedge \neg P_j(o))$	I (agent a) believe that object o has the property P_i and does not have the property P_j.
(3)	$B^a(\neg P_i(o) \wedge P_j(o))$	I (agent a) believe that object o does not have the property P_i and has the property P_j.
(4)	$B^a(\neg P_i(o) \wedge \neg P_j(o))$	I (agent a) believe that object o does not have the property P_i and the property P_j.
(5)	$B^a(P_i(o) \underline{\vee} P_j(o))$	I (agent a) believe that object o has either the property P_i or the property P_j.
(6)	$B^a(P_i(o) \vee P_j(o))$	I (agent a) believe that object o has the property P_i or the property P_j.

the cognitive agent as true, if and only if this formula is satisfied by the overall state of agent encapsulated knowledge. Such an approach to understanding satisfaction relation is alternative to the commonly known extensional definition of the satisfaction formulas accepted within Tarskian theory of truth [7]. In this sense the algorithm realises an original way of semantic and interpreted language generation. The aim of this paper is to investigate the level of formula grounding by measure of semantic closeness of the messages, which are generated individually by the agents using the algorithm for the choice of relevant messages. It is assumed that these formulas are syntactically identical. The question is: if the semantic of these formulas is also identical. We want to compare the states of agents knowledge which let them generate such formulas. As a result of this comparison the set of agents with the maximal cognitive capacity to observe the states of objects and maximal similarity of knowledge states to the median state of knowledge of all the agents is computed.

2 Basic Notions

The state of the external world, recognised by the agent $a \in A$ at the particular time point t_n is represented in its body as a base profile and is given as:

$$BP^a(t_n) = \langle O, P^+_1(t_n), P^-_1(t_n), P^\pm_1(t_n), ..., P^+_K(t_n), P^-_K(t_n), P^\pm_K(t_n) \rangle$$

Remark 1. $O = \{o_1, o_2, ..., o_S\}$ and each o denotes a unique cognitive representation of a particular object of the external world W.

Remark 2. For $i = 1, 2, ..., K$, $P^+_i(t_n) \subseteq O$ and $P^-_i(t_n) \subseteq O$. For each $o \in O$ the relation $o \in P^+_i(t_n)$ holds if and only if the agent has perceived that this object o possesses atomic property P_i. For each $o \in O$ the relation $o \in P^-_i(t_n)$ holds if and only if the agent has perceived that this object o does not posses the atomic property P_i.

Remark 3. The set of objects $P^{\pm}_i(t_k) = O/(P^+_i(t_k) \cup P^-_i(t_k))$ is called the area of agent's incompetence related to the property P_i. For each $o \in O$ the relation $o \in P^{\pm}_i(t_k)$ holds if and only if the agent could not perceive the state of this object o in relation to the property P_i.

The overall state of perceptions, stored by the agent $a \in A$, is given as a temporal data base consisted of the set of base profiles:

$$KS^a(t_c) = \{BP^a(t_n): t_n \in T \text{ and } t_n \leq t_c\},$$

where t_n, t_c are time points.

3   The Idea of an Algorithm for the Choice of Relevant Messages

In this section the idea of an algorithm for the choice of relevant semantic messages is explained. Understanding the essence of this algorithm is a key point in understanding the method for the measure semantic correlation between messages, discussed in Section 4. It is assumed, that at the time point t_c the state of an object $o \in O$ in relation to the properties P_i and P_j for at least two agents is not known and can not be verified by sensing:

$$\forall\ PB^a(t_c) \in KS^a(t_c).\ (o \in P^{\pm}_i(t_c) \text{ and } o \in P^{\pm}_j(t_c)) \text{ for } a \in A' \subseteq A, \text{card}(A') > 1,\ i,j \in K \qquad (1)$$

where A' denotes the set of agents, that can not observe the state of an object o in relation to the properties P_i and P_j.

All experiences, internally stored by each agent, represented by $KS^a(t_c)$, $a \in A'$, are the source of a meaning for the external formulas. The algorithm for the choice of relevant semantic messages is consisted of five steps [9], [10], [11]:

Step1. Message oriented classification of perceptions. The procedure for applying all stored perceptions uses a simple classification of empirical content in $KS(t_k)$. In particular, the following classes of base profiles are taken into account:

a) $C^a_1(t_c) = \{BP^a(t_n): a \in A', t_n \leq t_c, BP^a(t_n) \in KS^a(t_c) \text{ and both } o \in P^+_i(t_n) \text{ and } o \in P^+_j(t_n) \text{ hold for } BP^a(t_n)\}$

Obviously, this class of data consists of all base profiles stored up to the time point t_c, in which the object o has been found by the agent a as having both properties P^+_i and P^+_j.

b) $C^a_2(t_c) = \{BP^a(t_n): a \in A', t_n \leq t_c, BP^a(t_n) \in KS^a(t_c) \text{ and both } o \in P^+_i(t_n) \text{ and } o \in P^-_j(t_n) \text{ hold for } BP^a(t_n)\}$

c) $C^a_3(t_c) = \{BP^a(t_n): a \in A', t_n \leq t_c, BP^a(t_n) \in KS^a(t_c) \text{ and both } o \in P^-_i(t_n) \text{ and } o \in P^+_j(t_n) \text{ hold for } BP^a(t_n)\}$

d) $C^a_4(t_c) = \{BP^a(t_n): a \in A', t_n \leq t_c, BP^a(t_n) \in KS^a(t_c) \text{ and both } o \in P^-_i(t_n) \text{ and } o \in P^-_j(t_n) \text{ hold for } BP^a(t_n)\}$

Interpretations for $C^a_2(t_c)$, $C^a_3(t_c)$ and $C^a_4(t_c)$ are similar to $C^a_1(t_c)$.

The importance of $C^a_1(t_c)$, $C^a_2(t_c)$, $C^a_3(t_c)$ and $C^a_4(t_c)$ for the choice of relevant messages results from the semantic relations given in Table 2 (see also [5], [8]).

Table 2. Semantic correlation between external formulas and classes of perceptions

$B^a(P_i(o) \wedge P_j(o))$ is related to the content of $C^a_1(t_c)$
$B^a(P_i(o) \wedge \neg P_j(o))$ is related to the content of $C^a_2(t_c)$
$B^a(\neg P_i(o) \wedge P_j(o))$ is related to the content of $C^a_3(t_c)$
$B^a(\neg P_i(o) \wedge \neg P_j(o))$ is related to the content of $C^a_4(t_c)$
$B^a(P_i(o) \underline{\vee} P_j(o))$ is related to the content of $C^a_5(t_c) = C^a_2(t_c) \cup C^a_3(t_c)$
$B^a(P_i(o) \vee P_j(o))$ is related to the content of $C^a_6(t_c) = C^a_1(t_c) \cup C^a_2(t_c) \cup C^a_3(t_c)$

Step 2. A unique representative $CBP^a_m(t_c)$, m=1,2,3,4 of all perceptions stored in $C^a_m(t_c)$ is computed. The structure $CPB^a_m(t_c)$ is given as a consensus profile and interpreted as the agreement on the overall knowledge stored in a particular class $C^a_m(t_c)$. Obviously, from the formal point of view the consensus $CBP^a_m(t_c)$ from $C^a_m(t_c) \subseteq KS(t_c)$, m=*1,2,3,4* is given as:

$$CBP^a_m(t_c) = <O, cP^+_1(t_n), cP^-_1(t_n), cP^{\pm}_1(t_n), ..., cP^+_K(t_n), cP^-_K(t_n), cP^{\pm}_K(t_n)>$$

On this stage of algorithm the knowledge representation in the form of the base profiles is transformed to the single profile.

In order to determine $CPB^a_m(t_c)$ some rational requirements need to be fulfilled. Examples of the sets of requirements, postulates and algorithms based on these postulates are given in [6].

Step 3. For each m=1,2,3,4 the agent a computes the distance d^a_m between $CBP^a_m(t_c)$ and the current profile $BP^a(t_c)$. This distance reflects the numerical similarity between each set $C^a_m(t_c)$, m=1,2,3,4 and the current base profile $PB^a_m(t_c)$. An original approach to computing this type of measure is given in [9].

Step 4. The agent a computes a choice function values $V^a_m(X)$, m=1,...,6. Each of these values is derived from a subset of $\{d_1, d_2, d_3, d_4\}$ relevant to a particular message from Table 1. The rules for determining decision values are:

$V^a_1(X)$ is derived for $B^a(P_i(o) \wedge P_j(o))$ from the set of distances X=$\{d_1\}$

$V^a_2(X)$ is derived for $B^a(P_i(o) \wedge \neg P_j(o))$ from the set of distances X=$\{d_2\}$

$V^a_3(X)$ is derived for $B^a(\neg P_i(o) \wedge P_j(o))$ from the set of distances X=$\{d_3\}$

$V^a_4(X)$ is derived for $B^a(\neg P_i(o) \wedge \neg P_j(o))$ from the set of distances X=$\{d_4\}$

$V^a_5(X)$ is derived for $B^a(P_i(o) \underline{\vee} P_j(o))$ from the set of distances X=$\{d_2, d_3\}$

$V^a_6(X)$ is derived for $B^a(P_i(o) \vee P_j(o))$ from the set of distances X=$\{d_1, d_2, d_3\}$

Obviously, the choice of the subset of $\{d_1, d_2, d_3, d_4\}$ is strictly related to this part of $KS^a(t_c)$ to which a considered message is related. The choice function and some requirements, which should be fulfilled by this function are discussed in details in [12].

Step 5. The message is chosen by the agent a as externally appropriate, for which the decision value $V^a_m(X)$ is maximal.

4 The Semantic Correlation between Messages

As a result of applying an algorithm for the choice of relevant semantic messages, described in Section 3, each agent $a \in A' \subseteq A$ generates one belief formula about the current state of an object $o \in O$. It is assumed that all of these formulas are syntactically identical. Now we want to find out if the semantics of these formulas are also the same. It is assumed that from the agent's point of view stored perceptions are the original source of any meaning accessible for cognitive processes of symbol grounding. The assumptions about the meaning of formulas is closed to the phenomenological approach to analysis of meaning in which the meaning of any language symbol can ever be reduced by the cognitive agent to reflections of basic empirical data. Each formula is treated as an external reflection of agent's internal knowledge state. In other words, the only meaning that can ever be assigned by the cognitive agent to language symbols (formulas) comes out of the content of remembered experiences [11]. In this context of meaning it might be possible that one belief formula for two agents means completely different. In this paper we propose the method for the measure semantic correlation between messages. This method is applied to the algorithm for finding the set R of agents with the maximal cognitive capacity to observe the states of objects and maximal similarity of knowledge state to the median state of knowledge. In this algorithm called *the algorithm for finding the set of knowledge richest and reliable agents* (KRRA), the consensus profiles $CPB^a_m(tc)$, $a \in A'$, $m \in \{1,2,3,4\}$ for classes of perception $C^a_m(t_c)$ semantically correlated with external formula (see Table 2) are taken into account.

4.1 The Idea of KRRA

Let idea of an algorithm KRRA can be described as follows:

Step 1. Determine the set $CPB^{A'}_r(t_c)$, $r \in \{1,2,...,6\}$ of consensus profiles semantically related with the external formulas generated by the agents from the set A'. Each formula has its tag name (see Table 1) and depending on the external formula generated by each agent $a \in A'$ the proper set of base profiles is determined. In particular, the following sets of base profiles are taken into account:

a) $CPB^{A'}_1(t_c) = \{CBP^a_1(t_c): a \in A'\}$
b) $CPB^{A'}_2(t_c) = \{CBP^a_2(t_c): a \in A'\}$
c) $CPB^{A'}_3(t_c) = \{CBP^a_3(t_c): a \in A'\}$
d) $CPB^{A'}_4(t_c) = \{CBP^a_4(t_c): a \in A'\}$
e) $CPB^{A'}_5(t_c) = \{CBP^a_k(t_c): a \in A'$ and $k=2,3\}$
f) $CPB^{A'}_6(t_c) = \{CBP^a_k(t_c): a \in A'$ and $k=1,2,3\}$

Step 2. For the set $CPB^{A'}_r(t_c)$, $r \in \{1,2,...,6\}$ compute the median profile $Med^{A'}_r(t_c)$, $r \in \{1,2,...,6\}$. Median profile is interpreted as the agreement of the overall knowledge stored in a particular set $CPB^{A'}_r(t_c)$ and has the same structure as consensus profile $CPB^a_m(t_c)$, $m \in \{1,2,3,4\}$ and consists of the following sets: $medP^+_i(t_n)$, $medP^-_i(t_n)$, medP$^{\pm}_i$(t$_n$), i∈{1,2,...,K}. The distance between median profile Med$^{A'}_r$(t$_c$) and each consensus profile CPB^a_m(t$_c$)∈CPB$^{A'}_r$(t$_c$), a∈A', must be minimal. For the defined distance function d(Med$^{A'}_r$(t$_c$),CPB^a_m(t$_c$)) holds [13]:

$$\sum_{CPB\,^a_m(t_c)\in CPB\,^{A'}_r(t_c)} d\left(Med\,^{A'}_r(t_c), CPB\,^a_m(t_c) \right) = Min \tag{2}$$

where Min is the minimum of all d(Med$^{A'}_r$(t$_c$),CPB^a_m(t$_c$))∈D$_{Universe}$, for Med$^{A'}_r$(t$_c$)∈Med$_{Universe}$, r∈{1,2,...6} and CPB^a_m(t$_c$)∈CPB$^{A'}_r$(t$_c$), m∈{1,...,4}.Med$_{Universe}$ denotes the universe of all median profiles, D$_{Universe}$ is the universe of all distance function values between each median profile from the set Med$_{Universe}$ and each consensus profile form the set CPB$^{A'}_m$(t$_c$).

Step 3. For each consensus profile CPB^a_m(t$_c$)∈CPB$^{A'}_r$(t$_c$), r∈{1,2,...,6}, compute the value D$^{MC}_a$=card(MC$_a$):

$$MC_a = \left(medP^+_i(t_c) \backslash cP^+_i(t_c) \right) \cup \left(medP^-_i(t_c) \backslash cP^-_i(t_c) \right) \tag{3}$$

where medP^+_i(t$_c$), medP^-_i(t$_c$) ∈Med$^{A'}_r$(t$_c$) and cP^+_i(t$_c$), cP^-_i(t$_c$) ∈CPB^a_m(t$_c$), m=1,...,4. The aim of this step is to find the consensus profile closest to the median profile.

Step 4. For each consensus profile CPB^a_m(t$_c$)∈CPB$^{A'}_r$(t$_c$), r∈{1,2,...,6}, compute the value D$^{CM}_a$=card(CM$_a$), where the set CM$_a$ is equal:

$$CM_a = \left(cP^+_i(t_c) \backslash medP^+_i(t_c) \right) \cup \left(cP^-_i(t_c) \backslash medP^-_i(t_c) \right) \tag{4}$$

Step 5. For each consensus profile CPB^a_m(t$_c$)∈CPB$^{A'}_r$(t$_c$) compute the value of coefficient α$_a$, a∈A' and chose those agents as the most knowledge richest and reliable for which the coefficient α$_a$ is maximal.

$$\alpha_a = \frac{D^{MC}_a + 1.}{D^{CM}_a + 1} \tag{5}$$

Remark 4. Two factors have an influence on the value of coefficient α$_a$. The closure of an particular consensus profile CPB^a_m(t$_c$)∈CPB$^{A'}_r$(t$_c$) to the median profile Med$^{A'}_r$(t$_c$), r∈{1,2,...,6} (Step 3 of KRRA)and the information power of a particular consensus profile CPB^a_m(t$_c$) understood by means of the cardinality of objects, which states are known for an agent a∈A' in this consensus profile (Step 4 of KRRA). Let us note that the value D^{MC}a should be minimal. It means that the agent with the cognitive range closed to cognitive range of other agents' is needed. On the other hand the value D^{CM}a should be maximal. In this case the scope of agent's cognition is taken into account. The important is that only known states of objects are taken into account. As a consequence the objects from agents' area incompetence are omitted.

4.2 Computational Example

Let the external world W consist of atom objects $O=\{o_1,o_2,o_2,...o_{11}\}$. At the time point t_{12} the state of an object o_1 in relation to the properties P_1 and P_2 is not known for the agents $A'=\{a_1, a_2, a_5, a_7\}\subseteq A=\{o_1,o_2,...,o_{20}\}$. After applying the algorithm for the choice of relevant semantic messages the agents generate the messages $B=\{B^a(P_i(o)\wedge P_j(o)):$ a=1,2,5,7\}. Let apply the KRRA algorithm:

Step 1. The set $CPB^{A'}_1(t_{12}) =\{CBP^1_1(t_{12}),\ CBP^2_1(t_{12})\ CBP^5_1(t_{12})\ CBP^7_1(t_{12})\}$, where:
$CBP^1_1(t_{12})=\{\{o_1,o_2\},\{o_3,o_4,o_5\},\{o_6,o_7,o_8,o_9,o_{10}\};\{o_1,o_3,o_4\},\{o_5,o_7,o_9\},\{o_2,o_6,o_8,o_{10}\}\}$
$CBP^2_1(t_{12})=\{\{o_1,o_2,o_7\},\{o_3,o_4,o_8,o_{10}\},\{o_5,o_6,o_9\};\{o_1,o_3,o_4,o_6\},\{o_5,o_7,o_9\},\{o_2,o_8,o_{10}\}\}$
$CBP^5_1(t_{12})=\{\{o_1\},\{o_3,o_8,o_9\},\{o_2,o_4,o_5,o_6,o_7,o_{10}\};\{o_1,o_3,o_4\},\{o_6,o_7,o_9\},\{o_2,o_5,o_8,o_{10}\}\}$
$CBP^7_1(t_{12})=\{\{o_1,o_2\},\{o_3,o_4,o_6\},\{o_5,o_7,o_8,o_9,o_{10}\};\{o_1,o_3\},\{o_5,o_7,o_{10}\},\{o_2,o_4,o_6,o_8,o_9\}\}$

Step 2. The median profile for the set $CPB^{A'}_1(t_{12})$ is equal:
$Med^{A'}_1(t_{12})=\{\{o_1,o_2\},\{o_3,o_4\},\{o_5,o_6,o_7,o_8,o_9,o_{10}\}; \{o_1,o_3,o_4\},\{o_5,o_7,o_9\},\{o_2,o_6,o_8,o_{10}\}\}$

Step 3. For each consensus profile $CPB^a_1(t_{12})\in CPB^{A'}_1(t_{12})$ D^{MC}_a is equal: $D^{MC}_1=0$, $D^{MC}_2=0$, $D^{MC}_5=3$, $D^{MC}_7=1$.

Step 4. For each consensus profile $CPB^a_1(t_{12})\in CPB^{A'}_1(t_{12})$ D^{CM}_a is equal: $D^{CM}_1=1$, $D^{CM}_2=4$, $D^{CM}_5=3$, $D^{CM}_7=2$.

Step 5. The values of coefficient α_a are: $\alpha_1=2$, $\alpha_2=5$, $\alpha_5=1$, $\alpha_7=1.5$. The most knowledge richest and reliable is agent a_2.

5   Conclusions

Although the symbol grounding problem has been studied for quite a long time, there is still a lack of mathematical models for simplest languages' grounding [11].

The aim of this paper was to investigate the level of grounding the belief formulas generated individually by the agents in multiagent system after applying the algorithm for the choice of relevant messages. The semantic closeness of the messages was investigated. A method for computing the set of agents with the maximal cognitive capacity to observe the states of objects and maximal similarity of knowledge states to the median state of knowledge of all agents in a system was presented.

References

1. Daniłowicz, C., Nguyen, N. T.: Consensus–based Partition in the Space of Ordered Partitions. Pattern Recognition, 3, (1988) 269-273
2. Dennett, D.C.: Kinds of Minds. Basic Books, New York (1996)

3. Fauconier, G.: Mappings in Thought and Language. Cambridge University Press, Cambridge (1997)
4. Harnad, S.:The Symbol Grounding Problem. Physica, 42, 335-236
5. Katarzyniak, R.: Defining the Relation between Intentional and Implementation–oriented Levels of Agents Models. In: Proc. of ISAT'2000, Szklarska Poręba, (2000), 208-216
6. Katarzyniak, R., Nguyen, N.T.: Reconciling Inconsistent Profiles of Agents' Knowledge States in Distributed Multiagent Systems Using Consensus Methods. Systems Science, Vol. 26 No. 4, (2000) 93-119.
7. Katarzyniak, R., Pieczyńska-Kuchtiak, A.: Formal Modeling of the Semantics for Communication Languages in Systems of Believable Agents. In: Proc. of ISAT'2001, Szklarska Poręba, (2001), 174-181
8. Katarzyniak, R., Pieczyńska-Kuchtiak, A.: Intentional Semantics for Logic Disjunctions, Alternatives and Cognitive agent's Belief. In: Proc. of the 14th International Conference on System Science, Wrocław, Poland, (2001), 370-382
9. Katarzyniak, R., Pieczyńska-Kuchtiak, A.: A Consensus Based Algorithm for Grounding Belief formulas in Internally Stored Perceptions. Neural Network World, 5, (2002) 671-682
10. Katarzyniak, R., Pieczyńska-Kuchtiak, A.: Distance Measure Between Cognitive Agent's Stored Perceptions. In: Proc. of IASTED International Conference on Modelling, Identification and Control, MIC'2002, Innsbruck, Austria (2002) 517-522
11. Katarzyniak, R., Pieczyńska-Kuchtiak, A.: Grounding Languages in Cognitive Agents and Robots. In: Proc. of Sixteenth International Conference on System Engineering, Coventry (2003) 332-337
12. Pieczyńska-Kuchtiak, A.: A Decision Function in the Algorithm for the Choice of Semantic Messages. In: Proc. of Information Systems Architecture and Technology, Poland (2002)
13. Nguyen, N.T.: Consensus System for Solving Conflicts in Distributed Systems. Information Sciences, 147, (2002) 91-122

Modelling Intelligent Virtual Agent Skills
with Human-Like Senses

Pilar Herrero and Angélica de Antonio

Facultad de Informática. Universidad Politécnica de Madrid.
Campus de Montegancedo S/N.
28.660 Boadilla del Monte. Madrid. Spain
{pherrero,angelica}@fi.upm.es

Abstract. The research work presented in this paper represents a significant advance, specifically in the area of perception, within the multidisciplinary field of Intelligent Virtual Agents (IVAs) and multi-Intelligent Virtual Agent Systems (mIVAS) .Within the cognitive research area there are some studies underwriting that human perception can be understood as a first level of an "awareness model". Bearing in mind these researches, we have developed a human-like perceptual model based on one of the most successful awareness models in Computer Supported Cooperative Work (CSCW), called the Spatial Model of Interaction (SMI), which has been applied to Collaborative Virtual Environments (CVEs). This perceptual model extends the key concepts of the SMI introducing some human-like factors typical from human being perception as well as it makes a reinterpretation with the aim of using them as the key concepts of a IVA's human-like perceptual model.

1 Introduction

Nowadays, virtual environments often incorporate human-like embodied virtual agents with varying degrees of intelligence, getting what we call Intelligent Virtual Agents (IVAs).

An IVA may simply evolve in its environment or it may interact with this environment or even communicate with other IVAs or humans, but, in order to make this interaction possible, an IVA has to be aware of its environment. This awareness can be understood as the result of an IVA perceiving its surroundings and therefore those object/agent which are in the environment and nearby to the observer IVA.

The research that we present in this paper is precisely oriented towards endowing IVAs with perceptual mechanisms that allow them to be "realistically *aware*" of their surroundings. We propose a perceptual model, which seeks to introduce more coherence between IVA perception and human being perception. This will increment the psychological "*coherence*" between the real life and the virtual environment experience. This coherence is especially important in order to simulate realistic situations as, for example, military training, where soldiers must be trained for living and surviving risky situations. A useful training would involve endowing soldier agents with a human-like perceptual model, so that they would react to the same *stimuli* as a human soldier. Agents lacking this perceptual model could react in a non-

M. Bubak et al. (Eds.): ICCS 2004, LNCS 3038, pp. 575–582, 2004.
© Springer-Verlag Berlin Heidelberg 2004

realistic way, hearing or seeing things that are too far away or hidden behind an object. The perceptual model we propose in this paper introduces these restrictions inside the agent's perceptual model with the aim of reflecting more faithfully a human-like perception.

Having in mind that the physical perception can be understood as the first level of an "awareness model" [2] the first goal for our research was to select a model of awareness which could be valid for our purposes. As the "Spatial Model of Interaction" (SMI) [1] used the properties of the space to get knowledge of the environment, it was based on a set of key awareness concepts – which could be extended to introduce some human-like factors - and it had been tested with successful results in CSCW multi-user environments, this model had the essential qualifications for our purposes and we selected it in order to develop the perceptual model we propose in this paper.

2 Key Concepts in the Perceptual Model

As we mentioned in previous sections, the key concepts of our Perceptual Model are based on the main concepts of a CSCW awareness model known as *The Spatial Model of Interaction (SMI)* [1].

2.1 Key Concepts in the SMI

The spatial model, as its name suggests, uses the properties of space as the basis for mediating interaction. It was proposed as a way to control the flow of information of the environment in CVEs (Collaborative Virtual Environments). It allows objects in a virtual world to govern their interaction through some key concepts: awareness, focus, nimbus and boundaries. In this paper we are going to concentrate on the three first, leaving the boundaries out of the paper's scope.

Focus. "The more an object is within your focus the more aware you are of it". The focus concept has been implemented in the SMI as an "ideal" triangle limited by the object's aura. In the same way, it is possible to represent the observed object's projection in a particular medium, this area is called *nimbus*: "The more an object is within your nimbus the more aware it is of you". The implementations of these concepts –focus and nimbus- in the SMI didn't have in mind human aspects, thus reducing the level of coherence between the real and the virtual agent behaviour. *Awareness* between objects in a given medium is manipulated via *focus* and *nimbus*, controlling interaction between objects.

2.2 Making the Perceptual Model More Human-Like

There are many factors that contribute to our ability as humans to perceive an object, some of which are directly working on the mental processes, being not easily modelled or reproduced in a virtual world. We have selected the two most useful senses: sight and hearing and we have analysed some human perception key concepts to determine which of them could be introduced in our agent's perceptual model. These concepts, selected for being the more representative of human visual and hearing perception, are [3,4]:

- Sense Acuity: It is a measure of the sense's ability to resolve fine details In a visual medium, it is known as Visual Acuity. In a hearing medium, it is know as Auditory Acuity There are two different kinds of auditory acuity: spatial acuity and frequency acuity. Both acuities are inter-related.
- Sense Transition Region: It is the interval in the space between perfect and null perception. This factor plays an important role in a visual medium where it is known as Lateral Vision. The Lateral Vision corresponds to the visual perception towards the extremes of the visual focus. In a hearing medium this concept can be understood as the cone in the space known as Cone of confusion. The cone of confusion is a cone extending outwards from each ear. Sound events that originate from a point in this cone is subject to ambiguity.
- Filters: Allowing the selection, from all the objects in an extensive focus, of only those that the agent is especially interested in.
- Directivity of sound: This factor , associated to a hearing medium, represents the directional characteristic of the sound source propagation.

2.3   Reinterpreting the SMI's Key Concepts

Neither the SMI nor its implementations considered aspects of human perception. Thus, if the SMI were applied just as it was defined by Benford, the level of coherence between real and virtual agent behaviour would be minimum. We have decided to identify the factors concerning human-like perception which provide more realistic perception, and introduce them into the SMI. In this section, we are going to describe how the key concepts defining the SMI have been modified to introduce these human factors.

Focus. Benford introduced the focus concept in 1993 as *"The more an object is within your focus the more aware you are of it"* [1]. This concept meant that the observing object's interest for each particular medium could be delimited. According to this definition, the focus notion is the area within which the agent is perceiving the environment. We have analysed how sensitive perception works in humans beings, and, from this analysis, we have decided to select some factors that should have an effect on the physical area delimiting the observing object's interest.

Visual Focus. In a visual medium, these factors are the *Sense Acuity* and the *Sense Transition Region*. Starting from the focus concept in the spatial model, and bearing in mind previous implementations, where focus was implemented as a triangle, we have defined a new mathematical function to represent the human-like visual focus concept.

This mathematical function describes the focus as a double cone centered on the agent's eye and which length is limited by agent's visual resolution acuity. Each of these two cones is dependent on the opening angle, which will be the agent's foveal angle of vision in the foreground region and the agent's lateral angle of vision in the lateral region of perception.

Auditory Focus. A human listener can hear a sound arriving to him in two different ways: combining information from the two ears, *"Binaural Hearing"*, or taking information from one ear or from each ear independently, *"Monaural Hearing"* .Modern psychoacoustic research has turned its attention to binaural hearing [4] and, for this reason, we are also going to focus on this kind of hearing along this research.

Sound waves usually travel out in all directions from the source of sound, with an amplitude that depends on the direction and distance from the source. We have represented the IVA's hearing focus concept by an sphere circumscribing the agent's head.

Nimbus. Benford introduced the nimbus concept in 1993 as *"The more an object is within your nimbus the more aware it is of you"* [1]. This concept meant that the observed object's projection for each particular medium could be delimited. The nimbus concept, as defined in the Spatial Model of Interaction, has always been implemented as a circumference in both visual and hearing media. The radius of this circumference has an "ideal" infinite value, although, in practice, it is limited by the object's aura. Just as with the above-mentioned focus concept, the nimbus concept in the Spatial Model of Interaction does not consider any human factors, thus hypothetically reducing the level of coherence between real and virtual agent behaviour.

Visual Nimbus. In a visual medium, we are going to represent the nimbus of an object as an ellipsoid or a sphere, depending on the conic by which the object is circumscribed, centred on its geometrical centre. The way of practically determining which conic has to be associated with each object in the environment is to look for the bounding box that has been associated to this object in the environment. If the bounding box is a rectangle, we will approximate the nimbus as an ellipsoid; if the bounding box is a circle, then we will approximate the nimbus as a sphere. The nimbus radius, or its eccentricity if it is an ellipsoid, will depend on two factors: the object's shape and the furthest distance at which a human being would be able to distinguish the object. This distance is determined by visual acuity, which depends on the object's size; thus, indirectly, the nimbus conic will depend on the object's size as well.

Auditory Nimbus. In a hearing medium, the nimbus delimits the physical area of projection of a sound source for a particular medium. Before modelling the nimbus shape, it is important to analyse the way of projecting the sound, but also the factors that can have an effect on the physical projection of the sound.

We start from the assumption that the sound is propagated in the medium by a spherical wavefront, but even if this occurs, it could happen that the sound amplitude, and therefore its intensity, weren't the same in all the directions. For this reason, in this model we interpret the nimbus concept as the region within which the sound source is projected with the same intensity.

Starting from this interpretation of nimbus we can determine which factors have an influence on this concept and its representation within an environment. From all these factors, in this research we have just considered directivity of sound, living the rest of the factors, as for example, the presence of non-linear effects or the homogeneity of the medium, for future research and extensions of this work.

We have also centered our research in the projection of human voice. The directivity pattern associated to the human voice is a cardioid representing the sound pressure level (SPL) - in decibels (dB) - versus horizontal angle and being the direction of speaker's mouth located at 0 degrees.

3 Clarity of Perception

In the previous sections we have introduced the key concepts of our perceptual model. In this section we are going to concentrate on how to use these concepts in order to endow IVAs with a human-like perceptual model.

In order to make this possible, we have selected an appropriate agent architecture in which to introduce this perceptual model. From all the possible architectures, we have selected a vertical layered architecture which is composed by three different blocks: Perception, Central Processing and Action. We have also proposed several modules to form the agent's perception block being one of them the *Sensitive Perception*. The *Sensitive Perception* module simulates the typical perceptual process by which organisms receive sensations from the environment, depending on some relevant sensorial concepts. More details of this design and its implementation are given in [3,4]. The IVA's sensitive perception module will calculate what we have called *Clarity of Perception*, taking into account the focus of the IVA and the nimbus of those objects that are surrounding this agent.

Clarity of Perception is a measurement of the ability to perceive an object (or a sound) inside the agent's visual focus, as well as the clearness of this perception. Once an agent's focus intersects with an object's nimbus, this agent is aware of this object and therefore the sensitive perception module will calculate the clarity of perception for this object.

Following the research conducted by Levi et al. [5,6] in a visual medium and Shinn-Cunningham [7] in a hearing medium, we propose several mathematical function to describe the variation in the clarity of perception with the eye/ear-object/sound distance. More details are given in [3,4].

4 Perceptual Model Implementation

This model has been implemented in Visual C++ as a library. This library has been integrated with MASSIM_AGENT. MASSIM_AGENT was the prototype system built using the MASSIVE-3 – a CVE system - and the SIM_AGENT – a toolkit for developing agents. MASSIM_AGENT was the result of a collaboration established between the Mixed Reality Laboratory (MRL) at the University of Nottingham and the Universidad Politécnica de Madrid (UPM). More details of its implementation are given in [3,4].

5 Some Scenarios for Human-Like Perception

In order to prove the usefulness of the proposed perception model, lets consider that, as it was previously mentioned, mIVAS can be used to simulate risky situations, as for example, a world-wide war, where the soldiers' training plays a very important role. In this kind of systems, soldiers can be trained for living and surviving the worse real-life situations. To get a useful training, it is important to endow soldier agents with a human-like perception model. Different scenarios and situations can be raised where human-like perception plays a very important role. In this section we are going to describe some of them.

5.1 Visual Perception

Lets imagine that a soldier agent is at the battlefield. He is placed at a physical position given by the co-ordinates $(x,y,z)= (1,0,0)$ in the space, in meters, with an orientation of 90° related to the x axis of co-ordinates. This soldier is endowed with a visual acuity (in Snellen notation) equivalent to 20/20 and his foreground angle of vision is $\theta=30°$ while his lateral angle of vision is $\theta'=65°$. Lets also imagine that a fighter plane, a Focke-Wulf BMW whose size is (length x wide x height)=(20,15,7), in meters, appears in the air space. Introducing all these values in the implemented visual perceptual model, we get the foreground and lateral soldier's cone of vision. In the same way, we get the nimbus geometry associated to the plane, which in this case is an ellipsoid, and the plane's nimbus. In the same way, the perceptual model calculates the maximum distance of resolution for an object of the plane's size (D_m), which in this case is 64.20 m. When the plane is placed at co-ordinates $(x,y,z)=(1,25,0)$, in meters, away from the soldier, it is possible to perceive the object in the foreground area of perception. The clarity of perception in this area gets the maximum normalised value 1. The soldier can perceive most of the plane's details. When the plane is placed at co-ordinates $(x,y,z)=(-40,25,25)$, in meters, away from the soldier, it is possible to perceive the object in the lateral area of perception. The clarity of perception in this area is 0.5. The soldier can perceive just a few details of the plane. The soldier will probably perceive an object but he will not be able to identify its physical details. As the plane is placed in the lateral area of perception, the details that this soldier can get from the plain will be associated to its movement. When the plane is placed at co-ordinates $(x,y,z)=(15,70,25)$, in meters, away from the soldier, it is possible to perceive the object in the foreground area of perception. The clarity of perception in this point is 0.033 (very low). Maybe the soldier can perceive the plane's shape and its movement but this soldier can make a mistake confusing the coming plane with a friendly plane instead of recognising it as the hostile plane that it really is.

5.2 Auditory Perception

Let us imagine that a soldier agent (A) is in the countryside, keeping watch. He is placed at a position $(x,y,z)=(0,0,0)$ in the space, in metres. The soldier azimuth angle is 80 degrees and his elevation angle is 87.5 degrees. At this moment, a couple of people approach him and one of them starts speaking when at co-ordinates

Fig. 1. The war scenario in MASSIM_AGENT

$(x,y,z)=(1,2,0)$. As they are having a moderate conversation, the intensity of the sound that this person is emitting is 60dB. In this case, the intensity of the sound that is reaching soldier agent A, and which is calculated by our perceptual model, is 11.22 dB. Then the method that implements this perceptual model determines that this intensity of sound is between the standard thresholds of hearing and, therefore, the agent can hear it. However, the clarity of perception with which this agent can hear this sound is *Very_Low*. Soldier A can hear a sound, but he cannot understand what they are speaking about. If the same couple continue walking and the same soldier continues speaking with the same intensity, when he reaches co-ordinates $(x,y,z)=(0.50,0.75,0)$ the perceptual model determines that the intensity of the sound that is reaching soldier agent A is 42.63 dB, determining that the sound can be heard by soldier agent A because it is between the standard thresholds of hearing, and calculating that this sound can be heard by soldier A with a clarity of perception *Medium*. Soldier A can perceive something of the couple's conversation but not too much. If they continue walking under the same conditions, when the speaker reaches co-ordinates $(x,y,z)=(0.50,0.50,0)$ the perceptual model determines that the intensity of the sound that is reaching soldier agent A is 51.77 dB, returning a clarity of perception *High*. The soldier A can perceive the couple's conversation.

6 Conclusions

We have developed a human-like perceptual model for IVAs based on one of the most successful awareness models in Computer Supported Cooperative Work (CSCW), called the Spatial Model of Interaction (SMI) [1]. This perceptual model extends the key concepts of the SMI introducing some factors typical from human being perception as well as it makes a reinterpretation of the key concepts and introduces some new concepts – as the Clarity of Perception (CP)- with the aim of using them as the key concepts of an IVA's human-like perceptual model.

Acknowledgements. The work presented in this paper has been supported by the Communication Research Group (CRG), led by Steve Benford and Chris Greenhalgh at the School of Computer Science and Information Technology in the University of Nottingham, in UK.

References

1. Benford, S.D., and Fahlén, L.E. *A spatial Model of Interaction in Large Virtual Environments*. Proc. Third European Conference on Computer Supported Cooperative Work (ECSCW'93), Milano, Italy. Kluwer Academic Publishers, 1993.
2. Endsley. M., *Design and evaluation for situation awareness enhancement*. Proceedings of Human Factors Society and Annual Meeting, volume 1, 1988.
3. Herrero P., De Antonio A., Benford S., Greenhalgh C., *A Hearing Perceptual Model for Intelligent Virtual Agents*. Proceedings of the Second International Joint Conference on Autonomous Agents and Multiagent Systems, Melbourne, Australia, July, 2003.
4. Herrero P., De Antonio A. *Keeping Watch: Intelligent Virtual Agents Reflecting Human-Like Perception in Cooperative Information Systems*. The Eleventh International Conference on Cooperative Information Systems (CoopIS 2003). Catania, Sicily. Italy, 2003.
5. Levi, D.M., Hariharan, S. & Klein, S.A. *Suppressive and Facilitatory Spatial Interactions in Peripheral Vision: Peripheral Crowding is neither size invariant nor simple contrast masking*. Journal of Vision, 2, 167-177. 2002.
    http://www.journalofvision.org/2/2/3/
6. Levi, D.M., Klein, S.A. & Hariharan, S. *Suppressive and Facilitatory Spatial Interactions in Foveal Vision: Foveal Crowding is simple contrast masking*. Journal of Vision, 2, 140-166. 2002. http://journalofvision.org/2/2/2/
7. Shinn-Cunningham, BG. *Distance cues for virtual auditory space* Proceedings of the IEEE 2000 International Symposium on Multimedia Information Processing. Sydney, Australia. December 2000.

Reuse of Organisational Experience Harnessing Software Agents

Krzysztof Krawczyk[1], Marta Majewska[1], Mariusz Dziewierz[1], Renata Słota[2], Zoltan Balogh[3], Jacek Kitowski[1,2], and Simon Lambert[4]

[1] ACC CYFRONET-AGH, ul. Nawojki 11, 30-950 Cracow, Poland
{krafcoo,mmajew,aristot}@icsr.agh.edu.pl
[2] Institute of Computer Science, AGH-UST, al. Mickiewicza 30, Cracow, Poland
{rena,kito}@agh.edu.pl
[3] Institute of Informatics, SAS, Dubravska cesta 9, 845 07 Bratislava, Slovakia
balogh.ui@savba.sk
[4] CCLRC Rutherford Appleton Laboratory, Chilton, Didcot, OX11 0QX, UK
S.C.Lambert@rl.ac.uk

Abstract. This paper presents the application of agent technology for managing employee's experience within an organisation. The proposed approach is discussed on the basis of the software platform developed within the scope of the EU IST Pellucid project. The system is intended to provide universal assistance for organisationally mobile employees. Its functionality is realized by a set of agents. The focus of this paper is on reuse of organisational experience, which is transmitted by active hints. The experience is structured using ontologies.

1 Introduction

An agent-based experience management system is being developed within the scope of the EU IST Pellucid project (IST-2001-34519) [1], as a response to needs formulated by many organisations to reuse their employees' experience. The Pellucid system is aimed at departments of organisations that are characterized by a high level of staff rotation which results in reduced efficiency and consequent losses. The Pellucid system allows the capturing of employees' experience and reusing it to assist other employees in a similar position in the future. It also assists a mobile employee to reuse his knowledge at a new position within the organisation. The paper discusses the use of agents for experience management in the Pellucid platform. Certain aspects of experience management have been already described in [2,3,4,5].

In Section 2 the state of the art is sketched. Afterwards, the experience definition for the Pellucid platform is given in Section 3. The experience manifestation, presented to the employees, called active hints, is specified too. Then, the foundations of a process of the experience adaptation to the employee's needs are presented (in Section 3) together with the roles of agents participating in it (shown in Section 4). Next, Section 5 presents a detailed description of the way the agents perform activation of the active hints. Conclusions arising from

M. Bubak et al. (Eds.): ICCS 2004, LNCS 3038, pp. 583–590, 2004.
© Springer-Verlag Berlin Heidelberg 2004

the application of agents to experience management are presented in Section 7, together with some insight into future work.

2   State of the Art

There are some projects which apply agent technology to knowledge management environments. AMKM [6] is a framework, which uses agent concepts to analyse and model organisations and as a result to provide a reusable architecture to build knowledge management systems. The AMKM framework is intended to provide a common, uniform description of knowledge items for users performing different knowledge intensive tasks. Next, DIAMS is a multi-agent system which helps users to access, collect, organise and exchange information [7]. The personal agents are exploited to provide their owners with well-organised information collections, as well as friendly information management utilities. The CoMMA project [8] implements a corporate memory management framework based on agent technology, that captures, stores and disseminates knowledge to support users' work with documents. Ontologies are exploited to annotate documents in the corporate memory.

In general the scientific community tries to support employees by providing a service called Organisational Memory [9,10]. Such a memory is treated as the collector of all kinds of organisational knowledge and is specially designed for reuse in the future. The DECOR [11], FRODO[12], KnowMore[13] projects deal with such knowledge repositories. The heterogenous nature of knowledge resources implies the use of agents for managing the content of such a memory.

3   Experience Definition and Representation

It is assumed that an organisation follows a formalised and well established business administration process to reach its goals. The process could be supported by a Workflow Management System (WfMS). WfMS specifies in detail successive activities to be performed. Most of the activities are knowledge-intensive – they require knowledge of a miscellaneous nature to be performed correctly. Such activities require from the employees broad knowledge about how to proceed in given circumstances. The Pellucid working definition of experience uses such knowledge. Experience is a kind of knowledge concerning an action which has to be performed in a particular context.

It seems obvious that experience requires human intelligence in both creation and reuse. For the purposes of the experience management system an adequate representation of experience has to be found. Ontologies are the natural way to represent knowledge.

On ontology is a specification of conceptualisation, which is readable both by human and computer systems [14]. The experience could be expressed with the help of ontology. In the frame of Pellucid operation the experience is represented by a set of active hints. The active hint is a representation of a piece of experience. According to the context of the employee's work selected hints

Fig. 1. Active hint definition

are presented to him spontaneously. The elements of the active hint, shown in Fig. 1, are: an action to be performed, resources involved during performing the action, a context of the employee work and justification for recommendation of the action. The justification represents possible motivation for the employee to follow the hint. During the operation the Pellucid system tracks all actions taken by employees as well as working context, and stores relevant instances of the ontologies. These instances define the employee's experience representation. Pellucid ontologies are implemented in DAML+OIL [15] and manipulated by the Java-based Jena library[16].

Profiles of Employees. Experience is a highly subjective matter in both creation and reuse. The decision whether an active hint should be shown to the employee has to take into account factors arising not only from the scope of work characteristic for a given position but also from the employee profile. The active hints which are going to be proposed to the employee have to fit the employee's needs. The expectations of the employee strongly depend on experience and skills he already possesses. According to the above criterion several types of employees are distinguished:

- Novice – a totally inexperienced employee (e.g. a recent graduate).
- New-hired or Relocated Employee – an employee who has got already some practical skills needed at the given position acquired during earlier work for another employer.
- Relocated Employee – an employee who has got already some practical skills needed at the given position acquired during earlier work in this organisation.
- Experienced Employee – an employee who has worked at the given position in the organisation for a long time (for the Pellucid system he is an experience provider).

The active hints should be adapted to the level of employee's experience – the experienced user should not be bothered with hints concerning low level skills. The adaptation of active hints also has to consider that employees' experience evolves and new skills are developed over time. Moreover, the employee's profile is defined by the employee's history of work in the organisation and with the Pellucid system.

Pellucid Formula. One of the most important issues of the Pellucid experience management system is the choice of experience which should be presented to the user. On the one hand there is knowledge about the role and the job the employee is assigned to. On the other hand there is experience which the employee has already acquired. The above relation could be formalised as follows:

$$E_G = E_R - E_A, \tag{1}$$

where E_R is experience required, E_A is experience available due to the employee's previous activities and E_G is the experience gap between required and available ones. The task of the Pellucid system is to present a subset of E_G to the employee on the basis of E_R resulting from description of his current job and role and due to E_A derived from the employee's profile.

4   Experience Management Agents

Based on Eq. 1 the areas of responsibility for elements of the system can be defined. The agent technology fits the Pellucid system features well. Cooperating, intelligent, autonomous agents can be exploited to achieve the overall goals of Pellucid, i.e, reuse of experience to assist the employees.

As mentioned before, the system should find the experience gap for an employee accomplishing a task within a performed role. Therefore, two main components are introduced : Role Agent (RA) and Personal Assistant Agent (PAA). The purpose of RA is to find experience required for a particular role (E_R). PAA is responsible for finding experience of a particular employee (E_A). The required experience is not only related to the role of the employee but it also depends on the task the employee is performing. So, Task Agent (TA) is introduced to find out what part of experience is required for accomplishment of a particular task. PAA cooperates with RA and TA agents to learn about the experience gathered by the employee while playing a role. All these agents are able to find experience necessary for the employee.

Experience gathered by the agents is stored with the help of a special component called Organisational Memory (OM).

The nature of work in any organisation requires access to information resources, such as documents, e-mails, databases, etc. It is necessary to ensure fast and uniform access to information resources and allow extraction of the information and knowledge from them, since the experience is not only just knowledge of how to perform a required task, but also often knowledge extracted from organisational assets, i.e. documents and data. This task is performed by Information Search Agent (ISA), being responsible for accessing document repositories, other external resources and searching using full-text indexing, document similarity strategies and ontology searching.

To provide active assistance the system has to track employees' activities and the context which they are working in. Therefore, two agents, Monitoring Agent (MA) and Environment Tracking Agent (ETA) are introduced respectively.

In Fig. 2 the components involved in E_G computation are presented.

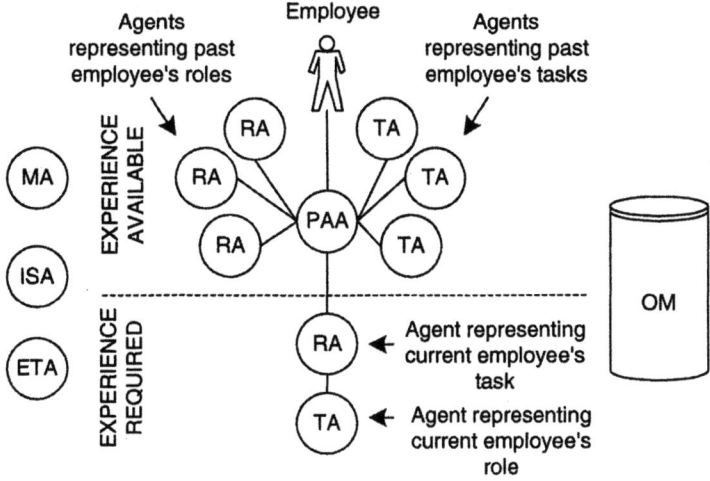

Fig. 2. Pellucid agents

5 Activation of Active Hints

Several methods could be invented to deal with Eq. 1. A method, which allows finding active hints to be presented to the employee, is proposed below.

The factors influencing activation and deactivation of active hints can be twofold – can derive from both the required and available experience. A set of factors that influence selection of the active hints can be defined. Beneath there are some examples of the critical factors:

- The employee has already been working on an activity with a similar context.
- Skills of the employee refer to the skills represented by the active hint.
- Employee's actions have been used to instantiate the active hint.

The approach can be formalised as follows. Thus, E is a set of employees, H is a set of active hints and F is a set of functions for computing different factors:

$$E = \{e_1, e_2, \ldots, e_n\}, \quad \text{where} \quad n \in \mathbf{N},$$
$$H = \{h_1, h_2, \ldots, h_m\}, \quad \text{where} \quad m \in \mathbf{N},$$
$$F = \{f_1, f_2, \ldots, f_l\}, \quad \text{where} \quad l \in \mathbf{N}.$$

The activation function is

$$A(f_1, f_2, \ldots, f_l) \in (0, 1).$$

where $f_i(e_j, h_k)$, is the factor function.

The activation function takes all factor functions f_i for a given active hint h_k for a particular employee e_j and returns a normalised rating value. The activation function is computed for each active hint. The active hints are rated according to the value of the activation function - the greater the value, the more relevant is the active hint for the particular situation.

Table 1. Activation function computation for a particular employee

e_j	h_1	h_2	$\ldots$	h_m
f_1	$f_1(e_j, h_1)$	$f_1(e_j, h_2)$	$\ldots$	$f_1(e_j, h_m)$
f_2	$f_2(e_j, h_1)$	$f_2(e_j, h_2)$	$\ldots$	$f_2(e_j, h_m)$
$\ldots$	$\ldots$	$\ldots$	$\ldots$	$\ldots$
f_l	$f_l(e_j, h_1)$	$f_l(e_j, h_2)$	$\ldots$	$f_l(e_j, h_m)$
A	$A(f_1, f_2, \ldots, f_l)$	$A(f_1, f_2, \ldots, f_l)$	$\ldots$	$A(f_1, f_2, \ldots, f_l)$

For each employee the following array should be computed:
The activation function exploits factor functions to determine if an active hint should be presented to the employee; for instance in the simplest case the activation function could be a sum or a weighted sum of all factors.

6 Scenario

The activation function is computed using software agents technology. Each agent is responsible for computing one or more factors, but to merge them all collaboration and communication between agents is necessary.

At the beginning of Pellucid's operation in the organisation, the organisational memory will be fed with the initial active hints. During regular work the system will acquire the knowledge or experience and transform it into new active hints with the help of a knowledge engineer. The graphical user interface for active hints creation is also proposed but targeted mainly to advanced users. So the system memory will grow continuously. As presented above the selection of active hints could involve many factors. We will show how subsequent factors limit the number of active hints to be shown.

We assume that two different employees are in the same working context. The system should extract only these active hints which match to the context of our two employees. By matching working context the position in the workflow tracking/management system is considered. Then the domain specific properties are regarded. They should be similar to those specified in the context of the active hint.

The user context and the problem context are as follows.

- User context in workflow system. It is a part of Working Context concerning the position of employee in business process. Active hint is candidate for activation if its context is similar to this one.

$$ProcessContext(e_j) \sim ProcessContext(h_k) \qquad (2)$$

- Problem context. These are all properties describing the current problem domain. Active hint could be activated if the problem context is similar to the problem in which the active hint is valid.

$$ProblemContext(e_j) \sim ProblemContext(h_k) \qquad (3)$$

Afterwards, the employee profile is taken into account. The participation in the similar tasks increases the level of expertise which in turn results in lower amount of active hints shown to the user. The active hints that recommend an action that the user is familiar with (he always performs it when dealing with such a task) are also not suggested. If dealing with a novice all active hints matching the current situation are shown. However, the skilled new hire employee also require experience assistance, but Pellucid does not contain any information about his work history. Such an employee should be classified with the level of experience and from the other side each active hint should have the type of employee it is valid for. The experience classification could be very simple (e.g. novice, basic, intermediate, advanced) or it could be a more detailed one with the description of several skills possessed by the employee. Some examples of skills are: language skills, documenting skills, cooperation skills, negotiation skills etc. Each skill might be described with the skill level: novice, basic, etc. Also any active hint could be characterised with the skills it concerned. Thus the context in workflow and the problem the user is solving determine the experience required for the current situation while the user profile represents the experience available.

7   Conclusions

The area of experience management systems is naturally related to agent technology. The management of experience often entails ontologies for knowledge and experience representation as well as reasoning for creation, adaptation and maintenance of the experience. The usage of agents within the Pellucid system has turned out to be profitable and effective. Reuse of experience manifested by spontaneous appearance of active hints is well suited to employees' expectations.

Acknowledgments. Thanks are due to European Commission and participants of the Pellucid project, IST-2001-34519. AGH-UST grant is also acknowledged.

References

1. Pellucid – A Platform for Organisationally Mobile Public Employees, EU IST-2001-34519 Project, 2002, http://www.sadiel.es/Europa/pellucid/.
2. Słota, R., Majewska, M., Dziewierz, M., Krawczyk, K., Laclavik, M., Balogh, Z., Hluchý, L., Kitowski, J., Lambert, S., "Ontology assisted access to document repositories in public sector organizations", Proc. of 5th Int. PPAM 2003 Conf., Sept. 7-10, 2003, Częstochowa, Poland.
3. Laclavik, M., Balogh, A., Hluchý, L., Słota, R., Krawczyk, K. and Dziewierz, M., "Distributed Knowledge Management based on Software Agents and Ontology", Proc. of 5th Int. PPAM 2003 Conf., Sept. 7-10, 2003, Częstochowa, Poland.
4. Lambert, S., Stringa, S., Vianno, G., Kitowski, J., Słota, R., Krawczyk, K., Dziewierz, M., Delaître, S., Oroz, M.B., Gomez, A.C., Hluchý, L., Balogh, Z., Laclavik, M., Fassone, M., Contursi, V., "Knowledge management for organisationally mobile public employees", Proc. of 4th IFIP Int. Working Conf. on Knowledge

Management in Electronic Government, KMGov 2003, Rhodes, May 26-28, 2003, Lecture Notes in Computer Science (Lecture Notes on Artificial Intelligence) no. 2645, Springer, 2003, pp. 203-212.

5. Kitowski, J., Krawczyk, K., Majewska, M., Dziewierz, M., Słota, R., Lambert, S., Alvaro, A., Miles, A., Hluchý, L., Balogh, Z., Laclavik, M., Delaître, S., Vianno, G., Stringa, S., Ferrentino, P., "Model of Experience for Public Organisations with Staff Mobility", submitted to 5th Working Conf. on Knowledge Management in Electronic Government, May 17-18, 2004, Kerms, Austria.

6. Dignum, V., "An Overview of Agents in Knowledge Management", http://www.cs.uu.nl/~virginia/amkm.pdf.

7. Chen, J.R., Wolfe, S.R., Wragg, S.D., "A Distributed Multi-Agent System for Collaborative Information Management and Sharing", Proc. of 9th ACM International Conference on Information and Knowledge Management, November 6-11, 2000, McLean, VA, USA.

8. CoMMA project, http://www.si.fr.atosorigin.com/sophia/comma/Htm/HomePage.htm.

9. Abecker, A., Bernardi, A., Hinkelmann, K., Kühn, O., Sintek, M.,"Toward a technology for organizational memories", IEEE Intelligent Systems, 13(3), pp. 40-48.

10. Reimer, U., Margelisch, A., Staudt, M., 2000, "EULE: A Knowledge-Based System to Support Business Processes", Knowledge-based Systems Journal, Elsevier, Vol. 13 No. 5. pp. 261-269.

11. DECOR: Delivery of Context-Sensitive Organizational Knowledge, http://www.dfki.uni-kl.de/decor/deldec/D1-Final.pdf.

12. Abecker, A., Bernardi, A., van Elst, L., Lauer, A., Maus, H., Schwarz, S., Sintek, M., FRODO: A framework for distributed organizational memories. Milestone M1: Requirements analysis and system architecture", D-01-01, 2001, DFKI GmbH.

13. Abecker, A., Bernardi, A., Hinkelmann, K., Kühn, O., Sintek, M. "Context-aware, proactive delivery of task-specific knowledge: The KnowMore project", International Journal on Information Systems Frontiers, 2000, 2(3/4), pp. 139-162.

14. Uschold, M., and Gruninger, M., "Ontologies: Principles, Methods and Applications", Knowledge Engineering Review, 11(2), June 1996, pp. 93-137.

15. DAML – DARPA Agent Markup Language, http://www.daml.org/.

16. Jena – A Semantic Web Framework for Java, http://jena.sourceforge.net/.

The Construction and Analysis of Agent Fault-Tolerance Model Based on π-Calculus

Yichuan Jiang[1], Zhengyou Xia[2], Yiping Zhong[1], and Shiyong Zhang[1]

[1] Department of Computing & Information Technology, Fudan University
Shanghai 200433, P.R.China
http://www.cit.fudan.edu.cn
[2] Department of computer, NanJing University of Aeronautics and Astronautics
Nanjing 210043, P.R.China
http://ice.nuaa.edu.cn/academic/4.php
{jiangyichuan,zhengyou_xia}@yahoo.com.cn,
{ypzhong,szhang}@fudan.edu.cn

Abstract. Agent replication and majority voting is a typical method to realize agent fault-tolerance. However, with such method, many agent replicas are produced in agent execution, which may cost much network and time resource. The paper constructs a novel agent migration fault-tolerance model based on integrity verification (AMFIV), which can reduce the complexity degree of agent communication and agent replicas amount so that network and time resource can be much saved. At last, the paper makes analysis for AMFIV based on π-calculus. The π-calculus analysis result proves that the novel model provided by the paper is correct and valid.

1 Introduction

Mobile agent technology can support agent migration among hosts and make network application more flexible and effective. However, mobile agent system may also bring out such problem: when there are malicious hosts, how to protect mobile agents against them? This is the **Problem of Malicious Host** [1].

To solve the Problem of Malicious Host, there have been some works, such as Time Limited Blackbox [1], Reference States [2], Cryptographic Traces [3], Authentication and State Appraisal [4], and some other solutions which adopted the measures of cryptography, digital signature and trusted environment [5], etc.

The above researches have made very effect for solving the Problem of Malicious Host. However, they often focus on the prevention and detection of the problem, and not cope with how to keep the mobile agent system uninterruptedly operating well when the Problem of Malicious takes place. Aiming at such situation, the concept of **agent fault-tolerance** was presented. Among the relative researches, [6] is the representative one in which the measure of agent replication and majority voting was adopted.

Now we introduce the typical relative work on agent fault-tolerance-**replicated agent migration computation with majority voting (RAMMV)** in [6], as shown in Fig.1.

M. Bubak et al. (Eds.): ICCS 2004, LNCS 3038, pp. 591–598, 2004.
© Springer-Verlag Berlin Heidelberg 2004

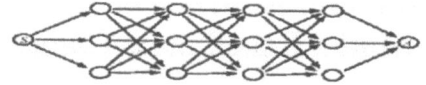

Fig. 1. Replicated agent migration computation with majority voting

In RAMMV, a node p in stage i takes as its input the majority of the inputs it receives from the nodes comprising stage i-1. And, p sends its output to all of the nodes that it determines comprising stage i+1 [6]. The voting at each stage makes it possible for the computation to heal by limiting the impact of the faulty host in one stage on hosts in subsequent stages. More precisely, it is possible to tolerate faulty values from a minority of the replicas in each stage.

However, such model is not feasible in practice, mainly as the model requests that all agent replicas should keep alive until the end of agent migration and assures that replicated hosts fail independently [5], and the large numbers of agent replicas may cost much network and host resource. Otherwise, the result voting among the replicas of agent can also cost much resource and time.

To resolve the deficiency of the method in [6] and other relative works, we suggest a novel agent migration fault-tolerance model based on integrity verification (AMFIV). The new model suggested by us can reduce the cost of resource and time very much.

The rest of the paper is organized as follows. Section 2 presented the novel agent migration fault-tolerance model-AMFIV. Section 3 makes analysis for the model based on π-calculus. Then the paper concludes in Section 4.

2 A Novel Agent Migration Fault-Tolerance Model (AMFIV)

2.1 Illustration of AMFIV

To solve the Problem of Malicious Host, we presented a novel agent migration fault-tolerance model based on integrity verification called AMFIV. We can see the trace example of AMFIV in Fig 2.

Fig 2 can be explained as following: agent at stage i runs on $host_i$ and selects a node with the highest priority as the next host to migrate which can be denoted as $host_{i+1}(0)$; then the agent spawns a replica which migrates to $host_{i+1}(0)$; agent replica runs on $host_{i+1}(0)$, after running its integrity is verified by $host_i$; if the integrity verification result is ok, then the agent on $host_{i+1}(0)$ spawns a replica to migrate to $host_{i+2}(0)$, and the agent on $host_i$ is terminated; otherwise $host_{i+1}(0)$ is a malicious one, then the agent on $host_i$ re-selects another host with the second priority as the next one to migrate which can be denoted as $host_{i+1}(1)$, and the model will execute the operations as the same as above operations. If $host_{i+1}(1)$ is also malicious, then the agent on $host_i$ will re-select another host $host_{i+1}(2)$ with the third priority as the next one to migrate...until there exists a normal host to migrate or there don't exist any other adjacent nodes to select. If $host_i$ hasn't any other adjacent nodes, then the agent on $host_i$ returns to $host_{i-1}$, and selects another node as $host_i(1)$.

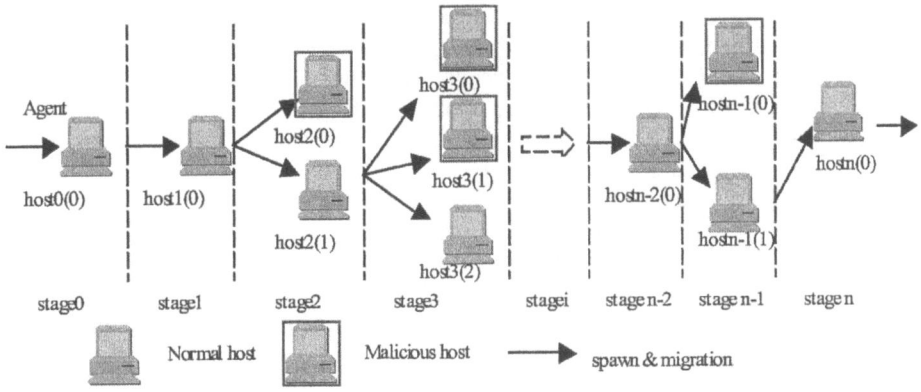

Fig. 2. The agent migration trace example of AMFIV

From Fig 2, we can see that agent needn't to produce replica at every migration step. In the AMFIV model, firstly agent migrates according to linear trace, only when the agent integrity is damaged by a malicious host then a new path is re-selected. But RAMMV model requires that at every migration step the replicas should be produced. Otherwise, AMFIV model limits the fault-tolerance problem to be solved in single hop, which avoid the multi steps accumulative problem.

Let the number of agent migration steps is n, and the number of standby nodes in every step is m, obviously the complexity of agent migration communication degrees in RAMMV is $O(n*m^2)$, and the one in AMFIV is $O(n*m)$. So AMFIV reduces the complexity degrees of agent migration communication from cube level to square level. Therefore, in AMFIV the network load can be reduced much accordingly.

On the amount of replicas produced, the average complexity in RAMMV is $O(n*m)$, but in AMFIV only under the worst situation, i.e. in every step the first m-1 nodes are all malicious, the complexity can reach $O(n*m)$. Obviously, the worst situation seldom takes place in practice, so AMFIV can also reduce the amount of agent replicas.

2.2 The Verification of Agent Integrity in AMFIV

The agent integrity includes the integrity of agent code, data and state. Here we discuss how to make agent code and data integrity verification in AMFIV model. In our verification protocol, we suppose the hosts have a shared key.

▶ The Sub-module of Agent Code Integrity Verification

After the agent runs on $host_{i+1}$, we make code integrity verification to detect that whether the agent code is damaged by $host_{i+1}$.

The agent code integrity verification protocol is explained as follows: $(K_{i,i+1}(x)$ denotes that encrypting x with the key shared by $host_i$ and $host_{i+1}$).

- A). $host_i \rightarrow host_{i+1} : i, R_i, K_{i,i+1}(t_i)$;
- B). $host_{i+1} \rightarrow host_i: R_{i+1}, K_{i,i+1}(R_i, t_{i+1})$; /*$R_i$ denotes the request message sent by $host_i$, and R_{i+1} denotes the request message sent by $host_{i+1}$ */

- C). $host_i \rightarrow host_{i+1}$: $K_{i,i+1}(R_{i+1})$; /*A), B), C) denote the identification authentication between $host_i$ and $host_{i+1}$*/
- D). $host_{i+1} \rightarrow host_i$: $K_{i,i+1}(hash(Code_{i+1}||t_{i+1}))$; /* $host_{i+1}$ sends the hash value of the agent code on $host_{i+1}$ with time stamp to $host_i$ */
- E). $host_i$: Check:
  compute $hash(Code_i||t_{i+1})$;
     if $hash(Code_i||t_{i+1}) == hash(Code_{i+1}||t_{i+1})$
      then Agent code integrity is ok;
      else Agent code integrity isn't ok.

/* $host_i$ computes the hash value of the agent code on itself, then compares it with the hash value returned by $host_{i+1}$ and judge if the agent code integrity is ok.*/

Analysis for the protocol: since the agent code shouldn't be changed in migration, so if some malicious hosts change the agent code, then the hash value of code should be different and can be detected. Since $host_{i+1}$ don't know the $hash(Code||t_{i+1})$ computed by $host_i$, so it can't forge $hash(Code||t_{i+1})$. If $host_{i+1}$ makes any change to the agent code, the hash value returned is different from the one computed by $host_i$, so the change can be detected. Therefore, the protocol is secure and correct.

▶ **The Sub-module of Agent Data Integrity Verification**
On the base of the work of [7], we design the sub-module of agent data integrity verification in AMFIV.

Thereinafter D_i signifies the data collected by the agent on $host_i$, and AD_i signifies the list of data collected by the agent from $host_0$ to $host_i$ accumulatively.

- Stage 0: $Host_0$ generates a secret number C_0, then computes $C_1 = hash(C_0)$, and passes C_1 to the agent, now $AD_0 = \{\}$;
- Agent encrypts C_1, then migrates to $host_1$;
- On $host_1$: C_1 can be obtained by decryption, agent collects data D_1, $AD_1 = AD_0 \cup D_1$, computes the data proof $PROOF_1 = hash(D_1, C_1)$, $C_2 = hash(C_1)$;
- On $host_i$ ($1 \le i \le n-1$): C_i can be obtained by decryption, Agent collects data D_i, $AD_i = AD_{i-1} \cup D_i$, computes the data proof $PROOF_i = hash(D_i, C_i, PROOF_{i-1})$, $C_{i+1} = hash(C_i)$; then Agent encrypts C_{i+1} and passes it with AD_i, $PROOF_i$ together to $host_{i+1}$.
- The protocol that $host_i$ verifies the agent data integrity after running on $host_{i+1}$ is shown as follows.
    - A). $host_i \rightarrow host_{i+1}$: i, R_i, $K_{i,i+1}(t_i)$;
    - B). $host_{i+1} \rightarrow host_i$: R_{i+1}, $K_{i,i+1}(R_i, t_{i+1})$;
    - C). $host_i \rightarrow host_{i+1}$: $K_{i,i+1}(R_{i+1})$; /*Similar to the protocol of agent code integrity verification, the A), B), C) are used for identification authentication between $host_i$ and $host_{i+1}$ */
    - D). $host_{i+1} \rightarrow host_i$: C_{i+p} AD_{i+p} $PROOF_{i+p}$; /* $host_{i+1}$ passes the agent data information to $host_i$ */
    - E). $host_i$: Computes $proof_{i+1} = hash(AD_{i+1} - AD_p \ hash(C_i), PROOF_i)$; /* Computes $proof_{i+1}$ on $host_i$ */
    - F). $host_i$: if $(proof_{i+1} == PROOF_{i+1})$ and $(C_{i+1} == hash(C_i))$
          then agent data integrity is ok;
          else agent data integrity isn't ok.

Analysis for the protocol: since $C_{i+1}=$ hash(C_i), so $host_{i+1}$ can't obtain C_i from C_{i+1}, and $host_{i+1}$ can't obtain $C_j(j<i+1)$; $host_{i+1}$ doesn't know $C_j(j<i+1)$, and can't modify $D_j(j<i+1)$, so it can't forge PROOF. Therefore the protocol is secure. Obviously, if the original data of agent is damaged by $host_{i+1}$, then $proof_{i+1}$ isn't equal to $PROOF_{i+1}$, so the damage of data integrity can be detected, therefore the protocol is correct. Obviously, we can see that the protocol can only detect any tampering of the data collected before $host_{i+1}$, and can't detect whether $host_{i+1}$ collects dirty data. Therefore, the protocol only guarantees the integrity of validly collected data.

3   Make Analysis Based on π-Calculus

Now we will make analysis to AMFIV based on π-calculus.

In our π-calculus model, the channels used are seen in Table 1.

Table 1. List of Channel Name in the π-Calculus Model

Channel	Sender	Receiver	Message
next	$host_i$	$host_i$	cha: denotes the next host to migrate
cha	$host_i$	$host_{i+1}$	Agent replica (includes data, code, etc.)
chr	$host_{i+1}$	$host_i$	The agent running result on $host_{i+1}$
chv_1	$host_i$	$host_i$	The verification result of agent integrity
chv_2	$host_i$	$host_{i+1}$	The verification result of agent integrity

We can define the π-calculus model of $host_i$ as Formula (1)[1]:

$$
\begin{aligned}
host_i &\stackrel{def}{=} next(cha).\overline{cha}(RUN(agent_i)) \mid chr(x).(let\ (code_{i+1}, data_{i+1}) \\
&= x\ in\ (\overline{chv_1}(VALIDATE(code_{i+1}, data_{i+1})). \\
&\overline{chv_2}(VALIDATE(code_{i+1}, data_{i+1})))) \\
&\mid chv_1(y).([y = true].TERMINATE(agent_i) + \\
&[y = false].(\overline{next}(SELECT(i+1)).host_i))
\end{aligned}
\tag{1}
$$

The Formula (1) is explained as following: from channel *next* $host_i$ obtains the $host_{i+1}$ which is denoted as channel *cha*, then spawns a replica of $agent_i$ after running on $host_i$, and migrates the replica through channel *cha* to $host_{i+1}$; From channel *chr* $host_i$ obtains the running result of $agent_{i+1}$ on $host_{i+1}$, and makes verification (VALIDATE) for its code and data integrity, then passes the verification result to channels chv_1, chv_2; From channel chv_1 $host_i$ obtains the verification result, if it is true the agent on $host_i$ is terminated, or else $host_i$ should re-select a new node to migrate, and passes the new node to channel *next*, then repeats all the acts of the model.

---

[1] A pair splitting process *let (x, y)=M in P* behaves as *P[N/x][L/y]* if term M is the pair (N,L), and otherwise it is stuck.

We can define the π-Calculus model of $host_{i+1}$ as Formula (2):

$$host_{i+1} \overset{def}{=} cha(agent_{i+1}).\overline{chr}(RUN(agent_{i+1}))|$$
$$chv_2(z).([z = true].GETMASTER + [z = false].ISOLATED) \qquad (2)$$

The Formula (2) is explained as following: from channel cha, $host_{i+1}$ receives the agent replica and runs it, and then passes the result (data part and code part) back to $host_i$; From channel chv_2 $host_{i+1}$ receives the verification result, if the verification result is true then $host_{i+1}$ gets the control power and agent can migrate further, or else $host_{i+1}$ is a malicious one and it should be isolated.

Therefore, we can define the π-Calculus model of AMFIV as Formula (3):

$$AMFIV \overset{def}{=} (vnext, cha, chr, chv_1, chv_2)(host_i \mid host_{i+1})$$

$$\equiv (vnext, cha, chr, chv_1, chv_2)(next(cha).\overline{cha}(RUN(agent_i))|$$

$$chr(x).(let\ (code_{i+1}, data_{i+1}) = x\ in\ (\overline{chv_1}$$

$$(VALIDATE(code_{i+1}, data_{i+1})).\overline{chv_2}(VALIDATE(code_{i+1}, data_{i+1})))) \qquad (3)$$

$$|chv_1(y).([y = true].TERMINATE(agent_i) + [y = false].$$

$$\overline{next}(SELECT(i+1).host_i))|$$

$$cha(agent_{i+1}).\overline{chr}(RUN(agent_{i+1}))| chv_2(z).([z = true].$$

$$GETMASTER + [z = false].ISOLATED))$$

Now we can use the π-calculus simulate the execution of AMFIV.

$$AMFIV \xrightarrow{next(cha)} (vnext, cha, chr, chv_1, chv_2)(\overline{cha}(RUN(agent_i))|$$

$$chr(x).(let\ (code_{i+1}, data_{i+1}) = x\ in\ (\overline{chv_1}$$

$$(VALIDATE(code_{i+1}, data_{i+1})).\overline{chv_2}(VALIDATE(code_{i+1}, data_{i+1})))) $$

$$|chv_1(y).([y = true].TERMINATE(agent_i) + [y = false]. \qquad (4)$$

$$(\overline{next}(SELECT(i+1)).host_i))|$$

$$cha(agent_{i+1}).\overline{chr}(RUN(agent_{i+1}))|$$

$$chv_2(z).([z = true].GETMASTER + [z = false].ISOLATED))$$

$$\overset{\tau}{\longrightarrow}(vnext, chr, chv_1, chv_2(chr(x).(let \ (code_{i+1}, data_{i+1}) = x \ in$$

$$(\overline{chv_1}(VALIDATE(code_{i+1}, data_{i+1})).\overline{chv_2}$$

$$(VALIDATE(code_{i+1}, data_{i+1})))) \mid chv_1(y).([y = true]$$

$$.TERMINATE(agent_i) + [y = false].(\overline{next(SELECT(i+1))}.host_i)) \mid$$

$$\overline{chr}(RUN(agent_{i+1})) \mid chv_2(z).([z = true]$$

$$.GETMASTER + [z = false].ISOLATED))$$

$$\overset{\tau}{\longrightarrow}(vnext, chv_1, chv_2)(let \ (code_{i+1}, data_{i+1}) = RUN(agent_{i+1})$$

$$in \ (\overline{chv_1}(VALIDATE(code_{i+1}, data_{i+1})).\overline{chv_2}(VALIDATE$$

$$(code_{i+1}, data_{i+1})))) \mid chv_1(y).([y = true].TERMINATE(agent_i)$$

$$+ [y = false].(\overline{next(SELECT(i+1)}.host_i)) \mid$$

$$chv_2(z).([z = true].GETMASTER + [z = false].ISOLATED))$$

$$\overset{\tau}{\longrightarrow}(vnext)(([VALIDATE(code_{i+1}, data_{i+1}) = true].$$

$$TERMINATE(agent_i) +$$

$$(VALIDATE(code_{i+1}, data_{i+1}) = false].(\overline{next(SELECT(i+1)}.host_i))$$

$$\mid ([VALIDATE(code_{i+1}, data_{i+1}) = true].GETMASTER +$$

$$[VALIDATE(code_{i+1}, data_{i+1}) = false].ISOLATED))$$

$$\begin{cases} \xrightarrow{[VALIDATE(code_{i+1}, data_{i+1})=true]} TERMINATE(agent_i) \mid GETMASTER \\ \xrightarrow{[VALIDATE(code_{i+1}, data_{i+1})=false]} (vnext_1)(\overline{next}(SELECT(i+1)).host_i \mid \\ \qquad\qquad ISLOATED) \end{cases}$$

From above we can see that: if the integrity of agent code and data is ok, the ultimate result is $TERMINATE(agent_i) \mid GETMASTER$, so the agent on $host_i$ is terminated, $host_{i+1}$ gets the control power, and agent migrates according to a linear trace; if the integrity of agent code and data is damaged by $host_{i+1}$, the ultimate result is $(vnext_1)(\overline{next_1}(SELECT(i+1)).host_i \mid ISOLATED)$, so $host_i$ re-selects another node as the next one to migrate, and repeats the acts of the model, and $host_{i+1}$ is isolated.

Therefore, from the above simulation result of the π-Calculus model of AMFIV, we can see that AMFIV is correct.

4 Conclusion

In this paper, aiming at the deficiency of other typical agent fault-tolerance models, we suggested a novel agent migration fault-tolerance model based on integrity verification called AMFIV. Comparing to other agent fault-tolerance models, our model can reduce the complexity degree of agent communication and agent replicas amount. The π-calculus simulation validation results prove that AMFIV is correct and efficient.

References

1. Fritz Hohl: Time Limited Blackbox Security: Protecting Mobile Agents from Malicious Hosts. Mobile Agents and Security, Giovanni Vigna (ed.), Springer-Verlag, 1998, pp. 92–113
2. Fritz Hohl: A Protocol to Detect Malicious Hosts Attacks by Using Reference States. http://elib.uni-stuttgart.de/opus/volltexte/2000/583/
3. Giovanni Vigna: Cryptographic Traces for Mobile Agents. Mobile Agents and Security, Giovanni Vigna (ed.), Springer-Verlag, 1998, pp. 137–153
4. William M. Farmer, Joshua D. Guttma, and Vipin Swarup: Security for Mobile Agents: Authentication and State Appraisal. In: Proceedings of the Fourth European Symposium on Research in Computer Security, 1996
5. Bennet S. Yee: A Sanctuary for Mobile Agents. In: DARPA Workshop on Foundations for Secure Mobile Cde, 1997. http://www.cs.ucsd.edu/~bsy/pub/sanctuary.ps
6. Fred B. Schneider: Towards Fault-tolerant and Secure Agentry. Invited paper. In: Proceedings of 11th International Workshop on Distributed Algorithms, Saarbrücken, Germany, Sept 1997
7. Paolo Maggi and Riccardo Sisto: Experiments on Formal Verification of Mobile Agent Data Integrity Properties. www.labic.disco.unimib.it/woa2002/papers/15.pdf
8. R. Milner: The Polyadic π-Calculus: a Tutorial. In. F.L. Bauer, W. Braueer, and H. Schwichtenberg (eds.), Logic and Algebra for Sepcification. Berlin: Springer-Verlag, 1993, pp. 203–246.

REMARK – Reusable Agent-Based Experience Management and Recommender Framework[*]

Zoltan Balogh[1], Michal Laclavik[1], Ladislav Hluchy[1], Ivana Budinska[1], and Krzysztof Krawczyk[2]

[1] Institute of Informatics, SAS, Dubravska cesta 9, Bratislava 84237, Slovakia
{balogh.ui,laclavik.ui,hluchy.ui,utrrbudi}@savba.sk
[2] ACC CYFRONET AGH, Nawojki 11, 30-950 Cracow, Poland
krafcoo@icsr.agh.edu.pl

Abstract. In this paper we introduce an advanced experience management and recommendation framework. The framework exploits software agent technology to ensure distributed and robust functioning. In design and implementation we use state of the art methodologies, technologies and tools. Ontology is used for modeling and describing knowledge. The infrastructure is able to provide relevant recommendations to a requester entity based on recorded experiences from the past depending on the current context of the environment. Such infrastructure is suitable for environments where instant recommendation is required in a given situation. A concrete example implementation of this framework is introduced in the public administration organization where employees need knowledge according to their current work context which is represented by activities in a workflow process.

1 Introduction

There are many experience management (EM) frameworks but each of them is usually liaison to a concrete application area. Having a general reusable infrastructure for EM where final implementation would not require redesign of the whole system but only customization and maintenance of a knowledge model is highly demanded. In this paper we describe such experience management framework which employs software agent technology. Our aim was to make our framework integrable with existing software so we do not create a barrier for information flows.

This introduction is followed by a section in which we give an overview of relevant experience management systems and sketch the motivation of our work (Section 2). The core framework is described in Section 3. CommonKADS is a methodology for developing and maintaining knowledge management (KM) systems. We discuss the use of CommonKADS methodology for developing systems based on our framework is Section 4. A project called Pellucid [8,9,10] being

---

[*] This work was supported by EC Project Pellucid 5FP RTD IST-2001-34519 and Slovak Scientific Grant VEGA 2/3132/23.

M. Bubak et al. (Eds.): ICCS 2004, LNCS 3038, pp. 599–606, 2004.
© Springer-Verlag Berlin Heidelberg 2004

developed in scope of the 5th EU FP is introduced in section 5 as a verification of our framework. We conclude (in Section 6) with account of further possible implementation areas for the framework.

2   Motivation and State of the Art

The goal of developing an EM system is to create a knowledge repository that contains problem solving experiences for a complex problem domain that changes over time. There have been proposed several systems which we examine in this section.

HOMER [1] is a generic experience management architecture and a set of related tools that stores the experience of the help-desk operators in a case base and enable them to access, reuse, and extend the knowledge in a natural and straightforward manner. HOMER has been developed mainly for realizing help-desk and self-service experience management applications. An object-oriented approach is used to model the experiences in HOMER. The system is implemented utilizing pure client-server architecture. Each request must be processed by the server. Server is therefore the critical point of the overall architecture what is typical for purely centralized architectures.

In [2] Lacher and Koch propose a theoretical reference agent-based knowledge management framework called Knowledge Management Information Technology (KM IT). Authors identified the advantages to employ software agents in KM system. The framework is built of agencies made up of several agents each. The agencies represent the organizational knowledge entities and manage the respective information objects. Also a so called shared ontology is proposed used for consistent communication among agencies as a basis for the user profile in the user agency and to specify information requests and replies. Even though KM IT proposes to use agents and ontologies in EM system design, the whole framework is purely theoretical with no implementation or verification work.

In [3] authors describe a development of an Intranet-based knowledge management framework called KnowledgePoint in a consulting firm. The paper further introduces a simple ontology model of organizational knowledge. The article is rather a case study than a scientific article. Anyhow very important implementation problems are raised here such as security or managerial issues in the organization. Scientists often neglect such problems whereat those must be considered during the system design.

There are also other KM systems but they usually deal only with a concrete knowledge model design and implementation. Such projects are CoMMA [4],DECOR [5], FRODO [6] or KnowMore [7].

All these above described systems address only a specific problem domain. That makes these systems too specialized and therefore inappropriate as candidates for generic EM framework. We aimed at developing a flexible and adaptable platform for enhancing EM in organizations. The framework is a set of methods, tools and reusable components that may be applied to develop further EM applications.

3   Framework Description

Herein we describe the component architecture for advanced experience management. The main goal was to create a general infrastructure for EM. The infrastructure is primarily intended for use for human users, but can be easily applied to serve computational clients as well. Henceforward we will refer to any entity which is served from the framework as client. The overall framework is on the figure bellow.

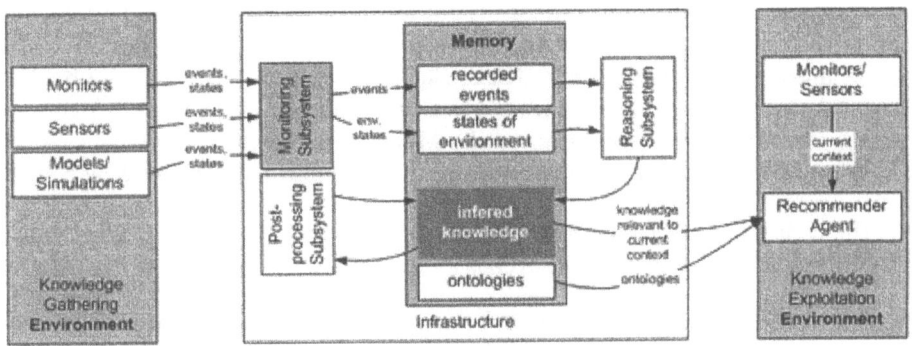

Fig. 1. The REMARK framework

Processing of knowledge in the infrastructure is fulfilled in three stages which are typical for the EM domain: capture, capitalization and reuse. In the capture phase knowledge is recorded. Additional reasoning and data mining is performed in the capitalization phase where the inferred knowledge is produced. Recorded and inferred knowledge is then used to assist users in the reuse phase. The core of the framework is the Processing Infrastructure (PI). PI has the following components:

- Corporate Memory (CM),
- Monitoring Subsystem (MSS),
- Reasoning Subsystem (RSS),
- Post-processing Subsystem (PPSS),
- Recommender Subsystem (RSS).

There are also monitors and sensors which are part of the framework but are not executed inside the PI. Monitors/sensors need to be executed on the application side with enabled active environment sensing and monitoring interfaces. Software agents have been chosen for the overall implementation of the framework. The main advantage of using agent technology is the distributed nature of agents, their well established communication infrastructure based on ontology and execution autonomy and mobility (required mainly for sensors and monitors).

All the information stored in scope of the framework is modeled by the ontology, what is simply the object model of the real world. The ontology is extensible, what makes also the framework open. We have identify the need for:

- a Generic Ontology which is usable for any application and
- a Domain-Specific Ontology which is specific for concrete application domain.

The Knowledge Gathering Environment (KGE) was separated from the Knowledge Exploitation Environment (KEE) in the schema. KGE is simply the environment in which the knowledge is gathered. KEE is monitored and depending on the context concrete experience is recommended to the client. KGE can be but must not be identical with KEE. The knowledge is gathered from KGE by Sensor Agents (SA) or directly from a running model or simulation. SA communicates the captured information to the Monitoring Agent (MA) which records the information to the CM. The structure of information monitored by SA is:

- the context of selected properties of the KGE and
- the events performed in the KGE.

Both context and events must be modeled by ontology. In simpler cases context can be modeled as part of the event concept and therefore only events need to be modeled. In most cases events and context will share some common general and domain-specific concepts.

Stored contexts and events are processed by the Reasoning Agent (RE) which generates active experiences in various forms. These experiences are stored as inferred knowledge in the CM. The structure of possible recommendations presented to the client is pre-modeled. The set of pre-modeled recommendations can be extended according to application needs.

SA are used also to monitor the context of the KEE. SA communicates the monitored information to the Recommender Agent (RCA). If RCA finds knowledge which is valuable for the client in the current context it will return the knowledge to client. RCA may utilize Task Agents (TA) for concrete experience retrieval from the CM. The recommendation is handed over to Personal Assistant Agent (PAA) in a case when the client is a human user.

4   Development Methodology

CommonKADS [11] methodology is standard for developing knowledge management systems and can be very well used for creating model for our experience management framework. It contains of several standard model describing a problem domain:

- Organizational Model, which describes organization structure, its goals, and also problem domain of concrete application;
- Task Model describes tasks preformed in organization;

- Agent Model describes actors involved in organization and it can be humans or computer systems involved in performing tasks;
- Knowledge and Communication Model are created from 3 models above and describes knowledge and communication between actors and
- Design Model comes from Knowledge and Communication model and describes how system should be implemented.

Knowledge model is basically ontology which describes entities (users, problems, application/domain related terms, resources) and its relation in the organization. In our framework we utilize DAML+OIL [15] based on XML/RDF for interfacing with existing systems and also for processing of knowledge results.

5   Framework Implementation

The described framework was verified in the Pellucid (IST-2001-34519) EU project [8]. The aim of the Pellucid platform is to provide assistance for organizationally mobile employees. Capture and capitalization of employee's knowledge in public organizations was the main interest in the project. Workflow process is used as a work context for knowledge creation, capitalization and reuse. It means that each event or action performed in an organization is stored in respect to the current workflow process status. Stored and capitalized knowledge is then recommended to other employees as they work on particular tasks in the process.

Knowledge ontology was created according to the application. First we have created an Organizational Model (Fig 2).

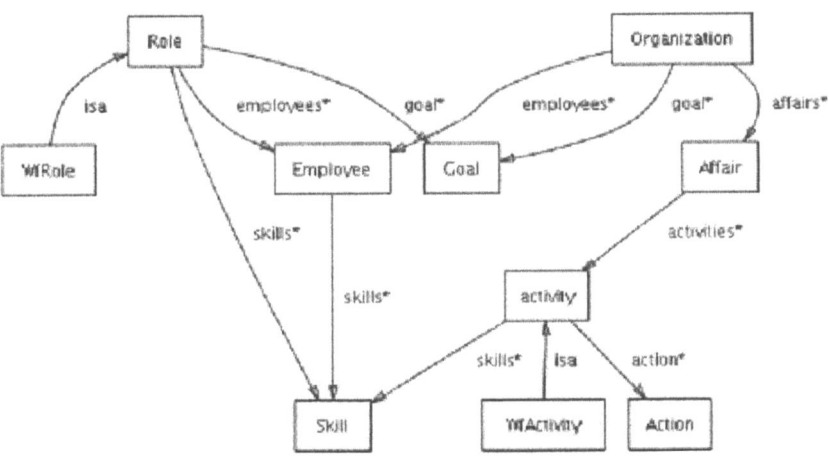

Fig. 2. Organizational Model of the Pellucid Framework

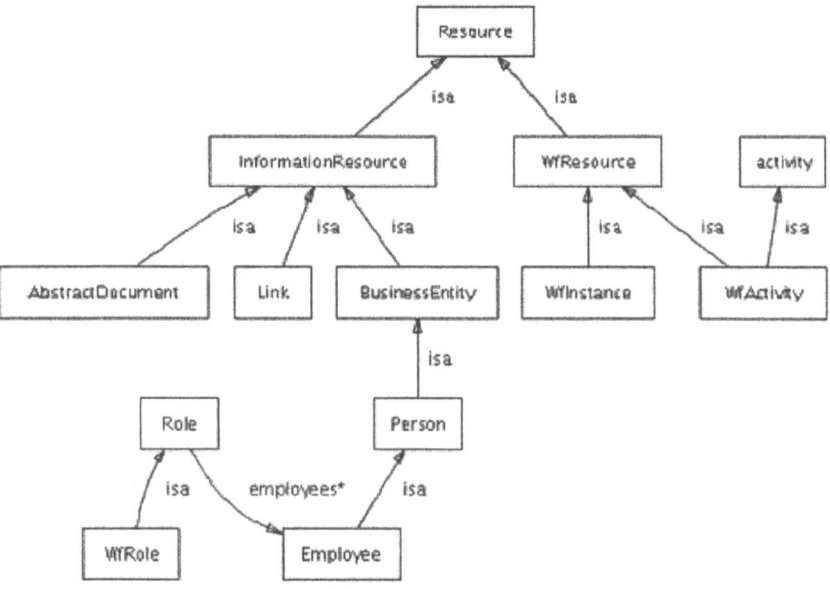

Fig. 3. Model of Resources in the Pellucid System

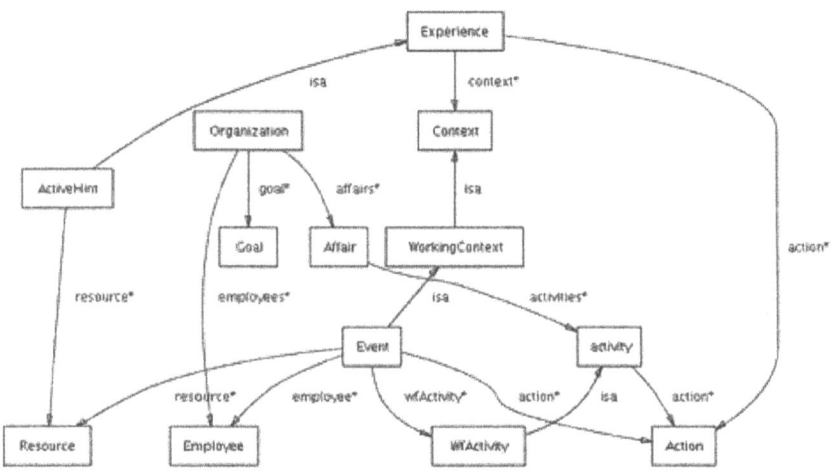

Fig. 4. Ontology Model in Pellucid

In scope of the Organizational Model we have also identified roles of the actors involved in organization together with their descriptions. As another important model all relevant resources in the organizations were modeled (Fig 3).

All the described ontologies create the generic ontology for any workflow driven application in public sector organizations. There are also three pilot sites were the Pellucid platform is installed. For all of these pilot sites a domain-specific ontology had to be created.

Active Hints (AH) were created as examples of inferred knowledge. It was defined that AH is composed according to the following formula:

```
[ACTIVE_HINT] = [WORK_CONTEXT] + [ACTION] + [RESOURCE] +
                [EXPLANATION]
```

The matching of the proper AH is then driven by the following condition:

```
If [ACTIVE_HINT].[WORK_CONTEXT] = [CURRENT].[WORK_CONTEXT]
Then Display [ACTIVE_HINT]
```

It is a simple rule which matches any AH from the CM which matches the users current work context. All the concepts from the formulas are modeled in the ontologies (see Fig. 3). Personal Assistant Agent was created to communicate the Active Hints returned from Pellucid to the end-user.

The ontology on Fig. 4 is the Design Model of the Pellucid platform according to the CommonKADS methodology. Eeach important ontology is included thus creating the overal model of the application domain.

In our framework we utilize DAML+OIL language for interfacing with existing systems, for interfacing among agents and also for processing of knowledge results. For system implementation we use the HP Jena library [12] which supports DAML+OIL and RDF models. Jena library also includes the RDQL query language and specialized query engine which is useful for reasoning and querying knowledge in the CM. We use JADE [13,14] as the agent implementational platform.

6 Conclusion

Our experience management framework can be used not only for public administrational organizations, but also for other application areas. One of possible applications is military combat strategy recommendation system where the knowledge base is build from real world combat situations or virtual combat simulations. Context of the situation can be represented for example by geographical location of the combat troop. The recommendations can be transmitted to objects from satellites to wireless devices. Analogical research and development project is being prepared in conjunction with the Slovak Military Academy as a verification of our framework. Disease diagnosis and cure recommendation system is another application area this once from medical environment. In such case the context can be represented by patient's health condition, patient's physiological measures and patient's medical record. There are further possibilities for the exploitation of our framework including recommender system for consultation companies or help desk applications. In the future we plan to continue developing our framework for the above mentioned application areas.

References

1. Goker, M. and T. Roth-Berghofer (1999). The development and utilization of case-based help-desk support system HOMER. Engineering Applications of Artificial Intelligence 12(6), 665-680.
2. Lacher S. and Koch M. (2000). An Agent-based Knowledge Management Framework.
3. Sarkar R. and Bandyopadhyay S. Developing an Intranet-based Knowledge Management Framework in a Consulting Firm: A Conceptual Model and its Implementation.
4. CoMMA project, http://www.si.fr.atosorigin.com/sophia/comma/
5. DECOR: Delivery of Context-Sensitive Organizational Knowledge, project outline at http://www.dfki.uni-kl.de/decor/deldec/D1-Final.pdf.
6. Abecker, A., Bernardi, A., van Elst, L., Lauer, A., Maus, H., Schwarz, S., Sintek, M. (2001) "FRODO: A framework for distributed organizational memories. Milestone M1: Requirements analysis and system architecture", DFKI Document D-01-01, DFKI GmbH.
7. Abecker, A., Bernardi, A., Hinkelmann, K., Kühn, O., Sintek, M. "Context-aware, proactive delivery of task-specific knowledge: The KnowMore project", International Journal on Information Systems Frontiers, 2000, 2(3/4), pp. 139-162.
8. Pellucid – A Platform for Organisationally Mobile Public Employees, EU IST-2001-34519 Project, 2002, http://www.sadiel.es/Europa/pellucid/.
9. M. Laclavik, Z. Balogh, L. Hluchy, G. T. Nguyen, I. Budinska, T. T. Dang: Pellucid Agent Architecture for Administration Based Processes, IAWTIC 2003, Vienna (2003).
10. R. Slota, M. Majewska, M. Dziewierz, K Krawczyk, M. Laclavik, Z. Balogh, L. Hluchy, J. Kitowski, S. Lambert: Ontology Assisted Access to Document Repositories for Public Sector Organizations. PPAM Conference (2003).
11. CommonKADS.
12. Jena (A Semantic Web Framework for Java) Homepage, http://jena.sourceforge.net/.
13. JADE (Java Agent DEvelopment Framework) Homepage, http://agentcities.cs.bath.ac.uk/docs/jade/.
14. Giovani Caire (2002). JADE Tutorial Application-defined Content Languages and Ontol-ogy, http://jade.cselt.it/.
15. DAML (DARPA Agent Markup Language) Homepage, http://www.daml.org/.

Behavior Based Detection of Unfavorable Resources

Krzysztof Cetnarowicz[1] and Gabriel Rojek[2]

[1] Institute of Computer Science,
AGH University of Science and Technology,
Al. Mickiewicza, 30 30-059 Kraków, Poland
cetnar@agh.edu.pl
[2] Department of Computer Science in Industry,
AGH University of Science and Technology,
Al. Mickiewicza 30, 30-059 Kraków, Poland
rojek@agh.edu.pl

Abstract. This article considers a problem of security in a computer systems in context of some mechanisms which act in societies and which mechanisms are seen as being useful to assure security in a computer system. A new approach to the security problem is discussed, which refers to identifying computer resource (e.g. agent, program) on the basis of this resource behaviour (actions which this resource undertakes). Mechanisms are presented, which may enable to recognize and dismiss resources undesirable or harmful in the computer system on the basis of behavior observation. Proposed mechanisms were tested in simulations of the computer system under DoS attack. The results of tests are presented and discussed.

1 Introduction

Present computer systems have become more and more complex and related to each other. Large nets, as well as almost all devices connected to the Internet, are open systems. It is impossible to have full knowledge about topology or current state of such systems even for the owner, or an administrator of a computer system [5]. Agent technology makes possible the full flow of resources among open computer systems. Autonomic agents can yet freely migrate in the net without the knowledge of the owner or an administrator. Agents can also execute their tasks without anybody's knowledge. These tasks could be useful or destructive for the system on which an agent operate. An agent which migrate in an open system could be equally desirable or undesirable in a computer system. This ambiguousness causes problems with the use of artificial immune systems which makes possible to distinguish two classes of resources: self or nonself [3,4,6,7, 8]. In the case of migrating agent in open system, the origin of supply (self / nonself) does not play an essential part in the process of assuring security [2].

M. Bubak et al. (Eds.): ICCS 2004, LNCS 3038, pp. 607–614, 2004.
© Springer-Verlag Berlin Heidelberg 2004

2 Social Approach to Security Problem

In the introduction mentioned facts and trends lead to the formulation of a new foundation of solution of computer's security problem. Attacks have generally decentralized nature and harmful resources (which cause these attacks) should be seen as in some way autonomous agents. In order to oppose against such threats, the computer system should be seen as a multiagent system, which consists of:

- "good" (desirable) resources (e.g. agents, programs),
- "bad" (undesirable) resources,

as it was proposed in [2].

Taking into consideration the above–mentioned problems with security systems, which are centralized and operate on fragments of code (look for known signatures), a way is proposed, how the division "good" / "bad" should be made. It could be stated, that:

- decentralized security system could be realized in the form of functions, mechanisms or algorithms that are build in as big amount of components of protected system as possible;
- division "good" / "bad" could be based on observation and evaluation of behavior of components of the protected system.

Authors formulating above mentioned paradigms where inspired by some of ethically–social mechanisms that act in human societies. That mechanisms prevent from misuses in societies in the way to enable secure functioning of an individual in the environment of others individuals. Some of individuals in society could be prejudicial for others individuals, what is main analogy to the society of agents, programs and other resources in a computer system.

An individual in a society seems trustworthy if behavior of this individual could be observed by other individuals in a society and this behavior is evaluated by majority as good and secure [9,10]. The decision about trustworthy of an individual are in society made in the decentralized way — all individuals in a society make own decisions which form one decision of the society. The decisions undertaken by individuals in the societies are made on the ground of observation and evaluation of behavior (and / or intentions) of the individual which decisions concerns.

2.1 Decentralization of Security System

Decentralization paradigm could be realized in multiagent algorithms by means of equipping all agents (agents that exist in the protected system) with some additional goals, tasks and mechanisms. Those goals, tasks and mechanism are named a division profile and should be designed to assure security for agents and the computer system, which those agents assemble. So the agents will execute tasks they have been created for and simultaneously will execute tasks connected with security of all agents in the system and / or the computer system. The name "division profile" is inspired by M–agent architecture which could be used to describe an agent (M–agent architecture was introduced among others in [1]).

2.2 Observation and Evaluation of Behavior

Undertaking actions by an acting agent should be seen as objects. Those objects create a sequence which could be registered by agents observing that acting agent. Registered objects–actions could be processed in order to qualify whether it is a "good" or "bad" acting agent. It should be mentioned, that the quoted notions of "good" and "bad" do not have absolute meaning. "Good" resource is a desirable resource for a definite computer system in which evaluation takes place. "Bad" resource is an undesirable resource for a given system, although it can happen that it would be greatly desirable in a different computer system.

3 Division Profile

Division profile is a class of agent activity whose goal is to observe others agents in society and possible other elements of the environment. Those observations should be made in order to distinguish individuals whose behavior is unfavorable or incorrect ("bad") for the observer. Such distinguished "bad" individuals should be adequately treated (e.g. convicted, avoided, liquidated) which should also be formed by a division profile. In the case of a multiagent system, it is possible to equip every agent in the system with division profile mechanisms, so the security is assured by all agents existed in the system. Division profile is defined as:

$$a_d = (M_d, Q_d, S_d)$$

where M_d is a set of division states m_d of agent a, Q_d is a configuration of goals q_d of agent's a division profile, S_d is a configuration of strategies s_d of agent's a division profile.

3.1 Division State

Division state m_d of agent a is represented as a vector:

$$m_d = (m_d^1, m_d^2, ..., m_d^{j-1}, m_d^j)$$

where j is the number of neighboring agents; neighboring agents are agents which are visible for agent a (including itself, if all agents in the system are visible for agent a, j is equal to the number of all existing agents in the system); m_d^k is the factor subordinated to neighboring agent number k; this factor can be a number from any range and it indicates whether the agent number k is "good" or "bad" (low number indicates "good", high number indicates "bad").

3.2 How to Fix the Division State

To fix division state (or to distinguish between "bad" and "good" individuals) some mechanisms of the immune system can be used. Fixing of division state is inspired by immunological mechanisms — generation of T cells in the immune system. This article does not present an artificial immune system, but some immunological mechanisms are used in a part of presented solutions. Immunological mechanisms should operate on actions made by observed agents. This approach is opposite to the one proposed in e.g. [3,12] in which immunological mechanisms operate on the resource's structure.

Immunological intruders detection in the computer environment has to be done on the basis of certain characteristic structures. These structures in the case of behavior observation can be chains of actions performed by an observed agent. These chains are of the settled length l, so one chain contains l objects, which present undertaken actions by observed agent (one object represents one action). We should define the way how agent a will recognize (notice, but not estimate) actions undertaken by neighbors. It is possible to store all actions undertaken by agents in the environment of agent computer system. The action stored should be accompanied by the notion by whom a particular action has been undertaken. This method presumes the mediation of the environment and / or resources of the environment in the process of recognizing undertaken actions.

Detectors. The method of fixing the division state refers to the mechanism of immune system. Once detector's set is generated, this detector's set is used to find "bad" among presented sequences of action–objects.

In order to generate a set of detectors R, own collection W should be specified. This collection includes correct, "good" sequences of action–objects. This collection W should consist of action–object sequences of length l, which is undertaken by the agent–observer. This is correct, because of the assumption that actions which the agent undertakes are evaluated as "good" by him. Presuming there are stored h last actions undertaken by every agent, own collection W will contain $h - l + 1$ elements.

Algorithm of Detectors Generation. The algorithm of detectors generation refers to the negative selection — the method in which T–lymphocytes are generated. From set R_0 of generated sequences of length l those reacting with any sequence from collection W are rejected. Set R_0 contains every possible sequence (but it is also possible to use a set of random generated sequences). Sequence reaction means that elements of those sequences are the same. Sequences from set R_0 which will pass such a negative selection create a set of detectors R.

Behavior Estimation of Neighboring Agents. First stage is a neighbor observation during which actions (and order of those actions) executed by neighboring agents are remembered. Those remembered actions create sequence N of presumed length h. After the next stage of detectors generation, generated detectors are used to find "bad", unfavorable agents. Every subsequence n of length l of sequence N is compared with every detector r from set R, as it is shown in Fig. 1. If sequence n and r match, it means finding "bad", unfavorable actions. Sequence

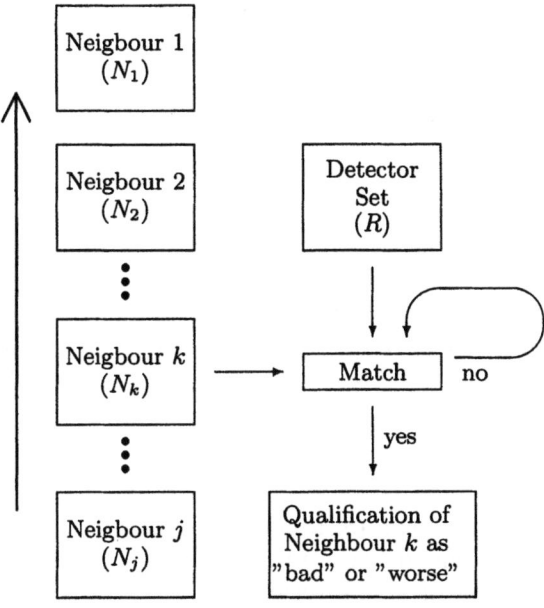

Fig. 1. Process of behavior estimation of neighboring agents, process presented for agent number k

matching means that the elements of the sequences compared are the same. The number of matches for every observed agent should be counted. On this basis behavior estimation is made — division state $m_d = (m_d^1, m_d^2, ..., m_d^{j-1}, m_d^j)$ of agent-observer is modified to the $m_{d'} = (m_d^{1'}, m_d^{2'}, ..., m_d^{j-1'}, m_d^{j'})$, where j is the number of agents in the environment, $m_d^{k'}$ is assigned to the number of counted matches for agent number k.

3.3   Configuration of Goals of Agent's Division Profile

The way neighboring agents are treated is described by Q_d — configuration of goals q_d of agent's division profile. Configuration of goals of an agent is constant (however it is possible to design such a system, which in is possible the goal's adaptation). In the system described the configuration of goals consists only from one goal — liquidation neighboring agent (or agents) number k, if $m_d^k = \max(m_d^1, m_d^2, ..., m_d^{j-1}, m_d^j)$.

3.4   Configuration of Strategies of Agent's Division Profile

Actions, which should be undertaken by agent a in order to treat agent number k in the way described by the configuration of goal, are specified by S_d — the configuration of strategies s_d of agent's division profile. The configuration of strategies of the agent is constant and in the system described the configuration of strategies consists only of one goal: if the goal is to liquidate agent number k, a demand of deleting agent number k is send to the environment (coefficient o_d equal to the m_d^k is attributed to this demand).

This configuration of strategies presumes an intervention of system's environment in the liquidation of the agent. In the systems described the environment calculates the sum of coefficients for every agent separately attributed to demands and liquidates all agents which have the maximum sum of coefficients and this sum is larger than constant OU. Periodically, after a constant time period, the calculated sums of coefficients are set to 0. Constant coefficient OU is introduced in order to get tolerance for behavior that is evaluated as "bad" in a short time, or is evaluated as "bad" by a small amount of agents.

4   Experiment

In the computer system there are some operation which must be executed in couples, for example: open and close a file, connection request and disconnection request. There are a lot of attack techniques which consist in doing only one from a couples (or trios...) of obligatory operations (for example so–called SYN flood attack [11]). There is simulated a system with two types of agents:

- good agents which perform some operations in couples (e.g. open, open, open, close, open, close, close, close);
- bad agents which perform only one from a couples of some operations (e.g. open, open, open, open, open, open, open, open).

In the simulation there is no possibility of distinguishing the type of an agent on the basis of the agent's structure. So the only possibility is to observe the agent's behavior and process the actions observed (actions–objects) to distinguish whether the agent is good or bad.

4.1   Simulation — Unsecured Computer System

First a case was simulated, in which only good agents exist in the computer system — initially there are 80 good agents in the system, which do not have any security mechanisms. Next a case was simulated, in which good and bad unsecured agents exist in the computer system — initially there are 64 good agents and 16 bad agents, all agents do not have any security mechanisms. The system in those two cases was simulated to 300 time periods and there were 10 simulations performed. Diagram in Fig. 2 shows the average numbers of agents in separate time periods.

If there are not any bad agents in the simulated system, all good agents can exist without any disturbance. The existence of bad agents in the system causes problems with executing tasks by good agents, which die after some time periods. Bad agents still remain in the system, which is blocked by those bad agents.

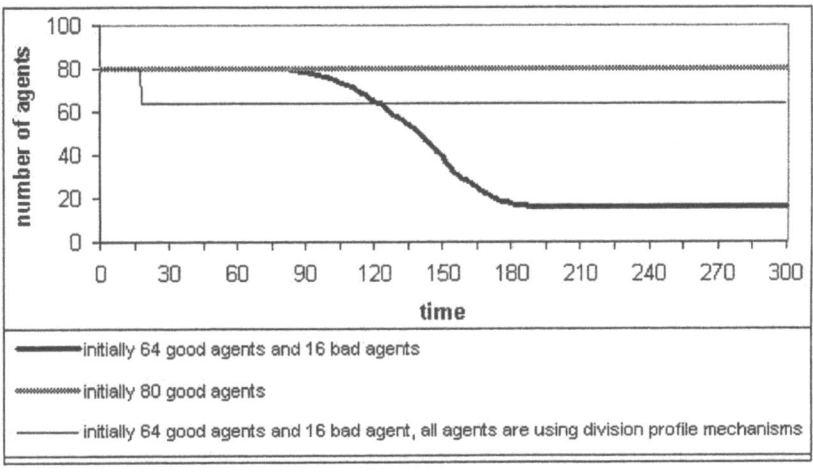

Fig. 2. Number of agents in separate time periods

4.2   Simulation — Secured Computer System

A case was simulated, in which good and bad secured agents exist in the computer system — initially there are 64 good agents and 16 bad agents. All agents
in the system were equipped with the division profile security mechanisms. The
system was simulated to 300 time periods and there were 10 simulations performed. Diagram in Fig. 2 shows the average numbers of agents in separate time
periods.

In the environment there are last 18 actions stored undertaken by every agent.
After 18 actions have been undertaken by every agent, detectors are constructed
of length $l = 5$. Agents use their division profile mechanisms to calculate which
neighboring agent they want to eliminate. Agent demand to eliminate these
neighbors which have the maximum of detector's matchings. Agents present their
demands to the environment with the number of matchings. The environment
counts matchings in the demands presented and eliminates agents as it was
proposed in the description of division profile mechanisms. The constant OU is
set up to 480.

After detectors were constructed, bad agents were distinguished due to the
division profile mechanisms. In the same time period, recognized agents were
deleted, what makes it possible for agents belonging to good type to function
freely.

5   Conclusion

This paper presents a discussion of security problem in the computer system.
Security solutions which create new security paradigms were proposed:

- equip all system resources (e. g. agents, programs) with security mechanisms,
- security mechanisms should be based on activity observation rather than looking for some fragments of code (signatures),
- design the environment of computer system in such a way so as to support security mechanisms with which system's resources are equipped.

In this paper security mechanisms with immunological approach were presented which fulfill the said security paradigms. All these security mechanisms were called a division profile. The conception presented was simulated and the obtained results confirm the effectiveness of this solution. The simulation enables to anticipate how the described mechanisms will function in the real world of computer systems.

Security mechanisms designed on the grounds of the conception presented have such advantages as detection of previously unseen danger activities, detection based on activity observation and decentralized detection.

References

1. Cetnarowicz K., Nawarecki E., Żabińska M.: M–agent Architecture and its Application to the Agent Oriented Technology. Proc. of the DAIMAS'97, St. Petersburg (1997)
2. Cetnarowicz K., Rojek G., Werszowiec-Plazowski J., Suwara M.: Utilization of Ethical and Social Mechanisms in Maintenance of Computer Resources' Security. Proc. of the Agent Day 2002, Belfort (2002)
3. Forrest S., Perelson A. S., Allen L., Cherukuri R.: Self-nonself Discrimination in a Computer. In Proc. of the 1994 IEEE Symposium on Research in Security and Privacy, IEEE Computer Society Press, Los Alamitos (1994) 202–212
4. Forrest S., Perelson A. S., Allen L., Cherukuri R.: A Change–detection Algoritm Inspired by the Immune System. IEEE Transactions on software Engineering, IEEE Computer Society Press, Los Alamitos (1995)
5. Gibbs W. W.: Jak przetrwać w niebezpiecznym świecie? Świat nauki, Wydawnictwo Prószyńska i s-ka (2002)
6. Hofmeyr S. A., Forrest S.: Architecture for an Artificial Immune System. Evolutionary Computation, vol. 7, No. 1 (2002) 45–68
7. Kim J., Bentley P.: The Human Immune System and Network Intrusion Detection. 7th European Congress on Intelligent Techniques and Soft Computing (EUFIT '99) Aachen September 13-19 (1999)
8. Kim J., Bentley P.: Negative Selection within an Artificial Immune System for Network Intrusion Detection. The 14th Annial Fall symposium of the Korean Information Processing Society, Seoul October 13-14 (2000)
9. Ossowska M.: Normy moralne. Wydawnictwo Naukowe PWN, Warszawa (2000)
10. Ricken F.: Etyka Ogólna. Wydawnictwo ANTYK — Marek Derewiecki, Kety (2001)
11. Schetina E., Green K., Carlson J.: Bezpieczeństwo w sieci. Wydawnictwo HELION, Gliwice (2002)
12. Wierzchoń S. T.: Sztuczne systemy immunologiczne: teoria i zastosowania. Akademicka Oficyna Wydawnicza Exit, Warszawa (2001)

Policy Modeling in Four Agent Economy

Adam Woźniak

Institute of Control and Computation Engineering
Warsaw University of Technology,
ul. Nowowiejska 15/19, 00-665 Warszawa, Poland
wozniak@ia.pw.edu.pl

Abstract. This paper proposes a simple model of policy game played by four main agents of economy: profit maximizing firms active on competitive market, monopolistic trade union (aggregated wage setters), government and central bank. The interactions between agents are described by the single period aggregate demand and supply equations. We also adopt assumption of bounded rationality of agents. After reduction of firms as active agent, the resulting three agents decision system is modeled as Stackelberg game with central bank as leader and trade union and government as composite follower, aggregated by means of Nash bargaining solution. The simulation of policy game is provided.

1 Introduction

Classical wisdom holds that, activity of conservative central bank, that is a central bank (CB) which places a greater weight on reduction of inflation than society, reduces equilibrium inflation with no (or very low) cost in terms of output. However, at the end of nineties, papers modeling decision process in the economy as policy game played within multiagent system, questioned desirability of conservative CB. For example V. Guzzo and A. Velasco concluded with proposition that decisions of central bank that pay little attention to price stabilization, so called populist CB, give better result because they lead to full employment, high output and low inflation in equilibrium [1], [2].

More complicated models of decision process used in the papers start with multiagent situation, but at the end describe the interactions among two agents only – aggregate wage setters (trade union) and central bank. The objective functions of agents are chosen as quadratic and economy is modeled using Keynesian aggregate demand – aggregate supply framework. Within these models, the case for a populist central bank can be made when aggregated wage setters are inflation averse Stackelberg leader with respect to a central bank that controls inflation [3].

However, G. Ciccarone and E. Marchetti in their recent paper pointed out that hypothesis that workers' organizations are interested, besides real wages and employment, in inflation *per se* is difficult to prove [4]. The fulfillment of second assumption that trade unions act as leader, that is announce their decision in advance knowing the response function of central bank is also problematical. Third, the assumption that agents in modeled economy are perfectly rational and optimize *quadratic functions has rather* mathematical than case study research origin. This all is

M. Bubak et al. (Eds.): ICCS 2004, LNCS 3038, pp. 615–622, 2004.
© Springer-Verlag Berlin Heidelberg 2004

crucial because the presented results are not robust to removal of any from the above assumptions.

This paper aims at presentation of simple, but free from the above weakness, model of policy game played by four main agents of economy: profit maximizing firms active on competitive market, monopolistic trade union (aggregated wage setters), government and central bank. Their decision variables are the level of employment, L, the nominal wage, W, the budget deficit, B, and the supply of money, M, respectively. When the decisions are taken, the economy 'produces' in equilibrium state output, Y, and rate of inflation, Π.

The interactions between agents will be described by the single period aggregate demand and supply equations. We also adopt, to some extend, assumption of bounded rationality of agents.

2 Interactions among Agents Constituting an Economy

First, we assume that output (production), Y, is described by short-run production function

$$Y = L^{\gamma} K^{1-\gamma},$$

where L is labor (level of employment), K – capital, and $\gamma \in (0,1)$.

The first agent in our model is an aggregated one – all production firms active on competitive market. As mentioned, this agent always tends to maximize his (aggregate) profit

$$Z = Y - L(W/P),$$

where P is given price level, choosing level of employment (labor demand) L. Therefore, his decision problem is as follows

$$\texttt{find}\ L^{\circ} = \arg \max{}_{L}[L^{\gamma} K^{1-\gamma} - L(W/P)]. \tag{1}$$

The solution of this problem is easy to obtain by differentiation and is the following

$$L^{\circ} = \left(\gamma \frac{P}{W}\right)^{\frac{1}{1-\gamma}} K$$

and takes the form of response function to external, for this agent, variables P and W. The above equation is nonlinear and troublesome in use, so as is customary, we take natural logs and convert it into linear one

$$l^{\circ} = -\frac{1}{1-\gamma}(w - p) + \frac{\ln \gamma}{1-\gamma} + k \tag{2}$$

where small letters denote logs, e.g., $k = \ln K$.

The assertion that firms are perfect rational and their behavior in any circumstances can be described by their best response (2) opens possibility to eliminate from further consideration activity of first agent described by decision problem (1) and using 'stiff' equation of labor demand (2) instead. Consequently the number of agents is reduced by one and we have the following explicit equation relating output (in logs) to wage and price level

$$y = \gamma l^{\circ} + (1-\gamma)k = -\frac{\gamma}{1-\gamma}(w-p) + \frac{\gamma}{1-\gamma}\ln\gamma + k.$$

In the sequel for simplicity, we shorthand the constant term to $y_c = \frac{\gamma}{1-\gamma}\ln\gamma + k$, what leads to the following, so called, aggregate supply (AS) equation

$$y = -\frac{\gamma}{1-\gamma}(w-p) + y_c. \tag{3}$$

It describes the aggregate supply of output by competitive profit-maximizing firms as decreasing when the wage is growing and increasing with the price level.

We assume that budget deficit B is covered using monetary base; therefore, exchange equation has the form

$$M + B = PY.$$

When we define $D = B/M$ the above equation changes to $M(1 + D) = PY$ or in logs

$$y = m - p + \ln(1 + D).$$

Because D is generally less than 0.2, we can linearize this equation to

$$y = m + D - p. \tag{4}$$

The equation (4) is called aggregate demand (AD) equation. It shows by making aggregate demand for output dependent upon money balance, the traditional inverse relationship, for a given money supply, between demand for output and price level.

As we remember, trade union, government and central bank are remaining agents in modeled economy. The level of inflation and unemployment occurs as the result of their decisions concerning levels of nominal wage, budget deficit and money supply. Therefore, we have to transform slightly our description and eventually add equation describing unemployment.

The inflation rate equals $\Pi = (P - P_{-1})/P_{-1} = P/P_{-1} - 1$. By assuming some "prior" level of prices, P_{-1}, we may talk of inflation and current prices interchangeably, cf. [5]. For convenience, we assume $P_{-1} = 1$, so

$$\Pi = P - 1 = \exp(p) - 1.$$

Bearing above in mind, we can now transform equation (3) and (4) to the following form

$$y = \gamma(m + D - w) + (1 - \gamma)y_c \tag{5}$$

$$p = (1 - \gamma)(m + D) + \gamma w - (1 - \gamma)y_c. \tag{6}$$

The above equations describe how output, eq. (5), and inflation, eq. (6), depend on money supply, budget deficit and nominal wage.

Now, we will present equation, which describes unemployment. Let N be a given level of labor force. When first agent (production firms) is rational, employment rate is equal $E = L^{\circ}/N$, in logs $e = l^{\circ} - n$. Under our assumptions it gives

$$e = \frac{1}{\gamma}y + \ln\gamma - \frac{1-\gamma}{\gamma}y_c - n. \tag{7}$$

Equation (7) means that unemployment rate $U = 1 - E$ decreases when output increases, so maximizing output means minimizing unemployment.

3 Primal Decision Problems of Agents

Basing on the above discussion of interactions in the economy, we can now define primal (isolated) decision problems of the remaining, active agents: trade union, government and central bank.

Usually one assumes that agents seek to optimize their objective functions, quadratic in variables, cf. [6]. This means that tacitly it is postulated, as in classical economics, that agents always succeed in choosing the best solution optimizing their functions. We adopted this assertion when concerned production firms on competitive market. Now we weaken it and will follow the way of H. Simon and assume that remaining agents interpret results as either satisfactory or unsatisfactory and that there exists an aspiration level (threshold) constituting the boundary between satisfactory and unsatisfactory. The agent, instead of seeking for the best alternative, looks for a good one only [7]. In other words, we assume that rationality of trade union, government and central bank is bounded and they are satisficing decision makers.

The classical thinking of organized labor asserts that the increase of wages is the main interest of trade union although it dislikes increase of unemployment. Although it is problematical, we assume that to some extend trade unions take into account requirement of keeping inflation on proper level.

Because we will state decision problem of trade union as maximization and keeping the above in mind we form the objective function of this agent as a weighted sum of real wage W/P, and terms measuring threshold violation

$$TU_p = (w - p) + \alpha_1 \min(0, y - y_{TU}) - \alpha_2 \max(0, \Pi - \Pi_{TU}), \tag{8}$$

$0 < \alpha_2 < \alpha_1 < 1$, $y_{TU} = \gamma(e_D + n - \ln\gamma) + (1-\gamma)y_c$, $E_D = 1 - U_D$.
The signs of terms describing threshold violation are selected in such a way that first term is smaller than zero when output is smaller than its threshold y_{TU} calculated by trade union basing on accepted level of unemployment, U_D, and second – is smaller than zero when inflation is larger than threshold Π_{TU}.

The decision problem of trade union is now the following

$$\text{find } w^\circ = \arg\max_{w \geq 0}[TU = (1-\gamma)(w - m - B(m)/\exp(m)) +$$
$$+ \alpha_1 \min(0, \gamma(m + B(m)/\exp(m) - w) + (1-\gamma)y_c - y_{TU}) + \tag{9}$$
$$- \alpha_2 \max(0, \exp((1-\gamma)(m + B(m)/\exp(m)) + \gamma w - (1-\gamma)y_c) - 1 - \Pi_{TU})].$$

In the above problem constraint on minimal wage had to be added. For simplicity, we assumed that the minimal wage, W_{min}, equals 1.
Similar considerations lead to objective function of government

$$G_p = \beta_1 \min(0, B - B_G) + \beta_2 \min(0, y - y_G) - \beta_3 \max(0, \Pi - \Pi_G), \quad \beta_1, \beta_2, \beta_3 > 0 \tag{10}$$

and the decision problem of this agent

find $B° = \arg \max_B [G = \beta_1 \min (0, B - B_G) +$
$+ \beta_2 \min(0, \gamma(m + B/\exp(m)) - w) + (1 - \gamma)y_c - y_G) +$
$+ \beta_3 \max (0, \exp ((1 - \gamma)(m + B/\exp (m)) + \gamma w - (1 - \gamma)y_c) - 1 - \Pi_G)].$ (11)

Ending modeling of agents' interests, we adopt as the objective function of the last agent (central bank) the function depending on terms measuring inflation and output (unemployment) target missing

$$CB_p = - \max (0, \Pi - \Pi_{CB}) + \delta \min (0, y - y_{CB}), \delta > 0.$$ (12)

The decision problem of central bank is stated as

find $m = \arg \max_m [CB =$
$= - \max (0, \exp((1 - \gamma)(m + B(m)/\exp(m)) + \gamma w - (1 - \gamma)y_c) - 1 - \Pi_{CB}) +$ (13)
$+ \delta \min (0, \gamma(m + B(m)/\exp(m) - w) + (1 - \gamma)y_c - y_{CB})].$

This completes descriptive part of our model – description of agents' decision problems with their decision instruments and interactions joining them. Now we must model the rules (protocol) regularizing behavior of agents. In other words, we must describe cooperation-coordination mechanism in economy. We assume that central bank is independent in his decisions and acts first announcing chosen supply of money. Knowing level of this external variable trade union and government negotiate level of wage and budget deficit. As a result, equilibrium determined by aggregate demand and aggregate supply equations (3), (4) gives employment, output and inflation.

4 Cooperation-Coordination Mechanism

The assumption about trade union – government negotiations is typical. As mechanism of negotiations, we propose concession mechanism derived by F. Zeuthen in his book devoted to this topic [8]. We adopted this mechanism because of two reasons. From one side, it is a dynamic adjustment procedure modeling real negotiations. From the other side, starting from the work of Harsanyi [9], it is known that this procedure converges to axiomatic Nash bargaining solution. We are not interested in presentation details of Zeuthen process here, short description can be found, e.g. in [10]. As we will see, convergence to relatively easy computable Nash bargaining solution significantly simplifies description of central bank operation rules.

On the first gland, central bank right to the first move gives him some kind of supremacy. But from the hierarchical games theory it is known that, so called leader, has supremacy only when he has appropriate information about the other agents, followers, giving him also information priority [11].

When we assume that presented model, trade union and government primal decision problems included, is known to the central bank, together with information that his partners tend in negotiation to the Nash bargaining solution, we can say that he has this priority. Both above assumptions seem quite reasonable.

The bestowal of right to the first move and information priority on central bank enables to model his decision situation as Stackelberg game with bank as leader and trade union and government as composite follower, aggregated by means of Nash bargaining solution.

The essential tool in analyzing Stackelberg game is response mapping of follower. Let (w^N, B^N) denotes Nash bargaining solution to union – government negotiation. For given level of money supply m, having information priority, central bank is able to compute $r(m) = (w^N(m), B^N(m))$ solving the following nonlinear optimization problem (cf. [12])

$$r(m) = (w^N(m), B^N(m)) = (TU^{-1}(m, \cdot, \cdot)(s_1(m)), G^{-1}(m, \cdot, \cdot)(s_2(m))),$$

$$(s_1(m), s_2(m)) = \arg\max_{(s_1, s_2) \in S_d(m)} (s_1 - s_1^d)(s_2 - s_2^d) \tag{14}$$

where

$$S_d(m) = S(m) 3 \{(s_1, s_2) \mid s_i \geq s_i^d, \ i = 1, 2\},$$

$$S(m) = \{(s_1, s_2) \mid (\exists \ w \geq 0)(\exists \ B)(s_1 = TU(m, w, B) \land s_2 = G(m, w, B))\}$$

and (s_1^d, s_2^d) is known *status quo* (disagreement) point.

We recall that TU and G denote functions maximized in trade union (9) and government (11) decision problem, respectively. Of course, it is impossible to solve the problem (14) analytically, but for given set $\{m_j\}$ of money supplies, central bank is able to compute set of responses $\{r(m_j)\}$ and basing on both sets, to estimate response function[1]

$$m \to r(m) = (w^N(m), B^N(m)).$$

Now the decision problem of central bank takes the form

find $m^o = \arg\max_m [CBS(m) =$

$$- \max (0, \exp((1 - \gamma)(m + B^N(m)/\exp(m)) + \gamma w^N(m) - (1 - \gamma)y_c) - 1 - \Pi_{CB}) +$$
$$+ \delta \min (0, \gamma(m + B^N(m)/\exp(m) - w^N(m)) + (1 - \gamma)y_c - y_{CB})]. \tag{15}$$

That is the reformulation of problem (13) using defined response function.

It is worth noting, that central bank knows only estimate of function $m \to CBS(m)$ maximized in his decision problem, so called composite objective function. However, as we remember, the problem of central bank was stated in fact, not as optimization but satisficing one (assumption about bounded rationality). So somewhat blurred description of composite objective function do not prevent this agent from taking suitable, satisfactory decision, which will give desired proper level of inflation and output.

5 Simulation Result

Now, we present simulation results of modeling hypothetical closed economy with described four agents. We adopt the following values of parameters:

- for production function: $\gamma = 0.7$, $y_c = 6.3$ ($K = 1251.7$);
- for primal decision problem of trade union: weight of output $\alpha_1 = 0.6$, output threshold $y_{TU} = 6.55$, weight of inflation $\alpha_2 = 0.3$, inflation threshold $\Pi_{TU} = 0.1$ (rather modern trade union);

---

[1] Equation (14) defines function, only when we assume that $(w^N(m), B^N(m))$ is unique for every m.

- for primal decision problem of government: deficit threshold $B_G = 40$, weight of deficit $\beta_1 = 10$, output threshold $y_G = 6.5$, weight of output $\beta_2 = 0.6$, weight of inflation $\beta_3 = 0.7$, inflation threshold $\Pi_G = 0.08$;
- for primal decision problem of central bank: inflation threshold $\Pi_{CB} = 0.05$, output threshold $y_{CB} = 6.3$, weight of output $\delta = 0.2$ (rather conservative central bank).

The *status quo* point was calculated basing on Nash equilibrium in noncooperative game of trade union and government: $(s_1^d, s_2^d) = (-0.235211, -0.159484)$.

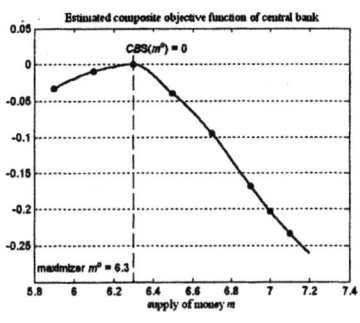

Fig. 1.

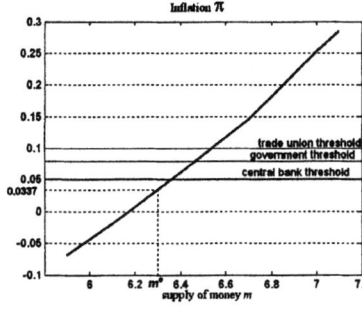

Fig. 2.

The shape of estimated composite function $CBS(\cdot)$ is shown in the Fig 1. Its maximal value equals zero and is realized by supply of money $m^\circ = 6.3$. The resulted equilibrium state of modeled economy is presented above and in the following figures. The goals of central bank are met – inflation is below and output is above threshold. However, economic development is below trade union and government expectations: $Y^\circ = 588.0$, compared to thresholds $Y_{TU} = 699.2$ and $Y_G = 665.1$. Therefore, employment is also below expectations ($L^\circ = 425.4$ compared to $L_{TU} = 544.8$ and $L_G = 507.3$). Separate analysis has showed that for given values of parameters the maximal attainable value of the real wage is slightly greater than one. As a consequence obtained value of real wage $(W/P)^\circ = 0.97$ is satisfactory.

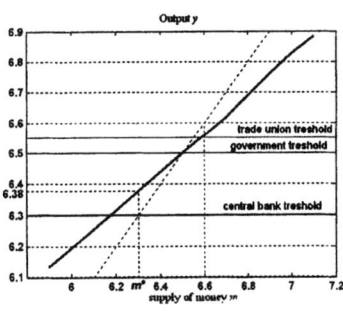

Fig. 3.

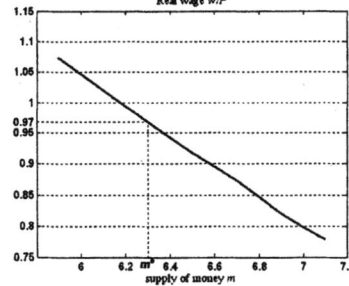

Fig. 4.

To complete analysis of simulation result, let us note that for $m = 6.4$, inflation is 0.06 which is still below government and trade union expectations, and economy gives output $Y = 626.2$ (compared to 588.0), employment $L = 465.4$ (compared to

425.4, increase by 9.4%) and real wage $(W/P) = 0.94$. It means that the adopted objective function of central bank with small weight of term assigned to output, characterizes, as we anticipated, conservative money authority.

6 Conclusions

The presented simulation result is the first attempt at using described model. It comes off well. The further research will be concentrated first, on examining of different parameter combination influence on result. Next, after tuning model, different literature hypotheses about connections between preference of agents modeled by their objective functions and outcomes will be checked.

References

1. Guzzo, V., Velasco, A.: The case for a populist central banker. European Economic Review, 43 (1999) 1317 - 1344
2. Guzzo, V., Velasco, A.: Revisiting the case for a populist central banker: A comment. European Economic Review, 46 (2002) 613 - 621
3. Jerger, J.: How strong is the case for a populist central banker? A note. European Economic Review, 46 (2002) 623 – 632
4. Ciccarone, G., Marchetti, E.: Trade unions' objectives and inflation. Public Economics Department, University of Rome *La Sapienza*, unpublished paper (2002)
5. Cubitt, R.P.: Corporatism, monetary policy and macroeconomic performance: a simple game theoretic analysis. Scandinavian Journal of Economics, 97 (1995) 245 – 259
6. Acocella, N., Di Bartolomeo, G.: Non-neutrality of monetary policy in policy games. Public Economics Department, University of Rome *La Sapienza*, Working Paper n. 49 (2002)
7. Simon, H.A.: A behavioral model of rational choice. Quarterly Journal of Economics, 69 (1955) 99 - 118
8. Zeuthen, F.: Problems of Monopoly and Economic Warfare. Routledge & Kegan, London (1930)
9. Harsanyi, J.C.: Approaches to the bargaining problem before and after the theory of games. Econometrica, 24 (1956) 144 - 156
10. Ren, Z., Anumba, C.J., Ugwu, O.O.: The development of a multi-agent system for construction claims negotiation. Advances in Engineering Software, 34 (2003) 683 - 696
11. Germeyer, Yu.B.: Igry s neprotivopolozhnymi interesami (Games with Non-antagonistic Interests). Nauka, Moskva (1976)
12. Nash, J.F.: The bargaining problem. Econometrica, 18 (1950) 155 - 162

Multi-agent System for Irregular Parallel Genetic Computations

J. Momot, K. Kosacki, M. Grochowski, P. Uhruski, and R. Schaefer

Institute of Computer Science, Jagiellonian University, Kraków, Poland
{momot,kosacki,grochows,uhruski,schaefer}@ii.uj.edu.pl

Abstract. The paper presents the multi-agent, parallel computing system (MAS) composed of a platform of software servers and a set of computing agents. The generic actions of agents and the system government are so designed that it can perform irregular concurrent genetic computations in heterogeneous computer network with a number of computation nodes and connection topology varying in time. The effectiveness of MAS solution is discussed in terms of average migration and communication overheads. Additionally, the MAS system with autonomous, diffusion-based scheduling is compared with low-level distributed implementation, which utilizes the centralized greedy scheduling algorithm.

1 Introduction

The multi-agent system (MAS) seems to be an attractive way to overcome transparency and scalability problems in distributed computing systems. Moreover MAS's are well suited to maintain loosely-coupled, locally synchronized parallel tasks due to the autonomous activity of agents (there is no overhead for global synchronization) and low cost of local communication among the agents located close together. Distributed computing systems composed of mobile tasks have recently been intensively developed (see e.g. PYRAMID in NASA [5]). The advantages of scheduling by agent migration have also been proved in [4,6]. A particular case of such loosely-coupled parallel computations is multi-deme, parallel genetic algorithms. They are an advantageous tool for solving difficult global optimization problems, especially in case of problems with many local extrema (see e.g. [1]). The Hierarchical Genetic Strategy (HGS) introduced by Kołodziej and Schaefer [2] constitutes one of their instances. The main idea of the HGS is to run a set of dependent evolutionary processes, called *branches*, in parallel. The dependency relation has a tree structure with the restricted number of levels m. The processes of lower order (close to the root of the structure) represent a chaotic search with low accuracy. They detect the promising regions on the optimization landscape, in which more accurate process of higher order are activated. Populations evolving in different processes can contain individuals which represent the solution (the phenotype) with different precision. This precision can be achieved by binary genotypes of different length or by different phenotype scaling. The strategy starts with the process of the lowest order 1, called

M. Bubak et al. (Eds.): ICCS 2004, LNCS 3038, pp. 623–630, 2004.
© Springer-Verlag Berlin Heidelberg 2004

the root. After the fixed number of evolution epochs the best adapted individual is selected. We call this procedure a *metaepoch* of the fixed period. After every metaepoch a new process of order 2 can be activated. This procedure is called a *sprouting operation* (SO). Sprouting can be generalized in some way to HGS branches of higher order up to $m - 1$. Sprouting is performed conditionally according to the outcome of the *branch comparison operation* (BCO). BCO can be also used to reduce branches of the same degree that checks the same region of the optimization landscape. Details of both the SO and BCO depend strongly upon the implementation of the HGS.

The first HGS implementation [2] utilizes the Simple Genetic Algorithm (SGA) as the basic mechanism of evolution in every process. The next implementation of the HGS_RN was obtained by using real-number encoding, normal mutation, and the simple arithmetic crossover instead of SGA (see [3]).

2   Autonomous Agent's Platform

The paper depicts an agent approach to the HGS application, which utilizes the MAS platform (see [4] and references inside). The MAS platform is composed of software servers statically allocated on computer nodes that perform information, migration, and hibernation policies for mobile computing units called agents. The MAS computing application is composed of intelligent, mobile agents that wrap the computational tasks. They can be dynamically created and destroyed. Each of them, with partial autonomy, can decide on its current location. The computing application is highly transparent with respect to the hardware platform, the number of computers and the configuration and addressing of the computer network. In order to facilitate agent building, which should combine the computational and scheduling purposes, *Smart Solid* architecture was introduced. Each Smart Solid Agent is represented by the pair $A = (S, T)$ where T is the computational task executed by the agent, including all data required for computation, S is shell responsible for the agent logic required to maintain the computational task, including the communication capabilities and scheduling mechanism. To perform scheduling the shell S enforces on computational task T the following functionalities:

- T has to be able to denominate the current requirement for computational power and RAM,
- T must allow pausing and continuing of its computation (pausing is needed for the hibernating task in case of agent migration or partitioning, and continuation is needed to restore the paused job),
- T must allow partitioned into two subtasks $T \rightarrow \{T_1, T_2\}$.

The shell S can perform the following actions: (a-1) execute T; (a-2) pause T; (a-3) continue T; (a-4) denominate own load requirement; (a-5) compute gradient, and check the migration condition; (a-6) partition $T \rightarrow \{T_1, T_2\}$ and create $A_i = (S_i, T_i), i = 1, 2$; (a-7) migrate; (a-8) disappear. These actions allow A to accomplish two goals: (G-1) - perform computation of carried task, (G-2) - find a better execution environment. The easiest possibility to get (G-1) is to execute (a-1). A more extended decision algorithm fitted to the HGS implementation will be presented in the next section. If a new agent is created and also when the shell recognizes a significant change of load on the local servers, the agent then checks if it should continue to realize (G-1). If no, then it passes to (G-2) by using the scheduling mechanism based on the diffusion phenomenon. Briefly depicting that mechanism we can say that the shell queries task requirements for resources (action (a-4)) and computes load concentration on the local server and nearest surrounding servers (action (a-5)). To achieve those values the shell communicates with both the computational task and the local server on which the agent is located. That information is necessary to compute agent *binding energy* on the server and local load concentration gradient. The gradient specifies a possible direction for migration (destination server). If the binding energy on the destination server exceeds the binding energy on the current server more than the predefined threshold, then the agent migrates (action (a-7)). The agent can keep migrating until it finds sufficient resources. If agent A is too large to migrate, then the actions (a-6) and (a-8) are performed, and both produced agents A_1 and A_2 will start scheduling independently.

3   HGS Agents

We concentrated on the agent-oriented project of the HGS that searches global extrema of the objective function. The HGS produces a tree with nodes being demes, and the edges being created when the sprouting operation occurs. A tree is created by the stochastic process, thus its size (number of demes) might be different in various runs of the algorithm. The HGS is concurrent in its nature, because computations performed in each branch do not have much influence on each other. The only point where two demes meet is when a deme sprouts. The information passed to produce a new deme is very small - only one individual. If the HGS is synchronized regularly (as it was done in [2]), all demes stop after each metaepoch, which enables us to relatively easy employ operations of *branch reduction* (BR) and *conditional sprouting* (CS) in those points of global synchronization. CS searches all the demes of one level and checks, using BCO, whether the deme that is to be sprouted will not be too close to any of them. BR requires a comparison of each pair of the demes on the same level and, if they are searching the same area, reducing them to a single deme. These operations are complex and very time consuming. Furthermore, in the case of implementation in distributed environment, they require more communication.

The architectural idea that naturally springs to one's mind is to put one deme into an agent and let it do the calculations. However, the time needed to process one deme is small in relation to the time needed to create and maintain an agent. Thus we have decided to construct an agent in such a way that it can contain a limited number of demes. When demes sprout, children (new demes) are stored in the same agent until there is no room for them. If this occurs, a new agent with all demes that could not be stored is created.

Instead of CS we introduce *local conditional sprouting* (LCS), which behaves in exactly the same way as CS but only within one agent. This means that before creating another deme it is checked whether there are any other demes of the same level that are very close to this new deme. If there are no such cases, then we create a new deme, otherwise we don't. Another mechanism is the 'killing agent'. It walks around the platform asking other agents for data about demes computed by them and finally reduce all demes that are scanning the same region to a single one. So this agent would actually perform an operation similar to BR and also reduce redundant demes that were sprouted due to a lack of global CS.

We use a very restrictive stop condition for the demes of higher levels. A deme finishes when either it has computed a maximal fixed number of metaepochs or it is stopped when it did not make enough progress in the best fitness.

Following the idea given above we have implemented the HGS as a set of Smart Solid Agents. In T (computational task space of an agent) we store a chart of active demes ("populations") and a chart of sprouted demes ("sprouted"). The agent can denominate the requirements for computational power by estimating the upper bound of a number of metaepochs to be computed by all the active populations. T can be paused after the metaepoch for each contained deme is finished. Its activity, when trying to achieve (G-2), can be described by the code below:

```
if (thereIsBetterEnvironment()) {
    computeGradient(); // (a-5)
    if (thereIsSufficientEnvironment()) {
        pause(); migrate(); continue(); // (a-2), (a-7), (a-3)
    } else {
        // create two agents, both with half of the populations of old agent
        partition T → {T₁,T₂} and create Aⱼ = (Sⱼ,Tⱼ), j = 1,2; // (a-6)
        disappear(); // (a-8)
    }
}
```

The following code is activated when the agent tries to achieve (G-1).

```
execute T; // (a-1)
if (mustDoPartitioning) {
    // create two new agents first with "parents" second with "children"
    partition T → {T₁,T₂} and create Aⱼ = (Sⱼ,Tⱼ), j = 1,2; // (a-6)
}
disappear(); // (a-8)
```

The functioning of task T is described in the code below:

```
do {
    for (int i = 0; i < populationsCount; i++) {
        if (populations[i].endOfComputing()) {
            kill(populations[i]);
        } else {
```

```
        popualtions[i].metaepoch();
        if (populations[i].canSprout()) {
            sprouted.insert(populations[i].sprout());
        }
    }
}
if (isPlaceForSprouted()) { storeSprouted(); }
else {
    mustDoPartition = true;
    return;
}
} while (thereAreAnyLivingPopulations())
storeResults(); // returns results to the Requester unit
```

A partition is done polymorphously; it acts differently depending on the state of the "mustDoPartition" flag (see code below).

```
if (mustDoPartition) {
    T₁ ← T; T₂ ← sprouted;
} else {
    T₁ ← first half of T; T₂ ← second half of T;
}
```

Aside from Smart Solid Agents dynamically created and destroyed, the HGS application contains a single requester unit that sends the first computing agent on the platform. All the results computed by the agents are sent to the requester which stores them and then chooses the best.

4 Experiments

The MAS-HGS implementation was tested with sample inputs to check its runtime properties. The experiments were conducted within a network of 45 PC machines connected by a TCP/IP protocol based network. Machines ranged from PIII, 256Mb RAM up to dual PIV 1Gb RAM machines and worked under Linux and MS Windows operating systems.

Table 1. The average results of HGS computations with diffusive scheduling

Objective	Agents amount		Execution times [sec]			Parallel	Over-
function	Total	Average	Migration	Communication	Computation	time [sec]	head %
1	2	3	4	5	6	7	8
Rastrigin	193,0	42,7	358,9	141,7	7923,0	187,1	5,94
Griewangk	183,7	22,0	168,55	110,95	6421,3	288,2	4,17
Schwefel	163,0	40,4	288,6	138,0	22259,4	558,1	1,88

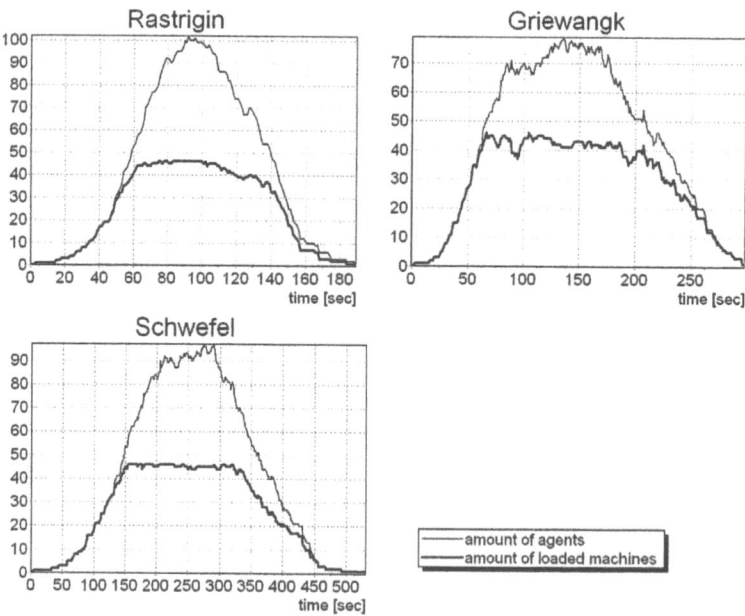

Fig. 1. The dynamics of HGS computing agents.

We performed computations for three well known global optimization bench-marks: 20-dimensional Rastrigin, 10-dimensional Griewangk, and 10-dimensional Schwefel. Each row of Table 1 contains values averaged over 10 runs performed for the particular objective. The Total Agents amount (column 2) is the number of agents produced in the whole run, while Average (column 3) is the mean number of active agents during each run for each objective. Columns 4-6 contain sums of all agents' migration, communication, and computation times. Column 7 shows the average parallel time, which includes migration and communication, measured by the requester. It is a good approximation of the mean wall clock time period of the total experiment. The computation time, stored in column 6, can be greater then the serial time for the whole computation, because significant negative synergy may occur among agents allocated on the same computer. The last column, Overhead %, shows the fraction of execution times that agents spend on migration and communication.

Figure 1 presents the amount of actively computing agents varying during the experiment runtime. We chose runs whose parameters were closest to the mean ones presented for each objective in the Table 1. A high agent dynamic is caused by the HGS strategy that sprouts many demes in the first phase of the search and then demes are reduced to ones that search close to the local extrema. Sprouting is stochastic so it is impossible to predict when the particular agent will be partitioned and, in consequence, to plan the optimal a'priori allocation for each new agent.

Table 2. Speedup of HGS computations for the Griewangk objective with Round-Robin and Diffusive Scheduling. Selected experiments with largest (a) and smallest (b) amounts of demes are shown. Section (c) presents mean values for all conducted experiments (10).

	Serial time [sec]	Diffusive Scheduling			Round-Robin	
		Total agents amount	Parallel time [sec]	Speedup	Parallel time [sec]	Speedup
(a)	6891	244	371	18,57412	318	21,66981
	6766	299	299	22,62876	334	20,25749
(b)	3387	122	374	9,05615	163	20,77914
	2687	94	228	11,78509	131	20,51145
(c)	**4221,9**	**183,7**	**288,2**	**14,44104**	**204,9**	**20,66099**

The mean number of active agents (see column 3 in Table 1) is lower while the maximum is about two times greater than the number of processors (see Figure 1), so the upper limit of speedup was temporarily activated.

The irregularity of parallelism in the agent-oriented application may be measured as the ratio between the maximum and the mean number of agents active during the whole computation. This ratio reaches the maximum value for the Griewangk benchmark. In this case we performed a detailed speedup comparison with a low level message passing (using fast RMI communication) distributed application with a predefined number of fixed PC nodes, which utilizes centralized Round-Robin (RR) scheduling performed by the master unit. The RR is one of the well known greedy policies that can handle cases with randomly appearing tasks. It is currently implemented as follows: each new sprouted deme is registered by the master unit using the message sent by the process holding the parent deme. Next, the master unit introduces the message to the process on a selected machine, pressing it to start computation for the new deme. Both messages contain only deme parameters (deme branch degree, standard deviation of mutation, etc.) and the seed individual (see [3]). Each deme sends the report with the computation result to the master after the stop condition is satisfied.

The appearance of the tasks is totally random, therefore a low-level distributed application with fast explicit communication and RR scheduling is close to the optimal solution in the case of clusters and LANs dedicated to parallel computation. This solution may be treated as a reference point for the fastest solution towards multi-agent solution.

5 Conclusions

- Multi-agent systems (MAS) are well suited to irregular parallel computations because the number of processing units (agents) may be dynamically adapted to the degree of concurrency of the algorithm. Moreover they allow utulizing of a multi purpose, heterogeneous computer network with a number of computation nodes and connection topology varying in time.

- The flexibility of the presented MAS solution was obtained by both architectural and implementation properties. We used Java language and CORBA framework services due to their transparency and support for object migration (serialization). The MAS is composed of a network of software servers and a set of Smart Solid Agents. Each one of the agents is a pair $A = (S, T)$, where T stands for task space and S is a shell that provides all services related to scheduling and tasks' communication. Diffusion scheduling ensures proper agent location in the dynamic network environment.
- Preparing a parallel computing application of this kind we have to specify only the task space T of agent classes. However, we are burdened by specifying all details associated with explicit low-level distributed programming.
- Smart Solid Architecture used to implement the HGS imposes coarse-grained parallelism. Grain size may be easily configured by changing agent capacity (maximum number of demes).
- Total overhead caused by the use of Agent paradigm is low in comparison with the computation time (about 5%, see Table 1). Comparison of this solution with a fine-grained, fast low-level message passing application on dedicated workstation cluster shows moderate losses of the average speedup (see Table 2 section (c)). Presented solution becomes faster for larger problems, even winning with Round Robin (see Table 2 section (a)).
- Speedup computed in all the tests is far from linear. It is caused not only by migration and communication overheads, but mainly by HGS irregularity. Though a mean number of agents is lower than the number of PCs, the maximum number is much higher.

References

1. Cantu-Paz E.: *Efficient and Accurate Parallel Genetic Algorithms*. Kluwer 2000.
2. Schaefer R., Kołodziej J.: Genetic search reinforced by the population hierarchy. in *De Jong K. A., Poli R., Rowe J. E. eds. Foundations of Genetic Algorithms 7* Morgan Kaufman Publisher 2003, pp. 383-399.
3. Wierzba B., Semczuk A., Kołodziej J., Schaefer R.: Hierarchical Genetic Strategy with real number encoding. *Proc. of the 6th Conf. on Evolutionary Algorithms and Global Optimization* Łagów Lubuski 2003, Wydawnictwa Politechniki Warszawskiej 2003, pp. 231-237.
4. Grochowski M., Schaefer R., Uhruski P.: Diffusion Based Scheduling in the Agent-Oriented Computing Systems. Accepted to *LNCS*, Springer 2003.
5. Norton Ch.: "PYRAMID: An Object-Oriented Library for Parallel Unstructured Adaptive Mesh Refinement" accepted to *LNCS*, Springer 2001.
6. Luque E., Ripoll A., Cortés A., Margalef T.: A distributed diffusion method for dynamic load balancing on parallel computers. *Proc. of EUROMICRO Workshop on Parallel and Distributed Processing*, San Remo, Italy, January 1995. IEEE CS Press.

Strategy Extraction for Mobile Embedded Control Systems Apply the Multi-agent Technology

Vilém Srovnal[1], Bohumil Horák[1], Radim Bernatík[1], and Václav Snášel[2]

[1] Department of measurement and control, FEECS, VŠB-Technical University of Ostrava,
17.listopadu 15,CZ-708 33 Ostrava-Poruba, Czech Republic
vilem.srovnal@vsb.cz
2 Department of computer science, FEECS, VŠB-Technical University of Ostrava,
17.listopadu 15,CZ-708 33 Ostrava-Poruba, Czech Republic
vaclav.snasel@vsb.cz

Abstract. Mobile embedded systems belong among the typical applications of the distributed systems control in real time. An example of a mobile control system is the robotic system. The proposal and realization of such distributed control system represents a demanding and complex task of real time control. In the process of robot soccer game application extensive data is accumulated. A reduction of such data is a possible win in game strategy. SVD decomposition of matrixes is used in data reduction.

1 Introduction

A typical example of a distributed control system with embedded systems is the proposal of the control system of mobile robots for the task robot-soccer game. The selection of this game for a laboratory task was the motivation both for students and for the teachers as well because this was a question of proposing and realizing a complicated multidisciplinary task, which can be divided into a whole number of partial tasks (the evaluation of visual information and processing of an image, the hardware and software implementation of a distributed control system, wireless data transmission and processing of information and the control of robots). For the development of the game strategy itself the multi-agents method is considered with the use of opponent strategy extraction algorithms. The results are used as a control action database for the decision agent.

Embedded systems are represented by up to 11 own and 11 opponent autonomous mobile robots. The core of an embedded control system is a digital signal processor Motorola - DSP56F805. PWM output of the signal processor are connected to a pair of power H-bridge circuits, which supply a pair of DC drives with integrated pulse encoders. For communication the communication module is used with the control IC Nordic nRF2401, which ensures communication with a higher level of the control system.

The higher level of control system is represented by a personal computer. In the PC the signal is entered, which represents the picture of a scene with robots scanned with an above the playground placed CCD camera. At the output is connected radio line which transmits commands for all own mobile robots.

M. Bubak et al. (Eds.): ICCS 2004, LNCS 3038, pp. 631–637, 2004.
© Springer-Verlag Berlin Heidelberg 2004

The software part of a distributed control system is realized by decision making and executive agents. The algorithm of agents cooperation was proposed with the control agent on a higher level. The algorithms for agents realized in robots are the same. The control agent determines the required behavior of the whole control system as the response to the dynamic behavior of robots and to the one's own global strategy in the task and knowledge about the last situations, which are saved in the database of the scene. The agent on a higher level controls the other agents [5].

The separate task is the transformation which converts the digital picture into the object coordinates (robots and ball in the task of robot soccer) which are saved in the database of the scene [4]. This database is common for all agents in the control system. Each agent sees actual the whole scene and is capable of controlling its behavior in a qualified way. The basic characteristic of a control algorithm of a subordinate agent is the independence on the number of decision making agents for robots on the playground.

Both agent teams (one's own and the opponent's) have a common goal, to score a goal and not to have any against. For successful assertion of one's own game strategy the extraction and knowledge of an opponent's game strategy is very important. From object coordinates of the picture scene strategy extraction algorithms are created from the opponent's game strategy database.

2   Control Agent

The control agent (Fig. 1) goals (score the goal into an opponent's goal, defend one's own goal area against the opponent players) can be achieved by the correct selection of cooperating agents (division of tasks among agents). The decision making agents are capable of selecting the correct tasks for themselves and further to select the executive agents.

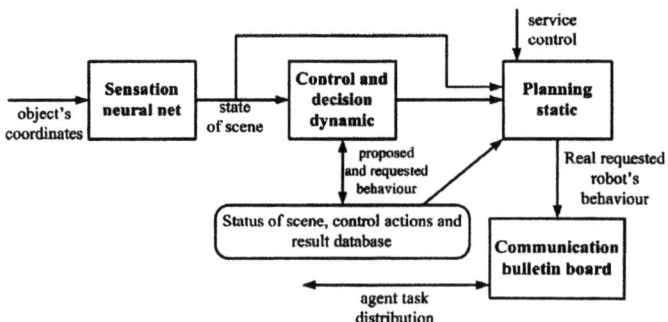

Fig. 1. Control agent scheme

The sensation module contains the neuron network which evaluates the actual state of the scene. The output vector (state of scene) does not describe the position of robots but gives the information about the capability of robots to move into places with critical situations. The neuron network proposed is suitably capable to correctly evaluate any situation.

In the case that the dynamic control and decision making system will not find the solution for a certain critical situation, then it will try to find the partial solution from several situations. The controller selects the relevant tasks for the agents for these situations. The controller saves or updates the information about the scene state and control in the relevant database which can be used for static control.

3 Decision Agent

The main task of the decision agent (Fig. 2) is to schedule the relevant task. The agent will schedule the tasks in relation to the environment and the internal state of the relevant robot. The module of perception provides the information about the environment.

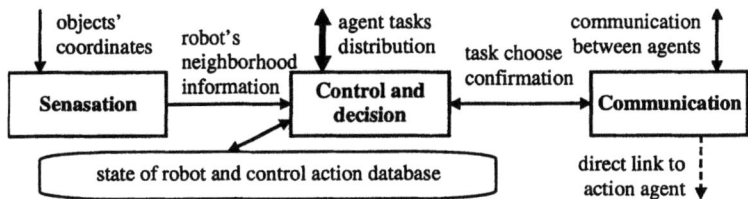

Fig. 2. Decision agent scheme

In the case that more than one agent will select the same task the decision agent must evaluate the probability of the action's success. The agent with the highest probability of success will schedule the actual task. The task selected is handed over to the executive agent which realizes it. The executive agent will receive also the information from cooperating agents for optimization of the robot's movement.

4 Action Agent

The activity of the action agent (Fig. 3) is simple. The agent moves with the robot from an actual position to a new position. The information concerning the actual position is obtained from the sensation module.

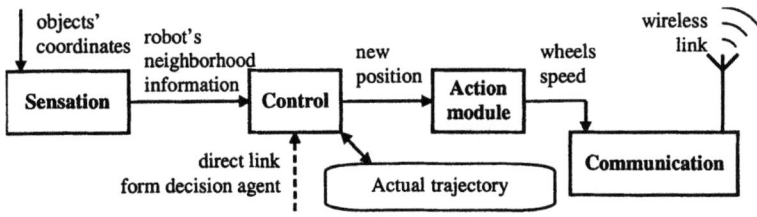

Fig. 3. Action agent scheme

A new position for further moment (for instance another frame of the camera) is evaluated by the control module. The new position is calculated on the basis of a pre-calculated trajectory. The advantage of pre-calculated trajectories is the possibility of the realization of the task of controlling position and time which is useful in avoiding other robots.

After completion of the trajectory calculation the agent will determine the position for next frame (new position) and will transmit it to the action module

5 Strategy Extraction Method

One's own and an opponent's robots create a very dynamically changed environment. This environment is scanned by a CCD camera with a sample frequency (in the present time) up to 50 fps. The picture sample with objects coordinates before processing is demonstrated in Fig.4.

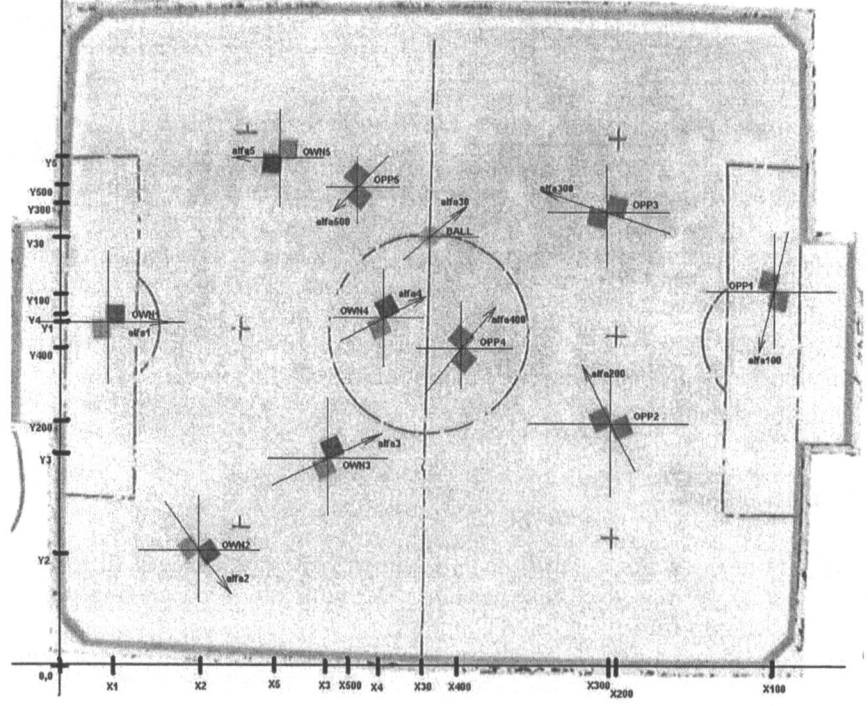

Fig. 4. Picture sample from CCD camera – playground with coordinates of mobile robots

The neuronal net of a control agent in a sensation module process the picture signal and encoded information (position, orientation) saved in one of output vectors. The time sequence of these vectors builds a very wide matrix (up to 420000 vectors - processed samples in the time of a game). This matrix inputs in the proper extraction of the game strategy process by using latent semantic analysis (LSA).

LSA is a statistical model of word usage that permits comparisons of the semantic similarity between pieces of textual information. It was originally designed to improve the effectiveness of information retrieval methods by performing retrieval based on the derived "semantic" content of words in a query, as opposed to performing direct word matching. LSA was used for extracting semantics in many other situations, see [1], [3], [7].

In this paper, LSA is used as a tool for the strategy extraction problem. In our approach, a game is coded as a game matrix.

We can extract vector V_t in time t from game:

$$V_t = \{ \quad t, X_{1;} Y_{1;} \alpha_{1;}$$
$$X_{2;} Y_{2;} \alpha_{2;}$$
$$X_{3;} Y_{3;} \alpha_{3;}$$
$$X_{100;} Y_{100;} \alpha_{100;} \tag{1}$$
$$X_{200;} Y_{200;} \alpha_{200;}$$
$$X_{300;} Y_{300;} \alpha_{00;}$$
$$X_{30;} Y_{30;} \alpha_{30} \quad \}$$

Where are:

$\quad X_i$ is x coordinate of one's own robot i
$\quad Y_i$ is y coordinate of one's own robot i
$\quad \alpha_i$ is angle of orientation of one's own robot i

$\quad X_{i00}$ is x coordinate of an opponent's robot i
$\quad Y_{i00}$ is y coordinate of an opponent's robot i
$\quad \alpha_{00}$ is angle of orientation of an opponent's robot i

$\quad X_{30}$ is x coordinate of the ball
$\quad Y_{30}$ is x coordinate of the ball
$\quad \alpha_{30}$ is angle of motion α ball

A game matrix (in next GM) we can define in the following way:

$$GM = (V_0^T, V_1^T, ..., V_{n1}^T, V_n^T) \tag{2}$$

The results in [1], [7], [8], [9] and [6] indicate that LSA can perform matching based on semantic content. The game matrix is analyzed by LSA and semantic information about game is obtained. This semantic information can be interpreted as a strategy. This strategy is use for for agent management see [10] and [11].

LSA is based on singular value decomposition (SVD). The SVD is commonly used in solution of unconstrained linear least squares problems, matrix rank estimation, and canonical correlation analysis [8]. The singular value decomposition takes a rectangular $m \times n$ matrix A and calculates three matrices U, S, and V. S is a diagonal $m \times n$ matrix (the same dimensions as A). U and V are *unitary* matrices with sizes $m \times m$ and $n \times n$ respectively. The matrices are related by the equation $A = USV^T$. Diagonal elements matrix S are called the singular value of the matrix A. Informally, we can say that greater singular value correspond to greater semantic information.

The algorithms of an opponent's game strategy extraction proceeds parallelly and independently at the main control process. Exploits of vector (matrix) representation *of picture analysis results in going to the extract general elements of an opponent's* game strategy. Knowledge of one's own and an opponent's strategy allows the building of an opponent's game strategy dependent on the decision rules for task scheduling for individual action agents. These rules are gradually saved in the decision rules database by the decision agent. In this way the possibility of temper game state is increased.

6 Software Development

The technology of agents enables the solving of the proposal and implementation of the distributed embedded control system. The embedded systems are used for controlling of mobile robots. For generation of the source code of DSP Motorola processor Metrowerks CodeWarrior can be used. It enables the simulation in real-time and simulation of the hardware.

For development of control software of the distributed system is proposed by the unified modeling language (UML). UML has the corresponding output for servicing the requirements and analysis of models on a higher level of abstraction but it is also connected to the detailed proposal of the conception as threads, multitasking, resource control, etc. [2], [3].

7 Conclusion

The algorithm of the control system should be proposed in a such way so that it would ensure the requirements for the immediate response of control, so that the system with robots would be controlled in real-time. That is why, it is very important so that the algorithm for critical speed would be optimized. The system response should be shorter than the time between two frames from a camera. In the event that this limit is exceeded, the frame is cut out and the control quality is decreased.

The main possibilities of algorithm adjustment are as follows:

- Dynamic control in the control and decision module of a control agent.
- The control and decision modules and communication protocol of the decision agents.
- The strategy of planning in the control model of the action agent.
- Extraction of an opponent's game strategy and using the extraction results for decision rules generation as a part of the rules decision database of a decision agent.

It is necessary to know that the system response should take a shorter time than the time between the frames from the PAL movie camera, e.g. 20 ms. If this limit is exceeded, the frame is dropped and the control quality decreases.

Parallel processing of extraction algorithms of an opponent's game strategy allows adding and refining information of opponent game strategy. This way expands the rule database of a decision agent for decision making within the bounds of one's own game strategy given by a control agent.

Acknowledgement. The Ministry of Education of Czech Republic supplied the results of the project CEZ: J17/98:272400013 with subvention. The Grant Agency of Czech Republic supplied the results of the project 102/02/1032 with subvention.

References

1. Berry, M. W., Browne, M.: Understanding Search Engines: Mathematical Modeling and Text Retrieval. SIAM Book Series: Software, Environments, and Tools, (June 1999), ISBN: 0-89871-437-0.
2. Adam G.K., Grant E. and Adam K.: Qualitative Modelling and Control of Industrial Processes, in Proceedings of the IASTED International Conference on Modelling and Simulation (MS'2000), Acta Press, ISBN: 0-88986-284-2, pages 477-482, May 15-17, 2000, Pittsburgh, Pennsylvania, USA.
3. Douglas B.P.: The UML for system engineering, I-Logix Rhapsody. White-paper, USA 2000
4. Bernatik,R., Horak,B., Kovar,P.: Quick image recognize algorithms. In: Proceeding International workshop Robot-Multi-Agent-Systems R-MAS 2001. VSB Ostrava 2001, Czech Republic, ISBN 80-7078-901-8 p.53-58
5. Srovnal V., Pavliska,A.: Robot Control Using UML and Multi-agent System. In: Proceeding 6th World Multiconference SCI 2002. Orlando, Florida, USA, ISBN 980-07-8150-1, p.306-311
6. Húsek D., Frolov A. A., Řezanková H., Snášel V.: Application of Hopfield-like Neural Networks to Nonlinear Factorization. COMPSTAT 2002, Proceedings in Computational Statistics (Eds.: Härdle W., Rönz B.), 177-182. Physica-Verlag, Heidelberg 2002. ISBN 3-7908-1517-9.
7. Berry, M. W. (Ed.): Survey of Text Mining: Clustering Classification, and Retrieval. Springer Verlag 2003.
8. Praks P., Dvorský J., Snášel V.: Latent Semantic Indexing for Image Retrieval Systems. SIAM Conference on Applied Linear Algebra (LA03) The College of William and Mary, Williamsburg, U.S.A. 2003.
9. Praks P., Dvorský J., Snášel V., Černohorský J.: On SVD-free Latent Semantic Indexing for Image Retrieval for application in a hard industrial environment. IEEE International Conference on Industrial Technology - ICIT'03, Maribor 2003.
10. Smid, J., Obitko, M., Snášel, V.: Communicating Agents and Property-Based Types versus Objects. Sofsem MatfyzPress 2004.
11. Obitko, M., Snášel, V.: Ontology Repository in Multi-Agent System. IASTED, International Conference on Artificial Intelligence and Applications (AIA 2004), Innsbruck, Austria.

Multi-agent Environment for Dynamic Transport Planning and Scheduling

Jaroslaw Kozlak[1,2], Jean-Charles Créput[1], Vincent Hilaire[1], and Abder Koukam[1]

[1] UTBM Systems and Transportation Laboratory,
University of Technology of Belfort-Montbéliard, Belfort, France
[2] Department of Computer Science, AGH University of Science and Technology,
Al. Mickiewicza 30, Kraków, Poland
kozlak@agh.edu.pl,
{Jean-Charles.Creput,Vincent.Hilaire,Abder.Koukam}@utbm.fr

Abstract. Nowadays, transport requests of the society are becoming more and more common and important. Computer systems may be a useful tool for transport companies. This paper is aimed at presenting a system which simulates the functioning of a transport company. The theoretical problem, which needs to be solved, is a dynamic Pickup and Delivery Problem with Time Windows and capacity constraints (PDPTW). The nature of the problem, its distribution and the possibility of using a lot of autonomous planning modules, predestines to apply a multi-agent approach. Multi-agent approach facilitates the introduction of elements which do not appear in classical PDPTW as company organisation, different strategies of requests acceptation by different vehicles or communication among vehicles.

Keywords: Multi-agent systems, dynamic transport planning and scheduling.

1 Introduction

Nowadays, the importance and commonness of transport requests in the society increases. The transport services offered are more and more advanced and better designed to meet the needs of users. Important parts of costs of many goods are transport costs. The size of the market of transport services and the sales volume and derived profits intensify a competition. Therefore, a reduction of costs, better adaptation of strategy to the demand as well as better planning and scheduling of the use of available resources are important for transport companies. Computer systems may be a useful tool for transport companies. On the one hand, they may support a rapid creation of effective transport plans and schedules. On the other, they enable simulation research leading to the correct selection of company organization, and their resources like vehicles, their capacities, and their location on the depots. We present a type of transport problem which have important practical applications: the Dynamic Pickup and Delivery Problem with Time Windows (PDPTW).

Each request is described by two locations: one for pickup and one for delivery and two time windows: the time window when a pickup operation can be started and the one when a delivery operation may be started. Both the request points: pickup and delivery places should be visited by the same vehicle in proper order. There is also a

M. Bubak et al. (Eds.): ICCS 2004, LNCS 3038, pp. 638–645, 2004.
© Springer-Verlag Berlin Heidelberg 2004

possibility of introducing an additional capacity constraint: each request is characterised by a load, and each vehicle has a maximum capacity which cannot be exceeded by the total load.

PDPTW is worth exploring because of its practical applications, like the transport of elderly and handicapped people, shared taxi or microbus company, sealift and airlift, school bus routing and scheduling. Additionally, PDPTW is a problem relatively weakly explored, especially in comparison to VRPTW.

The goal of this paper is to present a system which simulates the functioning of a transport company. The basic problem, which needs to be solved, is a dynamic version pf Pickup and Delivery problem with Time Windows and capacity constraints, where requests arrive during system running. The nature of the problems - their distribution and possibility of applying a lot of autonomous planning modules - predestines to use a multi-agent approach. Multi-agent approach facilitates to consider aspects which do not appear in classical PDPTW as company organisation, different strategies of requests acceptation by different vehicles or communication among vehicles.

The structure of the paper is as follows: Section 2 contains research overview, we put the emphasis on the heuristic methods to solve pickup and delivery problem and description of multi-agent systems for transport planning and scheduling. In Section 3, main features and advantages of our approach are given. In Section 4, the model of our system is described. Section 5 presents goals and configurations of performed experiences as well as obtained results. Section 6 concludes and presents plans of future works.

2 Research Overview

2.1 Heuristic Methods of Plan and Schedule Construction

An overview of optimal and heuristic methods of pickup and delivery problem solving may be found in [3, 4, 14]. An optimal, strict solution - because of its calculation complexity - is possible to obtain only for small size problems. Heuristic solutions for PDPTW are performed using methods like tabu search [10, 11, 15] simulated annealing [11], squeaky wheel optimization with local search [12].

Very few papers focus on dynamic aspects of PDPTW problem. Usually, the algorithms are similar to the ones used for static problems, and it is assumed that incoming of new requests interrupts the optimization function which is then restarted with a new set of requests. In [7], the authors describe a method for determining times of requests generation on the basis on Poisson distribution. Space is divided into regions, and each of them has its own probability that requests pickup or delivery points will be located on it. Basing upon these parameters, the points of pickup and delivery are determined. Times in the time windows are calculated in a way to make possible that the vehicle returns to depot before the end of the simulation time. For that purpose, the times of driving between pickup and delivery points and between delivery points and a depot are taken into consideration. In [13] authors describe REBUS system, which solves real-time dynamic PDPTW. The system was designed for transporting elderly or disabled persons in Copenhagen.

2.2 Multi-agent Systems

Multi-agent approach to transport problems enables the analysis of complex, heterogeneous and dynamic systems, and focuses mostly on complex cargo shipping problems (sometimes taking into consideration transshipments and transport multi-modality [2]) but some systems are also targeted at transport on demand problem [8].

In [1], a method of simulated trading was presented and applied for solving vehicle routing problem. Several processes are executed concurrently and each of them represents one route. There is also one process which manages the stock of unattached requests. The idea is based on buying and selling requests by processes representing routes. The process tries to sell requests which have bad influence especially on its cost function, and to buy these which do not increase it appreciably. A graph, representing the flow of requests between routes, is built and serves for the creation of feasible and efficient plans (trading matching).

MARS [6] is a system realized to simulate planning and scheduling for a society of shipping companies using multi-agent approach. The agents represent transportation companies and trucks. Protocol Contract Net [17] is used to assign requests to particular vehicles (company is a manager and truck-contractor). Simulated trading procedure is used for dynamic re-planning (if because of unexpected events – the traffic jams – the vehicle is not able to realize a request) or for optimization of the current solution by searching to a better one.

On the basis of experience, gathered using simulated trading and MARS, a distributed multi-agent system for shipping company planning and scheduling - TeleTruck [2] was realized and practically tested. The architecture of TeleTruck is based on the concept of holonic agents. In this system, there exists different kinds of agent-components (driver, truck, trailer, container), which possesses and manages specific type of resources. This approach gives a possibility of considering different aspects of transport problem (storing, transportation, management of cargos, drivers, trucks etc.).

3 System Aims and Methodology

Our goal is to create a system which makes possible the simulation of transport company. Transport requests, realized by a company, should suit dynamic PDPTW with capacity constraints. It is assumed that only feasible requests are accepted for performance. If the company is not able to realize a request respecting all constraints needed, such a request should be rejected. Besides, the system should simulate the network of communication connections (routes) and manage the features of scheduling process for each vehicle. A vehicle should have its own conditions of acceptance and estimation of a request.

Agent approach makes possible the development of systems consisting of many autonomous elements which are able to create plans, choose actions for realisation in a certain way, to perform its goals. Because of different locations of vehicles and of the many independent decisions they have to take, the transport planning and scheduling problem may be considered as an example of a problem with a distributed domain, which is very suitable for multi-agent approach [5].

Agent approach makes possible to take into consideration autonomous , goal-driven agents, which represent a company and vehicles. Each agent-vehicle manages its route. An agent estimates a request taking into consideration if it is feasible, which payment it will obtain and what expenses it will bear. A transport network may be represented as a directed graph. Agents may be equipped with message communication modules which gives a possibility of direct interactions between vehicles and drivers.

4 System Model

A model of multi-agent system for transport planning is composed of the following elements:
- *environment*: it represents a transport network and it is implemented through a graph describing communication connections,
- *several kinds of agents: customer agent* - responsible for the generation of transport requests, *agent-company* - representing a transport company and *agent-vehicle* - representing a vehicle of transport company, like mini-bus, bus, taxi etc.

4.1 Specification of System Organisation

We base our work on the RIO framework previously defined in [9]. This framework is based on three interrelated concepts : role, interaction and organization. Roles are generic behaviors. These behaviors can interact mutually according to interaction pattern. Such a pattern with groups generic behaviors and their interactions constitutes an organization. Organizations are thus descriptions of coordinated structures. Coordination occurs between roles as and when interaction takes place.

In this context, an agent is only specified as an active communicative entity which plays roles. In fact agents instantiate an organization (roles and interactions) when they exhibit behaviors defined by the organization's roles and when they interact following the organization interactions. An agent may instantiate dynamically one or more roles and a role may be instantiated by one or more agents.

The organization of the MAS is described in Fig. 1. There are two organizations: one specifying the Contract Net protocol and one specifying the interactions between clients and pick-up and delivery service providers. The latter organization specifies

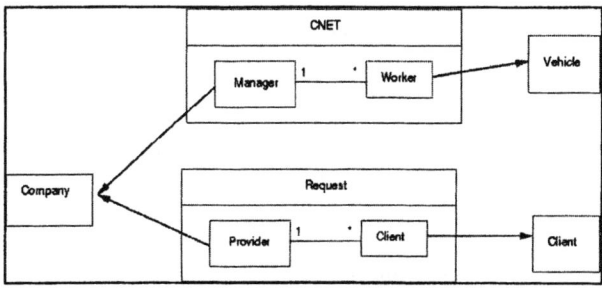

Fig. 1. System organization

two roles : Provider and Client. The Client role represents virtual clients that send request of pick-up and delivery to a virtual Provider that accept these requests and answer with offers. The Provider role is played by the Company agent. The former organization is composed of two roles : Manager and Worker. This organization specifies a contract net based negotiation between a Manager which propose pick-up and delivery requests and some Worker which are able to realize these requests. The Manager role is played by the Company agent and Worker role are played by Vehicle agents.

4.2 Model Elements Description

Environment. The transport network is represented as a directed graph $TN(N, E)$, where N is a set of nodes and E is a set of arcs, respectively. Nodes represent the locations. They may be either a start or destination point of the goods transfer. With each node V_i are associated the following parameters: a pair (x_i, y_i), where x_i and y_i are coordinates on the map, numerical value describing the probability that this node will appear in the transport request as pickup or delivery point (w_i), list of requests waiting for pickup from this node (pl_i). Each arc E_i is described by time period tp_i, which informs how much time the vehicle needs to traverse it. Value tp_i expresses also the cost of travel carried out by a vehicle.

The following types of agents exist in the system: agent-customer, agent-vehicle, agent-company.

Agent-customer. Agent-customer is responsible for generation of random events. The methods of events generation is inspired by [7]. Agent Customer sends them later to agent-company. Agent Customer AC is described by a list of periods of requests generations (prg_i), each period is characterized by: γ_i - coefficient of Poisson distribution describing the frequency of request generation, β_i - probability that time window of pickup starts just after request arrival, δ – an expected size of time windows.

Agent-company. Agent-company – is responsible for request reception from agent-customer and its scheduling to agent-vehicles. Agent-company AC_i is represented as a tuple (g, c, s, f, RL), where: g – gains (incomes-costs), c – costs, s – an average level of satisfaction of clients, which should be guaranteed, f – a function of choice of agent-vehicle, which will perform the request, RL – a list of received requests with information concerning the state of their realization (received, rejected, scheduled to agent, pickup performed, delivery performed).

Agent-vehicle. Agent-vehicle represents vehicle. It moves among nodes in the transport network and possesses plans of its routes and schedules of request realization. To obtain realization of request, it sends offers to agent-company. An Agent-vehicle AV_i is defined as a tuple $(loc_i, g_i, c_i, cap_i, pass_i, LRN_i, R_i, LO_i)$, where: loc_i - current location (node, direction, percentage of traversed arc), g_i - current gain, c_i - current costs, cap_i - maximal capacity, $pass_i$ - number of passengers, LRN_i - list of request nodes, R_i - list of nodes (route), LO_i - list of embarked orders (after pickup and before delivery).

Agents-vehicles plan their routes so as to visit all nodes of accepted transport requests. Each agents-vehicle uses two forms of route representation: *list of request nodes* and *list of nodes*. The list of request nodes is composed of pickup and delivery points for each accepted request ordered according to vehicle time of visit (thus each request is represented in the list by two points. On the basis of request nodes, the route composed of all traversed nodes (list of nodes) is constructed

5 Experiments

The goal of performed experiments is the choice of optimum company features (number of vehicles, capacity of vehicles) for the given topology of communication network as well as the quantity and features of requests. In particular, we will examine configurations with different time windows size and request frequencies.

In the experiments performed, the transport network consisting of 25 nodes and 60 arcs were used. Average length of arc is 7 minutes and 25 seconds.

Quantities of vehicles, their capacities as well as sizes of time windows were changing during experiments.The total simulation time was divided into 5 intervals (4 simulation hours each) with different frequency of request arrival. Their Poison distribution λ parameters were equal to 0.55, 0.70, 0.1, 0.4, 0.1, as in one of the experiments described in [7], and the sixth supplementary period of 1 hour, aimed at allowing the realization of the last requests . The average value of generated requests was equal to 360.

The following diagrams present the percentages of performed requests in relation to the quantity of the vehicles used and their capacities. In Fig. 2a are shown the results obtained for wide time windows (end point is calculated using even distribution, when the last possible time is the simulation end), whereas the ones presented in Fig.2b. concern narrow ones (equal 10 minutes each).

Basing upon Fig. 2a one can notice that an increase of rate of request realization of vehicle is relatively small. For this configuration of requests and transport networks, the vehicles rarely have a total load higher then 50, a stronger constraint is the

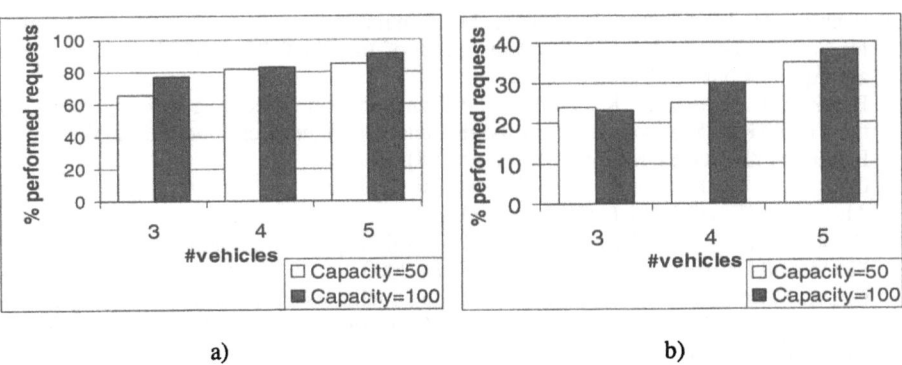

a)   b)

Fig. 2. Request realization in relation with vehicles quantity and capacity for a problem with a) wide time windows, b) thin time windows

participation of a vehicle in the realization of the requests previously submitted. In the case of wide time window, the increase in the quantity of vehicles improves the percentage of the performed requests.

For narrow time windows (Fig2b), the percentage of the request performed for the same number of vehicles and their capacities, is clearly lower then for wide ones. For the number of vehicles in the period of values 3-6, the difference is equal to 2-3. Similarly as in the case of wide windows, increase of the percentage of performed requests when the maximal capacity of vehicles is two times higher, is relatively small.

6 Conclusions

In this paper we have presented a simulation environment to solve dynamic transport problems of type PDPTW. It seems to us that the main advantages of the proposed approach to the construction of systems to transport planning is the possibility of adding emergencies, flexibility of configuration and change of features of particular vehicles, as well as taking the structure of transport organization and its policy into consideration during the optimization process.

We intend to enrich the multi-agent environment by introducing the following elements to the system, which may consist of taking into consideration several cooperating transport companies, allowing transport requests to arrive from different sources, also directly to a vehicle which is located in a node where pickup is to be realized and developing direct interactions among vehicles and request exchange between them. The another step is the implementation of the model of multi-agent systems presented in the paper by using a standard multi-agent platform, and through the development of a decentralised multi-agent system so as to improve the global computation calculation performance.

References

1. Bachem, A., Hochstattler, W. Malich, M.: Simulated Trading A New Parallel Approach for Solving Vehicle Routing Problems. Proc. of Int. Conf. "Parallel Computing: Trends and Applications", 1994
2. Burckert, H.-J., Fischer, K., Vierke., G. : Transportation scheduling with holonic MAS - the TELETRUCK approach. Third International Conference on Practical Applications of Intelligent Agents and Multiagents (PAAM 98), 1998
3. Cordeau, J.-F., Laporte G.: The Dial-a-Ride Problem: Variants, Modeling Issues and Algorithms. Les Cahiers du GERAD, 2002
4. Desaulniers, G., Desrosiers, J., Erdmann, A., Solomon, M.M., Soumis, F.: The VRP with Pickup and Delivery. *Les Cahiers du GERAD*, 2000
5. Ferber, J.: Les systèmes multi-agents: Vers une intelligence collective, InterEditions, 1995
6. Fischer, K., Müller, J. P., Pischel, M.: Cooperative Transportation Scheduling: an Application Domain for DAI. *Applied Artificial Intelligence*, vol.10, 1996
7. Gendreau, A., Guertin, F. , Potvin, J.Y., Séguin, R. :Neighborhood search heuristics for a dynamic vehicle dispatching problem with pick-ups and deliveries. *Rapport technique CRT-98-10*, Université de Montréal (1998)

8.   Gruer, P., Hilaire, V., Kozlak, J., Koukam, A.: A multi-agent approach to modelling and simulation of transport on demand problem. J. Sołdek , L. Drobiazgiewicz (eds) : Artificial Intelligence and Security in Computing Systems, The Kluwer Int. Series In Eng. And Computer Science - 752, 2003

9.   Hilaire, V.: Vers une approche de specification, de prototypage et de verification de Systèmes Multi-Agents. PhD thesis, UTBM, 2000

10. Lau H.C., Liang Z.: Pickup and Delivery with Time Windows : Algorithms and Test Case Generation. Proceeedings of 13th IEEE International Conference on Tools with Artificial Intelligence (ICTAI'01), Dallas, USA, 2001

11. Li, H., Lim A.: A Metaheuristic for the Pickup and Delivery Problem with Time Windows. in *Proceedings of 13th IEEE International Conference on Tools with Artificial Intelligence (ICTAI'01)*, Dallas, USA, 2001

12. Lim, H. , Lim, A., Rodrigues, B.: Solving the Pick up and Delivery Problem using ''Squeaky Wheel'' Optimization with Local Search. Proceedings of American Conference on Information Systems, AMCIS 2002, Dallas, USA

13. Madsen, O.B.G., Ravn, H.F., Rygaard, J.M.: A heuristic algorithm for a dial-a-ride problem with time windows, multiple capacities, and multiple objectives. *Annals of Operations Research* 60 (1995)

14. Mitrowic-Minic, S.: Pickup and Delivery Problem with Time Windows: A Survey. SFU CMPT TR. 1998-12, `ftp://fas.sfu.ca/pub/cs/techreports/1998`

15. Nanry, W.P., Barnes, J.W.: Solving the pickup and delivery problem with time windows using reactive tabu search. Transportation Research Part B 34, Elsevier Science Ltd 2000,

16. Potvin, J. Y., Bengio, S.: The vehicle routing problem with time windows - Part II: Genetic search., INFORMS Journal on Computing 8 (1996)

Agent-Based Models and Platforms for Parallel Evolutionary Algorithms*

Marek Kisiel-Dorohinicki

Institute of Computer Science
AGH University of Science and Technology, Kraków, Poland
doroh@agh.edu.pl

Abstract. The goal of the paper is to provide an overview of classical and agent-based models of parallel evolutionary algorithms. Agent approach reveals possibilities of unification of various models and thus allows for the development of platforms supporting the implementation of different PEA variants. Design considerations based on *AgWorld* and *Ant.NET* projects conclude the paper.

Keywords: Parallel evolutionary algorithms, evolution in multi-agent systems.

1 Introduction

Today *evolutionary algorithms* (EA) are used for more and more complex problems requiring large populations and long computational time (cf. e.g. [2]). Parallel implementations seem to be a promising answer to this problem, especially that evolutionary processes are highly parallel by nature. What is more, it turns out that some parallel models of evolutionary computation are able to provide even better solutions than comparably sized classical evolutionary algorithms — considering not only the quality of obtained solutions and convergence rate, but first of all the convergence reliability [4].

In the first part of the paper classical models of *parallel evolutionary algorithms* (PEA) are discussed. Then agent approach is proposed as a means to develop a unified model covering different variants of PEA. Two agent-based architectures of distributed evolutionary computation illustrate this idea. The paper ends with a short presentation of *AgWorld* and *Ant.NET* projects that aid realisation of both classical and agent-based models of PEA.

2 Classical Models of Parallel Evolutionary Algorithms

Evolutionary algorithms, as an abstraction of natural evolutionary processes, are apparently easy to parallelize and many models of their parallel implementations have been proposed [1,4]. The standard approach (sometimes called a *global parallelisation*) consists in distributing selected steps of the sequential algorithm among several processing

---

* This work was partially sponsored by State Committee for Scientific Research (KBN) grant no. 4 T11C 027 22.

M. Bubak et al. (Eds.): ICCS 2004, LNCS 3038, pp. 646–653, 2004.
© Springer-Verlag Berlin Heidelberg 2004

units. In this case a population is unstructured (*panmictic*) and both selection and mating are global (performed over the whole population).

Decomposition approaches are characterised by non-global selection/mating and introduce some spatial structure of a population. In a *coarse-grained* PEA (also known as *regional* or *multiple-deme* model) a population is divided into several subpopulations (regions, demes). In this model selection and mating are limited to individuals inhabiting one region and a migration operator is used to move (copy) selected individuals from one region to another. In a *fine-grained* PEA (also called a *cellular* model) a population is divided into a large number of small subpopulations with some neighbourhood structure. Here selection and mating are performed in the local neighbourhood (overlapping subpopulations). It is even possible to have only one individual in each subpopulation (this is sometimes called a *massively* parallel evolutionary algorithm).

And finally there are also methods which utilise some combination of the models described above, or even a combination of several instances the same model but with different parameters (*hybrid* PEAs).

2.1 Global Parallelisation

In PEA with a global population selected steps of the sequential algorithm (mostly evaluation of individuals) are implemented in a distributed or multiprocessor environment.

In a *master-slave* model (fig. 1) one processing unit (*master*) is responsible for management of the algorithm and distribution of tasks to other processing units (*slaves*). This often consists in sending selected individuals to slaves to perform some operations (e.g. calculate fitness). In a *sequential selection* model a master processing unit waits for finishing computation in all slave nodes so there is a clear distinction between successive generations. This makes the implementation simple, yet the cost of idle time of slaves is a potential bottleneck of this method. In a *parallel selection* model a master processing unit does not wait for all slaves but when one finishes the work it is immediately allocated a new task. In this case selection may be done in several variants, for example in form of a tournament (so one needs to know only a subset of fitness values).

Conversely, in a system with *shared memory* all processors have access to all individuals. This approach has all advantages of the master-slave one without its handicap, yet requires some kind of control over possibly simultaneous operations on individuals.

2.2 Regional Models

Regional models introduce coarse-grained spatial structure of a population, suitable for implementation in a distributed architecture. Each node has its own (sub)population and runs a separate thread of evolution (fig. 2), thus conceptually subpopulations 'live' on geographically separated regions [9]. A new operator – *migration* – controls the process of exchanging individuals between regions. The model is usually described by a few parameters: a number of regions, a number of individuals in each region, as well as migration topology, rate/interval and a strategy of choosing individuals to migrate.

Migration topology describes how individuals migrate from one region to another. This often depends on software architecture and the most common are hypercube, ring,

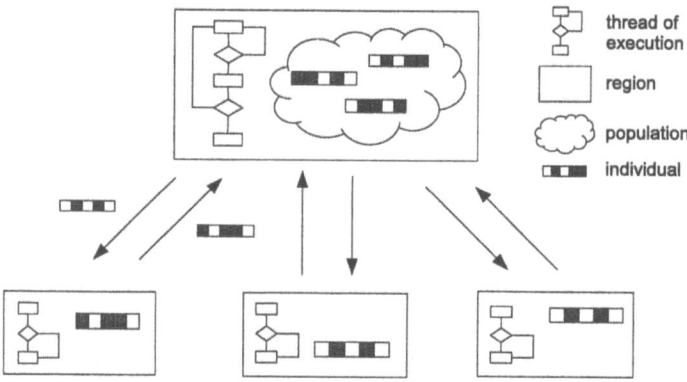

Fig. 1. A master-slave model of a globally parallel PEA

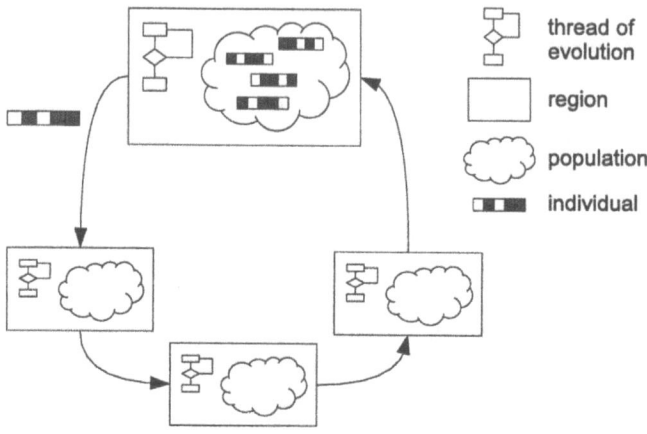

Fig. 2. A regional PEA – arrows indicate possible migration in stepping stone topology

or k-clique. In an *island* model individuals can migrate to any other subpopulation, while in a *stepping stone* model individuals can migrate only to neighbouring region(s). Migration rate and interval denote how many and how often individuals migrate. Of course migration rate should be greater if migration interval is longer. Typically the best individuals are chosen for migration and immigrants replace the worst individuals in a destination region. Other possibility is that immigrants replace the most similar individuals (e.g. using Hamming distance as a measure) or just replace emigrants.

2.3   Cellular Models

In cellular PEA a population has a fine-grained structure with neighbourhood relation defined [10]. In this case local selection/mating is performed over neighbouring individuals — subpopulations are not geographically separated but rather overlap and consist

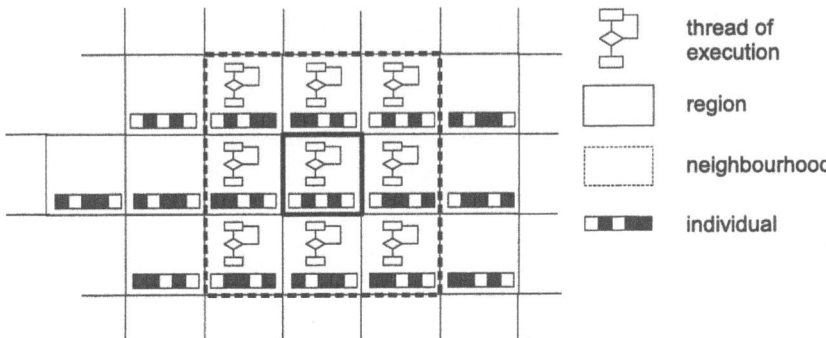

Fig. 3. A cellular PEA with local neighbourhood of central individual marked

of a few individuals (even only one individual). Usually the neighbourhood structure is a grid (each individual has a fixed number of neighbours) with a torus topology. A parameter is needed to describe a neighbourhood radius, which defines a direct neighbourhood of an individual where selection and mating is performed.

This model is strongly related to massively parallel computing. In such cases usually each processor contains one individual and is responsible for calculating its fitness and executing genetic operators. Neighbourhood topology is then determined by MPC architecture.

3   Agents for Parallel Evolutionary Computation

Designing complex evolutionary computation systems (e.g. consisting of a huge number of individuals or with spatially structured hybrid subpopulations) requires considering several aspects, that strongly affect both ease of building/modifying/tuning and computational properties (mainly efficiency). Recognizing common structure and identifying core functionality of such systems allows for development of universal software tools and libraries (platforms) facilitating their realisation. It seems that agent-based architectures may help a lot in this difficult task.

3.1   PEA **Common Organisation**

It seems that all (or almost all) classical and hybrid models of parallel evolutionary algorithms may be modelled using the following entities:

Environment – represents the whole system, main logical or physical aggregating unit used to directly manage regions, flocks or even individuals.

Region – plays role of a distribution unit, immobile entity, which may contain different kinds resources and may be used to directly manage flocks or individuals.

Flock – a unit of migration, mobile aggregating entity, which may be used to manage individuals.

Individual – an evolution unit and thus a basic element of the computation.

This structural model with four levels of organisation may be easily illustrated as a hierarchy of elements. Of course not every level of organisation is present in each particular technique or its purpose may be different, e.g.:

- in a *coarse-grained* PEA only **Regions** and **Individuals** are present,
- in a *fine-grained* PEA there is a geographically-structured **Environment**, and **Individuals**.

The hierarchical structure of hybrid PEAs can be organised in several ways revealing different possibilities of parallelisation.

3.2 Agent Systems and Evolutionary Computation

During the last decade the idea of an intelligent autonomous agent – a software entity, which is situated in some environment and autonomously acts on it so as to satisfy its own goals – gains more and more interest, and thus software agents are used in various domains [6]. At first sight evolutionary computation and agent approach seem to have nothing in common – the former is a search and optimisation technique, while the latter is a general concept of modelling decentralised systems. Yet one may notice that evolutionary processes are decentralised by nature and indeed multi-agent systems turn out to be a perfect tool for modelling them [7].

For each classical model of PEA alone the use of agents seems hardly advantageous – as an autonomous active entity, an agent may be identified with a region (coarse-grained PEA) or cell (fine-grained PEA). But considering a platform covering a wide range of different PEA models, agents may allow for a great deal of unification in the modelling scheme, no matter whether they would represent individuals or populations of them [3]. Depending on this decision two different agent-based architectures of evolutionary computation systems may be distinguished: in an evolutionary multi-agent system an agent represents a single individual, in a flock-based one an agent manages a group of individuals inhabiting one region (a flock). Of course in both approaches it is possible to realise classical models of PEA, but their strengths are manifested only for more complicated cases, e.g. for hybrid soft computing systems (cf. [8]).

3.3 Evolutionary Multi-agent Systems

In EMAS phenomena of *death* and *reproduction*, crucial for existing of inheritance and selection – the main components of evolution process, are modelled as agent actions:

- action of death results in the elimination of the agent from the system,
- action of reproduction is simply the production of a new agent from its parent(s).

Inheritance is accomplished by an appropriate definition of reproduction, like in classical evolutionary algorithms. Core properties of the agent (genotype) are inherited from its parent(s) – with the use of mutation and recombination. Besides, the agent may possess some knowledge acquired during its life, which is not inherited. Both the inherited and acquired information determines the behaviour of the agent in the system (phenotype).

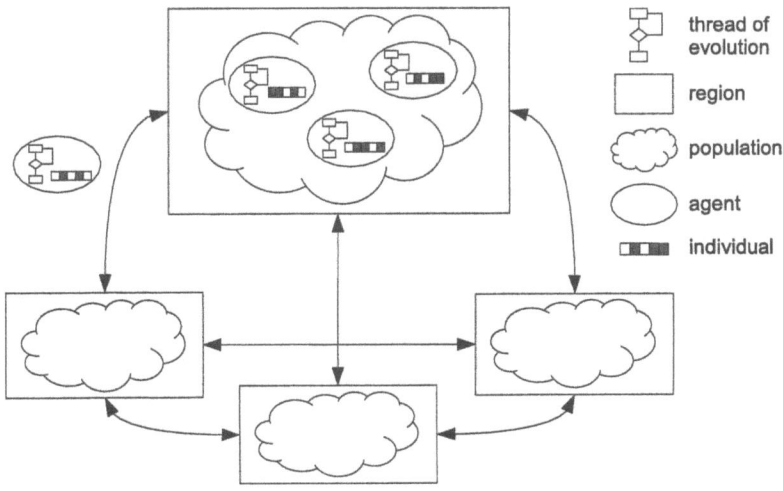

Fig. 4. Evolutionary multi-agent system

Selection is the most important and most difficult element of the model of evolution employed in EMAS [7]. This is due to assumed lack of global knowledge (which makes it impossible to evaluate all individuals at the same time) and autonomy of agents (which causes that reproduction is achieved asynchronously). The proposed principle of selection is based on the existence of non-renewable resource called *life energy*. The energy is gained and lost when the agent executes actions in the environment. Increase in energy is a reward for 'good' behaviour of the agent, decrease – a penalty for 'bad' behaviour (of course which behaviour is considered 'good' or 'bad' depends on the particular problem to be solved). At the same time the level of energy determines actions the agent is able to execute. In particular low energy level should increase possibility of death and high energy level should increase possibility of reproduction.

3.4   Flock-Based Multi-agent Model of Evolutionary Computation

A flock-based multi-agent system (FMAS) extends an island model of PEA providing additional organisational level. Subpopulations on the islands are divided into flocks, where independent processes of evolution (e.g. some classical sequential EA) are managed by agents. In such an architecture it is possible to distinguish two levels of migration, just like in *dual individual* distributed genetic algorithm [5]:

- exchange of individuals between flocks on one island,
- migration of flocks between islands.

Also merging of flocks containing similar individuals or dividing of flocks with large diversity allows for dynamic changes of population structure to possibly well reflect the problem to be solved (the shape of fitness function, e.g. location of local extrema). (Self-)management of agents-flocks may be realised with the use of non-renewable resources, just like selection in EMAS.

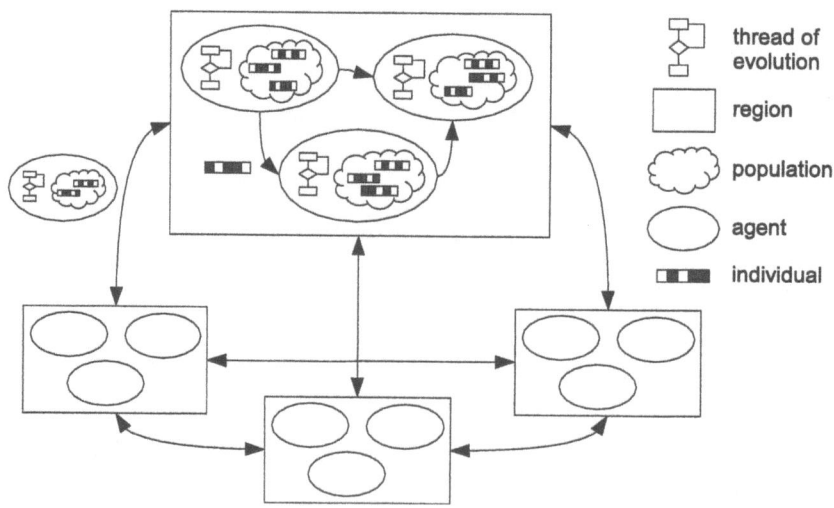

Fig. 5. Flock-based approach to multi-agent evolutionary computation

3.5   Agent-Based Platforms Supporting PEA Implementation

Based on the shortly presented models two software frameworks (platforms) were designed so as to aid realisation of various PEA applications in different domains [3]. Both platforms have similar logical organisation, yet differ in the technology used. The *AgWorld* project is based on parallel computation paradigm (PVM – Parallel Virtual Machine), whereas the *Ant.NET* project utilises software components (.NET) technology. Of course each platform (and technology) has its advantages and shortcomings.

Using *AgWorld* platform efficient parallel computation systems can be created, where groups of agents are placed in different tasks that can be run on individual computing nodes. Large systems, especially in a coarse-grained structure, can benefit much from using PVM this way.

Ant.NET proves useful for implementing applications requiring the use of sophisticated structures and dependencies between agents. The proposed architecture assumes that different aspects of the system (e.g. different selection or reproduction mechanisms) are implemented as software components that can be easily switched to change the behaviour of the whole system, or even a part of the system. Thus *Ant.NET* should become a good choice for implementing rather simulation systems, while efficient parallel computation requires more effort on .NET platform.

4   Concluding Remarks

Parallelisation of evolutionary algorithms is a conceptually easy task, since evolution is a parallel process by nature. Thus one may find lots of different parallel implementations of evolutionary computation paradigm, but still a unified model of parallel evolution and a universal platform allowing for comparison of all models is lacking. In the paper agent approach was proposed as a possible solution to this problem – providing a unified model of evolution that became a base for development of two software platforms: *AgWorld* and *Ant.NET*.

Agent-based evolutionary computation also provides a more complex model of evolution and thus closer to its natural prototype. It should enable the following:

- local selection allows for intensive exploration of the search space, which is similar to classical parallel evolutionary algorithms,
- activity of an agent (individual phenotype in EMAS or subpopulation behaviour in FMAS) depends on its interaction with the environment,
- self-adaptation of the population size is possible when appropriate selection mechanisms are used.

However it still remains an open question whether agent-based models are more powerful then the these of classical (parallel) evolutionary algorithms considering their computational properties. This would be a subject of further research.

References

1. P. Adamidis. Parallel evolutionary algorithms: a review. In *Proc. of the 4th Hellenic-European Conference on Computer Mathematics and its Applications*, 1998.
2. T. Bäck, D. B. Fogel, and Z. Michalewicz, editors. *Handbook of Evolutionary Computation*. Institute of Physics Publishing and Oxford University Press, 1997.
3. A. Byrski, L. Siwik, and M. Kisiel-Dorohinicki. Designing population-structured evolutionary computation systems. In T. Burczyński, W. Cholewa, and W. Moczulski, editors, *Methods of Artificial Intelligence (AI-METH 2003)*. Silesian Univ. of Technology, Gliwice, Poland, 2003.
4. E. Cantú-Paz. A summary of research on parallel genetic algorithms. *IlliGAL Report No. 95007. University of Illinois*, 1995.
5. T. Hiroyasu, M. Miki, M. Hamasaki, and Y. Tabimura. A new model of distributed genetic algorithm for cluster systems: Dual individual DGA. In *Proc. of the Third International Symposium on High Performance Computing*, 2000.
6. N. R. Jennings, K. Sycara, and M. Wooldridge. A roadmap of agent research and development. *Journal of Autonomous Agents and Multi-Agent Systems*, 1(1):7–38, 1998.
7. M. Kisiel-Dorohinicki. Agent-oriented model of simulated evolution. In W. I. Grosky and F. Plasil, editors, *SofSem 2002: Theory and Practice of Informatics*, Lecture Notes in Computer Science. Springer-Verlag, 2002.
8. M. Kisiel-Dorohinicki, G. Dobrowolski, and E. Nawarecki. Agent populations as computational intelligence. In L. Rutkowski and J. Kacprzyk, editors, *Neural Networks and Soft Computing*, Advances in Soft Computing. Physica-Verlag, 2003.
9. W. Martin, J. Lienig, and J. Cohoon. *Island (migration) models: evolutionary algorithms based on punctuated equilibria*, chapter C6.3. In Bäck et al. [2], 1997.
10. C. Pettey. *Diffusion (cellular) models*, chapter C6.4. In Bäck et al. [2], 1997.

A Co-evolutionary Multi-agent System for Multi-modal Function Optimization

Rafał Dreżewski

Department of Computer Science
AGH University of Science and Technology, Kraków, Poland
`drezew@agh.edu.pl`

Abstract. *Niching methods* for evolutionary algorithms are used in order to locate all desired peaks of multi-modal landscape. Co-evolutionary techniques are aimed at overcoming limited adaptive capacity of evolutionary algorithms resulting from the loss of useful diversity of population. In this paper the idea of *niching co-evolutionary multi-agent system* (NCoEMAS) is introduced. In such a system the niche formation phenomena occurs within one of the preexisting species as a result of co-evolutionary interactions. Also, results from runs of NCoEMAS against Rastrigin function and the comparison to other niching techniques are presented.

1 Introduction

Terms *Evolutionary Computation (EC)* and *Evolutionary Algorithms (EAs)* cover a wide range of global search and optimization techniques based on analogies to natural evolution. Evolutionary algorithms (EAs) have demonstrated in practice efficiency and robustness as global optimization techniques. However, they often suffer from premature loss of population diversity what results in premature convergence and may lead to locating local optima instead of a global one. What is more, both the experiments and analysis show that for multi-modal problem landscapes a simple EA will inevitably locate a single solution [10]. If we are interested in finding multiple solutions of comparable fitness, some multi-modal function optimization techniques should be used. *Niching methods* for EAs [10] are aimed at forming and stably maintaining niches (species) throughout the search process, thereby allowing to identify most of desired peaks of multi-modal landscape.

The loss of diversity also limits the adaptive capacities of EAs in dynamic environments. Co-evolutionary techniques are aimed at improving adaptive capacities and introducing open-ended evolution into EAs [11].

This paper introduces the idea of *niching co-evolutionary multi-agent system (NCoEMAS)*, which opens new possibilities of modeling biological speciation mechanisms based on predator-prey and host-parasite co-evolution, sexual preferences, competition for limited resources, and geographical isolation. Also, results from runs of sample NCoEMAS against Rastrigin test function are presented and the comparison to other niching techniques is made.

M. Bubak et al. (Eds.): ICCS 2004, LNCS 3038, pp. 654–661, 2004.
© Springer-Verlag Berlin Heidelberg 2004

2   Niching Techniques

Various mechanisms have been proposed to form and stably maintain species throughout the search process. Most of these techniques allow niche formation through the implementation of *crowding, fitness sharing* or some modifications of these mechanisms.

In the *crowding* (CR) technique [5] each generation, a proportion of the population G (*generation gap*) is selected for reproduction. For each offspring CF (*crowding factor*) individuals are selected at random. The most similar individual, according to a similarity metric, is then replaced by the offspring.

Mahfoud developed niching mechanism called *deterministic crowding* (DC) [9]. In his technique children are directly compared to their parents. Parent is replaced only if the competing child has higher fitness.

Fitness sharing (FS) was first introduced by Holland and further developed by Goldberg and Richardson [7]. In sharing technique each individual is considered to be the center of a niche with radius σ_{sh}. Fitness of each individual is reduced for every other individual, which lives in its niche, in a proportion to their similarity. The reduced fitness of an individual i is given by $f'_i = \frac{f_i}{m_i}$, where f_i is its raw fitness and m_i is the niche count. The niche count is given by $m_i = \sum_{j=1}^{n} sh(d_{ij})$, where d_{ij} is the distance between individual i and individual j, determined by a similarity metric. The sharing function is given by

$$sh(d) = \begin{cases} 1 - (\frac{d}{\sigma_{sh}})^\alpha & \text{if } d < \sigma_{sh}, \\ 0 & \text{otherwise,} \end{cases} \tag{1}$$

where α is a constant that regulates the shape of the sharing function (it is commonly set to 1).

Co-evolutionary shared niching (CSN) technique was developed by Goldberg and Wang [8]. Their technique was inspired by the economic model of *monopolistic competition*. The customer population is the usual population of candidate solutions. The businessman population evolve to obtain largest payoff (best cover the peaks in multi-modal domain). Customer c is served by businessman b if b is the nearest businessman according to some similarity measure (Hamming distance of binary strings is used). The modified customer fitness is $f'(c) = \left.\frac{f(c)}{m_{b,t}}\right|_{c \in C_{b,t}}$, where $C_{b,t}$ is the set of customers that are served by businessman b at generation t, and $m_{b,t} = \|C_{b,t}\|$ is the number of customers that businessman b serves at generation t. The modified businessman fitness is $\phi(b) = \sum_{c \in C_{b,t}} f(c)$.

3   Previous Research in Co-evolutionary Algorithms

In classical EAs each individual in the population is considered to be a potential solution of the problem being solved. The fitness of each individual depends only on how well it solves the problem. Selection pressure causes that better fit individuals have the greater chance to survive and/or reproduce and the less fit ones have the smaller chance.

In co-evolutionary systems the fitness of each individual depends not only on the quality of solution to the given problem but also on other individuals' fitness. As the result of ongoing research many co-evolutionary techniques have been proposed. Generally, each of these techniques belongs to one of two classes: "Competitive Fitness Functions" (CFF) or multi-population [11].

In CFF based systems two (or more) individuals compete in a game and their "Competitive Fitness Functions" are calculated based on their relative performance in that game [4]. Each time step given individual competes with different opponents, so its fitness value varies. Because in such systems an individual's fitness depends on other individuals' fitness, they are co-evolutionary in nature.

The second group consists of systems that use multiple populations. In such systems a problem is decomposed into sub-problems and each of them is then solved by different EA [12]. Each individual is evaluated within a group of randomly chosen individuals coming from different sub-populations. Its fitness value depends on how well the group solved the problem and on how well the individual assisted in the solution.

Although co-evolutionary techniques are aimed at overcoming limited adaptive capacity of evolutionary algorithms resulting from the loss of useful diversity of population, they are not very often applied in the field of multi-modal optimization. In fact, to our best knowledge, only one niching technique based on co-evolution was developed (CSN).

In the following sections we will present the idea of co-evolution realized in multi-agent system, which allows us to define many co-evolutionary interactions that exist in nature.

4   The Idea of Co-evolutionary Multi-agent Systems

The main idea of *evolutionary multi-agent system (EMAS)* is the modeling of evolution process in multi-agent system (MAS) [3]. *Co-evolutionary multi-agent system (CoEMAS)* allows co-evolution of several species of agents [6]. CoEMAS can be applied, for example, to multi-objective optimization and multi-modal function optimization (*niching co-evolutionary multi-agent system — NCoEMAS*).

In CoEMAS several (usually two) different species co-evolve. One of them represents solutions. The goal of the second species is to cooperate (or compete) with the first one in order to force the population of solutions to locate Pareto frontier or proportionally populate and stably maintain niches in multi-modal domain.

It seems that CoEMAS is especially suited for modeling different co-evolutionary interactions (resource competition, predator-prey and host-parasite co-evolution, sexual preferences, etc.)

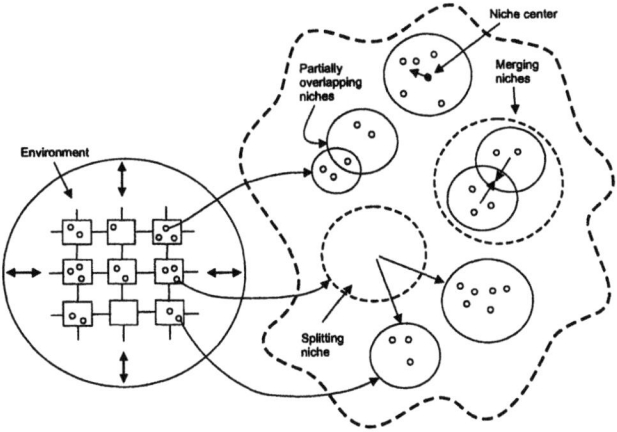

Fig. 1. Niching co-evolutionary multi-agent system used in experiments

5   Sample Niching Co-evolutionary Multi-agent System

In figure 1 sample co-evolutionary multi-agent system for multi-modal optimization is presented. The topography of environment, in which individuals live, is graph with every node (place) connected with its four neighbors. Within the environment two co-evolving species: niches and solutions live. There exist resource in the environment which is given to the niches and then distributed between solutions, that live within each niche, proportionally to their fitness. The more solutions live within the niche the more resource is given to it. Every action (such as migration or reproduction) of individual costs some resource.

Niches can migrate within the environment and all solutions live within niches. Each time step every solution searches for the niche that is located on the same peak. Modified version of hill-valley function [13] is used in order to check if two individuals are located on the same peak. If there are no niches located on the same peak, solution creates new niche, which genotype is the copy of its own genotype (niche is split into two niches). Then each solution searches its niche for the reproduction partner. Reproduction takes place only when individuals have enough amount of resource. The genotypes of all individuals are real-valued vectors. Intermediate recombination [2] and mutation with self-adaptation [1] are used for solutions and special mutation for niches. Each time step the niche's genotype is mutated in such a way, that the resulting genotype is the center of gravity of solutions that belong to the niche (fitness of each individual serves as a weight value). Niches can merge if they are at the same place and are located on the same peak in multi-modal domain.

The system was applied to multi-modal function optimization and then compared to other niching techniques.

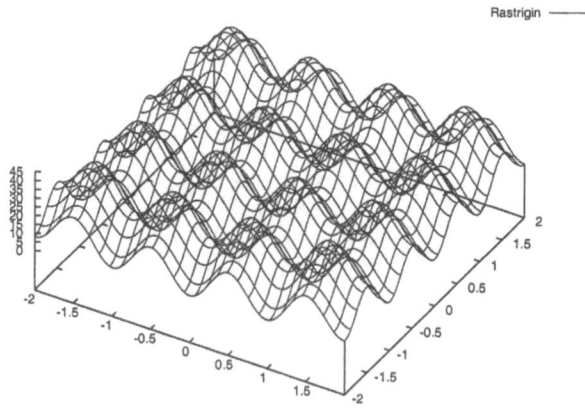

Fig. 2. Rastrigin function

6   Simulation Experiments

First simulation experiments were aimed at testing if NCoEMAS described in previous section is able to detect and stably maintain most of peaks in multi-modal domain throughout the search process. Also, the comparison to other niching techniques such as DC and FS, and EMAS was made.

6.1   Test Function

In all experiments Rastrigin function was used as the test fitness landscape (see fig. 2). This is multi-modal function commonly used in studies of niching methods. Rastrigin function used in experiments is given by

$$10*n + \sum_{i=1}^{n}(x_i^2 - 10*\cos(2*\pi*x_i)) \quad x_i \in [-2.0, 2.0] \text{ for } i = 1, \dots, n \quad (2)$$

where n is the number of dimensions ($n = 2$ in all experiments). The function has 16 maxima for $x_1, x_2 \in [-2.0, 2.0]$.

6.2   Results

In this section the results from runs of NCoEMAS against Rastrigin function are presented. Also the comparison to other niching techniques is made.

Figures 3 and 4 show the location of NCoEMAS individuals in fitness landscape during the typical simulation. At the beginning there are only 5 niches (represented with black circles) and 5 solutions (represented with dots) with identical genotypes as niches. It can be seen that as the simulation goes on the

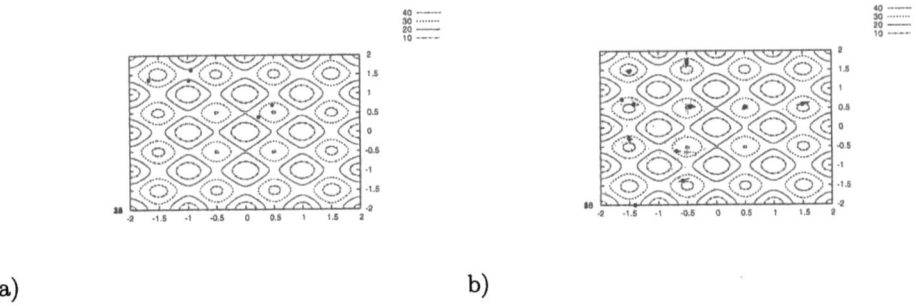

a)   b)

Fig. 3. The location of individuals in NCoEMAS during the 0th (a) and 100th (b) simulation step

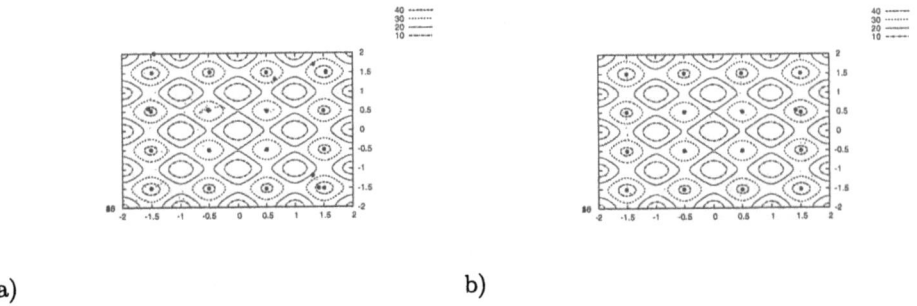

a)   b)

Fig. 4. The location of individuals in NCoEMAS during the 1000th (a) and 5000th (b) simulation step

individuals reproduce and locate themselves near the centers of peaks in multi-modal domain. What is more the subpopulations are stable, and do not disappear throughout the simulation. The population size self-adapts to the number of peaks in multi-modal domain.

Figure 5 shows the average number of located peaks from 10 simulations. The peak was classified as located when there was at least one individual closer than 0.05 to that peak. The experiments was made for four techniques: CoEMAS, EMAS, DC and FS. Each experiment was carried out with the different values of most important parameters (like initial population size, σ_{sh} for FS, etc.)

CoEMAS stood relatively well when compared to other techniques. On the average, it stably maintained about 15 peaks starting from the smallest population of all techniques (between 5 and 10 individuals). DC started from 300–500 individuals, quickly located the same number of peaks but there was quite strong tendency to lose almost all peaks during the rest part of simulation. FS technique used almost the same initial population size as DC and detected and stably maintained about 8–9 peaks on the average. Simple EMAS was not able to stably populate more than one peak, although it started from much bigger population

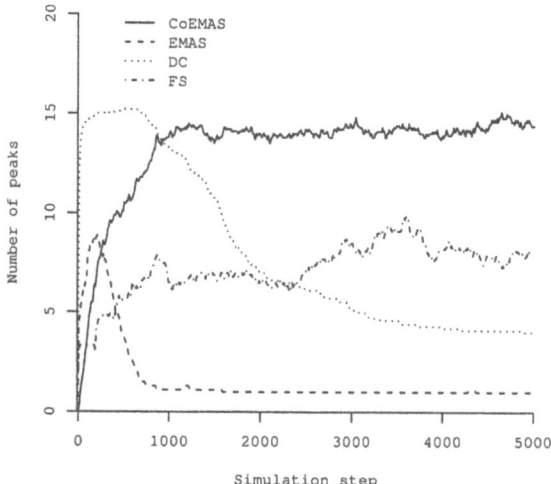

Fig. 5. The average number of detected peaks from 10 simulations

of about 150 individuals. It turned out that in case of multi-modal landscape it works just like simple EA.

To sum up, simple EMAS can not be applied to multi-modal function optimization without introducing special mechanisms such as co-evolution. FS and DC have some limitations as niching techniques, for example DC has the strong tendency to lose peaks during the simulation. The fact of relatively poor performance of DC was also observed in other works [14]. CoEMAS is valid and promising niching technique but still more research is needed.

7 Concluding Remarks

The idea of *co-evolutionary multi-agent system (CoEMAS)* allows us to model many ecological co-evolutionary interactions between species such as resource competition, predator-prey and host-parasite co-evolution, sexual preferences, etc.

In this paper sample CoEMAS with two co-evolving species: niches and solutions was presented. This system was applied to multi-modal function optimization. It properly detected and maintained most of the peaks in multi-modal fitness landscape and, as presented preliminary results show, has proved to be the valid and promising niching technique. What is more, it turned out that presented system was able to detect and stably maintain more peaks of Rastrigin function than other classical niching techniques.

Future research will include more detailed comparison to other niching techniques, CoEMAS based on the mechanisms of predator-prey or host-parasite co-evolution and sexual preferences. Also the parallel implementation of CoEMAS using MPI is included in future research plans.

References

1. T. Bäck, D. Fogel, D. Whitley, and P. Angeline. Mutation. In T. Bäck, D. Fogel, and Z. Michalewicz, editors, *Handbook of Evolutionary Computation*. IOP Publishing and Oxford University Press, 1997.
2. L. Booker, D. Fogel, D. Whitley, and P. Angeline. Recombination. In T. Bäck, D. Fogel, and Z. Michalewicz, editors, *Handbook of Evolutionary Computation*. IOP Publishing and Oxford University Press, 1997.
3. K. Cetnarowicz, M. Kisiel-Dorohinicki, and E. Nawarecki. The application of evolution process in multi-agent world to the prediction system. In *Proc. of the 2nd Int. Conf. on Multi-Agent Systems — ICMAS'96*, Osaka, Japan, 1996. AAAI Press.
4. P. J. Darwen and X. Yao. On evolving robust strategies for iterated prisoner's dilemma. *Lecture Notes in Computer Science*, 956, 1995.
5. K. A. De Jong. *An analysis of the behavior of a class of genetic adaptive systems*. PhD thesis, University of Michigan, Ann Arbor, Michigan, USA, 1975.
6. R. Dreżewski. A model of co-evolution in multi-agent system. In V. Mařík, J. Müller, and M. Pěchouček, editors, *Multi-Agent Systems and Applications III*, number 2691 in LNAI, pages 314–323, Berlin, Heidelberg, 2003. Springer-Verlag.
7. D. E. Goldberg and J. Richardson. Genetic algorithms with sharing for multimodal function optimization. In J. J. Grefenstette, editor, *Proc. of the 2nd Int. Conf. on Genetic Algorithms*, Hillsdale, NJ, 1987. Lawrence Erlbaum Associates.
8. D. E. Goldberg and L. Wang. Adaptive niching via coevolutionary sharing. Technical Report IlliGAL 97007, Illinois Genetic Algorithms Laboratory, University of Illinois at Urbana-Champaign, Urbana, IL, USA, 1997.
9. S. W. Mahfoud. Crowding and preselection revisited. In R. Manner and B. Manderick, editors, *Parallel Problem Solving From Nature, 2*, Amsterdam, 1992. Elsevier Science Publishers (North Holland).
10. S. W. Mahfoud. *Niching methods for genetic algorithms*. PhD thesis, University of Illinois at Urbana-Champaign, Urbana, IL, USA, 1995.
11. J. Morrison and F. Oppacher. A general model of co-evolution for genetic algorithms. In *Int. Conf. on Artificial Neural Networks and Genetic Algorithms ICANNGA 99*, 1999.
12. M. A. Potter and K. A. De Jong. Cooperative coevolution: An architecture for evolving coadapted subcomponents. *Evolutionary Computation*, 8(1), 2000.
13. R. K. Ursem. Multinational evolutionary algorithms. In P. J. Angeline, Z. Michalewicz, M. Schoenauer, X. Yao, and A. Zalzala, editors, *Proceedings of the Congress on Evolutionary Computation*, volume 3, pages 1633–1640, Mayflower Hotel, Washington D.C., USA, 6-9 1999. IEEE Press.
14. J.-P. Watson. A performance assessment of modern niching methods for parameter optimization problems. In W. Banzhaf, J. Daida, A. E. Eiben, M. H. Garzon, V. Honavar, M. Jakiela, and R. E. Smith, editors, *Proceedings of the Genetic and Evolutionary Computation Conference*, volume 1, pages 702–709, Orlando, Florida, USA, 13-17 1999. Morgan Kaufmann.

MIX
Papier aus verantwortungsvollen Quellen
Paper from responsible sources
FSC® C105338

If you have any concerns about our products,
you can contact us on
ProductSafety@springernature.com

In case Publisher is established outside the EU,
the EU authorized representative is:
**Springer Nature Customer Service Center GmbH
Europaplatz 3, 69115 Heidelberg, Germany**

Printed by Libri Plureos GmbH
in Hamburg, Germany

Lecture Notes in Computer Science 3038

Commenced Publication in 1973
Founding and Former Series Editors:
Gerhard Goos, Juris Hartmanis, and Jan van Leeuwen

Editorial Board

Takeo Kanade
Carnegie Mellon University, Pittsburgh, PA, USA
Josef Kittler
University of Surrey, Guildford, UK
Jon M. Kleinberg
Cornell University, Ithaca, NY, USA
Friedemann Mattern
ETH Zurich, Switzerland
John C. Mitchell
Stanford University, CA, USA
Moni Naor
Weizmann Institute of Science, Rehovot, Israel
Oscar Nierstrasz
University of Bern, Switzerland
C. Pandu Rangan
Indian Institute of Technology, Madras, India
Bernhard Steffen
University of Dortmund, Germany
Madhu Sudan
Massachusetts Institute of Technology, MA, USA
Demetri Terzopoulos
New York University, NY, USA
Doug Tygar
University of California, Berkeley, CA, USA
Moshe Y. Vardi
Rice University, Houston, TX, USA
Gerhard Weikum
Max-Planck Institute of Computer Science, Saarbruecken, Germany

Springer-Verlag Berlin Heidelberg GmbH

Marian Bubak   Geert Dick van Albada
Peter M.A. Sloot   Jack J. Dongarra (Eds.)

Computational
Science - ICCS 2004

4th International Conference
Kraków, Poland, June 6-9, 2004
Proceedings, Part III

Springer

Volume Editors

Marian Bubak
AGH University of Science and Technology
Institute of Computer Science and Academic Computer Center CYFRONET
Mickiewicza 30, 30-059 Kraków, Poland
E-mail: bubak@uci.agh.edu.pl

Geert Dick van Albada
Peter M.A. Sloot
University of Amsterdam, Informatics Institute, Section Computational Science
Kruislaan 403, 1098 SJ Amsterdam, The Netherlands
E-mail: {dick,sloot}@science.uva.nl

Jack J. Dongarra
University of Tennessee, Computer Science Department
1122 Volunteer Blvd, Knoxville, TN 37996-3450, USA
E-mail: dongarra@cs.utk.edu

Library of Congress Control Number: Applied for

CR Subject Classification (1998): D, F, G, H, I, J, C.2-3

ISSN 0302-9743
ISBN 978-3-540-22116-6   ISBN 978-3-540-24688-6 (eBook)
DOI 10.1007/978-3-540-24688-6

This work is subject to copyright. All rights are reserved, whether the whole or part of the material is concerned, specifically the rights of translation, reprinting, re-use of illustrations, recitation, broadcasting, reproduction on microfilms or in any other way, and storage in data banks. Duplication of this publication or parts thereof is permitted only under the provisions of the German Copyright Law of September 9, 1965, in its current version, and permission for use must always be obtained from Springer-Verlag. Violations are liable to prosecution under the German Copyright Law.

springeronline.com

© Springer-Verlag Berlin Heidelberg 2004
Originally published by Springer-Verlag Berlin Heidelberg New York in 2004.

Typesetting: Camera-ready by author, data conversion by PTP-Berlin, Protago-TeX-Production GmbH
Printed on acid-free paper   SPIN: 11009337   06/3142   5 4 3 2 1 0

Preface

The International Conference on Computational Science (ICCS 2004) held in Kraków, Poland, June 6–9, 2004, was a follow-up to the highly successful ICCS 2003 held at two locations, in Melbourne, Australia and St. Petersburg, Russia; ICCS 2002 in Amsterdam, The Netherlands; and ICCS 2001 in San Francisco, USA.

As computational science is still evolving in its quest for subjects of investigation and efficient methods, ICCS 2004 was devised as a forum for scientists from mathematics and computer science, as the basic computing disciplines and application areas, interested in advanced computational methods for physics, chemistry, life sciences, engineering, arts and humanities, as well as computer system vendors and software developers. The main objective of this conference was to discuss problems and solutions in all areas, to identify new issues, to shape future directions of research, and to help users apply various advanced computational techniques. The event harvested recent developments in computational grids and next generation computing systems, tools, advanced numerical methods, data-driven systems, and novel application fields, such as complex systems, finance, econo-physics and population evolution.

Keynote lectures were delivered by David Abramson and Alexander V. Bogdanov, *From ICCS 2003 to ICCS 2004 – Personal Overview of Recent Advances in Computational Science*; Iain Duff, *Combining Direct and Iterative Methods for the Solution of Large Sparse Systems in Different Application Areas*; Chris Johnson, *Computational Multi-field Visualization*; John G. Michopoulos, *On the Pathology of High Performance Computing*; David De Roure, *Semantic Grid*; and Vaidy Sunderam, *True Grid: What Makes a Grid Special and Different?* In addition, three invited lectures were delivered by representatives of leading computer system vendors, namely: Frank Baetke from Hewlett Packard, Eng Lim Goh from SGI, and David Harper from the Intel Corporation.

Four tutorials extended the program of the conference: Paweł Płaszczak and Krzysztof Wilk, *Practical Introduction to Grid and Grid Services*; Grzegorz Młynarczyk, *Software Engineering Methods for Computational Science*; the *CrossGrid Tutorial* by the CYFRONET CG team; and the Intel tutorial.

We would like to thank all keynote, invited and tutorial speakers for their interesting and inspiring talks.

Aside of plenary lectures, the conference included 12 parallel oral sessions and 3 poster sessions. Ever since the first meeting in San Francisco, ICCS has attracted an increasing number of more researchers involved in the challenging field of computational science. For ICCS 2004, we received 489 contributions for the main track and 534 contributions for 41 originally-proposed workshops. Of these submissions, 117 were accepted for oral presentations and 117 for posters in the main track, while 328 papers were accepted for presentations at 30 workshops. This selection was possible thanks to the hard work of the Program

Committee members and 477 reviewers. The author index contains 1395 names, and almost 560 persons from 44 countries and all continents attended the conference: 337 participants from Europe, 129 from Asia, 62 from North America, 13 from South America, 11 from Australia, and 2 from Africa.

The ICCS 2004 proceedings consists of four volumes, the first two volumes, LNCS 3036 and 3037 contain the contributions presented in the main track, while volumes 3038 and 3039 contain the papers accepted for the workshops. Parts I and III are mostly related to pure computer science, while Parts II and IV are related to various computational research areas. For the first time, the ICCS proceedings are also available on CD. We would like to thank Springer-Verlag for their fruitful collaboration. During the conference the best papers from the main track and workshops as well as the best posters were nominated and presented on the ICCS 2004 Website. We hope that the ICCS 2004 proceedings will serve as a major intellectual resource for computational science researchers, pushing back the boundaries of this field. A number of papers will also be published as special issues of selected journals.

We owe thanks to all workshop organizers and members of the Program Committee for their diligent work, which ensured the very high quality of the event. We also wish to specifically acknowledge the collaboration of the following colleagues who organized their workshops for the third time: Nicoletta Del Buono (New Numerical Methods) Andres Iglesias (Computer Graphics), Dieter Kranzlmueller (Tools for Program Development and Analysis), Youngsong Mun (Modeling and Simulation in Supercomputing and Telecommunications).

We would like to express our gratitude to Prof. Ryszard Tadeusiewicz, Rector of the AGH University of Science and Technology, as well as to Prof. Marian Noga, Prof. Kazimierz Jeleń, Dr. Jan Kulka and Prof. Krzysztof Zieliński, for their personal involvement. We are indebted to all the members of the Local Organizing Committee for their enthusiastic work towards the success of ICCS 2004, and to numerous colleagues from ACC CYFRONET AGH and the Institute of Computer Science for their help in editing the proceedings and organizing the event. We very much appreciate the help of the Computer Science and Computational Physics students during the conference. We owe thanks to the ICCS 2004 sponsors: Hewlett-Packard, Intel, IBM, SGI and ATM, SUN Microsystems, Polish Airlines LOT, ACC CYFRONET AGH, the Institute of Computer Science AGH, the Polish Ministry for Scientific Research and Information Technology, and Springer-Verlag for their generous support.

We wholeheartedly invite you to once again visit the ICCS 2004 Website (http://www.cyfronet.krakow.pl/iccs2004/), to recall the atmosphere of those June days in Kraków.

June 2004   Marian Bubak, Scientific Chair 2004
                                                   on behalf of the co-editors:
                                                      G. Dick van Albada
                                                        Peter M.A. Sloot
                                                        Jack J. Dongarra

Organization

ICCS 2004 was organized by the Academic Computer Centre CYFRONET AGH University of Science and Technology (Kraków, Poland) in cooperation with the Institute of Computer Science AGH, the University of Amsterdam (The Netherlands) and the University of Tennessee (USA).

All the members of the Local Organizing Committee are the staff members of CYFRONET and/or ICS. The conference took place at the premises of the Faculty of Physics and Nuclear Techniques AGH and at the Institute of Computer Science AGH.

Conference Chairs

Scientific Chair – Marian Bubak (Institute of Computer Science and ACC CYFRONET AGH, Poland)
Workshop Chair – Dick van Albada (University of Amsterdam, The Netherlands)
Overall Chair – Peter M.A. Sloot (University of Amsterdam, The Netherlands)
Overall Co-chair – Jack Dongarra (University of Tennessee, USA)

Local Organizing Committee

Marian Noga
Marian Bubak
Zofia Mosurska
Maria Stawiarska
Milena Zając
Mietek Pilipczuk
Karol Frańczak
Aleksander Kusznir

Program Committee

Jemal Abawajy (Carleton University, Canada)
David Abramson (Monash University, Australia)
Dick van Albada (University of Amsterdam, The Netherlands)
Vassil Alexandrov (University of Reading, UK)
Srinivas Aluru (Iowa State University, USA)
David A. Bader (University of New Mexico, USA)

J.A. Rod Blais (University of Calgary, Canada)
Alexander Bogdanov (Institute for High Performance Computing and Information Systems, Russia)
Peter Brezany (University of Vienna, Austria)
Marian Bubak (Institute of Computer Science and CYFRONET AGH, Poland)
Rajkumar Buyya (University of Melbourne, Australia)
Bastien Chopard (University of Geneva, Switzerland)
Paul Coddington (University of Adelaide, Australia)
Toni Cortes (Universitat Politècnica de Catalunya, Spain)
Yiannis Cotronis (University of Athens, Greece)
Jose C. Cunha (New University of Lisbon, Portugal)
Brian D'Auriol (University of Texas at El Paso, USA)
Federic Desprez (INRIA, France)
Tom Dhaene (University of Antwerp, Belgium)
Hassan Diab (American University of Beirut, Lebanon)
Beniamino Di Martino (Second University of Naples, Italy)
Jack Dongarra (University of Tennessee, USA)
Robert A. Evarestov (SPbSU, Russia)
Marina Gavrilova (University of Calgary, Canada)
Michael Gerndt (Technical University of Munich, Germany)
Yuriy Gorbachev (Institute for High Performance Computing and Information Systems, Russia)
Andrzej Goscinski (Deakin University, Australia)
Ladislav Hluchy (Slovak Academy of Sciences, Slovakia)
Alfons Hoekstra (University of Amsterdam, The Netherlands)
Hai Jin (Huazhong University of Science and Technology, ROC)
Peter Kacsuk (MTA SZTAKI Research Institute, Hungary)
Jacek Kitowski (AGH University of Science and Technology, Poland)
Dieter Kranzlmüller (Johannes Kepler University Linz, Austria)
Domenico Laforenza (Italian National Research Council, Italy)
Antonio Lagana (Università di Perugia, Italy)
Francis Lau (University of Hong Kong, ROC)
Bogdan Lesyng (ICM Warszawa, Poland)
Thomas Ludwig (Ruprecht-Karls-Universität Heidelberg, Germany)
Emilio Luque (Universitat Autònoma de Barcelona, Spain)
Michael Mascagni (Florida State University, USA)
Edward Moreno (Euripides Foundation of Marilia, Brazil)
Jiri Nedoma (Institute of Computer Science AS CR, Czech Republic)
Genri Norman (Russian Academy of Sciences, Russia)
Stephan Olariu (Old Dominion University, USA)
Salvatore Orlando (University of Venice, Italy)
Marcin Paprzycki (Oklahoma State University, USA)
Ron Perrott (Queen's University of Belfast, UK)
Richard Ramaroson (ONERA, France)
Rosemary Renaut (Arizona State University, USA)

Alistair Rendell (Australian National University, Australia)
Paul Roe (Queensland University of Technology, Australia)
Hong Shen (Japan Advanced Institute of Science and Technology, Japan)
Dale Shires (U.S. Army Research Laboratory, USA)
Peter M.A. Sloot (University of Amsterdam, The Netherlands)
Gunther Stuer (University of Antwerp, Belgium)
Vaidy Sunderam (Emory University, USA)
Boleslaw Szymanski (Rensselaer Polytechnic Institute, USA)
Ryszard Tadeusiewicz (AGH University of Science and Technology, Poland)
Pavel Tvrdik (Czech Technical University, Czech Republic)
Putchong Uthayopas (Kasetsart University, Thailand)
Jesus Vigo-Aguiar (University of Salamanca, Spain)
Jens Volkert (University of Linz, Austria)
Koichi Wada (University of Tsukuba, Japan)
Jerzy Wasniewski (Technical University of Denmark, Denmark)
Greg Watson (Los Alamos National Laboratory, USA)
Jan Węglarz (Poznań University of Technology, Poland)
Roland Wismüller (LRR-TUM, Germany)
Roman Wyrzykowski (Technical University of Częstochowa, Poland)
Jinchao Xu (Pennsylvania State University, USA)
Yong Xue (Chinese Academy of Sciences, ROC)
Xiaodong Zhang (College of William and Mary, USA)
Alexander Zhmakin (Soft-Impact Ltd, Russia)
Krzysztof Zieliński (Institute of Computer Science and CYFRONET AGH, Poland)
Zahari Zlatev (National Environmental Research Institute, Denmark)
Albert Zomaya (University of Sydney, Australia)
Elena Zudilova (University of Amsterdam, The Netherlands)

Reviewers

Abawajy, J.H.	Aluru, S.	Balogh, Z.
Abe, S.	Anglano, C.	Bang, Y.C.
Abramson, D.	Archibald, R.	Baraglia, R.
Adali, S.	Arenas, A.	Barron, J.
Adcock, M.	Astalos, J.	Baumgartner, F.
Adriaansen, T.	Ayani, R.	Becakaert, P.
Ahn, G.	Ayyub, S.	Belleman, R.G.
Ahn, S.J.	Babik, M.	Bentes, C.
Albada, G.D. van	Bader, D.A.	Bernardo Filho, O.
Albuquerque, P.	Bajaj, C.	Beyls, K.
Alda, W.	Baker, M.	Blais, J.A.R.
Alexandrov, V.	Baliś, B.	Boada, I.
Alt, M.	Balk, I.	Bode, A.

Bogdanov, A.
Bollapragada, R.
Boukhanovsky, A.
Brandes, T.
Brezany, P.
Britanak, V.
Bronsvoort, W.
Brunst, H.
Bubak, M.
Budinska, I.
Buono, N. Del
Buyya, R.
Cai, W.
Cai, Y.
Cannataro, M.
Carbonell, N.
Carle, G.
Caron, E.
Carothers, C.
Castiello, C.
Chan, P.
Chassin-de-
  Kergommeaux, J.
Chaudet, C.
Chaves, J.C.
Chen, L.
Chen, Z.
Cheng, B.
Cheng, X.
Cheung, B.W.L.
Chin, S.
Cho, H.
Choi, Y.S.
Choo, H.S.
Chopard, B.
Chuang, J.H.
Chung, R.
Chung, S.T.
Coddington, P.
Coeurjolly, D.
Congiusta, A.
Coppola, M.
Corral, A.
Cortes, T.
Cotronis, Y.

Cramer, H.S.M.
Cunha, J.C.
Danilowicz, C.
D'Auriol, B.
Degtyarev, A.
Denazis, S.
Derntl, M.
Desprez, F.
Devendeville, L.
Dew, R.
Dhaene, T.
Dhoedt, B.
D'Hollander, E.
Diab, H.
Dokken, T.
Dongarra, J.
Donnelly, D.
Donnelly, W.
Dorogovtsev, S.
Duda, J.
Dudek-Dyduch, E.
Dufourd, J.F.
Dumitriu, L.
Duplaga, M.
Dupuis, A.
Dzwinel, W.
Embrechts, M.J.
Emiris, I.
Emrich, S.J.
Enticott, C.
Evangelos, F.
Evarestov, R.A.
Fagni, T.
Faik, J.
Fang, W.J.
Farin, G.
Fernandez, M.
Filho, B.O.
Fisher-Gewirtzman, D.
Floros, E.
Fogel, J.
Foukia, N.
Frankovic, B.
Fuehrlinger, K.
Funika, W.

Gabriel, E.
Gagliardi, F.
Galis, A.
Galvez, A.
Gao, X.S.
Garstecki, L.
Gatial, E.
Gava, F.
Gavidia, D.P.
Gavras, A.
Gavrilova, M.
Gelb, A.
Gerasimov, V.
Gerndt, M.
Getov, V.
Geusebroek, J.M.
Giang, T.
Gilbert, M.
Glasner, C.
Gobbert, M.K.
Gonzalez-Vega, L.
Gorbachev, Y.E.
Goscinski, A.M.
Goscinski, W.
Gourhant, Y.
Gualandris, A.
Guo, H.
Ha, R.
Habala, O.
Habib, A.
Halada, L.
Hawick, K.
He, K.
Heinzlreiter, P.
Heyfitch, V.
Hisley, D.M.
Hluchy, L.
Ho, R.S.C.
Ho, T.
Hobbs, M.
Hoekstra, A.
Hoffmann, C.
Holena, M.
Hong, C.S.
Hong, I.

Hong, S.
Horan, P.
Hu, S.M.
Huh, E.N.
Hutchins, M.
Huynh, J.
Hwang, I.S.
Hwang, J.
Iacono, M.
Iglesias, A.
Ingram, D.
Jakulin, A.
Janciak, I.
Janecek, J.
Janglova, D.
Janicki, A.
Jin, H.
Jost, G.
Juhola, A.
Kacsuk, P.
Kalousis, A.
Kalyanaraman, A.
Kang, M.G.
Karagiorgos, G.
Karaivanova, A.
Karl, W.
Karypis, G.
Katarzyniak, R.
Kelley, T.
Kelly, W.
Kennedy, E.
Kereku, E.
Kergommeaux, J.C. De
Kim, B.
Kim, C.H.
Kim, D.S.
Kim, D.Y.
Kim, M.
Kim, M.J.
Kim, T.W.
Kitowski, J.
Klein, C.
Ko, P.
Kokoszka, P.
Kolingerova, I.

Kommineni, J.
Korczak, J.J.
Korkhov, V.
Kou, G.
Kouniakis, C.
Kranzlmüller, D.
Krzhizhianovskaya, V.V.
Kuo, T.W.
Kurka, G.
Kurniawan, D.
Kurzyniec, D.
Laclavik, M.
Laforenza, D.
Lagan, A.
Lagana, A.
Lamehamedi, H.
Larrabeiti, D.
Latt, J.
Lau, F.
Lee, H.G.
Lee, M.
Lee, S.
Lee, S.S.
Lee, S.Y.
Lefevre, L.
Leone, P.
Lesyng, B.
Leszczynski, J.
Leymann, F.
Li, T.
Lindner, P.
Logan, B.
Lopes, G.P.
Lorencz, R.
Low, M.Y.H.
Ludwig, T.
Luethi, J.
Lukac, R.
Luksch, P.
Luque, E.
Mairandres, M.
Malawski, M.
Malony, A.
Malyshkin, V.E.
Maniatty, W.A.

Marconi, S.
Mareev, V.
Margalef, T.
Marrone, S.
Martino, B. Di
Marzolla, M.
Mascagni, M.
Mayer, M.
Medeiros, P.
Meer, H. De
Meyer, N.
Miller, B.
Miyaji, C.
Modave, F.
Mohr, B.
Monterde, J.
Moore, S.
Moreno, E.
Moscato, F.
Mourelle, L.M.
Mueller, M.S.
Mun, Y.
Na, W.S.
Nagel, W.E.
Nanni, M.
Narayanan, M.
Nasri, A.
Nau, B.
Nedjah, N.
Nedoma, J.
Negoita, C.
Neumann, L.
Nguyen, G.T.
Nguyen, N.T.
Norman, G.
Olariu, S.
Orlando, S.
Orley, S.
Otero, C.
Owen, J.
Palus, H.
Paprzycki, M.
Park, N.J.
Patten, C.
Peachey, T.C.

Peluso, R.
Peng, Y.
Perales, F.
Perrott, R.
Petit, F.
Petit, G.H.
Pfluger, P.
Philippe, L.
Platen, E.
Plemenos, D.
Pllana, S.
Polak, M.
Polak, N.
Politi, T.
Pooley, D.
Popov, E.V.
Puppin, D.
Qut, P.R.
Rachev, S.
Rajko, S.
Rak, M.
Ramaroson, R.
Ras, I.
Rathmayer, S.
Raz, D.
Recio, T.
Reichel, L.
Renaut, R.
Rendell, A.
Richta, K.
Robert, Y.
Rodgers, G.
Rodionov, A.S.
Roe, P.
Ronsse, M.
Ruder, K.S.
Ruede, U.
Rycerz, K.
Sanchez-Reyes, J.
Sarfraz, M.
Sbert, M.
Scarpa, M.
Schabanel, N.
Scharf, E.
Scharinger, J.

Schaubschlaeger, C.
Schmidt, A.
Scholz, S.B.
Schreiber, A.
Seal, S.K.
Seinstra, F.J.
Seron, F.
Serrat, J.
Shamonin, D.P.
Sheldon, F.
Shen, H.
Shende, S.
Shentu, Z.
Shi, Y.
Shin, H.Y.
Shires, D.
Shoshmina, I.
Shrikhande, N.
Silvestri, C.
Silvestri, F.
Simeoni, M.
Simo, B.
Simonov, N.
Siu, P.
Slizik, P.
Slominski, L.
Sloot, P.M.A.
Slota, R.
Smetek, M.
Smith, G.
Smolka, B.
Sneeuw, N.
Snoek, C.
Sobaniec, C.
Sobecki, J.
Sofroniou, M.
Sole, R.
Soofi, M.
Sosnov, A.
Sourin, A.
Spaletta, G.
Spiegl, E.
Stapor, K.
Stuer, G.
Suarez Rivero, J.P.

Sunderam, V.
Suzuki, H.
Szatzschneider, W.
Szczepanski, M.
Szirmay-Kalos, L.
Szymanski, B.
Tadeusiewicz, R.
Tadic, B.
Talia, D.
Tan, G.
Taylor, S.J.E.
Teixeira, J.C.
Telelis, O.A.
Teo, Y.M
Teresco, J.
Teyssiere, G.
Thalmann, D.
Theodoropoulos, G.
Theoharis, T.
Thurner, S.
Tirado-Ramos, A.
Tisserand, A.
Toda, K.
Tonellotto, N.
Torelli, L.
Torenvliet, L.
Tran, V.D.
Truong, H.L.
Tsang, K.
Tse, K.L.
Tvrdik, P.
Tzevelekas, L.
Uthayopas, P.
Valencia, P.
Vassilakis, C.
Vaughan, F.
Vazquez, P.P.
Venticinque, S.
Vigo-Aguiar, J.
Vivien, F.
Volkert, J.
Wada, K.
Walter, M.
Wasniewski, J.
Wasserbauer, A.

Watson, G.	Xiao, Y.	Zhang, J.W.
Wawrzyniak, D.	Xu, J.	Zhang, N.X.L.
Weglarz, J.	Xue, Y.	Zhang, X.
Weidendorfer, J.	Yahyapour, R.	Zhao, L.
Weispfenning, W.	Yan, N.	Zhmakin, A.I.
Wendelborn, A.L.	Yang, K.	Zhu, W.Z.
Weron, R.	Yener, B.	Zieliński, K.
Wismüller, R.	Yoo, S.M.	Zlatev, Z.
Wojciechowski, K.	Yu, J.H.	Zomaya, A.
Wolf, F.	Yu, Z.C.H.	Zudilova, E.V.
Worring, M.	Zara, J.	
Wyrzykowski, R.	Zatevakhin, M.A.	

Workshops Organizers

Programming Grids and Metasystems

V. Sunderam (Emory University, USA)
D. Kurzyniec (Emory University, USA)
V. Getov (University of Westminster, UK)
M. Malawski (Institute of Computer Science and CYFRONET AGH, Poland)

Active and Programmable Grids Architectures and Components

C. Anglano (Università del Piemonte Orientale, Italy)
F. Baumgartner (University of Bern, Switzerland)
G. Carle (Tubingen University, Germany)
X. Cheng (Institute of Computing Technology, Chinese Academy of Science, ROC)
K. Chen (Institut Galilée, Université Paris 13, France)
S. Denazis (Hitachi Europe, France)
B. Dhoedt (University of Gent, Belgium)
W. Donnelly (Waterford Institute of Technology, Ireland)
A. Galis (University College London, UK)
A. Gavras (Eurescom, Germany)
F. Gagliardi (CERN, Switzerland)
Y. Gourhant (France Telecom, France)
M. Gilbert (European Microsoft Innovation Center, Microsoft Corporation, Germany)
A. Juhola (VTT, Finland)
C. Klein (Siemens, Germany)
D. Larrabeiti (University Carlos III, Spain)
L. Lefevre (INRIA, France)
F. Leymann (IBM, Germany)
H. de Meer (University of Passau, Germany)
G. H. Petit (Alcatel, Belgium)

J. Serrat (Universitat Politècnica de Catalunya, Spain)
E. Scharf (QMUL, UK)
K. Skala (Ruder Boskoviç Institute, Croatia)
N. Shrikhande (European Microsoft Innovation Center, Microsoft
Corporation, Germany)
M. Solarski (FhG FOKUS, Germany)
D. Raz (Technion Institute of Technology, Israel)
K. Zieliński (AGH University of Science and Technology, Poland)
R. Yahyapour (University Dortmund, Germany)
K. Yang (University of Essex, UK)

Next Generation Computing

E.-N. John Huh (Seoul Women's University, Korea)

Practical Aspects of High-Level Parallel Programming (PAPP 2004)

F. Loulergue (Laboratory of Algorithms, Complexity and Logic,
University of Paris Val de Marne, France)

Parallel Input/Output Management Techniques (PIOMT 2004)

J. H. Abawajy (Carleton University, School of Computer Science, Canada)

OpenMP for Large Scale Applications

B. Chapman (University of Houston, USA)

Tools for Program Development and Analysis in Computational Science

D. Kranzlmüller (Johannes Kepler University Linz, Austria)
R. Wismüller (TU München, Germany)
A. Bode (Technische Universität München, Germany)
J. Volkert (Johannes Kepler University Linz, Austria)

Modern Technologies for Web-Based Adaptive Systems

N. Thanh Nguyen (Wrocław University of Technology, Poland)
J. Sobecki (Wrocław University of Technology, Poland)

Agent Day 2004 – Intelligent Agents in Computing Systems

E. Nawarecki (AGH University of Science and Technology, Poland)
K. Cetnarowicz (AGH University of Science and Technology, Poland)
G. Dobrowolski (AGH University of Science and Technology, Poland)
R. Schaefer (Jagiellonian University, Poland)
S. Ambroszkiewicz (Polish Academy of Sciences, Warsaw, Poland)
A. Koukam (Université de Belfort-Montbeliard, France)
V. Srovnal (VSB Technical University of Ostrava, Czech Republic)
C. Cotta (Universidad de Málaga, Spain)
S. Raczynski (Universidad Panamericana, Mexico)

Dynamic Data Driven Application Systems

F. Darema (NSF/CISE, USA)

HLA-Based Distributed Simulation on the Grid

S. J. Turner (Nanyang Technological University, Singapore)

Interactive Visualisation and Interaction Technologies

E. Zudilova (University of Amsterdam, The Netherlands)
T. Adriaansen (CSIRO, ICT Centre, Australia)

Computational Modeling of Transport on Networks

B. Tadic (Jozef Stefan Institute, Slovenia)
S. Thurner (Universität Wien, Austria)

Modeling and Simulation in Supercomputing and Telecommunications

Y. Mun (Soongsil University, Korea)

QoS Routing

H. Choo (Sungkyunkwan University, Korea)

Evolvable Hardware

N. Nedjah (State University of Rio de Janeiro, Brazil)
L. de Macedo Mourelle (State University of Rio de Janeiro, Brazil)

Advanced Methods of Digital Image Processing

B. Smolka (Silesian University of Technology, Laboratory of Multimedia Communication, Poland)

Computer Graphics and Geometric Modelling (CGGM 2004)

A. Iglesias Prieto (University of Cantabria, Spain)

Computer Algebra Systems and Applications (CASA 2004)

A. Iglesias Prieto (University of Cantabria, Spain)
A. Galvez (University of Cantabria, Spain)

New Numerical Methods for DEs: Applications to Linear Algebra, Control and Engineering

N. Del Buono (University of Bari, Italy)
L. Lopez (University of Bari, Italy)

Parallel Monte Carlo Algorithms for Diverse Applications in a Distributed Setting

V. N. Alexandrov (University of Reading, UK)
A. Karaivanova (Bulgarian Academy of Sciences, Bulgaria)
I. Dimov (Bulgarian Academy of Sciences, Bulgaria)

Modelling and Simulation of Multi-physics Multi-scale Systems

V. Krzhizhanovskaya (University of Amsterdam, The Netherlands)
B. Chopard (University of Geneva, CUI, Switzerland)
Y. Gorbachev (St. Petersburg State Polytechnical University, Russia)

Gene, Genome and Population Evolution

S. Cebrat (University of Wrocław, Poland)
D. Stauffer (Cologne University, Germany)
A. Maksymowicz (AGH University of Science and Technology, Poland)

Computational Methods in Finance and Insurance

A. Janicki (University of Wrocław, Poland)
J.J. Korczak (University Louis Pasteur, Strasbourg, France)

Computational Economics and Finance

X. Deng (City University of Hong Kong, Hong Kong)
S. Wang (Chinese Academy of Sciences, ROC)
Y. Shi (University of Nebraska at Omaha, USA)

GeoComputation

Y. Xue (Chinese Academy of Sciences, ROC)
C. Yarotsos (University of Athens, Greece)

Simulation and Modeling of 3D Integrated Circuits

I. Balk (R3Logic Inc., USA)

Computational Modeling and Simulation on Biomechanical Engineering

Y.H. Kim (Kyung Hee University, Korea)

Information Technologies Enhancing Health Care Delivery

M. Duplaga (Jagiellonian University Medical College, Poland)
D. Ingram (University College London, UK)
K. Zieliński (AGH University of Science and Technology, Poland)

Computing in Science and Engineering Academic Programs

D. Donnelly (Siena College, USA)

Sponsoring Institutions

Hewlett-Packard
Intel
SGI
ATM
SUN Microsystems
IBM
Polish Airlines LOT
ACC CYFRONET AGH
Institute of Computer Science AGH
Polish Ministry of Scientific Research and Information Technology
Springer-Verlag

Table of Contents – Part III

Workshop on Programming Grids and Metasystems

Workshop on First International Workshop on Active and Programmable Grids Architectures and Components

Workshop on Next Generation Computing

Workshop on Practical Aspects of High-Level Parallel Programming (PAPP 2004)

Workshop on Parallel Input/Output Management Techniques (PIOMT04)

Workshop on OpenMP for Large Scale Applications

Workshop on Tools for Program Development and Analysis in Computational Science

Workshop on Modern Technologies for Web-Based Adaptive Systems

Workshop on Agent Day 2004 – Intelligent Agents in Computing Systems

Workshop on Dynamic Data Driven Applications Systems

Workshop on HLA-Based Distributed Simulation on the Grid

Workshop on Interactive Visualisation and Interaction Technologies

Workshop on Computational Modeling of Transport on Networks

Workshop on Modeling and Simulation in Supercomputing and Telecommunications

Workshop on QoS Routing

Workshop on Evolvable Hardware

Table of Contents – Part I

Track on Parallel and Distributed Computing

Track on Grid Computing

Track on Models and Algorithms

Track on Data Mining and Data Bases

Track on Networking

Poster Papers

Table of Contents – Part II

Track on Numerical Algorithms

Track on Finite Element Method

Track on Neural Networks

Track on Applications

Poster Papers

Table of Contents – Part IV

Workshop on Computer Graphics and Geometric Modelling (CGGM 2004)

Workshop on Computer Algebra Systems and Applications (CASA 2004)

Workshop on New Numerical Methods for DEs: Applications to Linear Algebra, Control and Engineering

Workshop on Parallel Monte Carlo Algorithms for Diverse Applications in a Distributed Setting

Workshop on Modelling and Simulation of Multi-physics Multi-scale Systems

Workshop on Gene, Genome, and Population Evolution

Workshop on Computational Methods in Finance and Insurance

Workshop on Computational Economics and Finance

Workshop on GeoComputation

Workshop on Simulation and Modeling of 3D Integrated Circuits

Workshop on Computational Modeling and Simulation on Biomechanical Engineering

Workshop on Information Technologies Enhancing Health Care Delivery

Workshop on Computing in Science and Engineering Academic Programs

Dynamic Data Driven Applications Systems: A New Paradigm for Application Simulations and Measurements

Frederica Darema

National Science Foundation, Arlington VA 22230, USA
darema@nsf.gov

Abstract. Dynamic Data Driven Application Systems (DDDAS) entails the ability to incorporate additional data into an executing application - these data can be archival or collected on-line; and in reverse, the ability of applications to dynamically steer the measurement process. The paradigm offers the promise of improving modeling methods, and augmenting the analysis and prediction capabilities of application simulations and the effectiveness of measurement systems. This presents the potential to transform the way science and engineering are done, and induce a major impact in the way many functions in our society are conducted, such as manufacturing, commerce, hazard management, medicine. Enabling this synergistic feedback and control-loop between application simulations and measurements requires novel application modeling approaches and frameworks, algorithms tolerant to perturbations from dynamic data injection and steering, and systems software to support the dynamic environments of concern here. Recent advances in complex applications, the advent of grid computing and of sensor systems, are some of the technologies that make it timely to embark in developing DDDAS capabilities. Research and development of such technologies requires synergistic multidisciplinary collaboration in the applications, algorithms, software systems, and measurements systems areas, and involving researchers in basic sciences, engineering, and computer sciences. The rest of the papers in the proceedings of this workshop provide examples of ongoing research developing DDDAS technologies within the context of specific and important application areas.

1 Introduction

Accurate analysis and prediction of the behavior of a complicated system is difficult. Today, applications and systems simulations are fairly complex, but still are still lacking the ability to accurately describe such systems. Even elaborate complex models of systems produce simulations that diverge from or fail to predict the real behaviors of those systems. This situation is accentuated in cases where real-time dynamic conditions exist. Application simulations that can dynamically incorporate new data, archival or from on-line measurements of the actual systems, offer the promise of more accurate analysis, more accurate predictions, more precise controls, and more reliable outcomes. These capabilities are fostered with the Dynamic Data Driven Applications Systems (DDDAS) paradigm. DDDAS can be viewed as a methodology to counterbalance incompleteness in model and capability to enhance

M. Bubak et al. (Eds.): ICCS 2004, LNCS 3038, pp. 662–669, 2004.
© Springer-Verlag Berlin Heidelberg 2004

the application models by imparting additional information into the model as at runtime additional data are used to selectively enhance or refine the original model.

In addition DDDAS paradigm fosters the ability of an application simulation to control and guide the measurement processes, thus creating more effective measurement processes. Such capabilities are highly desirable in cases where measurements are difficult to perform, expensive or time-critical. The ability to guide the measurement process and selectively focus on a subset of the measurement space can result in more efficient and effective measurements, which can desirable in reducing cost, collection time, or improving on the quality of data collected.

The author conceived the idea of this paradigm in the early 80's while doing radiation transport (Monte Carlo and Discrete Ordinates methods) computations for simulations and measurements relating to oil exploration [1]. These computations were quite expensive, executing in batch mode for many hours/days on state-of-the-art supercomputers of that time. The original motivation for this idea was the desire to speed the computations by using additional experimental data in selective places of the computation, to improve the accuracy of statistical sampling and speedup convergence in regions of interest, and reversely to control the measurement process, which for oil exploration is time-critical and expensive. That immediately led to the broader scope of such a paradigm, namely using the capability of injecting data as more broadly for refining the modeling analysis and prediction capabilities of applications. However at that time these computations were routinely run in batch mode. Even technologies such as user interactivity were just emerging, and even user interactivity is still a large step behind the kinds dynamic interfaces needed between the measurement system and the executing application in DDDAS environments. Similarly significant limiting factors were the then state of the art in complex modeling of systems, technologies for dynamic composition of applications, dynamic runtime support, and numeric and non-numeric algorithms. Some preliminary analysis of the opportunities and challenges were discussed in [2], in the context of the Gedanken Laboratory. While in the 80's many of the relevant technologies to enable this paradigm were at a stage of being non-existent or prohibitively inadequate, recent advances in the late 80's, the 90's make DDDAS more feasible to pursue now than ever before. In addition while the initial inspiration for this paradigm were engineering/scientific kinds of applications, later experiences [3] with manufacturing process and other management applications showed that there is significant relevance of DDDAS to these classes of applications also. The term Dynamic Data Driven Applications Systems (DDDAS) was coined around the time the author led the organizing effort of the NSF Workshop [4] in March of 2000, which provided a forum of discussion these ideas with broader research community, as well as addressing in more detail the challenges and the opportunities.

Manufacturing process controls, resource management, weather and climate prediction, and traffic management, systems engineering, geo-exploration, bio-sensoring are examples of areas that can benefit from the DDDAS paradigm. The vision of DDDAS, goes beyond the current concepts of real-time control, in terms of the concept of interaction of the application simulation with the measurement system, the range of applications, the streamed input data and the scope of systemic approach to addressing the ensuing challenges: assessment of data quality, incorporation of uncertainty, ability to combine different kinds of data taken at different times, application algorithms with stable response to the streamed data, distributed systems (for sensing, for computation and for output, and for data storage), development of effective interfaces of applications with measurement processes, supporting the execution of such applications with dynamic resource requirements.

2 Research Challenges

This new paradigm creates a rich set of new challenges and new class of problems for the applications, algorithms, and systems researchers to address. Such challenges include: advances at the applications level for enabling this kind of dynamic feedback and coupling with measurements; advances in complex, multi-modal models, and ability to combine dynamically such models as is dictated by the injected data, advances in the applications algorithms for the algorithms to be amenable to perturbations by the dynamic data inputs and enhanced capabilities for handling uncertainties in input data; new technology in the computer software systems areas to support such environments where the applications requirements are dynamic, depending on the injected data.

A number of these research and technology issues were addressed in a workshop [3] organized by NSF to examine the technical challenges and research areas that need to be fostered to enable such capabilities. Representative examples of applications were addressed at this workshop to illustrate the potential impact that this the DDDAS paradigm can have. To enable such capabilities multidisciplinary research and technology development is needed in the four component areas: applications, in mathematical algorithms, in systems software, and in measurements.

In terms of the applications challenges, DDDAS requires ability in the application simulations to accept data at execution time and be dynamically steered by such data. To handle dynamically injected data, requires application models that describe the application system at different levels of detail, and the ability to dynamically invoke such models as needed by the input data. New application modeling capabilities are necessary, including: multi-modal methods of describing the system at hand; modeling multiple levels of system detail; model enhancements for dynamic resolution; dynamic data assimilation; ability to incorporate asynchronously collected data. Methods and frameworks enabling dynamic composition of applications depending on the streamed data, namely: dynamic selection of models, based on input data; application model interfaces. Mathematical and computational models of multiscale systems often involve a hierarchy of models; how to pass information between the levels of the hierarchy is not yet well understood.

Enabling DDDAS requires advances in the existing mathematical algorithms, or development new mathematical algorithms, that will be stable and have robust convergence properties under dynamic data inputs. The of mathematical and statistical aspects include: methods to assess the quality and propagation of measurement errors and of uncertainty, in both stochastic and deterministic systems, and methods to incorporate data of variable quality, such as for example in cases where there is need to combine data (measured or computed) taken at different spatial or temporal scales is a significant problem. Need to address the challenge of small sample sizes, incomplete data, and extreme events. Differences between computational results and external data can be viewed as a measure for controlling the course of the simulation. Here all the issues of uncertainty, sensitivity, combination, scatter, quality, propagation, and hierarchy arise again in attempting a systematic control of the simulation. In addition there is need to address issues of varying discretization schemes, depending on the streamed data, and develop the underlying algorithms with the ability to interpolate or extrapolate between grids of different granularity.

In terms of systems software technology and infrastructure needed to support the dynamic computational, communication and data requirements of DDDAS applications, the interfaces with measurement systems, supporting fault-tolerance and quality-of-service guarantees to ensure streamed data and resource availability. Research advances in this area include systems software that will support the execution of applications whose requirements are dynamic depending on the streamed data. In addition new systems software approaches supporting interfaces of applications software with measurement systems, including sensor systems, and systems software to manage the underlying computational grid resources and measurement systems. Such technologies include: complex application frameworks, dynamic application execution support environments; support for dynamic selection of application components. Multiresolution capabilities require the ability dynamically select application components, embodying algorithms suitable for the kinds of solution approaches depending on the streamed data, and depending on the underlying resources. Dynamic computing requirements and matching dynamic resource requirements: DDDAS will employ heterogeneous platform environments such as embedded sensors for data-collection, distributed high-performance simulations environments and special-purpose platforms for pre- and post processing of data, and visualization.

In terms of data and measurements, there is need for developing interfaces to physical devices (including sensor systems) and application/measurement interfaces together with new capabilities for managing measurement systems, sensors and actuators which are additional "resources" in the computational grid, and as such, resource discovery and allocation of sensors and actuators becomes an important issue. Combined application and measurement data would require new approaches for data management systems supporting different naming schemas or ontologies, information services, or information views.

3 Why Now

Why now is the time for developing such capabilities? DDDAS is a powerful and new paradigm, requiring advances on several technologies. Over the recent years there has been progress in a number of technology areas that makes the realization of DDDAS possible. These include advances in hardware platforms' computing speeds, the advent of grid computing and sensor networks, applications and algorithms for parallel and distributed platforms, computational steering and visualization, computing, networking, sensors and data collection, and systems software technologies. These are some of the recent advances that will be exploited to enable the new paradigm of DDDAS. It is necessary however to further develop these technologies with enhanced capabilities to advance the state of the art in DDDAS. This background work make this a timely endeavor to pursue, as they provide the basis to springboard and enable the new capabilities discussed here.

Dynamic Data Driven Application Systems (DDDAS) is an opportunity to formulate application simulation models and methods that incorporate/integrate dynamically measurement data, algorithms, system tools, and mathematical and statistical advances to produce effective simulation systems for important applications across the range of science and engineering disciplines.

4 Overview of Selected Research Efforts Presented in This Workshop

The rest of the papers presented in this workshop represent ongoing research projects each addressing a number of aspects of the technical challenges mentioned above. With the exception of [18], all other projects [5-17, 19-25] are predominantly funded by NSF ITR program [26]. Several papers address the aspect of application simulation enhanced by additional data incorporated into the computation and simulation guided and controlled measurements. While many of the papers are developing new capabilities at the application and algorithm levels, several of the papers focus in the underlying software technology to support the development of the complex DDDAS applications, their dynamic runtime support requirements and interfaces to control measurement systems. Several papers provide examples of simulation controlled measurement processes. Most of the work on mathematical algorithms tolerant and stable to dynamically streamed data is done in the context and together with the development of the application models. In all cases the mathematical algorithms and systems software development is done in the context of specific applications driving and validating the developed technologies.

The projects discussed in [5, 11, 12, 24] include capabilities of application guided control of measurements. In [5] the project employs dense networks of high resolution radars to spatially resolve and improve location accuracy of tornadoes by overcoming earth induced blockage and optimizing data resources in response to multiple and sometimes conflicting needs. The paper [11] presents development of robust, efficient and scalable forward and inverse 3D wave propagation solvers and integrated steerable field measurements to improve techniques for forecasting of ground motion during future earthquakes. The project in [12] discusses development of a prototype hardware and software system to integrate dynamically empirical fire spread models with turbulence models representing the fire generated winds, together with up-to-date aerial survey and thermal sensor data and other information related to a wildfire in progress, and to provide prediction of the course of the wildfire in visual form and provide tools for predicting the outcome of firefighting strategies. The project in [24] provides an example of the underlying hardware and software platform developed for the case of neural modeling interacting with real-time data streams, by enabling real-time decoding of neural data streams and allowing neuronal models to interact with living nervous systems, and enabling integrated interactive control in the analysis of the neuronal function.

Projects such as those in [8 , 14, 15, 16] are dealing with appliying DDDAS for exnhanced structural integrity and physical systems design capabilities. In [8] analysis of structural integrity aspects of a vessel, using sensor data encoding strain-field measurements and other loading conditions, induced for example from an explosion, and which are streamed into the structural analysis program, or input to an aeroelastic model where embedded sensors provide datastreams of free-stream pressure, density and Mach numbers in simulating airplane wing flutter. In addition to the capabilities in the application level, the project is developing an environment for a multi-physics computationally implementable framework in the form of a Multidisciplinary Problem Solving Environment. In [14] development of state-of-the-art, efficient and robust methods for producing data enhanced, reduced order models of large state-space systems, and applied to structural impact response simulations for events such as for example the Pentagon plane crash on 9/11/2001. In [15] Data Driven Design optimization methodology for multi-objective optimization of engineering design of a submerged inlet is it will occur in fluid flow duct through which air is delivered to a turbojet, turbofan or ramjet. In [16] MRI measurement enhanced computer simulations to enable reliable and efficient determination of modeling functions for patients with reconstructed ears.

Papers [5, 6, 7, 9, 10, 11] focus on environments and crisis management applications, such as [5,6] for weather related phenomena such tornado studies and track prediction. Like [5]. also [6] Weather Surveillance radar Systems w continuous data streaming can lead to improved ability to discriminate tornadic from non-tornadic events. In [7] physical and biogeochemical ocean dynamics can be highly variable and involve interactions on multiple scales. For improved accuracy and efficiency in forecasting the structures and parameter models must evolve and respond dynamically to new data injected into the executing prediction system. In [11] as mentioned above develops enhanced methods for seismic modeling and detection. Papers [9, 10] deal with applications for subsurface and ambient atmosphere pollutant modeling respectively. In [9] are described numerical procedures for performing multi-scale interpolation of data to allow continuously updating simulation computations with sensor data streamed into the computation and shows that the methods significantly improve the prediction results for applications such as contaminant tracking in subsurface environments. Paper [10] describes continuous assimilation of data in the modeling of aerosol dynamics for more accurate tracking of chemical pollutant in ambient environments.

Other crisis management applications like fire propagation in enclosed environments, like buildings and submarimes, and in ambient environments are discussed respectively in [13, 12]. In [13], agent-based and data-driven simulation of fire propagation dynamics are modeled, and these enhanced modeling approaches can provide more accurate information to firefighting personnel and effective response.

Paper [22] discusses the development of applications for oil recovery, and the corresponding application development and runtime support framework. In addition the project is also using this framework for the development of enhanced medical applications, such as MRI (virtual microscope). These modeling methods and underlying systems software framework that this project has developed enables autonomic oil production management, by implementing dynamic data driven modeling ideas and ability to support the execution of the ensuing complex dynamic models and data in Grid environments.

Other application areas include [18] for transportation systems management ability to effectively analyze and manage heavy road traffic by relating dynamically the observed dynamics in speed and density of vehicles and driver behaviors and reactions. In the area of human stress recognition [23] applies the DDDAS paradigm to develop new techniques for image-based stress recognition using model-based dynamic face tracking system, thus enhancing the capability to detect human stress an application of which is in security.

Examples of mathematical algorithms aspects are addressed in papers [17, 25]. In [17] this work is developing a generalized feature mining framework to deal with analysis of evolutionary data (in the case study of CFD computations used in designing vehicles), and improve on the methods of response surfaces used to describe performance of the system (in this case vehicles) as functions of various geometrical design parameters, or sensor data measuring for example air speed and drag. In [25] are discussed efficient methods for calculating the effect on the fidelity of the computed results (as represented by error bars in the computed entities) which results from the uncertainty errors in the input data. In enabling DDDAS systems, where data with their corresponding uncertainties have to be incorporated dynamically into the computation, it is essential to have capabilities of efficiently calculating the error propagation.

Underlying infrastructure technologies that are being developed are discussed in the papers [19, 20, 13, 21, 22]. The SAMAS framework [19] allows multilevel agent-based simulations to incorporate dynamically data from different levels of abstraction of the multilevel models, and in particular allow multitemporal abstractions, an important aspect of enabling DDDAS capabilities requires the ability the integrate multimodal, multispatial and multitemporal data and models, and the ability to ensure the execution of such simulations with guaranteed quality of service, such as for example in time-critical applications. In [20] are explored techniques to allow the capability of dynamic composition of simulation components required to support DDDAS environments. The systems software technologies developed include abstractions of such components, syntactic definition of their interfaces, and their role in the semantic definition of the entire system, including both measurement and simulation modeling aspects. In [21] a new web services framework is discussed that is better suited for the DDDAS application environments.

5 Summary

The DDDAS paradigm has the potential to transform the way science and engineering are done, and induce a major beneficial impact in the way many functions in our society are conducted, such as manufacturing, commerce, transportation, hazard prediction/management, ecology and medicine, to name a few. To achieve such goals, fundamental, multidisciplinary and collaborative research is needed and develop technologies spanning applications, mathematical algorithms, and systems software, aimed at creating DDDAS capabilities. The remainder of the papers in these proceedings present efforts in these research and technology areas.

References

1. Schlumberger-Doll Research; www.slb.com
2. Parallel Applications and the Gedanken Laboratory, F. Darema; Conference of the Society of Engineering Sciences, StFe, NM, Oct 22-24, 1990.

3. On the Parallel Characteristics of Engineering/Scientific and Commercial Applications: differences, similarities and future outlook", F. Darema, invited paper, in the book: Parallel Commercial Processing, ed. J. A. Keane, UK, 1994 and 1996.

4. NSF Workshop, March 2000; www.cise.nsf.gov/dddas

5. Distributed Collaborative Adaptive Sensing for Hazardous Weather Detection, tracking and Predicting; Brotzge, Chandresakar, Droegemeier, Kurose, McLaughlin, Philips, Preston, Sekelsky; Proceedings ICCS'04

6. Rule-Based Support Vector Machine Classifiers Applied to Tornado Prediction; Trafalis, Santosa, Richman; Proceedings ICCS'04

7. Adaptive Coupled Physical and Biogeochemical Ocean Predictions: A Conceptual Basis; Lermusiaux, EvangelinosTian, Haley Jr., McCarthy, Patrikalakis, Robinson, Schmidt; Proceedings ICCS'04

8. Dynamic-Data-Driven Real-Time Computational Mechanics Environment; Michopoulos, Farhat, Houstis; Proceedings ICCS'04

9. A Note on Data-Driven Contaminant Simulation; Douglas, Shannon, Efendiev, Ewing, Ginting, Lazarov, Cole, Jones, Johnson, Simpson; Proceedings ICCS'04

10. Computaional Aspects of Data Assimilation for Aerosol Dynamics; Sandu, Liao, Carmichael, Henze, Seinfeld, Chai, Daescu; Proceedings ICCS'04

11. A Framework for Online Inversion-Based 3D Site Characterization; Akcelic, Bielak, Biros, Epanomeritakis, Ghattas, Kallivokas, Kim; Proceedings ICCS'04

12. A Note on Dynamic Data driven Wildfire Modeling; Mandel, Chen, Franca, Johns, Puhalskii, Coen, Douglas, Kremens, Vodacek, Zhao; Proceedings ICCS'04

13. Agent-based Simulation of Data-Driven Fire Propagation Dynamics; Michopoulos, Tsompanopoulou, Houstis, Joshi; Proceedings ICCS'04

14. Model Ruduction of Large-Scale Dynamical Systems; Antoulas, Sorensen, Gallivan, Van Dooren, Grama, Hoffman, Sameh; Proceedings ICCS'04

15. Data Driven Design Optimization Methodology Development and Application; Zhao, Knight, Taskinoglu, Jovanovic; Proceedings ICCS'04

16. A Dynamic Data Driven Computational Infrastructure for Reliable Computer Simulations; Oden, Browne, Babuska, Bajaj, Demkowicz, Gray, Bass, Feng, Prudhomme, Nobile, Tempone; Proceedings ICCS'04

17. *Improvements to Response-Surface Based Vehicle Design using a Feature-Centric Approach*; Thompson, Parthasarathy, Machiraju, Lawrence; Proceedings ICCS'04

18. An Application for the Virtual traffic laboratory: calibrating speed dependence on heavy traffic (a demonstration of a study in a data driven traffic analysis); Visser, Zoetebier, Yakali, Hertzberger; Proceedings ICCS'04

19. SAMAS: Scalable Architecture for Multi-resolution Agent-based Simulation; Chaturvedi, Chi, Mehta, Dolk; Proceedings ICCS'04

20. Simulation Coercion Applied to Multiagent DDDAS; Loitiere, Brogan, Reynolds; Proceedings ICCS'04

21. O'SOAP – A Web Services Framework for DDDAS Applications; Pingali, Stoghill; Proceedings ICCS'04

22. Application of Grid-enabled Technologies for Solving Optimization Problems in Data-Driven Reservoir Studies; Parashar, Klie, Catalyurek, Kurc, Matossian, Saltz, Wheeler; Proceedings ICCS'04

23. Image-Based Stress Recognition Using a Model-Based Dynamic Face Tracking System; Metaxas, Venkataraman, Vogler; Proceedings ICCS'04

24. Developing a Data Driven System for Computational Neuroscience; Snider, Zhu; Proceedings ICCS'04

25. Karhunen-Loeve Representation of Periodic Second-Order Autoregressive Processes; Lucor, Su, Karniadakis; Proceedings ICCS'04

26. NSF Information Technology Research (ITR) Program (1999-2004)

Distributed Collaborative Adaptive Sensing for Hazardous Weather Detection, Tracking, and Predicting*

J. Brotzge[1], V. Chandresakar[2], K. Droegemeier[1], J. Kurose[3],
D. McLaughlin[4], B. Philips[4], M. Preston[4], and S. Sekelsky[4]

[1] Center for Analysis and Prediction of Storms, University of Oklahoma,
100 East Boyd Norman, OK 73019-1012,
{jbrotzge,kkd}@ou.edu
[2] Dept. Electrical & Computer Engineering Colorado State University,
Fort Collins, CO 80523-1373,
chandra@engr.colostate.edu
[3] Dept. Computer Science, University Massachusetts,
Amherst MA 01003
kurose@cs.umass.edu
[4] Dept. Electrical and Computer Engineering, University Massachusetts,
Amherst MA 01003
{mclaughlin,bphilips,mpreston,sekelsky}@ecs.umass.edu

Abstract. A new data-driven approach to atmospheric sensing and detecting/ predicting hazardous atmospheric phenomena is presented. Dense networks of small high-resolution radars are deployed with sufficient density to spatially resolve tornadoes and other dangerous storm events and overcome the earth curvature-induced blockage that limits today's ground-radar networks. A distributed computation infrastructure manages both the scanning of the radar beams and the flow of data processing by dynamically optimizing system resources in response to multiple, conflicting end-user needs. In this paper, we provide a high-level overview of a system architecture embodying this new approach towards sensing, detection and prediction. We describe the system's data rates, and overview various modes in which the system can operate.

1 Introduction

Current approaches for observing the atmosphere are based upon a paradigm of widely separated, geographically fixed sensors that, by necessity, operate independent of the phenomena being observed and of the sometimes disparate needs of multiple end users. This is true of the WSR-88D Next Generation Doppler Radar (NEXRAD) system, which consists of 141 ground-based radars and serves as the cornerstone of the weather and storm-sensing system in the United States. This system has

---

* This work was supported by a grant from the Engineering Research Centers program of the National Science Foundation under cooperative agreement EEC-0313747. Any opinions, findings, and conclusions or recommendations expressed in this material are those of the authors and do not necessarily reflect the views of the National Science Foundation.

M. Bubak et al. (Eds.): ICCS 2004, LNCS 3038, pp. 670–677, 2004.
© Springer-Verlag Berlin Heidelberg 2004

tremendous capabilities to observe many types of weather phenomena, yet it remains *fundamentally constrained* in sensitivity and spatial resolution owing to the very long operating range (hundreds of km) over which the radars operate. Systems such as this are unable to view much of the lower atmosphere because the radar beam overshoots approximately the lowest 1 km due to the curvature of the Earth. These systems are also unable to spatially resolve many tornadoes and other damage-causing phenomena that have spatial scales smaller than the 2-4 km resolution achieved by the radar beams at long ranges.

This paper introduces a new, transforming paradigm for atmospheric sensing based on Distributed Collaborative Adaptive Sensing (DCAS) networks designed to overcome the fundamental limitations of current approaches to sensing and predicting atmospheric hazards. *Distributed* refers to the use of large numbers of small, solid-state radars, spaced appropriately, to overcome blockage due to the Earth's curvature, resolution degradation caused by spreading of radar beams at long ranges, and the large temporal sampling intervals that result from today's use of mechanically scanned antennas. In addition to providing high-resolution sampling throughout the entire troposphere, this distributed concept lends itself to the efficient utilization of low-power solid-state radars. These radars operate *collaboratively*, via coordinated targeting of multiple radar beams, based on atmospheric and hydrologic analysis tools (detection, tracking, and predicting algorithms). This enables the critical resources of the sensing system, such as radiated power, antenna beam positions, and data communications bandwidth, to be allocated to enable sampling and data acquisition in specific regions of the atmosphere where particular threats exist. *Adaptive* refers to the ability of these radars and the associated computing and communications infrastructure to rapidly reconfigure in response to changing conditions in a manner that optimizes response to competing end user demands. For example, such a system could track tornadoes for public warning while simultaneously collecting information on the parent storm and providing quantitative precipitation estimates for input to hydrologic prediction models. The system is thus driven by the data needs of the end-users as shown in Figure 1. The very first test bed of the project employing four radars will not use phased arrays: it will use off-the-shelf dishes moved around using motorized pedestals, and that system will be up and running in Spring 2005. There is underway a technology push for the project to make a phased array version of the system. This version will replace the individual dishes with phased array panels. Since there won't be any moving parts, the beams will be able to move around much quicker in response to data requests. That system will be fielded in the Spring of 2006 and will have 9 phased array radars arranged in a cooperative network.

In October of 2003, the National Science Foundation created an Engineering Research Center (ERC) among the University of Massachusetts (lead university), University of Oklahoma, Colorado State University, and the University of Puerto Rico at Mayaguez, and a consortium of industrial partners to lay the fundamental and technological foundations for DCAS and to investigate the practicability of this new paradigm for storm sensing and prediction. Called the Center for Collaborative Adaptive Sensing of the Atmosphere (CASA) [2], this ERC will create a series of test-beds that enable exploration of the design-space of DCAS using fielded hardware and software and enable proof-of-concept demonstration experiments involving specific end-users. This paper describes the architecture and data requirements for NetRad, which is the first of a series of system-level test-beds to be created and deployed by this center.

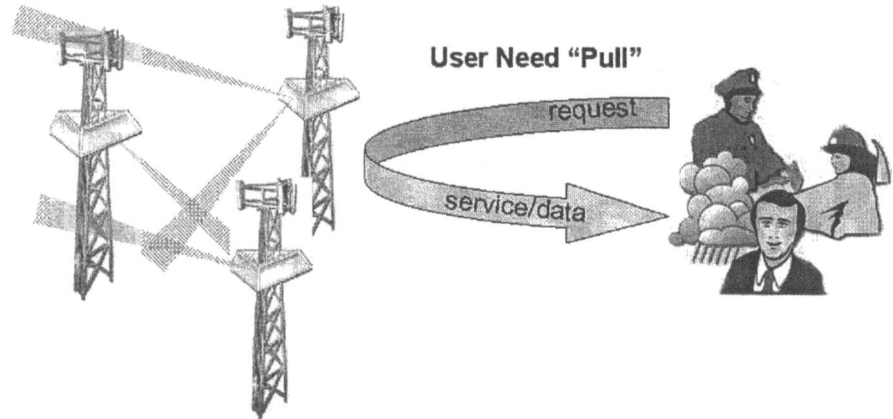

Fig. 1. Distributed Collaborative Adaptive Sensing System (DCAS) Driven by End-User Data Needs

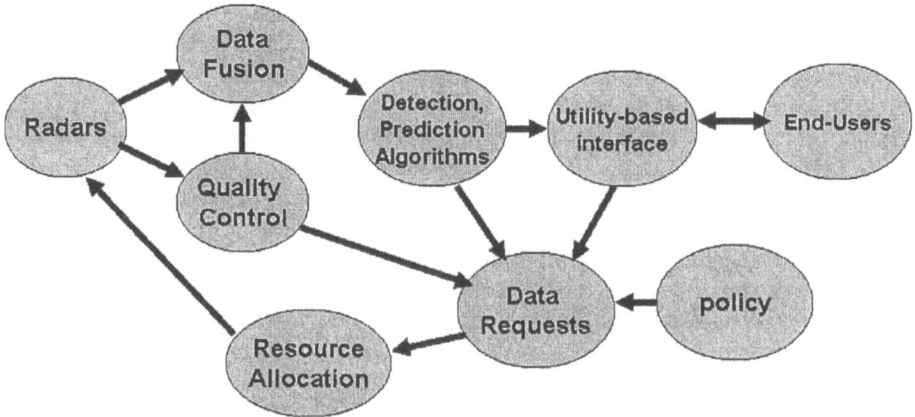

Fig. 2. DCAS System Functional View

2 DCAS Data Handling Architecture

High-level functional views of the NetRad architecture are shown in Figures 2 and 3. Users, either directly, or through meteorological algorithms, request stored and live radar data from the system. Competing resource requests (e.g., for radar beam scheduling) are mediated by the resource control algorithms in order to optimize overall utility. Data is returned from the radars through the data store, and to the requesting algorithms and users.

Figure 3 shows that data is at the very heart of the NetRad system. A **distributed data storage** facility is currently being designed that supports *(i)* querying, *(ii)* reading, and *(iii)* writing of data. The data store acts as an intermediary between data producers and consumers. The data producers include the radars and QC algorithms

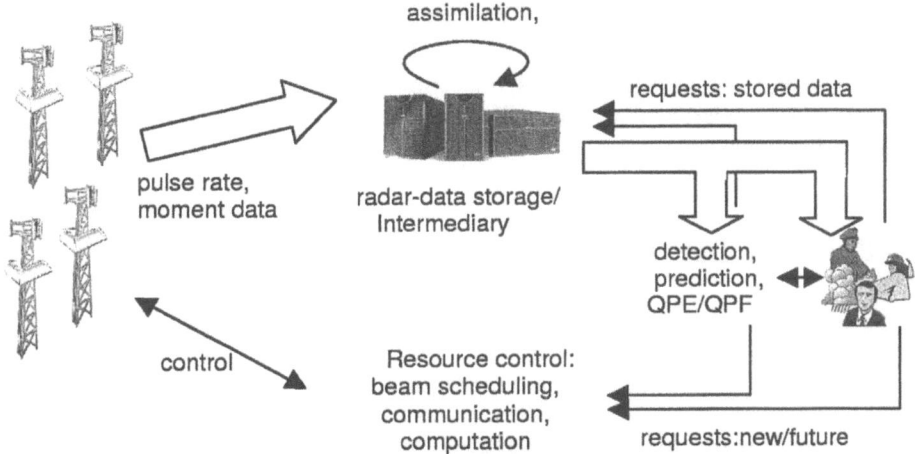

Fig. 3. Overview of NetRad architecture

that create QC-enhanced data; the data consumers include end-users who want direct access to retrieved data, hazardous weather detection and prediction algorithms, quantitative precipitation estimation and prediction algorithms, and quality control algorithms that read data in order to create new QC-enhanced data. The use of a data storage facility to provide a level of indirection [4] (i.e., act as an intermediary) between data producers and data consumers provides many benefits in terms of system structure. For example, with this level of indirection, it is easy for multiple data consumers to receive the same data, and to dynamically add new data consumers for the data stream. Similarly, a late arriving request for data that is currently being streamed to an existing consumer can be satisfied by adding the new consumer to the ongoing multicast data stream, and initiating a second patching stream that allows the new consumer to receive the missed data.

The interface to the data store will provide for read, write, and query [1] operations.

- **Data store write interface.** Radar observations are written into the data store. Similarly outputs from the QC (Quality Control) algorithms (e.g., to perform dealiasing) are written into the data store. Each stored data set will have a data provenance indicating the data source characteristics, and a history of all processing that has been performed to produce the data set.
- **Data store query interface.** The data store will provide a query interface over the globally stored data. An individual query will return a "handle" to data matching the query, as well as the provenance information stored with the data. The handle can then be used directly to read the data, if desired. Two basic types of queries that will be supported. The first type of query specifies a spatial region and a time interval; a handle to all matched data is returned. The second type of query is one that specifies a condition on the *value of data*; a handle to all matched data is again returned.
- **Data store read interface.** A handle is presented to the data store to read data associated with the query. The data store provides a pipe abstraction that *allows data to be read as they are being written.* This allows a data consumer

to receive data in real-time, i.e., without waiting for it to first be stored in the data store.

We note that while the actual storage location of data may be contained in the data pedigree, data readers and writers will typically read/write/query data in a storage-location-blind manner.

3 Per-radar Data Rates

In this section we describe the radar data being produced, as well as the overall rates associated with this data. A NetRad radar operating in "surveillance mode" (surveying the volume of atmosphere above the radar and out to a radius of 30 km) will sense roughly over an area with a radius of 30km, to a height of 3 km. This sensed volume is divided into unit volumes (referred to as *voxels*) of approximately 500m by 500m by 100m. These dimensions are set by the combination of the radar's two-degree antenna beamwidth and the 1 uSEC width of the transmitted pulse. A NetRad radar will thus sample approximately 350K voxels every 30 seconds.

The following six quantities are estimated per voxel every 30 seconds. These values are often referred to as *moment data* (referring to moments of the Doppler Spectrum of the received signals), as they are averaged over a number of radar pulses transmitted in a given beam position.

- **Reflectivity (Z):** This is a measure of the amount of the backscattered signal returned to a radar due to scatterers (e.g., raindrops and hail) in the voxel being sampled. It is proportional to the volumetric radar cross section of the observation volume and proportional to the sixth power of the diameter of the ensemble of hydrometeors in the observation volume.
- **Mean Doppler Velocity (V).** This is the mean of the Doppler velocity spectrum, indicating the average radial velocity of scatterers in the voxel.
- **Doppler Spectrum Width (W).** This is a measure of the spectral spread and often approximated by the standard deviation of a Gaussian shaped model for the Doppler spectrum.
- **Differential reflectivity (Z_{dr}).** NetRad will be deploying polarimetric radars, transmitting a signal that has equal powers at horizontal and vertical states Z_{dr} is the ratio of the power returned by the horizontally and vertically polarized pulses and is a measure of the deviation from a spherical shape, of the scatterers.
- **Correlation coefficient (ρ_{hv}).** This is a measure of the correlation between the horizontally and vertically polarized returns.
- **Differential phase (Φ_{DP}).** This is a measure of the difference in phase between the horizontally and vertically polarized returns, dominated by the propagation phase difference between two polarization states.

Given that the 350K voxels each produce six four-byte moment data (Z,V,W, Z_{dr}, ρ_{hv}, Φ_{DP}) every 30 seconds, the moment data rate is slightly over 2 Mbps. A small amount of additional overhead (including timing, radar operating parameters, and packetization overhead) will increase this value slightly.

Certain meteorological algorithms will require "raw" (unaveraged) radar data – per-pulse data corresponding to the received signals (amplitude and phase in the horizontal and vertical polarization directions, for 8 bytes total per point) received in response to each of the N pulses for a given voxel. The data rate for this *pulse-rate data* is thus approximately 100 Mbps.

We note that these requirements are for uncompressed data. Studies of compression of WSR-88D NEXRAD moment data (which is of much coarser scale than NetRad data) indicates that compression ratios between 1:6 and 1:10 can be achieved [3]. The compression ratios of NetRad moment and pulse data, which result from sensing the atmosphere at a much finer scale than NEXRAD, remain an open question.

In summary then, the per-radar data rates are approximately 2 Mbps for moment data, and 100 Mbps for pulse rate data. Each radar in the network will contribute this amount of data.

4 Adaptive Sensing through Data Mining

Once the moment data have been collected in the distributed data storage facility, the data are ready to be 'mined'. Data mining describes the process by which observations and/or gridded analyses/forecasts are interrogated to extract useful information.

A number of algorithms are included within the NetRad architecture for the detection of hazardous weather phenomena. Which algorithm is applied and when is dependent upon the scanning mode of the radar. Five scanning modes are established: Severe Storm Anticipation, Tornado Acquisition, Tornado Pinpointing, Quantitative Precipitation Estimation, and Data Assimilation and Prediction. Thus, through data mining processes, the radar scanning strategy adapts based upon the development of hazardous weather events.

During the *Severe Storm Anticipation* mode, the radar scanning strategy consists of general survey volume scans in search of any hazardous weather development. A Storm-Cell algorithm determines storm-cell boundaries and other radar observed thunderstorm characteristics. A Boundary Identification algorithm determines discontinuities that may trigger further storm growth and/or lead to tornadogenesis. Other algorithms search for shear and mid-level convergence as possible precursors for tornado development.

If a tornado precursor is identified, one or more nearby radars will switch to *Tornado Acquisition* mode. During this mode, the radar(s) will only focus on the areas of interest while collaborative radars continue to survey for general thunderstorm development. Meanwhile, several additional detection algorithms now begin to search for actual tornado signatures. Reflectivity and vorticity features, such as hook echoes and regions of circulation, are determined. From these features, once a tornado is identified, then *Tornado Pinpointing* mode is activated.

Tornado Pinpointing mode takes advantage of the collaborative radar system by triangulation of the beams to isolate the location of a tornado to within 100 m. Tornado triangulation merges data from multiple nodes to pinpoint the exact location (and possible trajectory) of the tornado(s).

Heavy rainfall often is an overlooked hazard in thunderstorms. Once storm cells are detected, the *Quantitative Precipitation Estimation* (QPE) mode is activated.

Scanning strategies will commence that allow rainfall fields to be mapped in real-time. The *Data Assimilation and Prediction* mode is also activated as storm cells are identified.

Assimilation of the NetRad data into a gridded 3-D volume analysis provides a critical advantage for improved detection and forecast performance: a means to combine NetRad data with external data sources such as NEXRAD Doppler radar, geostationary satellite, surface and upper-air observations, and operational numerical analysis and prediction grids. Assimilation describes the process by which atmospheric observations are combined with prior forecast information to yield a physically complete, dynamically consistent 3-dimensional state of the atmosphere, and includes such techniques as 3D- and 4D-variational analysis, ensemble Kalman filtering, single-Doppler velocity retrieval, and simple adjoints.

Such real-time assimilation of NetRad data provides the analyses for forecasts of hydrometeors, 3-D winds, and thermodynamic quantities, on spatial scales of order 1 km, for use in nowcasting and forecasting severe storm threats including but not limited to strong surface winds, hail, heavy precipitation, and tornadic potential. Nowcasting involves the use of meteorological information to generate 0-1 hour forecasts which rely heavily upon statistical techniques. One- to three-hour forecasts are made using much more advanced numerical weather prediction models, such as the Advanced Regional Prediction System (ARPS) and the Weather Research Forecast (WRF) models.

As a tornado or other important weather feature is identified, the corresponding coordinates (latitude and longitude) are sent to the **Distributed Resource Control Component** (DRCC). The DRCC converts the tornado coordinates into the appropriate polar coordinates (elevation and azimuth). The DRCC then adjusts scan strategies among the nodes. The exact scan strategy is based, in part, on the prioritization needs as defined by the End Users. These scan strategy commands are then sent back to the individual nodes.

5 Summary

The DCAS paradigm described in this paper is a dynamic data-driven system that samples the atmosphere when and where end-user needs are greatest. The concept has the potential to dramatically improve our ability to observe the lower troposphere, and to vastly improve our ability to observe, understand, and predict severe storms, floods, and other atmospheric and airborne hazards. Many knowledge and technology barriers need to be overcome prior to creating practicable DCAS systems, however. Among these barriers are the need to define an architecture and an appropriate interface to end-users, the lack of small, low-cost solid-state radars, the need to define optimization approaches and policies to adaptively allocate resources in a quantifiably optimal fashion. The NetRad system described at high level in this paper is but the first of a series of system-level test-beds that will be created by the CASA Engineering Research Center to explore the DCAS design space and instantiate the concept in the field.

References

1. J. Annis, Y. Zhao, J. Voeckler, M. Wilde, S. Kent, I. Foster, "Applying Chimera Virtual Data Concepts to Cluster Finding in the Sloan Sky Survey." *Supercomputing 2002 (SC2002)*, November 2002.
2. Center for Collaborative Adaptive Sensing of the Atmosphere, www.casa.umass.edu
3. S.D. Smith, K. Kelleher, Lakshmivarahan, "Compression of NEXRAD (WSR-88D) Radar Data Using Burrows-Wheeler Algorithm," *Proc.18th IIPS*, Orlando FL, Amer. Meteor. Soc., 133-135.
4. I. Stoica, D. Adkins, S. Zhuang, S. Shenker, S. Surana, "Internet Indirection Infrastructure," *Proc. ACM Sigcomm 2002*.

Rule-Based Support Vector Machine Classifiers Applied to Tornado Prediction

Theodore B. Trafalis[1], Budi Santosa[1], and Michael B. Richman[2]

[1] School of Industrial Engineering, The University of Oklahoma
202 W. Boyd, CEC 124, Norman, OK 73019
{ttrafalis,bsant}@ou.edu
[2] School of Meteorology, The University of Oklahoma
100 E. Boyd, SEC 1310, Norman, OK 73019
mrichman@ou.edu

Abstract. A rule-based Support Vector Machine (SVM) classifier is applied to tornado prediction. Twenty rules based on the National Severe Storms Laboratory's mesoscale detection algorithm are used along with SVM to develop a hybrid forecast system for the discrimination of tornadic from non-tornadic events. The use of the Weather Surveillance Radar 1998 Doppler data, with continuous data streaming in every six minutes, presents a source for a dynamic data driven application system. Scientific inquiries based on these data are useful for dynamic data driven application systems (DDDAS). Sensitivity analysis is performed by changing the threshold values of the rules. Numerical results show that the optimal hybrid model outperforms the direct application of SVM by 12.7 percent.

1 Introduction

Rule-based classification methods have shown promise in physical systems applications [1]. One builds a rule-based model by incorporating prior information. In the case of Support Vector Machines (SVMs), prior knowledge is incorporated into the model as additional constraints in the form of polyhedral rule sets in the input space of the given data. These rule sets are supposed to belong to one of two categories into which all the data are divided [2, 3].

Tornado forecasting is an active area of research in the meteorological community [4, 5]. State-of-the-science weather radar scans volumes of the atmosphere, producing a large amount of information that is updated every 5 to 6 minutes. Scientific inquiries based on these data are useful for Dynamic Data Driven Applications Systems (DDDAS). Once the data are collected, they are quickly processed by algorithms that look for signatures of tornadoes in near-real time, since an extra minute of lead-time can save lives. The dynamic nature of DDDAS problems requires us to address the time dependency or real time nature of the applications. Certain applications (e.g., tornado formation) require real time response to observations from data. Typically, in the prediction of severe weather potential, data from observations taken hours previous to the formation are used and these are not updated with real data as they become available. Incorporating new dynamically injected data is a fundamental

M. Bubak et al. (Eds.): ICCS 2004, LNCS 3038, pp. 678–684, 2004.
© Springer-Verlag Berlin Heidelberg 2004

change in the design. The use of the Weather Surveillance Radar 1998 Doppler (WSR-88D) data, with continuous data streaming in every six minutes, presents a source for data driven simulations. One of the severe weather detection algorithms, created by the National Severe Storms Laboratory (NSSL) and in use at the WSR-88D, is the Mesocyclone Detection Algorithm (MDA) [4]. This dynamic algorithm uses the data stream outputs of the WSR-88D and is designed to detect storm–circulations associated with regions of rotation in thunderstorms. The MDA is used by meteorologists as one input in their decision to issue tornado warnings. Recent work by Trafalis et al. [4, 5] has shown that SVMs applied to the MDA offer a promising role in improved tornado classification. We present a novel approach by incorporating rules into SVMs of the MDA attributes as they stream in just prior to tornado formation. These rule based sets classify the data into one of three categories, tornado, non-tornado and unclassified. Thus, the rules partition the input space into regions for which we know, with a high degree of certainty, the label of points located in those regions. Our approach is different from [3] in the sense that the rules are combined with SVM in a sequential approach. This paper is organized as follows. In section 2, the data description is given. In section 3, we provide a description of rule-based SVM classifiers. Section 4 describes the experimentation procedure. Section 5 provides computational results and, in section 6, analysis and conclusions are provided.

2 Data

The MDA data set used for this research is based on the outputs from the WSR-88D radar that is collected just prior to the formation of a pre-tornadic circulation. Any circulation detected on a particular volume scan of the radar can be associated with a report of a tornado. In the severe weather database, supplied by NSSL, there is a label for tornado ground truth that is based on temporal and spatial proximity. If there is a tornado reported between the beginning and ending of the volume scan, and the report is within a reasonable distance of a circulation detection, then the ground truth value is flagged. If a circulation detection falls within the prediction "time window" of -20 to +6 minutes of the ground truth report duration, then the ground truth value is also flagged. The key idea behind these timings is to determine whether a circulation will produce a tornado within the next 20 minutes, a suitable lead time for advanced severe weather warnings by the National Weather Service. Owing to the autocorrelation in the MDA attributes, a sampling strategy is used to minimize serial correlation. These sampled data are divided into independent training and testing sets, with 749 and 618 observations, respectively.

3 Rule-Based Support Vector Machine Classifiers

3.1 Rule Generation

In this work, we consider a rule-based approach of a decision tree type as shown in Fig. 1. Nodes in the decision tree involve testing a particular attribute. The test at a

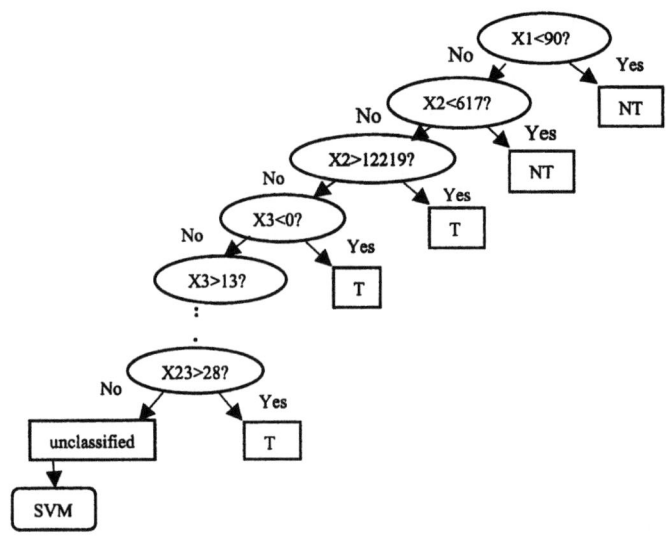

Fig. 1. Tree diagram of rule generation and decisions. Ovals represent nodes, squares represents leaves

node compares an attribute value with a constant threshold. Leaf nodes give a classification that applies to all instances that reach the leaf. When a leaf is reached, the instance is classified according to the class assigned to the leaf. Note that the output of the last node referring to the unclassified category becomes an input to the SVM that provides the final label to the unclassified cases.

There were 23 MDA attributes available for discriminating tornadoes from non-tornadoes [4]. For each attribute, we considered the corresponding probability distribution function for tornado and non-tornado cases arising from the training data. The selection of the threshold for each rule was based on eliminating misclassification by investigating if the minimum for a non-tornado case had a value less than the minimum for a tornado case for a specific attribute. If such a condition holds, then a region unique to non-tornadoes is found.

Similarly, if the maximum for a non-tornado case had a value less than the maximum for a tornado case, for a specific attribute, a region unique to tornado cases is found. Of the 23 attributes, only 20 were found to be useful for rule generation. The thresholds used for tornado and non-tornado discrimination are shown in Table 1.

3.2 Support Vector Machines (SVMs)

Given a set of data points $\{(x_i, y_i), i = 1,..., \ell\}$ with $x_i \in \Re^n$ and $y_i = \pm 1$, the SVM finds a classifier that separates the two classes of points with maximum margin separation (Fig. 2). The SVM formulation can be written as follows [6],

$$\min_{w,b,\eta} C \sum_{i=1}^{\ell} \eta_i + \tfrac{1}{2} \| w \|^2 \tag{1}$$

$$st \quad y_i(wx + b) + \eta_i \geq 1 \qquad \eta_i \geq 0 \quad i = 1,...\ell$$

Table 1. Threshold values for each MDA attribute. See [4] for description of attributes

Non-tornado thresholds	Tornado thresholds
if x1 < 90, then non-tornado	
if x2 < 617, then non-tornado	if x2 > 12219, then tornado
if x3 < 0, then non-tornado	if x3 > 13, then tornado
if x4 < 813, then non-tornado	
if x5 < 1091, then non-tornado	
if x6 < 124, then non-tornado	
if x7 < 6, then non-tornado	
if x8 < 10, then non-tornado	
if x9 < 122, then non-tornado	
if x10 <2, then non-tornado	if x10 > 77, then tornado
	if x11 > 83, then tornado
if x12 < 106, the non-tornado	
if x13 < 3, then non-tornado	
if x14 < 11, then non-tornado	
if x15 < 122, then non-tornado	
if x16 < 106, then non-tornado	
if x17 < 617, then non-tornado	
	if x18 > 113, then tornado
	if x22 > 26, then tornado
	if x23 > 28, then tornado

where C is a parameter to be chosen by the user that controls misclassifications, w is referring to the vector perpendicular to the separating hyperplane, η_i refers to the misclassification error variables and b is the bias of the separating hyperplane. A larger C corresponds to assigning a larger penalty to errors. Introducing positive Lagrange multipliers α_i, to the inequality constraints in model (1) we obtain the following dual formulation:

$$\min_{\alpha} \frac{1}{2} \sum_{i=1}^{\ell} \sum_{j=1}^{\ell} y_i y_j \alpha_i \alpha_j x_i x_j - \sum_{i=1}^{\ell} \alpha_i$$

$$st \ \sum_{i=1}^{\ell} y_i \alpha_i = 0 , \qquad\qquad (2)$$

$$0 \leq \alpha_i \leq C \qquad i = 1,...\ell$$

The solution of the primal problem is then given by $w = \Sigma_i \alpha_i y_i x_i$ where w is the vector that is perpendicular to the separating hyperplane. The free coefficient b can be found from the relation $\alpha_i (y_i (w x_i + b) - 1) = 0$, for any i such that α_i is not zero. The use of a kernel function allows the SVM to operate efficiently in nonlinear high-dimensional feature space [7].

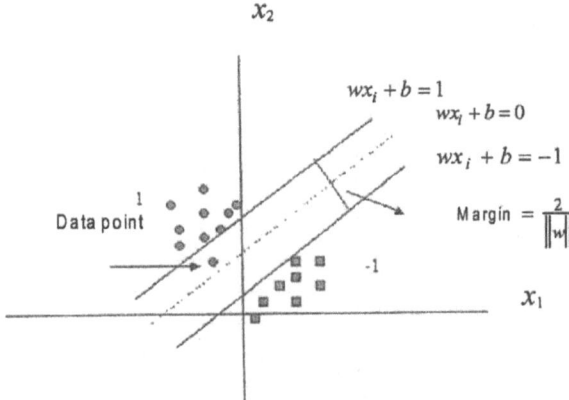

Fig. 2. The geometric illustration of SVM

4 Experiments

In our experimentation, the data are split into training and testing sets. The testing set is sampled independently five times. The first set of experiments is performed by using SVM only on the five testing samples. The total misclassification error is computed as the average of the misclassification error of each sample. The second set of experiments is performed by extracting the rules from the training data and applying those rules in the testing phase. Based on the rules, each testing sample is divided into three different sets: non-tornado, unclassified, and tornado. In the testing phase, those observations not classified by the rules are used as inputs to SVM. The SVM is trained on the training set then tested on five different unclassified samples. For each testing set, the misclassification error for the non-tornado rules set, SVM and tornado rules set are computed. The OSU SVM Classifier Matlab Toolbox [8] was used to run experiments of SVM.

5 Computational Results

The results of the experiments are presented in Fig. 3 and Table 2. The values in the table are misclassification error rates for non-tornado and tornado and SVM components of the total hybrid system. After initial experimentation, it was noted that the rules components of the system had a lower error rate than the SVM component of the system. Accordingly, altering the rules to admit additional cases was considered by creating a multiplier for the threshold values in Table 1. This multiplier controls the level of threshold values (e.g., in Table 1, for attribute 1, the original threshold, 90, corresponds to multiplier 1 and 1.05 times 90 equals 94.5 and this value admits additional observations into the non-tornado category). Table 2 and Fig. 3 illustrate the sensitivity of misclassification error with respect to the threshold. Table 3 shows the misclassification error for SVM for each testing sample and the average of the five samples.

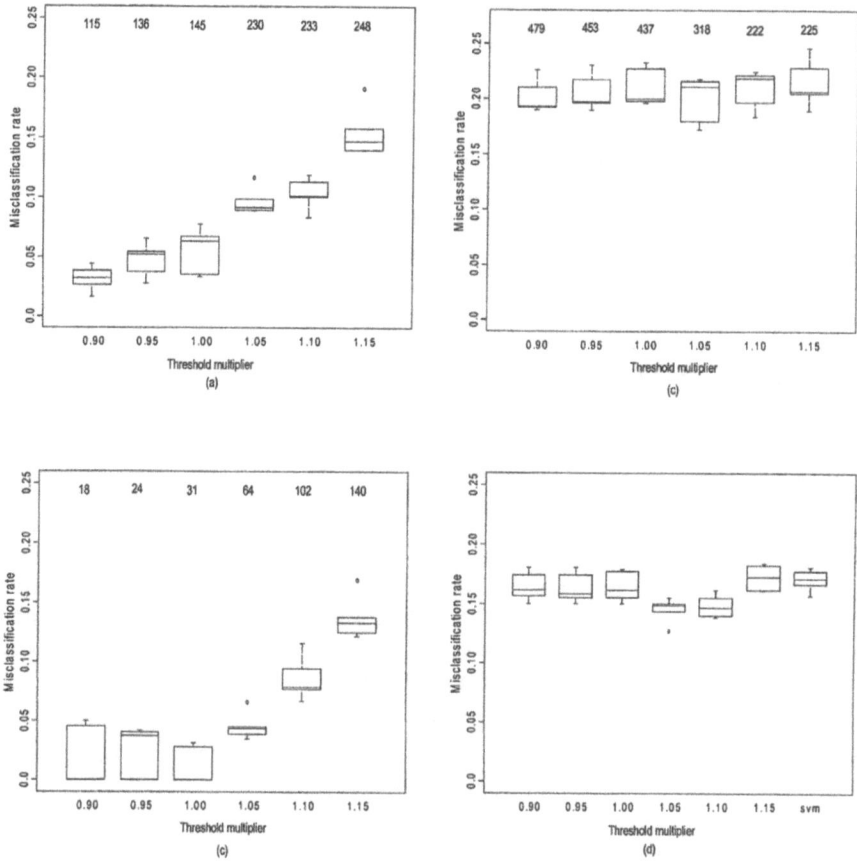

Fig. 3. Boxplots of misclassification error due to (a) non-tornado rules set, (b) SVM, (c) tornado rules set and (d) total hybrid system. Threshold multipliers are shown on X-axis and the numbers of cases classified are shown above the boxplots in (a), (b) and (c)

Table 2. Misclassification error of the hybrid system components and total system

Multiplier	0.90	0.95	1.00	1.05	1.10	1.15
Non-tornado rules	0.0191	0.0237	0.0174	0.0454	0.086	0.1373
Tornado rules	0.0316	0.0474	0.0550	0.0968	0.1032	0.1550
SVM	0.2024	0.2063	0.2110	0.1997	0.2093	0.2154
Total system	0.1648	0.1638	0.1648	0.1449	0.1485	0.1725

Table 3. Misclassification error for SVM

	Sample 1	Sample 2	Sample 3	Sample 4	Sample 5	Average
SVM	0.1664	0.1778	0.1713	0.1811	0.1566	0.1706

6 Analysis and Conclusions

Tables 2 and 3, show that the best misclassification error for the hybrid model (0.1449) is 12.7% lower than the one for the model based solely on SVM (0.1706). The reason for the total system improvement can be seen in Figure 3a,c and Table 2, where the non-tornado rules, based on the threshold given in Table 1, have a mean error rate of 0.0174. Similarly, the tornado rules have a mean error rate of 0.055 at the same multiplier. In contrast, the SVM component has an error rate of 0.211. The behavior of the rules, as seen in Fig. 3 a, c is interesting as the misclassification rate is remarkably low (approximately 5 percent) for threshold multipliers of 0.90 to 1.00. The trade-off is that fewer observations are classified as tornadoes or non-tornadoes. As the threshold multipliers increase to 1.05 and beyond, the misclassification error increases considerably to approximately 15 percent indicating a poorer discrimination between tornadoes, and non-tornadoes. In contrast, the SVM, based on unclassified data (Fig. 3c), is insensitive to the threshold multiplier.

Therefore, given the lower rates in the non-tornado and tornado rules, it is logical to create a hybrid system to capitalize on the disparity in error rate. By admitting additional observations into the leaves of the decision tree prior to sending the remaining observations to the SVM, the optimal system is found. This occurs at a multiplier of 1.05 times the threshold values in Table 1. Experiments are planned for injection of information from successive volume scans to assess additional predictive capability in a constantly updating form of DDDAS.

Acknowledgements. This work has been supported by NSF grant EIA-0205628.

References

1. Mitchell, T.M., Machine Learning, McGraw-Hill, New York, 1997
2. Fung, G.M., Mangasarian, O.L., Shavlik, J.W.: Knowledge-based Support Vector Machines Classifiers, Data Mining Institute. Technical Report 01-09. Computer Sciences Department, University of Wisconsin (2001)
3. Fung, G.M., Mangasarian, O.L., Shavlik, J.W.: Knowledge-based Nonlinear Kernel Classifiers. Data Mining Institute Technical Report 03-02. Computer Sciences Department, University of Wisconsin (2003)
4. Trafalis, T.B., Santosa B., Richman, M.B.: Tornado Detection with Kernel-based Methods. In: Dagli, C.H., Buczak, A.L., Ghosh, J., Embrechts, M., Ersoy, O. (eds.): Intelligent Engineering Systems Through Artificial Neural Networks. ASME Press, Vol. 13 (2003) 677-682
5. Trafalis, T.B., Santosa B., Richman, M.B.: Tornado Detection with Kernel-Based Classifiers From WSR-D88 Radar. Submitted to: Darema, F. (ed.) Dynamic Data Driven Application Systems, Kluwer (2004)
6. Haykin, S.: Neural Networks: A Comprehensive foundation, 2nd edition, Prentice-Hall, Upper Saddle River New Jersey (1999)
7. Schölkopf, B., Smola, A.: Learning with Kernels. MIT Press, Cambridge Massachusetts (2002)
8. Junshui, M., Zhao, Y. Ahalt, S.: OSU SVM Classifier Matlab Toolbox. Available at http://eewww.eng.ohio-state.edu/~maj/osu_SVM/

Adaptive Coupled Physical and Biogeochemical Ocean Predictions: A Conceptual Basis

P.F.J. Lermusiaux[1], C. Evangelinos[2], R. Tian[1], P.J. Haley[1], J.J. McCarthy[1], N.M. Patrikalakis[2], A.R. Robinson[1], and H. Schmidt[2]

[1] Harvard University, Cambridge, MA 02138, U.S.A.
[2] Massachusetts Institute of Technology, Cambridge, MA 02139, U.S.A.

Abstract. Physical and biogeochemical ocean dynamics can be intermittent and highly variable, and involve interactions on multiple scales. In general, the oceanic fields, processes and interactions that matter thus vary in time and space. For efficient forecasting, the structures and parameters of models must evolve and respond dynamically to new data injected into the executing prediction system. The conceptual basis of this adaptive modeling and corresponding computational scheme is the subject of this presentation. Specifically, we discuss the process of adaptive modeling for coupled physical and biogeochemical ocean models. The adaptivity is introduced within an interdisciplinary prediction system. Model-data misfits and data assimilation schemes are used to provide feedback from measurements to applications and modify the runtime behavior of the prediction system. Illustrative examples in Massachusetts Bay and Monterey Bay are presented to highlight ongoing progress.

1 Introduction

A team of scientists and engineers is collaborating to contribute to the advancement of interdisciplinary ocean science and forecasting through an effective union of ocean sciences and information technologies, focusing on adaptive modeling and adaptive sampling for coastal physical-biogeochemical-acoustical processes. Accurate ocean predictions and optimized rapid responses are essential for many oceanic applications, including fisheries, pollution control, hazard management, and maritime and naval operations. Because of multiple uncertainties in oceanic measurements and dynamical models, these capabilities require assimilation of physical, biogeochemical and acoustical data into dynamical models. Importantly, the coastal environment can be highly variable and intermittent on multiple scales, and oceanic variables, parameters and interactions that matter vary in time and space. Thus, efficient models must evolve during predictions. This evolution occurs as new data is injected and assimilated into the prediction system. The data-model comparisons of the data assimilation process then also involve a direct feedback of the data to the models, forming a dynamic data-driven application system (DDDAS [1]). The focus of this paper is to present and illustrate a conceptual basis for such adaptive ocean modeling and prediction.

M. Bubak et al. (Eds.): ICCS 2004, LNCS 3038, pp. 685–692, 2004.
© Springer-Verlag Berlin Heidelberg 2004

The advances in oceanic numerical models and data assimilation (DA, [2]) schemes of the last decade have given rise to interdisciplinary Ocean Observing and Prediction Systems (e.g. [3]) that are used in operational settings. The next generation of such systems will advance the interaction between simulation and measurement to a new level where the forecast application changes its runtime behavior to adapt to new measurements. Importantly, the data assimilation community [4] is starting to recognize the importance of this adaptation, from the correction of model biases to the multi-model data assimilation and automated evolution of model structures as a function of model-data misfits.

Automated objective adaptive modeling allows the optimal use of approximate models for rapidly evolving ocean dynamics. Presently, a model quantity is said to be adaptive if its formulation, classically assumed constant, is made variable as a function of data values. Both structural as well as parametric adaptation are possible. Physical adaptive modeling includes regime transition (e.g., well-mixed to stratified) and evolving turbulent mixing parameterizations. Biogeochemical adaptive modeling includes variations of biological assemblages with time and space (e.g., variable zooplankton dynamics, summer to fall phytoplankton populations, etc) and evolving biogeochemical rates and ratios. This is especially important because biogeochemical modeling is in its infancy and model uncertainties are very large. The adaptive component also greatly facilitates quantitative comparisons of competing physical or biogeochemical models, thus ultimately leading to better scientific understanding.

In what follows, Section 2 outlines several properties of ocean forecasting and DA schemes that are relevant for adaptive modeling. Section 3 describes the conceptual basis of our implementation of adaptivity. Section 4 illustrates some initial progress toward such dynamic data-driven systems. Section 5 concludes.

2 Adaptive Ocean Predictions and Data Assimilation

The presently employed interdisciplinary nowcasting and forecasting system is the Harvard Ocean Prediction System (HOPS, [3]). The DA scheme is Error Subspace Statistical Estimation (ESSE, [5]). The high level architecture involves the component encapsulation of binaries using XML [6]. HOPS is a portable and generic system that simulates the 4D ocean dynamics. It has been applied to many regions [3] and has provided accurate operational forecasts. With adaptive modeling, the data assimilated by HOPS corrects not only field estimates but also the forecast model itself, leading to a dynamic system. ESSE combines forecasts and data based on their errors. It captures the uncertainties that matter by focusing on the largest error modes. These dominant errors are evolved by an ensemble (Monte-Carlo) approach. The ESSE predictions of uncertainties in the forecast are central to adaptive modeling based on Bayesian system identification.

To carry-out adaptive ocean modeling based on dynamic data-driven schemes, one must account for specific properties of ocean predictions and data assimilation. These include: forecasting timelines, data transfers and processing, measurement models to link data to predictive models, dynamical adjustments, and multivariate error covariances. These properties are discussed next.

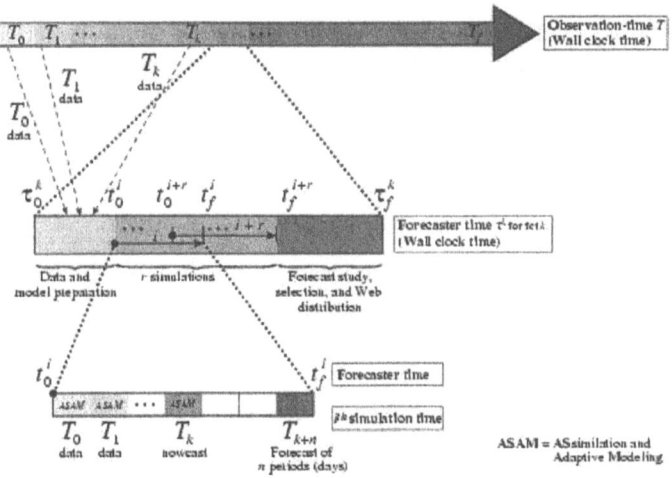

Fig. 1. Forecasting timelines. Top row: "Observation" or "ocean" time T during which measurements are made and the real phenomena occur. Middle row: "Forecaster" time τ^k during which the kth forecasting procedure and tasks are started and finished. Bottom row: "i^{th} simulation" time t^i which covers portions of the real "ocean" time for each simulation. Multiple simulations are usually distributed on several computers, including ensembles of forecasts for uncertainty predictions (ESSE).

An important consideration for real-time adaptive modeling relates to the different times involved in ocean forecasting: the observation time, forecaster time and simulation time (Fig. 1). New observations are made available in batches (Fig. 1, first row) during periods T_k, from the start of the experiment (T_0) up to the final time (T_f). During the experiment, for each prediction k (Fig. 1, zoom in middle row), the forecaster repeats a set of tasks (from τ_0^k to τ_f^k). These tasks include the processing of the currently available data and model (from τ_0^k to t_0^i), the computation of $r + 1$ data-driven forecast simulations (from t_0^i to t_f^{i+r}), and the study, selection and web-distribution of the best forecasts (from t_f^{i+r} to τ_f^k). Within these forecast computations, a specific forecast simulation i (Fig. 1, zoom in bottom row) is executed during t_0^i to t_f^i and associated to a "simulation time". For example, the ith simulation starts with the assimilation and adaptive modeling based on observations T_0, then integrates the dynamic model with data assimilation and adaptive modeling based on observations T_1, etc., up to the last observation period T_k which corresponds to the nowcast. After T_k, there are no new data available and the simulation enters the forecasting period proper, up to the last prediction time T_{k+n}.

Data-driven ocean applications involve measurement models which link the measured variables to the model state variables. These measurement models can be complex, e.g. for linking the measured acoustic travel times to the simulated fields of zooplankton concentrations and temperature. In addition, many of the state variables are usually not observed and inferring the non-observed variables by dynamical adjustment (e.g. [7]) prior to data assimilation is often crucial.

This reduces unphysical data shocks/adjustments that would otherwise occur after assimilation. With a fully multivariate DA scheme such as ESSE, forecast error covariances can infer the non-observed variables from the observed ones but it usually remains helpful to impose weak-constraint dynamical adjustments.

3   Implementing Adaptivity

We have been considering the following cases in our design thus far (Fig. 2):

1. running a single, adaptive interdisciplinary model,
2. running a single physical model, coupled to a set of competing biological models whose parameters are adapted,
3. running competing interdisciplinary models and adapting their parameters.

In all cases, the basis of the adaptation are the misfits between model estimates and data. When misfits are large, models are adapted. In the latter two cases, models can be rejected when seen as too inadequate. In the context of ESSE, expected bounds on misfits are given by the forecast uncertainties. Importantly, the computation of uncertainties for adaptive modeling can be based on small-size ESSE ensembles, one for each candidate models. This is feasible because error variances often converge faster than covariances. The adaptation triggers are provided to the software through external files, regularly checked at runtime and updated when data-forecast misfits warrant a specific model structure/parameter modification.

For example, in the first case (Fig. 2, left), the code may modify the carbon-to-chlorophyll ratio, increase the number of state variables (by adding meso-zooplankton) or alter the form of a source term. The latter two structural adaptations are cleanly implemented by using C function pointers to choose between Fortran modules. Physical and biological adaptation must be dynamically compatible but don't need to be computationally concurrent. In the second case (Fig. 2, right), the forecasts of competing biogeochemical models are compared based on their respective data-forecast misfits and the best biological functional forms selected for each time period T_i (Fig. 1). The third case is a combination of the former two.

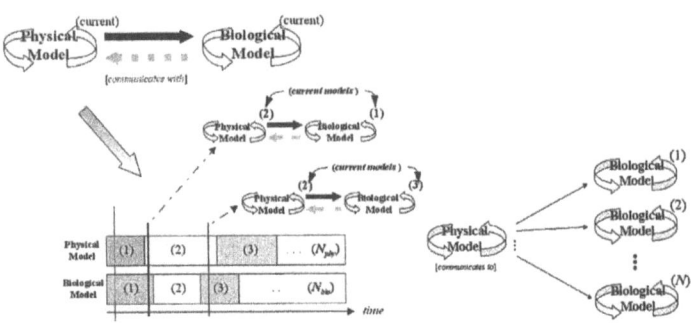

Fig. 2. Approaches for adaptive coupled physical-biogeochemical modeling

We are developing all three cases using a model of concurrent processes directly coupled (using MPI); we are considering the use of a more flexible coupling framework [8] for the future. Several computational challenges involved in the implementation of adaptivity for coupled physical-biological forecasts remain to be researched. They include the optimum operational choice between cases (i) to (iii), the impact of multiple observing and forecasting timelines, the interaction between parameter and structural adaptivity, the details of the quantitative schemes which trigger the adaptive switch/behavior, the relationships between adaptivity and consistent model re-initializations and data assimilation, the different computational meshes, and finally, the issues of communications costs and load imbalance in many-to-one scenarios.

4   Initial Progress and Prospectus

Three initial results are now presented. The first two are experiments which exercise the concept of adaptive modeling but do not yet utilize the whole dynamic data-driven computational system. The last result summarizes the development of a new adaptive biogeochemical computational model.

4.1   Biogeochemical Adaptive Modeling in Massachusetts Bay

The first example corresponds to dynamic data-driven predictions of coupled physical and biological dynamics in Massachusetts Bay during June 2001, as part of the Assessment of Skill for Coastal Ocean Transients (ASCOT-01) experiment. The goals of ASCOT-01 were to enhance the efficiency and extend the scope of nowcasting and forecasting of oceanic fields for Coastal Predictive Skill Experiments and Rapid Environmental Assessments. The limited physical dimensions of the Bay allowed a relatively comprehensive sampling and forecasting of transient conditions. The focus was on coastal ocean responses to wind events, including upwellings and subsequent advections of nutrients and organisms. The predictions illustrated by Fig. 3 were carried-out with the Harvard coupled models, initialization procedure and data assimilation schemes (see [7]).

In June (end of the Spring bloom), the ecosystem is usually in a different state than it is in Aug-Sept (summer/summer-to-fall transition). For June 2001,

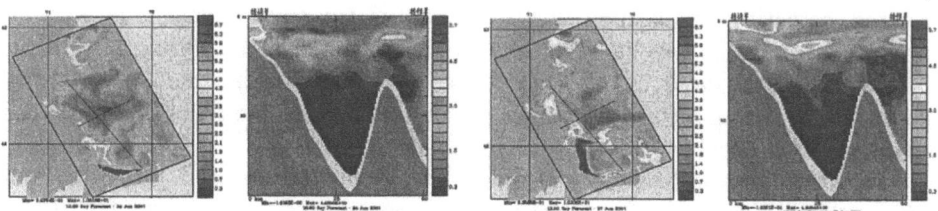

Fig. 3. Adaptive biogeochemical modeling during ASCOT-01 in Mass. Bay, June 2001. Two left panels: June 24 nowcast of surface Chl (mg/m^3) and corresponding west-east cross-section. Two right panels: June 27 forecast. All fields are after adaptation.

important state variables were the 4D fields of phytoplankton, zooplankton, detritus, nitrate, ammonium and chlorophyll-a (e.g. Fig 4), as in Aug-Sept 1998 [7]. However, model parameters needed to be modified. The parameters whose values where adapted to the June 2001 conditions included light attenuation scales, photosynthetic parameters, nutrient uptake constants, phytoplankton and zooplankton mortality rates, zooplankton grazing parameters, settling/sinking velocities for phytoplankton/detritus, and re-mineralization timescales.

Once the experiment started, model parameters were adapted to the new data collected, using the concept of DDDAS. The techniques utilized are illustrated on Fig 3 (right side, cases 2 and 3). Firstly, a set of physical-biological predictions was executed with a single physical model but different biological model parameters. The adaptive biological parameters were the: photosynthetic parameters, half saturation constants for ammonium/nitrate uptakes, zooplankton grazing parameters and zooplankton mortality rates. Each member of the set of predictions was then compared to the new data collected. The biological parameters of the coupled prediction which was the closest to this new data (within error bounds) were ultimately selected as the best parameters. This process was repeated several times during the experiment. Secondly, the representation of the Charles River outflow from Boston Harbor was adapted. This was necessary because the real-time data likely indicated that the outflow was larger in June 2001 than in August 1998. The physical strength of the simulated nitrate outflow was therefore increased. A sample of the nowcast and forecast chlorophyll concentration fields after adaptation are shown on Fig. 3. Note the effects of the southwesterly winds during the June 23-27 evolution (advection of Chl plume and increase in sub-surface maxima). Importantly, the adaptive modeling allowed the 2-3 day forecast of the Chl field to be better than persistence (assume oceanic conditions are constant) by about 5 to 20 percent.

4.2   Physical Adaptive Modeling in Monterey Bay

The second example corresponds to the Autonomous Ocean Sampling Network (AOSN-II) field experiment in the Monterey Bay region. It utilized a plethora of remote and in-situ sensors and platforms including multiple satellite images, drifters, gliders, moorings, AUV and ship-based data (www.mbari.org/aosn). These data were assimilated into numerical models and daily predictions of the ocean fields and uncertainties were issued. The different sensors provided correlated, sometimes redundant, measurements of the environment. Another important component was the multiple predictive models. In addition, multiple ESSE runs of the same model were carried-out, including stochastic forcings to represent model uncertainties. These ensembles of ESSE simulations produced plausible dynamical scenarios for the region.

Prior to the experiment, model parameters were calibrated to historical conditions judged to be similar to the conditions expected in August 2003. Once the experiment started, several parameters of the physical ocean model were adapted to the new 2003 data. This adaptation involved the parameterization of the transfer of atmospheric fluxes to the upper-layers of the sea. As shown on

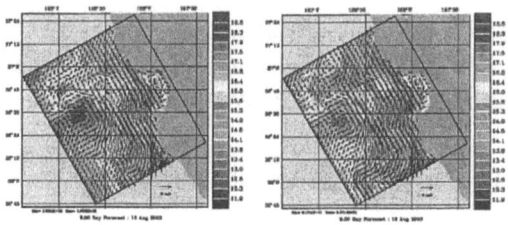

Fig. 4. Adaptive physical modeling during AOSN-II in Monterey Bay, August 2003: surface T on 16 Aug 2003. Left: before adaptation. Right: after adaptation.

Fig. 4, the new values for wind mixing clearly modified surface properties and improved the temperature fields and ocean currents (not shown).

4.3 Generalized Adaptive Biogeochemical Ocean Model

Oceanic states evolve and go through transitions. Efficient predictive models must have the same behavior. This is especially important for marine ecosystems. To automate the dynamic switch of biogeochemical parameters and structures, a new generalized biogeochemical ocean model is being developed [9].

This generalized model is fully modular and flexible (Fig. 5). It has been scientifically constructed based on a serious study of all possible functional groups and parameterizations for coastal ecosystems such as Massachusetts Bay and Monterey Bay (Sects. 4.1,4.2). A general set of state variables and of mathematical structures representing their interactions was selected, based on importance, completeness, efficiency and accuracy. This led to a generalized model (Fig. 5) with the following functional groups of state variables: nutrients (N_i), phytoplankton (P_i), zooplankton (Z_i), detritus (D_i), dissolved organic matter (DOM_i), bacteria (B_i) and auxiliary variables (A_i). Within each functional group, the number of state variables varies from 1 to n. The parameterizations of the interactions among state variables is also variable. Changes in their number at each trophic level can result in automatic changes in these parameterizations.

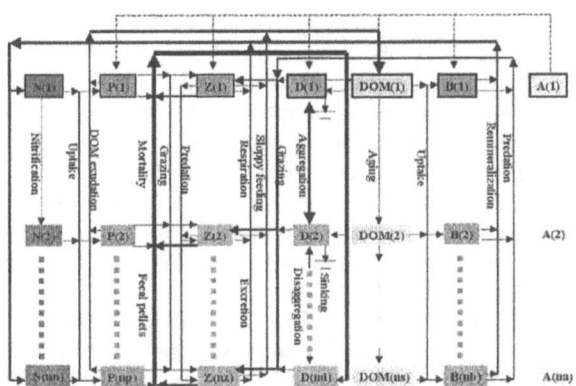

Fig. 5. Generalized biogeochemical model. Thicker arrows imply more processes.

With this flexibility, the generalized biogeochemical model can adapt to different ecosystems, scientific objectives and available measurements. Details of its properties and implementation are given in [9].

5   Conclusion and Future Research

Adaptive modeling based on DDDAS will become essential for interdisciplinary ocean predictions. The present manuscript provides a conceptual basis and illustrative examples for this adaptation. Opportunities and challenges in adaptive ocean modeling abound over a wide spectrum of needs, from coastal ocean science to global climate predictions. Adaptive ocean modeling must be anticipated to accelerate progress in fundamental research and to enable rapid operational predictions otherwise not possible.

Acknowledgments. We thank W.G. Leslie, the ASCOT-01 and AOSN-II teams, and M. Armstrong. This project was funded in part from NSF/ITR (grant EIA-0121263), ONR (grants N00014-01-1-0771, N00014-02-1-0989 and N00014-97-1-0239) and DoC (NOAA via MIT Sea Grant, grant NA86RG0074).

References

1. Darema, F., et al.: NSF sponsored workshop on dynamic data driven application systems. Technical report, National Science Foundation (2000)
   http://www.cise.nsf.gov/cns/darema/dd_das/dd_das_work_shop_rprt.pdf.
2. Robinson, A., Lermusiaux, P.: Data Assimilation in Models. In: Encyclopedia of Ocean Sciences. Academic Press Ltd., London (2001) 623–634
3. Robinson, A., et al.: Harvard Ocean Prediction System (HOPS) (2001-2004)
   http://oceans.deas.harvard.edu/HOPS/HOPS.html.
4. Stammer, D., Rizzoli, P., Carton, J., Cummings, J., Lermusiaux, P., Moore, A.: A U.S. plan for sustained data assimilation in ocean research (2004)
   http://www.atmos.umd.edu/~carton/dameeting.
5. Lermusiaux, P., Robinson, A., Haley, P., Leslie, W.: Advanced interdisciplinary data assimilation: Filtering and smoothing via Error Subspace Statistical Estimation. In: The OCEANS 2002 MTS/IEEE, Holland Publications (2002) 795–802
6. Evangelinos, C., Chang, R., Lermusiaux, P., Patrikalakis, N.: Rapid real-time interdisciplinary ocean forecasting using adaptive sampling and adaptive modeling and legacy codes: Component encapsulation using XML. In Sloot, P., et al., eds.: ICCS 2003. Volume 2660 of LNCS., Springer (2003) 375–384
7. Besiktepe, S., Lermusiaux, P., Robinson, A.: Coupled physical and biochemical data driven simulations of Massachusetts Bay in late summer: real-time and post-cruise data assimilation. J. of Marine Systems 40 (2003) 171–212. M. Gregoire, P. Brasseur and P.F.J. Lermusiaux (Eds.).
8. Larson, J.W., Jacob, R.L., Foster, I., Guo, J.: The Model Coupling Toolkit. Technical Report ANL/CGC-007-0401, Argonne National Laboratory (2001)
9. Tian, R., Lermusiaux, P., McCarthy, J., Robinson, A.: The Harvard generalized biogeochemical ocean model. Technical report, Harvard University DEAS (2004)

Dynamic-Data-Driven Real-Time Computational Mechanics Environment

John Michopoulos[1], Charbel Farhat[2], and Elias Houstis[3,4]

[1] Special Projects Group, Code 6303, U.S. Naval Research Laboratory, U.S.A.
john.michopoulos@nrl.navy.mil
[2] Dept. of Aerospace Engineering Sciences, University of Colorado at Boulder, U.S.A
charbel.farhat@colorado.edu
[3] Computer Sciences Department, Purdue University, U.S.A.
[4] Dept. of Comp. Eng. and Telecommunications, University of Thessaly, Greece
enh@cs.purdue.edu

Abstract. The proliferation of sensor networks in various areas of technology has enabled real-time behavioral monitoring of various physical systems in various length and time scales. The opportunity to use these data dynamically for improving speed, accuracy, and general performance of predictive behavior modeling simulation is of paramount importance. The present paper identifies enabling modeling methods and computational strategies that are critical for achieving real-time simulation response of very large and complex systems. It also discusses our choices of these technologies in the context of sample multidisciplinary computational mechanics applications.

1 Introduction

The main objective of the effort described in this paper is to establish and use a strategy for selecting, generating, improving, and applying methodologies capable of enabling data-driven real-time simulation of large, complex, multi-field, multi-domain physical systems. This paper reports on the initial success in achieving this goal, under the additional fundamental requirement that sensor networks will be providing live data originating from the actual physical systems that are simulated by the exercise of their corresponding computational models.

This effort is a part of a larger effort of developing a Data Driven Environment for Multiphysics Applications (DDEMA) [1,2,3,4], as a computationally implementable framework in the form of a Multidisciplinary Problem Solving Environment (MPSE). Although DDEMA's scope also entails utilizing dynamic data for adaptive modeling or model selection, this paper only discusses data usage for adaptive simulation, steering, and short future behavioral prediction assuming that the model is known and well established. This model approximates the behavior of a system that belongs in the continuum mechanics engineering domain.

The rest of the paper is organized as follows. In Section 2, we describe the context and required enabling technologies achieving this objective. In Section

M. Bubak et al. (Eds.): ICCS 2004, LNCS 3038, pp. 693–700, 2004.
© Springer-Verlag Berlin Heidelberg 2004

3, we specify the technologies we selected for DDEMA: precomputed solution synthesis and model reduction, along with multidimensional computational parallelism. Finally, Section 4 presents two preliminary sample applications and concludes this paper.

2   Modeling Methods and Computational Strategies for Improving Simulation Performance

Discussion in this section is limited only to the systemic modeling methodologies that capture system-behavior representation, and the computational strategies of their usage. We do not intend to discuss computational, networking, infrastructure and process mobility strategies and methodologies. However, it should be recognized that these can also have a considerable effect on the performance of a simulation environment. This task has been discussed in [1,2,3,4].

To achieve the main objective as described in the introduction, one has first to define the context of the problem at hand. Figure 1 provides a generic description of this context. It shows a schematic representation of the dataflow relationship among the actual physical system and the corresponding sensor network, the computational model, and the associated simulation environment along with the simulated system. In addition, the data-flow paths have been labeled to allow the distinction of both the real-time solving (RTS) and the precomputed solving and real-time solution synthesis (PS-RTSS) modes of simulation usage. All components in this figure are generic and do not express the architecture of a particular system or technology.

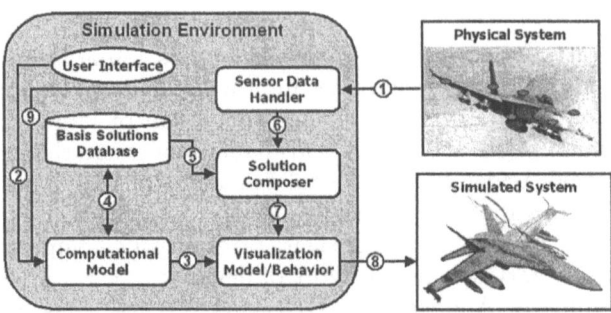

Fig. 1. Relationship among physical system, sensor data, and computational infrastructure for behavioral simulation.

As shown in Fig. 1, path $[2 - 4 - 3 - 8]$ and path $[1 - 9 - 3 - 8]$ represent the RTS mode, while path $[2 - 4] \wedge [(5, 6) - 7 - 8]$ corresponds to the PS-RTSS mode. The latter path consists of the conjunction of the precomputed solving part represented by the sub-path $[2 - 4]$, and the real-time solution synthesis part represented by the sub-path $[(5, 6) \wedge 7 - 8]$. The path syntax used here has been devised to avoid the complications of temporal logic syntax while allowing the ability to capture synchronicity and asynchronicity. Paths in regular parentheses separated by a coma represent synchronous parallel-in-time dataflow, while square brackets are used to denote complete paths, and ∧ is used to denote asynchronous logical conjunction of paths.

The two-way dataflow between the *Basis Solutions Database* and *Computational Model* in Fig.1 underscores the complexity of the relationship that can exist between these two modules. Indeed, computational models generate in general a basis database, but such a database can also be exploited to construct, for example, reduced-order models (ROM). Furthermore, a key property of the *Computational Model* module is that it may include inside it many levels of recursive application of the entire *Simulation Environment* operating in PS-RTSS mode. The difference from the current incarnation of the *Computational Model* module is that for each level below, a ROM participates in the corresponding instance of the *Computational Model* module.

Dynamic-data can improve the simulation via a number of various manners. These include the updating of computational models and the reduction of simulation uncertainty to enhance simulation validity, and the enabling of original ideas for parallel processing in the time-domain and of the construction of reduced-order models to improve computational efficiency.

Thus, it is important to select, create, and utilize methodologies and strategies that can inherently take advantage of real-time data. The most important aspect of what is referred to as real-time simulation is that the time between the time of user action (or the arrival of new data into the system), and the time the system responds back with the corresponding adjustment of behavioral simulation, has to be as small as possible. Therefore, the common requirement for all processes involved in the simulation is that they have to be as fast as possible. Here we examine how the modeling methods and the computational strategies can be combined for achieving real-time simulation response. Thus, a distinction between a computational strategy and a modeling methodology needs to be drawn here. A computational strategy refers to the abstract methodology employed for optimal usage of the computational and networking infrastructure. A modeling methodology refers to the particular behavior model (usually in equational form) and its corresponding computational implementation.

A detailed survey of behavior-modeling methods capable of enabling real- or near-real-time simulation lies well outside the scope of the present paper. However, before selecting, inventing, improving, or implementing a method, it is reasonable to expect that certain common attributes have been identified and a taxonomic classification has been constructed.

We have identified 36 behavior modeling methods to date that lend themselves to accelerated systemic behavior computations. Our list [5] is continuously growing, thus not allowing us to include it in the present paper. We have also created a classification taxonomy based on certain attributes of these methods, and have established four classes of selection criteria based on required attributes.

The first classification level contains the *Dimensional Reduction Methods* (DRM); all those which do not fall in this category are the *Rest*. The common characteristic of all DRMs that classifies them as such is that they all attempt to project some quantifiable feature (state-variable field and space- or time-domain of applicability) of the problem to a discrete lower-dimensional space. They subsequently recover the global systemic behavior via appropriate reconstruction from the reduced projections.

The DRMs are further subdivided into the *Reduced Basis Behavior Decomposition Methods* (RBBDM), and the *Reduced Domain Decomposition Methods* (RDDM). The common characteristic of all RBBDMs is that they often represent behavioral field variables as a composition of sub-behaviors defined in lower-dimensional spaces, mostly through the use of specialized basis functions. Similarly, the common characteristic of all RDDMs is that they decompose the domain space of application of behavior to subdomains, and subsequently reconstruct systemic behavior from the composition of the subdomains.

RBBDMs are model reduction methods subdivided further into those which are physics-based, and those which are physics-blind. Examples of physics-based RBBDMs are methods such as Proper Orthogonal Decomposition (POD), Hankel-norm approximation, balanced truncation, Singular perturbation, Cross grammian methods, Principal Components, Karhunen-Loeve, Multiresolution Green's Function, Hierarchical Multiresolution, Differential Geometry Components, Examples of physics-blind RBBDMs are methods such as Multilayered Perceptrons, Radial Basis Networks, Local Polynomials (Splines etc.), Neural Networks, Polynomial Chaos, Support Vector Machines, etc.

Examples of RDDMs are methods like the partial realization, Padé methods, multi-resolution geometry patches, general domain decomposition and the recently-developed PITA [6].

Examples in the category of all of the *Rest* methods are Recursive Kalman Filtering Estimation, Capacitance Matrix, Reaction and Lagrangian Constraint Optimization, and Equationless Lifting Operator.

All these methods have a great potential for enabling real-time computations, but are often reliable only for time-invariant systems. Dynamic data enables their adaptation, when the system simulated is time-varying, using interpolation and/or sensitivity approaches. Certain common desired attributes of all of the above methods can be used to construct selection criteria. We marked [5] all methods according to the following performance criteria: capability for real-time simulations in terms of both RTS and PS-RTSS, capability for modeling uncertainty, capability for adaptive model formation, and finally capability for parallel implementation.

3   Data-Driven Acceleration Choices of DDEMA

According to the behavioral model classification of the previous section, we have limited ourselves to those methods which satisfy only the first and last criteria of our selection strategy. The dynamic-data-driven aspect of the system to be simulated is relevant to the applied methods and strategies both trivially and non-trivially, depending on our particular choices. For example, the RTS strategy can utilize the dynamic data to alter initial and/or boundary conditions, to construct or update a ROM, and to provide seed values to a PARAREAL (parallel real-time) solution method [7] or PITA (parallel implicit time-integrator algorithm) [6] that parallelizes in the time-domain the dynamic solution of a ROM problem. It should be noted that for a ROM, a time-parallel solution approach is more efficient than a conventional degree-of-freedom-parallel approach (also known as space-parallel approach) because by definition a ROM contains few degrees of freedom and therefore can benefit only to a certain extent from parallel processing at the degree-of-freedom level.

The PS-RTSS strategy utilizes the data only at the synthesis stage in their capacity to select proportions of precomputed solutions. It is important to remember that PS-RTSS is a computational strategy and not a modeling method. It still requires utilizing a behavioral model of the system. It uses this model to compute and store in a database behavioral solutions corresponding to the bases of low-dimensional space parameterizations of the generalized input as it is exercised by the actions of the user- or the sensor-data. At the time of real-time usage of the simulator, the *Solution Composer* module takes the sensor data or/and the user's actions and synthesizes a solution that corresponds to these choices, based on the stored database content. The solution synthesis is done as a postprocessing of elementary solutions, and therefore involves evaluative computation — and not solving computation which is only performed in the precomputation stage. This strategy uses models in an off-line mode prior to the real-time simulation period, and therefore is more insensitive to the efficiency of the computational method that implements the behavioral model of the system to be simulated. Clearly, all possible computational methods for implementing the systemic behavior are acceptable in this strategy. However, it is reasonable to expect that the more efficient a method is, the more it contributes to the creation of an efficient system that is inexpensive to use from an overall perspective, regardless of the distinction between the precomputation solving and the real-time computation modes.

4   Sample Applications

4.1   PS-RTSS of an Underwater Composite Vessel

Preliminary results of this strategy have been obtained for the case of a cylindrical I-beam ring-stiffened composite pressure vessel with semi-spherical steel end-caps (see Fig. 2). The model of this vessel represents one of the components of a submarine-attached "Dry-Deck-Shelter" (DDS) utilized in underwater naval missions. Embedded sensors provide datastreams that encode strain-field measurements used to assess generalized loading conditions on the vessel in terms of three basis loading cases that span the loading space [8]. The first is the external pressure that is proportional to the depth of the submerged cylinder. The other two basis cases are bending moments about axis-x L_x and axis-y L_y (see

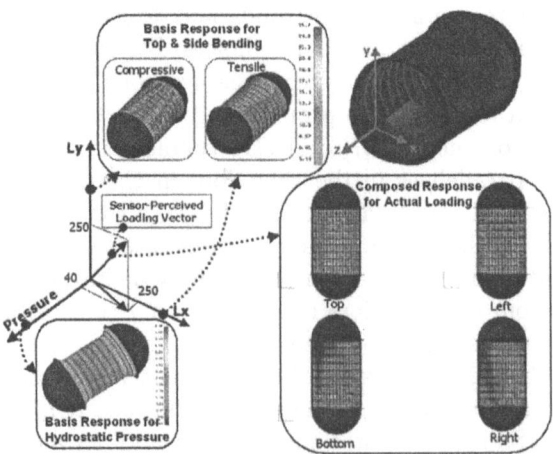

Fig. 2. Material softening distributions from precomputed basis solutions and sensor-controlled reconstruction for an underwater composite pressure vessel, under conditions of hydrostatic pressure and explosion-induced horizontal and vertical bending.

Fig.2). They can correspond to maneuvering inertial loads, or to the quasi-static bending effect of a pressure wave induced by a remote underwater explosion.

In precomputation mode, the simulator computes and stores into the database the material softening and other state-variable field distributions over the entire geometry of the structure loaded by each one of the three loading cases. This was done by finite element analysis (FEA) of the associated model for the three distinct loading conditions. The results are shown in Fig. 2. For the bending cases, both the tensile and compressive sides of the vessels are shown. The difference between the tension and compression sides of the vessel is due to the nonlinear constitutive response of the composite material that was predetermined by a mechatronically driven automated approach [9,10].

In the real-time mode, our simulator received strain measurement input from at least three sensors [8]. A hypothetical data report from these three sensors determined that the applied loading condition on the structure is defined by the loading vector $(40P, 250L_x, 250L_y)$ as shown in Fig. 2. The intrinsic power of this strategy is that the simulator does not have to run an additional FEA to determine the corresponding state variable field distributions and pay the solution cost. The *Solution Composer* module described in Fig. 1 multiplies the basis solutions stored in the database by the coefficients of the components of the loading vector bases and sums the results in just one pass. The visualization results through the *Visualization* module are instantaneous and satisfy the technical needs of the user. In this case, the user may be the submarine commander, in which case the simulator is on-board the structure it simulates and can be used as data-driven computational decision support system [2,3,4].

4.2 Combined RTS/PS-RTSS of the Aeroelastic Behavior of the AGARD Wing 445.6

Preliminary results of this hybrid strategy have been obtained for a flutter analysis of the AGARD Wing 445.6. This wing is an AGARD standard aeroelastic configuration with a 45 degrees quarter-chord sweep angle, a panel aspect ratio of 1.65, a taper ratio of 0.66, and a NACA 65A004 airfoil section. The model selected here is the so-called 2.5-ft weakened model 3 whose measured modal frequencies and wind-tunnel flutter test results are reported in [11], and for which a full-order aeroelastic computational model with 178,758 degrees of freedom is described in [12]. Embedded sensors provide datastreams that encode free-stream pressure, density, and Mach number.

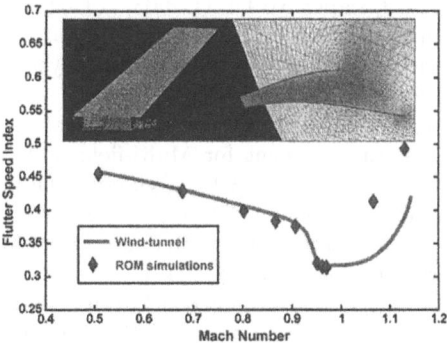

Fig. 3. AGARD Wing 445.6: flutter speed indices simulated by a hybrid RTS/PS-RTSS strategy.

In a first precomputation mode, the simulator used the POD approach described in [13] to generate and store in the *Basis Solutions Database* four aeroelastic ROMs with 200 degrees of freedom each, at the following four free-stream Mach numbers: $M_\infty = 0.499$, $M_\infty = 0.901$, $M_\infty = 0.957$, and $M_\infty = 1.141$. Then, it considered a scenario of flight acceleration from the subsonic regime ($M_\infty = 0.499$) to the supersonic regime ($M_\infty = 1.141$). During this acceleration, it received eight requests from the Mach number sensor for predicting the flutter speed in order to clear the target flight envelope. Four of these requests were at the Mach numbers mentioned above, and four others at $M_\infty = 0.678$, $M_\infty = 0.954$, $M_\infty = 0.960$, and $M_\infty = 1.072$. For each of the latter free-stream Mach numbers, the *Solution Composer* generated in near-real-time a corresponding new ROM by interpolating the angles between the subspaces of the stored ROMs [13], and predicted in real-time the corresponding flutter speed index. The visualization results through the *Visualization* module are reported in Fig. 3 and compared with the flutter speed index curve obtained from wind-tunnel test data [11]. The aeroelastic results generated by this RTS/PS-RTSS hybrid strategy are shown to be in good agreements with the test data.

Acknowledgments. The authors acknowledge the support by the National Science Foundation under grant ITR-0205663.

References

1. Michopoulos, J., Tsompanopoulou, P., Houstis, E., Rice, J., Farhat, C., Lesoinne, M., Lechenault, F., DDEMA: A Data Driven Environment for Multiphysics Applications,in: Proceedings of International Conference of Computational Science - ICCS'03, Sloot, P.M.A., et al. (Eds.) Melbourne Australia, June 2-4, LNCS 2660, Part IV, Springer-Verlag, Haidelberg, (2003) 309-318
2. Michopoulos, J., Tsompanopoulou, P., Houstis, E., Rice, J., Farhat, C., Lesoinne, M., Lechenault, F., Design Architecture of a Data Driven Environment for Multiphysics Applications, in: Proceedings of DETC'03, ASME DETC2003/CIE Chicago IL, Sept. 2-6 2003, Paper No DETC2003/CIE-48268, (2003).
3. Michopoulos, J., Tsompanopoulou, P., Houstis, E., Farhat, C., Lesoinne, M., Rice, J., Joshi, A., On a Data Driven Environment for Multiphysics Applications, Future Generation Computer Systems, in-print (2004).
4. Michopoulos, J., Tsompanopoulou, P., Houstis, E., Farhat, C., Lesoinne, M., Rice, Design of a Data-Driven Environment for Multi-field and Multi-Domain Applications, book chapter in Darema, F. (ed.), Dynamic Data Driven Applications Systems, Kluwer Academic Publishers, Netherlands, in-print (2004).
5. DDEMA-group, Taxonomic Classification of Reduced Order Models, available from http://ddema.colorado.edu/romtaxonomy, (2004).
6. Farhat, C., Chandesris, M., Time-Decomposed Parallel Time-Integrators: Theory and Feasibility Studies for Fluid, Structure, and Fluid-Structure Applications, Internat. J. Numer. Meths., Engrg. 58, (2003) 1397-1434.
7. Lions, J. L., Maday, Y., Turinici, G., Résolution d'EDP par un Schéma en Temps "Pararéel", C. R. Acad. Sci. Paris, Serie I Math. 332, (2001) 661-668.
8. Michopoulos, J.G., Mast, P.W., Badaliance, R., Wolock, I., Health Monitoring of smart structures by the use of dissipated energy, ASME proc. 93 WAM on Adaptive structures and material systems, G.P. Carman/E. Garcia, eds., ASME, AD-Vol. 35, (1993) 457-462.
9. Michopoulos, J., Computational and Mechatronic Automation of Multiphysics Research for Structural and Material Systems, Invited paper in "Recent advances in Composite Materials" in honor of S.A. Paipetis, by Kluwer Academic publishing (2003) 9-21.
10. Mast, P., Nash, G., Michopoulos, J., Thomas, R., Wolock, I., Badaliance, R., Characterization of strain-induced damage in composites based on the dissipated energy density: Part I. Basic scheme and formulation, J of Theor. Appl. Fract. Mech., 22, (1995) 71-96
11. Yates, E., AGARD Standard Aeroelastic Configuration for Dynamic Response, Candidate Configuration I. – Wing 445.6, NASA TM-100492, 1987.
12. Lesoinne, M., Farhat, C., A Higher-Order Subiteration Free Staggered Algorithm for Nonlinear Transient Aeroelastic Problems, AIAA Journal 36, No. 9, (1998) 1754-1756.
13. Lieu, T., Lesoine, M., Parameter Adaptation of Reduced Order Models for Three-Dimensional Flutter Analysis, 42nd AIAA Aerospace Sciences Meeting and Exhibit, Reno, NV, (2004).

A Note on Data-Driven Contaminant Simulation

Craig C. Douglas[1,2], Chad E. Shannon[1], Yalchin Efendiev[3], Richard Ewing[3],
Victor Ginting[3], Raytcho Lazarov[3], Martin J. Cole[4], Greg Jones[4],
Chris R. Johnson[4], and Jennifer Simpson[4]

[1] University of Kentucky, Department of Computer Science, 325 McVey Hall,
Lexington, KY 40506-0045, USA
{craig.douglas,ceshan0}@uky.edu
[2] Yale University, Department of Computer Science, P.O. Box 208285
New Haven, CT 06520-8285, USA
douglas-craig@cs.yale.edu
[3] Texas A&M University, ISC, College Station, TX, USA
{efendiev,ginting,lazarov}@math.tamu.edu, richard-ewing@tamu.edu
[4] Scientific Computing and Imaging Institute, University of Utah, Salt Lake City,
UT, USA
{gjones,mjc}@sci.utah.edu{crj,simpson}@cs.utah.edu

Abstract. In this paper we introduce a numerical procedure for performing dynamic data driven simulations (DDDAS). The main ingredient of our simulation is the multiscale interpolation technique that maps the sensor data into the solution space. We test our method on various synthetic examples. In particular we show that frequent updating of the sensor data in the simulations can significantly improve the prediction results and thus important for applications. The frequency of sensor data updating in the simulations is related to streaming capabilities and addressed within DDDAS framework. A further extension of our approach using local inversion is also discussed.

1 Introduction

Dynamic data driven simulations are important for many practical applications. Consider an extreme example of a disaster scenario in which a major waste spill occurs in a subsurface near a clean water aquifer. Sensors can now be used to measure where the contamination is, where the contaminant is going to go, and to monitor the environmental impact of the spill. One of the objectives of dynamic data driven simulations is to incorporate the sensor data into the real time simulations. Many important issues are involved in DDDAS for this application (see [1,2]).

Subsurface formations typically exhibit heterogeneities over a wide range of length scales while the sensors are usually located at sparse locations and sparse data from these discrete points in a domain is broadcasted. Since the sensor data usually contains noise it can be imposed either as a *hard* or a *soft* constraint. To incorporate the sensor data into the simulations, we introduce a multiscale interpolation operator. This is done in the context of general nonlinear parabolic

M. Bubak et al. (Eds.): ICCS 2004, LNCS 3038, pp. 701–708, 2004.
© Springer-Verlag Berlin Heidelberg 2004

operators that include many subsurface processes. The main idea of this interpolation is that we do not alter the heterogeneities of the random field that drives the contaminant. Instead, based on the sensor data, we rescale the solution in a manner that it preserves the heterogeneities. The main idea of this rescaling is the use of local problems. This interpolation technique fits nicely with a new multiscale framework for solving nonlinear partial differential equations. The combination of the interpolation and multiscale frameworks provides a robust and fast simulation technique that is employed in the present paper.

We have tested our method on some synthetic examples by considering both linear and nonlinear cases. We compare the numerical results for simulations that employ both updating the data at sensor location and the simulations that do not update the locations. Our numerical studies bear out that simulations that do not use the update or use it very infrequently produces large errors. Finally, we discuss the extension of the approach that uses some inversion of the local data.

In a real field, the sensors are typically in wells at different depths. A nice, tensor product mesh does not fit the reality of sensor locations. We are forced to work with unstructured, quite coarse grids if we want the sensors to be nodes on a mesh. The sensor data comes in the form of telemetry. The sensors can provide information at specified times, when specific conditions occur, be polled, or any number of combinations. We have developed a virtual telemetry package that is now part of SCIRun [1,2,3,4] so that we can simulate a field while computing remotely, as is typical in a real field simulation.

2   Data Driven Simulations

In this section we discuss the mapping of the sensor data to the finite dimensional space where the solution is calculated. This procedure is nontrivial in general since the solution space usually has high dimension while the sensors are located only at few locations. To demonstrate this in Fig. 1 we schematically plot a gray scale image of a 121×121 heterogeneous field with exponential variogram and the correlation lengths $l_x = 0.2$ and $l_x = 0.01$. Sensors in Fig. 1 are marked by an X. Due to uncertainties in the data, the random fields used for the simulations and the true field can differ significantly. Thus, we must be able to reconcile the streamed data from the sensors with that produced by our simulations.

Our simplified approach presented in this paper consists of passing the sensor data to the simulations and its use for the next time step simulations. Since the sensor data represents the solution only at few coarse locations we have to modify the solution conditioned to this data. We call this step *multiscale interpolation*. It consists of mapping the sensor data to the solution space. Before discussing the mapping, we demonstrate the main idea of our general approach handling dynamic data. It is depicted in Fig. 2 and consists of the following: At each time step the sensor data is received by the simulator. We can treat the data either as a hard or soft constraint. The latter means the data contains some noise and need not be imposed exactly.

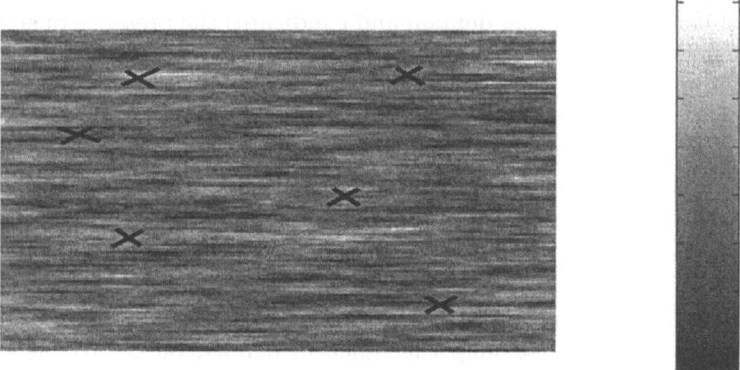

Fig. 1. Permeability field with spherical variogram and correlation lengths $l_x = 0.2$, $l_z = 0.02$. The locations of the sensors are marked.

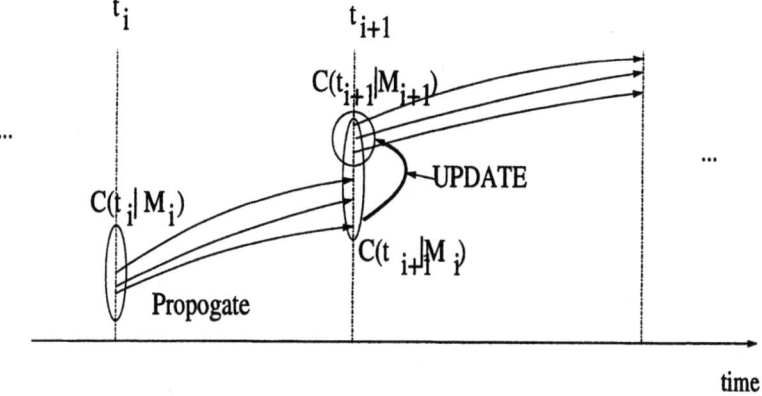

Fig. 2. Schematic description of data driven simulations. $C(t_i|M_i)$ designates the quantity of interest, such as the concentration, at time t_i, conditioned to some data M_i. In our basic methodology we consider M_i to be the sensor data. In the extension of our basic method the local heterogeneous field is also updated at each time based on sensor readings.

Consider the hard constraint case. At the beginning of each time step the sensor data needs to be mapped to the discrete solution space. This is performed using our DDDAS mapping operator, whose main goal is *not* to alter the heterogeneous field, i.e., at each time we update the data while not seeking the error source.

The proposed mapping for the sensor data is very general and applicable to various classes of equations. We consider general nonlinear parabolic equations of the form

$$D_t u_\epsilon = div(a_\epsilon(x, t, u_\epsilon, D_x u_\epsilon)) + a_{0,\epsilon}(x, t, u_\epsilon, D_x u_\epsilon), \text{ in } Q_0 \times [0, T], \quad (1)$$

where Q_0 refers to the spatial domain and ϵ indicates the presence of small scale heterogeneities. Equation (1) includes various physical processes that occur in subsurfaces.

Assume the domain is divided into a coarse grid such that the sensor points are nodal points of the coarse grid. Note that we do not require all nodal points to be sensor locations. Further denote by S^h the space of piecewise linear functions on this partition,

$$S_h = \{v_h \in C^0(\overline{Q_0}) : \text{the restriction } v_h \text{ is linear for each triangle } K \in \Pi_h\}.$$

Note that the K's are the coarse elements.

Our objective now is to map the function defined on S^h to the fine grid that represents the heterogeneities. This grid is obtained from *a priori* information about the field using geostatistical packages. Denote by the operator E the mapping from the coarse dimensional space into the fine dimensional space:

$$E : S^h \to V^h_\epsilon,$$

which is constructed as follows: For each element in $u_h \in S^h$ at a given time t_n we construct a space-time function $u_{\epsilon,h}(x,t)$ in $K \times [t_n, t_{n+1}]$ that satisfies

$$D_t u_{\epsilon,h}(x,t) = div(a_\epsilon(x,t,\eta, D_x u_{\epsilon,h})) \qquad (2)$$

in each coarse element K, where η is the average of u_h. $u_{\epsilon,h}(x,t)$ and is calculated by solving (2) on the fine grid, and thus is a fine scale function.

To complete the construction of E we need to set boundary and initial conditions for (2). We can set different boundary and initial conditions, giving rise to different maps that only differ from each other slightly. The main underlying property of our map is that it is constructed as a solution of local problems. The latter guarantees that the solution is consistent with prescribed heterogeneities.

In our numerical simulations we take the boundary and initial condition for the local problems to be linear with prescribed nodal values. The nodal values are obtained from the sensor data, if available. If the sensor data is unavailable at some location we use the values obtained from the simulations at previous time. Note that we cannot impose the values of the solutions directly at some locations since it can cause artificial discontinuities in the solution. See [5] for mathematical aspects of this interpolation operator, including convergence properties.

Once the solution at time $t = t_n$ is computed its values with sensor data at the sensor locations can be compared. After changing the values of the solution we interpolate it to the fine grid and use it for the next time step. At the last step we use a multiscale approach which is computationally efficient. In particular, the solution at the next time step is calculated based on

$$\int_{Q_0} (u_h(x,t_{n+1}) - u_h(x,t_n))v_h dx + \sum_K \int_{t_n}^{t_{n+1}} \int_K ((a_\epsilon(x,t,\eta, D_x u_{\epsilon,h}), D_x v_h) +$$

$$a_{0,\epsilon}(x,t,\eta, D_x u_{\epsilon,h})v_h)dxdt = \int_{t_n}^{t_{n+1}} \int_{Q_0} f v_h dx dt. \qquad (3)$$

Recall that Q_0 refers to the spatial domain and the K's are the coarse elements. This approach, combined with the interpolation technique, has great CPU advantages over just a fine scale calculations (see [5]).

3 Numerical Examples

We now present numerical results that demonstrate the accuracy and limitations of our proposed method. We have explored both linear and nonlinear heterogeneous diffusion cases. Due to space limitations here, we only provide a representative example. We will provide more examples and discussions in a future work.

The systems we consider are intended to represent cross sections (in the x-z plane) of the subsurface. For that reason we take the system length in x (L_x) to be five times the length in z (L_z). All of the fine grid fields used in this study are 121×121 realizations of prescribed overall variance (σ^2) and correlation structure. The fields were generated using GSLIB algorithms [6] with an exponential covariance model. In the examples below, we specify the dimensionless correlation lengths l_x and l_z, where each correlation length is nondimensionalized by the system length in the corresponding direction. For example, $l_x = 0.3$ means that the actual correlation length of the permeability field is $0.3L_x$.

In our calculations, we consider (1) with a fixed concentration at the inlet edge of the model $(x = 0)$ and a fixed concentration at the outlet edge $(x = L_x)$ of the model. The top and bottom boundaries are closed to flow. Initially zero contaminant concentration is assumed in the domain. For comparison purposes most results are presented in terms of cross sectional averages and l_2 norm errors. The computations are carried out until the final time $t = 0.1$ with a different frequency of updating.

We have tested both linear and nonlinear problems and observed similar numerical results. Here we will only present numerical results for the nonlinear case which represents simplified Richards' equation. Consider

$$D_t u_\epsilon = div(a_\epsilon(x, u_\epsilon)D_x u_\epsilon),$$

where $a_\epsilon(x, \eta) = k_\epsilon(x)/(1+\eta)^{\alpha_\epsilon(x)}$. $k_\epsilon(x) = \exp(\beta_\epsilon(x))$ is chosen such that $\beta_\epsilon(x)$ is a realization of a random field with the exponential variogram and with some correlation structure. $\alpha_\epsilon(x)$ is chosen such that $\alpha_\epsilon(x) = k_\epsilon(x) + const$ with the spatial average of 2. For the numerical examples we will only specify $\beta_\epsilon(x)$.

In all of the figures, solid line designates the true solution, dotted line designates the solution obtained using our simulations with some number of updates, and the dashed line designates the solution that has not been updated.

For the first example we consider the true field to be a random field with exponential variogram and with $l_x = 0.2$, $l_x = 0.02$, and $\sigma = 1$. For our simulations we use the same field with $\sigma = 2$. In Fig. 3 we compare the average solutions for the case with 20 updating, i.e., the simulated solution is updated 20 times during the course of the simulations. Fig. 3 demonstrates that the predictions that do not use the sensor data perform very poorly. The l_2-norm of the difference

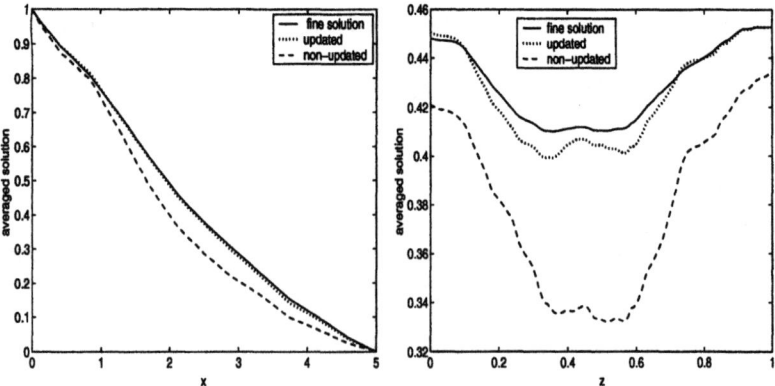

Fig. 3. Comparisons of the average solutions across x and z directions for nonlinear case. The true field is a realization of a random field with $l_x = 0.2$, $l_z = 0.01$, $\sigma = 1$, while the random field used for the simulations has $\sigma = 2$.

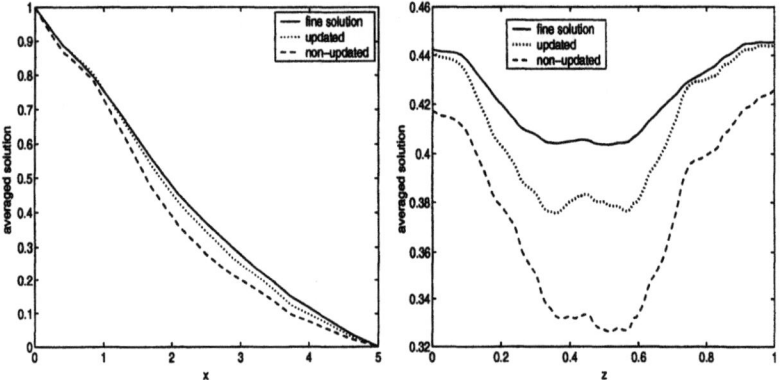

Fig. 4. Comparisons of the average solutions across x and z directions for nonlinear case. The true field is a realization of a random field with $l_x = 0.2$, $l_z = 0.01$, $\sigma = 1$, while random field used for the simulations has $\sigma = 2$.

between the solutions for the data that are frequently updated is 2.5% while the data that is not updated is 14%.

In our next example we ran the same case with less frequent updating. In particular, we use 4 updates during the simulations. The results are depicted in Fig. 4. The l_2-errors of the solution that is updated is 5.7% while the l_2 error of the non-updated solution is still 14%.

Finally we have considered a case where three different heterogeneous fields with exponential variograms are used with different probabilities. In particular the true field consists of $l_x = 0.1$, $l_z = 0.02$ with probability 0.1, $l_x = 0.2$, $l_z = 0.02$ with probability 0.8, $l_x = 0.4$, $l_z = 0.02$ with probability 0.1. For all these fields we take $\sigma = 1$. For our simulations we use these random fields with $\sigma = 2$ and with different probabilities. In particular, we take the field with

$l_x = 0.1$, $l_z = 0.02$ with probability 0.3, $l_x = 0.2$, $l_z = 0.02$ with probability 0.4, $l_x = 0.4$, $l_z = 0.02$ with probability 0.3. This is a typical field scenario when one does not know the probability distributions of the random fields. Our numerical results showed that the l_2 error between true and updated solution is 5%, while the error between the true solution and non-updated solution is 9%. The numerical results with frequent updates demonstrated higher accuracy.

4   Fixing Error Sources Using Inversion

So far we have presented a numerical method that incorporates the sensor data from sparse locations into the simulations. Currently we are working on the possible extension of the methodology that involves some kind of inversion during the simulations.

In the methodology described previously we only correct the solution without fixing the error source. One of the error sources is the incorrect heterogeneous diffusion field.

Consider the linear heterogeneous diffusion model $D_t u_\epsilon = div(a_\epsilon(x) D_x u_\epsilon)$. Our objective is to compare the misfit of the data from the sensors and the data from our simulator. If this misfit is larger than a certain threshold for some sensor points we perform an inversion in the neighborhood of the sensor point by modifying the diffusion field $a_\epsilon(x)$ in the neighborhood of the sensor point. By considering the neighborhood of the sensor location as a coarse block we determine the effective diffusivity, a^*_{sim}, corresponding to the existing heterogeneous diffusivity field. This quantity is determined such that the local problem with a single diffusion coefficient, a^*_{sim}, will give the same response as the underlying fine heterogeneous diffusion field.

Our next goal is to find an a^*_{true} that will give us the same response as the one from the sensor data. This is done using some classical inversion, such as parameter estimation. Since we only have to fit one parameter this problem can be easily handled without extensive computations.

Now we can impose a^*_{true} both as a soft as well as a hard constraint. For the hard constraint we must rescale the local heterogeneous field based on the ratio a^*_{sim}/a^*_{true}. Note that the hard constraint does not change the structure of the local heterogeneities and can be easily implemented since it only requires local rescaling.

For the soft constraint we use Bayesian framework (see [7]). Assume the a priori distribution for the local coarse heterogeneous field is given. Our objective is to generate fine-scale random diffusivity fields conditioned on the coarse-scale data. Denote a^c and a^f to be coarse and fine scale diffusivity fields respectively. Using Bayes theorem we obtain

$$P(a^f|a^c) \propto P(a^f)P(a^c|a^f), \tag{4}$$

where $P(a^f)$ is the probability distribution for the fine-scale field. In (4) $P(a^f)$ is a priori distribution of the fine-scale field which is prescribed.

To impose the coarse-scale data as a soft constraint we take $a^c = J(a^f) + \epsilon$, where J is a local upscaling operator and ϵ is a random noise, $\epsilon \sim N(0, \sigma^2)$. The local upscaling operator involves the solution of the local partial differential equations similar to (2). A soft constraint assumes that the coarse scale information is not accurate. Note that letting $\sigma \to 0$ we get a hard constraint.

The realizations from the posterior distribution (4) is generated using Markov Chain Monte Carlo (MCMC) [8]. This approach is known to be very general and can handle complex posterior distributions. The main idea of MCMC is to use a Markov Chain with a specified stationary distribution. The random drawing from the target distribution can be accomplished by a sequence of draws from full conditional distribution. Since the full conditionals have non-explicit form involving local problems Metropolis-Hastings algorithm is used.

We have tested our methodology with both soft and hard constraint on simple examples using Markov Random Fields as prior distributions for fine scale fields. Our numerical results indicated very good correspondence between the true and simulated diffusion fields on the coarse level. Further research into this area is warranted.

References

1. Douglas, C.C., Efendiev, Y., Ewing, R., Lazarov, R., Cole, M.R., Johnson, C.R., Jones, G.: Virtual telemetry middleware for DDDAS. Computational Sciences - ICCS 2003, P. M. A. Sllot, D. Abramson, J. J. Dongarra, A. Y. Zomaya, and Yu. E. Gorbachev (eds.) **4** (2003) 279–288
2. Douglas, C.C., Shannon, C.E., Efendiev, Y., Ewing, R., Lazarov, R., Cole, M.R., Johnson, C.R., Jones, G., Simpson, J.: Virtual telemetry middleware for DDDAS. In Darema, F., ed.: Dynamic Data-Driven Application Systems. Kluwer, Amsterdam (2004)
3. Johnson, C., Parker, S.: The scirun parallel scientific computing problem solving environment. In: Ninth SIAM Conference on Parallel Processing forScientific Computing, Philadelphia, SIAM (1999)
4. SCIRun: A Scientific Computing Problem Solving Environment. Scientific Computing and Imaging Institute (SCI), http://software.sci.utah.edu/scirun.html, 2002.
5. Efendiev, Y., Pankov, A.: Numerical homogenization of nonlinear random parabolic operators. SIAM Multiscale Modeling and Simulation (to appear) Available at http://www.math.tamu.edu/~yalchin.efendiev/ep-num-hom-parab.ps.
6. Deutsch, C.V., Journel, A.G.: GSLIB: Geostatistical software library and user's guide, 2nd edition. Oxford University Press, New York (1998)
7. Lee, S.H., Malallah, A., Datta-Gupta, A., Higdon, D.: Multiscale data integration using markov random fields. SPE Reservoir Evaluation and Engineering (2002)
8. Gilks, W., Richardson, S., Spegelhalter, D.: Markov Cain Monte Carlo in Practice. Chapman and Hall/CRC, London (1996)

Computational Aspects of Data Assimilation for Aerosol Dynamics*

A. Sandu[1], W. Liao[1], G.R. Carmichael[2], D. Henze[3], J.H. Seinfeld[3], T. Chai[2], and D. Daescu[4]

[1] Department of Computer Science Virginia Polytechnic Institute and State University, Blacksburg, VA 24060. {sandu,liao}@cs.vt.edu
[2] Center for Global and Regional Environmental Research, University of Iowa, Iowa City, IA 52240. {gcarmich,tchai}@cgrer.uiowa.edu
[3] Department of Chemical Engineering, California Institute of Technology, Pasadena, CA 91152. {daven,seinfeld}@caltech.edu
[4] Department of Mathematics and Statistics, Portland State University, Portland, OR 97207. daescu@pdx.edu

Abstract. In this paper we discuss the algorithmic tools needed for data assimilation for aerosol dynamics. Continuous and discrete adjoints of the aerosol dynamic equation are considered, as well as sensitivity coefficients with respect to the coagulation kernel, the growth rate, and emission and deposition coefficients. Numerical experiments performed in the twin experiment framework for a single component model problem show that initial distributions and the dynamic parameters can be recovered from time series of observations of particle size distributions.

1 Introduction

As our fundamental understanding of atmospheric particles and their transformations advances, novel computational tools are needed to integrate observational data and models together to provide the best, physically consistent estimate of the evolving state of the atmosphere. Such an analysis state better defines the spatial and temporal fields of key gas and particle phase chemical components in relation to their sources and sinks. Assimilation of chemical information is only now beginning in air quality/chemistry arenas [1,7,8,10,11], but offers the same motivations as those realized in the field of meteorology. Assimilation techniques can be utilized to produce three-dimensional, time varying optimal representations of the particle distributions in the atmosphere, that are consistent with the observed physical and chemical states.

Forward modeling of aerosols predicts the evolution of particle size distributions given the known parameters of the evolution (coagulation kernel, growth rate, emission rates and deposition velocities) as well as the initial size distribution. Numerous numerical methods have been proposed in the literature for

* The authors thank the National Science Foundation for supporting this work through the award NSF ITR AP&IM 0205198. The work of A. Sandu was also partially supported by the award NSF CAREER ACI 0093139.

M. Bubak et al. (Eds.): ICCS 2004, LNCS 3038, pp. 709–716, 2004.
© Springer-Verlag Berlin Heidelberg 2004

solving particle dynamics, e.g. [4,6] etc. The *inverse modeling* problem consists of recovering the initial or emitted size distribution and the parameters of evolution given information about the time evolution of the system, for example periodic measurements of the number, surface, mass, or volume density.

In this paper we discuss the computational tools needed for inverse modeling of aerosol dynamics, including continuous and discrete adjoints of the dynamic equation are derived, as well as sensitivity coefficients with respect to the coagulation kernel, the growth rate, and emission and deposition coefficients. A detailed derivation is given in [12,13]. At this time we do not treat chemical and thermodynamic transformations, and consider only space independent (box) models. The methods discussed here are a first step toward the goal of performing data assimilation for comprehensive three-dimensional particle, chemistry, and transport models.

2   Forward Modeling of Particle Dynamics

2.1   Continuous Formulation

We consider the particle dynamic equation in number density formulation, with particle volume the independent variable. The size distribution function (number density) of a family of particles will be denoted by $n(v, t)$; the number of particles per unit volume of air with the volume between v and $v + dv$ is $n(v, t)dv$. This describes completely a population of single-component particles. The aerosol population undergoes physical and chemical transformations (Seinfeld and Pandis, [14]) which change the number density in time:

$$\frac{\partial n(v, t)}{\partial t} = -\frac{\partial}{\partial v} \left[I(v, t) \, n(v, t) \right] \tag{1}$$

$$+ \frac{1}{2} \int_0^v \beta_{v-v', v'} \, n(v', t) \, n(v - v', t) \, dv'$$

$$- n(v, t) \int_0^\infty \beta_{v, v'} \, n(v', t) dv' + S(v, t) - L(v, t) \, n(v, t) \, ,$$

$$n(v, t = t^o) = n^o(v) \, , \quad n(v = 0, t) = 0 \, .$$

In this equation $I(v, t) = dv/dt$ is the rate of particle growth (e.g., due to condensation, evaporation, deposition and sublimation), $\beta_{v, v'}$ is the coagulation kernel, $S(v, t)$ is any source of particles of volume v (e.g., nucleation and emissions) and $L(v, t)$ is the first-order rate of removal of particles of volume v (e.g., by deposition).

2.2   Discrete Formulation

Here we use a discrete formulation of (1) based on the piecewise polynomial approach as described by Sandu [9] and (for simplicity of the presentation) on the Forward Euler time stepping scheme. The finite dimensional approximation of the number distribution $n(v, t)$ is taken in the space spanned by the set $\{\phi_i\}_{1 \leq i \leq s}$ of piecewise polynomial basis functions. The equation (1) is discretized as

$$n^{k+1} = n^k + \Delta t \left\{ G(t^k)n^k + F(n^k) + A^{-1} B\left(n^k, n^k\right) + S(t^k) - L(t^k)n^k \right\}. \quad (2)$$

The growth terms are discretized using a Discontinuous Galerkin approach where $F(n^k)$ is the Godunov flux difference and

$$G_{i,j}(t) = \int_0^\infty I(v, t)\phi_i(v)\phi_j'(v) \, dv \ .$$

The coagulation terms are discretized using a collocation approach [9] with collocation points $\{c_j\}_{1 \leq i \leq s}$, such that

$$A_{i,j} = \phi_j(c_i) \ , \quad \{B(n,n)\}_i = n^T B^i \, n \ ,$$

$$B_{i,k}^j = \frac{1}{2} \int_0^{c_j} \beta_{v,c_j-v}\phi_i(c_j)\phi_k(c_j - v) \, dv - \int_0^\infty \beta_{v,c_j-v}\phi_i(v)\phi_k(c_j - v) \, dv \ .$$

3   The Inverse Modeling Problem

Variational methods (e.g., 4D-Var) provide an optimal control approach to the data assimilation problem. The optimal analysis state is obtained through a minimization process to provide the best fit to the background estimate and to all observational data available in the assimilation window.

To be specific assume that observations $y^1 \ldots y^N$ are available at the discrete times $t^1 \ldots t^N$. The observed quantities are functions of the model state, $y(t) = h\big(n(v,t)\big)$. Observations can be particle number or mass densities, optical properties etc. The initial value of the solution n^o and/or the problem parameters p (i.e. I, β, S and L) are not known accurately, and data assimilation uses the information from observations to better estimate them.

A cost functional that measures the mismatch between model predictions and observations is

$$J(n^o, p) = \frac{1}{2}\,(n^o - n_B^o)^T\,R_0^{-1}\,(n^o - n_B^o) + \frac{1}{2}\,(p - p_B)^T\,B^{-1}\,(p - p_B) \quad (3)$$

$$+ \frac{1}{2}\sum_{k=1}^N (y^k - h(n^k))^T\,R_k^{-1}\,(y^k - h(n^k)) \ .$$

The first two terms penalize the departure of the initial solution and parameters from the apriori estimates (n_B^o and p_B), and are called background terms. The matrices R_0 and B are error covariances associated with background terms and R_k are error covariances associated with observations. The parameters n^o and p, which initially have values n_B^o and p_B, are adjusted such that the value of the functional is minimized. The minimization procedure needs the value of J and of its gradients with respect to the model parameters $\nabla_{n^o} J$ and $\nabla_p J$. For given values of (n^o, p) the functional value is obtained by a forward integration of the forward model, while its derivatives can be efficiently obtained through adjoint modeling as explained below.

3.1   Continuous Adjoints of the Dynamic Equation

In the continuous formulation one derives the adjoint of the exact dynamic equation (1), to obtain the derivative of the exact solution. The *adjoint equation* of the tangent linear model of (1) for the cost functional (3) is defined on each time interval (t^{k-1}, t^k) between consecutive observations for $k = N, N-1, \dots, 1$ by

$$\frac{\partial \lambda(v,t)}{\partial t} = -\frac{\partial \lambda(v,t)}{\partial v} I(v,t) - \int_v^\infty \lambda(v',t)\beta_{v,v'-v}n(v'-v,t)\, dv' \tag{4}$$

$$+\lambda(v,t)\int_0^\infty \beta_{v,v'}n(v',t)\, dv' + \int_0^\infty \lambda(v',t)n(v',t)\beta_{v',v}\, dv'$$

$$+\lambda(v,t)L(v,t)\,, \qquad t^{k-1}+\epsilon \leq t \leq t^k - \epsilon \quad (\epsilon \to 0)$$

$$\lambda\left(v,t^k-\epsilon\right) = \lambda\left(v,t^k+\epsilon\right) + R_k^{-1}\left(y^k - h(n^k)\right)\,,$$

$$\lambda\left(v,t^N+\epsilon\right) = 0\,, \quad \lambda(v,t^o) = \lambda(v,t^o+\epsilon) + R_0^{-1}\left(n^o - n_B^o\right)\,,$$

$$\lambda(v=0,t) = 0\,, \quad \lambda(v=\infty,t) = 0\,.$$

A complete derivation of the adjoint equations and the sensitivity relations is given in [12]. The gradients of the cost functional (3) with respect to forward model parameters are [12]:

$$\nabla_{n^o}\mathcal{J} = \lambda(v,t^o)\,, \quad \nabla_{I(v,t)}\mathcal{J} = \frac{\partial \lambda(v,t)}{\partial v}n(v,t)\,,$$

$$\nabla_{\beta_{v,v'}}\mathcal{J} = \int_{t_0}^T \left(-\lambda(v,t) + \frac{1}{2}\lambda(v+v',t)\right) n(v,t)n(v',t)\, dt$$

$$\nabla_{S(v,t)}\mathcal{J} = \lambda(v,t)\,, \quad \nabla_{L(v,t)}\mathcal{J} = -\lambda(v,t)n(v,t)\,.$$

All the above derivatives of the cost functional can be obtained by one integration of the forward model (1) during which the state $n(v,t)$ is saved, followed by a single backwards in time integration of the adjoint model (4). When solving the data assimilation problem each optimization iteration requires one evaluation of the cost functional and one evaluation of its gradient, i.e. one forward and one backward integration.

3.2   Discrete Adjoints of the Dynamic Equation

In the discrete approach the numerical discretization (2) of the the particle dynamic equation is considered to be the forward model. This is a pragmatic view, as only the numerical model is in fact available for analysis. The adjoint of the discrete model (2) is formulated and solved. The approach amounts to computing the derivatives of the numerical solution, rather than approximating the derivatives of the exact solution. Clearly, the formulation of the discrete adjoint equation depends not only on the dynamic equation, but also on the numerical method used to solve it.

Direct Approach. Taking the adjoint of the discrete equation (2) leads to a method to propagate the adjoint variables backwards in time [12]. The adjoint equation of (2) reads

$$\lambda^{k-1} = \lambda^k + \Delta t \ \{G^T(t^{k-1}) + F_n^T(n^{k-1}) + [(B+B^T) \times n^{k-1}] A^{-T} - L(t^{k-1})\} \cdot \lambda^k \ (6)$$
$$+ R_k^{-1} \left(y^k - h(n^k) \right), \qquad k = N, N-1 \dots 1$$
$$\lambda^N = 0, \qquad \lambda^o \longleftarrow \lambda^o + R_0^{-1} \left(n^o - n_B^o \right). \tag{7}$$

The adjoint variable at t^o gives the gradient of the cost functional (3) with respect to the initial distribution, $\lambda^0 = \nabla_{n^o} J$. Note that this is the derivative of the numerical solution as used in the definition of (3), as opposed to the continuous adjoint formulation (5a), where λ^0 defines the derivative of the continuous solution.

In practice the continuous forward model (1) is solved numerically, and so is the continuous adjoint equation (4). Therefore the continuous adjoint approach is in practice a hybrid approach. The operations of numerical discretization and adjoint do not commute in general, and consequently the numerical solution of (4) is different from the discrete adjoint (6). For data assimilation problems one needs the derivative of the numerical solution, i.e. the discrete adjoints are in principle preferred. For sensitivity studies using the adjoint method one wants to approximate the sensitivities of the continuous model, and the continuous adjoint seems more appropriate.

Automatic Differentiation. Given a program that implements the forward model, automatic differentiation builds a new, augmented program, that computes the analytical derivatives (accurate up to machine precision) along with the original program [3]. Different modes of automatic differentiation can produce the tangent linear model or the discrete adjoint model. In this paper we use the automatic differentiation tool TAMC [15] of Giering and Kaminski to generate the discrete adjoint derivatives.

Automatic differentiation only requires (an implementation of) the original forward model. It constructs gradients for complicated forward models, and even models which contain legacy code. Next, automatic differentiation produces derivatives of the numerical solution which are appropriate for optimization purposes.

4   Numerical Results

The Test Problem. For the numerical experiments we consider the test problem from [2]. With N_t the total initial number of particles, V_m the mean initial volume, the constant coagulation kernel $\beta_{v,w} = \beta_0$, and the linear growth rate $I(v) = \sigma_o v$, the problem admits the analytical solution

$$n^A(v,t) = \frac{4 N_t}{V_m (N_t \beta_o t + 2)^2} \cdot \exp \left(\frac{-2v \exp(\sigma_o t)}{V_m (N_t \beta_o t + 2)} - \sigma_o t \right).$$

We solve the dynamic equation for $\beta_o = 2.166 \times 10^{-6}$ cm^{3}h^{-1}particles^{-1}, $\sigma_o = 0.02$ h^{-1}, $N_t = 10^4$ particles, $V_m = 0.03$ μm^3. The values are chosen such that coagulation and growth have effects of comparable magnitude. The size range is truncated to $V_{min} = 10^{-3}$ μm^3, $V_{max} = 1\mu m^3$. A piecewise linear discretization with 8 log-uniform bins is employed. The time interval is $[t_0 = 0, T = 48]$ hours, and the time step $\Delta t = 6$ minutes. The actual implementation is done in the (equivalent) volume concentration density formulation.

The experiments are carried out in the twin experiment framework. A run with the reference values for initial conditions, coagulation kernel, and growth rate is used to generate hourly pseudo-observations $\{y_1 \ldots y_M\}$ of the number density, or a function of the number density.

The model is re-run with perturbed values of the initial conditions, coagulation kernel, and growth rate. The cost functional (3) is defined without background terms, and with observation covariance matrices equal to identity $R_k = I$. The lack of background terms is justified by our apriori knowledge of the fact that the initial guesses are wrong, while the observations are correct. The minimization algorithm is LBFGS [5], and the adjoint gradients are generated by TAMC. During optimization we prescribe the following bounds for the growth and coagulation parameters: $0.0 \leq \sigma \leq 0.5$ and $0.0 \leq \beta \leq 10^{-4}$.

As a measure of accuracy in the optimization process we consider the RMS difference between the reference and the optimized solutions:

$$RMS = \sqrt{\frac{1}{ndof+2} \sum_{j=1}^{ndof} \left(\frac{n_{ref}^o[j]-n_{opt}^o[j]}{n_{ref}^o[j]}\right)^2 + \left(\frac{\sigma_{ref}-\sigma_{opt}}{\sigma_{ref}}\right)^2 + \left(\frac{\beta_{ref}-\beta_{opt}}{\beta_{ref}}\right)^2}$$

In the data assimilation experiments the initial guesses are obtained by significant perturbations of the reference values, with $\beta_p = 5\beta_o$ and $\sigma_p = 25\sigma_o$.

Sensitivity Analysis. Adjoint calculations are a useful tool for sensitivity analysis. A widely used measure of the relative change in the functional due to relative changes in the input parameters is given by the logarithmic sensitivities

$$s = \partial \log J\left(t^F\right) \ / \ \partial \log V\left(t^o\right) \ .$$

Figure 1 shows the logarithmic sensitivity of the mean volume densities in bins 4 and 8 at the end of the 48 hours simulation interval with respect to the initial mean volume densities (left) and slopes (right). For both cases relative changes in the bin slopes have a considerably smaller influence on the final result than relative changes in the mean densities.

Data Assimilation. We first consider a complete set of hourly observations, i.e. all parameters of the solution (mean concentrations and slopes in each bin) are observed once every simulation hour. The test problem consists of the evolution under both coagulation and growth.

In Figure 2 the results for recovering simultaneously the initial distribution, β and σ are shown. This experiment is challenging for the inversion procedure since perturbations now affect not only the initial distribution, but the dynamic

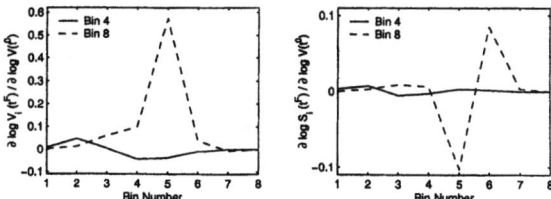

Fig. 1. The logarithmic sensitivities of the mean volume densities in bins 4 and 8 w.r.t. the initial distribution bin averages (left) and bin slopes (right).

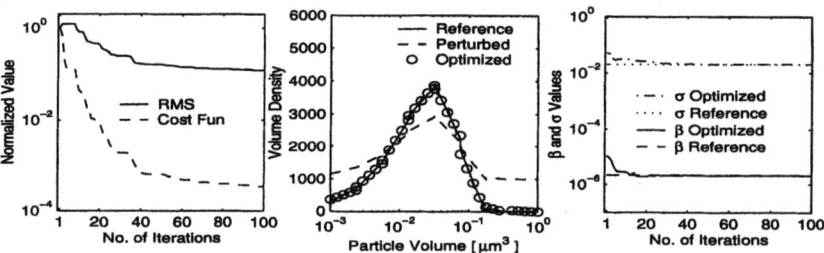

Fig. 2. Results for recovering V^o, β and σ for coagulation and growth from a complete set of hourly observations.

equation itself (through β and σ). The left Figure 2 panel shows the decrease of the cost function and RMS error with the number of iterations. The central panel shows the exact, reference, perturbed, and optimized distributions at the initial time. The optimized distribution is visually identical to the reference one. The right panel shows the reference and optimized β and σ, which are recovered accurately. Recovering only the initial distribution or only the parameters σ, β is easier and in both cases the minimization procedure converges in a small number of iterations.

Additional data assimilation experiments were carried out and the following conclusions were drawn: (1) The recovery procedure works even if observations are provided in only some of the bins. However less frequent observations in all bins are to be preferred to frequent but partial observations; (2) Observations of total surface, number, or volume density lead to a good recovery of densities only in the large size bins; (3) For the growth equation the optimization converges faster with the continuous adjoints, as compared to discrete adjoints; (4) The optimization performs similarly with discrete adjoints obtained by the direct and the automatic implementation.

5 Conclusions and Future Work

In this paper we have discussed the algorithmic tools needed for inverse aerosol modeling. Continuous and discrete adjoints of the integro-differential particle dynamic equation have been derived, together with formulas for sensitivity coefficients with respect to the coagulation kernel, the growth rate, and emission and deposition coefficients. Numerical tests were carried out using a single component particle dynamics model problem. From hourly measurements of the particle size one can recover the initial distribution as well as the parameters of the model.

The overall conclusion is that variational data assimilation is a feasible approach for particle dynamics. Discrete adjoints can be obtained easily through automatic differentiation, while continuous adjoints are a useful alternative.

A testing of the computational tools developed in this paper for a realistic growth problem of three-species aerosols is presented in [13]. Future work will extend the inverse modeling of aerosols to include chemical and thermodynamic processes. The techniques developed will be used ultimately to perform data assimilation in full three-dimensional models.

References

1. H. Elbern, H. Schmidt and A. Ebel. Variational data assimilation for tropospheric chemistry modeling. Journal of Geophysical Research 102(D13):15967–15985, 1997.
2. F.M. Gelbard and J.H. Seinfeld. Coagulation and growth of a multicomponent aerosol. *Journal Of Colloid and Interface Science* 63(3):472-479, 1996.
3. Andreas Griewank. Evaluating Derivatives: Principles and Techniques of Algorithmic Differentiation Frontiers in Applied Mathematics, SIAM, 2000.
4. M.Z. Jacobson. *Fundamentals of atmospheric modeling.* Cambridge University Press, 1999.
5. J. Nocedal. Department of Electrical & Computer Engineering, NorthWestern University, http://www.ece.northwestern.edu/ nocedal/lbfgs.html
6. A. Sandu and C. T. Borden. A Framework for the Numerical Treatment of Aerosol Dynamics. *Applied Numerical Mathematics* 45:475–497, 2003.
7. A. Sandu, D. Daescu, and G.R. Carmichael. Direct and Adjoint Sensitivity Analysis of Chemical Kinetic Systems with KPP: I – Theory and Software Tools. *Atmospheric Environment* 37:5083-5096, 2003.
8. D. Daescu, A. Sandu, and G.R. Carmichael. "Direct and Adjoint Sensitivity Analysis of Chemical Kinetic Systems with KPP: II – Validation and Numerical Experiments", *Atmospheric Environment* 37:5097-5114, 2003.
9. A. Sandu. Piecewise Polynomial Solutions of Aerosol Dynamics. Submitted, 2004.
10. A. Sandu, D.N. Daescu, T. Chai, G.R. Carmichael, J.H. Seinfeld, P.G. Hess, and T.L. Anderson. Computational Aspects of 4D-Var Chemical Data Assimilation in Atmospheric Models. F. Darema (ed.), Dynamic Data Driven Applications Systems, Kluwer Academic Publishers, 2004.
11. A. Sandu, D.N. Daescu, G.R. Carmichael, and T. Chai. Adjoint Sensitivity Analysis of Regional Air Quality Models. Submitted, 2003.
12. A. Sandu, W. Liao, G.R. Carmichael, D. Henze, and J.H. Seinfeld. Inverse Modeling of Aerosol Dynamics Using Adjoints – Theoretical and Numerical Considerations. Submitted, 2004.
13. D. Henze, J.H. Seinfeld, A. Sandu, W. Liao, and G.R. Carmichael. Inverse Modeling of Aerosol Dynamics. Submitted, 2004.
14. J.H. Seinfeld and S.N. Pandis. *Atmospheric chemistry and physics. From air pollution to climate change.* John Wiley & Sons, Inc., 1997.
15. Ralf Giering. Tangent Linear and Adjoint Model Compiler, http://www.autodiff.com/tamc.

A Framework for Online Inversion-Based 3D Site Characterization

Volkan Akçelik[1], Jacobo Bielak[1], George Biros[2], Ioannis Epanomeritakis[1], Omar Ghattas[1], Loukas F. Kallivokas[3], and Eui Joong Kim[4]

[1] Carnegie Mellon University, Pittsburgh PA, USA,
{volkan,jbielak,ike,omar}@cmu.edu
[2] University of Pennsylvania, Philadelphia PA, USA, biros@seas.upenn.edu
[3] University of Texas at Austin, Austin TX, USA, loukas@mail.utexas.edu
[4] Duke University, Durham NC, USA, ekim@duke.edu

Abstract. Our goal is to develop the capability for characterizing the three-dimensional geological structure and mechanical properties of individual sites and complete basins in earthquake-prone regions. Toward this end we present a framework that integrates in situ field testing, observations of earthquake ground motion, and inversion-based modeling.

1   Geotechnical Site Characterization

An important first step for forecasting strong ground motion during future earthquakes in seismically-active regions is to characterize the three-dimensional mechanical properties and geological structure of sites and basins within those regions. Characterizing a site refers to our ability to reconstruct as faithfully as possible the soil profile in terms of a limited set of material parameters, such as shear and compressional wave velocities, density, attenuation, and the slow velocity for poroelastic media.

Geological and geotechnical materials, soil and rock, impact the performance of the built environment during earthquakes. They play a critical role in the generation of ground motion, and, consequently, on determining the spatial extent and severity of damage during earthquakes. Yet, they are not well-investigated, even though they are the most variable and least controlled of all materials in the built environment. Since soils cannot be accessed easily, their properties can be inferred only indirectly. Currently, geomaterials are characterized with essentially the same general testing methods that were used 25 years ago. These methods rely on testing a small number of specimens in the laboratory and conducting a limited number of small-strain field tests. There is a critical need to advance beyond current methods to reduce the level of uncertainty that currently exists in the estimation of geological and geotechnical material properties.

A variety of techniques for reconstruction of earth properties from noninvasive field tests have been pursued, notably within the gas and oil exploration communities. However, the goals of our work are distinctly different, both in terms of the nature of the problem (e.g. complete material profile reconstruction

M. Bubak et al. (Eds.): ICCS 2004, LNCS 3038, pp. 717–724, 2004.
© Springer-Verlag Berlin Heidelberg 2004

vs. estimates of material contrast) and in the models employed (e.g. all elastic waves vs. just acoustic waves).

Reconstruction of the soil model results in a time-dependent or time-harmonic wave propagation inverse problem. Solution of this inverse problem represents an enormous challenge from the theoretical, algorithmic, and computational points of view. An added challenge is the lack of a systematic approach to in situ measurements. Such measurements are often decoupled from the needs of the computational process, due to cost considerations, equipment limitations, or the adoption of ad-hoc and simplified analysis procedures.

With current test equipment, the volume of soil that can be excited from a single location is somewhat limited because the maximum loads that can be applied are restricted and the response amplitude decays exponentially with distance, frequency, amount of soil damping, and slowness of the propagating waves. The advent of large-scale test equipment makes it possible to apply much larger loads over a wide range of frequencies, and thus excite a larger volume of soil from a single location than has been possible till now.

To effectively extract the desired information from the field test data, we need robust, efficient, and scalable forward and inverse three-dimensional wave propagation solvers. We have developed such methods and fine-tuned them for the analysis of earthquake ground motion in large basins. Our forward and inverse modeling methods are overviewed in Sections 2 and 3, respectively. Finally, in Section 4, we present an on-line framework for local site characterization that integrates steerable field experiments with inverse wave propagation.

2   Forward Elastic Wave Propagation Modeling

Our forward elastic wave propagation simulations are based on wavelength-adaptive mesh algorithms, which overcome many of the obstacles related to the wide range of length and time scales that characterize wave propagation problems through heterogeneous media [1,2,3,4,5,6,7]. In highly heterogeneous media such as sedimentary basins, seismic wavelengths vary significantly, and wavelength-adaptive meshes result in a tremendous reduction in the number of grid points compared to uniform meshes (e.g. a factor of 2000 in the Los Angeles Basin). Our code includes octree-based trilinear hexahedral elements and local dense element-based data structures, which permit wave propagation simulations to substantially greater resolutions than heretofore possible.

We have validated our code using the Southern California Earthquake Center (SCEC) LA Basin model and an idealized model of the 1994 Northridge earthquake. To illustrate the spatial variation of the 1994 Northridge earthquake ground motion, Fig. 1 presents snapshots at different times of an animation of the wave propagation through the basin. The left part of the figure shows a plan view and cross-section of the basin, as defined by the distribution of shear wave velocity. The projections of the fault and hypocenter are also shown. The directivity of the ground motion along strike from the epicenter and the concentration of motion near the fault corners are response patterns of the actual earthquake

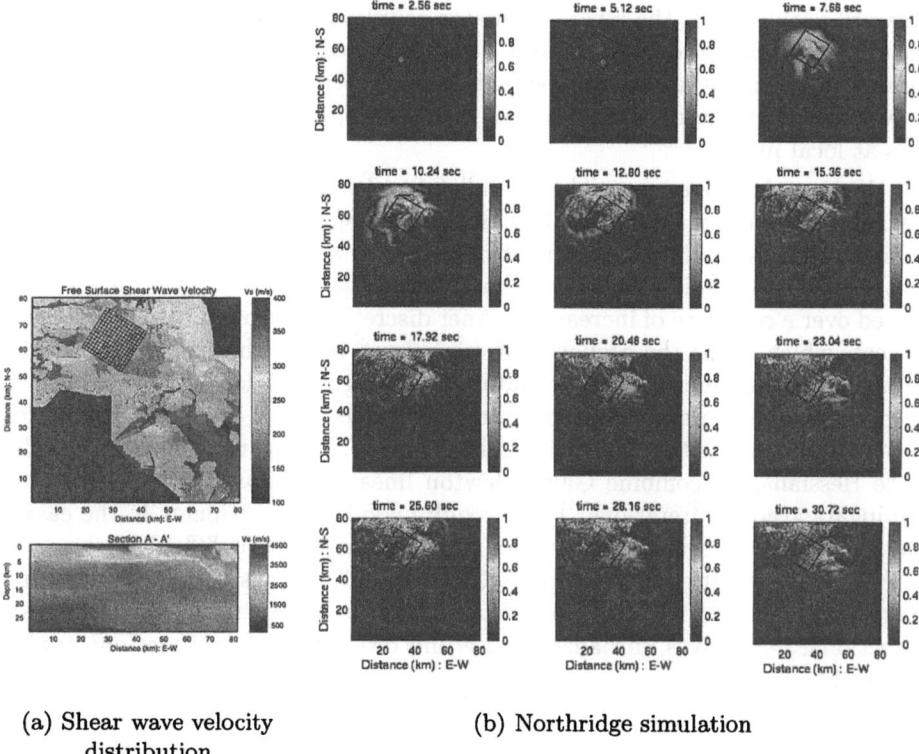

(a) Shear wave velocity
    distribution

(b) Northridge simulation

Fig. 1. Basin profile and basin response snapshots

that are reproduced in the simulation. Other effects, however, such as strong
ground motion in the southeast portion of the San Fernando Valley (Sherman
Oaks), in the Santa Monica area, and La Cienega, are not. This discrepancy is
due, in part, to the simplified nature of the source, but also because the current
SCEC model does not yet provide the necessary fidelity.

3   Inverse Elastic Wave Propagation Modeling

The discrepancy between observations and simulation illustrated in the previous
section underscores the need to invert for basin material properties, either from
records of past earthquakes or from observations of ground motion generated
by seismic testing equipment. The inverse problem is formulated as a nonlinear
optimization problem, with an objective function that represents a norm misfit
between observations and predictions, and constraints in the form of the seismic
wave propagation initial-boundary value problem. The inverse wave propagation
problem is significantly more difficult to solve than the corresponding forward
problem, because (1) forward solution is just a subproblem of the inverse problem, (2) the inverse problem is often ill-posed despite the well-posedness of the

forward problem, (3) the inverse operator couples the entire time history of the system's response and leads to large dense matrices, and (4) the objective function is often highly oscillatory, with basins of attraction that shrink with the wavelength of the propagating waves, thus entrapping most optimization methods at local minima.

We have developed a scalable, parallel algorithm for the inverse wave propagation problem that addresses these difficulties [8]. To treat multiple local minima, we employ multiscale continuation, in which a sequence of initially convex, but increasingly oscillatory, approximations to the objective function are minimized over a sequence of increasingly finer discretizations of state and parameter spaces, which keeps the sequence of minimizers within the basin of attraction of the global minimum. Total variation (TV) regularization is used to address ill-posedness and preserve sharp material interfaces such as between geological layers. To overcome the difficulties of large, dense, expensive to construct, indefinite Hessians, we combine Gauss-Newton linearization with matrix-free inner conjugate gradient iterations. These require good preconditioners in the case of the TV regularization operator, which presents considerable difficulties due to its strong heterogeneity and anisotropy. Our preconditioner is based on the limited memory BFGS algorithm to approximate curvature information using changes in directional derivatives, initialized by second order stationary iterations applied to the TV operator.

Here we provide evidence of the feasibility of this approach for finely-parameterized models. We illustrate first with a two-dimensional sedimentary basin undergoing antiplane motion. The seismic model under consideration comprises a portion of the vertical cross-section of the Los Angeles basin in Fig. 1, as shown at the bottom of the right frame in Fig. 2a, where the trace of a strike-slip fault and the hypocenter of an idealized source are identifiable. In our numerical experiments, we use waveforms synthesized from the target model on the free

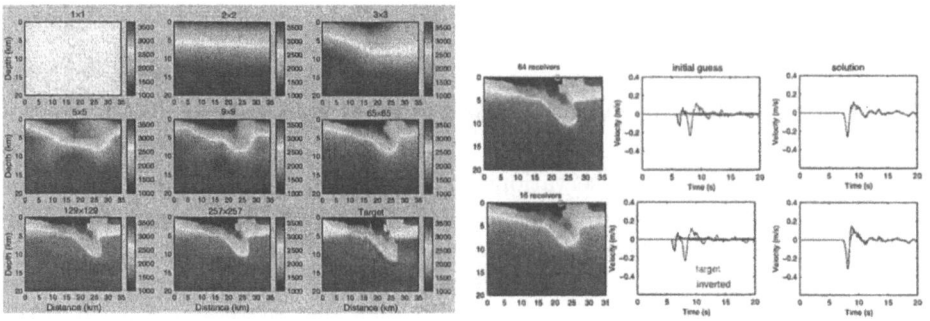

(a) Stages of multiscale inversion   (b) Effect of receiver density

Fig. 2. Stages of the multiscale inversion, starting from a homogeneous guess (*left*); comparison of wave velocity profiles and time histories (*red*: target; *blue*: inverted) obtained using different numbers of receivers (*right*)

surface as pseudo-observed data. We assume that the source, defined by its rupture velocity and slip function, is known, and invert for the shear wave velocity distribution of the geological structure. We have assumed that the material is lossless and that the density distribution is known. Five percent random noise has been added artificially to the synthetic seismogram. Fig. 2a shows a sequence of inverted material models, corresponding to increasingly finer inversion grids. Inversion was performed with 64 observers distributed uniformly on the free surface. The high fidelity of the results is noteworthy. Fig. 2b compares results for 64 and 16 receivers. As expected, the resolution of the inverted model for 16 receivers is not as great as for 64 receivers, yet it still closely approximates the target model. The synthetics in Fig. 2b show the velocity at one non-receiver location for the two models, both for the initial guess and for the inverted solution.

Corresponding results for a three-dimensional simulation of the San Fernando Valley are shown in Figs. 3 and 4. The material in this model is linearly elastic and lossless, with density and Lamé parameter λ fixed. We invert for the shear modulus using a grid of 15×15 receivers placed on the free surface and uniformly distributed throughout the valley. The source is a point-source double-couple located at a $16\,\mathrm{km}$ depth. Fig. 3 displays sections of the distribution of target and inverted shear wave velocity. The right panel shows the N-S component of

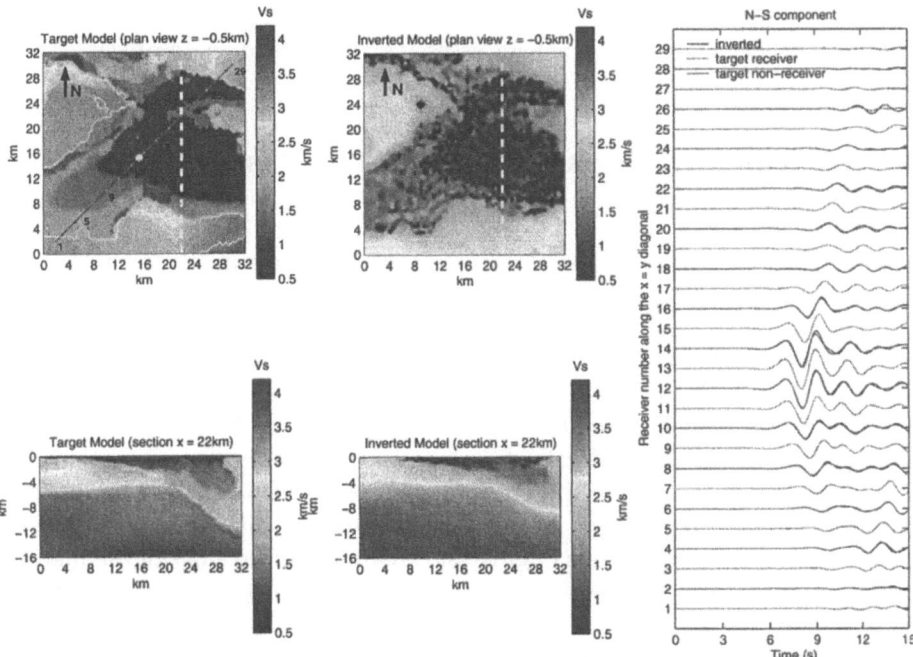

Fig. 3. LA Basin 3D wave-based inversion: target and computed material profiles; *top row*: plan view at $500m$ depth; *bottom row*: section view

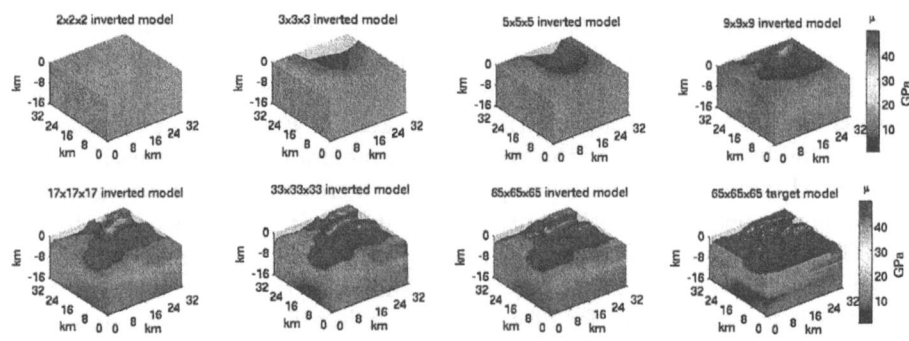

Fig. 4. Convergence of computational to target model via inversion grid refinement

the particle velocity at the points located on the diagonal line shown in the top left panel. Here again the high fidelity of the results is notable; major features of the target model are captured by the inverted model, and the agreement of the target and inverted waveforms is excellent, even at the non-receiver locations. Figure 4 illustrates the different levels of the multiscale algorithm. In this figure we show 12 GPa isosurfaces of the shear modulus. The similarity between the finest-grid inverted model and the target model is evident by comparing the last two images. Current efforts are aimed at extending the method to incorporate inversion for attenuation properties and nonlinear constitutive parameters of the soil, both of which are important for assessing response of geotechnical systems under strong ground motion.

4   A Framework for Online Inversion-Based 3D Site Characterization

The integration of field tests with seismic inversion presents some unique challenges and opportunities, both from the computational as well as experimental point of view. Our goal is to close the loop between experiment and simulation in the context of characterizing individual sites and basins. Toward this end we are pursuing an online framework for site characterization that integrates inversion-based modeling and steerable field experiments. The idea is to repeatedly: (1) excite the soil and collect surface and downhole ground motion data at a particular field location, (2) transmit the data back to high-end supercomputing facilities, (3) perform inversion to predict the distribution of material properties, (4) identify regions of greatest remaining uncertainty in the material properties, and (5) provide guidance on where the next set of field tests should be conducted.

The first key issue will be to turn around the inversion calculations in a time frame appropriate for the field experiments. Getting data to and from the compute nodes will not be a problem; two-way satellite connection will be able to keep up with the sensor ground motion data that is sent, and the material model that is received, on a daily basis. On the other hand, the run times needed to compute the 3D inversions in the previous section (on the order of a day on 512 processors) would be prohibitive in the online context. However, there are several improvements and simplifications we can make. First, we will invert for significantly smaller regions than the San Fernando Valley example considered above, typically just a square kilometer surrounding the test site. Second, whereas the inversions presented above began with homogeneous initial guesses, here we will incorporate priors based on an *offline phase* in which records of past earthquakes are inverted for an initial model of the basin. With the incorporation of a prior, we can expect faster convergence to the optimum. Third, ongoing improvements to the Hessian preconditioner are expected to yield a significant speedup in iterations. With these improvements, we can expect to comfortably turn around an inversion overnight on a reasonable number of processors.

The second key issue is the ability to invert the material model for observations associated with *multiple* excitations generated by the field equipment for a particular location. These multiple sources provide a richer dataset for inversion, and thus result in better resolution of the reconstructed soil model. However, this entails a large increase in necessary computational resources, since at each iteration of the inversion algorithm, a forward model must be solved for *each* independent excitation, and several dozen such excitations may be available. Fortunately, there is considerable coarse granularity available in the algorithm: all of the forward wave propagation simulations can proceed independently in parallel, at each inversion iteration. Therefore, by exploiting multiple, possibly distributed, groups of processors, we can effect inversion for multiple sources within the same clocktime as a single source—perhaps faster, since the inverse problem is better conditioned with the more copious available data, and the number of iterations should be fewer. The processor groups can be tightly coupled, as for example in the PSC AlphaCluster, or else loosely coupled across multiple sites of the NSF NEESGrid/TeraGrid. The inversion algorithm is highly latency tolerant—the coordination among the multiple forward simulations is in the form of broadcasts of material model corrections—done only after the forward (and adjoint) simulations are completed. Thus we can harvest cycles on processor groups wherever they are available.

The third key issue is the ability to steer the experimental equipment towards regions of high uncertainty in basin properties. The opportunity for steering occurs when the inversion is completed, the material model (of the entire basin) updated to reflect the recently acquired information, and the model transmitted back to the field to help direct the next round of field tests. Uncertainty in the estimates of the material parameters, as indicated by their variances, can be used to steer the equipment towards regions where conducting new tests is most profitable. Evaluating the variances exactly is intractable for such large-scale problems, but through the use of improved preconditioners we may be able to generate sufficiently good estimates of the important components of the Hessian, since the misfit term is typically of very low rank. The main task is to balance quality of the estimates against speed in obtaining them, while keeping sight of the goal of on-line experimental steering.

Acknowledgements. This work was supported by the National Science Foundation's Knowledge and Distributed Intelligence (KDI) and Information Technology Research (ITR) programs (through grants CMS-9980063, ACI-0121667, and ITR-0122464) and the Department of Energy's Scientific Discovery through Advanced Computation (SciDAC) program through the Terascale Optimal PDE Simulations (TOPS) Center. Computing resources on the HP Alpha-Cluster system at the Pittsburgh Supercomputing Center were provided under NSF/AAB/PSC award BCS020001P.

References

[1] Bao H., Bielak J., Ghattas O., Kallivokas L.F., O'Hallaron D.R., Shewchuk J.R., and Xu J. Large-scale Simulation of Elastic Wave Propagation in Heterogeneous Media on Parallel Computers. *Computer Methods in Applied Mechanics and Engineering*, 152(1–2):85–102, January 1998.

[2] Bielak J., Xu J., and Ghattas O. Earthquake ground motion and structural response in alluvial valleys. *Journal of Geotechnical and Geoenvironmental Engineering*, 125:413–423, 1999.

[3] Bielak J., Loukakis K., Hisada Y., and Yoshimura C. Domain reduction method for three-dimensional earthquake modeling in localized regions, part I: Theory. *Bulletin of the Seismological Society of America*, 93:817–824, 2003.

[4] Yoshimura C., Bielak J., Hisada Y., and Fernandez A. Domain reduction method for three-dimensional earthquake modeling in localized regions, part II: Verification and applications. *Bulletin of the Seismological Society of America*, 93:825–840, 2003.

[5] Kim E., Bielak J., Ghattas O., and Wang J. Octree-based finite element method for large-scale earthquake ground motion modeling in heterogeneous basins. In *Eos Trans. AGU, 83(47)*, Fall Meet. Suppl., Abstract S12B-1221, 2002.

[6] Volkan Akcelik, Jacobo Bielak, George Biros, Ioannis Epanomeritakis, Antonio Fernandez, Omar Ghattas, Eui Joong Kim, Julio Lopez, David O'Hallaron, Tiankai Tu, and John Urbanic. High resolution forward and inverse earthquake modeling on terascale computers. In *Proceedings of SC2003*, Phoenix, 2003. ACM/IEEE.

[7] E. Kim, J. Bielak, and O. Ghattas. Large-scale Northridge Earthquake simulation using octree-based multiresolution mesh method. In *Proceedings of the 16th ASCE Engineering Mechanics Conference*, Seattle, Washington, July 2003.

[8] Akcelik V., Biros G., and Ghattas O. Parallel multiscale Gauss-Newton-Krylov methods for inverse wave propagation. In *Proceedings of IEEE/ACM SC2002 Conference*, Baltimore, MD, November 2002. SC2002 Best Technical Paper Award.

A Note on Dynamic Data Driven Wildfire Modeling

J. Mandel[1], M. Chen[1], L.P. Franca[1], C. Johns[1], A. Puhalskii[1], J.L. Coen[2],
C.C. Douglas[3], R. Kremens[4], A. Vodacek[4], and W. Zhao[5]

[1] University of Colorado Denver, Denver, CO 80217-3364, USA
[2] National Center for Atmospheric Research, Boulder, CO 80307-3000, USA
[3] University of Kentucky, Lexington, KY 40506-0045, USA
[4] Rochester Institute of Technology, Rochester, NY 14623-5603, USA
[5] Texas A&M University, College Station, TX 77843-1112, USA

Abstract. A proposed system for real-time modeling of wildfires is described.
The system involves numerical weather and fire prediction, automated data
acquisition from Internet sources, and input from aerial photographs and
sensors. The system will be controlled by a non-Gaussian ensemble filter
capable of assimilating out-of-order data. The computational model will run on
remote supercomputers, with visualization on PDAs in the field connected to
the Internet via a satellite.

1 Introduction

Today, there exists virtually no capability to predict the spread of wildfires. In 2000
alone (the worst fire season in over 50 years), over 90,000 wildfires cost an estimated
$10 billion in suppression and lost resources. Current field tools for diagnosing exp-
ected fire behavior are simple algorithms that can be run on simple pocket calculators.
Researchers and fire managers alike envision a future when they can rely on complex
simulations of the interactions of fire, weather, and fuel, driven by remote sensing
data of fire location and land surface properties. *And* get the results as easy to
understand animations on a small laptop or PDA, have the computer incorporate all
information as soon as it becomes available, and assess the likelihood of possible
scenarios. This is how computers work in the imagination of sci-fi movie directors.
This project is to build a prototype as the first step to make it a reality.

This note is based on [24], where further details can be found.

2 Overview of the Proposed System

The objective of this project is to develop a prototype hardware and software system
to integrate available information related to a wildfire in progress, and to provide a
numerical prediction of the wildfire in a visual form, including tools to predict the
outcome of firefighting strategies. The proposed system will have the following main
components:

M. Bubak et al. (Eds.): ICCS 2004, LNCS 3038, pp. 725–731, 2004.
© Springer-Verlag Berlin Heidelberg 2004

- Numerical coupled atmosphere/fire model
- Data acquisition (measurements)
  - From Internet: maps (GIS), aggregated fire information, weather
  - Field information: aerial photos, sensors
- Visualization and user interface
- Dynamic Data Assimilation control module
- Guaranteed secure communication infrastructure

The numerical model accepts data in a mesh format. The Dynamic Data Assimilation control module will call the numerical model to execute multiple simulations, extract data from the state of the numerical model to be compared with the measurements and modify the state of the numerical model to match the measurements. The visualization and user interface module will display the results of the simulation and support user input to control alternative firefighting scenarios.

The numerical model will run on one or more remote supercomputers, while the visualization and user interface will run on PDAs in the field. Software agents and search engines will provide internet data, while networked sensors and cameras in airplanes will provide the field data. The field internet connection will be implemented using a local wireless network, bridged to the internet by a consumer-grade cheap broadband satellite connection and satellite data phones.

3 Weather and Fire Model

NCAR's *coupled atmosphere-fire model* is described in detail in [9, 12, 13, 14]. A *3D atmospheric prediction model* [7, 8, 10, 11] has been coupled with an *empirical fire spread model* [1, 2, 29, 30] such that the heat fluxes from the fire feed back to the atmosphere to produce fire winds, while the atmospheric winds drive the fire propagation. This wildfire simulation model can thus represent the complex interactions between a fire and local weather.

The existing numerical model is a legacy FORTRAN code that took a quarter of century and substantial scientific expertise to develop. We are proceeding in these steps: encapsulate execution of one time step as a subroutine; define and enforce software interfaces; upgrade the fire model from empirical to differential equations based; and speed up the model using techniques such as OpenMP and multigrid.

The overall system will run many instances of the model simultaneously; we expect that each instance of the model will run on a single CPU or an SMP node.

Our next step up from an empirical model of fire propagation is a model based on a reaction-convection-diffusion equation describing simplified fire physics with additional stochastic terms, which also model spotting (secondary fires started by flying embers). For related models, see [4, 16, 22]. There are two coupled equation: one for the temperature, derived from the balance of heat fluxes, and one equation for the rate of consumption of the fuel. The temperature influences the burn rate, which in turn determines the term for the heat flux generated by the burning in the equation for the temperature. Added stochastic terms are a noise expressing the uncertainties of the model, and flight of burning embers, modeled by replacing (with a certain probability, dependent on the temperature and the wind) the temperature a point by the temperature at another randomly selected point.

Anticipated developments of the fire model include several species of fuel to model different types of fire (grass, brush, crown fire), spotting terms with delay to account for the time it takes for the fire started by the flying embers to become visible on the computational grid scale, and of course coupling with the NCAR atmosphere model.

4 Data Acquisition

Maps, sometimes including fuel information, are available from GIS files maintained by public agencies as well as private companies. We are working on translating the GIS format files into meshed data suitable for input into the model. Raw as well as assimilated weather data are readily available from numerous sources on the Internet, including NOAAPORT broadcast, MesoWest weather stations, and the Rapid Update Cycle (RUC) weather system by NOAA. Aggregated fire information is available from the GeoMAC project at USGS. The challenge here is to develop intelligent converters of the information, which can deal with the constantly changing nature of Internet sites.

Thermal and near infrared images obtained from a manned or unmanned airborne platform is perhaps one of the best means of tracking the advance of wildland fires [28, 32]. Infrared radiation will pass through smoke relatively unattenuated, thus providing a signal of the exact fire location.

To obtain the geographic coordinates for the ground locations of individual pixels in an image that may correspond to the fire front or hotspot, the location and the direction that the camera was pointing at the time of image capture can be obtained by a combination of GPS readings for aircraft position and 3-axis inertial measurement unit data to determine the pointing direction of the camera. Terrain induced distortion can be corrected by parallax measurements on overlapping stereo images, or from given knowledge of the terrain and the camera pointing direction and altitude [25]. While topographic maps exist for much of the U.S., one very promising data set for the U.S. and much of the rest of the world is from the Shuttle Radar Topography Mission [27].

Networked, autonomous environmental detectors may be placed in the vicinity of a fire for point measurements of fire and weather information [21]. For geolocation, the units can either be built with a GPS capability or a less expensive approach is to plan for a GPS position to be taken at the exact location of the deployment using an external GPS unit and uploading the position data to the detector memory.

5 Dynamic Data Assimilation

An important feature of DDDAS is that the model is running all the time and it incorporates new data as soon as it arrives. Also, in this application, uncertainty is dominant because important processes are not modeled, there are measurement and other errors in the data, the system is heavily nonlinear and ill-posed, and there are multiple possible outcomes. This type of problem is a natural candidate for sequential Bayesian filtering (sequential because the data is incorporated sequentially as it arrives rather than all at once). The state of the system at any time period is represented by a collection of physical variables and parameters of interest, usually at mesh points. To be able to inject data arriving out of sequence, we will work with the *time-state vector x*, which will contain snapshots of system states at different points in time. The knowledge of the time-state of the system is represented in the model as a *probability density function p(x)*. The model will represent the probability distribution using an ensemble of time-state vectors $x_1, \ldots, x_n$. Thus, the number of the system states maintained by the model will be equal to the number of time snapshots saved for updating out of order data multiplied by the ensemble size, a product that will easily reach the thousands. Each of these system states will be advanced in time via separate simulations and substantial parallel supercomputing power will be required.

Sequential filtering consists of successive updating of the model state using data via Bayes theorem. The current state of the model is called the *prior* or the forecast probability density $p^f(x)$. The arriving data consists of measurements y along with an information how the measured quantities are derived from the system state x, and information about the distribution of measurement errors. That is, the additional information provided by the data to the model is represented by the vector y and a conditional probability density function p(y|x). The update is incorporated into the model using the Bayes theorem, resulting in the *posterior* or *analysis* probability density

$$p^a(x) = \frac{p(y \mid x) p^f(x)}{\int p(y \mid \xi) p^f(\xi) d\xi}, \tag{1}$$

which then becomes the new state of the model. The model states continue advancing in time, with a corresponding advancement of the probability density, until new data arrives. The system states for times within the time stamps of the data are system estimation. The system states beyond the time stamp of the data constitute a prediction. Note that advancing the model in time and injecting the data into the model is decoupled.

Clearly, the crucial question is how the probability distributions are represented in the model. The simplest case is to assume normal distributions for the states x and the conditional distributions p(y|x) and that the observations y depend linearly upon the states. These assumptions give the well-known Kalman filter formulas [20]. In particular, the posterior is also normal with the mean equal to the solution x^a of the least squares problem

$$\min_{x^a} \left\{ (x^a - x^f)^T \Sigma^{-1} (x^a - x^f) + (y - Hx^a)^T R^{-1} (y - Hx^a) \right\}. \tag{2}$$

Ensemble Kalman filters [17, 18, 19, 31] represent a normal distribution by a sample, which avoids manipulation of covariance matrix Σ, replacing it with sample covariance estimate. Thus, the ensemble approach is particularly useful in problems where the state vector x or the observation vector y are large. However, existing methods do not support out of order data, because they operate on only one time snapshot, and extensions of ensemble filters to the time-state model described here have to be worked out. Further, if the time evolution function f is nonlinear or the observation matrix H in (2) is replaced by a nonlinear function, the probability distribution $p(x)$ is not expected to be normal. Particle filters [3, 15] can approximate any probability distribution, not only normal, by a sample of vectors $x_1,\ldots,x_n$, with corresponding weights $w_1,\ldots,w_n$; then approximately $x \sim D$, where the probability of any event ω is calculated by

$$D(\omega) = \sum_{x_i \in \omega} w_i . \qquad (3)$$

Time-state models have been used in processing out-of-order data by particle filters [23]. But the approximation (3) is rather crude and, consequently, particle filters generally require large samples, which is not practical in our application. For this reason, we are investigating data assimilation by ensemble filters based on Gaussian mixtures [5, 6], where the probability distributions are approximated by a weighted mixture of normal distributions, such as

$$x \sim N(\mu_i, \Sigma_i) \text{ with probability } p_i . \qquad (4)$$

To assimilate data that depends on system state in a highly nonlinear way, such as fire perimeters from aerial photographs, one of the options we are investigating is replacing the least squares problem (2) by the nonlinear optimization problem

$$\min_{x^a}\left\{(x^a - x^f)^T \Sigma^{-1}(x^a - x^f) + (y - h(x^a))^T R^{-1}(y - h(x^a))\right\}, \qquad (5)$$

which may have multiple local minima. Alternatively, we consider updating the model state by minimizing some probabilities that the solution is off by more than a given amount, e.g.,

$$\min_{x^a}\left\{P(\neg | x^a - x^f | < \varepsilon_x \ \vee \ \neg | y - h(x^a) | < \varepsilon_y)\right\}, \qquad (6)$$

where the inequalities are understood term by term, and ε_x, ε_y are vectors of tolerances. This type of threshold estimates, related to Bahadur efficiency [26], could be quite important in fire problems, because of the nature of fires. Unlike in the case when the distribution of x is normal, here a least squares estimate may say little about whether or not the ignition temperature in a given region has been reached.

The Dynamic Data Assimilation module needs also to steer the data acquisition. In terms of the Bayesian update, one method of steering would be to select the measurements to minimize the variance of the posterior distribution. In the linear Gaussian case, this becomes an optimization problem for the observation matrix H under the constraints of what measurements can be made.

Finally, guaranteed secure communication [33] delivers messages within a preset delay with a given probability close to one. This means that loss of data or loss of communication with ensemble members could be, in principle, handled as a part of the stochastic framework, and the Dynamic Data Assimilation module should take advantage of that.

Acknowledgments. This research has been supported in part by the National Science Foundation under grants ACI-0325314, ACI-0324989, ACI-0324988, ACI-0324876, and ACI-0324910.

References

1. Albini, F. A.: PROGRAM BURNUP: A simulation model of the burningof large woody natural fuels. Final Report on Research Grant INT-92754-GR by U.S.F.S. to Montana State Univ., Mechanical Engineering Dept. (1994)
2. Anderson, H.: Aids to determining fuel models for estimating fire behavior. USDA Forest Service, Intermountain Forest and Range Experiment Station, Report INT-122 (1982)
3. Arulampalam, M., Maskell, S., Gordon, N., Clapp, T.: A tutorial on particle filters for online nonlinear/non-Gaussian Bayesian tracking, *IEEE Transactions on Signal Processing*, 50 (2002) 174-188
4. Asensio, M. I., Ferragut, L.: On a wildland fire model with radiation, *Int. J. Numer. Meth. Engrg*, 54 (2002) 137-157
5. Bengtsson, T., Snyder, C., Nychka, D.: A nonlinear filter that extends to high dimensional systems. *J. of Geophys. Res. - Atmosphere, in review* (2003)
6. Chen , R., Liu, J. S.: Mixture Kalman filters, *J. of the Royal Statistical Society: Series B*, 62 (2000) 493-508
7. Clark, T. L.: A small-scale dynamic model using a terrain-following coordinate transformation, *J. of Comp. Phys.*, 24 (1977) 186-215
8. Clark, T. L.: Numerical simulations with a three-dimensional cloud model: lateral boundary condition experiments and multi-cellular severe storm simulations, *J. Atmos. Sci.*, 36 (1979) 2191-2215
9. Clark, T. L., Coen, J., Latham, D.: Description of a coupled atmosphere-fire model, *Intl. J. Wildland Fire*, 13, in print (2004)
10. Clark, T. L., Hall, W. D.: Multi-domain simulations of the time dependent Navier- Stokes equations: Benchmark error analysis of some nesting procedures, *J. of Comp. Phys.*, 92 (1991) 456-481
11. Clark, T. L., Hall, W. D.: On the design of smooth, conservative vertical grids for interactive grid nesting with stretching, *J. Appl. Meteor.*, 35 (1996) 1040-1046
12. Clark, T. L., Jenkins, M. A., Coen, J., Packham, D.: A coupled atmospheric-fire model: Convective feedback on fire line dynamics, *J. Appl. Meteor*, 35 (1996) 875-901
13. Clark, T. L., Jenkins, M. A., Coen, J., Packham, D.: A coupled atmospheric-fire model: Convective Froude number and dynamic fingering, *Intl. J. of Wildland Fire*, 6 (1996) 177-190
14. Coen, J. L., Clark, T. L., Latham, D.: Coupled atmosphere-fire model simulations in various fuel types in complex terrain, in 4th. Symp. *Fire and Forest Meteor. Amer. Meteor. Soc., Reno*, Nov. 13-15 (2001) 39-42
15. Doucet, A., Freitas, N. de, Gordon, N. (ed.): *Sequential Monte Carlo in Practice, Springer.*
16. Dupuy, J. L. and Larini, M. (1999). Fire spread through a porous forest fuel bed: A radiative and convective model including fire-induced flow effects, *Intl. J. of Wildland Fire*, 9 (2001) 155-172.
17. Evensen, G.: Sequential data assimilation with nonlinear quasi-geostrophic model using Monte Carlo methods to forecast error statistics, *J. Geophys. Res.*,99 (C5) (1994) 143-162
18. Evensen, G.: The ensemble Kalman filter: Theoretical formulation and practical implementtation. http://www.nersc.no/geir (2003)
19. Houtekamer, P., Mitchell, H. L.: Data assimilation using an ensemble Kalman filter technique, *Monthly Weather Review*, 126 (1998) 796-811
20. Jazwinski, A. H.: Stochastic processes and filtering theory, Academic Press, New York (1970)
21. Kremens, R., Faulring, J., Gallagher, A.,Seema, A., Vodacek, A.: Autonomous field-deployable wildland fire sensors, *Intl. J. of Wildland Fire*, 12 (2003) 237-244
22. Linn, R. Reisner, J.,Colman, J., Winterkamp, J.: Studying wildfire behavior using FIRETEC, *Int. J. of Wildland Fire*, 11 (2002) 233-246

23. Mallick, M., Kirubarajan, T., Arulampalam, S.: Out-of-sequence measurement processing for tracking ground target using particle filters, *in Aerospace Conference Proceedings*, vol. 4, *IEEE* (2002) 4-1809-4-1818

24. Mandel, J., Chen, M., Franca, L.P., Johns, C., Puhalskii, A., Coen, J.L., Douglas, C.C., Kremens, R., Vodacek, A., Zhao, W.: Dynamic data driven wildfire modeling, *F. Darema (ed.), Dynamic Data Driven Applications Systems, Klumer Academic Publishers* (2004)

25. Mostafa, M., Hutton, J., Lithopoulos, E.: Covariance propagation in GPS/IMU - directly georeferenced frame imagery, in Proceedings of the Asian Conference on Remote Sensing 2000, Taipei, Center for Space; Remote Sensing Research, National Central University and Chinese Taipei Society of Photogrammetry and Remote Sensing (2000)

26. Puhalskii, A., Spokoiny, V.: On large-deviation efficiency in statistical inference, *Bernoulli*, 4 (1998) 203-272

27. Rabus, B. Eineder, M. Roth, A., Bamler, R.: The shuttle radar topography mission - a new class of digital elevation models acquired by spaceborne radar, *Photogrammetric Engineering and Remote Sensing*, 57 (2003) 241-262

28. Radke, L. R., Clark, T.L., Coen, J. L., Walther, C., Lockwood, R. N., Riggin, P., Brass, J., Higgans, R.: The wildfire experiment (WiFE): Observations with airborne remote sensors, *Canadian J. Remote Sensing*, 26 (2000) 406-417

29. Richards, G. D.: The properties of elliptical wildfire growth for time dependent fuel and meteorological conditions, *Combust. Sci. and Tech.*, 95 (1994) 357-383

30. Rothermel, R. C.: A mathematical model for predicting fire spread in wildland fires. USDA Forest Service Research Paper INT-115 (1972)

31. Tippett, M. K., Anderson, J. L., Bishop, C. H., Hamill, T. M., Whitaker, J. S.: Ensemble square root filters, *Monthly Weather Review*, 131 (2003) 1485-1490

32. Vodacek, A., Kremens, R. L., Fordham, A. J., VanGorden, S. C., Luisi, D., Schott, J. R.: Remote optical detection of biomass burning using a potassium emission signature, *Intl. J. of Remote Sensing*, 13 (2002) 2721-2726

33. Zhao, W.: Challenges in design and implementation of middlewares for real-time systems, *J. of Real-Time Systems*, 20 (2001) 1-2

Agent-Based Simulation of Data-Driven Fire Propagation Dynamics

John Michopoulos[1], Panagiota Tsompanopoulou[2], Elias Houstis[3,2], and Anupam Joshi[4]

[1] Special Projects Group, Code 6303, U.S. Naval Research Laboratory, U.S.A.
john.michopoulos@nrl.navy.mil
[2] Dept. of Comp. Eng. and Telecommunications, University of Thessaly, Greece
yota@inf.uth.gr
[3] Computer Sciences Department, Purdue University, U.S.A.
enh@cs.purdue.edu
[4] Dept. of Comp. Sci. and Electr. Eng., U. of Maryland Baltimore County, U.S.A.
joshi@cs.umbc.edu

Abstract. Real world problems such as fire propagation prediction, can often be considered as a compositional combination of multiple, simple but coupled subproblems corresponding to analytical and computational behavior models of the systems involved in multiple domains of action. The existence of various computational resources (legacy codes, middleware, libraries, etc.) that solve and simulate the subproblems successfully, the coupling methodologies, the increasing and distributed computer power (GRID etc.) and the polymorphism and plurality of the available technologies for distributed mobile computing, such as Agent Platforms, motivated the implementation of multidisciplinary problem solving environments (MPSE) to overcome the difficulties of their integration and utilization. In this paper we present the onset of the development of computational infrastructure for the simulation of fire propagation in multiple domains using agent platforms as an informal validation of our data-driven environment for multi-physics applications (DDEMA).

1 Introduction

The main objective of this effort is to provide informal validation to the general framework of DDEMA [1,2,3,4], for the area of fire-environment qualitative assessment and evolutionary fire propagation prediction based on dynamic sensor data. To achieve this objective we have targeted the development of a system that will be a multidisciplinary problem solving environment (MPSE), and will provide simulation based behavior prediction for monitoring situational assessment and decision support required by fire fighting teams. We will be referring to it as the Data-Driven Fire Hazard Simulator (DDFHS).

The context, general and specific requirements, architecture and meta-archite-cture of the DDEMA framework has been described already elsewhere [1,2,3,4].

M. Bubak et al. (Eds.): ICCS 2004, LNCS 3038, pp. 732–739, 2004.
© Springer-Verlag Berlin Heidelberg 2004

Development of DDFHS in the context of the above referenced main objective requires that it should be producible by utilizing the resources of the DDEMA framework as a meta-meta-system and that it will demonstrate two main features. The first feature is the software encapsulation of multiphysics data-driven model selection. The second feature is heavy- and light-weight computational solution implementations and comparison for solving of multi domain coupling problems (i.e. multiple rooms building with embedded sensor network under fire conditions).

The rest of the paper is organized as follows. In Section 2, we describe the fire dynamics in terms of two models of various complexities. In Section 3, the description of DDEMA's multilayered implementation for creating DDFHS is presented. Finally, conclusions follow in Section 4.

2   Modelling of Fire Dynamics

For the sake of demonstrating dynamic model selection capability we are employing two modelling formulations of significantly different complexity. They are the partial and the complete fire propagation models referring them with the acronyms (PFPM) and (CFPM) respectively. The system should be able to select which of the two models will actually be utilized for the simulation based on user specification. Our selection of these two models is by no means unique and many other formulations could be used in their place.

2.1   Partial Fire Propagation Model

For this case we are considering a time shifted term-selective sequential application of the convention-conduction partial differential equation (PDE):

$$\frac{\partial T}{\partial t} = -\beta_i \frac{\partial T}{\partial x_i} + \frac{1}{\rho C}\nabla(k_i\frac{\partial T}{\partial x_i}) - \frac{\alpha}{\rho C}(T_0 - T) + f, \tag{1}$$

where T is the temperature, β_i are convection coefficients, ρ is the density of the medium, C is the heat capacity, k_i are conductivity coefficients, α is the thermal expansion coefficient, T_0 is the initial temperature and f is the heat production. The second term, in the right-hand side, is the convective term, while the third term is the conductive term. When the convective term is missing the obtained PDE is elliptic, while when the conductive term is missing the resulting PDE is hyberbolic, and this suggests different solution approaches.

The solution of this heat evolution PDE will produce the approximate temperature distribution field in the solution domain(s) and will become the input of the embedded sensor network. This formulation represents a gross-simplification of the actual case since most conservation laws have been ignored from a thermodynamic modelling perspective.

2.2 Complete Fire Propagation Model

For this case we are considering one of the most complete formulations for reactive flaw dynamics as they are relevant to fire propagation dynamics. All thermodynamic and conservation laws have been incorporated and they yield the following set of PDEs [5]:

$$\frac{\partial \rho}{\partial t} + \nabla \cdot \rho \mathbf{u} = 0, \tag{2}$$

$$\frac{\partial}{\partial t}(\rho Y_l) + \nabla \cdot \rho Y_l \mathbf{u} = \nabla \cdot \rho D_l \nabla Y_l + \dot{m}_l''', \tag{3}$$

$$\rho \left(\frac{\partial \mathbf{u}}{\partial t} + (\mathbf{u} \cdot \nabla)\mathbf{u} \right) + \nabla p = \rho \mathbf{g} + \mathbf{f} + \nabla \cdot \tau, \tag{4}$$

$$\frac{\partial}{\partial t}(\rho h) + \nabla \cdot \rho h \mathbf{u} = \frac{Dp}{Dt} - \nabla \cdot \mathbf{q_r} + \nabla \cdot k\nabla T + \sum_l \nabla \cdot h_l \rho D_l \nabla Y_l, \tag{5}$$

where the most important variables are $\rho, \mathbf{u}, T, \mathbf{q_r}, Y_l, M_l, p$ and p_0 and are density of the fluid, velocity vector, temperature, radiative heat flux vector, mass fraction of each species, molecular weight of each gas species, pressure and background pressure respectively, while $\mathbf{f}, \mathbf{g}$ and τ are external forces, acceleration of gravity and viscous stress tensor respectively.

Equation (2), expresses the conservation of total mass, Eq. (3) expresses the conservation of mass for each species participating in the combustion process, Eq. (4) expresses the conservation of momentum, and finally Eq. (5) expresses the conservation of energy according to [5].

This coupled system of PDEs yields an approximate form of the generalized Navier-Stokes equations for reactive flow application with low Mach number. This approximation involves exclusion of acoustic waves while it allows for large variations of temperature and density [6]. This endows the equations with an elliptic character that is consistent with low speed, thermal convective reactive processes valid for our interests.

3 DDEMA Implementation for the Case of DDFHS

3.1 Fire Scenario Specification

To avoid providing a lengthy specification of user requirements that falls outside the sizing scope of this paper, we present a common scenario for various fire-fighting situations. This scenario will be equivalent to a qualitative specification of the involved user (fire-fighter) requirements. The development of DDFHS will be addressing the needs of all participants of this scenario over the selected represented hardware/network topology. Figure 1 depicts a schematic diagram of this scenario.

According to the scenario a room is assumed as the space defining the area of interest where a fire may brake out. The room is equipped with a stationary or moving network of sensors with various degrees of clustering aggregation [7]. There is a base-station farm that is capable of collecting the sensor

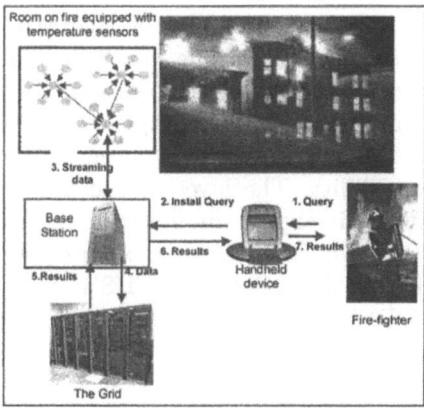

Fig. 1. Fire scenario for micro-future behavioral prediction of fire dynamic propagation upon actions and conditions specified by the fire fighter.

and/or cluster-head output data dynamically. In the beginningof this project we are considering these data to be temperature values at the associated sensor locations. They can also be coefficients corresponding to the interpolation polynomials representing the temperature distribution fields in the domain of the room. This implies that the sensors may extend from non-intelligent thermocouples, to very intelligent sensors with associated embedded microprocessors for local computation of the necessary coefficients. A farm of high performance computing (HPC) systems can be considered as the appropriate GRID replication resource. When a fire brakes out in the room and the fire-fighter arrives, then he/she uses his portable digital assistant (PDA) or PDA-enabled mobile phone or other handheld device to issue queries to the system to help him assess the current situation (based on the sensor outputs) and more importantly to ask 'what-if' questions of the type: *Is it safe to open a particular door in the next 1-2 minutes, to gain access in the room or not?*, or, *What are the risk and the confidence levels that the fire will not propagate through the door when the door is opened in the next 1-2 minutes?* Clearly these type of questions can only be answered when a sensor data-driven simulation of the fire dynamics is continually adjusted and steered by the fire-fighter's queries (usually actions translate to boundary conditions modifications) and the sensor data.

The flow of data is denoted by the vectors connecting the various resources in Fig. 1, and their sequential order by their numerical labels.

The software implementation as described below, is capable of responding to all requirements associated with the above mentioned scenario.

Since the design of DDEMA has been discussed from various perspectives [3, 4] here we will not repeat the detailed description of three-layered architecture. Only a short description will be provided for the scope present paper.

On the top (closer to the user and developer) is the *surfaceware* layer that implements the user interface functionality of DDEMA at the stage solution development time. The second layer is the *middleware* and serves as an interface between the surfaceware and the layer below it. Its most important activity is to allow process migration within multiple virtual machines for increased redundancy, control of the processes at the layer below and very lightweight hardware integration capability. The layer at the bottom, *deepware*, contains the legacy codes, the symbolic algebra modules, and specific codes for coupling multi-domain PDE written for DDEMA and execute them on the available resources. As an instantiation of DDEMA, DDFHS's implementation will inherit this architecture from DDEMA. Separate descriptions of these layers are going to be provided below as detailed as they are relevant to the DDFHS.

3.2   Implementation of Surfaceware

The top level of DDEMA and its validation instantiations (such as the DDFHS) is based on two modes, the application design mode and the application use mode. In the first mode the designer will describe the actual application architecture in terms of data flow and message passing diagrams. The use of Ptolemy-II from UC Berkeley [8] enable us to convert the visual representation of the application to components suitable for the meta-meta-system presented in [4]. In the application use mode, the user initiates the procedure for the automatic generation of necessary java source code or bytecode that implements the described application at runtime. The execution of the code will provide the user a stand alone application, which will perform the activities associated with the corresponding requirements. In most of the cases the actors (by Ptolemy-II) encapsulate existing code or agent-wrapped code that calls the legacy codes which simulate the specific application.

3.3   Implementation of Middleware

To select a suitable agent platform for our middleware we have conducted a comparison study based on knockout and performance criteria that will be published independently. The systems that surfaced as the three leaders of this study are the Java Agent DEvelopment (JADE) framework [9], GRASSHOPPER [10] and ProActive [11]. The first two of them are FIPA [12] compatible and therefore they can be integrated to a common but heterogeneous, integration.

For our initial implementation we exclusively selected JADE, driven mainly, by the fact that it has been ported to a very wide variety of hardware platforms extending from handheld devices to HPC clusters.

A typical surfaceware view of the middleware implementation of a multi-domain (corresponding to multiple rooms) fire propagation simulator is expressed by the actor-based data-flaw network in Fig. 2 where Ptolemy-II Vergil editor has been utilized. The actors are wrappers for agents that are wrappers for legacy codes that implement the Finite Element Tear and Interconnect method. Other domain compositions methods will also be considered.

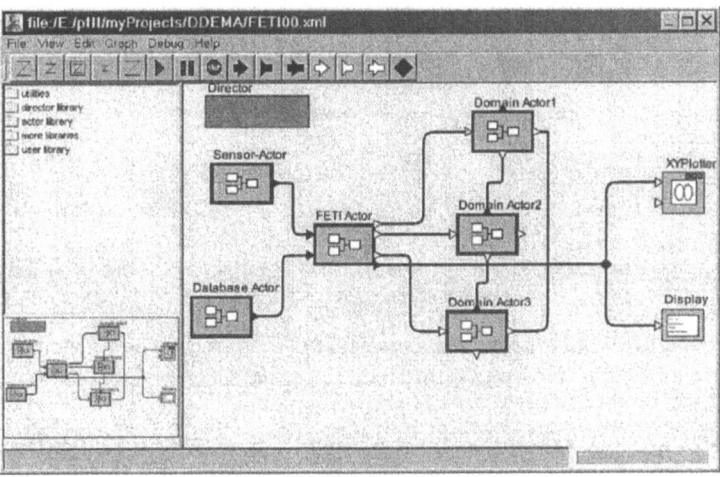

Fig. 2. Ptolemy-II surfaceware view of JADE middleware agents as a data-flaw diagram implementing a multi-domain FETI-based simulator

3.4 Implementation of Deepware

The deepware of DDEMA consists of two major components:

(1) legacy codes capable of implementing the solution of PDEs that express behavioral models of fire propagation dynamics such as those expressed by in Eqs. (1) or (2)-(5).

(2) intermediate codes and scripts responsible for data preparation, conversion and integration.

There are many codes capable of handling the numerical solution of the PDE encapsulated models given in Section 2. We have decided to use the general purpose code freeFEM+ [13] for solving Eq. (1) for the case of PFPM and NISTS's FDS [5] for the system of Eqs. (2)-(5) that corresponds to the CFPM. Preliminary results from our use of freeFEM+ are shown in Fig 3.

The major issue that has developed for the case FDS are the legacy code conversion and wrapping difficulties. Working with the main module of FDS we faced the well-known problem of calling a FORTRAN routines from a JAVA agent, because FDS is implemented in FORTRAN 90 and the agent platform we use is written in JAVA. This procedure could have been done *automatically* by converting the FORTRAN code to JAVA directly using tools like f2j, f2c and f90toC, however, various expressions of problematic behavior in each case prohibited the conversion. The tools f2j and f2c work for FORTRAN 77, and FDS code is in FORTRAN 90, while even in the latest version of f90toC important modules (e.g. I/O, vector subscripts, memory allocation) are not yet implemented. Since the code of FDS contains approximately 35 Klines, manual conversion to JAVA is as very tedious endeavor and has therefore been excluded. Thus, we use our experience from the implementation of GasTurbnLab [14] and we convert the main FORTRAN program to a subroutine which we call from a

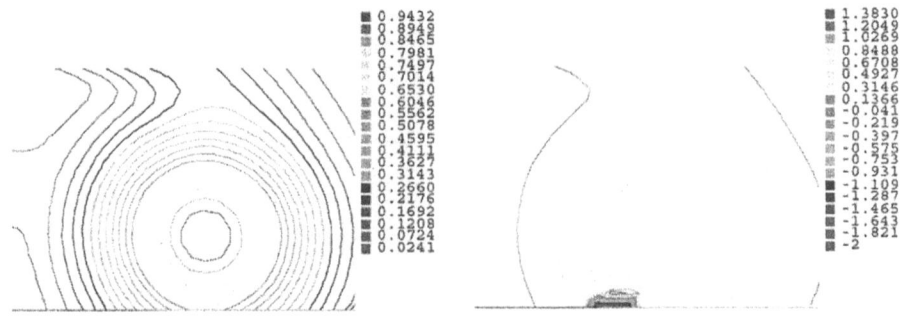

Fig. 3. Typical freeFEM+ produced simulation of 2D temperature distribution due to fire propagation before (left) and after (right) opening the door of the modelled room.

C/++ function. These C/++ functions are wrapped using the automated Java wrapping technology of JACAW [16](which utilizes Java Native Interface [15]). This ensures the capability of our legacy code functionality to be available for JADE agents and Ptolemy-II actors.

4   Conclusions

In this paper we presented the first steps towards utilizing DDEMA's infrastructure to develop DDFHS. We presented two potential modelling approaches for the fire propagation dynamics. A fire-fighting scenario was used to explain certain aspects of the software's implementation to be used for this effort. Efficient and accurate legacy codes are used to simulate the fire dynamics, while new code has to be written for the agentization of them and their coupling with surfaceware actors, based on previously used techniques.

In the subsequent steps of the implementation, we will continue with the legacy code wrapping and implementation of compositional solutions. Finally, we shall remain aware of new technologies relative to our project and we will adapt our approach as needed.

Acknowledgements. The authors acknowledge the support by the National Science Foundation under grants ITR-0205663 and EIA-0203958.

References

1. Michopoulos, J., Tsompanopoulou, P., Houstis, E., Rice, J., Farhat, C., Lesoinne, M., Lechenault, F., DDEMA: A Data Driven Environment for Multiphysics Applications,in: Proceedings of International Conference of Computational Science - ICCS'03, Sloot, P.M.A., et al. (Eds.) Melbourne Australia, June 2-4, LNCS 2660, Part IV, Springer-Verlag, Haidelberg, (2003) 309-318.

2. Michopoulos, J., Tsompanopoulou, P., Houstis, E., Rice, J., Farhat, C., Lesoinne, M., Lechenault, F., Design Architecture of a Data Driven Environment for Multiphysics Applications, in: Proceedings of DETC'03, ASME DETC2003/CIE Chicago IL, Sept. 2-6 2003, Paper No DETC2003/CIE-48268, (2003).
3. Michopoulos, J., Tsompanopoulou, P., Houstis, E., Farhat, C., Lesoinne, M., Rice, J., Joshi, A., On a Data Driven Environment for Multiphysics Applications, Future Generation Computer Systems, in-print (2004).
4. J. Michopoulos, J., Tsompanopoulou, P., Houstis, E., Farhat, C., Lesoinne, M., Rice, Design of a Data-Driven Environment for Multi-field and Multi-Domain Applications, book chapter in Darema, F. (ed.), Dynamic Data Driven Applications Systems, Kluwer Academic Publishers, Netherlands, in-print (2004).
5. McGrattan, K.B., Baum, H.R., Rehm, R.G., Hamins, A., Forney, G.P., Fire Dynamics Simulator - Technical Reference Guide, National Institute of Standards and Technology, Gaithersburg, MD., NISTIR 6467, January 2000.
6. Rehm, R.G., Baum, H.R., The Equations of Motion for Thermally Driven, Buoyant Flows. J. of Research of the NBS, 83:297–308, (1978).
7. Hingne, V., Joshi, A., Houstis, E., Michopoulos, J., On the Grid and Sensor Networks, Fourth International Workshop on Grid Computing November 17–17, 2003 Phoenix, Arizona (2003).
8. Davis II, J., Hylands, C., Kienhuis, B., Lee, E.A., Liu, J., Liu, X., Muliadi, L., Neuendorffer, S., Tsay, J., Vogel, B., Xiong, Y.: Heterogeneous Concurrent Modeling and Design in Java, Memorandum UCB/ERL M01/12, EECS, University of California, Berkeley, CA USA 94720 March 15, 2001, http://ptolemy.eecs.berkeley.edu/ptolemyII/ .
9. The JADE Project Home Page. http://jade.cselt.it
10. The Grasshopper Agent Platform, IKV++ GmbH, Kurfurstendamm 173-174, D-10707 Berlin, Germany. http://www.ikv.de
11. Caromel, D., Klauser, W., Vayssiere, J.: Towards Seamless Computing and Metacomputing in Java, pp. 1043–1061 in Concurrency Practice and Experience, September-November 1998, 10(11–13). Editor Fox, G.C., Published by Wiley & Sons, Ltd., http://www-sop.inria.fr/oasis/proactive/ .
12. FIPA 2000 Specifications. Available at http://www.fipa.org .
13. Bernardi, D., Hecht, F., Ohtsuka, K., Pironneau, O., freeFEM+, a finite element software to handle several meshes. (1999), downloadable from http://www.frefem.org .
14. Fleeter, S., Houstis, E.N., Rice, J.R., Zhou, C., Catlin, A.: GasTurbnLab: A Problem Solving Environment for Simulating Gas Turbines. Proc. 16th IMACS World Congress, (2000) No 104-5.
15. Java Native Interface Specification. http://web2.java.sun.com/products/jdk/1.1/docs/guide/jni.
16. JACAW: A Java-C Automatic Wrapper Tool, http://www.wesc.ac.uk/projects/jacaw .

Model Reduction of Large-Scale Dynamical Systems

A. Antoulas[1], D. Sorensen[2], K.A. Gallivan[3], P. Van Dooren[4],
A. Grama[5], C. Hoffmann[5], and A. Sameh[5]

[1] Department of Electrical Engineering, Rice University, Houston, TX
[2] Department of Computational and Applied Math, Rice University, Houston, TX
[3] School of Computational Science, and Information Technology, Florida State
University, Tallahassee, FL
[4] Department of Mathematical Engineering, Catholic University of Louvain,
Louvain-la-Neuve, BELGIUM
[5] Department of Computer Sciences, Purdue University, W. Lafayette, IN

Abstract. Simulation and control are two critical elements of Dynamic
Data-Driven Application Systems (DDDAS). Simulation of dynamical
systems such as weather phenomena, when augmented with real-time
data, can yield precise forecasts. In other applications such as structural
control, the presence of real-time data relating to system state can enable
robust active control. In each case, there is an ever increasing need for
improved accuracy, which leads to models of higher complexity. The basic motivation for system approximation is the need, in many instances,
for a simplified model of a dynamical system, which captures the main
features of the original complex model. This need arises from limited
computational capability, accuracy of measured data, and storage capacity. The simplified model may then be used in place of the original
complex model, either for *simulation and prediction*, or *active control*. As
sensor networks and embedded processors proliferate our environment,
technologies for such approximations and real-time control emerge as
the next major technical challenge. This paper outlines the state of the
art and outstanding challenges in the development of efficient and robust methods for producing reduced order models of large state-space
systems.

1 Introduction

Many physical processes in science and engineering are modeled accurately using finite dimensional dynamical systems. Examples include weather simulation,
molecular dynamic simulations (e.g., modeling of bio-molecules and their identification), structural dynamics (e.g., flex models of the international space station, and structural response of high-rise buildings to wind and earthquakes),
electronic circuit simulation, semiconductor device manufacturing (e.g., chemical vapor deposition reactors), and simulation and control of micro-electro-
mechanical (MEMS) devices. An important subclass of these processes can be effectively monitored to gather data in support of simulation, diagnosis, prognosis,

M. Bubak et al. (Eds.): ICCS 2004, LNCS 3038, pp. 740–747, 2004.
© Springer-Verlag Berlin Heidelberg 2004

prediction, and control. This class of dynamic data-driven application systems (DDDAS) pose challenging problems ranging from data gathering, assimilation, effective incorporation into simulations, and real-time control.

In the vast majority of applications where control can be affected, the original system is augmented with a second dynamical system called a controller. In general, the controller has the same complexity as the system to be controlled. Since in large-scale applications of interest, one often aims at real-time control; reduced complexity controllers are required. One possible solution to building low-complexity controllers is to design such controllers based on reduced order models of the original dynamical systems. In yet other systems in which real-time prognostics and prediction are required, constraints on compute power, memory, communication bandwidth, and available data might necessitate the use of such reduced-order models. This paper outlines the state-of-the-art in model reduction, discusses outstanding challenges, and presents possible solution strategies for addressing these challenges.

Model reduction seeks to replace a large-scale system of differential or difference equations by a system of substantially lower dimensions that has nearly the same response characteristics. Two main themes can be identified among several methodologies: (a) *balancing based methods*, and (b) *moment matching methods*. Balanced approximation methods are built upon a family of ideas with close connection to the singular value decomposition. These methods preserve stability and allow for global error bounds. Available algorithms for these methods, however, are not suitable for large-scale problems since they have been developed mainly for dense matrix computations. Moment matching methods are based primarily on Padé-like approximations, which, for large-scale problems, have led naturally to the use of Krylov and rational Krylov subspace projection methods. While these moment matching schemes enjoy greater efficiency for large-scale problems, maintaining stability in the reduced order model cannot be guaranteed. Consequently, their use can be problematic at times. Moreover, no a priori global error bounds exist for moment matching schemes.

A current research trend aims at combining these two approaches by deriving iterative methods that incorporate the desirable properties of both of these classes. This paper addresses several important unresolved issues in model reduction of large-scale Linear Time-Invariant (LTI) systems, and extending them to a new class of problems that require adaptive models. It also addresses large-scale structured problems that are either time-varying, or which require iterative updating of the initial reduced models to obtain better approximation properties.

2   Motivating Data Driven Control Applications

The problems addressed in this paper focus on reduced models for dynamic data-driven systems, which are characterized by model adaptivity and need for maintaining or exploiting a specific type of structure. This can be either due to time variations in the model itself, or due to the fact that the model to be approximated varies at each iterative step of some design loop.

Large-scale mechanical systems: Consider simulating a structure that is discretely modeled by a large scale finite element formulation such as a long-span bridge, a tall building, or even a car windshield which is subject to various external forces. In tall buildings, for example, these forces could be the result of strong ground motion or wind. Typically, the model is given by second-order systems of equations of the form:

$$\mathbf{M}(\omega, t)\ddot{\mathbf{x}}(t) + \mathbf{D}(\omega, t)\dot{\mathbf{x}}(t) + \mathbf{K}(\omega, t)\mathbf{x}(t) = \mathbf{f}(t) + \mathbf{B}\mathbf{u}(t), \quad \mathbf{y}(t) = \mathbf{C}\mathbf{x}(t). \quad (1)$$

Here, $\mathbf{M}(\omega, t)$ is the mass matrix, $\mathbf{D}(\omega, t)$ is the damping matrix, and $\mathbf{K}(\omega, t)$ is the stiffness matrix of the system. All three matrices are assumed to be frequency-dependent and possibly time-dependent. For large-scale structures, the state vector $\mathbf{x}(t)$ is of order N, which can reach in the tens of millions. The forcing function $\mathbf{f}(t)$ represents wind and earthquake effects and is of the same dimension. We assume that the structure under consideration is densely instrumented by networked sensing and actuator elements. Further, assuming that a sensor-actuator complex (SAC) is monolithically integrated with a strut system containing controllable dampers that can change the stiffness characteristics of the structure in milliseconds, the control function $\mathbf{u}(t)$, an m-dimensional vector, represents the action that can be affected through the smart strut system. Finally, the p-dimensional vector $\mathbf{y}(t)$ represents the signals collected via the sensor network. The dimension N of the state vector $\mathbf{x}(t)$, is much larger than m and p. The objective of the model reduction problem is to produce a model of the structure that possesses the "essential" properties of the full-order model. Such a reduced order model can then be used for the design of a reduced-order controller to affect real-time control. The reduced-order model may be used to predict the onset of failure, and the reduced-order controller can be used for real-time control of the structure so as to mitigate failure.

There are several preliminary problems that must be solved before we realize an appropriate reduced order model. These include: (i) computing the optimal location of the sensors and actuators, where we could start from a nominal model with $\mathbf{M}$, $\mathbf{D}$, and $\mathbf{K}$ constant, (ii) improve the dynamical model once the sensor/actuator placements are determined, using validated simulations (via physical laboratory models of the structure) or via measurements performed on the actual structure, if possible, (iii) design a closed-loop controller which is derived from the reduced-order model and, finally, (iv) adapt the reduced order model to the closed-loop configuration of the system (this will change its dynamics and therefore, also its reduced order model). In each of these steps it is useful to have simplified (i.e. reduced-order) models on which the most time consuming tasks can be performed in reasonable time.

3   Balanced Approximation Methods

Projection methods have provided effective model reduction algorithms for constructing reduced-order models in the large scale setting. Given a continuous-time state space system:

$$\dot{\mathbf{x}}(t) = \mathbf{A}\mathbf{x}(t) + \mathbf{B}\mathbf{u}(t), \quad \mathbf{y}(t) = \mathbf{C}\mathbf{x}(t), \quad (2)$$

with input $\mathbf{u}(t) \in \Re^m$, state $\mathbf{x}(t) \in \Re^N$ and output $\mathbf{y}(t) \in \Re^p$, one defines a *projected* model of reduced order $n \ll N$ for a reduced state $\hat{\mathbf{x}}(t) \in \Re^n$:

$$\dot{\hat{\mathbf{x}}}(t) = \hat{\mathbf{A}}\hat{\mathbf{x}}(t) + \hat{\mathbf{B}}\mathbf{u}(t), \quad \hat{\mathbf{y}}(t) = \hat{\mathbf{C}}\hat{\mathbf{x}}(t), \tag{3}$$

where

$$\mathbf{I}_n = \mathbf{W}^T\mathbf{V}, \hat{\mathbf{A}} = \mathbf{W}^T\mathbf{A}\mathbf{V}, \hat{\mathbf{B}} = \mathbf{W}^T\mathbf{B}, \hat{\mathbf{C}} = \mathbf{C}\mathbf{V}. \tag{4}$$

The widely used balanced truncation technique constructs the "dominant" spaces $\mathbf{W}$ and $\mathbf{V}$ such that

$$\mathcal{P}\mathbf{W} = \mathbf{V}\Sigma_+, \quad \mathcal{Q}\mathbf{V} = \mathbf{W}\Sigma_+, \quad \mathbf{W}^T\mathbf{V} = \mathbf{I}_n \tag{5}$$

where Σ_+ is symmetric, positive definite and usually diagonal, and $\mathcal{P}$, $\mathcal{Q}$ are the controllability and observability gramians, respectively. It follows that $\mathcal{P}\mathcal{Q}\mathbf{V} = \mathbf{V}\Sigma_+^2$ and $\mathcal{Q}\mathcal{P}\mathbf{W} = \mathbf{W}\Sigma_+^2$ and hence Σ_+^2 should contain the largest eigenvalues of $\mathcal{P}\mathcal{Q}$ or $\mathcal{Q}\mathcal{P}$. The gramians are non-singular if and only if the system $\{\mathbf{A},\mathbf{B},\mathbf{C}\}$ is controllable and observable. However, the eigenspaces of (5) essentially yields a projected system that keeps those states that are the most observable and controllable simultaneously. This is related to the fact that for a given state $\hat{\mathbf{x}}$, $\hat{\mathbf{x}}^T\mathcal{P}^{-1}\hat{\mathbf{x}}$ can be viewed as the input energy needed to steer the system from the zero state to $\hat{\mathbf{x}}$, and $\hat{\mathbf{x}}^T\mathcal{Q}\hat{\mathbf{x}}$ can be viewed as the energy of the output resulting from the initial state $\hat{\mathbf{x}}$ (see [5] for a more rigorous discussion).

Efficient algorithms for the large-scale continuous time case have been derived by exploiting the fact that the gramians can be obtained from the Lyapunov equations

$$\mathbf{A}\mathcal{P} + \mathcal{P}\mathbf{A}^T + \mathbf{B}\mathbf{B}^T = \mathbf{0}, \quad \mathbf{A}^T\mathcal{Q} + \mathcal{Q}\mathbf{A} + \mathbf{C}^T\mathbf{C} = \mathbf{0}.$$

and in the discrete-time case from the solution of the Stein equations :

$$\mathbf{A}\mathcal{P}\mathbf{A}^T - \mathcal{P} + \mathbf{B}\mathbf{B}^T = \mathbf{0}, \quad \mathbf{A}^T\mathcal{Q}\mathbf{A} - \mathcal{Q} + \mathbf{C}^T\mathbf{C} = \mathbf{0}.$$

For example, in [4,6] an iteration called AISIAD was developed. The algorithm alternates between two coupled Sylvester equations that are approximations to the projected equations

$$\mathbf{A}(\mathcal{P}\mathbf{W}) + (\mathcal{P}\mathbf{W})\mathbf{H}^T + \mathbf{B}(\mathbf{B}^T\mathbf{W}) = \mathcal{P}\mathbf{F}_w,$$
$$\mathbf{A}^T(\mathcal{Q}\mathbf{V}) + (\mathcal{Q}\mathbf{V})\mathbf{H} + \mathbf{C}^T(\mathbf{C}\mathbf{V}) = \mathcal{Q}\mathbf{F}_v,$$

where $\mathbf{F}_w = (\mathbf{I} - \mathbf{W}\mathbf{V}^T)\mathbf{A}^T\mathbf{W}$ and $\mathbf{F}_v = (\mathbf{I} - \mathbf{V}\mathbf{W}^T)\mathbf{A}\mathbf{V}$. Given a current guess $\mathbf{W}, \mathbf{V}$, the algorithm proceeds by determining $\mathbf{Z}$ by solving $\mathbf{A}\mathbf{Z} + \mathbf{Z}\mathbf{H}^T + \mathbf{B}(\mathbf{B}^T\mathbf{W}) = \mathbf{0}$, where $\mathbf{H} = \mathbf{W}^T\mathbf{A}\mathbf{V}$ and $\mathbf{V} \leftarrow \mathbf{V}_+$ is obtained from $\mathbf{Z}$. $\mathbf{Y}$ is then determined by solving $\mathbf{A}^T\mathbf{Y} + \mathbf{Y}\mathbf{H}_+ + \mathbf{C}^T(\mathbf{C}\mathbf{V}) = \mathbf{0}$ where $\mathbf{H}_+ = \mathbf{W}^T\mathbf{A}\mathbf{V}_+$, and $\mathbf{W} \leftarrow \mathbf{W}_+$ is obtained from $\mathbf{Y}$ to maintain relations (5).

The AISIAD iteration is effective, but no convergence results are known. These results may be potentially derived from the fact that one can show (see

[2,3]) that $\mathbf{V}$ and $\mathbf{W}$ will span the dominant invariant subspace of the *positive definite generalized eigenvalue problems* $(\lambda \mathcal{P}^{-1} - \mathcal{Q})$ and $(\lambda \mathcal{Q}^{-1} - \mathcal{P})$ if and only if they satisfy the respective "Tracemin" conditions:

$$\min \ trace \ \mathbf{V}^T \mathcal{P}^{-1} \mathbf{V}, \quad s.t. \ \mathbf{V}^T \mathcal{Q} \mathbf{V} = \mathbf{I}_n,$$
$$\min \ trace \ \mathbf{W}^T \mathcal{Q}^{-1} \mathbf{W}, \quad s.t. \ \mathbf{W}^T \mathcal{P} \mathbf{W} = \mathbf{I}_n.$$

4   Model Reduction for Second Order Systems

Models of mechanical systems are often of the type (1), where $\mathbf{M} = \mathbf{M}^T, \mathbf{D} = \mathbf{D}^T$ and $\mathbf{K} = \mathbf{K}^T$ are respectively the mass, damping and stiffness matrices, since this represents the equation of motion of the system. In the Laplace domain, the *characteristic polynomial matrix* $\mathbf{P}(s) := \mathbf{M}s^2 + \mathbf{D}s + \mathbf{K}$ and the transfer function $\mathbf{G}(s) = \mathbf{C}\mathbf{P}^{-1}(s)\mathbf{B}$ appear and the zeros of $\det(\mathbf{P}(s))$ are also known as the *characteristic frequencies or poles* of the system. Using an extended state $\xi = \begin{bmatrix} \mathbf{x}^T & \dot{\mathbf{x}}^T \end{bmatrix}^T$, this can be linearized to a generalized state-space model $\{\mathcal{E}, \mathcal{A}, \mathcal{B}, \mathcal{C}\}$:

$$\underbrace{\begin{bmatrix} \mathbf{D} & \mathbf{M} \\ \mathbf{M} & \mathbf{0} \end{bmatrix}}_{\mathcal{E}} \dot{\xi} = \underbrace{\begin{bmatrix} -\mathbf{K} & \mathbf{0} \\ \mathbf{0} & \mathbf{M} \end{bmatrix}}_{\mathcal{A}} \xi + \underbrace{\begin{bmatrix} \mathbf{B} \\ \mathbf{0} \end{bmatrix}}_{\mathcal{B}} \mathbf{u}, \quad y = \underbrace{\begin{bmatrix} \mathbf{C} & \mathbf{0} \end{bmatrix}}_{\mathcal{C}} \xi. \tag{6}$$

Since $\mathbf{M}$ is invertible, we can also transform this model to the standard state-space form $\{\mathcal{E}^{-1}\mathcal{A}, \mathcal{E}^{-1}\mathcal{B}, \mathcal{C}\}$, where $\mathcal{E}^{-1}\mathcal{A} = \begin{bmatrix} \mathbf{0} & \mathbf{I} \\ -\mathbf{M}^{-1}\mathbf{K} & -\mathbf{M}^{-1}\mathbf{D} \end{bmatrix}$ and $\mathcal{E}^{-1}\mathcal{B} = \begin{bmatrix} \mathbf{0} \\ \mathbf{M}^{-1}\mathbf{B} \end{bmatrix}$. The controllability gramian $\mathcal{P}$ and observability gramian $\mathcal{Q}$ of the state-space model $\{\mathcal{E}^{-1}\mathcal{A}, \mathcal{E}^{-1}\mathcal{B}, \mathcal{C}\}$ can be computed via the solution of the Lyapunov equations :

$$\mathcal{E}^{-1}\mathcal{A}\mathcal{P} + \mathcal{P}\mathcal{A}^T\mathcal{E}^{-T} + \mathcal{E}^{-1}\mathcal{B}\mathcal{B}^T\mathcal{E}^{-T} = 0, \quad \mathcal{Q}\mathcal{E}^{-1}\mathcal{A} + \mathcal{A}^T\mathcal{E}^{-T}\mathcal{Q} + \mathcal{C}^T\mathcal{C} = 0. \tag{7}$$

Moreover, if $\mathbf{C} = \mathbf{B}^T$, the transfer function and model (6) are symmetric and the gramians of the state-space model are defined in terms of the solution $\mathcal{G}$ of a single generalized Lyapunov equation :

$$\mathcal{A}\mathcal{G}\mathcal{E} + \mathcal{E}\mathcal{G}\mathcal{A} + \mathcal{B}\mathcal{B}^T = 0, \quad \mathcal{P} = \mathcal{G}, \quad \mathcal{Q} = \mathcal{E}\mathcal{G}\mathcal{E}. \tag{8}$$

Most model reduction methods use a projection to build the reduced-order model: given a generalized state-space model $\{\mathcal{E}, \mathcal{A}, \mathcal{B}, \mathcal{C}\}$, the reduced-order model is given by $\{\mathbf{Z}^T\mathcal{E}\mathbf{V}, \mathbf{Z}^T\mathcal{A}\mathbf{V}, \mathbf{Z}^T\mathcal{B}, \mathcal{C}\mathbf{V}\}$, where $\mathbf{Z}$ and $\mathbf{V}$ are matrices of dimension $2N \times k$, with k the order of the reduced system. The widely used balanced truncation technique constructs $\mathbf{V}$ and $\mathbf{Z} := \mathcal{E}^{-1}\mathbf{W}$ such that (5) is satisfied with Σ_+^2 the $k \times k$ matrix containing the largest eigenvalues of $\mathcal{P}\mathcal{Q}$.

This technique cannot be applied directly to a second order system since, in general, the resulting reduced order system is no longer a second order system.

There is a need for *adaptive* techniques to find reduced order models of second order type. The idea is to choose $k = 2n$ and to restrict the projection matrices $\mathbf{V}$ and $\mathbf{W}$ to have a block diagonal form, where each block is $N \times n$:

$$\mathbf{W} = \begin{bmatrix} \mathbf{W}_{11} & 0 \\ 0 & \mathbf{W}_{22} \end{bmatrix}, \quad \mathbf{V} = \begin{bmatrix} \mathbf{V}_{11} & 0 \\ 0 & \mathbf{V}_{22} \end{bmatrix}. \tag{9}$$

which turns out to be a *sufficient* condition for obtaining an appropriate reduced order model. Notice that for the symmetric case we have automatically $\mathbf{V} = \mathbf{W}$. If we want to impose a block diagonal form, we can, for example, relax the equality conditions in (6).

In this context, the following technical challenges must be addressed: (i) use a relaxation that can be dealt with using Tracemin like optimization methods, such as:

$$\min \ trace \ \mathbf{V}_{11}^T [\mathcal{P}^{-1}]_{11} \mathbf{V}_{11} + \mathbf{V}_{22}^T [\mathcal{P}^{-1}]_{22} \mathbf{V}_{22}; \tag{10}$$
$$s.t. \ \mathbf{V}_{11}^T [\mathcal{Q}]_{11} \mathbf{V}_{11} = \mathbf{I}_n, \ \mathbf{V}_{22}^T [\mathcal{Q}]_{22} \mathbf{V}_{22} = \mathbf{I}_n$$

(ii) define alternative gramians of dimension $N \times N$ for second order systems and analyze their use for model reduction, and (iii) analyze modal approximation based on the polynomial matrix $\mathbf{P}(s)$. One then computes the right and left eigenvectors $\mathbf{P}(\lambda_i)\mathbf{x}_i = 0$ and $\mathbf{y}_i^T \mathbf{P}(\lambda_i) = 0$ via the linearized eigenvalue problem :

$$\begin{bmatrix} \lambda_i \mathbf{D} + \mathbf{K} & \lambda_i \mathbf{M} \\ \lambda_i \mathbf{M} & -\mathbf{M} \end{bmatrix} \begin{bmatrix} \mathbf{x}_i \\ \lambda_i \mathbf{x}_i \end{bmatrix} = \mathbf{0},$$

$$[\mathbf{y}_i^T \ \lambda_i \mathbf{y}_i^T] \begin{bmatrix} \lambda_i \mathbf{D} + \mathbf{K} & \lambda_i \mathbf{M} \\ \lambda_i \mathbf{M} & -\mathbf{M} \end{bmatrix} = \mathbf{0},$$

and selects n of them to form the projection matrices $\mathbf{V}_{11}$ and $\mathbf{W}_{11}$.

5   Time Variance and Adaptivity

Time variance and adaptivity can be either due to actual time variation of the parameters of the system, or due to an external iteration used to adaptively tune the reduced order system. We first explain how to cope with general time-varying systems and show how these methods can also be used for time invariant or for periodic systems. We then discuss systems that are iteratively adapted within an outer tuning loop.

5.1   Iterative Schemes

Reduced order stabilization of a large plant $\mathbf{G}(s)$ amounts to constructing a small order controller $\hat{\mathbf{K}}(s)$ such that the closed loop transfer function $\phi(\mathbf{G}, \hat{\mathbf{K}}) =$

$G(s)(I + \hat{K}(s)G(s))^{-1}$ is stable and such that at the same time some performance criterion is optimized.

Most often a crude low-order (PI) controller is available initially. In order to come up with a simple controller which minimizes an appropriate norm of the transfer function between the inputs and the tracking/focusing errors respectively, an *iterative procedure* is used. The point of the iteration is to successively reduce the order of the high order system until the corresponding optimal controller has low enough order as well as the specified performance properties. Thus, at each reduction step, performance is checked by means of estimating a certain relative error. The rationale of this approach is that the controller should be designed for a small order plant $\hat{G}_i(s)$ using dense matrix techniques. However, this controller will only have a good performance for the original plant $G(s)$ if the approximation $\hat{G}_i(s)$ is obtained using the closed loop error $\|\phi(G, \hat{K}_i) - \phi(\hat{G}_i, \hat{K}_i)\|$, where $\phi(G, K) := G(s)(I + K(s)G(s))^{-1}$, hence the need for an outer iteration. An important ingredient for making such an outer iteration tractable is contained in our previous work on the construction of stabilizing controllers using invariant subspace ideas and Riccati iterations for large sparse system models [1].

In this context, one needs to analyze the convergence of the closed loop iteration and find global/local convergence strategies These ideas can be applied to finding reduced order models of plants where certain properties are imposed, such as stability, passivity, or second order model constraints. Finally, this problem can be embedded in a model reduction problem with time-varying coefficients. Therefore it is possible to take advantage of this fact in order to compute the new plant/controller by *updating* the previous ones.

6   Simulation and Validation Support

An important aspect of model reduction and control is to develop a comprehensive environment for validation and refinement. Instrumenting full-scale systems with control mechanisms can be an expensive proposition. For this reason, control techniques must be extensively tested in a simulation environment prior to scale model testing. We have developed an extensive simulation environment for complex structures and demonstrated the power of this environment in the context of several structural impact response simulations (the tragic Pentagon crash on 9/11 is illustrated in Figure 1).

With a view to understanding the nature, extent, and cause of structural damage, we have used our simulation environment to perform high-fidelity crash simulations on powerful parallel computers. A single simulation instance illustrated in Figure 1 with one million nodes (3 d.o.f. per finite element node) over 0.25 seconds of real time takes over 68 hours on a dedicated 8 processor IBM Regatta SMP. A coarser model with 300K nodes over 0.2s of real time takes approximately 20 hours of simulation time. While we are also working on improving the underlying solvers in our simulation, these computational requirements provide strong motivation for effective error-bounded model reduction. In this

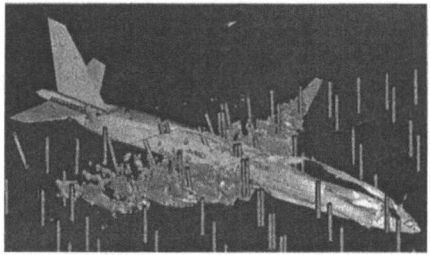

(a) Aerial view of damage to the
Pentagon building.

(b) Simulated aircraft impact on
RC columns in the building.

Fig. 1. High fidelity simulation of the Pentagon crash has yielded significant insights into structural properties of reinforced concrete and design of resilient buildings.

context, development of effective preconditioners and iterative solvers, meshing and mesh adaptation techniques, parallel algorithms, and software development and integration pose overarching technical challenges.

7   Concluding Remarks

Model reduction is an extremely important aspect of dynamic data-driven application systems, particularly in resource-constrained simulation and control environments. While considerable advances have been made in model reduction, simulation, and control, significant challenges remain. In conjunction with the research activity in areas of sensor networks, actuation mechanisms, and ad-hoc wireless networks, techniques for model reduction and control hold the key to fundamental advances in future generation systems.

References

1. X. Rao, K. Gallivan, and P. Van Dooren. Efficient stabilization of large scale dynamical systems. In *Proceedings IEEE Conference on Computer Aided Control Systems Design*, 2000.
2. A. Sameh and Z. Tong. The trace minimization method for the symmetric generalized eigenvalue problem. *Journal of Computational and Applied Mathematics*, 2000.
3. A. H. Sameh and J. A. Wisniewski. A trace minimization algorithm for the generalized eigenvalue problem. *SIAM Journal of Numerical Analysis*, 1982.
4. Y. K. Zhou. *Numerical Methods for Large Scale Matrix Equations with Applications in LTI System Model Reduction*. PhD thesis, Department of Computational and Applied Mathematics, Rice University, 2002.
5. Y. K. Zhou, J. C. Doyle, and K. Glover. *Robust and Optimal Control.* Prentice Hall, 1995.
6. Y. K. Zhou and D. C. Sorensen. Approximate implicit subspace iteration with alternating directions for lti system model reduction. Technical report, Rice University, 2002.

Data Driven Design Optimization Methodology Development and Application

H. Zhao, D. Knight, E. Taskinoglu, and V. Jovanovic

Dept of Mechanical and Aerospace Engineering
Rutgers - The State University of New Jersey
New Brunswick, NJ 08903
knight@soemail.rutgers.edu

Abstract. The Data Driven Design Optimization Methodology (DDDOM) is a Dynamic Data Driven Application System (DDDAS) developed for engineering design optimization. The DDDOM synergizes experiment and simulation in a concurrent integrated software system to achieve better designs in a shorter time. The data obtained from experiment and simulation dynamically guide and redirect the design optimization process in real or near-real time, and therefore realize the full potential of the Dynamic Data Driven Applications Systems concept. This paper describes the DDDOM software system developed using the Perl and Perl/Tk programming languages. The paper also presents the first results of the application of the DDDOM to the multi-objective design optimization of a submerged subsonic inlet at zero angle of attack and sideslip.

1 Introduction

A Dynamic Data Driven Application System (DDDAS) is an application which can interact with experiment and/or simulation in real-time, building a bridge between them and guiding them dynamically [1]. DDDAS is an important area of current research in Information Technology (IT) with a wide range of engineering and scientific applications [2,3,4]. Engineering design optimization can benefit significantly from the DDDAS concept whereby the concurrent integration of remote real-time input from high performance computing resources (e.g., NSF Teragrid [5]) and experiments will achieve the best designs. D. Knight, G. Elliott, Y. Jaluria, N. Langrana and K. Rasheed proposed the Data Driven Design Optimization Methodology (DDDOM) [6] for Multi-criteria Design Optimization (MDO) [7] as a Dynamic Data Driven Application System. DDDOM therefore differs from the conventional, sequential engineering design approach. DDDOM utilizes advanced experimental technology (*e.g.*, Rapid Prototyping) and computational technology (*e.g.*, parallel processing, grid computing [5]). The DDDOM software system under development is shown in Fig. 1. It is comprised of six elements: User Interface, Controller, Optimizer, Experiment, Surrogate Model (SM) and Simulation. Further details are presented in Knight *et al* [8].

M. Bubak et al. (Eds.): ICCS 2004, LNCS 3038, pp. 748–755, 2004.
© Springer-Verlag Berlin Heidelberg 2004

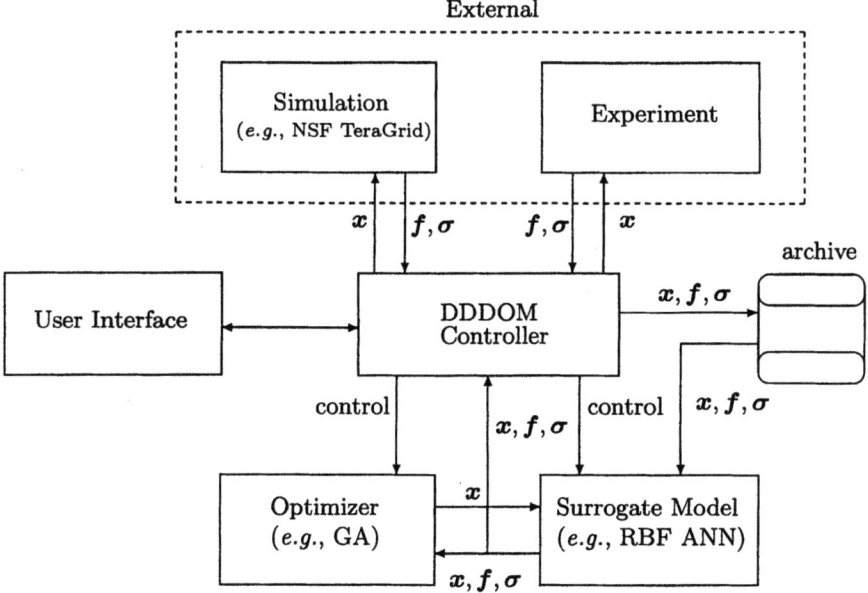

Fig. 1. Data Driven Design Optimization Methodology (DDDOM)

2   The DDDOM Controller

The DDDOM Controller is the central element in the DDDOM and controls the overall operations of the DDDOM cycle. Its basic function is to initiate and monitor the design optimization, and to direct the simulations and experiments needed to build and update the surrogate models. The functions are illustrated in Fig. 1 and are described below.

Control of Optimizer: The Controller initiates the optimization and monitors the progress of the optimization. The Controller monitors each design, denoted by its vector of design variables x, generated by the optimizer and the resultant vector of objective functions f and uncertainty σ generated by the Surrogate Model (SM). The Controller determines when new data is needed for training the SM and initiates new simulations and identifies new experiments needed.

Control of Surrogate Model: The Controller manages the state of the SM between training and modelling. It decides when the SM must be retrained using data stored on the archive.

Control of Simulation: The Controller determines the design x to be simulated and stores the resultant objective function f and uncertainty σ on the archive.

Control of Experiments: The Controller determines the design x to be examined by experiment and requests the experiment to be performed. In certain cases, the experiment may be conducted automatically.

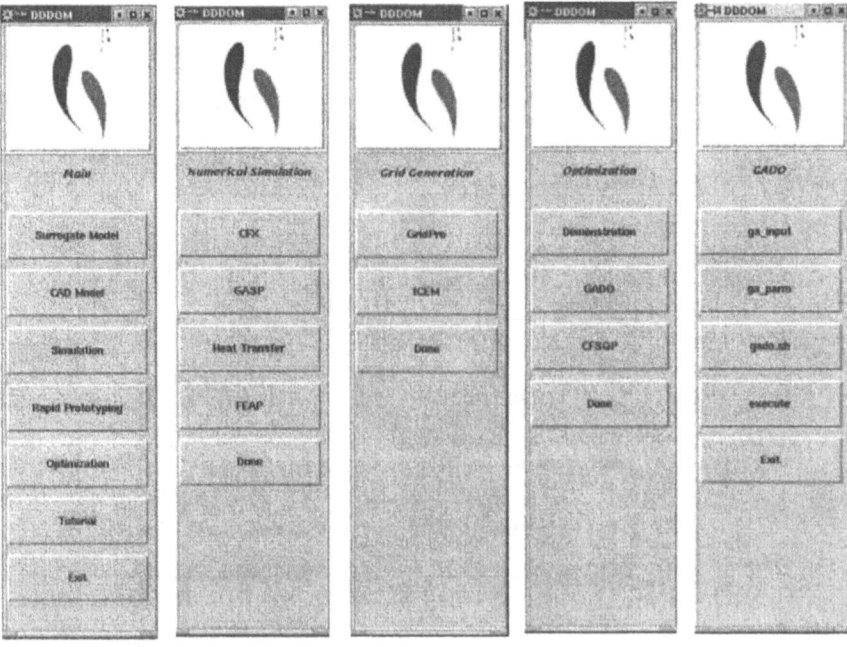

Fig. 2. Main panel for DDDOM controller

The DDDOM Controller must deal with executable codes, link different files and directories, and connect to databases. The User Interface is the access to the DDDOM Controller. The User Interface must be closely combined with the DDDOM Controller. It must be user-friendly, block-structured and well documented. The Controller is written in Perl and Perl/Tk [9,10,11]. The main panel of the DDDOM User Interface is shown in Fig. 2.

3   Multi-objective Design Optimization

3.1   Problem Statement

The DDDOM software system is applied to the multi-objective design optimization of a submerged inlet. A generic submerged inlet is shown in Fig. 3. Basically, it is a fluid flow duct through which air is delivered to a turbojet, turbofan or ramjet. The main objective for an inlet designer is to provide a high quality air flow, $i.e.$, the optimum levels of total pressure and flow uniformity which are vital for engine operability and stability. Because of the curved shape, inertial forces create a swirl in the inlet leading to flow non-uniformity. The specific design objectives are to minimize the Distortion Coefficient (DC) and Swirl Coefficient

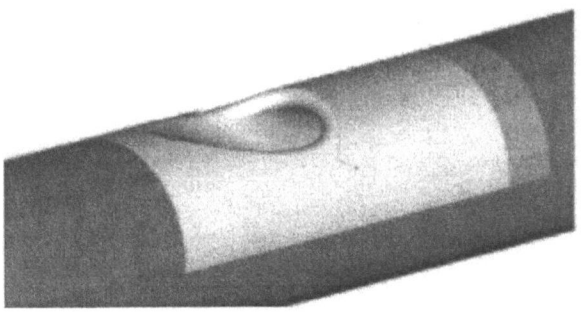

Fig. 3. Generic inlet

(SC) at the engine face (for definition, see [12]). Since the DC and SC are competing design objectives, the optimization problem is multi-objective and the solution is the Pareto Set of optimal designs.

This problem was selected to demonstrate the capability of the DDDOM to achieve an optimal set of designs (the Pareto Set) through concurrent integration of experiment and simulation while synergizing the respective advantages of each. For example, the computational cost of each simulation is independent of the inlet geometry, angle of attack or sideslip, and requires hundreds of total cpu hours. Since the simulation code is parallelized, the typical turnaround time for a simulation is six hours on the cluster at the Rutgers School of Engineering. The time to set up the experiment (including fabrication of the inlet model) is typically one day. However, changes in the inlet angle of attack or sideslip can be accomplished in a matter of minutes in the experiment. Thus, experiment and simulation can be synergized to achieve a more effective and efficient design optimization whereby the simulation is used to explore changes in the inlet geometry, and the experiment is used to explore changes in the inlet angle of attack and sideslip.

We present the results of the first stage of the multi-objective design optimization of a submerged inlet using experiment and simulation. In this stage, we conduct a design optimization of the inlet at zero angle of attack and sideslip using simulation alone, and then compare the predicted results for DC with the experiment. This establishes a baseline for calibration of the simulation against the experimental results which is an important element in the DDDOM. In the second stage (currently in progress), we will synergize the experiment and simulation in a concurrent manner to obtain the Pareto Set of designs for the case of non-zero angle of attack and sideslip.

Trade studies have shown [12,13] that the inclusion of a fin inside the inlet provides a significant improvement in flow quality (Fig. 4). The design variables are therefore taken to be the length, width and angle of attack of the fin (Fig. 4(a)).

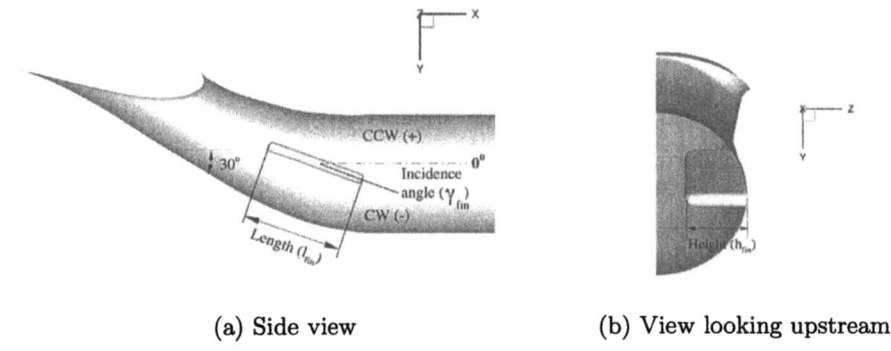

(a) Side view   (b) View looking upstream

Fig. 4. Design variables for a fin type of deformation

3.2   Design Optimization

The design optimization incorporates five software components linked together in an automated loop built using Perl/Tk. The components are the Optimizer (CF-SQP), Geometry Modeller (Pro/Engineer [14]), Mesh Generator (GridPro [15]), Flow Solver (GASPex [16]) and Auxiliary Program (to compute DC and SC). The Optimizer selects a specific geometry for the inlet. The geometry is passed to the Geometry Modeler (Pro/Engineer) which constructs a Computer Aided Design model of the inlet including the surface alteration in the form of a fin. The inlet surface definition is provided to the Mesh Generator (GridPro) which creates the grid in the flow domain. The grid is provided to the flow solver (GASPex) which solves the Reynolds-averaged compressible time dependent Navier Stokes equations in three dimensions. The computed flowfield is provided to the Auxiliary Program which calculates the objective functions DC and SC. The objective functions are provided to the Optimizer which then selects a new design. The loop continues, until the set of optimal designs (Pareto Set) is obtained. The entire process is controlled using Perl/Tk and the software components are executed on distributed systems based upon the licensing arrangements.

The set of designs examined in the optimization and the Pareto Set are shown in Fig. 5. The Pareto Set is disjoint and is characterized by two discrete sets of designs denoted A and B (circled). The designs in set A and set B show fundamentally different flow behavior. Set A designs are characterized by high swirl coefficient SC and low distortion coefficient DC, while Set B designs have low swirl coefficient SC and a higher distortion coefficient DC. Both sets of designs have improved values of both DC and SC compared to the baseline inlet without the fins [12].

3.3   Experiment

The experimental portion [12] of the inlet optimization problem is conducted in the subsonic wind tunnel located in the Low Speed Wind Tunnel Labora-

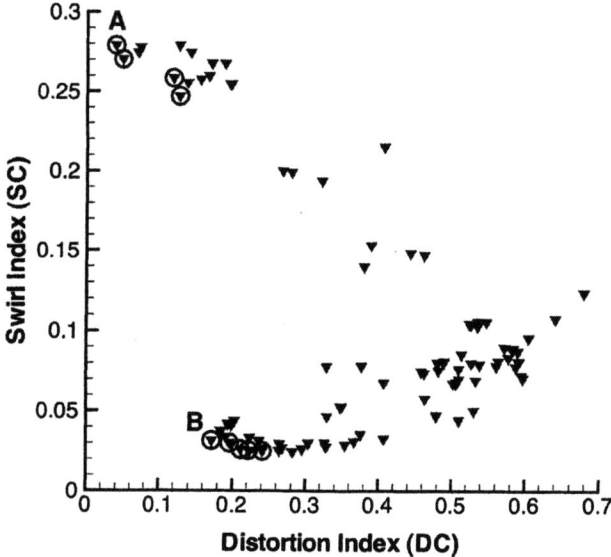

Fig. 5. Set of designs including Pareto Set (circled)

tory at Rutgers University (Fig. 6). A stainless steel model has been fabricated and installed into the test section of the subsonic wind tunnel. The model was designed with a removable submerged inlet (Fig. 7), which is fabricated using the Stratasys FDM Model 3000 rapid prototyping system at Rutgers University. Solid modeling of the submerged inlet is performed in ProEngineer using coordinates obtained from the computations. A STereo-Lithography (STL) file, which is a 3D faceted geometric model, is created in ProEngineer.

A comparison of the simulation results for one of the designs in Set B with the experimental measurements was performed. Compared with the baseline inlet, the distortion coefficient is improved (*i.e.*, reduced) in both the computation and the experiment. Specifically, the DC is reduced from 0.48 (baseline inlet) to 0.08 (optimum inlet) in the simulation and from 0.27 (baseline inlet) to 0.22 (optimum inlet) in the experiment. However, there is a significant difference between the computed and experimental DC values for both the baseline and optimal inlets which is attributable to the limitations of the accuracy of the simulation (specifically, the turbulence model used in the simulations). This is a common occurrence in such complex flow configurations, and will be addressed in future work (see below).

4   Future Work

The second stage of our research on optimal design of subsonic submerged inlets will focus on incorporating the experiment into the design optimization using

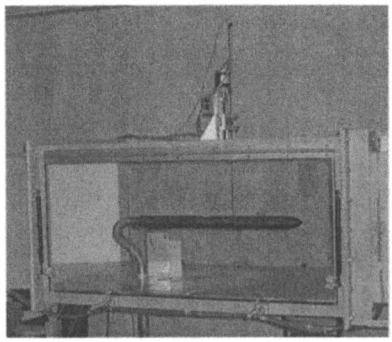

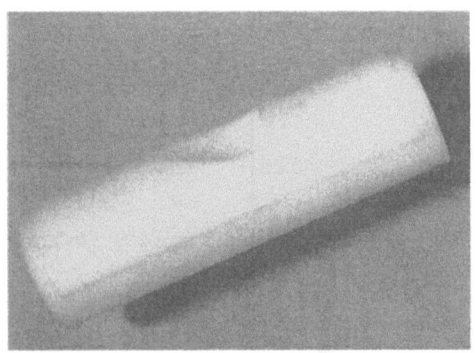

Fig. 6. Experimental facility   **Fig. 7.** Rapid prototyping model

the DDDOM software system. Two specific issues will be addressed. First, the experimental results will be used to extend the parametric range of the design optimization to non-zero angles of attack and sideslip. Evaluation of the distortion coefficient at nonzero angles of attack and sideslip is more readily (and quickly) achieved by experiment than by simulation. Thus, the experiment (rather than the simulation) will dynamically drive the optimization at nonzero angles of attack and sideslip. This demonstrates the enhancement offered by the Data Driven Design Applications System (DDDAS) concept. Second, we will incorporate the effects of uncertainty in the simulation (as defined by the results presented herein) into the DDDOM.

5   Conclusions

Engineering design is an example of a Dynamic Data Driven Application System wherein data from experiment and simulation can be synergized to dynamically drive the optimization process to achieve better designs in a shorter time. A Data Driven Design Optimization Methodology (DDDOM) is described which incorporates experiment and simulation in a real-time, synergistic manner. The DDDOM software system is developed using Perl and Perl/Tk. The DDDOM is applied to the multi-objective design optimization for a submerged inlet. The first stage results are presented. The Pareto Set of optimal designs at zero angle of attack and yaw are obtained, and the results compared with experiment. The uncertainty in the simulation (due to the limitations of the turbulence model) is identified. The second stage of research will extend the parametric space to nonzero angles of attack and yaw using the experimental data (rather than the simulation results) to drive the optimization.

Acknowledgments. The research is sponsored by the US National Science Foundation under grant CTS-0121058. The program managers are Drs. Frederica Darema, C. F. Chen and Michael Plesniak.

References

1. http://www.cise.nsf.gov/eia/dddas
2. https://www.fastlane.nsf.gov/servlet/showaward?award=0121177.
3. https://www.fastlane.nsf.gov/servlet/showaward?award=0121667.
4. https://www.fastlane.nsf.gov/servlet/showaward?award=0121263.
5. http://www.ncsa.uiuc.edu/About/TeraGrid
6. Knight D., Elliott G., Jaluria Y., Langrana N., and Rasheed K. Automated Optimal Design Using Concurrent Integrated Experiment and Simulation. AIAA Paper No. 2002-5636, AIAA/ISSMO Symposium on Multidisciplinary Analysis and Optimization, Atlanta, GA (2002).
7. Deb, K. Multi-Objective Optimization using Evolutionary Algorithms. John Wiley & Sons, New York (2001).
8. Knight D., Zhao H., Icoz T., and Jaluria Y. Data Driven Design Optimization Methodology. F. Darema (ed.) Dynamic Data Driven Applications Systems. Kluwer Academic Publishers, Dordrecht, The Netherlands (2004), to appear.
9. Wall L., Christiansen T., and Orwant J. Programming Perl. O'Reilly and Associates, Sebastopol, CA (2000).
10. Christiansen T., and Torkington N. Perl Cookbook. O'Reilly and Associates, Sebastopol, CA (1998).
11. Lidie S., and Walsh N. Mastering Perl/Tk. O'Reilly and Associates, Sebastopol, CA (2002).
12. E.S. Taskinoglu, V. Jovanovic, D. Knight. AIAA 2004-25 Multi-objective Design Optimization and Experimental Measurements for a Submerged Inlet. 42nd AIAA Aerospace Sciences Meeting and Exhibit, Reno, Nevada (2004).
13. E. Taskinoglu, D. Knight. Design Optimization for Subsonic Submerged Inlets. *AIAA Paper 2003-1247*, Jan, 2003.
14. Parametric Technology Corporation. *Pro/Engineer User's Guide. (1996)*.
15. Program Development Corporation. *GridPro/az3000 User's Guide and Reference Manual*. White Plain, New York (1998).
16. Aerosoft Inc. *General Aerodynamic Simulation Program User Manual. (1996)*.

A Dynamic Data Driven Computational Infrastructure for Reliable Computer Simulations

J.T. Oden[1], J.C. Browne[1], I. Babuška[1], C. Bajaj[1], L.F. Demkowicz[1],
L. Gray[2], J. Bass[1], Y. Feng[1], S. Prudhomme[1], F. Nobile[1], and R. Tempone[1]

[1] Institute for Computational Engineering and Sciences,
The University of Texas at Austin, U.S.A.
[2] Department of Otolaryngology - Head and Neck Surgery, University of Texas
Medical School, University of Texas Health Science Center at Houston, U.S.A.

Abstract. The paper presents an initial study on designing a Dynamic Data Driven (DDD) computational environment to enable reliable and efficient determination of Head Related Transfer Functions (HRTF) for patients with reconstructed ears. The determination involves a synthesis of coordinated experiments and large scale, parallel computer simulations. Time limits (one working day), cost of MRI scans, and parallelism in experimental and computer simulations impose the need for a DDD enabling fast and reliable results.

1 Introduction: Dynamic Data Driven Application Systems

Traditional application simulations are conducted with static data inputs. In context of Finite Element (FE) simulations, this means defining geometry of the domain of computations, physical coefficients and parameters and source and boundary data. Once the data are specified, the FE codes are run, possibly in an adaptive loop, to assure discretization error control. Any change in data implies restarting the codes from the beginning. In a Dynamic Data Driven Application System [14], an initial set of data is deposited in a depository, but, during the simulation process that follows, the data may evolve or be updated as a result of new measurements, or a feedback mechanism resulting from comparing simulations with experiments. A classical example is a weather prediction system where current simulation results may be updated as new measurements become available "online". Incoming extra information may be incomplete for simply restarting the simulations and more sophisticated "matching mechanisms" are necessary.

Verification and Validation (V and V). The systematic study of reliability of computer simulations of physical phenomena is generally divided into two basic processes, Verification and Validation (V and V). As a scientific discipline, V and V is in its very early stages of development. Verification is the process of determining the accuracy with which a given mathematical model of physics is

M. Bubak et al. (Eds.): ICCS 2004, LNCS 3038, pp. 756–763, 2004.
© Springer-Verlag Berlin Heidelberg 2004

solved; validation is the process of determining that the mathematical model represents the actual physical system with sufficient accuracy. Here verification does not include software engineering issues of the logical correctness of the program which implements the simulations (this is so-called *code* verification). Verification is thus concerned with estimating and controlling numerical approximation error. In recent years, significant progress has been made in this area. The idea of developing an adaptive control system for implementing V and V was made possible by the discovery of the concept of hierarchical modeling and "goal-oriented" adaptive methods based on a posteriori estimates of modeling error [7] and approximation error [1,2]. These error estimates techniques are crucial to the success of any adaptive control paradigm that must react to dynamic data management, and to adapting the models used in simulations.

Uncertainties in Data. Another source of error relevant to validation is due to the randomness of the material and geometry data. There are several approaches available to quantify this error. One direct approach is provided by the perturbation method described in [6]. If it is possible to develop statistical information on the material data to determine the *probability density function* (PDF), the material coefficients can be represented as a sum of a deterministic average and a stochastic perturbation. A similar decomposition holds then for the solution to the problem. Bounds on the stochastic perturbation of the solution can then be calculated and used as an additional measure of the modeling . The second approach is based on the theory of stochastic functions characterized by Karhunen-Loève expansion (see [4]). Similar methodologies can be used for estimating uncertainties in geometry data, although the problem is tackled in the literature much less frequently.

This paper presents an initial study on designing a Dynamic Data Driven (DDD) computational environment to enable reliable and efficient determination of Head Related Transfer Functions[1] (HRTF) for patients with reconstructed ears. The proposed project brings together specialists from Otolaryngology Research Laboratory at the University of Texas Houston Health Science Center, and researchers at the Institute for Computational Engineering and Sciences, at the University of Texas at Austin.

2   Analysis of the Effect of the Exterior Geometry of the Human Head on Human Hearing

Historically, many scientific investigations of important physical systems have been purely experimental, out of necessity since the complexities of the systems have been beyond the capability of numerical simulation. An illustrative example of such a system is the human auditory system. A purely empirical approach has placed limitations upon determining basic mechanisms of the sense of hearing, as

---

[1] *Pressure on the ear drum* expressed as a function of frequency and direction of an incoming plane wave

well as the development of hearing-aid technology for the hearing-impaired, and hearing protection for those working in noisy environments. Application of experimental tools to characterize human hearing have tended to be idiosyncratic to the subject, expensive and time-consuming in application, and dependent upon subjective feedback of the subject being tested. The latter issue is particularly problematic in extremely young patients who lack the ability to fully or productively participate in many testing procedures [5,13].

We describe how a new approach based on DDDAS in which the numerical modeling capabilities of geometry reconstruction and visualization schemes, and parallel *hp* boundary element method, developed at ICES, are linked to experimental data obtained at the UT Health Science Center in Houston. The project includes means for controlling modeling and discretization errors. While the technique has a myriad of potential application areas, initial studies are planned to apply the technique to the investigation of the effect of external ear reconstruction in patients born with only one external ear. The sounds that stimulate the ear, and the ability of a subject to localize sound, is dependent in large part on the filtering and amplification characteristics of the external ear, as well as the brain's ability to interpret the person's own HRTF. The work enables the study of sensitivity on the variation of geometry from one external ear, to a surgically-constructed ear, to a normally formed ear and the effects of the variation upon the resultant HRTF. Potential for application of the coupled modeling and experimental technique to other hearing-related issues, and the pursuit of sponsoring of future research are also under investigation.

3   Geometry Reconstruction from MRI Scans

In many applications, including the current one, the geometry is too complicated to be modeled by standard CAD systems, and only a linear scan obtained by MRI or CT techniques is available. In the CAD supported case, generation of higher order meshes is possible simply by interpolating the surface parametrization provided by the CAD modeler. In the case of scans, the generation of higher order meshes is not straightforward, since, *a priori*, only a linear triangulation is available. A number of different procedures are available that will produce higher order meshes from linear triangulations, including element clustering, decimation schemes, etc. A C^1 surface reconstruction has many advantages from the viewpoint of scattering problems in acoustics, including the elimination element clustering around artificial C^0 edges, and a lowering of the order of singularities in the integral kernels. One approach for generating a C^1 representation from a linear triangulation has been given by Bajaj, Xu, [3].

The C^1 geometry reconstruction, done within a *Geometrical Modeling Package*, a small CAD system developed at ICES that interfaces with the boundary element codes, is discussed next, and allows for an initial mesh generation as well automatic geometry updates during mesh refinements.

4   Modeling of Human Head Acoustics

The acoustics of the human head is modeled with a comparatively simple linear acoustics model which, in the frequency domain, reduces to solving the Helmholtz equation in the domain exterior to the human head (but including the ear channel), accompanied by Sommerfeld radiation condition at infinity, and impedance boundary conditions. The material data are represented by impedance, usually assumed to be piecewise constant, with different values for the head, ear channel, and the ear drum. The values may depend (mildly) on the patient, i.e. the impedance of the head may reflect the amount of patient's hair.

5   Parallel *hp*-Adaptive Boundary Elements

Major components of the parallel *hp* BEM code developed to solve the acoustics problem, are given below.

Burton-Miller Integral Equation (BMIE). The Helmholtz equation is replaced with an equivalent Burton-Miller integral equation defined on the surface of the head, including the ear channel and the ear drum. The formulation is uniformly stable in frequency. The integral equation is formulated in a variational form, allowing for reduction of singularities to weak singularities only. This facilitates and accelerates the integration of element matrices.

hp-discretizations. The integral equation, in its variational form, is discretized then using Galerkin approximation and *hp* finite element meshes in which element size h and polynomial order p may be varied locally to minimize the error per given problem size. If done properly, the *hp* discretizations have the potential of delivering exponential convergence and superaccurate results.

Goal-oriented adaptivity. The adaptivity is driven by minimizing the error in a specific quantity of interest, in our case - the average pressure on the ear drum. The algorithm requires solving simultaneously both the original and *dual* problems [11]. The error in pressure is then represented in terms of a functional involving both solutions, and serves as a starting point to determine optimal *hp* mesh refinements. The algorithm produces a sequence of meshes of increasing size and complexity with corresponding solutions, i.e. the pressure distribution on the head. The problem involves *multiple loads* as one mesh serves all (typically sixty) directions of the incoming plane wave. The problem can thus be very time consuming, and even with a parallel implementation, attaining a 1 percent accuracy, may require at this point many CPU hours. During the mesh optimization process, the code deposits intermediate meshes and corresponding (multiple) solutions on the disk from which the results may be accessed and visualized.

Parallel implementation. The parallel implementation has been done within PLAPACK [10]. This includes integration of element matrices, a dense linear solver based on LU decomposition, and a posteriori error estimation done in parallel. The problem does not involve any domain decomposition; a copy of data structure supporting the geometry, mesh and solution, is maintained on each processor.

Error control. At present, the code includes only an a-posteriori error control of the discretization error. Neither modeling nor geometry induces errors are being estimated. Sensitivity of the solution with respect to the geometry is critical in the case of frequencies corresponding to resonating modes (concha or ear channel modes).

6   A DDDAS for Modeling Acoustics of Human Head

A DDD adaptive control system is under study in which scanned images of external ears (both premature and reconstructed) is to be provided from remote repository at the UT Health Sciences Center at Houston. The team in Austin constructs mathematical models of sound propagation into the ear canals of specific patients. The computed HRTF's are then to be reported back to Houston to allow for a comparison with those obtained experimentally.

Existing computational methodology are being upgraded to a fully dynamic system enabling a continuous exchange of data between Houston and Austin, and a critical comparison of the computational and experimental results. A typical patient visit to the Houston facility lasts one day (eight working hours), and the targeted mode of operation is to complete the patient analysis within that time.

A schematic mode of operation is presented in Fig. 1. All software including visualization will be run in a client/server mode, with the actual codes being executed at ICES platforms. A common depository will be established with a new directory opened for each new patient. At any time during the operation, the depository may include the following files.

MRIscan1. The first MRI scan of the patient head and ear channel.
MRIscan2. The second MRI scan, done later, if the geometry reconstruction code fails to deliver acceptable results.
head_geometry. Patient head reconstructed geometry, based on one or two MRI scans.
Results of computation. A sequence of optimal meshes with corresponding pressure distributions and computed HRTFs.
Error estimates. Results of sensitivity analysis providing error bounds for obtained numerical results (for each mesh and each frequency), as well as possible results of comparison of measured and computed HRTFs.

The following is a possible scenario. A patient arrives at the Houston facility early in the morning, the first MRI scan ($1,000 each) is made and deposited in

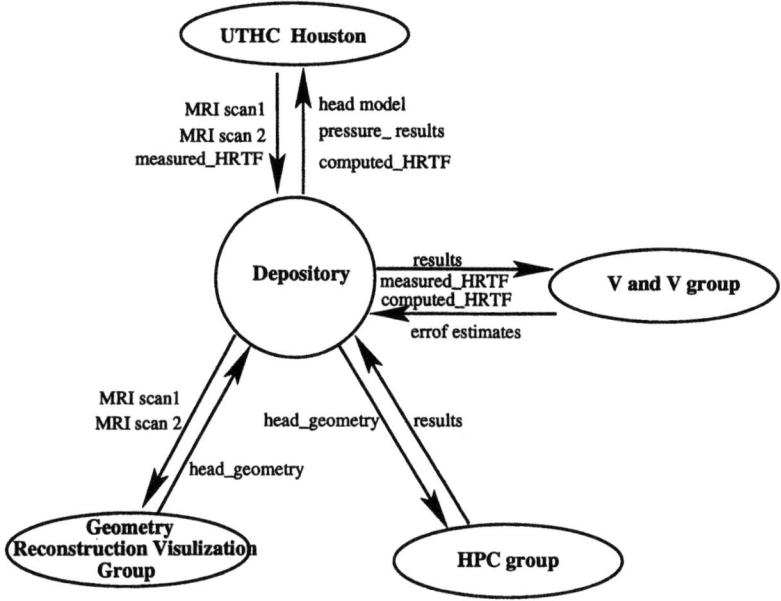

Fig. 1. A schematic idea of the proposed DDD computational environment

the depository. The initiation of a new patient case includes also specifying the corresponding impedance data.

The geometry of the patient is reconstructed in the form of a fine, unstructured linear mesh and stored in the depository. The reality of the geometry reconstruction schemes is that, in presence of noisy data, they produce many artifacts. Such discrepancies prove to be critical in highly accurate computer simulation of acoustics resulting in erroneous results. The model has to be "cleaned" interactively. The corrections will be done in communication between Houston and the geometry reconstruction group at Austin. The process may result in the necessity of taking another MRI scan and producing a better resolution geometry representation based on both sets of data.

The initial, "dirty" image is still sufficient to start the solution process and begin the mesh optimization process. As new, cleaner data arrive in the depository, the geometry input will be upgraded in the ongoing numerical simulations. One should emphasize that several [2] copies of the code will be executed in parallel.

The moment first meshes are generated and results are computed and stored in the depository, the Validation and Verification group begins analyzing sensitivity of the results with respect to geometry and impedance data, as well as more thorough (and expensive) discretization error estimates are computed (initial, rough estimates are provided by mesh optimization algorithms).

In meantime, the team in Houston completes determining HRTFs experimentally and the results are communicated to the V&V group for a final, quantitative

---

[2] The HRTFs are usually determined for 6-10 different frequencies

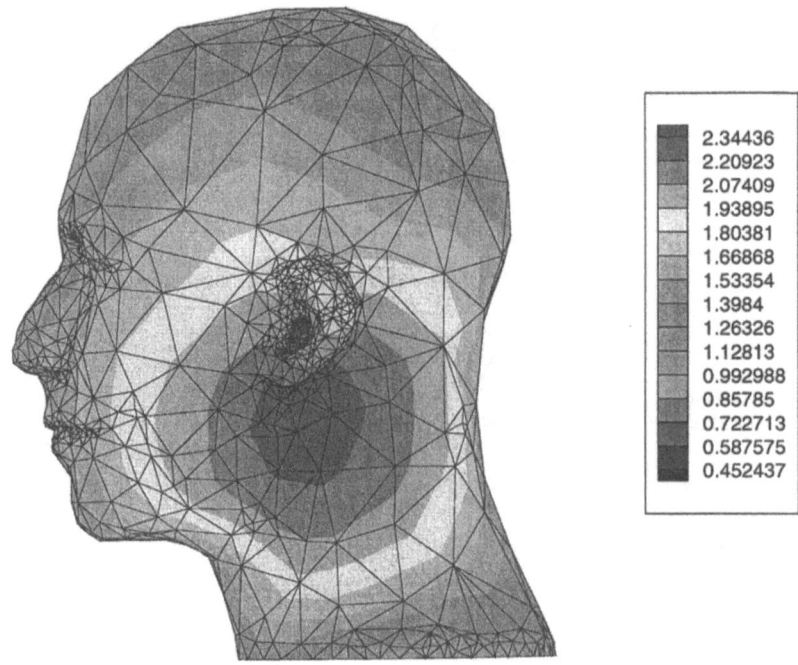

Fig. 2. Pressure distribution on a model of human head

error analysis. Dependently upon results of the sensitivity analysis, the group at Houston may decide to change the impedance data, as well.

A new methodology for comparing the experiments with simulations is needed. This will involve comparing not only the measured and computed HRTF's but also designing experiments that may verify the computed data in an "inverse mode". For example, we can test the accuracy of vertical and horizontal sound localization using established "auditory virtual reality" techniques. To do this we use the calculated HRTF's from reconstructed ears to make special stereo sounds and determine if localization under headphones is accurate.

7   Conclusions

The project described is an excellent example of an important DDDAS. The system builds on collaborations of medical researchers, computer scientists, computational scientists, engineers, and mathematicians in Houston and Austin. On top of necessary improvements of existing methodologies, the main challenges are:

- to build an environment allowing for a dynamic data transfer and simultaneous visualization of results by the three involved parties (experiment, geometry reconstruction and visualization, modeling), and,

- to develop algorithms for estimating sensitivity of the simulated results with respect to reconstructed geometry and material data,
- to establish a mathematically sound methodology that would allow for comparing the experiments and simulations with a feedback and a dynamic modification of data on both sides.

Acknowledgment. The authors gratefully acknowledge support of the NSF under grant 0205181.

References

1. Ainsworth, M., and Oden, J.T.: A Posteriori Error Estimation in Finite Element Analysis. John Wiley & Sons, New York, (2000)
2. Babuška, I., Strouboulis, T.: Finite Element Method and its Reliability. Oxford Univ. Press (2001)
3. Bajaj, C., Chen, J., and Xu, G.: Modeling with cubic A-patches, ACM Transactions on Graphics, **14** (2) (1995) 103–133.
4. Deb, D.K., Babuška, I., Oden, J.T.: Solution of Stochastic Partial Differential Equations Using Galerkin Finite Element Techniques. Comput. Methods Appl. Mech. Eng. **190** (2001), 6359–6372
5. Jahrsdoerfer, R.A., Yeakley, J.W., Aguilar, E.A., Cole, R.A., and Gray, L.C.: A grading system for the selection of patients with congenital aural atresia. American Journal of Otolology **13** (1992) 6–12
6. Kleiber, M., Hien, T.D.: The Stochastic Finite Element Method: Basic Perturbation Technique and Computer Implementation. John Wiley & Sons, (1992)
7. Oden, J.T., Prudhomme, S., Estimation of modeling error in computational mechanics. J. Comput. Phys., **182** (2002), 496–515
8. Oden, J.T., Babuska, I., Nobile, F., Feng, Y., and Tempone, R.: Theory and methodology for estimation and control of errors due to modeling, approximiation, and uncertainty. Computer Methods in Applied Mechanics and Engineering, MAFELAP 2003 Special issue (to appear)
9. Oden, J.T., Browne, J.C., Babuska, I., Liechti K.M., Demkowicz, L., Bass J., Feng, Y., Prudhomme, S., Nobile, F., Tempone, R.: A computational infrastructure for reliable computer simulations. In F. Darema (ed.) Dynamic Data Driven Application Systems, Kluver Academic Publishers (to appear)
10. van de Geijn, R.A.: Using PLAPACK, The MIT Press (1997)
11. Walsh, T., Demkowicz, ,L.: *hp* Boundary element modeling of the external human auditory system - goal oriented adaptivity with multiple load vectors, Computer Methods in Applied Mechanics and Engineering, **192** (2003) 125–146
12. Walsh, T., Demkowicz, L, Charles, R.: Boundary element modeling of the external human auditory system, JASA (to appear)
13. Wilmington, D., and Gray, L., Jahrsdoerfer, R.: Binaural processing after corrected congenital unilateral conductive hearing loss. Hearing Research, **74** (1994) 99–114
14. NSF Workshop (2000) http://www.cise.nsf.gov/dddas

Improvements to Response-Surface Based Vehicle Design Using a Feature-Centric Approach

David Thompson[1], Srinivasan Parthasarathy[2], Raghu Machiraju[2], and Scott Lawrence[3]

[1] Department of Aerospace Engineering and Computational Simulation and Design Center, Mississippi State University, MS
dst@erc.msstate.edu
[2] Computer and Information Science, The Ohio State University, Columbus, OH
{raghu,srini}@cis.ohio-state.edu
[3] Systems Analysis Branch, Ames Research Center, NASA, Moffett Field, CA
Scott.L.Lawrence@nasa.gov

Abstract. In this paper, we present our vision for a framework to facilitate computationally-based aerospace vehicle design by improving the quality of the response surfaces that can be developed for a given cost. The response surfaces are developed using computational fluid dynamics (CFD) techniques of varying fidelity. We propose to improve the quality of a given response surface by exploiting the relationships between the response surface and the flow features that evolve in response to changes in the design parameters. The underlying technology, **generalized feature mining**, is employed to locate and characterize features as well as provide explanations for feature-feature and feature-vehicle interactions. We briefly describe the components of our framework and outline two different strategies to improve the quality of a response surface. We also highlight ongoing efforts.

1 Introduction

Design studies for advanced aerospace vehicles are typically performed using a suite of computational fluid dynamics (CFD) simulation tools. An integral component of these design studies is the development of response surfaces describing vehicle performance characteristics as functions of the various geometrical design parameters. Essentially, the response surfaces provide a method for abstracting the results of a simulation [1]. A response surface may be generated using tools of differing fidelity, i.e., variable complexity modeling [1]. Because of various factors associated with the simulations, e.g., incomplete convergence of iterative methods, round-off errors, and truncation errors [2], numerical noise may be present in the response surfaces. When such noise occurs, the optimization procedure may be slow to converge or converge to an erroneous local extrema in the design space. This can be partially obviated through the use of response surface modeling [1]. The response surfaces typically include force and moment components and consist of the outputs from a series of CFD simulations. Continuous approximations to the response surfaces are obtained through a surface fitting process. These continuous approximations are then employed

M. Bubak et al. (Eds.): ICCS 2004, LNCS 3038, pp. 764–770, 2004.
© Springer-Verlag Berlin Heidelberg 2004

in a design algorithm to determine a geometrical parameter set that results in a set of vehicle characteristics that satisfy specified criteria.

Response surfaces can also be employed to optimize vehicle trajectories. In this context, the design parameters include vehicle attitude, control surface deflections, etc. Knowledge about the presence and location of flow features would provide guidance on judicious use of the available analysis tools as well as provide an understanding of the level of confidence that can be placed on the computational results in various regions of the parameter space.

Unfortunately, the quality of a given response surface is difficult to ascertain. This is primarily due to the fact that development of analysis techniques for CFD data has not kept pace with our ability to generate massive data sets. To address the difficulties associated with analyzing evolutionary data generated by large-scale simulations of complex phenomena, we are developing a framework that we term **generalized feature mining**. In the context of this effort, i.e., aerospace vehicle design, we are interested in how flow features, which influence the performance of the vehicle, evolve in response to changes in the design parameters. Hence, the features, which may be static features in the sense that they do not evolve over time, do evolve as the design parameters are modified.

1.1 Motivation

We can formalize the notion of response surfaces by letting $\mathbf{a}=\{a_1,\ldots,a_n\}$ be the set of n parameters that are considered in the design of a vehicle. These typically include the geometrical characteristics of the configuration of interest. Let $\mathbf{f}(\mathbf{a})=\{\psi_1(\mathbf{a}),\ldots,\psi_m(\mathbf{a})\}$ be m quantities of interest such as force and moment components. Each of the $\psi_i(\mathbf{a})$ form a response surface. Vehicle design attempts to find a set of design parameters $\mathbf{a}$ which yields a desirable vector $\mathbf{f}(\mathbf{a})$ for certain operational conditions. These response surfaces are obtained by computing the flow field and the desired quantities ψ_i for a given configuration $\mathbf{a}$. It should be noted that some response surfaces, such as configuration gross weight, depend only on the geometry and material properties and are independent of the flow simulation.

What is lacking from the approaches described above is a true "cause and effect" perspective for vehicle design. Changes in the response surfaces $\psi_i(\mathbf{a})$ are correlated only to changes in the design parameter vector $\mathbf{a}$. In our approach, we seek **explanations** for the behavior of a response surface ψ_i. Desirable conditions may occur in due to the presence (or absence) of specific flow phenomena. Understanding the properties of these phenomena and their interactions with the vehicle or other flow features provides more **insight** than correlations with the design parameters. We believe that these insights will lead to more cost effective methods for estimating the response surface, thereby reducing design cost.

1.2 Outline of Paper

In Section 2 we describe the basic components of our feature-centric design framework. Section 3 describes two scenarios which can exploit generalized feature mining in a design context. In Section 4, we describe related work and in Section 5 we describe on-going work in to develop generalized feature mining technologies. Finally, we include a summary of the paper in Section 6.

2 Components of Generalized Feature Mining

The generalized feature mining framework will accept as input a set of CFD simulation results **S** and the corresponding set of designs **A** that produced these results. Additional inputs include a definition of the material properties and a specification of flight and environmental conditions. There are four components that must be developed to accomplish our stated objective:

- **Extraction of the feature set:** The simulation results **S** will be processed to extract features **F**. Shape descriptors will be used to characterize the features. Here, a shape descriptor for a given feature is an abstract representation of the feature including its geometrical and dynamical characteristics.
- **Explanation of feature-feature and feature-vehicle interactions:** In addition to shape descriptors, generalized feature mining will generate explanations **E** correlated with changes in the design **a** through the evolution of the feature set **F**.
- **Characterization of a response surface:** Once a response surface **R** is generated, it will be characterized qualitatively and quantitatively using shape descriptors **F** and explanations **E**.
- **Estimation of the response surface:** New points on the response surface will be estimated based on the response surface characterization. This will facilitate determination of whether an additional computation needs to be performed to locally 1) refine the response surface or 2) improve its accuracy user a higher fidelity method.

3 Application to Aerospace Vehicle Design

We envision that our framework may be employed for aerospace vehicle design via two different **design-of-experiment** paradigms. Both approaches are designed to improve the quality of a response surface that can be generated for a given cost resulting in a higher confidence design. We describe them below:

- The quality of a response surface may be improved locally through the use of higher fidelity methods. An initial response surface is developed using lower fidelity (less expensive) computational tools. Our framework is then employed to identify regions of the response surface whose associated feature catalogs, i.e., flow features, feature-feature interactions, and feature-vehicle interactions, suggest that a higher fidelity simulation tool should be employed. The cost of generating the response surface is decreased since the number of expensive simulations needed to adequately define the response surface is reduced.

- The quality of a response surface may be improved locally through refinement. An initial response surface is developed using a relatively coarse sampling. Our framework is then employed to identify regions of the response surface whose associated feature catalogs suggest are under-sampled. Additional simulations of appropriate fidelity may be performed at the identified design points. The cost of generating the response surface is again decreased since the number of expensive simulations needed to adequately define the response surface is reduced.

We also suggest that it may be possible to employ the two different paradigms in tandem to adaptively generate a high quality response surface for a reduced cost.

4 Related Work

In this section we report on work related to the current effort. A timely book on scientific and engineering data mining techniques edited by Grossman, Kamath, Kegelmeyer, Kumar, and Namburu [3] discusses many issues relevant to this effort. Kamath provides a beautiful overview of the area along with some of the inherent challenges (Chapter 1). Marusic and his associates (Chapter 13) and Han and his associates (Chapter 14) consider the problem of mining computational fluid dynamics simulations using standard data mining techniques in conjunction with basic physical modeling of entities or features. The work by the authors Machiraju, Thompson and colleagues (Chapter 15) on the EVITA project is a precursor to the current effort. Han and colleagues (Chapter 25), Sekhar and colleagues (Chapter 26), and Lesage and Pace (Chapter 24) report on the usefulness of spatial data analysis and survey such techniques in the context of engineering and scientific applications. Additionally techniques for mining astronomical data (Chapters 5 and 6) and earth science observations (Chapter 10) are also reported in this book. However, few of the above methods actually account for the structural or physical properties of the data. Also, none of them address the application of these techniques to design.

Yip and Zhao [4] proposed a framework called **Spatial Aggregation** (SA). They relied on segmentation (or spatial aggregation) to cluster both physical and abstract entities and constructed imagistic solvers to gain insights about physical phenomena. Our paradigms for feature mining have close relationship to the SA framework. The work by Ester and colleagues [5] on trend detection in spatial databases and spatial database primitives are also quite relevant to our work on spatial associations. The work by Graves and colleagues [6] is also relevant but focused mainly on feature extraction and mining of image data (satellite). It should be noted that the ADAM framework is extensible and it may be possible to incorporate many of the techniques described here.

5 Ongoing Work

In this section we describe our preliminary efforts in data mining and effort towards realizing a framework for generalized feature mining. Our combined preliminary

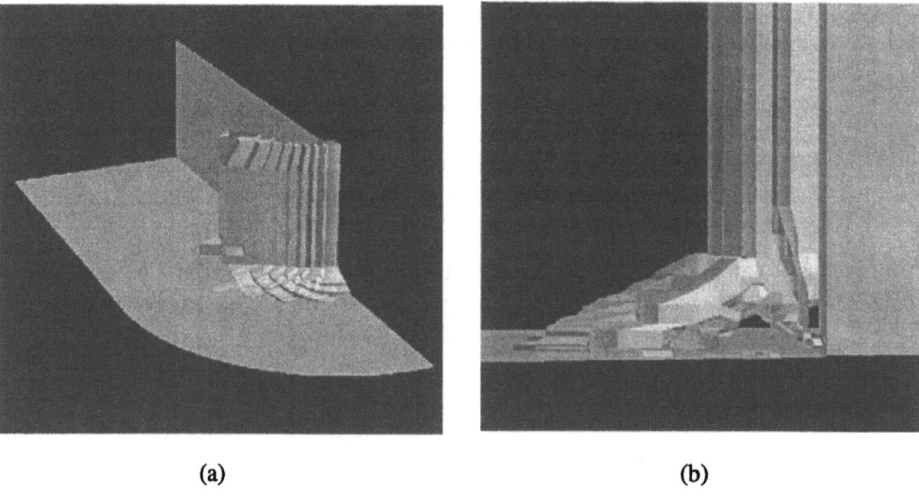

(a)   (b)

Fig. 1. Point Classification Techniques (a) The Blunt Fin Example (b) A zoomed version

work on feature mining has been reported in the literature [7-9]. It should be noted that our work spans two application domains, namely CFD and molecular dynamics. Because of the paucity of space and vehicle design context, we focus on work applicable to CFD simulation data.

We first describe two distinct feature mining paradigms. The common thread is that both are bottom-up feature constructions based on underlying physically based criteria. They consist essentially of similar steps; however, the order of the steps is different. The first paradigm, based on **point classification**, is suitable for features that can be defined and detected with local operators such as shocks and defects. The second paradigm, based on an aggregate **classification**, is suitable for features with a more global influence such as vortices. More details can be found in [8-9].

As an example of the **point classification paradigm**, we consider shock wave detection in flow fields. A shock is an abrupt compression wave that may occur in fluid flows when the velocity of the fluid exceeds the local speed of sound. The properties of shocks are can be exploited to develop highly discriminating shock detection algorithms [10]. Figure 1(a) shows a detached oblique shock that wraps around the blunt fin in the standard blunt fin/flat plate flow field solution. Figure 1(b) shows the symmetry plane which intersects a lambda-shock.

We recently developed vortex detection and verification algorithms that exploit the aggregate classification paradigm [11,12]. For the local detection operator, a combinatorial labeling scheme is employed. Our detection algorithm labels the velocity vectors at the grid points and identifies grid cells that are likely to contain the vortex core. Our technique then segments candidate vortex core regions by aggregating points identified from the detection step. We then classify (or verify) these candidate core regions based on the existence of swirling streamlines surrounding them. In Figure 2 we show the vortex cores and a detailed view of the vortical flow near the fin/plate intersection in the left and center images, respectively. The **aggregate classification** paradigm identifies individual points as being probable candidate points in a feature and then aggregates them. The verification algorithm

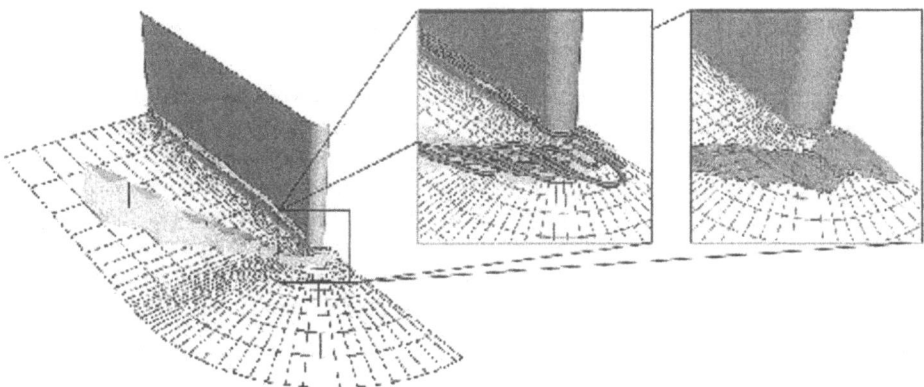

Fig. 2. Aggregate Classification Techniques. We use the Blunt Fin example again. The first inset shows the verification algorithm at work. The second inset shows the modeling of the swirling region as a collection of frusta.

then is applied to the aggregate using physics-based non-local criteria to determine whether the aggregate is actually a feature. In this case, the binary feature classification is applied to an aggregate of points. Features that are regional in nature, such as a vortex, benefit from the aggregate classification approach.

Having extracted the features, we now describe the next step: *characterizing them using geometric shape attributes.* The swirling region of a vortex can be characterized using a sequence of elliptical frusta (a conical frusta with two ends as ellipses) [8]. Starting from the upstream extent of the vortex core, the modeling process seeds a set of swirling streamlines surrounding the core region. Each elliptical frustum is oriented along the longest segment of the vortex core that does not curve by more than a user-specified amount. We illustrate the adaptive and hierarchical shape characterization process in the right image of Figure 2. The elliptical shape of the primary vortex as well as its curved core provide visual validation of the choice of the geometric descriptors.

We now describe an ongoing data mining effort that has, to date, been employed for the discovery of substructures in large molecules. Motifminer, is a novel scalable toolkit to efficiently discover frequently occuring spatial and structural relationships in scientific data [13]. The technique relies on range pruning (limits the search for viable spatial relationships) and candidate pruning (similar to candidate pruning in association rule mining but with a spatial twist) for pruning the search space of possible frequent structures. We also rely on fuzzy recursive hashing, a technique akin to geometric hashing, for rapid matching of frequent spatial relationships (to determine frequency of occurrence).

In addition feature characterization, much work is being conducted towards tracking of these swirling regions. Also, work has begun toward determining spatial associations and using the gained knowledge to glean explanations about the underlying phenomena.

6 Summary

We have described our vision for a feature-centric framework to improve simulation-based aerospace vehicle design. We proposed two different strategies, both based on generalized feature mining, to improve the quality of response surfaces employed in the design process by exploiting the relationships between the response surface and the flow features that evolve in response to changes in the design parameters. Also included were brief discussions of the various components of our framework and current efforts to realize these components. We believe our approach may prove to be a viable cost effective alternative to current automated design techniques.

References

1. Giunta, A., Golividov, O., Knill, D. L., Grossman, B., Haftka, B., Mason, W., Watson, L.: Multidisciplinary Design Optimization of Advanced Aircraft Configurations. 15th International Conference on Numerical Methods in Fluid Dynamics, Monterey, CA, June 1996.
2. Baker, C., Grossman, B., Haftka, R., Mason, W., Watson, T.: High-Speed Civil Transport Design Space Exploration Using Aerodynamic Response Surface Approximations, J. Aircraft, 39(2):215–220, 2002.
3. Grossman, R., Kamath, C., Kegelmeyer, P., Kumar, V., Namburu, R.: Data Mining for Scientific and Engineering Applications. Kluwer Academic Press, 2001.
4. K. Yip and F. Zhao: Spatial Aggregation: Theory and Applications. J. of Artificial Intelligence Research, 5:1–26, Aug 1996.
5. Ester, R., Frommelt, A., Kriegel, H., Sander, A.: Algorithms for characterization and trend detection in spatial databases. KDD, 1998.
6. Ramachandran, R., Conover, H., Graves, S., Keiser, K.: Algorithm development and mining (ADAM) system for earth science applications. Second Conference on Artificial Intelligence, 80th American Meteorogical Society Annual Meeting, 2000.
7. Machiraju, R., Parthasarathy, S., Thompson, D. S., Wikins, J., Gatlin, B., Choy, T. S., Richie, D., Jiang, M., Mehta, S., Coatney, M., Barr, S., Hazzard, K.: Mining Complex Evolutionary Phenomena. In H. Kargupta , editor, Data Mining for Scientific and Engineering Applications. MIT Press, 2003.
8. Jiang, M., Choy, T. S., Mehta, S., Coatney, M., Barr, S., Hazzard, K., Richie, D., Parthasarathy, S., Machiraju, R., Thompson, D., Wilkins, J., Gatlin, B.: Feature Mining Algorithms for Scientific Data. In Proceedings of SIAM Data Mining Conference, pages 13–24, May 2003.
9. Thompson, D., Machiraju, R., Jiang, M., Nair, J., Craciun, G., Venkata, S.: Physics-Based Feature Mining for Large Data Exploration. In IEEE Computing in Science and Engineering, 4(4):22–30, July-August 2002.
10. Lovely, D., Haimes, R.: Shock Detection from Computational Fluid Dynamics Results. AIAA Paper 99-3285, AIAA 14th Computational Fluid Dynamics Conference, June 1999.
11. Jiang, M., Machiraju, R., Thompson, D.: Novel Approach to Vortex Core Region Detection, In Proceedings of Joint Eurographics-IEEE TCVG Symposium on Visualization, pages 217–225, May 2002.
12. Jiang, M., Machiraju, R., Thompson, D.: Geometric Verification of Swirling Features in Flow Fields. In Proceedings of IEEE Visualization '02, pages 307–314, October 2002.
13. Parthasarathy, P., Coatney, M.: Efficient discovery of common substructures in acromolecules. In Proceedings of IEEE International Conference on Data Mining, 2002.

An Experiment for the Virtual Traffic Laboratory: Calibrating Speed Dependency on Heavy Traffic

A Demonstration of a Study in a Data Driven Traffic Analysis

Arnoud Visser, Joost Zoetebier, Hakan Yakali, and Bob Hertzberger

Informatics Institute, University of Amsterdam

Abstract. In this paper we introduce an application for the Virtual Traffic Laboratory. We have seamlessly integrated the analyses of aggregated information from simulation and measurements in a Matlab environment, in which one can concentrate on finding the dependencies of the different parameters, select subsets in the measurements, and extrapolate the measurements via simulation. Available aggregated information is directly displayed and new aggregate information, produced in the background, is displayed as soon as it is available.

1 Introduction

Our ability to regulate and manage the traffic on our road-infrastructure, essential for the economic welfare of a country, relies on an accurate understanding of the dynamics of such system. Recent studies have shown very complex structures in the traffic flow [1], [2]. This state is called the synchronized state, which has to be distinguished from a free flowing and congested state. The difficulty to understand the dynamics originates from the difficulty to relate the observed dynamics in speed and density to the underlying dynamics of the drivers behaviors, and the changes therein as function of the circumstances and driver motivation [3].

Simulations play an essential role in evaluating different aspects of the dynamics of traffic systems. As in most application areas, the available computing power is the determining factor with respect to the level of detail that can be simulated [4] and, consequently, lack of it leads to more abstract models [5]. To be able to afford more detailed situations, we looked how we could use the resources provided by for instance Condor[6], or the Grid [7].

Simulation and real world experimentation both generate huge amount of data. Much of the effort in the computer sciences groups is directed into giving scientists smooth access to storage and visualization resources; the so called middle-ware on top of the grid-technology. Yet, for a scientist seamless integration of the information from simulated data and measurements is the most important issue, the so called data-driven approach (see for instance [8]).

M. Bubak et al. (Eds.): ICCS 2004, LNCS 3038, pp. 771–778, 2004.
© Springer-Verlag Berlin Heidelberg 2004

Our department participated in the Grid-based Virtual Laboratory Amsterdam (VLAM-G) [9]. VLAM-G had as goal to hide resource access details from scientific users, and to allow scientific programmers to build scientific portals. These portals give access to the user interfaces for scientific studies: combinations of information gathering, processing, visualization, interpretation and documentation. Typical applications can be found in Astronomy, Earth Observation and High Energy Physics, Medical Diagnosis and Imaging, Food- and Bio-Informatics, as bundled in the 'Virtual Laboratory for e-science' [10].

In this article we show our experience with building our Virtual Traffic Laboratory as a data driven experimentation environment. This experience can be used as input for the future development of the Virtual Laboratory on other application domains.

2   VLAM-G Architecture

The Scientist is the person that actually is performing the studies. In a study often the same steps are repeated, as for instance testing a hypothesis on a certain dataset. Some steps can be quite time-consuming, so the Scientist can log-out from this study, prepare another study, and come back to inspect the intermediate results and perform another step of the study.

So, when the Scientist starts working with VLAM-G, there is support in the form of the information management system VIMCO and the run time system RTS [11]. VIMCO archives study, module and experiment descriptions [12], together with the application specific databases. The RTS takes care of scheduling, instantiating and monitoring the computational modules of an experiment. It makes thereby extensive use of Globus services, the actual standard in Grid computing.

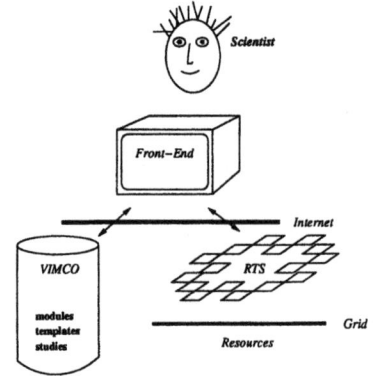

Fig. 1. The different systems for a study

3   The Front-End

The Scientist can use the front-end that is optimal for its domain. For complex system engineering, as traffic systems, we favor the Matlab environment. So, we have coupled a prototype of the RTS [13] with the Matlab environment. Here the RTS is used for the heavy computational tasks, while the Matlab environment is used for analysis and visualization of the results.

To be able to demonstrate the possibilities of Matlab as front-end, we have implemented a gateway routine, which allows the user to load VLAM-G modules, couple them, configure them and start them. We have hidden our gateway routine inside a user-defined block of Simulink. Simulink is an interactive, graphical, tool for modeling, simulating, and analyzing dynamic systems.

Further we used the Condor system to start up a cluster of jobs on several resources, including the Globus-resources at our site. The Simulink system was used to monitor the progress of the jobs, by monitoring the database where the results of the analysis and simulation are stored.

In the following figure one can see an example of such Simulink model. The first block generates a trigger when new data is available in the database, the second block then queries the database and updates a figure with the new data.

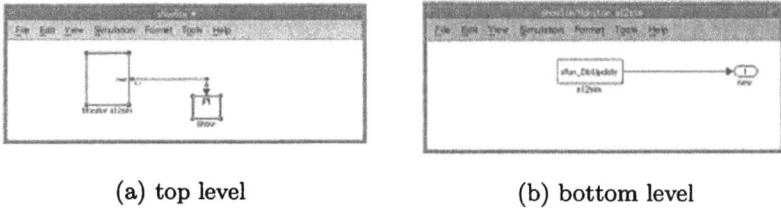

(a) top level   (b) bottom level

Fig. 2. The ShowSim Monitor in Simulink

Simulink is a part of the Matlab suite. It has an extensive library of predefined blocks, which perform operations on their input and generate output. In that sense they are comparable with the modules of VLAM-G. The difference is that these operations are performed as threads of the current Matlab process, on the current Machine, while at VLAM-G the modules are processes on remote resources. The VLAM-G concept of re-usable modules is perfect for the initial processing, analyzing and cleaning of the data, while the library of functions that Simulink has to offer is perfect to perform some final filtering of the results before they are visualized to the Scientist.

4   The Application

Traffic flow on the Dutch highway A12 is investigated for a wide variety of circumstances in the years 1999-2001. The location was especially selected for the absence of disturbing effects like nearby curvature, entrees or exits. This location has the unique characteristic that, although the flow of traffic is high, traffic jams are very sporadically seen. In this sense it is a unique measurement point to gather experimental facts to understand the microscopic structures in synchronized traffic states ([2]), which was not reported outside of Germany yet.

Previous research ([1]) has shown that three types of synchronized traffic can be distinguished:

- i stationary and homogeneous states, where the average speed and flow are nearly constant for several minutes
- ii stationary, non-homogeneous states, where the average speed is constant for a lane, but the flow noticeably changes.
- iii non-stationary and non-homogeneous states, where both average speed and flow change abruptly.

In addition, it is found that transitions from these states to free flowing traffic are rare, but between synchronized states are frequent.

This means that for understanding the microscopic structures in synchronized traffic states the relations between several aggregates of single vehicle measurements have to be made. Important aggregate measurements are for instance average speed, average flow, average density, headway distribution and speed difference distribution. The dynamics of these one-minute aggregates over 5-10 minutes periods are important for a correct identification of the state.

To facilitate the analysis of aggregate measurements over time we designed the following architecture:

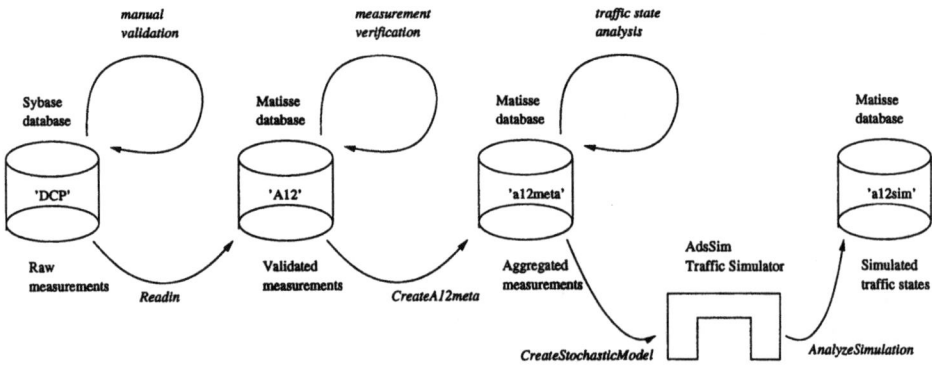

Fig. 3. The measurement analysis architecture

Along the A12 there was a relational database from Sybase that collected the measurements from two independent measurement systems. One system was based on inductive loops in the road, the other on an optical system on a gantry above the road. Although both were quality systems, some discrepancies occur between measurements due to different physical principles. Video recordings were used to manually decide the ground truth when the measurements were not clear.

After this validation process, the measurements were converted to an object oriented database from Matisse. This database was used to verify the quality of the measurement systems themselves. While the manual validation process was used to get the overall statistics of errors in the measurements, the object oriented database was used to analyze the circumstances of the measurement errors. Several hypothesis of underlying failure processes were raised, as for instance the characteristics of trailers that had higher changes to be characterized as an independent passage.

The validated measurements were used to generate the statistics that characterize the traffic flow. Different measurements-periods could be combined based on different criteria, for instance 'Flowing'/'Congestion', 'High-Density'/'Low-Density', weather conditions, fraction of heavy traffic, etc, etc. The right combination of criteria results in candidate traffic flow states. The statistics that are important to characterize the microscopic structure of the traffic flow are not only averages and standard deviations, but require for non-Gaussian distributions fits of complex probability density functions. Generating such statistics typically takes 20 minutes on a UltraSPARC-II workstation, which makes it worthwhile to store the results a separate database.

An example of such analysis is given in figure 4, where the average speed is given as a function of the flow (as percentage of the maximum flow) and the fraction of lorries (as percentage of the number of passages).

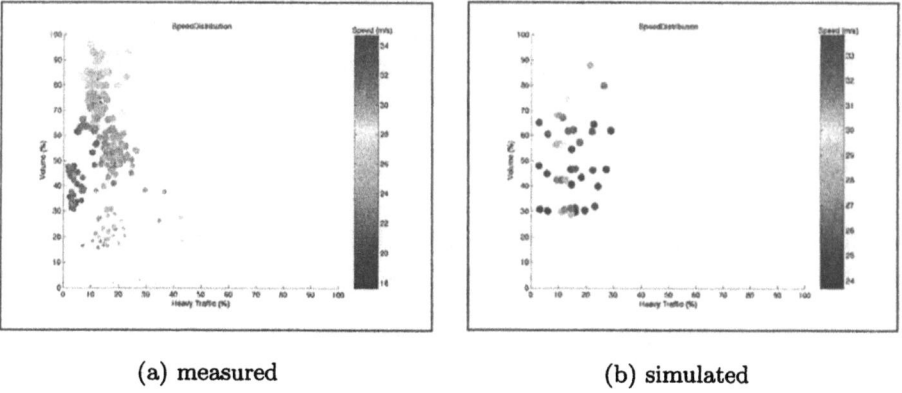

(a) measured   (b) simulated

Fig. 4. The average speed as function of the flow and the fraction heavy traffic

The average speed is indicated with a colorcode, red (top of the bar) indicates high speeds, blue (bottom of the bar) indicates low speeds. Each point indicates an aggregate over longer period (30-60 minutes), which are typically equivalent with a few thousand passages.

Combinations of measurement-periods that showed the same patterns on their aggregated traffic flow measurements over time were candidate traffic flow states. These aggregated measurements could be translated into the parameters of a microscopic traffic simulator, AdsSim [14], which is based on the microscopic Mixic model [15].

The characteristics of the simulated data were aggregated in the same way as the real data, and the resulting dynamics were compared to the original dynamics, to see if the model was complete (see figure 4). As one can see, the simulated points (each representing 5000 passages) are more homogeneous spread over the spectrum because one can ask for certain combination of parameters. Yet, the results are less to be trusted when one has to extrapolate far from actual measured parameter-combinations space. For instance, the average speed is unexpectedly high for Heavy Traffic percentages above 30%. This is due to the fact that this type traffic was only seen for low Volumes, when the fast lane is only sporadically used (with high speeds). When the flow increases, the fast lane is used more often, which gives this lane more influence on the average speed, and the speed-characteristics of this flow are extrapolated from the sporadic passages at low Volumes.

The *CreateA12meta* and *CreateStochasticModel* were originally Matlab-functions with more than 1000 lines of code. We converted those functions to standalone programs, which made it possible to run those functions in the background with the aid of the Condor software[6]. The latest versions of this software even make it possible to add Grid-resources to the pool with the *glide-in* technique [16].

The major advantage of our approach was to store all these meta-data in separate databases. This made it possible to start from the Matlab commandline a number of daemons, implemented in Simulink, which constantly monitor those databases and update the diagrams when new analysis results are ready.

5   Discussion

We have chosen this application, because of the complexity of both the measurement analysis and the traffic flow model. For instance, the Mixic model has 68 parameters in its traffic flow model [15], and most parameters are described as functions of single vehicle data such as lane, speed and headway. For AdsSim this resulted in 585 variables that can be adjusted to a specific traffic condition. Compare this with the 150 keywords in the standard application in molecular dynamics [17] in the UniCore environment [18].

To be able to calibrate such a model for a certain traffic state, the Scientist needs to be able to select characteristic subsets in the bulk of measurements, and visualize the dynamics of the aggregates in different ways. It is no problem that it takes some time to generate aggregates, as long as the Scientist is able to switch fast between diagrams of parameters and their dependencies as soon as the aggregates are ready. Storing the analysis results in a database solves this problem.

Typically, the Scientist can concentrate on a few diagrams (say three) at a time. The Scientist sees a certain pattern in the dependency, and can select measurements and simulation to add extra points to the diagram, to look if the pattern holds.

While processes at the background fill in the missing data-points, the Scientist starts the visualization of other dependencies, till an unexpected pattern appears. At that moment other subsets in the measurements become important. This means that new analysis has started up, and the decision has to be made to stop the current analysis.

In most cases this decision is negative, because the analysis of the previous characteristic is often quite far, and the Scientist wants to be sure how the unexpected pattern looks for that complete subset, even as there is a subset identified that should show the effect more clearly.

The key of our approach is that we don't see the Scientist as a user that repeats the same analysis repeatedly on the different datasets, but is an expert analyzing the problem from different viewpoints. These viewpoints are not known beforehand, and slowly shifting. The Matlab environment allows full control of a dataset, and facilitates different ways to search, fit and display dependencies. At the moment an interesting viewpoint is found, additional datapoints can be generated in the background, with an interface a high throughput system like Condor. The results are automatically displayed by monitoring the databases with meta-data via Simulink.

This approach differs from the approach of for instance [8], where Matlab is seen as inappropriate due to license policies and speed issues. By using high throughput systems in the background speed is no longer an issue. With its license the Matlab environment provides the Scientist directly a rich set of graphics and solvers, without the need to construct this functionality from home-made modules. Yet, both approaches do not exclude each other. In our view the programming effort can concentrate on often used algorithms, and optimize these algorithms into modules that can be executed in the background, while Matlab is used for direct analysis and visualization.

6   Conclusions

In this article we have introduced an experiment for the Virtual Traffic Laboratory. To aid the scientist, analysis results are stored in databases with aggregated data. This allows to repeatedly display the results from different viewpoints, where the scientist does not have to worry that too rigorous filtering will force him to do the aggregation again.

New aggregate data can be generated by exploring a dependency by performing new analysis on sets selected on different parameter-combinations in the background. This analysis can be performed seamlessly on both real data and simulated data. New data can be automatically displayed by adding monitors to the databases.

References

1. B.S. Kerner and H. Rehborn, "Experimental properties of complexity in traffic flow", Physical Review E, Vol. 53, No. 5, May 1996.
2. L. Neubert, et al., "Single-vehicle data of highway traffic: A statistical analysis", Physical Review E, Vol. 60, No. 6, December 1999.
3. Th. Jörgensohn, M. Irmscher, H.-P. Willumeit, "Modelling Human Behaviour, a Must for Human Centred Automation in Transport Systems?", Proc. BASYS 2000, Berlin, Germany, September 27-29, 2000.
4. K. Nagel, M.Rickert, "Dynamic traffic assignment on parallel computers in TRANSIMS", in: Future Generation Computer Systems, vol. 17, 2001, pp.637-648.
5. A. Visser et al. "An hierarchical view on modelling the reliability of a DSRC-link for ETC applications", IEEE Transactions on Intelligent Transportation Systems, Vol. 3: No. 2, June 2002.
6. Douglas Thain, Todd Tannenbaum, and Miron Livny, "Condor and the Grid", in Grid Computing: Making The Global Infrastructure a Reality, John Wiley, 2003. ISBN: 0-470-85319-0
7. I. Foster, C. Kesselman, "The Grid: Blueprint for a New Computing Infrastructure", Morgan Kaufmann, 1999.
8. Craig C. Douglas, "Virtual Telemetry for Dynamic Data-Driven Application Simulations"; Proc. ICCS 2003, Melbourne, Australia, Lecture Notes in Computer Science 2657. p. 279-288, Springer-Verlag 2003.
9. H. Afsarmanesh, et al. "VLAM: A Grid-Based virtual laboratory", Scientific Programming (IOS Press), Special Issue on Grid Computing, Vol. 10, No. 2, p. 173-181, 2002.
10. http://www.vl-e.nl
11. A. Belloum et al. "The VL Abstract Machine: A Data and Process Handling System on the Grid". Proc. HPCN Europe 2001, 2001
12. E. C. Kaletas, H. Afsarmanesh, and L. O. Hertzberger. "Modelling Multi-Disciplinary Scientific Experiments and Information". In Proc. ISCIS'03, 2003.
13. B. van Halderen, "Virtual Laboratory Abstract Machine Model for Module Writers", Internal Design Document, July 2000.
    See $http://www.dutchgrid.nl/VLAM - G/colla/proto/berry\_running/$
14. A. Visser et al.
15. Tampére, C. and Vlist, M. van der, "A Random Traffic Generator for Microscopic Simulation", Proceedings 78th TRB Annual Meeting, Januari 1999, Washington DC, USA.
16. James Frey et al. "Condor-G: A Computation Management Agent for Multi-Institutional Grids", Proceedings of the Tenth IEEE Symposium on High Performance Distributed Computing (HPDC10) San Francisco, California, August 7-9, 2001.
17. W. Andreoni and A. Curioni "New Advances in Chemistry and Material Science with CPMD and Parallel Computing", Parallel Computing 26, p. 819, 2000 .
18. Dietmar W. Erwin and David F. Snelling, "UNICORE: A Grid Computing Environment", in Lecture Notes in Computer Science 2150, p. 825-839, Springer-Verlag Berlin Heidelberg, 2001.

SAMAS: Scalable Architecture for Multi-resolution Agent-Based Simulation*

Alok Chaturvedi[1], Jie Chi[1], Shailendra Mehta[1], and Daniel Dolk[2]

[1] Purdue University, West Lafayette, IN, 47907, USA
{alok,chij,mehta}@mgmt.purdue.edu
[2] Naval Postgraduate School, Monterey, CA, 93943, USA
drdolk@nps.navy.mil

Abstract. Large scale agent-based simulation has become the focus of great interest from both academy and industry in resent years. It has been shown an effective tool for understanding a variety of complex systems, such as market economics, war games, and epidemic propagation models. As the systems of interest grow in complexity, it is often desirable to have different categories of artificial agents that execute tasks on different time scales. With the added complexity, the scalability of a simulation environment becomes a crucial measure of its ability in coping with the complexity of the underlying system. In this paper, we present the design of SAMAS, a highly scalable architecture for multi-resolution agent-based simulation. SAMAS is a dynamic data driven application system (DDDAS) that adopts an efficient architecture that allows large number of agents to operate on a wide range of time scales. It uses gossiping, an efficient broadcasting communication model for maintaining the overall consistency of the simulation environment. We demonstrate the effectiveness of this communication model using experimental results obtained from simulation.

1 Introduction

Large-scale agent-based simulation has become an effective modeling platform for understanding a variety of highly complex systems and phenomena such as market economics, war games, and epidemic propagation [1]. What is becoming increasingly clear from the deployment of such environments is that they are powerful media for integrating models of widely differing aggregation and granularity. This multi-resolution property of agent-based simulations has already been demonstrated in the spatial dimensions and in the emergence of multi-agent systems, which support a diversity of agents playing different roles and exhibiting different behaviors.

What has been slower in forthcoming is the support of temporal multi-resolution models. Specifically, as systems and their associated models grow in complexity, it becomes increasingly desirable and necessary to have different categories of artificial agents that execute tasks on different time scales. Consider, for example, a simple network in which an agent that simulates network routers needs to operate on the scale of milliseconds while an agent that simulates a mail server, which depends on

---

* This research is funded by the National Science Foundation grant # ACI-0325846.

M. Bubak et al. (Eds.): ICCS 2004, LNCS 3038, pp. 779–788, 2004.
© Springer-Verlag Berlin Heidelberg 2004

the routers to get mail, need only operate on the scale of minutes. Such requirements present great challenges to conventional agent-based systems in terms of temporal scalability. These systems typically dictate a single time resolution based on the requirement of the agents running at the highest resolution (lowest granularity, as in the router agents in the above example). Essentially, such systems now run on a single time resolution. Agents are either idle most of the time, causing considerable waste of resources, or they have to be explicitly tracked and activated by the system at different time intervals by the simulation environment, causing excessive overhead for the system. While these strategies are simple to implement, and work well for systems of largely uniform agents, they do not scale to a large number of diverse agents operating on a wide range of time resolutions. In this paper, we propose the design of SAMAS, a dynamic data driven application system (DDDAS) for multi-resolution simulation that uses the gossiping communication model to allow agents to maintain consistency across different time resolutions. Consistency between layers is maintained through a "sample and synchronize" process. In this process, an agent running asynchronously at a higher level can dynamically sample the lower level, which may be another agent (human or artificial), system, component, instrument, and/or sensor, and adjust its parameters.

We approach this problem by first discussing in general terms the power of multi-resolution, agent-based simulation to integrate organizational processes and models at many levels of scope and time (Section 2). We then focus our attention on the problem of temporal multi-resolution and discuss a preliminary design of SAMAS (Section 3.2), which relies upon a gossiping algorithm to coordinate agents operating in different time domains (Section 4). Using simulation, we provide results showing the effectiveness of the gossiping algorithm as a communication model for maintaining global consistency of agents in a simulation environment (Section 5). We summarize our results within the larger context of multi-resolution systems, and briefly present a plan for continuing this venue of research.

2   Multi-resolution, Agent-Based Simulation in Support of Organizational Model Integration

Agent-based simulation with robust multi-resolution capabilities offers the potential for coalescing the entire portfolio of processes and models upon which an organization relies for policy and decision-making. Figure 1 shows a conceptual notion of an integrated simulation environment, which takes a layered approach to organizational models. In this framework, models exist at different levels of organization with the Organizational Model being the highest level in terms of generality. The Organizational layer drives the Business Process layer which drives the Workflow layer which, in turn, depends, as all the models do, upon the underlying Infrastructure layer. Each of these layers can be modeled and then solved for different scenarios using the SEAS [2, 3] agent-based simulation engine. The resulting goal is to be able to tie these layers together in an overarching virtual environment, which captures the overall organizational behavior to an acceptable degree of verisimilitude.

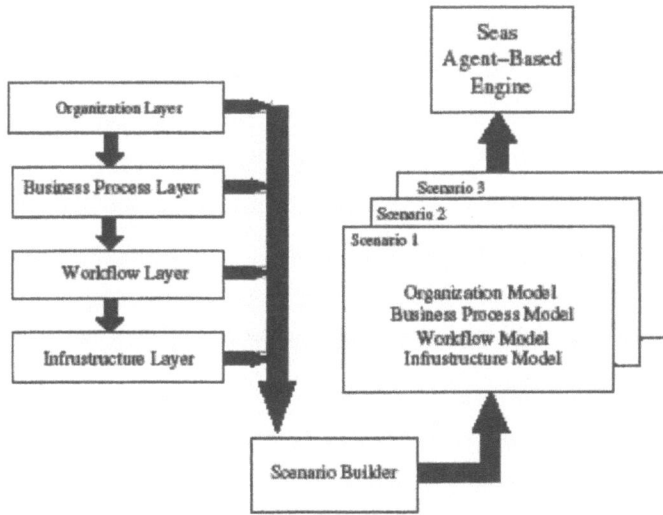

Fig. 1. Integration of Multi-resolution Organizational Processes and Models

One of the critical problems in effecting the kind of integration envisioned in Figure 1 is how to handle the time dimension in simulations. The timeframes for organizational processes vary widely in scope. At the strategic level, the focus may be on the next few months or years, at the tactical level days or weeks, and at the operational level, minutes or hours. This disparity in time resolution has a large impact on an organization's modeling processes, which tend to focus model development on a particular domain within a specific timeframe, for example. Although this may be locally efficient, it is often globally suboptimal. The fragmentation of models by domain and timeframe interferes with an organization's ability to integrate and reuse these models, which then can result in the inability to plan and operate to the fullest extent possible. We address the problem of time resolution below by showing an architecture in which agents existing at different time levels in an integrated multi-level model can communicate effectively with one another without consuming inappropriate amounts of overhead.

3   SAMAS Architecture

3.1   Naïve Design for Multi-resolution Agent-Based Simulation

In most of the complex systems that we study, there are thousands, if not millions, of input elements that may affect the outcome of the system in any number of different ways. Agent-based simulation attempts to capture this complexity by using a large number of artificial agents, each of which plays the role of one or more of the elements in the real system. While existing agent-based simulation systems, given a sufficient number of agents, can effectively mimic the behavior of real world systems, they generally do not effectively take into consideration the temporal relationship

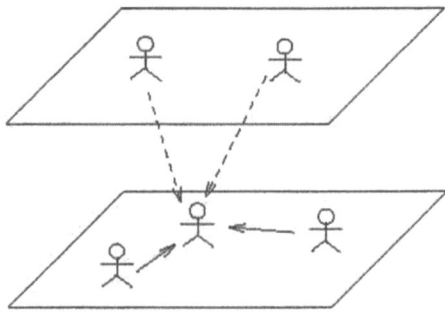

Fig. 2. Multi-resolution simulation based on naïve design

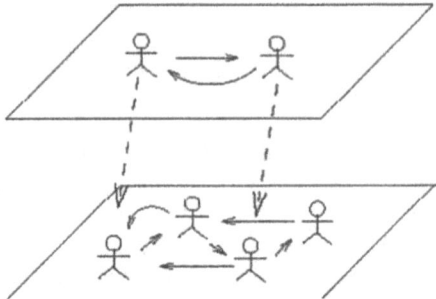

Fig. 3. Multi-Resolution Agent-based Simulation. Solid lines represent gossiping among agents at the same level. Dashed lines represent sampling among agents at adjacent levels.

among the different agents. As a result, the entire simulation environment uses one global clock time interval for the execution of all agents in the systems. This design, however, does not scale well to large number of heterogeneous agents and may lead to gross inefficiencies in simulation execution. In many sufficiently complex systems, there is often a hierarchy of different types of agents that need to operate on different time scales. The agents that operate at a lower resolution (slower) often have to rely on agents that operate at higher time resolution (faster) for information. To implement such a scenario using the naïve design, we have to assign a central location, e.g. an agent, at the lower resolution level to gather all the information needed by agents at the upper resolution level. The agents at the upper level periodically gather information from the designated party. This scenario is shown in Figure 2. There are two major drawbacks associated with this design:

1. Since there is only one global clock, most of the time the agents on the upper level are doing nothing but checking for information in the lower level. These activities waste the resources in the simulation environment.

2. The agent that is responsible for gathering information at the lower level becomes the single point of contention as the number of agents increase.

3.2 SAMAS Design

The design of SAMAS is motivated by the need to build an agent-based simulation system that is capable of scaling up to millions of artificial agents running at multiple time resolutions. The SAMAS simulation environment no long requires a single uniform global clock. Instead, SAMAadopts the multi-resolution time interval concept that enables agents to operate at anytime resolution. As shown in Figure 3, the agents on the same level maintain global state information by using gossiping. The gossiping communication model adapted in SAMAS works as the simple steps shown in Figure 4. Essentially, the approach enables the agents at the higher levels operate asynchronously by sampling the agents at lower level and removes the point of contention in previous design.

Gossiping allows the maintenance of global information without imposing aggressive network bandwidth requirement on the underlying systems. By maintaining the consistent global information among all agents at the same level of time resolution, agents at the higher level can periodically contact any one of the agents at the lower level to obtain necessary information. The agents can safely assume with a high level of confidence that this acquired information, resulting from gossiping communication among agents, truly reflects the actual state information within a bounded degree of error.

---

1. **Initialize**, for every agent i, the set $A_i \leftarrow \{i\}, \forall 0 < i < n$

2. **Loop** until $A_i = \{0, 1, ..., n\}, \forall 0 < i < n$

3.   Agent i randomly chooses a set agents, $S, i \notin S, |S| = N_s < n$ and send A_i to each of them

4.   Agent i randomly receives a set $R, |R| \leq N_R < n$ of messages addressed to it, so that $A_i = A_i \cup A_j, \forall A_j \in R$

5. **End loop**

Fig. 4. Outline of the Gossiping Algorithm Used in Simulation

4 Related Work

The gossiping problem, which is also known as all-to-all broadcast, involves each node in a network of n nodes sending a unique piece of information to the rest of the n − 1 nodes. The gossiping problem has long been the subject of research in various contexts [4]. Researchers investigate a number of different metrics for the solution of the problem. For a complete network of n nodes, the lower bound on time complexity (the number of communication steps) is established to be O(log n) [5] (all log are of base 2 in this paper unless otherwise specified). The minimum number of calls needed for the operation is proved to be 2n − 4 for n > 4 [6]. Czumaj et. al.[7] shows the trade-off relationship between the time and cost of the gossip communication model, where the cost is measured by either the number of calls or the number of links used in the communication. Various strategies have been purposed to perform gossiping on different network architectures with different lower bounds on the performance metrics [8, 9]. In particular, these works show that gossiping can be carried out with desirable performance properties on hypercube architecture. Krumme and Cybenko [10] further show that gossiping requires at least $\log_\rho N$, where ρ is the golden ratio, communication steps under the weakest communication model(H1), in which the links are half duplex and a node can only communicate with one other node at each communication step. Gossiping algorithms are at the heart of many fundamental parallel operations, such as matrix transposition and fast fourier transformation [8]. In SAMAS, we implement a simple randomized gossiping algorithm. The details of the operations are discussed in detail in Section 3.2

5 Experimental Results

Experiments have been conducted using simulation to measure the performance of the gossip algorithm on a network of agents exhibiting various sizes. We define the following parameters used in the simulation.

1. Number of Communication Rounds is the number of steps it takes to finish the all-to-all broadcast. This is the performance metric measured in the experiments.

2. The number of messages to send (NS) is the upper bound on the number of message an agent can send at each communication round. It represents the outbound bandwidth of an agent.

3. The number of messages to receive (NR) is the upper bound on the number of messages of an agent can receive at each communication round. It represents the inbound bandwidth of an agent.

When the number of agent is small, some irregularities are observed (the zigzag in Figure 5). These irregularities are attributed to the randomness nature of the simulation. As the number of agent in the network increase, the run time converges nicely to a continuous surface (Figure 6 and 8).

This performance metric is measured in three different agent network sizes of 128, 256, and 512 agents respectively. We assume overlay networks, which means the networks are fully connected. The simulation is carried out using the procedure shown in Figure 4. We devote the remainder of this section to the discussions and observations of the simulation results.

The number of communication rounds is associated with the run time of the gossiping algorithm. We demonstrate the number of communication round as a function of the outbound bandwidth and the inbound bandwidth of agents in the network (Figures 5 6 and 8). For illustrative purposes, Figures 6 and 7 provide alternative views of the same data in both 3D and 2D.In all three networks with different sizes, we observe the following patterns:

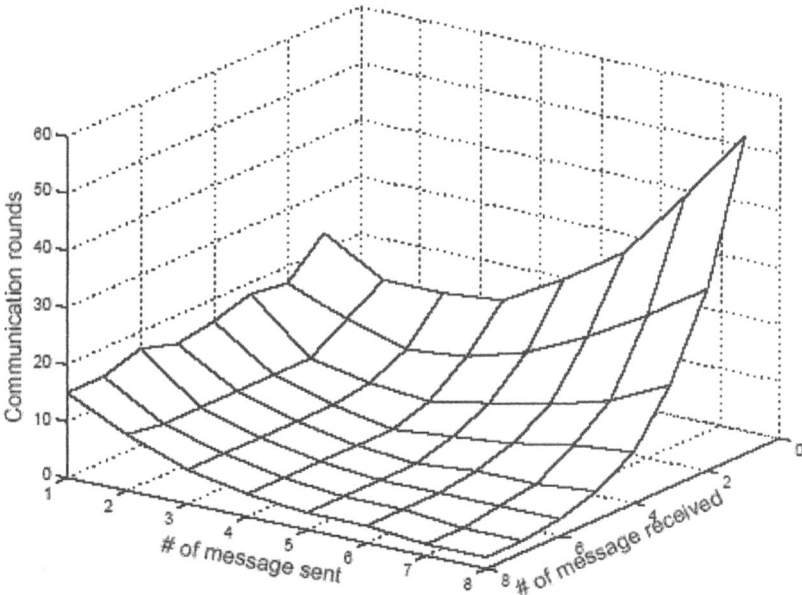

Fig. 5. Network size=128. Number of communication rounds as a function of both the number of messages sent and received. Both number of messages sent and received axis are log-scaled

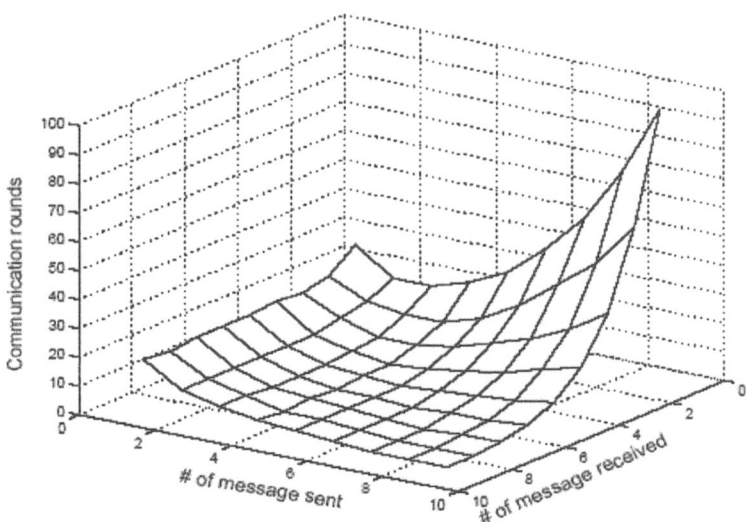

Fig. 6. Network size=256. Number of communication rounds as a function of both the number of messages sent and received. Both number of messages sent and received axis are log-scaled

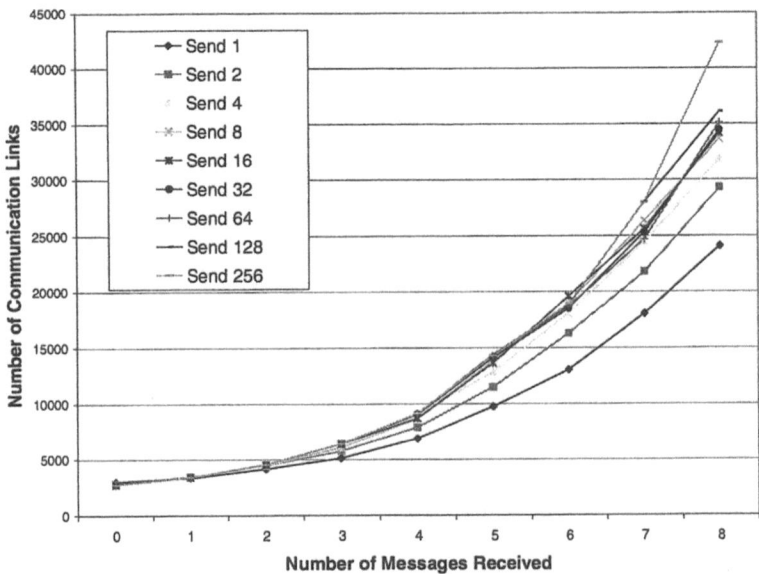

Fig. 7. Two dimensional view of the data shown in Figure 6

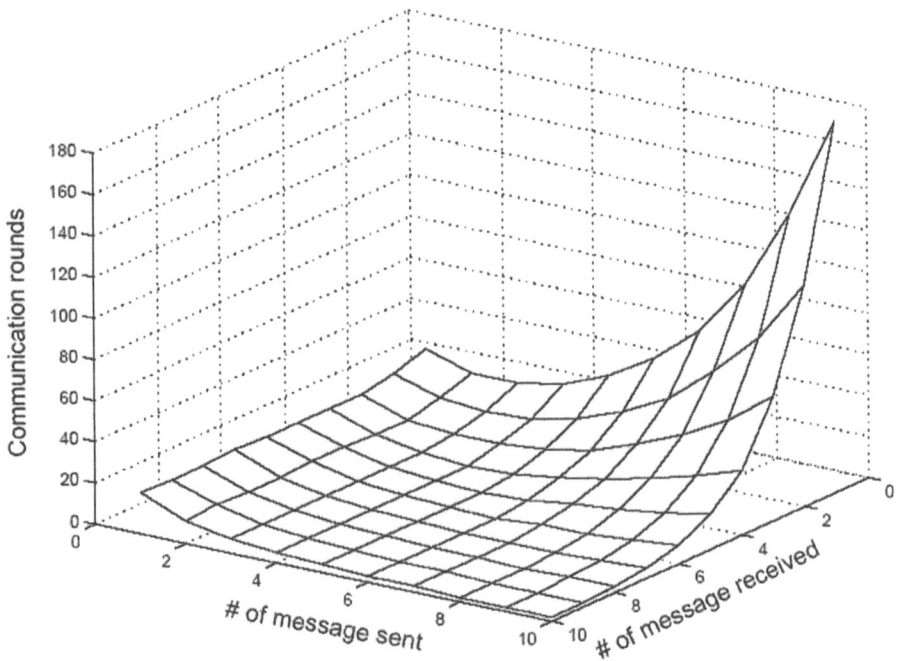

Fig. 8. Network size=512. Number of communication rounds as a function of both the number of messages sent and received. Both number of messages sent and received axis are log-scaled

- Only relatively small network bandwidth is required to perform gossip optimally, or very close to optimal. The optimality referred to here is O(log n) steps, where n is number of agents. In fact, the simulation on average finishes in 7 steps when agents both send and receive 4 messages in each step in $2^7=128$ agent case, and similarly in other cases.

- A network of agents that has more balanced inbound and outbound bandwidth results in better performance.

- The performance is more sensitive to inbound bandwidth. This asymmetric property can be demonstrated by observing the large increase in time when the agents are allowed to send large number of messages but only receive a small number of them, whereas the inverse is not true.

Our empirical results indicate that the randomized gossiping algorithm adopted in SAMAS can operate at, or very close, to the optimal level shown in previous theoretical works [5,11]. In the context of the multi-resolution simulation model presented in this paper, these results demonstrate that the gossiping algorithm can be effectively used for the purpose of maintaining global state information across agents on the same time interval. For example, if there are 1000 agents of type I operating on a 10 millisecond time interval and several type II agentsoperating on a 1secondtime interval, the worst case would require type I agents about 10 steps (100 milliseconds) to update any global information needed by type II agents. This translates into a minimum refresh rate of 10 times/interval for the type II agents, which ensures that type II agents can almost always get the most updated information.

6 Conclusion and Future Work

In this paper, we have discussed the importance of multi-resolution simulation for integrating models and processes at all levels of an organization independent of time granularity. We have proposed a preliminary design for SAMAS, a highly scalable dynamic data driven application system (DDDAS) for multi-resolution, agent-based simulation. At the core of SAMAS, we use a gossiping algorithm to efficiently maintain global information among the agents, which operate at different time resolutions. We demonstrate through simulation results that gossiping can be used to implement an architecture that allows large numbers of agents to operate on a wide range of time scales. Should our future experiments with large-scale simulations confirm the feasibility of the SAMAS gossiping approach, we will have taken a significant step in creating a virtual environment for integrating and coordinating mission critical organizational models.

References

1. Chaturvedi, A., Mehta, S., Drnevich, P., "Live and computational experimentation in bio-terrorism response". In *Dynamic Data Driven Applications Systems*, Kluwer Academic Publishers, (2004)

2. Chaturvedi, Mehta, Dolk, Ayer, "Artificial labor market", *European Journal of Operations Research*, (2003)
3. Chaturvedi, A.R., Choubey, A.K., Roan, J.S., "Active replication and update of content for electronic commerce". *International Journal of Electronic Commerce*, 5, (2003)
4. Hedetniemi, S.M., Hedetniemi, T., Liestman, A.L., "A survey of gossiping and broadcasting in communication networks", *NETWORKS*, 18, (1988), 319–349
5. Bavelas, A. "Communication patterns in task-oriented groups", *J. Acoust. Soc. Amer.*, 22 (1950), 271–282
6. Baker, B., Shostak, R., "Gossips and telephones", *Discrete Mathematics*, 2 (1972), 191–193
7. Czumaj, A., Gasieniec, L., Pelc, A., "Time and cost trade-offs in gossiping", *SIAM Journal on Discrete Mathematics*, 11 (1998), 400–413
8. Grama, A., Kumar, V., Gupta, A., Karypis, G. "Introduction to Parallel Computing, An: Design and Analysis of Algorithms", 2nd Edition. Addison-Wesley (2003)
9. Krumme, D.W., "Fast gossiping for the hypercube", *SIAM J. Comput.* 21 (1992) 365–380
10. Krumme, D.W., Cybenko, G., Venkataraman, K.N. "Gossiping in minimal time", *SIAM J. Comput.*, 21 (1992) 111–139
11. Landau, H. "The distribution of completion times for random communications in a task-oriented group", *Bull. Math.Biophys.*, 16 (1954) 187–201

Simulation Coercion Applied to Multiagent DDDAS

Yannick Loitière, David Brogan, and Paul Reynolds

Computer Science Department
University of Virginia, Charlottesville, VA 22901, USA
{ycl2r,dbrogan,pfr}@virginia.edu

Abstract. The unpredictable run-time configurations of dynamic, data-driven application systems require flexible simulation components that can adapt to changes in the number of interacting components, the syntactic definition of their interfaces, and their role in the semantic definition of the entire system. Simulation coercion provides one solution to this problem through a human-controlled mix of semi-automated analysis and optimization that transforms a simulation to meet a new set of requirements posed by dynamic data streams. This paper presents an example of one such coercion tool that uses off-line experimentation and similarity-based lookup functions to transform a simulation to a reusable abstract form that extends a static feedback control algorithm to a dynamic, data-driven version that capitalizes on extended run-time data to improve performance.

1 Introduction

Dynamic, data-driven application systems (DDDAS) emphasize the run-time flexibility of data/control systems dynamically composed from complex combinations of sensors, networks, computational resources, simulation software, and human-in-the-loop interaction. Dynamic composition of DDDAS changes the number of interacting components, the syntactic definition of their interfaces, and their role in the semantic description of the entire system. These unpredictable run-time reconfigurations pose significant challenges to the straightforward reuse of complex scientific simulations for DDDAS.

For simulation designers and developers, anticipation of and provision for all possible contexts and uses of a simulation is generally unattainable. Except under very constrained circumstances, reusing simulations without design or code modification has proven elusive. For example, a simulation may be unable to perform its desired role in a DDDAS due to decisions made as early as when an underlying scientific model was selected for its design or as late as when run-time bindings to memory structures capped the maximum amount of data permissible. Despite the challenges of building reusable simulations, additional research is justified by the many forms of benefits: employing a program in an alternative context, composing a program with others to create a new system with greater objectives, combining a program with another to broaden the levels of

M. Bubak et al. (Eds.): ICCS 2004, LNCS 3038, pp. 789–796, 2004.
© Springer-Verlag Berlin Heidelberg 2004

abstraction represented, and modifying a DDDAS to exhibit such properties as better performance or robustness. We are developing an approach to simulation transformation, COERCE, that tackles the immediate challenges of making simulations work more effectively in existing dynamic, data-driven applications and pursues the long-term challenges of specifying how today's simulations should be designed such that they can easily adapt to the unpredictable needs of tomorrow's DDDAS.

COERCE has the potential to increase the flexibility of simulation components comprising DDDAS. COERCE consists of two elements: coercion and coercibility. Coercion represents a process of transforming a simulation, using a human-controlled mix of semi-automated analysis and optimization, to meet a different set of requirements than those for which the simulation was originally designed. Through coercion, preexisting (legacy) simulations can be transformed to adapt to new or missing data sources, to change fidelity, and to change their roles in the larger data-driven system. Coercibility is a simulation property whereby designer knowledge is documented within the simulation to support future coercion. In ongoing research related to that presented here, coercibility techniques are being used to define necessary simulation metadata, including mechanisms for capturing and representing this data, so that coercion can more easily perform automated analysis and simulation transformation.

Simulation coercion and coercibility are ambitious goals that will immensely benefit the DDDAS enabling technology requirements. Not only is there great breadth in the contributing fields within computer science (software engineering, programming languages, computational science, etc.), but contributions to these goals will come from mathematics (optimization, sensitivity analysis, modeling) as well as from science and engineering. This paper explores one technique of many that must be studied in the context of simulation coercion: simulation transformation through off-line experimentation and similarity-based lookup functions. A simulation of robotic soccer serves as a complex, low-level simulation that must be transformed to interact with a high-level strategic planning algorithm. The transformation process we develop for this problem is data driven and is intended to be applicable to many similar transformations that arise at run time in dynamic applications.

2   Background

In its current form, the process of transforming a simulation to adapt to dynamic run-time data streams is primarily a manual process due to the complex logical interactions and engineering constraints that dictate a simulation's form and function. Research in the simulation reuse community demonstrates that code standardization, functional abstraction, and procedural documentation are tools that facilitate the reuse process. These techniques share a common belief that human creativity and insight are among the scarcest resources, but they differ in how and where they bring human skills to bear on critical issues. A growing number of research projects are demonstrating how human-guided, data-centric abstraction processes can re-represent simulation performance to suit new run-time needs.

Davis and Bigelow [2] argue that an abstraction mechanism called motivated metamodels can support simulation transformation. The code that defines a simulation is abstracted to numerical and functional models that generate the same simulation output. Consider, for example, how the stochastic models of queuing theory can replicate the automobile flow rates of a high-resolution traffic simulation. Not only is the queuing model a simplified form of the original simulation, it also provides an intuitive way to be reused for scenarios where the mean automobile arrival rate is varied (perhaps a capability not easily generated through the modification of the multiagent simulation directly). No general-purpose process exists to create motivated metamodels, but the authors outline the opportunities and risks of their use.

Grzeszczuk et al. [4] introduce a specific example of simulation abstraction with their Neuroanimator. Their technique demonstrates that a computationally intensive rigid body simulator can be substituted with a trained neural network. This technique requires repeated execution of a simulation under different initial conditions in order to accumulate the database of performance data required for training, but this offline investment is recouped with improved run-time performance. Although no trace of the scientific logic behind rigid body dynamics remains, the neural network representation executes more quickly and, because it is analytically differentiable, it enables the use of gradient descent search in run-time optimal controllers.

Additional exploration of simulation transformation has been conducted by those studying simulation coercion. Through code modification, Drewry et al. [3] demonstrate a semi-automated process that uses user-guided numerical optimization to retune simulation performance to operate in conditions different from those for which it was originally designed. Waziruddin et al. [6] further formalize simulation coercion as a formal process by constructing a formal language describing its execution and reasoning about provable outcomes in dynamic scenarios. Carnahan et al. [1] contribute to the specification of the coercion process by outlining the roles of simulationists and subject-matter experts and by producing a suite of software tools to support their collaborations. The systems produced through these research efforts are suited to run-time transformation, but currently lack a unifying theory of system transformation that serves the broad community of simulation and designers in DDDAS.

3   Simulation Transformation

This paper studies a transformation strategy for addressing misalignments between data sources. For example, suppose a DDDAS user desires to compose two simulations such that the output of one becomes the input of another. Figure 1 depicts the data relationship between the two simulations. In particular, note that the exact data required in the input vector i_2 must be contained within the output vector o_1 in order for the computational sequence to complete in a seman-

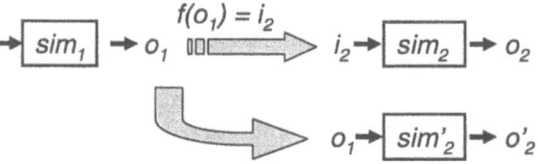

Fig. 1. In the top row, the output from simulation$_1$ aligns with the input to sim_2 to support simulation composition. If no mapping from o_1 to i_2 exists, sim_2 is mapped to a new version, sim_2' that accepts o_1 as input and generates output, o_2'.

tically meaningful way. If the exact data required for i_2 are not in o_1, a mapping function could be created to synthesize the required data from o_1: $f(o_1) = i_2$. With the assistance of a domain expert, such mapping functions may be feasible solutions. If the data in o_1 represents sensor data in the form of units/second and the data in i_2 requires units/minute, a mapping function could simply scale the output data by 60 to create the appropriate input data. Frequently, however, the mismatch between a simulation's input and the available data sources will be more significant.

To integrate two simulations with irreconcilable misalignments between their input and output data, we perform more substantial mappings. Figure 1 depicts a scenario where two simulations must work in a series such that the input to the two-simulation sequence, i_1 cannot be changed and the output from the sequence, o_2, is fixed as well. Because no effective mapping function from o_1 to i_2 exists, we instead consider mapping sim_2 to a new version, sim_2' such that the input to sim_2' is o_1 and the output, o_2', matches the requirements met by o_2. The method we use to accomplish this simulation transformation is data driven.

Our data-driven simulation transformation technique exercises sim_2 in its native environment in order to record its behavior for future abstraction. sim_2 is executed with many input instances of the form i_2. The specific instances of i_2 used during this exploration stage are guided by a user or the run-time execution to reflect conditions the DDDAS will subsequently encounter. Because the accuracy of sim_2' depends on the data generated during this exploration, the sampling of the space represented by i_2 is very important. If the simulation is well behaved in some areas of state space and chaotic or nonlinear in others, the data sampling must capture these behaviors and not the potentially infinite variety of less-interesting ones.

The exploration stage produces a database consisting of mappings from i_2 to o_2. These are point samples of the true functional relationship, $sim_2(i_2) = o_2$, encoded by the second simulation. Because the goal of this transformation process is to produce an alternative functional relationship, $sim_2'(o_1) = o_2$, we must create sim_2'. As an initial step, the data representing i_2 in the database is mapped into a format, i_2', that matches the requirements of output data, o_1. This mapping algorithm is typically easier to describe than its inverse. The transformed i_2' input data is compatible with the o_1 output data from sim_1 and we thus have a database of point samples that demonstrate the mapping sim_2 applies to an input derived from o_1. To be most useful, sim_2' must be more than a database lookup function and must be converted to a tuned numerical model (neural network, radial basis functions, wavelets) possessing the ability to interpolate and extrapolate the data samples, in a manner semantically consistent with the behavior of sim_2'.

We note that this parameter-mapping approach is inadequate for many cases of simulation incompatibility. For example, no amount of parameter mapping can create new degrees of freedom in sim_2' or force the frequency of data output from sim_1 to increase in order to meet sim_2's temporal requirements. The simulations will have to adjust to the run-time data properties. Furthermore semantic incompatibilities between simulations require more substantial alterations to unite their data paths. These more complex simulation transformations are currently being studied by the COERCE community.

4   Application to Soccer

We will demonstrate simulation transformation in a DDDAS created to implement a physical simulation of robotic soccer. Simulated robotic soccer teams are showcased each year in the RoboCup competition. Recent RoboCup games demonstrate a high degree of human-like intelligence by the autonomous players. Each team is composed of eleven physically simulated players, all of which are controlled during the game by an autonomous control algorithm that specifies unique player behaviors such as protecting the ball or kicking the ball up the field. The conventional player control algorithms make little use of run-time data about their opponents' behaviors, primarily because each player only has access to local information about the opponent-player locations. We seek to improve the performance of one of these conventional controllers by infusing additional data at run time that contains the locations of all the players.

We propose to improve performance by using the additional run-time data to generate predictions about future player positions. The existing soccer simulator specifies where players will be after one timestep, but it cannot accurately predict player positions further into the future. Due to the aforementioned complexity of transforming a simulation through source code modification, we interface the existing soccer simulation with its data-driven version using our data mapping approach.

The run-time data utilized by the data-driven version of the soccer simulator is obtained from a virtual camera simulation (sim_1 from our earlier example). The virtual camera is observing a simulated soccer game from a bird's-eye position above the field and outputting virtual snapshots (o_1). The soccer simulator (sim_2) must adapt to this new data and improve the quality of its agents' actions (o_2) by specifying where the players will be after one timestep and predicting their locations further into the future.

The DDDAS is a composition of sim_1 and sim_2 and their data are incompatible. The input to sim_2 is the position, orientation, and velocity of all 22 players and the ball as well as additional state variables representing each player's energy level, perceptual model, and locomotion (sprinting, turning, stopping) abilities. The output from sim_1 is a low-resolution image created by quantizing the soccer playing field and rendering colored 2-D Gaussians centered about the 22 players and the ball. Each grid cell of the quantized soccer field is represented by a pixel of the image. The final image of the field is rendered by assigning players of Team A, Team B, and the ball the colors red, green, and blue respectively. The amount of red, green, and blue each entity contributes to a pixel is summed according to the distance between the player/ball and the pixel center.

Because sim_2 cannot utilize the data generated by sim_1 in a straightforward manner, we pursue mapping sim_2 to sim_2' according to the method described in the previous section. After the mapping, sim_2' will be able to utilize the run-time data from sim_1 and output predictions of the players' future positions. To create the mapping, the positions of the soccer players are stored after every timestep of a simulated soccer game. Each intermediate game state serves as the simulation output of the previous timestep and the simulation input for the next timestep. This accumulated database of game performance provides the training data required to tune sim_2'. Instead of providing an exact substitution of sim_2's functionality, the transformed sim_2' we wish to build for this example will map the image of player positions output by sim_1 at one moment to a new state many timesteps in the future in order to provide the desired predictive ability.

The output from sim_1 is semantically incompatible with the input required by sim_2. Because we are unable to transform the output from sim_1 to match the format required by sim_2, we instead transform sim_2 so it accepts the output of sim_1 as input. The database of game states that we created serves to characterize the behavior of sim_2 and we map each state to a format matching the output of sim_1 by using the camera simulation to create the corresponding image for each. Instead of storing data from sim_2's native data format, the database now describes sim_2's execution behavior as a mapping from one image of game state to subsequent images.

4.1 Building sim_2'

The image database represents the performance of sim_2' in specific circumstances, but to be useful at run time in the DDDAS, sim_2' must produce correct outputs for inputs that are not in the database. For a unique input, i, sim_2' produces output in two stages: a matching algorithm uses a similarity metric to first compare i to all images in the database and then a prediction algorithm constructs o_2' as a function of i's similarity to the database frames and their outputs. Using this process, the transformed simulation, sim_2', will regenerate the mappings contained within the database when provided with identical inputs and will approximate the mapping for other inputs.

All the images stored in the database are represented by $\mathcal{I}$. Each image of $\mathcal{I}$ is an element of the set of all images, $\mathbf{I}$. The ordering of the image sequences in the database is preserved to maintain time-dependent frame correlations. The matching algorithm compares the camera output image to all the images in the

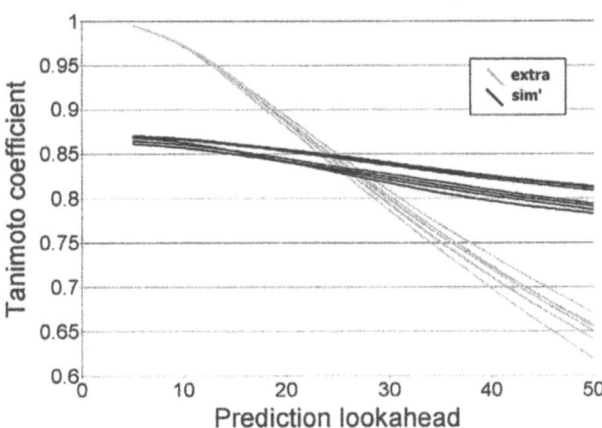

Fig. 2. This graph plots the prediction performance of the data-driven predictor (bold) and the kinematic extrapolator (grey) for seven experiments. The line for each of the seven experiments is generated by plotting the prediction lookahead value (δt) against the average of the similarity measure across all images in that test game log. A perfect score of 1.0 indicates an exact match between the predicted images and the actual future image from the test data.

database and evaluates the similarity of each pair by assigning a match value in the range $[0, 1]$. The similarity $\mathcal{T}_{pq}$ between two images $p, q \in \mathbf{I}$ is measured by computing the Tanimoto coefficient [5] between the vector representation of each image:

$$\mathcal{T}_{pq} = \frac{\vec{P} \cdot \vec{Q}}{\vec{P} \cdot \vec{P} + \vec{Q} \cdot \vec{Q} - \vec{P} \cdot \vec{Q}} \tag{1}$$

The measured Tanimoto coefficients are converted into weights, w, that map exactly to $[0, 1]$ in order to better distinguish the most similar images and to further reduce the contribution of low-similarity matches:

$$u_{pq} = \mathcal{T}_{pq} - \min_{q}(\mathcal{T}_{pq}) \tag{2}$$

$$w_{pq} = \left(\frac{u_{pq}}{\max_{q}(u_{pq})}\right)^2 \tag{3}$$

After all images in the database have been assigned a weight, the output image can be computed through a weighted average of all the images.

4.2 Experimental Results

To evaluate the effectiveness of our data-driven prediction technique, we used sim'_2 to predict the game state δt timesteps in the future for every frame of a new soccer game. We constructed sim'_2 from the 6,000 state vectors of player positions obtained during the execution of one simulated soccer game. We use the Tanimoto similarity measure to compare this predicted output to the actual state to determine the accuracy of the transformed simulation. As a reference technique, we constructed a baseline kinematic extrapolator that computes an output image based on the velocities and positions of the player and ball. Because the simulated soccer players are physically simulated, the kinematic extrapolator should perform well for small δt values, but it will fail to predict the accelerations and changes in direction caused by the dynamic nature of the game.

Figure 2 demonstrates the comparison of our data-driven predictor to the kinematic extrapolator in seven different experiments. The length and width of the quantized soccer field image in each is 40 and the number of frames simulated by the two systems, δt, ranges from five to 50. Although the kinematic extrapolator performs better for small δt values, its performance degrades rapidly as δt increases and the data-driven predictor consistently outperforms it for values of δt greater than 26.

5   Conclusion

We have presented a semi-automated method for transforming – coercing – simulations to meet the new requirements that arise at run time in dynamic, data-driven applications. We have discussed an example from RoboCup where we have applied ideas from COERCE to effect data alignment and capitalize on the infusion of new data. We have experienced some encouraging success in the transformation of our position-predicting simulation, as we have reported here. Further, our study of simulation transformation has provided not only another example demonstrating the viability of COERCE-like transformations in data-driven applications, but it has also provided us with insights into how to further develop COERCE technology.

References

1. J. Carnahan, P. Reynolds, and D. Brogan. Semi-automated abstraction, coercion, and composition of simulations. In *Interservice/Industry Training, Simulation, and Education Conference*, 2003.
2. P. Davis and J. Bigelow. Motivated metamodels. In *Proceedings of the 2002 PerMIS Workshop*, 2002.
3. D. Drewry, P. Reynolds, and W. Emmanuel. An optimization-based multi-resolution simulation methodology. In *Winter Simulation Conference*, 2002.
4. R. Grzeszczuk, D. Terzopoulos, and G. Hinton. Neuroanimator: Fast neural network emulation and control of physics-based models. In *Proceedings of SIGGRAPH '98*, pages 9–20. ACM Press, July 1998.
5. T. Tanimoto. Internal report. In *IBM Technical Report Series*, November, 1957.
6. S. Waziruddin, D. Brogan, and P. Reynolds. The process for coercing simulations. In *Fall Simulation Interoperability Workshop*, 2003.

O'SOAP – A Web Services Framework for DDDAS Applications*

Keshav Pingali and Paul Stodghill

Department of Computer Science
Cornell University, Ithaca, NY 14853, USA
{pingali,stodghil}@cs.cornell.edu

Abstract. Because of the continued development of web services protocols and the apparent convergence of Grid services with web services, it is becoming evident that web services will be an important enabling technology for future computational science applications. This is especially true for Dynamic Data-Driven Application Systems (DDDAS's). In this paper, we argue that the current systems for web services development are ill-suited for DDDAS applications. We describe O'SOAP, a new framework for web service applications that addresses the specific needs of computation science. We discuss a multi-physics simulation developed using O'SOAP and show that O'SOAP is able to deliver excellent performance for a range of problem sizes.

1 Introduction

There are certain classes of Dynamic Data-Driven Application Systems (DDDAS) that are, by their very nature, distributed systems. Some obvious examples include applications that incorporate geographically distributed instruments, such as VLA radio telescopes or sensor nets. Other, less obvious, examples include large-scale loosely-coupled applications[1] developed by multi-institutional teams. Elsewhere [2,19], we have described the Adaptive Software Project (ASP) [11], a multi-institutional project that is developing multi-disciplinary simulation systems. What is unique about our systems is that they have been deployed as a geographically distributed set of application components. There are many advantages to this over the conventional approach of building monolithic applications. First, component developers only have to deploy and maintain their codes on a single local platform. This saves the developer time and enables codes with intellectual property constraints to be used by other project members. Second, the loosely-coupled nature of distributed components facilitates their reuse in the development of new simulations. It also enables their simultaneous use in any number of research projects.

What has made this approach possible is the standardization of protocols for the interoperability of distributed components. In particular, our infrastructure is based on

---

* This research is partially supported by NSF grants EIA-9726388, EIA-9972853, and ACIR-0085969.

[1] We consider an application to be loosely-coupled if its components communication infrequently, as opposed to tightly-coupled, in which communication is frequent, or embarrassingly parallel, in which communication is absent.

M. Bubak et al. (Eds.): ICCS 2004, LNCS 3038, pp. 797–804, 2004.
© Springer-Verlag Berlin Heidelberg 2004

the standard web service protocols (i.e., XML [21], SOAP [10], and WSDL[4]). While original developed for business applications, we have found that web services are ideally suited for building computations science applications as well[2].

Because these standardization efforts have led to the development of many interoperable systems for the deployment and use of web services, we believe that many future computational science and DDDAS applications will use web services to varying degrees. Such web services could be particularly useful for DDDAS applications, where runtime dynamic invocation of components is required based on additional injected data, proximity of data repositories, or when the simulation is used to control measurement processes.

As a result, one of the key technologies that will lead to the wide spread deployment of future DDDAS applications will be web services frameworks that make it relatively easy to build distributed component-based applications. Hence, it is critical that computational scientists be able to prototype DDDAS applications without an enormous development effort.

In Section 2, we will discuss existing web services systems and why they do not directly meet the needs of DDDAS application programmers. In Section 3, we will discuss O'SOAP, a new web services framework that is designed to address these needs. In Section 4, we discuss some performance result using O'SOAP. In Section 5, we discuss our conclusions and future directions of O'SOAP development.

2   Motivation

Many systems have been developed for deploying web services. These range from large systems, such Apache Axis [8], Microsoft .NET [5] and Globus [7] to more modest frameworks like, SOAP::Lite [12], SOAPpy [14] and GSOAP [6]. Unfortunately, these systems present a relatively high entry point for DDDAS application developers.

Let's consider what is required to deploy an existing application using these systems. First, the developer must write code to interface the application with a web services framework. While this code is often short, it presents a learning curve that can discourage computational scientists from experimenting with DDDAS systems.

The second difficulty is that there are many issues that arise in deploying an existing application in a web services environment that do not arise in the traditional interactive environment. As a result, the DDDAS application developer must consider:

- Generating WSDL - WSDL is the means for documenting a web service's interface. Some web service frameworks provide tools for generating WSDL documents automatically, but many require that the developer write these documents by hand.
- Data management - Data sets in computational science applications vary greatly in size. While small data sets can be included in the SOAP envelopes, other mechanisms must be used for larger data sets. Also, the developer must manage intermediate and result files that are generated by the application.

---

[2] Since the emerging standards for Grid computing [20,9] are based upon the basic web services protocols, we consider Grid services to be part of the more general class of web services.

- Asynchronous interactions - SOAP over HTTP is essentially a synchronous protocol. That is, the client sends a SOAP request to the server and then waits to receive a SOAP response. However, many computational science applications can take a very long time to execute, which can result in the remote client timing out before receiving the results. This must be considered when writing the code for interfacing with the application.
- Authentication, Authorization and Accounting (AAA) - The developer will certainly wish to restrict which remote users are able to use the web service.
- Job scheduling - Very often, the machine that is hosting the web service is not the same as that on which the application will run. Very often, the web service will have to interact with a job scheduling system in order to run the application.
- Performance - High performance is an important consideration for many computational science applications. It is likely more so for DDDAS applications.

The existing tools offer a blank slate for the programmer. This enables the experienced and knowledgeable web services developer to write efficient and robust solutions for each application. For the novice web service developer, this presents a tremendous hurdle that will only be tackled if absolutely necessary.

To summarize, the very general nature of existing web and Grid service tools makes deploying web services a very costly undertaking for a novice DDDAS application developer. This cost makes it unlikely that computational scientists will try to build DDDAS systems unless there is an absolute need to do so. What is needed is a new web services framework that is designed to address the needs of the DDDAS application developer. Ideally, this framework would enable a computational scientist to deploy new and existing applications as web services with little or no interfacing code. This framework must also address the considerations listed above.

3   Overview of O'SOAP

O'SOAP [17] is a web services framework that is designed to enable a non-expert to quickly deploy and use legacy applications as fully functional web services without sacrificing performance. The primary benefits of O'SOAP over other frameworks is the manner in which it builds upon the basic SOAP protocol to enable efficient interactions between distributed scientific components.

3.1   Deploying Applications as Distributed Components

On the server side, O'SOAP enables existing, command-line oriented applications to be made into web services without any modification. The user only needs to write a small CGI script that calls O'SOAP server-side applications. Placed in the appropriate directory on a web server, this script will execute when the client accesses its URL. An example of such a script is shown in Figure 1.

The oids_server program, which is provided by the O'SOAP framework, processes the client's SOAP request. The -n, -N, and -U parameters specify the short name, full name, and namespace, respectively, of the web service. What appears after -- is a template of the command line that is to be used to run the legacy program, add.sh.

```
#! /bin/bash

oids_server \
    -n arithmetic-test -U urn:test -N 'Arithmetic Server' \
    -- ./add.sh '[in val x:int]' '[in val y:int]' \
        '>' '[out file result:int]'
```

Fig. 1. Sample O'SOAP Server

The text that appears within [...] describes the arguments to the legacy program. Each argument specification includes at least four properties,

- The directionality of the parameter, i.e., "in", "out", or "in_out".
- Whether the parameter value should appear directly on the command line ("val") or whether the parameter value should be placed in a file whose name appears on the command line ("file").
- The name of the parameter, e.g., "x", "y" and "result".
- The type of the parameter value, i.e., "int", "float", "string", "raw" (arbitrary binary file), "xml" (a structured XML file).

A component implemented using O'SOAP will expose a number of methods, discussed below, that can be invoked using the SOAP protocol. O'SOAP also automatically generates a WSDL document that describes these methods, their arguments, and additional binding information.

On the client-side, O'SOAP provides two tools for accessing remote web services. The `osoap_tool` program provides a command-line interface to remote web services. In addition, the `wsdl2ml` program generates stub code for invoking web services from O'Caml [15] programs.

To summarize, O'SOAP is a framework that hides most of the details of the SOAP protocol from the client and server programs. With this in place, we can now discuss how the interactions between the clients and servers can be organized to support distributed computational science applications.

3.2 Asynchronous Interactions

As previously mentioned, SOAP over HTTP was designed for synchronous communication. To accommodate long running computational science applications, O'SOAP's server-side programs provide basic job management by exposing a number of methods to the client. The "spawn" method invokes the application on the server and returns a job id to the client. The client can then pass this job id as the argument to the "running" method to discover whether or not the application has finished execution. Once completed, the client uses the "results" method to retrieve the results. There are additional methods, such as "kill", for remotely managing the application process.

Since the server is able to generate a response for these methods almost immediately, the synchronous SOAP protocol can be used for such method invocations. Also, since a new network connection is established for each method invocation, detached execution and fault recovery are possible without additional system support (e.g., to re-establish network connections).

3.3   Support for Small and Large Data Sizes

In computational science applications, data set sizes can vary greatly. Small data sets can be included within the SOAP envelope that is passed between the client and the server. This eliminates the need for a second round of communication to retrieve the data. However, there are several reasons why embedding large data sets in SOAP envelopes is problematic. One reason that has been observed by others [3,16] is that translating binary data into ASCII for inclusion in the SOAP envelope can add a large overhead to a system. The second reason is that many SOAP implementations have preset limits on the size of SOAP envelopes. Many of our data sets exceed these limits.

For these reasons, O'SOAP enables data sets to be separated from the SOAP request and response envelopes. If a data set is included, it is encoded using XML or Base64 (called, "pass by value"). If it is not included, then a URL to the data set is included (called "pass by reference"). Furthermore, O'SOAP enables clients and servers to dynamically specify whether a data set will be passed by value or reference.

O'SOAP manages a pool of disk space that is used for storing data sets downloaded from the client and data sets generated by the application that will be accessed remotely. O'SOAP currently supports the HTTP, FTP, and SMTP protocols, and we have plans to provide support for IBP [13].

4   Performance

In the previous section, we discussed how O'SOAP generates WSDL automatically, provides mechanisms for transferring large data sets, and enables asynchronous interactions with long running applications. In this section, we will discuss the performance of O'SOAP for a real application. The Pipe Problem application simulates an idealized segment of a rocket engine modeled after actual NASA experimental spacecraft hardware. The object is a curved, cooled pipe segment that transmits a chemically-reacting, high-pressure, high-velocity gas through the inner, large diameter passage, and a cooling fluid through the outer array of smaller diameter passages. The curve in the pipe segment causes a non-uniform flow field that creates steady-state but non-uniform temperature and pressure distributions on the inner passage surface. These temperature and pressure distributions couple with non-uniform thermomechanical stress and deformation fields within the pipe segment. In turn, the thermomechanical fields act on an initial crack-like defect in the pipe wall, causing this defect to propagate.

The components of this system were deployed on servers at Cornell Computer Science and the Engineering Research Center at Mississippi State University. All components were deployed using O'SOAP, except for one, which used SOAP::Clean [18,2], a predecessor of O'SOAP. All clients were developed using O'SOAP.

To understand how increasing the problem size changes the performance of our system, we ran experiments for three different sizes of the Pipe Problem. The sizes of the meshes for the solid and interior volumes of the Pipe are shown in Table 1.

Table 2 shows the total running time, in seconds, for the Pipe Problem application. The column labeled "Local runtime" shows the running time when each component it is executed directly, without using the web services infrastructure. These times correspond

Table 1. Pipe Problem Sizes

| Problem | Solid Mesh | | | Interior Mesh | | |
Size	vertices	triangles	tet's	vertices	tri's/quad's	tet's/prisms
1	4,835	4,979	22,045	19,242	3,065	38,220
2	16,832	10,322	83,609	41,216	5,232	85,183
3	54,849	21,127	289,500	79,407	9,074	170,179

Table 2. Pipe Problem Runtimes

Problem Size	Local runtime (secs.)	CU Client runtime (secs.)	overhead	UAB Client runtime (secs.)	overhead
1	1630.89	1719.44	5.43%	1695.24	3.95%
2	5593.66	5776.55	3.27%	5745.78	2.72%
3	22202.73	22901.39	3.15%	22222.49	0.09%

to the performance of a monolithic application and overheads are measured relative to these times. The columns labeled "CU Client" and "UAB Client" show the running times when the client is run on different machines than the components. The "CU Client" client runs on a machine at Cornell on the same LAN as the Cornell server, and the "UAB Client" client runs on a machine at the University of Alabama at Birmingham. Overall, the total overhead for both clients falls as the problem size increases. The overhead for the largest problem size is 3.2% and 0.1% for the "CU Client" and "UAB Client" clients, respectively.

These results are in marked contrast to the studies in the literature [3,16] that concluded that the use of SOAP and XML adds enormous overhead to computational science applications. Our conclusion is that the organization of a distributed simulation system makes more of a difference to its performance than the underlying web services infrastructure. Tightly-coupled applications appear to perform poorly for large problem sizes, even when a highly optimized web services infrastructure is used. However, loosely-coupled applications, such as ours, appear to perform very well.

A more complete description of the setup, results, and analysis of these experiments can be found in [19].

5   Conclusions and Future Work

In this paper, we have described O'SOAP, a framework that enables legacy applications to be deployed as feature-rich web services without any interface programming. O'SOAP provides automatic WSDL generation, mechanisms for efficient data set transport, and asynchronous client-server interactions. This makes it ideally suited for computational scientists that are not web services programmers to develop distributed component-based and DDDAS applications.

Just as importantly, our experiments have demonstrated that applications can be developed using O'SOAP-based web services without sacrificing performance. In fact, we have shown that, for the Pipe Problem application, the overhead introduced by the

O'SOAP framework decreases as the problem size increases. While there will certainly be some DDDAS applications for which O'SOAP is not appropriate, we believe that O'SOAP is perfectly suited for a very large class of DDDAS applications.

While we are greatly encouraged by the features and performance of O'SOAP to date, there are a number of improvements that we plan to make. First, we need to address the two remaining items from the list of requirements given in Section 2. We have implemented WS-Security [1] in SOAP::Clean, the predecessor of O'SOAP; it remains to fold this code into O'SOAP. Also, while O'SOAP has hooks for interfacing with conventional job submission systems, it remains to provide the interfaces for commonly used systems.

One other direction in which O'SOAP can be developed is to expand support for commonly used protocols. For instance, the SOAP standard specifies a means of sending envelops using SMTP (i.e., email). While this transport mechanism is unlikely to deliver the same level of performance as HTTP or TCP, it dramatically lowers the entry point for the DDDAS application developer, because it enables applications to be deployed without the need for any system administrator support. This will make it much easier to develop prototype or "one-off" DDDAS applications.

The Open Grid Services Infrastructure (OGSI) specification [20] has been developed by the Global Grid Forum (GGF) to define basic low-level protocols that Grid services will need to interoperate. OGSI is currently supported by a number of Grid toolkits. IBM, HP and the Globus Project have recently proposed the Web Services Resource Framework (WSRF), a new set of Grid protocols designed to be more compatible with existing web services specifications. At the moment, the evolution of the Grid protocols is a little murky, but once it becomes clearer, O'SOAP can be extended to support the protocols the GGF adopts.

Last, but not least, there are a number of ways in which the performance of O'SOAP can be improved. One way would be to enable O'SOAP-based web services to be deployed as servlets (e.g., using `mod_ocaml` and Apache). While servlets require much more system administrator involvement and support, there are certainly some DDDAS application developers who are willing to incur this cost in return for the performance improvements.

References

1. Bob Atkinson et al. Web services security (WS-Security), version 1.0. Available at `http://www-106.ibm.com/developerworks/webservices/library/ws-secure/`, April 5 2002.
2. Paul Chew, Nikos Chrisochoides, S. Gopalsamy, Gerd Heber, Tony Ingraffea, Edward Luke, Joaquim Neto, Keshav Pingali, Alan Shih, Bharat Soni, Paul Stodghill, David Thompson, Steve Vavasis, and Paul Wawrzynek. Computational science simulations based on web services. In *International Conference on Computational Science 2003*, June 2003.
3. Kenneth Chiu, Madhusudhan Govindaraju, and Randall Bramley. Investigating the limits of soap performance for scientific computing. In *Proceedings of the Eleventh IEEE International Symposium on High Performance Distributed Computing (HPDC'02)*, July 2002.
4. Erik Christensen, Francisco Curbera, Greg Meredith, and Sanjiva Weerawarana. Web services description language (wsdl) 1.1. Available at `http://www.w3.org/TR/wsdl`, March 15 2001.

5. Microsoft Corporation. Microsoft .NET. Accessed February 11, 2003.
6. Robert A. Van Engelen and Kyle A. Gallivan. The gSOAP toolkit for web services and peer-to-peer computing networks. In *2nd IEEE/ACM International Symposium on Cluster Computing and the Grid (CCGRID'02)*, page 128, Berlin, Germany, May 21 – 24 2002.
7. I. Foster and C. Kesselman. The globus project: A status report. In *IPPS/SPDP '98 Heterogeneous Computing Workshop*, pages 4–18, 1998.
8. The Apache Foundation. Webservices - axis. http://ws.apache.org/axis/.
9. Globus Alliance. The WS-Resource framework. Available at http://www.globus.org/wsrf/, January 24 2004.
10. Martin Gudgin, Marc Hadley, Noah Mendelsohn, Jean-Jacques Moreau, and Henrik Frystyk Nielsen. Soap version 1.2 part 1: Messaging framework. Available at http://www.w3.org/TR/SOAP/, June 24 2003.
11. The itr/acs adaptive software project for field-driven simulation. Available at http://www.asp.cornell.edu/.
12. Paul Kulchenko. Web services for perl (soap::lite, xmlrpc::lite, and uddi::lite). Accessed on June 3, 2003.
13. James S. Plank, Micah Beck, Wael R. Elwasif, Terry Moore, Martin Swany, and Rich Wolski. The internet backplane protocol: Storage in the network. In *NetStore99: The Network Storage Symposium*, Seattle, WA, USA, 1999.
14. Python web services. http://pywebsvcs.sourceforge.net/.
15. Didier Rémy and Jérôme Vouillon. Objective ML: An effective object-oriented extension to ML. *In Theory And Practice of Objects Systems*, 4(1):27–50, 1998.
16. Satoshi Shirasuna, Hidemoto Nakada, Satoshi Matsuoka, and Satoshi Sekiguchi. Evaluating web services based implementations of gridrpc. In *Proceedings of the Eleventh IEEE International Symposium on High Performance Distributed Computing (HPDC'02)*, 2002.
17. Paul Stodghill. O'SOAP - a web services framework in O'Caml. http://www.asp.cornell.edu/osoap/.
18. Paul Stodghill. SOAP::Clean, a Perl module for exposing legacy applications as web services. Accessed February 11, 2003.
19. Paul Stodghill, Rob Cronin, Keshav Pingali, and Gerd Heber. Performance analysis of the pipe problem, a multi-physics simulation based on web services. Computing and Information Science Technical Report TR2004-1929, Cornell University, Ithaca, New York 14853, February 16 2004.
20. Steve Tuecke et al. Open grid services infrastructure (OGSI) version 1.0. Available at https://forge.gridforum.org/projects/ogsi-wg/document/Final_OGSI_Specification_V1.0/en/1, June 27 2003.
21. World Wide Web Consortium. Extensible markup language (xml) 1.0 (second edition). W3C Recommendation, October 6 2000.

Application of Grid-Enabled Technologies for Solving Optimization Problems in Data-Driven Reservoir Studies[*]

Manish Parashar[1], Hector Klie[2], Umit Catalyurek[3], Tahsin Kurc[3],
Vincent Matossian[1], Joel Saltz[3], and Mary F. Wheeler[2]

[1] Dept. of Electrical & Computer Engineering,
Rutgers, The State University of New Jersey, New Jersey, USA
{parashar,vincentm}@caip.rutgers.edu
[2] CSM, ICES, The University of Texas at Austin, Texas, USA
{klie,mfw}@ices.utexas.edu
[3] Dept. of Biomedical Informatics, The Ohio State University, Ohio, USA
{umit,kurc,jsaltz}@bmi.osu.edu

Abstract. This paper presents use of numerical simulations coupled with optimization techniques in reservoir modeling and production optimization. We describe three main components of an autonomic oil production management framework. This framework implements a dynamic, data-driven approach and enables execution in a Grid environment for large scale optimization formulations in reservoir modeling.

1 Introduction

The ultimate goal of reservoir simulation is to generate both good estimates of reservoir parameters and reliable predictions of oil production to optimize return on investment from a given reservoir. The objective function $f(w, s)$ can be seen in terms of a performance measure depending on a vector of decision variables w (e.g., well location indices) and on a vector of uncontrollable conditions s (e.g.,rock porosity values, oil and water saturation). Function f usually represents a mismatch between observed and computed values (history matching) or a economical model based on the amount of oil produced or displaced. In either case, since sampling locations in the field are sparse and the amount of information is scarce, solutions to f are plagued with sources of error and uncertainity[1]. This means that any objective function is limited to a partial knowledge of reality.

---

[*] This work is partly supported under the National Science Foundation under Grants ACI-9619020 (UC Subcontract 10152408), ANI-0330612, EIA-0121177, SBR-9873326, EIA-0121523, ACI-0203846, ACI-0130437, ACI-9982087, NPACI 10181410, ACI 9984357, EIA 0103674 and EIA- 0120934, Lawrence Livermore National Laboratory under Grant B517095 (UC Subcontract 10184497), Ohio Board of Regents BRTTC BRTT02-0003, and DOE DE-FG03-99ER2537.
[1] Note, that in order to simplify the discussion we have omitted the use of constrains which are also common in this setting: fixed budget, allowable locations, predefined surface facilities, to name a few.

M. Bubak et al. (Eds.): ICCS 2004, LNCS 3038, pp. 805–812, 2004.
© Springer-Verlag Berlin Heidelberg 2004

Black oil and more complex compositional, geomechanical, thermal, and chemical models can be used as forecasting tools in both day-to-day operational management of production facilities and long term field development planning. However, little use has been made of reservoir simulations coupled with systematic optimization techniques. The main advantage of applying these mathematical tools to decision-making process is that they are less restricted by human imagination than conventional case-by-case comparisons. A key issue is to come up with reliable prediction models, despite the inherent uncertainty and scales involve in all subsurface measurements, that operate by searching a large space of oil production and reservoir parameters.

One of the main obstacles to the application of optimization techniques coupled with a reservoir simulator is the computational time required to complete simulations of complex, large scale reservoir models. Optimization strategies normally evaluate hundreds or even thousands of scenarios (each representing a simulation run) in the course of searching for the optimal solution to a given management question. This process is extremely time-consuming and data-intensive [5, 8] and can easily overwhelm local computational capacity at any single institution. This approach is further hampered by the need to navigate multi-terabyte datasets from simulations and field measurements.

Grid computing is rapidly emerging as the dominant paradigm for large-scale parallel and distributed computing. A key contribution of Grid computing is the potential for seamless aggregations of and interactions among computing, data and information resources, which is enabling a new generation of scientific and engineering applications that are self-optimizing and dynamic data driven. However, achieving this goal requires a service-oriented Grid infrastructure that leverages standardized protocols and services in accessing hardware, software, and information resources [4,8].

In a previous work, we described a suite of tools and middleware that enable analysis of large, distributed collections of simulation datasets [13]. In this paper, we present an infrastructure for solving optimization problems in data-driven reservoir simulations in the Grid. The infrastructure builds on 3 key components; a computational engine consisting of a simulation framework (IPARS) and optimization services, middleware for distributed data querying and subsetting (STORM), and an autonomic Grid middleware (Discover) for service composition, execution, and collaboration. We describe each of these components and their application in autonomic data-driven management of the oil production process [9].

2   Computational Grid Components

2.1   The Integrated Parallel Accurate Reservoir Simulator (IPARS)

IPARS represents a new approach to a parallel reservoir simulator development, emphasizing modularity of code portability to many platforms, and ease of integration with other software. It provides a set of computational features such as memory management for general geometric grids, portable parallel communication, state-of-the-art non-linear and linear solvers, keyword input and output for visualization. There are currently several models in IPARS, including multi-phase gas-oil-water, air-water and one-phase flow, compositional, geomechanical and reactive transport models. The framework supports both the use of IMPES (implicit pressure explicit saturations) and fully implicit formulations. A key feature of IPARS is that it allows the definition of different numerical and physical models in different blocks of the domain (i.e., multi-numeric, multiphysics and multiblock capabilities). A more technical description of IPARS with further applications can be found in [1].

2.2 Optimization Algorithms

Very Fast Simulated Annealing(VFSA). This algorithm is a simulated annealing variant the speedups the process by allowing a larger sampling at the beginning and a much narrower sampling at its latest stages. This is achieved by the use of a Cauchy like distribution. The second appealing feature is that each model parameter can have its own cooling schedule and model space sampling schemes. This allows selective control of the parameters and the use of *a priori* information (e.g., [14])

Simultaneous Perturbation Stochastic Algorithm (SPSA). The novelty of the SPSA is the underlying derivative approximation that requires only two (for the gradient) or four (for the Hessian matrix) evaluations of the loss function regardless of the dimension of the optimization problem. In other words, it does not require full gradient function information. This feature allows for a significant decrease in the cost of optimization, specially in problems with a large number of decision parameters to be inverted. This algorithm is suitable for noisy measurements of the objective function and the search for a global optimizer (e.g., [15]).

Gradient based. These methods essentially use the approximated gradient of the response surface to derive a search direction. Along the search direction a better point is located based on the response values. Different ways for generating the search direction result in different methods. Newton and quasi-Newton methods[3] and finite-difference stochastic approximation (FESA) methods [15] are representative examples.

Hybrid approaches. These methods are based on the coupling of either the VFSA or the SPSA methods with any of the gradient based methods. This allows to improve the overall convergence of the optimization procedure in the vicinity of the desired solution.

3   Querying and Subsetting of Distributed Data: STORM

STORM (a.k.a. GridDB-Lite) [11] is a services-oriented middleware that is designed to provide basic database support for 1) *selection of the data of interest:* The data of interest is selected based on attribute values or ranges of values, and can involve user-defined filtering operations. 2) *transfer of data from storage nodes to compute nodes for processing:* After the data of interest has been selected, it can be transferred from storage systems to processor memories for processing by a potentially parallel data analysis program.

STORM supports data select and data transfer operations on scientific datasets through an object-relational database model. With an object-relational view of scientific datasets, the data access structure of an application can be thought of as a *SELECT* operation as shown in Figure 1. The $<$ *Expression* $>$ statement can contain operations on ranges of values and joins between two or more datasets. *Filter* allows implementation of user-defined operations that are difficult to express with simple comparison operations.

Datasets generated in scientific applications are usually stored as a set of flat files. STORM services provide support to create a view of data files in the form of virtual tables using application specific *extraction* objects. An extraction object is implemented by an application developer and returns an ordered list of attribute values for a data element in the dataset, thus effectively creating a virtual table. The analysis program can be a data parallel program. The distribution of tuples in a parallel program can be represented as a *distributed array*, where each array entry stores a tuple. This abstraction is incorporated into our model by the *GROUP-BY-PROCESSOR* operation in the query formulation. *ComputeAttribute* is another user-defined function that generates the attribute value on which the selected tuples are grouped together based on the application specific partitioning of tuples.

SELECT $<$ *Attributes* $>$
    FROM $Dataset_1, Dataset_2, ..., Dataset_n$
    WHERE $<$ *Expression* $>$ AND $Filter(<$ *Attributes* $>)$
    GROUP-BY-PROCESSOR $ComputeAttribute(<$ *Attributes* $>)$

Fig. 1. Formulation of data retrieval steps as an object-relational database query.

STORM has been developed using a component-based framework, called DataCutter [2], which enables execution of application data processing components in a distributed environment. Using the DataCutter runtime system, STORM implements several optimizations to reduce the execution time of queries: **Distributed Execution of Filtering Operations.** Both data and task parallelism can be employed to execute user-defined filtering operations in a distributed manner. If a select expression contains multiple user-defined filters, a network of filters can be formed and executed on a distributed collection of machines. **Parallel Data Transfer.** Data is transferred from multiple data sources to multiple destination processors by STORM data mover components. Data movers can be instantiated on multiple storage units and destination processors to achieve parallelism during data transfer.

4   An Autonomic Grid Middleware for Oil Reservoir Optimization

Discover [7] enables a seamless access to and peer-to-peer integration of applications, services and resources on the Grid. The middleware substrate integrates Discover collaboratory services with the Grid services provided by the Globus Toolkit using the CORBA Commodity Grid (CORBACoG) Kit [12]. It also integrates the Pawn peer-to-peer messaging substrate [9]. Pawn enables decentralized (peer) services and applications to interact and coordinate over wide area networks. Finally, Discover/DIOS [10] distributed object infrastructure that enables development and management of interaction objects and applications, encapsulating sensors and actuators, and a hierarchical control network. DIOS also allows the dynamic definition and deployment of policies and rules to monitor and control the behavior applications and/or application services in an autonomic manner [6]. Detailed descriptions of the design, implementation and evaluation of Discover components can be found in [7,9,10,6].

5   Putting It Together: Data-Driven Oil Production Optimization

Oil production optimization process involves (1) the use of an integrated multiphysics/multi-block reservoir model and several numerical optimization algorithms (global, local and hybrid approaches) executed on distributed computing systems in the Grid; (2) distributed data archives that store historical, experimental (e.g., data from sensors embedded in the field) and observed data; (3) Grid services that provide secure and coordinated access to the resources required by the simulations; (4) external services that provide data relevant to optimization of oil production or of the economic profit such as current oil market prices, and (5) the actions of scientists, engineers and other experts, in the field, the laboratory, and in management offices.

In this process, item 1 is implemented by the IPARS framework. Both forward modeling (comparison of the performance of different reservoir geostatistical parameter scenarios) and inverse modeling (searching for the optimal decision parameters) for solving optimization problems in reservoir management can greatly benefit from integration and analysis of simulation, historical, and experimental data (item 2). Common analysis scenarios in optimization problems in reservoir simulations involve economic model assessment as well as technical evaluation of changing reservoir properties (e.g., amount of bypassed oil, concentration of oil and water) [13]. In a Grid environment, data analysis programs need to access data subsets on distributed storage systems. This need is addressed by STORM. An example query for exploring regions of bypassed oil in one or more simulation datasets is given in Figure 2. The Discover autonomic Grid middleware implements the support for items 3, 4, and 5. We now discuss the use of Discover to enable oil reservoir optimization [8].

The overall application scenario is illustrated in Figure 3. The peer components involved include: IPARS providing sophisticated simulation components that encapsulate complex mathematical models of the physical interaction in the subsurface, and execute on distributed computing systems on the Grid; IPARS Factory responsible for configuring IPARS simulations, executing them on resources on the Grid and managing their execution; Optimization Service (e.g. very fast simulated annealing); and Economic Modeling Service that uses IPARS

SELECT $R.Cell_x, R.Cell_y, R.Cell_z, R.Id, R.Time$
 FROM $Realization_1, Realization_2, ..., Realization_n$
 WHERE $T_{start} <= R.Time$ AND $R.Time <= T_{end}$
  AND $R.SOIL > SOIL_{tol}$
  AND $Speed(R.V_{oil,x}, R.V_{oil,y}, R.V_{oil,z}) < Speed_{tol}$
 GROUP-BY-PROCESSOR $Partition(R.Id, R.Time)$

Fig. 2. An example query for analysis of data in oil reservoir management studies: "Retrieve all the mesh cells, from simulations $Realization_1, ..., Realization_n$, which contain bypassed oil (which is defined as cells, in which oil saturation is greater than user-defined oil saturation threshold, $SOIL_{tol}$, and oil speed is less than user-defined speed threshold."

simulation outputs and current market parameters (oil prices, costs, etc.) to compute estimated revenues for a particular reservoir configuration.

These entities need to dynamically discover and interact with one another as peers to achieve the overall application objectives. Figure 3 illustrates the key interactions involved: (1) the experts use the portals to interact with the Discover middle-ware and the Globus Grid services to discover and allocate appropriate resource, and to deploy the IPARS Factory, Optimization Service and Economic model peers. (2) The IPARS Factory discovers and interacts with the Optimization Service peer to configure and initialize it. (3) The experts interact with the IPARS Factory and Optimization Service to define application configuration parameters. (4) The IPARS Factory then interacts with the Discover middle-ware to discover and allocate resources and to configure and execute IPARS simulations. (5) The IPARS simulation now interacts with the economic model to determine current revenues, and discovers and interacts with the Optimization Service when it needs optimization. (6) The Optimization Service provides IPARS Factory with optimized well information, which then (7) launches new IPARS simulations with update parameters. (8) Experts can at anytime discover, collaboratively monitor and interactively steer IPARS simulations, configure the other services and drive the scientific discovery process. Once the optimal well parameters are determined, the IPARS Factory configures and deploys a production IPARS run.

Figure 4 shows the convergence history for the optimization of well location using the VFSA optimization service, to maximize profits for a given economical revenue objective function. The well positions plot (on the left) shows the oil field and the positions of the wells. The black circles represent fixed injeciton wells and the well at the bottom most part of the plot is a fixed production well. The plot also shows the sequence of well position guesses for the other production well returned by the VFSA service (shown by the lines connecting the light squares), and the corresponding normalized cost value (plot on the left).

6   Conclusions

In this paper, we presented an infrastructure and its components to support autonomic oil production management process. Use of this infrastructure to im-

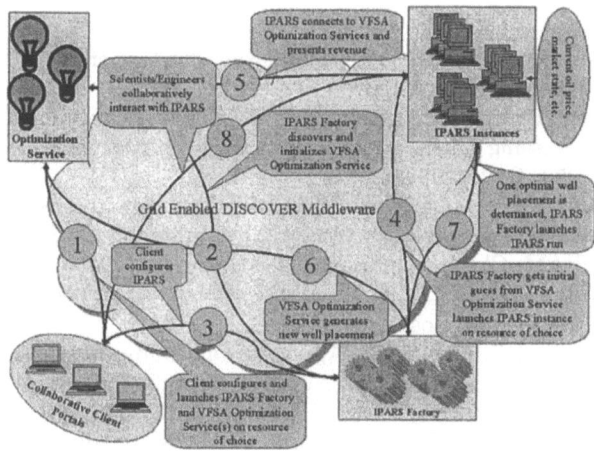

Fig. 3. Autonomous oil reservoir optimization using decentralized services.

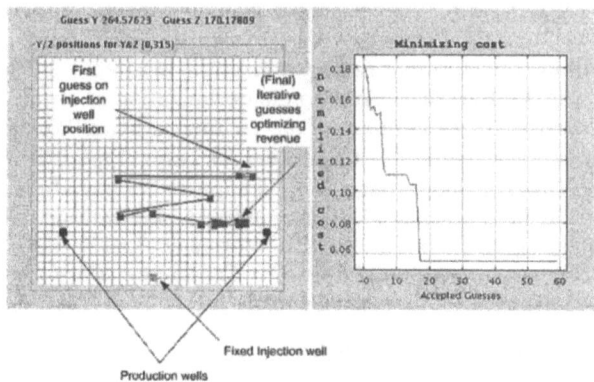

Fig. 4. Convergence history for the optimal well placement in the Grid.

plement Grid-enabled data-driven application support can aid in gaining better understanding of subsurface properties and decision variables. With a better understanding of these properties and variables, engineers and geoscientists can implement optimized oil production scenarios. We believe autonomic oil production management strategies combined with Grid-enabled data and parameter space exploration technologies can lower infrastructure costs and change the economics of productivity maximization.

Acknowledgment. We would like to thank Ryan Martino and Małgorzata Peszyńska for their help in executing the VFSA-based experiments.

References

1. IPARS: Integrated Parallel Reservoir Simulator. The University of Texas at Austin, http://www.ices.utexas.edu/CSM.
2. M. D. Beynon, T. Kurc, U. Catalyurek, C. Chang, A. Sussman, and J. Saltz. Distributed processing of very large datasets with DataCutter. *Parallel Computing*, 27(11):1457–1478, Oct 2001.
3. J. Dennnis and R. Schnabel. *Numerical Methods for Unconstrained Minimization and nonlinear equations*. Prentice Hall, Englewood Cliffs, New York, 1983. reprinted by SIAM, Classics in Applied Mathematics, 1996.
4. I. Foster and C. Kesselman. *The Grid: Blueprint for a new computing infrastructure*. Morgan Kaufmann, Los Altos, CA, 1999.
5. V. M. Johnson and L. L. Rogers. Applying soft computing methods to improve the computational tractability of a subsurface simulation-optimization problem. *Journal of Petroleum Science and Engineering*, 29:153–175, 2001.
6. H. Liu and M. Parashar. Dios++: A framework for rule-based autonomic management of distributed scientific applications. In H. Kosch, L. Boszormenyi, and H. Hellwagner, editors, *Proceedings of the 9th International Euro-Par Conference (Euro-Par 2003)*, volume 2790 of *Lecture Notes in Computer Science*, pages 66–73. Springer-Verlag, August 2003.
7. V. Mann and M. Parashar. Engineering an Interoperable Computational Collaboratory on the Grid. *Grid Computing Environments. Special Issue of Concurrency and Computations: Practice and Experience*, 14(13-15):1569 – 1593, 2002.
8. V. Matossian and M. Parashar. Autonomic optimization of an oil reservoir using decentralized services. In *Proceedings of the 1st International Workshop on Heterogeneous and Adaptive Computing– Challenges for Large Applications in Distributed Environments (CLADE 2003)*, pages 2–9. Computer Society Press, June 2003.
9. V. Matossian and M. Parashar. Enabling peer-to-peer interactions for scientific applications on the grid. In H. Kosch, L. Boszormenyi, and H. Hellwagner, editors, *Proceedings of the 9th International Euro-Par Conference (Euro-Par 2003)*, volume 2790 of *Lecture Notes in Computer Science*, pages 1240–1247. Springer-Verlag, August 2003.
10. R. Muralidhar and M. Parashar. A Distributed Object Infrastructure for Interaction and Steering. *Special Issue - Euro-Par 2001, Concurrency and Computation: Practice and Experience*, 15(10):957–977, 2003.
11. S. Narayanan, T. Kurc, U. Catalyurek, X. Zhang, and J. Saltz. Applying database support for large scale data driven science in distributed environments. In *Proceedings of the Fourth International Workshop on Grid Computing (Grid 2003)*, pages 141–148, Phoenix, Arizona, Nov 2003.
12. M. Parashar, G. von Laszewski, S. Verma, J. Gawor, K. Keahey, and N. Rehn. A CORBA Commodity Grid Kit. *Grid Computing Environments. Special Issue of Concurrency and Computations: Practice and Experience*, 14(13-15):1057–1074, 2002.
13. J. Saltz and et.al. Driving scientific applications by data in distributed environments. In *Dynamic Data Driven Application Systems Workshop, held jointly with ICCS 2003*, Melbourne, Australia, June 2003.
14. M. Sen and P. Stoffa. *Global Optimization Methods in Geophysical Inversion*. Advances in Exploration Geophysics 4, series editor: A.J. Berkhout. Elsevier, 1995.
15. J. C. Spall. *Introduction to stochastic search and optimization:estimation, simulation and control*. John Wiley & Sons, Inc., Publication, New Jersey, 2003.

Image-Based Stress Recognition Using a Model-Based Dynamic Face Tracking System

Dimitris Metaxas, Sundara Venkataraman, and Christian Vogler

Center for Computational Biomedicine, Imaging and Modeling (CBIM)
Department of Computer Science
Rutgers University
dnm@cs.rutgers.edu
http://www.cs.rutgers.edu/~dnm

Abstract. Stress recognition from facial image sequences is a subject that has not received much attention although it is an important problem for a host of applications such as security and human-computer interaction. This class of problems and the related software are instances of Dynamic Data Driven Application Systems (DDDAS). This paper presents a method to detect stress from dynamic facial image sequences. The image sequences consist of people subjected to various psychological tests that induce high and low stress situations. We use a model-based tracking system to obtain the deformations of different parts of the face (eyebrows, lips, mouth) in a parameterized form. We train a Hidden Markov Model system using these parameters for stressed and unstressed situations and use this trained system to do recognition of high and low stress situations for an unlabelled video sequence. Hidden Markov Models (HMMs) are an effective tool to model the temporal dependence of the facial movements. The main contribution of this paper is a novel method of stress detection from image sequences of a person's face.

1 Introduction

Stress detection of humans has been a well researched topic in the area of speech signal processing [Steeneken, Hansen 99], while very little attention has been paid to recognizing stress from faces. Recognizing stress from faces could complement speech-based techniques and also help in understanding recognition of emotions. The challenges in this domain is that the data from each person are continuous and dynamic and each person expresses stress differently. Therefore the recognition of stress and the associated development of the necessary software is a DDDAS. In the next two sections we illustrate the data collection procedure and the algorithms that will be used for training/recognition.

1.1 Overview of the System

The data used for our experiments were obtained from a psychological study at the University of Pennsylvania. The subjects of the study were put through a battery of

M. Bubak et al. (Eds.): ICCS 2004, LNCS 3038, pp. 813–821, 2004.
© Springer-Verlag Berlin Heidelberg 2004

tests that induce high and low stress situations. The subjects were videotaped as they took the tests.

A generic model of the face is fitted to the subjects' face and the tracking system is run on the face image sequence with this initial fit of the model in the first frame. The tracking system does a statistical cue integration from computer vision primitives such as edges, point trackers and optical flow. The face model incorporates some parametric deformations that give jaw, eyebrow and basic lip movements. The face tracker gives values for these parametric deformations as a result of the tracking. These are the parameters we will use to learn the movements that correspond to different stress situations. The learning methods will be trained on these parameters and the parameters of a sequence from an unknown stress situation are tested against the learned system to classify a given sequence as a high or low stress condition.

1.2 Learning and Recognition

We will evaluate different learning approaches to train the system for low and high stress conditions. The tracking result will give us parameters which account for the rigid transformations between the face movements and also deformations of the mouth, eyebrow etc. for stress conditions. The recognition will be done using Hidden Markov Models (HMMs). HMMs were chosen since they have been used to model temporal dependence very effectively in American Sign Language (ASL) recognition. Also, we will use the boosting approach to enhance the learning methods and avoid overfitting.

2 Deformable Model Tracking

2.1 Deformable Models

The face tracking system uses a face model that deforms according to movement of a given subject's face. So the shape, position and orientation of the model surface can change. These changes are controlled by a set of n parameters $\mathbf{q}$. For every point i on the surface of the model, there is a function F_i that takes the deformation parameters and finds

$$p_i = F_i(\mathbf{q}) \tag{1}$$

where p_i is the position of the point in the world frame [Metaxas 97].

In addition, computer vision applications, such as deformable model tracking, require the first order derivatives, so we restrict F_i to the class of functions for which the first order derivative exists everywhere with respect to $\mathbf{q}$. This derivative is the Jacobian J_i, where

$$J_i = \begin{bmatrix} | & & | \\ \frac{\partial p_i}{\partial q_1} & \cdots\cdots & \frac{\partial p_i}{\partial q_n} \\ | & & | \end{bmatrix} \tag{2}$$

Each column of the Jacobian J_i is the gradient of p_i with respect to the parameter q_i.

2.2 Fitting and Tracking

In principle, there exists a clean and straightforward approach to track deformable model parameters across image sequences. Low-level computer vision algorithms generate desired 2D displacements on selected points on the model, that is differences between where the points are currently according to the deformable model and where they should be according to measurements from the image. These displacements, also called 'image forces', are then converted to n-dimensional displacement f_g in the parameter space, called the *generalized force* and used as a force in the first-order massless Lagrangian system :

$$\dot{q} = f_g + F_{internal}(q) \tag{3}$$

where $F_{internal}(q)$ is the result of internal forces of the model (i.e. elasticity of the model, preset). We integrate this system with the classical Euler integration procedure, which eventually yields a fixed point, where $f_g = 0$. This fixed point corresponds to the desired new position of the model.

In order to use the system of (1), we have to accumulate all the 2D image forces from the computer vision algorithms into f_g. First, we convert each image force f_i on a point p_i into a generalized force f_{gi} in parameter space, which describes the effect that single displacement at point p_i has on all the parameters. Obtaining a generalized force fg then simply consists of summing up all f_{gi} :

$$f_g = \sum_i f_{gi} \quad \text{where} \quad f_{gi} = \sum_i B_i^T f_i \tag{4}$$

and

$$B_i = \left.\frac{\partial \mathbf{Proj}}{\partial p}\right|_{pi} J_i \tag{5}$$

B_i is the projection of the Jacobian J_i from world coordinates to image coordinates via the projection matrix **Proj** at point p_i.

Generating the generalized force this way works fine as long as all the image forces come from the same cue (diff. algorithms on the same image). When there are multiple cues from multiple vision algorithms, combining the cues becomes a hard problem. In order to effectively combine them statistically we will need to know the distributions of the individual cues ahead of time, but it is hard to estimate these distributions beforehand.

We choose the framework of affine regions [Goldenstein et al. 2001, 2003] to estimate the distributions of the cues within small regions and apply the equivalent of the central limit theorem to these affine regions to make a Gaussian approximation. Then we use maximum likelihood estimation to get the final generalized force.

The face model itself was made from a publicly available geometric model of the head, available from the University of Washington as part of [Pighin et al.99]. A face mask was cut out of this original model and we obtained a static model with 1,100 nodes and 2000 faces. Then, parameters and associated regions are defined for the raising and lowering of eyebrows, for the smiling and stretching of the mouth, for the opening of the jaw as well as the rigid transformation parameters for the model frame.

2.3 Improvements of the Face Model

In this version of the system we have added asymmetric deformations for the eyebrows and the mouth region i.e. the left and right eyebrows, the left and right ends of the lips of the mouth are no longer tied together. This is essential since one of the major indicators of stress is asymmetric lip and eyebrow movements and the original framework did not support that.

The deformation parameters all put together form about 14 parameters. The tracking results from a particular video sequence will give us these 14 parameters for the model for each time instance. We perform these tracking experiments on various subjects and use the parameters we obtain to train the HMMs.

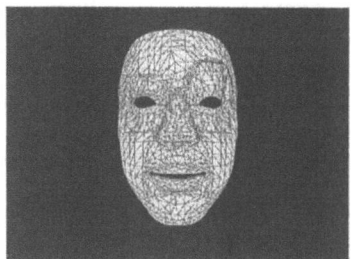 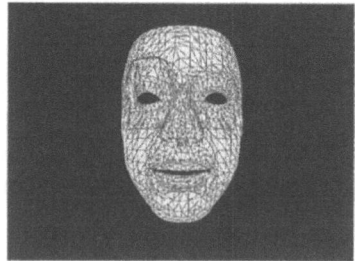

Fig. 1. Left eyebrow movement in the model and right eyebrow movement in the model

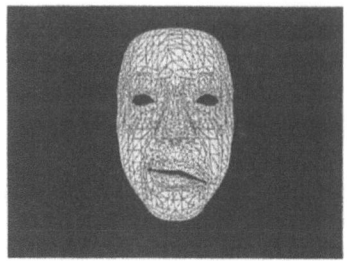 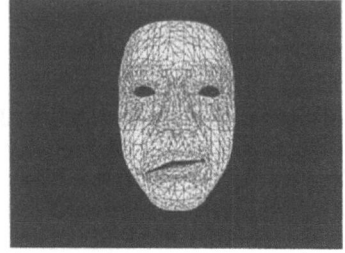

Fig. 2. Left and right Risorius movement in the model

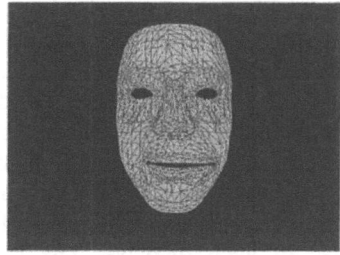 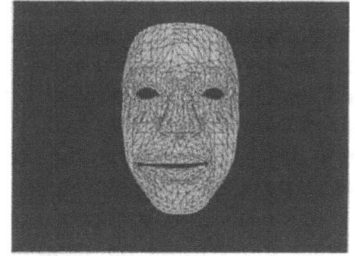

Fig. 3. Left and right lip stretching in the model

3   Stress Recognition

The computational detection of a human undergoing a stress reaction can be broken up into two stages. The first stage consists of recognizing the possible individual displays of stress response in the human face such as eye movements and blinking, various negative facial expressions. The second stage consists of accumulating the information collected in the first stage and deciding whether these displays occur frequently enough to classify as a stress response.

From an abstract point of view, the first stage corresponds to the task of detecting specific patterns in a time-varying data signal. Depending on the specific task and the pattern that we are looking for, the signal often simply consists of the deformable model parameter vectors that we estimate during the face tracking process. For instance, in order to detect rapid head movements, we are interested in the rigid body component of the parameter vector. The orientation, position, and derivatives thereof contain all the necessary information for this particular pattern. Likewise, in order to detect specific negative facial expressions, we are interested in the nonrigid deformation component of the parameter vector, which controls the eyebrow and mouth movements, and nearby regions.

Eye blinking is slightly more complicated to handle, because the deformable model does not contain any parameters that control it directly. However, the output of the tracking algorithm does contain information on the location of the eyes in the human face at any given point in time. The eyes themselves are represented as holes in the deformable model, delimited by the region that is formed by a set of nodes. Because from the deformable model parameters the position of these nodes can be deduced, it is possible to project their positions into image space, and thus find out which region in the video frame corresponds to the eyes. We then use a grayscale level averaging method on the region in the video frame to determine the degree, to which the eyes are opened or closed - uniform grayscale levels indicate that the eyelids are covering the irises, whereas more diverse grayscale levels indicate that the irises are visible. Just like the model parameters, the degree of openness of the eyes can thus be quantified in a few numbers, which is important for the subsequent training of the recognition algorithm.

Detecting stress patterns in a time-varying signal is very similar to other well-known activity recognition tasks, such as gesture recognition, and sign language recognition. For these tasks, hidden Markov models (HMMs) have been shown to be highly suitable, for several reasons: First, the HMM recognition algorithm (Viterbi decoding) is able to segment the data signal into its constituent components implicitly, so it is not necessary to concern ourselves with the often extremely difficult problem of segmenting a data signal explicitly. Second, the state-based nature of HMMs is a natural match for the task of recognizing signals over a period of time. Third, the statistical nature of HMMs makes them ideal for recognizing tasks that exhibit an inherent degree of variation; for example, due to motor limitations, humans generally do not perform the same movement twice in exactly the same way, even if they intend to. To make HMM-based recognition of potentially stress-related displays work, we first train the HMMs on hand-labeled examples of such displays. The labeled examples include information on the starting and ending frame of the display, as well as the class into which it belongs: a specific type of negative facial expression, rapid head movement, eye blinking, and so on. Then, during the recognition phase, the HMM algorithm detects from the tracking data which ones of these types of displays occur in the video, and when; that is, at which frames.

4 Hidden Markov Models

4.1 Background

Hidden Markov Models (HMMs) have been a very effective tool in capturing temporal dependencies in data and fitting them to models. They have been applied very effectively to the problem of American Sign Language (ASL) recognition and to speech recognition in particular with reasonable commercial success.

Hidden Markov models are a type of statistical model embedded in a Bayesian framework. In their simplest form, an HMM λ consists of a set of N states $S_1, S_2, ...S_N$. At regularly spaced discrete time intervals, the system transitions from state S_i to S_j with probability a_{ij}. The probability of the system initially starting in state S_i is Π_i.

Each state S_i generates output $O \in \Omega$, which is distributed according to a probability distribution function $b_i(O) = P\{\text{Output is } O \,|\text{System is in } S_i\}$. In most recognition applications $b_i(O)$ is actually a mixture of Gaussian densities.

In most applications, the HMMs correspond to specific instances of a situation (in our case, different high stress situations). Then the recognition problem is reduced to find the most likely state sequence through the network. So we would like to find a state sequence $Q = Q_1,....,Q_T$ over an output sequence $O = O_1,.....,O_T$ of T frames, such that $P(Q,O|\lambda)$ is maximized. So we have

$$\delta_t(i) = \max_{Q1,...,Qt-1} P(Q_1 Q_2 \cdots Q_t = S_i, O \,|\, \lambda) \tag{6}$$

and by induction

$$\delta_{t+1}(i) = b_i(O_{t+1}) \cdot \max_{1 \le j \le N} \{\delta_t(j)\, a_{ji}\} \tag{7}$$

$$P(Q,O \,|\, \lambda) = \max_{1 \le i \le N} \{\delta_T(i)\}$$

The Viterbi algorithm computes this state sequence in $O(N^2T)$ time, where N is the number of states in the HMM network. The Viterbi algorithm implicitly segments the observation into parts as it computes the path through the network of HMMs.

A more detailed introduction and description of algorithms and inference in HMMs is described in [Rabiner 89].

5   Experiments and Results

The data for the experiments were collected at the University of Pennsylvania NSBRI center. The subjects of the experiment were put through the neurobehavioral test battery (NTB), while being videotaped. The NTB tests consist of two sessions. Session I : The subjects perform the 'Stroop word-color inference task (Stroop). This requires the subject to filter out meaningful linguistic information whereby subjects must respond with the printed color name rather than the ink color name. During the task it is difficult to ignore these conflicting cues and automatic expectations that are associated with impulse control. The Psychomotor Vigilance Task (PVT) is a simple, high-signal-load reaction time test designed to evaluate the ability to sustain attention and respond in a timely manner to salient signals [Dinges 85]. The probed recall memory (PRM) test controls report bias and evaluates free working memory [Dinges 93]. Descending subtraction task (DST) requires the subject to perform serial subtractions of varying degrees of difficulty. Visual motor task (VMT) requires subjects to remember and replicate positions of a continually increasing sequence of flashing blocks and Digit symbol substitution task (DSST) assesses cognitive speed and accuracy trade-offs.

Session II : Serial addition subtraction task (SAST) assesses cognitive throughput (speed and accuracy trade-offs),. Synthetic workload (SYNW) task is a multi-cognitive task comprising four tasks completed simultaneously on a split screen, including probed memory, visual monitoring and simple auditory reaction time. Meter reading task (MRT) is a numerical memory and recall task; Working memory task (WMT) requires the subjects to determine whether the target stimulus is the same or different from a previously displayed cue stimulus; Logical reasoning task (LRT) involves attention resources, decision-making and response selection; Haylings sentence completion task (HSC) involves subjects completing a sentence with a single word that is either congruous or incongruous with the overall meaning of the sentence.

The Stroop, HSC and DST are verbal tasks, while PVT, PRM, VMT, SAST, SYNW, MRT, WMT, LRT and DSST are non-verbal tasks, requiring computer inputs as responses.

Workload demands are initially low, and increase across the performance bout, so that in the second half subjects are performing under high workload demand, designed to induce high behavioral distress. Periodically throughout the test bout, onscreen feedback is provided. During the low workload periods, this feedback is positive. As the workload demands increase, the feedback is negative. The feedback is identical for all subjects, and is independent of performance levels. Additionally, during the high workload portions of the test bout, uncontrolled and unexpected "computer failures" occur that are experimenter generated. The obtained video sequences were classified into high and low stress conditions and blind video sequences on which testing needs to be done.

The following figures show examples of high and low stress situations from the tests. The images are normalized i.e. the inverse of the rigid transformations inferred from the deformable model tracking are applied to the face model to produce a frontal face with the image from the video sequence texture mapped onto this model.

From the sequences that were analyzed, we found that some of the major indicators of high stress were eyebrow movements, asymmetric lip deformations and baring of teeth. The tracking results from such sequences were used as input to the HMM learner.

The experiments were conducted on 25 datasets in all with approximately one half of the data being high stress sequences and the other half low stress sequences.

We used two ergodic HMMs one for low and one for high stress conditions. The feature vector we used for training the HMMs were the face asymmetries (difference between the left and right eyebrows, Risorius and lip stretching deformation parameters. The training and testing was independent of the subjects who were part of the experiment.

Recognition was performed with different amounts of data splits. A 75% - 25% data split between training and test data respectively gave us results where all the 6 datasets in the test sets were correctly identified as low/high stress conditions. A 50% - 50% split between the training and test data sets gave us results where 12 out of the 13 samples were correctly identified as low/high stress conditions.

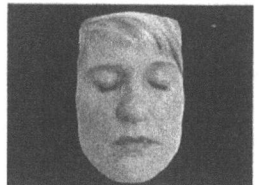 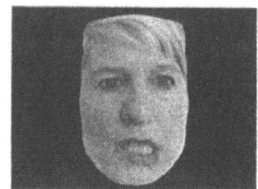 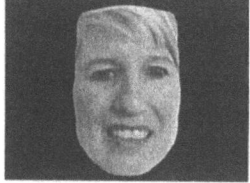

Fig. 4. (a) Low stress condition, (b) high stress condition (asymmetric lip deformation), and (c) high stress condition (baring of teeth)

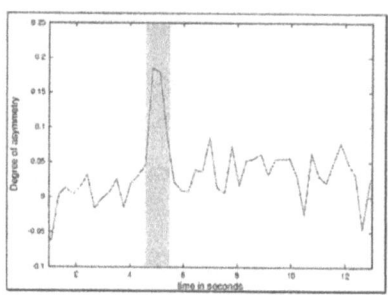 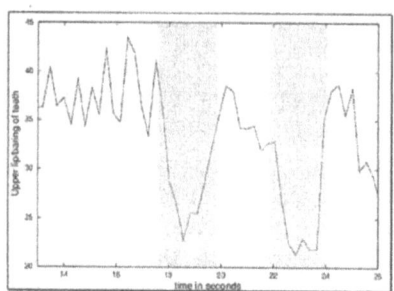

Fig. 5. Asymmetry in lips (indicator of high stress) and baring of teeth (indicator of high stress)

6 Conclusions

In this paper we have presented a DDDAS for the recognition of stress from facial expressions. Our method is based on the use of deformable models and HMMs that can deal with the dynamically varying data and variances in the expression of stress among people.

Acknowldedgements. This work has been funded by an NSF-ITR 0313184 to the first author.

References

[1] Steeneken, H. J. M., Hansen, J. H. L. (1999). *Speech under stress conditions: Overview of the effect on speech production and on system performance.* IEEE International Conference on Acoustics, Speech and Signal Processing.

[2] Goldenstein, S., Vogler C., Metaxas D.N., (2001). *Affine arithmetic based estimation of cue distributions in deformable model.*

[3] Goldenstein, S., Vogler C., Metaxas D.N., (2003). *Statistical Cue Integration in DAG Deformable Models.* IEEE Transactions on Pattern Analysis and Machine Intelligence, July 2003.

[4] Vogler, C., Metaxas, D. N. (1999). *Parallel Hidden Markov Models for American Sign Language recognition.* International Conference on Computer vision, Kerkyra, Greece.

[5] Pighin, F., Szeliski, R., Salesin, D., (1999). *Resynthesizing Facial animation through 3D Model-based tracking.* International Conference on computer vision.

[6] Rabiner, L. R., (1989). *A tutorial on hidden Markov models and selected applications in speech recognition.* Proceedings of the IEEE, Vol. 77.

[7] Metaxas, D. N., (1997). *Physics-based deformable models: Applications to computer vision, graphics and medical imaging.* Kluwer Academic Press.

Developing a Data Driven System for Computational Neuroscience

Ross Snider and Yongming Zhu

Montana State University, Bozeman MT 59717, USA

Abstract. A data driven system implies the need to integrate data acquisition and signal processing into the same system that will interact with this information. This can be done with general purpose processors (PCs), digital signal processors (DSPs), or more recently with field programmable gate arrays (FPGAs). In a computational neuroscience system that will interact with neural data recorded in real-time, classifying action potentials, commonly referred to as spike sorting, is an important step in this process. A comparison was made between using a PC, DSPs, and FPGAs to train a spike sorting system using Gaussian Mixture Models. The results show that FPGAs can significantly outperformed PCs or DSPs by embedding algorithms directly in hardware.

1 Introduction

A data driven system is being developed for computational neuroscience that will be able to process an arbitrary number of real-time data streams and provide a platform for neural modeling that can interact with these real-time data streams. The platform will be used to aid the discovery process where neural encoding schemes through which sensory information is represented and transmitted within a nervous system will be uncovered. The goal of the system is to enable real-time decoding of neural information streams and allow neuronal models to interact with living simple nervous systems. This will enable the integration of experimental and theoretical neuroscience. Allowing experimental perturbation of neural signals while in transit between peripheral and central processing stages will provide an unprecedented degree of interactive control in the analysis of neural function, and could lead to major insights into the biological basis of neural computation.

Integrating data acquisition, signal processing, and neural modeling requires an examination of the system architecture that is most suitable for this endeavor. This examination was done in the context of spike sorting, which is a necessary step in the acquisition and processing of neural signals. The spike sorting algorithm used for this study was based on Gaussian mixture models that have been found useful in signal processing applications, such as image processing, speech signal processing, and pattern recognition [1,2]. The hardware platforms compared were the desktop PC, a parallel digital signal processing (DSP) implementation, and the use of field programmable gate arrays (FPGAs).

M. Bubak et al. (Eds.): ICCS 2004, LNCS 3038, pp. 822–826, 2004.
© Springer-Verlag Berlin Heidelberg 2004

2   Gaussian Mixture Model

Spike sorting is the classification of neural action potential waveforms. The probability that an action potential x belongs to neuron o_i is given by Bayes' rule:

$$p(o_i|x) = \frac{p(x|o_i)p(o_i)}{\sum_{k=1}^{M} p(x|o_k)p(o_k)} \tag{1}$$

where the Gaussian component density $p(x|o_i)$ is given by

$$p(x|o_i) = \frac{1}{(2\pi)^{R/2}|\Sigma_i|^{1/2}} e^{-\frac{1}{2}(x-\mu_i)^T \Sigma_i^{-1}(x-\mu_i)} \tag{2}$$

and μ_i is the mean vector waveform and Σ_i is the covariance matrix of the Gaussian model o_i.

2.1   Expectation Maximization Algorithm

The parameters of the Gaussian mixture model were found via the Expectation-Maximization (EM) algorithm. The EM algorithm is a general method of finding the maximum-likelihood estimate of the parameters of an underlying distribution from a given data set. Details of this iterative parameter estimation technique can be found in [3]. A log version of the EM algorithm was used to deal with underflow problems.

3   PC Implementation

For comparison purposes we implemented the training of the EM algorithm in Matlab on various PCs. We used version 6.5 Release 13 of Matlab since it significantly increases the computational speed of Matlab as compared to prior versions. The spike sorting process can be divided into two phases: the model training phase which is computationally expensive and the spike classification phase which is much faster. The test results for estimating the parameters of 5 Gaussian components from 720 160-dimensional waveform vectors are shown in table 1.

Table 1. Training time of EM algorithm using Matlab on PCs. Times are in seconds

	Pentium III 1.2 GHz[1]	AMD 1.37 GHz[1]	AMD Dual 1.2 GHz[2]	Pentium IV 3.3 GHz[3]	Average	Best
Training Time	8.33	3.42	1.45	1.31	3.22	1.31

1 The Pentium III 1.2GHz and AMD 1.37GHz systems both had 512 MB DDRAM

2 The Dual AMD 1.2GHz system had 4 GB DDRAMM

3 The Pentium IV 3.3GHz system has 1GB DDRAM.

The best performance for training was 1.31 seconds on the 3.3 GHz Pentium IV system. Classification performance was 0.05 seconds per waveform vector which meets real time performance since typical neural spike duration is around 2 ms. However, the training process ran much slower than the classification process and needs to be accelerated to minimize delays in an experimental setup.

4   Parallel DSP Implementation

To speed up the training, the first approach we tried was to implement the EM algorithm on a parallel DSP system consisting of 4 floating-point DSPs. Digital signal processors have become more powerful with the increase in speed and size of on-chip memory. Furthermore, some modern DSPs are optimized for multiprocessing. The Analog Devices ADSP-21160M DSP has two types of integrated multiprocessing support, which are the link ports and a cluster bus. We used Bittware's Hammerhead PCI board with 4 ADSP-21160M DSPs to speed up the training algorithm. We used the profile command in Matlab to analyze the EM algorithm and found that 70% of the execution time was spent in two areas. These were calculating $p(x|o_i)$ and updating the means and covariance matrices. We wrote a parallel version of the EM to take advantage of the multiprocessing capability of the board. The results are shown in table 2.

Table 2. Training time of EM algorithm using Parallel DSP board. Times are in seconds

Data Transfer Method	Single DSP	Four DSPs
No Semaphores	N/A	stopped at 7000
Semaphores	14.5	13.5
DMA transfer	15.9	3.4

Not using semaphores in data transfers resulted in bus contention that significantly lowered performance. Using semaphores to transfer data eliminated the bus contention but there was hardly any performance gain when going from one DSP to 4 DSPs. The code was then written to take advantage of the on board DMA controllers and this resulted in a nearly linear speedup. The 4 DSP solution appears to be faster than 4 times the single DSP solution. This is because all the data could be stored in internal memory with 4 DSPs and not with the single DSP solution.

The parallel DSP method ended up being slower than the best PC implementation. The reason for this was that the clock speed of DSP was much less than the high end PC. The clock speed of the ADSP 21160 ran at 80MHz in order to get single cycle multiplies, while the clock speed of the Pentium IV ran at 3.3 GHz, which is heavily pipelined. Thus, the high-end PC was about 41 times faster than the DSP than in terms of clock speed. Even so, the DSP performance was only 11 times slower, highlighting the architectural advantage that the DSPs have in terms of multiply and accumulate operations.

5   FPGA Implementation

Field Programmable Gate Arrays (FPGAs) are reconfigurable devices where algorithms can be embedded directly in hardware. The advantage with using FPGAs is there can be hundreds of embedded multipliers running in parallel tied to custom logic. We targeted Xilinx's Virtex II FPGA that had the following resources (table 3).

Table 3. Virtex II XC2V3000 Resources

System Gates	CLB Slices	Multipliers	Block RAMs
3M	14336	96	96

One drawback of using FPGAs to implementation complex algorithms like the EM algorithm is the long time it can take to design and optimize the algorithms for internal FPGA resources. We used a recently developed tool, AccelFPGA [4], to shorten the design time. AccelFPGA can compile Matlab code directly into VHDL code. By using AccelFPGA, we could focus on optimizing the algorithm in Matlab without dealing with low level hardware details inside the FPGA. The design cycle of the FPGA implementation using this method can be reduced significantly. The FPGA implementation through AccelFPGA is not as optimal as directly coding the algorithms in VHDL by hand. However, for a prototype design, AccelFPGA can provide a time-efficient FPGA solution with reasonably good performance. The performance of the FPGA implementation is shown in table 4.

Table 4. Training time of EM algorithm using FPGA

FPGA Execution Time (sec)	Speedup relative to best PC	Speedup relative to best DSP implementation
0.08	16.4	42.5

Using AccelFPGA v1.6 and Xilinx ISE v5.2.02i

The FPGA implementation was 16.4 times faster than the fastest PC and 42.5 times faster than the best parallel DSP implementation. It should be noted that FPGAs are best for fixed-point processing and that the embedded multipliers are 18x18 bits. Thus the optimal use of FPGAs require the conversion from floating-point to fixed-point. This is ideally done in Matlab using the fixed-point toolbox where simulations can be made examining the effects of reduced precision. Given that the algorithm can be mapped effectively to fixed-point, the next step is to add compiler directives for AccelFPGA. These directives can specify the usage of FPGA's embedded hardware resources such as BlockRAM and embedded multiplier blocks. The third step is the creation of the RTL model in VHDL and the associated testbenches which are automatically created by the AccelFPGA compiler. The fourth step is to synthesize the VHDL models using a logic synthesis tool. The gate-level netlist is then simulated to ensure functional correctness against the system specification. Finally the gate-level netlist is placed and routed by the place-and-route appropriate for the FPGAs used in the implementation. The design used 42 of the 96 embedded multipliers (42%), 58 of the 96 BlockRAMs (60%), and 6378 of the 14336 logic slices (44%).

One advantage of AccelFPGA is that bit-true simulations can be done directly in the Matlab environment. Testbenches are automatically generated along with the simulations. You can use these testbenches later in the hardware simulation and compare the results with the Matlab simulation. As a result, it is very easy to know whether your hardware implementation is correct or not.

Since Matlab is a high level language, the compiled implementation will typically no be as good as direct VHDL coding. However, by using proper directives, you can specify the parallelism of your algorithm and maximize the usage of the on-chip hardware resources. Hence, you can get a reasonably good performance out of your implementation.

The structure of the FPGAs is optimized for high-speed fixed-point addition, subtraction and multiplication. However, a number of other math operations such as division or exponentiation have to be implemented. This was done via lookup tables using BlockRAMs.

6   Conclusion

It appears that the use of FPGAs are ideal for use in data driven systems. Not only are they suitable for digital signal processing applications where incoming data streams need to be processed in real-time, they can also be used to implement sophisticated algorithms such as the EM algorithm.

References

1. Yang, M.H., Ahuja, N.: Gaussian Mixture Model for Human Skin Color and Its Application in Image and Video Databases. Proc. of the SPIE, (1999) 3635
2. Reynolds, D.A.: Speaker identification and verification using Gaussian mixture speaker models. Speech Communication, 17 (1995)
3. Redner, R.A., Walker, H.F.: Mixture Densities, Maximum Likelihood and EM Algorithm. SIAM Review, 26 (1984) 195-239.
4. AccelFPGA User's Manual. V1.7. www.accelchip.com, (2003)

Karhunen–Loeve Representation of Periodic Second-Order Autoregressive Processes

Didier Lucor, Chau-Hsing Su, and George Em Karniadakis

Division of Applied Mathematics, Brown University, Providence, RI 02912, USA.

Abstract. In dynamic data driven applications modeling accurately the uncertainty of various inputs is a key step of the process. In this paper, we first review the basics of the Karhunen-Loève decomposition as a means for representing stochastic inputs. Then, we derive explicit expressions of one-dimensional covariance kernels associated with periodic spatial second-order autoregressive processes. We also construct numerically those kernels by employing the Karhunen-Loève expansion and making use of Fourier representation in order to solve efficiently the associated eigenvalue problem. Convergence and accuracy of the numerical procedure are checked by comparing the covariance kernels obtained from the Karhunen-Loève expansions against theoretical solutions.

1 Introduction

In many application of dynamic data driven systems there is a significant degree of uncertainty associated with the various inputs due to inaccuracy of measurements, gappy data, or incomplete knowledge. We can incorporate this uncertainty directly into the models by formulating stochastic algorithms, but a significant first step is to represent accurately and efficiently all stochastic inputs. The Karhunen-Loève (KL) expansion is a very powerful tool for representing stationary and non-stationary random processes with explicitly known covariance functions [1]. In a few cases, the eigen-solutions of the covariance function can be obtained analytically but in most cases they have to be obtained numerically [2]. The KL representation is optimal as both the deterministic basis functions and the corresponding random coefficients are orthogonal. If the random process is assumed stationary and the ratio between the length of the domain and the correlation length is infinite or periodic boundary conditions are imposed, the KL representation becomes a *spectral* expansion with uncorrelated random coefficients [3].

One of the simplest and most used random processes is the first-order Markov process, which relates to the Brownian motion of small particles and the diffusion phenomenon. It is a unilateral type of scheme extended only in one direction. The covariance kernel associated with that one-dimensional first-order autoregressive process takes an exponential form [2]. However, realistic models of random series *in space* require autoregressive schemes with dependence in *all* directions. In some cases, it has been shown that schemes of bilateral type in one dimension can be effectively reduced to a unilateral one [4].

M. Bubak et al. (Eds.): ICCS 2004, LNCS 3038, pp. 827–834, 2004.
© Springer-Verlag Berlin Heidelberg 2004

In the current work, we derive explicit expressions of one-dimensional co-variance kernels associated with periodic second-order autoregressive processes. We then represent numerically those kernels using a KL representation where we make use of the Fourier series representation in order to compute efficiently the associated eigenvalue problem. Convergence and accuracy of the numerical method are checked against theoretical solutions for the kernels.

2   The Karhunen-Loève Expansion

The KL expansion is based on the spectral expansion of the covariance function $c(t_1, t_2)$ of the process $v(t, \theta)$, where t_1 and t_2 are two temporal coordinates and time is from a to b. The KL expansion then takes the following form:

$$v(t, \theta) = \bar{v}(t) + \sigma_v \sum_{i=1}^{\infty} \sqrt{\lambda_i} \psi_i(t) \xi_i(\theta), \tag{1}$$

where $\bar{v}(t)$ denotes the *mean* of the random process, σ_v denotes the *standard deviation* of the process, and $\xi_i(\theta)$ is a set of independent random variables with a given random distribution; they form an orthonormal random vector (throughout this paper, we will use the symbol ξ to denote a random variable with zero mean and unit variance.) Obviously, truncated expansions will be used in practice. Also, $\psi_i(t)$ and λ_i are the eigenfunctions and eigenvalues of the covariance function, respectively, i.e.,

$$\int_a^b c(t_1, t_2) \psi_i(t_2) dt_2 = \lambda_i \psi_i(t_1). \tag{2}$$

The KL expansion minimizes the mean-square error resulting from a finite-term representation of the process [2]. Its use, however, is limited as the covariance function of the solution process is not known *a priori*. Nevertheless, the KL expansion still provides a powerful means for representing *input* random processes when the covariance structure is known. Some numerical experiments regarding the accuracy of truncated KL representation for different covariance kernels have been reported in [5,6]. The rate of convergence of the KL expansion is closely related to the smoothness of the covariance kernel and to the ratio between the length of process T and the correlation length A. For instance, for the particular case of a first-order Markov process, given by $c(t_1, t_2) = \sigma_v^2 e^{-|t_2 - t_1|/A}$, an upper bound for the relative *error in variance* ϵ of the process represented by its KL expansion is such that:

$$\epsilon \leq \frac{4}{\pi^2} \frac{1}{n} \frac{T}{A} \approx 0.4053 \frac{1}{n} \frac{T}{A}. \tag{3}$$

We see that it depends on the ratio $\frac{T}{A}$, where T is the length of the time domain, and it is inversely proportional to the number of retained terms. The dependence in $(\frac{1}{n})$ is related to the rate of convergence of the Fourier decomposition for this particular kernel, which has a discontinuity in its first derivative at the origin [6].

3   Fourier Series Representation of the Integral Equation

If the random process is *stationary* with respect to t, then the covariance satisfies:

$$c(t, t_1) = \langle v(t)v(t_1) \rangle = c(t - t_1). \tag{4}$$

The corresponding integral equation (2) for the eigenfunctions and eigenvalues used in the KL representation can be cast into a spectral form by taking a Fourier series representation of the kernel function $c(t - t_1)$ and the eigenfunctions $\psi(t)$. The Fourier series representation provides a natural way of solving the problem if it is periodic, i.e. both the kernel function c and the eigenfunctions ψ are periodic functions with period T. In this case, the integral equation can be solved exactly.

With both $\psi(t)$ and $c(t)$ being periodic functions of period T, we represent them in terms of their Fourier series as follows:

$$\psi(t) = \frac{\psi_{co}}{2} + \sum_{n=1}^{\infty} \left[\psi_{cn} \cos \frac{2n\pi}{T}(t - a) + \psi_{sn} \sin \frac{2n\pi}{T}(t - a) \right], \tag{5}$$

$$c(t) = \frac{c_{co}}{2} + \sum_{n=1}^{\infty} \left[c_{cn} \cos \frac{2n\pi}{T}t + c_{sn} \sin \frac{2n\pi}{T}t \right], \tag{6}$$

with

$$\psi_{cn} = \frac{2}{T} \int_a^b dt\psi(t) \cos \frac{2n\pi}{T}(t - a), \quad \psi_{sn} = \frac{1}{T} \int_a^b dt\psi(t) \sin \frac{2n\pi}{T}(t - a) \tag{7}$$

$$c_{cn} = \frac{2}{T} \int_{-T/2}^{T/2} dtc(t) \cos \frac{2n\pi}{T}t, \quad c_{sn} = \frac{2}{T} \int_{-T/2}^{T/2} dtc(t) \sin \frac{2n\pi}{T}t \tag{8}$$

In most applications $c(t)$ is an even function and thus $c_{sn} = 0$. Substituting equations (5-6) into equation (2), we obtain:

$$\left(\lambda - \frac{c_{co}}{2}T \right) \frac{\psi_{co}}{2} + \sum_{n=1}^{\infty} \left(\lambda - \frac{c_{cn}}{2}T \right) \left[\psi_{cn} \cos \frac{2n\pi}{T}(t - a) + \psi_{sn} \sin \frac{2n\pi}{2}(t - a) \right] = 0. \tag{9}$$

We derive the set of eigenvalues and eigenfunctions as:

1. $\lambda = \lambda_0 = \frac{T}{2}c_{co}$, $\psi_{co} \neq 0$, $\psi_{cn} = \psi_{sn} = 0$ for $n \neq 0$ and $\psi^{(0)} = \frac{1}{\sqrt{T}}$.
2. $\lambda = \lambda_n = \frac{T}{2}c_{cn}$, $\psi_{cn} \neq 0$, and $\psi_{sn} \neq 0, \psi_{cm} = \psi_{sm} = 0$ for $m \neq n$, and

$$\psi_1^{(n)}(t_1) = \sqrt{\frac{2}{T}} \cos \frac{2n\pi}{T}(t_1 - a), \quad \psi_2^{(n)}(t_1) = \sqrt{\frac{2}{T}} \sin \frac{2n\pi}{T}(t_1 - a). \tag{10}$$

The KL representation of the periodic random process $v(t, \xi)$ is:

$$v(t, \xi) = \sqrt{\frac{\lambda_o}{T}} \xi_{co} + \sqrt{\frac{2}{T}} \sum_{n=1}^{\infty} \sqrt{\lambda_n} \left[\xi_{cn} \cos \frac{2n\pi}{T}(t - a) + \xi_{sn} \sin \frac{2n\pi}{T}(t - a) \right],$$

$$\tag{11}$$

where

$$\lambda_n = \frac{T}{2}c_{cn} = \int_{-T/2}^{T/2} c(t)\cos\frac{2n\pi}{T}t\,dt = 2\int_0^{T/2} c(t)\cos\frac{2n\pi}{T}t\,dt. \qquad (12)$$

Given a covariance function $c(t)$, one can solve these integrals using appropriate numerical quadrature methods. The corresponding covariance function which can be obtained directly from equation (11):

$$\langle v(t,\xi)v(t_1,\xi)\rangle = \frac{c_{co}}{2} + \sum_{n=1}^{\infty} c_{cn}\cos\frac{2n\pi}{T}(t-t_1) = c(t-t_1), \qquad (13)$$

is just the Fourier series representation of the covariance function c.

It is worthwhile to note that this approach can also be used as an approximation for non-periodic cases if the domain T of the problem is much larger than the correlation length A. The range of the functions $\psi(t)$ and $\psi(t_1)$ is (a,b) while that of $c(t-t_1)$ is $(-T,T)$ where $T = b - a$. In order to represent the extended functions as Fourier series, i.e., we need to take the period to be $2T$. However, if $c(t-t_1)$ has a compact support, or if it is different from zero only for $|t-t_1| \le A$ for some finite A, then in cases where $A \ll T$, the relative error in using the period being T rather than $2T$ is of the order of A/T. In the literature, this approximation is sometime refered to as *spectral approximation* [7,8].

Notations and more detailed derivations for Section 3 can be found in the work of Lucor [9].

4   Second-Order Autoregressive Processes

For *space series*, a *bilateral* autoregression takes the form:

$$v_i = \gamma v_{i-1} + \delta v_{i+1} + \epsilon_i \quad \text{or}$$

$$v_i = \frac{b}{2}(v_{i-1} + v_{i+1}) + af\xi_i, \qquad (14)$$

where it is intuitively clear that γ and δ cannot be too large. Let us denote by L the spatial length of the periodic domain and take $L = 1$ without loss of generality. We discretize the domain with $2m + 1$ equidistant points, so that we have: $m = L/2\Delta x$. The expression for v_i, due to the homogeneity of the system based on the assumed periodicity, reduces to:

$$v_i = \sum_{j=0}^{2m-1} \alpha_j \xi_{i+j}.$$

The coefficients are such that:

$$\alpha_j = \begin{cases} \frac{af}{1-\frac{b^2}{D_m}} = 1 & \text{for } j = 0 \\[2ex] \alpha_{j-1}/D_{m-i+1} & \text{for } 0 < j \le i \text{ with } D_1 = 1;\ D_k = 2 - b^2/D_{k-1};\ 1 < k \le m \\[2ex] 0 & \text{for } j > i \end{cases}$$

Because we take $\alpha_0 = 1$, the variance of the process is conserved and is constant for each point in the domain, so we have: $af = 1 - b^2/D_m$. The covariance $\langle v_i v_j \rangle$ between two points of the grid is obtained from the dynamical system:

$$\langle v_i v_j \rangle = \sum_{j=0}^{2m-1} \alpha_j \alpha_{j+i-k}. \tag{15}$$

Let us define the mode number q that represents the number of waves that one could fit within the domain L; the corresponding wave number is $k = 2\pi q/L$. There exists an analogy between a second-order regressive dynamical system of type (14) and the discrete version of a non-homogeneous second-order ordinary equation [10]:

$$\frac{v_{i+1} - 2v_i + v_{i-1}}{\Delta x^2} \pm k^2 v_i = F(x_i). \tag{16}$$

Indeed, the system of equations (14) can be transformed into equation (16) by setting $b = 1/(1 \pm \frac{1}{2}(k\Delta x)^2)$ and $af\xi_i = -F(x_i)\Delta x^2/(2 + (k\Delta x)^2)$. This gives in the continuous limit of Δx going to zero:

$$b = e^{\pm \frac{1}{2}(k\Delta x)^2} = e^{\pm \frac{1}{2}(\frac{2\pi q}{L}\Delta x)^2} \tag{17}$$

The wave number k has the dimension of the inverse of a correlation length A, $A = 1/k$. The coefficient b tends to unity as Δx (grid size) tends to zero. The plus or minus signs in equation (17) indicate that b can approach unity from above or below. This distinction gives totally different results for the covariance kernel form.

If $b = e^{-\frac{1}{2}(k\Delta x)^2} < 1$, which corresponds to equation (16) with a negative sign, and under the assumption that the domain is periodic, the covariance of the solution is:

$$\langle v(x), v(y) \rangle = \frac{1}{4k^2 \sinh^2 \frac{kL}{2}} \int_0^L \int_0^L \langle\, F(x_1)F(x_2)\rangle \cosh k\Big(|x - x_1| - \frac{L}{2}\Big)dx_1$$

$$\cosh k\Big(|y - x_2| - \frac{L}{2}\Big)dx_2. \tag{18}$$

If the variance of the forcing function is conserved for each point of the domain, i.e. $\langle F(x)F(y)\rangle = \sigma_F^2 \delta(x - y)$, it becomes:

$$\langle v(x), v(y) \rangle = \frac{\sigma_F^2}{4k^2 \sinh^2 \frac{kL}{2}} \int_0^L \cosh k\Big(|x - x_1| - \frac{L}{2}\Big) \cosh k\Big(|y - x_1| - \frac{L}{2}\Big)dx_1. \tag{19}$$

The normalized covariance kernel becomes in this case:

$$c(x, y) = \frac{1}{2\alpha + \sinh(2\alpha)}\Big[2\alpha + \sinh(2\alpha) + \alpha|x - y|(\cosh(2\alpha) - 1)\Big]\cosh(\alpha|x - y|)$$

$$+ \Big[1 - \cosh(2\alpha) - \alpha|x - y|\sinh(2\alpha)\Big]\sinh(\alpha|x - y|) \quad ; \quad \alpha = q\pi \tag{20}$$

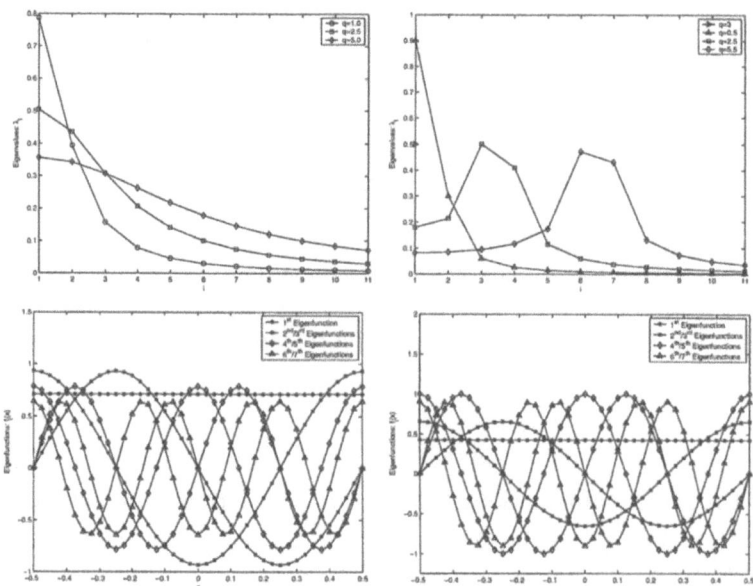

Fig. 1. Left column: **Kernel1**. Right column: **Kernel2**. Top row: Eigenvalues λ_i for various values of the correlation length, here given by q, with $q = L/(2\pi A)$ and $L = 1$. Bottom row: First four-pairs eigenfunctions $f_i(x)$ with $|x| \leq L/2$ and $L = 1$ and $q = 2.5$.

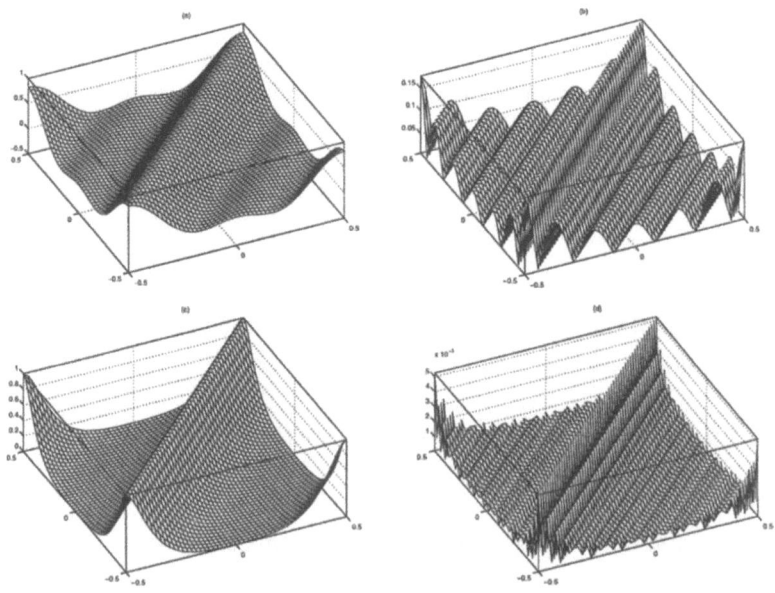

Fig. 2. Approximation of Covariance **Kernel1** with $q = 2.5$; $k = 5\pi$. (a): $n = 5$-term KL approximation (b): $n = 5$-term absolute error (c): $n = 21$-term KL approximation (d): $n = 21$-term absolute error.

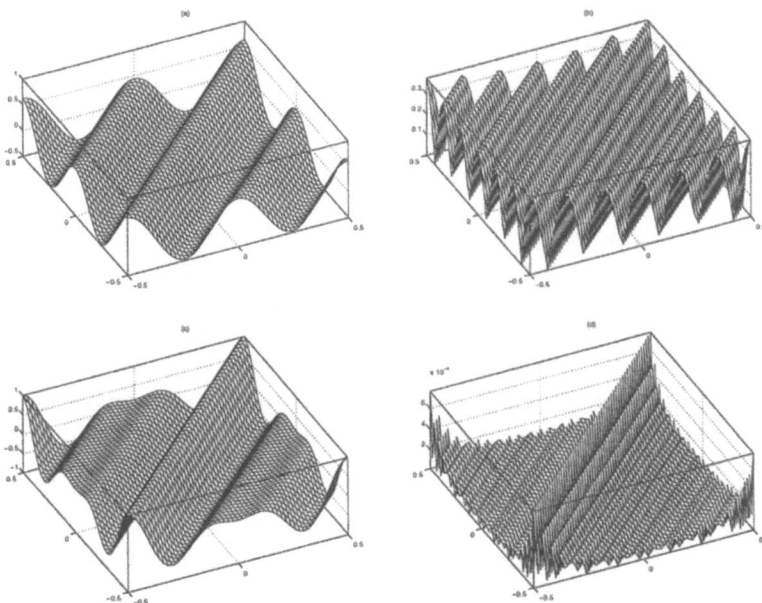

Fig. 3. Approximation of Covariance **Kernel2** with $q = 2.5$; $k = 5\pi$, (a): $n = 5$-term KL approximation (b): $n = 5$-term absolute error (c): $n = 21$-term KL approximation (d): $n = 21$-term absolute error.

In the following, we will refer to the covariance kernel given by equation (20) as **Kernel1**. We show a close approximation to this kernel in Figure 2. This kernel is periodic, as expected, and the decay rate is controlled by the correlation length value. The kernel decays faster for smaller correlation lengths.

Similarly and under the same assumptions as in the previous case, if $b = e^{\frac{1}{2}(k\Delta x)^2} > 1$, which corresponds to equation (16) with a positive sign, it can be shown that the covariance kernel is similar to **Kernel1** and becomes:

$$c(x, y) = \frac{1}{2\alpha + \sin(2\alpha)}\left[2\alpha + \sin(2\alpha) + \alpha|x - y|(\cos(2\alpha) - 1)\right]\cos(\alpha|x - y|)$$
$$+\left[1 - \cos(2\alpha) + \alpha|x - y|\sin(2\alpha)\right]\sin(\alpha|x - y|). \tag{21}$$

In the following, we will refer to the covariance kernel given by equation (21) as **Kernel2**. We show a close approximation to this kernel in Figure 3. This kernel decays faster for smaller correlation lengths. Interestingly, if q is an integer, the covariance kernel given by equation (21) simply becomes $c(x, y) = \cos(\alpha|x - y|)$. Since all those kernels are periodic, it is possible to employ the Fourier decomposition method of the integral equation described in the previous paragraph, in order to build a KL representation of the corresponding processes. In this case, the time dimension is treated as a spatial dimension without loss of generality. We obtain the eigenvalues and eigenfunctions of the kernels from equations (10) and (7-8). The integrals of equations (12) for **Kernel1** and

Kernel2 are computed numerically using numerical quadratures, except for the case of **Kernel2** with q being an integer. In that case, we can directly derive exact expressions for the eigenvalues and eigenfunctions of the kernel. We only have one double eigenvalue corresponding to two eigenfunctions:

$$\lambda_0 = \lambda_1 = \frac{L}{2} \, ; \quad \psi^{(0)} = \sqrt{\frac{2}{L}} \cos(kx) \quad \text{and} \quad \psi^{(1)} = \sqrt{\frac{2}{L}} \sin(kx) \qquad (22)$$

Figure 1-top shows the eigenvalues for **Kernel1** (left) and for **Kernel2** (right) for different correlation lengths A. Note that the smaller the value of A, the more contribution is from terms associated with smaller eigenvalues.

Figure 1-bottom shows the first seven eigenfunctions for **Kernel1** (left) and **Kernel2** (right) for $q = 2.5$. As seen in equations (10), each eigenvalue λ_n is associated with a pair of eigenfunctions $\psi_1^{(n)}$ and $\psi_2^{(n)}$ corresponding to a cosine and sine contributions, respectively. Figures 2 and 3 show the 5-term and 21-term KL approximations of **Kernel1** and **Kernel2** for $q = 2.5$ and the associated errors. We notice that the kernels obtained from the KL representations tend to the corresponding exact kernels as the number of terms n increases. For both cases with identical mode number q, the 21-term approximation does much better than the 5-term approximation. It is interesting to notice that the convergence is not monotonic, and the maximum absolute error is about one order of magnitude lower for **Kernel2** than for **Kernel1** for the 21-term approximation, see Figures 2-(d) and 3-(d).

Acknowledgements. This work was supported by the DDDAS program of the National Science Foundation under the supervision of Dr. Frederica Darema.

References

1. M. Loève. *Probability Theory, Fourth edition.* Springer-Verlag, 1977.
2. R.G. Ghanem and P. Spanos. *Stochastic Finite Elements: a Spectral Approach.* Springer-Verlag, 1991.
3. H. Stark and W.J. Woods. *Probability, Random Process, and Estimation Theory for Engineers.* Prentice-Hall, Inc., Englewood Cliffs, New Jersey, 1994.
4. P. Whittle. On stationary processes in the plane. *Biometrika,* 41:434–449, 1954.
5. S.P. Huang, S.T. Quek, and K.K. Phoon. Convergence study of the truncated Karhunen-Loeve expansion for simulation of stochastic processes. *Int. J. Numer. Methods Eng.,* 52:1029–1043, 2001.
6. D. Lucor, C.-H. Su, and G.E. Karniadakis. Generalized polynomial chaos and random oscillators. *Int. J. Numer. Methods Eng.,* in press, 2004.
7. M. Shinozuka and G. Deodatis. Simulation of the stochastic process by spectral representation. *ASME Appl. Mech. Rev.,* 44(4):29–53, 1991.
8. M. Grigoriu. On the spectral representation method in simulation. *Probabilistic Engineering Mechanics,* 8(2):75–90, 1993.
9. D. Lucor. *Generalized Polynomial Chaos: Applications to Random Oscillators and Flow-Structure Interactions.* PhD thesis, Brown University, 2004.
10. C.H. Su and D. Lucor. Derivations of covariance kernels from dynamical systems and PDEs. In *ICOSAHOM-04, Brown University, Providence USA, June 21-25,* 2004.

Using Web Services to Integrate Heterogeneous Simulations in a Grid Environment

J. Mark Pullen[1], Ryan Brunton[2], Don Brutzman[3], David Drake[2], Michael Hieb[4],
Katherine L. Morse[2], and Andreas Tolk[5]

[1] Department of Computer Science & C3I Center, George Mason University, Fairfax,
Virginia, 22030, USA
mpullen@gmu.edu
[2] Science Applications International Corporation (SAIC), 10260 Campus Point Drive,
San Diego, CA 92121, USA
{RYAN.P.BRUNTON,DAVID.L.DRAKE-2,KATHERINE.L.MORSE}@saic.com
[3] MOVES Institute, Naval Postgraduate School, Monterey, California 93943 , USA
brutzman@nps.navy.mil
[4] Alion Science & Technology, S &T, 1901 N. Beauregard St. Alexandria,
Virginia 22311, USA
Michael.R.Hieb@us.army.mil
[5] VMASC, Old Dominion University, Norfolk, Virginia 23529, USA
atolk@odu.edu

Abstract. The distributed information technologies collectively known as Web services recently have demonstrated powerful capabilities for scalable interoperation of heterogeneous software across a wide variety of networked platforms. This approach supports a rapid integration cycle and shows promise for ultimately supporting automatic composability of services using discovery via registries. This paper presents a rationale for extending Web services to distributed simulation environments, including the High Level Architecture (HLA), together with a description and examples of the integration methodology used to develop significant prototype implementations. A logical next step is combining the power of Grid computing with Web services to facilitate rapid integration in a demanding computation and database access environment. This combination, which has been called Grid services, is an emerging research area with challenging problems to be faced in bringing Web services and Grid computing together effectively.

1 Introduction

The distributed information technologies collectively known as Web services recently have been shown to offer powerful capabilities for scalable interoperation of heterogeneous software across a wide variety of networked platforms. A particular appeal of this approach is the rapid integration cycle it provides, which shows promise for supporting automatic composability of services ultimately by using discovery on Web service registries. The authors have championed extension of Web services to distributed simulation environments, including the HLA, under the name Extensible Modeling and Simulation Framework (XMSF)[1]. We present below our

M. Bubak et al. (Eds.): ICCS 2004, LNCS 3038, pp. 835–847, 2004.
© Springer-Verlag Berlin Heidelberg 2004

rationale for the viability of the approach and a description of the integration methodology we have used to develop significant prototype implementations. XMSF has the potential to become a unifying framework for heterogeneous distributed simulation involving a wide variety of languages, operating systems, and hardware platforms. Interoperation of non-HLA simulations with HLA federations and other software systems such as military command and control is an attractive possibility.

We are now engaged in expanding the XMSF paradigm to address a wider range of simulation issues. An overarching approach to system design using the Model Driven Architecture (MDA) in the context of web service implementations is proving to be a foundational standard, as presented in Section 3. Connecting high-performance computational resources to diverse physics-based 3D visualizations is increasingly important. An example is shown in Section 4.3 where Sonar is visualized by computation and rendering of sonar-ping propagation via a Web service. XMSF has the potential to become a unifier for heterogeneous distributed simulation involving a wide variety of languages, operating systems, and hardware platforms. As the scale and complexity of such systems increases, the required computational, database, and networking resources soon will outstrip the collections of networked workstations we have used to create our prototypes. A logical extension is to bring together Web services with Grid computing, as addressed in the final section of this paper.

2   Web Services Definition

The fundamental idea behind Web services is to integrate software applications as services. This concept is based on a defined set of technologies, supported by open industry standards, that work together to facilitate interoperability among heterogeneous systems, whether within an organization or across the Internet. In other words, Web services enable applications to communicate with other applications using open standards. This approach has tremendous potential to build bridges between "stove-piped" legacy systems, that is, systems that were designed without open lateral interfaces to each other. Fundamentally, Web services provide an approach to distributed computing with application resources provided over networks using standard technologies. Because they are based on standard interfaces, Web services can communicate even if they are running on different operating systems and are written in different languages. The standards are widely supported by industry and have been applied successfully in a wide range of different domains. For this reason, Web services provide an excellent approach for building distributed applications that must incorporate diverse systems over a network.

The Web services framework provides a set of operations, modular and independent applications that can be published, discovered, and invoked by using industry standard protocols described below. The resulting distributed computing model represents the interaction between pairs of programs, rather than the interaction between programs and user. An equivalent statement is that Web services are discrete Web-based applications that interact dynamically with other Web services. In order to make this happen, several sub-functions are necessary:

Self-description of the service functionality
Publishing the service descriptions using a standardized format

Locating the service with the required functionality
Establishing message communications with the service
Requesting the required data to initiate the service
Exchanging data with other Web services, including delivering the results.

Protocols most commonly used for these purposes are the Extensible Markup Language (XML), Simple Object Access Protocol (SOAP), Web Service Description Language (WSDL), and Universal Distribution Discovery and Interoperability (UDDI).

2.1 Messaging for Web Services

All web service implementations consist of three core components: a consumer, a provider, and an optional service registry. The consumer application locates providers either by querying a registry service or by accessing a known service endpoint. Service requests are encoded using an XML vocabulary understood by both the consumer and the provider, encapsulated in a SOAP envelope, and sent to the provider using Internet protocols. The message is decoded, acted on, and the response encoded and returned in a parallel process. While the consumer/provider abstraction is an effective descriptive simplification, the roles are not always so straightforward. SOAP envelope headers can contain routing information that allows messages to be propagated in a chain of connected web services, each node modifying the payload and acting as both a provider for the previous node and a consumer for the succeeding node. Alternatively, in a system consisting of two bidirectional nodes like the Web Enabled RTI, each node is both a provider and a consumer. Figure 1 illustrates these relationships.

While the most common implementation of Web services sends messages in SOAP over the Hypertext Transfer Protocol (HTTP) used for webpages, the formal W3C definition is far more robust and flexible:

A Web service is a software system identified by a URI [RFC 2396], whose public interfaces and bindings are defined and described using XML. Its definition can be discovered by other software systems. These systems may then interact with the Web service in a manner prescribed by its definition, using XML based messages conveyed by Internet protocols.[2].

Given this definition, Web services bindings are not constrained by the traditional limitations of HTTP based Web services. For example, as a request-response based protocol, the HTTP is inherently limited to operations that take place synchronously within a brief connection time. Protocols such as Simple Mail Transfer Protocol (SMTP) and the Blocks Extensible Exchange Protocol (BEEP) allow for asynchronous message processing and response. We will consider here the most general form of Web service.

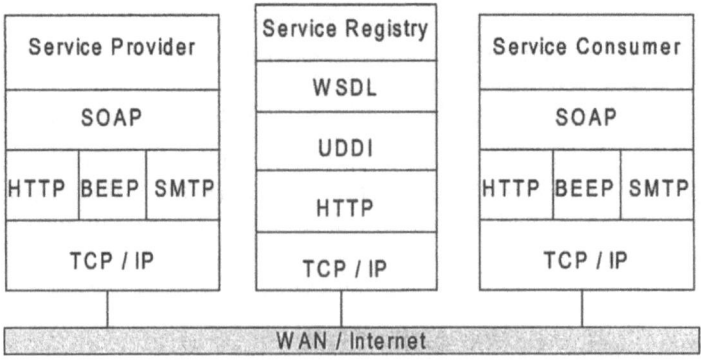

Fig. 1. Web Services Protocol Stack

2.2 Web Services and the HLA-RTI

The High Level Architecture (HLA) for simulation, as defined in IEEE Standard 1516 [3], provides a structure and set of services for networked interoperation of simulation software. The interoperation strategy is based on defining a Federation Object Model (FOM) that is common to all members of a simulation federation and providing a Run Time Infrastructure (RTI) that supports a standard set of communication services among the federates. We have proposed framework (XMSF) that expands the concepts of the HLA into the realm of Web services. In the most general case, we envision each simulation providing one or more Web services that provide the data needed to interoperate with other simulations. The rules of the HLA restrict federates to sharing FOM data via the federation's RTI, but do not preclude sharing data with other software or other federations by means of a Web service offered through the RTI.

Moreover, interoperability of software via Web services depends on creating a common vocabulary (in effect, an ontology) that is shared by two or more software applications and expressed as an XML tagset. This arrangement can be seen as a superset of the FOM, which also can be expressed as an XML tagset under IEEE standard 1516. Thus we see that the stage is already set for adapting Web services to support HLA-compliant distributed simulations. Benefits might occur in two ways:

- The RTI supporting an HLA federation can be constructed from commercial Web service components, reducing cost to support the HLA because the cost of commercial components is amortized over a larger number of users.
- HLA federations can interoperate with other software applications supported by Web services, using as a tagset a mapping between the federation's FOM and whatever ontology is shared with the other applications. This approach offers considerable promise for interoperation of military command, control, communications, computing and intelligence (C4I) software whose sponsors have not been willing to adopt the HLA but are beginning to use via Web services as an interoperability solution.

3 Composing Simulation Systems Using the MDA Techniques

For the XMSF approach to be viable, we must show how it can bridge the gap between HLA-compliant federations and other software systems and also the gap that arises when simulations are supported by different underlying infrastructures, such as traditional systems and the Grid computing approach. To achieve this will require a combination of a technical approach and an overarching conceptual approach, allowing composition and orchestration of the participating systems. While Web services are sufficient on the technical level, Web services alone are insufficient to achieve this at a conceptual level; therefore XMSF advocates applying techniques from the Model Driven Architecture of the Object Management Group (OMG) [4] to facilitate design of meaningful interoperability among distributed and individually implemented components.. Applying the MDA approach can provide benefits in structuring functionality in a cross-platform, cross-language way using well-defined patterns for model based system design and re-engineering. Conscientious use of well-defined design patterns, as exemplified by MDA principles, provides a basis on which predictably interoperable and composable software components can be assembled. This holds true both within complex applications and across interconnected applications running across the Internet.

The Web service vision is that services will work together seamlessly because they are developed to the same standards for self-description, publication, location, communication, invocation, and data exchange capabilities. As all the standards concerned are open, the technologies chosen for Web services are inherently neutral to compatibility issues that exist between programming languages, middleware solutions, and operating platforms. As a result, applications using Web services can dynamically locate and use necessary functionality, whether available locally or from across the Internet.

However, the existence of the appropriate standards is not sufficient to ensure consistency and meaningful interoperability in the distributed simulation environment. As currently defined, the Web services family of standards does not support the composition of agile and dynamic simulation components. Meaningful implementation of interoperability among simulation systems on this level requires composability of the underlying conceptual models. The HLA and related approaches provide the mechanics of interoperation, but they fall short when aiming at composability [5][6]. Conceptual problems cannot be solved with only implementation-driven solutions, whether they take the form of the HLA or of Web services. To succeed, Web services for distributed heterogeneous simulation must be embedded into a larger context ensuring "meaningful" interoperability. Applying the techniques for software design and re-engineering within a distributed software project can help to ensure composability and orchestration on the conceptual level [4].

Previously the OMG has developed the Common Object Request Broker Architecture (CORBA) standard and related technologies. The MDA approach encompasses the established standards and standard development procedures of OMG and raises them to a new level. The MDA methods ensure that components can be described in a common way, and that the processes of composing these components as well as orchestrating them in a common composition are commonly understood and consistent. The main objective of the MDA is the ability to derive software from a stable model as the underlying infrastructure shifts over time. To achieve this, the model of the application is captured in an implementation and platform-independent language. The specification of this core model is based on the established OMG standards Unified Modeling Language (UML), Meta-Object Facility (MOF), and the Common Warehouse Metamodel (CWM). The result is a Platform Independent Model (PIM) that captures the concepts of the application rather than its implementation. These PIMs are essential to ensure "meaningful" interoperability, as they capture the concepts and intents of the applications while the other elements the technical connectivity between the components on the communication level. When applied for system design,the PIM is transformed into a Platform Specific Model (PSM) that can be used for CORBA applications, HLA compliant applications, or Web service applications following standardized mapping rules. When applied in the context of re-engineering, following similar inverse mapping rules a PIM can be established as a formal representation of the conceptual model of the component. This PIM can be integrated easily into the formal description of the mission space comprising all PIMs, and hence the concepts and functionality, of all participating components. Therefore, the MDA not only complements the Web services above the implementation level by introducing a common language for conceptual modeling with the PIM, but also supports the migration of legacy components into web enabled components reusable for distributed heterogeneous simulation in future applications [7].

In summary, XMSF is a Web service oriented implementation of the conceptual ideas captured by the MDA. This approach allows legacy systems implemented in heterogeneous information technology environments to migrate toward a common simulation framework, implementing a common conceptual schema within a heterogeneous infrastructure. Thus XMSF not only has the potential to bridge the gap between HLA federations and other systems, but also can facilitate bridging gaps among various hardware and operating system solutions,. We will make the case below that this facilitates Grid applications as well as traditional applications.

4   Implementing Web Services for Simulation

Implementing XMSF requires a shared ontology for data exchange, which takes the form of an XML tagset. As described below, the process of defining this ontology varies considerably according to the organizational environment involved and the complexity of the data to be defined. The process is facilitated considerably by the nature of the Web service model, which involves each software system making its own data available on its own terms. Nevertheless, where the goal is a very broad level of interoperability as in the Command and Control Information Exchange Data Model (C2IEDM) based ontology described below, the process of definition can be quite protracted, as with an HLA FOM. After the tagset has been defined, the strength of the open standards for message-based data exchange becomes evident. We report here on three prototypes, each of which demonstrated rapid, successful integration of existing software.

4.1   XMSF DCEE Viewer

The Distributed Continuous Experimentation Environment (DCEE), managed by the Experimentation Directorate of the U.S. Joint Forces Command (J9/JFCOM), has established a framework of common terminology for the information to be exchanged between components using an enhancement of the Real-time Platform Reference (RPR) FOM. Although the DCEE presently uses HLA as the technical backbone, the concept is open for extensions to emerging solutions. Use of XML, standardized tagsets, Web services, and Internet technology is part of the general concept of DCEE.

Under a project sponsored by JFCOM, we demonstrated the benefits of XMSF in the DCEE with the XMSF DCEE Viewer (XDV). XDV is a web-based, low-cost viewer for DCEE based events in the Joint Semi Automated Forces (JSAF) and relation simulations. The concept for XDV is simple: every eligible stakeholder interested in observing execution of the ongoing experiment can connect to the HLA federation and use the viewer software to follow the actual experiment. The necessary software was installed on widely available computer workstations connected via Internet protocols, allowing eligible stakeholders to follow the experiment from wherever they were located. Key components enabling rapid implementation were the Web Enabled RTI (WE RTI) [8] developed by SAIC and the map graphics packages developed by the Virginia Modeling, Simulation and Analysis Center (VMASC) for visualization as used on command and control maps (unit based viewer) and by the Naval Postgraduate School (NPS) for visualization on the platform level as used in emerging environments (entity based viewer).

Figure 2 illustrates the logical architecture of the XDV. The XDV was tested successfully in all possible four permutations of installation the viewer and server locations at VMASC and JFCOM.

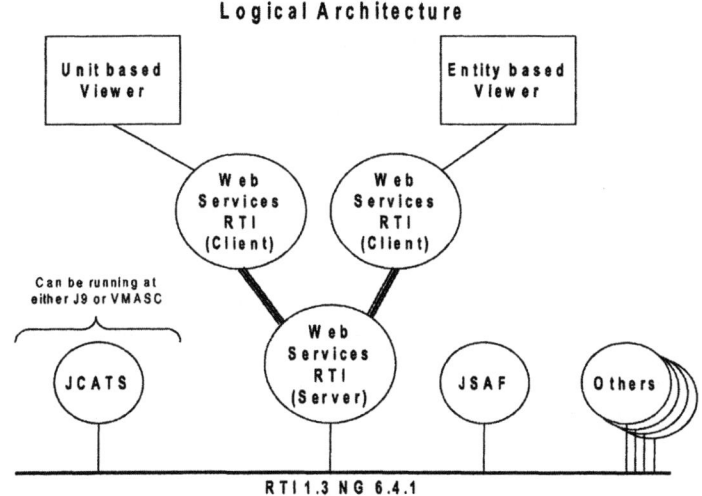

Fig. 2. XDV Logical Architecture

A key future goal is to operationalize the viewer. While the prototype was successful within the limits of its requirements, it is still a prototype, i.e. not robust enough or sufficiently documented to ensure easy distribution, installation, and use by a broad user base. Operationalizing the viewer will entail applying a standard software engineering. We are also investigating the introduction of Area Of Interest Management (AOIM) to support compartmentalization of data and bandwidth use reduction. This work also will be applicable to the Joint National Training Capability (JNTC) currently under development by the Training Directorate of JFCOM.

4.2 Extensible Battle Management Language

Military Command and Control (C2) communication in a network centric environment is postulated to be very data intensive. However, the commander's intent, orders and directives, which is the most critical C2 information, does not currently flow as data. It is communicated as "free text" elements within messages or as stand-alone files. Its vocabulary is found in doctrinal task lists and manuals, but it lacks clearly delineated rules governing its use (semantics and syntax). It is riddled with ambiguity and overlapping definitions. As such, it is incapable of transitioning to the full range of automation that the Department of Defense is implementing. It will not support either digitized C2 or decision support based upon integrated modeling and simulation. While the current formats are suitable for interpersonal communication, they are inadequate for use with simulations, or for future forces that have robotic components. As commanders increasingly rely upon simulation-based decision aids (and therefore use their C2 devices to control simulations) a solution for this "free text" problem must be found. Battle Management Language (BML) was developed as a solution to this problem.

The Battle Management Language (BML) is defined as the unambiguous language used to: 1) command and control forces and equipment conducting military operations and, 2) provide for situational awareness and a shared, common operational picture [9][10]. It can be seen as a representation of a digitized commander's intent to be used for real troops, for simulated troops, and for future robotic forces. The U.S. Army developed a prototype of BML demonstrating the ability to generate an actual National Training Center (NTC) Brigade Operations Order in an advanced Army C4I Planning System, CAPES (Combined Arms Planning and Execution System) and have it executed in widely used Army Simulation, the OneSAF Test Bed (OTB) [11]. The BML prototype used a database containing C4I data, called the Multi Source Data Base (MSDB) to interoperate between the various components in the prototype.

A major achievement of BML was to adapt representations of command and control so that they are directly derived from doctrine and are expressed in a flexible syntax. This provides a means to link the BML (terminology and symbology) directly to their doctrinal source; and it allows operational forces to use their own information systems both to interact with supporting simulations to conduct rigorous, realistic training and support mission rehearsals, and, in the future, to support an expedited military decision making process.

We used the XMSF approach to transform BML to a Web enabled or Extensible Battle Management Language (XBML)[12]. Figure 3 shows Phase One of XBML, which distributed the Army BML Prototype by implementing interfaces to the MSDB

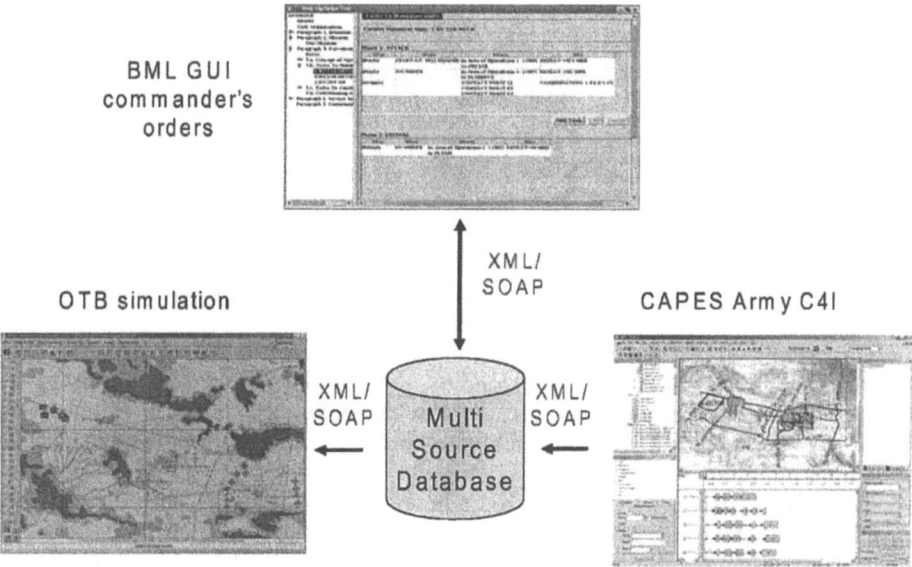

Fig. 3. XBML Testbed project converted to Web services in 3 months

based upon SOAP and XML. This allowed different components to be distributed geographically, and opened the interfaces so more C4I and Simulation nodes could join in an event. The new capability was introduced in less than three months, demonstrating that it is straightforward to rework proprietary interfaces with XMSF standards. Our next goal is to extend XBML into a Joint (involving more than one US military service) and coalition solution, based on open standards. The XML tag set as well as the MSDB of this version is based on the NATO standard, the C2IEDM[13] that will be extended by applying the standards promoted by XMSF. The goal of XBML is to demonstrate a methodology for developing standard doctrinal terms and allowing these to be accessed as Web services.

XBML addresses a longstanding problem in military modeling and simulation: C4I system to simulation interoperability. It does so in two areas:

Development of a Methodology for Easily Integrating Simulations within C4I Systems: The methodology used in "opening" the interfaces in XBML will be documented such that it can be easily reapplied by other members of the community easily, therefore becoming the basis for a standardized way to deal with this challenge.

Defining a Live/C4I Ontology for Simulations: As with other XMSF applications, a common tagset is needed for interoperation. Using the C2IEDM and its standardized procedures for extensions of the model, XBML is contributing to the ontology definitions as the basis of a broadly accepted international standard.

4.3 Sonar Visualization

The NPS Sonar Visualization Project merges physics-based sonar-modeling algorithms, 3D graphics visualization tools and Web-based technologies to provide the military with relevant real-time sonar analysis. Tactical decision aids in this project utilize Web-based Extensible 3D (X3D) models for composable rendering together with Web Services messaging and XML Schema-Based Compression (XSBC) for reliable transmission.

Environmental effects are processed in real-time by distributed sonar servers. Largescale environmental datasets maintained on supercomputers also can be accessed via Web services using XML SOAP messaging.

X3D visualization schemes provide intuitive and useful ways to describe multipath sonar propagation results to USW operators. Animated X3D results are shown in a deployable real-time tactical application. For example, collision detection between sensor raylets and scene geometry aids in evaluation of bottom interactions, surface reflections, and threat contacts.

The project also is developing a common XML tagset to describe acoustic, environmental and sonar path data. This will allow a deployed client to access varying levels of environmental data and processing power, as available, to support specific mission needs.

5 Web Services and Grid

We stated previously that the most common implementation of Web services consists of a SOAP messaging framework using HTTP as the transport mechanism, and that the formal definition of a web service is far more flexible. We now introduce a definition of Grid computing for purposes of comparison:

> A parallel, distributed system composed of heterogeneous resources located in different places and belonging to different administrative domains connected over a network using open standards and protocols [14].

Using this definition, Grid computing can be seen as a natural evolution of distributed computing technologies such as RMI and CORBA (see Figure 4). Where traditional distributed computing technologies tightly coupled the remote resource location to the client stub and often required end-to-end control of the network, Grid computing has moved to a more robust and transparent architecture, allowing resources to discover each other over wide area networks without control of the infrastructure. A lesson also has been learned from the mistakes of those previous proprietary technologies and the positive example provided by Web services. By using open standards and protocols, Grid computing solutions gain the one thing that is necessary for their success: ubiquity of peer resources.

Further comparing the definitions of Grid computing and Web services, we find that while the two both describe distributed computing technologies, they describe them in largely orthogonal ways. The Web services definition focuses on the use of XML to describe both service interfaces and the communication messaging format. Grid computing focuses on the system architecture, leaving the particulars of protocols and message formats unspecified. With complimentary goals and orthogonal requirements it is unsurprising that Grid computing and Web services have been merged into coherent distributed systems. These solutions, known as Grid services, are Grid computing implementations which use XML to describe their interfaces and encode messages, and open internet protocols for communication. Examples of service implementations that aim to meet this ideal are JXTA, a peer-to-peer architecture supported by Sun Microsystems, and the Open Grid Service Architecture (OGSA), the Web services interface from Globus, whose toolkit is the defacto standard for Grid computing [14][15].

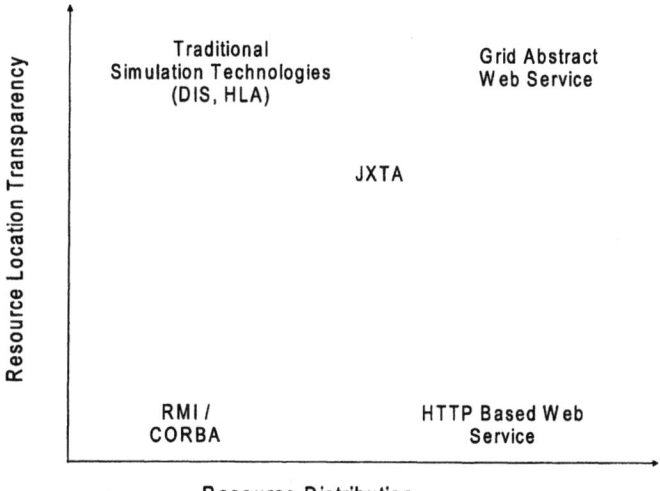

Fig. 4. Comparison of Distributed Computing Paradigms

Having established that Grid computing and distributed HLA based simulations can both be implemented using Web services, we need only discover common ground between Grid computing and the HLA to create a foundation for a Grid services RTI implementation. This common ground exists in the HLA specification: federates exist as simulation resources with complete transparency to each other. While current RTI implementations limit the distribution of federates, this is an implementation decision and not a requirement of the HLA. The set of objects and interactions available to a federation are defined in its FOM, and object and ownership management allow instances to be shared and divested between running resources without regard to federate location. Synchronization issues can be managed through the judicious use of the time management functions. One shortcoming of the HLA that could prevent a Grid services implementation is the lack of an explicit authentication mechanism. Any federate can join a federation, send any interactions it publishes, receive any attribute updates or interactions it subscribes to, and update any attributes for owned object. In addition, updates and interactions do not carry any explicit identification of their origin, leaving no recourse for application accounting within the specification. This does not necessarily preclude implementing the HLA over Grid services, but it would place severe constraints on other simultaneous uses of the underlying Grid.

6 Conclusions

The XMSF project is only two years old and already has made major progress by defining the role of Web services for interoperation of simulations and demonstrating their effectiveness in multiple projects. The rapid success of these efforts and lack of any major problems in their implementation is highly encouraging. XMSF is able to support interoperation among disparate software systems, including HLA federations, on heterogeneous platforms. The principles of Web services are also applicable in the Grid computing environment where they promise to allow further expansion of the range of distributed simulation possibilities, via Grid services.

Acknowledgement. Work described in this paper was supported in part by the US Defense Modeling and Simulation Office (DMSO) and JFCOM.

References

1. Donald Brutzman, Michael Zyda, J. Mark Pullen and Katherine L. Morse: Extensible Modeling and Simulation Framework (XMSF): Challenges for Web-Based Modeling and Simulation, Findings and Recommendations Report of the XMSF Technical Challenges Workshop and Strategic Opportunities Symposium, October 2002, http://www.movesinstitute.org/xmsf/XmsfWorkshopSymposiumReportOctober2002.pdf
2. W3C (2002) Web Services Architecture Requirements, November 14 2002, Online documents at http://www.w3.org/TR/wsa-reqs.
3. IEEE 1516, Standard for Modeling & Simulation - High Level Architecture.
4. Object Management Group (OMG) website: Model Driven Architecture (MDA); http://www.omg.org/mda/
5. Andreas Tolk, James Muguira: The Levels of Conceptual Interoperability Model (LCIM). Paper 03F-SIW-007; *Proceedings of the IEEE Fall 2003 Simulation Interoperability Workshop*, 2003; http://www.odu.edu/engr/vmasc/TolkPublications.shtml
6. Mikel Petty *et al*: Findings and Recommendations from the 2003 Composable Mission Space Environments Workshop, Paper 04S-SIW-050, *Proceedings of the IEEE 2004 Spring Simulation Interoperability Workshop*
7. Andreas Tolk: Avoiding another Green Elephant, A Proposal for the Next Generation HLA based on the Model Driven Architecture. Paper 02F-SIW-004, *Proceedings of the IEEE 2002 Fall Simulation Interoperability Workshop*, 2002; http://www.odu.edu/engr/vmasc/TolkPublications.shtml
8. Katherine L. Morse, Ryan Brunton and David Drake: Web Enabling an RTI – an XMSF Profile, Paper 03E-SIW-046, *Proceedings of the IEEE 2003 European Simulation Interoperability Workshop*, 2003.
9. Scott Carey, Martin Kleiner, Michael Hieb, and Richard Brown: Standardizing Battle Management Language - A Vital Move Towards the Army Transformation, Paper 01FSIW-067, *Proceedings of the IEEE 2001 Fall Simulation Interoperability Workshop*, 2001.
10. Scott Carey, Martin Kleiner, Michael Hieb, and Richard Brown: Standardizing Battle Management Language - Facilitating Coalition Interoperability, Paper 02E-SIW-005, *Proceedings of the IEEE 2002 European Simulation Interoperability Workshop*, 2002.
11. William Sudnikovich, Michael Hieb, Martin Kleiner: Developing the Army's Battle Management Language Prototype Environment, Paper 04S-SIW-115, *Proceedings of the IEEE 2004 Spring Simulation Interoperability Workshop*, 2004.

12. Michael Hieb, William Sudnikovich, Andreas Tolk, and J. Mark Pullen: Developing Battle Management Language into a Web Service, Paper 04S-SIW-113, *Proceedings of the IEEE 2004 Spring Simulation Interoperability Workshop*, 2004.
13. Multilateral Interoperability Programme (MIP): The C2 Information Exchange Data Model (C2IEDM). 20 November 2003, approved by MIP in Greding, Germany, http://www.mipsite.org/MIP_Specifications/Baseline_2.0/C2IEDMC2_Information_Exchange_Data_Model/t
14. Wikipedia: Grid computing, January 4 2004, Online documents at http://en2.wikipedia.org/wiki/Grid_computing.
15. Rajkumar Buyya: Grid Computing Info Centre, January 29, 2004, Online documents at http://www.gridcomputing.com.

Support for Effective and Fault Tolerant Execution of HLA-Based Applications in the OGSA Framework*

Katarzyna Rycerz[1], Marian Bubak[1,2], Maciej Malawski[1], and Peter Sloot[3]

[1] Institute of Computer Science, AGH, al. Mickiewicza 30, 30-059 Krakow, Poland
[2] Academic Computer Centre – CYFRONET, Nawojki 11, 30-950 Kraków, Poland
[3] Faculty of Sciences, Section of Computational Science, University of Amsterdam
Kruislaan 403, 1098 SJ Amsterdam, The Netherlands
phone: (+48 12) 617 39 64, fax: (+48 12) 633 80 54
{kzajac,bubak,malawski}@uci.agh.edu.pl, sloot@science.uva.nl

Abstract. The execution of High Level Architecture (HLA) distributed interactive simulations in an unreliable Grid environment requires efficient brokering of resources, which is an important part of the Grid Services framework supporting execution of such simulations. This paper presents the overall architecture of the framework with emphasis on services supplying the Broker Service with information about application performance on a Wide Area Network (WAN). For that purpose, a benchmark interaction-simulation-visualization schema is designed basing on CrossGrid medical application architecture [1,10].

Keywords: Interactive simulation, Grid computing, HLA, OGSA

1 Introduction

Distributed interactive applications often require extensive computing resources. The Grid is a promising concept for solving this problem as it offers a possibility to access resources which are not centrally controlled and are under different administrative policies.

Interactive applications built with an implementation of the HLA standard [9] allow for merging geographically-distributed parts of simulations into a coherent whole. However, applying HLA to the Grid environment on an "as is" basis is not enough. HLA-based simulations require a certain quality of service in the underlying infrastructure. First, HLA is based on the assumption that the environment it executes on is faultless, while the Grid environment is inherently unreliable and requires fault tolerance mechanisms. Secondly, the Grid facilitates access to computing resources and makes them more transparent to the user.

---

* This research is partly funded by the European Commission IST-2001-32243 Project "CrossGrid" and the Polish State Committee for Scientific Research, SPUB-M 112/E-356/SPB/5.PR UE/DZ224/2002-2004.

M. Bubak et al. (Eds.): ICCS 2004, LNCS 3038, pp. 848–855, 2004.
© Springer-Verlag Berlin Heidelberg 2004

Currently, creating a distributed application based on HLA requires manual setup and configuration. HLA limitations are discussed in [15,16].

[15] identifies different levels on which it is possible to integrate HLA applications with Grid solutions. On the uppermost layer, we can consider HLA components as available resources that can be used to run HLA–based application on the Grid. In the case of the Department of Defense Runtime Infrastructure (DoD RTI) implemenation of HLA, we can treat RTIExec and RTI libraries as resources that are needed to execute applications dynamically linked with the RTI library. The lower layers of porting HLA concepts to the Grid would depend on implementing HLA communication using Grid technologies (i.e. using GridFTP optimizations[8] for data transfer[15]) or redisigning HLA service (such as time or data management) implementations [14].

We have decided to focus on the uppermost level, which involves considering HLA as a resource and propose a system to support transparent and faultless execution of HLA-based applications in a Grid environment. The aim of this research is to design and build a system supporting interactive applications in the Grid environment and at the same time preserve backward compatibility with the HLA standard. As an example we use the medical application developed in the CrossGrid project [2]. In this application, connections between blood flow simulation, visualization and interaction are provided by means of Interactive Simulation System Studio (ISS-Studio) agents middleware [17] that, in turn, uses HLA in its communication layer.

In our previous paper [16] we have presented the overall architecture of the Grid Service framework for supporting execution of HLA-based applications. In this paper we present a more advanced concept of system architecture. We focus on the part of the framework responsible for benchmarks, which supports decisions regarding application setup and migration. In Section 2 we describe related work in the area of integrating simulation standards with Web services and the Grid. In Section 3 we present the overall system design. In Section 4 we focus on the part of the system responsible for supporting the broker in selecting the appropriate HLA Services for distributed interactive applications. We propose Grid Services that manage HLA benchmarks according to the model of application distribution. We base our measurements on CrossGrid [2] medical application requirements. Benchmark results are presented in Section 5. Section 6 concludes the paper.

2   Related Work

The modeling and simulation community has come to realize that currently, there does not exist a widely-accepted standard which could handle both modeling and simulation mechanisms, such as the HLA [9] and Web/IT standards [13]. Such was the motivation behind the Extensible Modelling and Simulation Framework (XMSF) project [14], whose aim is to develop a common standard for Web and DOD modeling and simulation.

A parallel research effort is going on within the Commercial Off-The-Shelf Simulation (COTS) Package Interoperability Forum (HLA-CSPIF) [3]. As there is no real standard use pattern for the High Level Architecture within the context of this area, the goal of the Forum is to create standards through Simulation Interoperability Standards Organization (SISO) [11] that will facilitate the interoperation of COTS simulation packages and thus present users of such packages with the benefits of distributed simulations enjoyed by other modeling and simulation communities.

Then there is research related to supporting management of HLA–distributed simulations. Cai et al. [12] implemented a load management system for running large-scale HLA simulations in the Grid environment basing on GT2. They also present a framework where the modeler can design parallel and distributed simulations with no knowledge of HLA/RTI [12]. The authors hope that in the future the framework design should facilitate easy checkpointing and federate migration for dynamic load balancing.

Creating a scalable and fault-tolerant framework for HLA–based applications is also the goal of the Swedish Defence Research Agency (FOI), where the Decentralized Resource Management System (DRMS) is being developed[6]. DRMS is a JXTA-based peer-to-peer system for execution of HLA (High Level Architecture) federations in a decentralized environment.

3   Overview of Framework Design

In our previous paper [16] we presented the overall architecture of the Grid Service framework for supporting execution of HLA-based applications. In particular, we discussed an experimental HLA–speaking service that manages execution of a federate as described below. In this work we focus on the part of the framework that supports the *Broker Service* in making decisions on which HLA–speaking service to choose. Our proposed architecture involves a framework of Open Grid Services Architecture (OGSA) Grid services [7] that cooperate to set up distributed interactive applications. Within our framework, presented in Fig.1, each site providing HLA to the Grid community sets up the following *HLA Management Services*:

The *HLA–Speaking Service*, responsible for setting up, running and directly managing user federate code on its site and also acting as an interface between other services and the federate. The service starts execution of federate code on its site, saves the application state and restores it from the checkpoint file on *Migration Service* requests. A prototype of the *HLA–Speaking Service* was described in [16].

Benchmark Services that perform HLA benchmarks according to various scenarios. In this paper we concentrate on the simple interaction–simulation–visualization scenario described in Section 4.

Benchmark Analyzer Services analyze HLA benchmarks results according to *Broker Service* needs. These services analyze data produced by the *Benchmark*

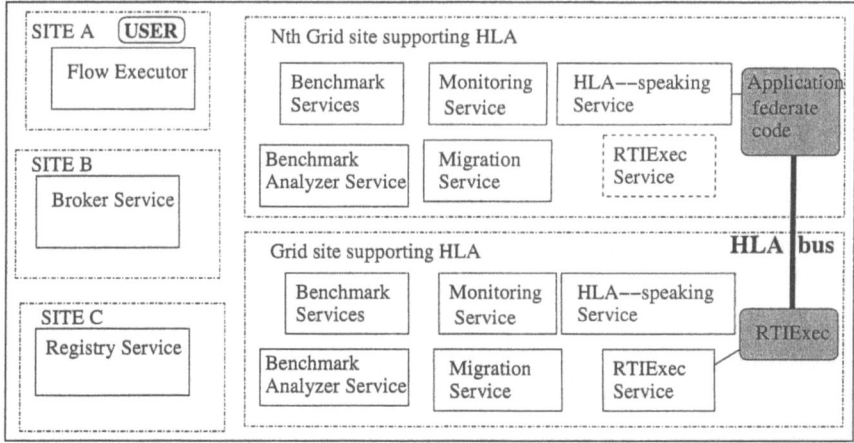

Fig. 1. Grid Services Framework for managing HLA–based application

Service online (by asking it to perform measurements) or offline (by analyzing data from benchmarks stored in a database).

The *Monitoring Service* monitors performance of current federation.

The *Migration Service* orchestrates migration of HLA federates from its site to other sites basing on *Broker Service* information (see below).

The *RTIExec Service* (optional)– manages execution of the RTIExec process.

We also assume the presence of the *Registry Service* that stores locations of possible *HLA Management Services* described above and can be used as a database for finding appropriate services.

Additionally, the user has to have access to the following services:

The *Broker Service* that finds appropriate HLA–speaking Services for a given execution scenario basing on *Benchmark Analyzer Services*.

The *Flow Setup* service that sets up the application for the user and supplies appropriate HLA–speaking services with application code that is already compiled and dynamically linked with the HLA RTI library.

The goal of our system is to facilitate the usage of these distributed resources (such as HLA RTI library) in robust manner regardless of the Grid's unpredictable behavior.

4   Proposal of Support for Broker Decisions

Broker Service algorithms are crucial for performance of distributed applications. The broker has to make decisions not only during application setup, but also during migration of its parts. One important problem is performance of the distributed application on WANs. It is obvious that components, between which *communication is intensive*, should be placed within the same high performance network. A problem arises if there are not enough resources on one site. In that

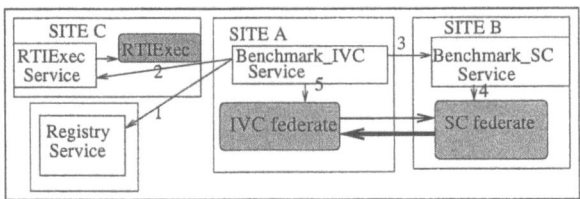

Fig. 2. Performing HLA Interaction-Simulation-Visualization Benchmarks in the Grid Services Framework

case some intelligent decisions must be made on how to effectively distribute parts of applications between Grid nodes.

In this paper we focus on the part of the framework that supplies the *Broker Service* with information about communication performance of the application. In order to make brokerage decisions, the system needs to monitor hosts and the network infrastructure. Although various monitoring tools are currently being developed in the CrossGrid project[2], it is extremely difficult to monitor WAN connections. One possible solution is to perform on-line HLA-based benchmarks and make decisions about migration basing on their results.

In general, benchmarks should be suited to the application model of distribution. In this work, we focus on the scenario present in the CrossGrid biomedical application [1]. This application consists of three components : interaction, simulation and visualization. While simulation is being conducted, the user can change its parameters using the interaction module (sending small messages from interaction to simulation) and see results through visualization (simulation sends large updates to visualization). According to this scenario, our benchmark consists of two components: an integrated Interaction-Visualization Component (IVC) and a Simulation Component (SC). IVC sends small messages that are human interactive requests for change of SC parameters. SC responses are large messages that present the altered state of the simulation to the user. Certainly, this scenario is suitable only for some class of applications. In the future we are going to extend the set of benchmarks (Section 6).

The HLA benchmark federation consists of two federates, acting as IVC and SC respectively. IVC sends SC requests in which it asks for certain amounts of data and measures the time between sending the request and receiving a response. Messages are sent through updates of HLA data objects [9]. The benchmark is managed within the Grid Services framework by the *Benchmark_IVC Service* and the *Benchmark_SC Service*, that start, control and measure performance of IVC and SC respectively.

The task of the *Benchmark_SC Service* is to start and control SC federates. This is done by invoking their **Start** operation. The *Benchmark_IVC Service* takes from the *Registry Service* a list of available Grid Service Handlers (GSH) (step1 in Fig.2) of *Benchmark_SC Services* and *RTIexec* services. Then, it can perform operations within the *Benchmark_IVC PortType* listed in Tab.1.

Table 1. Operations of the *Benchmark_IVC PortType*.

check – makes requests to a specific *Benchmark_SC Service* indicated by Grid Service Handler (GSH)(step 2 in Fig.2). To perform this operation, *Benchmark_IVC Service* chooses the *RTIExec Service* and creates a federation. Then, it invokes the **start** operation of the *Benchmark_SC PortType* of the *Benchmark_IVC Service* (step 3) that sets up the SC part of the benchmark by connecting it to the same federation(step 4). Following that, the *Benchmark_IVC Service* starts its IVC federate (step 5) and performs measurements as described above
scan – performs **check** operations on the selected list of sites. If there is no response until a specific timeout, the *Benchmark_IVC Service* proceeds to the next site.
forscan – performs the **scan** operation from start time to stop time for each time interval.

5   Implementation and Results

The structure of our benchmarks was based on the CrossGrid medical application requirements [10]. In the original application, for a medium resolution of the computational mesh used in the simulation, the uncompressed update message from visualization to simulation is about 600 MB in size. For larger resolutions, the uncompressed update can reach several gigabytes. Obviously, compression will be used for reducing the volume of update messages.

Our measures confirm what can be expected of the Grid environment in conditions similar to those described above. Our current implementation consists of the *Benchmark_IVC Service* – **check** and **scan** operations, as well as the *Benchmark_SC Service* – **start** operation. Measurements were performed on the DutchGrid DAS2 [5] coupled with the Krakow ACC Cyfronet [4] infrastructure. The internal bandwidth between DAS2 nodes was 10Gbps and the bandwidth between DAS2 nodes and Cyfronet was 100Mbps. We set up the *Benchmark_IVC Service* at Vrije University in Amsterdam and performed the **scan** operation for connections between all other sites of DAS2 as indicated in Tab.2. The RTIExec process was run in Krakow, at ACC Cyfronet. The first column presents the measured time of the **check** operation, which consisted of setting up the federation in the OGSA framework as described above and sending actual data (10 requests (small message) and 10 responses (400MB each)). Each response was in fact di-

Table 2. Benchmark results (in seconds).

	check operation (incl. 10 x send) time		actual 10 x send time	
	avr	δ	avr	δ
A'dam(Vrije)–Leiden	695	10.5	656	10.7
A'dam(Vrije)–A'dam (UvA)	450	13.5	413	15.1
A'dam(Vrije)–Delft	289	9.9	252	8.2
A'dam(Vrije)–Utrecht	377	8.9	340	9.1
Total **scan** operation	1811.2	30.9		

vided into 1MB chunks of data due to memory allocation issues (RTI requires four times more memory then the size of the actual sending/receiving buffer). In the next column we show time of ten request-response operations made by pure HLA object updates. In the last row of the table we show the total duration of the **scan** operation. The measurements were taken outside working hours form 9-10 p.m., to avoid additional traffic. They give us an idea of what to expect from application performance in a simple interaction–simulation scenario and helps decide if it is good enough for running interactive applications (for some applications this can be satisfactory, for others – less so). The results also show that the benchmarks can provide support for decisions on where to put application components (the location of the interaction and visualization component is fixed due to human presence constraints, but the simulation component can be placed – for instance – in Delft according to benchmark results). Some tuning of benchmark parameters is necessary depending on whether they are performed on– or off–line.

6   Conclusions and Future Work

In this paper we describe the framework for management of HLA-based interactive simulations in the Grid environment. While in [16] we describe the migration support part of the framework, this paper focuses on support for the *Broker Service* that will help in choosing between HLA Services on the Grid not only during setup of the HLA-based application, but also while performing migration decisions under two different scenarios - failure or bad performance. We are currently working on the following issues:

Broker Decisions and Benchmark Analyzer. While designing *Broker Service* architecture and algorithms, it is also necessary to have a service that will analyze benchmark results according to specific needs of the broker.

Various Benchmark scenarios. The benchmarking case described here is based on a simple Interaction-Simulation-Visualization scenario, however we are working on extending it to other scenarios.

Migration Service - periodic checkpointing. Periodic checkpointing will allow saving application state before a failure and restoring it afterwords. Also, a decision should be made where to back up checkpoint files in order to use them in the case of a failure.

Single HLA-speaking service – many federates. In general, an HLA-speaking service should be able to manage many federates on its site. We are working on extending our experimental implementation to cover that case.

Monitoring Service As described in Section 3, there is a need for a service that will monitor performance of the current application to supply *Migration Service* with information on whether and when migration is needed.

Acknowledgments. The authors wish to thank Dick van Albada, Alfredo Tirado-Ramos and Zhiming Zhao for valuable discussions about OGSA and HLA, and Piotr Nowakowski for his remarks.

References

1. Bubak, M., Malawski, M., Zajac, K.: Architecture of the Grid for Interactive Applications, in: Sloot, P. M. A., et al. (Eds.), Proceedings of Computational Science - ICCS 2003, International Conference Melbourne, Australia and St. Petersburg, Russia, June 2003, vol. I, no. 2657, Lecture Notes in Computer Science, Springer, 2003, pp. 207-213.
2. Crossgrid project web page. http://www.crossgrid.org/
3. HLA CSPIF home page: http://www.cspif.com
4. ACK Cyfronet http://www.cyfronet.krakow.pl
5. Distributed Ascii Supercomputer http://www.cs.vu.nl/das2/
6. Eklof, M., Sparf, M., Moradi, F: HLA Federation Management in a Peer-to-Peer Environment in In Proceedings of the 03 European Simulation Interoperability Workshop, organized by the Simulation Interoperability Standardization Organization. Stockholm, 2003. http://www.sisostds.org/
7. Foster I, Kesselman C., Nick J., Tuecke S.: The Physiology of the Grid: An Open Grid Services Architecture for Distributed Systems Integration. Open Grid Service Infrastructure WG, Global Grid Forum, June 22, 2002.
8. Globus project homepage http://www.globus.org
9. HLA specification, http://www.sisostds.org/stdsdev/hla/
10. Private communication with Robert Bellman (available upon request).
11. Simulation Interoperability Standards Organization's (SISO) http://www.sisostds.org/
12. Yuan, Z., Cai, W. and Low, M.Y.H.: A Framework for Executing Parallel Simulation using RTI published in: in: S.J. Turner and S.J.E. Taylor, editors, Proceedings Seventh IEEE International Symposium on Distributed Simulation and Real Time Applications (DS-RT 2003),pp. 20-28 IEEE Computer Society, Delft, The Netherlands, October 2003.
13. Web services, http://www.w3.org/2002/ws/
14. XMSF project home site. http://www.movesinstitute.org/xmsf/xmsf.html
15. Zajac, K., Tirado-Ramos A., Bubak, M., Malawski, M., Sloot, P.: Grid Services for HLA-based Distributed Simulation Frameworks, to appear in Proceedings of First European Across Grids Conference, Santiago de Compostela, Spain, Springer-Verlag, Heidelberg, February 2003.
16. Zajac, K., Bubak, M., Malawski, M., Sloot, P.: Towards a Grid Management System for HLA–based Interactive Simulations, in: S.J. Turner and S.J.E. Taylor, editors, Proceedings Seventh IEEE International Symposium on Distributed Simulation and Real Time Applications (DS-RT 2003), pp. 4-11. IEEE Computer Society, Delft, The Netherlands, October 2003.
17. Zhao, Z., van Albada, G. D., Tirado-Ramos, A., Zajac, K., Sloot, P. M. A., ISS-Studio: A Prototype for a User-Friendly Tool for Designing Interactive Experiments in Problem Solving Environments, in: Sloot, P. M. A., et al. (Eds.), Proceedings of Computational Science - ICCS 2003, International Conference Melbourne, Australia and St. Petersburg, Russia, June 2003, vol. I, no. 2657, Lecture Notes in Computer Science, Springer, 2003, pp. 679-688.

Federate Migration in HLA-Based Simulation

Zijing Yuan[1], Wentong Cai[1], Malcolm Yoke Hean Low[2], and
Stephen J. Turner[1]

[1] School of Computer Engineering
Nanyang Technological University, Singapore 639798
{p144328830,aswtcai,assjturner}@ntu.edu.sg
[2] Singapore Institute of Manufacturing Technology
71 Nanyang Drive, Singapore 638075
yhlow@simtech.a-star.edu.sg

Abstract. The High Level Architecture promotes simulation interoperability and reusability, making it an ideal candidate to model large-scale systems. However, a large-scale simulation running in a distributed environment is often affected by the imbalance of load level at different computing hosts. Migrating processes from heavily-loaded hosts to less-loaded hosts can overcome this shortcoming. We have previously developed a SimKernel framework to execute HLA-based simulations in the Grid environment with migration support as a prominent design feature. In this paper, we will introduce a transparent migration protocol for SimKernel-based federates that minimizes migration overhead.

Keywords: HLA-based simulation, federate migration, load management, Grid computing

1   Introduction

The High Level Architecture (HLA) was developed by the U.S. Defence Modeling and Simulation Office (DMSO) to provide simulation interoperability and reusability across all types of simulations and was adopted as an IEEE standard [1]. The Runtime Infrastructure (RTI) is an implementation of the HLA standard that provides the platform for simulation.

In HLA, a simulation is called a federation and a simulation component is called a federate. Federates communicate with their peers by sending interactions or updating object attributes. The interaction is the formal HLA definition of a transient message of parameter compounds. In this paper, we will use interaction, event and message interchangeably. Federates do not communicate directly with each other and all communication is administrated by the RTI based on each individual federate's interest.

While HLA/RTI aims to provide a software platform for simulation interoperability, the Grid [2] provides an ideal hardware infrastructure for the high performance computing community. There has been research effort to integrate the HLA simulations with the Grid environment to solve large-scale compute-intensive problems [3,4,5,6].

M. Bubak et al. (Eds.): ICCS 2004, LNCS 3038, pp. 856–864, 2004.
© Springer-Verlag Berlin Heidelberg 2004

Large-scale simulations running in a distributed environment are often affected by the imbalance of dynamic load level at individual participating hosts, and this leads to poor performance. Load balancing is a technique that can improve hardware utilization and shorten the execution time. To achieve a balanced load, a process is migrated from a heavily-loaded host to a less-loaded one.

However, process migration generally incurs a large overhead mainly due to poor migration protocol design. This is especially true with HLA-based simulations where not only analyzing and extracting the federate execution state requires tremendous effort, but also migration coordination may undesirably halt non-migrating federates in the simulation. Hence, it would be particularly beneficial to minimize the undesirable large migration overhead.

We have previously developed a SimKernel framework for modelers to design and deploy parallel and distributed simulations in the Grid environment using HLA/RTI. In this paper, we will introduce a migration mechanism based on SimKernel that could minimize migration overhead.

2 Related Works

Federate migration can be achieved at various levels. Obviously, general purpose process migration schemes could also be used to migrate HLA federates. In this paper, however, we focus only on application level federate migration.

The easiest approach to migrate a federate is to utilize the HLA standard interfaces: *federationSave* and *federationRestore*. The drawback is apparent: federation wide synchronization is required. Another side-effect is that all non-migrating federates are required to participate in the federation save and restore process for every migration request. As seen in [5], the migration overhead increases almost proportionally with the number of federates in the simulation.

Other implementations generally adopt the checkpoint-and-restore approach. In both [3,4], the migrating federate's essential state is checkpointed and uploaded to an FTP server. The restarted federate will reclaim the state information from the FTP server and perform a restoration. These implementations introduce further latency since communicating with the FTP server is more time consuming.

Hence, minimizing the migration latency would be of great interest. As migration overhead stems mainly from synchronization and communication with a third party (such as FTP), mechanisms avoiding these issues would be desirable. Such algorithms exist in other migration studies. An interesting *freeze free* algorithm for general purpose process migration was proposed by Roush [7]. In this algorithm, the source host receives messages before the communication links are transferred. Any message arriving while the communication links are in transit will be held temporarily and will be sent to the destination when the links are ready at the new host. Message receipt is only delayed while the communication links are in transit. This greatly reduces process freeze time since non-migrating processes are not involved in the migration. The *migration mailbox* is another approach [8] where a predefined address called a migration mailbox receives messages for the migrating process. After the process is successfully migrated, it will retrieve the messages from the mailbox and inform other processes to send messages directly to it. The shadow object approach used in avatar migration [9] also achieves the same target. In this approach, each server monitors an interest area and neighboring servers' interest areas overlap in a migration region. An avatar is maintained by a server. When an avatar moves into the migration region, a shadow object is created on the neighboring server. Once the avatar is out of the original server's scope, the shadow object is upgraded to an avatar on the neighboring server and the old avatar at the original server is destroyed.

We adopt a similar approach to the shadow object concept. However, we need to ensure that the restarted federate should be identical to the original federate when a successful migration is achieved. The main concern is to ensure that no event is lost or duplicated. Federate cloning addresses the same problem, and relies on event history logging [10] to ensure the integrity. The approach requires logging every event and may result in a large overhead. We will use an alternative method for message integrity enforcement.

3   Federate Migration

We proposed a *SimKernel* framework for easy development of parallel and distributed simulation using RTI and for deployment of simulation in the Grid environment in [11]. The framework consists of three major components, a federate execution model named SimKernel, a GUI that allows modelers to specify essential simulation information at process level and an automatic code generation tool that translates the modeler's design into executable Java codes for deployment. The SimKernel framework is designed with the following characteristics:

- Simulation design is allowed to be specified at the Logical Process (LP) level.
- Each federate is abstracted to a main simulation loop with event queues inQ and $outQ$ holding incoming and outgoing events respectively (Figure 1).
- All federates adopt the same execution pattern.
- Event interest of a particular federate is specified in a configuration file.
- Event processing detail is defined in a user-defined *consume*() routine.
- Each federate is identified by a unique literal name at the LP level.

These features facilitate easy federate migration at a higher abstraction level. As the SimKernel is application independent, information transferred at migration time is drastically reduced. In the aspect of code transfer, if the standard library of the SimKernel framework is placed at the destination hosts, the migrating federate can be dynamically reconstructed at the destination with the LP specifications and the received events restored.

Migrating a federate, or process in general, requires the transfer of the program executable and essential execution state. With reference to the features identified above, migrating a SimKernel-based federate can reduce the size of the program executable to be transferred at migration time since the SimKernel code library is application independent. Essential execution state includes any

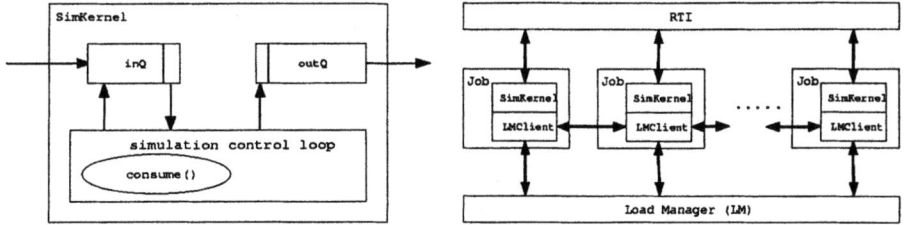

Fig. 1. SimKernel Federate Model   **Fig. 2.** Migration Architecture

local variables and events in both *inQ* and *outQ*. At migration time, the events in *outQ* will be delivered to the RTI and thus will not be transferred. Hence, only the *inQ* information and the local attributes need to be transferred to the destination. This also reduces the amount of data to be transferred. In this section, we will describe the system architecture and protocol to migrate SimKernel-based federates. In the following text, we will refer to the federate before a particular migration operation as the original federate or migrating federate and the federate created at the destination for migration purpose as the restarting federate.

3.1   Architecture

Our simulation execution support system consists of two subsystems, namely simulation subsystem and load management subsystem. A SimKernel component performs the simulation activity using the HLA/RTI and the Load Manager (LM), with the LMClients at each individual hosts, performs the load management activity (Figure 2).

The LM determines the destination host for the federate to be migrated. The LMClients at the source and destination hosts will communicate until a successful migration is achieved.

The LMClient at each host performs three major tasks. First, it monitors the load level at the host and reports the information to the LM. The information will be used by the LM to determine the federate and hosts involved in migration. Second, on receiving a migration request from the LM, the LMClient will migrate the selected federate using the protocol described in the next subsection. Third, the LMClient will create, suspend, communicate with, and destroy the federate when necessary.

To support migration, the SimKernel main simulation loop is adapted to a state-based process model (Algorithm 1). A set of valid states is illustrated in Figure 3. Most state names are self-explanatory except that state "collected" means that the execution state information is already extracted and transferred to the destination. State "joined" means that the restarting federate has joined the federation execution and has successfully subscribed all events of its interests. State transition is described in the migration protocol. The state "restarting" is not shown in the loop because a restarting federate will join the federation, subscribe and publish its event interests and set the state to "joined" before executing the main simulation loop.

Algorithm 1 SimKernel Main Simulation Loop

```
while (notEndOfSimulation()){
    switch(fedStatus) {
        case running:     processEvent(); // identical to main loop in [11]
                          break;
        case suspended:   flushOutgoingEvents();
                          waitForExecutionStateRequest();
                          performFlushQueueRequest();
                          collectStateInfo();
                          setFedState("collected"); break;
        case collected:   waitForSuccessfulMigrationAck();
                          setFedState("terminating"); break;
        case joined:      waitForExecStateInfo();
                          setFedState("restoring"); break;
        case restoring:   performFlushQueueRequest();
                          checkAndRemoveDuplicates();
                          restoreStateInfo();
                          setFedState("running"); break;
        case terminating: break;
        default:          System.out.println("invalid state.");
    } }
```

3.2 Migration Protocol

Our migration protocol (Figure 4) begins with the LM issuing a migration request to the LMClients at both source and destination hosts. The LMClient at the source host will set the federate state to "suspended". After the current event if any is processed, the migrating federate sends out all outgoing events in its *outQ*. Then, the federate waits for the LMClient at the destination host to request transfer of execution state. The LMClient will request execution state from the federate only when its peer at the migration destination sends a "requestInformation" request. The LMClient at the destination host will create a new instance of the federate with state "restarting" upon receiving the migration request from LM. The *restarting* federate will proceed to join the federation execution and subscribe and publish any event of its interests with the same configuration of the original federate. After the new federate successfully completes the event subscription (i.e., "joined"), the new federate starts to receive messages of its interest. Note that the new federate is identical to the original one. After the new federate subscribes to the events, both federates will receive messages from the same set of federates. The LMClient at the destination will send "requestInformation" to its peer at the source host when the restarting federate reaches the "joined" state.

When the migrating federate is instructed to collect execution state, it first invokes *flushQueueRequest()* with the parameter of its current logical time, causing the RTI to deliver all messages by calling its *receivedInteraction()* callback regardless of time management constraints. Received events will be stored in the federate's *inQ*.

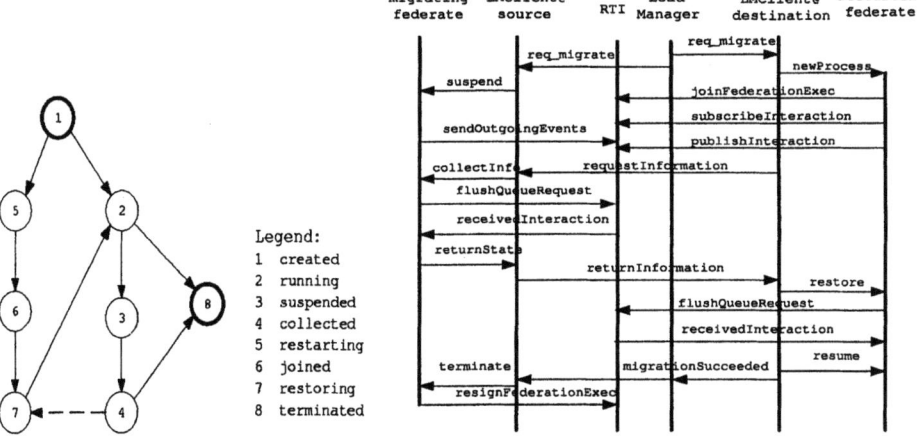

Fig. 3. Federate States   **Fig. 4.** Federate Migration Protocol

Upon the completion of flush queue request, the migrating federate encodes its internal execution state including events in the *inQ* and local attributes in the attribute table into a formatted string and in the meantime, sets a flag to indicate that execution state data is ready for migration. The federate state is also set to "collected". The LMClient at the host periodically checks the federate until the flag is set and starts to transfer the information to its peer at the destination host. The "collected" migrating federate will set its state to "terminating" after its *migrationSucceededFlag* is set by the LMClient at the host on receiving a "migrationSucceeded" message. Later the information transferred to the destination host is restored by the *restarting* federate.

The *restarting* federate begins restoration after it receives the state information. A dynamic class loading technique [12] is used to reconstruct event objects from the string specification transferred from the source host. Reconstructed event objects are inserted to *inQ*. Subsequently, the *restarting* federate invokes *flushQueueRequest()* with parameter of its current logical time to obtain all events sent by RTI since it registered its event interests. When restoring the received information, duplicates are removed.

When the restarting federate has successfully restored its execution state, normal simulation execution resumes. The LMClient at the destination will notify the LM and its peer at the source that the federate is successfully migrated.

Note that the LMClient at each host is regularly updating the load level information to the LM. When an LMClient fails to do so, the host is assumed inactive and not eligible for migration. If the selected destination host is down after the migration decision is made, no socket channel to the host can be successfully created and the LM has to select another destination for migration.

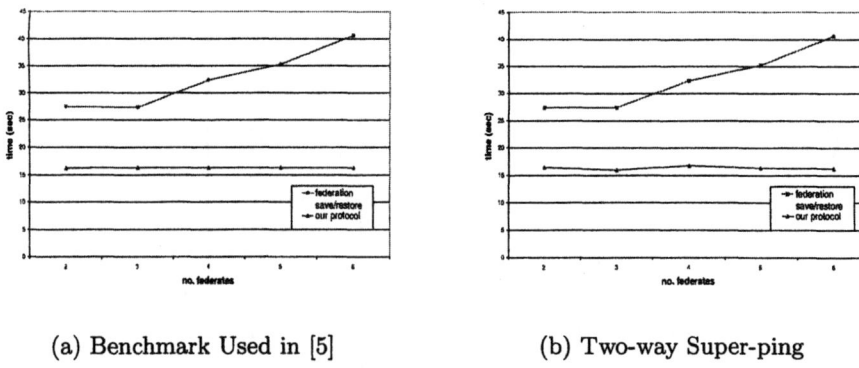

(a) Benchmark Used in [5]   (b) Two-way Super-ping

Fig. 5. Migration Latency Comparison

4   Implementation Results and Discussion

To evaluate the protocol's performance, experiments were carried out in a LAN environment using 2.20GHz Pentium 4 PCs with 1GB RAM. Two test cases were investigated. The first case implemented a two way super-ping, with the total number of nodes (federates) varied from 2 to 6. The second case was identical to the one used in [5], where a federate publishes events, and all other federates subscribe and consume the received events. In both setups, migration requests were issued by the LM every 30 seconds and the event processing time was simulated using random delay. Similar results were obtained in both cases. The migration latency using our protocol was plotted in Figure 5 in comparison to the federation save/restore approach provided by the HLA/RTI standard interface.

Unlike the federation save/restore approach, the time taken for the migration process using our protocol remains constant with increasing number of federates. Migration overhead spans from the time when the original federate is suspended to the time when the migrated federate resumes normal execution. In comparison to other migration approaches, our protocol results in reduced migration overhead due to the following factors:

No explicit federation-wide synchronization is required. Federate migration that employs federation-wide synchronization suffers from poor performance since federates not involved in the migration must also be synchronized.

No communication with third party is required. In our design, migrating a federate requires only peer to peer communication between the source and the destination hosts. This greatly reduces the migration time.

Our framework for simulation design is completely transparent to modelers. Modelers only need to specify at LP level the LPs in the simulation, the events between LPs and the processing details of each event. The design is translated into Java codes by our automatic code generator. This allows modelers to concentrate on the simulation model rather than low level implementation.

Our protocol further benefits the non-migrating federates with complete migration transparency. During the entire migration period, non-migrating federates that interact with the migrating federate continue to send and receive events. These federates have no knowledge whether the federate is processing a message or migrating to another host.

5   Conclusion

In this paper, we have presented a migration protocol for federates based on our SimKernel framework. The protocol adopts a shadow-object like model and reduces the influence on other non-migrating federates to the minimum. Our protocol also achieves complete transparency and reduces overhead without violating the simulation constraints. Although the protocol reduces migration overhead, it still needs to be improved to guarantee complete message consistency. There is a potential problem of message loss if the network is heavily congested. We are working on an improved algorithm to guarantee complete message consistency by providing a counter for each interaction class and verifying the continuity of the counter values. The improved algorithm will be presented in our future publication.

References

1. IEEE: P 1516, Standard for Modeling and Simulation (M&S) High Level Architecture (HLA) - IEEE Framework and Rules (1998)
2. Foster, I., Kesselman, C., Tuecke, S.: The Anatomy of the Grid: Enabling Scalable Virtual Organizations. International J. Supercomputer Applications **15**(3) (2001)
3. Cai, W., Turner, S.J., Zhao, H.: A Load Management System for Running HLA-based Distributed Simulations over the Grid. In: Proceedings of the Sixth IEEE International Symposium on Distributed Simulation and Real Time Applications (DS-RT '02). (2002) 7–14
4. Lüthi, J., Großmann, S.: The Resource Sharing System: Dynamic Federate Mapping for HLA-based Distributed Simulation. In: Proceedings of Parallel and Distributed Simulation, IEEE (2001) 91–98
5. Zając, K., Bubak, M., Malawski, M., Sloot, P.: Towards a Grid Management System for HLA-based Interactive Simulations. In: Proceedings of the Seventh IEEE International Symposium on Distributed Simulation and Real Time Applications (DS-RT '03), Delft, The Netherlands (2003) 4–11
6. Zając, K., Tirado-Ramos, A., Zhao, Z., Sloot, P., Bubak, M.: Grid Services for HLA-based Distributed Simulation Frameworks. In: Proceedings of the First European Across Grids Conference. (2003)
7. Roush, E.T.: The Freeze Free Algorithm for Process Migration. Technical Report UIUCDCS-R-95-1924, UIUC (1995) Available online at http://www.cs.uiuc.edu/Dienst/UI/2.0/Describe/ncstrl.uiuc_cs/UIUCDCS-R-95-1924.
8. Heymann, E., Tinetti, F., Luque, E.: Preserving Message Integrity in Dynamic Process Migration. In: Proceedings of Euromicro Workshop on Parallel and Distributed Processing (PDP-98). (1998) 373–381

9. Huang, J., Du, Y., Wang, C.: Design of the Server Cluster to Support Avatar Migration. In: Proceedings of The IEEE Virtual Reality 2003 Conference (IEEE-VR2003), Los Angeles, USA (2003) 7-14

10. Chen, D., Turner, S.J., Gan, B.P., Cai, W., Wei, J.: A Decoupled Federate Architecture for Distributed Simulation Cloning. In: Proceedings of the 15th European Simulation Symposium (ESS 2003), Delft, The Netherlands (2003) 131-140

11. Yuan, Z., Cai, W., Low, M.Y.H.: A Framework for Executing Parallel Simulation using RTI. In: Proceedings of the Seventh IEEE International Symposium on Distributed Simulation and Real Time Applications (DS-RT '03), Delft, The Netherlands (2003) 12-19

12. Liang, S., Bracha, G.: Dynamic Class Loading in the Java Virtual Machine. In: Conference on Object-oriented Programming, Systems, Languages, and Applications (OOPSLA'98). (1998) 36-44

FT-RSS: A Flexible Framework for Fault Tolerant HLA Federations

Johannes Lüthi[1] and Steffen Großmann[2]

[1] University of Applied Sciences FHS KufsteinTirol
Business Informatics, Kufstein, Austria
Johannes.Luethi@fh-kufstein.ac.at
[2] German Army Signal School
Feldafing, Germany
steffengrossmann@bundeswehr.org

Abstract. The absence of fault tolerance mechanisms is a significant deficit of most current distributed simulation in general and of simulation systems based on the high level architecture (HLA) in particular. Depending on failure assumptions, dependability needs, and requirements of the simulation application, a choice of different mechanisms for error detection and error processing may be applied. In this paper we propose a framework for the configuration and integration of fault tolerance mechanisms into HLA federates and federations. The administration and execution of fault tolerant federations is supported by the so-called fault tolerant resource sharing system based on a ressource sharing system previously developed by the authors.

Keywords: Dependable distributed simulation, fault tolerance, high level architecture, resource management

1 Introduction

In its current state, no formal failure model is included in the High Level Architecture (HLA) [1]. Services of the runtime infrastructure (RTI) such as e.g. the save/restore services may be used to implement fault tolerant federations. However, failure detection as well as failure processing and all corresponding management acivities are left to the federation developer. Only little effort has been made in combining *fault tolerance* (FT) mechanisms with distributed simulation. Exceptions include the work of Agrawal and Agre proposing replicated objects in time warp [2], and the optimistic fault tolerant simulation approach presented by Damani and Garg [3]. An overview of possibilities for fault tolerant distributed simulation can be found in [4]. Anyway, these approaches have only restricted applicability to distributed simulation systems based on the HLA.

There is a wide spectrum of dependability enhancing mechanism that may be integrated into distributed simulation. However, there is also a wide spectrum of different types and application domains for HLA-based distributed simulation. Depending on the specific dependability, realtime, repeatability, and other requirements, and depending on the failure assumptions for the federates and the

M. Bubak et al. (Eds.): ICCS 2004, LNCS 3038, pp. 865–872, 2004.
© Springer-Verlag Berlin Heidelberg 2004

computing nodes, an appropriate choice of FT mechanisms has to be used. Moreover, not every federate of an HLA federation will have the same requirements with respect to dependability and fault tolerance. In this paper, we present a framework for the development, administration, and execution of fault tolerant HLA simulations. The proposed framework is based on the *resource sharing system* (RSS), presented in [5]. Hence it is called *fault tolerant RSS* (FT-RSS). The framework and its components allow the individual configuration of FT mechanisms to be included in federates and federations. In the design and implementation of a fault tolerant federation, the developers are supported by an FT configuration tool. Furthermore, in its final state the FT-RSS will provide code modules and interfaces for error detection and error processing functions and associated HLA interactions. During the execution of a distributed HLA simulation using the FT-RSS, the manager and client components of the FT-RSS are responsible for the enforcement of FT mechanisms.

The paper is organized as follows. In the next section, the general concept and architecture of the framework is presented. The configuration options for HLA federates are discussed in Section 3. Federation configuration is considered in Section 4. The work is summarized in Section 5.

2   Concept for Configurable Fault Tolerance

2.1   Introducing the Framework

The objective of the proposed framework is to support a flexible configuration of HLA federates and federations with respect to fault tolerance mechanisms. The three main components of the proposed framework are the *FT configurator*, a set of *FT code modules and interfaces*, and the *fault tolerant resource sharing system* (FT-RSS):

- During the design and implementation of a federation, FT requirements determine the choice of error detection and error processing[1] mechanisms for the federates. The configuration tool can be used to select the appropriate options for these mechanisms. The developers of the federates are supported with code segments and interfaces corresponding to the selected FT mechanisms. The developer is responsible to integrate these mechanisms into the HLA federates.
- Depending on the available computing resources, prior to simulation runs at the target environment, the migration configuration of an FT HLA federation is specified. For that purpose, again the configuration tool can be used.
- The federation with integrated FT mechanisms can be run using the FT-RSS environment. An FT monitor can be used to visualize detected errors and initiated error processing mechanisms.

---

[1] In our terminology, error processing includes error recovery, error passivation, and error compensation.

2.2   Architecture of the FT-RSS

The proposed architecture of the runtime environment for configurable fault tolerant simulation of HLA federations is based on the *Resource Sharing System (RSS)* presented in [5]. The RSS is designed to allow users of workstations to control the participation of their computers in an HLA-based distributed simulation. In the original RSS architecture, the *RSS manager* keeps track of participating workstations and simulation federates that are registered in the federation. *RSS clients* at the workstations allow the users to connect or disconnect their workstation to the RSS. Whenever a node connects or disconnects to the RSS, federates may be migrated. The migration operations are controlled by the manager. Migration specific communication between the manager and simulation federates is performed via a special *communication federate*. This way, non-RTI communication of the simulation federates can be avoided. A separate ftp server is used as a persistent storage system. This may not be the most efficient solution for that purpose. However, this way dependability issues of that aspect need not be considered in the FT-RSS itself. In order to provide a runtime environment for configurable fault tolerant distributed HLA simulation, the architecture of the RSS is extended in several ways (see Fig. 1).

- A configuration file generated by the configurator *FT Config* tells the FT-RSS manager the fault tolerance configuration of every federate as well as the configuration of the whole federation.
- The RSS manager is extended to control error detection, error processing, and migration mechanisms according to the configuration.
- The functionality of the clients is extended to support checks required for error detection and methods for error processing. In contrast to the RSS, this may also include non-RTI-based communication with the federates. Since no simulation specific information is exchanged this is not a violation of the corresponding HLA rule.
- An *FT monitor* is introduced to allow observation of error/failure behavior of the simulation federation and measures taken by the FT-RSS.

3   Federate Configuration

The configuration of FT options can be divided into two types: (a) aspects that can be configured for individual federates and (b) FT configuration of the whole federation. In this section, the configuration options for FT properties of single simulation federates are discussed. Three classes of options are considered: configuration of migration possibilities, configuration of error detection mechanisms, and configuration of error processing methods.

3.1   Migration Configuration

For some error processing methods, federates have to be migrated to another node. There are many situations where a configuration of migration options

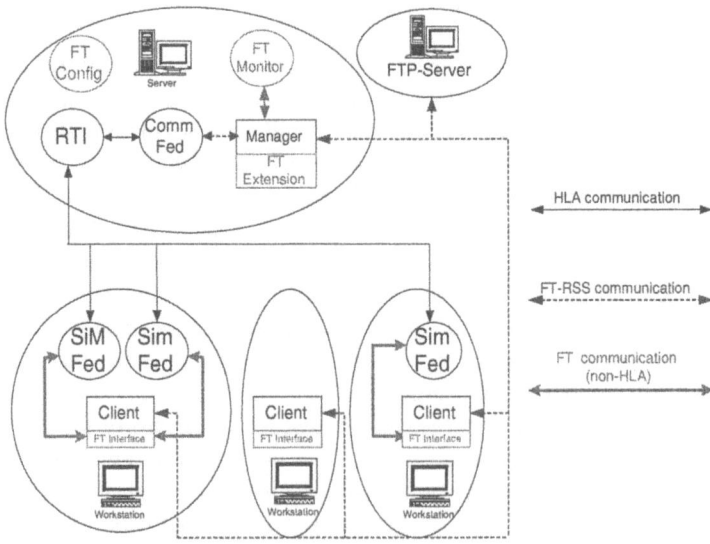

Fig. 1. Architecture of the fault tolerant resource sharing system (FT-RSS).

is required. Consider for example interactive simulation: only nodes that are within reasonable distance and equipped with appropriate I/O devices may be considered for migrating a federate with human interaction. More restrictions can be imposed by heterogeneous hardware and operating systems (OS). In the proposed FT configurator, three means to specify the migration behavior of a simulation federate are available:

– In a *node pool*, IP addresses of nodes the federate may be migrated to are listed. Also address ranges and sub-networks may be specified.
– A list of *restrictions* – again provided as addresses, host names and/or address ranges or sub-networks – can be used to prevent the federate from being migrated to certain computers.
– In a list of *OS assignments*, the operating systems the federate can be executed on can be specified.

3.2   Error Detection Configuration

In the federate configuration, only detection of errors in simulation federates are taken into account. The detection of errors in computing nodes are considered in the federation configuration (see Section 4). A choice of four basic options is considered for error detection (see Fig. 2):

– *No error detection*: if a federate is assumed to be highly dependable or some reasons make the use of error detection mechanisms for that federate not practicable, a federate can be excluded from error detection. Note that the error *processing* behavior may still be configured for that federate because the configuration may be used in the case of failures in computing nodes.

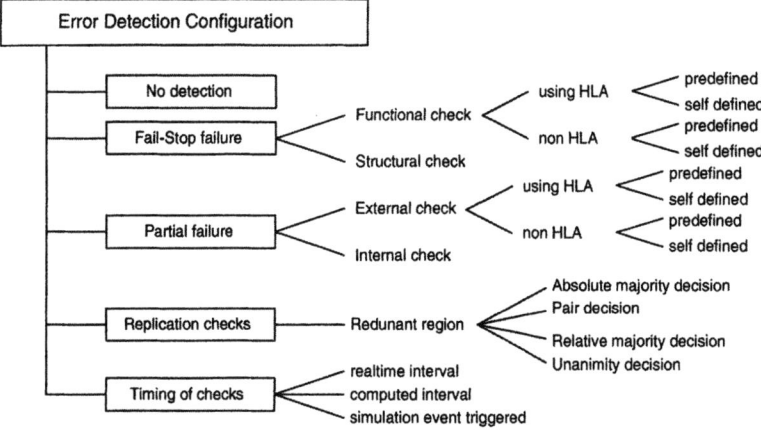

Fig. 2. Federate configuration options for error detection.

- *Fail-Stop failures*: for the detection of fail-stop failures *functional checks* or *structural checks* may be applied. In a functional check, a prespecified function of the federate is called at certain times. If the expected function results are provided by the federate, it is assumed to be running fault-free. The function used in a functional check may be an HLA interaction or it may be called by the RSS-client directly. For both cases, predefined functions may be used that are provided in the set of *FT code modules and interfaces*. As an alternative, the developer of the federate may specify the function to be used. For HLA internal function checks, the communication between the manager (triggering such a check) and the federate is performed via the communication federate.

  Structural checks are triggered by the manager and are performed by the local FT-RSS client. They use structural information about the federate (e.g., task lists provided by the OS) to check whether the process associated with the federate is still active. Since functional checks can be performed outside the federate, they can be provided by the FT-RSS. No additional implementation effort by the simulation developer is required.

- *Partial failures*: if failures in separate parts of the federate are considered in the failure assumption, such partial failures can be detected by appropriate functional or structural checks. However, since a partition of a federate into several failure regions is typically internal to that federate (for example separate threads that may fail independently), structural checks for partial failures have to be performed internally by the federate (for example by examination of a thread list). External functional checks to test for partial failures offer the same options as functional checks for fail-stop failures.

- *Replication checks*: replication can serve as a sufficient tool for both, error detection and (if more than two replicates exist) error processing. A number of federates can be grouped as a redundant region. A choice of options for

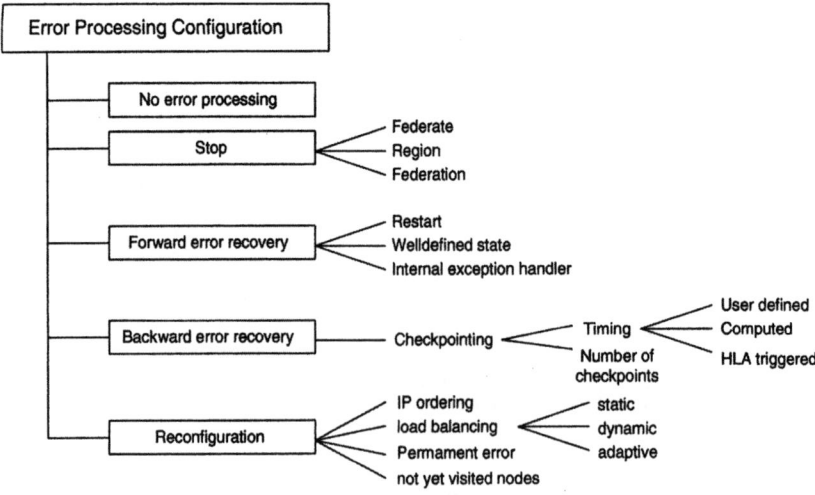

Fig. 3. Federate configuration options for error processing.

error detection is available in the configurator: absolute and relative majority decision, pair decision, and unanimity decision.

The FT-RSS manager is responsible to trigger checks for fail-stop failures and external checks for partial fail-stop failures. In these cases, timing of the checks must be specified. Three options are proposed for that purpose: A fixed realtime interval, an adaptive checkpointing interval, and checkpoints triggered by prespecified simulation events.

3.3   Error Processing Configuration

Various options are also available for error processing (see Fig. 3):

- *No error processing*: errors may be detected but no FT mechanism is applied.
- *Stop*: a failed federate is stopped without further FT mechanisms. Three variants are available: (a) only the failed federate may be stopped. This may for example be useful if the federate of just one trainee in a joined training simulation fails and it is reasonable that all others can continue the training program. (b) a sub-model consisting of several federates may be stopped. (c) the whole federation may be stopped.
- *Forward error recovery*: depending on the functionality of the federate, it may not be necessary to use a saved checkpoint to restart a failed federate. This class of methods is referred to as forward error recovery. The options in this case are: (a) restarting the federate with its initial state, (b) using a specified valid state, and (c) evoking an internal exception handler within the federate to obtain a valid state. In any case, the responsibility to consider time management conflicts with forward error recovery is left to the federate/federation developer.

– *Backward error recovery*: restarting a federate using a previously saved state is referred to as backward recovery. For backward recovery the checkpointing behavior has to be specified. Timing for checkpoints can be chosen to be (a) a realtime interval, (b) automatically computed using an underlying failure model, or (c) triggered by certain simulation events. Solving or avoiding time management conflicts is a crucial problem when using backward recovery. In strictly optimistic distributed simulation, backward recovery can be implemented by deploying timewarp and its anti-message mechanism [3]. However, locally decreasing simulation time is not an available HLA time management service.

If forward or backward recovery is chosen for error processing, the federate has to be restarted after a failure. As a default, every failure is assumed to be transient. Thus, a failed federate is restarted at the same node. However, in the error processing configuration it can be specified that every failure is assumed to be permanent. This means that in a failure situation the federate is migrated to another computing node before restart. As an additional option it can be specified that in the case of failure-driven migration, a federate has to be migrated to a node it has not yet been executed on in the current run.

4 Federation Configuration

In addition to configuration options for HLA federates, some aspects concerning FT behavior of the whole federation have also to be configured. The HLA definitions do not specify the behavior of federates and of the RTI in failure situations. For example, some RTI implementations (e.g., DMSO's RTI NG [6]) fail to work properly after a fail-stop failure of one of the federates whereas other RTI implementations (e.g., portable RTI [7]) are able to treat a failed federate as if it had resigned from the federation. Such differences have to be considered in the mechanisms used by the FT-RSS. Thus, an RTI in a list of versions available for the FT-RSS has to be specified in the federation configuration.

Also hardware failures of participating computing resources should be detected and processed (here, only fail-stop failures are considered). In the federation configuration, options for functional and/or structural checks for hardware failures as well as associated timing behavior can be specified. When an error of a computing node is processed, every federate running on that node is treated as if a failure of the federate had occurred.

Detection and processing of RTI fail-stop failures is also included in the federation configuration. Structural as well as functional checks can be applied for error detection. Stopping and backward recovery may be used for error processing of RTI failures. A consistent internal state of the RTI after a failure can only be reached with the HLA service *federation restore*.

There are several reasons why error processing measures required in a fault tolerant distributed simulation using the FT-RSS may fail. For example a node that conforms to a necessary migration operation or resources (e.g. files) required for an error processing measure may not be available. For such situations it is important to include an *exception handling* mechanism in the FT-RSS. A basic feature of FT-RSS exception handling is to report the situation to the simulation user and to stop the simulation run to avoid propagation of incorrect or misleading simulation results.

5   Summary and Conclusions

Although failed simulation runs may be safety critical and costly, currently most distributed simulation systems restrict fault tolerance measures to the use of dependable hardware. The requirements with respect to dependability, interactivity, repeatability, and timing behavior varies not only from one federation to another but even among the federates of a single federation. An appropriate selection of mechanisms to support fault tolerance has to be implemented. In order to achieve a flexible FT configuration of HLA-based simulations we propose a framework including a *fault tolerant resource sharing system* (FT-RSS). The FT-RSS allows the configuration of migration, error detection, and error processing for individual federates and for the federation. Prototypes of the configuration tool and the FT-RSS runtime environment have been implemented. Future work includes the implementation of FT federations, extension of the FT-RSS implementation with respect to the number of FT methods, and research in the direction of replication approaches for HLA federates.

References

1. Dahmann, J.S.: The high level architecture and beyond: Technology challenges. In: Proc. 13th Workshop on Parallel and Distributed Simulation (PADS '99, Atlanta, GA, USA), IEEE CS Press (1999) 64–70
2. Agrawal, D., Agre, J.R.: Replicated objects in time warp simulations. In: Proc. 1992 Winter Simulation Conference, SCS (1992) 657–664
3. Damani, O.P., Garg, V.K.: Fault-tolerant distributed simulation. In: Proc. 12th Workshop on Parallel and Distributed Simulation (PADS '98, Banff, Alberta, Canada), IEEE CS Press (1998) 38–45.
4. Lüthi, J., Berchtold, C.: Concepts for dependable distributed discrete event simulation. In: Proc. Int. European Simulation Multi-Conference (ESM 2000, Ghent, Belgium), SCS (2000) 59–66
5. Lüthi, J., Großmann, S.: The resource sharing system: Dynamic federate mapping for HLA-based distributed simulation. In: Proc. 15th Workshop on Parallel and Distributed Simulation (PADS 2001, Lake Arrowhead, CA, USA). IEEE CS Press (2001) 91–98
6. DMSO: RTI Next Generation Programmer's Guide Version 3.2. Defense Modeling and Simulation Office (DMSO), Alexandria, VA, USA (2000)
7. Pitch Kunskapsutveckling AB. http://www.pitch.se. validated: December 2003.

Design and Implementation of GPDS*

Tae-Dong Lee, Seung-Hun Yoo, and Chang-Sung Jeong**

Department of Electronics Engineering Graduate School,Korea University
{lyadlove,friendyu}@snoopy.korea.ac.kr, csjeong@charlie.korea.ac.kr

Abstract. In this paper, we describes the design and implementation of Grid-based Parallel and Distributed Simulation environment(GPDS). GPDS not only addresses the problems that it is difficult for parallel and distributed application to achieve its expected performance, because of some obstacles such as deficient computing powers, weakness in fault and security problem, but also supports scalability using Grid technologies. GPDS supports a 3-tier architecture which consists of clients at front end, interaction servers at the middle, and a network of computing resources at back-end including DataBase, which provides three services: Automatic Distribution Service, Dynamic Migration Service and Security Service, designed by UML-based diagrams such like class diagram and interaction diagram. The GPDS has been implemented as Grid Agent(GA) and Simulation Agent(SA) using C++. The object-oriented design and implementation of GA and SA in GPDS provides users with modification, extensibility, flexibility through abstraction, encapsulation and inheritance.

1 Introduction

Parallel and distributed simulation (PADS) is concerned with issues introduced by distributing the execution of a discrete event simulation program over multiple computers. In paper [1], we described the problems with PADS of performance and deficient computing power, weak in fault and security problem. for solving these problems, the paper suggested the three services: Automatic Distribution Service, Dynamic Migration Service, and Security Service. The three services provide both the supply of computing resources and robustness of the system which PADS does not provide. The GPDS is composed of two agents, Grid Agent (GA) and Simulation Agent (SA). The GA has fundamental functions of resource broker using Globus toolkit[5]. It accomplishes three major services, which include automatic distribution, dynamic migration, and security. Automatic distribution service makes parallel and distributed system have the strong extensibility to utilize abundant resources. Also, dynamic migration enables the

---

* This work has been supported by KOSEF and KIPA-Information Technology Research Center, University research program by Ministry of Information & Communication, and Brain Korea 21 projects in 2004.
** Corresponding author.

M. Bubak et al. (Eds.): ICCS 2004, LNCS 3038, pp. 873–880, 2004.
© Springer-Verlag Berlin Heidelberg 2004

fault tolerance of whole system as well as the improvement of overall performance. The SA provides several modules which assist the PADS to achieve its objective. It is responsible for communication, monitoring, and dynamic configuration for parallel and distributed architecture, and manages available servers via the communication with the GA. GA and SA enable clients to transparently perform a large-scale object-oriented simulation by automatically distributing the relevant simulation objects among the computing resources while supporting scalability and fault tolerance by load balancing and dynamic migration schemes

In this paper, we describe design, implementation and performance evaluation of GPDS. It shows how grid-based parallel and distributed simulation improves the existing simulation environment. In Sect.2, we illustrate the design of GPDS including architecture and UML-based diagrams such like class diagram and interaction diagrams. In the next section, we depict the implementation of GPDS and experimental results. At last, we conclude in Sect. 4.

2   Design of GPDS

2.1   System Architecture

The system architecture of GPDS is 3-tier architecture based on client-server model shown in Fig.1(a). Here, the server indicates the broad range which is enclosed by virtual organization (VO). The hosts within the VO can be servers of the GPDS. All processes of the server side are transparent to the client. The client only sees the results from the server side. The client in client tier delivers an executable and standard input files of required simulation to the GPDS Manager in Server tier. The GA creates a process to allocate simulations to remote hosts and the SA. Note that there is a simulation process in the SA. This simulation process is used to accomplish dynamic migration service. The SA manages the DBs in Database tier while cooperating with the GA. Each simulation process within remote hosts and the SA joins to simulation-specific middleware to execute the PADS. The result of simulation is returned to the client by the SA.

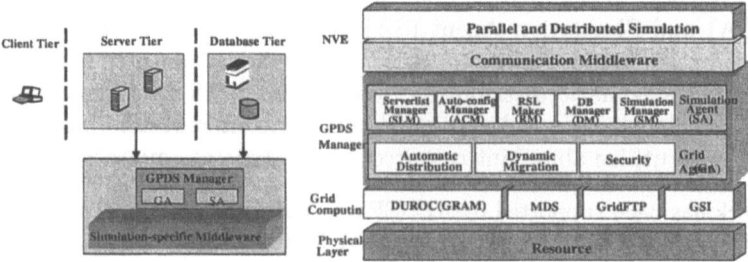

Fig. 1. (a) 3-tier architecture of GPDS (b) Layered structure of GPDS

Moreover, the characteristics in the architecture of GPDS can be grasped by observing its layered structure. The layer structure of GPDS is shown in Fig.1(b). The GCE(Grid Computing Environment) accomplishes the management of resources. In the GPDS, powerful modules of the Globus toolkit are used on the GCE. The GCE comprises of four modules: GRAM which allocate and manage the job in the remote hosts, MDS which provides information services, GridFTP which is used to access and transfer files, and GSI which enables authentication via single sign-on using a proxy. The NVE(Networked Virtual Environment) consists of application and middleware to communicate efficiently. In the application layer, the logic of a simulation is performed through the interaction of entities. Each application joins to a corresponding middleware. The middleware layer provides the communication of entities, interest management, data filtering, and time management required to achieve stable and efficient simulation. In brief, the GPDS means the NVE over the GCE. The GPDS Manager is an intermediate layer between NVE and GCE. It is in the charge of a bridge between both layers. As mentioned earlier, GPDS Manager is composed of Grid Agent and Simulation Agent. Agent is an identifiable computational entity that automates some aspect of task performance or decision making to benefit a human entity.

2.2   Class Diagram

Figure 2 shows a class diagram which uses a facade pattern which provides a unified interface to a set of interfaces in a subsystem. The facade pattern offers the benefits to shield clients from subsystem components, promote weak coupling between the subsystem and its clients. In Fig. 2, CGPDS is a facade for GPDS system, CGA for Grid-based classes and CSA for simulation-based classes. Each applications approach to agent classes through CGPDS. Also, the agents can be extended through aggregation to CGPDS. This object-oriented architecture provides extensibility of agents. CGA manages and controls the Grid-based classes, and CSA does simulation-specific classes. The SA includes five classes: ServerlistManager(CSLM), RSLMaker(CRSLMaker), Auto-configuration Manager(CACM), Simulation Manager(CSM) and DB Manager(CDBM). The CSM makes the list of resources available in corresponding virtual organization(VO). The number and performance of available hosts have great effect on the configuration of the PADS. This severlist of available resources is periodically updated

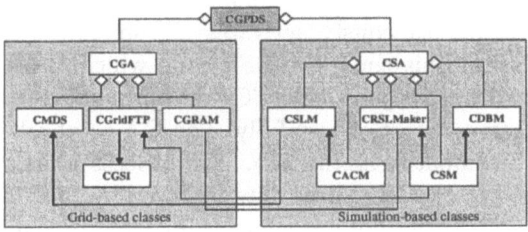

Fig. 2. Class diagram of GPDS

and referenced. The CRSLMaker dynamically creates a RSL code to meet the status of simulation and the requirements of the GA. The CACM automatically makes configuration files to provide information needed to initiate the PADS, according to the serverlist. The CSM has three missions. First, it establishes a connection to the client. Second, it periodically received and monitored simulation data from one simulation process within the SA, and delivers them to the CDBM. Third, the simulation data is returned to the client as simulation results by CSM. Lastly, the CDBM stores the simulation data of each host periodically. The stored data is applied to the recovery of the fault.

2.3 Interaction Diagram

Automatic distribution service means that GPDS Manager can automatically create the PADS by transferring and executing solicitated executable and standard input files on new host. Dynamic migration strategy means that the computing which has been proceeding on one host can be transferred to and continuously performed on another host. This service is used to achieve two purposes in GPDS. First goal is the fault tolerance of the whole system, and second is the improvement of performance. Each service composes of 3 steps.

Interaction Diagram for Automatic Distribution Service. Figure 3(a) shows the interaction diagram for Automatic Distribution Service. In (1) step, client connects to GPDS. The CSM of SA establishes a connection with the client. The client submits the job and sends indispensable files to GPDS. (2) step accomplishes the preparation for creating remote servers. It consists of four stages, that is, server list production, configuration, storage, and transmission. In server list production stage, the CSLM in SA makes out server list which includes available resources, using the metadata of hosts registered to GIIS through CMDS. In next stage, the CACM automatically modifies an initial configuration into several configuration files corresponding to parallel and distributed architecture by considering the number of available hosts and reflecting all information which is required to initiate remote servers. The CRSLMaker automatically generates the required RSL codes to use Globus. In the storage stage, the initial configuration data of each remote server is saved in DB by CDBManager. At last, the program files and configuration file is sent to remote hosts through CGridFTP service using CGSI by GA in the transmission stage. In (3) step, The GA forks one process to execute remote hosts. Through CGRAM service of Globus, the GA simultaneously activates the simulation process of remote servers and the SA. At this time, the RSL code which CRSLMaker created is used to submit simultaneous jobs. The CGRAM provides the barrier that guarantees all remote servers start successfully. The GA establishes the flag so that the process of SA can identify its mission. This process plays a role in delivering periodic simulation data to SA. These data is stored by CDBManager, transmitted to the client by CSM. Remote servers are initiated by reading each assigned configuration file, and periodically deliver own computation to the simulation process of the SA through simulation-specific middleware.

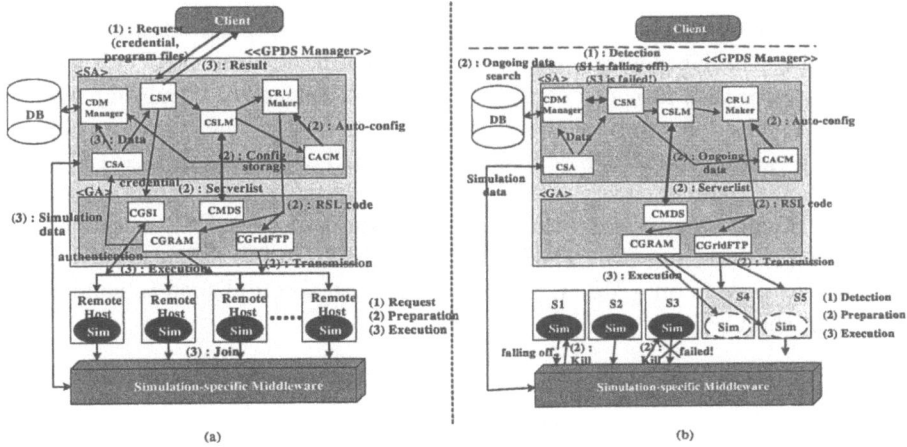

Fig. 3. Interaction diagram for (a)Automatic Distribution Service (b) Dynamic Migration Service

Interaction Diagram for Dynamic Migration Service. Figure 3(b) shows the interaction diagram for Dynamic Migration Service. In first step, The CSM of SA employs the timeout mechanism to detect the fault of remote servers. In Fig. 2(b), S1, S2, and S3 are remote hosts on the execution step of automatic distribution service. The CSM perceives the fault of S3 because GA regularly retrieves the current status of servers who are the members of GIIS through the metadata of CMDS. By analyzing this information, the GA can look up the better resource or recognize the degradation of current slave servers. We assume that the performance of S1 is falling off and S5 is founded as better server in Fig. 2(b). In second step, the CGPDS needs to prepare the creation of new remote server. The CSM retrieves ongoing data of the target server in DB through CDBManager, which manages the simulation data per remote host. Following mechanisms are the same as the storage and transmission stages of the preparation step in automatic distribution strategy. In Fig. 2(b), after the CSM searches for the ongoing data of S1 or S3, configuration file is made out using this data by the Auto-configuration Manager and transmitted to S5 or S4 through the GridFTP service by GA. The third step is same as the third step of Automatic Distribution Service.

3   Implementation of GPDS

For an experiment of the GPDS system, we implemented parallel and distributed war game simulation where distributed forces are operated by movement, detection, and close combat. The HLA/RTI is used as simulation-specific middleware to provide stable communication and interoperability. High Level Architecture (HLA) was developed by the Defense Modeling and Simulation Office (DMSO) to provide a common architecture that facilitates simulation interoperability and reusability across all classes of simulations [8]. HLA has been adopted as IEEE standard 1516 in September 2000. The Runtime Infrastructure (RTI) is a collection of software that provides commonly required services to simulation systems using HLA. The information about forces is supplied in the configuration file, and distributed by the GPDS Manager. The GPDS uses the RTI NGv1.3 [8] and Globus v2.4. Client is constructed on windows based system, while servers are based on Linux. Our implementation is accomplished on 4 PCs as clients, and 10 clusters(5 : Pentium IV 1.7GHz, 5 : Pentium III 1.0GHz Dual) and one 486 computer as servers on a VO. Our experiments are accomplished to confirm key services of the GPDS.

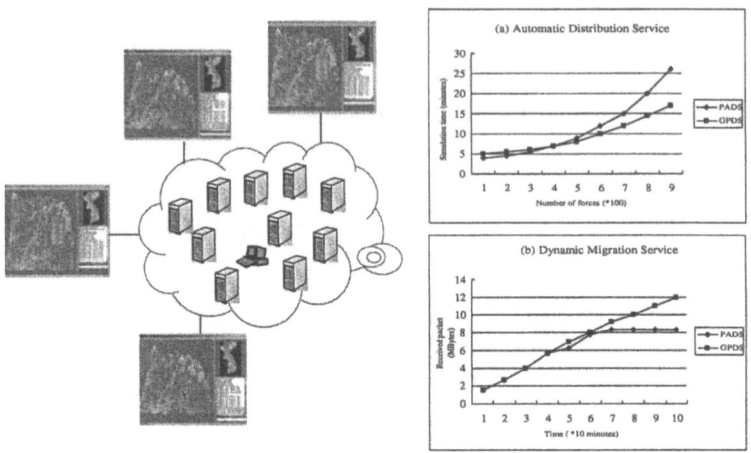

Fig. 4. Experimental results : (a) Simulation time according to the Increase of forces, (b)Accumulated received packets updated by 600 forces per 30 second

First experiment is for the automatic distribution service. We organizes a PADS which has five servers(we assume that 486 computer is included as server, because of the limitation in local condition), and the GPDS which has a VO of 11 servers(10 clusters and 486 PC). Then, we estimated the complete time of simulation as the number of forces increases. As we expect, the resource selection of the GPDS did not choose the 486 computer. In Fig.4(a),the GPDS is superior to the PADS as the scale of simulation is increasing,although the time consumption of initialization have an effect on the state of small forces. The GPDS can utilize abundant computing power and adapt for various environment,as well as provide convenient user interface.

To verify the dynamic migration service, we comprised second experiment. In this test,we measured accumulated received packets updated by 600 forces per 30 second. One packet has the size of 100 bytes. In 70 minutes, we intentionally made a failure on one server. DB Manager stores the information related to federation

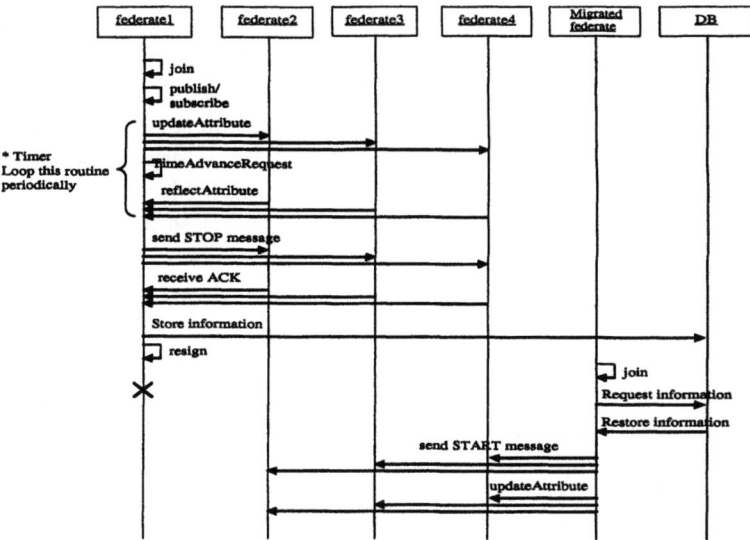

Fig. 5. Sequence diagram for second experiment

like LBTS and object information periodically, and then the application resigns from the federation. The application sends the stop message to federates before resignation. GA gathers the information of MDS and SA sends the application to a selected host. The application sends the restart message to all federates and receives the data stored before failure. As shown Fig.4(b), the GPDS can fulfill its mission after the failure, while the PADS is halted. Fig.5 shows the sequence diagram for second experiment.

4   Conclusion and Future Work

The paper has described the design and implementation of GPDS. GPDS not only addresses the problems that it is difficult for parallel and distributed application to achieve its expected performance, because of some obstacles such as deficient computing powers, weakness in fault and security problem, but also supports scalability using Grid technologies. GPDS supports a 3-tier architecture which consists of clients at front end, interaction servers at the middle, and a network of computing resources at back-end including DataBase, which provides three services: Automatic Distribution Service, Dynamic Migration Service and Security Service, describing the design by UML-based class diagram and interaction diagrams. Grid and simulation agents in the interaction server enable client to transparently perform a large-scale object-oriented simulation by automatically distributing the relevant simulation objects among the computing resources while supporting scalability and fault tolerance by load balancing and dynamic migration schemes.

As for future work, GT2.4 are being replaced by GT3 which is implemented by Java, and GPDS must be changed based on GT3 using Java. We will develop web-based GPDS in future using Java. Also, in the paper we did not describe HLA-specific issues concerning the design of migration service for HLA components. We have developed RTI [9] according to HLA interface specification and are developing RTI implementation (RTI-G) using Grid components. We will submit the RTI-G related works, and then we will describe the HLA-specific issues in detail.

References

1. C.H. Kim, T.D. Lee, C.S. Jeong, "Grid-based Parallel and Distributed Simulation Environment" 7th international conference PaCT2003 Nizhni Novogorod Russia, Proceedings LNCS pp. 503–508, September 2003
2. www.globus.org
3. I. Foster, C. Kesselman, G. Tsudik, S. Tuecke, "A Security Architecture for Computational Grids," Proc. 5th ACM Conference on Computer and Communications Security Conference, pp. 83–92, 1998.
4. I. Foster, C. Kesselman, S. Tuecke, "The Anatomy of the Grid: Enabling Scalable Virtual Organizations," International J. Supercomputer Applications, 15(3), 2001.
5. I. Foster, C. Kesselman, "Globus: A Metacomputing Infrastructure Toolkit," Intl J. Supercomputer Applications, 11(2):115–128, 1997.
6. K. Czajkowski, I. Foster, "Grid Information Services for Distributed Resource Sharing," Proceedings of the Tenth IEEE International Symposium on High-Performance Distributed Computing (HPDC-10), IEEE Press, August 2001.
7. J. Dahmann, R.M. Fujimoto, R.M. Weatherly, "The DoD high level architecture: an update," Winter Simulation Conference Proceedings of the 30th conference on Winter simulation Washington, D.C., United States Pages: 797–804, 1998
8. U.S. Department of Defense(DMSO), "High Level Architecture Run-Time Infrastructure (RTI) Programmer's Guide Version 1.3," http://hla.dmso.mil, 1998.
9. T.D. Lee, C.S. Jeong, "Object-oriented Design of RTI using Design Patterns," 9th international conference OOIS2003 Geneva Switzerland Proceedings LNCS 2817 pp. 329–333, 2003

HLA_AGENT: Distributed Simulation of Agent-Based Systems with HLA

Michael Lees[1], Brian Logan[1], Ton Oguara[2], and Georgios Theodoropoulos[2]

[1] School of Computer Science and IT
University of Nottingham, UK
{mhl,bsl}@cs.nott.ac.uk
[2] School of Computer Science
University of Birmingham, UK
{txo,gkt}@cs.bham.ac.uk

Abstract. In this paper we describe HLA_AGENT, a tool for the distributed simulation of agent-based systems, which integrates the SIM_AGENT agent toolkit and the High Level Architecture (HLA) simulator interoperability framework. Using a simple Tileworld scenario as an example, we show how the HLA can be used to flexibly distribute a SIM_AGENT simulation with different agents being simulated on different machines. The distribution is transparent in the sense that the existing SIM_AGENT code runs unmodified and the agents are unaware that other parts of the simulation are running remotely. We present some preliminary experimental results which illustrate the performance of HLA_AGENT on a Linux cluster running a distributed version of Tileworld and compare this with the original (non-distributed) SIM_AGENT version.

1 Introduction

Simulation has traditionally played an important role in agent research and a wide range of simulators and testbeds have been developed to support the design and analysis of agent architectures and systems [1,2,3,4,5,6]. However no one simulator or testbed is, or can be, appropriate to all agents and environments, and demonstrating that a particular result holds across a range of agent architectures and environments often requires using a number of different systems. Moreover, the computational requirements of simulations of many multi-agent systems far exceed the capabilities of a single computer. Each agent is typically a complex system in its own right (e.g., with sensing, planning, inference etc. capabilities), requiring considerable computational resources, and many agents may be required to investigate the behaviour of the system as a whole or even the behaviour of a single agent.

In this paper we present an approach to agent simulation which addresses both interoperability and scalability issues. We describe HLA_AGENT, a tool for the distributed simulation of agent-based systems, which integrates an existing agent toolkit, SIM_AGENT, and the High Level Architecture (HLA) [7]. Simulations developed using HLA_AGENT are capable of inter-operating with other HLA-compliant simulators and the objects and agents in the simulation can be flexibly distributed across multiple computers so as to make best use of available computing resources. The distribution is transparent to the

M. Bubak et al. (Eds.): ICCS 2004, LNCS 3038, pp. 881–888, 2004.
© Springer-Verlag Berlin Heidelberg 2004

user simulation and symmetric in the sense that no additional management federates are required.

2   An Overview of SIM_AGENT

SIM_AGENT is an architecture-neutral toolkit originally developed to support the exploration of alternative agent architectures [4,8][1]. In SIM_AGENT, an agent consists of a collection of modules, e.g., perception, problem-solving, planning, communication etc. Groups of modules can execute either sequentially or concurrently and at different rates. Each module is implemented as a collection of condition-action rules in a high-level rule-based language. The rules match against data held in the agent's database, which holds the agent's model of its environment, its goal and plans etc. In addition, each agent also has some public data which is "visible" to other agents.

SIM_AGENT can be used both as a sequential, centralised, time-driven simulator for multi-agent systems, e.g., to simulate software agents in an Internet environment or physical agents and their environment, and as an agent implementation language, e.g., for software agents or the controller for a physical robot.

The toolkit is implemented in Pop-11, an AI programming language similar to Lisp, but with an Algol-like syntax. It defines two basic classes, sim_object and sim_agent, which can be extended (subclassed) to define the objects and agents required for a particular simulation scenario. The sim_object class is the foundation of all SIM_AGENT simulations: it provides slots (fields or instance variables) for the object's name, internal database, sensors, and rules together with slots which determine how many processing cycles each module will be allocated at each timestep and so on. The sim_agent class is a subclass of sim_object which provides simple message based communication primitives.

As an example, we briefly outline the design and implementation of a simple SIM_AGENT simulation, SIM_TILEWORLD. Tileworld is a well established testbed for agents [2,9]. It consists of an environment consisting of tiles, holes and obstacles, and one or more agents whose goal is to score as many points as possible by pushing tiles to fill in the holes. The environment is dynamic: tiles, holes and obstacles appear and disappear at rates controlled by the experimenter.

SIM_TILEWORLD defines three subclasses of the SIM_AGENT base class sim_object to represent holes, tiles and obstacles, together with two subclasses of sim_agent to represent the environment and the agents. The subclasses define additional slots to hold the relevant simulation attributes, e.g., the position of tiles, holes and obstacles, the types of tiles, the depth of holes, the tiles being carried by the agent etc. By convention, external data is held in slots, while internal data (such as which hole the agent intends to fill next) is held in the agent's database.

The simulation consists of two or more active objects (the environment and the agent(s)) and a variable number of passive objects (the tiles, holes and obstacles). At simulation startup, instances of the environment and agent classes are created and passed to the SIM_AGENT scheduler. At each simulation cycle, the environment agent causes tiles,

---

[1] See http://www.cs.bham.ac.uk/~axs/cog_affect/sim_agent.html

obstacles and holes to be created and deleted according to user-defined probabilities. The scheduler then runs the Tileworld agents which perceive the new environment and run their rules on their internal databases updated with any new sense data. Each agent chooses an external action to perform at this cycle (e.g., moving to or pushing a tile) which is queued for execution at the end of the cycle. The cycle then repeats.

3   Distributing a SIM_AGENT Simulation with HLA

There are two distinct ways in which SIM_AGENT might use the facilities offered by the HLA. The first, which we call the *distribution* of SIM_AGENT, involves using HLA to distribute the agents and objects comprising a SIM_AGENT simulation across a number of federates. The second, which we call *inter-operation*, involves using HLA to integrate SIM_AGENT with other simulators. In this paper we concentrate on the former, namely distributing an existing SIM_AGENT simulation using SIM_TILEWORLD as an example. Based on the SIM_TILEWORLD implementation outlined in section 2, we chose to split the simulation into $n + 1$ federates, corresponding to n agent federates and the environment federate respectively.

In the distributed implementation of SIM_TILEWORLD, the communication between the agent and environment federates is performed via the objects in the FOM, through the creation, deletion and updating of attributes.[2] The FOM consists of two main subclasses: *Agent* and *Object*, with the *Object* class having *Tiles*, *Holes* and *Obstacles* as subclasses. The classes and attributes in the FOM are mapped in a straightforward way onto the classes and slots used by SIM_AGENT. For example, the *depth* attribute of the *Tile* class in the FOM maps to the depth slot of the sim_tile class in SIM_AGENT.

HLA requires federates to own attribute instances before they can update their value. In HLA_AGENT we use ownership management to manage conflicts between actions proposed by agents simulated by different federates. For example, two (or more) agents may try to push the same tile at the same timestep. Once the tile has been moved by one agent, subsequent moves should become invalid, as the tile is no longer at the position at which it was initially perceived. If the agents are simulated by the same agent federate such action conflicts can be handled in the normal way, e.g., we can arrange for each action to check that its preconditions still hold before performing the update and otherwise abort the action. However, this is not feasible in a distributed setting, since external actions are queued by SIM_AGENT for execution at the end of the current cycle. We therefore extend current practice in SIM_AGENT and require that attribute updates be *mutually exclusive*. Ownership of a mutually exclusive attribute can only transferred at most once per simulation cycle, and a federate relinquishes ownership of an attribute only if it has not already been updated at the current cycle. (If multiple attributes are updated by the same agent action, we require that federates acquire ownership of the attributes in a fixed order to avoid deadlock.) For example, if two agents running on different federates try to move a given tile at the same cycle, whichever agent's action is processed first will acquire ownership of the tile and succeed, while the other will be denied ownership and fail.

---

[2] In this example, SIM_AGENT's inter-agent message based communication is not used. Message passing also handled by the RTI, using interactions.

SIM_AGENT is a centralised, time-driven system in which the simulation advances in timesteps or cycles. We therefore synchronise the federation at the beginning of each cycle, by making all federates both time-regulating and time-constrained. This ensures that the federates proceed in a timestep fashion, alternating between performing their external actions and perceiving changes.

4   Extending SIM_AGENT

In this section we briefly sketch the extensions necessary to the SIM_AGENT toolkit to allow an existing SIM_AGENT simulation to be distributed using the HLA. Together, the extensions constitute a new library which we call HLA_AGENT.

In what follows, we assume that we have an existing SIM_AGENT simulation (e.g., SIM_TILEWORLD) that we want to distribute by placing disjoint subsets of the objects and agents comprising the simulation on different federates. Each federate corresponds to a single SIM_AGENT process and is responsible both for simulating the local objects forming its own part of the global simulation, and for maintaining *proxy* objects which represent objects of interest being simulated by other federates. Each federate may be initialised with part of the total model or all federates can run the same basic simulation code and use additional information supplied by the user to determine which objects are to be simulated locally. For example, in SIM_TILEWORLD we may wish to simulate the agent(s) on one federate and the environment on another.

The overall organisation of HLA_AGENT is similar to other HLA simulators. Each SIM_AGENT federate requires two ambassadors: an RTI Ambassador which handles calls to the RTI and a Federate Ambassador that handles callbacks from the RTI. Calls to the RTI are processed asynchronously in a separate thread. However, for simplicity, we have chosen to queue callbacks from the RTI to the Federate Ambassador for processing at the end of each simulation cycle. SIM_AGENT has the ability to call external C functions. We have therefore adopted the reference implementation of the RTI written in C++ developed by DMSO, and defined C wrappers for the RTI and Federate Ambassador methods needed for the implementation. We use Pop-11's simple serialisation mechanism to handle translation of SIM_AGENT data structures to and from the byte strings required by the RTI. All RTI calls and processing of Federate Ambassador callbacks can therefore be handled from HLA_AGENT as though we have an implementation of the RTI written in Pop-11.

To distribute a simulation, the user must define the classes and attributes that constitute the Federation Object Model and, for each federate, provide a mapping between the classes and attributes in the FOM and the SIM_AGENT classes and slots to be simulated on that federate. If the user simulation is partitioned so that each federate only creates instances of those objects and agents it is responsible for simulating, then no additional user-level code is required. In the case in which all federates use the same simulation code, the user must define a procedure which is used to determine whether an object should be simulated on the current federate. The user therefore has the option of partitioning the simulation into appropriate subsets for each federate, thereby minimising the number of proxy objects created by each federate at simulation startup, or allowing all federates to create a proxy for all non-local objects in the simulation. For very large simulations, the latter approach may entail an unacceptable performance penalty, but has the advantage that distributed and non-distributed simulations can use identical code.

5   Experimental Results

To evaluate the robustness and performance of HLA_AGENT we implemented a distributed version of SIM_TILEWORLD using HLA_AGENT and compared its performance with the original, non-distributed SIM_AGENT version.

The hardware platform used for our experiments is a Linux cluster, comprising 64 2.6 GHz Xeon processors each with 512KB cache (32 dual nodes) interconnected by a standard 100Mbps fast Ethernet switch. Our test environment is a Tileworld 50 units by 50 units in size with an object creation probability (for tiles, holes and obstacles) of 1.0 and an average object lifetime of 100 cycles. The Tileworld initially contains 100 tiles, 100 holes and 100 obstacles and the number of agents in the Tileworld ranges from 1 to 64. In the current SIM_TILEWORLD federation, the environment is simulated by a single federate while the agents are distributed in one or more federates over the nodes of the cluster[3]. The results obtained represent averages over 5 runs of 100 SIM_AGENT cycles.

We would expect to see speedup from distribution in cases where the CPU load dominates the communication overhead entailed by distribution. We therefore investigated two scenarios: simple reactive Tileworld agents with minimal CPU requirements and deliberative Tileworld agents which use an A^* based planner to plan optimal routes to tiles and holes in their environment. The planner was modified to incorporate a variable "deliberation penalty" for each plan generated. In the experiments reported below this was arbitrarily set at 10ms per plan.

For comparison, Figure 1(a) shows the total elapsed time when executing 1, 2, 4, 8, 16, 32 and 64 reactive and deliberative SIM_TILEWORLD agents and their environment on a single cluster node using SIM_AGENT and a single HLA_AGENT federate. This gives an indication of the overhead inherent in the HLA_AGENT library itself independent of any communication overhead entailed by distribution. As can be seen, the curves for SIM_AGENT and HLA_AGENT are quite similar, with the HLA overhead diminishing with increasing CPU load. For example, with 64 reactive agents the HLA introduces a significant overhead. In SIM_AGENT, the average elapsed time per cycle is 0.145 seconds compared to 0.216 seconds with HLA_AGENT, giving a total overhead for the HLA of approximately 54%. For agents which intrinsically require more CPU and/or larger numbers of agents, the overhead is proportionately smaller. With 64 deliberative agents, the average elapsed time per cycle is 0.522 seconds with SIM_AGENT and 0.524 seconds with HLA_AGENT, giving a total overhead for the HLA of just 0.4%.

We also investigated the effect of distributing the Tileworld agents across varying numbers of federates. Figure 1(b) shows a breakdown of the total elapsed time for an agent federate when distributing 64 reactive and deliberative agents over 1, 2, 4, 8 and 16 nodes of the cluster.[4] In each case, the environment was simulated by a single environment federate running on its own cluster node. As expected, the elapsed time

---

[3] For our experiments, only one processor was used in each node.

[4] Unfortunately, it was not possible to obtain exclusive access to all the nodes in the cluster for our experiments.

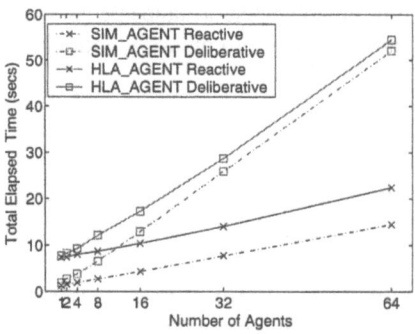
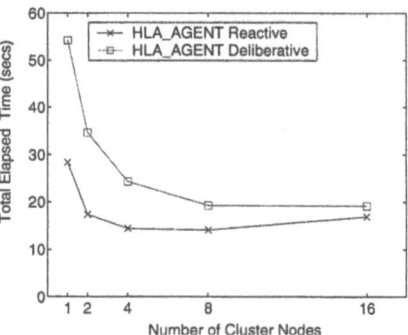

(a) Total elapsed times for 1-64 Reactive and Deliberative agents in SIM_AGENT and HLA_AGENT (single federate).

(b) Total elapsed times for an Agent Federate (64 Reactive and Deliberative Agents distributed over 1-16 nodes).

Fig. 1. Elapsed Times

drops with increasing distribution, and with 4 nodes the elapsed time is comparable to the non-distributed case for the reactive agents.[5] For the more computation-bound deliberative agents a greater speedup is achieved, and with four nodes the elapsed time is approximately half that of the non-distributed case. However in both the reactive and deliberative cases, as the number of nodes (and hence the communication overhead) increases, the gain from each additional node declines. For example, with a single agent federate we would expect to see at least 128 attribute updates per cycle (since agents update their x and y positions every cycle). With 16 agent federates, the environment federate still receives 128 attribute updates per cycle, but in addition each agent federate receives at least 120 updates from the 60 agents on the other agent federates. As a result, the number of callbacks processed by the RTI in each cycle grows from 128 with 1 agent federate to 2048 with 16 agent federates.

In addition, without load balancing, the speedup that can be obtained is limited by the elapsed time for the slowest federate. An analysis of the the total cycle elapsed times for the simulation phase of HLA_AGENT (i.e., the time required to run the user simulation plus object registration and deletion, attribute ownership transfer requests and queueing attribute updates for propagation at the end of the user simulation cycle) shows that with more than 4 agent federates, the simulation phase time for the environment federate is greater than that for any single agent federate. Prior to this point, the environment federate spends part of each cycle waiting for the agent federate(s) to complete their simulation phase, and after this point agent federates spend part of each cycle waiting for the environment federate. With 8 agent federates, the elapsed time of the environment federate forms a lower bound on the elapsed time for the federation as a whole, and further speedup can only be obtained by distributing the environment across multiple federates. Without

---

[5] This represents a significant improvement on results reported previously [10], where 16 nodes were required to obtain speedup with 64 reactive agents. We believe the increase in performance is largely attributable to a change in the ticking strategy adopted. In the experiments reported in this paper, we used the no argument version of `tick`.

the communication overhead of distribution, we would therefore expect the total elapsed time to reach a minimum between 4 and 8 agent federates and thereafter remain constant.

Although preliminary, our experiments show that, even with relatively lightweight agents like the Tileworld agents, we can get speedup by distributing agent federates across multiple cluster nodes. However with increasing distribution the broadcast communication overhead starts to offset the reduction in simulation elapsed time, limiting the speedup which can be achieved. Together with the lower bound on elapsed time set by the environment federate, this means that for reactive agents the elapsed time with 16 nodes is actually greater than that for 8 nodes. With 64 deliberative agents, which intrinsically require greater (though still fairly modest) CPU, we continue to see small improvements in overall elapsed time up to 16 nodes.

6 Summary

In this paper, we presented HLA_AGENT, an HLA-compliant version of the SIM_AGENT agent toolkit. We showed how HLA_AGENT can be used to distribute an existing SIM_AGENT simulation with different agents being simulated by different federates and briefly outlined the changes necessary to the SIM_AGENT toolkit to allow integration with the HLA. The integration of SIM_AGENT and HLA is transparent in the sense that an existing SIM_AGENT user simulation runs unmodified, and symmetric in the sense that no additional management federates are required. In addition, the allocation of agents to federates can be easily configured to make best use of available computing resources.

Preliminary results from a simple Tileworld simulation show that we can obtain speedup by distributing agents and federates across multiple nodes in a cluster. While further work is required to analyse the RTI overhead and characterise the performance of HLA_AGENT with different kinds of agents and environments, it is already clear that the speedup obtained depends on the initial allocation of agents to federates. If this results in unbalanced loads, the slowest federate will constrain the overall rate of federation execution. It should be relatively straightforward to implement a simple form of code migration to support coarse grain load balancing by swapping the execution of a locally simulated object with its proxy on another, less heavily loaded, federate. We also plan to investigate the performance implications of distributing the simulation across multiple (geographically dispersed) clusters. Together, these extensions will form the first step towards a GRID-enabled HLA_AGENT.

Another area for future work is *inter-operation*, using HLA to integrate SIM_AGENT with other simulators. This would allow the investigation of different agent architectures and environments using different simulators in a straightforward way. In a related project, we are currently developing an HLA-compliant version of the RePast agent simulator [11], which will form part of a combined HLA_AGENT /RePast federation.

Acknowledgements. We would like to thank the members of the Centre for Scientific Computing at the University of Warwick, and in particular Matt Ismail, for facilitating access to the Task Farm cluster. This work is part of the PDES-MAS project[6] and is supported by EPSRC research grant No. GR/R45338/01.

---

[6] http://www.cs.bham.ac.uk/research/pdesmas

References

1. Durfee, E.H., Montgomery, T.A.: MICE: A flexible testbed for intelligent coordination experiements. In: Proceedings of the Ninth Distributed Artificial Intelligence Workshop. (1989) 25–40
2. Pollack, M.E., Ringuette, M.: Introducing the Tileworld: Experimentally evaluating agent architectures. In: National Conference on Artificial Intelligence. (1990) 183–189
3. Atkin, S.M., Westbrook, D.L., Cohen, P.R., Jorstad., G.D.: AFS and HAC: Domain general agent simulation and control. In Baxter, J., Logan, B., eds.: Software Tools for Developing Agents: Papers from the 1998 Workshop, AAAI Press (1998) 89–96 Technical Report WS-98–10.
4. Sloman, A., Poli, R.: SIM_AGENT: A toolkit for exploring agent designs. In Wooldridge, M., Mueller, J., Tambe, M., eds.: Intelligent Agents II: Agent Theories Architectures and Languages (ATAL-95). Springer–Verlag (1996) 392–407
5. Anderson, J.: A generic distributed simulation system for intelligent agent design and evaluation. In Sarjoughian, H.S., Cellier, F.E., Marefat, M.M., Rozenblit, J.W., eds.: Proceedings of the Tenth Conference on AI, Simulation and Planning, AIS-2000, Society for Computer Simulation International (2000) 36–44
6. Schattenberg, B., Uhrmacher, A.M.: Planning agents in JAMES. Proceedings of the IEEE **89** (2001) 158–173
7. Kuhl, F., Weatherly, R., Dahmann, J.: Creating Computer Simulation Systems: An Introduction to the High Level Architecture. Prentice Hall (1999)
8. Sloman, A., Logan, B.: Building cognitively rich agents using the SIM_AGENT toolkit. Communications of the ACM **42** (1999) 71–77
9. Ephrati, E., Pollack, M., Ur, S.: Deriving multi-agent coordination through filtering strategies. In Mellish, C., ed.: Proceedings of the Fourteenth International Joint Conference on Artificial Intelligence, San Francisco, Morgan Kaufmann (1995) 679–685
10. Lees, M., Logan, B., Oguara, T., Theodoropoulos, G.: Simulating agent-based systems with HLA: The case of SIM_AGENT – Part II. In: Proceedings of the 2003 European Simulation Interoperability Workshop, European Office of Aerospace R&D, Simulation Interoperability Standards Organisation and Society for Computer Simulation International (2003)
11. Minson, R., Theodoropoulos, G.: Distributing RePast agent-based simulations with HLA. In: Proceedings of the 2004 European Simulation Interoperability Workshop, Edinburgh, Simulation Interoperability Standards Organisation and Society for Computer Simulation International (2004) (to appear).

FedGrid: An HLA Approach to Federating Grids

Son Vuong[1], Xiaojuan Cai[1], Juan Li[1], Sukanta Pramanik[1],
Duncan Suttles[2], and Reggie Chen[2]

[1] Computer Science Department, University of British Columbia
Vancouver, BC V6T 1Z4 Canada
{vuong,xjcai,juanli,pramanik}@cs.ubc.ca
[2] Magnetar Games Corp,Vancouver, BC Canada
{duncan,reggie}@magnetargames.com

Abstract. Research on Grids has received considerable attention in recent years. Whereas existing work on connecting the grids primarily focused on grid services, in our work we propose a unified approach, so-called *FederationGrid* (or *FedGrid* for short), which integrates both virtual organization (VO) and grid services, thus enabling formation of virtual communities on top of grids. FedGrid is based on the standard High Level Architecture (HLA) that enjoys the advantages of high-level information modeling, including re-usability of software components and real-time performance of the overall systems. In addition, FedGrid is a fractal grid, comprised of hierarchical HLA federations, e.g. *Realms* and *Lobby*, which supports well the concept of virtual communities and scalability.

Keywords: Grid, Federation, HLA, RTI, Grid Computing, Grid Services, Grid Access, Virtual Organization, Community Grid, Collaborative Computing.

1 Introduction

Grid Computing [1-2] has made a rapid progress in both concept and evaluation in recent years, which is evident in several large-scale applications in high-energy physics, earth science, and other areas [3-6]. It glues a great amount of distributed recourses and powers together to allow for the provisioning of ultra extra time and resources to solve problems previously intractable and accelerate computing and process time. It also allows widely distributed organizations or branches of organizations to create a virtual organization to cooperate and share resources. The need for virtual organization support in the grids has been increasing as it allows and attracts more and more grid users to form the grids' virtual communities; hence enabling interactive and collaborative distributed applications to run in the grid environment. Our work focuses on federating grids via the High Level Architecture (HLA) [7], the US DoD Standard which has become an IEEE Standard defined for modeling and execution of interoperable distributed simulations. HLA has experienced tremendous growth over the past few years, expanding its uses in various distributed application areas such as multi-user games and distance learning.

The multidimensional advantages of grid computing have led to many new Grid projects in both scientific research and enterprise domains. However, since Grids are

M. Bubak et al. (Eds.): ICCS 2004, LNCS 3038, pp. 889–896, 2004.
© Springer-Verlag Berlin Heidelberg 2004

owned and controlled by specific groups, and are built out of differing substrate technologies with perhaps incompatible communication systems, it is inevitable that Grids will be established as islands of resource collections. Therefore, although continued support of Open Grid Services Architecture (OGSA) and Open Grid Services Infrastructure (OGSI) is essential, interoperability standards are also needed to support the federation of distributed Grids developed by distributed team. The federation of grids is essential to deal with a range of important problems such as unified   interface,   interoperable/federated   services,   sharing   information, communication across boundaries, and so on.

There has been existing work that addresses the problem of federating different types of grids, e.g. the Narada brokering [8], but such work mainly focuses on the core lower-level service, rather than the distributed object-based service level. Besides offering the message-based service at a higher-level than Narada, FedGrid via HLA can be implemented as a distributed P2P service rather than a client-server service, a fact which can enable rapid and widespread use of the technology on the Internet. By using HLA and the special virtual community supporting federations, so-called *Realms*, FedGrid can provide all benefits of Narada, and at the same time it can federate grids from both the perspectives of higher-level virtual organizations and grid service federations.

This paper is organized into five sections as follows. In Section 2 we briefly introduce HLA/RTI and existing work on how to federate grids and manage interactive grid applications. In Section 3, we present our proposal of the FedGrid framework. Finally, in Section 4 we describes some major work in progress and for future research on FedGrid and offer some concluding remarks.

2 Background and Related Works

The Grid concept embodies a vision of a global computing and communications infrastructure providing transparent, flexible and pervasive access to distributed resources.

2.1 Virtual Organization and OGSA

In effect, Grid enables resource sharing among dynamic collections of autonomous entities that can be individuals and institutions, so-called virtual organizations [9], which are governed by definite rules for authorization, authentication, resource access, etc. These distributed resources across heterogeneous virtual organizations, developed inevitably by distributed teams, need some inter-grid protocols to ensure interoperability. The OGSA (Open Grid Service Architecture) [10] addresses this issue by aligning grid technology with Web Service [11] technologies. OGSA defines a set of standard conventions and WSDL (Web Service Description Language) interfaces from which all Grid Services are implemented. This core set of consistent interfaces control the semantics of the Grid Service instance: its creation, naming, lifetime, communication and thereby facilitating the construction of higher-order services.

2.2 HLA/RTI and Distributed Interactive Simulation

While OGSA is still evolving, there are already existing solutions for distributed simulation systems such as HLA [7] that can be used as underlying framework for grids. The High Level Architecture (HLA) is a U.S. DoD-wide initiative to provide an architecture for support of interoperability and reusability of all types of models for Distributed Interactive Simulations. HLA has been adopted by IEEE as an open standard, the IEEE 1516. HLA models individual simulation participant as a federate and all federates in a simulation forms a federation. Federates interact with the federation through the interface between federates called Run-Time Infrastructure (RTI). The Key components of HLA are:

- *HLA rules*: principles and conventions that must be followed to achieve proper interaction among federates in a federation.
- *Interface specification*: describes RTI services and interfaces, and identify "callback" functions each federate must provide to accept the information from RTI.
- *Object Model Template (OMT)*: describes allowed structure of a Federate Object Model (FOM). It provides a standardized framework for describing HLA object models. It defines the format and syntax (but not content) of HLA object models. OMT is a meta-model for all FOMs. Its main components are:

Some well-known implementations of HLA/RTI include the DoD DMSO, the Pitch's portable RTI, and more recently Magnetar's RTI implementation in .NET over Directplay, called *Chronos*, and the UBC's RTI implementation in Java over JXTA, called *Galaxy*.

2.3 HLA Approach to Grid

There has been research attempts to use grid services in support of HLA functions. Zajac *et al.* [12, 13] adopts a three-level approach to migrate HLA concept into Grid. In the first step they address the issue of HLA Federate Discovery mechanism by publishing them as a Grid Service. The second step deals with Grid-based data communication protocols like Globus GridFTP [14] as an alternative to RTI communication. Their final step encapsulates the RTI library within the Grid Service. Similarly, the CrossGrid project [15] considers HLA from the perspective of the Grid hosting an HLA Federation. In contrast, our FedGrid approach makes use of HLA in support of scalable grid services and their integration in a unified information model. In our FedGrid approach, HLA is used to form an integral part of the global-scale Grid architecture and to contribute to enhanced Grid functionality, scalability and performance.

3 System Architecture

From both the information modeling point of view and the fact that it can be implemented as a P2P service, HLA is deemed to be the ideal way to federate the virtual organizations to support interactive distributed applications. The scalable data distribution mechanism via HLA also provides benefits to the FedGrid approach. The OGSI Open Grid Services Infrastructure takes a step in the direction of HLA by supporting service data variables which hold service state and may be exposed through a client side proxy. FedGrid incorporates this OGSI concept while proposing a major extension through the mechanism of a federate proxy "*Realm Engine*" which exposes a federated object model to clients. The benefits of this extension are considerable, as listed below and explained in the next 2 subsections.

1. Uniform information modeling of grid services through the HLA OMT standard.
2. Ability to easily create real-time updates between Realm peers using the HLA runtime infrastructure (RTI).
3. Proven integration technology for federating heterogeneous grid services.
4. Provide a mechanism for composing basic object models BOM [16] into aggregate models.

3.1 Overall Framework

FedGrid uses High Level Architecture (HLA) modeling and simulation standards to implement a super-scalable grid supporting real-time collaborative applications. The FedGrid is organized as hierarchical HLA federations, which form "Realms" defined by Federated Object Models (FOM). *Realms* correspond to the grid concept of virtual organization and are federation communities. For example, both the *chess realm* and the massive multiplayer online games *(MMOG) realm* require the support of a game lobby and repositories of resources or game digital assets. The HLA Run Time Infrastructure (RTI) can provide very advanced P2P services used in the Federation Grid's three-tiered network architecture consisting of core, super-peer, and peer federations. This is a view of federation grid from the perspective of application network building. Figure 1 shows this network architecture.

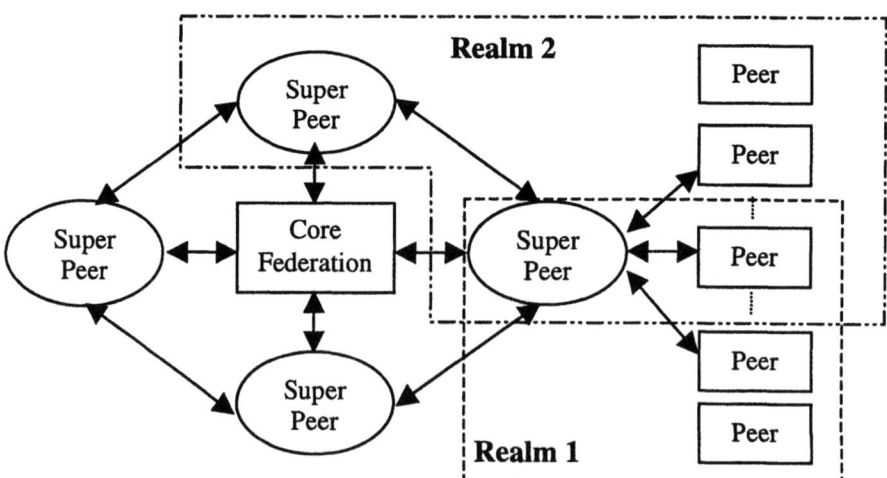

Fig. 1. FedGrid Network Architecture

The *core federation* provides authorization and discovery services. The *super-peers* are *Realm* service provider and they manage each federation community, while the peers are the entry point to the entire network. The super-peers, typically powerful servers, can be viewed to constitute an overlay network. *Realms* are virtual communities that constitutes of a set of peers and super-peers which abide by some definite FOM. They can also overlap as is shown in Figure 1.

Federation Grid further leverages HLA by introducing a single process federation (SPF) of *Agents* architecture for software *Engines* and through federation of engines into *Systems*. As is illustrated in Figure2, a System Federation is essentially a federation of Systems each of which is again a federation of Engines communicating using an internal RTI. The Engines are designed as an SPF of Agents and the Agents themselves talk via a blackboard. The detailed structure layering is described in the next section.

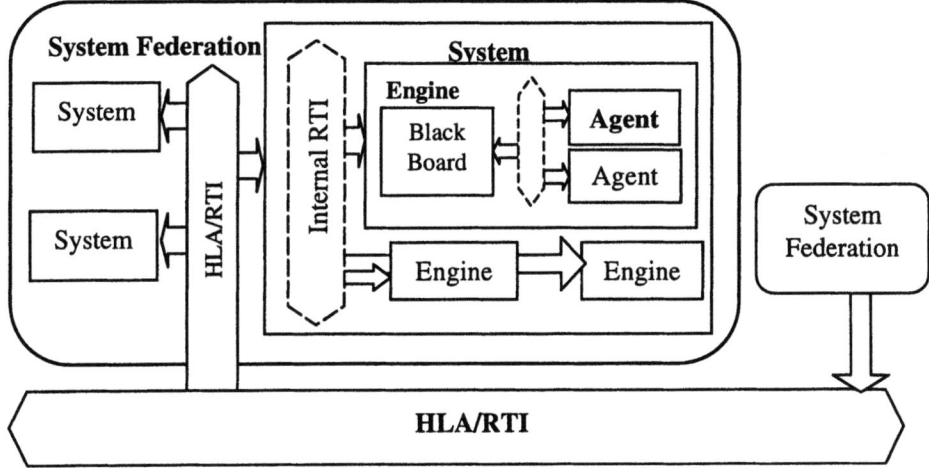

Fig. 2. Hierarchy of Federations

3.2 Federation Grid Layers

In the context of FedGrid organization, *"Realms"* are defined as collections of Virtual Organizations, which share a common information model. The basic idea of the realm model is illustrated in Figure 3. From the layer structuring point of view, *Engines*, which are single process federations (SPF) of *Agents* (software module), constitute the lowest rung in a HLA hierarchical federation model, which can be federated up to form *Systems*. Systems can in turn be federated to form a System Federation or so-called *Federation Community*. A FedGrid is formed by a collection of Federation Communities. A FedGrid can also be viewed as global virtual grid, analogous a fractal grid, that comprises an integration of grids within an application domain. These hierarchical federations are described as follows:

3.2.1 *Engine* as an Internal Federation

The HLA use is extended to include a SPF model for complex software *"Engines"* composed of plugin *"Agent"* federates. SPF acts like an internal bus for those agents. The idea is shown in Figure 3. The SPF agent paradigm is a promising avenue in defining re-usable AI component technology thereby reducing the cost of developing new engines. The re-use strategy also creates the basis for a true Meta product line technology. For this internal federation, there is an internal FOM for all the agents to follow. For example, in the realm engine, there are different messenger adapter like IM, Jabber, and different directory service adapter to LDAP server, X500 directory etc., and all these agents will use the "realm" Federate Object Models (FOM). Realms correspond to the grid concept of virtual organization and are federation communities.

Federation Grid engines are implemented using MAGNETAR (Metaprogrammable AGent NETwork ARchitecture) [17]. The structure of a MAGNETAR engine is a single process federation, which uses an internal HLA infrastructure engine (that acts as a Blackboard) to provide communication between *"agent"* plugin components. The first MAGNETAR engine is *"Chronos"*, a full HLA RTI implementation. Further development of MAGNETAR will include generic logic components, which can script behavior by driving the engines agents.

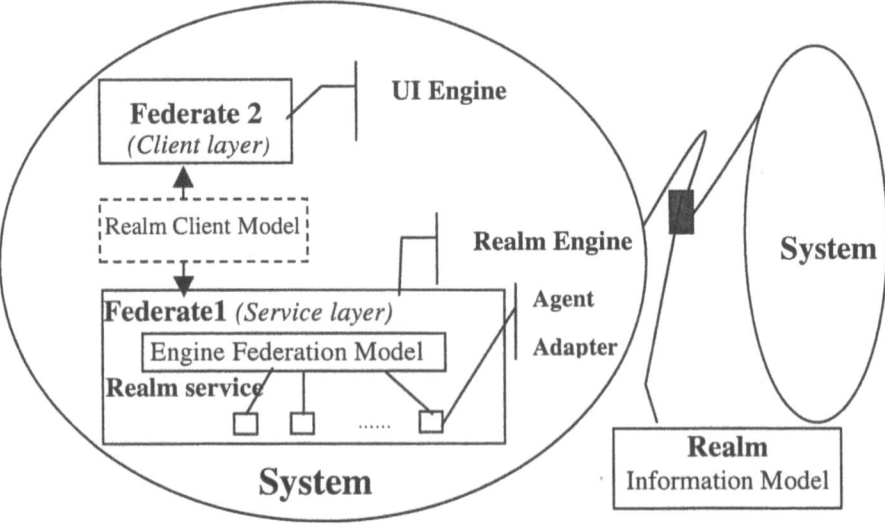

Fig. 3. The Realm Federation Model

3.2.2 *System* as a Federation of Engines

A *System* is a federation of engines. For example, in a game system, we need different types of engines, such as *realm* engine, *game* engine, *realm user interface* engine, network engine and so on, to work together to achieve the overall game functions.

Figure 3 above shows a system federation to illustrate the main idea of FedGrid. The UI engine and the realm engine follow the FOM of engine/service model, so that the realm could provide a unified interface to the client, no matter how various the services that the realm engine supports might be. The existing grid service (even from different types of grids) can be implemented as a grid service adapter (shown as an agent/adaptor in Figure 3) in the realm engine so as to provide a uniform interface to the user.

3.2.3 Federation Communities

To form a virtual organization or even a global virtual grid, systems need to interact with each other. These systems will form a high-level federation, called a *Federation Community*. An example is a *Realm federation* for a realtime interactive application environment in which we need to provide realtime directory information for purposes such as effective game match making. In this case, the systems will exchange realtime directory information through the HLA Realm federation.

4 Work in Progress and Concluding Remarks

We present a unified approach to federate grids for supporting global-scale interactive applications over virtual organizations. In FedGrid, HLA forms an integral part of the Grid architecture and contributes to significantly improve the Grid functionality, reusability, scalability and performance. We have developed a framework and partially experimented with the FedGrid approach by implementing the underlying components. FedGrid can find its applications in a variety of global-scale integrated grid services such as **Community Grid, Content Grid** and **Computing Grid**. We have developed two working versions of RTI/HLA engines and some working version of the game lobby and repository. We are in the process of integrating them into a distributed chess prototype, including a *ChessGrid* system, to demonstrate the concept of FedGrid. Concurrently, we are working on a Web-based Flash FedGrid prototype, which is motivated by the increasing popularity of *Flash Games* and the fact that distributed web simulations has become increasingly important, falling in line with the XMSF initiative [18]. FedGrid is intended provide a complete XMSF-based solution. The popularity of "Flash Games" can be seen by noting that the Google search for "Flash Games" returns with 769000 hits as compared to 20,100 hits for "Wireless Games", a supposedly very hot area, i.e. less than 3% of the hits for "Flash Games".

The Flash FedGrid prototype includes a Flash-based game lobby with community service called *Flash Tryst* which implements the complete FedGrid Realm model for massive multiplayer game applications. Such prototypes can be adapted for interactive e-learning and other collaborative distributed applications. The Realm engine will interface with one or more P2P instant messaging services and provide a federated object model to the applications. Flash Tryst consists of a Flash client and a .Net Realm engine with *Chronos*, a P2P RTI engine developed by Magnetar, as the communication infrastructure between them. Another ongoing project is to build a Federated Lobby prototype to prove the concept of FedGrid. Yet another interesting project is to consider the issue of QoS and performance in HLA [19] and therefore in FedGrid.

The FedGrid approach is deemed unique and powerful as it promises to provide a unified information model for federating existing grids in an efficient, scalable and seamless manner. The FedGrid work presented in this paper serves to highlight some preliminary ideas to encourage further open concerted development in this direction.

While FedGrid offers a promising unified global scale approach to gluing (federating) existing grids together within its own information model, it can also be viewed as complementary to the OGSA approach as FedGrid aims at federating grids at the higher (information model) level whereas OGSA focuses on integrating the grids at one level below (below the HLA level). In fact, OGSA grids or other existing grids can be integrated into the FedGrid environment by simply developing the appropriate agent adaptors for the respective grids. Further research and experiment will be necessary to fully demonstrate and validate the FedGrid approach.

Acknowledgments. The work in this paper is supported in part by the joint grants from the Canadian Institute for Telecommunication Research (CITR) Network of Centres of Excellence, the Advanced Systems Institute of B.C., Telus Mobility, Proface INDE, and Magnetar Games. The authors also wish to thank Aleks Mandic, Mandi Paugh, and Ryan Levere from Magnetar Games for the fruitful discussions on HLA and RTI implementation.

References

1. Foster, I. And Kesselman, C. (eds.), *The Grid: Blueprint for a New Computing Infrastructure,* Morgan Kaufman, 1999.
2. Foster, I., *What is the Grid? A three checkpoints list,* GridToday Daily News and Information for the Global Community, July 22, 2002.
3. The Grid for UK Particle Physics, http://www.gridpp.ac.uk
4. The GriPhyN Project, http://www.griphyn.org/
5. NEES Earthquake Engineering Grid, http://www.neesgrid.org/
6. SCEC Earthquake Science Grid http://www.scec.org
7. IEEE 1516 Standard for Modeling and Simulation (M&S) High Level Architecture (HLA) - Federate Interface Specification 2000, ISBN 0-7381-2622-5, http://shop.ieee.org/store/product.asp?prodno=SS94883
8. Narada Brokering, http://www.naradabrokering.org/
9. Foster, I., Kesselman, C. and Tuecke, S., *The Anatomy of the Grid: Enabling Scalable Virtual Organizations,* International J. Supercomputer Applications, *15(3), 2001.*
10. Foster, I., Kesselman, C., Nick, J.M. and Tuecke, S., *The Physiology of the Grid: An Open Grid Services Architecture for Distributed Systems Integration,* Open Grid Service Infrastructure WG, Global Grid Forum, June 22, 2002.
11. Web Services, http://www.w3.org/2002/ws/
12. Zajac, K., Tirrado-Ramos, A., Zhao, Z., Sloot, P. and Bubak, M., *Grid Services for HLA based Distribution Systems Framework*
13. Tirrado-Ramos, A., Zajac, K., Zhao, Z., Sloot, P. Albada, D.V. and Bubak, M., *Experimental Grid Access for Dynamic Discovery and Data Transfer in Distributed Interactive Simulation Systems*
14. Allcock, W., Bester, J., Bresnahan, J., Chervenak, A., Liming, L., Meder, S. and Tuecke. S., *GridFTP Protocol Specification,* GGF GridFTP Working Group Document, September 2002.
15. CrossGrid Project, http://www.eu-crossgrid.org/
16. Base Object Models, http://www.boms.info
17. GameXML, http://www.gamexml.org
18. Web Enabled Modeling and Simulation Network, http://www.websim.net/
19. Hui Zhao, *"HLA Real-Time Extension",* Concurrency and Computation: Practice and Experience Journal. (Selected by the IEEE DS-RT 2001 Workshop). ISSN: 1532-0626.

Do Colors Affect Our Recognition Memory for Haptic Rough Surfaces?

Zhaowu Luo and Atsumi Imamiya

Department of Computer Science and Media Engineering, University of Yamamashi,
Takeda 4-3-11, Kofu, Yamanashi Prefecture, 400-8511, Japan
{luozhwu,imamiya}@hci.media.yamanashi.ac.jp

Abstract. Haptic (tactile) interaction is a promising approach to be used for human computer interfaces. But how various kinds of haptic effects are learned and memorized, and what factors might influence user performance, remain unanswered. Little is known about the recognition memory for haptic information. The authors used a recognition memory paradigm to study the influence of color information on recognition memory for haptic rough surfaces. Participants' performance is less confident at exposure duration of 2s than at other three durations. The performance shows little difference as long as rough surfaces are presented and queried in the same color. However, Experiment 2 revealed that the influence of colors on haptic memory is sensitive to color presentation and query conditions. Our results can be used as guidelines for haptic interfaces for selecting colors.

1 Introduction

Current mainstream human-computer interaction (HCI) is visual-information-centered [1]: information displayed on computer screens accounts for an extremely large proportion of that available to users.

Among new interaction modes feasible from a technical perspective, haptic interaction (i.e., forces transmitted back to human hand or fingers in a way that mimics the sensation of touching real objects, using specialized systems) is a promising one. Compared with visual and auditory interaction, it has a unique bi-directional nature. As we touch and manipulate objects, we simultaneously change their state and receive information about them [2].

For more than 50 years, the influence of verbal (for example, spatial language) or visual (for example, color) information on memory performance has been studied. The effect of spatial language on recognition memory for spatial scenes was studied [3], finding that spatial language influences the encoding and memory of spatial relation presented visually in pictures. In another study, the color's contribution to recognition memory for natural scenes was investigated [4], finding that colors enhance an individual's visual recognition memory.

Most studies to date have been concentrated on memory for visual and/or verbal information ([5], [6]), using visual or verbal information as the to-be-remembered material in recognition memory tests, although there are few exceptions in olfactory ([7], [8]) and haptic memory ([9], [10]). But to our big surprise, less or little is known

M. Bubak et al. (Eds.): ICCS 2004, LNCS 3038, pp. 897–904, 2004.
© Springer-Verlag Berlin Heidelberg 2004

about the influence of colors on the recognition memory for haptic information (for example, roughness), considering that neural correlates of both encoding and retrieval vary with the to-be-remembered material ([11]).

On the other hand, how haptic effects are learned, memorized and later used for interaction with computers is a rather complex process. Identifying factors making significant contribution to the process is important. Through a haptic interface designed for virtual reality applications, such as remote surgery, users usually can both see and feel objects within a single workspace. In such cases, the color of objects may be changed from one view to another, as the operation procedure proceeds. Can we intentionally select certain color for the objects within each view in such a way that the user's performance can be strengthened, thus raising the efficiency of haptic interfaces and reducing the users' workload?

To answer this question, we conducted a series of recognition memory experiments in which participant's haptic recognition memory for rough surfaces was tested. In our experiment task, both the processing of colors and haptic memory systems of the brain are involved. Experimental apparatus, colors and computer-generated rough surfaces used for experiments are described in Section 2, followed by two experiments and a discussion of the results in Section 3. Finally, conclusions are drawn in Section 4.

2   General Method

2.1 Experiment Apparatus

The hardware setup, as shown in Figure 1, consists of a PHANToM (Model: Premium EW) from SenseAble Technologies, a dual Pentium III PC operating on the Windows 2000 Professional platform, and a wireless liquid crystal stereo shutter eyewear (Model: CrystalEye 3) from StereoGraphics. This model of PHANToM has a maximum stiffness of around 1100N s /m and a workspace of 19.5cm × 27.0cm × 37.5cm. The shutter eyewear was used to enable participants to see the 3D experimental environment.

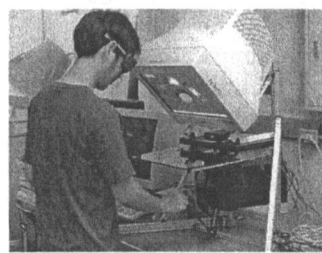
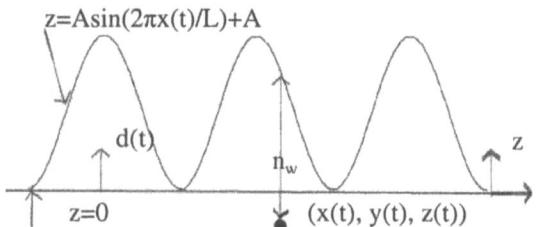

Fig. 1. The hardware setup for experiments

Fig. 2. An illustration of the textured surface and its associated variables

2.2 Stimuli

2.2.1 Colors
In order to investigate colors' influence on recognition memory for rough surfaces, we need to create a condition in which color information can be reduced to the minimum so that participants' performance in colored conditions can be compared with it. We use RGB values of (0.3, 0.3, 0.31) to paint all rough surfaces for this condition, considering the fact that RGB values of the default background of a scene created by Reachin API are RGB(0.3, 0.3, 0.3). The resultant visual effect is that surfaces almost integrate with the background of the experimental environment, with their profiles being marginally discernible. But the color information available is reduced to the minimum.

In addition to the (almost) gray values of RGB(0.3, 0.3,0.31) simulating Black and White (B&W) images being viewed in other studies, three types of colors are used as visual stimuli, and namely: RGB(1, 0, 0) for Red, RGB(1, 1, 0) for Yellow and RGB(0, 1, 0) for Green. The colors are fully (100%) saturated.

2.2.2 Haptic Rough (Textured) Surface
In order to create a touchable, rough surface, used is a one-dimensional sinusoidal grating superimposed on an underlying box. The sinusoidal grating is described by $z=A\sin(2\pi x(t)/L)+A$, where A and L are the amplitude and the spatial wavelength, respectively (see Figure 2).

We use the first method $F_1(t)$ in Choi' s study [12] for texture rendering. The force $F(t)$ generated can be calculated as follows:

$$d(t) = \begin{cases} 0, & z(t)<0 \\ z(t) -A\sin(2\pi x(t)/L) -A, & z(t)\geq 0 \end{cases}$$
$$F(t) = Kd(t)n_w,$$

where K is the stiffness of the surface, n_w is the normal vector of the surface of the underlying box, $(x(t), y(t), z(t))$ are the coordinates of the stylus at time t, $d(t)$ is the penetration depth of the stylus into the textured surface at time t.

For all the experiments described in this paper, amplitude A is fixed at 0.4mm. Since our PHANToM has a fairly larger value of stiffness K than that of Choi's [12], it is fixed at 450 Newton seconds/meter. To create six different levels of roughness in the query phase (described in Section 3), the six wavelengths Li (L1, L2, L3, L4, L5 and L6) used are 1.0, 2.0, 4.0, 8.0, 16.0 and 32.0mm, respectively. They are divided into two groups, and namely: L1, L2, L3 for one group, and L4, L5, L6 for the other group.

3 Experiments

In the experiments, there are two phases, and namely: presentation phase and query phase. In the presentation phase, participants are asked to explore three rough target surfaces (either L1, L2, L3 or L4, L5, L6, in randomized order) successively, with a 3-second interval between successive explorations. In the query phase immediately following it, the three rough target surfaces are randomly mixed with another three rough non-target surfaces (either L4, L5, L6 or L1, L2, L3). Participants are then asked to explore the resultant, randomly-positioned, six rough surfaces and give an answer as to whether they have explored each of them in the presentation phase, setting the push button to the right of each surface to either "Yes" or "No".

Figure 3 shows the display layout for the presentation phase. In it there are one big box (rough surface) and seven push buttons. The surface is such that when it is explored continuously, three kinds of roughness can be felt successively, with each being presented for an exposure duration of 8 seconds, with a 3-second interval (changed from a rough surface to a smooth one) between the successive presentations. After three kinds of roughness are presented, the surface remains smooth forever, and the query phase will be entered.

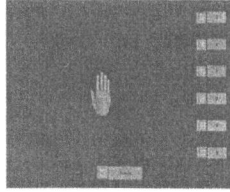   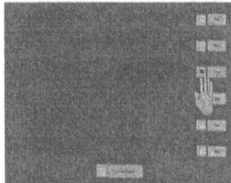

Fig. 3. Display layout for presentation phase   **Fig. 4.** Display layout for query phase

In query phase (see Figure 4 for the display layout), three target and three non-target surfaces are randomly mixed together. Participants are asked to explore each of the six surfaces and indicate whether he or she has explored it in the presentation phase by giving a "Yes" or "No" response. Twenty participants, 12 males and 8 females, aged from 18 to 38, took part in Experiments 1 and 2. All participants had normal or corrected-to-normal visual acuity and normal color vision.

3.1 Experiment 1: Exposure Duration

In Experiment 1, the rough surfaces are presented and queried in the same color condition, and namely: they are presented and queried in either (1) Gray (AA); (2) Green (GG); (3) Red (RR); or (4) Yellow (YY). Four experiments are conducted, each for one condition. Each of the three rough target surfaces (L1, L2, L3 or L4, L5, L6) is presented for either 2s, 4s, 8s, or 16s, with a 3-second interval between successive presentations. Twenty participants took part in Experiment 1, and they were randomly divided into two groups of 10 persons each to counterbalance the performance.

RESULTS: The results for Experiment 1 are shown in Figures 5 and 6.

The mean hit rate, and the mean value of the d', averaged across 20 participants and the 4 conditions are plotted (y-axis) against exposure duration of the rough surfaces during the presentation phase (x-axis), respectively. Here, and in all the following data figures, the error bars plotted correspond to plus or minus one standard error.

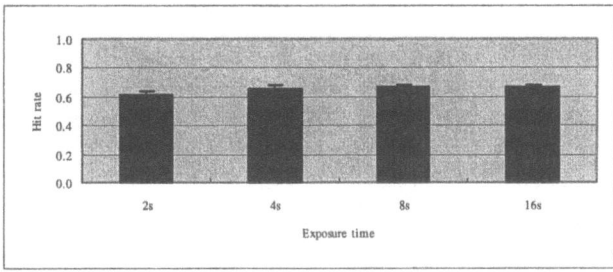

Fig. 5. Hit rate as a function of exposure duration of the surfaces in presentation phase

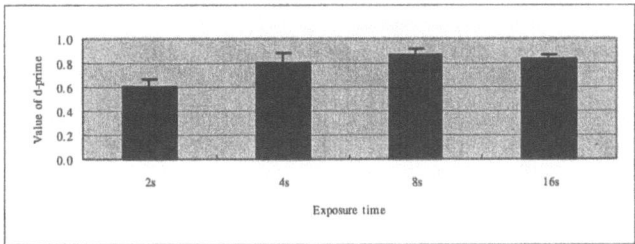

Fig. 6. Value of d' as a function of exposure duration in the presentation phase

DISCUSSION: As Figure 5 shows, the performance in terms of averaged hit rate did not change very much as the exposure duration is varied. The effect of exposure duration is typical for transfer to short-term memory: it appears to be relatively insensitive to speed of presentation [13]. Analysis of variance (ANOVA) in terms of hit rate did not show significant differences between exposure durations, between paired conditions (paired combination among AA, GG, RR and YY), or between GG, RR, YY combined and AA. In other words, the performance changes little as long as surfaces are presented and queried in the same color. Participants' performance is less confident at exposure duration of 2s than at 4s, 8s and 16s.

3.2 Experiment 2: Presentation and Query Conditions

The results of experiment 3 in [5] revealed that (recognition) performance for visual images worsened when they were presented in color and tested in black and white, or vice versa. Therefore, we set up the following six conditions in Experiment 2: (1) AG: rough surfaces presented in gray, but queried in Green; (2) AR: those presented in gray, but queried in Red; (3) AY: those presented in gray, but queried in Yellow; (4) GA: those presented in Green, but queried in gray; (5) RA: those presented in Red, but queried in gray; (4) YA: those presented in Yellow, but queried in gray. Exposure time for each of three rough surfaces in Experiment 2 is set as 8s. The same twenty participants took part in Experiment 2.

RESULTS: Analysis of data in Experiment 2 also included the data for AA condition at an exposure time of 8s in Experiment 1. Since there is no significant difference in false alarm rate between paired presentation and query conditions, only hit rate is analyzed. The results are shown in the (a), (b), (c) and (d) of Figure 7.

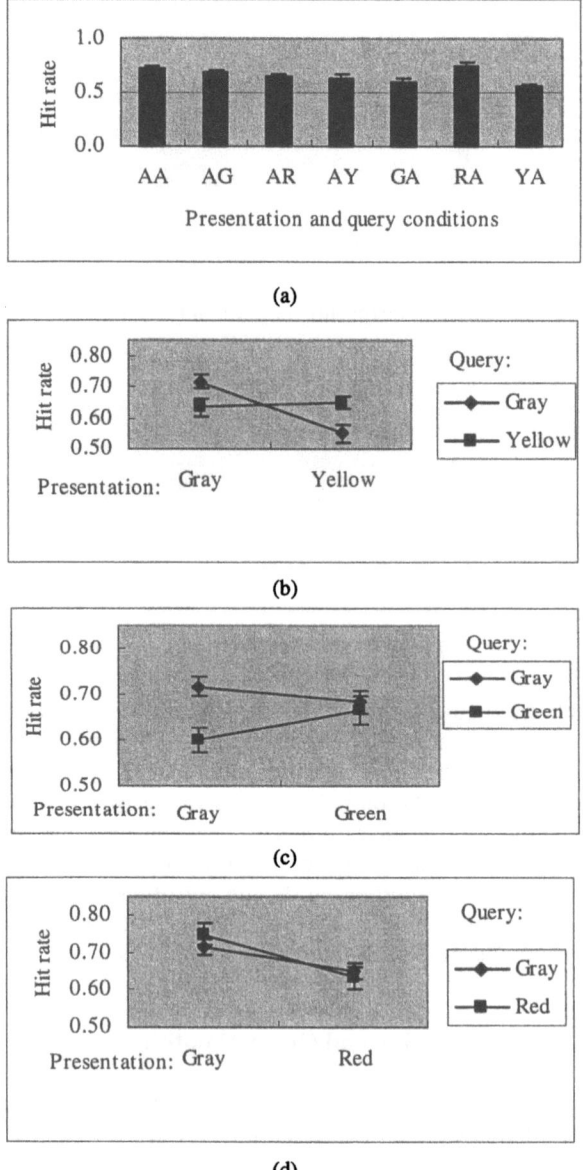

Fig. 7. Results of Experiment 2. (a) Hit rate for presentation and query conditions. (b), (c) and (d): the x-axis plots whether rough surfaces were presented in gray or colors (green, red or yellow); the y-axis plots the hit rate for surfaces queried in gray or colors (green, red or yellow).

DISCUSSION: Our experimental task involves two processes: a learning process in presentation process, and a retrieval process in the query process. We control the color available to learning and retrieval systems, because neuroimaging studies revealed a possible involvement of visual cortex in haptic processing ([14], [15]).

From Experiment 2, we found that participants' performance is worse for (1) rough surfaces presented in gray and queried in Yellow than those presented and queried in gray; and (2) those presented in Yellow and queried in gray than those presented and queried in Yellow, as shown in Figure 7(b). This finding is similar to that of the Experiment 3 in [4]. However, we found that participants' performance is slightly better for (3) rough surfaces presented in gray and queried in Red than for those presented and queried in gray, and (4) those presented in Red and queried in gray than those present and queried in Red, as shown in Figure 7(d).

The result for Green is different. When rough surfaces are presented in gray and queried in Green, or when they are presented and queried in Green, the performance is worse than when they are presented and queried in gray, than when they are presented in Green and queried in gray, respectively, as shown in Figure 7(c).

Our results are in part in line with one theory according to which the probability of recall of an item is a direct function of the similarity between the recall situation and the original learning environment [16], [17]. But the cases for RA, GR and YR color presentation and query conditions in our experiment did not result in lower hit rate.

The performance patterns are also what the encoding specificity principle predicts [18]. But colors' influence on haptic memory seems to be much more complex and versatile.

4 Conclusions

A recognition memory paradigm was used to study the influence of colors (Green, Yellow and Red) on recognition memory for haptic rough surfaces based on two experiments. Participants' performance is less confident at exposure duration of 2s than at other 3 durations. The performance shows little difference as long as rough surfaces are presented and queried in the same color.

Experiment 2 revealed that the influence of colors on haptic memory is sensitive to presentation and query conditions: RA condition results in significantly better performance than GA condition. The former could be used to improve user performance, whereas the latter should not be used in haptic interfaces whenever possible. Our results can be used as guidelines for haptic interface design for selecting colors.

Acknowledgements. We would like to thank Professors XiaoYang Mao, Kenji Ozawa, Ryutaro Ohbuchi and Kentaro Go of Yamanashi University for helpful discussions and comments. We are very grateful for Dr. Li Jing, of the National University of Singapore, who supported us with the statistical analysis of our experiment data.

This research was supported in part by the Japan Society for the Promotion of Science, by the Telecommunications Advancement Organization of Japan, and by the Research Institute of Electrical Communication of Tohoku University awarded to A.Imamiya.

References

1. Luo, Z., Imamiya, A.: How do colors influence the haptic perception of textured surfaces? Univ Access Inf. Soc. 2(2) (2003) 160-172
2. Salisbury, J. K.: Making graphics physically tangible. Commun ACM. 42(8) (1999) 75-81
3. Feist, M.I., Gentner, D.: An influence of spatial language on recognition memory for spatial scenes. Proc. of the Twenty-Third Annual Conference of the Cognitive Science Society, London, UK, (2001) 279-284
4. Wichmann, F.A., Sharpe, L.T., Gegenfurtner, K.R.: The contributions of color to recognition memory for natural scenes. Journal of Experimental Psychology: Learning, Memory, and Cognition. 28(3) (2002) 509-520
5. Gegenfurtner, K.R., Rieger, J.: Sensory and cognitive contributions of color to the recognition of natural scenes. Current Biology, 10 (2000) 805–808
6. Miller, G.A.: The magical number seven, plus or minus two: some limits on our capacity for processing information. The Psychological Review, 63 (1956) 81-97
7. Dade, L.A., Jones-Gotman, M., Zatorre, R.J., Evans, A.C.: Human brain function during order encoding and recognition. A PET activation study, Ann. N. Y. Acad. Sci. 855 (1998) 572-574
8. Morrot, G.: The color of ordors. Brain and Language, 79(2) (2001) 309-320
9. Bonda, E., Petrides, M., Evans, A.: Neural systems for tactual memories. J. Neurophysiol. 75 (1996) 1730-1737
10. Lepage, M., Randal, A., Tulving, E.: Transperceptual encoding and retrieval processes in memory: a PET study of visual and haptic objects. NeuroImage, 14 (2001) 572-584
11. Kelley, W.M., Miezin, F.M., McDermott, K.B., et al.: Hemispheric specialization in human dorsal frontal cortex and medial temporal lobe for verbal and nonverbal memory encoding. Neuron, 20 (1998) 927-936
12. Choi, S., Tan, H.Z.: A parameter space for perceptually stable haptic texture rendering. Proceedings of the Fifth PHANToM User Group (PUG) Workshop, Aspen, Colorado, (2000) 31-34
13. Wilson, R.A. Keil R.C. (Ed.): The MIT encyclopedia of the cognitive science. Cambridge, Massachusetts, London, England, The MIT Press (1999)
14. Deibert, E., Kraut, M., Krement, S., Hart, J. J.: Neural pathways in tactile objection recognition. Neurology, 52 (1999) 1413-1417
15. Zangaladze, A., Epstein, C. M., Grafton, S. T., Sathian, K.: Involvement of visual cortex in tactile discrimination of orientation. Nature, 401 (1999) 587-590
16. Hollingworth, H. L.: Psychology: its fact and principles. New York, Applet (1928)
17. Melton, A.W. Implications of short-term memory for a general theory of memory. Journal of Verbal Learning and Verbal Behavior, 2 (1963) 1-21
18. Tulving, E., Thomson, D.M.: Encoding specificity and retrieval processes in episodic memory. Psychological Review, 80(5), (1973) 352-373

Enhancing Human Computer Interaction in Networked Hapto-Acoustic Virtual Reality Environments on the CeNTIE Network

Tony Adriaansen[1], Alex Krumm-Heller[1], and Chris Gunn[2]

[1] CSIRO, ICT Centre,
P O Box 76, Epping, 1710, NSW, Australia.
http://www.ict.csiro.au
[2] CSIRO, ICT Centre
Building 108 North Road, ANU campus,
Acton, 2601, ACT, Australia

Abstract. We discuss some recent advances made to our Virtual Reality system which includes touch feedback in a networked environment. Some issues which affect the realisation and implementation of the Hapto-Acoustic Virtual Reality Environment and human–computer interaction (HCI) are mentioned, as well as networking issues. Our system is called a "haptic workbench" which combines 3D graphics visualisation, touch (haptics) and audio interaction with a computer generated model. The aim is to produce a semi-immersive 3D environment, experienced in real time by the user. Networking two or more systems makes it possible to collaborate in the virtual space and achieve a sense of perceived virtual presence. Recent additions to the system in the form of real time video with audio is demonstrated and shown to enhance the level of virtual presence in the networked environment. Such face to face communication with the remote user in addition to haptic and audio collaboration with the virtual model adds to the sense of presence between users. Two methods of sending video over the internet are compared and our example implementation is described. These additions make it possible to add value in learning scenarios and opens up the possibility of increasing participation during a training session.

1 Introduction

Our virtual environment is known as a "haptic workbench"[1] which combines a Virtual Reality (VR) system with the sense of touch. Other typical VR systems enable user interaction by tracking and monitoring a user's head or body position[2], altering the displayed output based on the updated orientation and position. Such systems may use a heads-up display worn by the user, be displayed on a monitor or projected onto a screen for applications where a wide angle view is necessary like flight simulator or similar. The haptic workbench uses a semi-immersive virtual environment where the interactive space is the volume occupied by the hands and parts of the arms when seated, unlike other systems. This space is behind a mirror and contains a PHANToM haptic feedback device. Such an arrangement obviates the need for user position tracking, but navigating around the virtual scene requires an extra input. We use a 3D

M. Bubak et al. (Eds.): ICCS 2004, LNCS 3038, pp. 905–912, 2004.
© Springer-Verlag Berlin Heidelberg 2004

space-ball mouse that can be directly controlled by the user. This means the user is stationary and the scene is rotated in space for alternate viewpoints. The user therefore has direct control of both viewpoint position and zoom, and is not constrained by machine dependent measurements from a tracking device.

Human interaction with the virtual world requires adequate tactile, visual and audio feedback so that the user's perception of the world is as natural and intuitive as possible. Ideally the user should be focussed on manipulating the virtual object and performing the interactive task rather than being concerned with the intricacies or limitations of the machine. Since there is some interaction in human perception between the senses, the response of each of these should not be considered in isolation. Rather, factors relating to a unified view leading to perception must be taken into account when designing a VR system, to ensure maximum useability.

Networking haptic workbenches provides a shared environment where two or more users can interact with a single model. The Australian CeNTIE (Centre for Networking Technologies in the Information Economy) network* is a high speed research network which allows user-perceived real-time communication between systems possible and makes remote collaboration a reality. CeNTIE is made up of a consortium of Australian companies and research institutions, dedicated to research utilising high bandwidth networks. The aim of our research is to enable a sense of presence with the remote user during collaborative interaction. Adding real-time video as well as audio to the system enhances the sense of presence between users. In effect the users may collaborate on two levels. Firstly, together they may interact with and manipulate the virtual model. Secondly, and concurrent with model interaction, they have direct face to face communication available. Factors affecting real time video and audio on the network are discussed and our implementation is described.

2   System Overview

The haptic workbench system enables human interaction using the senses of sight, sound and touch. 3D stereo vision is made possible with the aid of LCD shutter glasses and by using a graphics card where the left and right image fields are alternately displayed. The graphics card generates a left/right frame signal which allows synchronisation with the 3D shutter glasses. The virtual 3D object appears projected through the mirror. By co-locating a force feedback PHANToM device behind the mirror the user perceives a 3D virtual object without occlusion which can be touched as shown in Fig 1.

The haptic tool is a PHANToM premium 1.5 with an effective workspace size of 19.5cm x 27cm x 37.5cm. This has 6 degrees of freedom in position sensing (x, y, z, roll, pitch, yaw) and 3 degrees of freedom in force feedback (x, y, z). It is capable of supplying a continuous force of 1.4N, a maximum force of 8.5N, updates at a frequency of 1KHz and has a position accuracy of 0.03mm. The system was originally developed on a Silicon Graphics computer system and has recently been ported to a PC with dual 2.4Ghz processors, 512KB level 2 cache, 4GB RAM, a dual head graphics card and monitors able to support 120Hz refresh rate for stereo video.

---

* Details on CeNTIE can be found at http://www.centie.net.au/

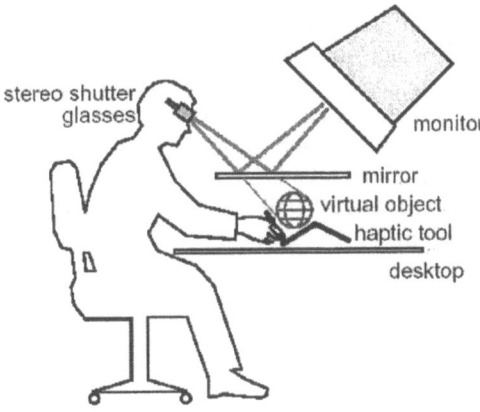

stereo shutter
glasses

monitor

mirror
virtual object
haptic tool
desktop

Fig. 1. Configuration of the Haptic Workbench

3   Human Computer Interaction

Human sensory response and computer/machine performance dictate the users
perception and interaction with the virtual model. Not surprisingly, most VR systems
focus on visual cues as the sense of sight generally tends to dominate relative to sound
or touch when all three are present, even though some studies have shown auditory
dominance[3] over visual. However, some studies have shown visual processing to be
instrumental in tactile perception [4].

In terms of spatial resolution, a typical monitor is capable of displaying about
2,000 pixels in either dimension, or 5pixels/mm for a 53cm display, enough for
relatively smooth visual perception of graphics. In contrast, the spatial resolution
capability for control of fingertip position is about 2mm[5] or about 20 times less
sensitive, while fingertip displacement touch detection resolution is much more
sensitive at about 1 micron.

For video dynamics, the relatively slow video frame update rate of 100-120Hz for
3D stereo has been found to be perceptually flicker free when used with liquid crystal
shutter glasses. This together with scene refresh rates of about 30Hz[6] for the
movement of virtual objects in the 3D space gives the illusion of smoothly varying
object motion. Conversely, the dynamics of human touch perception require much
faster update rates and frequencies up to 1KHz can be sensed[7]. However,
combining the finger touch perception response time with (arm-hand-finger) position-
feedback control means that response rates of between 1-10Hz are adequate for touch
dynamics[8].

The synchronisation of audio with both video and haptics is important, delays of
generally less than 200ms of audio with respect to video is required for concurrent
perception. Good audio quality relies on ensuring no appreciable gaps in transmission.
Poor audio/video synchronisation in some television news broadcasts and audio
communication drop-out gaps when using mobile phones are common everyday
examples highlighting the importance of synchronisation and missing data.

The points outlined above show that improving human interaction and perception
with the virtual world relies on matching human characteristics and expectations to

machine capabilities. As the available computer and hardware resources are limited, some shortcuts are necessary in order to develop a system which responds in perceived real time. Briefly, the haptic workbench splits the model into two parts, one for haptic interaction and the other for video, before re-combination. The visual model has high spatial resolution and relatively low update rate while the haptic model has low spatial resolution and a high update rate. In this way the user perceives a continuous real-time experience in all three modes of interaction with the virtual model and when synchronised gives the illusion of seeing, touching and hearing interaction with a single entity.

4 Networking

Networking systems makes a collaborative environment possible where both users can interact with the same model in perceived real time.

To successfully realise a networked virtual environment the issue of communication delay (latency) needs to be addressed. For haptics, work by Matsumoto et al.[9] shows that a latency of greater than 60ms prevents the effective use of haptics across a network. For audio communication, normal telephony has a goal of less than 250ms latency, a greater delay causes people to speak at the same time and "infringe on each other's words"[10]. An adequate network then must have sufficient bandwidth, low latency and quality of service capabilities in order to provide a workable system. The CeNTIE network is a high speed layer 2 switched research network, which uses no routing and therefore reduces latency compared to routing networks. The core network is a dedicated fibre-optic research network which now stretches across Australia and links Sydney, Brisbane, Canberra, Melbourne, Adelaide and Perth, and environs, stretching about 4,000 km east to west and covering some 1,500 km on the east coast. CeNTIE is currently configured for carrying traffic at speeds of 1GBit/s and up to 10GBit/s bandwidth has been demonstated. The CeNTIE network makes collaborative communication possible and fulfils bandwidth and latency requirements. The system uses both TCP/IP and UDP protocols to transfer model data to fulfill real-time perception.

For adequate face to face user communication in addition to the real time model collaboration, and for the same reasons outlined above, there should ideally be adequate resolution, no missing video frames and synchronisation between video and audio streams.

5 Adding Real Time Video and Audio

Increasing the perception of being present between users working collaboratively will make user interaction more useful and appealing. This can be achieved by the addition of real time video and audio streams to the system.

Enabling both video and voice communication with the remote user while simultaneously interacting with the virtual model creates a more engaging and natural communication environment compared to not seeing the remote person. This enhances the experience of both users and is of real benefit in aiding the understanding between users during collaborative tasks. In the same way that moving from purely written communication to speech enables a user to convey extra meaning by intonation and emphasis rather than just content, adding video enables better communication via means such as facial expression. Real time video also enables context dependent communication and provides an understanding of what is happening around the remote person. Creating the enhanced environment also makes it possible to involve others by having additional displays, while incurring no extra computational overhead.

6 Comparison of Two Methods

We examined two methods of implementing video and audio conferencing; Digital Video over IP (DV) and MPEG2. DV is a recent video format that offers broadcast quality audio and video, created by a consortium of 60 manufacturers and is gaining in popularity due to agreements with the major players.

DV employs a lossy compression algorithm, based on Discrete Cosine Transform and Huffman coding and provides 5:1 compression of the video image. It has a lower compression ratio compared with MPEG2 and uses a purely intra-frame compression technique. As MPEG2 does intra-frame compression as well as inter-frame compression, latency is higher than for DV. For our application, some data loss in compression/de-compression is tolerable. Table 1 compares the two de/compression algorithms and although DV does not have as high compression values, we chose DV due to lower latency and cost.

Table 1. Comparison of DV and MPEG2

	DV	MPEG2
Compression	5:1	between 8:1 and 30:1
Compression Type	Intra frame	Both Intra and Inter frame
Bitrate	Fixed:~30Mbit/s	Scalable: 2-15 Mbit/s
HW Encoder Cost	~$600	~$2000-$5,000

By using software that divides each DV video frame into discrete parts we can put this data into packets and send over an IP network such as CeNTIE.

7 Implementation of DV

DV data is sent over the Internet using Real Time Protocol (RTP), a protocol for real time applications. Our implementation is based on work by Ogawa et al.[11]. The sending sequence is: Camera receives the DV stream, encapsulates DV packet and transmits to the IP network. The receiving sequence is: data received from network, DV frame reconstructed, attach header, decode and display. With the overhead of encapsulation into IP packets, the data rate is slightly more than 28 Mbit/second. A 100baseT (CAT-5) network cable is able to carry four streams, two in each direction. DV over IP can be considered as an elementary stream implemented as a library that can be added to any application. We can manipulate this stream to offer the services required for each application. For example altering the frame-rate allows us to alter bandwidth but image quality is retained.

8 Ergonomics

Having a stationary user seated at the haptic workbench allows us to add the additional monitoring hardware like a fixed camera and enables positioning of a display not constrained by the orientation or position of the user. In other VR systems where the user's head moves around this would not be possible without elaborate camera motion control.

In order to maintain a realistic viewpoint of each user, the camera should ideally be positioned at the centre of the display so it appears users are facing each other. It is intuitive and natural to look directly at the person on the monitor rather than at an off-axis camera. However, placing a camera at the front center of the display would partly obscure the display. Space constraints also influence the type and positioning of the extra hardware. A compromise using a very small camera mounted close to a small LCD display minimises the off-axis viewpoint position while allowing the user to work without altering head position.

Audio echo cancellation also turns out to be important to avoid audio feedback. We use a special echo canceller which compares outgoing audio with incoming and performs automatic subtraction which almost entirely eliminates round trip echo.

9 Results

Combined Network Traffic Load
Preliminary testing of the system shows network traffic loads of just over 30Mb/s for combined haptic workbench interaction together with real time video and audio, with no user perceived delay. Similar "current" and "maximal" values for incoming and outgoing traffic are shown in green and blue in figure 2 below.

The system on the CeNTIE network is able to sustain real time activity continuously. The above 3 hour period shows typical data throughput with no loss of continuity, no gaps in data transfer or loss of perceived real time interaction. The resulting implementation of the enhanced workbench is shown in figure 3.

The camera is centrally mounted and aimed directly at the seated user so that the field of view captures a head and shoulders image while the monitor size and position was chosen so that each user's viewpoint is as close as possible to the on-axis position within the confines of the space available. This arrangement eliminates the need for turning away from the virtual world while performing collaborative interaction and simultaneously allows face to face communication. A surgical model can be seen reflected in the mirror as well as the remote user video image with the haptic device partly obscured below.

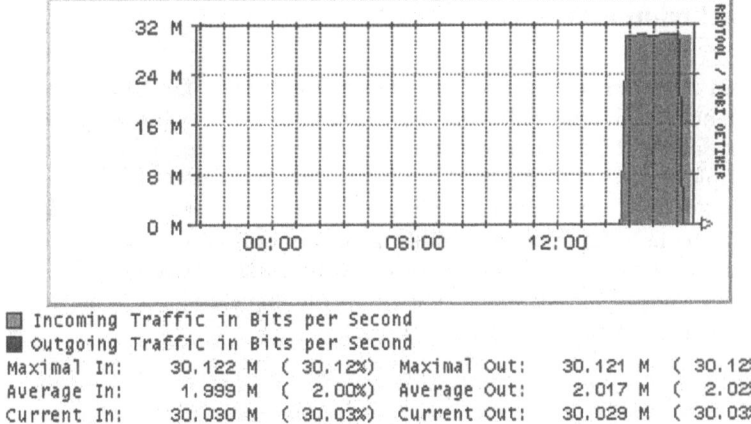

Incoming Traffic in Bits per Second
Outgoing Traffic in Bits per Second

Maximal In:	30.122 M	(30.12%)	Maximal Out:	30.121 M	(30.12%)
Average In:	1.999 M	(2.00%)	Average Out:	2.017 M	(2.02%)
Current In:	30.030 M	(30.03%)	Current Out:	30.029 M	(30.03%)

Fig. 2. Combined network traffic load (ignore average values)

Fig. 3. Enhanced haptic workbench showing small CCD camera (top center), LCD monitor (top right of center) and microphone (right foreground).

Preliminary results on two students and one experienced user confirm increased useability and user performance on a task initiated by the remote user. While it is still too early to generalise, all three showed an improvement in understanding a task when the real time video and audio was present. Objective measurements are planned for the near future and initial comparisons will be made on task completion times. In terms of synchronisation, the real-time perception parameters mentioned in section 3 have been achieved, resulting in a perceptually smooth interaction of haptics, video and audio.

10   Conclusion

The system has been operational for only a few weeks, however preliminary tests show that both real time collaborative interaction with a virtual world as well as concurrent face to face communication between users is possible. We also confirm that the implementation of the extra hardware does not detract from the collaborative VR task but in fact enhances the experience. The extra hardware is sufficiently well laid out for the users to feel comfortable in face to face communication and results in substantial improvements in interaction, while not detracting from concurrent collaborative model interaction. The aim of enhancing the networked haptic workbench has been demonstrated.

References

1. Stevenson,D.R., Smith,K.A., McLaughlin,J.P., Gunn,C.J., Veldkamp,J,P., and Dixon,M.J. "Haptic Workbench: A multi-sensory virtual environment". In Stereoscopic Displays and Virtual Reality Systems VI, proceedings of SPIE Vol. 3639, 1999. Pages 356–366.
2. LaViola,J.J.Jr., Feliz,D.A., Keefe,D.F., Zeleznik,R.C. "Hands-Free Multi-Scale Navigation in Virtual Environments". Proceedings of 2001 ACM Symposium on Interactive 3D Graphics, pp 9-16, March 2001.
3. Repp,B.H. and Penel,A. "Auditory Dominance in Temporal Processing: New Evidence From Synchronization With Simultaneous Visual and Auditory Sequences", Journal of Experimental Psychology-Human Perception and Performance, 2002, Vol. 28, No. 5, 1085–1099
4. Zangaladze,A., Epstein,C.M., Grafton,S.T., Sathian,K., Nature, Volume 401, 7 October 1999, pp587-590.
5. Durlach,N.I., Delhorne,L.A., Wong,A., Ko,W.Y., Rabinowitz,W.M. and Hollerbach,J. Manual discrimination and identification of length by the finger span method. Perception and Psychophysics, 1989, 46(1), 29-38.
6. Chen,W., Towles,H., Nyland,L., Welch,G. and Fuchs,H. "Toward a Compelling Sensation of Telepresence: Demonstrating a portal to a distant (static) office" in Proceedings of IEEE Visualization 2000, October 2000.
7. Bolanowski,S.J., Gescheider,G.A., Verrillo,R.T., Checkosky,C.M. "Four channels mediate the mechanical aspects of touch", J. Acoustic Society of America, 84 (5), November 1988, pp. 1680-1694.
8. Brooks,T.L. Telerobotic response requirements. Proceedings of the IEEE International Conference on Systems, Man, and Cybernetics. Los Angeles, California, 1990, pp 113-120.
9. Matsumoto,S., Fukuda,I., Morino,H., Hikichi,K., Sezaki,K. and Yasuda,Y. "The Influences of network Issues on Haptic Collaboration in Shared Virtual Environments". Fifth Phantom Users' Group Workshop, 2000.
10. DeFanti,T.A., Brown,M.D. and Stevens,R. "Virtual Reality Over High-Speed Networks". IEEE Computer Graphics & Applications, Vol. 16, No. 4, July 1996, pp. 42-43.
11. Ogawa,A., Kobayashi,K., Sugiura,K., Nakamura,O., Murai.J. "Design and Implementation of DV based video over RTP", Packet Video Workshop 2000.

Collaborative Integration of Speech and 3D Gesture for Map-Based Applications

Andrea Corradini

Center for Human-Computer Communication
Oregon Health & Science University, Portland OR 97239, USA
andrea@cse.ogi.edu
http://www.cse.ogi.edu/CHCC

Abstract. QuickSet [6] is a multimodal system that gives users the capability to create and control map-based collaborative interactive simulations by supporting the simultaneous input from speech and pen gestures. In this paper, we report on the augmentation of the graphical pen input enabling the drawings to be formed by 3D hand movements. While pen and mouse can still be used for ink generation drawing can also occur with natural human pointing. To that extent, we use the hand to define a line in space, and consider its possible intersection point with a virtual paper that needs to be determined by the operator as a limited plane surface in the three dimensional space at the begin of the interaction session. The entire system can be seen as a collaborative body-centered alternative to the traditional mouse-, pen-, or keyboard-based multimodal graphical programs. Its potential applications include battlefield or crisis management, tele-medicine and other types of collaborative decision-making during which users can also be mobile.

1 Introduction and Related Work

Over the last decade, computers have moved from being simple data storage devices and calculating machines to become an assistant to the everyday' lives of many people. Web surfing, e-mailing and online activities made computers even extend to vehicle for human socialization, communication, and information exchange. However, for a few simple reasons computers are still far away from becoming an essential and indispensable component of our society.

In the first place, it is very difficult for computers to gain a sociological acceptance unless they will be able to seamlessly improve ordinary human activities. To that extent, before becoming essential, computers will have to understand, interact and dialogue with the real world around them the way people do. The interface for such computers will not be a menu, a mouse, a keyboard but instead a combination of speech, gestures, context and emotions. Human beings should not be required to adapt to technology but vice versa [4]. Moreover, "...*the personal computer has become a little too personal...*" [14]. In many organizations, people work in collaboration with each other by cooperating as a group. When much of an individual's work is team-related, software needs also to support interaction between users [11]. Yet, most current software is designed for the solely interaction between the user and computer.

M. Bubak et al. (Eds.): ICCS 2004, LNCS 3038, pp. 913–920, 2004.
© Springer-Verlag Berlin Heidelberg 2004

Most of the work that people do requires some degree of communication and coordination with others. Unfortunately, the development of technology to support teamwork has proven to be a considerable challenge in practice. To achieve successful computer-mediated implementations, successful designs require: 1) social psychological insights into group processes and organizational coordination, 2) computer science insights into mechanisms to coordinate, share, communicate and organize information, and 3) Human-Computer Interaction (HCI) design insights. Current multimodal systems for Computer Supported Collaborative Work (CSCW) primarily focus on the last two factors. They typically consist of an electronic whiteboard and a video/audio teleconferencing system and maintain data manipulation disjoint from communication.

In the Speech, Text, Image and Multimedia Advanced Technology Effort (Stimulate) project [12], researchers used a combination of prototype technologies to study how people interact with computers employing sight, sound, force-feedback and touch as multiple modes of communicating. In a military application, once a user tell the computer to create a camp, the computer places the new camp where the user's eyes are pointed or where the user points with the glove and then respond that it has created the camp. Users can work together from different locations through a standard Internet connection or other type of network. QuickTurn [13] is a DARPA-funded project aiming to build a collaborative environment for multi-source intelligence analysis. It provides support for multimodal speech/gesture/mouse interaction, access to mediated databases, maps, image tools and the capability to share information between system components and system users in several collaboration modalities. In [3], a map-based application for travel planning domain has been proposed. It runs on a handheld PDA, can access several external data sources (e.g. the web) to compensate for power and data storage limitednesses of PDAs, and makes use of a synergistic combination of speech, gesture and handwriting modalities. Users can draw graphical gestures or write natural language sentences on a map display and issue spoken utterances. These inputs or their combination are deployed to give information about objects of interest, such as hotels, restaurants etc. Rasa [20] is a multimodal environment with visual perceptual input, which augments paper-based tools for military command and control. With Rasa users make use of familiar tools and procedures, creating automatic couplings between physical objects (i.e. maps and Post-it notes) and their counterparts in the digital world. This in turn, allows the users to manipulate paper artifacts, and accordingly control digital objects in the command and control systems.

All of these systems perform interactive collaborative tasks while employing speech and pen gesture recognition. However, what distinguishes our system from those is its extension, rather than replacement, to human gesture recognition and free hand movements (i.e. without the need of a pen or keyboard) to input system commands. To this extent, we built up on top of QuickSet [6,19], a prototype system that was developed at our Department after extensive research efforts in the area of multiagent architectures and multimodal collaborative HCI.

2 Augmenting QuickSet for CSCW and Pervasive Computing

The idea behind the augmentation of QuickSet is based on a vision. Imagine one single command center connected with several single remote operators each keeping track of the events on a different area of a common monitored area [24]. The remote operators provide local information for the central decision center to be evaluated. They use voice commands, 3D gestures and pen marks to add multimodal annotations regarding their local environment on a shared map and to collaborate with the intelligence specialists at the operative center as the local situation evolves. In the event of command center failure, any remote operator can take over and become command center while the interaction with the other operators continues. Users are allowed to move about and concentrate on different aspects/displays of the interface at will, without having to concern themselves with proximity to the host system's.

QuickSet is a collaborative distributed system that runs on several devices ranging from desktops, wireless hand-held PCs to wall-screen display and tablets. Users are provided with a shared map of the area or region on which a simulation is to take place. By using a pen to draw and simultaneously issuing speech commands, each user can lay down entities on the map at any desired location and give them objectives. Digital ink, view and tools are shared among the participants to the simulation session and ink can be combined with speech to produce a multimodal command.

Collaboration capabilities are ensured by a component-based software architecture that is able to distribute tasks over the users. It consists of a collection of agents, which communicate through the Adaptive Agent Architecture [18] that extends the Open Agent Architecture [5]. Any agent registers with a blackboard both the requests for actions it is put into charge and the kind of messages it is interested in. Agents communicate by passing Prolog-type ASCII strings via TCP/IP to the facilitator. This ladder then, unifies messages with agents' capabilities and forwards it to other agents according to the registration information. Collaboration takes place anytime at least two agents produce or support common messages.

There is empirical evidence of both user preference for and task performance advantages of multimodal interfaces compared to single mode interfaces in both general [22] and map-based applications [23]. Speech is the foremost modality, yet gestures often support it in interpersonal communication. Especially when spatial content is involved, information can be conveyed more easily and precisely using gesture rather than speech. Electronic pens convey a wide range of meaningful information with a few simple strokes. Gestural drawings are particularly important when abbreviations, symbols or non-grammatical patterns are required to supplement the meaning of spoken language input. However, the user must always know to which interface the pen device is connected, and switch pens when changing interfaces. This does not allow for pervasive and transparent interaction, as the user gets mobile.

We have extended QuickSet to accept 'ink' input from 3D hand movements independent of any pen device (see Fig. 1). To give input to our hand drawing system, the user attaches four Flock of Birds (FOB) [1] magnetic field tracker sensors; one on the top head, one on the upper arm to register the position and orientation of the humerus, one on the lower arm for the wrist position and lower arm orientation, and finally one on the top of hand (see [9] for details). The data from the FOB are delivered at a frequency of approximately 50Hz via serial lines, one for each sensor, and is processed in real time by a single SGI Octane machine. Because the FOB devices uses a magnetic field that is affected by metallic objects, and the laboratory is constructed of steel-reinforced concrete, the data from the sensors is often distorted. As a result, the data is processed with a median filter to partially eliminate noise.

To facilitate our extended interface we have created a geometrical system to deal with relationships between entities in our setting. Such relationships are between both manually-entered screen positions (for drawing and visualization) and the perceptually-sensed current hand position and orientation (for recognition of certain meaningful 3D hand movements). This geometrical system is responsible for representing appropriately the information conveyed through hand interaction and automatically selecting the display to which the user's gesture pertains in the multi-display setting.

The system needs to know what display regions the users' 3D gestures can pertain to. Therefore, before the system is started for the first time, the system deployers have to manually set the regions the users will be able to paint in. The chosen regions will typically be a wall screen, a tablet or a computer screen on which the shared map or maps have been projected (see Fig. 1). The deployer accomplishes this by pointing at three of the vertices of the chosen rectangle for each painting region. However, since this procedure must be done in the 3D space, the deployer has to gesture at each of the vertices from two different positions. The two different vectors are triangulated to select a point as the vertex. In 3D space, two lines will generally not have an intersection, so we use the point of minimum distance from both lines.

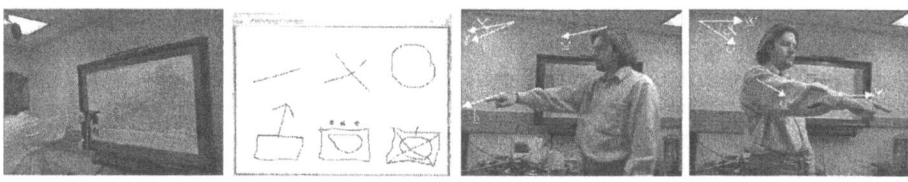

Fig. 1. *(from left to right)* Operating on a wall screen from a distance of approximately 2m; free-hand military symbols with increasing drawing complexity; imaginary head/wrist and upperarm/wrist pointing direction at gestural stroke

3   Multimodal Input

3.1   Gestural Input

Currently, the system supports the recognition of four kinds of gestures: pointing gestures, hand twisting about the index finger, rotating the hand about the wrist, and pushing with the palm up or down. Recognition is based on a body model [9] we use to track human movements and a set of rules for those movements, that extend the rules in [8]. These rules were derived from an evaluation of characteristic patterns we identified by analyzing sensor profiles of the movements underlying the various gestures.

Table 1. Average upper-/lower-arm angle (left) and head/wrist angle (right) for each subject (bars 1-10) along with the average of these values (bar 11). Blue bars have been determined using all collected data, red bars after dropping 10% (both higher and lower 5%) of the data

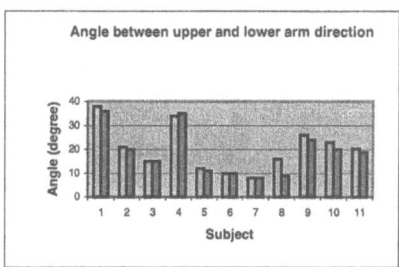 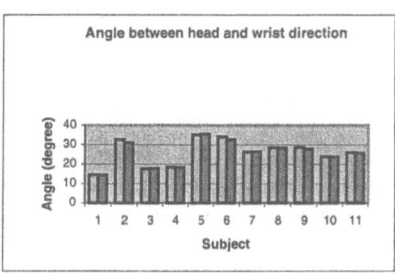

An Empirical Study. Pointing is probably the most common kind of gesture and human beings can easily recognize it. It is an intentional behavior controlled both by the pointer's eyes and by muscular sense (proprioception), which aims at directing the listener's visual attention to either an object or a direction [10,17,21]. Pointing gesture recognition by computers is much more difficult and in fact to date there is no system able to reliably recognize such class of gestures under any conditions.

We conducted an empirical experiment to expand the set of rules, we used in a previous gesture recognizer [8], wrt the geometrical relationships within our body model while a pointing gesture occurs. We invited ten subjects to both point at and describe any object in our laboratory. They were attached the four FOB sensors and then instructed to ignore anyone in the laboratory. Furthermore, they were required to keep the whole body in a kind of 'frozen' state at the end of the stroke for any pointing gesture performed (see Fig. 1). Once in that position, by averaging the FOB reports, we determined the pointing direction, the direction of the upper arm, that of the wrist and the direction of the eyes (see Table 1). The subjects were requested to perform one hundred pointing at as many objects or entities within the room, on the floor or the ceiling. Objects or entities were allowed to be pointed at more than once.

Deictic Gestures. Based on the above empirical study and on the partial analysis of collected gesture data [9], we characterize a pointing gesture as: 1) a hand movement between two stationary states, 2) lasting between 0.5 and 2.4 seconds, 3) whose dynamic patterns are characterized by a smooth and steady increase/decrease in the spatial component values, 4) whose head direction and pointing direction (see Fig. 1) form an angle below a heuristically determined threshold, and 5) whose pointing direction and the direction determined by the upper arm (see Fig. 1) forms an angle below some certain threshold. The first condition is quite artificial as it was introduced by us, rather than extrapolated from the data, for facilitating data stream segmentation. In this way, start and stroke phases of the pointing gestures can be used to trigger pen up/down events while for ink production. The fourth condition about the angular value between pointing and head direction, implicitly assumes that head orientation is a reliable indicator of the direction of the user's attention. Estimating where a person is looking at based on his solely head orientation is a plausible simplification [25] used to determine the focus of attention of the user without having to perform eye gaze tracking. The experimental study reported in the previous subsection provides a useful estimate for the threshold used for checking this condition and the fifth one as well.

In natural human pointing behavior, the hand and index finger are used to define a line in space, roughly passing through the base and the tip of the index finger. Normally, this line does not lie in any target plane, but it may intersect one at some point. When a user interacts with the system on any shared map, it is this point of intersection that we aim to recover from within the FOB's transmitter coordinate system. A pointing gesture phase is dealt with only if the imaginary line described by the sensor in the space intersects some of the virtual papers. Since in the context of ink production any recognition errors lead frequently to incorrect messages we implemented a substitute way to trigger pen down, and pen up events using a PinchGlove [2]. At the begin of the interaction, the user can choose which method to use. Drawings are passed on from the remote system to the central command and control for recognition.

Rotational Gesture. A similar rule-based analysis of hand twisting, hand rotating and hand pushing can be given using the quaternion components provided by the sensor.

All three kinds of gesture are very similar rotational movements, each one occurring about different orthogonal axes. To recognize such gestures, we analyze the hand rotation information using the quaternion components provided by the sensor. We characterize a rotational gesture as a hand movement for which: 1) the gesture takes place between two stationary states, 2) it is a rotation, 3) the unit vector along the axis of rotation is constant over the movement, 4) the rotation is smooth and takes place about the axis of rotation by at least N (a tunable parameter) degrees, 5) the upper arm does not move during the rotation, and 6) the hand moves during the rotation.

According the axis of rotation considered, each of the three rotational gestures can be recognized. While in the current system hand pushing and hand rotating are not attached to any command, a hand twisting gesture is required anytime the user wishes to pan over the map. In such case, the pointing direction with respect to the center of the virtual paper determines the direction of the panning. An additional twisting gesture makes the system to switch back to normal mode.

3.2 Speech and Gesture Integration

Voice is an essential component of natural interaction. Spoken language allows operators to keep their visual focus on the map environment, leaving their hands free for further operations.

For speech recognition, we use Dragon 4.0, a commercial off-the-shelf product. Dragon is a Microsoft SAPI 4.0 compliant, speaker independent speech engine. Any user can immediately interact via voice without having to train the system for his/her voice. Spoken utterances are sensed either by microphones in the room or by (wireless) microphones that users wear. Sensed utterances are sent to the speech recognition engine, which receives the audio stream and produces an n-best list of textual transcripts, constrained by a grammar supplied to the speech engine upon startup. These are parsed into meaningful constituents by the natural language processing agent, yielding an n-best list of alternative interpretations ranked by their associated speech recognition probabilities. Speech recognition operates in a click-to-speak microphone mode, i.e. the microphone is activated when a pen down event is trigger.

Concerning modality fusion, we exploit the original version of QuickSet that already provides temporal integration of speech and gesture. These modes are constrained to either have overlapping time intervals or to have the speech signal onset occur within a time window of up to 4 seconds [23] following the end of the gesture. The multimodal integrator determines which combinations of speech and gesture interpretations can produce an actual command, using the approach of [16]. The basic principle is that of typed feature structure unification [15], which is derived from term unification in logic programming languages. Unification rules out inconsistent information, while fusing redundant and complementary information via binding of logical variables that are values of 'matching' attributes. The matching process also depends on a type hierarchy. A set of multimodal grammar rules specify, for this task, which of these speech and gesture interpretations should be unified to result in a command.

4 Conclusion and Future Directions

This report describes a working implementation of a 3D hand gesture recognizer to extend the existing digital-ink input capabilities of a fully functional real-time, multimodal speech/pen-architecture. The architecture presented is flexible and easily extensible to provide more functionality. It exploits QuickSet's capabilities for recognition and fusion of distinct input modes while maintaining a history of the collaboration.

While drawing using free hand movements allows for non-proximity and transparency to the interface, creating detailed drawings is not easy as human pointing is not accurate [7]. Executing detailed drawings takes training and practice on the user's part, and relies on a precise calibration of both the tracking device and the geometrical model. The accuracy of drawings decreases with both increasing symbol complexity (see Fig. 1) and increasing distance from the virtual paper, thus this might cause lower recognition rates. We use the PinchGlove to signal pen-up/pen-down. As speech recognition in non click-to-speak mode in QuickSet becomes more reliable, such pen-up/pen-down gestures could also be entered by voice. Alternatively PinchGlove signaling could be replaced by defining a specific hand shape to trigger these events as an extension to our current 3D gesture recognition system.

For right now, hand rotation and pushing gesture are not attached to any command because of the lack of a natural intuitive mapping between these movements and any entity behavior on the map. For instance, while an entity can be put on, moved about, and removed from the map, it cannot be given an orientation (e.g. to visually indicate the direction of movement of a unit on the map) i.e. it is always displayed using the same symbol. Once such an entity will be given an orientation, it will be natural and straightforward to attach e.g. a hand rotation to a 2D rotation of the entity on the map.

Acknowledgments. This research has been supported by the ONR grants N00014-99-1-0377, N00014-99-1-0380 and N00014-02-1-0038. We are thankful to Phil Cohen, Ed Kaiser, and Richard Wesson for programming support and fruitful suggestions.

References

1. http://www.ascension-tech.com
2. http://www.fakespace.com
3. Cheyer A., and Julia, L., "Multimodal Maps: An Agent-based Approach", Multimodal Human-Computer Comm., Lecture Notes in AI #1374, Springer Verlag, 1998, 111-121
4. Coen, M.H., "The Future of Human-Computer Interaction or How I learned to stop worrying and love My Intelligent Room", *IEEE Intelligent Systems*, March/April 1999
5. Cohen, P.R., et al., "An Open Agent Architecture", Working Notes of the AAAI Spring Symposium on Software Agents, Stanford, CA, March 21-22, 1994, 1-8
6. Cohen, P.R., et al., "QuickSet: Multimodal Interaction for Distributed Applications", Proceeding of the 5th International Multimedia Conference, ACM Press, 1997, 31-40
7. Corradini, A., and Cohen, P.R., "Multimodal speech-gesture interface for hands-free painting on virtual paper using partial recurrent neural networks for gesture recognition", Proc. of the Int'l Joint Conf. on Artificial Neural Networks, Honolulu (HI), 2002, 2293-2298
8. Corradini A., et al., "A Map-based System Using Speech and 3D Gestures for Pervasive Computing", Proc. IEEE Int'l Conf. on Multimodal Interfaces, 2002, 191-196
9. Corradini A., and Cohen P.R., "On the Relationships Among Speech, Gestures, and Object Manipulation in Virtual Environments: Initial Evidence", Proc. of Int'l CLASS W'shop on Natural, Intelligent and Effective Interaction in Multimodal Dialogue Sys., 2002,.52-61
10. Efron D., "Gesture, Race and Culture", Mouton and Co., 1972
11. Ellis, C.A., Gibbs, S., and Rein, G., "Groupware: Some Issues and Experiences", *Communications of the ACM*, 34(1):39-58, 1991
12. Flanagan, J., et al., "NSF – STIMULATE: Synergistic Multimodal Communication in Collaborative Multiuser Environments", Annual Report, 1998
13. Holland, R., "QuickTurn: Advanced Interfaces for the Imagery Analyst", DARPA/ITO Intelligent Collaboration & Visualization Program PI Meeting, Dallas, TX, Oct. 10, 1996
14. Johansen, R., "Groupware: Computer Support for Business Teams'', The Free Press, 1988
15. Johnston, M., et al., "Unification-based multimodal integration", Proceedings of the 35th Annual Meeting of the Association for Computational Linguistics, Madrid, Spain, 1997
16. Johnston M., "Multimodal Language Processing", Proceedings of the 5th International Conference on Spoken Language Processing, Sydney, Australia, 1998
17. Kendon A., "The Biological Foundations of Gestures: Motor and Semiotic Aspects", Lawrence Erlbaum Associates, 1986
18. Kumar, S., et al., "The Adaptive Agent Architecture: Achieving Fault-Tolerance Using Persistent Broker Teams", Proc. of 4th Int'l Conf. on Multi-Agent Systems, 2000, 159-166
19. McGee, D.R., Cohen, P.R, "Exploring Handheld, Agent-based Multimodal Collaboration", Proceedings of the Workshop on Handheld Collaboration, Seattle, WA, 1998
20. McGee, D. R., et al., "A Visual Modality for the Augmentation of Paper", Proceedings of the Workshop on Perceptive User Interfaces, ACM Press: Orlando, FL, Nov. 14-16, 2001
21. McNeill, D., "Language and Gesture: Window into Thought and Action", David McNeill, editor, Cambridge: Cambridge University Press, 2000
22. Mellor, B.A., et al., "Evaluating Automatic Speech Recognition as a Component of Multi-Input Human-Computer Interface", Proc. of Int'l Conf. on Spoken Lang. Processing, 1996
23. Oviatt, S.L., "The Multimodal Interfaces for Dynamic Interactive Maps", Proc. of the Conference on Human-Factors in Computing Systems, ACM Press, 1996, 95-102
24. Sharma, R., et al., "Speech-Gesture Driven Multimodal Interfaces for Crisis Managment", Proceedings of IEEE, special issue on Multimodal Human-Computer Interface, 2003
25. Stiefelhagen, R., "Tracking Focus of Attention in Meetings", Proc. IEEE Int'l Conference on Multimodal Interfaces, Pittsburgh, PA, USA, October 14-16, 2002, 273-280

Mobile Augmented Reality Support for Architects Based on Feature Tracking Techniques

Michael Bang Nielsen[1], Gunnar Kramp[2], and Kaj Grønbæk[1]

[1] Department of Computer Science, Aarhus University
Åbogade 34, 8200 Århus N, Denmark.
{bang,kgronbak}@daimi.au.dk
[2] Aarhus School of Architecture
Nørreport, 8000 Århus, Denmark.
gunnar.kramp@a-aarhus.dk

Abstract. This paper presents a mobile Augmented Reality (AR) system called the SitePack supporting architects in visualizing 3D models in real-time on site. We describe how vision based feature tracking techniques can help architects making decisions on site concerning visual impact assessment. The AR system consists only of a tablet PC, a web cam and a custom software component. No preparation of the site is required and accurate initialization, which has not been addressed by previous papers on real-time feature tracking, is achieved by a combination of automatic and manual techniques.

1 Introduction

The AR system described in this paper has been developed in the WorkSPACE project, which is an EU IST project focusing on developing interactive working environments for architects. Among other things the project has developed a 3D system called Topos, which is collaborative spatial environment for sharing documents and 3D models. Topos has been extended with various kinds of Augmented Reality components to support integration of the architects' documents and models into the real physical world.

Augmented Reality research [1][11][12], focus on linking digital information and virtual objects to physical objects and places. Typical applications are to link and display digital annotations on top of objects and places by means of some identifying code. Examples are Cybercodes [13] and ARToolkit [2] allowing information to be linked to objects visually tagged with a two-dimensional bar-code label, which is interpreted by a camera-based reader. Several mobile augmented reality systems exist. The MARS prototype [8] is an example of an AR system applying very precise GPS and see-through glasses to provide historical information superimposed onto a campus visitor's view of Columbia University's campus. Another example is the use of a Rekimoto's CyberCode-based Navicam, which recognizes visual tags and superimposes digital text, images, and 3D models on top of the visual tag (the CyberCode). A final example is the Archeoguide system [15], which is a vision based system geared to mix models of old buildings and findings into real-time streams of video of the archeological site. The system is based on registration of reference photos in fixed positions.

M. Bubak et al. (Eds.): ICCS 2004, LNCS 3038, pp. 921–928, 2004.
© Springer-Verlag Berlin Heidelberg 2004

Our AR system called the SitePack is a mobile system combining GPS tracking for location based photo capture and annotation and vision based feature tracking for real-time visualizations.

Section 2 describes the design and implementation of the SitePack software. Section 3 shows how the SitePack was used by landscape architects in evaluating a design proposal on a building site. Section 4 compares the SitePack system to similar systems. Finally, Section 5 concludes the paper.

2   The SitePack Design

On site, architects need to make assessments of the visual impact of their design proposals. This can be facilitated by overlaying the architects' 3D model on a real-time video stream on site. Our mobile AR system, the SitePack, is depicted in figure 1. It runs the Topos 3D environment that supports a number of tracking technologies. In [5] we discussed how GPS and digital compass are sufficiently accurate when supporting collaboration, collecting materials and doing annotations. Here we describe the vision based feature tracker we use for on-site real-time visualizations.

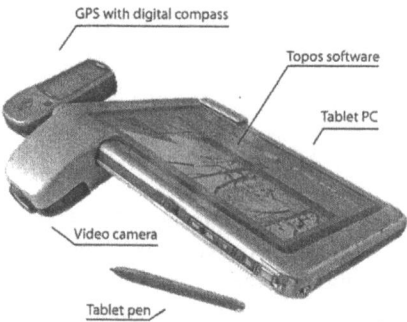

Fig. 1. The SitePack appliance is assembled from off-the-shelf hardware components, consisting of a tablet PC with a pen input device, 1GHZ processor, 512 MB RAM, and a 16MB graphics card. The external devices are a 640 x 480 pixel web cam integrated in a bracket mount, and a GPS with digital compass.

2.1   Requirements from the Architectural Domain

The challenges inherent in supporting the visual impact assessment task of architects imply a number of key requirements for the tracking software. In particular it must:
- be robust to a dynamic environment, e.g. occlusions and changing weather.
- require minimal preparation of and intrusion on the environment.
- not be subject to significant drift (i.e. where the 3D overlay drifts from its true position over time), as the system may be in use over a fair amount of time.
- work in a wide range of environments not necessarily exhibiting regular structures, e.g. a building site may be mostly mud.
- be precise enough to assess design proposals based on the overlays produced.

- require minimal calibration of tracking devices.
- work both indoor and outdoor enabling maximal mobility, i.e. it should be possible to move around while tracking and to use the tracker at many different places.
- run in real-time and be able to handle abrupt camera motions.

Furthermore the tracking technology chosen must lead to a reasonably priced SitePack which in addition must be easy to carry and operate.

No existing tracking technology complies with all of these requirements. GPS technology has been used by numerous AR systems, but it is not possible to obtain the required precision unless combined with RTK GPS which restricts the mobility[7]. Using marker based approaches[10] on site is unpractical because of occlusions, the number of markers needed, marker registration in the same coordinate system etc.

Our work extends and combines previous work on real-time vision based feature tracking[9][14][15] to meet most of and relax some of the requirements listed above. Furthermore, previous work on real-time feature tracking has not addressed the issue of how to accurately initialize the feature tracker on site in real world scenarios, e.g. properly aligning a 3D model with the real world.

2.2 Initial Registration

All vision based approaches based on feature tracking must address the important problem of initially aligning the coordinate system of the model with the real world, as only the relative change in pose is computed by the tracker.

Our vision based tracker currently supports two ways of aligning a 3D model with the real world. Both methods require that some sort of correspondence between the 3D model and the real world be available. For this reason they cannot be applied in totally unstructured environments such as e.g. a field with nothing but mud.

Once the user is at the location from where he wishes to view the virtual 3D model, he points the SitePack in the desired direction and presses a button to initialize the tracker. This causes the video in the background of the Topos application to freeze and the 3D model is shown on top of the background as depicted in figure 2. To align the 3D model with the background video, the user can then mark a number of point correspondences between the background video and the 3D model. If the points are coplanar at least four points are specified [10], otherwise we use a simple method that requires at least six points[6], although better methods exist. When the correspondences have been marked, the 3D model is shown correctly aligned on top of the video stream, as shown in figure 3. In order to move the camera whilst maintaining the virtual overlay, the user simply points the SitePack in the initial direction, starts the tracker and subsequently moves the SitePack. This registration method has been applied with great success in indoor scenarios. However, consider the courtyard building site in figure 4. Most often the 3D model to be superimposed on the video stream does not contain significant texture detail in order to make sufficient point correspondences between the model and the video. Furthermore, identifiable features such as corners are often hidden by mud, tools or other equipment thus making the use of point correspondences alone non-feasible. For this reason we offer a second method for doing the initial registration, based on identifying directions between the 3D model and the background video.

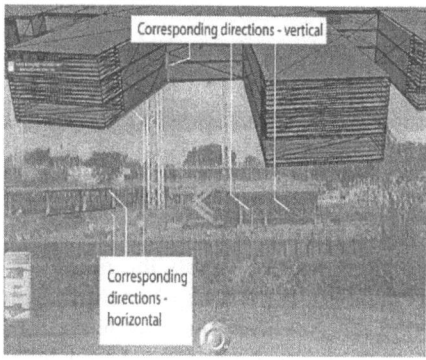 

Fig. 2. Shows recovering the orientation of a 3D model by specifying corresponding lines on the background image and the model.

Fig. 3. Shows the result of the initial registration using corresponding lines for orientation recovery and manual recovery of the position.

To recover the orientation of the camera, the user selects two orthogonal directions on the 3D model, drawn with an overlaying wire frame representation during the course of this registration method (figure 2). The two identical directions must be indicated on the video background. This is done by specifying vanishing points [6]. To specify a vanishing point at least two lines must be drawn. If the lines specified are detected to intersect at infinity, which is the case for the vertical directions in figure 2, the direction is recovered by exploiting that it must be parallel to the image plane of the camera.

Using these two direction correspondences, the orientation is easily computed. The position of the camera can be recovered by manually moving the camera using the manipulators supported by Topos. It could also be done by specifying two point correspondences between the 3D model and the background video. However, recovering the orientation manually is very cumbersome.

One could attempt to make this initial registration process automatic, however due to the limited, dynamic and often complex structures on a building site, this may not always be possible. For indoor scenarios, however, it is more likely to work.

Using the combination of automatic and manual techniques described above requires some skill of the user in addition to knowledge of both the construction site and the 3D model. However, once these are in place, the process is fast and easy.

The initialization processes described above requires the camera to be calibrated. If this is not done beforehand, a sensible approximation that works well in practice can easily be computed on site by using four rectangular points [4] or specifying the vanishing points of three orthogonal directions [6].

2.3 Homography Computation from Reference Images

The relative transformation of the camera between two images is computed from the homography[6] relating the two, and this homography is in turn computed from a number of feature correspondences. Our core real-time vision based tracker is similar to the methods in [6][9][14] that can be consulted for pseudo-code. Our method differs from previous work in the way reference images are captured and used to determine the transformation of the camera in each frame.

Each time a transformation is computed, an error is introduced. If the tracker computes the current camera pose based solely on the transformation in the previous image and the transformation between the current and previous image, this error will accumulate and thus cause undesirably drift[4][14]. Instead our tracker is based on relating the current image to a number of reference images, ie. computing the transformation between the current image and one of several reference images for which the transformation is known. The method described by Prince et al. [9] similarly relates the current image to a single reference image or reference mosaic, both of which must be captured in advance. Our method allows any number of reference images like the method of Stricker et al.[15]. However, contrary to the methods above, our reference images are registered automatically or on demand on site while tracking, thus avoiding the need to prepare the scene in advance whilst maintaining the extendible nature of the tracker.

The camera pose for each reference image is known as it is computed when the reference image is captured. Determining which reference image is to be used to compute the transformation in the current image is based simply on the maximal 2D overlap of each of the reference images with the current image. This is very fast, contrary to the reference image selection strategy of Stricker et al.[15]. The overlap is based on the homography computed in the previous frame and it is possible to do this 2D comparison as we are estimating a homography, i.e. a 2D projective transformation. The image used for the initial registration as described above is used as the first reference image. Additional reference images are needed if the overlaps between the current image and the existing reference images are too small which will cause the tracker to become unstable. Registration of a reference image can be done either manually by pressing a button (the camera pose for the reference image is then computed by the tracker) or automatically by a simple monitoring process that measures and ensures a minimal overlap between the current image and any reference image.

The camera pose computed for a reference image is contaminated by a small error. When subsequently relating the current image to such a reference image, the error will propagate to the pose estimated for the current image. However, since a small error is only introduced when registering a reference image and when relating the current image to the reference image, the error in the pose for the current image will be insignificant in most practical applications and significant drift will not occur.

Each time a camera transformation is estimated it is accompanied by an estimate of its quality. If this quality is below a certain threshold, the transformation is ignored and the one estimated in the previous frame is used instead. This has the practical consequence that if the tracking is lost due to e.g. abrupt camera motions, the camera can simply be pointed approximately in the direction in which the tracking was lost, and the model will jump into place.

3 Using the SitePack for Zone of Visual Impact Assessment

We have tried out the SitePack together with a group of landscape architects from the Scottish DLP, our user partner in the WorkSPACE project. During the last year they have been working on a site outside Edinburgh, covering an area of 78 acres.

Fig. 4. Courtyard building site.   **Fig. 5.** Shows the revised 3D model superimposed on the video in the courtyard garden.

One of the many challenges for the involved constructors and architects on a building site this size, is to communicate and visualize proposals to the contractors and the building owner. At the time of testing of the SitePack on site, the architects were working on a proposition plan for a courtyard garden, see figure 4. The proposition plan was to be carried out in a full scale mock up, so the building owner could get an impression of what the finished courtyard would be like. Grass, hedges, trees etc. were to be planted full scale in the courtyard during a week, and then removed afterwards costing a considerable amount of money, time and inconvenience for the construction workers.

Thus the landscape architects were interested in using the SitePack to visualize their 3D proposition plan of the garden on site. Primarily to facilitate evaluation of their own proposals before presentation, but also focusing on the possibility of using the SitePack in the future instead of full scale mock-ups.

During the test of the SitePack in the courtyard garden, a very direct result of the visualization was that the height of the hedges had to be reconsidered. From the viewpoint of the SitePack, chosen to be that of a standing person looking out of the window from the bottom of the courtyard, the superimposed 3D model showed that some of the hedges would actually block the view to the hills in the background.

Another issue of interest offered by overlaying the 3D model on the courtyard vista was the visualization of the visual impact of the finished three story building. When standing in the courtyard site there was an impression of semi-openness, due to the fact that only the first story of the building was partly finished. Watching the 3D overlay on the SitePack in the courtyard illustrated how the finished building would alter the courtyard impression into a feeling of enclosure and privacy.

In this scenario the initial registration was done on site by recovering the orientation from vanishing points and manually recovering the position. No preparation was required. Even though the illumination changed and construction workers moved around on the site, the tracking remained sufficiently robust throughout the session. This would have been harder using previous techniques such as eg. a marker based approach described in section 3.1.

4 Experiences and Related Work

During the landscape architects' evaluation of the SitePack on site, a number of issues were revealed. Tablet PC hardware needs to become more powerful and the screen technology is not currently designed for outdoor use which means that screen visibility is reduced. However, compared to previous work such as MARS [7] and ArcheoGuide [15] the SitePack is much more mobile than the equipment required by those systems.

On site, the collaboration between the landscape architects was facilitated by the screen on the SitePack. They were able to directly discuss the impact of the 3D model proposition by sharing the same picture and focusing the camera towards views of interest and importance. Yet they could still sense the place by being there.

The vision based tracker proved to be sufficiently robust to occlusions and changes in illumination during a session. Contrary to other vision based trackers it does not require preparation of or intrusion on the site [9][10][15]. By combining the strengths of previous real-time feature tracking techniques [9][14] and using either an automatic or manual strategy for on-site registration of reference images we have obtained a feature tracker that is both extendible and does not suffer significantly from drift when used over time, contrary to previous work. Both of these features proved invaluable on site. If the camera was moved too fast or the video image contained too much motion blur, the tracker sometimes failed to estimate the transformation, however, as described previously, abrupt camera motions are handled in a gentle way.

The tracker only works for rotations or when tracking planar structures. This reduced the mobility of SitePack with respect to real-time overlays. However, most often the architects desired to remain in a certain position over a period of time to discuss a proposal anyway. Contrary to previous work on real-time feature tracking, we also described how to quickly and accurately register the camera to initialize the tracker at a new position directly on site in real world scenarios. This increased the mobility otherwise restricted by the homography-based vision tracker, although a tracker that allows full mobility while tracking by recovering 6 DOF is of course desirable. Prince et al. [9] describe how registration of the camera can be done by placing a marker in the first frame. However, in real life settings this is difficult to do precisely, as the position of the marker must be known in relation to the model and be placed exactly at this position in the real world. The main problems we experienced using the tracker was the high frequency jitter caused mainly by the noisy grained pictures captured by the web cam. This can be improved both algorithmically and by using a better camera. Another problem encountered on site was that if a pan of the SitePack was not close enough to a rotation, the tracker got confused by the parallax introduced in the movement of the tracked features. This was mainly a problem if the features being tracked were too close to the camera and can be solved by placing the web cam on a tripod, at the cost of some mobility.

5 Concluding Remarks

The paper has discussed an architect's SitePack appliance which features vision based feature tracking to support architectural work on site. We have described a successful example of use and the experiences gained. What distinguishes the SitePack software from previous systems is that tracking is extendible with insignificant drift and requires no preparation of the site – the site per se provides sufficient context for locating the user with overlay models precisely. Accurate initialization, which has not been addressed by previous work on real-time feature tracking, is achieved by a novel combination of automatic and manual techniques.

Acknowledgements. This work has been funded by the EU IST project WorkSPACE [www.daimi.au.dk/workspace/]. We would like to thank the entire WorkSPACE group for work on the Topos prototype and many fruitful discussions.

References

[1]   Azuma R.T. A Survey of Augmented Reality. In Presence: Teleoperators and Virtual Environments 6, 4 (August 1997), 355-385.

[2]   Billinghurst, M., Kato, H. & Poupyrev, I. (2001). The MagicBook - Moving Seamlessly between Reality and Virtuality. Computer Graphics and Applications, 21(3), 2-4.

[3]   Billinghurst, M., Kato, H., Kiyokawa, K., Belcher, D., Poupyrev, I. Experiments with Face to Face Collaborative AR Interfaces. Virtual Reality Journal, Vol 4, No. 2, 2002.

[4]   Gilles,S., Fitzgibbon,A.W., and Zisserman,A. Markerless Tracking using Planar Structures in the Scene. In *Proc. International Symposium on Augmented Reality*, Oct. 2000

[5]   Grønbæk, K., Vestergaard, P. P., & Ørbæk, P. (2002). Towards Geo-Spatial Hypermedia: Concepts and Prototype Implementation. In Proc. of HyperText 2002. Maryland: ACM.

[6]   Hartley R. & Zisserman A., Multiple View Geometry, Cambridge Univ. Press, 2000.

[7]   Höllerer, T. et al. Exploring MARS: Developing Indoor and Outdoor User Interfaces to a Mobile Augmented Reality System, Computers & Graphics 23(6), pp 779-785.

[8]   Höllerer, T., Feiner, S., and Pavlik, J. Situated Documentaries: Embedding Multimedia Presentations in the Real World. Proc. ISWC '99 (Third Int. Symp. on Wearable Computers), San Francisco, CA, October 18-19, 1999, pp. 79-86

[9]   J.D.Prince S., Xu Ke, Cheok A.D, Augmented Reality Camera Tracking with Homographies, *IEEE Computer Graphics and Applications, special issue on Tracking* , November-December 2002

[10]   Kato H. and Billinghurst M, Marker Tracking and HMD Calibration for a Video-based Augmented Reality Conferencing System, Proceedings of the 2nd IEEE and ACM International Workshop on AR '99.

[11]   Mackay, W. Augmented Reality: Linking real and virtual worlds. A new paradigm for interacting with computers. In: Proceedings of AVI'98, ACM Conference on Advanced Visual Interfaces, ACM Press, New York, 1998.

[12]   Milgram, P. & Kishino, F.A. "Taxonomy of Mixed Reality Visual Displays," Institute of Electronics, Information, and Communication Engineers Trans. Information and Systems (IECE special issue on networked reality), vol. E77-D, no. 12, 1994, pp.1321-1329.

[13]   Rekimoto, J. & Saitoh, M. Augmented Surfaces: A spatially Continuous Work Space for Hybrid Computing Environments. In: Proceedings of CHI'99 Conference on Human Factors in Computing Systems (Pittsburgh, Pennsylvania USA) ACM/SIGCHI, New York, NY, 1999, pp. 378-385

[14]   Simon G. and Berger M.-O., Pose Estimation for Planar Structures, *IEEE Computer Graphics and Applications, special issue on Tracking* , November-December 2002

[15]   Stricker, D. Tracking with Reference Images: A Real-Time and Markerless Tracking Solution for Out-Door Augmented Reality Applications, VAST 01

User Interface Design for a Navigation and Communication System in the Automotive World

Olaf Preißner

Manager Design,
Harman/Becker Automotive Systems GmbH
Raiffeisenstr. 34
70794 Filderstadt, Germany
Tel: +49 / (0) 711 / 90 77 0 – 551
opreissner@harmanbecker.com

Abstract. What makes a "smart graphic" and what distinguishes it from other design or graphic variants? Many existing designs for navigation and communication systems in high-end cars are not so attractive, don't feel the user emotional and are not easy to use. In contrast, smart graphics have the advantage that it entice the user to interact with the display, the systems or the application and appealing to his aesthetic expectations while maintaining high quality appearance. It is totally user-friendly and uncomplicated thus avoiding misapplication. Smart graphic uses real time execution of all functions ensuring simple and fast interaction. A limiting factor of performance is the capacity of the computer systems so a graphic has to adapt to the characteristics of the medium used. In the end, it also must satisfy high design expectations and technical features. The aim of this paper is to investigate the potential of 'smart graphics' in the design of user interfaces for automotive navigation and communication systems. A comparison will be made with other design and graphic variants to fully grasp the additional value of 'smart graphics'. Our experience has taught us that a smart graphic integrates all of these aspects while optimally combining aesthetic qualities, user specific conditions and the environment it is used in (e.g. in a vehicle). These insights are of great importance to arrive at interface designs that fully adhere to technical, aesthetic, emotional and usability requirements.

1   The Driver Information and Communication System

Driver information and communication systems must strictly comply with these characteristics to effectively display accessible information in the vehicle. Harman/Becker has developed such a system together with an interdisciplinary team of developers, designers and ergonomic scientists.

It is taken for granted that almost every car today has a radio, CD player and navigation system. This trend is steadily continuing. Increasingly, services which are used in the office or at home, such as Internet or email, are being transferred and modified specifically to fit into the unique environment of a car. The aim is to offer a logical and structured integration of the desired functions through the help of display technology. *Complex driver information and communication systems combine*

M. Bubak et al. (Eds.): ICCS 2004, LNCS 3038, pp. 929–935, 2004.
© Springer-Verlag Berlin Heidelberg 2004

Fig. 1. New driver information system with speech recognition

individual components such as radio, CD, navigation, telephone into a single integrated unit, enabling the user to interact using logical interface with high quality displays. As a general principle, all functions of the various components in an automobile are handled by a small number of controls require the driver's close attention to operate and are often difficult to see. This restricts the overall interaction with the system.

In order to make driver information systems for vehicles safe and comfortable the technology of speech recognition is a very important factor (Fig. 1). Using voice commands a driver doesn't have to look away from the road or take his hand off the steering wheel to access the navigation system or the car telephone. Additionally, by using speech for system in and output new options can be integrated such as listening to incoming emails while driving. Unfortunately, the technology in speech recognition systems hasn't been fully perfected and therefore, it can't be used alone in operating driver information systems. As mentioned above, complex situations are created when competing demands are placed on a driver of an automobile, these demands can only be resolved with an optimally configured smart graphic. The graphic design "user interface" is therefore devised by integrating the following aspects:

2 Demands of the Medium, the Surroundings, and the Graphic Realization

Operating a vehicle in traffic always demands high concentration and quick reactions by the driver. This means crucial information, which should be taken in parallel to driving, must be quickly recognized, understood, and processed.

Even complex components and system applications like navigation, air conditioning or the Internet in the future need to be operated using simple input elements. The frequently deeply branched out and involved menu structures are necessary to access all functions of the system, poses a unique challenge.

While driving it is important to localise all instruments on the cockpit quickly and easily. The better the display is placed in the cockpit the easier it is for the driver to read the indicated information. The driver must find the desired information in split seconds. Because of the better resolution in the middle retina the information can be better identified then the information in the lateral field of view, when the display is placed centrally placed in the cockpit [1].

Unfortunately, only in exceptional cases is it possible to place ideally the display. The packaging in the car such as the climate channels in the cockpit, are limit the positioning of the display. Since the location cannot always meet optimal ergonomic demands, the information must also be absorbed from a limited viewing angle or from a considerable distance.

The "Ergonomic aspects of in-vehicle visual presentation of information" from the International Organisation for standardization recommends specific types and sizes of fonts to optimize the readability during driving [2]. Also extreme lighting situations, during the day or at night, can drastically reduce readability of the display. Varied lighting conditions greatly affect the display contrast resulting in poor readability of the display.

3 Demands of the Driver

These restrictions conflict with the main desires of the driver for ease of operation and elegant design. The user expects a graphic design that follows logical rules, integrates and reflects his environment, expectations and thought processes as much as possible, that also encourages him to interact. Operating must be made to be simple and intuitive; the driver shouldn't have to concentrate when he asks for information in a timely manner.

Since such a system is built mainly into upper class automobiles, the design of the user interfaces must complement the high aesthetic standard of the rest of the car's interior that the driver is accustomed to.

The top priority for developing a graphic concept is to follow the design guide lines that the customer is used to having. The design of the interface must convey this message, such as representing the company's philosophy, and fitting into the respective vehicle.

Also, different drivers have distinct patterns of habit and action. It is therefore already essential to differentiate the users' specifics in the conceptional phase and the design of the interface. As an example we can consider the group of older traffic participants. The more complex the information is designed and the more tasks are need to resolved at the same time, the more inhibited is the information processing by older people [3]. We can assume that older people have a general, old-age worsening visual acuity [4]. Examinations by B. Bell, E. Wolf and C.D. Bernholz [5] show, for example a decreasing of the depth perception between the age of 40 and 50. Furthermore, J.G. Gilbert [6] was found out that there is a reduction in sensitivity to the color spectrum of about 24 % between the ages 60-70. The reading precisions from displays could be strong assuaged by designing the screens adequate. This kind of flexibility in operating the system can be achieved by a design that the user can tailor to his specific desires.

4 The Design of the System's Graphic Display

The expectations of the driver as well as the demands of the system have to combine using a smart graphic in order to satisfy the above mentioned needs.

The exterior and interior design create the individual statement of a vehicle type. In the interior of the vehicle the active communication surfaces are an important part of this statement even though they only take up a relatively small part of the interior. It is therefore essential to adhere to the established guidelines and to unify them.

Fig. 2. One hardkey for each function permitts a flat menu hierarchy

In order to simplify the menu guide, flat menu hierarchies are being developed that demonstrate clearly separated menu points and distribution of the various functions. The contents of the individual categories is dependent on the different components in the vehicle and the needs of the user, such as the division of the main menu into radio, audio, navigation, telephone, information and address book (Fig. 2).

It is essential that the menu structure is consistent and well organized. Only in this way intuitive operation is possible and therefore an effortless orientation of the driver is guaranteed. The need for an easily understandable graphic is required to make up for limiting the main menu and greatly reducing structure logic.

To enable intuitive use, the designed form of the user elements is embody into the design of the graphic display of the user interface. This system uses a rotary input device with a push function. That means that the input elements are reflected in the design and give the driver the feeling that he initiates actions directly in the display. If a user element is turned to the right the cursor on the screen also moves to the right. Basically, the direction of movement of the user element and the screen can't conflict, instead they should correspond with each other.

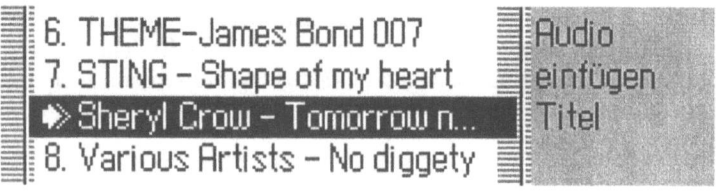

Fig. 3. Different cursor while storing a station

Next to the formal design, another deciding factor is the color-coding of the different main menus. Using bright and different color choices create easy orientation even under unfavourable viewing conditions. The user knows immediately which menu he is viewing. This is the reason why the cursor, used to mark the actual position of the input element, is identified clearly by color and significant contrast with the background making it immediately visible no matter what the lighting conditions are. The cursor is simply the navigation tool, marking the current position on the screen and is the reference point for the next action of the user. The cursor can change into different forms that express different statements, informing the user of the next possible step (Fig. 3).

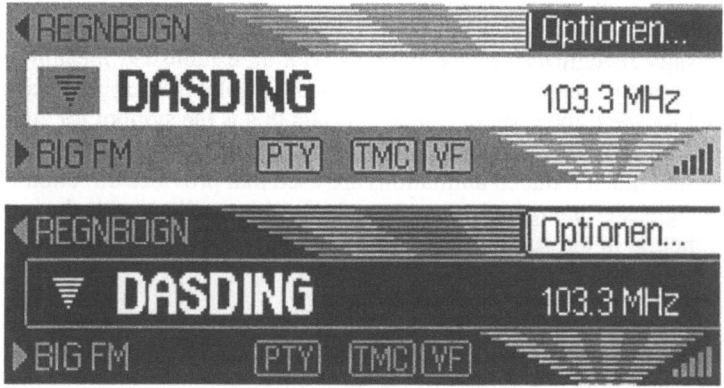

Fig. 4. Day and night design

The manipulation of light and shadow, such as use of contrast values, enable optimal readability even in unfavourable light conditions like the sun shining through the back window. One option in this area is to create a day and night design (Fig. 4). Depending on the time of day and variations in brightness an optimal graphic display is offered to the driver, which is adapted to the specific characteristics of the incoming light conditions.

During daylight a brighter screen is considered more comfortable to read. Since the eyes aren't strained by uneven lighting conditions, less fatigue is observed that with a dark screen during daylight. Uncomfortable light reflections are also avoided by using a bright screen background. At night darker colors are more appropriate so the driver isn't blinded. The contrasts can be reduced in the night screen design since a screen designed for daylight tends to be too bright under dark conditions.

Next to a rich contrast display, the typographic on the display play an important role. Due to the more difficult sensory environment encountered in a vehicle, the font used in the display has to adhere to stringent rules in order to achieve the primary goal of optimal readability. Important criteria for the suitable typographic are the font type, cut, size and running width. Fonts without serifs (Arial, Helvetia, Univers, Frutiger) are preferred over those having serifs (like Times and Bodoni). Regular or bold font cut is best suited, while cursive or condensed fonts should be avoided. The font size depends on a number of factors such as viewing distance, form, contour or contrast. Having chosen an optimal font type for form, contour and contrast a font size of at least 5 mm for capital letters is required when viewing from a 70 cm distance. Smaller font sizes are acceptable for non-critical messages or status indications, which do not need to be viewed while driving. Besides the criteria of readability the font should also reflect the corporate identity of the brand and accent the appearance.

A decisive factor is also the differentiation between an information display, which shows status information and that of active display elements that are triggered when initiating a certain action. Menu levels are set in theme-related backgrounds and therefore, the function lying behind it visualized. For totally new functions a new symbolic system must be developed that intuitively makes sense without having to refer to an operator's manual.

In order to be able to use the symbolic system successfully it has to be absolutely indisputable. Icons have to always be recognizable and readable and to the greatest degree adhere to the common standard or DIN norms. The meaning of the icons has to be universally recognized. Many symbols with an established meaning in Europe may be interpreted differently by other cultures. Example: In Europe the analog clock is the symbol for the dimension of "time", in Asia this is the symbol for "death".

In order to be able to understand the statically displayed information with minimal effort, small and much reduced animations are used that provides direction to the user, thus permitting easier navigation. An example for this application is the animation of the repeat function in the audio menu (Fig 5).

Fig. 5. Animated MP3 repeat function

The arrangement of the graphic elements in a linear fashion throughout all hierarchy levels is the basic requirement for all screens of the system. Even though the layout configuration is strictly fixed, the arrangement of information along with text fields must be flexible enough in order to conform to other language variants requiring a different number of symbols. Here, the system will look for the smallest common denominator in flexibility between the number of the symbols and retaining the page arrangement (Fig. 6)

Fig. 6. Flexible text fields for different languages

A test with 13 people showed improved user interaction with the interface design discussed above. A formative usability test of the interaction principles and concept has been carried out. In the study 13 people, 8 male and 5 female, took part. The test persons were between 22 and 44 years old. Nine test persons hadn't had experience with navigation systems, the other 4 had used navigation systems before. The tasks comprised often used applications like radio, telephone, audio and navigation without any guidance or support. This new system compared with a similar driver information system with a smaller display, (Becker Online Pro) was the starting point for the benchmark. The analysis of the subjective evaluation shows a decided preference for the new system with the smart graphics. The key points were: positive overall impression, aesthetic design, attractive color, good readability from a distance, clearly arranged screen design, further there were easy to use, not complicated and generally user friendly.

The uniqueness of the above described graphic is that it not only considers the specific factors of the environment it is being used in but places great emphasis on aesthetic expectations and being user friendly. All these unique characteristics and distinctions make for a optimally sized graphic that is logical and to the point, but at the same time not boring or unimaginative - just a smart graphic.

References

1. Hollwich, F.: Augenheilkunde, Thieme, Stuttgart 1988
2. ISO TC22/SC13/WG8 "TICS M.M.I". Road vehicles – Traffic information and control systems – ergonomics aspects of in-vehicle visual presentation of information.
3. Welford, A.T.: Aging and human skills, Greenwood Press, Connecticut, 1975
4. Cohen,A.S.: Sensorik und ihre altersabhängige Variation. In U. Tränkle, 1994
5. Bell, B., Wolf, E. & Bernholz, C.D.: Depth perception as a function of age, Aging and Human Development, (77-88), 1972
6. Gilbert,J.G.: Age changes in colour matching. Journal of Gerontology, 210-215, 1957
7. Apple Computer Inc, Human Interface Guidelines, The Apple Desktop Interface, Addison-Wesley Publishing Company, Bonn Paris, 1987
8. Bühl Achim: Cyber Society, Mythos und Realität der Informationsgesellschaft, PapyRossa Verlags GmbH & Co. KG, Köln, 1996
9. Brugger, Ralf: Professionelle Bildgestaltung in der 3D Computergrafik, Addison-Wesley, Bonn, 1995
10. Cyrus Dominik Khazaeli: Typo und Layout, Vom Zeilenfall zum Screendesign, Rowohlt Taschenbuch Verlag GmbH, 1995
11. Hannemann Jörg, Thüring Manfred: Designing User Interfaces for Hypermedia, Springer Verlag, Berlin Heidelberg, 1995
12. Helander M.: Handbook of Human-Computer Interaction, Elsevier Science Publishers, Amsterdam, 1988
13. Kieselbach Ralf: The drive to design, av-Ed., Stuttgart 1998
14. Schneiderman B.: Designing the User Interface, Strategies for Effective Human-Computer Interaction, Addison-Wesley, 1992
15. Stary Christian: Interaktive System; Software-Entwicklung und Software-Ergonomie, Vieweg Verlag, 1996
16. Stary, Joachim: Visualisieren, ein Studien- und Praxisbuch, Cornelsen Scriptor, Berlin, 1997
17. Human Factors Design Guidlines for Advanced Traveler Information Systems (ATIS) and Commercial Vehicle Operations (CVO), U.S. Department of Transportation, Federal Highway Administration, Report FHWA-RD-98-057, 1998

Target Selection in Augmented Reality Worlds

Jamie Sands, Shaun W. Lawson, and David Benyon

Human Computer Interaction Group,
School of Computing, Napier University,
10 Colinton Road, Edinburgh,
United Kingdom, EH10 5DT.
{j.sands,s.lawson,d.benyon}@napier.ac.uk

Abstract. This paper describes our work in developing the human-computer interface of an augmented reality (AR) system that has been designed to facilitate the accurate selection of real world locations in 3D space using 3D cursors. We detail the results of an experiment undertaken to highlight the benefit of certain features integrated into the design of a 3D cursor and the interface employed to manipulate the cursor. We conclude that object selection is aided by adding additional synthetic depth cues to the cursor and also by using constraints on the number of degrees of freedom exhibited by the cursor.

1 Introduction

Augmented reality (AR) is the term used when fusing real world environments with context-aware computer generated information. The underlying idea is that a human's task, or experience is augmented in some way by the use of overlaid computer graphics, text or even audio. The potential applications of AR are numerous and wide ranging and readers are directed to a review of recent advances in the field [1] for a summary of these. Despite the wealth of potential applications, AR is still a relatively infant technology with many issues still poorly understood. From an engineering point of view, for example, it can be said that the ultimate goal of AR research is to attain perfect dynamic registration between the real and virtual worlds. This is indeed the focus of much current research, including some of our own [2]. However, the problems of registration have tended to over-shadow many other important issues including the physical, physiological and psychological (cognitive) limitations of using AR systems. Indeed, detailed studies of Human Computer Interaction (HCI) with AR systems have appeared infrequently in the literature. Above all, the issue of usability of AR systems and the associated problem of applying augmented reality to a task to do *work*, has received very little attention.

Whilst common problems between AR and Virtual Environments/Reality (VE/VR) exist (selection of input/output devices, choice of tools and metaphors for instance) we believe that AR presents an entirely new set of problems to the HCI community. These problems stem from the fact that an AR display contains, by definition, an amount of real world content and, in many cases, very little prior information is known about this content. This novel setting, unlike VR, combines an environment and objects that conform to the physical laws of the real world and virtual overlaid

M. Bubak et al. (Eds.): ICCS 2004, LNCS 3038, pp. 936–945, 2004.
© Springer-Verlag Berlin Heidelberg 2004

Fig. 1. Panoramic picture of the surface of Mars, landscapes like this could be remotely mapped from Earth using 3D AR cursors.

graphics that may not. The accurate manipulation of virtual objects with respect to the real world therefore represents a problem not currently addressed by VR work.

Consider a user interacting with a visual display of a VE - an interface designer can make use of the fact that the graphics system can detect, via the underlying projection model employed, when a user positions a selection tool over the top of an object. This allows the designer to highlight the object to let the user know that it has been selected. Additionally, the selection tool in a VR system can become occluded by or will occlude objects in the environment and is subject to the same rendering rules (shadowing, shading and coloring) as all other objects within the environment. When selecting a real object in an AR display none the above factors hold true: a selection tool will simply occlude all objects that lie 'behind it' no matter their relative positions in 3D space. The selection tool is also rendered according to an approximation, at best, of the real world lighting conditions. The end result is an interface that conveys poorly constructed, or even conflicting, depth cues to the user. Couple these problems with the already well known issues and deficiencies of typical interface devices such as Head Mounted Displays (HMD's) and 3D mice then this results in an interactive system that is potentially very difficult to use.

Despite these inherent problems research into AR systems is a worthwhile pursuit. In comparison to VR, AR has an almost limitless scope for application. Whereas Brooks suggests that VR has only two main domains of application, training/simulation and the presentation/demonstration of environments or objects [3], AR applications include anything where an advantage is gained by combining overlaid graphics and the real world. In his 1997 and 2001 surveys of AR Azuma presents a good synopsis of many areas of proposed applications [1,4].

An essential element of successful AR systems is the ability for the real and virtual to appear to co-exist in the same space. An area of particular interest is the accurate alignment of 3D cursors with the real world in order to measure distances. One particular application that could utilize the accurate positioning of augmented 3D cursors is the mapping of remote inaccessible or inhospitable environments such as the surface of Mars, Figure 1. 3D AR systems offer a repeatable method of accurately examining such environments without the need for the presence of a human user and can conducted 'offline'. Users can measure and map these environments from a remote location using only the received stereo images, thus eliminating the need for complex and expensive mapping technology.

In this paper we describe our efforts to produce a set of general guidelines, independent of problems associated with system variables such as display devices, to aid AR researchers in the development of selection tools such as virtual cursors. In particular we describe recent work in addressing the accuracy of object selection via the addition of artificial depth cues in a stereoscopic display. Our experiment was designed to assess some aspects of the system chosen to manipulate our virtual cursor. We conclude by stating the findings of the work, which have confirmed a number of well-known HCI principles. We also describe our current efforts to move our users away from the desktop and into a walking, wearable AR environment.

2   The Importance of Object Selection in AR

Using virtual tools to select real objects is a task common to numerous proposed AR applications. The accurate selection of chosen targets precludes the ability to label objects, people and buildings, and the measurement of sizes and distances. Much of our own work has concentrated on the development of AR tools for the exploration of remote environments using tele-operated mobile vehicles [2,5,6]. These tasks require a human operator to select points in 3D space to a very high precision. However, given that the mobile vehicle can be accurately tracked, the technique is repeatable and could, potentially, be scaled to allow full mapping and characterization of complete remote scenes – the inside of buildings or planetary landscapes for instance.

Similar work to our own has been undertaken at the University of Toronto [7] with the primary objective being to determine whether subjects can accurately mark the location of real points in an unprepared space. This work, like our own features either a stereoscopic desktop display or HMD coupled with a 3D (often 6 degrees of freedom (DoF)) desktop mouse. Experiments in this area are notoriously difficult to repeat across research groups since numerous systematic sources of error are inherently present in any AR system. The sources of these errors include the calibration strategy employed, the tracking of any devices used, and distortions and flaws in the chosen display system.

Another related piece of work is the *Tinmith-Metro* system [8] under development at the University of South Australia. The developers of *Tinmith-Metro* have incorporated a glove-based interface into a walking AR system so that new virtual objects can be created. This can allow a subject either to completely model the existing world, or to create new augmented parts to real worlds. This system allows the user to walk about the environment whilst interacting with it at a very involved level.

2.1   Borrowing Object Slection Tools from VES

It is clear from the above discussion that the selection of objects will comprise an important set of interaction techniques in a wide range of future AR applications. However – how best to implement such interactions is, at present, unclear. The manual selection of objects in a 3D *virtual* environment (VE) is a well-studied HCI problem with many solutions having being proposed in the literature. Poupyrev et al, for instance, give taxonomy of object selection techniques in [9]. They classify relevant techniques as using either virtual hand or virtual pointer metaphors. Many such methods can work in VEs because the entire environment is known, whereas in AR the real element of the environment is often not known (we use the term *unprepared*). Therefore virtual pointer methods, such as ray casting for instance, are inappropriate as they rely on knowing when a ray intersects a known object. An equivalent ray in a mixed reality environment would simply sit on top of all objects within the display.

Our 3D cursor is manipulated by the user until its "hot-spot" (for instance the tip of a cone) appears, to the subject, to coincide with the position of an object. In pure VE's, 3D cursors are simply another virtual object in the environment – they are subject to the same collision detection, occlusion, shadowing and rendering rules as all other objects. In an unprepared AR environment however a virtual 3D cursor is subject to none of these rules –a cursor will occlude all real objects in the world no matter of their relative position. The end result of this is that users have difficulty in positioning a 3D cursor in an AR scene – largely, we believe, because the depth information available to the user exhibits conflicting cues. This remains true even when stereoscopic displays are used to present the mixed world to the user. Other factors such as the general awkwardness of 3D stereoscopic displays and interface devices such as 3D mice also serve to make cursor manipulation a difficult task.

3 Experimental Conditions and Procedure

In order to determine the effect of a number of variables in a typical object selection task using a 3D cursor we have recently conducted a series of experiments using our stereoscopic video see-through AR system [2]. Our experiments were designed to assess the following:

- The accuracy of subjects' ability to position a 3D cursor in an unprepared mixed reality environment with no other extra information present to the user
- Whether the use of a synthetic shadow cast by the virtual cursor improved subjects accuracy in positioning of the cursor
- Whether the overlay of textual information regarding the cursors whereabouts in the world improved subjects accuracy in positioning of the cursor
- Whether adding user selectable constraints to the number of degrees of freedom (DoF) available when manipulating the cursor improved subjects' accuracy in positioning of the cursor
- Whether the presence of an object of known size within the display improved subjects' accuracy in positioning of the cursor
- Whether there were any correlations between subjects' spatial ability (measured using a general purpose mental object rotation test) and cursor positioning accuracy.

An early version of our AR system has been described in detail in [2]. Our most recent system features two fixed color CCD cameras in a parallel configuration gazing at a real scene containing a number of targets in known real world locations. The left and right analogue PAL camera signals are overlaid with corresponding graphics streams (using analogue luma-keying devices) and presented to a user in a remote room via a stereoscopic HMD (a 5dt device). Usually, the only overlaid computer generated objects are a 3D cursor and textual information shown in a fixed location in the display. A static calibration of the system is performed as described in [2].

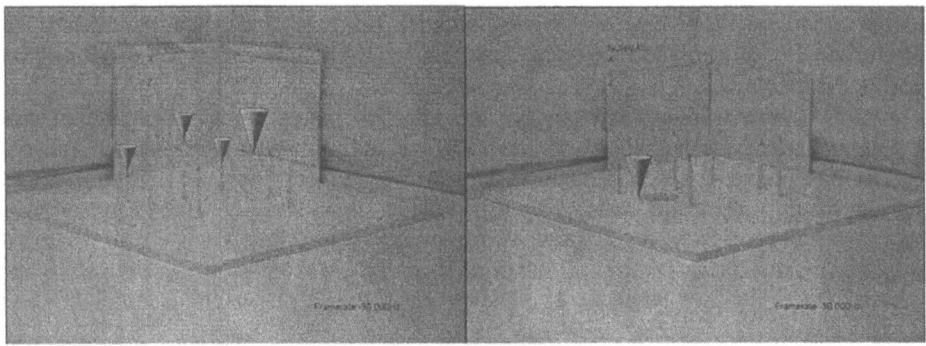

Fig. 2. Screenshots of base condition display containing 3 target cones and the shadow condition display.

Due to the luma-keying method employed in the overlaying of virtual graphics a blank, white scene, lit using diffuse lighting was employed to avoid areas of conflicting shadow and varying degrees of brightness. The blank scene also kept the depth information available to the user at a minimum.

The purpose of our experiments was to investigate subjects' accuracy in aligning a 3D virtual cursor with a number of real objects (targets) in a stereoscopic display. The targets consisted of the tips of a number of upturned sharpened dowels mounted on a rigid base (see Figure 2). Each individual experiment required the subject to drop a virtual marker at six such targets using a Spacemouse device to manipulate the cursor. A Spacemouse interface was chosen as it naturally offered 3DoF to subjects in a familiar desktop setting; the experimenters also had previous experience programming Spacemouse interface devices. Subjects' errors in the alignment tasks, in millimeters, were measured during eight experimental conditions using a within subjects repeated measures procedure. The eight conditions are described in Table 1.

The experimental conditions described in Table 1 aim to uncover which additional synthetic depth cue provides the user with the most useful extra information. These particular cues were chosen as each could be added to the 3D cursor without affecting the cursors movement or the users view of the cursors hotspot. They were also chosen as they did not appear to be application specific and therefore could be easily applied in any AR system requiring the accurate positioning of a cursor. The cues can be separated into two broad groups: cues that assist positioning by indicating a ground intercept and those that assist the user by offering scalar information. Figure 2 illustrates the correct shadow condition showing the indication of the 'ground' level.

Fifty-six subjects (thirty six male and twenty female) aged between 19 and 54 years took part in the experiment. No prior experience was necessary. No subjects reported blindness in either eye although visual acuity varied between subjects.

At the start of each experiment each subject was briefed on the task they were about to perform, shown how to use the Spacemouse interface and allowed to practice moving the cursor. Subjects were made aware that the tasks would not to be timed. Once subjects confirmed they understood the experiment and could satisfactorily move the cursor they were instructed on how to use the stereoscopic HMD. Subjects were given the choice of having the text displayed to their left or right eye, and were also given the choice to wear or remove their spectacles as they saw fit. It was also

Table 1. List of experimental conditions

Condition Name	Description
Base condition	In this condition the alignment task was undertaken using a simple upturned cone cursor as shown in Figure 1.
Non-constrained 3D motion	In all other conditions including the base condition, subjects were able to constrain the motion of the cursor by switching off motion in x, y and z directions. However in this condition, the cursor moves freely in all DoF.
Co-ordinates display	In this condition a world centered (not user centered) text overlay indicated the absolute cursor position to the user in the upper right corner of the display.
Correct Shadow	This condition incorporated a 'token' virtual shadow cast by the cone cursor onto the experimental tabletop plane. The shadow position conformed to the real light source that was present in the real world (to the upper left of the subject's viewpoint). The shadow moved and behaved, as would a real shadow except that it could not, of course, be occluded by real world objects. If the subject positioned the cursor under the surface of the table (at a position of y <=0), then the shadow disappeared.
Crosshairs Cursor	The Crosshairs condition replaced the upturned cone of the base condition with a 3D crosshairs with a central dot indicating the 'hot spot'. The spars of the crosshairs were created along the lines of the x, y and z dimension axis.
Incorrect Shadow	Same as the Correct Shadow condition except that the shadow appeared in the 'wrong' location considering the position of real light source i.e. it appeared on the opposite side.
Scale	In this condition an additional object – a scale of known size - was added along the perceived x and z-axis. The scale comprised a series of black and white bands each 1cm apart.
Textured cursor	In this condition the cone cursor was enhanced by applying a texture to its surface, thus adding texture gradient cues.

stated that subjects could re-adjust the focus and position of the HMD as often as they wanted and at any time during the experiment sessions.

Each subject was then asked to perform the base condition; this condition was always performed first for a number of reasons. The base condition served to further familiarize the subject with the task prior to the more complex experimental conditions. As the base condition was conducted as a baseline measurement, against which the other conditions would be compared, this condition had to be devoid of any other information cues. Performing this condition first meant that no residual information could be carried over from the tasks containing extra cues. After completing the base condition subjects completed the six experimental conditions in a randomized order. The final condition was always the non-constrained 3D-motion condition; this allowed subjects the maximum possible practice with moving the cursor in 3D before conducting this condition. Overall the time taken to complete instruction and experimentation was around 40 minutes with each task taking around 4 minutes each.

4 Experimental Results and Analysis

Table 2 shows the mean errors, standard deviations and variance for each of our experimental conditions. One of our primary objectives was to investigate the effect of additional depth information on alignment accuracy -therefore our analysis compares subject's performance in each condition with the base condition where no extra depth information was presented. Therefore the benefit of a single depth cue can be examined. As can be seen from Table 2 only the non-constrained 3D-motion condition has a higher mean, or performed worse, than the base condition, all the additional stimulus conditions performed better than base condition. Inspection of the standard deviations also shows that all conditions other than the 3D-movement condition had relatively similar degrees of deviation from the mean.

Table 2. Summary of results.

Condition Name	Mean Error /mm	Std./mm	Variance /mm
Base condition	12.41	8.47	71.65
Un-constrained 3D motion	46.82	34.65	1200.89
Co-ordinates display	8.38	6.74	45.37
Correct Shadow	6.97	6.09	37.13
Crosshairs Cursor	12.29	8.06	64.90
Incorrect Shadow	8.46	9.63	92.77
Scale	10.44	8.83	77.93
Textured cursor	9.25	7.05	49.66

A within subjects repeated measures analysis of variance (ANOVA) was conducted to determine any significant differences between the experimental conditions. Due to the highly varied nature of the data obtained and performance of subjects it was decided to remove all the outlying and extreme values. During testing, observations showed some subjects were unable to complete all the tasks in a similar fashion. As these discrepancies generally did not indicate an inability to complete the tasks correctly but instead, a loss of concentration or a misunderstanding it was decided that these results were not indicative of the overall performance. These values were replaced by the mean value, in total 37 results were replaced over all conditions. Removal of these values did not affect the levels of significance of any of the results. Statistical analysis found a significant effect of stimulus type (i.e. between conditions): $F_{(7,55)} = 51.224$, $p < 0.001$. It was concluded that there exists a significant difference between the means of the experimental conditions.

Table 3. Significance results for the between condition paired t-tests.

Condition paired with Base Condition	Difference in Mean Errors /mm	Significance result for paired test
Non-constrained 3D motion	34.41	**t(55) = 7.38, p<0.007**
Co-ordinates display	-4.04	**t(55) = -3.39, p<0.007**
Correct Shadow	-5.45	**t(55) = -4.55, p<0.007**
Crosshairs Cursor	-.124	t(55) = -0.09, p>0.007
Incorrect Shadow	-3.96	t(55) = -2.43, p>0.007
Scale	-1.98	t(55) = -1.33, p>0.007
Textured cursor	-3.17	t(55) = -2.51, p>0.007

Further analysis was then performed to localize the differences among the individual conditional means. As the aim of the experiments was to investigate the performance of additional depth information in comparison to the base condition, post-hoc comparisons were only conducted for all conditions versus the base condition to determine where significant differences existed. Seven paired sample t-tests were carried out on all possible combinations of other condition versus the base condition. For the t-tests the p value was adjusted for Bonferroni correction therefore p<0.007 was required to achieve significance between the means of the conditions. Table 3 shows the results of the seven t-tests performed.

Table 3 shows that only the 3D movement condition, the co-ordinates display condition and the correct shadow condition versus the base condition reached the required p<0.007 significance level. The mean difference for the 3D movement condition is significantly higher than the base condition; therefore the 3D movement condition mean is significantly less accurate than the base condition. However the Co-ordinate and Correct shadow conditions are significantly lower than the base condition and are therefore significantly more accurate than the base condition.

5 Discussion

The primary aim of this experiment was to investigate the beneficial affect of the presentation of additional depth information in a simple 3D AR target alignment task. Our initial hypothesis was that *any* additional depth information would aid a users performance over conditions without such depth information. Our results support this view; however they also suggest that certain depth cues are more helpful than others. In particular, co-ordinate information about the absolute position of the cursor and the addition of a correctly oriented shadow proved to significantly improve performance.

It is hypothesized that these results indicate the ability is for these experimental conditions to indicate the zero plane (when Y=0), thus affording the user useful information about the positional height of the cursor relative to its starting position. The benefit of this knowledge may be accentuated in our experiments as the Y=0 plane was situated on the tabletop, as were the bases of the target objects. Users' judgments must still therefore be made as an evaluation of the cursor position depending on the users understanding of the origins position. Enabling the user to easily see when they have returned to this position, or the zero plane, increases a user's understanding of where the cursor is in relation to the perceived world.

Another significant finding highlights the benefit of control of individual dimensions of movement. All experimental conditions apart from the non-constrained 3D-movement condition allowed the user to switch on/off the individual x, y and z DoFs. Comparison of the base and non-constrained 3D-movement condition indicates the effect of the differing control techniques. As shown in Table 3, the base condition significantly outperforms the non-constrained 3D-movement condition. It can therefore be seen that having control over the individual x, y and z dimensions significantly increases the accuracy of performance during an alignment task. It is assumed that when in control of all dimensions; subjects are unable to distinguish possible errors made and therefore unable to reconcile them, therefore any increase in accuracy is due to the ability of users to limit errors to a single DoF.

6 Conclusions and Future Work

The results presented here show that many factors have an influence on the accuracy of object alignment, or selection, tasks in stereoscopic AR displays. Further work is required to determine the optimal combination of these influential factors. However, our experiments indicate that using additional depth cues such as simulated cursor shadows and text overlays of cursor positions significantly enhance subject accuracy in object selection. Additionally, allowing subjects to constrain the motion of the cursor also increases accuracy.

The results presented here have ramifications not only for the development of AR but any technology in which objects within a 3D world are required to be selected. In VE, interface techniques such as ray casting are designed to select entire objects, these techniques would have difficulty with the selection of more accurate or complex objects or parts of objects. Future desktops computers will inevitably utilize 3D space to cope with the ever-increasing multitasking workload of the user. Some 3D desktops have already been proposed [10].

Our experimental results also reinforce two well-known principles of human-computer interaction, namely feedback and control (see [11] for example). The non-constrained condition is worse than any other condition because users do not have adequate control whereas the three most accurate conditions provide control – either visually as in the shadow condition, or conceptually as the y co-ordinate shows zero in the co-ordinates display. As a general finding, we should conclude that developers of devices for use in AR worlds should provide better facilities for allowing users to take control and for providing suitable feedback as to position in the AR world.

It should also be noted that, of course, the use of desktop devices, such as the Spacemouse might not be suitable for manipulating cursors in wearable or mobile AR systems. Previously described systems of this kind have made use of hand-held joystick or glove type devices to perform basic object selection and menu driven commands. Although much work has been done in glove-based interfaces for VE interaction it is unclear whether such devices are appropriate for use in AR systems. Potentially new ways of implementing easy, fast and natural ways of object selection for walkable or wearable AR applications include eye movement analysis and even brain computer interface (BCI) techniques. Some work in eye-movement analysis for non-cursor based selection of objects has already been described for VE's [12] – it is of interest to discover whether such techniques map easily to AR systems. The performance of BCI systems remains, at present, very limited – though it is intriguing to note that a good proportion of BCI work has been aimed at cursor manipulation (see [13] for instance). To-date such work has been limited to cursor control in one or two dimensions in purely computer generated environments– again it will be interesting to discover whether this can be extended to three dimensions and then, in turn, to AR environments.

Acknowledgments. This work has been funded by the UK Engineering and Physical Sciences Research Council (EPSRC) under Grant No. GR/R25309/01.

References

1. Azuma, R.; Baillot, Y.; Behringer, R.; Feiner, S.; Julier, S. and MacIntyre, B. Recent Advances in Augmented Reality. *IEEE Computer Graphics and Applications*, 25(6) (2001), 24-35.
2. Lawson, S.W., Pretlove, J.R.G., Wheeler, A.C., Parker, G.A. Augmented Reality as a Tool to Aid the Telerobotic Exploration and Characterization of Remote Environments. *Presence: Teleoperators and Virtual Environments*, 11(4) (2002), 352-367.
3. Brooks, Jr., F.P., What's Real About Virtual Reality? *IEEE Computer Graphics and Applications*, 19(6) (1999) 16-27.
4. Azuma, R.; A survey of augmented reality, *Presence: Teleoperators and Virtual Environments*, 6(4), (1997), 355-386,
5. Sands, J., Lawson, S.W. Towards a usable stereoscopic augmented reality interface for the manipulation of virtual cursors. To be presented at *ISMAR 2003, 2nd IEEE and ACM International Symposium on Mixed and Augmented Reality*, Oct. 2003, Tokyo, Japan.
6. Pretlove, J.R., Lawson, S.W. Integrating Augmented Reality and Telepresence for Telerobotics in Hostile Environments, *Proc First IEEE International Workshop on Augmented Reality (IWAR'98)* A.K. Peters (1998) 101-107.
7. Milgram P., and Drascic, D. Perceptual Effects in Aligning Virtual and Real Objects in Augmented Reality Displays. *Proc. Human Factors and Ergonomics Society 41st Annual Meeting* (1997).
8. Piekarski, W., and Thomas, B.H. Tinmith-Metro: New Outdoor Techniques for Creating City Models with an Augmented Reality Wearable Computer. *Proc. 5th Int. Symposium on Wearable Computers*, (2001) 31-38.
9. Poupyrev, I., Weghorst, S., Billinghurst, M., Ichikawa, T. Egocentric object manipulation in virtual environments: empirical evaluation of interaction techniques. *Computer Graphics Forum*, 17(3) (1998) 41-52.
10. Robertson, G.C.; van Dantzich, M.; Robbins, D.C.; Czerwinski, M.; Hinckley, K.; Risden, K.; Thiel, D.; and Gorokhovsky, V. The Task Gallery: a 3D window manager, Proceedings of the SIGCHI conference on Human factors in computing systems, The Hague, The Netherlands. April, 2000, 494-501
11. Benyon, D. R., Turner, P. and Turner, S. *Designing Interactive Systems*. Addison-Wesley (2004).
12. Tanriverdi, V. and Jacob, R.J.K. Interacting with Eye Movements in Virtual Environments. *Proc. CHI 2000* ACM Press (2000) 265-272.
13. Black, M. J., Bienenstock, E., Donoghue, J. P., Serruya, M., Wu, W., Gao, Y. Connecting brains with machines: The neural control of 2D cursor movement. *Proc. 1st International IEEE/EMBS Conference on Neural Engineering*, (2003) 580-583.

Towards Believable Behavior Generation for Embodied Conversational Agents

Andrea Corradini[1], Morgan Fredriksson[2], Manish Mehta[1], Jurgen Königsmann[2],
Niels Ole Bernsen[1], and Lasse Johannesson[2]

[1] Natural Interactive Systems Laboratory
University of Southern Denmark, 5230 Odense M, Denmark
{andrea,manish,nob}@nis.sdu.dk
http://www.nis.sdu.dk/staff
[2] Liquid Media AB
Skaanegatan 101, 11635 Stockholm, Sweden
{morgan,jurgen,lasse}@liquid.se
http://www.liquid.se

Abstract. This paper reports on the generation of coordinated multimodal output for the NICE (Natural Interactive Communication for Edutainment) system [1]. In its first prototype, the system allows for fun and experientially rich interaction between primarily 10 to 18 years old human users and 3D-embodied fairy tale author H.C. Andersen in his study. User input consists of domain-oriented spoken conversation combined with 2D input gesture, entered via a mouse-compatible device. The animated character can move about and interact with his environment as well as communicate with the user through spoken conversation and non-verbal gesture, body posture, facial expression and gaze. The described approach aims to make the virtual agent's appearance, voice, actions, and communicative behavior convey the impression of a character with human-like behavior, emotions, relevant domain knowledge, and a distinct personality. We propose an approach to multimodal output generation, which exploits a richly parameterized semantic instruction from the conversation manager and splits the instruction into synchronized text instructions to the text-to-speech synthesizer, and behavioral instructions to the animated character. Based on the implemented version of this approach, we are in the process of creating a behavior sub-system that combines the described multimodal output instructions with parameters representing the current emotional state of the character, producing animations that express emotional state through speech and non-verbal behavior.

1 Introduction

CASA (Computers Are Social Actors) studies [2] have shown that humans automatically and unconsciously tend to treat computers and other new media as real social entities. When interacting with computer systems that are able to display attitude and personality, people find them polite, extrovert, etc. Humans respond to these affective stimuli and traits in the same way as they do to people in relation to the politeness, extroversion, or other psycho-social phenomena they have perceived. These findings

M. Bubak et al. (Eds.): ICCS 2004, LNCS 3038, pp. 946–953, 2004.
© Springer-Verlag Berlin Heidelberg 2004

advocate the Media Equation: media = real life, which essentially postulates that interaction between humans and machines tends to be social in nature, i.e. can be viewed as a specialization of the anthropomorphic tendency of human beings. In this light, it has been argued that Human-Computer Interaction (HCI) is fundamentally social and thus that social rules governing human-human communication apply to HCI as well. Consequently, the user interface of interactive technologies can be improved by leveraging human expectations of natural human social features.

Supported by these trends, and by today's advances in human language technologies and increasingly powerful computer graphics techniques that are making available a broader set of tools to create more flexible, human-centered and adaptive user interfaces, there has been a thrust towards designing agent based interfaces which exhibit human-like behavior and appearance. These interfaces, termed embodied conversational agents (ECAs) [3], aim both to use and to realize cues inherently peculiar to human-human communication, such as sense of presence, mixed initiative and non-verbal behaviors to hold up their end of the dialogue with the user.

We have implemented a domain-oriented (non-task-oriented [4]) conversation system that allows users to interact in a natural, fun, and experientially rich manner with an embodied conversational character impersonating the Danish fairytale writer Hans Christian Andersen (HCA) within a graphical 3D environment. The system accepts spoken and 2D gestural input entered via a mouse-compatible device, recognizes, interprets, and fuses these modalities with context information, and eventually generates an appropriate coherent behavior in response to the user input.

In this paper, we report on our approach to generating coordinated multimodal output with the aim of making the virtual agent's appearance, voice, actions, and communicative behavior believable, i.e. conveying the impression of a character with human-like behavior, emotions, relevant domain knowledge, and a distinct personality, entertaining, and instructive, i.e. providing true historical information, to the user.

2 Multimodal Output Generation

2.1 Related Work

Other researchers have developed embodied multimodal agents that produce natural conversation in task-oriented applications [3].

Rea [18] plays the role of a real estate salesperson; she uses gaze, head movements and facial expressions for functions such as turn-taking, emphasis and greetings. The verbal and non-verbal communication aspect of Rea is limited to task oriented dialogue in which the agent interacts with users to determine their needs, shows them around virtual properties, and attempts to sell them a house. Baldi [19] is an animated conversational agent, who helps students accurately produce expressive speech. The interactive system's curriculum development software lets teachers and students customize class work. The animated agent communication is limited to facial movements that are synchronized to its audible speech. Greta [20] responds to user queries by speaking and exhibiting gaze, head movements, and facial expression. The conversation simulated is restricted to a limited task.

Also in the game industry there have been attempts to provide the user with different modalities in order to provide a rich interaction experience. Black and White [21] is a strategy game developed by Lionhead Studios Ltd., which involves drawing symbols on the screen to cast miracles such as a rain spell to increase food production. The shape of the symbols ranges from arrows and swirls to letters and numbers. As the user advances in the game, the symbols she has to draw become more complex, making it harder to cast them quickly during gameplay. Arx Fatalis [22] is a first-person role-playing game that allows the player to draw burning runes in mid-air using the mouse. A series of these gestures combine to create powerful magic spells that will protect the player or empower him to defeat his enemies and pursue his quest. These games provide unimodal user interaction yet there have been rather few attempts to enhance user experience with multimodal input modalities. The same goes for synchronization of verbal and non-verbal output from the player and non-player character.

2.2 Response Generator

The HCA character module is always in one of three output states. While in the Communicative Function (CF) output state, the character shows in his behavior that he is aware of being spoken to or addressed by the user. A Non-Communicative Action (NCA) output state refers to the situation in which HCA is not engaged in conversation with a user. The agent is in a Communicative Action (CA) output state anytime he takes the turn in producing outputs that serve as a vehicle for the on-going conversation. CA output is the character's actual conversational contribution in the form of verbal realization of one or more speech acts within a single turn, physical actions within the graphical environment, emotion display, gaze and gestural behaviors. Responses to, and questions for, the user, observations, confirmations and acknowledgements, and meta-communication, such as clarification questions, are produced by HCA when in this state.

Similarly to non-verbal output categorization in [5], we have employed a common strategy to produce output behaviors for both NCA and CF states while we have developed a response generator module to explicitly treat CA state output. For NCA and CF states, we have defined approximately 30 and 15 behaviors, respectively. CF states are generated from a pre-defined subset of elementary behaviors. Finite-state machines whose states are also elementary behaviors are instead utilized to generate NCA states due to the indefinitely long sequence of actions required while in such state. Most of these behaviors are non-verbal, yet a few have also a non-speech audio component, e.g., for playing footsteps sound when the character is moving about in his study. NCA behaviors are typically state-of-the-mind expressions, such as being idle or thinking, and physical actions, such as picking up an object or dropping one. CF behaviors are feedback gestures, e.g., nodding, and posture changes, e.g., clasping hands. Since the CF and NCA output repertoire is fixed at design-time, its realization does not need to involve the response generator. CF and NCA behaviors are randomly selected at certain time intervals and sent directly to the graphic engine for realization.

As concerns CA output, giving HCA a richer persona through emotion and personality modeling is not sufficient to creating a life-like character [6,7]. To increase the character's believability [8], it also has to be able to display human-like behavior by combining non-verbal (gesture, facial expression, pose and gaze) and verbal (speech) output in a consistent way [9]. Coherence is a general rule in social human-human communication [10]. Hence, following the Media Equation, we can expect this rule to

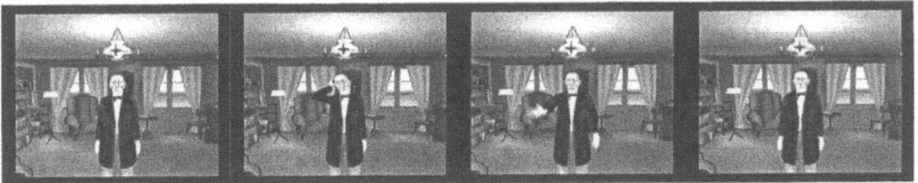

Fig. 1. *(left to right)* HCA performing a DONT_KNOW animation while in the CA output state

apply to human-machine interfaces as well. Furthermore, as people dislike interacting with individuals who behave incoherently [11,2], consistency may also contribute to increasing user effectiveness and satisfaction.

Our response generator links emotional patterns and character animation in terms of speech and non-verbal communication. It performs in real-time and enables generation of a comprehensive representation of communicative actions which can be rendered via speech synthesis and by the animation component (see Fig. 1).

In its technical implementation, the response generator receives from the character module a parameterized semantic instruction composed of input values, text-to-speech (TTS) references, and/or references to non-verbal behaviors. The TTS references are used to retrieve text template output with embedded start and end tags for non-verbal behaviors (bookmarks) that are stored in a form such as: *"I [g0] don't remember exactly [/g0] when I wrote {FAIRYTALE}"*. In this example, elements within square brackets starting with numbered 'g' letters represent onset and offset of non-specified non-verbal parts of the template. Elements within curly parentheses, like *FAIRYTALE,* are placeholders for variable values to be filled in using input value information delivered by the natural language understanding module [12] during a conversational turn. 'The Princess and the Pea' and 'The Little Mermaid' are two possible values in the present example (see [12] for details on the structure of conversation in NICE). Both verbal variable values, as was just mentioned, and non-verbal behavioral elements are initially uninstantiated in the sense that they need to be retrieved at run-time. This approach allows for a high degree of flexibility as the binding of non-verbal behavior to speech occurs at run-time rather than being hard-coded, enabling a sentence to be synthesized at different times with, e.g., different accompanying gestures.

In order to provide timing information for speech and gesture during rendering, non-verbal behavioral elements are made up of two sets of tags to indicate their start and end, respectively. Thus, in the example above, tags *[g0]* and *[/g0]* indicate that a single movement has to co-occur with uttering the spoken text *'don't remember exactly'* around which they are wrapped. Any non-verbal behavior can be attached to the gestural behavior *[g0]* while uttering the short text.

Once non-verbal-behavior tags have been processed and variable values inserted into the templates, one surface language string results. This string is sent to the speech synthesizer, which synthesizes the verbal output and, whenever it meets a bookmark, sends a message to the response generator. The response generator creates an XML representation of the non-verbal element and sends it to the animation engine that takes care of the graphics output. Fig. 1 shows a series of snapshots of the animation generated with the text template used as example, while the XML representation of the non-verbal element in this case simply looks like:

```
<play>
    <animation>
            <name>DONT_KNOW</name>
            <startTime>0</startTime>
            <stopTime></stopTime>
            <data>H, 0; E, 200; J, 500; Sil, 800</data>
    </animation>
</play>
```

The item `<name>` indicates the name of the animation to play, among those that are loaded in memory upon start-up by the rendering application. We refer to these animations as elementary or primitive animations. The `<data>` tag is used to specify additional animation-related information while `<startTime>` provides timing information for the start of the animation. The optional item `<stopTime>` is not being used because we did not want to explicitly set the duration of the animation.

In addition to elementary animations, more complex non-verbal behaviors can be created, assigned a name, and stored by the response generator. To create a new animation, its behavior must be defined in terms of existing animations. For example, assuming the existence of an animation called PUFF, a new composite animation GIVE_UP_FRUSTRATED can be created using PUFF and DONT_KNOW. Because the rendering engine recognizes only elementary animations, to play the new animation, the response generator has to create as many different XML representation strings as the number of primitive animations used in the description of the new one, i.e. two for GIVE_UP_FRUSTRATED. Sequentiality, parallelism and partial overlapping of existing animations to create the new animation can be tuned by setting appropriate values for the temporal items in the XML representation. So, GIVE_UP_FRUSTRATED can occur either as sequence of DONT_KNOW and PUFF, or vice versa, or by having these two running simultaneously, etc.

In the our first prototype, we use approximately 300 spoken utterance templates, many of which are no-variable stories to be told by HCA, and 100 different non-verbal behavior primitives that we identified after analyzing a 4-hour video recording of a human actor performing as HCA in a children theater in the city of Odense, Denmark.

2.3 Graphical Engine and Animations

The rendering module consists of a number of different subsystems that take care of different aspects of the animation process. These are the following.

Visual System. Handles everything related to on-screen visualization, such as rendering of the 3D-environment and the characters as well as output of 2D-text.

XML-based Entity Factory. A set of XML files loaded at start-up to specify which components, such as characters, animations and objects, are included and how they are configured.

Input Handler and Object Tracker. Handle the user's keyboard and mouse input. For each input device position, the object tracker keeps track of the objects in the scenery. This is important as some objects in the scene are 'active', i.e. can be manipulated, e.g. picked up, dropped, or pointed at, by the character.

Fig. 2. HCA's study: *(left)* general view, *(right)* skinned mesh of the HCA character

Network Handler. For TCP-IP or UDP socket communication with other modules.

Navigation System. Handles the movement of characters, stops them from walking through walls or colliding with objects in the environment.

3D Sound System. Handles playback of environmental sounds as well as character speech; the sound can be global or positioned.

Scheduler. Manages events like synchronization of animations, collisions checking, rendering etc. For example, it is used to synchronize the start of animations at a specific time and checking for collisions every Nth frame while rendering is done at every frame. Events in the scheduler are either framebased or time-based.

Animation and Camera System. Handle the updating of characters (currently we only have the HCA character, yet more will be added soon) and other animated objects and keep track of all cameras (currently there are five views) used by the application. The camera decides which part of the 3D scene is rendered. To animate a three-dimensional character, we change its position, scale and orientation at different points in time. The animation system uses three scalar values to represent position and scale, and quaternion values to represent orientation. A character is built upon a hierarchy of elementary parts, referred to as frames, where each single frame represents a bone in the character. The hierarchy of frames, together with a textured polygon mesh and skin weighting information, is represented as a skinned mesh (Fig. 2). The skinning information specifies the influence a frame has on its mesh. To avoid breaking up the mesh while animating the characters, we use vertex blending. The root frame contains a transformation matrix relative to the world space. An animation that affects the root node affects the whole scene while one that affects a leaf node does not affect any other node. Hence, the frame hierarchy gives the system the ability of playing single animations in parallel for different parts of the body to obtain complex animations.

Each animation is given a priority index. For example, let us assume that we set the priorities of animation IDLE to 0, WALK to 5 and NOD to 50. The WALK animation, which affects all nodes in the frame hierarchy, would replace the IDLE animation completely while NOD would only affect the nodes from the neck and down (Fig. 2). NOD with its higher priority overloads only the relevant nodes while WALK or IDLE affects the rest of the hierarchy. The WALK animation overloads all nodes of the IDLE animation. The result is a walking man nodding his head.

The animation system works as a sequencer, it receives network commands and schedules animation events via the scheduler system. The animations can be started in parallel, having, e.g., the same start time. The graphical engine uses its own methods for memory allocation to facilitate the tracing of memory usage and leaks.

3 Conclusions and Future Work

Most of the information exchanged in human face-to-face communication takes place over the non-verbal channel. Therefore, when it comes to developing embodied conversational agents, non-verbal representation is at least as important as verbal representation. Unfortunately, defining an exhaustive representation that encapsulates the attributes necessary for ECA behavior remains an unsolved problem.

In this paper, we have proposed how to build multimodal output generation for an ECA system that supports agent's believability in displaying full-body human-like behavior. Our system attains its educational goal by providing correct factual information, both visually, via a variety of non-verbal behaviors, notably gesture, body-posture, gaze and emotion, and orally via spoken utterances. Any inconsistency between verbal and non-verbal components may deceive and mislead the user [13]. Thus, coordination between output modalities is a very important step toward this goal. Incoherent output realization is counterproductive as it undermines the user's learning process rather than reinforcing it. The system attains its entertainment goal as well. Our character is lifelike, reproduces the human physics in detail, and performs non-verbal behaviors in an exaggerated manner as this has been proven to convey emotions more efficiently and directly than regular performance (in fact, caricaturists take advantage of this), making interaction a fun experience. Embodiment enhances entertainment and effective user engagement [14]. A preliminary system test, we recently run with eighteen 10 to 18 years old kids, seems to support these facts.

Several XML-based languages have been proposed to specify human communicative behavior [15] but none of these fully capture the non-verbal information conveyed by humans. In our approach, we use a high-level XML formalism to describe the overt form of non-verbal communication by utilizing a parameterized library of a few hand-crafted behaviors that can be flexibly and easily expanded to include new elements for modeling additional behaviors, emotions and moods. The main limitation of our approach is the inherent impossibility to fine-control the motions of single body parts. However, with the capability to combine a set of few elementary behaviors we can cover a large variety of movements, thus reaching a compromise between the number of pre-defined behaviors and the complexity to generate others from them.

The system is undergoing continuous improvement. We plan to deal with animation co-articulation to allow smooth rendering of an animation that is started when another one is not yet concluded. We also plan to focus on the auditory realization of communicative intention. Providing HCA's spoken utterances with stress and intonation will be a first step. Linking facial expression and prosody to further personalize the character's communication of emphasis and topic will be the next step. Moreover, in the effort to improve character believability and, potentially, the perceived quality of spoken segments [16,17], we plan to synchronize speech and lip movements using phoneme-to-viseme mapping.

Acknowledgement. We gratefully acknowledge the support from the EU Human Language Technologies programme under Contract no. IST-2001-35293.

References

1. http://www.niceproject.com
2. Reeves, B., and Nass, C.: The Media Equation: how people treat computers, televisions and new media like real people and places. Cambridge, Cambridge University Press, 1996
3. Cassell, J., Sullivan, J., Prevost, S., and Churchill, E. (eds.): Embodied conversational agents. Cambridge, MA: MIT Press, 2000
4. Bernsen, N.O., Dybkjær, H., and Dybkjær, L.: Designing Interactive Speech Systems. From First Ideas to User Testing. London:Springer Verlag, 1998
5. Beskow J., Edlund, J., and Nordstrand, M.: A model for generalised multi-modal conversation system output applied to an animated talking head. In: Minker, W., et al. (eds.): Spoken Multimodal Human-Computer Conversation in Mobile Envs, Kluwer Academic, 2004
6. Argyle, M.: Bodily Communication. 2nd edition, London and NYC: Methuen & Co., 1986
7. Knapp, M.L.: Non-verbal Communication in Human Interaction. 2nd edition, Holt, Rinehart and Winston Inc., New York City, 1978
8. Loyall, A.B.: Believable Agents: Building Interactive Personalities. PhD thesis, Tech Report CMU-CS-97-123, Carnegie Mellon University, 1997
9. Picard, R.: Affective Computing. MIT Press, 1997
10. Fiske, S.T., and Taylor, S.E.: Social Cognition. New York, McGraw Hill, 1991
11. Nass, C., Isbister, K., and Lee, E.-J.: Truth is beauty: Researching embodied conversational agents. In Cassell, J., et al (eds.):Embodied conversational agents. MIT Press, 374-402, 2000
12. Bernsen, N.O., Charfuelàn, M., Corradini, A., et al.: First Prototype of Conversational H.C. Andersen. In: Proc. of ACM Int'l Working Conf. on Advanced Visual Interfaces, 2004
13. Ekman P., and Friesen, W.V.: Nonverbal leakage and clues to deception. Psychiatry 32, 1969, 88-95
14. Koda, T., and Maes, P.: Agents with faces: The effects of personification of agents. Proceedings of Human-Computer Interaction, London, UK, 1996, 239-245
15. http://www.vhml.org/workshops/AAMAS/papers.html
16. Massaro, D.W., and Cohen, M.: Speech perception in perceivers with hearing loss: Synergy of multiple modalities. Jou. of Speech, Language, and Hearing Res., 42, 1999, 21-41
17. McGurk, H., and MacDonald, J.: Hearing lips and seeing voices. Nature 264, 1976, 746-748
18. Casell, J., Bickmore, J., Billinghurst, M., Campbell L., Chang K., Vilhjalmsson H., and Yan H.: Embodiment in conversational interfaces: Rea In: Proc. of CHI 99, 1999, 520-527
19. Massaro, D.W., Bosseler, A., and Light, J.: Development and Evaluation of a Computer-Animated Tutor for Language and Vocabulary Learning. 15th Int'l Congress of Phonetic Sciences, Barcelona, Spain, 2003
20. Pelachaud, C., Carofiglio, V., De Carolis, B., de Rosis, F., and Poggi, I.: Embodied Contextual Agent in Information Delivering Application, First International Joint Conference on Autonomous Agents & Multi-Agent Systems, Bologna, Italy, 2002
21. http://www.blackandwhite.ea.com/
22. http://www.arxfatalis-online.com/

A Performance Analysis of Movement Patterns

Corina Sas[1], Gregory O'Hare[1], and Ronan Reilly[2]

[1] University College Dublin, Belfield, Dublin 4, Ireland
[2] National University of Ireland, Maynooth, Co. Kildare, Ireland

Abstract. This study investigates the differences in movement patterns followed by users navigating within a virtual environment. The analysis has been carried out between two groups of users, identified on the basis of their performance on a search task. Results indicate significant differences between efficient and inefficient navigators' trajectories. They are related to rotational, translational and localised-landmarks behaviour. These findings are discussed in the light of theoretical outcomes provided by environmental psychology.

1 Introduction

An understanding of how people explore an indoor virtual space can provide not only theoretical contributions but can also be harnessed within practical applications. Designing flexible Virtual Environments (VEs) which are able to adapt themselves in order to support user navigation is one of the most promising application fields. Designing adaptive VEs for navigation support necessitates sensitivity to differing types of users. Such adaptive VEs should be able to discriminate between different groups of users, who require different accommodations. These groups of users differ not only in their performance on spatial tasks, but also in their spatial behaviour. Another significant aspect in the design of adaptive VEs for navigation support is the identification of ways to accommodate individual differences in navigational patterns.

This paper focuses on the analysis of users' spatial behaviour, as reflected in their trajectory paths. Movement paths allow an online and unobtrusive identification of user groups. Trajectory classification offers potential in performing such kinds of identification [6]. The analysis of movement paths also allows the implicit extraction of navigational patterns embedded in trajectory paths.

2 Study Design

The experiments have been carried out within a desktop VE [3], which due to its tractable characteristics permitted us to record the users' positions and headings at each moment in time. Adopting a physical world metaphor the VE consists of a virtual multi-story building where each one of the levels contains three rooms. Its projection has a rectangular shape of 16×27 virtual metres. The ECHOES environment comprises a virtual multi-story building, each one of

M. Bubak et al. (Eds.): ICCS 2004, LNCS 3038, pp. 954–961, 2004.
© Springer-Verlag Berlin Heidelberg 2004

Fig. 1. Virtual Training Room   **Fig. 2.** Virtual Library

the levels containing several rooms: conference room, training room (Figure 1), library (Figure 2), lobby etc.

There is no predefined set of paths, such as halls or corridors which would limit the user choice of movements. The users can move freely in the space, freedom limited only by the walls and objects located within the spatial layout. Users can navigate in terms of moving forwards, backwards or rotating, through the use of the directional keys. They merely use the mouse for selecting a new floor on the panel located in the virtual lift.

The study involved three phases: familiarisation, exploration and performance measurement. Initially, users were allowed to become accustomed with the VE and to learn movement control. After this, they were asked to perform an exploration task. The exploration task within the virtual building lasts for approximately 25 minutes. After the completion of this task, during which participants acquired (implicitly) spatial knowledge related to the VE, they were tested. Users were placed on the third level and asked to find a particular room located on the ground floor of the virtual building (the library). The time needed to accomplish this task acted as an indicator of the level of spatial knowledge acquired within the VE: the shorter the search time, the better the spatial knowledge [5]. According to the time required for the search task, users have been identified as *low spatial users*, when they needed significantly longer time to find the library (Mean = 49 seconds), or *high spatial users* who found the library straight away (Mean = 7 seconds). Within this paper, the terms of low versus high spatial users are related to this particular outcome and they also capture the dichotomy between poor versus good or inefficient versus efficient navigators.

The sample consisted of 32 students from the Department of Computer Science in University College Dublin and volunteers were paid for their participation.

The following sections focus on aspects supporting differentiation between efficient and inefficient navigators' trajectories. They are related to rotational, translational and localised-landmarks user behaviour during navigation.

2.1   Rotational Behaviour

Given the significance of rotations along a movement path, we analysed these in order to identify differences which discriminate efficient from inefficient spatial

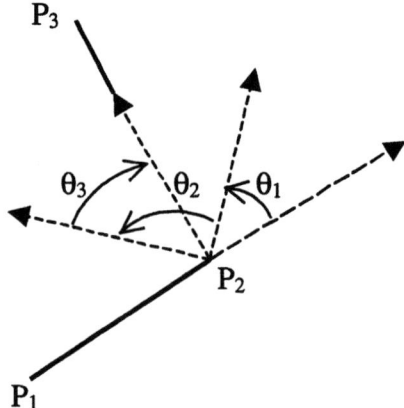

Fig. 3. Observation Angle

behaviour. Each action performed by the user is either a rotation or a translation. Rotations receive increased attention, since they represent those joints, where trajectory segments are articulated at different angles. The location where rotation is performed and the rotation angle provide valuable information about the user's orientation and user's intentions.

Each angle performed at a particular location, possibly consisting of one or more successive rotations, has been computed and averaged for each user. This angle we entitle *observation angle* since it allows the user to observe the environment through the increased view field it facilitates. Figure 3 presents a sample of user's behaviour, where P_1, P_2 and P_3 represent three consecutive positions in time, while θ_1, θ_2 and θ_3 represent there consecutive changes of heading, carried out at location P_2.

The observation angle was computed as follows:

$$\text{Observation angle} = \sum_{i=1}^{k} |\theta_i| \, , \tag{1}$$

where k is the number of successive rotations (in our example $k = 3$).

A t-test suggests significant difference between the values of these angles corresponding to rotations performed by efficient versus inefficient users. Thus, low spatial users perform significantly higher changes of heading (Mean above $65°$) compared to high spatial users (Mean below $45°$) ($t(12) = 1.92$, $p < 0.05$).

When the sign of the rotation angle has been also considered, another angle indicator was obtained. It represents the angle between two adjacent segments of trajectory which suggests a change of direction between two translations. Such an angle is referred to as *moving angle* and differs from the previous indicator when users change the rotation direction within a set of consecutive rotations. Such a *moving angle* between consecutive translations performed at different headings is equal or smaller than the *observation angle*. Figure 4 presents the

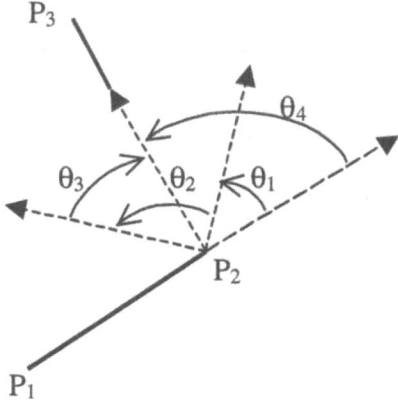

Fig. 4. Moving Angle

same sample of user's behaviour as that depicted in Figure 3, where the moving angle is the angle θ_4.

Findings suggest that low spatial users performed significantly higher changes of heading (Mean above 50°) when measured as *moving angles*, compared to high spatial users (Mean below 35°) ($t(12) = 1.83$, $p < 0.05$). Looking at the distribution of these angles, it appeared that low spatial users performed significantly more rotations higher than 90° (61.2%), compared to high spatial users (38.8%), (continuity correction: $\chi^2(1) = 20.71$, $p < 0.05$).

The analysis of spatial behaviour with respect to the number of rotations and translations within the entire set of trajectories, revealed another interesting finding. On average per trajectory, high spatial users performed a significantly higher number of rotations (Mean $= 12.07$) than low spatial users (Mean $= 10.36$), ($t(11178) = 10.98$, $p < 0.001$). When this analysis was restricted to the numbers of *successive rotational steps* performed between two translations, it appeared that high spatial users performed significantly less such kinds of consecutive rotations (Mean $= 2.23$), compared to low spatial users (Mean $= 3.01$), ($t(10947) = 9.93$, $p < 0.001$). Such successive rotational steps compose a larger change of heading performed in a given location.

2.2 Translational Behaviour

High spatial users performed also significantly less consecutive translations (Mean $= 1.60$) than low spatial users (Mean $= 1.80$), ($t(11514) = 3.92$, $p < 0.001$). High spatial users performed also significantly more translations per trajectory (Mean $= 11.97$) compared to low spatial users (Mean $= 10.26$), ($t(11192) = 10.90$, $p < 0.001$). In addition, the length of each straight line segment, namely the distance covered through consecutive translations, was measured along each trajectory. Such segments have been obtained as a result of consecutive translations performed by the user. A t-test indicated significant differences with respect to the average length of the straight line segments between high spatial users (Mean = 2.94) and low spatial users (Mean = 3.82) ($t(12) = 2.49, p < 0.05$).

2.3   Behaviour Around Landmarks

An important aspect related to translations refers to behaviour around landmarks. Are there any differences in the way efficient and inefficient navigators visit or revisit landmarks, in the way they navigate along a given landmark, or to what distance from the landmarks of interest they navigate? These are the kind of questions that this section tries to answer. Without being statistically significant, low spatial users visit less landmarks (Mean = 11.25) then high spatial users (mean = 16.27) ($t(12) = 1.50, p > 0.05$).

An important outcome suggests that low spatial users revisited significantly more time the same landmarks (Mean = 6.53), as opposed to high spatial users (Mean = 3.60), ($t(12) = 2.95, p < 0.05$). Without reaching significance, other findings suggest that high spatial users visited more rooms (Mean = 9) than low spatial users (Mean = 6.93). Findings also indicate that high spatial users move along the nearest landmarks, significantly longer (Mean = 20.88 events), compared to low spatial users (Mean = 14.31 events) ($t(12) = 1.97, p < 0.05$). In doing so it seems that such landmarks act as some form of attractor. T-test analysis indicates that low spatial users move along the landmarks of interest at significantly longer distances (Mean = 1.92), compared to high spatial users (Mean = 1.68), ($t(12) = 1.95, p < 0.05$). This implies that the search performed by high spatial users is more systematic, focusing thoroughly on one landmark at the time. Once a particular landmark acts as an attractor and a complete search has been performed in its vicinity, the need for revisiting it decreases significantly.

There is no significant difference between trajectories performed by high versus low spatial users in terms of the trajectory length. Given this result, it would be interesting to investigating whether the same relationship holds with respect to the area covered by high and low spatial users.

3   Discussion

Results of this study suggest that the poor trajectories are characterised by more turns at sharp angles, more changes of direction and significantly more straight line segments. The key outcome of this study is that compared to inefficient spatial behaviour, the efficient spatial behaviour can be interpreted as an example of the *Minimum Energy Principle*. The minimum energy principle expresses itself through those procedural and strategic rules which enable users to achieve the goals with a minimum investment of resources, both temporal and cognitive [7].

The efficiency of spatial behaviour performed by high spatial users is reflected also in the fact that these users try to grasp the entire picture, rather than focusing on its details. Thus, they visit each level of the building, usually in a systematic manner, then they visit more rooms and within the rooms, more landmarks than low spatial users. These users are interested in acquiring a larger and more synthetic view of the virtual building, rather than an analytic one. Such a top down approach in exploring the virtual building is complemented by a thorough search in the vicinity of each landmark of interest.

In addition, the way in which high spatial users search around landmarks differs from that employed by low spatial users. The same holistic approach employed at a macro level can be identified: high spatial users perform a thorough search around the landmarks which are considered of interest. They perform significantly more translations and rotations around these landmarks, at shorter distances and accordingly, they revisit these landmarks less.

Spatial orientation refers to users' knowledge of position and orientation within the VE [1,2]. Through their behaviour, high spatial users succeeded in maintaining a better *spatial orientation*. Spatial orientation refers to users' knowledge of position and orientation within the VE [1,2]. High spatial users usually maintain the same main direction or steady orientation, only smoothly altered through small turns. They avoid both long straight movements and large changes of direction, thus maintaining the course of movement.

On the other hand, trajectories performed by low spatial users contain more changes of directions of great angles. These users change their direction not only more often but also more drastically, with serious negative impact on the spatial orientation.

The rotational behaviour of low spatial users presents several major limitations. These users rotate more often at angles greater than 90°, for both observing and changing the movement orientation. They perform significantly more consecutive rotations which dramatically increase the risk of disorientation.

Findings suggest that high spatial users carry out significantly more rotations and translations along a trajectory (on average 12 as opposed to 10), but significantly less such consecutive events (on average 2 rotations as opposed to 3). The actions of high spatial users seem to be more frequent, evenly distributed in time, and much smaller (both in changing the heading and changing the location). This is an important outcome which explains how the efficient navigators are more successful in maintaining their orientation and subsequently acquire better spatial knowledge. Within Table 1 we have codified a set of rules which may effectively be used to differentiate between efficient navigators and inefficient navigators within VEs.

4   Conclusions

This paper investigates individual differences in navigating in VEs. Findings indicate that there are individual differences in navigational patterns, which are related to performance on spatial tasks, or in other words with users' spatial abilities.

Efficient navigators maintain the movement direction and avoid sharp angles. Low spatial users significantly violate the Minimum Energy Principle which ad-

Table 1. Low Level Navigational Rules

Efficient Rules	Inefficient Rules
Rotations	
More small rotations	Fewer small rotations
Fewer great rotations	More great rotations (sharp angles)
Fewer changes of heading ($> 90°$)	More changes of heading ($> 90°$)
More rotations on average per trajectory (12)	Fewer rotations on average per trajectory (10)
Fewer successive rotations in average per trajectory (2)	More successive rotations in average per trajectory (3)
Observational angle	
Small changes of heading ($< 45°$)	High changes of heading ($> 65°$)
Moving angle	
Small changes of heading ($< 35°$)	High changes of heading ($> 50°$)
Translations	
More translations in average per trajectory (12)	Fewer translations in average per trajectory (10)
Fewer successive translations in average per trajectory (1.6)	More successive translations in average per trajectory (1.8)
Shorter segments covered by consecutive translations (3 virtual metres)	Longer segments covered by consecutive translations (4 virtual metres)
Behaviour around landmarks	
More rooms visited (9)	Fewer rooms visited (7)
More landmarks visited (16)	Fewer landmarks visited (11)
More movements near landmarks of interest (20 events)	Fewer movements near landmarks of interest (14 events)
Closer moves to the landmarks of interest (1.5 virtual metres)	Farther moves from the landmarks of interest (2 virtual metres)

vocates the reduction of needless expenditure of energy. They are inefficient and their frequent changes of direction increase the risk of disorientation.

Apart from offering a better understanding of spatial behaviour performed within VEs, in terms of individual differences in spatial knowledge acquisition, the study findings have a significant potential for practical applications. They can be used for elaborating design principles for building adaptive VEs, capable of providing real-time navigation support for low spatial users. This can be achieved through enabling low spatial users' access of the efficient navigational rules, governing behaviour of high spatial users. Thus, differentiating between efficient and inefficient spatial behaviours and their underlying strategies could be exploited through applications dedicated to improve spatial skills, such as adaptive VEs. On-going work is encapsulating such rules in the form of an intelligent navigational support agent, based upon a Belief Desire Intention (BDI) model. Such an agent, upon receipt of various observations from a listener agent, updates its set of beliefs (i.e. Belief(User, fewer_small_rotations)). Based upon such beliefs and combinations thereof the agent may adopt commitments to provide support for the navigational process. Examples of such may include degradation of the responsiveness of the mouse or movement keys, introduction of directional cues, or indeed the provision of an embodied navigation assistant in the form of an avatar within the VE [4].

References

1. Bowman, D., Davis, E., Badre, A., Hodges, L.: Maintaining spatial orientation during travel in an immersive virtual environment. Presence: Teleoperators and Virtual Environments 8(6) (1999) 618–631
2. Bowman, D.A.: Interaction Techniques for Common Tasks in Immersive Virtual Environments. PhD Thesis, Department of Computer Science, Georgia Institute of Technology (1999)
3. O'Hare, G.M.P., Sewell, K., Murphy, A. and Delahunty, T.: An agent based approach to managing collaborative work within both a virtual environment and virtual community. In: Proceedings of the Workshop on Intelligent Agents for Computer Supported Co-Operative Work: Technologies & Risks (2000)
4. Sas, C., O'Hare, G., Reilly, R.G.: Virtual environment trajectory analysis: A basis for navigational assistance and scene adaptivity. Future Generation Computer Systems. Special Issue on "Interaction and Visualisation Techniques for Problem Solving Environments"(2004) in press.
5. Sas, C., O'Hare, G.M.P.: Presence and individual differences in virtual environment: Usability study. In Sharp, H., Chalk, P., LePeuple, J., Rosbottom, J. (eds) Proceedings of 16th British HCI Conference volume 2, 50–53, London British Computer Society (2002)
6. Sas, C., O'Hare, G.M.P., Reilly, R.G.: On-line trajectory classification. Lecture Notes in Computer Science, Vol. 2659, 1035–1044, Springer-Verlag, Berlin, (2003)
7. Warnett, L., McGonigle, B.: Unsupervised navigation using an economy principle. In Prince, C.G., Demiris, Y., Marom, Y., Kozima, H., Balkenius, C. (eds) in Proceedings of the Second International Workshop on Epigenetic Robotics: Modeling Cognitive Development in Robotic Systems (2002)

On the Motivation and Attractiveness Scope of the Virtual Reality User Interface of an Educational Game

Maria Virvou, George Katsionis, and Konstantinos Manos

Department of Informatics, University of Piraeus,
80 Karaoli & Dimitriou Str., 18534 Piraeus, Greece
mvirvou@unipi.gr, {gkatsion,kman}@singular.gr

Abstract. Software games are very popular among children and adolescents and thus they have often been used in educational software research projects to increase motivation of students. However, before educational software games are designed to be targeted to real classroom students there are many questions to be answered concerning the scope of motivation and attractiveness of these games. This paper investigates the extent to which learning can be combined with pleasure and vice versa so that the end users of an educational virtual reality game may gain an enjoyable learning experience. For this reason the likeability of an educational virtual reality game interface has been evaluated both in the school environment and in the home environment of student-users.

1 Introduction

There is a fast growing industry of software games that are targeted to children and adolescents who are fascinated by these games. Such games are mainly created for pleasure. However, the attractiveness of software games has often been considered as a promising means for the creation of attractive educational software. Indeed, there are many researchers and educators that advocate the use of software games for the purposes of education. Papert [7] notes that software games teach children that some forms of learning are fast-paced, immensely compelling and rewarding whereas by comparison school strikes many young people as slow and boring. As a result, many researchers have developed games for educational purposes (e.g. [5], [1], [3]).

However, there are criticisms about the quality of the existing educational games. For example, Brody [2] points out that the marriage of education and game-like entertainment has produced some not-very-educational games and some not very-entertaining learning activities. Indeed, educational software games aim at serving two distinct aims, which are often conflicting each other: education and entertainment. Both these aims have to be obtained to a satisfactory extent; otherwise the existence of such software becomes pointless. If an educational game is not entertaining then it is neither motivating nor attractive. On the contrary the game environment may become distractive and educationally less effective than other kind of educational software. Thus the likeability of these games has to be examined before one may say that such software can be targeted to large masses of students and be included in school classrooms.

On the other hand, the prevailing preoccupation of students with computers at home is playing games. In contrast, games are not played at school. Mumtaz [6] points out that this has created an enormous gap between home and school

M. Bubak et al. (Eds.): ICCS 2004, LNCS 3038, pp. 962–969, 2004.
© Springer-Verlag Berlin Heidelberg 2004

perceptions and use of computers by students and that further research needs to examine how to close this gap. Indeed, it would be useful if this gap was closed. Students would have a more positive attitude towards computers at all times both in the school environment and at home. Moreover, education would not look so hard or boring to them. In this respect, educational games can provide a means for closing this gap. However, in order to say that an educational software game serves this purpose well, the likeability of the educational game has to be examined both in school and leisure time conditions.

In view of the above, we have examined the scope of motivation and attractiveness of a virtual reality educational game both in classroom and leisure time conditions. The game is called VR-ENGAGE [8] and teaches students geography.

2   Description of the Educational Game and Its VR Interface

VR-ENGAGE is an educational virtual reality game. The environment of VR-ENGAGE is similar to that of the popular game called "DOOM" [4], which has many virtual theme worlds with castles and dragons that the player has to navigate through and achieve the goal of reaching the exit. Similarly with DOOM, VR-ENGAGE has also many virtual worlds where the student has to navigate through. There are mediaeval castles in foreign lands, castles under the water, corridors and passages through the fire, temples hiding secrets, dungeons and dragons. The main similarity of VR-ENGAGE with computer games like DOOM lies in their use of a 3D-engine.

The story of VR-ENGAGE incorporates a lot of elements from adventure games. The ultimate goal of a player is to navigate through a virtual world and find the book of wisdom, which is hidden. To achieve the ultimate goal, the player has to be able to go through all the passages of the virtual world that are guarded by dragons and to obtain a score of points, which is higher than a predefined threshold. The total score is the sum of the points that the player has obtained by answering questions. In particular, while the player is navigating through the virtual world, s/he finds closed doors, which are guarded by dragons as illustrated in the example of Figure 1. A guard dragon poses a question to the player from the domain of geography. If players give a correct answer then they receive full points for this question and the dragon allows them to continue their way through the door, which leads them closer to the "book of wisdom".

As part of the adventure of the game the player may come across certain objects or animated agents. These objects or animated agents appear at random and give hints to students or guide them to tutoring places respectively. In tutoring places, students are encouraged to read a new part of the domain being taught.

The user interface of VR-ENGAGE involves the navigation of the player through the virtual worlds using the mouse and the keyboard. In case players are lost in the virtual world they may use a map, which is provided on line.

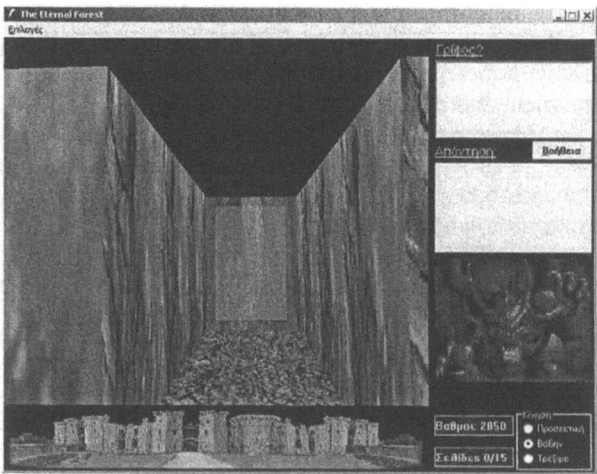

Fig. 1. Closed door in a virtual woods world

3 Aims, Settings, and Parts of the Empirical Study

For the purposes of finding out the extent to which students would like the educational game, we conducted an empirical study which consisted of two parts and involved students from schools. Both parts involved the same students. The first part was conducted in the environment of school classrooms whereas in the second part students had to use software in the environment of their own homes at leisure time.

More specifically, the empirical study involved in total 50 school children of 11-12 years old from 5 geography classes. The students that participated were selected from each geography class on the basis of their game-playing expertise. For this purpose, all the students of the five classes were interviewed concerning their experience in game playing. For example, they were asked to estimate how long they spend weekly on playing virtual reality computer games. Moreover, they were asked how long they have been familiar with such games, and they were requested to navigate for 5 minutes in the virtual worlds of the commercial game we provided them. As a result, the students were divided into three groups: experienced, intermediate and novice players. Then, some students from each group were selected at random to participate so that there was sufficient representation of all three categories. Indeed, among the participants there were 15 novice, 20 intermediate and 15 expert game players. The novice players were given a short training before they used the game on their own.

School children usually have a preconception of educational means as being totally different from entertainment. In this respect, the first part of the empirical study aimed at finding out how school children would react to an educational game in the settings of a real classroom where an entertaining aspect of education is rather unexpected. Therefore the first experiment took place in school-classrooms. The experiment aimed at estimating the likeability of the VR-ENGAGE interface in comparison to the likeability of educational software that had a simple user interface with no virtual reality and no game at all. This program was a simple window application with forms, menus, buttons, and edits which gave the student the opportunity to read questions and answer them. Also it provided help about the theory examined. The comparison was made in terms of the attractiveness of each of these applications for their end-users.

In particular, all students were asked to play the game at school for about 2 hours. Then, after the 2 hours that all students had used VR-ENGAGE, they were asked to use for another 2 hours an educational software application that had a simple user interface with no virtual reality and no game at all. Then, all the students were given the choice to use for 1 hour either of the two applications (VR-ENGAGE or the simple UI application) to repeat the lessons. We measured the time that the students had spent using each of the applications. Students who had spent more time with VR-ENGAGE were considered to have preferred it over the other kind of educational application. At a first glance, one might consider it obvious that school children would prefer VR-ENGAGE since it would be more amusing. However, in reality it could not be foreseen whether the students who were novice game-players would find playing the game amusing and whether the expert game-players would find VR-ENGAGE interesting enough to bother to play. Finally, students might find the game distractive and thus might prefer the other kind of software for their learning purposes.

The second part of the empirical study aimed at finding out the extent to which VR-ENGAGE could be used by children and adolescents at their leisure time. The underlying rationale of this part of the empirical study was to find out whether VR-ENGAGE could replace other computer games, which did not have any educational value, in the children's preferences for their entertainment. In this way, the children's game culture could be enriched with educationally beneficial games. Moreover, the educational game could be used both at work time and leisure time and thus would have a greater educational impact on children.

In view of the above aims, the likeability of the educational software game interface was compared to the likeability of a popular commercial game that had no educational content. Thus, after the 3 hours that students had spent with VR-ENGAGE in the classroom (2 hours of compulsory use and 1 hour where they had free choice) the same students were given the opportunity to play VR-ENGAGE at home during the weekend. They were also given a popular commercial game to play with. Students were encouraged to play both of these games for as much as they liked but it was also made clear that they did not have to play if they did not want to. After the weekend they were interviewed about how they liked VR-ENGAGE and were asked to state how many hours they had spent for each game.

For each phase of the empirical study we performed 4 t-tests to compare the time spent on VR-ENGAGE and on the respective application that VR-ENGAGE was compared to by: 1) Novice game-players (students), 2) Intermediate game-players (students), 3) Expert game-players (students) and 4) All the students.

4   Results of the Comparative Study in Classrooms

In the first phase of the empirical study that was conducted in classrooms we performed 4 t-tests to compare the time spent on VR-ENGAGE and on the application with the simple user interface.

The null hypothesis, H_0 was that there was no difference between the time spent for each of the two applications. The research hypothesis, H_1 was that there was a difference between the time spent for each of the two applications. A summary of the results of the 4 t-tests for correlated samples in classroom comparing the use of VR-ENGAGE and the application with the simple user interface is illustrated in Table 1 and Table 2. Table 1 illustrates the total minutes spent on each application by each group. Table 2 illustrates the t-values found and the critical values for each t-test. The t-value of each t-test is calculated by performing a t-test for correlated samples for the time spent on each of the applications [9]. The critical value for each t-test is the value taken from Table T for a one-tailed research hypothesis depending on the sample number.

Table 1. Time spent by each group on VR-ENGAGE and on the educational application with the simple UI

	Minutes spent on VR-ENGAGE	Minutes spent on SIMPLE UI	Sum of Differences
Novice	532	333	199
Intermediate	741	387	354
Expert	664	168	496
ALL	1937	888	1049

Table 2. T-test results for each group

	T Value	Critical Value
t-test for novice	1.99	1.76
t-test for intermediate	2.71	1.73
t-test for expert	4.26	1.76
t-test for all students	5.1	1.68

The results of the t-test for the novice game-players have shown that t, 1.99 is greater than the critical value 1.76. Thus, we can reject H_0 and accept H_1. Therefore novice students prefer to play VR-ENGAGE than to use the simple UI educational application for leisure time at school or to repeat lessons. However, we can see that the difference is not great. This is expected to some extent because novice users have more operating difficulties with a virtual reality system than experienced users and thus they may be put off from the use of the game for this reason. In the case of intermediate game-players t is 2.71, which is much greater than the critical value 1.73. Thus again we can reject H_0 and accept H_1. Therefore intermediate students strongly prefer to play VR-ENGAGE than use the simple UI application. In the case of expert game players t is 4.76 which is extremely greater than the critical value 1.76. Therefore experienced students almost all the time use VR-ENGAGE. Indeed, the fact that they are expert game players means that they are used to the culture of games and they like it a lot. These users were very pleased to see the games introduced in the classroom. The t-test for the total of game-players (students) showed that the t value is 5.10, which is extremely greater than the critical value 1.68. Therefore students prefer to play VR-ENGAGE than use the simple UI educational application.

5   Results of the Comparative Study During Users' Leisure Time

The second phase of the empirical study was based on students' interviews after they had used VR-ENGAGE and a commercial game at their homes. The students' answers to the questionnaires for the interviews were used for t-tests that aimed at showing what the comparison results were. However, the students' answers were also analysed to reveal how students had liked or disliked VR-ENGAGE irrespective of its comparison with a commercial game.

Similarly with the first phase, in the second phase, which was based on the students' likeness of the software when it was played at home, we performed 4 t-tests to compare the hours spent for each kind of software (VR-ENGAGE and a popular commercial game).

The null hypothesis, H_0 was that there was no difference between the time spent for each of the two applications. The research hypothesis, H_1 was that there was a difference between the time spent for each of the two applications. A summary of the results of the 4 t-tests for correlated samples between use of a Commercial Game or VR-Engage in leisure time at home is illustrated in Table 3 and Table 4. The t-value and the critical value of each t-test are calculated similarly to section 4 [9].

Table 3. Time spent by each group on VR-ENGAGE and the commercial game

	Hours spent on Commercial	Hours spent on VR-ENGAGE	Sum of Differences
Novice	32	28	4
Intermediate	68	41	27
Experienced	75	27	48
ALL	175	96	79

Table 4. T-test results for each group

	T Value	Critical Value
t-test for novice	0.56	1.76
t-test for intermediate	2.08	1.73
t-test for expert	4.05	1.76
t-test for all students	3.95	1.68

The results of the t-test for the novice game-players have shown that t, 0.56 is much smaller than the critical value 1.76 we can accept H_0. Therefore novice students do not prefer to play a commercial game more than VR-Engage for leisure time at home. This can be explained because novice users have operating difficulties with virtual reality systems. Thus, they do not enjoy a commercial game more than VR-ENGAGE because the commercial game has a more sophisticated virtual environment, which is more difficult for them to handle. Also the time spent by novice users was the smallest as compared to the other two groups of users as a consequence of these difficulties. The t-test for the intermediate student "game players" has shown that t, 2.08 is greater than the critical value 1.73; thus H_0 is rejected and H_1 is accepted. Therefore intermediate students prefer to play a commercial game more than VR-

Engage at home. The t-test for the experienced student "game players" has shown that the t value of 4.05 is extremely greater than the critical value 1.76. Thus we can reject H_0 and accept H_1. Therefore experienced students almost all the time played commercial games for leisure time at home. We can understand this because commercial games have a more sophisticated gaming environment than VR-ENGAGE and thus seem more amusing to experienced users who seek adventure and amusement. The t-test for all students resulted in a t-value of 3.95, which is extremely greater than the critical value 1.68. Thus, we can reject H_0 and accept H_1. Therefore students prefer to play a commercial game more than VR-Engage for leisure time at home.

The above t-tests' results were expected to some extent. Indeed, it was almost certain that the commercial game would be more appealing than VR-ENGAGE. This is so because commercial games have very sophisticated virtual environments and can be more challenging in terms of adventure since they do not have to care about educational content. However, we wanted to find out whether our VR-ENGAGE application was anywhere close to the commercial for each of the 4 categories. In this respect VR-ENGAGE needs a lot of improvement so that it may be more competitive to commercial games.

The results from this experiment were quite different from the first one. Since children were not given the game to work with it as an assignment, they considered it merely as a game similar to the commercial games they were familiar with. Therefore their judgment on it focused on the game environment. The students' interviews revealed many interesting comments about what they expected and what they would like. Most of the students (62%) pointed out that the game would be better if it had more virtual objects, more background sounds and more adventure. These comments came to a large extent from experienced game players rather than novice ones. This was due to the fact that most of them were familiar with commercial virtual reality games therefore they compared VR-ENGAGE with them and had higher expectations in this aspect. Some of the students (8%) criticised it for being non-violent. Again, this was probably due to the fact that the culture of commercial VR-games has penetrated the world of adolescents and children in a way that they expect all games to be similar even if this is not good for them. However, although most children had given comments for the enhancement of the entertaining aspect of the game, a very large percentage of them (84%) said that they would like to have this game at their homes and play it at their leisure time together with other computer games they had. This was a very encouraging result.

6 Conclusions

This paper has described and discussed the evaluation of an educational virtual reality game for geography, VR-ENGAGE in terms of the attractiveness and motivation scope that it may have on students. The results from the evaluation showed that students in classrooms would be quite happy to work with a computer game, which represents a more amusing teaching fashion than that of conventional educational software. On the other hand, during their leisure time students would prefer to play a popular commercial game instead of VR-ENGAGE. It was shown that the game environment of the educational game has to be very competitive with commercial games to make the most of its motivation and engagement effects on students. This is so because children are quite familiar with commercial games and therefore they have high expectations from the game environment.

The results from the evaluation have provided some important guidelines for the improvement of educational games. First, the virtual reality environment of VR-ENGAGE has to be enhanced, so that the game is more competitive with other commercial games. Moreover, there has to be more automatic help for the game, so that novice players can play more easily and thus have a more enjoyable experience.

References

1. Amory, A., Naicker, K., Vincent, J. & Claudia, A. (1998). "Computer Games as a Learning Resource." *Proceedings of ED-MEDIA, ED-TELECOM 98*, World Conference on Education Multimedia and Educational Telecommunications, 1, pp. 50-55.
2. Brody, H. (1993) "Video Games that Teach?". In *Technology Review*, November/December 1993, pp. 51-57.
3. Conati, C. & Zhou, X. (2002) " Modeling students' emotions from cognitive appraisal in educational games". In S. A. Cerri, G. Gouarderes and F. Paraguacu (Eds.) : *Intelligent Tutoring Systems 2002*, LNCS, 2363, pp. 944-954, Springer-Verlag Berlin Heidelberg 2002.
4. ID-software 1993.
5. Inkpen, K., Upitis, R., Klawe, M., Lawry, J., Anderson, A., Mutindi, N., Sedighian, K., Leroux, S. & Hsu, D. (1994). "We Have Never-Forgetful Flowers In Our Garden: Girl's Responses to Electronic Games". *Journal of Computers in Math and Science Teaching*, 13(4), pp. 383-403.
6. Mumtaz, S. (2001) " Children's enjoyment and perception of computer use in the home and the school" In *Computers & Education*, 36 (2001), pp. 347-362.
7. Papert, S. (1993). "The Children's Machine: Rethinking School in the Age of the Computers". Basic Books, New York, 1993.
8. Virvou, M., Manos, C., Katsionis, G. & Tourtoglou, K.(2002): "VR-ENGAGE: A Virtual Reality Educational Game that Incorporates Intelligence" In *Proceedings of IEEE International Conference on Advanced Learning Technologies (2002)*, Kazan, Russia, September 16-19, 2002.
9. Voelker, D.: Statistics, Wiley Publishing, Inc. New York 2001.

A Client-Server Engine for Parallel Computation of High-Resolution Planes

D.P. Gavidia*, E.V. Zudilova, and P.M.A. Sloot

Section Computational Science
Faculty of Science, University of Amsterdam
Kruislaan 403, 1098 SJ Amsterdam, the Netherlands
daniela@cs.vu.nl, {elenaz,sloot}@science.uva.nl

Abstract. The paper describes a visualization mechanism that permits fast extraction of high-resolution slices of interest from large biomedical datasets stored remotely. To provide the user with real-time interaction, a sub-sampling mechanism of the 3D images has been developed. The 3D sub-sampled view is detailed enough to provide a context and allow a user to orient him/herself. The result of clipping is a high-resolution plane, which can be explored independently from the 3D image. The clipping engine is built as a client-server application, where the computation of high-resolution planes is executed on a server-site in parallel. This approach results in remarkably good performance.

Keywords: Interactive visualization, clipping, data parallelism, client-server, VTK

1 Introduction

Biomedical data is data collected by applying imaging techniques [2] to an object of biological or medical origin. Information sources may be widely distributed, and the volume of datasets to be processed is usually extremely big. As a result, biomedical imaging requires more and more computational resources. Processing, visualization and integration of information from various sources play an increasingly important role in modern biomedicine [9].

Current 3D imaging systems have the capability to produce overwhelming amounts of data converted into visual representations in order to be analyzed. Specialists in medicine and biology rely on these visualizations to achieve a better understanding of the structures and relationships presented in their subjects of study [2].

The improvements in imaging techniques have led to the development of higher resolution devices resulting in ever increasing amounts of data being produced. The problem, though, is not the lack of techniques for extracting information from 3D datasets and visualizing them accordingly, but the shift that is necessary between the traditional practice in biomedicine of visualizing slides in two dimensions and doing so in 3D [7]. Even though 3D data is available, the traditional practice of analyzing

---

* The author is now a PhD-student at the Department of Computer Science, Vrije Universiteit (Amsterdam, the Netherlands).

M. Bubak et al. (Eds.): ICCS 2004, LNCS 3038, pp. 970–977, 2004.
© Springer-Verlag Berlin Heidelberg 2004

the information using 2D slices is still predominant. In order for such shift to occur, applications that use 3D data need to present it in a way that makes exploration and analyses intuitive.

Even though clipping techniques have been known for at least 15 years, they have been usually considered as an additional feature provided by a visualization package or a toolkit. However, clipping can also be an interactive visualization technique that permits to bring to the forefront the information that might be obscured in the density of a dataset, allowing a user to explore data intuitively. More importantly, clipping can help, for instance, in reducing the need for exploratory surgery as well as for dissection, by presenting both the 3D reconstruction of an object and a set of 2D slices, similar to normal scans that physicians usually use in their everyday practice. In this respect the 3D representation helps physicians to orient themselves within an object's geometry and 2D allows them to have a closer look into a familiar setting. The clearness and accuracy of the representation of a slice of interest for most real cases is more important than for the 3D representation of an object as a whole.

The research presented in this paper is focused on the development of a new visualization mechanism for the fast extraction of high-resolution clipping planes from a large datasets stored remotely, for example, on the GRID [9]. Section 2 describes the main design concept on which the clipping engine is based. The technical aspects of the implementation are given in section 3. Section 4 focuses on the parallel computations needed for the calculation of a high resolution plane done on the server site. The results showing the benefits of parallel processing are presented here as well. Conclusions and final discussions can be found in Section 5.

2 Design Concept

3D biomedical data is usually acquired by means of such imaging techniques as Computed Tomography (CT), Magnetic Resonance Imaging (MRI) or Positron Emission Tomography (PET). The choice of the imaging technique is determined by the structure or anomaly that needs to be observed, given that some techniques are better suited for certain cases than others.

Data produced by imaging techniques can vary from relatively small (e.g., 64 x 64 x 32 voxels) to extremely large (e.g., 2048 x 2048 x 1024 voxels). In the case of large datasets, when heavy processing is involved, the interactivity is usually hindered by the time required both for the data transfer from the remote repository and visualizing it on the user site.

One possible solution to reduce this delay is data compression, which provides good performance results (see for instance [3]) and is widely used. Other works (e.g. [1]) report on methods like sub-sampling for finding a balance between interactive performance and accuracy of the data presented in 3D, where interaction is possible with an acceptable delay and the result presented to a user is detailed enough to allow a thorough understanding of the data analyzed.

The potential users of a clipping engine are physicians who are used to work with 2D scanned images. 3D representation hardly plays a role for them, while 2D representation of a slice of interest is of the utmost importance. That is why it has been decided to use sub-sampling to provide a user with the real-time interaction capabilities.

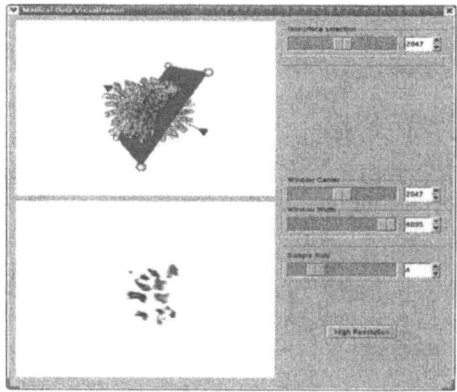

Fig. 1. Clipping engine: GUI screenshot

The goal of this project is to allow for visualization of data resulting from biomedical imaging by interacting with a 3D representation of a dataset. This is achieved through the use of iso-surfaces as a means of giving a user an idea of the content of the dataset and clipping planes that allow a more in-depth analysis of the characteristics of the data observed. Surface rendering has been selected for the visualization of a static object in 3D because it gives similar possibilities for data exploration as volume rendering but is less computational and time consuming.

The GUI of the application contains two viewers (see Fig. 1): one presents the 3D object and another one shows a high-resolution slice of interest, which has been generated as a result of the intersection of a 3D object with a clipping plane built by a user.

3   Communication Considerations

3.1   The Client-Server Model of the Application

Imaging devices can produce enormous amounts of data, which are frequently stored remotely in high-capacity storage facilities. The size of these data files makes it inconvenient or impossible to have them stored locally. This leads to the natural choice for building a clipping engine as a client-server application. Additionally, there are two other reasons to implement the clipping engine using a client-server model:

1. The computation of a high-resolution plane is expensive due to the need to find the points that make up the plane through interpolation. This can be quite time-consuming. Assuming that better computational resources are available at the server, doing this operation at the server-site might yield better performance.
2. If high-performance computational resources are available at the server (e.g., a parallel super computer), performance can be further improved by conducting the computations of a high-resolution plane in parallel on the server site.

The clipping engine functions as follows. A sub-sampled dataset is transferred to the client-site and used for further interactions. A user can manipulate the 3D representation of the data and choose the clipping plane in real-time. Once a selection has been made, the client communicates with the server to obtain the desired high-resolution slice of interest.

The client-server implementation relieves the client from having to store enormous data files and having a fast processor to compute planes. His/her main concern would then be to have a fast enough graphics workstation for visualization and rendering.

The flow of data between client and server occurs in an asynchronous manner, with the server sending subsets of the scanned data by request through Remote Method Invocations (RMI). According to the flow direction, the data can be defined as follows. *Control data* is the data sent from the client to the server, comprised of the requests for data made by the client, such as the definition of the desired clipping plane or the sample rate for a sub-sampled dataset. *Computed data* is the data sent from the server to the client. In general, it is a subset of the scanned data sent to the client by request.

Control data is limited to a few bytes since it is comprised of requests for data. On the other hand, computed data, being a subset of the scanned data, results in a significantly larger amount of information being sent.

3.2 Implementation

The system runs on Linux and Unix computers. For the purpose of visualization and image processing, the Visualization Toolkit (VTK) [8] has been chosen. Besides visualization algorithms provided by the VTK, support for parallelization and running remote processes has played an important part in the development of the application. The GUI was built using the QT GUI application framework. VTK is cleanly integrated into the QT interface through the use of the VtkQt library [10]. VtkQt is an extension of VTK developed at the Section Computational Science of the University of Amsterdam. It allows for the insertion of VTK windows in the QT interface and transparent use of VTK classes alongside Qt, simplifying the development.

The CT scans of the coral "Madracis mirabilis" used for experiments have been made in a joint project on modeling and analysis of growth and form of corals in collaboration with the Netherlands Institute for Sea Research and Cooperative Institute for Marine and Atmospheric Studies, USA [5].

3.3 Operations and Communication

Figure 2 gives a general overview of the way the client-server clipping engine operates. The communication between client and server is triggered by two events: a request for a sub-sampled dataset and a request for a clipping plane. To obtain a sub-sampled data set, a sample rate value is sent to the server, triggering the execution of a sampling routine and generation of the sub-sampled dataset. In a similar manner, the client can request a clipping plane. In this case, the plane definition - a point in the plane and the normal of the plane - is sent to the server, triggering a routine for extracting the plane. The result is then made available to the client.

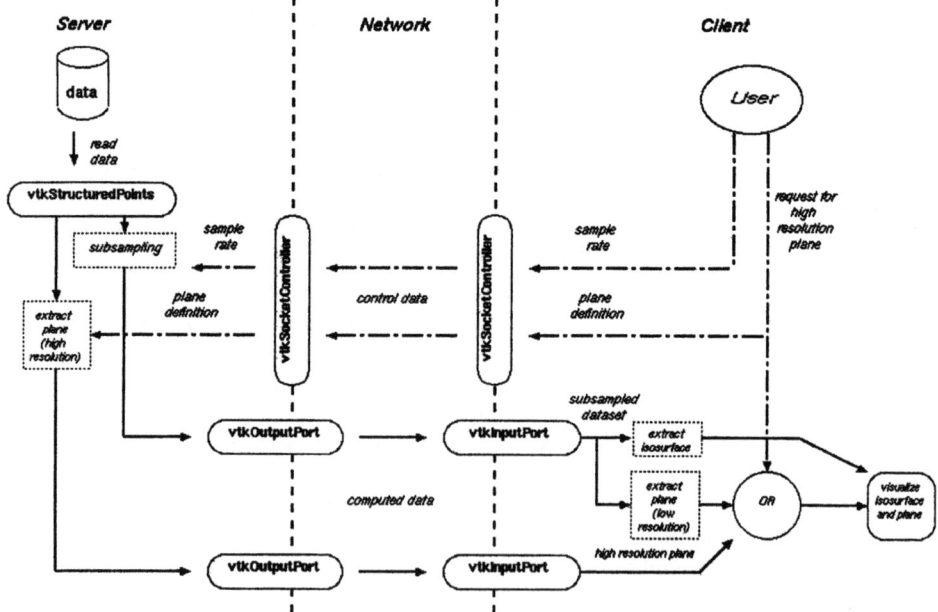

Fig. 2. Pipeline of the operation in the client-server application. The communication between client and server is handled by *vtkSocketController*, *vtkOutputPort* and *vtkInputPort*

The communication between client and server is handled by three VTK classes: *vtkSocketController*, *vtkOutputPort* and *vtkInputPort*, as illustrated in Fig. 2. The control data is sent from the client to the server through the *vtkSocketController*. After the necessary computations are performed on the server-site, the resulting datasets (computed data) are sent back to the client through the *vtkOutputPort* and *vtkInputPort* objects. These two classes allow for the construction of a pipeline that starts on the server-site and continues on the client-site.

Several RMI methods are written on the server and registered with the *vtkSocketController*. The *vtkSocketController* on the client connects to the one on the server using the name of the server and a designated port number. Once the connection is made, the client application triggers the RMIs to communicate changes in the parameters controlled by a user, namely sample rate for the dataset and definition of the clipping plane. The server then modifies these parameters in the server-site pipeline and makes the results available through the *vtkOutputPort* objects.

With the sub-sampled dataset, the client can generate iso-surfaces in accordance with the parameters set interactively by a user through the GUI. Iso-surfaces are rendered in a 3D render window along with the low-resolution clipping plane. A high-resolution slice of interest is generated only after a corresponding request is sent to the server. Then a high-resolution 2D image is calculated in a parallel on the server and rendered on the client-site.

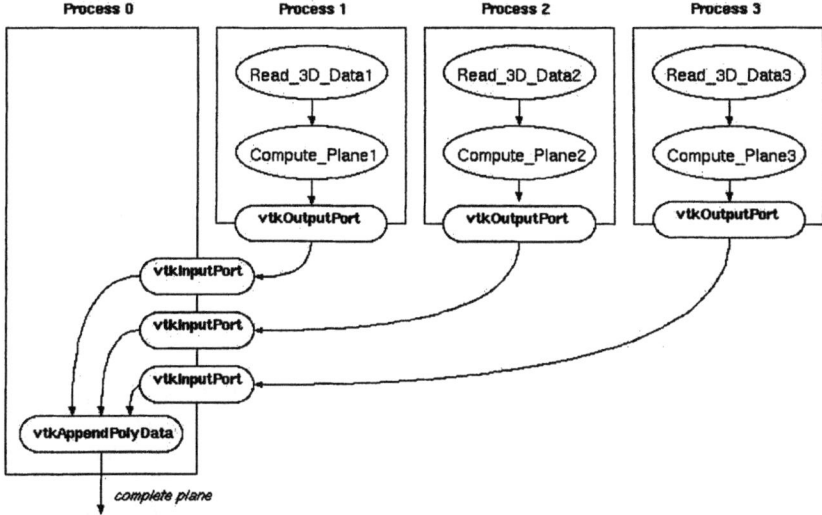

Fig. 3. Data parallelism. Each process, except one reserved for appending the results, computes part of a high-resolution plane. In the figure, the computation is done using 3 processes. Process 0 appends the results from all the processes together.

4 Parallel Data Processing

For the extraction of the arbitrary plane from the dataset, the class *vtkCutter* has been used. This class is a filter that cuts through a dataset. It creates a polygonal surface corresponding to the implicit function $F(x, y, z) = value(s)$. In this case, the implicit function is a plane function with the desired position and orientation. Cutting with an implicit function is time-consuming, which hampers interactive performance. Taking advantage of VTK's support for parallel computing, a parallel implementation of the plane computation is in order to speed up the process. VTK supports task, pipeline and data parallelism. Data parallelism is useful for processing extremely large datasets, not unlike the ones produced by medical [1].

In order to reduce the time required to obtain the high-resolution plane after its request by the client, the calculation of the plane has been done in parallel. The implementation uses multiple processors that communicate through MPI (see Fig. 3). Each process works on a different piece of the original dataset producing a section of a high-resolution plane. These results are appended together by process 0 using a *vtkAppendPolyData* filter, which has as inputs the sections of the plane calculated by each of the other parallel processes. The output is a single high-resolution plane.

VTK handles data parallelism through parallel streaming, where the data is partitioned into independent subsets using VTK's streaming data model [6]. The *vtkAppendPolyData* filter implements this functionality. Coding is simplified considerably since it is not needed to explicitly assign a different subset of data for each process to work on. Instead, in parallel streaming mode the *vtkAppendPolyData* filter asks for a different piece of the plane polydata from each of its inputs (the outputs of the other processes). When the update request is propagated up the pipeline

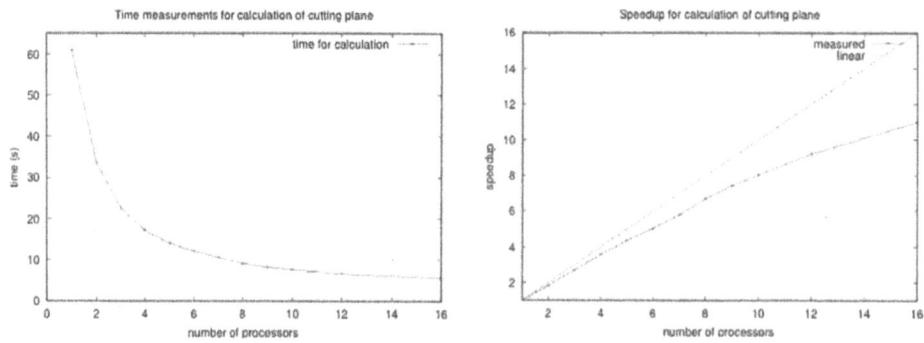

Fig. 4. Time measurements (left) and speedups (right) for the calculation of a high-resolution plane of dimensions 512x512 with 16 bits per sample when done in parallel. Process 0 was not taken into account when considering the number of processes running in parallel.

of each process, the request for a subset of polydata is translated into a request for an independent subset of the input dataset (i.e. the volumetric data).

Fig. 4 shows the time required to calculate the plane in parallel using an increasing number of parallel processes and the associated speedup, demonstrating significant improvements in performance. The parallel implementation of the high-resolution plane computation was tested on the Distributed ASCI Supercomputer 2 - DAS-2 [4].

5 Conclusions and Future Work

Interactivity and accuracy are main concerns for the exploration of the biomedical data. Sub-sampling has been used to make interactivity possible in real-time, but the 2D slices extracted from a sub-sampled dataset might not be detailed enough. It is desirable to use the totality of the data available for the extracted slice. However, the computation of a high-resolution plane can be very time-consuming when done on a single processor, crippling the performance. The parallel computations on the server-site ease this problem and speedup the process significantly.

In respect to the graphics hardware available on the client and the complexity of the dataset, a higher or lower sample rate would be required for the interaction in real-time. Currently, the determination of an appropriate sample rate is given to a user. Though finding the optimal sample rate falls beyond the scope of this project, implementing an intelligent way of doing this would certainly prove beneficial. Moreover, in cases when the client is able to handle the entire dataset without a noticeable effect towards the interaction capabilities, it would be reasonable to avoid sub-sampling and to work with the entire dataset on the client.

Acknowledgements. The authors would like to thank Dr. Jaap Kaandorp and Dr. Robert Belleman for their input to the project and Denis Shamonin for his technical assistance.

References

1. Ahrens J., Law C., Schroeder W., Martin K., Papka M.: A parallel approach for efficiently visualizing extremely large time-varying datasets. Tech. Rep. LAUR-00-1620, Los Alamos National Laboratory, 2000.
2. Belleman R.G. and Sloot P.M.A.: Simulated vascular reconstruction in a virtual operating theatre. Proceedings of the CARS2001 Conference, Elsevier Science B.V., Berlin, Germany, June 2001, pp. 938-944.
3. Belleman R.G., Shulakov R.M.: High performance distributed simulation for interactive simulated vascular reconstruction. Proceedings of The 2002 International Conference on Computational Science (ICCS 2002), Amsterdam, the Netherlands, April 2002. Series Lecture Notes in Computer Science (LNCS) volume 2331, ISBN 3-540-43594-8, Springer-Verlag.
4. Distributed ASCI Supercomputer 2 home page: http://www.cs.vu.nl/das2/.
5. Kaandorp J.A., Koopman E.A., Sloot P.M.A, Bak R.P.M., Vermeij M.J.A. and Lampmann L.E.H.: Simulation and analysis of flow patterns around the scleractinian coral Madracis mirabilis (Duchassaing and Michelotti). Phil. Trans. R. Soc. Lond. B DOI:10.1098/rstb.2003.1339, 2003.
6. Law C.C., Schroeder W.J., Martin K.M., Temkin J.: A multi-threaded streaming pipeline architecture for large structured datasets. Proceedings of Visualization'99 (October 1999), IEEE Computer Society Press.
7. Robb R.A.: Handbook of Medical Imaging: Processing and Analysis. Isaak N.Bankman, Academic Press, 2000, Ch. Three-Dimensional Visualization in Medicine and Biology, pp.685-712.
8. Schroeder W., Martin K., Lorensen L: The Visualization Toolkit: An Object-Oriented Approach to 3D Graphics. 2nd Edition. Prentice-Hall, Inc., 1997.
9. Sloot P.M.A., Albada van G.D., Zudilova E.V., Heinzlreiter P., Kranzlmüller D., Rosmanith H., Volkert J.: Grid-based Interactive Visualisation of Medical Images, Proc. of the 1st European HealthGrid Conference, Lyon, France, January 2003, pp.57-67.
10. VtkQt web-page: http://carol.wins.uva.nl/dshamoni/VtkQt

A Framework for 3D Polysensometric Comparative Visualization

Javed I. Khan, Xuebin Xu, and Yongbin Ma

Media Communications and Perceptual Engineering Research Laboratory
Department of Math & Computer Science
Kent State University
233 MSB, Kent, OH 44242
{javed,xxu1,yma}@kent.edu

Abstract. Typically any single sensor instrument suffers from physical/observation constraints. This paper discusses a generalized framework, called polysensometric visual information fusion framework (PVIF) that enables information from multiple sensors to be fused and compared to gain broader understanding of a target of observation. An automate software shell supporting comparative cognition has been developed to form 3D models based on the datasets from different sensors. This fusion framework not only provides an informatic engineering tool to overcome the constraints of individual sensor's observation scope but also provides a means where theoretical understanding surrounding a complex target can be mutually validated and incrementally enhanced by comparative cognition about the object of interest.

1 Introduction

3D visualization is becoming widespread and is being used in many fields. However, a relatively new area is comparative visualization. The methodology of correlating information from different experimental techniques to find the 3D structures improves scientists' understanding about the objects they work on. This has been recently demonstrated in a growth in interest in combining *CT* (computerized tomography) and *MRI* (magnetic resonance imaging) information [1,2]. Comparative cognition is an important and strong means to obtain new knowledge about any complex system. Multiple sets of data are collected, processed and compared in the process. The proper application of these data sets is however often quite a complex process and depends on the nature of the observation target, the field of research and the purpose of the research is, etc. The most straightforward usage is to combine the data sets to obtain a composite view about the interest object when these multiple sets of data are complementary to one another. Current research in the area mainly has produced very domain specific systems which fuses data from very specific modes. In our project we investigate and propose how a generalized fusion framework can be constructed. We propose an algorithmic model of multi-sensor information fusion that closely correspond the cognitive process of comparison based knowledge exploration and hypothesis refinement. As a proof of concept, along with it we have also developed a new automated system. It accepts sets of information from multiple techniques and

M. Bubak et al. (Eds.): ICCS 2004, LNCS 3038, pp. 978–985, 2004.
© Springer-Verlag Berlin Heidelberg 2004

combines them for 3D visualization. In the system, all sensor data are mapped into a unified information domain. The system can be easily extended to any new sensor instrument. Naturally the user knows the physical-chemical meaning of data. Thus s/he can insert domain specific intermediate data processing elements and yet reuse more generic or available information processing tools from library in a plug and play fashion. The actual framework works as an exploration shell for its user which allows experimentation. The paper briefly presents this new framework.

The following section first explains polysensometric visual information fusion framework. Section 3 and 4 then describe the exploration algorithms for inferring a 3D models and the architecture of the exploration shell. Finally section 5 presents a working example of polysensometric visualization which offers fusion of XPS *X-ray Photoelectron Spectroscopy* (XPS) and 3D *Laser Scanning Confocal Microscopy* (LSCM) data sets in the comparative study of liquid crystal film.

2 Polysensometric Visual Information Fusion (PVIF) Framework

The basic idea is illustrated in figure 1, in which we take a real physical-chemical experiment as an example. The proposed polysensometric visual information fusion model generates estimation of 3D structure of objects, based on multiple techniques and allows their systematic comparison and fusion. In this example, the data, collected from XPS and LSCM, might be modified by domain specific filters, as well as geometric transformations (such as rotation, cropping, zooming, tilting and data mapping, etc.). After being filtered, their corresponding 3D models are generated. Next, it allows these 3D models to be visualized as a composite model. It can further generate a fuller model by volume-filling algorithm set and measure the differences by using 3D comparison algorithm set. This process is extensible to additional sensor modalities. Overall the system allows domain specific as well as generic geometric algorithms, filters, and transformation algorithms to be easily infused in the processing pathway leading to the comparative 3D model.

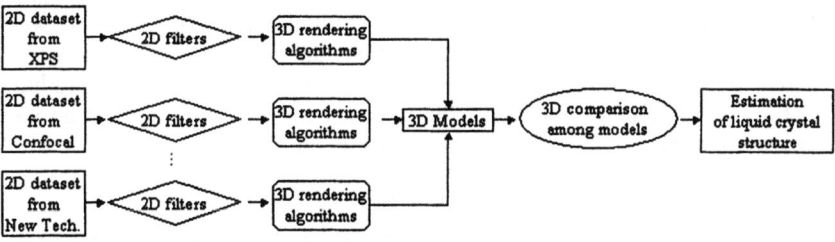

Fig. 1. Polysensometric visual information fusion model. In this figure, we derive the 3D models from sets of information, which generated by different experimental techniques, and compare these models.

In order to explain our information framework, we only use two abstract techniques A and B. At the same time, we use the following concepts and corresponding notations to facilitate explaining our visualization strategy: (i) *I* - the *total information about the experimental object*; (ii) Estimation *I* - estimation of *I* from experimental data; (iii) I^A - the information generated by technique A; (iv) I^B - the information generated by technique B; (v) F - framework for the real structure of experimental object; (vi) F^A - framework for the structure "seen" through technique A; (vii) F^B - framework for the structure "seen" through technique B; (viii) F^R - framework for the registration between the intersection of frameworks A and B, i.e. registration framework; (ix) F^V - visualization framework for F^R; (x) Transformations $T[i]$ - any transformation i applied on information set; (xi) Error E - the difference between two information sets;

In reality, for a complex target of observation we are never able to get the real structure of the target, total I under framework F. But we can approximate it as close as possible to obtain estimation I'. So our objective is to obtain estimation I and visualize it under F^V, which is close enough to F. The relations among these concepts and notations are: the total information, I, is corresponding to the total framework, F; I^A is information about I under its framework F^A; I^B is information about I under its framework F^B; F^A and F^B are registered towards F^R based on the intersection part between them; I_A^R is part of information I^A after registration under F^R; I_B^R is part of information I^B after registration under F^R; we use the union of transformed I^A and I^B as the estimation of I, I; I is visualized under F^V. The inferring algorithm is showed in figure 2 and the specific details will be covered in the next part of this paper.

3 Algorithms

The data fusion is achieved through a set of processing: space projection, registration, volumetric filling and comparison, etc. This section interprets those algorithms used to obtain reasonable data fusion results.

3.1 Sensometric to Geometric Projection Algorithms

The original input datasets are collected from some sensors. These sensors probably have different space coordinate systems and resolutions. In these cases, we need algorithms to project the sensor domain sample datasets to a common geometric space. After the original datasets processed by these algorithms, they should have the same spatial orientation and resolution and be ready for the following operations. The frequently used transformations are including zoom and tilt etc.

$$I * T[0] * ... * T[n] -> I' \tag{1}$$

In this formula, I is input dataset, I' is output dataset. $T[i](0 <= i <= n)$ are n+1 transformations applied on input dataset I.

3.2 Registration Algorithms

After we project the original sensometric datasets to geometric space, we need the automated registration algorithms to register the voxel in one spatial datasets with the corresponding voxel in the other spatial datasets. In our research, registration algorithms are divided into two groups: 2D registration and 3D registration. We use mutual information [4,5] or FFT [6] to align our datasets in registration framework, in which the common part of two datasets is normally 2D. Also, we use 3D registration to fuse two 3D estimation models, I_A^V and I_B^V, in visualization framework. Some 3D transformations are applied on two models, and difference between them is calculated. Difference metrics are discussed later in comparison algorithm. Here least error method is used to decide which transformation will be used to get the final estimation of the real model:

$$I_A^V = f_i^A[I^A]; \tag{2}$$

$$I_B^V = f_i^B[I^B]; \tag{3}$$

$$E' = \text{Diff}(I_A^V, I_B^V); \tag{4}$$

In these formula, f_i means ith transformation on dataset A, Diff means difference metric calculation between two parameters.

3.3 Comparison Algorithms

Comparison algorithms are used for inferring the final 3D model. It checks if the loop ending conditions have been met. There are many 2D whole-image level comparison algorithms, such as mean-absolute-error (MAE), root-mean-square-error (RMSE), peak signal-to-noise ratio (SNR) etc. In this example system, we used comparison algorithms that are extension of those used in 2D. The greater MAE and RMSE, the more different these two 3D models are; SNR is on the contrary; the closer p is to 1, the more correlated are the two sample images.

3.4 3D Model Inferring Algorithm

We use 3D model inferring algorithm to infer the final 3D visualization model for the object of interest (see figure 2). The algorithm is composed of two loops. We explain the algorithm by using two techniques, A and B, as an example. First, the input datasets from sensors generate 3D models of their own by using hypotheses H_a and H_b respectively. Then, these two 3D models are projected onto registration framework, F^R. In registration framework, the intersection parts of F^A and F^B are registered by applying a series of transformations. Least errors are used to get the optimum registration position. After registered, I_A and I_B are compared to verify the hypotheses at the very beginning. Least error principle is also used as the optimum means to obtain the best theoretic 3D model about the object of interest. At last, we use the union of I_A and I_B from the hypotheses with least error as the estimation of total information I.

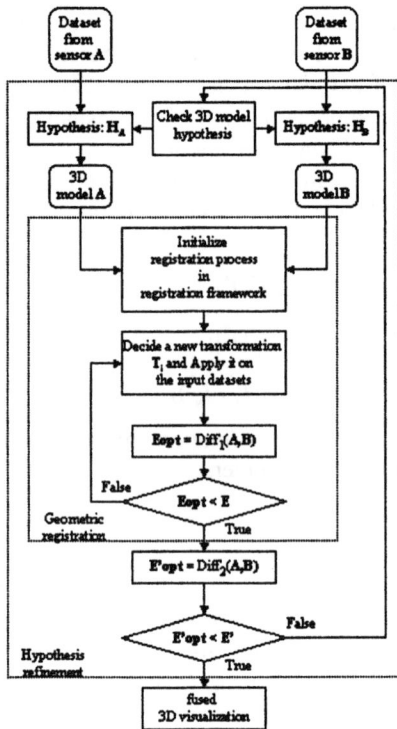

Fig. 2. "Pounding-and-inferring" algorithm for 3D model based on sets of information from two techniques A and B

4 System Architecture and Working Process

The architecture of our shell system is illustrated in figure 3. The shell is composed of such components as task controller, flowchart object store and some other functional parts, such as flowchart designer, data combiner, information verifier etc. Task controller works as a central control manager. It controls the state of our program and coordinates the work of the functional parts. Its work includes deletion of a flowchart object from the flowchart store, let the flowchart designer insert a flowchart object into the store according to the requirement of the user, injection of input data through the data combiner, declaration and definition of filters and algorithms through filter/algorithm combiner, verification of information in the flowchart object through verifer and driving the transformation engine to process the data following the controls in flowchart object and generating the final model. The flowchart object store is used to store all the information about data and operations what will be applied on the data. Information transformation engine drives the input data through the control of the flowchart and generate 3D results. Also, hypothesis refinement can be done by this component.

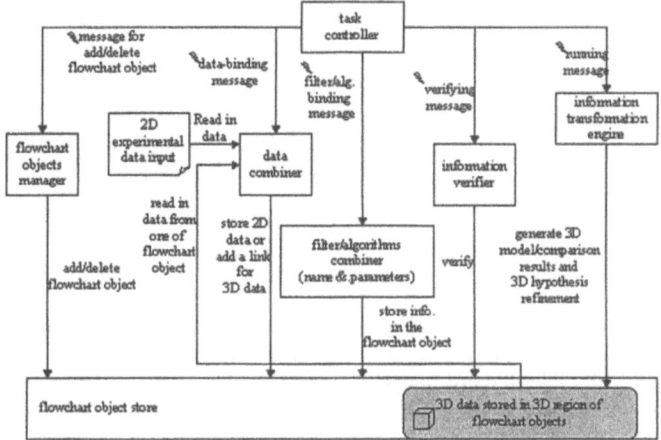

Fig. 3. System architecture

There are total 5 phases to run the whole process: design phase, name-binding phase, parameter-binding phase, verification phase and run phase. During *design phase*, a flowchart object can be designed by using flowchart designer. There are two kinds of flowchart object in the project: one is to take 2D experimental data as input and generate a 3D model; another is to take 3D models, which are generated by some flowchart object before, as input and generate the fusion 3D object or comparison between two 3D models. During *name Binding Phase*, we declare filters and algorithms. Another important thing in this phase is to bind experiment method and image data. During *parameter Binding Phase*, we set parameters for each filter and algorithm declared in the current flowchart object. During *verification Phase*, all those input information need to be verified before the data are to be transformed following the datapath defined by the flowchart object. During *run Phase*, a new 3D model will be created from 2D or 3D data. This model will be displayed in a separate window for further visualization. In this phase, user can also make parameters of filters and algorithms change on the fly in order to get a content result. These changes will be flush back to the flowchart aggregation component when the user pushes a button in the 3D model presentation window.

5   Case Study

We visualize the fused information about liquid crystal polymeric film, which was collected by using XPS and LSCM [3]. The polymeric film is 2 microns thick and composed of two non-equally distributed polymers: PVC and PMMA. In XPS, only top and bottom surface with 0.01 microns thick into film have contribution to top image and bottom image of 265 by 265 pixels respectively. The dimension of the shot area is 700 by 700 microns. We also know that the highest pixel intensity in the image corresponds to 35 percent PVC and the lowest pixel intensity is to 5 percent PVC. LSCM generates 32 images of 600 by 800 pixels in this example. The shot area is 300 by 400 microns. The first and last LSCM images are in the depth 0.2 microns to the

Fig. 4. Result images. *Left*: the stack of result images by combining XPS data and LSCM data in 3D presentation. We put the top image of XPS on the top of 1ˢᵗ image of LSCM and the bottom image of XPS under the last image of LSCM. *Middle*: the difference between theoretic model 1 and fusion model. MAE: 119.3, SNR: 129.8. *Right*: the difference between theoretic model 2 and fusion model. MAE = 87.3; SNR = 99.0

surface of the film. The highest pixel intensity in the fourth image corresponds to 70 percent PVC and the lowest pixel intensity of the same image is to 20 percent PVC. Also, we are ensured that the shot area in XPS is partially overlapped with that in LSCM. Now our task is to visualize the fusion information based on data from XPS and LSCM. Also, we let user choose theoretic model to fill space in between XPS top and bottom images and then compare them with fusion model. Thus, user can obtain a proposed theoretic model, which is closest to experimental data.

Our system is used to process the input XPS and LSCM datasets. First, we design two flowchart objects to process XPS and LSCM datasets respectively. These two flowchart objects should adjust the resolution, image dimension, concentration mapping, etc., in order to let them match for fusion. After 3D models of each sensor domain are generated, we fused them to form a combined 3D model. And we also use 3D in-filling algorithms to fill the space between XPS images to form several theoretic 3D models. Then comparison metrics are calculated and the best proposed 3D structure of material is decided.

The result of our example is illustrated in the attached figures. In figure 4, a fused 3D model of XPS and LSCM is illustrated. We put XPS top image on the top of the 1ˢᵗ image of LSCM image stack and XPS bottom image at the bottom of the last image of LSCM image stack to form a composite 3D model based on these two techniques. Also, we use linear insertion algorithm insert 30 images between top image and bottom image of XPS datasets. Thus both datasets have the same number of images. After using all filters, they also have the same resolution, dimension and been well registered. So we use our comparison metrics, MAE and SNR, to compare these two 3D models. The results are: MAE = 119.3; SNR = 129.8. The comparison results are illustrated in figure 4. We also use another different formula to fill the space in between the top and bottom image: $2*topPixel*x+botPixel*(1-x)$. Comparison between the new 3D XPS model and LSCM model is calculated: MAE = 87.3; SNR = 99.0. So according to MAE and SNR, the second in-filling function generates a result closer to LSCM. The results are illustrated in figure 4.

From the above case study, we see our system can generate a fused and theoretic 3D model based on a set of experimental data and user-defined algorithms. The user might propose different theoretic 3D models and compare these models with real experimental data. Thus the best theoretic model of chemical structure of the material of interest could be chosen based on the comparison metrics. This the framework can used to refine understanding about the chemical /physical process as well.

6   Conclusion

In this paper, we have proposed the Polysensometric Visual Information Fusion (PVIF) Model for estimation of total information of the object of interest based on partial information from multiple instruments. We have implemented an integrated system to automate fusion of the partial information and to compare different 3D models. Our system can be used as a tool for chemists and physicists to explore the structure of a material. There are still many complicated issues. Among these difficulties, 2D/3D registration algorithm itself is a challenge; how to find a good 3D comparison metric is also interesting. The effectiveness of various morphing and interpolation algorithms also changes when comparison is desired at 3D level. Speed of overall computation remains an ever important issue as it affects the mental process of perceptualization.

Acknowledgements. Dr. Julia Fulgum's analytical and surface chemistry research group at Kent State University, Department of Chemistry, and later at University of New Mexico, Chemical and Nuclear Engineering Department, has provided the XPS and Confocal datasets for the system. The work has been supported by National Science Foundation (NSF) Grant NSF-ITR-0113724.

References

1. Schiers C, Tiede U, Hohne KH 1989, Interactive 3D-registration of image volumes from different sources. In: Lemke HU et al. (eds), Computer Assisted Radiology (Proc. CAR '89), Springer, Berlin.
2. Hohne, K.H., Bomans, M, et al, 3D-Visualization of tomographic volume data using the generalized voxel-model, CH Volume Visualization workshop.
3. K. Artyushkova, "Application of Multitechnique Correlation And Multivariate Analysis To Heterogeneous Polymer Systems", dissertation, Kent State University, Dec., 2001.
4. William M. Wells, Paul Viola, Hideki Atsumi and Shin Nakajima, Multi-Modal Volume Registration by Maximization of Mutual Information, Medical Image Analysis, vol. 1, no. 1, pp. 35--51, March 1996.
5. M. E. Leventon and E. L. Grimson, Multi-Modal Volume Registration Using Joint Intensity Distributions, in Medical Image Computing and Computer-Assisted Intervention-MICCAI'98, ed. by W.M. Wells, A. Colchester, S. Delp, Cambridge, MA, Springer LNCS 1496, pp 1057.
6. P.E. Anuta, Spatial registration of multispectral and multitemporal digital imagery using fast fourier transform techniques. IEEE Trans. on Geoscience Electronics, GE-8(4):353--368, October 1970.

An Incremental Editor for Dynamic Hierarchical Drawing of Trees

David Workman[1], Margaret Bernard[2], and Steven Pothoven[3]

[1] University of Central Florida, School of EE and Computer Science
workman@cs.ucf.edu
[2] The University of the West Indies, Dept. Math. & Computer science
mbernard@fsa.uwi.tt
[3] IBM Corporation, Orlando, Florida

Abstract. We present an incremental tree editor based on algorithms for manipulating shape functions. The tree layout is hierarchical, left-to-right. Nodes of variable size and shape are supported. The paper presents algorithms for basic tree editing operations, including cut and paste. The layout algorithm for positioning child-subtrees rooted at a given parent is incrementally recomputed with each edit operation; it attempts to conserve the total display area allocated to child-subtrees while preserving the user's mental map. The runtime and space efficiency is good as a result of exploiting a specially designed Shape abstraction for encoding and manipulating the geometric boundaries of subtrees as monotonic step functions to determine their best placement. All tree operations, including loading, saving trees to files, and incremental cut and paste, are worst case O(N) in time, but typically cut and paste are $O(\log(N)^2)$, where N is the number of nodes.

1 Introduction

Effective techniques for displaying static graphs are well established but in many applications the data are dynamic and require special visualization techniques. Applications arise when the data are dynamically generated or where there is need to interact with and edit the drawing. Dynamic techniques are also used for displaying large graphs where the entire graph cannot fit on the screen. In this paper, we present an incremental tree editor, DW-tree, for dynamic hierarchical display of trees. Graphs that are trees are found in many applications including Software Engineering and Program Design, which is the genesis of this Dynamic Workbench DW-tree software [5, 1, 4]. DW-tree is an interactive editor; it allows users to interact with the drawing, changing nodes and subtrees. In redrawing the tree after user changes, the system reuses global layout information and only localized subtree data need to be updated. A full complement of incremental editing operations is supported, including changing node shape and size as well as cutting and pasting subtrees. The editor has the capability of loading (saving) a tree from (to) an external file. The tree editor uses Shape vectors for defining the boundary of tree or subtree objects. Incremental tree editing operations are based on algorithms for manipulating Shape vectors. The tree layout is hierarchical, left-to-right (horizontal). The DW-tree drawing algorithms

M. Bubak et al. (Eds.): ICCS 2004, LNCS 3038, pp. 986–995, 2004.
© Springer-Verlag Berlin Heidelberg 2004

attempt to conserve the total display area allocated to child-subtrees (area efficient) without appreciably distracting the user's mental continuity upon re-display. For a tree of size N nodes, the runtime efficiency is O(N) for load and save operations to an external file. For cut and paste operations at depth d, under reasonable assumptions, the runtime is $O(d^2)$, with $d = \log(N)$.

The remainder of the paper is organized as follows. First (Section 2), we present the principles for tree layout using the Shape abstraction. In section 3, we present key principles and algorithms that form the basis for the incremental editor operations and in Section 4 we discuss related work; we close with some concluding remarks in section 5.

2 Tree Layout Principles

In this section we present the layout design principles and definitions that provide the conceptual foundation and framework for the remainder of the paper. The display area of the editor defines a 2D coordinate system as depicted in Figure 1. Coordinate values along both axes are multiples of the width (*fduW*) and height (*fduH*) of the *Fundamental Display Unit* (*FDU*), the smallest unit of display space allocated in the horizontal and vertical directions, respectively.

Nodes can be of any size, but are assumed to occupy a display region bounded by a rectangle having width and height that are multiples of fduW and fduH, respectively. The absolute display coordinates of a node are associated with the upper left corner of its bounding rectangle. *Trees* are oriented left-to-right as shown in Figure 1, where the Parent node and its First Child always have the same vertical displacement. All children with the same parent node have the same relative horizontal displacement,

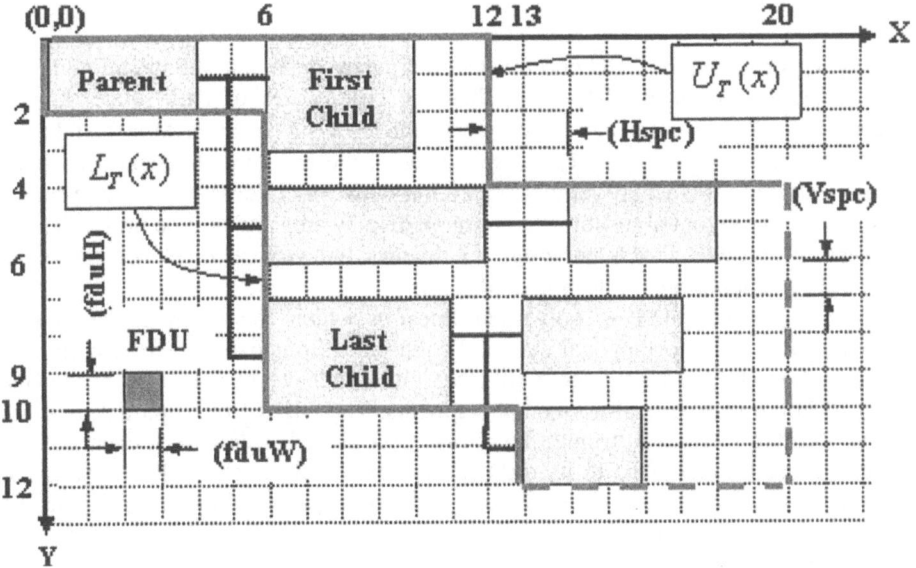

Fig. 1. Display Coordinate System,Tree Layout, and Bounding Shape Functions

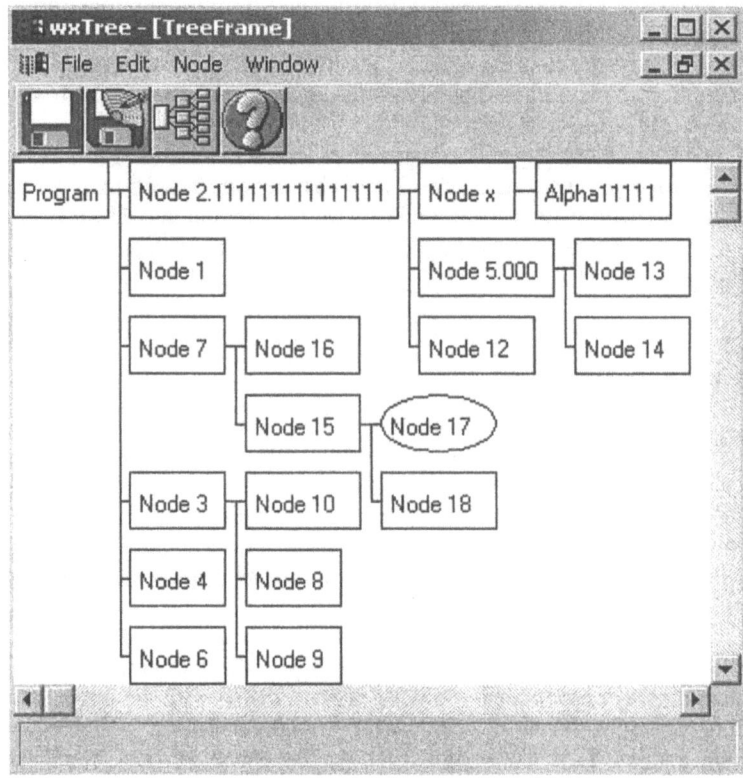

Fig. 2. Screen Image of DWtree Editor

Hspc. If (x,y) denotes the FDU coordinates of a parent node, and the parent node has width (*Pwidth*) and height (*Pheight*), then the coordinates of child node, k, will always be (x + *Pwidth* + *Hspc*, y + Δk), where for 1 ≤ k ≤ *Nchildren*, Δ1 = 0, and for k > 1, Δk ≥ Δ(k-1) + *Cheight*(k-1) + *Vspc*. *Vspc* is the minimum vertical separation between children. The actual vertical separation between children of the same parent is defined by our layout algorithms to conserve display area and will be presented in a later section entitled, Tree Operations. H denotes the vertical displacement of all children (H = ΔNchildren + *Cheight*Nchildren).

  Edges are represented as polygonal lines in which the segments are either horizontal or vertical (orthogonal standard). An edge from parent to first child is a horizontal line. Edges from parent to other children have exactly two bends and consist of horizontal-vertical-horizontal segments. The edges from a parent to all its children overlap to form a 'trunk'. To allow space for edge drawings, each node is allocated an additional Hspc to its right in the horizontal direction. The rectangular region of width *Hspc* and height H, located at coordinates (x + *Pwidth*, y), is used by the parent node to draw the edges to all its children. An example of a trees drawn by the DW-tree editor is illustrated by Figure 2.

2.1 Shape Functions

Shape functions are objects that define the outline of a tree or subtree. Figure 1 illustrates the concept for some tree, T. The solid upper line outlines the upper shape (U_T) of T, the solid lower line outlines the lower shape (L_T) of T. The subtree rooted at a given node has relative coordinates $(0, 0)$ and corresponds to the coordinates of its root node. Every such subtree, T, has a bounding rectangle of width W_T and height H_T; in Figure 1, $W_T = 20$ and $H_T = 12$. The width of a subtree always takes into account the horizontal spacing (Hspc) associated with the node(s) having maximum extent in the X direction.

The upper shape function, UT, is a step function of its independent variable, x. We represent UT as a vector of pairs: $UT = < (\delta x1, y1), \ldots (\delta xn, yn) >$, where $\delta xk \geq 1$ denotes the *length* of the kth *step*, $1 \leq k \leq n$, and $y1 = 0$. yk gives the value of the function in the kth step (interval). Along the X-axis the kth step defines a half-open interval [xk-1, xk) where $x_0 = 0$ and $x_k = \sum_{j=1}^{k} \delta x_j$. The *width of T* then becomes

$WT = xn$ – the right extreme of the last step interval of UT. Finally, we define *steps*(UT) $= \{x0, x1, \ldots, xn\}$, the set of interval end-points defined by UT, and *dom*(UT) $= [x0, WT]$, the closed interval defined by the extreme X-coordinates of the minimal bounding rectangle for T.

The lower shape function, LT, is defined in an analogous way with the roles of x and y reversed. That is, we represent LT as a step function with y as the independent variable. Specifically, for some $m \geq 1$, $LT = < (x1, \delta y1), \ldots (xm, \delta ym) >$. Along the Y-axis the kth step defines a half-open interval [yk-1, yk) where $y_0 = 0$ and

$y_k = \sum_{j=1}^{k} \delta y_j$. Thus the *height of T* becomes $HT = ym$ – the right extreme of the

last step interval of LT. And, we define steps(LT) $= \{ y0, y1, \ldots, ym\}$, the set of interval end-points defined by LT, and dom(LT)$=[y0, HT]$, the closed interval defined by the extreme Y-coordinates of the minimal bounding rectangle for T.

Because of our layout conventions, the key step in our layout algorithm requires computing the vertical separation of adjacent siblings in a given subtree. Since the lower shape function and upper shape function are based on different independent variables, we must convert the lower shape function to an equivalent step function with x as the independent variable. We let ΛT denote the lower shape function of T where x is the independent variable. $\Lambda T = < (\delta x1, y1), \ldots (\delta xm, ym) >$, where $\delta xk = (xk+1 - xk)$ and yk are computed from LT as defined above. However, because xm+1 is not defined in LT we compute $\delta xm = (WT - xm)$. As we will show later, it will always be the case that $WT > xm$, where xm is always taken from the last step interval of LT.

2.2 The Shape Algebra

Incremental tree editing operations are based on algorithms for manipulating shape functions. As the basis for these algorithms, we introduce a simple algebra for manipulating shape functions. Each will be defined for upper shape functions – analogous definitions apply to lower shape functions (LT and ΛT).

UShape(dx ,y) = an upper step function = $<(dx, y)>$. Similarly, *LShape(x, dy)* = $<(x, dy)>$. UShape and LShape are distinct *function types*.

Min(R,S) = Z, where *R,S and Z are UShape functions. Assume that dom(R) $\subseteq$ dom(S). Then dom(Z) = dom(S) and steps(Z) = steps(R) $\cup$ steps(S) for each x$\in$ steps(Z), Z(x) = min(R(x),S(x)), if x $\in$ dom(R) $\cap$ dom(S); Z(x) = S(x), if x $\in$ dom(S) – dom(R).*

Diff(R,S) = Z, where *R,S and Z are UShape functions. Then dom(Z) = dom(R) $\cap$ dom(S) and steps(Z) = (steps(R) $\cup$ steps(S)) $\cap$ dom(Z). Finally, for each x$\in$ steps(Z), Z(x) = R(x)-S(x).*

ScalarAdd(c, R) = ScalarAdd(R,c) = Z, where *R and Z are UShape functions and c is a scalar constant. Z(x) = R(x)+c, for all x$\in$ dom(Z) = dom(R). In effect, the scalar c is added to the y-component (dependent variable) of each step in a UShape, and added to the x-component of each step in an LShape. Note: steps(Z) = steps(R).*

Cat(R,S) = Z, where *R, S and Z are UShape functions. steps(Z) = { x | x$\in$ steps(R) or x = x'+W_R, for x'$\in$ steps(S) } – {W_R | y-value of the last step of R = y-value of the first step of S}, and dom(Z) = [0, W_R+W_S]; Z(x) = R(x), if x $\in$ dom(R); Z(x) = S(x'), if x = x'+ W_R, where x'$\in$ dom(S).*

MaxElt(R) = C, where *R is an UShape function. C = Max{ R(x) | x$\in$ dom(R)} = Max{ R(x) | x$\in$ steps(R)}.* For Min(R,S) the runtime is O(|R|+|S|), where |R| denotes the number of steps in R (analogously of S). For ScalarAdd(R), MaxElt(R) the runtime is O(|R|). For Cat(R, S) the runtime can be O(1) if linked structures are used to represent shape functions.

3 Tree Operations

Some of the most basic operations on trees are:

      (Op-1)   Create and edit a selected node.
      (Op-2)   Cut and paste a subtree.
      (Op-3)   Read (write) a tree from (to) an external file.
      (Op-4)   Select a node in a tree.

(Op 1). *Creating/editing a node* involves defining/changing the size of the bounding rectangle enclosing the node's display image. This means (re-)computing the shape functions for the node and then propagating these changes throughout the rest of the tree in which the node is embedded. As this is a special case of cutting/pasting a subtree, we describe how shape functions are defined for a single node and defer the rest of the discussion to the subtree cutting/pasting operation, Op-2. If R is a free-standing node with width (R.width) and height (R.height), then: UR = UShape (R.width + Hspc, 0) and LR = LShape(0, R.height). This is an O(1) operation.

(Op-2). *Cutting/Pasting a node.* Let S and T be fully defined free-standing trees. We consider the operation of <u>pasting S into T</u> at some position. The position in T is a node previously identified through a *select operation* (Op-4). Let P denote the selected node in T and let R = root(S) denote the root of the tree S. The paste operation requires a parameter that defines the new relationship P is to have with respect to R in the composite tree, that is: R can be an <u>upper/lower sibling</u> of P (assuming P ≠ root(T)); R can be the (new)(only) <u>first/last child</u> of P. We describe the case where R is to become a new lower sibling of P. The other variations differ only in minor details. Figure 3 illustrates the trees S and T before and after the paste operation, respectively. The relevant shape functions are presented in the accompanying table. There are three key sub-algorithms incorporated in the algorithmfor a complete Cut or Paste operation.

Algorithm 1 (Change Propagation): If P is the parent of some node (child subtree) whose shape changes, then the shape functions for P (for the subtree rooted at P) must be recomputed (See Algorithms 2 and 3). No additional change is necessary to the shape functions of any other child of P. Once the shape functions for P have been recomputed, then this algorithm is repeated at the parent of P, etc., terminating only when the root of the tree is reached. The worst case running time is O(N), where N is the number of nodes in a tree. The worst case would occur with trees where every node has a single child. However, if we assume a randomly chosen tree with N nodes, then the tree will tend to be balanced – each node has approximately the same number of children, say b. Thus Change Propagation will require O(dp) running time, where d denotes the depth in the tree where the first change occurs ($d \approx \log_b(N)$) and p denotes the worst case running time at each level on a path to the root. The running time at each level is essentially the running time of Algorithm 2 and 3 combined.

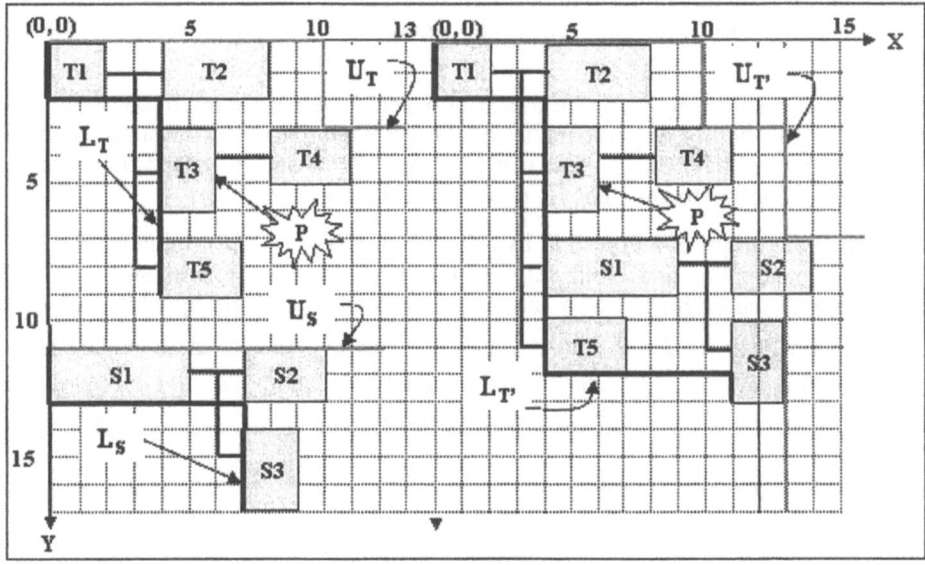

Fig. 3. Subtree Paste Operation

Tree (Figure 3)	Upper Shape (U)	Lower Shape (L)	Lower Shape (Λ)
T before paste	<(10, 0), (3,3)>	< (0,2), (4,7) >	< (4,2), (9,7) >
S before paste	<(12, 0)>	< (0,2), (7, 4) >	< (7,2), (5,4) >
T after paste	<(10, 0), (3,3), (3, 7)>	< (0,2), (4, 10), (11, 1) >	< (4,2), (7, 10), (5, 1) >

Algorithm 2 (Child Placement): The algorithm for (re-)positioning the child subtrees of a given root or parent node when a new child subtree is added, removed, or changed can be characterized as building a *forest of trees* relative to some given origin point – the *forest origin*. Figure 4 illustrates a Forest, F, with origin (XF, YF) and an arbitrary (child) subtree Ck with origin (Xk, Yk). The normal shape functions for Ck have their origin relative to the origin of Ck. These functions are depicted as Uk and Lk. The first step is to extend the shape functions for Ck so that they are relative to the origin of the Forest. This can be done by defining $U_k^F =$ ScalarAdd(Cat(UShape(Xk-XF,0),Uk), Yk – YF) and $L_k^F =$ Cat(LShape(0,Yk–YF), ScalarAdd(Lk, Xk-XF)). The extended shape functions, U_k^F and L_k^F, are computed by composing Cat() and ScalarAdd() in different orders but have the same form with the roles of X and Y reversed. This symmetry in form of the operations used to compute the upper and lower shape functions is one of the benefits of using the same basic representation for these functions, but with the roles of X and Y reversed. The running time for computing U_k^F and L_k^F is O($|U_k^F|$) and O($|L_k^F|$), respectively; that is, the number of steps in each of these functions. By extending the shape functions for individual trees within a Forest (Figure 4), we are able to define shape functions, UF and LF, for a Forest, F. The algorithm for constructing UF and LF can now be given.

(1) Let P be an existing node, the parent of a tree, T, we are about to construct. The tree that results will be a simple matter of placing P relative to the forest composed of its child subtrees, C1, C2, ..., Cn. If n = 0, then T = P and the shape functions are computed according to Op-1 above. If n > 0, continue with step (2).

(2) Define the origin of a Forest, F, by (XF, YF) = (XP+WP, YP). Initialize F = F1 = {C1} to contain the first child subtree, C1, by setting the coordinates of C1 to coincide with the origin of F. The algorithm proceeds iteratively by adding Ck+1 to Fk to obtain Fk+1,for $2 \le k \le n$. If we let L_F^k and U_F^k denote the bounding shape functions for the Forest, Fk, the Forest obtained after adding k child subtrees, then $L_F^1 = L1$ and $U_F^1 = U1$, the shape functions for C1. The iterative step is (3).

(3) For $2 \le k \le n$ do the following:

(3a) Set δ = Vspc + MaxElt(Diff(Λ_F^{k-1} ,Uk)), where Λ_F^{k-1} is the UShape function obtained from L_F^{k-1} and Uk is the UShape function associated with Ck.

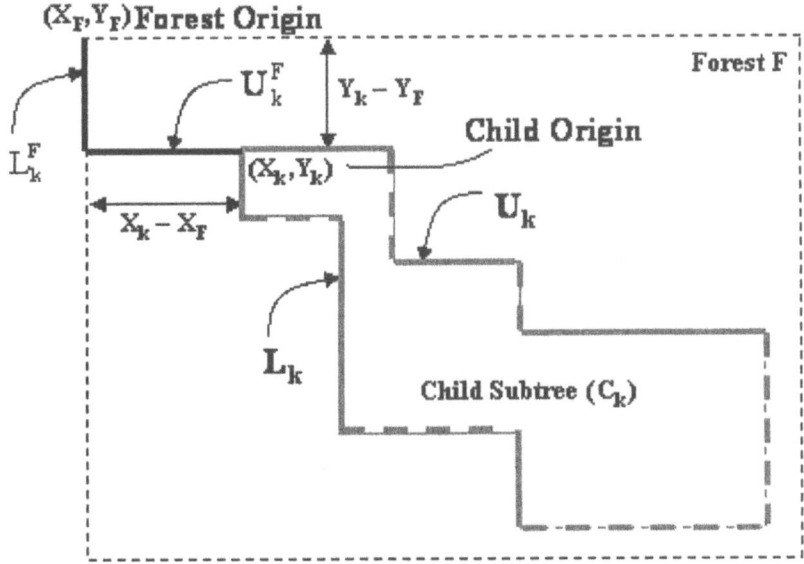

Fig. 4. Extending Shape Functions to a Forest Origin

(3b) Compute the origin (Xk, Yk) for Ck as follows: (Xk, Yk) = (XF, YF + δ). Add Ck to the Forest, F; that is, Fk = Fk-1 $\cup$ { Ck }.

(3c) Compute the new shape functions for Fk as follows: $U_F^k = \text{Min}(U_F^{k-1}, U_k^F)$ and $L_F^k = \text{Min}(L_F^{k-1}, L_k^F)$, where U_k^F and L_k^F are the extended shape functions defined by Ck at its new origin. Specifically recall, $U_k^F = \text{ScalarAdd}(\text{Cat}(\text{UShape}(Xk\text{-}XF,0),$ Uk), Yk – YF) and $L_k^F = \text{Cat}(\text{LShape}(0,Yk – YF), \text{ScalarAdd}(Lk, Xk\text{-}XF))$. In these expressions, Xk-XF = 0, and Yk – YF = δ. Thus U_k^F reduces to just ScalarAdd(Uk, δ) and L_k^F reduces to Cat(LShape(0,δ), Lk). Since the first step of Lk has a relative x-coordinate of 0, then the result of the Cat() operations simply increases the δy1 component of the first step by δ. Runtime Note: The running time for (3a) and (3c) is O($|U_k^F| + |L_k^F|$), thus the running time for (3) does not exceed O(2b*d') = O(d'), where d' is max(max{ $|U_k^F| | 2 \le k \le b$ },max{ $|L_k^F| | 1 \le k \le$ b-1 }), but under our assumption of a balanced tree, d' $\approx$ logb(N). These computations are traced for the Paste operation illustrated in Figure 3 and are summarized in Table 1 below; Vspc = 1, and (XF, YF) = (4,0) in these computations. Also, C3 is the subtree S in this scenario.

Cut operations use the same computations with slightly different preliminaries. If a subtree S, rooted at node R with parent P, is removed from T, then node R is unlinked from its parent P and its siblings (if any). The computations of Algorithm 2 are then applied to the remaining children of P. The computation defined Algorithm1 then propagates the change all the way to the root of the new tree.

Table 1. Computation of Bounding Shape Vectors and Child Placement

k	U_k	L_k	U_k^F	L_k^F	δ	U_F^k	L_F^k	Λ_F^k
1	<(6,0)>	<(0,2)>	<(6,0)>	<(0,2)>		<(6,0)>	<(0,2)>	<(6,2)>
2	<(9,0)>	<(0,3)>	<(9,3)>	<(0,6)>	3	<(6,0), (3,3)>	<(0,6)>	<(9,6)>
3	<(12, 0)>	<(0,2), (7,4)>	<(12, 7)>	<(0,9), (7,4)>	7	<(6,0), (3,3), (3,7)>	<(0,9), (7,4)>	<(7,9), (5,4)>
4	<(5, 0)>	<(0,2)>	<(5, 10)>	<(0,12)>	10	<(6,0), (3,3), (3,7)>	<(0,12), (7,1)>	<(7,12), (5,1)>

Algorithm 3 (Parent Update). Once the Forest of children has been computed as described in Algorithm 2, the shape functions for the entire subtree rooted at the Parent, P,must be updated. This implies that each node stores the shape functions for the node itself and also the subtree rooted at that node. Let T denote the subtree rooted at P and let F denote the Forest of children of P. Then UT = Cat(UP, UF) and LT = Min(LP, ScalarAdd(LF, WP)). The running time for Parent Update is O(|LF|) due to the Min() operation. The Cat() operation is O(1). By the analysis presented above for Algorithms 1, 2 and 3 it follows that Cut and Paste operations, under a balanced tree assumption, have a runtime cost of O(logb(N)2).

(Op-3). *Reading/Writing a tree to an external file.* By storing and incrementally maintaining the size of a subtree rooted at a given node, it is possible to write a tree to an external file without having to save the shape functions – when the tree is read back into memory, the shape functions can be incrementally reconstructed in O(1) time for each node of the tree. Thus, both write and read operations require O(N) time.

(Op-4). *Selecting a node.* The most basic tree operation is selecting a node to use as the basis for defining the above operations. Using the absolute display coordinates of a node selected on the screen, a simple recursive descent algorithm is used to locate the selected node in the tree structure. The algorithm uses a simple check to see if the selected point falls within the bounding shape functions of a given subtree. The worst case is O(N), but under a balanced tree assumption, selecting a node requires O(log(N)) operations.

4   Related Work

Early work on dynamic and incremental drawing of trees was given by Moen [3]. He provides a layout algorithm for constructing and drawing trees by maintaining subtree contours stored as bounding polygons. Moen's algorithm is similar to ours, differing essentially in two important respects. First, Moen's contours correspond to our step functions. However, Moen contours completely circumscribe a subtree, whereas our step functions bound essentially "half" the subtree; hence, Moen's contours are always greater in length than our step functions for a given subtree. Second, Moen's layout always positions the parent roughly at the midpoint of the forest of its children.

Our algorithm always places the parent next to its first child. Examples can be found where Moen's policy is more costly in display area than ours, and vice versa. This is an area requiring further investigation. Overall, the Big-O runtime complexity of the two approaches is the same under similar assumptions. Our algorithm, however, will generally run faster by a constant factor due to our parent placement policy and our more efficient choice of data structure for bounding subtree shapes.

Cohen et al [2] provide dynamic drawing algorithms for several classes of graphs, including upward drawings of rooted trees. Their tree layout is a layered drawing in which nodes of the same depth are drawn on the same horizontal line. The bounding box of a node is the rectangle bounding the node and its subtrees; subtrees are drawn such that bounding boxes do not overlap giving a drawing area $O(n^2)$. In our layout, the display space is used more efficiently as subtree shapes are defined by step functions and the child placement algorithm minimizes the vertical separation between children of the same parent to conserve display space.

5 Conclusion

In this paper, we have presented a new technique for an incremental tree editor based on algorithms for manipulating shape functions. The boundaries of subtree objects are defined as monotonic step functions and are manipulated to determine the best placement of the subtrees. The framework was applied to hierarchical tree drawings with a horizontal, left-to-right layout. The drawing system supports a full complement of editing operations, including changing node size and shape and insertion and deletion of subtrees. In redrawing the tree after user changes, the system reuses global layout information and only localized subtree data need to be updated. This ensures that the drawing is stable and incremental changes do not disrupt the user's sense of context. The algorithms conserve the total display area allocated to child-subtrees and their runtime and space efficiency is good with all tree operations being worst case $O(N)$ in time. The tree editor was described for a very specific layout of hierarchical trees. Our approach can be generalized to a family of Shaped-based algorithms where the position of the parent relative to children and relative positions of children to each other are based on layout parameters.

References

1. Arefi, F., Hughes, C., Workman, D., Automatically Generating Visual Syntax-Directed Editors, *Communications of the ACM*, Vol.33, No. 3, 1990.
2. Cohen, R., DiBattista, G., Tamassia, R., Tollis, I., Dynamic Graph Drawings:Trees, Series-Parallel Digraphs, and Planar ST-Digraphs, *SIAM Journal on Computing*, Vol.24, No. 5, 970-1001, 1995.
3. Moen, S., Drawing dynamic trees, IEEE Software, Vol. 7, 21-28, 1990
4. Pothoven, S., A Portable Class of Widgets For Grammar-Driven Graph Transformations, M.Sc. Thesis, University of Central Florida, 1996.
5. D. Workman, GRASP: A Software Development System using D-Charts, Software – Practice and Experience, Vol. 13, No. 1, pp. 17-32, 1983.

Using Indexed-Sequential Geometric Glyphs to Explore Visual Patterns

Jim Morey and Kamran Sedig

Cognitive Engineering Laboratory
Department of Computer Science
The University of Western Ontario, Canada
{jmorey,sedig}@uwo.ca

Abstract. This paper presents a visualization tool called PolygonR&D for exploring visual tiling patterns. To facilitate the exploration process, PolygonR&D uses dynamically-generated, interactive geometric glyph visualizations that intermediate reasoning between the sequential textual code and the parallel visual structure of the tilings. Sequential textual code generates indexed-sequential geometric glyphs. Not only does each glyph represent one procedure in the sequential code, but also a constituent element of the visual pattern. Users can reason with a sequence of glyphs to explore how tiling patterns are constructed. Alternatively, they can interact with glyphs to semantically unpack them. Glyphs also contain symbolic referents to other glyphs helping users see how all procedures work together to generate a tiling pattern. Experimenting with indexed-sequential glyphs in tools such as PolygonR&D can help us understand how to design interactive cognitive tools that support reciprocal reasoning between sentential and visual structures.

1 Introduction and Background

Mathematics has been described as the science of patterns [1]. Visual tilings are an example of mathematical patterns that are all around us [1, 2]. One of the best ways to investigate many mathematical concepts is to interact with their representations using computational cognitive tools—interactive tools that support and enhance cognition in the process of reasoning and experimentation [3]. Gaining insight into many ideas involves reasoning with multiple forms of representations of those ideas and interacting with those representations using different interaction styles and methods [3, 4, 5]. This is true of mathematical patterns. Due to their flexibility, malleable form, and dynamic nature, computational tools can easily present users with different representational forms of mathematical ideas and various interaction styles, allowing for different reasoning [3, 5]. In this paper, we are interested in investigating how to explore geometric tiling patterns using different representational forms.

Two forms of representation of geometric tilings include descriptive and visual [1, 2]. The first form is textual, sentential, and language-like; it is linear and sequential in nature; it linguistically describes how the tiling can be constructed. The second form is visual or diagrammatic; it is spatial and parallel in nature; it visually shows the structure of the tiling. These two forms of representation are informationally

M. Bubak et al. (Eds.): ICCS 2004, LNCS 3038, pp. 996–1003, 2004.
© Springer-Verlag Berlin Heidelberg 2004

equivalent, that is, they convey the same information. However, they are computationally non-equivalent, that is, they require different amounts and types of cognitive effort and can lead to different forms of knowledge [6]. These alternative forms of representation are complementary since they direct our attention to different aspects of their objects which we may otherwise overlook [4, 7]. The textual form communicates information about the logical sequence and ordering of the tilings, and the visual form conveys information about the geometry and structure of the tilings [6]. The textual form is cognitively processed in a sequential manner, whereas the visual form is cognitively processed in a more parallel and holistic manner.

This paper presents a visualization modeling tool called PolygonR&D, a programming environment that allows users to use a textual, procedural language to generate complex visual tiling patterns made of different types of polygons. Since tiling patterns are parallel visual structures, users can often find it difficult to reason about the constituent building blocks of such structures. Furthermore, reasoning back and forth between the sequential textual code and the parallel visual structure is not straightforward. Glyph, or iconic, visualizations can be used to facilitate comprehension of multi-dimensional data [8, 9]. These visualizations encode several dimensions of information, such as shape, size, and color, in a compact form to make these dimensions easily accessible to the users at a more perceptual level. Multidimensional glyphs have been used to encode different aspects of textual information. Such glyphs have proven to be easier and more efficient for users to process [10].

PolygonR&D uses dynamically-generated, interactive glyph visualizations to intermediate reasoning between the sequential textual code and the parallel visual structure of the tilings. Sequential textual code generates indexed-sequential geometric glyphs. Each glyph represents one procedure in the sequential code, as well as a constituent element of the visual pattern. Users can reason with a sequence of glyphs to explore how tiling patterns are constructed. Alternatively, they can interact with glyphs to semantically unpack them. Glyphs contain symbolic referents to other glyphs helping users see how all procedures combine to generate the tiling pattern.

2 Exploring Visual Tiling Patterns with PolygonR&D

PolygonR&D incorporates all the above representations: a sequential, textual representation, an intermediary, iconic representation, and a parallel, visuospatial representation. Fig. 1 shows how the environment of PolygonR&D[1] is separated into three panels each containing one of the representations of the tiling: Programming Panel (left) contains the sequential representation, Glyph Panel (centre) contains the intermediary representation, and Polygon Landscape Panel (right) contains the parallel spatial representation. Each of the panels is interactive: the Programming Panel allows for the standard text manipulations, the Glyph Panel allows for a semantic unpacking of the informationally-dense glyphs, and the Polygon Landscape Panel allows for interactive exploration of the execution of the programs.

---

[1] PolygonR&D is a java application that can be run in a browser with the Java plug-in http://www.csd.uwo.ca/~morey/CogEng/PolygonRnD.html

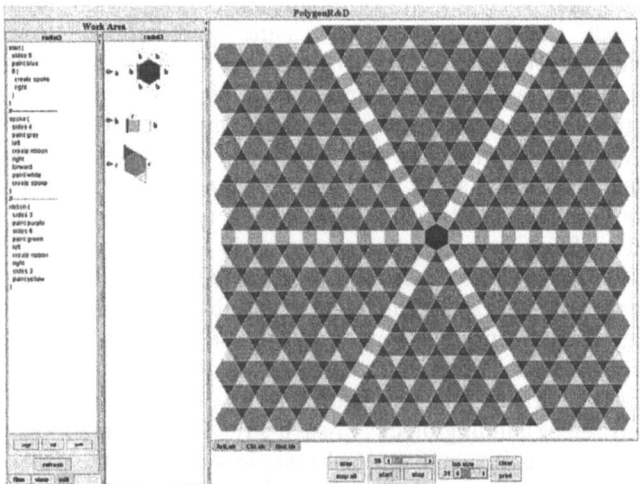

Fig. 1. A screen capture of the PolygonR&D

The sequential description of tilings is similar to the description of turtle movements in Logo except that the focus is on regular polygons rather than lines and the description allows for multiple turtle-like agents to perform the drawing[2]. The description is segmented into labeled procedures, paralleling the building blocks of the tiling. Table 1 shows the procedures that generate the tiling in Fig. 1, which is also present in its Programming Panel. The labels are used in the construction of new agents to reference the procedure the agent will perform. For instance, in the procedure labeled **start,** there is an instruction **create spoke**, which creates a new agent that performs the procedure labeled **spoke**.

Table 1. Example procedures associated with Fig. 1.

```
start {              spoke {             ribbon {
   sides 6              sides 4             sides 3
   paint blue           paint gray          paint purple
   6 {                  left                sides 6
     create spoke       create ribbon       paint green
     right              right               left
   }                    forward             create ribbon
}                       paint white         right
                        create spoke        sides 3
                     }                      paint yellow }
```

---

[2] The environment can be described as a 2D-Turing machine with multiple polygonal heads that write coloured polygon symbols.

2.1 Exploring Textual and Glyph Representations

A glyph for each procedure is constructed and put into the Glyph Panel. The glyph visually represents the effects of executing its corresponding procedure. It is constructed by executing the procedure with the exception that the **create** instructions are not executed, but instead tokens are placed to visually indicate where the unexecuted instruction occurs. For instance, the upper left corner of Fig. 2 shows the glyph associated with the procedure **start**. The hexagon represents a portion of the tiling while the six dotted squares represent six tokens each indicating spatial places where the unexecuted **create** instructions are called or connected. The tokens reference the procedure **spoke** in two ways: visually, by depicting its starting shape (a square), and symbolically, by referring to its index (**b**). The square is the first shape that appears in the procedure **spoke**. The index **b** is the index given to **spoke**'s glyph. The glyphs are indexed according to the order that they are used in the construction of the tiling. In the above example, **start** is the first procedure used, hence indexed as **a**; **spoke** is the second procedure used, hence indexed as **b**; and **ribbon** is the third procedure used, hence indexed as **c**. The short indices help make the glyphs visually compact.

The creation of the list of glyphs from the procedures is not difficult for the computer but the reasoning back and forth between the glyphs and code can be difficult for a programmer. To aid this reasoning process, the glyphs are made interactive. Each glyph can be unpacked so that the details of its construction can be viewed. The details are shown as a labeled sequence of icons. The label is the original name of the procedure. The list of icons corresponds to the instructions of the textual code. The arrow icons represent motion, the polygons represent the introduction of a new shape, and the coloured ovals represent the placement of a coloured landmark. The icon for the **create** instruction is a miniature of the glyph without the indices. Fig. 2 shows the unpacked versions of the glyphs from Fig. 1. The icons' instructions provide spatial cues that help visualize the construction of the glyph. The inclusion of the miniature glyphs allows for the linear scanning of the icons to be uninterrupted with a look up of a glyph on the list.

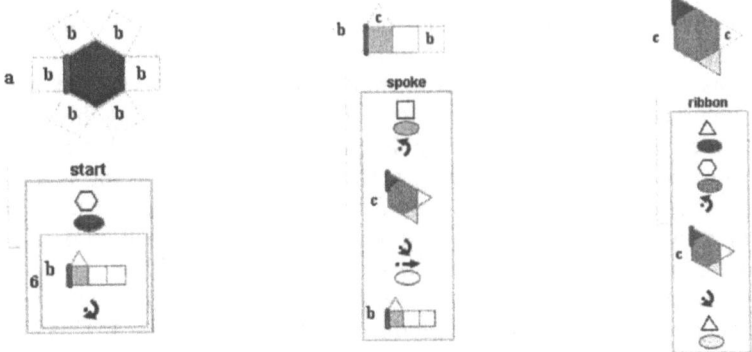

Fig. 2. Unpacked glyphs created from the code in Table 1.

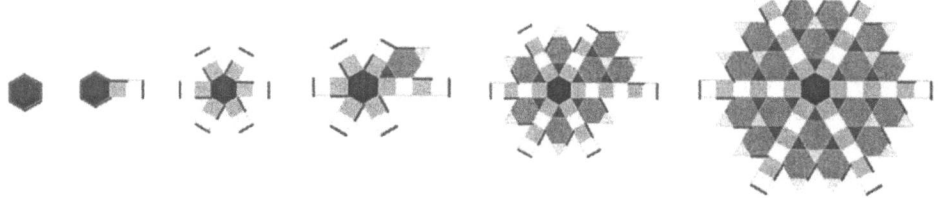

Fig. 3. A sequence of images created by the Polygon Landscape Panel during the execution of the code from Table 1.

2.2 Exploring Glyph and Visuospatial Tiling Representations

Reasoning between the list of glyphs and the tiling takes place in the spatial domain. The act of creating the tiling is a matter of starting with the first glyph and replacing the indexed tokens with their referenced glyphs. Mentally visualizing the tiling using the glyphs is a matter of visually chunking copies of the glyphs into one mental image. This task can sometimes be difficult due to the mental manipulation of a large number of glyphs. The key to comprehending the tiling is to recognize replacement patterns so that large numbers of glyphs can be chunked together. For instance, the above example has a replacement pattern that can be noted in glyph **b**. This glyph creates another **b** in such a way that a line of squares is created. Mentally visualizing this pattern can be difficult, but it is easy to see after viewing the animated construction of the tiling in the Polygon Landscape Panel. Fig. 3 shows a sequence of images depicting the gradual construction of the tiling in the polygon landscape. The line-of-squares feature in the tiling is prominent and relating this feature back to the glyphs is a matter of scanning the list for squares. Unlike the squares, which are only present in glyph **b**, the hexagons in the tiling could potentially come from either glyph **a** or glyph **b**. The colouring of the glyphs can aid in disambiguating the origins of polygons in the tiling. For instance, the central hexagon must come from glyph **a** (not glyph **c**) since it is the only glyph that contains a hexagon of that colour.

2.2.1 Single Glyph Tilings

By their constructions, glyphs represent local neighbourhoods of the tilings. When tilings are defined by one local neighbourhood, one glyph can represent the entire tiling. For instance, Fig. 4 shows an example with a single unpacked glyph that creates a tiling made of squares, hexagons, and triangles. Visualizing this tiling using the glyph can be supported by watching the gradual construction of the tiling in the polygon landscape as in Fig. 3. Creating a glyph to produce the tiling is a matter of constructing a neighbourhood, which can act as a building block of the tiling, and then relating it to its identical nearby neighbourhoods. The difficulty of predicting the outcome of a glyph can be offloaded to the Polygon Landscape Panel.

There is a many-to-one relationship between glyphs and tilings. The left image of Fig. 5 shows an alternative glyph that produces the tiling from Fig. 4. This alternative construction demonstrates another way of understanding the tiling as a network of paths between with hexagons that branch into three paths (Fig. 5) instead of six paths (Fig. 4). Exploring alternative constructions can aid in developing to deeper understandings of the tiling can be achieved. Concretely, an elaboration of a glyph

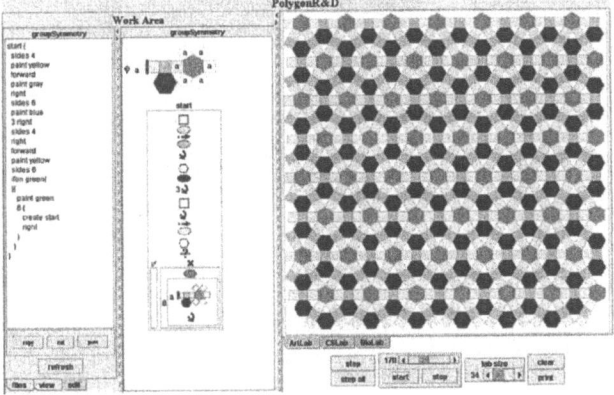

Fig. 4. An example of a single glyph tiling.

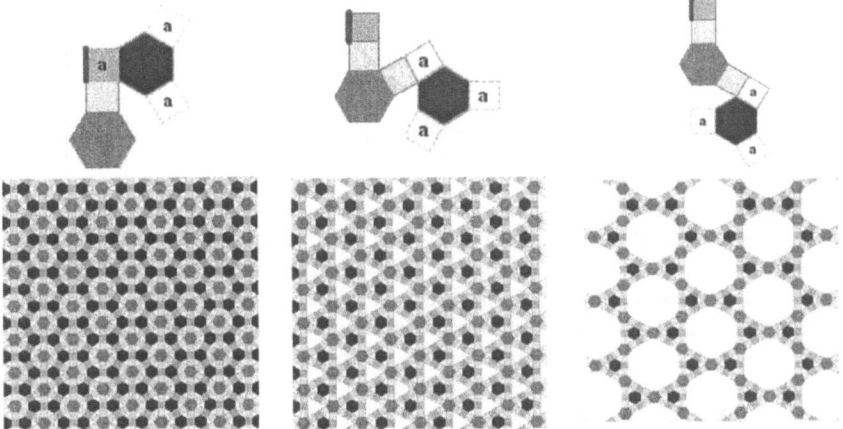

Fig. 5. Certain elaborations of the glyph on the left produce subsets on the tiling from Fig. 4

can expose the differences between alternative constructions of the same tiling. For instance, the second and third glyphs of Fig. 5 are elaborations of the first glyph. They result in tilings that difficult to create by elaborating the glyph from Fig. 4.

2.2.2 Negative Space

The tiling examples from Fig. 5 have features that may not be seen in the glyphs. For instance, the third example has large white rounded hexagons. PolygonR&D is restricted to using regular polygons as its basic shapes but the negative space, the gaps between the regular polygons, does not have this restriction. The creation of negative spaces often requires more experimentation in which the focus goes back and forth between the glyph construction and the resulting tiling. For example, Fig. 6 shows a complicated tiling designed by Kepler that incorporates five point stars and shapes with the outline of two fused decagons. The construction of this tiling was achieved in a piece-wise fashion where incomplete glyphs were used to partially construct the tiling. The partial tiling then helped to incrementally build and finish the glyphs.

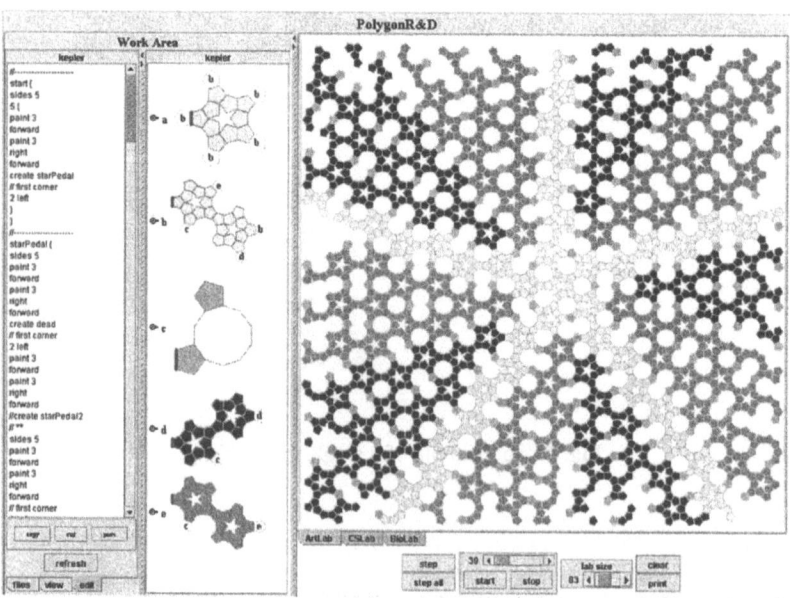

Fig. 6. A complicated tiling design by Kepler.

2.2.3 Multiple Glyph Tilings

Many tilings cannot be constructed from one local neighbourhood. Both the tilings from Fig. 1 and Fig. 6 are examples of this. These tilings are constructed by integrating a number of local neighbourhoods, which is reflected by their associated list of glyphs. Understanding how these local neighbourhoods are integrated seems to be facilitated by the glyph representation of the tiling's description. The glyph's visual description of the tiling often takes up less space than the textual description of the tiling. Notice in Fig. 6 that the complete set of glyphs that constructs Kepler's tiling is shown. At the same time, only a small portion of the textual description of the tiling is shown (one sixth, judging by the scroll bar). In this case, the amount of scrolling required when reasoning back and forth between the tiling and its glyphs is eliminated; this is not the case for the textual description.

Multiple glyphs can also aid in the construction of tilings that have only one local neighbourhood. One example would be if the defining neighbourhood were very large, in which case breaking down the neighbourhood into manageable sections may help organize its creation. A more important case occurs when the defining neighbourhood contains some smaller neighbourhoods that also repeat but do not define the tiling. In this way, repeating a section in the description is unnecessary.

3   Conclusions

PolygonR&D integrates index-sequential geometric glyphs to aid in bridging the text of the procedural description and the visuospatial pattern of the tiling. The transformation from a textual procedure to a glyph is supported by the glyph's interactive ability to be semantically unpacked into a list of icons. Each glyph is related to a section of the tiling. The sections connect and combine to form the tiling. The visual description of the list of glyph's interconnectedness is achieved through one-letter indices. The transformation from the glyph is supported by the interactive execution of the glyphs. Reasoning back and forth between the representations can aid in developing insight into the structure of tilings and how to construct them.

Experimenting with indexed-sequential glyphs in tools such as PolygonR&D can help us understand how to design interactive cognitive tools that support reciprocal reasoning between sentential and visual structures.

References

1. Devlin, K: Mathematics: The Science of Patterns. NY, Freeman and Company (2000)
2. Grunbaum, B, Shephard, GC: Tilings and Patterns: An Introduction. W.H. Freeman, New York (1989)
3. Sedig, K, Morey, J, Mercer, R, Wilson, W: Visualizing, interacting and experimenting with lattices using a diagrammatic representation. In G. Malcolm (Ed.), Multidisciplinary Approaches to Visual Representations and Interpretations. Elsevier Science (In press)
4. Peterson, D (ed.): Forms of representation. Intellect Books, Exeter, UK (1996)
5. Sedig, K, Rowhani, S, Morey, J, Liang, H: Application of information visualization techniques to the design of a mathematical mindtool: A usability study. Palgrave Macmillan Ltd. Information Visualization, (2003) 2(3): 142-16
6. Larkin, J, Simon, H: Why a diagram is (sometimes) worth 10000 words. In J. Glasgow, N. H. Narayanan, and B. Chandrasekaran (eds.), Diagrammatic reasoning: Cognitive and computational perspectives, The MIT Press (1995)
7. Myers, K, Konolige, K: Reasoning with analogical representations. In J. Glasgow, N. H. Narayanan, and B. Chandrasekaran (eds.), Diagrammatic reasoning: Cognitive and computational perspectives, The MIT Press (1995)
8. Post, FJ, van Walsum, T, Post, FH: Iconic techniques for feature visualization. In Proc. Visualization, IEEE Comp. Soc. Press (1995) 288-295
9. Ribarsky, W, Ayers, E, Eble, J, Mukherjea, S: Glyphmaker: Creating customized visualization of complex data. IEEE Computer (1994) 27(7):57-64
10. Spence R: Information Visualization. Addison-Wesley (2001)

Studying the Acceptance or Rejection of Newcomers in Virtual Environments

Pilar Herrero, Angélica de Antonio, and Javier Segovia

Facultad de Informática. Universidad Politécnica de Madrid.
Campus de Montegancedo S/N.
28.660 Boadilla del Monte. Madrid. Spain
{pherrero,angelica,fsegovia}@fi.upm.es

Abstract. In this paper, we present an experiment, called the "Old Man Experiment" carried out at the Universidad Politécnica de Madrid. This experiment pursued three objectives: to study the acceptance or rejection of newcomers, to study the field of "mixed reality" in more depth and to investigate perception. The "Old Man Experiment" was conducted as part of the Amusement Esprit Project 25197.

1 The Old Man Experiment

This is an interactive experiment, in which any spectator can take part. The experiment is based on a huge screen on which an avatar representing an Old Man will show up. The screen, large enough to represent the Old Man in full size (we used a 3m x 2m screen), must be located in a public space, and people must be allowed to get close to it. The system includes a hidden camera with two angles of vision (72° and 180°), to detect people moving and approaching. Our experiment was designed as it is showed in the Figure 1.

Fig. 1. The Old Man experiment

The Old Man was designed with *Alias Power Animator* on *Silicon Graphics O/2* stations, and the software was developed using *Direct3D and Microsoft Visual C++* tools. A *Frame Grabber SDK* and a *Data Translation DT3153* card were used to capture and observe the images.

M. Bubak et al. (Eds.): ICCS 2004, LNCS 3038, pp. 1004–1007, 2004.
© Springer-Verlag Berlin Heidelberg 2004

This is a two-scenario experiment. The first scenario shows the Old Man going for a walk around a beautiful green park near the Amusement Centre on a glorious evening, and the second scenario takes place in a square of the Amusement Centre. In both scenarios, the Old Man sits down on a bench and quietly starts to read a newspaper while a bird flutters by. It is then, while the Old Man is sitting on the bench (see Figure 2), that he starts to interact with his audience.

When people approach the bench, the Old Man will react. He is programmed to attract people or reject people depending on the situation. If anyone approaches the Old Man while he is reading his newspaper and gets too close, the Old Man could get a fright and drop the newspaper, shield himself with the newspaper or show some interest in the person. Some of the above-mentioned Old Man's gestures are illustrated in the Figure 2.

Fig. 2. The Old Man reactions

The behaviour of the Old Man is governed by an algorithm that contains two different situations. The first one, called attract situation, happens when the Old Man tries to catch the user's attention. In this scenario, the Old Man looks at the spectator and the application generates pre-programmed animations and gestures to catch the spectator's attention. In the second one, called rebuff situation, the Old Man rebuffs the spectator, trying to make him go away.

For the attract situation, we have defined different areas of reaction, as shown in Figure 3 and Table 1 bellow:

Table 1. Intervals in "Attract Situation"

Area	Range	The Old Man's Reaction
A	$(x1, x2), (x5, x6)$	Looks at the spectator
B	$(x2, x3), (x4, x5)$	Tries to catch spectator's attention
C	$(x3, x4)$	Watches the spectator's reaction

Where the values of the variables x_1, x_2, x_3, x_4, x_5 and x_6 depend on the angle of vision of the camera with which we are working. In this experiment, we located the origin of co-ordinates at the camera's position, the user's y-co-ordinate was a fixed value (y_1=1 metre) and the user's x-co-ordinate was x_1=-1, x_2=-0.75, x_3=-0.5, x_4=0.5, x_5=0.75 and x_6=1.

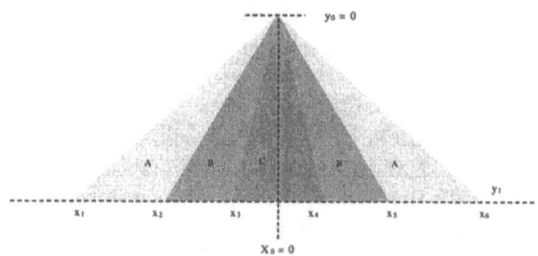

Fig. 3. Intervals in "Attract Situation"

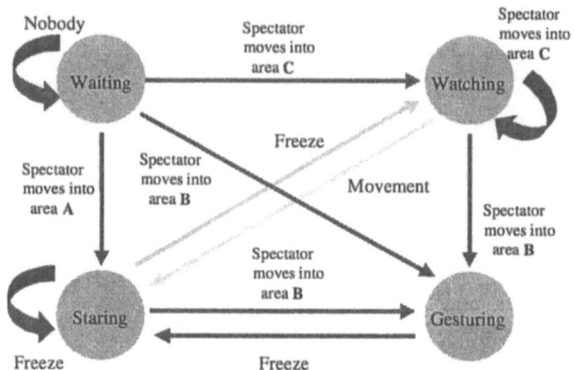

Fig. 4. State diagram in "Attract Situation"

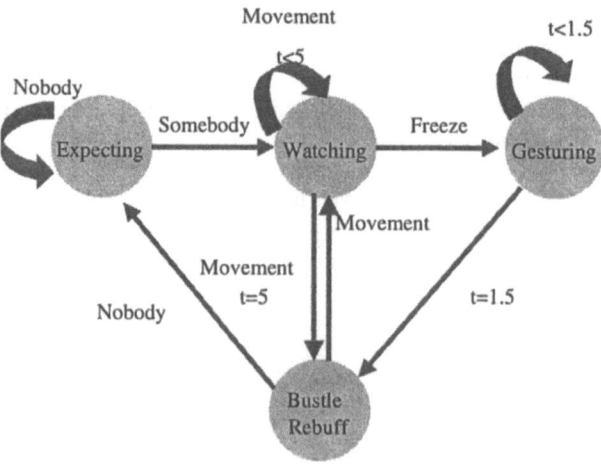

Fig. 5. State diagram in "Rebuff Situation"

According to the state diagram in the attract situation (Figure 4), the Old Man waits reading his newspaper until he perceives the presence of a spectator. Then, if the person is in area A, the Old Man looks at the spectator, if the person is in area B, the application issues gestures of attraction, and if the person is in area C, the Old

Man watches where the spectator goes. If the Old Man is watching the spectator and the spectator stops, the Old Man looks at him for a while. However, if he moves again, the Old Man will start to watch where he goes. If the Old Man is watching where the spectator is going and the spectator moves into area B, the application issues gestures of attraction. If the program issues gestures of attraction and the person stops, the Old Man will look at the spectator for a while.

As for the rebuff situation, the Old Man reacts according to the state diagram showed at the Figure 5. He waits reading his newspaper until he perceives the presence of a spectator. Then, the Old Man watches where the spectator is going for a while, after which the program issues a gesture of rejection. If the Old Man is watching the spectator and the spectator stops, the Old Man will look at him for a while. However, if the spectator stays still, the application will issue another gesture of rebuff. If the spectator moves while the program is in this state, the Old Man will again watch where the spectator goes until the spectator moves off.

2 Conclusions

In this paper we have presented an experiment, called the "Old Man Experiment" carried out at the Universidad Politécnica de Madrid. This experiment pursued three objectives: to study the acceptance or rejection of newcomers, to study the field of "mixed reality" in more depth and to investigate perception. This experiment was conducted as part of the Amusement Esprit Project 25197 [1,3] and it was running in parallel to another experiment, called *Run & Freeze* [2], which was also part of the Amusement project and which used the same system of perception for different purposes [4].

As a conclusion of this experiment, we realised the relevance of introducing a significant set of reactions to newcomers, how important the lateral area of perception is, and how user intentions and the Old Man's reactions are different in different areas. The "Old Man" was a very simple agent furnished with a simple model of perception - the camera's focus- and endowed with a very simple behaviour.

References

1. Herrero P., De Antonio A., Segovia J. *The Old Man Experiment*. I3net Annual Conference (I3AC'98). Nyborg. Dinamarca, 1998.
2. Herrero P., De Antonio A., Segovia J. *Run and Freeze*. I3net Annual Conference (I3AC'98). Nyborg. Dinamarca, Jun. 1998.
3. Herrero P. *Awareness of interaction and of other participants*. Technical Report. Amusement Esprit Project 25197, 1999
4. Herrero P. *A Human-Like Perceptual Model for Intelligent Virtual Agents* PhD Thesis. Universidad Politécnica de Madrid, June 2003.

Open Standard Based Visualization of Complex Internet Computing Systems

Seung S. Yang and Javed I. Khan

Media and Communications and Network Research Laboratory
Department of Computer Science
Kent State University
syang@cs.kent.edu, javed@kent.edu

Abstract. The emerging distributed internet computing paradigms envision involvement of large conglomeration of dynamic and internet-wide distributed computing resources. Visualization is an indispensable tool to confront the challenge of managing such complex systems. We present a monitor messaging framework for visualizing the lifecycle of a multiparty decentralized process. This design can be used as a blueprint for an open standard based visualization protocol for emerging internet computing systems.

1 Introduction

The Internet and particularly the Web is increasingly becoming a computation centric network. Emerging initiatives, ranging from scientific grids to application services networks, increasingly view network as an integrated platform for joint computing and communication rather than one only for communication. Research in topics such as composable services, active and programmable networking, etc. are pushing this envelop. In last four years, we have developed and experimented with a number of Ad-hoc active Internet Service Systems (made-to-order channel 1, transcoder channel 5, active prefetch-proxy, etc). Our experience with complex netcentric composable systems indicates that monitoring and management of complex distributed system is critical for enabling growth of complex distributed computing. In this paper we discuss a monitoring framework that we have successfully used in several of our systems Ad hoc Internet Service Systems (AISS). This infrastructure propagates the status information and control commands via multi-level decentralized agents and can be used to generate unified views of the overall operation of the services.

2 Design Considerations

Though traditional networking research has ignored visualization, but very recently there has been few pioneering works in the area of Grid visualization. Tierney et. al. [2] suggested an agent based monitoring system in Grid Environment. They use a direct connection between producer and consumer to reduce communication traffic. Waheed et. al [4] developed an monitoring infrastructure to share monitored data

M. Bubak et al. (Eds.): ICCS 2004, LNCS 3038, pp. 1008–1012, 2004.
© Springer-Verlag Berlin Heidelberg 2004

using common APIs. Another layer based visualization system was suggested by Bonnassieux et. al. [3]. They offer a flexible presentation layer for displaying monitoring data from a huge and heterogeneous environment. The framework we propose is particularly suitable for monitoring the lifecycle of loosely coupled and scalable complex multiparty active systems. Besides fulfilling the basic responsibility of process event reporting the design of our visualization framework offers the following distinguishing features. It allows sub-system to maintain its own status and control messages within it. A sub-system can further report its status and control messages to its upper system. Furthermore it supports several reporting modes to allow performance tuning. The time, type and content of messages all are decided initially by the system designer and can be overridden by the system operator or administrator at run time. A privileged user can freely controls and monitors the system status using a flexibly configurable multi-view visualization system from any authorized terminal.

3   Ad hoc Internet Service System Mode

3.1   Visualization of Control and Status Architecture

ABONE 6 is an operational network and provides an Internet wide network of routing as well as processing capable nodes. The software structure of ABONE node involves a native Node Operating System (NOS) and Execution Environments (EEs), which acts like a remote shell and provides a programming framework to ABONE applications. The ANETD is one of the developed EEs and allows users to obtain secured and controlled access to the ABONE resources. The AISS uses the ANENTD as its run time base infrastructure. However, the AISS is independent from the ANETD. AISS runs on our Virtual Switching Machine (VSM) EE which runs via ANETD. The Kent VSM can deploy and lunch authorized system components to build a service. [5]

Fig. 1 described the symbolic representation of visualization system architecture. The monitoring components of AISS are run on Kent VSMs. They are deployed and executed on Kent VSM as a part of a service construction. A Status Monitor (SM) processes status message of a sub-system. A SM stores status message structure descriptions and delivers or saves status messages of the sub-system. A Control Monitor (CM) handles control messages. A CM is added in a sub-system when the sub-system supports a control mechanism from outside of the system. Initiation and execution of a monitor is coordinated by sub-system management software.

3.2   Dynamic Message Binding and Interpretation

Based on the AISS design considerations, the visualization system should support 1) dynamic interpretation of the system status messages, 2) seamless navigation through layer abstraction and visualization of the given layer of a system, 3) uniform method of visualization at all levels. The meaning of a status message is represented in a well formed *status structure description* language and is gathered by a status monitoring

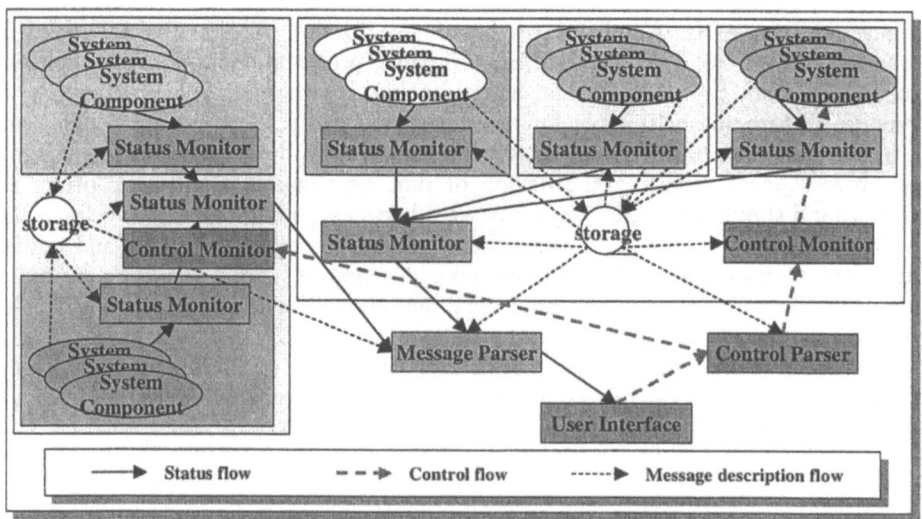

Fig. 1. Visualization System Architecture

system. When a new system component is developed, the descriptions of its status message structures and the descriptions of its state diagram are supplied together with the component. A status monitoring system and a visualization system dynamically binds and interprets the meaning of a status message with the given description.

The Visualizer integrates with a code server based hierarchical code server-based service deployment framework. Each system can have isomorphic sub-systems and/or code modules. Each time a system is installed (i.e. all of its sub-systems are launched) a hypothetical state-monitor is assumed to be concurrently instantiated. A set of messages are generated towards this state-monitor in sub-system's leader module. A visualization system can use a subset of the messages to present various perspectives on the system. The key challenge here is that these messages should carry enough information to identify it self with respect to the various perspective frameworks within which an active service operates along with the actual status information. Below we provide a try-partite identifier system. This message system encodes the fields in its messages (i) system identifier, (ii) subsystem module identifier, (iii) system state identifier, (iv) state execution count (v) system status, (vi) service instance identifier, (vii) service subsystem instance identifier, (viii) service instance status, (ix) service location instance identifier, (x) platform identifier (xi) platform status. The primary state identifier set i-iii is assigned by the programmer who has coded the active modules. This identifier set has to be *hierarchically unique* within a specific version of a specific software. The identifier set vi-vii are to be assigned are to be assigned by the active service administration system (such as EEs/ ANETDs) while installing and initializing instances of the service at each instantiating of loaded modules. Again, these identifier set has to be *hierarchically unique* within the service administration domain. The last identifier x is to be supplied by the active node owner. These are assigned when a node joins an active network domain. The status information iv and v is computed by the code modules at run time and thus its value is designed by the programmer. The service instance status information viii, if any, is passed on to the monitor messaging agents by the service administration local agent (such as node EE). The status information xi, if any, is set by the local node administrator during the period the service is running. The monitor messaging system collects and composes the messages prior to generating the messages. Messages can contain control flags to control the mode of reporting and even to filter the content to tune performance. The system allows three reporting modes (i) REAL-TIME, (ii) BATCH, (iii) TRACE-ONLY. In real-time mode the monitor messages are generated and sent when the code executes through the state points. In BATCH-ONLY mode the messages are generated at real-time but forwarded periodically in batch. The period is decided by a PERIOD field. The mode feature only modifies the time of sending the monitor messages but do not affect their content. Three flags are further used to negotiate filtering the three status fields in the messages. In every message sent by the monitor messaging agent the flags are set according to the current value of these flags. A set of control messages can be potentially sent in reverse direction to request change in these flags (and the PERIOD field).

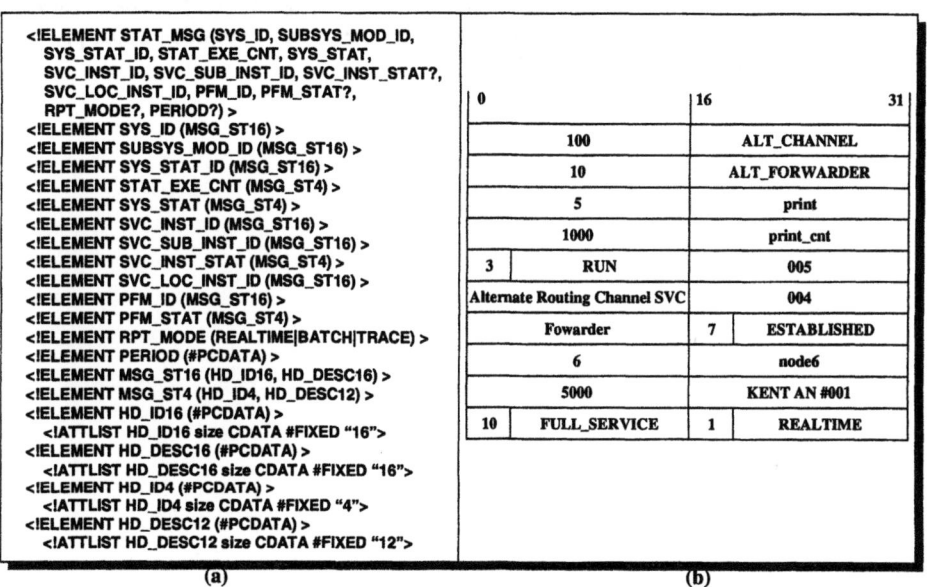

Fig. 2. A Status Message Structure and a Status Message

4 Conclusion

The suggested visualization system for Ad hoc Internet Service System gives intuitive status report and simple control using decentralized reporting for a multi-component multi-location internet service. Just like other IP based messaging (such as ICMP or TCP signaling, etc.) this monitor messaging should be wrapped with an authentication protocol. This overall design can be used as a blueprint for an open standard based visualization protocol for emerging internet computing systems. This work has been funded by the DARPA Grant F30602-99-1-0515 under its Active Network initiative.

References

1. Javed I. Khan, & S. S. Yang, Made-To-Order Custom Channels for Netcentric Applications over Active Network, Proc. Of the Conf. On Internet and Multimedia Systems and Applications, IMSA 2000, Nov 2000, Las Vegas, pp22-26.
2. B. Tierney, R. Aydt, D. Gunter, W. Smith, V. Taylor, R. Wolski, M. Swany, A Grid Monitoring Service Architecture, Global Grid Forum Performance Working Group, 2001
3. Bonnassieux F., Harakaly R., Primet P.: MapCenter: an Open GRID Status Visualization Tool, proceedings of ISCA 15th Int. Conf. on parallel and distributed computing systems, Louisville, Kentucky, USA, September 2002.
4. A. Waheed, W. Smith, J. George, J. Yan, An Infrastructure for Monitoring and Management in Computational Grids.
5. Seung S. Yang and Javed I. Khan, Delay and Jitter Minimization in Active Diffusion Computing, Int. Symp. on Applications and the Internet, Jan. 27-31 2003, Orlando, Florida.
6. Steve Berson, Bob Braden, Steve Dawson. Evolution of an Active Networks Testbed, Proceedings of the DARPA Active Networks Conference and Exposition 2002, pp. 446-465, San Francisco, CA, 29-30 May 2002.

General Conception of the Virtual Laboratory

Marcin Lawenda[1], Norbert Meyer[1], Tomasz Rajtar[1], Marcin Okoń[1],
Dominik Stokłosa[1], Maciej Stroiński[1], Lukasz Popenda[2],
Zofia Gdaniec[2], and Ryszard W. Adamiak[2]

[1] Poznań Supercomputing and Networking Center,
ul. Z.Noskowskiego 10, 61-704 Poznań, Poland,
lawenda@man.poznan.pl,
http://vlab.psnc.pl/
[2] Institute of Bioorganic Chemistry Polish Academy of Sciences,
ul. Z.Noskowskiego 12/14, 61-704 Poznań, Poland

Abstract. In the paper some theoretical considerations about virtual laboratory (VL) aspects are discussed. Possibilities of the VL system creation to control many laboratory apparatus in a remote way are considered by the authors. The main research goals on which the authors want to focus their attention are: laboratory framework and dynamic measurement scenarios.

1 Introduction

Virtual Laboratory (VL) is a heterogeneous, distributed environment, which allows scientists all over the world to work on a common group of projects [1]. This environment should allow conducting experiments with the usage of physical devices, doing simulation using the computational application, and communication between users working on the same topic. Similarly to typical laboratory tools and research techniques depending on the specific field of science, virtual laboratories can benefit from some collaboration techniques as tele-immersion, but they are not mandatory.

Research work of the authors is focused on virtual laboratory aspects [2] which have significant influence on creating a general system which will allow taking control on many different devices. In the virtual laboratory system there are many factors which are significant in our work. The most important of them are: laboratory framework and dynamic measurement scenarios.

The Virtual Laboratory system is developed in Poznań Supercomputing and Networking System in collaboration with the Institute of Bioorganic Chemistry and Radioastronomy Department of Nicolaus Copernicus University.

2 Virtual Laboratory Framework

In the virtual laboratory architecture we can distinguish four layers (see Fig. 1). The first of them is the Access layer. Originally it consists of tools which enable access to the laboratory resources and for presentation stored data. The most important modules are:

M. Bubak et al. (Eds.): ICCS 2004, LNCS 3038, pp. 1013–1016, 2004.
© Springer-Verlag Berlin Heidelberg 2004

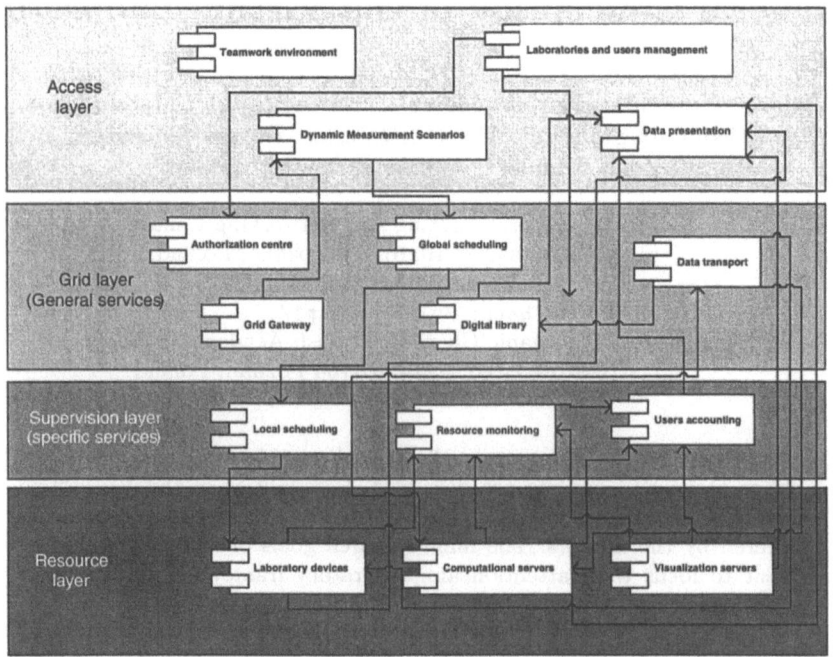

Fig. 1. The framework of virtual laboratory architecture

- Dynamic measurement scenarios - assure control and execution of the operations (experiments) chain, which is submitted by the user,
- Users and laboratories management - tools for creating new laboratories profiles, new users accounts, management of the access rights to devices, digital library and communication media,
- Data presentation tool - presentation experiment results collected in the data management system, on-line visualization tool and accounting data looking through,
- Users communication tools - allow users to communicate with one another

The second is the Grid layer where services which we can call general were accumulated. They are required in most laboratories instances. There are also connections with the outside grid services e.g. with Globus services [3]. The most important are:

- Authorization centre - granting (or not) access to many virtual laboratory resources, certificates management,
- Global scheduling - responsible for choosing the appropriate laboratory device and load balancing when possible,
- Digital library - storing experiments results and electronic publications,
- Data transport - responsible for download and upload data from/to destination machine,

- Grid Gateway - communication with the grid broker (e.g. GRMS [4]), sending computational tasks to the grid and receiving experiment tasks from the grid systems

On the supervision layer specific services are gathered. These services have to be implemented taking into consideration a given device and their specification. Usually it consists of:

- Local scheduling - tasks scheduling on a given device taking into consideration parameters and priority
- Resource monitoring - resources using control, tasks actual state control
- Users accounting - information about the used resources

Finally we can define the Resources layer which consists of devices to experiment execution and also the necessary software:

- Laboratory devices - laboratory apparatus and software for experiment execution
- Computational servers - also software for pre- and post-processing computation
- Visualization servers - also visualization software

3   Dynamic Measurement Scenarios

The experiment process execution in many types of laboratories consists of very similar stages. The given experimental process is often recurrent and must be executed many times by some parameters modification. In this situation the measurement scenarios conception seems to be very useful.

The conception of the dynamic measurement scenarios allows defining the process of an experiment in any way, from pre-processing, through executing the experiment, to the post-processing and visualization tasks. Users are also allowed to add their own module as a part of the scenario. Defining the measurement scenario allows to spare a lot of time during computation. The user does not have to wait for the end of a given process stage to submit another one. It is made automatically.

Initially, we divide the experiment execution process into four steps: preprocessing (data preparation to use as an input source), experiment (executed on real device), postprocessing (output data processing) and visualization.

This simple task's execution chain we call a static measurement scenario (SMS). It allows connecting particular stages only between themselves. Here, the execution path is specific and the user can not manipulate it.

To increase the possibility of the jobs scenario the dynamic measurement scenario (DMS) was defined. In the DMS model, besides the definition of the tasks execution sequence, we can also define some extra connections (e.g. loops, parallel connections), conditions on the connections and different lengths of the execution paths. In fact, DMS allows to define many SMS models in one scenario. Thanks to the possibility of the conditions defining on the connections paths the real path is determined during execution and can depend on computational results.

To properly define the dynamic measurement scenario the system has to have knowledge about connections which are enabled. In case of the laboratory type where the processing software is well known, it seems to be easy. The case will be more complex when we want to define the DMS model for a wider range of virtual laboratories.

To solve this problem we defined a special language which determines the following conditions: names of the connected applications, a condition (or conditions) connected with a given path, an additional condition which has to be met to pass a given path, e.g. when special conversion between application is needed and in the end a list of input or output files for applications.

An expertise from a particular discipline is necessary to define rules for DMS. It can be done by a laboratory administrator together with a domain expert.

4   Conclusions

The universal virtual laboratory framework allows to spare a lot of human energy and money during the development stage. It is possible to build many different types of laboratory systems which provide and adapt functionality to specified needs. Designers of a new laboratory have to focus their attention only on a few aspects connected with adaptation to a new data set, and a specific scheduling algorithm.

To proof our assumptions and as an exemplary implementation it is planned to make the NMR Spectrometer (Bruker Avance 600) and radiotelescope (diameter 32m) available to the Grid users [5].

References

1. Lawenda, M.: Laboratorium Wirtualne i Teleimersja. Poznań Supercomputing and Networking Center, RW nr 34/01, (2001), (in polish)
2. Lawenda, M., Meyer, N., Rajtar, T.: General framework for Virtual Laboratory. The 2nd Cracow Grid Workshop, Cracow, Poland, December 11 - 14, 2002
3. Adamiak, R. W., Gdaniec, Z., Lawenda, M., Meyer, N., Popenda, Ł., Stroiński, M., Zieliński, K.: Laboratorium Wirtualne w Środowisku Gridowym. Polish Optical Internet - PIONIER 2003, 9-11 April 2003, (in polish)
4. GridLab: A Grid Application Toolkit and Testbed http://www.gridlab.org/
5. Virtual Laboratory project http://vlab.psnc.pl/

Individual Differences in Virtual Environments

Corina Sas

University College Dublin, Belfield, Dublin 4, Ireland

Abstract. This paper summarises the results of several studies of individual differences among users navigating in virtual environments. These differences relate to performance of navigational tasks, and the degree of sense of presence experienced by the users. The individual differences addressed in this paper refer primarily to personality and demographic factors. The possibility of improving the design of virtual environments for a better accommodation of these differences is discussed.

1 Introduction

Individual differences is an umbrella term describing an entire field of research primarily involving psychology that focuses on aspects of behaviour that differentiate individuals from one another. Any attempt to accommodate individual differences should follow after they have been identified and after we have established the aspects with respect to which the individual differences emerged.

The work presented in this paper focuses on two major aspects which usually describe any interaction with a Virtual Environment (VE), such as navigation, and sense of presence.

Because of its prevalence, there is a tendency to take the navigational process for granted. However, navigation is a complex activity which requires seamless integration of several cognitive processes. The difficulties associated with navigation become especially obvious when it is performed in unfamiliar environments (i.e. one becomes lost). Due to their specific characteristics, VEs put additional demands on untrained users. In addition, within VEs the user set of actions is restricted, consisting mainly of navigation and locomotion, object selection, manipulation, modification and query [10].

Sense of presence is a psychological adjacent phenomenon experienced by users while (and not only) they interact with virtual reality systems.

This paper summarises the findings of some of our previous studies, based on which a set of factors that led to individual differences with respect to previously mentioned aspects, e.g. performance on spatial tasks and sense of presence, have been identified.

2 Individual Differences

Clinical studies involve an in–depth investigation of one or a limited number of subjects in order to highlight their individual unique profile. A more general

M. Bubak et al. (Eds.): ICCS 2004, LNCS 3038, pp. 1017–1024, 2004.
© Springer-Verlag Berlin Heidelberg 2004

approach to studying individual differences, and the one taken by this paper, strives to capture *dimensions* of individual differences rather than individual patterns. Accordingly, one can group subjects together, based on the commonly shared variance with respect to a particular aspect of their individuality. Such an approach leads to the identification of differences between groups of subjects (intra-group differences) as opposed to the differences among individuals within the same group (inter-group differences).

Identifying and consequently describing groups of individuals sharing common features is a preliminary stage in this kind of research. It leads to a descriptive taxonomy regarding the factors which trigger the individual differences. Attempts to highlight a causal relationship between these factors and other relevant aspects define another area of potential research in this field. The third area of research in this area is a theoretical one, aiming at explaining the structure and dynamics of individual differences [18].

3   Individual Differences in Virtual Environments

A significant amount of work has been carried out on individual differences in the area of Human Computer Interaction (HCI). The theoretical and empirical findings in fields like differential psychology and cognitive psychology present a great potential for the researchers in HCI [7]. Carroll's [2] work involved a hierarchical exploratory analysis of part of the studies in differential psychology. His findings regarding individual differences refer to general intelligence as a top level, which includes eight general abilities. Personality traits and cognitive style have been also considered of potential interest for HCI field.

Most of studies related to spatial navigation addressed the issue of individual differences with respect to navigation in abstract information space, such as hypermedia space or semantic space [4,3]. However, less research has been carried out in the area of spatial navigation within VE.

The basic goal of Virtual Environments (VEs) is to create a place for people to act [28]. An additional purpose, related primarily with the VEs developed for training spatial skills, addresses the users' ability to both learn and represent the spatial characteristics of such virtual spaces [29]. While considerable amount of work concerned the technological factors regarding the design of VEs, factors which might be related to the training effectiveness, little research has been done in the field of user characteristics. As Waller highlighted [29], this is unfortunate since individual differences are a major source of variation in performance in both real and virtual spatial tasks.

A major aspect related to VEs, which can greatly benefit from the insights into the psychology of individual differences, regards the possible transfer of skills from the VE to the real world. Since the inter-subjects variability in performance within a VE is higher than that for the analogous real tasks, it seems that knowledge acquired within a VE requires not only abilities for similar real tasks, but additional skills as well [29].

A major distinction among different types of VEs refers to immersive and non-immersive VEs. As opposed to non-immersive VEs, immersive VEs involve the restriction of users' senses in terms of their reference to the real world [8]. A desktop VE has been used in the studies summarised within this paper.

4   Study Design

The VE which has been used as an experimental testbed across our studies represents a training environment supporting technicians in the maintenance of engineering artefacts [17]. The ECHOES system consists of a virtual multi-storey building, each of the seven levels containing three rooms, e.g. lobby room, conference room, training room, library etc. Figures 1 and 2 presents a bird's eye view of the ground floor and first floor respectively.

Fig. 1. Bird's Eye View of Ground Floor

Fig. 2. Bird's Eye View of First Floor

The study involved three phases: familiarisation, exploration and performance measurement. Initially, users were allowed to become accustomed with the VE and to learn movement control. After this, they were asked to perform an exploration task within the virtual building which lasted for approximately 25 minutes. After the completion of this task, during which participants implicitly acquired spatial knowledge related to the VE, they were tested. Given the findings provided by studies on human spatial cognition, the strength of relationship between spatial abilities measured by psychometric tests and spatial knowledge acquisition from VE is arguable [29]. Therefore, in order to assess the level of spatial knowledge acquisition, we employed a search task. Users were placed on the third level and asked to find a particular room identified only through its name, and located on the ground floor.

The time needed to accomplish this task acted as an indicator of the level of spatial knowledge acquired within the VE. According to the time required for the search task, users have been identified as *low spatial users*, when they needed significantly longer time to find the library (Mean = 49 seconds), or *high spatial users* who found the library straight away (Mean = 7 seconds).

5   Individual Differences on Spatial Task Performance

The demographic factors which we considered of interest in understanding performances on the search task, were gender and prior computer games experience, since all participants were already familiar with computers. Findings indicate significant gender differences: male subjects outperformed female subjects by finding the room significantly quicker. These differences follow the line of findings in the area of spatial abilities that indicate better performances for males [11].

Understanding users' preferred manner of processing information opens a door towards their perception of the world, either physical or virtual. Several studies consider cognitive style and/or learning style important variable(s) potentially triggering individual differences in the way users perform [9]. However, little attention has been paid to the impact of personality cognitive style, defined in terms of Jung's personality types [13]. We conjectured that is might be a significant factor which could lead to differences on spatial task performance.

Without being statistically significant, findings suggest that individuals who are more extrovert, intuitive or thinking perform better in terms of task efficiency. Along the Extroversion–Introversion continuum, the extravert people are predominantly orientated towards the external world, thing which probably leads to an increased level of spatial awareness. On the Sensing–Intuition continuum, intuitive people look at the entire picture which emerges from the single parts, but also go beyond it. The search task is a complex one and its solution requires information perceived through our senses. However, this does not suffice and user has to perceive something which is beyond the immediate information. In order to be successful, one needs an internal representation of the spatial layout, or a so-called cognitive map. Most likely intuitive individuals are able to take the bird's eye view, as they would see the space from above. However, further work is required for validating this hypothesis. Along the Thinking–Feeling continuum, thinking type grounds his/her decision on logic and analysis. The need of thinking type to organise both things and ideas within its environment could prove beneficial in understanding it. Probably when both extrovert and thinking types are met within the same individual, he/she could achieve higher performance in building up the internal representation of the spatial layout through methodical coverage of the space.

6   Individual Differences on Navigational Pattern

Ford [9] investigated the individual differences in behavioural patterns of navigating in abstract VEs, triggered by cognitive style defined in terms of holist or serialist orientation. He showed that while holists like to have access to an overview of the underlying structure, serialists prefer keyword indices. As the best of our knowledge there have been no research on identifying patterns of human navigational behaviour within VEs.

Work carried out in one of our studies involved the use of self-organising maps for classifying users' trajectories [22]. This allowed identifying a set of five clusters, based on the shape of motion trajectory. These clusters helped differentiating study participants on the basis of efficiency of their navigational strategies. Clustering allowed the identification of *good* and *poor* motion trajectories and their associated characteristics, where *good* and *poor* have been determined in the light of users' performance.

Each cluster but one could prove beneficial for the performance of the search task, excepting one, which consisted of erratic trajectories, presenting lots of turns and straight line segments joined at sharp angles. Since more than 50% of trajectories composing this cluster belong to the subjects with the worst performance on the search task, we considered trajectories within this cluster as *poor* trajectories.

These findings indicate that there are individual differences in navigational patterns, which correlate with performance on spatial task, or in other words with users' spatial abilities. While high spatial users perform a good coverage of the space, by following a smooth path which resembles "going around the edge" feature, low spatial users navigate more erratically, performing more greater turns or crossovers. They seem to explore the environment in an unsystematic manner, which is reflected on navigation behaviour both on each level and across the levels. An analysis of these findings in the light of Ford's previously summarised results [9] determined us to conjecture that, through their pattern of navigation, high spatial users developed holistic strategies of exploration, while low spatial users seemed to be more serialistic in their approach.

7   Individual Differences on Experiencing Presence

One of the psychological phenomena experienced by users while they interact with virtual reality systems is sense of presence. It allows them *to be there* [23], to feel themselves immersed and moreover to perceive the virtual world as another world where they really exist. We defined presence as a psychological phenomenon, through which one's cognitive processes are oriented towards another world, either technologically mediated or imaginary, to such an extent that he or she experiences mentally the state of being (there), similar to one in the physical reality, together with an imperceptible shifting of focus of consciousness to the proximal stimulus located in that other world [19]. For measuring presence, we devised a questionnaire presented in Sas and O'Hare [21].

Within immersive VEs, most of the users become immersed, despite the intragroup variability based on cognitive factors. We conjecture that in this case, the presence is experienced because of the advanced technological aspects featured by these fully immersive systems. These technical issues are so impressive that they simply prevail over the cognitive determinants of presence. On the other hand, within non-immersive VEs whose technological infrastructure is less advanced, the user's experience of presence is mainly due not to the system characteristics, but rather to the associated human factors.

The potential set of personality factors which could impact upon sense of presence has been identified after a thorough review of presence literature [15,

12] accompanied by an approach in the psychology of hypnosis. Therefore, we choose to investigate the following factors: empathy, absorption, creative imagination, personality cognitive style and willingness to be transported in the VE. A presentation of each of these factors is briefly outlined below.

- *Empathy* was identified with a set of constructs associated with the responses of one individual to the experience of another [6]. It involves the ability to engage in the cognitive process of adopting another's psychological point of view, together with the capacity to experience affective reactions to the observed experience of others. In order to develop such capacities and exhibit empathic behaviour, one should be able to assume perceptual, cognitive and affective roles. Empathy was measured with Davis' Interspersonal Reactivity Index [6].
- *Absorption* construct elaborated by Tellegen [26] is defined as a state of "openness to experiencing, in the sense of readiness to undergo whatever experiential events, sensory or imaginal, that may occur, with a tendency to dwell on, rather than go beyond, the experiences themselves and the objects they represent". Absorption was measured with Tellegen Absorption Scale [27].
- *Creative Imagination* represents the ability to generate mental representations of objects, persons or events not immediately presented to the senses [24], was also considered to carry an impact upon presence [12]. This trait was measured with Barber and Wilson's Creative Imagination Scale [1].
- *Cognitive style.* Curry's Onion Model [5] proposes a hierarchical structure of cognitive styles, with the outermost layer referring to the individual's choice of learning environment, with the middle layer referring to information processing style and with the innermost layer consisting of cognitive personality style. Defined as the individual's tendency to assimilate information, cognitive personality style is an enduring and context-independent feature. Personality Cognitive Style was measured with Myers-Briggs Type Indicator (MBTI) [16].
- *Willingness to be transported into the virtual world.* The role of this factor can be better understood if one looks at it as a prerequisite for willingness to suspend disbelief, aspect often mentioned in relation to sense of presence. It seems to be a necessary condition for experiencing a high level of sense of presence [14,25] and ultimately enjoying a mediated experience of any kind (e.g. theatre, literature, television, film, VR).

With respect to the individual differences in experiencing sense of presence, findings indicate that individuals more absorbed ($t(30) = 2.10$, $p < 0.05$), more imaginative ($t(30) = 2.41$, $p < 0.05$), more empathic ($t(30) = 2.00$, $p < 0.05$) or more willing to be transported into the virtual world ($t(30) = 2.75$, $p < 0.01$), experienced a significantly greater level of sense of presence [21]. The way in which personality cognitive style shapes the experience of "being there" is indicated by the following results: individuals who are highly introvert ($t(28) = 1.68$, $p = 0.05$), sensitive ($t(28) = 1.47$, $p = 0.07$) or feeling type ($t(28) = 1.46$, $p = 0.07$) are more prone to experience a higher level of presence [20].

8   Conclusions

The results presented in this paper indicate the significant impact of some personality factors on spatial performance, navigational patterns and sense of presence experienced by users during their interaction with the system. These findings indicate that "one size fits all" is an obsolete concept when it comes to the design of VEs. Since the ultimate goal of studying individual differences is to enhance system usability, once they have been identified and their impact is significant, efforts should be made to accommodate the individual differences.

In this context, an important aspect to be considered is the purpose of the application. Whether the VE is designed to enable humans to exceed the barrier of our limited senses for allowing access to *inaccessible* spaces, to improve skills, to cure, or to entertain, greater attention should be paid to different aspects, and accordingly to the set of factors which could trigger remarkable individual differences on these aspects. However, the distinct goal of any VE could be placed on a continuum between two poles: best performance–most pleasure.

When, for instance, the VE is designed for medical applications, its main purpose is to enable surgeons to perform at least as well as they would perform in reality (with or without presence). In this case, particular attention should be given to introvert, sensitive and feeling types who experience a higher level of sense of presence but whose performance might be reduced. For example, a VE for games should be particularly designed to allow extrovert, intuitive and thinking type individuals not only to perform better in spatial tasks but to experience presence as well.

References

1. Barber, T.X., Wilson, S.C.: The Barber suggestibility scale and the creative imagination scale: Experimental and clinical applications. The American Journal of Clinical Hypnosis 21(2-3) (1979) 84–108
2. Carroll, J.B.: Human Cognitive Abilities: A Survey of Factor-Analytic Studies. Cambridge University Press, New York (1993)
3. Chen, C.: Individual differences in a spatial-semantic virtual environment. Journal of the American Society for Information Science 51(6) (2000) 529–542
4. Chen, S., Ford, N.: Towards adaptive information systems: Individual differences and hypermedia. Information Research 3(2) (1997)
5. Curry, L.: An organization of learning styles theory and constructs. ERIC Document (1983) 235-185
6. Davis, M.: Empathy: A social psychological approach. Westview Press, Oxford (1994)
7. Dillon, A., Watson, C.: User analysis in HCI: The hystorical lesson from individual differences research. International Journal of Human–Computer Studies 45(6) (1996) 619–637
8. Fernie, K. and Richards, J.D.: Creating and Using Virtual Reality: A Guide for the Arts and Humanities. URL: http://vads.ahds.ac.uk/guides/vr_guide/glossary.html (10.03.2003)

9. Ford, N.: Cognitive styles and virtual environments. Journal of American Society for Information Science 51(6) (2000) 543–557

10. Gabbard, J., Hix, D. Taxonomy of Usability Characteristics in Virtual Environments. Virginia Polytechnic Institute, Blackenburg (1997)

11. Goldstein, D., Haldane, D., Mitchell, C.: Sex differences in visual–spatial ability: The role of performance factors. Memory and Cognition 18(5) (1990) 546–550

12. Heeter, C.: Being there: The subjective experience of presence. Presence: Teleoperators and Virtual Environments 1(2) (1992) 262–271

13. Jung, C.G.: Psychological types. In Ress, L., McGuire, W. (eds) The Collected Works of C.G. Jung, Vol. 6 Princeton University Press Princeton (1971)

14. Laurel, B.: Computers as Theatre Addison-Wesley Longman Publishing Co., Inc, Boston (1993)

15. Lombard, M., Ditton, T.: At the heart of it all: The concept of presence. Journal of Computer–Mediated Communication 3(2) (1997)

16. Myers, I.B., McCaulley, M.H.: Manual: A Guide to the Development and Use of the MBTI. Consulting Psychologist Press, Palo Alto, (1998)

17. O'Hare, G.M.P., Sewell, K., Murphy, A., Delahunty, T.: An immersive training experience. In Brusilovsky, P., Stock, O., Strapparava, C. (eds) Lecture Notes in Computer Science Vol. 1892 179–188 Springer-Verlag, Berlin (2000)

18. Revelle, W.: Individual differences. In Kazdin, A. (ed) Encyclopedia of Psychology Oxford University Press (2000)

19. Sas, C. and G.M.P. O'Hare, G.M.P.: The presence equation: An investigation into cognitive factors underlying presence within non-immersive virtual environments. In 4th Annual International Workshop on Presence (2001)

20. Sas, C., O'Hare, G.M.P.: Impact of cognitive style upon sense of presence. In Jacko, J., Stephanidis, C. (eds) Proceedings of the 10th International Conference on Human–Computer Interaction 696–700 (2003)

21. Sas, C., O'Hare, G.M.P.: Presence equation: An investigation into cognitive factor underlying presence Presence: Teleoperators and Virtual Environments 12(5) (2003) in press.

22. Sas, C., O'Hare, G.M.P., Reilly, R.G.: On-line trajectory classification. In Lecture Notes in Computer Science Vol. 2659 1035–1044 Springer-Verlag, Berlin (2003)

23. Schloerb, D.W., Sheridan, T.B.: Experimental investigation of the relationship between subjective telepresence and performance in hand-eye tasks. Telemanipulator and Telepresence Technologies 62–73 (1995)

24. Singer, J.: Imagination. In Kazdin, A. (ed) Encyclopedia of Psychology Vol. 4 227–230 Oxford University Press, Oxford (2000)

25. Slater, M., Usoh, M.: Representations systems, perceptual position, and presence in immersive virtual environments. Presence: Teleoperators and Virtual Environments 2(3) (1993) 221–233

26. Tellegen, A.: Practicing the two disciplines of relaxation and enlightenment: Comment on "Role of the feedback signal in electromyography biofeedback: The relevance of attention" by Qualls and Sheehan. Journal of Experimental Psychology: General 110 (1981) 217–226

27. Tellegen, A.: Brief manual for the differential personality questionnaire. Unpublished manuscript. Department of Psychology, University of Minnesota, Minneapolis (1982)

28. Tromp, J.G.: Methodology of distributed CVE evaluations. Proceedings of UK VR SIG (1997)

29. Waller, D.: Individual differences in spatial learning from computer–simulated environments. Journal of Experimental Psychology: Applied, 8 (2000) 307–321

Ecological Strategies and Knowledge Mapping

Jose Bidarra[1] and Ana Dias[2]

[1] Universidade Aberta,
Rua da Escola Politecnica 147,
1269-001 Lisboa, Portugal
bidarra@univ-ab.pt
[2] TecMinho,
Universidade do Minho,
Campus de Azurem,
4800-058 Guimaraes, Portugal
anadias@tecminho.uminho.pt

Abstract. In this paper we discuss the concept of ecological strategy in relation to cognitive artefacts that reflect the expressive power of learners engaged in the development of significant knowledge paths. Our experiments show that learners strive to make sense of the fragmentary information surrounding them and in this endeavour they may benefit from the creation of knowledge maps, i.e. branched structures of knowledge that depart from a central node, a starting point or a homepage. In the course of creating these fractal hyperspaces on the Web, with integration of varied multimedia materials, the range of cognitive strategies used during the learning process becomes evident, both at individual and collective level. Perhaps ecological strategies may encourage a new research approach regarding the study of knowledge construction within complex hypermedia environments.

1 Introduction

Inspired by Resnick's *eThinking* [5], we discuss in this paper the need for a conceptual framework integrating multimedia hyperscapes [1], [4], with a new way of thinking and a cultural perspective that go beyond the traditional "clockwork approach" to reach a further "ecological approach". Ever since the 17th Century we have been presented with Newton's mechanical view of the universe, nonetheless, in the present day new ideas from ecology, ethology and evolution are influencing scientific research in many areas. Ecological strategies are very common in the biological world and share two common characteristics: **response to local conditions** and **adaptation to change**. These can also be described as decentralized, simple, flexible and robust problem solving strategies that transfer without difficulty to networked environments.

A network is, in its essence, a tissue of collaborations of many parts that can create a critical mass and reach certain states in an extremely effective way. Taking the WWW network, covering the whole planet, we can say that its complexity is so big that at the local level we do not perceive the global changes that are taking place. We just scratch on the surface and this has consequences for our lives. Perhaps we are at

M. Bubak et al. (Eds.): ICCS 2004, LNCS 3038, pp. 1025–1029, 2004.
© Springer-Verlag Berlin Heidelberg 2004

the threshold of a new culture, no doubt very rich in information, that forces us to read history in a non-linear way. In other words, sequential reasoning becomes difficult as we try to catch up and order the many fragments of information that reach us every hour, every minute, every second.

In this Web ecosystem we are becoming - perhaps our "selves" are becoming – digital like in character. Our identities, our communications, our work is being constructed by and through digital technology (e-mail, homepages, courseware, etc.). Any division between people and technology is becoming more and more artificial as we are constantly building hypermedia spaces that grow in an organic and generative way in the course of human interactions. This is the environment where learning occurs these days. For instance, when we have a large number of students connected through the network, we can create activities based on ecological principles, since these involve the participation of all the contributors and the combination of the best solutions to solve complex problems. Also, the development of new tools, software or digital artifacts, stimulates the creation of communities of practice (developers, experts, students, teachers), engaged in a construction process.

2 Ecological Learning Strategies

The creation of multimedia hyperscapes [1], [4], in essence the construction of networked or branching cognitive artefacts, tries to reflect the expressive power of learners in the development of significant knowledge paths. As people try to make sense of the fragmentary information surrounding them they tend to create branched structures of knowledge that depart from a single node, a starting point or a homepage. In the beginning there is something that triggers new thoughts, perhaps an unexpected question or a provoking idea.

In our experiments at Universidade Aberta the construction of knowledge is achieved through concept mapping and related cognitive activities. Since the Web makes possible new modes of collaboration and interaction, groups of students can work together very effectively on such tasks as gathering information, designing Web pages, discussing relevant topics, organizing events, writing assignments, making multimedia presentations, among other. In the course of creating fractal hyperspaces on the Web, with integration of varied multimedia materials, the range of cognitive strategies used by students during the learning process becomes evident, both at individual and collective level. As a result, assessment of portfolios turns out a straightforward task for the professor or trainer.

But reality is somehow more complex. Constructing knowledge by means of multimedia hyperscapes represents a shift from a very centralized didactic style, based on the teacher as the main source of knowledge, to a flexible networked learning environment where the student must be able to construct meaningful structures through connections to multiple sources of information. In particular, it implies ecological habits of thinking within a vast, complex and unpredictable environment. Immersed in a technological maze – the World Wide Web is often confusing - new modes of interaction seem to materialize. After studying the modes of hypermedia development used by students and teachers working as a team, relying on cognitive and concept mapping tools, we may understand the (varied) mechanisms used by learners to dominate the complexity of the vast available resources.

Furthermore, the Web may also be compared to natural ecosystems because it has no centralized decision-making device. Paradoxically, there is no apparent leadership to tell learners what to do but the ecologies of the Web make possible the use of information and knowledge to attain effective global results.

In this context is important to distinguish between knowledge and information; according to Salomon [6]:

- Information is discrete, knowledge is arranged in networks with meaningful connections between the nodes.
- Information can be transmitted as is; knowledge needs to be constructed as a web of meaningful connections.
- Information need not be contextualized; knowledge is always part of a context.
- Information requires clarity; the construction of knowledge is facilitated by ambiguity, conflict and uncertainty.
- Mastery of information can be demonstrated by its re-production; mastery of knowledge is demonstrated by its novel applications. (p. 4)

Eventually a greater problem emerges: how do we integrate interaction processes, authoring approaches and rich elements of hypermedia information in order to improve the effectiveness of educational environments? Are ecological strategies the answer to the mounting complexity of current learning systems?

3 Designing Knowledge Hyperscapes

Ecological learning strategies may be supported by the construction of multimedia hyperscapes; in this regard we must consider that Web design activities:

- introduce the notion of hyperlinked knowledge spaces enriched with digital multimedia (i. e. establishes the concept of a complex ecosystem);
- suggest a potential for the development of highly interactive constructivist learning environments (i. e. enables learner's response to given situations);
- permit the creation and growth of networked learning communities using synchronous and asynchronous communication devices (i. e. facilitates adaptation to new or changing conditions).

Within our learning activities at Universidade Aberta we are developing models that enable the construction of hyperscapes through knowledge mapping. This technique was developed for representing knowledge in graphs that constitute tree-like structures [2], [3]. These consist of nodes and links, where nodes represent concepts and links represent the relations between concepts. Concepts, and sometimes links, are labeled and may be categorized. The developing patterns of association and branching create fractal structures. Like clouds or trees, they form physical structures that do not possess a defined form; yet in fractal structures we can always describe other levels or scales of its structure where we always find the same basic elements (self-similarity). Fractal geometry establishes algorithms to describe/create fractal structures but these are not relevant at this time. Knowledge maps or hyperscapes may have the aspect of fractal structures (Fig. 1).

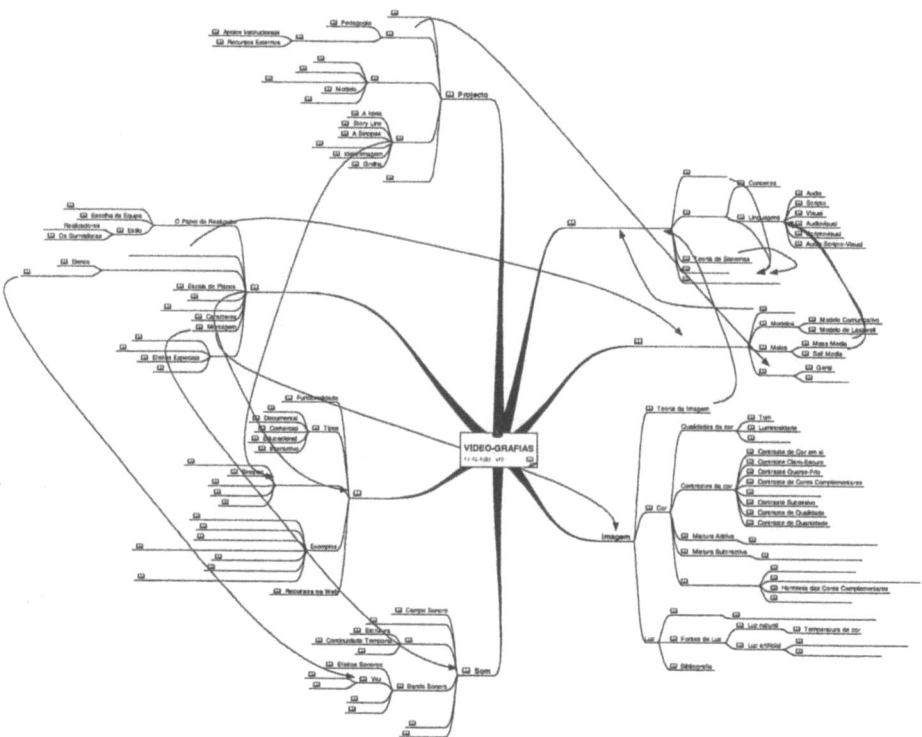

Fig. 1. Global visualization of a branched knowledge structure created with a mapping tool (MindManager™).

While recognizing the value of knowledge mapping, we were able to witness in our experiments that authoring with hypermedia tools is a rather difficult task for most people. Content creation, whatever the form and technology used, is a huge challenge for the average person and this explains why we have professional Web designers, movie makers, writers and musicians. Nevertheless, we believe that future scientists will also be artists – they will go beyond the "exact words" to reverberate across the culture shared with their audience. By establishing the right context we may have more success in recreating each scientist's own experience. The way to establish context, as we propose here, is through ecological thinking and the construction of multimedia hyperscapes.

4 Future Learning Environments

We have focused on the previous section on the creation of multimedia hyperscapes departing from an ecological thinking position. Basically, when we have a number of learners connected over the network we can create activities based on **ecological principles** that involve the evaluation of all the relevant information and the combination of the best solutions to solve complex problems within a suitable time.

The preliminary findings of our research suggest that activities leading to the mapping of concepts are effective to motivate students and increase the level of interaction with learning resources. We believe that any subject matter will be mastered more readily and more thoroughly when students become able to derive intrinsic rewards from dealing with complex domains. On the other hand, ecological ideas are possible to apply only within a set environment with a specific group of people trying to solve relevant problems, transforming information into knowledge.

Nonetheless, the construction of knowledge is a problem of conceptual development based on an evolutionary process with a rather unpredictable outcome. Implicit in this process are pro-active learning strategies, the collaboration with peers and other students and the adoption of a bold perspective concerning the problems to solve. In this context knowledge representations may take many forms. For instance, "chaotic" elements that enter the processes in creative activities (e.g. generation of ideas) have to be managed ad hoc, according to each learner's path and progression, to arrive at a final conceptual representation that is valid. This does not mean that a less professional approach is applied but that a great deal of the "authority" usually attributed to the teacher is not possible or desirable. This is a problematic notion that certainly points to a clear paradigm change in education.

Every time ecological strategies are chosen, decisions are based on emerging information, not on fixed or planned solutions. As the learner's environment changes, ecological strategies adjust and deliver new solutions adapted to new circumstances. Our challenge now is to find out how hypermedia mapping tools enable students to engage with their subject matter while working in communities of practice and adopting a symbolic world that will continue to provide curiosity and interest.

References

1. Bidarra, J., N. Guimarães, P. Kommers (2001). *Handling Hypermedia Complexity: Fractal Hyperscapes and Mind Mapping.* CINTEC 2001, Aveiro.
2. Buzan, T. (1995). The MindMap Book (2 ed.), BBC Books, London, UK.
3. Gaines, B. R., Shaw, M. L. G. (1995). *Concept Maps as Hypermedia Components.* URL: http://ksi.cpsc.ucalgary.ca/articles/ConceptMaps/CMa.html. Accessed February 1998)
4. Guimarães, N., T. Chambel, J. Bidarra (2000). From Cognitive Maps to Hypervideo: Supporting Flexible and Rich Learner-Centred Environments. *Interactive Multimedia Electronic Journal of Computer-Enhanced Learning* (Wake Forest University – USA). URL: http://imej.wfu.edu/articles/2000/2/03/index.asp. Accessed October 2000.
5. Resnick, M. (2002). *Thinking Like a Tree (and Other Forms of Ecological Thinking).* URL: http://llk.media.mit.edu/projects/eThinking/tree.html. Accessed June 2002.
6. Salomon, G. (2000). *It's Not Just the Tool, But the Educational Rationale that Counts.* Keynote Speech, ED-MEDIA 2000, Montreal.

Need for a Prescriptive Taxonomy of Interaction for Mathematical Cognitive Tools

Kamran Sedig

Cognitive Engineering Laboratory
The University of Western Ontario
London, Ontario, Canada
sedig@uwo.ca

Abstract. Cognitive tools encode and display information in different representational forms. Through interaction with these representations, people can make sense of and reason about the information. Mathematical cognitive tools can allow people to interact with mathematical ideas in ways and levels not hitherto possible. These tools can support reasoning and enhance the experience and understanding of mathematics. However, currently, there exists no prescriptive taxonomy of interaction with mathematical representations to guide designers of mathematical cognitive tools. Given different representations, cognitive activities, and users, such a taxonomy can help designers know what interaction techniques to use, when, and to what end. This paper discusses the need for the development of a comprehensive prescriptive taxonomy of interaction with visual representations used in mathematical cognitive tools. Additionally, the paper outlines some inter-related lines of action that can be followed to develop such a taxonomy.

1 Introduction and Background

Mathematical knowledge is an important asset in knowledge societies, where people work with their minds more than their bodies. Yet, due to the highly formal nature of mathematics, its abstract representational forms, and the means by which it is communicated, many people do not find mathematical ideas readily understandable, nor engaging. New computational technologies can help. Just as technological tools enhanced the physical reach and capabilities of people during the Industrial Revolution and helped increase their productivity, computational technologies can enhance our minds [1]. As Norman [2] states, "Without external aids, memory, thought, and reasoning are all constrained... The real powers come from devising external aids that enhance cognitive abilities... It is [external aids] that make us smart."

One category of external aids are cognitive tools or technologies, tools that people can use to support, guide, and enhance their mental activities and increase their productivity.[1] These tools allow for the encoding and displaying of information in differ

---

[1] Many terms have been used to signify cognitive tools: mindtools, instruments and technologies of the mind, information artifacts, mental gadgets, and so on [1, 2, 3]. This paper uses the term cognitive tools to refer to interactive computational devices, specifically interactive software applications.

M. Bubak et al. (Eds.): ICCS 2004, LNCS 3038, pp. 1030–1037, 2004.
© Springer-Verlag Berlin Heidelberg 2004

ent representational forms. Through interaction with these representations, different cognitive tasks can be performed, such as visualizing, analyzing, interpreting, meaning making, decision making, planning, problem solving, and learning [4]. These tools can be our partners in cognition, carrying some of the cognitive load and helping us perform mental activities of different degrees of complexity, some even impossible to accomplish without their aid [3, 4]. Examples of these tools include interactive visualization software to customize and observe climatological patterns, mind mapping tools to help externalize and organize thoughts and concepts, and online interactive mathematical applets to investigate how velocity and position graphs relate. Mathematical cognitive tools (MCTs) are a subset of cognitive tools, intended to support interaction with mathematical objects, concepts, and problems. By using appropriate, innovative human-computer interaction techniques, MCTs can allow people to explore mathematical ideas in ways and levels not hitherto possible, supporting reasoning and enhancing understanding of mathematics, and, hence, contributing to the creation of a more mathematically literate and productive society.

Two of the main components of MCTs are: representations and interaction.

1.1 Representations

External representations play a central role in the communication and understanding of mathematics. There are two broad types of representations: algebraic and visual. The former are made of formal, language-like symbols; the latter are made of image-based, graphical notations [5]. This paper is concerned with the latter: visual mathematical representations (VMRs). Research shows that human beings process and manage visual representations more efficiently than algebraic representations [5, 6].

Some examples of VMRs from different areas of mathematics include 3D and 4D geometric structures, statistical graphs, and tiling patterns. These graphical notations encode structural, relational, functional, and semantic properties of concepts, objects, patterns, problems, and ideas in mathematics. In other words, they embody the meaning that the images try to communicate. Effectiveness and appropriateness of an image depend to a large degree on how its users interpret it and make sense of its inherent meaning as well as how well it fits and supports the cognitive task [2, 5, 7]. Making sense of and reasoning with VMRs often depends on a process in which the user browses a visual image, poses 'what if' questions to it, and rearranges it until its meaning is gradually understood and then uses the representation for performing cognitive tasks [8]. This cognitive process of decoding and application can be mediated and supported through appropriate user interaction with VMRs.

1.2 Interaction

Interaction with a VMR has two aspects: a user acting upon the VMR through the intermediary of a human-computer interface, and the VMR communicating back through some form of reaction or response [9]. This addition of reaction, responsiveness, or interactivity to static visual representations enhances their communicative power and leads to some general benefits [7, 10, 11, 12]: supporting emergent understanding of encoded ideas; making mental manipulation of ideas easier; providing opportunities for experimentation, discovery, and hypothetical reasoning; supporting 'doing' rather than 'watching'; facilitating acquisition of qualitative insight into the nature of representations; supporting cognitive tracing and computational offloading by allowing users to think in partnership with a representation; and, serving as a coordinator between the internal mental models of users and an external representation.

Although, as discussed above, we know that interaction has general benefits, designing interaction to effectively mediate, support and enhance cognitive activities is challenging [9, 10]. It has been shown that using the wrong type of interaction can actually constrain human thinking and have negative, undesirable consequences [10, 13]. Just as users use different reasoning and thinking styles to approach different cognitive tasks [4], so they need different interaction techniques to mediate and support their reasoning with different VMRs [14]. Interaction technique and style play a crucial role in how users' minds are engaged with VMRs – that is, how they reason, what cognitive strategies they employ, how deeply they process the visual information, what knowledge they construct, and how well they learn [9, 13, 14, 15]. The types of VMRs used and the various given cognitive tasks for an MCT have a significant bearing on the types of interaction that are effective and should be used. Just as the effectiveness of a representation depends on whether it is appropriate for a given task, a given interaction technique should also fit the task, as it extends and mediates the representation in the process of cognitive activities.

1.3 Design Taxonomies

Being a young research area, much of the early research in cognitive technologies has been devoted to the building of domain-specific tools. This process of tool building has been needed for experimentation and has led to dozens of innovative interaction design techniques across different types of cognitive tools. When designing interaction for VMR-based MCTs, however, there exists no conceptual framework to guide the design. Often times, designers use techniques developed to support productivity tasks, which may not fit learning tasks. These tasks have different requirements and goals. In productivity tools, the goal is to reduce the user's cognitive effort; whereas in learning tasks, the goal is to engage users in mindful cognition [9, 15]. To address this type of problem in human-computer interaction, task and interface taxonomies are often developed [10]. A taxonomy abstracts, articulates, decomposes, and organizes a problem space, which, consequently, can facilitate analysis and understanding, provide opportunities for comparing different techniques, guide designers, and foster innovation and advancement [10]. There are two broad types of taxonomies: descriptive and prescriptive. Descriptive taxonomies characterize the decomposed elements of the problem space and provide examples of them. Prescriptive taxonomies, on the other hand, prescribe design rules and guidelines for the elements of the problem space and can provide best-practice examples of existing systems and techniques. Currently, there exist a number of taxonomies dealing with visualization and visual representations, interface styles, interactive externalizations, and interface tasks [3, 10, 12, 16, 17]. Though some of the terms and ideas used in these taxonomies are relevant to the objectives of designing MCTs, none of them prescriptively address how these tools should be designed to fit and support the cognitive tasks of the users.

Recently, Sedig and Sumner [9] have developed a descriptive taxonomy of interaction with VMRs. This taxonomy organizes and characterizes the different interaction techniques that can be used in MCTs. The techniques are organized and characterized according to their common features. The taxonomy consists of three primary interactions (conversing, manipulating, and navigating) and twelve secondary interactions (animating, annotating, chunking, constructing, cutting, filtering, fragmenting, probing, rearranging, repicturing, scoping, and searching). The taxonomy also identifies ten dimensions of interactivity (affordance, computational offloading, constraints, control, distance, feedback, flow, focus, participation, and scaffolding). Though a step in the right direction, this taxonomy is not prescriptive.

This paper proposes that there is a need for systematic research to develop a comprehensive prescriptive taxonomy of interaction for VMR-based MCTs. As online MCTs are now being delivered on the WWW as applets [17], and with the pervasiveness of the Internet, MCTs can virtually affect millions of people in terms of their application of mathematics in knowledge development and production. Since interaction is one of the central, distinguishing features of e-learning, a prescriptive taxonomy of interaction can guide the design and evaluation of not only MCTs, but also other types of cognitive tools. Mathematics has a rich repertoire of representational forms. As such, many interface concepts and interaction techniques developed for MCTs can transfer to cognitive tools embedding other knowledge domains.

2   A Proposed Strategy

This section proposes a strategy for conducting research to develop a prescriptive taxonomy of interaction and outlines some of the methodological actions needed to achieve the objectives of this type of research.

2.1   Problem Statement

Before discussing how a prescriptive taxonomy can be developed and what lines of action are needed for this research, this paper presents a problem statement to demonstrate some questions that may arise in the course of designing interaction for MCTs. The problem statement can be as follows:

> Given a VMR, a cognitive task or activity, and a prototypical user, what interaction technique(s) should be designed to render an MCT effective— that is, what technique(s) would support and enhance the user who is engaged with the cognitive task involving the VMR? How should interactivity (e.g., feedback or scaffolding) be operationalized to enhance the technique(s)?

Three scenarios are presented to situate the problem in real cases:

1) Chemists need to have good cognitive maps of the structures of different 3D lattices in order to reason about crystals. Given a tetrahedral lattice structure, the goal is to help chemistry students form an accurate cognitive map of the structure so they can mentally visualize it and reason about similar structures. What technique should be used to support the task of forming a cognitive map of a 3D

structure? Should the user 'walk' on the structure or 'fly' through and/or around it? Should the user annotate the structure by marking it with visual breadcrumbs?

2) Four-dimensional polytopes can have many and varied geometric components, making their visualized image complex, cluttered, and noisy. Given a rendered visualization of a 4D polytope, a mathematician must make sense of the image by identifying, locating, and distinguishing its faces and cells, which are some typical cognitive tasks performed on visualizations [16]. What interaction techniques support these cognitive tasks? Should the user interact with the VMR directly or not? Should the visualization be fragmentable?

3) Polyhedra solids can be transformed from one shape to another by truncating and/or augmenting their vertices and edges. Given a set of polyhedra solids, the goal is to explore and investigate how these transformations take place. Should these transformation processes be animated or should the user manipulate these VMRs to morph them? Should the interaction be continuous or discrete? Should the user interact with these VMRs directly or indirectly? What are the tradeoffs?

Without a framework within which to analyze and answer these questions, it is difficult to know how to design an MCT, the design remains arbitrary, and it is not easy to predict the impact the design may have on its users.

2.2 Towards a Prescriptive Taxonomy of Interaction

To address the types of questions posed in the above problem statement, we need to develop a prescriptive taxonomy of interaction with VMRs. Such a taxonomy should contain methods, rules, and guidelines for the design of MCTs in order to facilitate and support performance of given cognitive tasks by some prototypical users. These rules and guidelines should help with the analysis, prediction, and prescription of aspects of design as they pertain to mind-MCT interaction. The taxonomy should be tied to appropriate cognitive frameworks and provide a vocabulary that the designers can use when investigating and evaluating the implications of their design choices.

In order to develop such a taxonomy, several inter-related lines of action are proposed below. Although these lines of action are presented sequentially, aspects of them can overlap and take place simultaneously.

1) *Creation of test-bed MCTs.* In order for the taxonomy to cover a wide range of VMRs and cognitive tasks, it is necessary to have a varied and rich set of MCTs which instantiate different techniques and can be used as test beds. Our research group has already created several MCTs dealing with areas such as 2D transformation geometry, 2D and 3D lattices, 2D polygonal patterns, 3D solids, 4D complex polytope visualizations, state-transition diagrams, and visual mathematical programming. We are in the process of building more tools. However, what is needed is for research groups to collaborate to build more tools and share them with each other. We need a large repertoire of these test-bed MCTs.

2) *Designing and conducting of usability studies.* Interaction techniques used in the MCTs should be tested with quantitative and qualitative methods to assess their effectiveness and observe their patterns of use. Conducting such studies will allow for comparison of different techniques. As more studies are conducted, features of these techniques can be validated, abstracted, and generalized [1, 10]. This process will also help create profiles for prototypical users. There have been

a few such studies, and the results clearly suggest that interaction techniques affect learning for better or worse. There needs to be a more concerted and systematic way of conducting these studies. A descriptive taxonomy of interaction, such as the one developed by Sedig and Sumner [9], can provide a framework within which these studies are conducted.

3) *Development of rules and guidelines.* Results obtained from (2) should be analyzed to distill rules and guidelines for interaction with VMRs. Some important issues to consider will be: type of rules (e.g., if-then vs. procedural decision trees); representation of rules; and, level of granularity and degree of abstractness of rules. There are a few studies whose findings can be regarded as general, embryonic rules [13, 14, 15]. For instance, from one of the studies [13], the following rule can be distilled: if a VMR represents a concept (e.g., rotation), and the concept is made of sub-concepts encoded as different visual elements in the VMR (e.g., angle of rotation and center of rotation), and the cognitive task is to understand the concept, then gradual removal of the VMR's visual elements helps users take notice of its sub-concepts. Such rules and guidelines, however, should be further developed and validated to generalize to a wide set of user types and VMRs. This can be done by incorporating and evaluating a rule in the context of other MCTs so that the rule is refined and its validity increases.

4) *Organization of rules into taxonomy.* A three dimensional structure, such as a cube, can be used to help organize and onto which map the rules developed in (3). The axes of the cube would represent the three-tuple (user, cognitive task, VMR), and its cells would point to a set of rules prescribing which interaction techniques to use and would provide examples of best-practice MCTs that incorporate these techniques. For instance, an instantiation of the rule in (3) is embodied in an MCT that a designer can study as a practical application of that rule [5].

5) *Decomposition of taxonomy axes.* There exist classification schemes for visual representations [18], cognitive tasks [19, 20], and user types [19]. For instance, a structural classification of six types of visual representations along 10 characteristics has been developed [18], though not explicitly about VMRs. This and other classification schemes can be used as starting points for decomposing the taxonomy axes. Through evaluation and analysis, these schemes can gradually evolve and be refined. The strategy to be used for decomposing the axes in (4) can play an important role in the usability and effectiveness of the proposed taxonomy [10]. Issues to consider include type and level of granularity of decomposition. For instance, should decomposition be in terms of higher-level, conceptual tasks such as 'formation of cognitive maps', or should it be in terms of lower-level tasks such as 'locating an element', or should it be hierarchical, encompassing several levels? Furthermore, should the VMR axis be decomposed using broad categories or specific categories? For instance, the axis may have specific elements such as 3D lattices and 3D solids, or a broader category such as 3D structures.

6) *Validation of rules and taxonomy.* Validation of the rules and the taxonomy will be an important line of action. By necessity, the development and validation of a prescriptive taxonomy is evolutionary. However, there are some general steps that can increase the validity of such taxonomies [1]: 1) repeated evaluation of rules using reliable testing instruments, 2) designing and evaluating MCTs with and without the use of the taxonomy and comparing the results, and 3) having outside experts review, evaluate, and test the rules and the categories.

3 Summary

Interaction plays an important role in helping learners (or users) make sense of visual mathematical representations. Given different visual mathematical representations, cognitive tasks, and users, designers often need to use different interaction techniques. Therefore, they need to know what interaction techniques to use, when, and to what end. Using the wrong type of interaction can constrain human thinking and have negative, undesirable consequences for learners. Human-computer interaction designers use design taxonomies, particularly prescriptive ones, to facilitate analysis and understanding of design requirements, to provide opportunities for comparing different techniques, and to guide designers. Currently, there exist no design taxonomies to guide the design of interactive mathematical cognitive tools containing visual representations. This paper calls for the development of a comprehensive prescriptive taxonomy of interaction for mathematical cognitive tools.

To develop such a taxonomy, the paper proposes several inter-related lines of action. These lines of action are as follows: creating a large repertoire of mathematical cognitive tools that can be used as test-beds for experimenting with different design requirements; designing and conducting usability studies to investigate and compare different interaction techniques to find out how they support understanding of different mathematical concepts and to what end; developing general rules and guidelines for interaction design as a result of usability studies performed with a large set of test-bed tools; taxonomizing design rules and pointing designers to tools that embody best-practice examples of these prescriptive rules; and, as the rules and the taxonomy evolve, constantly evaluating them to increase their validity.

The scope of such a research agenda calls for collaboration among researchers in this area. As mathematics has a very rich repertoire of representational forms, many of the guidelines and interaction techniques developed in this area can transfer to other cognitive tools embedding other knowledge domains.

References

1. Jonassen, DH (ed.): Handbook of research for educational communications and technology. Simon & Schuster Macmillan (1996)
2. Norman, DA: Things that make us smart: Defining human attributes in the age of the machine. Addison Wesley (1993)
3. Beynon, B, Nehaniv, CL, Dautenhahn, K (eds.): Cognitive technology: Instruments of mind. Proceedings of International Cognitive Technology Conference. Springer-Verlag (2001)
4. Lajoie, S (ed.): Computers as cognitive tools. Lawrence Erlbaum Assoc. (2000)
5. Skemp, RR: The psychology of learning mathematics. Penguin Books (1986)
6. Card, SK, Mackinlay, JD, Shneiderman, B. (eds.): Readings in information visualization: Using vision to think. Morgan Kaufmann Publishers (1999)
7. Strothotte, T: Computational visualization: Graphics, abstraction, and interactivity. Springer-Verlag (1998)
8. Stylianou, D: On the interaction of visualization and analysis: The negotiation of a visual representation in expert problem solving. Journal of Mathematical Behavior 21: (2002) 303-317

9. Sedig, K, Sumner, M: Characterizing interaction with visual mathematical representations. International Journal of Computers for Mathematical Learning. Kluwer Academic Publishers (Under review)
10. Carroll, JM (ed.): Designing interaction: Psychology at the human-computer interface. Cambridge University Press (1991)
11. Hanson, A, Munzner, T, Francis, G: Interactive methods for visualizable geometry. IEEE Computer, (1994) 27(7): 73-83
12. Tweedie, LA: Characterizing interactive externalizations. In Proceedings of CHI '97, (1997) 375-382
13. Sedig, K, Klawe, M, Westrom, M: Role of interface manipulation style and scaffolding on cognition and concept learning in learnware. ACM Transactions on Computer-Human Interaction, (2001) 1(8), 34-59
14. Sedig, K, Rowhani, S, Morey, J, Liang, H: Application of information visualization techniques to the design of a mathematical mindtool: A usability study. Palgrave Macmillan Ltd. Information Visualization, (2003) 2(3): 142-160
15. Golightly, D: Harnessing the interface for domain learning. In Proceedings of CHI '96, (1996) 37-38
16. Keller PR, Keller MM: Visual cues: Practical data visualization. IEEE Computer Society (1993)
17. Gadanidis, G, Sedig, K, Liang, H: Designing online mathematical investigations. Journal of Computers in Mathematics and Science Teaching. Association for the Advancement of Computing in Education (In press)
18. Lohse, GL, Biolsi, K, Walker, N, Rueter, HH: A classification of visual representations. Communications of the ACM, (1994) 37(12): 36-49
19. Jonassen, DH, Tessmer, M & Hannum, WH: Task analysis methods for instructional design. Lawrence Erlbaum Associates (1999)
20. Schraagen, MC, Chipman, SF & Shalin, VL (eds.): Cognitive task analysis. Lawrence Erlbaum Assoc (2000)

Evolution of the Internet Map and Load Distribution

K.-I. Goh, B. Kahng*, and D. Kim

School of Physics and Center for Theoretical Physics, Seoul National University,
Seoul 151-747, Korea

Abstract. We track the evolutionary history of the Internet at the autonomous systems (ASes) level and provide the evidence that it can be described in the framework of the multiplicative stochastic process. It is found that the fluctuations arising in the process of diversifying connections of each node play an essential role in forming the *status quo* of the Internet. Extracting relevant parameters for the growth dynamics of the Internet topology, we are able to predict the connectivity (degree) exponent γ of the Internet AS map successfully. We also introduce a quantity called the load as the capacity of node needed for handling the communication traffic and study its distribution over the Internet across years. The load distribution follows a power law with the exponent $\delta \approx 2.0$ and the load at the hub scales with the network size as $\ell_h \sim N^{1.8}$.

1 Introduction

During recent years, the Internet has become one of the most influential media in our daily life, going beyond in its role as the basic infrastructure in the technological world. Explosive growth in the number of users and hence the amount of traffic poses a number of problems which are not only important in practice for, e.g., maintaining it free from any undesired congestion and malfunctioning, but also of theoretical interests as an interdisciplinary topic [1]. Such interests, also stimulated by other disciplines like biology, sociology, and statistical physics, have blossomed into a broader framework of network science [2,3,4]. The Internet is a primary example of complex networks. It consists of a large number of very heterogeneous units interconnected with various connection bandwidths, however, it is neither regular nor completely random. In their landmark paper, Faloutsos et al. [5] showed that the Internet at the autonomous systems (ASes) level is a scale-free (SF) network [6], meaning that it follows a power-law distribution

$$p_d(k) \sim k^{-\gamma} \tag{1}$$

in node degree k, the number of connections a node has. The degree exponent γ of the AS map is subsequently measured by a number of groups to be $\gamma \approx 2.1$.

Emergence of such power-law degree distribution calls for explanation and understanding of the basic mechanism underlying the growth of the Internet.

---

* Corresponding author (E-mail: kahng@phya.snu.ac.kr).

M. Bubak et al. (Eds.): ICCS 2004, LNCS 3038, pp. 1038–1045, 2004.
© Springer-Verlag Berlin Heidelberg 2004

Once revealed, it can be used to predict what the Internet will be like in the future, as well as how it has evolved into the present shape. In the first part of the paper, we will address this issue, showing that it can be described by a simple physical model, the multiplicative stochastic process. By extracting relevant parameters for the stochastic process from the time history of the AS map deposited in the Oregon route views project, we can predict the degree exponent of the Internet accurately.

The Internet is not a quiet object. Data packets are sent and received over it constantly, causing momentary local congestion from time to time. To avoid such undesired congestion, the capacity, or the bandwidth, of the routers should be as large as it can handle the traffic. In the second part of the paper, we will introduce a rough measure of such capacity, called the load and denoted as ℓ. The distribution of the load reflects the high level of heterogeneity of the Internet: It also follows a power law,

$$p_l(\ell) \sim \ell^{-\delta}, \tag{2}$$

with the load exponent δ. We will discuss the implication of the power-law behavior of the load distribution.

2   Internet Evolution as a Multiplicative Stochastic Process

The mechanism of the emergence of SF network is mostly captured by the Barabási-Albert (BA) model [7] which assumes the linear growth in numbers of nodes and links in time and the preferential attachment (PA) in establishing links from a new node to other previously existing ones. The PA means that the probability $\Pi_i(t)$ that a node i will receive a link from the new node created at time t is linearly proportional to its present degree $k_i(t)$, i.e., $\Pi_i(t) = k_i(t)/\sum_j k_j(t)$. The empirical evidence of the PA in the Internet has been reported [8,9]. As we will see, however, the assumption that the numbers of nodes and links increase linearly in time does not apply to the real situation of the Internet. Rather, the numbers of nodes and links increase exponentially but with different rates. Furthermore, the interconnections between nodes are being updated continually in the Internet, which was not incorporated in the original BA model.

Huberman and Adamic (HA) [10] proposed another scenario for SF networks. They argued that the fluctuation effect arising in the process of connecting and disconnecting links between nodes is an essential feature to describe the dynamics of the Internet topology correctly. In their model, the total number of nodes $N(t)$ increases exponentially with time as

$$N(t) = N(0)\exp(\alpha t). \tag{3}$$

Next, they assumed that the degree k_i at a node i evolves through the multiplicative stochastic process,

$$k_i(t+1) = k_i(t)[1 + \zeta_i(t+1)], \tag{4}$$

where $\zeta_i(t)$ is the growth rate of the degree k_i at time t, which fluctuates from time to time. Thus, one may divide the growth rate $\zeta_i(t)$ into two parts, $\zeta_i(t) = g_{0,i} + \xi_i(t)$, where $g_{0,i}$ is the mean value over time, and $\xi_i(t)$ the rest part, representing fluctuations over time. $\xi_i(t)$ is assumed to be a white noise satisfying $\langle \xi_i(t) \rangle = 0$ and $\langle \xi_i(t)\xi_j(t') \rangle = \sigma_{0,i}^2 \delta_{t,t'} \delta_{i,j}$, where $\sigma_{0,i}^2$ is the variance. Here $\langle \cdots \rangle$ is the sample average and $\delta_{a,b}$ is the Kronecker delta symbol. For later convenience, we denote the logarithm of the growth factor as $G_i(t) \equiv \ln[1+\zeta_i(t)]$. Then a simple application of the central limit theorem ensures that the probability distribution of $k_i(t)/k_i(t_0)$, t_0 being a reference time, follows the log-normal distribution for sufficiently large t. To get the degree distribution, one needs to collect all contributions from different ages τ_i, growth rates $g_{0,i}$, standard deviations $\sigma_{0,i}$ and initial degree $k_i(t_0)$. HA further assumed that ζ_i are identically distributed so that $g_{0,i} = g_0$ and $\sigma_{0,i} = \sigma_0$ for all i. Then the conditional probability for degree, $P_d(k, \tau \mid k_0)$, that $k_i(t_0+\tau) = k$, given $k_i(t_0) = k_0$ is given by

$$P_d(k, \tau \mid k_0) = \frac{1}{k\sqrt{2\pi\sigma_{\text{eff}}^2\tau}} \exp\left\{ -\frac{(\ln(k/k_0) - g_{\text{eff}}\tau)^2}{2\sigma_{\text{eff}}^2\tau} \right\}, \tag{5}$$

where $g_{\text{eff}} \equiv \langle G_i(t) \rangle$ and $\sigma_{\text{eff}}^2 \equiv \langle (G_i(t) - \langle G_i(t) \rangle)^2 \rangle$. g_{eff} and σ_{eff}^2 are related to g_0 and σ_0^2 as $g_{\text{eff}} \approx g_0 - \sigma_0^2/2$ and $\sigma_{\text{eff}}^2 \approx \sigma_0^2$, respectively [11]. Since the density of nodes with age τ is proportional to $\rho(\tau) \sim \exp(-\alpha\tau)$, the degree distribution collected over all ages becomes $p_d(k) = \int d\tau \rho(\tau) P_d(k, \tau \mid k_0) \sim k^{-\gamma}$, where the degree exponent γ is given in terms of the growth parameters as

$$\gamma = 1 - \frac{g_{\text{eff}}}{\sigma_{\text{eff}}^2} + \frac{\sqrt{g_{\text{eff}}^2 + 2\alpha\sigma_{\text{eff}}^2}}{\sigma_{\text{eff}}^2}. \tag{6}$$

In the next section, we will measure such parameters from the real evolutionary history of the Internet AS map and check if the HA scenario holds.

3   Growth Dynamics of Internet

A number of projects exist aiming to map the world-wide topology of the Internet. One such is the Route Views project initiated at the university of Oregon [12], the data of which are also archived at the National Laboratory of Applied Network Research (NLANR) [13]. Among the daily data from November 1997 to January 2000, we sample one AS map a month, with the total period of 26 months, and analyze them for various quantities. First we measure the growth rate of the number of ASes α. We also measure directly the growth rate of the number of links β, which can be crosschecked for consistency later. In Fig. 1, we show the total number of ASes $N(t)$ and the total number of links $L(t)$ as a function of time t. The straight line in log-linear plot means $N(t)$ and $L(t)$ indeed grows exponentially. The growth rates are determined to be $\alpha \approx 0.029$ and $\beta \approx 0.034$. We also find that the newly appeared AS would connect to only one or two existing ASes so that the average number of links the new AS establishes

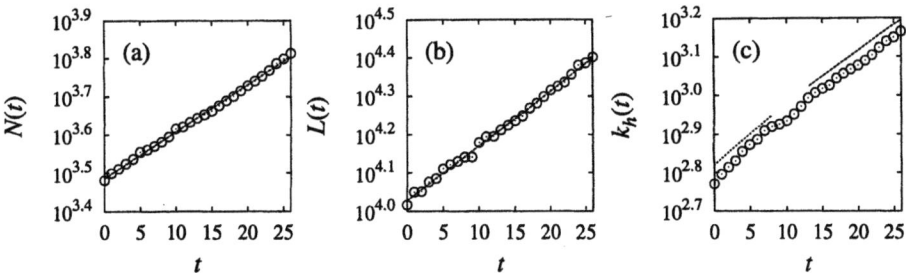

Fig. 1. The time evolution of the number of ASes $N(t)$ (a), the number of links between ASes $L(t)$ (b), and the degree of the hub $k_h(t)$ (c). Note that the ordinates in the figures are in logarithmic scale, indicating the exponential increase of corresponding quantities. The fitted line has a slope 0.029 in (a), 0.034 in (b), and 0.043 (0.030) for the dotted (dashed) one in (c).

is $k_{\text{new}} \approx 1.35$. Fig. 1(c) shows the growth of the degree of the hub, the node with the largest degree. It shows a change of growth rate around $t \approx 14$.

The measurement of g_0 and σ_0 is nontrivial due to the presence of large fluctuations. To this end, we measure the degree growth rate of a node i, $G_i(t)$, defined earlier as $G_i(t) \equiv \ln[1 + \zeta_i(t)] = \ln[k_i(t)/k_i(t-1)]$. To keep $G_i(t)$ well-defined for all t, we consider only the nodes existing for the entire time range $0 \leq t \leq 26$, the set composed of which is denoted by S hereafter. By the existence of a node we mean that its degree is nonzero, since we cannot identify an AS with no connection. For each i ($i \in S$), let $g_i = \langle G_i(t) \rangle_t$ and $\sigma_i^2 = \langle (G_i(t) - g_i)^2 \rangle_t$, where $\langle \cdots \rangle_t$ means the temporal average over the period $16 < t \leq 26$ ($T = 10$). If the HA scenario holds, the histogram of $\{g_i\}$ for all nodes would follow the Gaussian distribution with the mean g_{eff} and the variance σ_{eff}^2/T. We show such histogram in Fig. 2, the fit of which to the Gaussian gives the mean $\bar{g}$ as 0.016 and the standard deviation σ_g as 0.04. The measured values of $\{\sigma_i^2\}$ give the mean value $\overline{\sigma^2} \approx 0.017$.

It is most likely that $\bar{g}$ and $\overline{\sigma^2}$ would have a distribution over nodes. As HA assumed, however, we try to approximate the growth process by a single process whose effective mean growth rate and standard deviation are g_{eff} and σ_{eff}, respectively. Then the Eq. (5) should hold for all i and all t with a suitable choice of those parameters. For this purpose, we consider the distribution $P[k_i(t)/k_i(t_0)]$ in terms of the scaled variables x and y defined as

$$x \equiv \frac{\ln[k_i(t)/k_i(t_0)] - g_d(t - t_0)}{\sqrt{2\sigma_d^2(t - t_0)}}, \tag{7}$$

and

$$y \equiv P[k_i(t)/k_i(t_0)][k_i(t)/k_i(t_0)]\sqrt{2\pi\sigma_d^2(t - t_0)}, \tag{8}$$

where we set $t_0 = 0$ and g_d and σ_d are parameters to be chosen. From Eq. (5), with suitably chosen parameters g_d and σ_d, the distribution for different time

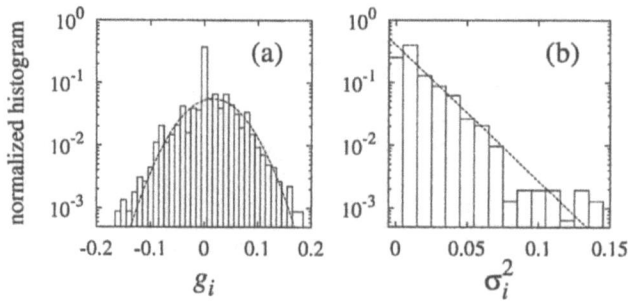

Fig. 2. The normalized histogram of g_i (a) and σ_i^2 (b). In (a), the data is fitted with a Gaussian with the mean 0.016 and the standard deviation 0.04. In (b), the data is fitted with an exponential decay $\exp(-x/x_c)$ with the characteristic scale $x_c \approx 0.02$. The measured value of the average is $\overline{\sigma^2} \approx 0.017$.

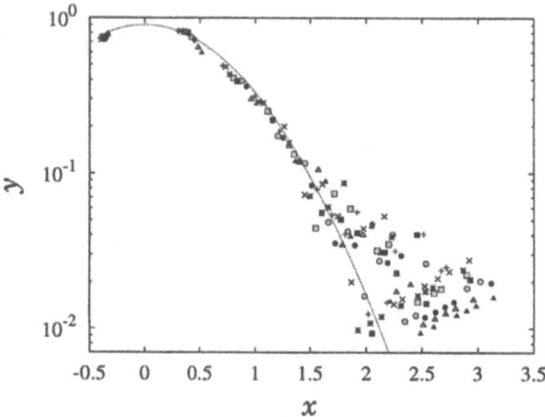

Fig. 3. Plot of $P[k_i(t)/k_i(t_0)]$ versus $k_i(t)/k_i(t_0)$ for different times $t > 16$ in terms of the scaled variables y and x defined in Eqs. (7) and (8). Larger deviations for large x are due to t being finite and are caused by the rare statistics.

t would collapse onto a single curve, $\ln y = -x^2$. We show such data in Fig. 3. The best collapse can be accomplished by choosing the parameters $g_d = 0.016$ and $\sigma_d = 0.14$, which should be identified with g_{eff} and σ_{eff}, respectively. The effective growth parameters found in this way are in accordance with the ones estimated before as $\overline{g} = 0.016$ and $\overline{\sigma^2} = 0.017$. As noted earlier, the consistency of estimated parameters can be checked as, for example, it should satisfy $\beta = \max(\alpha, g_{\text{eff}} + \sigma_{\text{eff}}^2)$, for which we have $\beta = 0.034$ and $g_{\text{eff}} + \sigma_{\text{eff}}^2 = 0.035$, being reasonably consistent with each other. Thus we conclude the parameters $g_{\text{eff}} = 0.016$ and $\sigma_{\text{eff}} = 0.14$ can be regarded as the effective parameters of the degree growth dynamics of the Internet AS map as a single process. Applying those values together with $\alpha = 0.029$ found earlier into Eq. (6), we found $\gamma \approx 2.1$, which is in excellent agreement with the directly measured ones.

4   Load Distribution of Internet

To a large extent, the Internet is the medium of communication. The continuous communication between hosts generates certain amount of data traffic. To make the best use of it, we have to avoid congestions, from which we suffer possible delays and the loss of information. What's worse, one doesn't know when and how much a host will generate the traffic. Absent is the central regulation in the Internet, hence each node should do its best for its own ends. To give a measure of such activity, we define the load ℓ_i as the amount of capacity or bandwidth that a node i can handle in unit time [14]. Not knowing the level of traffic, one assumes that every node sends a unit packet to everyone else in unit time. One further assumes that the packets are transfered from the source to the target only along the shortest paths between them, and divided evenly upon encountering any branching point. To be precise, let $\ell_i^{s \to t}$ be the amount of packet sent from s (source) to t (target) that passes through the node i. Then the load of a node i, ℓ_i, is the accumulated sum of $\ell_i^{s \to t}$ for all s and t,

$$\ell_i = \sum_{s \neq t} \ell_i^{s \to t}. \tag{9}$$

In other words, the load of a node i gives us the information how much the capacity of the node should be in order to maintain the whole network in a free-flow state. To calculate load on each node, we use the modified breath-first search algorithm introduced by Newman [15] and independently by Brandes [16], which can evaluate $\{\ell_i\}$ in time of order $\mathcal{O}(N^2)$ for sparse binary graphs.

For a number of SF networks in nature and society, the load distribution is also found to follow a power law, Eq. (2) [17]. The Internet AS map is no exception and the load exponent δ of the power law is estimated to be approximately $\delta \approx 2.0$ [17,18]. The power-law load distribution means that a few ASes should handle an extraordinarily large amount of load while most others should do only a little.

The load of a node is highly correlated with its degree. The Pearson correlation coefficient between the two quantities is as high as 0.98. This suggests a scaling relation between the load and the degree of a node as

$$\ell \sim k^\eta \tag{10}$$

and the scaling exponent η is estimated as $\eta = 1.06 \pm 0.03$ for January 2000 AS map (Fig. 4a). In fact, the exponent η depends on γ and δ as $\eta = (\gamma-1)/(\delta-1) \approx 1.1$, which is consistent with the direct measurement.

The time evolution of the load at each AS is also of interest. Practically, how the load scales with the total number of AS (the size of the AS map) is an important information for the network management. In Fig. 4b, we show $\ell_i(t)$ versus $N(t)$ for 5 ASes with the highest rank in degree, i.e., 5 ASes that have largest degrees at $t = 0$. The data of $\{\ell_i(t)\}$ shows large fluctuations in time. Interestingly, the fluctuation is moderate for the hub, implying that the connections of the hub is rather stable. The load at the hub is found to scale

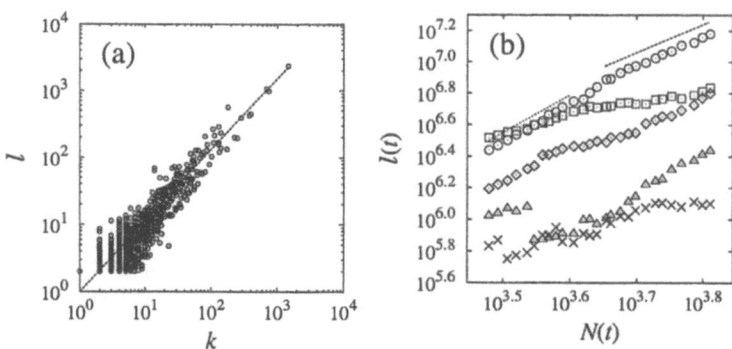

Fig. 4. (a) The scatter plot of the load versus the degree of each node for the AS map as of January 2000. The slope of the dashed line is 1.06, drawn for the eye. (b) Time evolution of the load versus $N(t)$ at the ASes of degree-rank 1 (O), 2 ($\square$), 3 ($\Diamond$), 4 ($\triangle$), and 5 ($\times$). The dashed line for larger N has slope 1.8, drawn for the eye.

with $N(t)$, $\ell_h(t) \sim N(t)^\mu$, but the scaling shows a crossover from $\mu \approx 2.4$ to $\mu \approx 1.8$ around $t \approx 14$, as it did for the degree.

5   Summary

We have studied the temporal evolution of the Internet AS map and showed that it can be described in the framework of multiplicative stochastic process for the degree growth dynamics. We measured the values of relevant parameters from the history of the AS map. With those values, the AS number growth rate $\alpha = 0.029$, the effective degree growth rate $g_{\text{eff}} = 0.019$, and its effective standard deviation $\sigma_{\text{eff}} = 0.14$, we were able to predict the degree exponent γ as $\gamma \approx 2.1$, which is in excellent agreement with the previously reported empirical values $\gamma_{\text{measured}} = 2.1 \sim 2.2$. Although it successfully accounts for the emergence of the scale-free characteristics of the Internet, the present description is by no means complete. More elaborated modeling [19,20,21] would improve our understanding of the evolutionary and organizational principle of the Internet and the research in this direction is highly called for.

In the second part of the paper, we introduced a quantity called the load. It can be thought of as the amount of traffic that a node should handle to keep the whole network away from the unwelcome congestion and maintain free-flow state, giving a measure of desired capacity of the nodes. The load distribution also follows a power law. The load and the degree of an AS are highly correlated with each other. The analysis of the temporal change of the load reveals that the load at the hub scales with the system size as $N^{1.8}$. Finally, we note that the contents of this article is in part overlapped with our previous studies published in [14,17,19].

References

1. Pastor-Satorras, R., Vespignani, A.: Evolution and Structure of the Internet. Cambridge University Press, Cambridge (2004)
2. Albert, R., Barabási, A.-L: Statistical mechanics of complex networks. Rev. Mod. Phys. **74** (2002) 47–97
3. Dorogovtsev, S. N., Mendes, J. F. F.: Evolution of Networks. Oxford University Press, Oxford (2003)
4. Newman, M. E. J.: The structure and function of complex networks. SIAM Rev. **45** (2003) 167–256
5. Faloutsos, M., Faloutsos, P., Faloutsos, C.: On power-law relationships in the Internet topology. Comput. Commun. Rev. **29** (1999) 251–262
6. Barabási, A.-L., Albert, R., Jeong, H.: Mean-field theory for scale-free random networks. Physica A **272** (1999) 173–187
7. Barabási, A.-L., Albert, R.: Emergence of scaling in random networks. Science **286** (1999) 509–512
8. Pastor-Satorras, R., Vázquez, A., Vespignani, A.: Dynamical and correlation properties of the Internet. Phys. Rev. Lett. **87** (2001) 258701
9. Jeong, H., Néda, Z., Barabási, A.-L.: Measuring preferential attachment for evolving networks. Europhys. Lett. **61** (2003) 567–572
10. Huberman, B. A., Adamic, L. A.: Evolutionary dynamics of the World Wide Web. e-print (http://arxiv.org/abs/cond-mat/9901071) (1999)
11. Gardiner, C. W.: Handbook of Stochastic Methods. Springer-Verlag, Berlin (1983)
12. Meyer, D.: University of Oregon Route Views Archive Project (http://archive.routeviews.org)
13. The NLANR project sponsored by the National Science Foundation (http://moat.nlanr.net)
14. Goh, K.-I., Kahng, B., Kim, D.: Universal behavior of load distribution in scale-free networks. Phys. Rev. Lett. **87** (2001) 278701
15. Newman, M. E. J.: Scientific collaboration networks. II. Shortest paths, weighted networks, and centrality. Phys. Rev. E **64** (2001) 016132
16. Brandes, U.: A faster algorithm for betweenness centrality. J. Math. Sociol. **25** (2001) 163–177
17. Goh, K.-I., Oh, E., Jeong, H., Kahng, B., Kim, D.: Classification of scale-free networks. Proc. Natl. Acad. Sci. USA **99** (2002) 12583–12588
18. Vázquez, A., Pastor-Satorras, R., Vespignani, A.: Large-scale topological and dynamical properties of the Internet. Phys. Rev. E **65** (2002) 066130
19. Goh, K.-I., Kahng, B., Kim, D.: Fluctuation-driven dynamics of the Internet topology. Phys. Rev. Lett. **88** (2002) 108701
20. Yook, S.-H., Jeong, H., Barabási, A.-L.: Modeling the Internet's large-scale topology. Proc. Natl. Acad. Sci. U.S.A. **99** (2002) 13382–13386
21. Rosvall, M., Sneppen, K.: Modeling dynamics of information networks. Phys. Rev. Lett. **91** (2003) 178701

Complex Network of Earthquakes

Sumiyoshi Abe[1] and Norikazu Suzuki[2]

[1] Institute of Physics, University of Tsukuba, Ibaraki 305-8571, Japan,
suabe@sf6.so-net.ne.jp
[2] College of Science and Technology, Nihon University, Chiba 274-8501, Japan
suzu@phys.ge.cst.nihon-u.ac.jp

Abstract. The complex evolving network is constructed from the seismic data taken in southern California, and its topological properties are studied. It is shown that the network associated with earthquakes is scale free. The value of the degrees of separation, i.e., the characteristic path length, is found to be small, between 2 and 3. The clustering coefficient is also calculated and is seen to be about 10 times larger than that in the case of the completely random network. These discoveries should play an important role in modeling earthquake phenomenon.

1 Introduction

Earthquake phenomenon has been attracting much attention from the viewpoint of science of complexity. [1]-[5]. Though seismicity has diverse physical aspects, some known empirical laws are remarkably simple. The Omori law for the temporal pattern of aftershocks [6] and the Gutenberg-Richter law for the relationship between frequency and magnitude [7] are celebrated classical examples. These are the power laws and represent the scale-free natures of seismicity.

In recent papers [8], [9], we have studied spatio-temporal complexity of seismicities in southern California and Japan based on nonextensive statistical mechanics [10]-[15], which is considered to be a consistent and unified framework for the statistical description of complex systems. We have found that both the spatial distance [8] and time interval [9] between two successive earthquakes are described extremely well by the so-called q-exponential distribution (see below) which maximizes the Tsallis entropy [10], [15] under appropriate constraints.

On the other hand, we have also discovered the nonextensive-statistical-mechanical element in the Internet time series of the packet round-trip time measured by performing the Ping experiment [16]-[19]. We have found that statistics of the sparseness time [16] obeys the q-exponential distribution. The sparseness time is the quantity analogous to the time interval between earthquakes. Thus, seismicity and the Internet exhibit a similar dynamical behavior. In addition, we have also found that the Omori and Gutenberg-Richter laws also hold for "Internetquakes" corresponding to the heavily congested states of the Internet [17]-[19]. This striking similarity suggests the existence of a deep root common in seismicity and the Internet. In this respect, the scale-free structure of topology

M. Bubak et al. (Eds.): ICCS 2004, LNCS 3038, pp. 1046–1053, 2004.
© Springer-Verlag Berlin Heidelberg 2004

of the Internet [20] is of central interest, and accordingly we are naturally led to examining a possible network structure underlying earthquake phenomenon.

Here, we propose the definition of the earthquake network and analyze its topological properties. Using the seismic data taken in southern California, we calculate the distribution of connectivities, the degree of separation (i.e., the characteristic path length), and the clustering coefficient of the earthquake network. We shall show that the distribution of connectivities decays as a power law, indicating that the earthquake network is scale free [21]-[24]. The degree of separation is found to take a small value between 2 and 3, exhibiting the small-world structure of the network [25]. We shall also show that the clustering coefficient is about 10 times larger than that in the case of the completely random network, implying the complex hierarchical structure [22], [23], [25].

The present paper is organized as follows. In Sect. 2, we explain how to construct the earthquake network from the seismic data. In Sect. 3, we discuss the scale-free nature of the earthquake network constructed from the data taken in southern California. An interpretation is given to the emergence of the scale-free nature in connection with the geological feature of aftershocks. In Sect. 4, we discuss the small-world and hierarchical properties. Sect. 5 is devoted to conclusion.

2   Earthquake Network

Our proposal for constructing the earthquake network is as follows. A geographical area under consideration is divided into a lot of small cubic cells. A cell is regarded as a vertex when earthquakes with any values of magnitude occurred therein. Two successive events define an edge between two vertices. In this way, the complex fault-fault interaction is replaced by this edge. If two successive earthquakes occur in the same cell, they form a loop. This procedure allows us to map the seismic data to an evolving network. This construction contains a unique parameter, which is the cell size. Since there are no *a priori* operational rules to determine the cell size, it is essential to examine the dependencies of the network properties on this parameter. Once the cell size is fixed, the earthquake network is unambiguously defined by the seismic data [26], [27]. The earthquake network contains a few special vertices associated with mainshocks. *Careful analysis of the seismic data shows that aftershocks associated with a mainshock tend to return to the locus of the mainshock geographically* and consequently a stronger mainshock tends to have the larger values of connectivities contributed by more aftershocks. Accordingly, the concept of preferential attachment [21]-[24] is realized by the existence of "hubs", the role of which is played by the mainshocks. This is schematically depicted in Fig. 1, in which the vertices, A and B, may be identified with the cells containing the mainshocks. This observation leads to the reasoning that the earthquake network may be scale-free and possess the small-world structure. In what follows, we shall show that it is indeed the case.

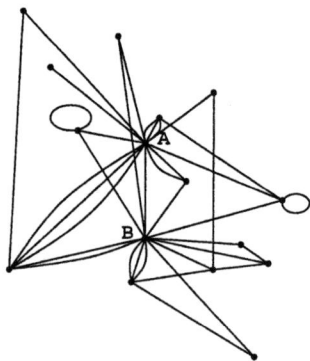

Fig. 1. A schematic description of the earthquake network. The vertices represent the cells in which earthquake occured and the edges replace the complex fault-fault interaction. A and B are the cells containing mainshocks and have large connectivities.

3   Scale-Free Nature of Earthquake Network

A key concept in the Barabási-Albert scale-free networks is the rule of preferential attachment. According to this rule, a newly created vertex is connected to the vertex v_i with connectivity k_i with probability

$$\Pi(k_i) = \frac{k_i + 1}{\sum_j (k_j + 1)}. \tag{1}$$

In [24], Albert and Barabási have discussed an exactly solvable model of an evolving random graph and have derived the analytic expression for the distribution of connectivities, $P(k)$, in the continuum limit. This solution is the Zipf-Mandelbrot-type distribution [28]

$$P(k) \sim (k + k_0)^{-\gamma} \tag{2}$$

with $\gamma > 1$ and $k_0 > 0$, which clearly decays as a power law. In the worked examples [22], the exponent γ ranges between 1.05 and 3.4.

It should be noted that, with identifications $q = 1 + 1/\gamma$ and $\kappa = (q - 1)/k_0$, $P(k)$ in Eq. (2) is reexpressed in terms of the q-exponential function as $P(k) \sim e_q(-k/\kappa)$, where $e_q(x) = (1 + (1 - q)x)_+^{1/(1-q)}$ with the notation $(a)_+ \equiv \max\{0, a\}$. This implies that the Albert-Barabási solution optimizes the Tsallis entropy [10], [15], $S_q[P] = (1-q)^{-1}(\int dk P^q(k) - 1)$, under the appropriate constraints on the normalization of $P(k)$ and the average number of edges in the continuum limit. Relevance of Tsallis nonextensive statistical mechanics [10]-[15] to a complex networks has recently been a noticed also in [29].

We have constructed the earthquake networks in the area of southern California by introducing two cell sizes, 10 km × 10 km × 10 km and 5 km × 5 km × 5 km. (We have examine these two cases, since as already mentioned there are no

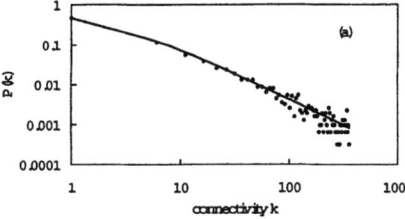 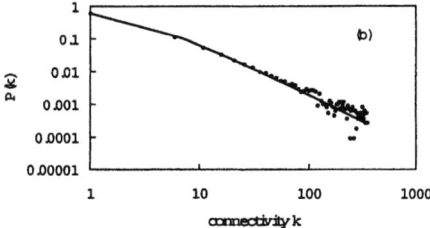

Fig. 2. The log-log plots of the distributions of connectivities. All quantities are dimensionless. The dots represent the observed data and the solid lines are drawn by using Eq. (2). (a) The cell size 10 km $\times$ 10 km $\times$ 10 km, $\gamma = 1.33$ (± 0.03), and $k_0 = 1.70$ (± 0.02). (b) The cell size 5 km $\times$ 5 km $\times$ 5 km, $\gamma = 1.61$ (± 0.03), and $k_0 = 2.04$ (± 0.02).

a priori operational rules to fix the cell size.) We have analyzed the earthquake catalog made available by the Southern California Earthquake Data Center (http://www.scecdc.scec.org/catalogs.html) covering the region $29°15.25'$N$-38°49.02'$N latitude and $113°09.00'$W $- 122°23.55'$W longitude with the maximal depth (of the foci of the observed earthquakes) 57.88 km in the period between $00:25:8.58$ on January 1, 1984 and $15:23:54.73$ on December 31, 2002. (We have taken this period since the data in 1983 are partially missing for a few months.) In Fig. 2, we present the plots of the distributions of connectivities. One appreciates that the data is well fitted by Eq.(2), and therefore the earthquake networks possess the scale-free nature. This may be interpreted as follows. The Gutenberg-Richter law, on the one hand, tells us that frequency of earthquakes with large values of moment decays as a power law. On the other hand, as already mentioned, aftershocks associated with a mainshock tend to be connected to the vertex of the mainshock, realizing preferential attachment. These imply that the scale-free nature of the distribution of connectivities may have its origin in the Gutenberg-Richter law.

We would also like to report the following further two findings regarding time evolution of the earthquake network. (i) As shown in Fig. 3, the factor, k_0, appearing in Eq. (2) is found to change in time, in contrast to the Albert-Barabási solution given in [24]. Monotonic increase of the value of k_0 is clearly observed there. (ii) *The value of the exponent, γ, is ascertained to remain constant in time according to evolution of the earthquake network.*

Concerning (i) above, we notice the following two points regarding consistency of time evolution with the principle of maximum Tsallis entropy. Firstly, the time-dependent problems have repeatedly been investigated within the maximum entropy principle in the literature. It is known that the maximum entropy principle is not limited to characterizing the equilibrium states in thermodynamics but can be used in much wider contexts of information theory and statistics. Secondly, as can be seen in Fig. 3, the rate of change of k_0 in the present case is less than 0.1 per a year, and therefore its time scale is much longer than that of microscopic evolution (typically about 20000 events per a year).

Closing this section, we stress the following point. We have also analyzed the data taken in Japan and have found same trends in the earthquake network [26].

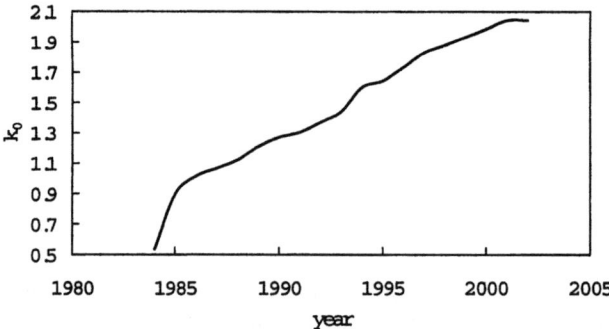

Fig. 3. Change of the value of k_0 according to evolution of the earthquake network. Here, the cell size employed is 5 km × 5 km × 5 km. The value of the exponent remains constant, $\gamma = 1.61$.

4 Small-World Property of Earthquake Network

The scale-free nature of the earthquake network implies that the network possesses the small-world and hierarchical structures. We discuss these properties in this section.

In Fig. 4, we present the degrees of separation, i.e., the characteristic path length, between an arbitrary pair of two vertices. Here, we examine the variation of the cell size from 5 km × 5 km × 5 km to 10 km × 10 km × 10 km by every 1 km. The values of the degree of separation have been calculated by random sampling of 60 pairs of vertices. The degrees tend to slightly decrease with respect to the cell size, as it should do. The values are typically in-between 2 and 3, very small, showing the small-world nature of the earthquake network.

Next we study the clustering coefficient, C, proposed by Watts and Strogatz [25]. The scale-free nature of the distribution of connectivities implies that the value of C is much larger than that in the case of the completely random network whose distribution of connectivities is Poissonian [21],[22],[30]. To evaluate C, it is essential to notice that loops attached to a single vertex should be removed and multiple edges between two vertices have to be identified with a single edge. For example, consider three vertices, v_1, v_2, and v_3. Suppose that they are originally connected as $v_1 \to v_2 \to v_1 \to v_2 \to v_2 \to v_3$. This should be identified with $v_1 \to v_2 \to v_3$ now.

The clustering coefficient is defined as follows. Let the v_i be the vertex which has $k_i - 1$ neighboring vertices. At most, $k_i(k_i - 1)/2$ edges can exist between them. Calculate $c_i \equiv$ (number of edges of v_i and its neighbors)$/[k_i(k_i - 1)/2]$. Then, the clustering coefficient is given by

$$C = \frac{1}{N}\sum_{i=1}^{N} c_i, \tag{3}$$

where N denotes the total number of vertices. In the case of the completely random network, this quantity is known to be expressed as follows [22], [23], [25]:

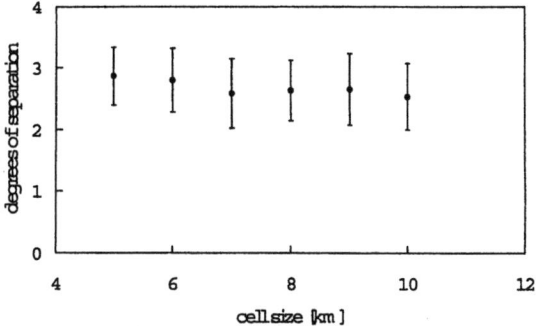

Fig. 4. The degrees of separation for various cell sizes. The values are 2.87 (±0.47) for 5 km $\times$ 5 km $\times$ 5 km ($N = 3434$), 2.80 (±0.51) for 6 km $\times$ 6 km $\times$ 6 km ($N = 2731$), 2.58 (±0.56) for 7 km $\times$ 7 km $\times$ 7 km ($N = 2165$), 2.63 (±0.49) for 8 km $\times$ 8 km $\times$ 8 km ($N = 1827$), 2.65 (±0.58) for 9 km $\times$ 9 km $\times$ 9 km ($N = 1582$), and 2.53 (±0.54) for 10 km $\times$ 10 km $\times$ 10 km ($N = 1344$), where N denotes the number of the vertices. The error bounds are given by the standard deviations for 60 pairs of vertices chosen at random.

$$C = C_{\text{random}} = \frac{<k>}{N} \ll 1, \tag{4}$$

where $<k>$ is the average connectivity. The point is that a complex network has the clustering coefficient which is much larger than C_{random} [25].

We have analyzed two subintervals of the seismic data in southern California mentioned before: (I) between 01:50:15.81 on January 30, 1992 and 05:48:10.93 on February 2, 1992, with 63 events, and (II) between 11:57:34.10 on June 28, 1992 and 20:48:22.61 on June 28, 1992, with 176 events. Seismicity in the period (I) is moderate, in contrast to the active period (II). The initial time of the period (II) is adjusted to be the event of the mainshock with M7.3 ($34°12.01'$N latitude, $116°26.20'$W longitude, and 0.97 km in depth), followed by a lot of aftershocks. This is why the period (II) is much shorter than (I). The cell size is taken to be 10 km $\times$ 10 km $\times$ 10 km for (I) and (II). Both of the corresponding earthquake networks have 50 vertices. The results are: (I) $C_{\text{actual}} = 0.680$ ($C_{\text{random}} = 0.046$), (II) $C_{\text{actual}} = 0.653$ ($C_{\text{random}} = 0.093$). Therefore, we conclude that compared to the completely random network, the clustering coefficient is about 10 times larger, highlighting the hierarchical property of the earthquake network.

Finally, we mention that employing the method of random sampling is due to the limitation of our computational power for combinatorial problems. However, through extensive tests, we confidently believe that the results reported here are correct ones.

5   Conclusion

We have constructed the complex evolving network for earthquakes in the area of southern California and have studied its topological properties. We have found that the earthquake networks have the scale-free nature in their distributions of connectivities and, thus, have presented a novel feature of seismicity as a complex critical phenomenon. We have presented an interpretation of this scale-free nature in conformity with the Gutenberg-Richter law. Then, we have studied the small-world and hierarchical properties of the earthquake network. We have shown that the degrees of separation between two vertices as the vertices take a small value between 2 and 3. We have also calculated the clustering coefficient and have found that its actual value is about 10 times larger than that in the case of the completely random network.

We believe that these new discoveries may play an important role in modeling earthquake phenomenon [31], [32] and shed light on an aspect of seismology from a novel viewpoints.

References

1. Bak, P., Tang, C.: Earthquakes as a Self-Organized Critical Phenomenon. J. Geophys. Res. **94** (1989) 15635-15637
2. Olami, Z., Feder, H.J.S., Christensen, K.: Self-Organized Criticality in a Continuous, Nonconservative Cellular Automaton Modeling Earthquakes. Phys. Rev. Lett. **68** (1992) 1244-1247
3. Huang, Y., Johansen, A., Lee, M.W., Saleur, H., Sornette, D.: Artifactual Log - Periodicity in Finite Size Data: Relevance for Earthquake Aftershocks.: J. Geophys. Res. **105** (B11) (2000) 25451-25471
4. Bak, P., Christensen, K., Danon, L., Scanlon, T.: Unified Scaling Law for Earthquakes. Phys. Rev. Lett. **88** (2002) 178501
5. Lise, S., Paczuski, M.: Nonconservative Earthquake Model of Self-Organized Criticality on a Random Graph. Phys. Rev. Lett. **88** (2002) 228301
6. Omori, F.: On the Aftershocks of Earthquakes. J. Coll. Sci. Imp. Univ. Tokyo **7** (1894) 111-216
7. Gutenberg, B., Richter, C.F.: Frequency of Earthquakes in California. Bull. Seism. Soc. Am. **34** (1944) 185-188
8. Abe, S., Suzuki, N.: Law for the Distance between Successive Earthquakes. J. Geophys. Res. **108** (B2) (2003) 2113
9. Abe, S., Suzuki, N.: Zipf-Mandelbrot Law for Time Intervals of Earthquakes. e-print cond-mat/0208344
10. Tsallis, C.: Possible Generalization of Boltzmann-Gibbs Statistics. J. Stat. Phys. **52** (1988) 479-487
11. Tsallis, C., Mendes, R.S., Plastino, A.R.: The Role of Constraints within Generalized Nonextensive Statistics. Physica A **261** (1998) 534-554
12. Abe, S., Okamoto, Y. (eds.): Nonextensive Statistical Mechanics and Its Applications. Springer-Verlag, Heidelberg (2001)
13. Kaniadakis, G., Lissia, M., Rapisarda, A. (eds.): Special Issue of Physica A **305** (2002)

14. Gell-Mann, M., Tsallis, C. (eds.): Nonextensive Entropy-Interdisciplinary Applications. Oxford University Press, Oxford (2003)
15. Abe, S.: Stability of Tsallis Entropy and Instabilities of Rényi and Normalized Tsallis Entropies: A Basis for q-Exponential Distributions. Phys. Rev. E **66** (2002) 046134
16. Abe, S., Suzuki, N.: Itineration of the Internet over Nonequilibrium Stationary States in Tsallis Statistics. Phys. Rev. E **67** (2003) 016106
17. Abe, S., Suzuki, N.: Omori's Law in the Internet Traffic. Europhys. Lett. **61** (2003) 852-855
18. Abe, S., Suzuki, N.: Gutenberg-Richter Law for Internetquakes. Physica A **319** (2003) 552-556
19. Abe, S., Suzuki, N.: Statistical Similarity between Internetquakes and Earthquakes. Physica D (to appear)
20. Faloutsos, M., Faloutsos, P., Faloutsos, C.: On Power-Law Relationships of the Internet Topology. ACM SIGCOMM Comput. Commun. Rev. **29** (1999) 251-262
21. Barabási, A.-L., Albert, R.: Emergence of Scaling in Random Networks. Science **286** (1999) 509-512
22. Albert, R., Barabási, A.-L.: Statistical Mechanics of Complex Network. Rev. Mod. Phys. **74** (2002) 47-97
23. Dorogovtsev, S.N., Mendes, J.F.F.: Evolution of Networks. Oxford University Press, Oxford (2003)
24. Albert, R., Barabási, A.-L.: Topology of Evolving Networks: Local Events and Universality. Phys. Rev. Lett. **85** (2000) 5234-5237
25. Watts, D.J., Strogatz, S.H.: Collective Dynamics of 'Small-World' Networks. Nature (London) **393** (1998) 440-442
26. Abe, S., Suzuki, N.: Scale-Free Network of Earthquakes. Europhys. Lett. **65** (2004) 581-586
27. Abe, S., Suzuki, N.: Small-World Structure of Earthquake Network. e-print cond-mat/0308208, Physica A (to appear)
28. Mandelbrot, B.B.: The Fractal Geometry of Nature. Freeman, San Francisco (1983)
29. Tadic, B., Thurner, S.: Information Super-Diffusion on Structured Networks. e-print cond-mat/0307670
30. Bollobás, B.: Random Graphs, 2nd edition. Cambridge University Press, Cambridge (2001)
31. Baiesi, M., Paczuski, M.: Scale Free Networks of Earthquakes and Aftershocks. e-print cond-mat/0309485
32. Peixoto, T.P., Prado, C.P.C.: Distribution of Epicenters in Olami-Feder-Christensen Model. e-print cond-mat/0310366, Phys. Rev. E (to appear)

Universal Features of Network Topology

K. Austin and G.J. Rodgers

Brunel University, Uxbridge, Middlesex UB8 3PH, U.K.

Abstract. Recent studies have revealed characteristic general features in the topology of real-world networks. We investigate the universality of mechanisms that result in the power-law behaviour of many real-world networks, paying particular attention to the Barabasi-Albert process of preferential attachment as the most successful. We introduce a variation on this theme where at each time step either a new vertex and edge is added to the network or a new edge is created between two existing vertices. This process retains a power-law degree distribution, while other variations destroy it. We also introduce alternative models which favour connections to vertices with high degree but by a different mechanism and find that one of the models displays behaviour that is compatible with a power-law degree distribution.

1 Introduction

A variety of real-world networks, such as communications networks, the World Wide Web and the Internet consist of many constituents that are connected by diverse interactions. Until recently, there was an absence of data on these large networks, making it difficult to extract reliable results about their topological features. However, as the accumulation of data becomes increasingly computerised, more information about large networks is becoming available. The studies that have been performed on these complex networks so far suggest that they share many common structural characteristics. Thus it has been proposed that there may be universal laws governing the structure and evolution of many diverse real-world networks.

Examples of such networks are the World Wide Web [1,2,3], the Internet [4], the graph of movie actors, [5], scientific collaboration networks, [6,7], citation networks [8,9] and the graph of long distance phone calls [10]. Many of the networks studied are found to exhibit power law degree distributions [11], $P(k) \sim k^{-\gamma}$, with γ mostly in the range $2 \leq \gamma \leq 3$. These findings have raised the important question of whether there is a universal mechanism by which the networks grow and self-organise that persistently results in the distributions found in real-world networks.

In this paper we introduce a mechanism that is not based on strict preferential attachment to suggest that the preferential attachment process may not be the only mechanism that results in the degree distributions observed in real-world networks.

M. Bubak et al. (Eds.): ICCS 2004, LNCS 3038, pp. 1054–1061, 2004.
© Springer-Verlag Berlin Heidelberg 2004

2   Preferential Attachment

The process of preferential attachment was first introduced by Barabasi and Albert [5]. Preferential attachment is the process whereby new vertices and edges are introduced to the network over time in such a way that vertices with more edges increase their degree more quickly. The simplest case of linear preferential attachment is given in Ref. [13], with a vertex of degree k receiving a new edge with probability $\Pi(k)$, given by $\Pi(k) \propto k$. At each time step, a new vertex and edge are added to the network and connect to a vertex already present. The probability that a new vertex will connect to an existing vertex of degree k, $\Pi(k)$, is given by $\Pi(k) = k/M_1(t)$. The moments $M_\alpha(t)$ are given by $M_\alpha(t) = \sum k^\alpha P_k(t)$, where $P_k(t)$ is the number of vertices of degree k at time t. The lower order moments, $M_0(t)$ and $M_1(t)$, give the total number of vertices in the network and the total degree of the network, respectively. The result of this model is a power law degree distribution, where the number of vertices of degree k at time t, $P_k(t)$, is given by $P_k(t) \sim tk^{-3}$ [5,13].

For non-linear preferential attachment, given by $\Pi(k) \propto k^\alpha$, where $\alpha > 0$ is a constant, then it is found that for $\alpha < 1$, $P_k(t)$ follows a stretched exponential and for $\alpha > 1$, a single vertex in the network links to all other vertices [13]. In Ref. [14], $\Pi(k)$ was measured for four large networks. This was acheived using the computerised data of network dynamics to monitor which vertices acquire new links as a function of the vertex degree. To avoid strong fluctuations, the accumulative distribution, $\kappa(k) = \int_0^k \Pi(k)dk$, was used instead of $\Pi(k)$. If $\Pi(k)$ follows non-linear preferential attachment, then $\kappa(k)$ should have the form $\kappa(k) \propto k^{\alpha+1}$. In Ref [14], the Internet and citation networks were both found to follow near linear preferential attachment, with $\alpha = 1.05$ and $\alpha = 0.95 \pm 0.1$, respectively. The scientific collaboration network and the actor network were found to follow sublinear preferential attachment, with $\alpha = 0.79 \pm 0.1$ and $\alpha = 0.81 \pm 0.1$, respectively. This sublinear behaviour implies that $P(k)$ should follow a stretched exponential [13], but the measured $P(k)$ indicate that a power law offers a better fit. The authors of Ref. [14] suggest that this is because external links to new vertices follow sub-linear preferential attachment, whereas internal links between existing vertices follow near linear preferential attachment. Since both the actor and science collaboration networks are dominated by internal links, the degree distribution exhibits the scale-free behaviour of linear preferential attachment.

Inspired by the above work, we introduce a model where at each time step a new edge is added with probability p or an edge and vertex with probability $1 - p$. Following the method of [13], the rate equation for $P_k(t)$ is found to be

$$\frac{dP_k(t)}{dt} = \frac{2p+1-p}{M_1(t)}[(k-1)P_{k-1} - kP_k] + (1-p)\delta_{k,1}. \tag{1}$$

The first term on the right hand side of this equation represents the creation of a vertex of degree k from a vertex of degree $k - 1$. The second term represents the creation of a vertex of degree $k + 1$ from a vertex of degree k. This happens with probability $2p$ if a new edge is added between existing vertices, and $1 - p$ if a new edge and vertex are added and link to an existing vertex. The final term represents the arrival of a new vertex of degree 1, which happens with probability $1 - p$. Following [13], we find the total number of edges in the network, $M_1(t)$, obeys $dM_1(t)/dt = 2$. Substituting $M_1(t) = 2t$ and $P_k(t) = tn_k$ into Eq.(1) yields the recursion relation

$$n_k = \frac{(k-1)}{k + [2/(1+p)]} n_{k-1} \qquad (2)$$

for $k > 1$. For large k, n_k has the solution $n_k \sim k^{-\gamma}$, where $\gamma = 1 + 2/(1+p)$. Since $0 \le p \le 1$, then $2 \le \gamma \le 3$ for large k, which is the range of the exponent found for many real-world networks.

We also consider the case where a new edge is introduced between existing vertices with linear preferential attachment, but the new vertices link to existing vertices with sublinear preferential attachment. The probability that an existing vertex will recieve a link to a new vertex is then given by $\Pi(k) \propto k^\alpha$, where $\alpha < 1$. We find that when $0 < \alpha < 1$ and $k \to \infty$, the degree distribution is dominated by the process whereby edges are added according to linear preferential attachment. A power-law degree distribution is obtained, $P_k(t) \sim tk^{-\gamma}$, with $\gamma = 1 + 1/p$, for large k, even when more edges are added according to the process with sub-linear preferential attachment, i.e. when $0 \le p < 0.5$. However, to yield $2 \le \gamma \le 3$ in agreement with experimental data [11], more edges must be added to the network from the process with linear preferential attachment, i.e. when $0.5 \le p \le 1$. This work suggests that real-world networks are dominated by the process of linear preferential attachment but that there may be other mechanisms at work.

Many other models based on linear preferential attachment have been proposed to include mechanisms such as the decay of edges, the "condensation" of edges and vertices with fitness [11]. The aim of these models is to maintain the scale-free nature found in real-world networks, whilst incorporating other mechanisms to model the different features that they exhibit.

3   Networks with Fitness

The different rates with which vertices acquire edges is often attributed to the ability of vertices to compete for new edges, called the vertex fitness [15]. In Ref. [16], models of growing random networks are studied in which the growth rates are dependent upon the vertex fitness.

Model A

In model A, a network is built by connecting new vertices to those already present in the network. The probability that an existing vertex will acquire a new edge is given by $\Pi(k, \eta) \propto (k-1) + \eta$, where k is the vertex degree and $\eta \ge 0$ is it's additive fitness chosen randomly from a probability distribution $f(\eta)$. For this model the degree distribution behaves as $P_k(\eta, t) \sim tk^{-\gamma}$, where $\gamma = <\eta> + 2$ and $<\eta>$ is the average additive fitness.

Model B

In model B, the network is built in the same way, but an existing vertex acquires a new edge with random additive fitness, η, random multiplicative fitness, ζ, and degree, k, with probability $\Pi(k, \eta, \zeta) \propto \zeta(k-1) + \eta$. η and ζ are initially chosen from a probability distribution $f(\eta, \zeta)$. After some analysis, it is simple to show that the degree distribution behaves as $P_k(\eta, \zeta, t) \sim t k^{-\gamma}$ for $k \to \infty$, where $\gamma = 1 + m/\zeta$, m is the reduced moment, given by $M_{11}(t) = mt$, and $P_k(t) = t n_k$. For model B, specific cases are considered as the general case cannot be solved explicitly.

The case where $f(\eta, \zeta) = 1$, $0 \le \eta \le 1$, and $0 \le \zeta \le 1$ is first considered. For this case, the reduced moment is given by $m = 1/(1 - e^{-2}) = 1.156$, which yields $n_k \sim k^{-\gamma}/\ln k$ with $\gamma = 1 + m = 2.156$.

For the case where $f(\eta, \zeta) = \delta(\zeta - \eta)$ with $0 \le \zeta \le 1$, then the degree distribution follows $n_k \sim k^{-\gamma}$ with $\gamma = 2.255$.

Finally, the case where $f(\eta, \zeta)$ takes the form $f(\eta, \zeta) = 6\zeta(1 - \zeta)\delta(\zeta - \eta)$ and $0 \le \zeta \le 1$ is solved to give the degree distribution $n_k \sim k^{-\gamma}/(\ln k)^2$ for $k \to \infty$, where $\gamma = 1 + m = 2.550$.

4 Extremal Networks

There are many variations of the BA model that preserve the scale-free nature of the model, such as networks where the vertices have random additive fitness [15, 16]. Other variations include processes such as aging and rewiring of edges [11]. However, it is important to note that the BA method of preferential attachment may not be the only mechanism to produce power law degree distributions. Here we present a model which favours connections to vertices with high degree, but by a different mechanism.

In this model, at each time step a set of $m(t)$ vertices are chosen from the $N(t)$ vertices already present in the network. The vertices to be in the set are chosen at random so that any vertex can appear in the set more than once. A new vertex is then connected to the vertex in the set with the highest degree. If two or more vertices have equal highest degree, one of them is chosen at random. With these rules, the degree, $L(t)$, of the hub vertex, the vertex with the largest degree, obeys

$$\frac{dL(t)}{dt} = 1 - \left[1 - \frac{1}{N(t)}\right]^{m(t)}. \tag{3}$$

The second term on the right hand side of this equation represents the probability of not choosing the vertex when $m(t)$ vertices are selected. Hence, as the hub vertex always gains another edge when selected, the rate of change of the degree of the hub vertex is one minus this probability. As $N(t) \sim t$ for large time we have

$$\frac{dL(t)}{dt} \sim 1 - exp\left[-\frac{m(t)}{t}\right]. \tag{4}$$

Model A $m(t) = m$

When $m(t)$ is time independent it is simple to use Eq.(4) to show that $L(t) \sim m \ln t$ for large t. We consider four different cases of this model in turn (i) $m = 1$, (ii) $m = 2$, (iii) $m > 2$ and (iv) $m \to \infty$.

For **(i)** $m = 1$ the model is just the random addition of an edge and vertex to a randomly selected vertex already in the network. All vertices are selected with equal probability, independent of degree. Consequently the number of vertices of degree k at time t, $P_k(t)$, obeys the equation

$$\frac{dP_k(t)}{dt} = \frac{1}{N(t)} [P_{k-1}(t) - P_k(t)] + \delta_{k,1}. \tag{5}$$

Following [13], Eq.(5) can be solved in the large time limit to yield $P_k(t) \propto t2^{-k}$. As one would expect from a network in which the vertices gain edges with a rate independent of their degree, this network has an exponential degree distribution.

For **(ii)** $m = 2$ we choose 2 vertices at random and join the incoming edge to the vertex with the larger degree. If they both have the same degree, one of them is chosen at random. The degree distribution $P_k(t)$ satisfies

$$\frac{dP_k(t)}{dt} = 2\frac{P_{k-1}(t)}{N^2(t)} \sum_{r=1}^{k-2} P_r(t) - 2\frac{P_k(t)}{N^2(t)} \sum_{r=1}^{k-1} P_r(t) + \frac{P_{k-1}^2(t)}{N^2(t)} - \frac{P_k^2(t)}{N^2(t)} + \delta_{k,1}. \tag{6}$$

From this equation it is a simple matter to check that $dN(t)/dt = dM_0(t)/dt = 1$ and $dM_1(t)/dt = 2$. Introducing $P_k(t) = tn_k$ and $N(t) = t$ for large t into Eq.(6) gives

$$n_k = 2n_{k-1} \sum_{r=1}^{k-2} n_r - 2n_k \sum_{r=1}^{k-1} n_r + n_{k-1}^2 - n_k^2 + \delta_{k,1}. \tag{7}$$

Fig. 1. is a plot of n_k against k obtained from a simulation of this model over 10^6 time steps with $m = 2$. We see that the data suggests that n_k has an exponential dependence on k. We can test this by inserting $n_k = (1 - e^{-\beta})e^{-\beta k}$ into Eq.(7). For large k this reveals $\beta = \ln(3/2) = 0.405...$, which is in good agreement with the simulation data, suggesting that $\beta = 0.4306...$ A similar value of β can be obtained by iterating Eq.(7) numerically.

For **(iii)** $m > 2$ we can show that the n_k approximately obey

$$n_k = mn_{k-1} \left[\sum_{r=1}^{k-2} n_r \right]^m - mn_k \left[\sum_{r=1}^{k-1} n_r \right]^m + \delta_{k,1} \tag{8}$$

which is obtained by dropping terms of order n_k^2 and higher. Inserting $n_k = (1 - e^{-\beta})e^{-\beta k}$ into this iteration for large k yields $\beta = \ln[(m+1)/m]$. Thus for all finite $m \geq 1$ we have an exponential degree distribution.

As **(iv)** $m \to \infty$ this exponential degree distribution breaks down and $\beta \to 0$. This is because a new edge is always added to the vertex with the highest degree in the network and the network develops a star geometry. More precisely, at time

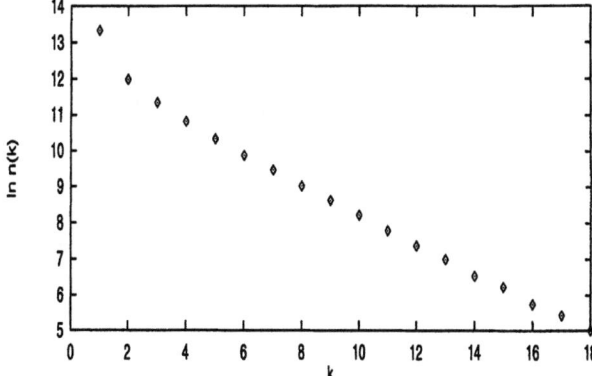

Fig. 1. The degree distribution when $m = 2$ vertices are chosen at random and the incoming edge is connected to the vertex with highest degree.

t it has a hub vertex with degree $N(t) - 1$ linked to $N(t) - 1$ vertices with degree 1. Consequently the degree distribution takes the bi-modal form

$$P_k(t) = \delta(k - N(t) + 1) + (N(t) - 1)\delta(k - 1). \tag{9}$$

Model B $m(t) = \mu N(t)$

In this network model a few hubs develop with very high degree and the vast majority of the the vertices have degree 1. This can be demonstrated by some simple analysis. For instance, putting $m(t) = \mu N(t) = \mu t$ for large t into Eq.(4) gives $L(t) \sim (1 - e^{-\mu})t$, which indicates that a proportion $1 - e^{-\mu}$ of the edges in the network are connected to the vertex of the highest degree. The number of vertices with degree 1, $P_1(t)$, obeys

$$\frac{dP_1(t)}{dt} = 1 - \left[\frac{P_1(t)}{t}\right]^{\mu t}. \tag{10}$$

By writing $P_1(t) = t + p(t)$ and substituting it into Eq.(10) we can solve the resulting differential equation for $p(t)$ in the large t limit to yield

$$P_1(t) = t - \frac{1}{\mu}\ln(t + c) \tag{11}$$

where c is a constant. Consequently the proportion of vertices with degree greater than 1 behaves as

$$\frac{t - P_1(t)}{t} \sim \frac{\ln(t + c)}{t} \sim 0 \qquad \text{as} \qquad t \to \infty. \tag{12}$$

Thus we see that for all $\mu > 0$ this network has $N_{hub} \sim 1/\mu$ hub vertices with degree $k_{hub} \sim \mu t$ and a much larger number, $N_1 \sim t$, of vertices with degree 1.

Model C $m(t) = N^{\nu}(t)$

Inserting $m(t) = N^{\nu}(t)$ for large t into Eq.(4) gives $L(t) \sim t^{\nu}$. Clearly for $\nu \geq 1$ the network develops a geometry like that of Model B above. For $0 < \nu < 1$ the precise geometry is more difficult to classify analytically. For this reason we have performed numerical simulations of this network for $0 < \nu < 1$. We found that the degree distribution P_k is compatible with a power-law distribution $P_k \sim k^{-\gamma}$ with for instance $\gamma = 1.45...$ when $\nu = 3/4$ and $\gamma = 1.46...$ when $\nu = 2/3$. This is clearer numerically for larger values of ν, because the closer that ν is to 1, the easier it is to obtain accurate data for a wide range of vertex degrees. A plot of the log-binned data for P_k against k for $\nu = 3/4$ and $\nu = 2/3$ after 5×10^6 time steps is given in Fig. 2. Plots of P_k against k for 10^6 and 2.5×10^6 time steps give the same exponents. The numerical work, and the work done on Models A and B, suggest that when we have $m(t) \sim N(t)^{\nu}$, $P_k \sim k^{-\gamma}$, with γ a decreasing function of ν. However, the fact that for some values of ν, $\gamma < 2$, suggests that this behaviour is destroyed for larger systems, and that P_k may follow some other functional form.

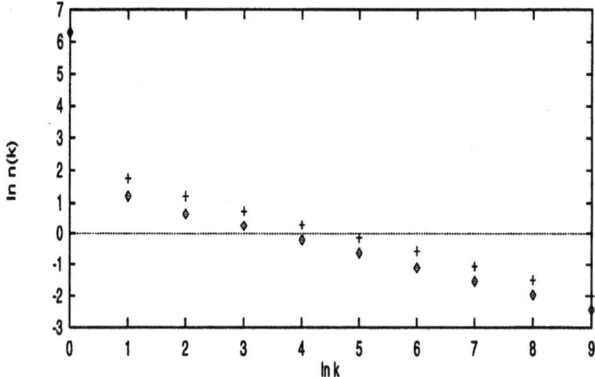

Fig. 2. The degree distribution when $m(t) = N^{\nu}(t)$ vertices are chosen at random and the incoming edge is connected to the vertex with highest degree; (a) $\nu = 3/4$ (*diamond*) ; (b) $\nu = 2/3$ (*cross*).

5   Discussion

Many real-world networks appear to exhibit power-law degree distributions, indicating that such networks may be classified according to their scale-free nature. The process of preferential attachment introduced by the BA model, along with it's variations, naturally results in these distributions. Thus it may be proposed that there are universal laws governing real-world networks and that the process of preferential attachment may prove to be one.

It seems unlikely, however, that all networks with power-law degree distributions grow according to the same mechanism. Indeed there are many variations of the BA model that preserve the scale-free nature of the model such as networks where the vertices have random additive fitness. Other variations include processes such as aging and rewiring of edges to mimic the behaviour of real world networks. This raises the question of whether preferential attachment is robust enough to incorporate different aspects of real-world networks whilst retaining their degree distributions. We have seen that the introduction of trivial, additive fitness retains power law behaviour, but multiplicative fitness destroys it.

Another important consideration is that what experimental studies indicate as power-law dependence is actually a linear fit to a comparatively narrow range on a log-log plot. It is very difficult for any experimental study to yield a functional form for the degree distribution because of strong fluctuations [11]. We introduced models that were not based on strict preferential attachment and found that one of the models has a degree distribution that is compatible with a power law, but may have a different functional form.

This work reveals important questions about the universality of complex networks: (i) Is preferential attachment robust enough to incorporate all of the diverse processes observed in real-world networks? (ii) Are there other processes that result in the degree distributions found in real-world networks?

References

1. R. Albert, H. Jeong and A. L. Barabasi, *Nature*, **401**, 130 (1999).
2. R. Kumar, P. Raghavan, S. Rajagopalan and A. Tomkins, *Proceedings of the 25th VLDB Conference, Edinburgh, Scotland, 7-10 Sep tember*, 639-650, (1999).
3. A Broder, R. Kumar, F. Maghoui, P. Raghavan, S. Rajagopalan, R. Stata, A Tomkins and J Wiener, *Proceedings of the 9th WWW Conference, Amsterdam, 15-19 May*, 309, (2000).
4. M. Faloutsos, P. Faloutsos and C. Faloutsos, *Comput. Commun. Rev.*, **29**, 251, (1999).
5. A.L. Barabasi and R. Albert, *Science*, **286**, 5 09 (1999).
6. M. E. J. Newman, *Phys. Rev. E.*, **64**, 016131, (2001).
7. M. E. J. Newman, *Phys. Rev. E.*, **64**, 016132 , (2001).
8. J. Laherre and D. Sornette, *Eur. Phys. J.*, **2**, 525, (1998).
9. S. Redner, *Eur. Phys. J.*, **4**, 131, (1998).
10. W. Aiello, F. Chung and L. Lu, *Proceedings of the Thi rty-Second Annual ACM Symposium on Theory of Computing*, 171-180, (2000)
11. S. N. Dorogovtsev and J. F. F. Mendes, *Adv. Phys.*, **51**, 1079, (2001).
12. R. Albert, H. Jeong and A. L. Barabasi, *Nature*, **4 06**, 378, (2000).
13. P. L. Krapivsky, S. Redner and F. Leyvraz, *Phys. Rev. Lett.*, **85**, 4629, (2000).
14. H. Jeong, Z. Neda and A. L. Barabasi, *Europhys. Let t.*, **61**, 567, (2003).
15. G. Bianconi and A. L. Barabasi, *Europhys. Lett.*, **54**, 436, (2001).
16. G. Ergun and G. J. Rodgers, *Physica A*, **303**, 26 1, (2002).

Network Brownian Motion: A New Method to Measure Vertex-Vertex Proximity and to Identify Communities and Subcommunities

Haijun Zhou and Reinhard Lipowsky

Max-Planck-Institute of Colloids and Interfaces, D-14424 Potsdam, Germany
{zhou,lipowsky}@mpikg-golm.mpg.de

Abstract. The networks considered here consist of sets of interconnected vertices, examples of which include social networks, technological networks, and biological networks. Two important issues are to measure the extent of proximity between vertices and to identify the community structure of a network. In this paper, the proximity index between two nearest-neighboring vertices of a network is measured by a biased Brownian particle which moves on the network. This proximity index integrates both the local and the global structural information of a given network, and it is used by an agglomerative hierarchical algorithm to identify the community structure of the network. This method is applied to several artificial or real-world networks and satisfying results are attained. Finding the proximity indices for all nearest-neighboring vertex pairs needs a computational time that scales as $O(N^3)$, with N being the total number of vertices in the network.

1 Introduction

Network models are necessary to understand the behavior of complex systems, such as a biological organism, an ecological system, a human society, or the Internet. A Network is composed of a set of vertices and a set of edges which connect these vertices. Many complex networks were constructed and studied in recent years, and the *small-world* and *scale-free* properties of real-world networks were discovered (for a review, see [1,2,3]).

As far as the dynamics of networks is considered, the concept of network Brownian motion (or random walks) has aroused some interest among statistical physicists [4,5,6,7,8,9,10]. For example, the diffusion constant on a small-world network was investigated in [4]; random walks were used in [5,6] to study network search problems and in [7,8] to study network traffic and congestion. In [9,10] a new approach based on Brownian motion was introduced, by which one can measure the extent of proximity between neighboring vertices of a given network and cluster vertices of this network into different communities and subcommunities.

The present work extends the basic idea of [9,10]. Intuitively, a community of a network should consist of a subset of vertices which are more "near" to each other than to vertices not included in this subset. We give a quantitative

M. Bubak et al. (Eds.): ICCS 2004, LNCS 3038, pp. 1062–1069, 2004.
© Springer-Verlag Berlin Heidelberg 2004

definition of proximity measure and show how to calculate its value efficiently. This proximity measure is based on biased Brownian motion on a network. We apply this proximity measure in identifying the community structure of several networks.

Section 2 introduces a class of biased network Brownian motions and define a vertex-vertex proximity index. Section 3 outlines the clustering algorithm *Netwalk* and shows its performance by application on random modular networks. Section 4 applies the *Netwalk* algorithm to several social and biological networks. We conclude this work in section 5 together with some discussion.

2   Biased Brownian Motion and Proximity Index

Consider a connected network of N vertices and M edges, with a weight matrix w. If there is no edge between vertex i and vertex j, $w_{ij} = 0$; if there is an edge in between, $w_{ij} \equiv w_{ji} > 0$ and its value corresponds to the interaction strength of this edge. In a social friendship network, for example, w_{ij} may be proportional to the frequency of contact between person i and person j. A Brownian particle moves on the network, and at each step it jumps from its present position, say i, to a nearest-neighboring position, say j. We assume that the jump probability P_{ij} has the form

$$P_{ij} = \frac{1}{K_i} w_{ij}(c_{ij} + 1)^\gamma, \tag{1}$$

where $K_i = \sum_k w_{ik}(c_{ik}+1)^\gamma$, and c_{ij} is the number of common nearest-neighbors of vertex i and vertex j. For $\gamma > 0$, the Brownian particle has greater probability at each vertex to jump to a nearest-neighboring vertex that shares more nearest-neighbors with the original vertex. Equation (1) thus defines a biased Brownian motion, with the degree of bias being controlled by the bias exponent γ. For $\gamma = 0$, eq. (1) reduces to the unbiased Brownian motion discussed in refs. [9,10].

For convenience, we introduce a generalized adjacency matrix $\mathbb{A}$ with matrix elements $\mathbb{A}_{ij} = \mathbb{A}_{ji} = w_{ij}(c_{ij} + 1)^\gamma$. In addition, we define a diagonal matrix $\mathbb{K}$ such that $\mathbb{K}_{ii} = K_i$. The transfer matrix P in eq. (1) is then given by $P = \mathbb{K}^{-1}\mathbb{A}$.

Suppose the Brownian particle is initially located at vertex i. The mean-first-passage-time d_{ij} [9] is the average number of steps the Brownian particle takes before it reaches vertex j (or return to i in the case of $i = j$) for the first time. This quantity is given by

$$d_{ij} = P_{ij} + \sum_{m=1}^{+\infty} (m+1) \sum_{k_1 \neq j; \ldots; k_m \neq j} P_{ik_1} P_{k_1 k_2} \ldots P_{k_m j}. \tag{2}$$

It has been shown in ref. [9] that d_{ij} is the solution of the linear equation

$$[I - B(j)] \begin{pmatrix} d_{1j} \\ \vdots \\ d_{Nj} \end{pmatrix} = \begin{pmatrix} 1 \\ \vdots \\ 1 \end{pmatrix}, \tag{3}$$

where $B(j)$ is the matrix formed by replacing the j-th column of matrix P with a column of zeros. Equation (3) seems to imply that N matrix inversion operations are needed to calculate the values of d_{ij} for all pairs i, j. This would lead to a computational time of $O(N^4)$. However, what we really need to know is the difference of mean-first-passage-times, $\Delta(i,j;k) = d_{ik} - d_{jk}$. In this article, we describe a method by which one can calculate all the $N^2(N-1)/2$ different $\Delta(i,j;k)$ values with a computational time of $O(N^3)$.

Equation (3) is equivalent to

$$
[\mathbb{K} - \mathbb{A}] \begin{pmatrix} d_{1j} \\ d_{2j} \\ \vdots \\ d_{Nj} \end{pmatrix} = \begin{pmatrix} K_1 - \mathbb{A}_{1j}d_{jj} \\ K_2 - \mathbb{A}_{2j}d_{jj} \\ \vdots \\ K_N - \mathbb{A}_{Nj}d_{jj} \end{pmatrix}.
\tag{4}
$$

Because the determinant of $\mathbb{K} - \mathbb{A}$ is zero, one cannot invert this matrix directly to find the solution of eq. (4). However, one can construct two $(N-1) \times (N-1)$ matrices $\mathbb{K}_r$ and $\mathbb{A}_r$ by removing the last rows and columns[1] of the matrices $\mathbb{K}$ and $\mathbb{A}$ [11]. This leads to

$$
d_{ij} = d_{Nj} + \sum_{l=1}^{N-1} \left(\frac{1}{\mathbb{K}_r - \mathbb{A}_r} \right)_{il} K_l - \frac{\mathrm{Tr}\mathbb{K}}{K_j} \sum_{l=1}^{N-1} \left(\frac{1}{\mathbb{K}_r - \mathbb{A}_r} \right)_{il} \mathbb{A}_{lj}, \quad (i < N). \tag{5}
$$

In deriving eq. (5) we have used the fact that d_{jj}, the average returning time, is independent of network topology, with $d_{jj} = \sum_{l=1}^{N} K_l/K_j = \mathrm{Tr}\mathbb{K}/K_j$ [12].

With eq. (5) one only needs to invert the matrix $\mathbb{K}_r - \mathbb{A}_r$ to obtain the values of all quantities $\Delta(i,j;k)$. The total computation time scales as $O(N^3)$.

For each nearest-neighboring pair of vertices i and j with $\omega_{ij} > 0$, we define the proximity index

$$
\Lambda(i,j) = \frac{\sqrt{\sum_{k \neq i,j} \Delta^2(i,j;k)}}{(N-2)}
\tag{6}
$$

in order to quantity the extent of proximity between i and j. If two nearest-neighboring vertices i and j belong to the same community, then the mean-first-passage-time d_{ik} from i to any another vertex k ($k \neq i, j$) will be approximately equal to that from j to k; in other words, the "coordinates" of the two vertices will be near to each other. Consequently, $\Lambda(i,j)$ will be small if i and j belong to the same community and large if they belong to different communities. The proximity index eq. (6) gives a quantitative measure of vertex-vertex proximity for a network that has no metric otherwise.

This proximity index is used in the *Netwalk* algorithm of the following section.

---

[1]   Actually one can remove an arbitrary row and an arbitrary column and the result is unchanged. Here for definiteness, we remove the N-th row and the N-th column.

Table 1. Number of misclassified vertices as a function of the between-community probability p. For each value of p, 100 random networks with 128 vertices are generated. The results obtained by using unbiased ($\gamma = 0$), linearly-biased ($\gamma = 1$) and squarely-biased ($\gamma = 2$) Brownian motions are compared.

p	$\gamma = 0$	$\gamma = 1$	$\gamma = 2$
0.3	0.43 ± 0.78	0.43 ± 0.70	0.62 ± 0.90
0.35	2.9 ± 2.9	1.96 ± 2.1	2.59 ± 2.4
0.4	13.0 ± 7.5	8.3 ± 5.4	10.2 ± 6.5
0.45	38.1 ± 14.6	26.5 ± 10.8	29.8 ± 10.6

3   *Netwalk* Algorithm

We exploit the proximity index to reveal the community structure of a network. The *Netwalk* algorithm works as follows:

1. Calculate the inverse of $\mathbb{K}_r - \mathbb{A}_r$.
2. Calculate the proximity index $\Lambda(i, j)$ for all nearest-neighboring pairs based on eq. (5) and eq. (6).
3. Initially, the network has N communities, each contains a single vertex. We define the proximity index between two communities α and β as

$$\Lambda_{\alpha,\beta} = \frac{1}{n_{\alpha,\beta}} \sum_{(i,j):\omega_{ij}>0,i\in\alpha,j\in\beta} \Lambda(i, j), \tag{7}$$

where the summation is over all edges (i, j) that connect communities α and β, and $n_{\alpha,\beta}$ is the total number of such edges. Merge the two communities with the lowest proximity index into a single community and then update the proximity index between this new community and all the other remaining communities that are connected to it. This merging process is continued until all the vertices are merged into a single community corresponding to the whole network.

4. Report the community structure and draw a dendrogram.

We tested the *Netwalk* algorithm on an ensemble of modular random networks. Each network in this ensemble has 128 vertices, 1024 edges and, hence, an average degree of 16 for each vertex. The vertices are divided into four communities or modules of size 32 each. The connection pattern of each network is random, except that each edge has a probability p to be between two different modules. For each value of p, 100 random networks have been generated and studied. The results are listed in Table 1. The performance of *Netwalk* is remarkable. For example, when $p = 0.4$, i.e., when each vertex has on average 6.4 between-community edges and 9.6 within-community edges, only about eight vertices (6% of all vertices) are misclassified by this algorithm using linearly-biased Brownian motion.

Table 1 also suggests that, the performance of the linearly-biased Brownian motion ($\gamma = 1$ in eq. (1)) is considerably superior to those of $\gamma = 0$ and $\gamma = 2$.

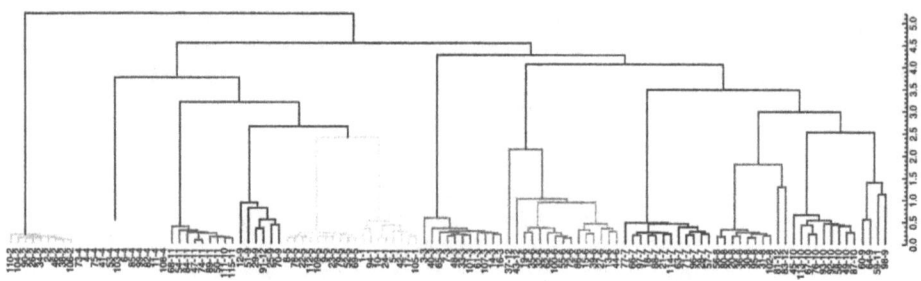

Fig. 1. Community structure of a football-team network. In the name pattern xx-yy, the number yy after the hyphen denotes the group identity of vertex xx according to information from other sources. The dendrogram is generated by P. Kleiweg's **den** algorithm (http://odur.let.rug.nl/ kleiweg/levenshtein/).

Indeed, we have observed that, for each generated random network, in most cases the number of misclassified vertices by using $\gamma = 1$ is less than that reported by using $\gamma = 0$ or $\gamma = 2$. Therefore, in our later applications, linearly-biased Brownian motion ($\gamma = 1$) will be used. (In general, one may also use non-integer values of γ but this has not been explored so far.)

4   Applications

We apply the *Netwalk* algorithm to several real-world networks in order to detect their community structures. In the following, we first discuss the results for two social networks and then for one biological network.

Karate club network of Zachary (1977).   The fission process of a social network was studied by Zachary in 1977 [13]. This network is composed of 34 members of a karate club, it has 77 weighted edges. It broke up into two parts because of a disagreement between the club's officer and its instructor. When applying our algorithm to this network, the two main communities identified by our algorithm are in full agreement with the actual fission pattern [13].

Football network of Girvan and Newman (2002).   The American football network collected by Girvan and Newman [14] contains 115 vertices (football teams) and 613 unweighted edges (matches). The community structure of this network is shown in fig. 1. Comparing the predicted pattern with the actual conference structure of these football teams, we see the following differences: (1) Conference 9 is divided into two parts by our algorithm. We have checked that there is no direct connection between the two parts of conference 9. (2) Vertex 111 is grouped into conference 11. We have checked that it has eight edges to conference 11 and only three edges to other conferences; similarly, we have checked that vertex 59 has stronger interaction with conference 9 than with any other conference. (3) Vertices in conference 12 are distributed into several conferences. We have also checked that there are very few direct interactions between the five members of this conference.

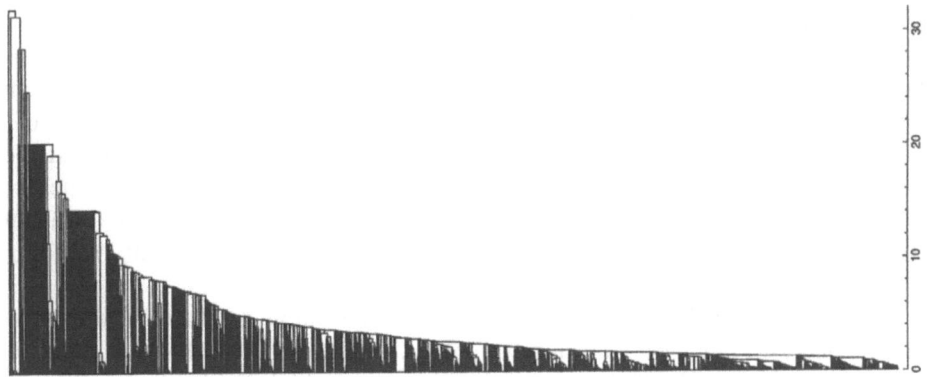

Fig. 2. Community structure of yeast's protein-protein interaction network.

Protein-protein interaction network of yeast *Saccharomyces cere-visiae*. The yeast protein-protein interaction network is constructed according to experimental data [15,16]. Each vertex of this network represents a protein, and each edge represents some physical interaction between the two involved proteins. The giant component of the reliable subset of this network contains 2406 proteins and 6117 unweighted edges (excluding self-connection) [15,16].

The community structure of this network as obtained via *Netwalk* is shown in fig. 2, which is strikingly different from those of the two social networks as described above. On the global scale, the protein-protein interaction network cannot be divided into two or more large communities of similar size. At each proximity level, the network has one giant community and many small communities, each of these small communities containing of the order of ten proteins. As the community-community proximity index is increased, these small communities are integrated into the giant community in a hierarchical order. This hierarchical organization of small modules seems to be a universal feature of biological networks. It is unlikely to be caused by a particular clustering method. Similar hierarchical patterns are observed in metabolic networks using different clustering methods [17,18]. We also investigated the community structure of the gene-regulation network of yeast [19] and found a similar hierarchical pattern. The construction principles underlying such hierarchical structures are still to be fully appreciated. It is plausible that such hierarchical structures contain information about the evolutionary history of the biological organisms (H. W. Peng and L. Yu, private communication).

Based on fig. 2 many communities of proteins can be obtained. To determine the best threshold value of the proximity index in dividing the network, one may calculate the network's modularity value [20] at each value of the proximity index. We found that for yeast's protein interaction network, the peak value of the modularity is 0.51, achieved by setting threshold proximity index to 1.20. We have checked that, the communities and subcommunities predicted by the *Netwalk* algorithm at this level of proximity index are composed of proteins that have similar cellular locations and are involved in similar biological processes.

5   Conclusion and Discussion

In this paper, we have discussed biased Brownian motion on networks. A quantitative measure for the degree of proximity between neighboring vertices of a network was described based on this concept of biased network Brownian motion. This proximity index integrates both the local and the global structural information of a given network. Based on this proximity measure, we have constructed a powerful algorithm, called *Netwalk*, of network community structure identification.

We have tested the performance of *Netwalk* on random modular networks and found good performance. The algorithm was then applied to two real-world social networks and to two biological networks. For the two biological networks, namely (i) the protein-protein interaction network and (ii) the inter-regulation network of transcription factors, the communities are organized in a hierarchical way. More work is needed to understand the evolutionary origin of this hierarchical organization and its biological significance.

The *Netwalk* algorithm includes a matrix inversion operation, and its computation time scales as $O(N^3)$, where N is the total number of vertices of the network of interest. For very large networks $N \gg 10^3$, it is impractical to calculate exactly the value of the proximity index as given by eq. (6). An approximate scheme is as follows. To calculation $\Lambda(i, j)$, one may first construct a subnetwork of (say) $N_s = 1000$ elements including vertices i, j and their nearest-neighbors, next-nearest-neighbors, etc., and all the edges between these elements. An estimate of $\Lambda(i, j)$ can then be obtained by applying eqs. (5) and (6) on this subnetwork. Because of the scale-free property of many real-world networks [2], we expect the value of the proximity index obtained by this method to be a good approximation of the exact value. If this approximate scheme is used, all the vertex-vertex proximity indices can be calculated in a computational time that scales linearly with the total number of edges.

For sparse networks of size $N < 10^3$, the *Netwalk* algorithm is comparable in computational time and performance with the graph-theoretical Girvan-Newman algorithm [14]. The advantages of the statistical-physics based algorithm are as follows: (i) It is applicable to weighted networks. Therefore it is able to uncover some structure even for a densely connected network, provided the edges of this network have different weights. (ii) It could be easily extended to very large graphs as discussed above. (iii) The local environment of each vertex is included by the bias coefficient γ in eq. (1).

We are confident that both the vertex-vertex proximity measure and the *Netwalk* algorithm described in this paper will find more applications in social and biological networked systems.

References

1. Strogatz, S. H.: Exploring complex networks. Nature **410** (2001) 268-276
2. Albert, R., Barabási, A.-L.: Statistical mechanics of complex networks. Rev. Mod. Phys. **74** (2002) 47-97

3. Dorogovtsev, S. N., Mendes, J. F. F.: Evolution of networks. Adv. Phys. **51** (2002) 1079-1187

4. Jespersen, S., Sokolov, I. M., Blumen, A.: Relaxation properties of small-world networks. Phys. Rev. E **62** (2000) 4405-4408

5. Tadic, B.: Adaptive random walks on the class of Web graphs. Eur. Phys. J. B **23** (2001) 221-228

6. Adamic, L. A., Lukose, R. M., Puniyani, A. R., Huberman, B. A.: Search in power-law networks. Phys. Rev. E **64** (2001) 046135

7. Guimera, R., Diaz-Guilera, A., Vega-Redondo, F., Cabrales, A., Arenas, A.: Optimal network topologies for local search with congestion. Phys. Rev. Lett. **89** (2002) 248701

8. Holme, P.: Congestion and centrality in traffic flow on complex networks. Adv. Compl. Sys. **6** (2003) 163-176

9. Zhou, H.: Network landscape from a Brownian particle's perspective. Phys. Rev. E **67** (2003) 041908

10. Zhou, H.: Distance, dissimilarity index, and network community structure. Phys. Rev. E **67** (2003) 061901

11. Newman, M. E. J.: A measure of betweenness centrality based on random walks. e-print: cond-mat/0309045 (2003)

12. Noh, J. D., Rieger, H.: Random walks on complex networks. e-print: cond-mat/0307719 (2003)

13. Zachary, W. W.: An information flow model for conflict and fission in small groups. J. Anthropol. Res. **33** (1977) 452-473

14. Girvan, M., Newman, M. E. J.: Community structure in social and biological networks. Proc. Natl. Acad. Sci. U.S.A. **99** (2002) 7821-7826

15. Xenarios, I., Salwinski, L., Duan, X. J., Higney, P., Kim, S. M., Eisenberg, D.: DIP, the database of interacting proteins: A research tool for studying cellular networks of protein interactions. Nucleic Acids Res. **30** (2002) 303-305

16. Deane, C. M., Salwinski, L., Xenarios, I., Eisenberg, D.: Protein interactions: Two methods for assessment of the reliability of high throughput observations. Mol. Cell. Proteomics **1** (2002) 349-356

17. Ravasz, E., Somera, A. L., Mongru, D. A., Oltvai, Z. N., Barabási, A.-L.: Hierarchical organization of modularity in metabolic networks. Science **297** (2002) 1551-1555

18. Holme, P., Huss, M., Jeong, H.: Subnetwork hierarchies of biochemical pathways. Bioinformatics **19** (2003) 532-538

19. Lee, T. I., Rinaldi, N. J., Robert, F., Odom, D. T., Bar-Joseph, Z., Gerber, G. K., Hannett, N. M., Harbison, C. T., Thompson, C. M., Simon, I., Zeitlinger, J., Jennings, E. G., Murray, H. L., Gordon, D. B., Ren, B., Wyrick, J. J., Tagne, J.-B., Volkert, T. L., Fraenkel, E., Gifford, D. K., Young, R. A.: Transcriptional regulatory networks in *Saccharomyces cerevisiae*. Science **298** (2002) 799-804

20. Newman, M. E. J., Girvan, M.: Finding and evaluating community structure in networks. e-print: cond-mat/0308217

Contagion Flow through Banking Networks

Michael Boss[1], Martin Summer[1], and Stefan Thurner[2]

[1] Oesterreichische Nationalbank, Otto-Wagner-Platz 3, A-1011 Wien, Austria*
[2] Complex Systems Research Group, HNO, Medical University of Vienna,
Währinger Gürtel 18-20, A-1090 Vienna, Austria
thurner@univie.ac.at

Abstract. Starting from an empirical analysis of the network structure of the Austrian inter-bank market, we study the flow of funds through the banking network following exogenous shocks to the system. These shocks are implemented by stochastic changes in variables like interest rates, exchange rates, etc. We demonstrate that the system is relatively stable in the sence that defaults of individual banks are unlikely to spread over the entire network. We study the contagion impact of all individual banks, meaning the number of banks which are driven into insolvency as a result of a single bank's default. We show that the vertex betweenness of individual banks is linearly related to their contagion impact.

1 Introduction

With the development of new ideas and tools of analysis, physics has lately strongly contributed to a functional understanding of the structure of complex real world networks. A key insight of this research has been the discovery of surprising structural similarities in seemingly very different networks, ranging from internet technology to cell biology. See [1] for an overview. Remarkably, many real world networks show power-law (scale free) degree distributions and feature a certain pattern of cliquishness, quantified by the *clustering coefficient*. Many also exhibit a so called *small world phenomenon*, meaning that the average shortest path between any two vertices ("degrees of separation") in the network can be surprisingly small. Maybe one of the most important contributions to recent network theory is an interpretation of these network parameters with respect to stability, robustness, and efficiency of an underlying system. From this perspective financial networks are a natural candidate to study. The financial world can be thought of as a set of intermediaries i.e. banks who interact with each other through financial transactions. These interactions are governed by a set of rules and regulations, and take place on an interaction graph of all connections between financial intermediaries. Financial crises i.e. the incapacity to finance businesses and industries have recently hit several countries all around the globe. These events have triggered a boom of papers on banking-crises, financial risk-analysis and numerous policy initiatives to find and understand the

---

* The views and findings of this paper are entirely those of the authors and do not necessarily represent the views of Oesterreichische Nationalbank.

weak-points of the financial system. One of the major concerns in these debates is the danger of so called *systemic risk*: the large scale breakdown of financial intermediation due to domino effects of insolvency [2,3]. The network of mutual credit relations between financial institutions is supposed to play a key role in the risk for contagious defaults.

In the past the theoretical economic literature on contagion [4,5] suggest network topologies that might be interesting to look at. In [4] it is suggested to study a complete graph of mutual liabilities. The properties of a banking system with this structure is then compared to properties of systems with non complete networks. In [5] a set of different network structures is studied. However, so far surprisingly little is known about the *actual* empirical network topology of mutual credit relations between financial institutions.

In a recent paper we have for the first time analyzed empirical data to reconstruct the banking network of Austria [6]. Here we characterized the interbank network by the liability (or exposure) matrix L. The entries L_{ij} are the liabilities bank i has towards bank j. L is a square matrix but not necessarily symmetric. We showed that the liability (L_{ij}) size distribution follows a power law, which can be understood as being driven by underlying size and wealth distributions of the banks, which show similar power exponents. We find that the interbank network shows – like many other realistic networks – power law dependencies in the degree distributions. We could show that different scaling exponents within the same distribution relate to different hierarchy levels in sub-networks (sectors) within the total network. The scaling exponents by the agricultural banks are very low, due to the hierarchical structure of this sector, while the other banks lead to scaling exponents of sizes also found in other complex real world networks. The interbank network shows a low clustering coefficient, a result that mirrors the analysis of community structure which shows a clear network pattern, where banks would first have links with their head institution, whereas these few head institutions hold links among each other. A consequence of this structure is that the interbank network is a small world with a very low "degree of separation" between any two nodes in the system.

2   Inter-bank Topology and Flow of Contagion

The knowledge of the detailed structure of the interbank topology enables us to use it as an input for a contagion-flow model [7], which is the main idea of this paper. Unlike previous research where we studied random flow on structured networks [8,9], here we follow the flow of payments uniquely determined by the structure of the liability matrix. We use this flow to perform stress tests to the system by artificially changing external financial parameters like interest rates, exchange rates etc., mimicking conditions and "global" events which are beyond the influence of banks. By doing this we can answer questions about the stability of the financial system with respect to external shocks, in particular, which banks are likely to default due to shocks, and which banks will drag other banks into default due to their mutual credit relations (contagion).

In the following we are looking for the bilateral liability matrix L of all (about $N = 900$) Austrian banks, the Central Bank (OeNB) and an aggregated foreign banking sector. Our data consists of 10 L matrices each representing liabilities for quarterly single month periods between 2000 and 2003. T obtain these data, we draw upon two major sources from the Austrian Central Bank: the Austrian bank balance sheet data base (MAUS) and the major loan register (GKE). The Austrian banking system has a sectoral organization due to historic reasons. Banks belong to one of seven sectors: savings banks (S), Raiffeisen (agricultural) banks (R), Volksbanken (VB), joint stock banks (JS), state mortgage banks (SM), housing construction savings and loan associations (HCL), and special purpose banks (SP). Banks have to break down their balance sheet reports on claims and liabilities with other banks according to the different banking sectors, Central Bank and foreign banks. This practice of reporting on balance interbank positions breaks the liability matrix L down to blocks of sub-matrices for the individual sectors. Banks with a head institution have to disclose their positions with the head institution, which gives additional information on L. Since many banks in the system hold interbank liabilities only with their head institutions, one can pin down many entries in the L matrix exactly. This information is combined with the data from the major loans register of OeNB. This register contains all interbank loans above a threshold of 360 000 Euro. This information provides us with a set of constraints and zero restrictions for individual entries L_{ij}. Up to this point one can obtain about 90% of the L-matrix entries exactly. For the rest we employ entropy maximization method [6]. The estimation problem can be set up as a standard convex optimization problem: Assume we have a total of K constraints. The column and row constraints take the form $\sum_{j=1}^{N} L_{ij} = b_i^r$ $\forall$ i and $\sum_{i=1}^{N} L_{ij} = b_j^c$ $\forall$ j with r denoting *row* and c denoting *column*. Constraints imposed by the knowledge about particular entries in L_{ij} are given by $b^l \leq L_{ij} \leq b^u$ for some i, j The aim is to find the matrix L that has the least discrepancy to some a priori matrix U with respect to the (generalized) cross entropy measure

$$C(L, U) = \sum_{i=1}^{N} \sum_{j=1}^{N} L_{ij} \ln \left(\frac{L_{ij}}{U_{ij}} \right) \quad . \tag{1}$$

U is the matrix which contains all known exact liability entries. For those entries (bank pairs) ij where we have no knowledge from Central Bank data, we set $U_{ij} = 1$. We use the convention that $L_{ij} = 0$ whenever $U_{ij} = 0$ and define $0 \ln(\frac{0}{0})$ to be 0. As a result we obtain a rather precise picture of the interbank relations at a particular point in time. Given L we find that the distribution of liabilities follows a power law for more than three decades with an exponent of -1.87 [6]. To extract the network topology from these data, for the present purposes we ignore directions and define an undirected but weighted adjacency matrix $A_{ij}^w = L_{ij} + L_{ji}$, which measures the gross interbank interaction, i.e. the total volume of liabilities and assets for each node. We next test the validity of our estimate of L, by computing the implied community structure and then comparing it to the known real bank-clusters, i.e. the sectors. There exist various

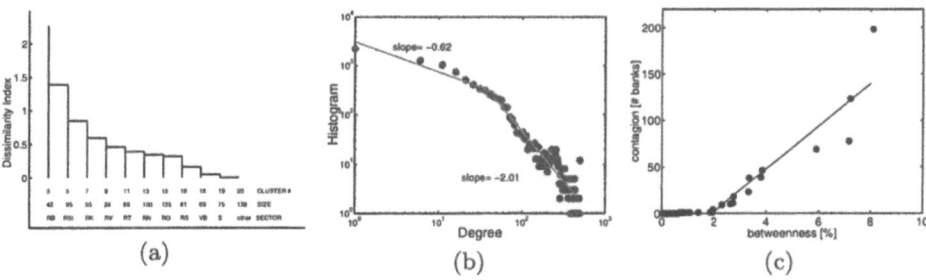

Fig. 1. (a) Community structure of the Austrian interbank market network from the same data (A_{clip}^w). The dissimilarity index is a measure of the "differentness" of the clusters. (b) Empirical degree distribution of the interbank connection network. Histograms are from aggregated data from the 10 datasets. (c) Contagion impact as a function of relative node betweenness for the L matrix. Below a value of $B(i)$ of 0.6 no contagion impact is found.

ways to find functional clusters within a given network. In [10] an algorithm was introduced which – while having at least the same performance rates as [11] – provides an additional measure for the differences of cluster, the so-called dissimilarity index. For analyzing our interbank network we apply this latter algorithm to the weighted adjacency matrix A^w. As the only preprocessing step we clip all entries in A^w above a level of 300 m Euro for numerical reasons, i.e. $A_{\text{clip}}^w = \min(A^w, 300m)$. The community structure obtained in this way can be compared to the actual community structure in the real world. The result for the community structure obtained from one representative data set is shown in Fig. 1 a. The algorithm associates banks with their corresponding sectors, like R, VB, and S. For banks which in reality are not structured in a strong hierarchical way, such as banks in the SP, JS, SM, HCL sectors, no significant community structure is expected. By the algorithm these banks are grouped together in a cluster called 'other'. The Raiffeisen sector, which has a sub-structure in the federal states, is further grouped into clusters which are clearly identified as R banks within one of the eight federal states (B,St,K,V,T,N,O,S). In Fig. 1 a these clusters are marked as e.g. 'RS', 'R' indicating the Raiffeisen sector, and 'S' the state of Salzburg. Overall, there were 31 mis-specifications into wrong clusters within the total $N = 883$ banks, which is a mis-specification rate of 3.5 %, demonstrating the quality of the dissimilarity algorithm and – more importantly – the quality of the entropy approach to reconstruct the matrix L.

Like many real world networks, the degree distribution of the interbank market follows a power law for the tail of graph A, Fig. 1 b. The exponent is $\gamma_{tail}(A) = 2.01$. We have checked that the distribution for the low degrees is almost entirely dominated by banks of the R sector. Typically in the R sector most small agricultural banks have links to their federal state head institution and very few contacts with other banks, leading to a strong hierarchical structure. This hierarchical structure is perfectly reflected by the small scaling exponents. Betweenness is a measure of centrality that considers the position of nodes in-between the shortest paths (geodesics) that link any other nodes of the network. Let g_{jk} be the number of geodesics linking nodes j and k. If all geodesics are equally likely to be chosen the probability of taking one of them is $1/g_{jk}$. The probability that a particular node i lies on the geodesics between j and k is denoted by g_{jik}. The betweenness B for node i is defined as the sum of these probabilities over all pairs of nodes not including node i. $B(i) = M \sum_{j,k} g_{jik}/g_{jk}$, M being some normalization constant. B has a minimum value of zero when i falls on no geodesics and a maximum at $(N-1)(N-2)/2$, which is the number of pairs of nodes not including i. We use a relative version, i.e. M is such that the sum of $B(i)$ over all N nodes adds up to 100%. Finally, the average path length in the (undirected) interbank connection network A is $\bar{\ell}(A) = 2.26 \pm 0.03$. A is unweighted, meaning $A_{ij} = 1$ for $A_{ij}^w \neq 0$ and $A_{ij} = 0$ else. From these results the Austrian interbank network looks like a very small world with about three degrees of separation. This result looks natural in the light of the community structure described earlier. The two and three tier organization with head institutions and sub-institutions apparently leads to short interbank distances via the upper tier of the banking system and thus to a low degree of separation.

The framework here is a model of a banking system based on a detailed description of the structure of interbank exposures L [7]. The model explains the feasible payment flows between banks endogenously from the given structure of interbank liabilities, net values of the banks arising from all other bank activities and an assumption about the resolution of insolvency for different random draws from a distribution of risk-factor changes, such as interest rate, foreign exchange rate and stockmarket changes, as well as changes in default frequencies for corporate loans. We expose the banks' financial positions apart from interbank relations to interest rate, exchange rate, stock market and business cycle shocks. For each state of the world, the network model uniquely determines endogenously actual, feasible interbank payment flows. Taking the feedback between banks from mutual credit exposures and mutual exposures to aggregate shocks explicitly into account we can calculate default frequencies of individual banks across states. The endogenously determined vector of feasible payments between banks also determines the recovery rates of banks with exposures to an insolvent counterparty. We are able to distinguish bank defaults that arise directly as a consequence of movements in the risk factors and defaults which arise indirectly because of contagion. The model therefore yields a decomposition into fundamental and contagious defaults. Risk scenarios are created by exposing those positions on the bank balance sheet that are not part of the interbank business to interest rate, exchange rate, stock market and loan loss shocks. In order to do so we undertake a historic simulation using market data, except for the loan losses where we employ a credit risk model. In the scenario part we use data from Datastream, the major loans register, as well as statistics of insolvency rates in various industry branches from the Austrian rating agency (Kreditschutzverband von 1870). For each scenario the estimated matrix of bilateral exposures L and the income positions determine via the network model a unique vector of feasible interbank payments and thus a pattern of insolvency. The basic idea is to determine the feasible flows of funds in the banking network that can occur

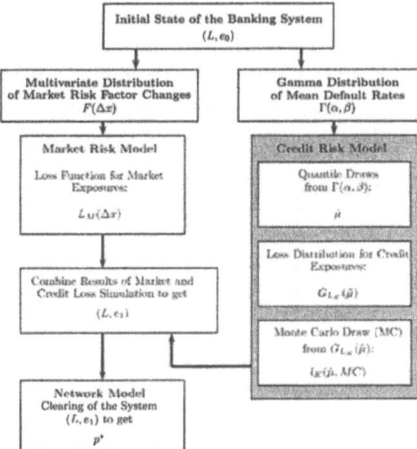

Fig. 2. Schematic diagram of the default flow model: At the initial state each bank is fully characterized by its wealth e_0 and its liabilities to other banks L. We expose each bank to stochastic shocks which model the market risk and the credit risk separately. The introduced risk changes the positions of the banks to new levels e_1. The crucial step is the clearing of the system where the structure of L becomes important. The result of the clearing is a vector p_i^* of payments bank i has to pay to the other banks in the system. If a component of this vector becomes less than the obligations d_i to pay to all the other banks $d_i = \sum_j L_{ij}$ bank i is insolvent.

after a realization of risk factor changes by a clearing procedure in which creditors that can not fulfill their payment promises (L) are proportionally rationed. One can show that a unique clearing vector always exists under mild regularity conditions [12]. The clearing vector can be found constructively as follows: for the given portfolio positions of banks e_i and the given risk factor changes we assume that every bank would pay its interbank liabilities as specified by L. If under this assumption one bank has a negative net value it is insolvent and its creditors receive a proportional claim on the remaining value of the insolvent bank. All banks with positive value are assumed to honor their promises fully. In a second iteration it can happen that under the new vector of feasible payment flows banks that were previously solvent are now insolvent because the value of their interbank claims is reduced impairing their capacity to pay. One can show that this procedure converges to a unique clearing payment vector p^* for the system as a whole. From the clearing vector found in this way one can directly read three pieces of information: first all banks that have a component in the vector that is smaller than the sum of their promises as specified by L are insolvent. The loss given default can be determined because it requires only a comparison between what has been promised (given by L) and what value can actually be payed (given by the clearing payment vector). Furthermore, insolvency cases can be distinguished by their occurrence in the clearing algorithm. Banks that are insolvent in the first round are fundamentally insolvent, others are insolvent because the value of interbank claims has been reduced by the insolvency of others. Their insolvency can therefore be interpreted as contagious. The analysis of

Table 1. Simulation results for probabilities of fundamental and contagious defaults. A fundamental default is due to losses arising from exposures to market and non-bank credit risk. Contagious defaults are triggered by the default of another bank that cannot fulfill its promises. The probability that only fundamental defaults occur is shown as well as the probability that fundamental and contagious defaults are observed.

Fundamental Defaults	No Contagion	Contagion	Total
0-10	93.38%	0.01%	93.39%
11 to 20	2.82%	0.40%	3.22%
21 to 30	0.11%	1.04%	1.15%
31 to 40	0.00%	0.40%	0.40%
41 to 50	0.00%	0.53%	0.53%
Total	96.31%	3.69%	100.00%

these data then allows us to assess the risk exposure – in particular for defaults – of all banks at a system level. The details of the model are described in [7].

3   Results and Conclusions

The given banking system is very stable and default events that could be classified as a "systemic crisis" are unlikely. We find that the mean default probability of an Austrian bank to be below one percent. Perhaps the most interesting finding is that only a small fraction of bank defaults can be interpreted as *contagious*. The vast majority of defaults is a direct consequence of macroeconomic shocks. More specifically, we find the median endogenous recovery rates to be 66%, and we show that the given banking system is quite stable to shocks from losses in foreign counterparty exposure and we find no clear evidence that the interbank market either increases correlations among banks or enables banks to diversify risk. Using our model as a simulation tool, we show that market share in the interbank market alone is not a good predictor of the relevance of a bank for the banking system in terms of contagion risk. Table 1 shows the decomposition of simulation scenarios with and without contagion following a fundamental default of a given number of banks. The simulation is run under the assumption that there is a recovery rate of 50% of loans to non banks and that the full value of an insolvent institution is transferred to the creditors. Clearing is done after the netting of interbank claims in L.

Finally, we ask the question of the impact of individual bank defaults on other banks. More specifically, if one particular bank becomes erased from the network, how many other banks become insolvent due to this event? We call this conditional *contagion impact* on default of a specific bank. This is similar in spirit to looking at avalanche distributions triggered by controlled toggling of grains in a sandpile. We observe the portfolio positions e_i of banks from our database. Instead of simulating risk factor changes that are applied to this portfolio and are then cleared in a second step we artificially eliminate the funds of each bank one at a time, clear the system, and count the number of contagious (induced) defaults. This is repeated N times so that each bank becomes removed, and all the others are present. We find that only 13 banks – when defaulting – drag more than one other bank into default. There are 16 banks which will cause one single default of one other bank. A natural guess would be to relate the contagion impact of a specific bank to its role in the network. Amongst many possible measures, we find that the betweenness of the defaulting bank is directly related to the contagion impact. This is shown in Fig. 1 c, where a linear relation between the betweenness $B(i)$ and the contagion impact is suggested for $B(i) > 2$.

In this work we combined the knowledge of the detailed structure of a real world banking network with an economic model, which allows to estimate the functional stability and robustness of the financial network. This model adds a dynamical component to the "static" liability matrix, which is nothing else but the promise for future financial flows between the banks. By stochastically varying external parameters of the system like interest rate shocks, we can follow the flow of financial transactions, and in particular can scan for defaults occurring in the system, due to insolvency. The results of this work is that the system seems to be relatively robust, and the probability for a contagious spread over the whole network is very small. However, there are several key banks (not including the Central bank), which upon their default, will lead to a considerable number of other banks defaulting as well. We showed that these key banks can be reliably identified by the vertex betweenness. We think that the existence of a threshold in the variable $B(i)$ in a "quasi" scale-free network, combined with complex but realistic contagion dynamics is a remarkable finding which is worthwhile to examine further.

References

1. Dorogovtsev, S.N. and Mendes, J.F.F.: Evolution of Networks: From Biological Nets to the Internet and WWW, Oxford University Press (2003)
2. Hartmann, P. and DeBandt, O.: Systemic Risk: An Overview. European Central Bank Working Paper (2000) 35
3. Summer, M.: Banking Regulation and Systemic Risk. Open Econ. Rev. **1** (2003) 43
4. Allen, F. and Gale, D.: Financial Contagion. J Polit. Econ. **108** (2000) 1
5. Thurner, S., Hanel, R., Pichler, S.: Risk trading, network topology, and banking regulation. Quantitative Finance **3** (2003) 306
6. Boss, M., Elsinger, H., Summer, M., Thurner, S.: The Network Topology of the Interbank Market. SFI working paper # 03-10-055; cond-mat/0309582 (2003)
7. Elsinger, H., Lehar, A., Summer, M.: Risk Assessment of Banking Systems. Austrian National Bank working paper 79 (2002).
8. Tadić, B., Thurner, S.: Information Super-Diffusion on Structured Networks, Physica A **332** (2004) 662
9. Tadić, B., Thurner, S., Rodgers, G. J.: Traffic on complex networks: Towards understanding global statistical properties from microscopic density fluctuations. Phys. Rev. E (2004) in press
10. Zhou, H.: Distance, dissimilarity index, and network community structure. Phys. Rev. E **67** (2003) 061901
11. Girvan, M. and Newman, M.E.J.: Proc. Natl. Acad. Sci. **99** (2002) 7831
12. L. Eisenberg and T. Noe, Systemic Risk in Financial Systems. Management Science **47** (2001) 236

Local Search with Congestion in Complex Communication Networks

Alex Arenas[1], Leon Danon[2], Albert Díaz-Guilera[2], and Roger Guimerà[3]

[1] Dept. Enginyeria Informàtica i Matemàtiques, Univ. Rovira i Virgili, 43006 Tarragona, Catalonia, Spain
aarenas@correu.urv.es
[2] Dept. Física Fonamental, Univ. de Barcelona, 08028, Barcelona, Catalonia, Spain
ldanon@ffn.ub.es,albert.diaz@ub.edu
[3] Dpt. Chemical Engineering, Northwestern Univ., Evanston, IL, USA
rguimera@northwestern.edu

Abstract. We present a formalism that is able to cope with search and congestion simultaneously. This formalism avoids the problem of simulating the dynamics of the search-communication process which turns out to be impracticable, specially close to the congestion point where search costs (time) diverge.

1   Introduction

In recent years, the study of static and dynamical properties of complex networks has received a lot of attention [1,2,3,4,5]. Complex networks appear in such diverse disciplines as sociology, biology, chemistry, physics or computer science. In particular, great effort has been exerted to understand the behavior of technologically based communication networks such as the Internet [6], the World Wide Web [7], or e-mail networks [8,9,10]. One of the common facts behind these structures is the short mean distance between nodes. Furthermore, these short paths can be found with strategies that do not precise a complete knowledge of the pattern of interactions between the nodes. Related to the search problem, when the network is facing a number of simultaneous processes, we find that the network can get collapsed because some of these problems are travelling through the same node, and this raises the problem of congestion. We introduce a formalism that enables to handle these two problems in a common framework by writing the dynamical properties of the search process in terms of the topological properties of the network [11].

First we calculate the average number of steps (search cost) needed to find a certain node in the network given the search algorithm and the topology of the network. The calculation is exact if the search algorithm is Markovian. Next, congestion is introduced assuming that the network is formed by nodes that behave like queues, meaning that are able to deliver a finite number of packets at each time step [12,13,14]. In this context, we are able (i) to calculate explicitly the point at which the arrival rate of packets leads to network collapse, in the

M. Bubak et al. (Eds.): ICCS 2004, LNCS 3038, pp. 1078–1085, 2004.
© Springer-Verlag Berlin Heidelberg 2004

sense that the average time needed to perform a search becomes unbounded, and (ii) to determine, below the point of collapse, how the average search time depends on the rate at which search process are started. In both cases, the relevant quantities are expressed in terms of the topology of the network and the search algorithm.

2   Search

Following the steps introduced in [11] let us consider a single *information packet* at node i whose destination is node k. The probability for the packet to go from i to a new node j in its next step is p_{ij}^k. In particular, $p_{kj}^k = 0 \ \forall j$ so that the packet is *removed* as soon as it arrives to its destination. This formulation is completely general, and the precise form of p_{ij}^k will depend on the search algorithm that will be discussed later on. In particular, when the search is Markovian, p_{ij}^k does not depend on previous positions of the packet. Using matrix notation we can define the matrices D^k whose elements D_{ij}^k are the average number of steps needed to go from i to j for a packet traveling towards k. D_{ij}^k are such that $d_{ik} = D_{ik}^k$. These matrices are related to the probabilities p_{ij}^k through the following expression [11]

$$D^k = \left[(I - p^k)^{-1} \right]^2 p^k, \tag{1}$$

where I is the identity matrix. In particular, the element D_{ik}^k is the average number of steps needed to get from i to k when using the search algorithm given by the set of matrices p^k. When the search algorithm has global knowledge of the structure of the network and the packets follow minimum paths between nodes, the effective distance will coincide with the topological minimum distance; otherwise, the effective distance between nodes will be, in general, larger than the topological minimum distance.

Finally, the average search cost in the network is

$$\bar{d} = \frac{\sum_{i,k} D_{ik}^k}{S(S-1)}, \tag{2}$$

where S is the number of nodes in the network. This expression allows to calculate exactly the average search cost performing simple matrix algebra. Note that simulation based calculation of this quantity would require, in principle, to generate an infinite amount of packets and let them travel from all possible origins to all possible destinations following all possible paths, which are in general arbitrarily long.

3   Load

When there are many simultaneous search process, the effective distance is not a good measure of performance since, even when the distance is small, accumulation of packets can generate long delays. Rather, the characteristic time, τ, needed to get from the origin to the destination is the right measure. According to Little's law of queuing theory[12], the characteristic time is proportional to the average total load of the network, $\overline{N}$. In our case, the total load is identified with the total number of floating packets, which is the algebraic sum of the single queues of the nodes. Thus minimizing τ is equivalent to minimizing $\overline{N}$. In the following, we show how to calculate the load of a network using only the p^k matrices as has been done for the case of no congestion.

First, we calculate the average number of times, b_{ij}^k, that a packet generated at i and with destination k passes through j,

$$b^k = \sum_{n=1}^{\infty} \left(p^k\right)^n = (I - p^k)^{-1} p^k. \tag{3}$$

Note that the elements b_{ij}^k are sums of probabilities but are not probabilities themselves.

The *effective* betweenness of node j, B_j, is defined as the sum over all possible origins and destinations of the packets, and represents the total number of packets that would pass through j if one packet would be generated at each node at each time step with destination to any other node:

$$B_j = \sum_{i,k} b_{ij}^k. \tag{4}$$

Again, as in the case of the effective distance, when the search algorithm is able to find the minimum paths between nodes, the effective betweenness will coincide with the *topological* betweenness, β_j, as usually defined [15,16].

4   Search and Congestion

Now consider the following general scenario. In a communication network, each node generates one packet at each time step with probability ρ independently of the rest of the nodes. The destination of each of these packets is randomly fixed at the moment of its creation. On the other hand, the nodes are queues that can store as many packets as needed but can deliver, on average, only a finite number of them at each time step—without lost of generality, we fix this number to 1. For low values of ρ the system reaches a steady state in which the total number of *floating* packets in the network $N(t)$ fluctuates around a finite value. As ρ increases, the system undergoes a continuous phase transition from this *free phase* to a *congested phase* in which $N(t) \propto t$ [14]. Right at the critical point, ρ_c, quantities such as $N(t)$ and the characteristic time diverge [17]. In the free phase, there is no accumulation at any node in the network and the number of packets that arrive to node j is, on average, $\rho B_j/(S-1)$. Therefore, a particular node will collapse when $\rho B_j/(S-1) > 1$ and the critical congestion point of the network will be

$$\rho_c = \frac{S-1}{B^*} \tag{5}$$

where B^* is the maximum effective betweenness in the network, that corresponds to the most central node.

To calculate the time average of the load of the network, $\overline{N}$, it is necessary to establish the behavior of the queues. In the general scenario proposed above, the arrival of packets to a given node j is a Poisson process with mean $\mu_j = \rho B_j/(S-1)$. Regarding the delivery of the packets, consider the following model. For a node j with ν_j packets stored in its queue, each packet jumps to the next node (chosen according to the algorithm defined through the matrices p^k) with probability $1/\nu_j$. In this model, the delivery of packets is also a Poisson process. In such a simple case in which both the arrival and the delivery are Poisson processes, queues are called M/M/1 in the computer science literature and the average size of the queues is given by [12]

$$\langle \nu_j \rangle = \frac{\mu_j}{1 - \mu_j} = \frac{\frac{\rho B_j}{S-1}}{1 - \frac{\rho B_j}{S-1}}. \tag{6}$$

The average load of the network $\overline{N}$ is

$$\overline{N} = \sum_{j=1}^{S} \langle \nu_j \rangle = \sum_{j=1}^{S} \frac{\frac{\rho B_j}{S-1}}{1 - \frac{\rho B_j}{S-1}}. \tag{7}$$

It is straightforward to extend the calculations to other types of queues. For instance, the queues used in [13] are such that one packet is delivered deterministically at each time step. These queues are called M/D/1 and the corresponding expression for the size of the queues is $\langle \nu_j \rangle = \mu_j^2/(1 - \mu_j)$. Moreover, it is worth noting that the fact that we are able to map the behavior of the nodes to that of M/M/1 queues implies that any conclusion that we are able to draw will be valid in general for any system of M/M/1 queues, and with slight modifications for other types of queues.

There are two interesting limiting cases of equation (7). When ρ is very small, $\langle \nu_j \rangle \approx \mu_j$ and taking into account that $\sum_j B_j = \sum_{i,k} d_{ik}^k$, one obtains

$$\overline{N} \approx \rho S \overline{d} \qquad \rho \to 0. \tag{8}$$

On the other hand, when ρ approaches ρ_c most of the load of the network comes from the most congested node, and therefore

$$\overline{N} \approx \frac{1}{1 - \frac{\rho B^*}{S-1}} \qquad \rho \to \rho_c, \tag{9}$$

where B^* is, as before, the betweenness of the most central node. The last two expressions suggest the following interesting problem: to minimize the load of a network it is necessary to minimize the effective distance between nodes if the amount of packets is small, but it is necessary to minimize the largest effective betweenness of the network if the amount of packets is large. The first is accomplished by a *star-like* centralized network, that is, a network with a few central nodes and all the others connected to them. Rather, the second is accomplished by a distributed, very decentralized, network in which all the nodes support a similar load. In [11] we checked that those are indeed the optimal structures by means of an extensive generalized simulated annealing [18]. This behavior is common to any system of queues provided that the communication depends only on the sender. In queues M/D/1, for example, equation (8) reads $\overline{N} \approx (\rho S \overline{d})^2$ (thus, minimization of $\overline{N}$ still implies minimization of $\overline{d}$) and equation (9) is unchanged.

5   Limitations of the Calculation and Bounds to Other Models

It is worth noting that there are only two assumptions in the calculations above. The first one has already been mentioned: the movement of the packets needs to be Markovian to define the jump probability matrices p^k. Although this is not strictly true in real communication networks — where packets are not allowed usually to go through a given node more than once — it can be seen as a first approximation [19,14,13]. The second assumption is that the jump probabilities p_{ij}^k do not depend on the congestion state of the network, although communication protocols sometimes try to avoid congested regions, and then $B_j = B_j(\rho)$. However, all the derivations above will still be true in a number of general situations, including situations in which the paths that the packets follow are unique, in which the routing tables are fixed, or situations in which the structure of the network is very homogeneous and thus the congestion of all the nodes is similar.

When these two assumptions are fulfilled the calculations are exact. For example, the calculation of ρ_c using equation (5) coincides exactly (within the simulation error) with simulations of the communication model where the communication only depends on the sender of the packet. Compared to situations in which packets avoid congested regions, equations (5)–(9) correspond to the worst case scenario and thus provide bounds to more realistic scenarios in which the search algorithm interactively avoids congestion. Consider, for example, ρ_c in the model presented in [14], where the communication depends not only on the sender but also on the availability of the receiver. The fact that the packets are sent with higher probability to less congested nodes implies that the flow is better balanced among nodes. Although the assumptions of the present calculation do not apply, one would expect that the value of ρ_c estimated analytically will be a lower bound to the real situation in which load is more balanced.

For an accurate description of the algorithm used in the computer simulations we need to specify the set of matrices p_{ij}^k. We have assumed that the movement of the packets is Markovian, this implies that the probabilities do not depend on the packet history. In a purely local search scenario, where the knowledge that the nodes have about the network structure is bounded, the nodes face the problem of forwarding the packets. In order to quantify this information we introduce the radius of knowledge as a measure the distance along network links that a node can identify the destination of a packet. Thus, when this radius is zero, packets travel completely at random until they reach its destination, and then we have

$$p_{ij}^k = (1 - \delta_{ik}) \frac{a_{ij}}{\sum_l a_{il}}. \tag{10}$$

where a_{ij} are the elements of the adjacency matrix of the network: $a_{ij} = 1$ if i and j are connected in the network and $a_{ij} = 0$ otherwise. This expression means that packets are uniformly distributed among neighbors unless they have reach their destination. The delta symbol ensures that $p_{kj}^k = 0 \; \forall j$ and the packet *disappears* from the network. When the radius of knowledge is 1, a node is able to recognize one of its neighbors as the destination and then the packet is sent to it; otherwise, the packet is just sent at random to one of the neighbors of the node. The corresponding p^k matrices are given by

$$p_{ij}^k = a_{ik} \delta_{jk} + (1 - a_{ik} - \delta_{ik}) \frac{a_{ij}}{\sum_l a_{il}}. \tag{11}$$

The first term corresponds to i and k being neighbors: then the packet will go to j if and only if $j = k$, i.e. the packet will be sent directly to the destination. The second term corresponds to i and k not being neighbors: in this case, j is chosen at random among the neighbors of i.

We have compared our predictions using a radius of knowledge equal to one with simulations of the communication model introduced in [14], where the communication between the nodes depends on the receiver's state as well, and packets are automatically delivered if a neighbor is its final destination. Figure 1 shows that our model is indeed a lower bound and, moreover, that the analytical estimation provides a good approximation to the simulated value.

This figure also confirms another expected and useful result. For a given size of the network and a given number of links, the most robust networks, that is those with higher ρ_c, are those with better balanced load. For these networks, the effect of avoiding congestion is less important and therefore the analytical estimation turns out to be more accurate.

6   Packet Dynamics

We have performed simulations of the communication model described in the previous sections for the structures shown in [11] to be optimal in the two regimes: for a low load, the star-like centralized one, and for a high load, the distributed one. We note that for a fixed number of nodes and links, the distributed network has a larger value of ρ_c than its centralized counterpart. In the case we are considering of 32 nodes and 128 links, the critical values of ρ_c are 0.256 and 0.162, respectively. Nevertheless, the dynamical behavior near the critical value of the load has to be very similar. In Fig. 1 we have plotted the standard deviation in the number of packets in the most connected nodes and in the total number of packets in the network.

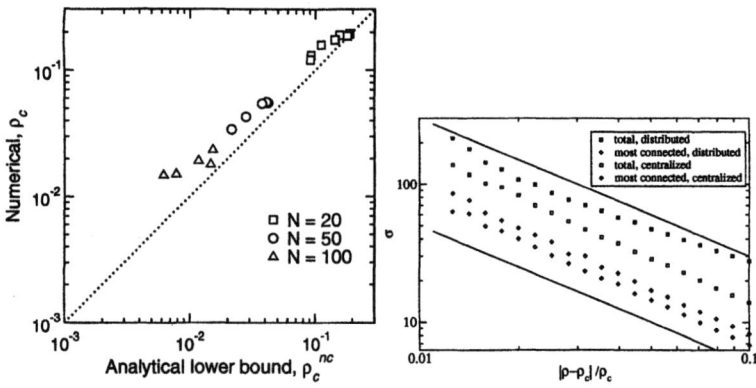

Fig. 1. Left: Comparison between the predictions of equation (5) for ρ_c with betweennesses computed according to (11) and the results obtained for the communication model discussed in [14]. The analytical value is a lower bound to the actual value. To keep the figure simple, we do not show results corresponding to the model discussed in the current work, but the points would lay exactly on the diagonal line, since all the assumptions of the calculation are fulfilled. Right: Standard deviation in (squares) total number of packets and (diamonds) number of packets in the most connected nodes. The straight lines have slope -1 and are guides to the eye.

In the type of queue we are considering, the variance in the number of packets in any node is given by [12]

$$\sigma_j = \frac{\sqrt{\mu_j}}{1 - \mu_j} = \frac{\sqrt{\frac{\rho B_j}{S-1}}}{1 - \frac{\rho B_j}{S-1}}. \tag{12}$$

Since the most connected nodes are those with the highest effective betweenness and its value determines the critical value of ρ, we expect that at these nodes

$$\sigma_{\text{m.c.n.}} \simeq \frac{1}{\rho_c - \rho}. \tag{13}$$

and this is precisely the behavior we infer from Fig. 1.

On the other hand, for the total number of packets, there a clear distinction between the two networks. In the centralized one, there are just a few nodes, three in our case, that carry most of the network load and hence the main contributions to the variance, making that the variance in the total number of packets follows the same law as the single most connected nodes. Contrary to this, for the distributed network there are much more nodes with a critical load and correlation are induced that may disrupt the single node behavior.

7   Conclusion

We have presented a single model that can handle simultaneously search and congestion in a general network. The dynamical properties of the network can be related to purely topological properties, in terms of the connectivity matrix and on the search algorithm. Close to the critical point where congestion develops the system presents an interesting behavior since the variance of the total number of packets cannot be easily related to the variance of the number of packets in the single nodes.

References

1. Watts, D.J., Strogatz, S.H.: Collective dynamics of 'small-world' networks. Nature, 393 (1998) 440.
2. Barabási, A.-L., Albert, R.: Emergenge of scaling in random networs. Science, 286 (1999) 509–512.
3. Amaral, L.A.N., Scala, A., Barthelemy, M., Stanley, H.E.: Classes of small-world networks. Proc. Nat. Acad. Sci. USA, 97 (2000) 11149–11152, 2000.
4. Albert, R., Barabási, A.-L.: Statistical mechanics of complex networks. Rev. Mod. Phys., 74 (2002) 47–97.
5. Dorogovtsev, S., Mendes, J.F.F.: Evolution of networks. Adv. Phys., 51 (2002) 1079.
6. Faloutsos, M., Faloutsos, P., Faloutsos, C.: On power-law relationships of the internet topology. Comp. Comm. Rev., 29 (1999) 251–262.
7. Albert, R., Jeong, H., Barabási, A.-L.: Diameter of the world-wide web. Nature, 401 (1999) 130.
8. Newman, M.E.J., Forrest, S., Balthrop, J.: Email networks and the spread of computer viruses. Phys. Rev. E, 66 (2002) 035101(R).
9. Ebel, H., Mielsch, L.I., Bornholdt, S.: Scale-free topology of e-mail networks. Phys. Rev. E, 66 (2002) 035103.
10. Guimera, R., Danon, L., Diaz-Guilera, A., Giralt, F., Arenas, A.: Self-similar community structure in a network of human interactions . Phys. Rev. E, 68 (2003) 065103(R).
11. Guimera, R., Diaz-Guilera, A., Vega-Redondo, F., Cabrales, A., Arenas, A.: Optimal network topologies for local search with congestion. Phys. Rev. Lett., 89(2002) 248701.
12. Allen, O.: *Probability, Statistics and Queueing Theory with Computer Science Application.* Academic Press, New York, 2nd edition, 1990.
13. Ohira, T., Sawatari, R.: Phase transition in a computer network traffic model. Phys. Rev. E, 58 (1998) 193.
14. Arenas, A., Diaz-Guilera, A., Guimera, R.: Communication in networks with hierarchical branching. Phys. Rev. Lett., 86 (2001) 3196–3199.
15. Freeman, L.C.: A set of measures of centrality based on betweenness. Sociometry, 40 (1977) 35–41.
16. Newman, M.E.J.: Scientific collaboration networks. ii. shortest paths, weighted networks, and centrality. Phys. Rev. E, 64 (2001) 016132.
17. Guimera, R., Arenas, A., Diaz-Guilera, A.: Communication and optimal hierarchical networks. Physica, 299A (2001) 247–252.
18. Tsallis, C., Stariolo, D.A.: *Annual Rev. Comp. Phys. II.* World Sci., Singapore, 1994. edited by D. Stauffer.
19. Sole, R.V., Valverde, S.:. Information transfer and phase transitions in a model of internet traffic. Physica, 289A (2001) 595–605.

Guided Search and Distribution of Information Flow on Complex Graphs

Bosiljka Tadić

Jožef Stefan Institute, Box 3000, 1001 Ljubljana, Slovenia
Bosiljka.Tadic ijs.si
http://phobos.ijs.si/~tadic/

Abstract. Within the numerical model proposed in [1] we study the flow of information on a graph with power-law organization of links and an emergent superstructure formed by two highly interconnected hubs. The local search algorithms which navigate transport at each node can use information in the range of zero, one, and two-nearest neighbourhood of the node. We show how the flow carried by a particular link is distributed over graph when range of guided search and posting rate of packets are varied. The probability density function of the flow is given by a universal log-normal law, which is determined by the overall packet density in the stationary traffic. However, the distribution becomes unstable when the traffic experiences temporary or permanent jamming.

1 Introduction

The complexity of structure and function of evolving networks is often expressed in emergence of topological communities [2,3,4] and functionally connected units, which are based on the linking properties of the network. The inhomogeneity and sparseness of complex networks effect dynamic processes that these networks support. Recent study of traffic on networks of diverse emergent topologies [1, 5] revels that functional efficiency of a network, however, is not determined by the network's topology alone, but it also depends on the microscopic rules of the dynamics. In particular, in the transport processes on networks the rules which are adapted to the locally changing networks' structure lead to more efficient network performance. In the case of random walk dynamics the influence of the structural diversity can be incorporated into search algorithms that navigate the walkers through the graph [6].

Inhomogeneity on the level of local connectivity of nodes makes possible to design variety of navigations rules on complex networks. The requirement for locality of the search is essential for costs reasons. In [1] we have studied two different search algorithms that are guided by locally encountered network structure. It has been found that (a) better guided search leads to better performance of the network; and (b) the same guided search algorithm performs better on networks with higher organizational complexity [1]. An example which illustrates these conclusions is the traffic of information packets on the Web-type graph [7]

M. Bubak et al. (Eds.): ICCS 2004, LNCS 3038, pp. 1086–1093, 2004.
© Springer-Verlag Berlin Heidelberg 2004

with the next-near-neighbour (*nnn-*) search. The network's output was found to be 40 times higher than when the rules of random diffusion are applied.

These results suggest that in artificially grown networks, although the topological complexity of the network is essential, the functional units can not be identified with topological communities independently on the nature of the dynamic process that the network performs. This motivates study of details of the network's activity as it performs a particular type of function. A detailed knowledge about the distribution of activity throughout the network may be then used as a starting point for restructuring of the network in order to improve its function at local levels. Another approach of network's restructuring to optimize its performance under global condition was suggested in [8].

Here we use the numerical model, which was recently introduced in [1] for information packet transport on structured graphs, to study the intensity of traffic carried along a particular link (flow) on the cyclic scale-free Webgraph. In particular, we determine how the distribution of flow depends on the radius of the area around visited node which is explored by the guided search rules, and on the packet density.

2   Network Topology and Search Algorithms

The network which we use here is represented by a directed graph (Web graph) grown with the microscopic dynamic rules which are proposed in [7] to model the evolution of the world-wide Web. The rules consist of preferential attachment and rewiring while the graph grows, which leads to an emergent structure with inhomogeneous scale-free ordering in both in-coming and out-going links and a number of closed cycles, as shown in Fig. 1. The statistical properties of the graph topology are studied in detail [7,9], in particular, the local and global connectivity, a giant component and clustering (number of triangles per node). A substantial degree of correlations between in- and out-link connectivity at local level was found [10] which is characteristic for dis-assortative [2] graph topologies. Further topological properties—betweenness of nodes and links—are studied by use of random walk methods [11,2]. An important feature of the Web graph, which makes it different from common scale-free graphs [12] is the emergence of two types of hub nodes—a node with high in-link connectivity and a node with high out-link connectivity—which are closely interlinked by a number of short triangles. The superstructure of interlinked nodes which is associated with the two hubs appears to play an important role in the transport processes on the graph [1,13].

The diffusion rules for *dense* traffic of packets on the graph are summarized as follows (see [1] for detailed description and [10] for details of the implementation of the numerical code):

Packets are created with a finite rate R at a node and each packet is assigned an address (destination) where it should be delivered. Both the origin node and destination node are selected randomly within the giant component of the graph. The packets are navigated through the nodes by one of the search rules given

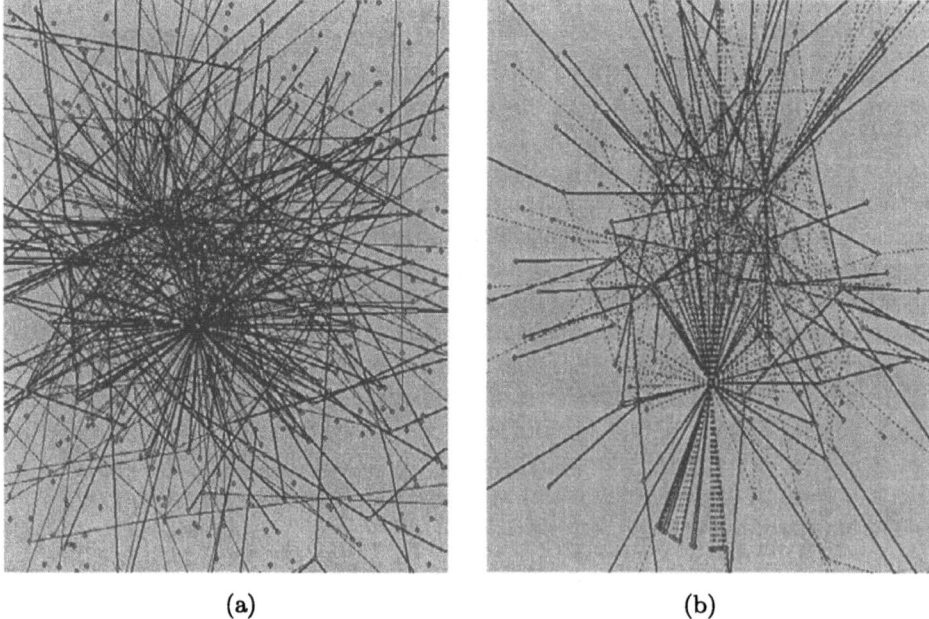

(a)   (b)

Fig. 1. (a) Webgraph with different widths of links representing four bins of flow: low, average, high, and very high flow in the case of *nnn*-search. (b) Only high flow links are displayed (solid lines), with the remaining graph connections shown by dotted lines.

below. The priority of move at a node is given according to the "last-in-first-out" (LIFO) queuing mechanism. When the packet arrives to its destination it is removed from the traffic.

The three local search algorithms which we consider here differ in the range of area where the search for the destination node is performed in the vicinity of the visited node. If the node just visited by the packet is not its destination node, the search performed around that node can be within

- **range zero**, when the packet proceeds by randomly selecting a neighbour node on the network;
- **range one**, where the search for the destination is asked among first neighbours of the visited node and, if found, packet is delivered to it, else it proceeds to a randomly selected neighbour;
- **range two**, where through all first neighbours the signaling to next-near-neighbours (*nnn*) of the visited node is done; if the message is returned by a neighbour of the destination node, then the packet is delivered to that (common) neighbour, else it proceeds to a randomly selected neighbour node.

The packet transport on the same network appears to be markedly different when different search algorithm is used. The efficiency of the transport measured by the transit times of packets increases with the range of search. The distributions of the transit times of packets in Fig. 2 show power-law dependences

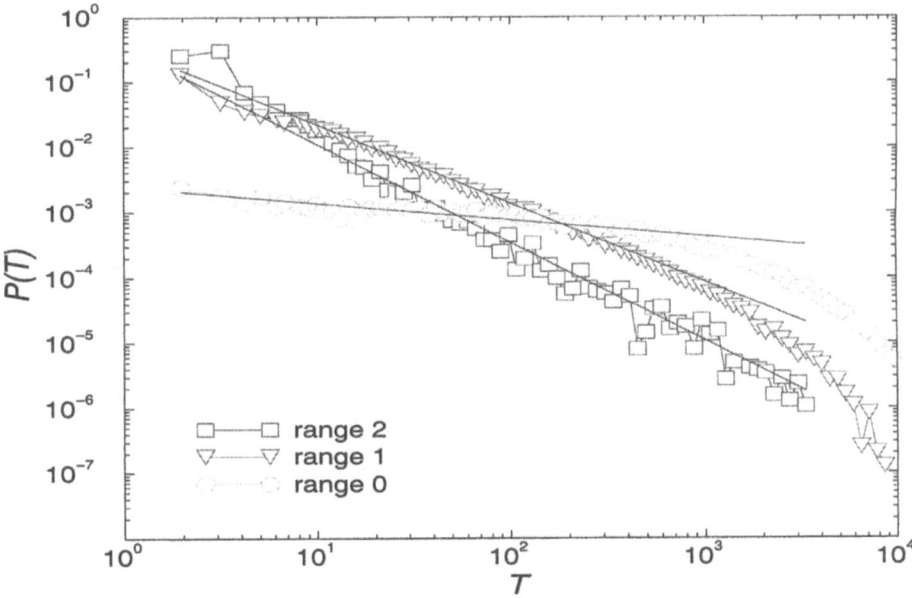

Fig. 2. Distribution of transit times for three algorithms of search in ranges 0, 1, and 2, and for fixed posting rate $R = 0.005$. The slopes of solid lines are: 1.5, 1.2, and 0.25. All data are averaged over many network realizations and log-binned.

with different slopes and cut-offs for different search algorithms. The maximum transit time increases in the case of less effective search, leading to increased accumulation of packets in the traffic (small output rate) and the network enters jamming regime at much lower posting rate, as compared to the advanced search algorithms. For shortest transit times $T = 1$, resulting in the case when selected pairs of nodes are at unit distance, the search rules with the range one and range two coincide.

3   Information Flow Distribution

The topological importance of a particular node on the graph is measured by betweenness, which is defined as the number of shortest paths between pairs on nodes on the graph that pass through that node. An analogous property, betweenness-centrality, is defined for links in the graph theory [14]. Here we consider a dynamically generalized betweenness-centrality called *flow* [14], by which we measure the importance of a particular link for the packet traffic on the graph. The actual paths of packets differ from topologically shortest paths, being subjected to navigation and queuing processes.

The flow of the traffic which is carried by a particular link is given by the number of packets transferred along that link within given transport process. In

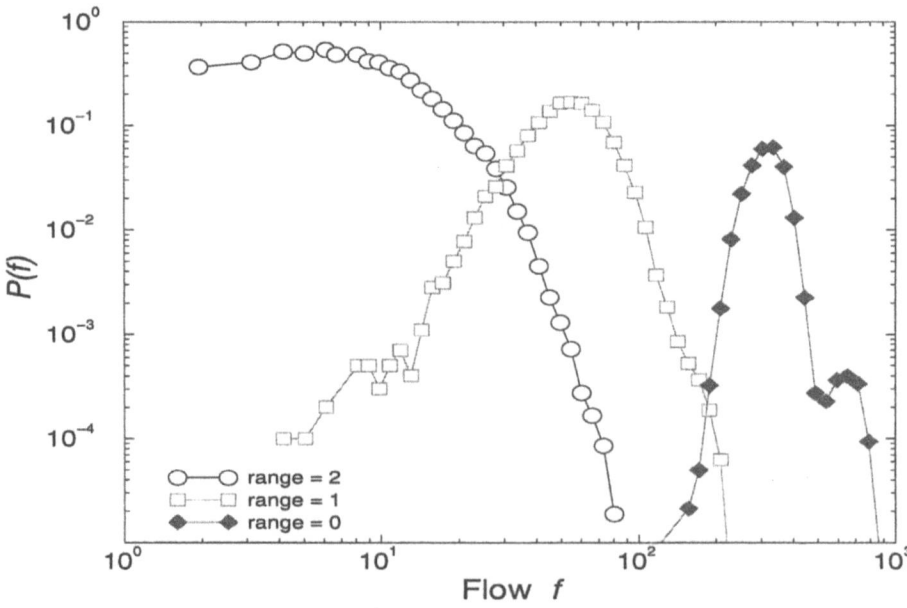

Fig. 3. Distribution of flow for three search algorithms of information range (right to left) zero, one, and two, and for fixed posting rate ($R = 0.005$) and time interval.

the packet transport between pairs of nodes on the graph we identify the exact path of a packet and update the flow on each link along the path. Depending on local network structure and the search algorithm, in the same time interval and fixed posting rate certain links carry much higher flow compared to the others. An example is shown in Fig. 1a: the intensity of flow on the links of the graph is expressed by the increased width of a link (within four bins). The results for the probability density function of flow along a link on the graph for different situations studied in this work are shown in Figs. 3 and 4 (averaged over network realizations and log-binned). The distribution appears to be close to a log-normal law (cf. Figs. 3 and 4).

Fig. 3 demonstrates the influence that the three search algorithms with different ranges of information have on the probability density of flow carried by a link for identical posting rates and time intervals. The efficient search algorithm (i.e., with the information range two) removes packets from the traffic quite quickly after their creation. Therefore, although certain links appear to carry quite large flow, the overall distribution is centered at low mean value. The location of links with heavy flow on the graph are shown in Fig. 1b for the case of the most effective search. On the other hand, in the case of less effective searches, packets remain on the network for much longer time, increasing the overall packet density. Consequently, the probability of large flow along a link will increase. The mean value of the distribution is shifted towards increasingly larger values when the range of the search algorithm is reduced (cf. Fig. 3).

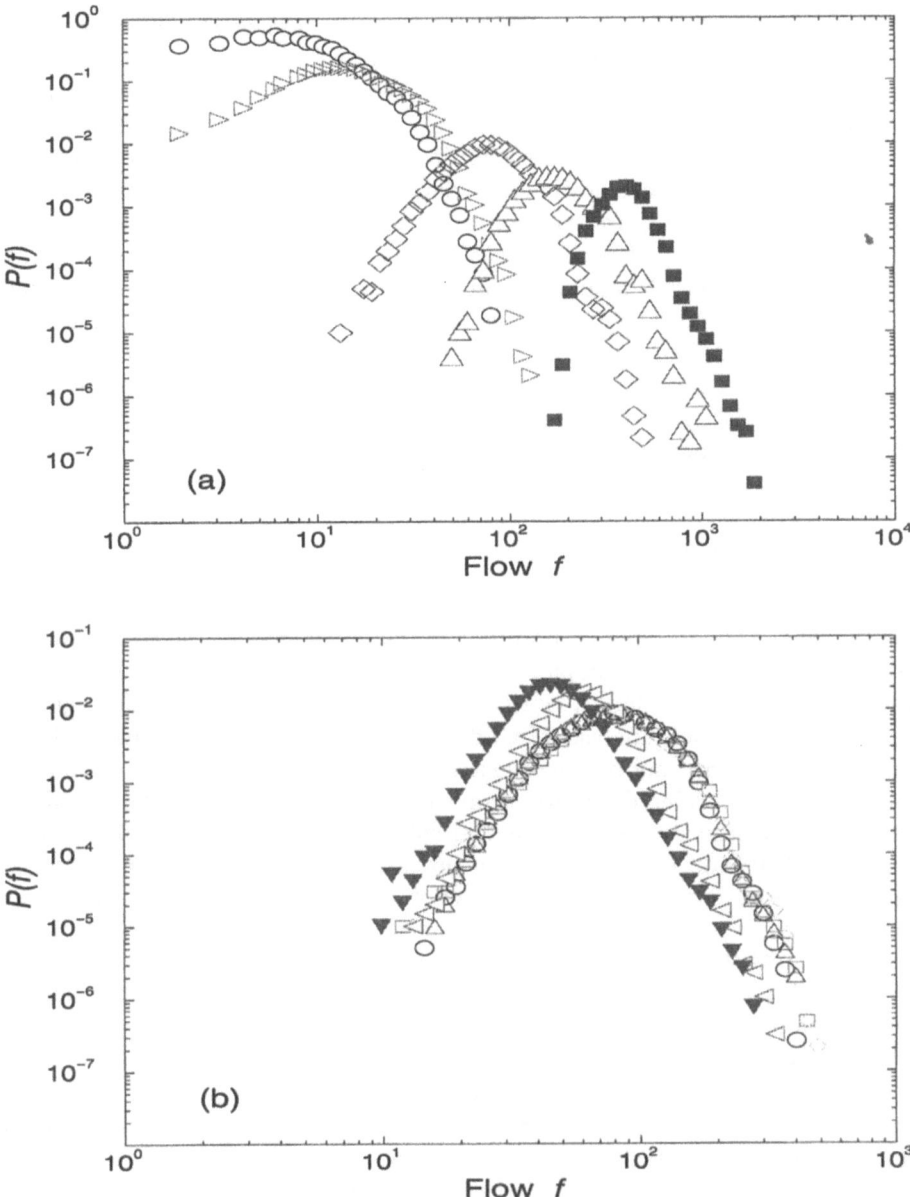

Fig. 4. Distribution of flow in the case of nnn-search (a) for different values of posting rate R =0.005, 0.01, 0.05, 0.1, and 0.3 (left to right) for fixed time (40000 timesteps), and (b) for fixed number of packets (2000 packets) in different time intervals, corresponding to stationary flow rates R = 0.005, 0.01,0.05, 0.1 (collapsed curves), and temporary and permanent jamming rates R = 0.3, and 0.4 (shifted curves).

Next we consider the flow distribution in the case of the efficient search algorithm (i.e., for the searched range in next-near-neighbourhood) for different density of packets. When the number of posted packets increases by increasing posting rate R within same time interval, the mean of the flow distribution moves towards higher values, as shown in Fig. 4a. In the conditions shown in Fig. 4a the traffic remains stationary, which can be checked by monitoring the work-load timeseries [13]. In order to further investigate this effect, we determine the flow distribution for the conditions when the time interval and posting rate vary so that the overall number of created packets remains fixed. The results are shown in Fig. 4b. For the posting rates where the traffic is stationary, the distributions for different posting rates collapse. Whereas, the flow distribution deviates from the universal curve for larger rates when the network experiences temporary "crisis" [13] which the system manages to clear out within long time intervals, or when permanent jamming occurs.

4   Conclusions

The topology of underlying graph plays an important role in the transport processes on networks. For example, the graph with high complexity, as the Web graph described in Section 2, can carry huge amount of traffic before getting jammed. The detailed analysis of traffic in terms of the transit time and flow statistics reveals the importance of search algorithm in the transport of information packets to specified destinations on the graph. In particular, the global efficiency of the traffic is directly correlated with the radius of the locally searched area around a node in the navigation process. The maximum transit time of packets decreases by a factor of two to ten when the range of search is increased from zero- to one- or two-near neighbours of a node. The distribution of flow along the links changes accordingly with the efficiency of the local search. Under the identical other conditions in the less effective local search the packets accumulation on the network, which leads to generally increased flow along each link. The mean of the flow distribution moves towards larger values.

On the global level, the probability distribution of flow appears to be close to a log-normal law. It is unstable with respect to random addition processes, which in the case of traffic is realized in the conditions of increased rate of packet creation. We find that for given (effective) search rule in the stationary traffic the distribution of flow along links of graph is universal when the overall number of created packets is fixed, independently on the time interval in which the packets are created. The observed deviations from the universal curve are correlated with the volatile fluctuations of the packet density (temporary jamming) or permanent jamming in the network. These results are in agreement with other findings based on queuing and network-load timeseries analysis [13], which suggest the occurrence of three distinct phases in the transport on complex graphs: stationary free flow, stationary flow with temporary jamming, and jammed traffic.

In the artificially grown networks the detailed knowledge of the flow distribution of packet traffic for a given search rule can be used to rebuild the network structure so that it performs that transport process with higher efficiency.

References

1. Tadić, B., Thurner, S.: Information Super-Diffusion on Structured Networks. Physica A **332** (2004) 566-584; cond-mat/0307670
2. Newman, M. E. J.: The structure and function of complex networks. SIAM Rev. **46** (2003) 167-225
3. Zhou, H.: Network landscape from a Brownian particle's perspective. Phys. Rev. E **67** (2003) 041908
4. Arenas, A., Danon, L., Diaz-Guilera, A., Gleiser, P.M., Guimerà, R.: Community analysis in social networks. Europ. Phys. J. B (in press).
5. Tadić, B., Rodgers, G.J.: Packet transport on scale-free networks. Advances in Complex Systems **5** (2002) 445-456
6. Tadić, B.: Adaptive random walks on the class of Web graphs. European Physical Journal B, **23** (2001) 221-228
7. Tadić, B.: Dynamics of directed graphs: the world-wide Web. Physica A **293** (2001) 273-284
8. Guimera, R., Diaz-Guilera, A., Vega-Redondo, F., Cabrales, A., Arenas, A.: Optimal network topologies for local search with congestion. Phys. Rev. Lett. **86** (2001) 3196-3199
9. Tadić, B.: Temporal fractal structures: origin of power laws in the world-wide Web. Physica A **314** (2002) 278-283
10. Tadić, B.: Modeling Traffic of Information Packets on Graphs with Complex Topology. Proceedings ICCS 2003, P. Sloot *et al.* Eds., Lecture Notes in Computer Science, Part I, **2567** (2003) 136-142, Springer (Berlin)
11. Tadić, B.: Exploring Complex Graphs by Random Walks. Proceedings of The Granada Seminar on Computational Physics "Modeling Complex Systems", P.L. Garrido and J. Marro (Eds.), AIP Conference Proceedings **661** (2002) 24-27
12. Albert, R., Barabasi A.-L.: Statistical Mechanics of Complex Networks. Reviews Mod. Phys. **74** (2002) 47-101
13. Tadić, B., Thurner, T., Rodgers, G. J.: Traffic on complex networks: Towards understanding global statistical properties from microscopic density fluctuations. Phys. Rev. E (in press).
14. Bollobás, B.: Modern Graph Theory. Springer (New York) 1998.

Network Topology in Immune System Shape Space

John Burns and Heather J. Ruskin

School of Computing,
Dublin City University,
Dublin 9, Ireland
{jburns,hruskin}@computing.dcu.ie
http://www.dcu.ie/computing/msc/index.html

Abstract. We consider the emergent network topology of an immune system shape space at the end of primary response. We extend the formalism of shape space in order to model the relationship between activated immune lymphocytes and stimulant antigen presentation cells by way of a graph consisting of a pair $G = (V, E)$ of sets. The vertex set V is the set of activated genotypes, while the edge set E connects such activated immune lymphocytes and stimulant antigen presentation cell in shape space. This paper shows how shape space graph edge weighting can be viewed, from the biological perspective, as the vigour with which an activated cytotoxic immune cell suppresses the infected antigen presentation cell which stimulated it. In this research, we also identify critical vertices (called α-vertices). These α-vertices act as *bridging* vertices in that they join subgraphs of unrelated immune response. As a consequence of this, such α-vertices ideally model immune cytotoxic lymphocyte *memory* cells. By representing memory cells as highly connected vertices, we show how such cells play a significant role in the elimination of pathogenic agents.

1 Introduction

In this paper we present results from recent work carried out to model the emergence of shape diversity within the immune system. Previously [1], we introduced a new process by which two formalisms, usually separately addressed, may be integrated. These formalisms are known as shape space and physical space. We highlighted a means by which localised dynamics effect global (or shape space) condition, and how global condition in turn may feed information down to local physical space. This approach is now further refined by treating shape space as a self-organising, dynamic network in 2-dimensional space. The system is considered to be exposed to a set of genetically varied pathogens in order to simulate normal human immune experience over a fixed period of time. We then study the cytotoxic lymphocyte activation patterns which emerge naturally in shape space. The results presented here show that, at the end of primary response, a network of activated cytotoxic lymphocytes and pathogen challengers emerges in shape space.

M. Bubak et al. (Eds.): ICCS 2004, LNCS 3038, pp. 1094–1101, 2004.
© Springer-Verlag Berlin Heidelberg 2004

The main contribution of this work is as follows: We present a means to model the genotype (or shape) space of the immune system as set of connected, directed and weighted subgraphs. These subgraphs, will, by way of the emergence of critical (or α-) vertices, merge over time. We show that disruption to α-vertex formation degrades immune response more severely than does the disruption of other (which we call β-) vertices. Disruption is likely to occur whenever a viral mutation is a factor, for example, Human Immune Virus (HIV) or Influenza. The means by which such graphs grow, and how *rewiring* of vertices improves response over time, is also investigated. This work demonstrates that edge weighting can be viewed, from the biological perspective, as the vigour with which an activated cytotoxic immune cell suppresses the infected antigen presentation cell which stimulated it. In shape space, this weighting is the distance (d) from α- and β- vertices to the stimulant pathogen (effectively, it is the length of the edge).

2   The Model

In this section we first review some important features of our previous work, and introduce new detail. For a full exposition of both shape and physical space, the reader is directed to [4]. The shape space formalism was introduced by [5] as a way to represent antibody-antigen binding dynamics. Further research refined this model, notably [6] and [7]. The features of cytotoxic lymphocyte (CTL) cells and antigen presentation (APC) cells which govern the dynamics of cell binding (known as the *antigenic determinant*), may be represented by N parameters. If the N parameters are combined into a vector, the antigenic determinant for each APC and each CTL can be considered as points within an N-dimensional Euclidean space of length L_{ss}. Some notational conventions are observed in the work which follows (where upper-case letters refer to shape space, and lower case, to physical space), and this convention is summarised as follows:

1. (CTL^+, ctl^+): activated cytotoxic lymphocyte cells which are ready to attack and remove infected antigen presentation cells. These cells are often referred to as *armed effectors*. The recirculation patterns of ctl^+ are different from ctl^-, in that ctl^+ will leave the lymphatic compartment and migrate to the location of infection detection. Alteration of recirculation patterns is a common feature of cellular immune response in healthy [8] and diseased [9] immune systems.
2. (APC^+, apc^+): infected antigen presentation cells (typically, dendritic cells) which, having engulfed a virus particle, has gone on to present characteristic viral peptide fragments on its surface.

At the start of each simulation, shape space is characterised by two (non-zero) subpopulations: CTL^- and APC^+, representing the number of precursor cytotoxic lymphocyte and active infected antigen presentation cell *genotypes* respectively. A further subpopulation (called CTL^+), arises once an APC^+ is detected by a CTL^-. CTL^+ are activated cytotoxic lymphocyte genotypes. The CTL^+

subpopulation level increases each time another detection occurs, such that the total CTL^- approaches 0. Denoting CTL and APC genotype vectors as $\mathbf{c}$ and $\mathbf{a}$ respectively, we further develop shape space as follows: Surrounding each $\mathbf{c}$ is a disc of radius ρ [1]. Any $\mathbf{a}$ located within this disc will be subject to a clearance pressure inversely proportional to the distance (d) between the $\mathbf{c}$ and $\mathbf{a}$ in shape space ($d = ||\mathbf{c} - \mathbf{a}||$). Our approach is to place into shape space an increasingly diverse set of antigen challenges and test the varying immune response.

Shape space may be further explored using graph theory [10] to model the relationship between APC^+ and CTL^+. In this approach, shape space is a graph consisting of a pair $G = (V, E)$ of sets satisfying $E \subseteq [V]^2$. The set of vertices V is made up of both CTL^+ and APC^+. Edges connect an APC^+ to the set of CTL^+ stimulated by its presence to become activated. Each G is directed and weighted. An initial set of vertices is introduced at time τ_0 based on model startup parameters. At any time $\tau_k > \tau_0$, a new vertex may be added in shape space with probability dependent both on an apc^+ and ctl^- being neighbours in physical space ($P(N)$) and the distance (d) between APC^+ and CTL^- in shape space is less than or equal to some threshold $\hat{\rho}$ ($P(d \leq \hat{\rho})$). The outcome of both events are independent of each other, so the probability of a new vertex being added is: $P(newvertex) = P(N)P(d \leq \hat{\rho})$ for any given $\{APC^+, CTL^-\}$ conjugate. A newly added vertex is designated CTL^+ (indicating it has been *recruited* from the CTL^- pool). Whenever a new vertex is added, an edge is added by joining the new vertex to the vertex which stimulated its activation. Thus the relationship represented by an edge between two vertices can be understood as: the CTL^+ *was recruited from* the CTL^- pool by the presence of the APC^+, and, the CTL^+ *acts against* the stimulant APC^+:

$$CTL^- \overset{APC^+}{\simeq} CTL^+ \tag{1}$$

and

$$CTL^+ \overset{attacks}{\simeq} APC^+ \tag{2}$$

respectively.

A new edge is added whenever a new vertex is, but also, whenever a new α-vertex appears. An α-vertex is one which although emerges in response to one individual APC^+ actually effects pressure on other APC^+. As shown in Fig. 1, the α-vertex acts against three APC^+ genotypes (as the APC^+ is the *median* vertex in each subgraph, we use the notation m_i). In so doing, it connects the otherwise unconnected subgraphs of Q, R and S. The importance of the α-vertex is clear: α-vertices are promiscuous, not only targeting the stimulant infected cell but also other APC^+ nearby in shape space (Fig. 1, the m_1, m_2 and m_3 vertices, respectively). The emergence of such vertices marks a diversity threshold of immune response (α_{crit}) which once reached, favours a full and healthy clearance of infected cells from the lymphatic compartment. Such α-vertices are unique in that they participate in Eqn.(2) without having first participated in Eqn.(1). Clearance pressure acts along directed edges which are defined by the triple:

---

[1] Clearly, with $N = 2$, the area of this disc is $\pi\rho^2$.

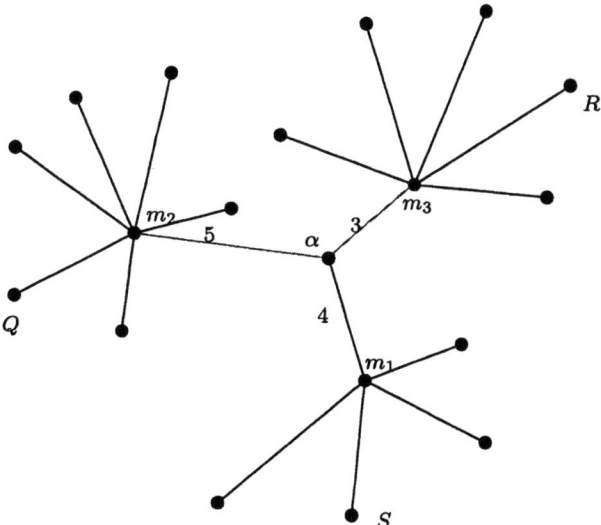

Fig. 1. Three subgraphs with median vertices (stimulant APC^+) at m_i. Leaves (β-vertices) represent activated cytotoxic lymphocyte genotypes (CTL^+). The α-vertex is also a CTL^+, but is one which though originally activated by m_1 now affects clearance pressure against both m_2 and m_3.

$$\vec{E} := \{(e, x, y) | e \in E; x, y \in V; e = xy\}$$

With respect to Fig. 1, the α-vertex clearance pressure acts in three directions (for $i = 1, .., 3$): $\vec{e}_i = (e_i, \alpha, m_i)$.

Rewiring of edges happens whenever a subpopulation of APC^+ disappears, to be replaced by a later APC^+. From the biological point of view, the disappearance of a subpopulation of APC^+ happens whenever the viral genotype challenge is completely eradicated. Rewiring can be viewed as a secondary immune response to some further antigenic challenge. Consistent with [11], we refer the APC^+ as the median vertex (m_i) in a subgraph S. Rewiring is a three step process as follows:

(i)   The median vertex m_i is deleted from the graph and the vertices connected to it become disconnected.

(ii) A new median vertex m_j ($i \neq j$) is introduced, representing a new viral infection in the system.

(iii) Each leaf [2] (or β-vertex) rewires to the new median m_j. Depending on the location of m_j in shape space, there is a probability that not all leaves will reconnect to the median vertex (due to $d > \hat{\rho}$) and such leaves will remain unconnected until a *more* central vertex appears. In biological terms, the disappearance of a stimulus APC^+ usually results in a gradual decline of the effector response, to some small, non-zero level which remains as a form of immune *memory*.

---

[2] A leaf is a vertex of $deg(1)$.

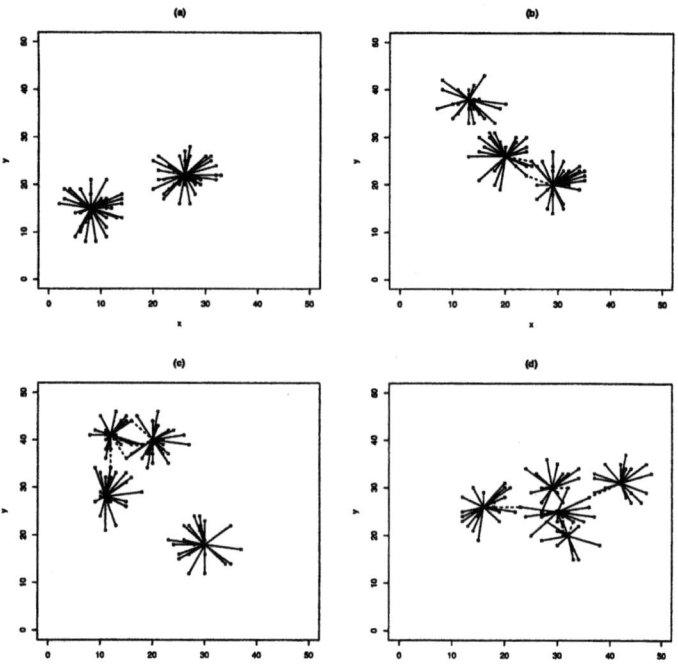

Fig. 2. Four representations of immune shape space at the end of exposure to five antigen genotypes ($\mathcal{R}_{1,..,4}$). The emergence of the α-indexes is indicated by the dashed edges appearing first when the system has been exposed to three antigens (b), with $\mathcal{R}_4$. At exposure to five antigens (d), with $\mathcal{R}_4$, the number of α-vertices is five, and all five of the subgraphs are linked to form one.

3 Results

Fig. 2 shows the state of shape space at the end of primary response. Progressive exposure to varied antigens (values drawn from $\mathcal{R}$) is shown from (a) to (d). In (d), the model has been exposed to 5 different and unrelated antigenic challenges, in much the same way that a maturing immune system would be at the end of $3-4$ years of development. The only parameter varied during simulation execution is $\mathcal{R}$. In (a) and (b), the central (or median) vertex is connected by a set of edges to leaves which appear in response to the prior appearance of the median vertices. Biologically, the infected antigen presentation cell, once recognised, triggers a process (known as clonal expansion) which eventually results in the immune system applying clearance pressure against the infected cell (and all cells presenting the same viral genetic material). In (b), when the system has been exposed to two infections, the subgraphs remain unconnected, indicating that no clearance pressure is applied cross-reactively, and there are no α-vertices. At the point of third antigen challenge, (c), there emerges two α-vertices (acting on the lower two subgraphs). The edges connecting the α-vertices are shown as

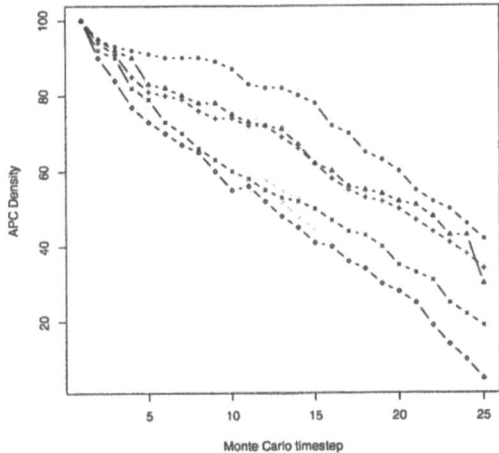

Fig. 3. Clearance rates of apc^+ for $\mathcal{R}_4$, under five different regimes of α-vertex disruption: 1 to 5 (respectively $\diamond$ $\times$ $+$ $\triangle$ and $\circ$), sampled 25 times during model execution and representing some 6 days of real-world time. Selective knock-out (or disruption due to pathogen mutation) of an α-vertex reduces the efficacy of infected cell detection and clearance. When all five vertices have been disrupted (shown by the circle), only $\approx 60\%$ of infected cell clearance takes place. In all cases, the β-vertex response remained healthy. In viral pathogens known to mutate slowly (influenza) or quickly (HIV), immune memory (our α-vertices) can often be less effective (as the viral target *drifts* in shape space). The figure shown here offers some insight into consequences of memory cell disruption.

a broken line. At this point, the third infection to appear has caused an activation pattern which results in cross-reactive pressure applied against it. This pressure (by way of the α-vertex) is significant, because it does not emerge over the normal time period of clonal expansion (usually some $3 - 5$ days). Rather, the pressure is applied *instantaneously* as the α-vertices exist *a priori* and are therefore primed in advance. When four infections have been experienced by the system (Fig. 2 (c)), three of the subgraphs have merged into one, by way of some 4 α-vertices. When the fifth unrelated infection is encountered by the system, a complex network of subgraphs, connected by some 5 α-vertices, has coalesced into one graph. Once the subgraphs merge, the dynamical immune process of up- and down regulation may be explained as follows:

(i)   The appearance of m_1 stimulates the development of both α- and β- vertices in R.
(ii) In turn, α-vertex acts to reduce (or down-regulate) m_2
(iii) The down-regulation of m_2 causes a down regulation in β-vertices of Q (as their source of stimulation declines)

And this process may be summarised thus:

$$\uparrow m_1 \Rightarrow \uparrow \{\alpha, \beta \in R\} \Rightarrow \downarrow m_2 \Rightarrow \downarrow \{\beta \in Q\}$$

Once $\alpha_{crit} = 5$ has been reached, the removal of any further infections which may arise is achieved by way of the rapid appearance of α-vertices, and rapid increase in edge density of the connected, weighted graph G in shape space. In Fig. 3, the relative importance of the α-vertices over the β-vertices is shown. This figure shows the model clearance rate of APC^+ from the lymphatic compartment during exposure to five infections. However, in each case, we have explored the effects of disrupting α-vertices. Disruption will occur whenever an APC^+ *drifts* from its original shape space coordinates $(\Delta x, \Delta y)$. Drift is likely to arise whenever an APC^+ mutates (for example, in the case of HIV). Disruption to one or two α-vertices does not seriously degrade clearance, but results in an average reduction of efficiency of $\sim 15\%$. This is intuitive: the importance of the α-vertex lies not in one individual but in the cumulative effect of all. Further disruption results in progressively worse clearance ability. Disruption of all five α-vertices reduces infected cell clearance by around 35%. We would not expect a total immune failure even under full disruption, as there still remains a healthy and effective β-vertex response. Each simulation is repeated 30 times, and the results are averaged. The results shown here were obtained from five separate model simulation runs (for $\mathcal{R}_4$), at each stage, a further α-vertex was suppressed and the edges connecting subgraphs consequentially did not emerge.

4   Discussion and Conclusions

In this research we have provided an outline of how the immune system shape space may be usefully extended to model the process by which infectious agents may be targeted by cells which have been primed in response to a previous and unrelated infection. Using an approach based on graph theory, we identified two qualitatively different vertex types: α and β. Although both vertex types form part of the cellular effector response, we have shown that an effective immune response depends largely on successful α-vertex activation for efficacious response, and only to a lesser extent, on β-vertices. We have seen how disruption to α-vertex activation results in a suppressed response characteristic of chronic infection. In the results presented we have proposed that α-vertices have strong biological equivalent: cytotoxic lymphocyte memory cells. Such cells, having been primed by way of previous immune challenge, require less time to respond, and, crucially, tend to be beneficially cross-reactive. This finding supports [12] and others. Disruption to α-vertices results in a significantly degraded pathogen clearance than disruption to β-vertices does. This supports the theory that during primary and secondary response, some cytotoxic genotypes are *more important than others* [12]. The network implication of α-vertices is that each edge connection formed from a median vertex to an α-vertex acts as a back-bone in joining disparate *subgraphs*. As these subgraphs connect, two related further questions arise (i) up and down regulation of competing CTL^+ *across subgraphs* and (ii) the emergence of a balanced graph[3]. We have already shown how up- and down- regulation may be explained by extending shape space as an evolving graph (or network).

References

1. Burns, J., Ruskin, H.: A Model of Immune Suppression and Repertoire Evolution. In: Sloot, P.M.A., Gorbachev, Y.E., (eds.): Lecture Notes in Computer Science, Vol. 2660. Springer-Verlag, Berlin Heidelberg New York (2003) 75-85
2. Albert, R., Barabasi, A.: Topology of Evolving Networks: Local Events and Universality. Phys. Rev. Lett. 85(24) (2000) 5234-5237
3. Buseyne, F., and Riviere, Y.: The flexibility of the TCR allows recognition of a large set of naturally occurring epitope variants by HIV-specific cytotoxic T lymphocytes Int. Immunol. (13) (1999) 941-950
4. Burns, J., Ruskin, H.: Viral Strain Diversity and Immune Response - a Computational Model. In: Hamza, M.H., (ed): Proceedings of the IASTED International Conference, Biomedical Engineering. ACTA Press (2003) 60-65
5. Perelson, A.S., Oster, G.F.: Theoretical Studies of Clonal Selection: Minimal Antibody Repertoire Size and Reliability of Self-Non-Self Discrimination. J. Theor. Biol. 81(4) (1979) 645-670
6. de Boer, R.J., Segel, L.A., Perelson, A.S.: Pattern formation in one- and two-dimensional shape-space models of the immune system. J. Theor. Biol. 155(3) (1992) 295-333
7. Smith, D., Forrest, S.: Deriving shape space parameters from immunological data. J. Theor. Biol. 189 (1997) 141-150.
8. Janeway, C.A., Travers, P., Walport, M., Capra, J.D.: Immunobiology. The Immune System in Health and Disease. 4th edn. Churchill-Livingston (1999)
9. Wick, D.: The Disappearing CD4+T Cells in HIV Infection: a Case of Over-stimulation?. J. Theor. Biol. 197 (1998) 507-516
10. Diestel, R.: Graph Theory. 2nd edn. Spinger (1997)
11. Wuchty, S., Stadler, P.: Centers of complex networks. J. Theor. Biol. 223 (2003) 45-53
12. Brehm, M.A., Pinto, A.K., Daniels, K.A., Schneck, J.P., Welsh, R.M., Selin, L.K.: T cell immunodominance and maintenance of memory regulated by unexpectedly cross-reactive pathogens Nat. Immunol. 3(7) (2002) 627-634

---

[3] A graph G is *balanced* if the maximum ratio of edges to vertices, taken over all subgraphs of G, occurs at G itself

An Evolutionary Approach to Pickup and Delivery Problem with Time Windows

Jean-Charles Créput[1], Abder Koukam[1], Jaroslaw Kozlak[1,2], and Jan Lukasik[1,2]

[1] University of Technology of Belfort-Montbeliard, Belfort, France
[2] AGH University of Science and Technology, Al. Mickiewicza 30, 30-059 Kraków, Poland
Jean-Charles.Creput@utbm.fr, kozlak@agh.edu.pl

Abstract. Recently, the quality and the diversity of transport services are more and more required. Moreover, in case of a great deal of services and selling goods, a significant part of price is transport cost. Thus, the design of models and applications which make possible efficient transport planning and scheduling becomes important. A great deal of real transport problems may be modelled by using Pickup and Delivery Problem with Time Windows (PDPTW) and capacity constraints, which is based on the realization of a set of transport requests by a fleet of vehicles with given capacities. Each request is described by pickup and delivery locations, time periods when pickup and delivery operations should be performed and needed load. Application of evolutionary approach has brought good results in case of another, simpler transport problem – the Vehicle Routing Problem with Time Windows (VRPTW). This paper is aimed at proposing a straightforward extension of VRPTW based heuristics for the PDPTW.

1 Introduction

Recently, the importance of passenger and cargo transport services increases. Taking into consideration a growing competition in the domain, companies make an effort to improve the quality of services and to offer new kinds of services, which are able to meet the needs of users.

A lot of transport services may be described by different kinds of vehicle routing problems such as vehicle routing problem (VRP), vehicle routing problem with time windows (VRPTW), pickup and delivery problem (PDP), pickup and delivery problem with time windows (PDPTW). These problems are defined by a set of known transport requests to be performed with the least resources available and time consumption, and with maximum satisfaction of customers. Time windows concern acceptable periods of realization of service in visited points. Maximal capacities of vehicles cannot be exceeded by overall load.

The routes of vehicles should start and finish at a given depot point. In VRPTW it is assumed that each transport request is described by one location which should be visited. The cargoes are being delivered either from starting depot to different destination locations or from different starting locations to one destination location. VRPTW has practical applications like post delivering, newspapers distribution or transport of employee to a workplace.

M. Bubak et al. (Eds.): ICCS 2004, LNCS 3038, pp. 1102–1108, 2004.
© Springer-Verlag Berlin Heidelberg 2004

In PDPTW it is assumed, that each transport request is described by two locations, which are to be visited by the same vehicle: pickup point and delivery point. PDPTW problem has a lot of possible applications [4, 9] such as the transport of elderly and handicapped people, shared taxi services, sealift and airlift, discharge of larvicide, school bus routing and scheduling.

The problem of finding a solution for VRPTW is better explored than for PDPTW case, there are publications concerning VRPTW subject since the seventies. Here, we focus on PDPTW with capacity constraints. Because evolutionary algorithms (EA) have made possible to obtain good solutions for similar problems like VRPTW, we try to extend their embedded specific heuristic designed for VRPTW to PDPTW in a straightforward way.

The organization of this paper is as follows: section 2 contains a state of the arts of recent attempts to solve PDPTW using modern meta-heuristics. Section 3 presents our evolutionary algorithm. Section 4 describes the experimental results obtained, whereas section 5 concludes.

2 Research Overview

The problem encountered while trying to solve PDPTW is the high computational complexity which causes the calculation time for non-trivial size problems to be very high.

A state of the arts of exact methods as well as heuristic solutions of PDPTW is presented in [3, 4, 10]. Most recent attempts to solve PDPTW problem use meta-heuristics based on tabu search. These approaches consist of two phases: in the first one some initial solutions are generated using heuristics of routes constructions and in the other, these solutions are improved through searching the solution space by introducing changes according to given rules [5, 6, 7, 8, 11].

From analysis of the related works we retain the following two classes:

Heuristic used for generation of initial solutions: The different kinds of algorithms based on insertion of requests to routes are applied. They differ by the method used to define the order of insertions. To obtain more balanced routes, the routes may be initialised by the seed-request – the one which is the farthest from depot point or which time-window of pickup starts first. Request points may be sorted by increasing angle $<AOS$ where S is location of request point, O the depot location and A a location which is used as a reference. For each route, only those for which pickup points are situated in an adequate bracket of values of $<AOS$ are analysed. Insertions place of request in a route may be selected in different ways. It may be the first feasible place found or the place where cost function increase, after insertion, is the lowest one.

Search in solution space: Some algorithms take into consideration the solutions which do not respect time and capacity constraints. It is worth noting that operations applied in the presented papers, despite different names, have similar algorithms and perform three kinds of actions: removing a request from a route and adding it into another route, exchanging pairs of pickup and delivery points among selected routes as well as changing order of visited points within one route.

We have not found solutions of PDPTW using evolutionary approaches, but EA provide solutions for VRPTW problem which are among the best ones [2]. Retaining the basic principles of solutions for PDPTW and starting from known heuristics for VRPTW we propose a new evolutionary algorithm for PDPTW.

3 System Model and Optimisation Algorithm

In this section, the evolutionary algorithm for solving PDPTW is described. It is composed of a method of representation of the individual, a fitness function expressing the quality of solution as well as a set of basic evolutionary operators. We try to develop the straightforward EA solution with the simples operators.

Each request point contains following information: type of point (pickup or delivery), times of arrival and departure, occupied vehicle capacity. Times of arrival and departure are defined by two values: min_T and max_T which define the maximum time for the vehicle to arrive at their request place and the minimum time unless it can not serve the request. If the vehicle arrives before min_T, it has to wait at the place.

3.1 Individual and Population

Individual in the evolutionary algorithm is a feasible solution of PDPTW, which consists of a list of routes. One route is associated with one vehicle. Each route consists of a sequence of request points (pickup and delivery points) which are visited by the given vehicle and of the following additional attributes: number of served requests, total travel time, total travel distance, total cost.

Population consists of PDPTW solutions which are modified by evolutionary operators to obtain the next generation of candidate solution.

3.2 Fitness Function

To estimate a quality of solution, we use the fitness function presented by the following equation:

$$F(x) = \frac{\alpha_1}{K} + \frac{\alpha_2}{\sum\limits_{k=1}^{K} D_k} + \frac{\alpha_3}{\sum\limits_{k=1}^{K} T_k}$$

where: K – number of vehicles, D_k – total travel distance of k-th vehicle, T_k – total travel time of k-th vehicle, α1, α2, α3 – weight parameters.

3.3 Construction of Initial Population of Solutions

The number of solutions in the initial population is equal to 50. We have used one single algorithmto create an initial generation of solutions. It is *Simple Random Insertion Algorithm*, defined as follows:

```
Let RS contains all requests to realize;
while RS not empty begin
    empty route VRᵢ for  vehicle vᵢ is created;
    number_of_attempts=number_of_elements(RS);
    attempt=0;
    repeat
        A request rⱼ is randomly selected from RS;
        Attempt to insert rⱼ into route VRᵢ;
        if insertion succeeded then
            remove rⱼ from RS;
    until (attempt = number_of_attempts)
end
```

3.4 Evolutionary Operators

In the following paragraphs, the principles of the *Reproduction, Mutation* and *Selection* operators are presented. Cross-over and mutation operators are inspired by those presented in [12], for VRPTW problem solving.

Reproduction. The individuals are selected to reproduce by a tournament method. An existing population is divided into groups and after that from each group the best individual is selected and added to the parent population. Two methods of creation of offspring from two parents are used: in the first one, an exchange of parts of two routes takes place, in the other, complete routes are exchanged. It is assumed, that after the reproduction operation only one offspring is created. These operators proposed in [12] for VRPTW are extended by considering a single request as a binary pickup and delivery pair.

Sequence-Based Crossover (SBX). The principle of this operator is an exchange of fragments of routes between two solutions. In order to do this, two parents are selected from the population using tournament method, then for each of these solutions, one route is selected. Lists consisting of requests served by each of vehicles are created and an index (within the range from 0 to up to size of shortest from among both the lists) is selected by random. The requests which are on positions lower or equal to the index value in first route or on position higher than index value in the second route are included into a new set of requests. On the basis of this set of requests a new route is constructed. If the creation of route is impossible then operation is rejected, otherwise a new solution is created. This new solution consists of the new route and the routes from replaced parent excluding the one which was changed. Then, the feasibility of the entire solution is examined, and, if necessary, a repair operation is performed. The unfeasibility takes place when:

- Not all the requests are served. In this case, absent requests are inserted into positions which gives best values of fitness function. If it is not possible, then operation is withdrawn and the offspring is abandoned.
- Some requests are served two times. If the same requests appear in the second parent and in the offspring, then they are removed from the parent, if they appear two times in the offspring than one of them is removed.

Route-Based Crossover (RBX). There is made a choice of two individuals from population, then for each of these two, one route is selected. Then, the selected routes are exchanged between individuals. As in the case of SBX operator, so as obtain feasible solution where each request occurs only once, the execution of repairing procedure may be necessary.

Mutation. Two different types of mutation operators are defined, extending the operators for VRPTW presented in [12].

One-level exchange (1M). The role of this mutation is trying to decrease the number of needed vehicles. An individual and a route are selected by random and then it is tried to move all requests served by this route to other routes in such a way, that the obtained solutions are feasible and total riding time of other routes is minimized.

Local search (LSM). The role of operation LSM is to improve the quality of a route by changing the order in which request points are visited.

One route is selected by random. For each requests from that route we try to find a better location inside that route, if there is such a place, we move a request to that place.

Selection. New generation is obtained after modification of previous one by crossover and mutation operators, adding best solution from previous generation or removing best existing or randomly selected solutions. Total number of individuals in the population cannot exceed the upper limit and be under the lower limit.

4   Results

The experiments were performed for the set of PDPTW benchmarks [1]. The main parameters of evolutionary algorithm were set as follows: size of population = 50, number of iterations = 500, weigh parameters of fitness function $\alpha_1=\alpha_2=\alpha_3=1$.

The Table 1. contains a list of evolutionary operators with probabilities of their applications for the given generation, and a frequency of attempts of execution for one generation.

Table. 1. Evolutionary operators

Operator	Probabiliy	Iterations	Details
Fitness	1	1	Vehicle number, Travel Time, Travel Distance are considered
PDP Cross Over	1	2	random selection of RBX or SBX, parents selected by tournament
LS PDP Mutation	0.3	3	Local Search for all individuals
M1 PDP Mutation	0.5	2	M1 for randomly selected individuals
Selection	1	1	add the 5 best individuals or remove randomly selected individuals

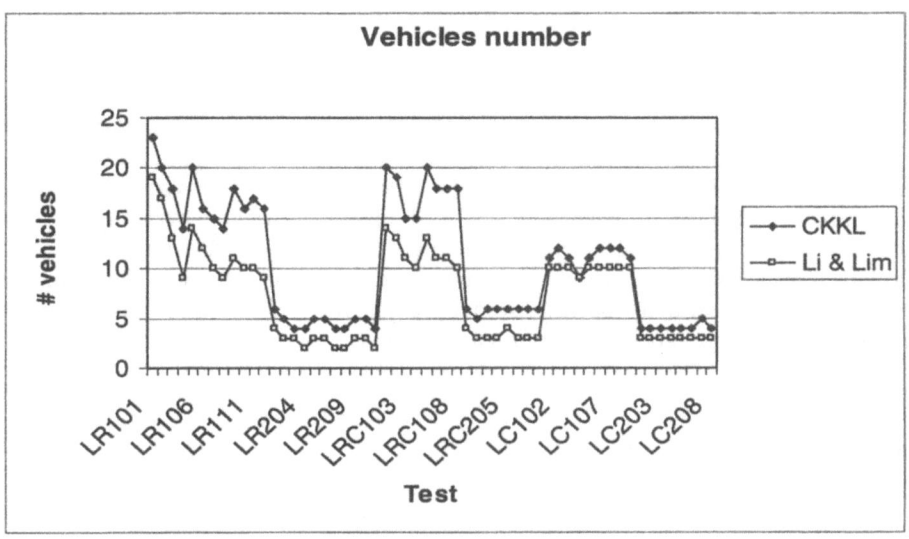

Fig. 1. Number of vehicles

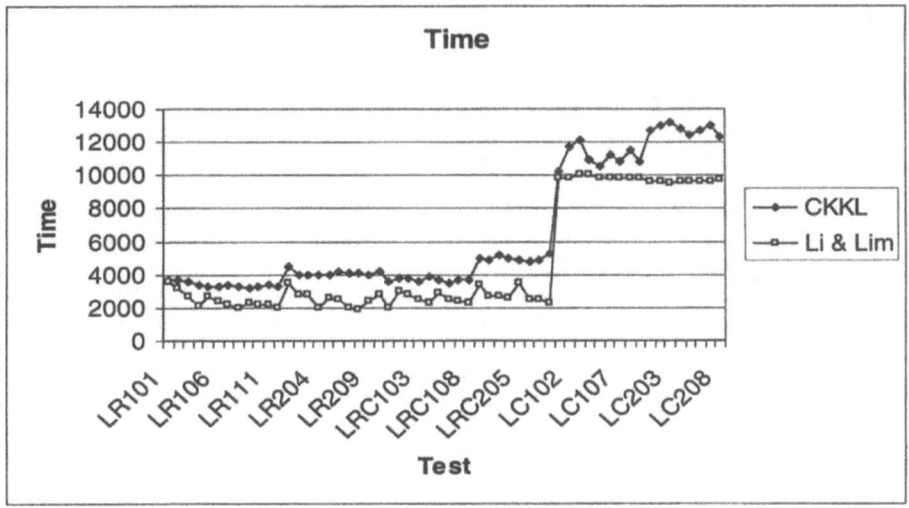

Fig. 2. Total travel times

On Fig. 1 and Fig. 2 are presented vehicle number and total travel time obtained by us [CKKL] and the ones accessible on WWW [1], obtained by Li & Lim [7]. On the X-axis the names of benchmarks are marked.

All requests are satisfied but the obtained results (number of used vehicles, total travel time, total distance) are usually worse about few dozens of percents than the best known ones.

5 Conclusions

In this article, an example of evolutionary approach to solve PDPTW was presented. It is a straightforward extension of VRPTW heuristic approach. Additionally, experiments and modifications including an adjustment of configuration parameters and some modifications of evolutionary operators are necessary. Results are suboptimal in regard to bests [Li & Lim] results, however we can see some concordance of the global shape of the plots with the [Li & Lim] results. It is explained by the simple procedure we have achieved in our approach. Thus, this first prototype would certainly be improved by adjustment of operator parameters and introduction of more problem specific knowledge.

References

1. Benchmarks – Vehicle Routing and Travelling Salesperson Problems.
    http://www.sintef.no/static/am/opti/projects/top/vrp/benchmarks.html,
    Updated; 23 April 2003.
2. Braysy, O.: Genetic Algorithms for the Vehicle Routing Problem with Time Windows. *Arpakanuus* 1, special issue on Bioinformatics and Genetic Algorithms, 2001.
3. Cordeau, J.-F., Laporte, G.: The Dial-a-Ride Problem: Variants, Modeling Issues and Algorithms. *Les Cahiers du GERAD*, 2002.
4. Desaulniers, G., Desrosiers, J., Erdmann, A., Solomon, M.M., Soumis, F.: The VRP with Pickup and Delivery. *Les Cahiers du GERAD*, 2000.
5. Gendreau, A., Guertin, F., Potvin, J.Y., and Séguin, R.: Neighborhood search heuristics for a dynamic vehicle dispatching problem with pick-ups and deliveries. *Rapport technique CRT-98-10*, Université de Montréal, 1998.
6. Lau, H.C., Liang, Z.: Pickup and Delivery with Time Windows : Algorithms and Test Case Generation. *Proceeedings of 13th IEEE International Conference on Tools with Artificial Intelligence (ICTAI'01)*, Dallas, USA, 2001.
7. Li H., Lim, A.:A Metaheuristic for the Pickup and Delivery Problem with Time Windows. *Proceedings of 13th IEEE International Conference on Tools with Artificial Intelligence (ICTAI'01)*, Dallas, USA, 2001.
8. Lim, H., Lim, A.,. Rodrigues, B.: Solving the Pick up and Delivery Problem using "Squeaky Wheel" Optimization with Local Search. *Proceedings of American Conference on Information Systems, AMCIS 2002*, Dallas, USA, 2002.
9. Madsen, O.B.G., Ravn, H.F., Rygaard, J.M.: A heuristic algorithm for a dial-a-ride problem with time windows, multiple capacities, and multiple objectives. *Annals of Operations Research* 60, 193-208, 1995.
10. Mitrowic-Minic, S.: Pickup and Delivery Problem with Time Windows: A Survey. SFU CMPT TR. 1998-12, ftp://fas.sfu.ca/pub/cs/techreports/1998, 1998.
11. Nanry, W.P., Barnes, J.W.: Solving the pickup and delivery problem with time windows using reactive tabu search. *Transportation Research* Part B 34, Elsevier Science Ltd, pages 107-121, 2000.
12. Potvin, J. Y., Bengio, S.: The vehicle routing problem with time windows - Part II: Genetic search. *INFORMS Journal on Computing* 8, pp. 165-172, 1996.

Automatic Extraction of Hierarchical Urban Networks: A Micro-Spatial Approach

Rui Carvalho[1] and Michael Batty[2]

[1] The Bartlett School of Graduate Studies
rui.carvalho@ucl.ac.uk,
[2] Centre for Advanced Spatial Analysis
University College London,
1-19 Torrington Place, London WC1E 6BT, UK
m.batty@ucl.ac.uk

Abstract. We present an image processing technique for the identification of 'axial lines' [1] from ridges in isovist fields first proposed by Rana [2,3]. These ridges are formed from the maximum diametric lengths of the individual isovists, sometimes called viewsheds, that make up the isovist fields [4]. We discuss current strengths and weaknesses of the method, and show how it can be implemented easily and effectively.

1 Axial Maps as Skeletons for Urban Morphology

Axial lines are used in 'space syntax' to simplify connections between spaces that make up an urban or architectural morphology. Usually they are defined manually by partitioning the space into the smallest number of largest convex subdivisions and defining these lines as those that link these spaces together. Subsequent analysis of the resulting set of lines (which is called an 'axial map') enables the relative nearness or accessibility of these lines to be computed. These can then form the basis for ranking the relative importance of the underlying spatial subdivisions and associating this with measures of urban intensity, density, or traffic flow [1,5,6]. Progress has been slow at generating these lines automatically. Lack of agreement on their definition and lack of awareness as to how similar problems have been treated in fields such as pattern recognition, robotics and computer vision have inhibited explorations of the problem and only very recently have there been any attempts to evolve methods for the automated generation of such lines [7,8,4].

One obvious advantage of a rigorous algorithmic definition of axial lines is the potential use of the computer to free humans from the tedious tracing of lines on large urban systems. Perhaps less obvious is the insight that mathematical procedures may bring about urban networks, and their context in the burgeoning body of research into the structure and function of complex networks [9,10]. Indeed, on one hand urban morphologies display a surprising degree of universality [11,12,13,14,15], but little is yet known about the relation between this observed universality and the transport and social networks embedded within

M. Bubak et al. (Eds.): ICCS 2004, LNCS 3038, pp. 1109–1116, 2004.
© Springer-Verlag Berlin Heidelberg 2004

urban space (but see [16,17]). On the other hand, axial maps are a substrate for human navigation and rigorous extraction of axial lines may substantiate the development of models for processes that take place on urban networks which range from issues covering the efficiency of navigation, and the vulnerability of network nodes and links to failure, attack and related crises.

Axial maps can be regarded as members of a larger family of axial representations (often called skeletons) of 2D images. There is a vast literature on this, originating with the work of Blum on the Medial Axis Transform (MAT) [18,19].

2   Axial Lines as Ridges on Isovist Fields

An isovist is the space defined around a point (or centroid) from which an object can move in any direction before it encounters some obstacle. We shall see that the paradigm shift from the set of maximal discs inside the object (as in the MAT) to the maximal straight line that can be fit inside its isovists holds a key to understanding what axial lines are.

As in 'space syntax', we simplify the problem by eliminating terrain elevation and associate each isovist centroid with a pair of horizontal coordinates (x, y) and a third coordinate - the length of the longest straight line across the isovist at each point which we define on the lattice as $\Delta_{i,j}^{\max}$. We extend previous work by Rana [3], where he noted that "the ridge lines give an indication of the disposition of the axial lines", by using a modification of the Medial Axis Transform [18, 19] and the Hough Transform [20]. The hypothesis states that all axial lines are ridges on the surface of $\Delta_{i,j}^{\max}$. The reader can absorb the concept by 'embodying' herself in the $\Delta_{i,j}^{\max}$ landscape: movement along the perpendicular direction to an axial line implies a decrease along the $\Delta_{i,j}^{\max}$ surface; and $\Delta_{i,j}^{\max}$ is an invariant, both along the axial line and along the ridge. The hypothesis goes further to predict that the converse is also true, i.e., that up to an issue of scale, all ridges on the $\Delta_{i,j}^{\max}$ landscape are axial lines.

Here we sample isovist fields by generating isovists for the set of points on a regular lattice [2,21,8,22]. Specifically, we are interested in the isovist field defined by the length of the longest straight line across the isovist at each mesh point, (i, j). This measure is denoted the maximum diametric length, $\Delta_{i,j}^{\max}$ [4], or the maximum of the sum of the length of the lines of sight in two opposite directions [8, p 204]. To simplify notation, we will prefer the former term.

First, we generate a Digital Elevation Model (DEM) [23] of the isovist field, where $\Delta_{i,j}^{\max}$ is associated with mesh point (i, j) [21,8]. Our algorithm detects ridges by extracting the strict maxima (i.e. a cell with value stricly greater than any of its nearest neighbours [24]) of the discrete DEM. Next, we use an image processing transformation (the Hough Transform) on a binary image containing the local maxima points which lets us rank the detected lines in the Hough parameter space. Finally, we invert the Hough transform to find the location of axial lines on the original image.

The process of using the HT to detect lines in an image involves the computation of the HT for the entire image, accumulating evidence in an array for

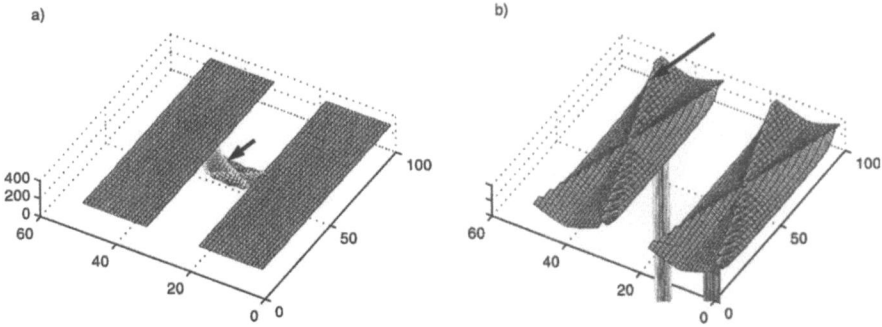

Fig. 1. (a) Plot of the Maximum Diametric Length ($\Delta_{i,j}^{\max}$) isovist field for an 'H' shapped open space structure. (b) Zoom-in (detail) of (a) showing the ridges on the longer arms of the 'H' shape. Arrows point to the ridges o nboth figures.

events by a voting (counting) scheme and searching the accumulator array for peaks which hold information of potential lines present in the input image. The peaks provide only the length of the normal to the line and the angle that the normal makes with the y-axis. They do not provide any information regarding the length, position or end points of the line segment in the image plane [25]. Our line detection algorithm starts by extracting the point that has the largest number of votes on parameter space, which corresponds to the line defined by the largest number of collinear local maxima of $\Delta_{i,j}^{\max}$, and proceeds by extracting lines in rank order of the number of their votes on parameter space. One of us [4] has previously proposed rank-order methods as a rigorous formulation of the procedure originally outlined of "first finding the longest straight line that can be drawn, then the second longest line and so on (...)" [1, p 99].

To test the hypothesis that axial lines are equivalent to ridges on the $\Delta_{i,j}^{\max}$ surface, we start with a simple geometric example: an 'H' shaped open space structure. As illustrated in Fig. 1, axial lines are equivalent to ridges for this simple geometric example, if extended until the borders on the open space. Indeed, one confirms this both in Fig. 1a) and Fig. 1b) by properly zooming-in the $\Delta_{i,j}^{\max}$ landscape. Next, we aim at developing a method to extract these ridges as lines by sampling. In Fig. 2a), we plot the local maxima of the discretized $\Delta_{i,j}^{\max}$ landscape, which are a discretized signature of the ridges on the $\Delta_{i,j}^{\max}$ continuous field. Figure 2b) is the Hough transform of Fig. 2a) where θ goes from $0°$ to $180°$ in increments of $1°$. The peaks on Fig. 2b) are the maxima in parameter space, (ρ, θ), which are ranked by height in Fig. 2c). Finally, the ranked maxima in parameter space are inverted onto the coordinates of the lines in the original space, yielding the detected lines which are plotted on Fig. 2d).

Having tested the hypothesis on a simple geometry, we repeat the procedure for the French town of Gassin —see Fig. 3. We have scanned the open space structure of Gassin [1, p 91] as a binary image and reduced the resolution of the scanned image to 300 dpi (see inset of Fig. 3). The resulting image has 171×300 points, and is read into a Matlab matrix.

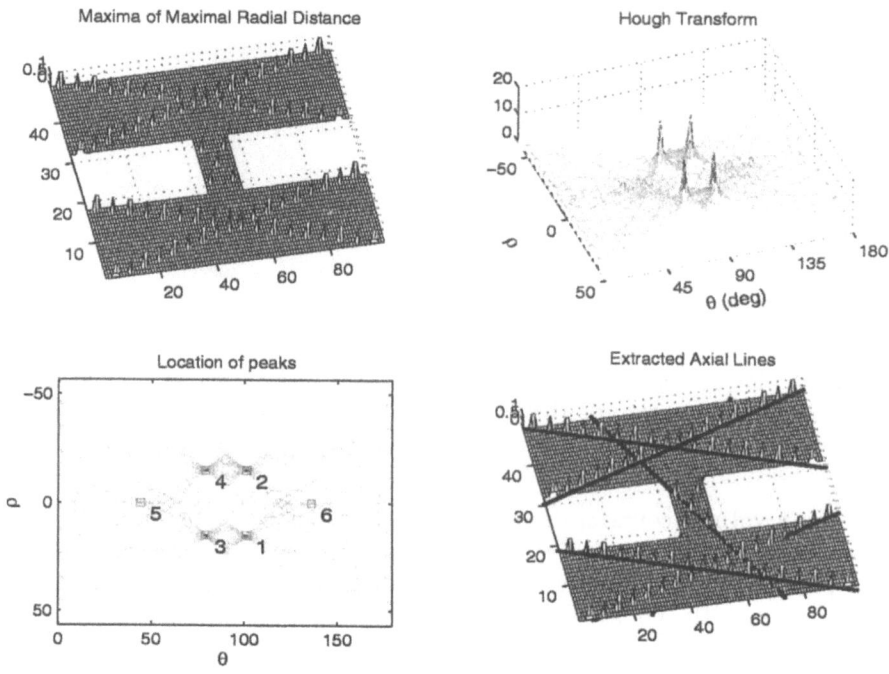

Fig. 2. (a) Local maxima of the Maximum Diametric Length ($\Delta_{i,j}^{\max}$) for the 'H' shaped structure in Fig. 1. (b) Hough transform of (a). (c) rank of the local maxima of the surface in (b). (d) The Hough transform is inverted and the 6 highest peaks in (c) define the axial lines shown.

Next we use a ray-tracing algorithm in Matlab (angle step=0.01°) to determine the $\Delta_{i,j}^{\max}$ measure for each point in the mesh that corresponds to open space. The landscape of $\Delta_{i,j}^{\max}$ is plot on Fig. 3. To extract the ridges on this landscape, we determine the local maxima. Next, we apply the Hough Transform, as in the 'H' shape example, and invert it to determine the 6 first axial lines (see Fig. 4). We should alert readers to the fact that as we have not imposed any boundary conditions on our definition of lines from the Hough Transform, three of these lines intersect building forms illustrating that the technique is identifying the dominant linear features in image space but ignoring any obstacles which interfere with the continuity of these linear features. We consider that this is a detail that can be addressed in subsequent development of the approach.

3   Where Do We Go from Here?

Most axial representations of images aim at a simplified representation of the original image, in graph form and without the loss of morphological information. Therefore, most shape graphs are invertible –a characteristic not shared with axial maps, as the original shape cannot be uniquely reconstructed from the

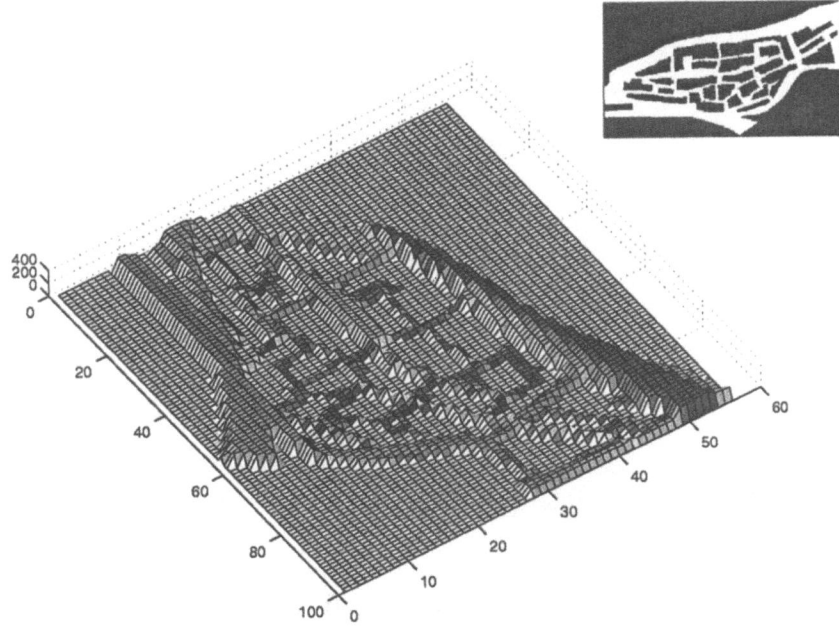

Fig. 3. Plot of the Maximum Diametric Length $(\Delta_{i,j}^{\max})$ isovist field for the town of Gassin. The inset shows the scanned image from [1]

latter. Also, metric information on the nodes length is often stored together with the nodes (the latter often being weighted by the former), whereas it is discharged in axial maps. On the other hand, most skeletonizations aim at a representation of shape as the human observer sees it and therefore aim mostly at small scale shapes (images), whereas the process of generating axial maps assumes that the observer is *immersed* in the shape and aims at the representation of large scale shapes (environments). Nevertheless, we have shown that the extraction of axial lines can be accomplished with methods very similar to those routinely employed in pattern recognition and computer vision (e.g. the Medial Axial Transform and the Hough Transform).

The hypothesis has successfully passed the test of extracting axial lines both for a simple geometry and for a classical case study in Space Syntax – the town of Gassin. Indeed, $l_{2,detected}, l_{3,detected}, l_{4,detected}, l_{5,detected}$ and $l_{6,detected}$ in Fig. 4 all match reasonably well lines originally drawn [1]. Differences between original and detected lines appear for $l_{3,original}$ and $l_{3,detected}$, where the mesh we used to detect lines was not fine enough to account for the detail of the geometry and the HT counts collinear points along a line that intersects buildings, and for $l_{5,original}$ and $l_{5,detected}$, where the original solution is clearly not the longest line through the space.

Fig. 4 highlights two fundamental issues. First, defining axial lines as the longest lines of sight may lead to unconnected lines on the urban periphery. The

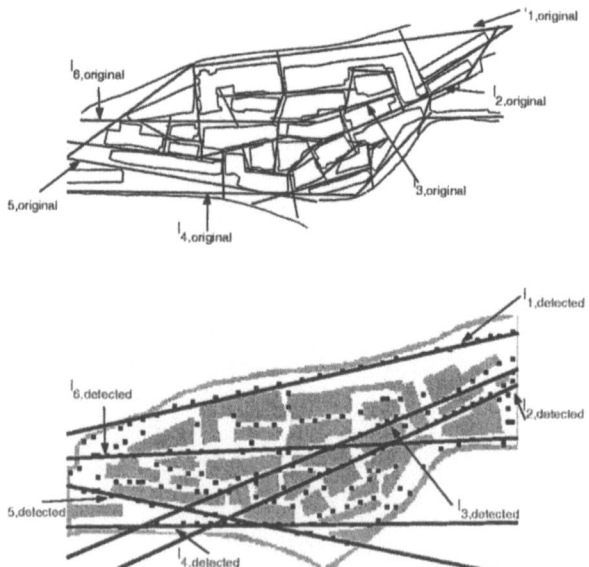

Fig. 4. (a) Axial lines for the town of Gassin [1]. (b) Local maxima of $\Delta_{i,j}^{\max}$ (squares) and lines detected by the proposed algorithm.

problem is quite evident with line $l_{1,original}$ in Fig. 4a) [1, p 91], where the solution to the longest line crossing the space is $l_{1,detected}$ —see Fig. 4b). Thus, the price to pay for a rigorous algorithm may be that not all expected connections are traced. The second problem is an issue of scale, as one could continue identifying more local ridges with increasing image resolution (see discussion in [4]). We believe that the problem is solved if the width of the narrowest street is selected as a threshold for the length of axial lines detected from ridges on isovist fields. Only lines with length higher than the threshold are extracted. We speculate that this satisfies almost always the condition that all possible links are established, but are aware that more lines may be extracted automatically than by human-processing. Again, this seems to be the price to pay for a rigorous algorithm.

By being purely local, our method gives a solution to the global problem of tracing axial maps in a time proportional to the number of mesh points. Thus, algorithm optimization is akin to local optimization (mesh placement and ray-tracing algorithm). Although most of the present effort has been in testing the hypothesis, it is obvious that regular grids are largely redundant. Indeed, much optimization could be accomplished by generating denser grids near points where the derivative of the boundary is away from zero (e.g. turns) to improve detection at the extremities of axial lines. Also, the algorithm could be improved by generating iterative solutions that would increase grid and angle sweep resolutions until a satisfactory solution would be reached or by parallelizing visibility analysis calculations [26].

Our approach to axial map extraction is preliminary as the HT detects only line parameters while axial lines are line segments. Nevertheless, there has been considerable research effort put into line segment detection in urban systems, generated mainly by the detection of road lane markers [27,28], and we are confident that further improvements involve only existing theory.

This note shows that global entities in urban morphology can be defined with a purely local approach. We have shown that there is no need to invoke the concept of convex space to define axial lines. By providing rigorous algorithms inspired by work in pattern recognition and computer vision, we have started to uncover problems implicit in the original definition (disconnected lines at boundary, scale issues), but have proposed working solutions to all of them which, we believe will engage other disciplines in the effort of gaining insight into urban morphology. Finally, we look with considerable optimism to the automatic extraction of axial lines and axial maps in the near future and believe that automatic processing of medium to large scale cities may be only a few years away from being implemented on desktop computers.

Acknowledgments. RC acknowledges generous financial support from Grant EPSRC GR/N21376/01 and is grateful to Bill Hillier and Alan Penn for valuable comments. The authors are indebted to Sanjay Rana both for the original idea that axial lines appear as ridges in the MDL isovist field and for using his Isovist Analyst Extension (see http://www.geog.le.ac.uk/sanjayrana/software_isovistanalyst.htm) to provide independent corroboration on the 'H' test problem.

References

1. Hillier, B., Hanson, J.: The Social Logic of Space. Cambridge University Press, Cambridge (1984)
2. Benedikt, M.: To take hold of space: isovists and isovist fields. Environ. Plan. B **6** (1979) 47–65
3. Rana, S.: Isovist analyst extension for arcview 3.2 (2002)
4. Batty, M., Rana, S.: The automatic definition and generation of axial lines and axial maps. Environ. Plan. B **31** (2004) forthcoming.
5. Hillier, B., Penn, A., Hanson, J., T, T.G., Xu, J.: Natural movement: or, configuration and attraction in urban pedestrian movement. Environ. Plan. B **20** (1993) 29–66
6. Peponis, J., Ross, C., Rashid, M.: The structure of urban space, movement and co-presence: The case of atlanta. Geoforum **28** (1997) 341–358
7. Peponis, J., Wineman, J., Bafna, S., Rashid, M., Kim, S.: On the generation of linear representations of spatial configuration. Environ. Plan. B **25** (1998) 559–576
8. Ratti, C.: Urban analysis for environmental prediction. Phd thesis, University of Cambridge (2001)
9. Albert, R., Barabási, A.L.: Statistical mechanics of complex networks. Rev. Mod. Phys. **74** (2002) 47–97
10. Newman, M.E.J.: The structure and function of complex networks. SIAM Rev. **45** (2003) 167–256
11. Batty, M., Longley, P.: Fractal Cities. Academic Press, San Diego, CA (1994)

12. Carvalho, R., Penn, A.: Scaling and universality in the micro-structure of urban space. Physica A **332** (2004) 539–547
13. Frankhauser, P.: La Fractalité des Structures Urbaines. Anthropos, Paris (1994)
14. Makse, H.A., Havlin, S., Stanley, H.E.: Modelling urban growth patterns. Nature **377** (1995) 608–612
15. Makse, H.A., Jr, J.S.A., Batty, M., Havlin, S., Stanley, H.E.: Modeling urban growth patterns with correlated percolation. Phys. Rev. E **58** (1998) 7054–7062
16. Latora, V., Marchiori, M.: Is the boston subway a small-world network? Physica A **314** (2002) 109–113
17. Chowell, G., Hyman, J.M., Eubank, S., Castillo-Chavez, C.: Scaling laws for the movement of people between locations in a large city. Phys. Rev. E **68** (2003) 066102
18. Blum, H.: Biological shape and visual science (part 1). J. Theor. Biol. **38** (1973) 205–287
19. Blum, H., Nagel, R.N.: Shape description using weighted symmetric features. Pattern Recogn. **10** (1978) 167–180
20. Illingworth, J., Kittler, J.: A survey of the hough transform. Computer Vision, Graphics, and Image Processing **44** (1988) 87–116
21. Batty, M.: Exploring isovist fields: space and shape in architectural and urban morphology. Environ. Plan. B **28** (2001) 123–150
22. Turner, A., Doxa, M., O'Sullivan, D., Penn, A.: From isovists to visibility graphs: a methodology for the analysis of architectural space. Environ. Plan. B **28** (2001) 103–121
23. Burrough, P.A., McDonnell, R.A.: Principles of Geographical Information Systems. Spatial Information Systems and Geostatistics. Oxford University Press (1998)
24. Niblack, C., Gibbons, P., Capson, D.: Generating skeletons and ceterlines from the distance transform. CVGIP: Graphical Models and Image Processing **54** (1992) 420–437
25. Gonzalez, R.C., Woods, R.E.: Digital Image Processing. Addison-Wesley (1992)
26. Mills, K., Fox, G., Heimbach, R.: Implementing an intervisibility analysis model on a parallel computing system. Computers & Geosciences **18** (1992) 1047–1054
27. Kamat-Sadekar, V., Ganesan, S.: Complete description of multiple line segments using the hough transform. Image and Vision Computing **16** (1998) 597–613
28. Pomerleau, D., Jochem, T.: Rapidly adapting machine vision for automated vehicle steering. IEEE Expert **11** (1996) 19–27

Design and Implementation of the Web-Based PSE *GridGate*

KyungWoo Kang, YunHee Kang, and KwangMoon Cho

Department of Computer and Communication Engineering,
Cheonan University, 115, Anseo-dong,
Cheonan 330-704, Choongnam, Republic of Korea
{kwkang,yhkang,ckmoon}@cheonan.ac.kr

Abstract. Grid-computing on networked computers is increasingly applied to solve a variety of large-scale computation problems. Several PSEs(Problem Solving Environment) are developed to provide the application programmers with computers which are available in wide environment. However, these systems do not supplies web-based interface or the real-time visualization facility. Web technology is becoming the general technology on the development of network applications, in particular, because its interface can be made platform independent. In this paper, we propose a web-based PSE for executing the parallel SPMD application written in MPI. Also, a web-based collaborative environment is developed with a real-time visualization technology.

1 Introduction

With the advance of network and software infrastructure, grid-computing technology on a cluster of heterogeneous computing resources becomes pervasive [1, 2,3,4]. Grid-computing describes a coordinated use of an assembly of distributed computers, which are linked by networks and are deployed by many kinds of softwares. The potential benefit of the grid-computing is that users can exploit a powerful monolithic virtual machine. However, construction of a grid-computing system has been a challenging task which involves broad spectra of technical issues. The major hurdles of grid-computing environments are due to the lack of tools to facilitate the development of parallel and distributed applications [5, 4]. Consider the basic operations that arise in the development cycle of such applications [6]:

1. Generation: source files
2. Transfer: source files and data files between computing resources.
3. Compile: each source file on each computing resource
4. Execution: execution file
5. Visualization: the result to be collected from computing resources after the finish of the execution

The steps including two, three and five can become quite time consuming as the number of resources increases because users must transfer many files between

M. Bubak et al. (Eds.): ICCS 2004, LNCS 3038, pp. 1117–1123, 2004.
© Springer-Verlag Berlin Heidelberg 2004

computers. Moreover, step five should wait for the completion of the previous step, the fourth step, which might require long time. In order to use the visualization service, the user needs to have an account on the machine on which the visualization system is installed. In this research, a web-based PSE(hereinafter referred to as *GridGate*) is developed for doing those five steps on one interface. For this purpose, this research will focus on more effective use of computing resources by making the systems more easily available and collaborative to the researchers.

This paper is organized as follows: Section 2 describes the system architecture of *GridGate*. Section 3 shows the experimental results on our testdbed. Section 4 concludes this paper and sketches some future works.

2   System Aarchitecture of *GridGate*

The simulation procedure in *GridGate* is schematically described in Fig. 1. The structure of *GridGate* consists of two major functional parts; grid portal and realtime visualization. The basic components for a computational run in *GridGate* are the user's source code for simulation (referred to as solver) and data files. It is assumed that the user prepares these files. The realtime visualization provides monitoring of the intermediate result of the assigned application by real-time graphical processing to a three-dimensional graph.

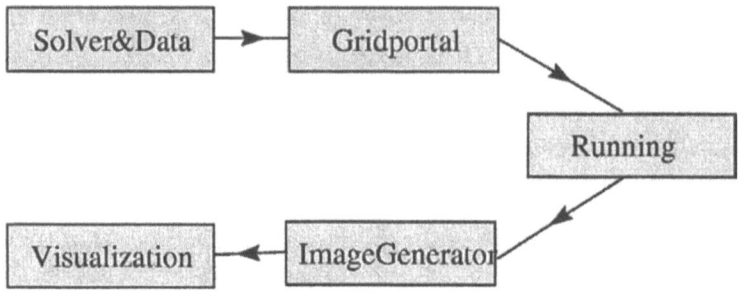

Fig. 1. System Flow of *GridGate*

2.1   Grid Portal

A grid portal is developed for easy and rapid development of SPMD applications on the grid environment. Our grid portal supplies the web-based functions of authentication, visualization of resource's state, distribution&compilation of user's job and spawning the job. These functions are shown in the following Fig. 2.

The basic components for a computational run in *GridGate* are the user's source code and data file for simulation. It is assumed that user prepares these files and the makefile to compile the source in each machine. The authentication step needs user's id and password one time. In this research, authentication step

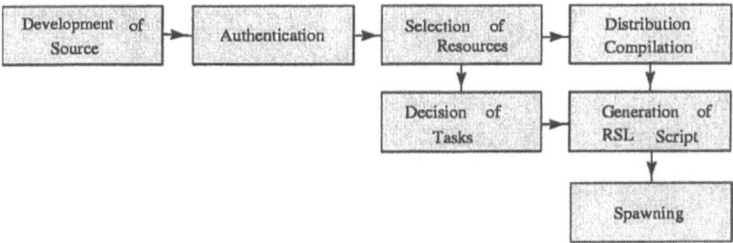

Fig. 2. Structure of grid portal

generates an user proxy using SWING[7]. In the next step, user could choose resources and the number of tasks according to the current information of the computing resources. The hardware information includes the name, architecture, CPU load and network traffic.

2.2 Realtime Visualization

Visualization should show the time dependence of a data set, the result of simulations, using several techniques. Simulations of fluid dynamics are examples of the applications that generate data sets in time. Static visualizations of dynamic information are sometimes ineffective and incorrect because they do not convey motion. Animations can help the user understand, particularly if the user is interested in the process of simulation. A traditional approach to the visualization is for the user to use the visualization system or a library to implement application specific visualizations. Many such visualization systems require that the user has an account on the machine on which the software is installed. For these reasons scientists often transfer the results of these visualizations by videotape or by converting a series of animation frames to a movie file(MPEG, GIF etc). The process of translating a visualization to the movie file eliminates user interaction with the visualization. Recent technological developments have created opportunities for new approaches for representing time-sensitive problems using web-based computer animation. More importantly, Java and the World Wide Web have provided a universal platform for building animations and visualizations. This web-based execution model solves many dissemination problems. Users can interact with animations without having an account on a particular machine. Fig. 3 shows the process of realtime visualization based on the web. Each task generates a series of intermediate files that could be merged into one file. Image generator gets the merged files as its input and generates the animation that represents its simulation. In this research, we use gnuplot as the image generator.

2.3 User Interface

A dedicated graphic user interface is designed for *GridGate*. This GUI is responsible for controlling the system and manipulation of a RSL(Resource Speci-

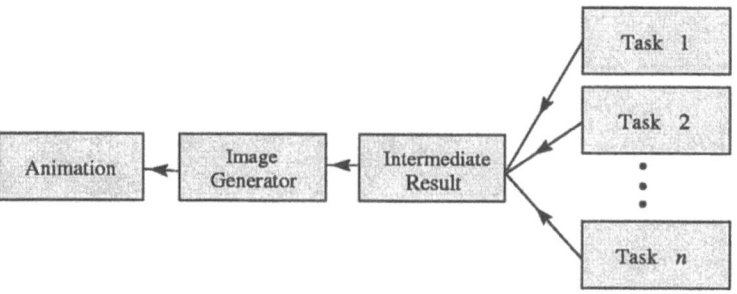

Fig. 3. The process of realtime visualization

fication Language) file. The RSL file is the important input for Globus as a text file which contains the information both of the resources and the solver required for running a grid-computing application. The resource information includes the host name, local scheduler, and working directory. The solver information describes data I/O as well as its name.

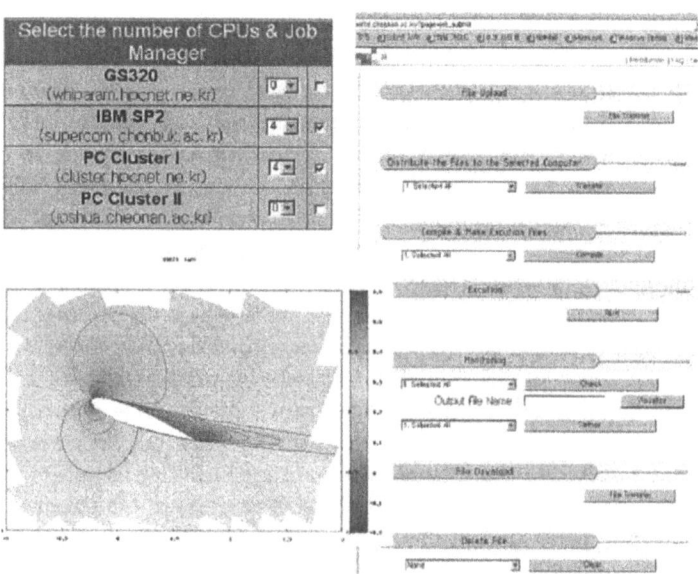

Fig. 4. User interfaces in *GridGate*

GridGate provides three major interfaces: list of available resources, panel for job-submit and visualization. The control panel supplies many buttons. A button "Run" as shown in Fig. 4 invokes the user's job according to the information prescribed in the RSL file that is automatically created by selecting the computing resources. A button "Visualize" opens the visualization interface, on which the status of execution of the solver is monitored by a real-time animation.

The other buttons, "Transfer" and "Compile" transfers the user's job to each resource and compiles the solver, respectively.

3 Experiment

3.1 Testbed for Grid-Computing

In this research, we established a testbed that consists of five parallel systems; Compaq HPC320, Compaq GS320, Linux-Cluster, two IBM SP2 machines. These systems achieve the different utilization from each other depending on the time. Eventually, our *GridGate* allows people to reach out and get the computational resources they need from their own desktop workstations and to monitor their intermediate results from anywhere. In this research, we developed *GridGate* System and used Globus Toolkits[3,4] and several job schedulers depending on the supercomputing resources as shown in Fig. 5. The *GridGate* System is the grid-computing tool that supplies GUI, job-distribution and remote compilation.

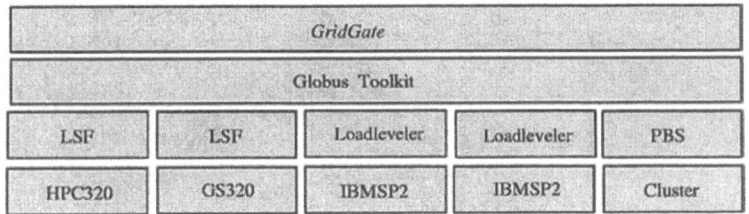

Fig. 5. Hierarchy of middlewares in our testbed

Fig. 6 shows the network configuration of our testbed. The supercomputers of KISTI are connected with 800 MBps HIPPI(High-Performance Parallel Interface) and are linked to the Korean R&D network; Kreonet and HPCnet. The HIPPI makes our grid faster because the latency time of the network is short. Two IBM SP2s are linked to 45 Mbps Kreonet/HPCnet

3.2 Experimental Results

Table 1 shows the number of processors on the experiments using a turbo machine fluid analysis program.

These experiments are conducted according to the number of processors when the size of computation is 28 blocks. The elapsed time is measured when the iteration reaches 100. We are not fully able to use the processors of the supercomputers because of the restriction of the management policy. Although we could obtain the speed-up according to the increase of processors, the communication time commanded the absolute majority.

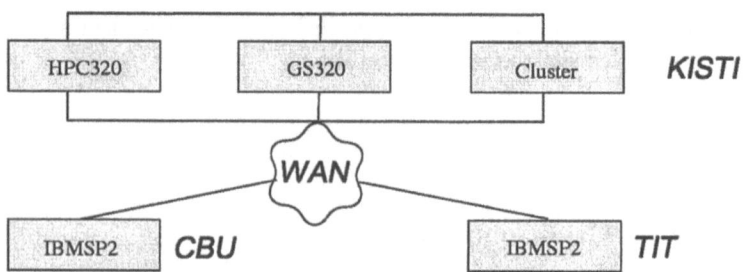

Fig. 6. Network configuration in our testbed

Table 1. The number of processors on the experiment

CPU #	CASE I		CASE II		CASE III			
	HPC320	GS320	IBM SP2	GS320	Cluster	IBM SP2	HPC320	GS320
1	1		1		1			
2	1	1	1	1				
4	2	2	2	2	1	1	1	1
7	3	4	2	5	1	1	2	3
14	4	10	2	12	1	2	4	7

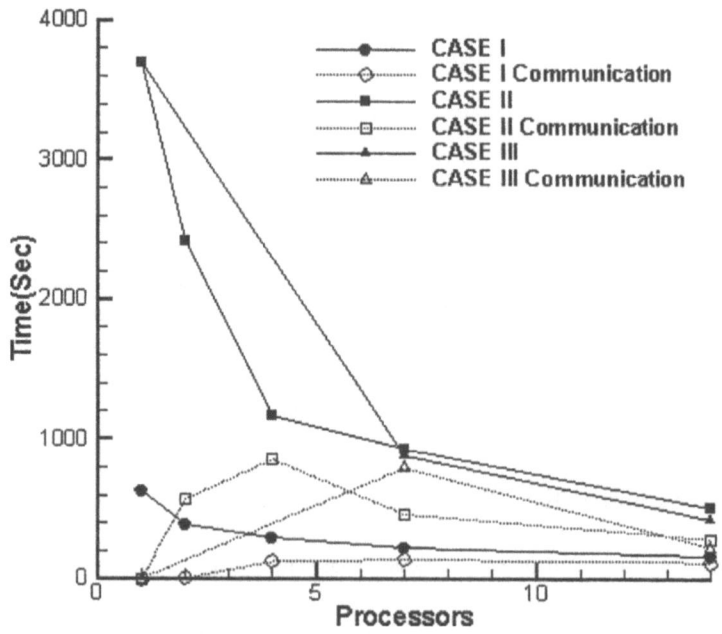

Fig. 7. Performance Analysis

1. CASE I (computation using HPC320, GS320) This experiment is conducted using two supercomputers connected with LAN. The result shows the speed-up in proportion to the number of processors in spite of the communication overhead.
2. CASE II (computation using IBM SP2, GS320) CASE II simulates the same code on the WAN environment between CNU and KISTI. In spite of WAN environment, the increase of processors makes the elapsed time reduced. In fact, we distribute the job on two supercomputers unequally in order to reduce communication data.
3. CASE III (computation using cluster, GS320, IBM SP2, HPC320) This case is similar to CASE II on the WAN environment and the unequal distribution of job.

4   Conclusion and Future Works

In this research, we implemented a grid toolkit named *GridGate* on several supercomputers connected with LAN or WAN. The users from any systems could submit jobs and have them transparently run on the grid environment. This would provide many benefits to the supercomputing centers, including utilization and to the users a convenient simulation environment. A major challenge for this research was providing a uniform software environment across the geographically distributed and diverse computational resources. To meet this challenge, we developed *GridGate* and used Globus Toolkit, which provides a variety of services including co-allocation, security, and parallel programming support.

References

1. V. Sunderam, "pvm: A framework for parallel distributed computing," *Concurrency: Practice and Experience*, vol. 2, December 1990.
2. C. K. I. Foster, "globus: A metacomputing infrastructure toolkit," *Intl. J. Supercomputer Applications*, vol. 11, no. 2, 1997.
3. I. Foster and C. Kesselman, *The Grid: Blueprint for a new Computing Infrastructure*. Morgan Kaufmann Publishers, Inc., 1998.
4. I. F. K. Czajkowski, S. Fitzgerald and C. Kesselman, "Grid information services for distributed resource sharing," in *Proceedings of the Tenth IEEE International Symposium on High-Performance Distributed Computing (HPDC-10)*, August 2001.
5. Bhatia and Burzevski, "webflow-a visual programming paradigm for web/java based corse grain distributed computing," *Concurrency: Practice and Experience*, vol. Java Special Issue, March 1997.
6. K. A. Hoffmann, *Computational Fluid Dynamics for Engineers*. Morgan Kaufmann Publishers, Inc., 1993.
7. "Creating new information providers," *MDS 2.1 GRIS Specification Document*, May 2002.

Performance Evaluation of ENUM Directory Service Design*

Hyewon K. Lee and Youngsong Mun

School of Computing, Soongsil University, Seoul, Korea
kerenlee@sunny.ssu.ac.kr, mun@computing.ssu.ac.kr

Abstract. ENUM (tElephone NUmbering Mapping) is a protocol that brings convergence between PSTN networks and IP networks using a unique worldwide E.164 telephone number. This mechanism couples two completely different environments with E.164 number and allows users to utilize IP based application services in PSTN networks and PSTN based application services in IP networks. We propose a new way to organize and handle ENUM name servers to improve the performance of name resolution process in ENUM based application service. We have built an ENUM based network model when NAPTR (Naming Authority PoinTeR) resource record is registered and managed by area code at the initial registration step.

1 Introduction

ENUM maps telephone numbers into domain name and provides a connection between IP network based services and PSTN based services [1]. This mechanism enables end-users to access web-sites, e-mail, fax, mobile-phone or instant message services with a single telephone number. ENUM protocol promises uniqueness, flexibility and convenience to both IP network and PSTN users; however, there is no sufficient proof that the performance of ENUM service is quite good to employ in Internet instead of using conventional DNS. ENUM clients get only specific NAPTR Resource Records (RRs) from name servers, not all NAPTR RRs, and further, NAPTR RRs joined to one E.164 number are too big to be cached in a local name server, so the availability of cached NAPTR RR should be lower than the availability of cached conventional RR in local name servers. Thus, people would have more a relatively long wait to get an appropriate response with ENUM service. In this paper, we discuss lookup service of ENUM, estimate response time and present some techniques to improve its performance.

2 Related Works

2.1 DNS and Name Lookup Service

IP based applications are indispensable to DNS [2], so the correct operation of DNS plays a prominent part in IP networks. Domain name is preferred to IP address because of its

---

* This work was done as a part of Information & Communication fundamental Technology Research Program supported by Ministry of Information & Communication in republic of Korea.

readability and writability; however, physical hardware cannot understand this, so DNS translates domain name into the logical IP address of device and provides information to most IP based applications. DNS offers the other service, *opposite mapping* from IP address to domain name.

DNS makes extensive use of caching to reduce server load, client latency and ultimately DNS traffic on the Internet. It is generally credited that cache is very efficacious even in changeable and variable IP environment because information in DNS database varies slowly, and small quantities of staleness are tolerable. Simply, the DNS caching design prefers availability above freshness [2][6].

2.2   ENUM Lookup Service

ENUM first transforms E.164 numbers into ENUM domain names and then uses the DNS-based architecture to access records from which URIs are derived. E.164 numbers is used to provide users with a number of addresses, including those used for phone, fax and email, at which the called party can be contacted.

3   Design of ENUM Directory Service

The ENUM delegation model is composed with three tiers: Tier 0 registry, Tier 1 registry, and Tier 2 name server provider. At first, Tier 0 corresponds to the ENUM root level. At this level, the ENUM architecture contains only one domain, called the ENUM root. The Tier 0 name servers contain records that point to ENUM Tier 1 name servers. Secondly, Tier 1 registry is an entity responsible for the whole management of the E.164 country code for ENUM. Name servers of Tier 1 contain records that point authoritative name servers for individual E.164 number or blocks of numbers in the country code. Finally, Tier 2's name servers contain NAPTR RRs with information for specific services. Name servers of Tier 2 responses to queries from other name servers or clients.

Hierarchical design around administratively delegated name spaces and aggressive use of caching are indispensable to the scalability of DNS [2] [6]. Contemporary DNS resolution makes use of caching, and the cache hit rate is up to 90 % depending on the starting state of database, so the response time of DNS query is independent with hop count. On the contrary, the response time of ENUM query is dependent on hop count because a local name server will not cache all of NAPTR RRs with E.164. The ENUM resolution will have excessively low cache hit rate.

The high cache hit rate is closely connected with scalability; so alternatively, we propose to separate Tier 2 name server's role according to local area numbers to improve ENUM performance. If Tier 2 name server represents one or more area codes, each local name server will pre-learn and cache Tier 2 name server's addresses corresponding area codes. More than two Tier 2 name server providers can represent a specific area; and moreover, one Tier 2 name server providers can represent more than one area, but their information should be mutually exclusive except caching.

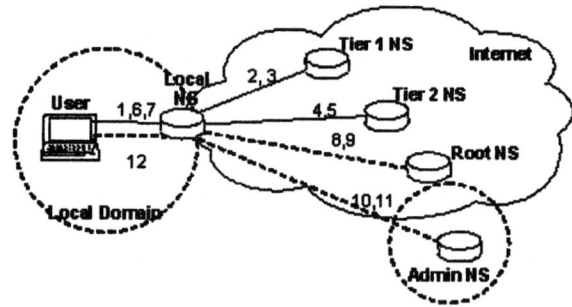

Fig. 1. Schematic topology of traced networks

4 Performance Evaluation

4.1 Modeling

The base assumptions for this experiment are as follows: domain name to query is restricted only to the .kr domain, so any query whose answer is not in the local name server's zone file or cache, is forwarded to the .kr root name server. A query is only generated in the SSU network, and the local name server is ns.ssu.ac.kr. Name server administrator designates the nearest .kr root name server. The administrative name server does not have any sub name server. Any current DNS name server can be the ENUM Tier 2 name server. Name server processing time for ENUM and DNS lookups is assmed to be same.

Once a local name server receives a query about NAPTR RR from a host, it first searches its cache. If the local name server finds appropriate NAPTR RRs for the query, it responds to the host, and the ENUM resolution is over. Otherwise, the local name server forwards the query to the Tier 1 name server, which then returns a referral response, composed of addresses of administrative Tier 2 name servers. The local name server then forwards the query to the Tier 2 name server again. Then, the Tier 2 name server builds response packet containing NAPTR RRs and send it back to the local name server. As the local name server receives a response, it caches and forwards the response to the host. Now, the ENUM client of the host receives NAPTR RR, and starts DNS name lookup process to map NAPTR RR to IP address. DNS lookup process is identical to ENUM lookup, except that the final response from an administrative name server is IP address, not NAPTR RRs. Fig. 1 illustrates the name lookup process. The solid line indicates the ENUM query process, while the dotted line indicates the conventional DNS query process. The expected response time of each step in name lookup process, x_i is as follows, where $i \leq 1$ then $m = i$, else $m = i + 1$:

$$x_i = 2\alpha \sum_{j=1}^{m} (n_j - 1) + m\beta + \gamma(i - 1) + 2 \sum_{j=1}^{m} n_j \delta_j, \tag{1}$$

where $n_i{}^1$ is a hop count in ith trace, α is router processing time, β is DNS name server table searching time, γ is DNS name server table caching time, δ is query propagation delay time between two hops, p_λ is probability related to ENUM resolution, and p_μ is probability related to original DNS resolution. Especially, p_{λ_1} is the probability that local name server has NAPTR RR corresponding to a query, p_{λ_2} is the probability that Tier 1 registry's name server gives address of Tier 2 name server corresponding to a query. Similarly, p_{μ_1} is probability that local name server can response to a query, and p_{μ_2} is the probability that root name server gives the address of administrative name server corresponding to a query.

Assume that x_λ is the response time of the ENUM resolution process, and x_μ is the response time of the DNS resolution process. The expectation of the response time for the whole resolution process using ENUM is the sum of each expectation.

4.2   The Data-Statistical Analysis

To evaluate the response time, the processing speed of root name server, the cache hit rate of local name server, and propagation time between two hops are measured via experiments. For router processing time, simply benchmark from Cisco [4] was used, and the router processing speed will be the reciprocal of throughput.

At first, for name server processing time, 5,000,000 virtual A records were used to get DNS query processing time of .kr root server on July 2003. From Table 1, the processing time of name server is 0.008 ms.

Table 1. Throughput of .kr root name server

DNS program	Zone file data	Query data	Check program
BIND 9.2.0	Virtual $5 * 10^6$ A records	Virtual $5 * 10^6$ A records in zone file	queryperf tool
Queries sent	Queries completed	Started at	Finished at
500,000	500,000	22:09:39	22:10:21

Secondly, an approximately one-day aged name server statistics from Soongsil university was used to get the response rate of the name server. The response rate (or cache hit rate) has a deep connection with how long name the server has been turned on, so we turned off and restarted the name server for the experiment. The response rate of name server will be the total queries responded by the name server to total queries from remote hosts.

Table 2[2] summarizes the traced statistics of ns.ssu.ac.kr. on October 2, 2003, and it shows that response rate of name server is 88.48%, so we assume the value of p_μ, as

---

[1] n_1 is always 1.

[2] Following explanation is extracted from [5]. Some unimportant are excluded, and for detailed information, see [5]. RQ is the count of queries received from relay. RR is the count of responses received from relay. RIQ is the count of inverse queries received from relay. RNXD is the count of no such domain answers received from relay. RFwdQ is the count of queries received from relay that needed further processing before they could be answered. RFwdR is the count of

Table 2. Name server statistics (ns.ssu.ac.kr)

Date	Place		Time	since boot	since reset	
Oct 2 2003	Soongsil U.		(in *sec*)	150269	150269	
RQ	RR	RIQ	RNXD	RFwdQ	RFwdR	RDUPQ
2745590	823370	0	439968	0	710946	22794
RDUPR	RFail	RFErr	RErr	RTCP	RAXFR	RLame
1334	1875	0	237	39866	0	18549
ROPTs	SSysQ	SAns	SFwdQ	SFwdR	SDupQ	SFail
0	436914	2402330	316407	710946	81456	13
SFErr	SErr	RNotNsQ	SNaAns	SNXD		
0	0	2352	131547	797155		

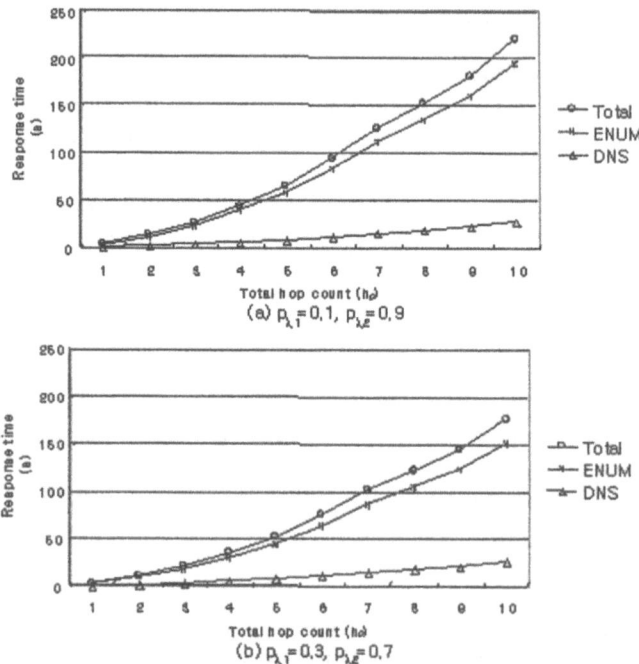

Fig. 2. DNS, ENUM, and total response time

duplicate queries from relay. RDupQ is the count of duplicate queries from relay. RDupR is the count of duplicate responses from relay. RFail is the count of server failure responses from relay. SSysQ is the count of system queries sent to relay. SAns is the count of answers sent to relay. SFwdQ is the count of queries that were sent to relay when the answer was not in the name server's zone or cache. SFwdR is the count of responses from some name server that were sent to relay. SDupQ is the count of duplicate queries sent to relay. SNaAns is the count of nonauthoritative answers sent to relay. SNXD is the count of no such domains answers sent to relay.

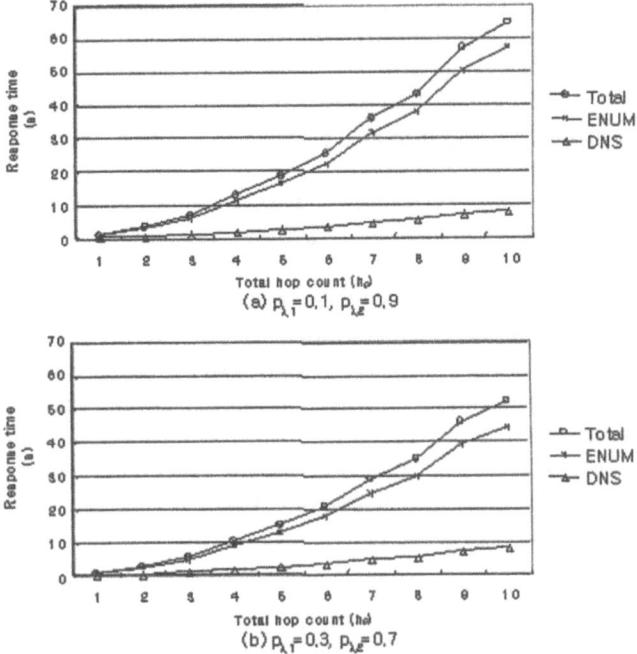

Fig. 3. DNS, ENUM, and total response time when NAPTR RRs are managed by area code

$p_{\mu_1} = 0.88$, and $p_{\mu_2} = 0.12$. In addition to it, the cache hit rate of local name server for ENUM query is supposed to be very low relatively to DNS query, and it will never get ahead of DNS's cache hit rate, because the number and size of NAPTR RR per E.164 number should be more and bigger than RR per IP, so we suspect that $p_{\mu_1} \ll p_{\mu_2}$.

The propagation time between hops is measured by transmitting 60,000 40-byte packets from one host to the first external router in SSU network on September 28, 2003. The average of propagation time, $\bar{X}$ is 43.30 ms, and σ is 48.98ms. With 95 % confidence level, the propagation time ranges from 42.91 to 43.70 ms. The average propagation delay time between two hops is assumed to be 43.2 ms. Propagation delay between hops follows an exponential distribution, and it is expressed as $\sigma_i = -\frac{\lg R}{\lambda}$, where R is random number between 0 and 1, and average propagation delay time, $\frac{1}{\lambda}$ is 43.2 ms.

Let us now return to formula (1). β and γ will be same. Router processing time, α,β and γ are negligible compared with other variables. Formula (1) and (2) can be revised as follows, where $i \leq 1$ then $m = i$, else $m = i + 1$:

$$x_i = 2 \sum_{j=1}^{m} n_j \delta_j. \tag{2}$$

$$E(X) = 2 \sum \sum_{j=1}^{m} (n_{\lambda_j} p_{\lambda_i} + n_{\mu_j} p_{\mu_i}) \delta_j. \tag{3}$$

As Fig. 2 indicate, DNS response time is insensitive to the increment of hop count and has no relevance at all to cache hit rate for ENUM. On the other hand, ENUM response time is very susceptible with increment of hop count and cache hit rate. As hop count increases from $2h_c$ to $5h_c$, ENUM response time at p_{λ_1} rapidly jumps from $8sec$ to $44sec$ approximately. As p_{λ_1} increases from 0.1 to 0.3, ENUM response time at $5h_c$ falls approximately from $57sec$ to $44sec$.

Fig. 2 proves clearly that ENUM lookup time is strongly deteriorated by increasing in the total hop count (h_c), and total response time is decided by ENUM response time, not by DNS response time.

To improve the total response time, it is inevitable to lessen total hop count, especially caused by ENUM lookup. Name server for ENUM is bound with geographical location, so it is difficult to lesson hop count, however, if we run name server to represent each local area as proposed, query would be forwarded to directly the appropriate administrative Tier 2 name server which possess NAPTR RRs corresponding E.164 number. Fig. 3 shows that response time rapidly falls off compared with Fig. 2. Under conditions that $p_{\lambda_1} = 0.3$ and $5h_c$, ENUM response time in two experiments is roughly estimated at $44sec$ and $12sec$ respectively.

5 Conclusions

ENUM provides connectivity to applications based on completely different communication infrastructures, with contemporary DNS. Simply, this mechanism enables PSTN users to communicate with other parties in no matter what environment they belong to and to access to resources on Internet.

In this paper, at first, we have analyzed original DNS query response time with priority given to propagation time delay between two nodes. Secondly, we have estimated ENUM query response time with the same condition of the first analysis. Results from two experiments lead us to the conclusion that total response time is much correlated with ENUM response time. We have proposed representative Tier 2 name server, so queries from local resolvers are seldom. For the most part, queries are directly forwarded to an appropriate administrative Tier 2 name server, and we have seen this idea promises better performance in the ENUM resolution process.

References

1. Faltstrom, P.: E.164 number and DNS. RFC 2916, IETF(2000)
2. Mockapetris, P., Dunlap, K.: Development of the Domain Name System. Computer Communication Review, Vol. 18, No.4, SIGCOMM, ACM(1988) pp123–133
3. Recommendation E.164, The international public telecommunication numbering plan-Supplement 2: Number Portability, ITU-T (1998)
4. Next-Generation Solutions for the service-enabled Internet, third Quarter 2001 PACKET, CISCO (2001)
5. Albitz, P., Liu, C.: DNS and BIND. 4th Ed, pp185–191, O'reilly (2001).
6. Jung, J., Sit, E., Balakrishnan, H., Morris, R.: DNS Performance and the Effectiveness of Caching. Internet Measurement Workshop, SIGCOMM ACM(2001)

A Single Thread Discrete Event Simulation Toolkit for Java: STSimJ

Wenguang Chen, Dingxing Wang, and Weimin Zheng

Tsinghua University, Beijing 100084, China

Abstract. Discrete event simulation is widely used in simulating complex systems. SimJava [6]is a popular java toolkit for discrete event simulation. However, SimJava employs multiple threads for the simulation process. The disadvantage of the multi-threaded approach is that the result of the simulation is not repeatable, because of the uncertainty introduced by multi-threads. In this paper, we propose a single thread discrete event simulation toolkit for Java, whose result is always repeatable.

1 Introduction

Discrete event simulation is one of the key method to validate and evaluate computer and telecom systems.

SIMULA[1] is a language designed for simulation. Hase++[4,3] is a simulation library for C++ which provides discrete process based simulation similar to SIMULA's class and libraries. In order to enable web browser based simulation, Java is also used for simulation. SimJava[6] is a popular java toolkit for discrete event simulation.

However, both Hase++ and SimJava used multiple threads or processes for the simulation process. The disadvantage of the multi-threaded approach is that the result of the simulation is not repeatable, because of the uncertainty introduced by multi-threads.

There're also other single thread simulation package, such as SMPL[5]. But SMPL is written in C difficult to be extended.

In this paper, we propose a single thread discrete event simulation toolkit for Java(STSimJ), which would always produce the same output for the same input.

2 STSimJ Overview

2.1 STSimJ Concepts

In this section, we illustrate some concepts used in STSimJ for system modelling..

- Entity: The concept of entity in STSimJ is similar to the entity in SimJava. A system consists of several entities.

M. Bubak et al. (Eds.): ICCS 2004, LNCS 3038, pp. 1131–1137, 2004.
© Springer-Verlag Berlin Heidelberg 2004

- State: At any given time, each entity must be in a certain state. The entity may have several possible states, and may change from one state to the other because of incoming events or internal logic.
- Transition: An entity may change its state from one to another. This process is called "Transition".
- Event: Events are the main reasons for entities to change their states. Events can be fired by either the entity itself or other entities.

2.2 STSimJ Java Library Overview

In STSimJ, there are mainly 6 classes in its library:

- STSEvent: A class for holding a generic event. Programmers can inherit this class to define their own event classes.
- STSEventList: A helper class to maintain the event queue of each entity.
- STSTransition: A class to record the information related to state transition.
- STSEntityState: It contains a method named eventHandler(), which is supposed to be overridden by real state classes to describe the entities behavior on the state.
- STSEntity: It contains member variables and methods for a generic entity, which include state management, transition management etc. The initialize() method of class STSEntity is supposed to be overridden by real entity classes to describe the events, states and transition information of the entity.
- STSSystem: It controls all entities, maintains and advances the simulation time.

In the next section, we are going to describe the usage of these classes in detail.

3   How to Simulate a System with STSimJ

In this section, we describe how to simulate a system with STSimJ.

3.1   Modelling the System with State Chart

Let's demonstrate the process with a simple example:

A system contains a sender and a receiver. The sender would send a message to the receiver, then waits for the receiver's response. After getting the response, it holds for 10 seconds and send a message to the receiver again. If it has sent 100 messages, it will exit.

The receiver is in idle state initially, after receiving a message from the sender, it holds itself for 1.234 seconds, then send a response message to the sender. If it received 100 messages, it will exit.

It's easy to draw the state chart[2] of the system. The Fig. 1 shows the state chart of the system. There are 2 entities in the system: sender and receiver. Each entity has 3 states: idle, send and receive. And there are 3 kinds of state transition illustrated in Fig. 1:

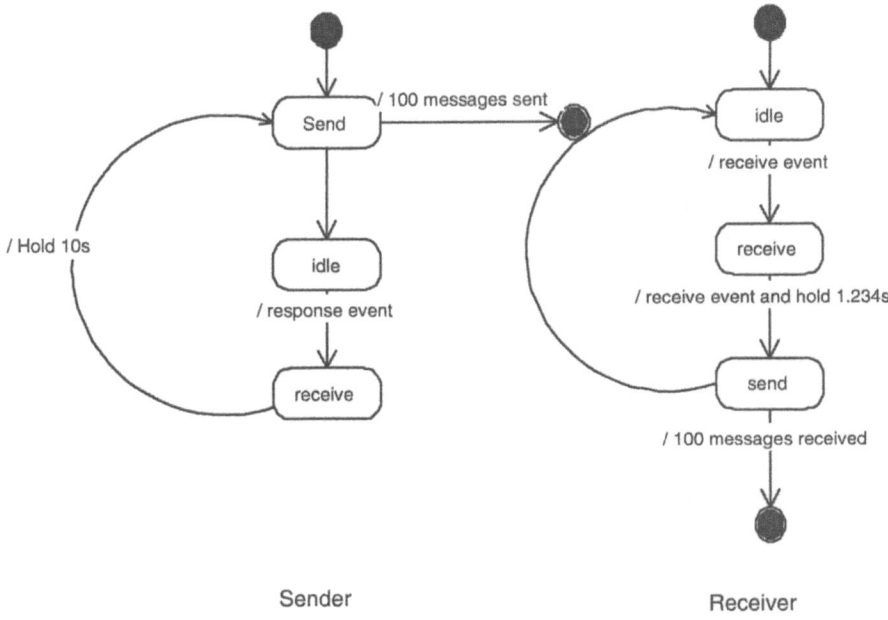

Fig. 1. State chart of the sender/receiver system

- From SENDER_SEND to SENDER_IDLE
  This is a immediate state transition, which does not involve any external events and time delay. The **sender** would enter SEND_IDLE states right after it sends out the message to the **receiver**. We call this kind of transition ITAA(Immediate Transition After Action).
- From SENDER_IDLE to SENDER_RECEIVE
  This is a blocked state transition. The transition would not happen until the **sender** received a message from the **receiver**. We call this kind of transition TBBE(Transition Blocked By Event).
- From SENDER_RECEIVE to SENDER_SEND
  This is a delayed state transition. The entity is held for the specified period and would not process any event. When the period passes, the entity would transit to the target state. We call this kind of transition DT(Delayed Transition).

3.2   Define the Behavior of Entities

Subclassing STSEntity to Define Entity Behavior
In order to define the behavior of a entity, we need to subclass the STSEntity and define states, events and transition by overriding its method initialize(). The definition of class **Sender** is as following:

```
public class Sender extends STSEntity{
  public Sender(String name) {
    super( name );
  }

  public void intialize() {
    //1. Define states
    STSEntityState senderIDLE =
                    new STSEntityState(this, "SENDER_IDLE");
    STSEntityState senderSEND =
                    new STSEntityState(this, "SENDER_SEND");
    STSEntityState senderRECEIVE =
                    new STSEntityState(this, "SENDER_RECEIVE");
    addState(senderIDLE);
    addState(senderSEND);
    addState(senderRECEIVE);

    //2. Add state transition table to entities
    STSEvent messageEvent =
                    new STSEvent(0.0, "MESSAGE", null, null);
    addStateTransition(senderIDLE, senderRECEIVE, messageEvent);
    addHoldStateTransition(senderRECEIVE, senderSEND, 10.0);

    //3. Set initial state
    setInitialState(senderSEND);
  }
}
```

The state definition is a simple step and does not need to be explained more. However, the definition of events and state transition is a little more tricky.

The class STSEntity have two methods to describe the state transition:

- addStateTransition()
  This method is used to specify the transition of TBBE type. In this sample, the statement addStateTransition(senderIDLE,senderRECEIVE, messageEvent) describes this transition:if the current state is senderIDLE, then it will wait for a MESSAGE event and change the entity state to senderRECEIVE.
- addHoldStateTransition()
  This method is used to specify the transition of DT type. In this sample, the statement addHoldStateTransition(senderRECEIVE,senderSEND, 10.0) describes this transition:if the current state is senderRECEIVE, then hold 10.0 seconds and enter the state senderSEND.

For the state transition type ITAA, we did not specify it here. Instead, it is specified in the eventHanlder() of state SENDER_SEND. See the next section.

Subclassing STSEntityState to Define State Event Handler

If the entity should perform some operations other than jump to another state or waiting for a event in a state, the operations should be specified in the simulation code. In this case, we should subclass STSEntityState and override its method eventHandler(). Besides, the transition of type ITAA should also be specified in the eventHandler() method.

```java
public class StateSenderSend extends STSEntityState {
  int msgCnt = 0;
  public StateSenderSend(STSEntity ofEntity, String stateName) {
    super(ofEntity, stateName);
  }
  public void eventHandler( ISTSEvent event) {
    if ( msgCnt ++ >= 100 ) {
      entity.setCurrentState("EXIT", null);
      return;
    }
    STSEvent evt = new STSEvent(STSSystem.getGlobalTime(),
                          "MESSAGE", entity.getName(), "hello" );
    STSSystem.sendEvent( "receiver", 0.5, evt );
    entity.setCurrentState("SENDER_IDLE", null);
  }
}
```

In this example, the operation performed is to send a MESSAGE event to the entity `receiver` by using the method STSSystem.sendEvent() . And the entity sender enter the state SENDER_IDLE right after the sendEvent() by calling STSEntity's method setCurrentState().

3.3   Put It All Together to Simulate!

Now we have defined the entities and their behavior. It's time to get them together:

```java
public class SampleSimulation {
  public static void main( String[] args ) {
    STSSystem.initialize();

    // Add sender entity into system
    STSEntity sender = new STSEntity( "sender" );
    STSEntity receiver = new STSEntity( "receiver" );
    STSSystem.add(sender);
    STSSystem.add(receiver);

    // Start to simulate
    STSSystem.run();
  }
}
```

Simply create and add them to the class STSSystem, it is ready to go. The
STSSystem.run() is a static method of the class STSSystem that would start
the simulation.

4   The Single Thread Simulation Kernel

The main feature of STSimJ is that it just use a single thread simulation kernel. All the situations that would require blocking, i.e. waiting for a event to
arrive and holding the system for a period etc., are defined as state transition in
STSimJ, and thus there's no need to block any simulation code in STSimJ. The
simulation kernel is like following:

```
public class STSSsytem {
  ...
  public static void run() {
    ...
    // Simulate
    boolean hasEvents = true;
    globalTime = advanceToNextEventTime();
    while ( hasEvents && globalTime < terminateTime ) {
      hasEvents = false;
      for (int i = 0; i < entityList.size(); i++) {
        STSEntity entity = (STSEntity) entityList.elementAt(i);
        hasEvents |= entity.run();
      }
    }
  }
}
```

The method advanceToNextEventTime() find the next event time from the
current event lists of all entities. globalTime is the global clock that can be
read by all entities. The STSEntity::run() method would handle the event
occurred at the time and/or transit states.

It should be noted that all entity.run()s are executed one by one and
there's no chance for them to execute simultaneously.

5   Conclusion and Future Work

In this paper, we proposed a single thread discrete event simulation toolkit for
Java:STSimJ. To use STSimJ, one can use the state chart to describe the system
to be simulated, then map the state chart to STSimJ code. We have illustrated
that with the frameworks and libraries provided by STSimJ, it is quite easy to
construct the simulation code.

In STSimJ, there's no blocked thread in the simulation code. All simulation
is done with synchronous call to entity method and event handler method. So

the whole simulation is completed within the single thread. Thus, it avoids the uncertainty of multi-thread simulation and is always repeatable.

In the future, we will integrate more features into STSimJ, which include statistics library and GUI libraries.

References

1. Birtwhistle, G.M., Dahl, O-J. Myhrhaug B. and Nygaard K.,Simula Begin, Academic Press, 1973
2. Fowler M. and Scott K., UML Distilled: A Brief Guide to the Standard Object Modeling Language(2^{nd}ed), Addison Wesley Longman, 2000
3. Howell, F.W. Hase++:a discrete event simulation library for C++.
   http://www.dcs.ed.ac.uk/home/fwh/hase++/hase++.html
4. Ibbett, R.N.,Heywood,P.E. and Howell,F.W., "Hase : a flexible toolset for computer architects." The Computer Journal, 2000
5. MacDougall M.H. Simulating computer systems:techniques and tools. The MIT Press, 1987
6. McNab, R. and Howell, F.W. Using Java for Discrete Event Simulation, Proceedings of Twelfth UK Computer and Telecommunications Performance Engineering Workshop (UKPEW),Edinburgh, 219-228,1996 38(10):755-764.

Routing and Wavelength Assignments in Optical WDM Networks with Maximum Quantity of Edge Disjoint Paths*

Hyunseung Choo[1] and Vladimir V. Shakhov[2]

[1] School of Information and Communication Engineering
Sungkyunkwan University, KOREA, choo@ece.skku.ac.kr
[2] Institute of Computational Mathematics and Mathematical Geophysics
Russian Academy of Sciences (Siberian Branch), shakhov@rav.sscc.ru

Abstract. In the present paper routing and wavelength assignment (RWA) in optical WDM networks is discussed. Previous techniques on based on the integer linear programming and graph coloring are complex and require extensive use of heuristics which makes them slow and sometimes practically not reasonable. Another approach employs the greedy algorithm for obtaining available edge disjoint paths. Even though it is fast, it produces a solution for any connection request which is far from the optimal utilization of wavelengths. We propose a novel algorithm which is based on the maximum flow to have the maximum quantity of edge disjoint paths. Comprehensive computer simulation shows that the proposed method outperforms previous ones significantly in terms of running time. Furthermore, it shows compatible or better performance comparing to others in number of wavelengths used.

1 Introduction

Among various research areas in optical networks, efficient source-destination routing and assigning a wavelength on the route selected are quite important for the network performance [1]. Their objectives include the minimization of the required number of wavelengths [2,3,4], the minimization of the blocking probability under fixed number of wavelengths [5], and the minimization of the network costs [6].

Routing and wavelength assignment (RWA) problem for the lightpaths of virtual topology in any WDM network is **NP** complete and many heuristic approaches [2,3,4,5,6] have been proposed in the literature. It has been studied for various criteria. One criterion of them is how to set up and release optical paths. Here we assume the static path assignment that optical paths are not released after once established. In static wavelength assignment the lightpaths and their routes are known in advance and we need to assign wavelengths to each lightpath such that the lightpath on a given link must have different wavelengths.

---

* This paper was supported in part by Brain Korea 21 and University ITRC project.

M. Bubak et al. (Eds.): ICCS 2004, LNCS 3038, pp. 1138–1145, 2004.
© Springer-Verlag Berlin Heidelberg 2004

That is practically more achievable with the current technology than the dynamic path assignment, where optical paths are randomly set up and released.

Another criterion is the existence of wavelength converters. Physical constraints in the RWA problem include wavelength channel spacing in a fiber link, the number of optical transceivers, and the number of wavelength converters. Actually they determine the number of wavelengths available on fiber links and thus the number of source-to-destination connections. Each wavelength is assumed to be assigned to a single channel and the bandwidth granularity of the channel is usually not considered for the simplicity of the problem. This paper investigates the case that no wavelength converters exist in the topology based on the static path assignment. It is predictable that the proposed scheme can easily apply to the case with wavelength converters. And the goal of this work is either to reduce the number of used wavelengths or to maintain the similar number with shorter running time. This can be achieved based on the maximum number of edge disjoint paths. The comprehensive computer simulation shows that the proposed scheme significantly outperforms other schemes in terms of the running time.

Basically there are two approaches for solving the RWA problem. The first approach is based on the *lpsolver* [7], which is a program to solve a linear program using the idea of multicommodity flows, to obtain the optimal solution for RWA. And the second one uses path selection based graph algorithms to minimize the number of wavelengths used heuristically. We briefly mention three popular RWA algorithms in the literature.

Lpsolver: Routing Heuristic Algorithm with Graph Coloring - Rounding heuristic algorithm based on lpsolver is used in lightpath routing and graph coloring is used in wavelength assignment phase [2]. *Graph Algorithm: BGA for EDP* - A simple edge disjoint paths scheme based on shortest path algorithm to identify available paths is used in [3]. Here authors investigate the performance of rounding heuristic algorithm and their scheme in terms of the number of wavelengths used and the actual running time. The result shows that the number of wavelengths used in two schemes is almost same, however, rounding heuristic algorithm consistently takes more time than that of *BGA for EDP*. Therefore we just compare the proposed scheme with *BGA for EDP* in our work. *Graph Algorithm: RWA with Lookup Table - LTB_RWA* is based on disjoint path set selection protocol (DPSP) [4,8]. Here authors investigate the performance of *BGA for EDP* and their scheme in terms of the number of wavelengths used and the running time. The simulation result shows that the number of wavelengths used in *BGA for EDP* is almost same. However, *LTB_RWA* yields better solutions for the running time.

2 The Proposed RWA Scheme

Let $G(V, E)$ denote a simple graph that models a network, where V is the vertex set and E is the edge set. Let $|V| = n$, i.e. a network contains n hosts. The demand set D is a set of pairs $\{(s_1, t_1), \ldots, (s_k, t_k)\}$, where $s_i, t_i \in V, i = 1, \ldots, k$. Here k is the total number of requests for connections, s_i is a host-sender, and t_i

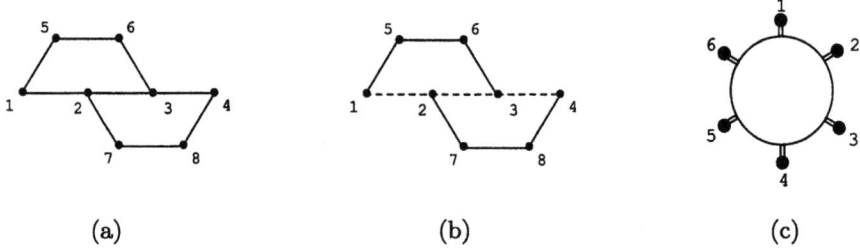

Fig. 1. The sample topologies for RWA

is a host-receiver. For the request (s_i, t_i) it is necessary to assign a wavelength λ_i and a lightpath $L_i \subseteq E$ (which means that edges in the lightpath are elements in E). Let us remark the lightpaths that are assigned the same wavelength *simultaneously* will not traverse the same physical link, i.e. these paths are edge disjoint. Let $l = max_i \lambda_i$. The aim of RWA problem is to minimize l under the condition $L_i \cap L_j = \emptyset, \forall i, j \in \{1, \ldots, k\}$. If the condition is false, then we say the assignment is not admissible.

If $s_1 = s_2 = \ldots = s_k = s$ and $t_1 = t_2 = \ldots t_k = t$, then the optimal solution for the problem is defined as k over the maximal number of edge disjoint paths between s and t, i.e. k divided by the *cardinality* of the *edge cut set* between s and t, $Card(s, t)$. i.e., *Optimal number of wavelengths* $= \lceil \frac{k}{Card(s,t)} \rceil$

Let us assume an optical network (refer to Fig. 1 (a)). Shortest path is assigned to the connection request. Let the demand set be $\{(1,4),(4,1),(2,3)\}$. First random selected request is (1,4). The shortest path between node 1 and 4 is $1 \rightarrow 2 \rightarrow 3 \rightarrow 4$. Hence the request (1,4) receives wavelength $\lambda = 1$ and uses path $1 \rightarrow 2 \rightarrow 3 \rightarrow 4$. We remove this path from graph and get unconnected graph. Paths for other connections do not exist (see Fig. 1 (b)). Now we return to the original graph and increase the number of wavelength, i.e. $\lambda = 2$. Let a second random selected request be (4,1). Shortest path is $4 \rightarrow 3 \rightarrow 2 \rightarrow 1$. Hence connection (4,1) uses this path with wavelength 2. We have no path for the demand (2,3). So we use wavelength 3 for this request. As a result, we use different wavelength for each connection, i.e. $max \lambda = 3$. However, if connection (1,4) uses path $1 \rightarrow 5 \rightarrow 6 \rightarrow 3 \rightarrow 4$, (4,1) uses $4 \rightarrow 8 \rightarrow 7 \rightarrow 2 \rightarrow 1$, (2,3) use $2 \rightarrow 3$, then $max \lambda = 1$.

As we all know, ring topology is one of important network architectures. Let a network be as in Fig. 1 (c). Let a demand set be $\{(1, 2), (6, 2)\}$. The graph diameter is 3. Hence the second request (6,2) cannot use a path $6 \rightarrow 5 \rightarrow 4 \rightarrow 3 \rightarrow 2$ because the length of this path is more than the diameter. The resulting assignment by the method in [3] is as follows. The request (1,2) uses path $1 \rightarrow 2$ and wavelength $\lambda = 1$. The request (6,2) uses path $6 \rightarrow 1 \rightarrow 2$ and wavelength $\lambda = 2$. However, only one wavelength can be enough if all edge disjoint paths are used.

Fig. 2 shows a strategic comparison between the conventional approach and the proposed one. The conventional approach is done by selecting routes for the set of connection requests. This will be repeated for the subsequent sets of requests. However the proposed approach finds maximum number of EDPs for all possible connections and stores them in a lookup table for lightpath assignments.

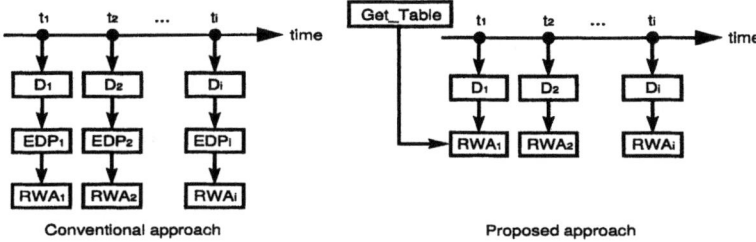

Fig. 2. Comparison of strategies between two approaches

To solve the RWA problem based on the table, just refer to the table without reselecting a route for the same connection request and assign a wavelength to the route as far as the network topology does not change. The optical network is a technology that is employed in core backbone networks. In general they are not changed frequently with fixed connection requests. Hence the proposed algorithm can be efficient in terms of time. If the same connection request occurs concurrently the probability that the proposed algorithm minimizes the number of wavelengths comparing to others is relatively high. In the simulation for the number of wavelength used, we generate random graphs and random demand sets. First we obtain a random graph for the physical topology by specifying the number of nodes in the graph and the probability of an edge existence (P_e) for any node pair. The demand set is also generated randomly by populating the matrix with the number of same requests, which is less than or equal to N_c, where the probability of the request is specified by P_l. After obtaining the table and we assign a path for a request of the demand set in order. Table 1 summarizes the notation used in this section.

Table 1. Notation

Variables	Meaning
G	The directed graph that models the network
V	The set of vertices
E	The set of edges
d	$\max(\text{diam}(G), \sqrt{|E|})$
τ	The demand set, $\tau = \{(s_1, t_1), \ldots, (s_k, t_k)\}$
P_l	The probability of request
P_e	The probability of edge existence between any node pairs
N_c	The multiplicity of a single request

In order to find a set of edge disjoint paths for effective wavelength assignment we solve the maximal flow problem from s to t in G, where any edge is undirected and all capacities are equal to one. Generally (s_i, t_i) and (s_j, t_j) for arbitrary i and j $(i \neq j)$ are different. Here we can get an optimal solution for RWA for the traffic between any couple of nodes as above. If the union of these solutions is acceptable, then the optimal common solution is found. Otherwise, some wavelengths assigned to edge common paths should be replaced by some other wavelength assignments. We do not necessarily get the optimal solution by this approach, furthermore the assignment is not trivial as in examples earlier, when $\lambda_i \neq \lambda_j, \forall i, j$.

As a network topology is usually fixed we offer to calculate a set of edge disjoint paths for all possible pairs in advance before investigation of the demand set. For this purpose the procedure **Get_Table**(G) has been implemented and used in the our method. By $P(s, t, i)$ denote the result of **Get_Table**(G), here $s = 1, 2 \ldots, n - 1$; $t = s + 1, s + 2, \ldots, n$; $i = 1, 2, \ldots, Card(s, t)$. Entire edge disjoint paths and their maximal quantity between nodes s and t can be found from the solution of maximal unit flow in G from s to t. Here the unit flow corresponds to one path. As we know, a straightforward implementation of the maximal unit flow algorithm runs in time $O(|E| \cdot Card(s, t))$.

The network topology can be altered due to link-host failures and network expansions. In this situation, we reconfigure the set of edge disjoint paths $P(s, t, i)$ and continue the RWA process. Now we provide the description of the method.

Algorithm **Max_EDP_RWA**
Get_Table(G)
while (the same G) **do**
   retrieve a new D
   $\lambda = 0$
   while $(D \neq \emptyset)$ **do**
     $\lambda = \lambda + 1$
     for $j = 1$ **to** k **do**
       if $\exists$ empty path $P(s_j, t_j, i)$ **then**
         Assign $P(s_j, t_j, i)$ and λ to (s_j, t_j)
         $D = D - (s_j, t_j)$
       end if
     end for
   end while
end while.

Usually network problems can be classified into two parts. The first part takes a serious view of path, and the other for taking a serious view of flow. Representative instance of the former is shortest path problem and the latter is source and destination pair. The mechanism that searches the maximum flow is based on the famous Ford-Fulkerson algorithm. The two sets of elements divided by an edge cut set are S and T, edge cut set is $e_1, \ldots, e_{Card(s,t)}$. **Get_Table**$(G)$ obtains maximum possible edge disjoint paths for each (s, t) pair based on the mechanism discussed above.

In Fig. 3 shows the graph of a network topology. And demand set is $\{(1,8), (4,6), (1,8), (6,9)\}$. So the reasonable value d is 4.1. First we make the Table through Get_Table(G). For each demand pair, we obtain the maximum number of EDPs. The demand pair (1,8) which is requested at first and third has

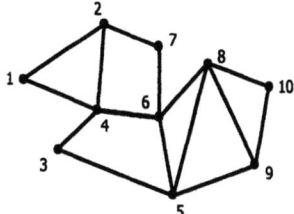

Fig. 3. Random Network Topology

$1\rightarrow2\rightarrow7\rightarrow6\rightarrow8$ and $1\rightarrow4\rightarrow3\rightarrow5\rightarrow8$. The second pair (4,6) has $4\rightarrow6$, $4\rightarrow2\rightarrow7\rightarrow6$, and $4\rightarrow3\rightarrow5\rightarrow6$. And the last demand (6,9) has $6\rightarrow8\rightarrow9$ and $6\rightarrow5\rightarrow9$. We select one at random $1\rightarrow2\rightarrow7\rightarrow6\rightarrow8$. For the second pair (4,6), assign the shortest EDP in the Table $4\rightarrow6$. And for the third one (1,8), we take the path $1\rightarrow4\rightarrow3\rightarrow5\rightarrow8$ because the other one is used for the first demand (1,8). At the last demand (6,9), assign the path $6\rightarrow5\rightarrow9$ since edge (6,8) is used for the first demand. The total number of wavelengths used is 1. We need 2 wavelengths for *BGA for EDP* based algorithm in this topology, i.e. $1\rightarrow4\rightarrow6\rightarrow8$ and $4\rightarrow3\rightarrow5\rightarrow6$ for wavelength 1, and $1\rightarrow4\rightarrow6\rightarrow8$ and $6\rightarrow5\rightarrow9$ for wavelength 2.

3   Performance Evaluation

We now describe some numerical results with which we compare the performance of the proposed scheme, the *LTB_RWA* algorithm [4], and the *BGA for
EDP* based *Greedy_EDP_RWA* described in [3]. We compare the number of wavelengths used and the running time. We use NSFNET for the running time as
done earlier and random graphs for the number of wavelengths used to test them
in various network topologies. The demand set is also generated randomly by
populating the $n \times n$ matrix with the number of requests, which is less than or
equal to N_c, where the probability of the request is specified by P_l. We simulate
1,000 times for each n, P_e, and P_l, and obtain the average number of wavelengths
used. Table 2 shows the number of wavelengths used for each combination of the
graph G and the demand set. Note that the proposed scheme (denoted as 'New'
in the Table 2) consistently requires the number of wavelengths which is quite
similar to those of *BGA for EDP* based *Greedy_EDP_RWA* (denoted as 'BGA')
and *LTB_RWA* (denoted as 'LTB').

We measure the running time for three algorithms on NSFNET with the
demand set as obtained earlier. We generate 100 to 1,000 demand sets separately
and take the average running time. Fig. 4 shows the average running time when
$P_l = 0.6$ and $N_c = 5$, and other P_l values present the similar trends. As a
result, the proposed scheme uses almost same number of wavelengths when the
multiplicity of a pair is close to 1, meanwhile it is better than other schemes
when the multiplicity becomes higher. Furthermore, the proposed scheme takes
significantly less running time than others.

Table 2. Performance comparison: the number of wavelengths

n	P_e	P_l	$N_c = 1$			$N_c = 3$			$N_c = 5$		
			BGA	LTB	New	BGA	LTB	New	BGA	LTB	New
10	0.6	0.4	2.02	2.09	2.1	5.44	5.38	5.39	8.81	9.04	8.63
10	0.6	0.6	2.69	2.78	2.75	8.1	7.66	7.86	13.22	13.27	12.74
10	0.6	0.8	3.49	3.54	3.53	10.55	10.34	10.2	17.34	17.69	16.7
10	0.8	0.4	1.25	1.33	1.34	3.6	3.56	3.57	5.88	5.86	5.74
10	0.8	0.6	1.92	1.94	1.94	5.24	5.11	5.11	8.6	8.68	8.26
10	0.8	0.8	2.15	2.17	2.18	6.8	6.68	6.6	11.44	11.52	10.93
14	0.6	0.4	1.99	2.06	2.05	5.36	5.23	5.2	8.95	8.99	8.58
14	0.6	0.6	2.74	2.79	2.8	8.03	7.53	7.68	13.19	13.18	12.5
14	0.6	0.8	3.44	3.42	3.46	10.37	10.05	9.88	17.34	17.6	16.21
14	0.8	0.4	1.22	1.31	1.34	3.68	3.64	3.6	6.05	6.0	5.82
14	0.8	0.6	1.97	1.99	1.98	5.33	5.21	5.19	8.95	8.96	8.55
14	0.8	0.8	2.2	2.23	2.27	7.1	6.84	6.79	11.87	11.89	11.16
18	0.6	0.4	2.06	2.03	2.06	5.45	5.21	5.33	8.9	8.82	8.42
18	0.6	0.6	2.67	2.84	2.73	7.86	7.53	7.5	13.08	13.09	12.24
18	0.6	0.8	3.38	3.0	3.38	10.48	9.85	9.81	17.88	17.57	16.61
18	0.8	0.4	1.15	1.21	1.26	3.73	3.74	3.67	6.09	6.16	5.86
18	0.8	0.6	1.97	2.0	1.98	5.47	5.27	5.26	9.12	9.13	8.58
18	0.8	0.8	2.28	2.24	2.35	7.27	6.98	6.91	12.2	12.17	11.4

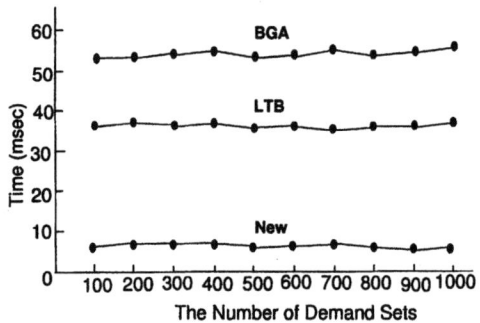

Fig. 4. The average running time (NSFNET, $P_l = 0.6$, and $N_c = 5$)

4   Conclusion

In this paper we present an RWA method in optical networks to minimize
the number of wavelengths needed with shorter running time than others. It
is studied in the static path assignment and without wavelength converters. The
method obtains the maximum edge disjoint paths for each and every possible
source-destination pair, and stores them in a table in advance. And for each demand pair of the demand set, we assign any available path from the table. The
table obtained is again directly used for other demand sets unless the network

topology changes. The comprehensive computer simulation results show that our scheme uses almost same number of wavelengths when the multiplicity of a pair is close to 1, meanwhile it is much better than others when the multiplicity becomes higher. Furthermore, the proposed scheme takes significantly less running time than others.

References

1. Zang, H., Jue, J.P., Mukherjee, B.: A review of routing and wavelength assignment approaches for wavelength-routed optical WDM networks. Optical Networks Magazine, Vol. 1. January (2000) 47-60
2. Banerjee, D., Mukherjee, B.: Practiacal Approach for Routing and Wavelength Assignment in Large Wavelength Routed Optical Networks. IEEE Jounal of Selection Areas Communication, Vol. 14 (1996) 903-908
3. Manohar, P., Manjunath, D., Shevgaonkar, R.K.: Routing and Wavelength Assignment in Optical Network From Edge Disjoint Path Algorithms. IEEE Communications Letters, Vol. 6. May (2002) 211-213
4. Kim, M.H., Choo, H., Yae, B.H., Lee, J.H.: A Practical RWA Algorithm Based on Lookup Table for Edge Disjoint Paths. Lecture Notes in Computer Science, Vol. 2713. (2003) 138-147
5. Swaminathan, M.D., Sivarajan, K.N.: Practical Routing and Wavelength Assignment Algorithms for All Optical Networks with Limited Wavelength Conversion. IEEE International Conference on Communications, Vol. 5 (2002) 2750-2755
6. Chlamtac, I., Farago, A., Zhang, T.: Lightpath (wavelength) routing in large WDM networks. IEEE IEEE Jounal of Selection Areas Communication, Vol. 14. June (1996) 909-913
7. http://only1.snu.ac.kr/software/lpsolver/
8. Papadimitratos, P., Haas, Z.J., Sirer, E.G.: Path Set Selection in Mobile Ad Hoc Networks. MobiHoc (2002)

Parallelism for Nested Loops with Non-uniform and Flow Dependences

Sam-Jin Jeong

Dept. of Information & Communication Engineering, Cheonan University, 115, Anseo-dong, Cheonan, Chungnam, 330-180, Korea.
sjjeong@cheonan.ac.kr

Abstract. Many methods are proposed in order to parallelize loops with non-uniform dependence, but most of such approaches perform poorly due to irregular and complex dependence constraints. This paper proposes an efficient method of tiling and transforming nested loops with non-uniform and flow dependences for maximizing parallelism. Our approach is based on the Convex Hull theory that has adequate information to handle non-uniform dependences, and also based on minimum dependence distance tiling, the unique set oriented partitioning, and three region partitioning methods. We will first show how to find the incrementing minimum dependence distance. Next, we will propose how to tile the iteration space efficiently according to the incrementing minimum dependence distance. Finally, we will show how to achieve more parallelism by loop interchanging and how to transform it into parallel loops. Comparison with some other methods shows more parallelism than other existing methods.

1 Introduction

Parallel processing is recognized as an important vehicle for the solution of many areas of computer applications. Most of the computing time is spent in loops in such applications. The existing parallelizing compilers can parallelize most of the loops with uniform dependences, but they do not satisfactorily handle loops with non-uniform dependences. Most of the time, the compiler leaves such loops running sequentially. Unfortunately, loops with non-uniform dependences are not so uncommon in the real world.

Several works have been done for loops with non-uniform dependences. All of the existing techniques do a good job for some particular types of loops, but show us a poor performance on some other types of loops.

Some techniques, based on Convex Hull theory [7] that has been proven to have enough information to handle non-uniform dependences, are the minimum dependence distance tiling method [5], [6], the unique set oriented partitioning method [4], and the three region partitioning [1], [3].

This paper will focus on parallelizing perfectly nested loops with non-uniform and flow dependences.

The rest of this paper is organized as follows. Section two describes our loop model, and reviews some fundamental concepts in non-uniform and flow dependence

M. Bubak et al. (Eds.): ICCS 2004, LNCS 3038, pp. 1146–1152, 2004.
© Springer-Verlag Berlin Heidelberg 2004

loop. Section three presents an improved tiling method for parallelization with nested loops with non-uniform and flow dependences. In this section, we show how to find the incrementing minimum dependence distances in the iteration space. Then, we discuss how to tile the iteration space efficiently according to the incrementing minimum dependence distance and how to achieve more parallelism by loop interchanging. Section four shows comparison with related works. Finally, we conclude in section five with the direction to enhance this work.

2 Data Dependence Analysis in Non-uniform and Flow Dependence Loop

The loop model considered in this paper is doubly nested loops with linearly coupled subscripts and both lower and upper bounds for loop variables should be known at compile time. The loop model has the form in Fig. 1.

$$
\begin{aligned}
&\text{do } i = l_1, u_1 \\
&\quad \text{do } j = l_2, u_2 \\
&\qquad A(a_{11}i + b_{11}j + c_{11},\ a_{12}i + b_{12}j + c_{12}) = \dots \\
&\qquad \dots = A(a_{21}i + b_{21}j + c_{21},\ a_{22}i + b_{22}j + c_{22}) \\
&\quad \text{enddo} \\
&\text{enddo}
\end{aligned}
$$

Fig. 1. A doubly nested loop model

The dependence distance function $d(i_1, j_1)$ in flow dependence loops gives the dependence distances $d_i(i_1, j_1)$ and $d_j(i_1, j_1)$ in dimensions i and j, respectively. For uniform dependence vector sets these distances are constant. But, for the non-uniform dependence sets these distances are linear functions of the loop indices. We can write these dependence distance functions in a general form as

$$
d(i_1, j_1) = (d_i(i_1, j_1),\ d_j(i_1, j_1))
$$
$$
d_i(i_1, j_1) = p_1 * i_1 + q_1 * j_1 + r_1
$$
$$
d_j(i_1, j_1) = p_2 * i_1 + q_2 * j_1 + r_2
$$

where p_i, q_i, and r_i are real values and i_1 and j_1 are integer variables of the iteration space.

The properties and theorems for tiling of nested loops with flow dependence can be described as follows.

Theorem 1. *If there is only flow dependence in the loop, DCH1 contains flow dependence tails and DCH2 contains flow dependence heads.*

Theorem 2. *If there is only flow dependence in the loop, then $d_i(x, y) = 0$ or $d_j(x, y) = 0$ does not pass through any DCH.*

If there exists only flow dependence in the loop, then $d_i(x_1, y_1) = 0$ or $d_j(x_1, y_1) = 0$ does not pass through any IDCH(Integer Dependence Convex Hull) because the IDCH is a subspace of DCH(Dependence Convex Hull) [5].

Theorem 3. *If there is only flow dependence in the loop, the minimum and maximum values of the dependence distance function $d(x_l, y_l)$ appear on the extreme points.*

Theorem 4. *If there is only flow dependence in the loop, the minimum dependence distance value d_{imin} is equal or greater than zero.*

From theorem 4, we know that when there is only flow dependence in the loop and d_{imin} is zero, d_{jmin} is greater than zero. In this case, since $d_j(x_l, y_l) = 0$ does not pass through the IDCH, the minimum value of $d_j(x_l, y_l)$, d_{jmin}, occurs at one of the extreme points.

Theorem 5 *If there is only flow dependence in the loop, the difference between the distance of a dependence and that of the next dependence, d_{inc}, is equal to or greater than zero.*

Thus, d_{inc} is equal to to or greater than zero when there is only flow dependence in the loop.

3   Improved Tiling Method

Cho and Lee [2] present a more general and powerful loop splitting method to enhance all parallelism on a single loop. The method uses more information from the loop such as increment factors, and the difference between the distance of dependence, and that of the next dependence. Cho and Lee [3] derive an efficient method for nested loops with simple scripts from enhancing [2].

The minimum dependence distance tiling method [6] presents an algorithm to convert the extreme points with real coordinates to the extreme points with integer coordinates. The method obtains an IDCH from a DCH. It can compute d_{imin}, the minimum value of the dependence distance function $d_i(i_l, j_l)$ and d_{jmin}, the minimum value of the dependence distance function $d_j(i_l, j_l)$ from the extreme points of the IDCH. The first minimum dependence distances d_{imin} and d_{jmin} are used to determine the uniform tile size in the iteration space.

3.1   Tiling Method by the Incrementing Minimum Dependence Distance

From theorem 5, when $p_1 > 0$ and $q_1 \geq 0$, we know that the difference between the distance of a dependence and that of the next dependence in loop with flow dependence, d_{inc}, is equal to or greater than zero.

For each i_l, d_{imin} is incremented as the value of i_l is incremented. So, the second d_{imin} is equal to or greater than the first one, and the third one is greater than the second one, and so on.

The improved tiling method for doubly nested loops with non-uniform and flow dependence is described as **Procedure Tiling_Method**, which is the algorithm of tiling loop by the incrementing minimum dependence distance as shown in Fig. 2.

This algorithm computes the incrementing minimum dependence distance, tiles the iteration space efficiently according to the incrementing minimum dependence distance, and transforms it into parallel loops.

Procedure Tiling_Method $(i_1, j_1, l_1, l_2, u_1, u_2, d_i(i_1, j_1))$

 i_1, j_1: i and j value for the source of the first minimum dependence in the loop computed by the extreme points of the IDCH

 l_1, l_2, u_1, u_2: the lower and upper bounds of outer loop and inner loop, respectively

 $d_i(i_1, j_1)$: the dependence distance function of the IDCH

begin

 Step 1: when the first source point, (i_1, j_1), is given, the first minimum dependence distance d_{imin} and first tile size are computed.

 Step 2: Next d_{imin} is computed. If (next sink point is greater than bound), Goto Step 4.

 Step 3: Next tile size is computed, and Goto Step 2.

 Step 4: the original loop is transformed into n parallel tiles.

end Tiling_Method.

Fig. 2. Algorithm of tiling loop by the incrementing minimum dependence distence

Example 1

do i = 1, 50
 do j = 1, 50
  A(3*i+1, 4*i+2*j+1) = ...
   ...= A(2*i-4, i+j-4)
 enddo
enddo

An example given in Example 1 illustrates the case that there is non-uniform and flow dependence. Fig. 3(a) shows CDCH(Complete Dependence Convex Hull) of Example 1. As the example, we can obtain the following results using the improved tiling method proposed in this section.

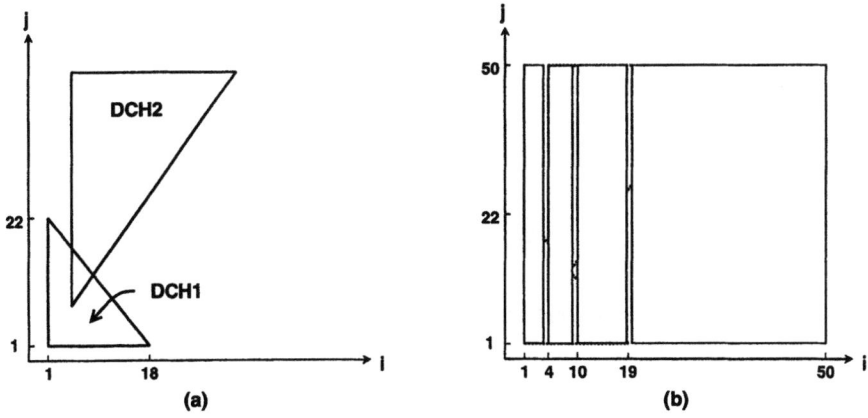

Fig. 3. (a) CDCH, (b) Tiling by minimum dependence distance in Example 1.

From the algorithm to compute a two-dimensional IDCH in [5], we can obtain the extreme points such as (1, 1), (1, 22), and (18, 1) as shown in Fig. 3(a). The first minimum value of $d_i(i, j_i)$ occurs at one of the extreme points. The i value for the source of the first dependence in the second tile is 4. The i value in the third tile is 10, and next values are 19, 31, and 49. Then, we can divide the iteration space by four tiles as shown in Fig. 3(b).

3.2 Loop Tiling Method Using Loop Interchanging

When there is only flow dependence in the loop, we can tile the iteration space into tiles with width = d_{imin} or width = d_{jmin}. In case $d_{jmin} > d_{imin}$, we can tile the iteration space into tiles with width = d_{jmin}.

In Example 1, because $d_j(i, j_i)$ (= $5/2*i + j + 5/2$) is greater than $d_i(i, j_i)$ (= $1/2*i + 5/2$), we can use an changed form of the example that the outer loop i and the inner loop j are interchanged as shown in Fig. 4.

```
do j = 1, 50
    do i = 1, 50
        A(3*i+1, 4*i+2*j+1) = ...
            ...= A(2*i-4, i+j-4)
    enddo
enddo
```

Fig. 4. Another form of Example 1 by loop interchanging.

If the upper limits of loop i and j are 100 by 100 as an example given in Fig. 4, the number of tiles for the original loop is six as shown in Fig. 5(a), and for the interchanged loop is five as shown in Fig. 5(b). When $d_{jmin} > d_{imin}$ in this loop, we can achieve greater parallelism by loop interchanging.

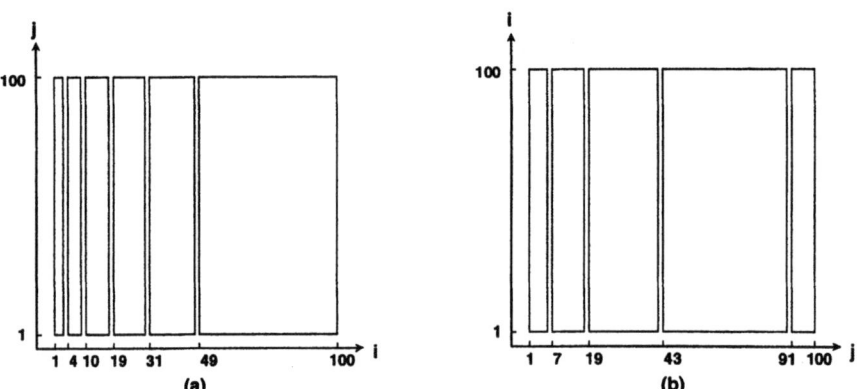

Fig. 5. (a) Tiling by the incrementing minimum dependence distance, (b) Tiling by Loop Interchanging in Example 1.

4 Performance Analysis

This section discusses the performance analysis of our proposed methods through the comparisons with related works theoretically.

Theoretical speedup for performance analysis can be computed as follows. Ignoring the synchronization, scheduling and variable renaming overheads, and assuming an unlimited number of processors, each partition can be executed in one time step. Hence, the total time of execution is equal to the number of parallel regions, N_p, plus the number of sequential iterations, N_s. Generally, speedup is represented by the ratio of total sequential execution time to the execution time on parallel computer system as follows:

$$Speedup = (N_i * N_j)/(N_p + N_s)$$
where N_i, N_j are the size of loop i, j, respectively

We will compare our proposed methods with the minimum dependence distance tiling method and the unique set oriented partitioning method as follows:

Let's consider the loop shown in Example 1. Fig. 3(a) shows original partitioning of Example 1. This example is the case that there is only flow dependence and DCH1 overlaps DCH2. Applying the unique set oriented partitioning to this loop illustrates case 2 of [4]. This method can divide the iteration space into four regions: three parallel regions, AREA1, AREA2 and AREA4, and one serial region, AREA3, as shown in Fig. 6. The speedup for this method is (100*100)/(3+44) = 212.8.

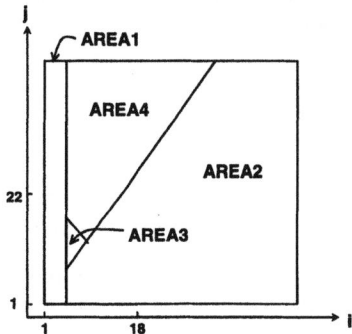

Fig. 6. Regions of the loop partitioned by the unique sets oriented partitioning in Example 1.

Applying the minimum dependence distance tiling method to this loop illustrates case 1 of this technique [5], which is the case that line $d_i(i, j) = 0$ does not pass through the IDCH. The minimum value of $d_i(i, j)$, d_{imin}, occurs at the extreme point (1, 1) and $d_{imin} = 3$. The space can be tiled with width = 3, thus 34 tiles are obtained. The speedup for this method is (100*100)/34 =294.

Let's apply our proposed method - the improved tiling method as given in section 3. This loop is tiled by six areas as shown in Fig. 5(a). The iterations within each area can be fully executed in parallel. So, the speedup for this method is (100*100)/6 = 1666.

Applying the loop interchanging method in this example, this loop is tiled by five areas as shown in Fig. 6(b). So, the speedup for this method is $(100*100)/5 = 2000$.

If the upper bounds of loop i, j are 1000 by 1000, the speedup for the original loop is $(1000*1000)/11 = 90909$, and the speedup for the interchanged loop is $(1000*1000)/8 = 125000$. Because $d_{jmin} > d_{imin}$ in this example, we can achieve more parallelism by loop interchanging.

5 Conclusions

In this paper, we have studied the problem of transforming nested loops with non-uniform and flow dependences to maximize parallelism.

When there is only flow dependence in the loop, we propose the improved tiling method. The minimum dependence distance tiling method tiles the iteration space by the first minimum dependence distance uniformly. Our proposed method, however, tiles the iteration space by minimum dependence distance values that are incremented as the value of i_j is incremented. Furthermore, when $d_{jmin} > d_{imin}$ in the given loop, loop parallelism can be improved by loop interchanging.

In comparison with some previous partitioning methods, the improved tiling method gives much better speedup than the minimum dependence distance tiling method and the unique set oriented partitioning method in the case that there is only flow dependence and DCH1 overlaps DCH2.

Our future research work is to develop a method for improving parallelization of higher dimensional nested loops.

References

1. A. A. Zaafrani and M. R. Ito, "Parallel region execution of loops with irregular dependences," in *Proceedings of the International Conference on Parallel Processing*, vol. II, (1994) 11-19
2. C. K. Cho, J. C. Shim, and M. H. Lee, "A loop transformation for maximizing parallelism from single loops with non-uniform dependences," in *Proceedings of High Performance Computing Asia '97*, (1997) 696-699
3. C. K. Cho and M. H. Lee, "A loop parallization method for nested loops with non-uniform dependences", in *Proceedings of the International Conference on Parallel and Distributed Systems*, (1997) 314-321
4. J. Ju and V. Chaudhary, "Unique sets oriented partitioning of nested loops with non-uniform dependences," in *Proceedings of International Conference on Parallel Processing*, vol. III, (1996) 45-52
5. S. Punyamurtula and V. Chaudhary, "Minimum dependence distance tiling of nested loops with non-uniform dependences," in *Proceedings of Symposium on Parallel and Distributed Processing*, (1994) 74-81
6. S. Punyamurtula, V. Chaudhary, J. Ju, and S. Roy, "Compile time partitioning of nested loop iteration spaces with non-uniform dependences," *Journal of Parallel Algorithms and Applications*, (1996)
7. T. Tzen and L. Ni, "Dependence uniformization: A loop parallelization technique," *IEEE Transactions on Parallel and Distributed Systems*, vol. 4, no. 5, (1993) 547-558

Comparison Based Diagnostics as a Probabilistic Deduction Problem

Balázs Polgár

Dept. of Measurement and Inf. Systems, BUTE
Magyar Tudósok krt. 2, Budapest, Hungary, H-1117,
polgar@mit.bme.hu

Abstract. A novel modeling approach for system-level diagnosis of multiprocessor systems has been introduced in previous publications. In this approach the diagnostic process is formulated as an optimization problem. The possible logical relations are identified between the different pieces of diagnostic information of the system and an optimal consistent combination of the relations is determined during the solution method. A part of the information is that, which can be observed at the outputs of the system. Another part is composed by hypotheses on the states of units. Relations between these information are described by consequence rules having probabilities assigned to them. These probabilities express the uncertainty of test results. The object is to draw back the set of observed information to a subset of hypotheses on unit states with the maximum likelihood, i.e. to determine the states of system units on the basis of the syndrome.

1 Introduction

Diagnostics is one of the major tools for assuring the reliability of complex systems in information technology. In such systems the test process is often implemented on system-level: main components are tested, results are collected, and based on this information the good or faulty state of each system-component is determined. This classification procedure is known as *diagnostic process*.

The early approaches employed oversimplified models, but these proved to be impractical, lately much effort has been put into extending the limitations of traditional models [1]. However, the presented solutions mostly concentrated on only one aspect of the problem.

Application of P-graph based modeling to system-level diagnosis [2] can provide a general framework that supports the solution of several different types of problems, that previously needed numerous different modeling approaches and solution algorithms. Furthermore, we have not only integrated existing solution methods, but proceeding from a more general base we have extended the set of solvable problems with new ones. The representational power of the model was illustrated in paper [3].

Another advantage of the P-graph models is that it takes into consideration more properties of the real system than previous diagnostic models. Therefore

M. Bubak et al. (Eds.): ICCS 2004, LNCS 3038, pp. 1153–1161, 2004.
© Springer-Verlag Berlin Heidelberg 2004

its diagnostic accuracy is also better. This means that it provides almost good diagnosis even when half of the processors are faulty [3], which is important for example in the field of wafer scale testing [4]. These favorable properties of the approach are achieved by considering the diagnostic system as a structured set of information called hypotheses with well-defined relations.

In part of the paper the adapted abstract model is presented in order to be able to reuse or apply it to other fields of diagnostics. The application for comparison based diagnostics and an example is also discussed.

2   Comparison Based System-Level Diagnostics in Multiprocessor Systems

System-level diagnosis considers the *replaceable units* of a system, and does not deal with the exact location of faults within these units. A *system* consists of an interconnected network of independent but cooperating *units* (typically processors, the elements of set **U**). These units can fail, therefore a diagnostic procedure is performed from time to time to ensure the correct behavior of the whole system. According to this the fault state of each unit is considered as either *good* or *faulty* (the domain of the states $D_{st} = \{g, f\}$ and the function $st(u)$ determines the actual state of unit $u \in U$). The collection of the fault states of every unit in the system is called the *fault pattern* [6].

A possible way to detect the failures of the units in multiprocessor systems is the comparison based diagnostics [7]. This is used for instance in the commercially available *APEMille supercomputer* [8] which was developed in collaboration by IEI-CNR of Rome and Pisa, and the DESY Zeuthen in Germany. A further possible application field of this model is the *wafer scale diagnosis* [9].

In the comparison based approach the system consists comparators between pairs of the homogeneous units. Both units perform the same test and the comparator compares the bit-sequence result. The set of comparators is denoted by **C** and the set of tests is denoted by $T \subset U \times C \times U$ (× denotes the Cartesian product of two sets), which has the elements (A,c,B), where units A and B are neighbors and have comparator c between them for comparing their test-sequence results. The so-called *syndrome* contains the results of the comparators, namely the information that 'the two units differ' or 'the two units operate similarly' denoted by *1* and *0*, respectively. Formally, $D_{tr} = \{0, 1\}$ is the domain of the test results and function $tr(t)$ determines the actual test result of test $t \in T$. The object is to determine the fault pattern on the basis of the syndrome.

If comparators are assumed to be fault free then the test result is always 0 if both unit is good, it is always 1 if one is good and the other is faulty and it can be either 0 or 1 if both is faulty. The behavior in the last case can be described by assigning probabilities to test results, see Table 1 [3].

The diagnostics is more complicated if possible faults of comparators should also be taken into consideration. The behavior of a faulty comparator can be described by the parameters $p_{C00}, p_{C01}, p_{C10}, p_{C11}$, where p_{Cxy} is the probability that a faulty comparator alters the expected test result from x to y (Table 2).

Table 1. Probabilities of test results depending on unit states if fault free comparators are assumed

State	State	Prob. of testres	
of A	of B	0	1
good	*good*	1	0
good	*faulty*	0	1
faulty	*good*	0	1
faulty	*faulty*	p_{d0}	p_{d1}

Table 2. Behavior of a comparator that can be faulty

State	Probability of altering			
of the	test result from x to y			
comp.	$0 \to 0$	$0 \to 1$	$1 \to 0$	$1 \to 1$
good	1	0	0	1
faulty	p_{C00}	p_{C01}	p_{C10}	p_{C11}

3   Basic Terminology and Its Application

In this section basic definitions and notations are presented. A part of it (for instance the concept of P-graphs, decision mappings and their relation) is taken directly from the field of PNS problems, introduced by Friedler et al. in [10]. Another part is taken from there with an adapted terminology and with some modified interpretation (for instance *hypotheses* instead of *materials*, *logical relations* instead of *operating units*). The model consists some additional properties, too (e.g. excluding hypotheses).

3.1   Hypotheses and Their Relations

Hypotheses are said to be a kind of statements or assumptions, that contain some information about the system, but the truth of which aren't necessarily known. Hypotheses can be for instance about the possible states of units or about the possible outcomes of tests. The probability that hypothesis h becomes true is determined by the function $p(h)$.

Considering the set **H** of hypotheses, logical relations between hypotheses can be defined as follows. The pair of $(\mathbf{h_p}, \mathbf{h_c})$, $\mathbf{h_p} \subset \mathbf{H}$, $\mathbf{h_c} \subset \mathbf{H}$ describes a *probabilistic consequence relation*, where $\mathbf{h_p}$ is the set of *premises*, $\mathbf{h_c}$ is the set of *consequences* and the *probability function* $p((\mathbf{h_p}, \mathbf{h_c}))$ determines the conditional probability that hypotheses in set $\mathbf{h_c}$ become true if hypotheses in set $\mathbf{h_p}$ are supposed to be true.

Function $\boldsymbol{exc} : \mathbf{H} \to 2^{\mathbf{H}}$ determines the *excluded sets of hypotheses* for the elements of **H**, that is for hypothesis h the set $exc(h)$ contains those hypotheses, which cannot be true if h is true (function $2^{\mathbf{H}}$ determines the *power set* of set **H**, i.e. the set of all subsets of the given set). With this function for instance that kind of constraints can be formalized, which assure the unique state of a unit.

Application in Comparison Based Diagnostics. Four sets of hypotheses can be distinguished in the diagnostic model described in section 2:

– hypotheses about the states of units

$$\mathbf{H_1} = \{A_g \mid A_g \equiv \text{'}st(A) = g\text{'} \equiv \text{'unit } A \text{ is good'}, A \in \mathbf{U}\}$$
$$\cup \{A_f \mid A_f \equiv \text{'}st(A) = f\text{'} \equiv \text{'unit } A \text{ is faulty'}, A \in \mathbf{U}\}$$

- hypotheses about the states of comparators

$$\mathbf{H_2} = \{c_g \mid c_g \equiv \text{'st}(c) = g\text{'}, c \in \mathbf{C}\} \cup \{c_f \mid c_f \equiv \text{'st}(c) = f\text{'}, c \in \mathbf{C}\}$$

- hypotheses about the comparison results if comparator faults are not taken into consideration

$$\mathbf{H_3} = \{AB_0 \mid AB_0 \equiv \text{'tr}((A, c, B)) = 0 \,\&\, st(c) = g\text{'} \equiv \text{'unit } A$$
$$\text{and unit } B \text{ are in the same state'}, A, B \in \mathbf{U}, c \in \mathbf{C} \text{ and } (A, c, B) \in \mathbf{T}\}$$
$$\cup \{AB_1 \mid AB_1 \equiv \text{'tr}((A, c, B)) = 1 \,\&\, st(c) = g\text{'} \equiv \text{'unit } A$$
$$\text{and unit } B \text{ are in different states'}, A, B \in \mathbf{U}, c \in \mathbf{C} \text{ and } (A, c, B) \in \mathbf{T}\}$$

- hypotheses about the comparison results if comparator faults can alter the real results

$$\mathbf{H_4} = \{AcB_0 \mid AcB_0 \equiv \text{'tr}((A, c, B)) = 0\text{'}, A, B \in \mathbf{U}, c \in \mathbf{C}, \text{ and } (A, c, B) \in \mathbf{T}\}$$
$$\cup \{AcB_1 \mid AcB_1 \equiv \text{'tr}((A, c, B)) = 1\text{'}, A, B \in \mathbf{U}, c \in \mathbf{C}, \text{ and } (A, c, B) \in \mathbf{T}\}$$

Consequence relations formalize the information given in Tables 1 and 2:

$$\mathbf{R_1} = \{(\{A_s, B_t\}, \{AB_x\}) \mid A_s, B_t \in \mathbf{H_1}, AB_x \in \mathbf{H_3}\} \qquad (1)$$

An element r of this set having probability $p(r)$ represents the relation '*if* A is in state s, B is in state t and there exists a fault free comparator between them *then* the test result is x with probability $p(r)$'. The probability function $p(\cdot)$ for these relations is determined by the parameters given in Table 1.

$$\mathbf{R_2} = \{(\{AB_x, c_s\}, \{AcB_y\}) \mid c_s \in \mathbf{H_2}, AB_x \in \mathbf{H_3}, AcB_y \in \mathbf{H_4}\} \qquad (2)$$

An element r of this set having probability $p(r)$ represents the relation '*if* the expected test result is x and comparator c is in state s *then* the real test result is y with probability $p(r)$'. The probability function $p(\cdot)$ for these relations is determined by the parameters given in Table 2.

The excluded set of hypotheses for a hypothesis $h \in \mathbf{H_1} \cup \mathbf{H_2}$ consists all hypotheses representing another assumption on the state of the given unit or comparator: $exc(A_s) = \{A_t \mid t \in \mathbf{D_{st}} \setminus \{s\}\}$.

Example. Let us consider a system with three units (A, B, C) and three comparators (c^1, c^2, c^3) according to Fig. 1, where both units and comparators can fail. The set of hypotheses is $\mathbf{H} = \mathbf{H_1} \cup \mathbf{H_2} \cup \mathbf{H_3} \cup \mathbf{H_4}$, where

$$\mathbf{H_1} = \{A_g, A_f, B_g, B_f, C_g, C_f\}$$
$$\mathbf{H_2} = \{c_g^1, c_f^1, c_g^2, c_f^2, c_g^3, c_f^3\}$$
$$\mathbf{H_3} = \{AB_0, AB_1, BC_0, BC_1, AC_0, AC_1\}$$
$$\mathbf{H_4} = \{Ac^1B_0, Ac^1B_1, Bc^2C_0, Bc^2C_1, Ac^3C_0, Ac^3C_1\}$$

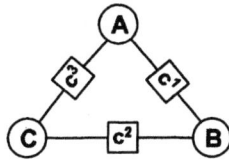

Fig. 1. System with units A, B, C and comparators c^1, c^2, c^3

The set of consequence relations is $\mathbf{R} = \mathbf{R_1} \cup \mathbf{R_2}$, where

$$\mathbf{R_1} = \{(\{A_g, B_g\}, \{AB_0\}), (\{A_g, B_f\}, \{AB_1\}), (\{A_f, B_g\}, \{AB_1\}),$$
$$(\{A_f, B_f\}, \{AB_0\}), (\{A_f, B_f\}, \{AB_1\}), \dots\}$$
$$\mathbf{R_2} = \{(\{AB_0, c_g^1\}, \{Ac^1 B_0\}), (\{AB_0, c_f^1\}, \{Ac^1 B_0\}), (\{AB_0, c_f^1\}, \{Ac^1 B_1\}), \dots\}$$

The excluded sets of hypotheses –where it is not empty– are according to the followings: $\boldsymbol{exc}(A_g) = \{A_f\}$, $\boldsymbol{exc}(A_f) = \{A_g\}$, $\boldsymbol{exc}(B_g) = \{B_f\} \dots$

3.2   Probabilistic Deduction Problem

The *probabilistic deduction problem* (PD problem, for short) based on set $\mathbf{H}$ of hypotheses is defined by the triplet $(\mathbf{H^i}, \mathbf{H^d}, \mathbf{R})$, where $\mathbf{H^i} \subset \mathbf{H}$ is the set of *initial hypotheses*, $\mathbf{H^d} \subset \mathbf{H}$ is the set of *hypotheses to be deduced* and $\mathbf{R} \subseteq 2^{\mathbf{H}} \times 2^{\mathbf{H}}$ is the set of *probabilistic consequence relations* between hypotheses. The object is to deduce the elements of set $\mathbf{H^d}$ from a subset of $\mathbf{H^i}$.

In the probabilistic deduction problem $(\mathbf{H^i}, \mathbf{H^d}, \mathbf{R})$ the base set of hypotheses $\mathbf{H^b}$ is the set of all hypotheses defined implicitly by the triplet $(\mathbf{H^i}, \mathbf{H^d}, \mathbf{R})$, i.e.

$$\mathbf{H^b} = \mathbf{H^i} \cup \mathbf{H^d} \cup \{h \mid h \in \mathbf{h_p} \cup \mathbf{h_c}, \ (\mathbf{h_p}, \mathbf{h_c}) \in \mathbf{R}\} \tag{3}$$

Application in Comparison Based Diagnostics. First the model consisting fault-free comparators is considered. Let set $\mathbf{H_1^d}$ be defined as that subset of $\mathbf{H_3}$ which corresponds to the syndrome determined by function tr: $\mathbf{H_1^d} = \{AB_x \mid x = tr((A, c, B)) \text{ and } AB_x \in \mathbf{H_3}\}$. Then the diagnostic task is formulated as the PD problem $(\mathbf{H_1}, \mathbf{H_1^d}, \mathbf{R_1})$.

If comparator faults are also included in the model, then the diagnostic task is formulated as PD problem $(\mathbf{H_1} \cup \mathbf{H_2}, \mathbf{H_2^d}, \mathbf{R_1} \cup \mathbf{R_2})$, where $\mathbf{H_2^d} = \{AcB_x \mid x = tr((A, c, B)) \text{ and } AcB_x \in \mathbf{H_4}\}$

The base set of hypotheses for the first model is $\mathbf{H_1^b} = \mathbf{H_1} \cup \mathbf{H_3}$, whereas for the second one it is $\mathbf{H_2^b} = \mathbf{H_1} \cup \mathbf{H_2} \cup \mathbf{H_3} \cup \mathbf{H_4}$.

Example (continued). Let's consider that the syndrome is *101*, i.e. $tr((A, c^1, B)) = 1$, $tr((B, c^2, C)) = 0$ and $tr((A, c^3, C)) = 1$. The set of hypotheses to be deduced for the example introduced above is $\mathbf{H_2^d} = \{Ac^1 B_1, Bc^2 C_0, Ac^3 C_1\}$.

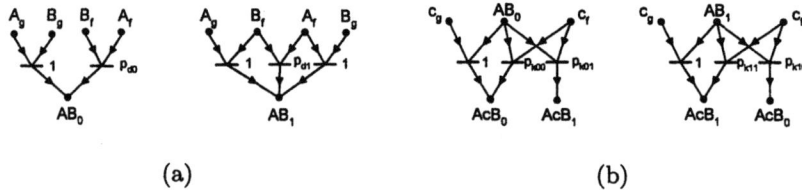

(a)   (b)

Fig. 2. P-graph model of a single comparator test

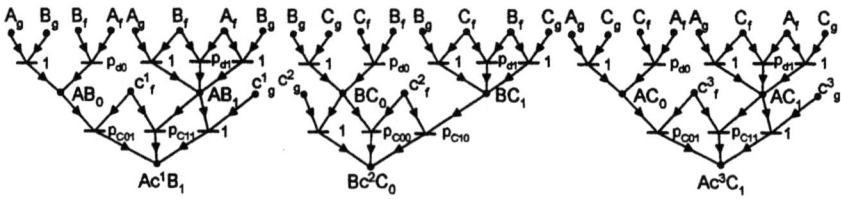

Fig. 3. P-graph model of the example (corresponding to the syndrome)

3.3   P-Graphs

For graphical representation of the probabilistic deduction problem the so-called P-graphs are used. Consider two finite sets $\mathbf{h}$ and $\mathbf{r}$ such that $\mathbf{r} \subseteq 2^{\mathbf{h}} \times 2^{\mathbf{h}}$. A *P-graph* is defined as a directed bipartite graph that is determined by the pair of $(\mathbf{h}, \mathbf{r})$ by the following way:

1. the *vertices* of the graph are the elements of the set $\mathbf{V} = \mathbf{h} \cup \mathbf{r}$; vertices belonging to set $\mathbf{h}$ are called *H-type* and are denoted by a dot, whereas those belonging to set $\mathbf{r}$ are called *R-type* and are denoted by a horizontal bar;
2. the *arcs* of the graph are the elements of set $\mathbf{E} = \mathbf{E_1} \cup \mathbf{E_2}$, where $\mathbf{E_1}$ and $\mathbf{E_2}$ are the sets of arcs from the premisses to the relations and from the relations to the consequences, respectively. I.e., $\mathbf{E_1} = \{(h, r) \mid r = (\mathbf{h_1}, \mathbf{h_2}) \in \mathbf{r} \text{ and } h \in \mathbf{h_1}\}$ and $\mathbf{E_2} = \{(r, h) \mid r = (\mathbf{h_1}, \mathbf{h_2}) \in \mathbf{r} \text{ and } h \in \mathbf{h_2}\}$.

Application in Comparison Based Diagnostics. The P-graph of a single comparison test between units A and B with comparator c can be seen on Fig. 2(a) if comparators are assumed to be fault-free. This P-graph should be extended with the graph on Fig. 2(b) if comparator faults are taken into consideration. The labels beside the H-type vertices are the names of the hypotheses, whereas beside the R-type vertices the probability of the relation can be seen.

Example (cont.). Part of the P-graph model of the hypotheses and relations defined previously, which corresponds to the syndrome can be seen on Fig. 3.

3.4 Deduction Structures

Consider the probabilistic deduction problem $(\mathbf{H^i,H^d,R})$ and its base set $\mathbf{H^b}$. Let $\mathbf{h \subseteq H^b}$ and $\mathbf{r \subseteq R}$ be two sets such that $\mathbf{r \subseteq 2^h \times 2^h}$. Then the structure consisting the set of hypotheses $\mathbf{h}$ and the set of relations $\mathbf{r}$ between them is called a *deduction structure of* $(\mathbf{H^i,H^d,R})$ and is formally determined by the P-graph $(\mathbf{h,r})$. A deduction structure given by a P-graph $(\mathbf{h,r})$ is called *consistent* if it does not consist hypotheses that exclude each other, i.e. $\mathbf{h \cap exc[h]} = \emptyset$.

Example (cont.). The P-graph on Fig. 3 and all of its subgraphs are deduction structures. The P-graphs on Fig. 4 are consistent deduction structures.

3.5 Solution Structures

The deduction structure given by P-graph $(\mathbf{h,r})$ is called a *solution structure* of the probabilistic deduction problem $(\mathbf{H^i,H^d,R})$ if

1. it is consistent,
2. it consist all hypotheses to be deduced, i.e. $\mathbf{H^d \subset h}$.
3. a vertex h of H-type has no arc pointing to it if and only if it corresponds to a hypothesis belonging to the set of initial hypotheses.
4. a vertex h of H-type has no arc starting from it if and only if it corresponds to a hypothesis belonging to the set of hypotheses to be deduced.
5. every vertex of R-type has at least one path leading to a vertex of H-type representing a hypotheses to be deduced.

Let $\mathbf{S(H^i,H^d,R)}$ denote the *set of all solution structure* of the probabilistic deduction problem $(\mathbf{H^i,H^d,R})$. The *weight function* $W(s)$ assigns a value to every solution structure $s \in \mathbf{S(H^i, H^d, R)}$. This function serves as the optimization objective function during the selection between solution structures.

Application in Comparison Based Diagnostics. The initial sets of hypotheses in solution structures represent the fault patterns that is compatible with the syndrome. The object is to select one of it, which is optimal according to a given criteria. This criteria is determined by the weight function, which can be for instance the conditional probabilities of the syndrome given the conditions of the possible fault patterns. In PD problem for a solution structure $(\mathbf{h_s, r_s}) \in \mathbf{S(H^i, H^d, R)}$, this is the probability of $\mathbf{P(H^d \mid H^i \cap h_s)} = \mathbf{P}(syndrome \mid fault\ pattern)$. In this case we are talking about maximum likelihood diagnostics, because that fault pattern is chosen by which the occurrence of the given syndrome has the maximal probability.

In comparison based diagnostics if fault-free comparators are assumed the former version of the weight function is determined according to the formula:

$$W((\mathbf{h_s, r_s})) = \mathbf{P(H^d \mid H^i \cap h_s)} = \bigcap_{h \in \mathbf{H^d}} \mathbf{P}(h \mid \mathbf{H^i \cap h_s}) =$$

$$= \bigcap_{h \in \mathbf{H^d};(\mathbf{h_p},\{h\}) \in \mathbf{r_s}} \mathbf{P}(h \mid \mathbf{h_p}) = \bigcap_{r \in \mathbf{r_s}} p(r) \tag{4}$$

By this definition, the solution structure having the maximum value of the weight function should be determined. It can be transformed into a minimization problem if necessary by taking the sums of the negative logarithms of the probabilities of relations as weight function.

Example. Two feasible solution structures s_1 and s_2 can be seen on Fig. 4(a) and Fig. 4(b). The weights of these structures are the products of the probabilities of the relations included. For s_1 it is 1, meaning that for the given fault pattern it is sure that the considered syndrome will be the result of the tests. The weight of s_2 is $p_{C01} \cdot p_{C01}$ meaning that for this fault pattern the syndrome arise with this probability. In maximum likelihood diagnostics that fault pattern is 'selected' for which the syndrome arise with the maximum likelihood. This means that if probability parameters in Tables 1 and 2 are smaller than 1 than the diagnosis will be the fault pattern according to s_1.

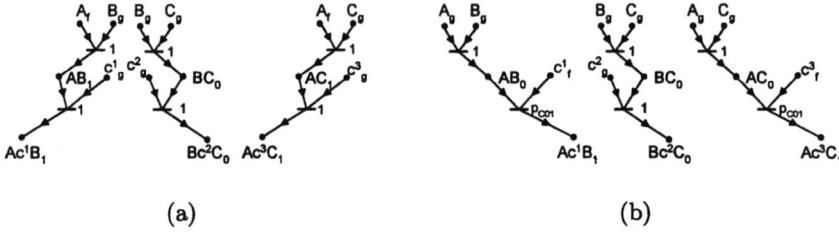

(a)   (b)

Fig. 4. Two feasible solution structures of the example

4   Conclusions

In this paper a formal description of the previously introduced model of P-graph based diagnostics is presented. In probabilistic deduction problems the relations of the information is modeled, which relations corresponds to the structure of the system and to the structure of the diagnostic procedure. Solving this model the most probable states of units can be determined on the basis of the syndrome. The concept of the model –formulating in general– is that knowing the relations between pieces of information and knowing the observed information it is possible to deduce for the hidden or embedded information with the maximum likelihood (an information is considered hidden or embedded if it cannot be observed directly). The PD problem can be solved either with a general solver for linear programming problems or with the adapted version of the solution algorithm of PNS problems developed by Friedler et al. [10]. The model and its solution algorithm have been already implemented in C++, and simulation results demonstrate its good diagnostic performance [3].

References

1. S. N. Maheshwari, S. L. Hakimi. On Models for Diagnosable Systems and Probabilistic Fault Diagnosis. *IEEE Trans. on Computers*, vol. C-25, pp. 228–236, 1976.
2. B. POLGÁR, SZ. NOVÁKI, A. PATARICZA, F. FRIEDLER. A Process-Graph Based Formulation of the Syndrome-Decoding Problem, In *4th Workshop on Design and Diagnostics of Electronic Circuits and Systems*, pp. 267–272, Hungary, 2001.
3. B. POLGÁR, T. BARTHA, E. SELÉNYI. On the Extension and Applicability of the P-Graph Modeling Paradigm to System-Level Diagnostic Problems, *invited and submitted to PDCP Special Issue of DAPSYS'2002 Workshop*.
4. S. CHESSA. Self-Diagnosis of Grid-Interconnected Systems, with Application to Self-Test of VLSI Wafers, *Ph.D. Thesis*, TD-2/99 Univ. of Pisa, Italy, March 1999.
5. F. FRIEDLER, L. T. FAN, AND B. IMREH. Process Network Synthesis: Problem Definition. *Networks*, 28(2), pp. 119–124, 1998.
6. M. BARBORAK, M. MALEK, AND A. DAHBURA. The Consensus Problem in Fault Tolerant Computing, *ACM Computing Surveys*, vol. 25, pp. 171–220, June 1993.
7. A. Dahbura, K. Sabnani, and L. King. The Comparison Approach to Multiprocessor Fault Diagnosis, *IEEE Trans. on Computers*, vol. C-36, pp. 373–378, 1987.
8. F. AGLIETTI, ET AL.. Self-Diagnosis of APEmille, *Proc. EDCC-2 Companion Workshop on Dependable Computing*, pp. 73–84, Silesian Technical University, Gliwice Poland, May 1996.
9. B. SALLAY, P. MAESTRINI, P. SANTI. Comparison-Based Wafer-Scale Diagnosis Tolerating Comparator Faults, *IEEE Journal on Computers and Digital Techniques*, 146(4), pp. 211–215, 1999.
10. F. FRIEDLER, K. TARJAN, Y. W. HUANG, L. T. FAN. Combinatorial Algorithms for Process Synthesis. *Comp. in Chemical Engineering*, vol. 16, pp. 313–320, 1992.

Dynamic Threshold for Monitor Systems on Grid Service Environments

E.N. Huh

Seoul Women's University, Division of Information and Communication, Seoul, Korea
huh@swu.ac.kr

Abstract. Grid technology requires use of geographically distributed multiple domain's resources. Resource monitoring services or tools consisting sensors or agents will run on many systems to find static resource information (such as architecture vendor, OS name and version, MIPS rate, memory size, CPU capacity, disk size, and NIC information) and dynamic resource information (CPU usage, network usage(bandwidth, latency), memory usage, etc.). Thus monitoring itself may cause system overhead. This paper proposes push based resource notification architecture on OGSI (Open Grid Service Infrastructure) and the dynamic threshold to measure monitoring events in accurate and with less overhead. By employing the new feature (dynamic threshold), we find out unnecessary system overhead is significantly reduced and accuracy of events is still acquired.

1 Introduction

In these days, the high-performance network technology enables geographically localized clusters to gather remote domain's clusters replacing supercomputing power with small cost. Remotely networked computers need to be cooperated to process huge amount of data in the area of bio-information technology (BT), space-technology (ST), environment-technology (ET), and nano-technology (NT) interfacing with electronic equipments. To enable information technology (IT) under the above areas, computational Grid is announced as a middleware system, which considers complex systems as a single system in point of a user. Thus, Grid is a virtual supercomputing infrastructure interacting with widely dispersed computing powers [1].

In order to effectively use the heterogeneous computing resources, such as storage devices, and various research equipments, these should be efficiently managed in Grid environments. Hence Grid is consisting of resource information service, resource brokering service, security service, resource management service, monitoring service, and data access service. The resource information service (GAIS-Grid Advanced Information Service) provides discovery and registration of resources. The brokering service functions an agent to provide higher level abstraction of resource information to user applications to be executed. The resource management service allocates and reserves resources. The security service using single sign on (SSO) based on public key infrastructure allows grid middleware components to access signed resources remotely. The monitoring service detects system changes and basic information, and delivers them to GRIS. The data access service manages distributed huge amount of

M. Bubak et al. (Eds.): ICCS 2004, LNCS 3038, pp. 1162–1169, 2004.
© Springer-Verlag Berlin Heidelberg 2004

data using meta-catalog for searching and replicas for duplicating data, respectively [2].

The monitoring system needs to collect the system information correctly while resource changes occur dynamically. The dynamic characteristics make resource allocator hard as delivered monitoring events are old under the situation. Thus the monitoring agents or sensors should run in real-time to provide accurate resource information of system events.

This paper presents a model of the dynamic update of Grid resource information service on GTK 3.0 using notification mechanism to measure monitoring events accurately during run-time. This paper is organized as follows: related studies are discussed in section 2; dynamic system change analysis is presented in section 3; and experiments are illustrated in section 4; and section 5 concludes this study.

2 Related Work

This section depicts Grid monitoring architecture from Grid Monitoring Architecture Working Group (GMA-WG) draft [3] and other monitoring tools such as iperf [4], ntop [5], tcpdump [6], topomon[8], and pingER [7] currently used for different purposes.

There are three components in monitoring architecture, directory service, consumer and producer. In order to describe and discover performance data on the Grid, a distributed directory service for publishing and searching must be available. The GMA directory service stores information about producers and consumers that accept requests. Functions of the directory service are Add, Update, Remove, and Search. A producer is any component that uses the producer interface to register its identification and basic information to the directory service and send events to a consumer. Functions of a producer defined in [3] are Locate Event, Locate Consumer, Register, Accept Query, Accept Subscribe, Accept Unsubscribe, Initiate Query, Initiate Subscribe, and Initiate Unsubscribe. A consumer locates resources from the directory service and requests events by query to the provider. Functions of a consumer are Locate Event, Locate Producer, Initiate Query, Initiate Subscribe, Initiate Unsubscribe, Register, Accept Query, Accept Subscribe, and Accept Unsubscribe. The overall mechanism among components is illustrated in Figure 1 [3].

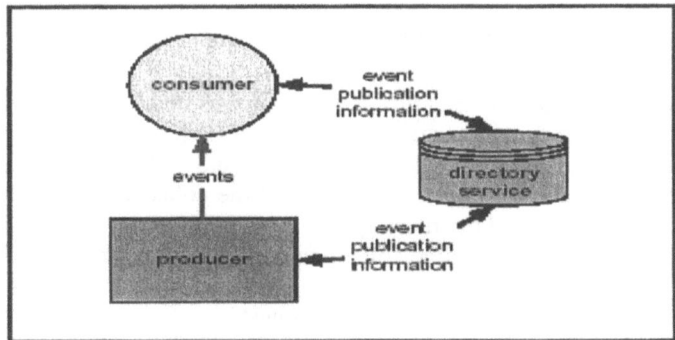

Fig. 1. Grid Monitoring Architecture

This architecture is designed for transmission of performance data, time-stamped events, and must define standard schema of events, which is extensible.

In [9], the monitoring mechanism for Information Power Grid (IPG) by NASA is explained shown in Figure 2. The service based monitoring consists of components, sensor, actuator and event service interface. It is designed for detection of the system behavior and control of QoS.

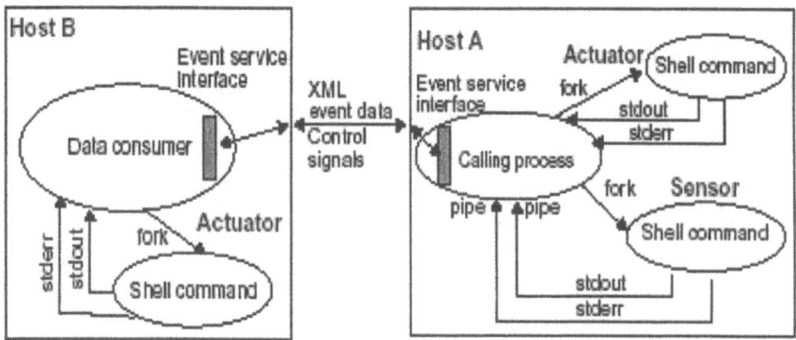

Fig. 2. IPG monitoring mechanism

The sensor is primary collecting tool and the actuator is pre-defined specific shell command (to collect events) by developer. And the event service provides standard interface to the requestor for transmission of inquires and results.

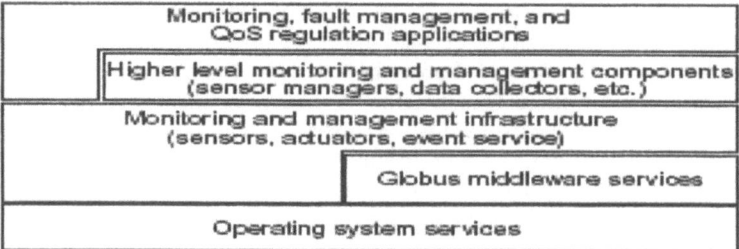

Fig. 3. Monitoring S/W layer

The monitoring S/W structure is designed on Globus middleware service [10] for security and monitoring event transmission defined in [11]. As shown in Figure 3, this focuses on the run-time environment to detect faults and QoS metrics on the external request (actuation). There is neither periodic real-time collection technique nor even overhead management in Grid monitoring tools.

Thus, this study supports real-time collection on Grid environments and proposes a model of the optimal and dynamic monitoring interval. Our sensors or agents are initiated to collect events by external and self requests.

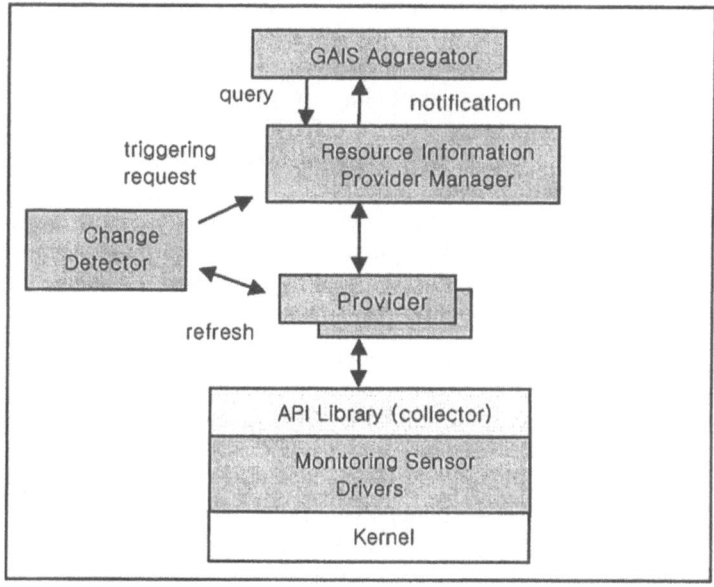

Fig. 4. Monitoring Architecture on OGSI using Notification

3 Proposed Monitoring Architecture and Dynamic Threshold

The resource of computing systems is changed very often during run-time as general processes and system processes (e.g., daemon) can be executed at any time. That is, we need a scheme for a threshold to determine if the significant system condition changes occur. For example, if the CPU is changed by 5% from the previous interval, then is it significant system change? Thus, we are proposing here additional technique, a *dynamic threshold*, to decide significant system changes at run-time to support precise resource information for future resource allocation.

Mainly, processes use CPU, memory, disk, and network resources, which are main types of resources that Grid resource allocation service uses. Analyzing client/server program's resource usage (DynBench [12] is used for benchmarking application set, which consumes resources dynamically.), a strong correlation among resource types is assessed that other resources are depending on CPU changes. Thus, we collect initially only CPU usage percentage event in real-time to decide overall system changes. If CPU change is significant, then collects other resource events. According to the step 5 at the previous section, the dynamic optimal interval for the next cycle is longer than the current optimal interval if CPU change ratio is not. Otherwise, the optimal interval becomes shorter.

Thus, it is hard to determine the significance of CPU change. For example, CPU is changed by 1% at every cycle for long periods, and suddenly 2% is changed at any cycle. Then can you say it is significant change? Assume another case that CPU change ratio by 5% at every cycle occurs for long periods, and 7% CPU change is detected at sudden. Then is it significant change? In order to find out the significance of condition change, we need to consider history of changes, current value, and difference from the previous.

The Figure 4 is the overall architecture proposed to update dynamically GAIS aggregator which is a global resource holder (Xindice XML DB is used). . The proc file system in Unix is widely used (e.g., NETSTAT , VMSTAT, TOP) to collect kernel and system information as well as process information. However, the event in proc file system is updated when process state is changed. Hence the accurate information of running processes is not guaranteed at the collecting time. Therefore, the device driver called "sensor" is employed in kernel space to collect more accurate events than other methods using directly reading the kernel variables. There exist 4 different types of providers we implement as follows:

✓ HostStaticInfoProvider : OS name, Vendor, clock speed, memory size, cache size, disk size, MIPS rate.
✓ HostDynamicInfoProvider: current CPU, CPU load info during 1, 5, and 15min., memory usage(buffered, swapped), uptime, disk usage, context switching time
✓ ProcessInfoProvider: CPU, memory usage, network usage, bandwidth, utilization
✓ NetworkInfoProvider: TCP parameters (loss, error, retransmit), utilization, network device usage (sent bytes, received. bytes, collision)
✓ PBSInfoProvider: Qinformation (name, queue-length) and Job status information

All of above events are formatted to XML for delivery to GAIS corresponding to the suggestion of Discovery and Monitoring Event Description (DAMED) Working Group [11].

Now, remain of this section describes how the dynamic threshold is derived for monitoring system, which requires les overhead and accuracy. Fortunately, Proportional, Integrated, Differential (PID) controller is designed to solve similar problems such as Quasar scheduler [13]. The PID controller is commonly used in water tank to keep proper water level automatically. Especially, in [14], PID controller is explained well as follows:

– The Proportional term causes a larger control action to be taken for a larger
– error
– The Integral term is added to the control action if the error has persisted for
– some time
– The Derivative term supplements the control action if the error is changing
– rapidly with time.

The formula for the desired level (denoted to $u(k)$) calculation by PID controller is described below:

$$u(k)= k_p \times e(k) + k_i \times (e(k) - e(k-1))$$

where, Kp, Ki, and Kd are parameters (tunable).

We consider that the steady state implies no changes of system resources. That is, $| e(k)-e(k-1) | = 0$, where $e(k) = R_{Max}-R_{obs}(k)$, R_{Max} is the maximum amount resource (e.g., for CPU, it is 100%), and $R_{obs}(k)$ is resource usage collected in current k^{th} interval. With steady state condition, we simplify PID controller formula by subtracting two states as follows:

$$\left| u(k) - u(k-1) \right|$$

$$= (k_p \times e(k) + k_i \times \sum_{j=1}^{k} e(j) + k_d \times (e(k) - e(k-1)))$$

$$- (k_p \times e(k-1) + k_i \times \sum_{j=1}^{k-1} e(j) + k_d \times (e(k) - e(k-1)))$$

$$= k_p \times (e(k) - e(k-1)) + k_i (\sum_{j=1}^{k} e(j) - \sum_{j=1}^{k-1} e(j))$$

$$\cong k_i \times e(k)$$

From the above simplification, |u(k)-u(k-1)| is less than equal to Ki * e(k) under steady state. Thus, if |u(k)-u(k-1)| is greater than Ki * e(k), the current state is not steady. From this fact, we bring the dynamic threshold, Ki * e(k). Therefore, the significant change of resource is determined if |u(k)-u(k-1)| > Ki * e(k).

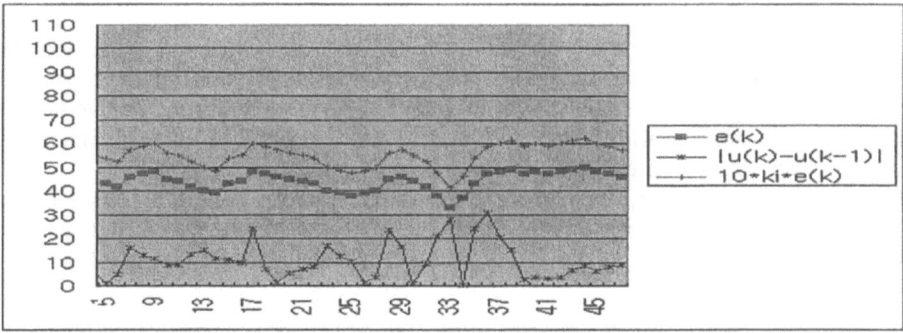

Fig. 6. Smooth system change scenario experiment

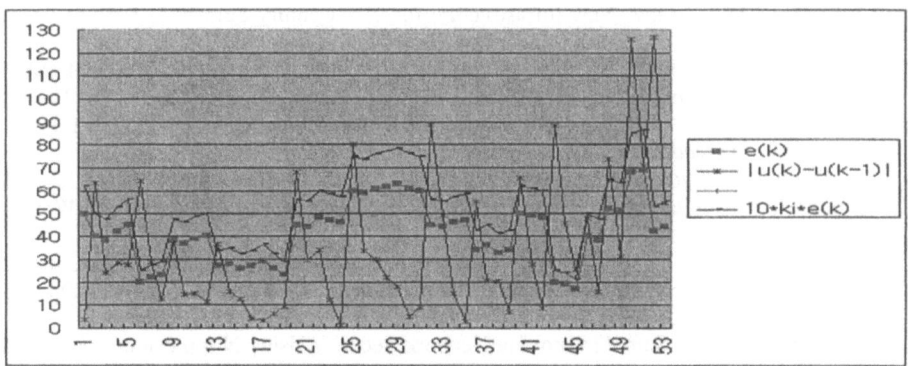

Fig. 7. The significant changes detection at the dynamic scenario

4 Experiments

In this section, two types of experiments are examined. The first experiment supports is done under the stable system. From the Figure 6, no significant change is detected. We change (add and drop) only by two percentage of CPU resource in this experiment.

The second experiment scenario for the dynamic system change is designed. There are CPU changes by 1% up to 15%. From the scenario, interesting result is illustrated in Figure 7.

There looks many significant changes expected but only 10 times out of 30 samples are significant from the Figure 7. When we look at the Table 1 carefully, the proposed scheme takes an influence of history trend and current changes.

From the two experiments, the results are very effective and our novel approach works well for the real-time Grid monitoring system on Grid Service based .

5 Conclusion

Grid will use many resources geographically dispersed. Grid will work together around world with a number of computers and equipments. Monitoring the resource is an essential component to enable better middleware services to users. The overhead of real-time monitoring tools might be tremendous without overhead management. This study provides many interesting factors in monitoring itself to researchers.

Our proposed scheme, the dynamic threshold, performs well in Grid Service environments, which reduces lots overhead and obtained accuracy of events from the experiments. And the complexity of our schemes is very small so that developers can implement easily.

References

[1]   I. Foster., "The Grid: A New Infrastructure for 21st Century Science.", *Physics Today*, 55(2):42-47 2002
[2]   Chervenak, E. Deelman, I. Foster, L. Guy, W. Hoschek, A. Iamnitchi, C. Kesselman, P. Kunst, M. Ripeanu, B, Schwartzkopf, H, Stockinger, K. Stockinger, B. Tierney, November 2002, "Giggle: A Framework for Constructing Scalable Replica Location Services.", *Proceedings of Supercomputing 2002 (SC2002)*
[3]   B. Tierney, R. Aydt, D. Gunter, W. Smith, M. Swany, V. Taylor, R. Wolski, "A Grid Monitoring Architecture", March 2000, *Global Grid Forum Performance Working Group*
[4]   Chung-Hsing Hsu, Ulrich Kremen, " IPERF : A Framework for Automatic Construction of Performance performance models", *Workshop on Profile and Feedback-Directed Compilation* (PFDC), Paris, France. October 1998.
[5]   Luca Deri, Stefano Suin, "Improving Network Security Using N top events", *raid 2000 Materials*.
[6]   tcpdump, http://www.tcpdump.org.
[7]   pingER, http://www.slac.stanford.edu/comp/net/wanmon/tutorial.html

[8]   Mathijs den Burger, Thilo Kielmann, and Henri E. Bal, TopoMon: A Monitoring Tool for Grid Network Topology, *Proceedings of ICCS 02.*

[9]   Tierney, B., B. Crowley, D. Gunter, M. Holding, J. Lee, M. Thompson "A Monitoring Sensor Management System for Grid Environments" *Proceedings of the IEEE High Performance Distributed Computing conference* (HPDC-9), August 2000, LBNL-45260

[10] A. Waheed, W. Smith, J. George, J. Yan. , "An Infrastructure for Monitoring and Management in Computational Grids", In Proceedings of the 2000 Conference on Languages, Compilers, and Runtime Systems.

[11] DAMED working group, https://forge.gridforum.org/projects/damed-wg

[12] Welch, L. R. and B. Shirazi., "DynBench: A Dynamic Real-Time Benchmark for Assessment of QoS and Resource Management Technology". *IEEE Real- time Application System*, 1999.

[13] Ashvin Goel, David Steere, Calton Pu, and Jonathan Walpole, "Adaptive Resource Management Via Modular Feedback Control", Submitted to HOTOS 1999

[14] PID Tutorial, http://www.engin.umich.edu /group /ctm /PID/PID.html

Multiuser CDMA Parameters Estimation by Particle Filter with Resampling Schemes

Jang-Sub Kim[1], Dong-Ryeol Shin[1], and Woo-Gon Chung[2]

[1] School of Information and Communication Engineering,
Sungkyunkwan University,
300 ChunChun-Dong, JangAn-Gu, Suwon, Korea 440-746
{jangsub,drshin}@ece.skku.ac.kr
[2] Computer Science Dept. CSU at Bakersfield, USA
wchung@csub.edu

Abstract. The joint estimation of linear and nonlinear state variables remains challenging, especially in multiuser communications applications where the state dimension is large and signal to noise ratio is low. In this paper, an efficient Particle Filter (PF) is developed to make estimates of nonlinear time delay parameters in the presence of non-Gaussian noise. The PF method has the advantage that the importance weights are easily evaluated and the importance density can be easily sampled. We propose a PF-based algorithm with Resampling schemes for the estimation of closely-spaced path delays and related coefficients in CDMA environments. Furthermore we present a number of resampling schemes, namely: Multinomial Resampling (MR), Residual Resampling (RR) and Minimum Variance Resamplings (MVR). The simulation results show that MR scheme outperforms the other selection schemes. We also show that it provides a more suitable method for tracking time-varying amplitudes and delays in CDMA communication systems than RR and MVR schemes.

1 Introduction

Nonlinearity in the measurement model has always been an obstacle for the reliability of the estimation in the Kalman filtering structure. Extended Kalman Filtering (EKF) approximates the a posteriori distribution to be Gaussian to simplify computation. Recently the unscented transform(UT) has been used in an EKF framework, and the resulting filter, known as the unscented Kalman filter, has been employed to tackle the nonlinearity and shown its effectiveness in terms of the divergence reduction or error propagation [1]. However, all the iterative solutions including Kalman filters appeared in the literature necessarily assume that the unknown parameters are Gaussian distributed.

In a CDMA environment which involves a joint channel coefficient and time-delay tracking techniques, non-Gaussian properties of unknown parameters are inherently taken. In this paper, the particle filter, recently introduced in the communication area, which does not require the Gaussian assumption is employed for such an exponentially distributed parameter like the propagation delay in the multipath environment [2]. The PF results in better performance in the parameter (including

M. Bubak et al. (Eds.): ICCS 2004, LNCS 3038, pp. 1170–1177, 2004.
© Springer-Verlag Berlin Heidelberg 2004

other channel coefficients) estimation [3]. We present PF with resampling techniques to get around degeneracy problem occurring in the PF without resampling schemes. Furthermore we demonstrate a number of resampling schemes, namely: Multinomial Resampling (MR), Residual Resampling (RR) and Minimum Variance Resamplings (MVR). The major contribution of this paper is to adapt the PF filter to the CDMA parameter estimation in the highly nonlinear environments to make easy and practical implementation by resampling techniques and to show performance comparisions among resampling techniques..

This paper is organized as follows. Section 2 introduces the signal and channel model that will be used throughout the paper and a description of problem formulation. Section 3 provides a description of PF and resampling schemes used for parameter estimation. The simulation and results of the performance from computer simulations are given in Section 4. Finally, Section 5 provides concluding remarks.

2   Problem Formulation

2.1   System and Channel Model

In DS spread-spectrum [2], the received baseband signal is given by

$$r(l) = \sum_{k=1}^{K}\sum_{i=1}^{M} c_{k,i}(l) d_{k,m_l} a_k (lT_s - m_l T_b - \tau_{k,i}(l)) + n(l) \tag{1}$$

where $\tau_{k,i}(l)$ and $c_{k,i}(lT_s) = \sqrt{P_k}\, f_{k,i}(lT_s) e^{j\phi_{k,i}(lT_s)}$ are delays and channel coefficient of i th multipaths associated with the k th user's channel, respectively. $a_k(t)$ and $d_{k,m}(l)$ are k th user's PN spreading code and binary data sequence, respectively. AWGN, $n(l)$, is assumed to have zero mean and variance of σ_n^2.

2.2   Problem Formulation

Given the received waveform samples, $r(l)$, the task is to obtain minimum variance estimates of the unknown parameters c_k and τ_k for $k = 1,2,\cdots,K$, given by

$$\hat{c}_k(l|l) = E\{c_k(l)\,|\,\mathbf{R}^l\} \qquad \hat{\tau}_k(l|l) = E\{\tau_k(l)\,|\,\mathbf{R}^l\} \tag{2}$$

where $\mathbf{R}^l$ $\{r(l), r(l-1),\cdots, r(0)\}$ is the set of received samples up to time lT_s.

Let the unknown parameters be represented by the $2K \times 1$ vector

$$\mathbf{x} = \begin{bmatrix} \mathbf{c} & \boldsymbol{\tau} \end{bmatrix}^T \tag{3}$$

where $\mathbf{c} = [c_1, c_2, \cdots, c_K]^T$ and $\boldsymbol{\tau} = [\tau_1, \tau_2, \cdots, \tau_K]^T$.

By [2], we can write the state model as

$$x(l+1) = F(l)x(l) + v(l) \tag{4}$$

where the state $x_l = x(l)$ and $F = diag\{F_c, F_\tau\}$ is $2K \times 2K$ augmented state transition matrix, $v = \begin{bmatrix} v_c^T & v_\tau^T \end{bmatrix}$ is $2K \times 1$ process noise vector with zero mean and covariance matrix $Q = diag\{Q_c, Q_\tau\}$, and $diag(\bullet)$ is diagonal matrix.

The scalar measurement model follows from the received signal of (1) by

$$z(l) = h(x(l)) + n(l) \tag{5}$$

where the measurement $y_l = y(l) = r(l)$, and

$$h(x(l)) = \sum_{k=1}^{K} \sum_{i=1}^{M} c_{k,i}(l) d_{k,m(l)} a_k (lT_s - m(l)T_b - \tau_{k,i}(l))$$

The scalar measurement $z(l)$ is a nonlinear function of the state $x(l)$. Hence, our goal is to find the $2K \times 1$ joint estimator $E\{x(l) | R^l\}$ with estimated error covariance

$$P = E\{[x(l) - \hat{x}(l | l)][x(l) - \hat{x}(l | l)]^T | R^l\} \tag{6}$$

3   Particle Filtering

3.1   Implementing the PF

A particle filter allows for a complete representation of the a posteriori distribution of the states, so that any statistical estimates, such as the mean, modes, kurtosis and variance, can be easily computed. They can therefore, deal with any nonlinearities or distributions. Particle filters rely on importance sampling and, as a result, require the design of proposal distributions that can approximate the a posteriori distribution reasonably well. The most common strategy is to sample from the probabilistic model of the states evolution. We are in a position to compute the particle filter algorithm [1]. The basic procedure to construct a particle filter is shown below. The detailed derivations are presented in [1] and [3]. The pseudo-code of a generic particle filter can now be presented.

1)   Initialization: $t = 0$

For $i = 1, \ldots, N$, draw the states $x_0^{(i)}$ from the prior $p(x_0)$.

2)   For $t = 1, 2, \ldots$

①   Importance sampling step
  - For $i = 1, \ldots, N$, sample $\hat{x}_t^{(i)} \sim q(x_t | x_{0:t-1}^{(i)}, y_{1:t})$
  - For $i = 1, \ldots, N$, evaluate the importance weights up to a normalizing constant

$$\mathbf{w}_t^{(i)} = \mathbf{w}_{t-1}^{(i)} \frac{p(y_t \mid \hat{\mathbf{x}}_t^{(i)}) p(\hat{\mathbf{x}}_t^{(i)} \mid \hat{\mathbf{x}}_{t-1}^{(i)})}{q(\hat{\mathbf{x}}_t^{(i)} \mid \mathbf{x}_{0:t-1}^{(i)}, y_{1:t})}$$

- For $i = 1, \ldots, N$, normalize the importance weights:

$$\tilde{\mathbf{w}}_t^{(i)} = \mathbf{w}_t^{(i)} / \sum_{j=1}^{N} \mathbf{w}_t^{(j)}$$

② Resampling

- Multiply/Suppress samples $\hat{\mathbf{x}}_{0:t}^{(i)}$ with high/low importance weights $\tilde{\mathbf{w}}_t^{(i)}$, respectively, to obtain N random samples $\mathbf{x}_{0:t}^{(i)}$ approximately distributed according to $p(\mathbf{x}_{0:t}^{(i)} \mid y_{1:t})$.

- For $i = 1, \ldots, N$, set $\mathbf{w}_t^{(i)} = \tilde{\mathbf{w}}_t^{(i)} = 1/N$

③ Output : The output of the algorithm is a set of samples that can be used to approximate the a posteriori distribution as follows

$$p(\mathbf{x}_{0:t} \mid y_{1:t}) = 1/N \sum_{i=1}^{N} \delta_{\mathbf{x}_{0:t}^{(i)}}(d\mathbf{x}_{0:t})$$

④ One obtains straightforwardly the following estimate of $E[g_t(\mathbf{x}_{0:t})]$

$$E[g_t(\mathbf{x}_{0:t})] = 1/N \sum_{i=1}^{N} g_t(\mathbf{x}_{0:t}^{(i)})$$

where the variables $\mathbf{w}_t$ is known as the normalized *importance weights* and y_t is measurement signal and the a posteriori density $p(\mathbf{x}_{0:t} \mid y_{1:t})$, where $\mathbf{x}_{0:t} = \{\mathbf{x}_0, \mathbf{x}_1, \cdots, \mathbf{x}_t\}$ and $y_{1:t} = \{y_1, y_2, \cdots, y_t\}$, constitutes the complete solution to the sequential estimation problem.

For example, letting $g_t(\mathbf{x}_{0:t}) = \mathbf{x}_{0:t}$ yields the optimal MMSE estimate $\hat{\mathbf{x}}_{0:t} = E[\mathbf{x}_{0:t} \mid y_{1:t}]$. The particles $\mathbf{x}_{0:t}^{(i)}$ are assumed to be independent and identically distributed (i.i.d.) for the approximation to hold. It is usually difficult to takes samples from given a posteriori distribution. Bayesian theory proves to result in another useful concept, the *proposal distribution*, $q(\mathbf{x}_{0:t} \mid y_{1:t})$, for the Monte Carlo Estimation calculation[1]. Here, the particle filters considered are three resampling schemes. The PF algorithm can be easily derived from the SIS algorithm by an appropriate choice of : (i) The importance density: $q(\mathbf{x}_t \mid \mathbf{x}_{k-1}^{(i)}, y_{1:t})$ is chosen to be the prior density $p(\mathbf{x}_t \mid \mathbf{x}_{t-1}^{(i)})$, and (ii) Resampling step: to be applied at every time index. The above choice of importance density implies that we need samples from $p(\mathbf{x}_k \mid \mathbf{x}_{k-1}^{(i)})$. Thus, the importance weights evaluate $\mathbf{w}_t^{(i)} = p(y_t \mid \mathbf{x}_t^{(i)})$. As the importance sampling density for the PF is independent of measurement y_t, the state space is explored without any knowledge of the observations. As resampling is applied at each iteration, this can result in rapid loss of diversity in particles. However, the PF method does have the advantage that the importance weights are easily evaluated and the importance density can be easily sampled.

3.2 Resampling Schemes

The particle filters rely on sequential importance sampling (SIS). The SIS algorithm discussed so far has a serious limitation: the variance of the importance weights increases stochastically over time. To avoid the degeneracy of the SIS simulation method, a resampling (selection) stage may be used to eliminate samples with low importance weights and multiply samples with high importance weights. It is possible to see an analogy to the steps in genetic algorithms [4]. A selection scheme associates to each particle $x_{0:t}^{(i)}$ a number of "children", say $N_i \in \mathbb{N}$, such that $\sum_{i=1}^{N} N_i = N$. Several selection schemes have been proposed in the literature. These schemes satisfy $E(N_i) = N\tilde{w}_t^{(i)}$ but their performance varies in terms of the variance of the particles $\mathrm{var}(N_i)$. Results in [5] indicate that the restriction $E(N_i) = N\tilde{w}_t^{(i)}$ is unnecessary to obtain convergence results. So it is possible to design biased but computationally inexpensive selection schemes. We will now present a number of selection or resampling schemes, namely: Multinomial Resampling (MR), Residual Resampling (RR) and Minimum Variance Resampling (MVR). We found that the specific choice of resampling scheme affects the performance of the particle filter.

3.2.1 Multinomial Resampling

Resampling involves mapping the Dirac random measure $\{x_{0:t}^{(i)}, \tilde{w}_t^{(i)}\}$ into an equally weighted random measure $\{x_{0:t}^{(j)}, N^{-1}\}$. This can be accomplished by sampling uniformly from the discrete set $\{x_{0:t}^{(i)}; i = 1, \cdots, N\}$ with probabilities $\{\tilde{w}_{0:t}^{(i)}; i = 1, \cdots, N\}$. Gordon [6] gave a mathematical proof of this result. After constructing the cumulative distribution of the discrete set, a uniformly drawn sampling index i is projected onto the distribution range and then onto the distribution domain. The intersection with the domain constitutes the new sample index j. That is, the vector $x_{0:t}^{(j)}$ is accepted as the new sample. Clearly, the vectors with the larger sampling weights will end up with more copies after the resampling process.

Sampling N times from the cumulative discrete distribution $\sum_{i=1}^{N} \tilde{w}_t^{(i)} \delta_{x_{0:t}^{(i)}}(dx_{0:t})$ is equivalent to drawing $\{N_i; i = 1, \cdots, N\}$ from a multinomial distribution with parameters N and $\tilde{w}_t^{(i)}$. As we are sampling from a multinomial distribution, the variance is $\mathrm{var}(N_i) = N\tilde{w}_t^{(i)}(1 - \tilde{w}_t^{(i)})$. As pointed out in [7], it is possible to design selection schemes with lower variance.

3.2.2 Residual Resampling

This procedure involves the following steps [4]. Firstly, set $\tilde{N}_i = \lfloor N\tilde{w}_t^{(i)} \rfloor$. Secondly, perform an SIR procedure to select the remaining $\overline{N}_t = N - \sum_{i=1}^{N} \tilde{N}_i$ samples with new weights $w_t'^{(i)} = \overline{N}_t^{-1}(\tilde{w}_t^{(i)} N - \tilde{N}_i)$. Finally, add the results to the current $\tilde{N}_i$. For this scheme, the variance $(\text{var}(N_i) = \overline{N}_i w_t'^{(i)} (1 - w_t'^{(i)}))$ is smaller than the one given by the SIR scheme. Moreover, this procedure is computationally cheaper.

3.2.3 Minimum Variance Resampling

This strategy includes the stratified/systematic sampling procedures introduced in [5]. One samples a set of N points U in the interval $[0, 1]$, each of the points a distance N^{-1} apart. The number of children N_i is taken to be the number of points that lie between $\sum_{j=1}^{i-1} \widetilde{w}_t^{(j)}$ and $\sum_{j=1}^{i} \widetilde{w}_t^{(j)}$. This strategy introduces a variance on N_i even smaller than the residual resampling scheme, $(\text{var}(N_i) = \overline{N}_i w_t'^{(i)} (1 - \overline{N}_i w_t'^{(i)}))$

4 Simulation

We now examine the performance of the PF for making parameter estimates for a multiuser detector. We compare the PF-based estimator with among three resampling schemes. The multipath coefficients of the channel can be generated using Clarke and Gans Fading Model [8, Chap 4], which provides taps with the appropriate distributions and near the correct tap autocorrelations, although the taps are somewhat correlated. For simplification purposes, we consider no multipath. For the state model, the augmented state transition matrix of (5) was chosen to be $\mathbf{F} = 0.999\mathbf{I}$. Also the process noise covariance matrix was $\mathbf{Q} = 0.001\mathbf{I}$.

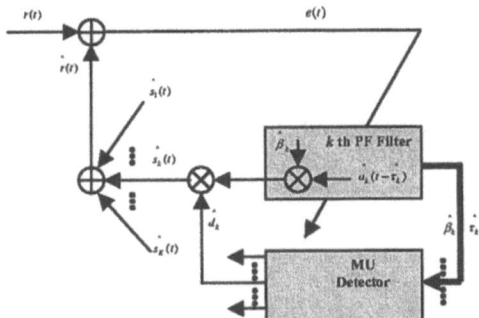

Fig. 1. Operation of multiuer parameter estimation receiver

We simulate a two-user scenario where the users' PN spreading codes are chosen from the set of Gold codes of length 31 and generated by the polynomials $x^5 + x^2 + 1$ and $x^5 + x^4 + x^3 + x^2 + 1$. The SNR (Signal-to-Noise Ratio) at the receiver of weaker user is 10dB. The Near-Far Ratio is 20dB. Oversampling factor (sample/chip) is 2. The SNR for the weaker user is set at 10dB. One aspect about using PF is that they require proper initialization. Depending on the problem, the initial guesses may need to be close to the correct value for convergence. For the simulation results, we assume such an initial estimator is used to start the tracking algorithm fairly close to the true values. Furthermore, we note that the data bits, $d_{k,m}$, are not included in the estimation process, but are assumed unknown a priori. In the simulations, we assume that the data bits are available from decision-directed adaptation, where the symbols $d_{k,m}$ are replaced by the decisions $\hat{d}_{k,m}$ shown in Fig. 1.

The tracker for a two-user system is simulated for a fading channel where the channel coefficients are time varying, but the delays remain constant. A simple channel model is assumed for each user with a single tap (flat fading). Furthermore it is assumed that each user is moving with a Doppler frequency of 200 Hz for User 1 and 300 Hz for User 2. The fading for in-phase and quadrature components of the the channel coefficients for each user was implemented by IFFT and normalized so that the average power is unity. The sampling time is taken as $T_c = 1/(1.2288Mbps \times 2)$ and The bit rate is assumed to be $1/T_b = 9600bps$ with processing gain 31.

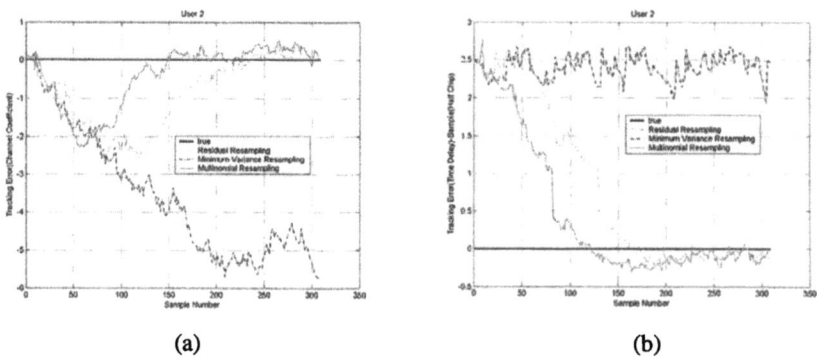

(a)   (b)

Fig. 2. Parameter estimation errors for (a) channel amplitudes and (b) time delays with a near-far ratio of 0dB

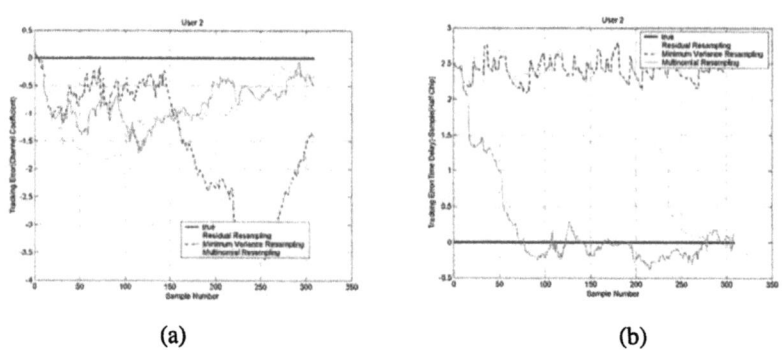

(a)   (b)

Fig. 3. Parameter estimation errors for (a) channel amplitude and (b) time delays with near-far ratio of 20dB

Fig. 2 shows the estimation error for channel coefficients and time delays with imperfect power controlled using the Multinomial Resampling, Residual Resampling and Minimum Variance Resampling, respectively. As the figure indicates, the estimator/tracker with MR and RR is able to accurately track the time-varying channel coefficients of each user, even for fast fading rates. But MVR is diverge. The results for the same set of assumptions, but with a near-far ratio of 20dB are shown in Fig. 3.

Again, the estimator is able to accurately converge to the correct values of the parameters for both the MR and RR. But MR algorithm indicates to converge faster RR.

It is seen that a user is able to accurately converge to the correct delays and channel coefficient for PF with MR and RR scheme. PF with MR scheme indicates to converge faster and it has smaller mean squared error (MSE).

5 Conclusions

We have presented a parameter estimator based on the RPF that are capable of estimating channel coefficients and time delays in MAI (near-far ratio = 20dB). The PF with MR has been demonstrated to have superior performance over RR and MVR scheme. PF with MR can provide a better alternative to nonlinear filtering than RR or MVR schemes since it possible to design selection schemes with lower variance. Computer simulations also show that it provides a more effective technique for tracking time-varying amplitudes and delays in CDMA communication systems than RR and MVR. Furthermore the PF with MR estimator is shown to have the ability to converge to the user's true coefficients and time delays for a near-far ratio of 20 dB.

References

1. Rudolph van der Merwe, Arnaud Doucet, Nando de Freitas, Eric Wan, "The Unscented Particle Filter," Oregon Graduate Institute Electrical and Computer Engineering, OR 97006, USA, August 16, 2000.
2. James J. Caffery, Wireless Location in CDMA Cellular Radio Systems, Kluwer Academic Publishers, 2000.
3. J.S. Kim and D.R. Shin, " Multiuser CDMA Parameters Estimation by Recursive Particle Filter", Multiuser CDMA Parameters Estimation by Recursive Particle Filtering(RPF), The 2003 Int'l Multiconference in Computer Science and Engineering, June 23~26, 2003, pages:444~447.
4. Higuchi, T. "Monte Carlo filter using the genetic algorithm operators," Journal of Statistical Computation and Simulation 59 , 1997, pages: 1-23.
5. Kitagawa, G. "Monte Carlo filter and smoother for non-Gaussian nonlinear state space models," Journal of Computational and Graphical Statistics, 1996, pages: 1-25.
6. Gordon, N. J. Bayesian Methods for Tracking, PhD thesis, Imperial College, University of London, 1994.
7. Liu, J. S. and Chen, R. "Sequential Monte Carlo methods for dynamic systems," Journal of the American Statistical Association, 1998, pages:1032-1044.
8. Theodore S. Rappaport, Wireless Communications, IEEE Press, 1996.

Routing, Wavelength Assignment in Optical Networks Using an Efficient and Fair EDP Algorithm

Pallavi Manohar and V. Sridhar*

Applied Research Group, Satyam Computer Services Ltd.,
14 Langford Avenue, Lalbagh Road, Bangalore 560 025 INDIA.
{pallavi_mm, sridhar}@satyam.com

Abstract. Routing and wavelength assignment (RWA) problem in wavelength routed optical networks is typically solved using a combination of integer programming and graph coloring. Such techniques are complex and make extensive use of heuristics. RWA with maximum edge disjoint path (EDP) using simple bounded greedy algorithm is shown to be as good as previously known solution method. In this paper, we present shortest path first greedy algorithm for maximum EDP with construction of path conflict graph which gives fair and efficient solution to the RWA problem in optical networks.

Keywords: Optical Networks, Routing and Wavelength Assignment, Edge Disjoint Path Algorithms.

1 Introduction

Routing and wavelength assignment (RWA) for the lightpaths of the virtual topology in wavelength routed optical networks is typically solved in two parts – lightpath routing, to determine the route of the lightpaths on the physical topology, and wavelength assignment, to assign the wavelengths to each lightpath in the virtual topology such that wavelength constraints are satisfied for each physical link. The RWA problem for a given physical network and virtual topology is known to be NP-complete. Further, routing and wavelength assignment problems are each known to be NP-complete [3] and many heuristic solutions have been proposed [7,8]. See [3,10,11] for a survey of the various solutions to the RWA problem. Lightpath routing is done by first selecting the lightpaths and then assigning routes to them. The order in which the lightpaths are assigned can be selected by different heuristic schemes such as random, fixed, longest-first, or shortest-first. Routes can be found by techniques such as shortest path algorithm, weighted shortest path, or $K-$shortest path algorithms. The route is selected from the available candidates using different schemes such as random, first-fit, or probability. The RWA problem is also solved by formulating it as an

---

* *Member, IEEE*

M. Bubak et al. (Eds.): ICCS 2004, LNCS 3038, pp. 1178–1184, 2004.
© Springer-Verlag Berlin Heidelberg 2004

integer linear program (ILP) optimization problem [9] with one of two objective functions – minimize the required number of wavelengths to establish the given set of lightpaths and maximize the number of established lightpaths with a fixed number of wavelengths. A hard version of the problem is when no wavelength conversion is used at the routing nodes of the physical network and this requires that the wavelength continuity constraint be satisfied, i.e., a lightpath should use the same wavelength on all the links in its path. Extensive and complex heuristics are used to solve the ILP. One solution considers a fractional relaxation with rounding techniques to obtain an integral solution and wavelength assignment with graph colouring method [1].

Solving the RWA problem using the mathematical programming is quite complex in view of the large number of variables and constraints, and complex problem size reduction heuristics. Further, it will not be known how close the solution is to the optimal. A simple and intuitive alternative method for RWA with performance comparable to that from mathematical programming techniques is given in [8]. In the next section we describe their method based on the simple greedy solution of a well known graph theory problem. Then, in Section 3 we present our solution for RWA using fair and efficient greedy solution for the same graph theory problem. We then give approximation bounds for the algorithm and conclude the paper in Section 5.

2   RWA Using Edge Disjoint Paths

In this section, we describe simple and intuitive solution to the RWA problem based on a well known graph theory problem [8]. Observe that the lightpaths that are assigned the same wavelength do not traverse through same physical link, i.e., these paths are *edge disjoint*. In graph theory, given a graph and a set of source-destination pairs and the requirement that a path be found for as many of the pairs as possible, finding the maximum set of edge disjoint paths is a well known problem and is formally defined as follows. Let $G = (V, E)$ be the graph of the physical network with no wavelength conversion, V the vertex set, and E the edge set. Connection request i is specified by a pair (s_i, t_i), $(s_i, t_i) \in V$. Let τ be the set of connections for whom edge disjoint paths need to be found, $\tau = \{(s_1, t_1), \ldots, (s_k, t_k)\}$. τ is said to be realizable in G if there exist mutually edge-disjoint paths $P_1, \ldots, P_k$ in G such that P_i has endpoints s_i and t_i. The *maximum edge disjoint paths problem* is to find a maximum size realizable subset of τ, given G and τ. Let the maximum size of a realizable subset of τ in G by $\alpha(\tau)$. It is easy to see that the maximum edge disjoint path (EDP) problem is a combinatorial optimization problem and is known to be NP-hard [5].

Given G and τ, let $\tau_1, \cdots \tau_i$ be partitions of τ such that partition is a solution to the maximum EDP problem on G, τ. Now the application to the RWA problem is follows. Given G and τ consider a solution of the maximum EDP problem that obtains a realizable set $\tau_1 = \alpha(G, \tau)$. Since the paths in τ_1 are edge disjoint they can be assigned the same wavelength, say λ_1. Now construct the set $\tau_1' = \tau - \tau_1$, i.e., τ_1' is the set of connections not contained in τ_1. An iteration of maximum

EDP on the G and τ_1' will give $\tau_2 = \alpha(G, \tau_1')$ and the paths in τ_2 can be assigned the same wavelength, say λ_2. This process can be repeated till τ_i' is empty for some i. The minimum such i will be denoted by $\chi(G, \tau)$. Finding $\chi(G, \tau)$ is the problem of minimizing the number of wavelengths in wavelength routed optical networks. It is easy to see that if the solution to the maximum EDP is optimum, then the optimum $\chi(G, \tau)$ can be obtained from above.

Simple greedy algorithm for maximum EDP is used in [8] and it is shown that it performs as good as that of more complex solution methods based on integer linear programming. They use bounded greedy algorithm (BGA) for the maximum EDP that was first described in [5]. Here, the order in which the lightpaths are selected is random. In this paper, we present improved greedy algorithm for maximum EDP. This uses shortest-first heuristics for the selection of lightpaths to be assigned and construction of path conflict graph [7] for assigning the routes from available candidate routes. In general, any network has the property that longer paths are likely to experience higher blocking than shorter ones. Consequently, the degree of fairness can be quantified by defining the unfairness factor as the ratio of the blocking probability on the longest path to that on the shortest path for a given RWA algorithm. Blocking of long lightpaths leaves more resources, such as wavelengths on the physical link in wavelength routed networks, available for short lightpaths. The problem of unfairness is more pronounced in networks without converters since finding long paths that satisfy the wavelength continuity constraint is more difficult than without this constraint. This motivates the use *shortest path first greedy* algorithm (SGA) [6] for RWA in wavelength routed optical networks. The first fit shortest path among the enumerated alternate shortest paths is selected in [8]. This may not give all possible disjoint paths in that partition. Basically this doesn't make use of the available alternate shortest paths which may lead to an inefficient solution. Further the algorithm can be improved with construction of *path conflict* graph from which a set of *desirable* routes for all (s_i, t_i) pairs in τ can be obtained. Thus, the route for each (s_i, t_i) pair is selected from this set. This efficiently packs disjoint paths for each partition and hence motivates its use for RWA especially in the case of reducing the blocking probability. In the following, we describe details of the solution of the RWA problem using the fair and efficient greedy algorithm to solve the maximum EDP.

3   RWA with Fair and Efficient Greedy EDP

For solving RWA using maximum EDP, in [8], a lightpath is selected randomly from traffic matrix and first fit shortest path is assigned to it. We improve this algorithm in two ways. First, by using SGA, that is, by selecting the lightpath from the traffic matrix which has minimum length shortest path. This implements fairness in routing for optical networks. The shorter paths are always assigned before the longer ones. Secondly, we make use of *alternate* shortest paths found for each (s_i, t_i) pair. Basically, we prune the set of routes further such that a route sharing minimum number of edges with routes for $(\tau - (s_i, t_i))$

is selected for each (s_i, t_i) pair. Then, from this *desirable* set, route is assigned. Note that it may not reduce the total number of wavelengths required in all the cases but it efficiently finds possible disjoint paths that are assigned the same wavelength. This improves reuse of the wavelength further. This *desirable* set of routes is obtained using path conflict graph as explained further in this section. We consider RWA for static case as selection of lightpath as well as route is performed offline while the algorithm in [8] can be applied to static and dynamic cases.

3.1 RWA with Shortest Path First Greedy EDP

In this section, we describe the details of the solution of the RWA problem using SGA to solve the maximum EDP. The inputs to the algorithm are the graph G of the physical network with no wavelength conversion and the lightpath set τ. The output of the algorithm is $\mathcal{P}$ which is the set of pairs $((s_i, t_i), P_i)$ where (s_i, t_i) is a routed lightpath and P_i is the physical path assigned to this lightpath. The algorithm, SGAforEDP(G, τ), operates as follows. (a) Select a lightpath (s_i, t_i) from the set τ, such that the shortest path P_i in G for this connection has minimum length among those of all requests in the set τ. If there are more than one candidates which can be selected, then choose one of them arbitrarily; (b) Find the shortest path P_i from *desired* set of routes (explained in Sect. 3.2) for this connection; (c) Add $((s_i, t_i), P_i)$ to the path set $\mathcal{P}$ and (s_i, t_i) to the set of routed lightpaths $\alpha(G, \tau)$ and delete the edges in G used by P_i; (d) Remove (s_i, t_i) from the set τ and repeat this while τ contains a request which can be routed in G. The algorithm is summarized in Fig. 1. Note that $\alpha(G, \tau)$ contains the lightpaths that get assigned the same wavelength.

Algorithm SGAforEDP(G, τ)
Begin
    $\alpha(G, \tau) = \phi$; $\mathcal{P}(G, \tau) = \phi$
    While (τ contains a request which can be routed in G) do
        If $((s_i, t_i)$ such that a shortest path P_i from
        s_i to t_i in G has minimum length among the
        requests in τ) then
            $\alpha(G, \tau) = \alpha(G, \tau) \bigcup (s_i, t_i)$
            $\mathcal{P}(G, \tau) = \mathcal{P}(G, \tau) \bigcup ((s_i, t_i), P_i)$
            $\tau = \tau - (s_i, t_i)$
            Delete edges in path P_i from G
        EndIf
        EndWhile
End

Fig. 1. Pseudo code for algorithm SGAforEDP

Each iteration of the SGAforEDP results in a set of lightpaths $\alpha(G, \tau)$ that can be assigned the same wavelength. And τ is the set of unassigned lightpaths. We run SGAforEDP with the physical topology graph G and the set of unassigned lightpaths to obtain the set of lightpaths that get assigned a distinct wavelength. This is repeated till all the lightpaths are assigned. The pseudo code is shown in Fig. 2.

```
Algorithm Greedy_EDP_RWA(G,τ)
Begin
     λ = 0
     While ( τ ≠ φ) do
          λ = λ + 1
          SGAforEDP(G,τ)
          Assign λ to all paths Pᵢ ∈ P
     EndWhile
End
```

Fig. 2. Pseudo code for the algorithm Greedy_EDP_RWA

3.2   Obtain Desired Set of Routes

The shortest path for a selected lightpath in Fig. 1 is chosen from *desired* set of routes. This set is obtained after pruning enumerated alternate shortest paths using path conflict graph [7]. This is a graph having k partitions each for (s_i, t_i) pair. The path conflict graph $G_P = (V_P, E_P)$ has V_P vertices and each vertex represents one route. For an (s_i, t_i) pair there are one or more candidate routes that is alternate shortest paths which implies $|P_{s_i,t_i}| \geq 1$. Thus there are $\sum_{i=1}^{k} |P_{s_i,t_i}|$ vertices in V_P. Because of construction of G_P, the vertices in V_P can be partitioned into k disjoint sets such that $V_P = V_{s_1,t_1} \bigcup \ldots \bigcup V_{s_k,t_k}$. Each partition corresponds to one (s_i, t_i) pair. For any $u, v \in V_P, (u, v) \in E_P$ if and only if two routes corresponding to u and v share at least one edge in $G = (V, E)$. Now the question is how to choose the *best* vertex that is route from each partition to obtain *desired* set of routes. We use the heuristic given in [7] which is as follows. First remove all edges whose vertices are in the same partition, then choose a vertex with the smallest degree as the *best* vertex and remove all other vertices in the same partition. Continue this to obtain such a vertex from each partition. This gives us the *desired* set of routes.

3.3   Limited Wavelengths

The RWA can also be solved with objective of maximizing the number of lightpaths that are routed using a fixed number of wavelengths, say L. This is similar to the objective function of Sivarajan and Ramaswami [9]. The Greedy_EDP_RWA algorithm defined above can be used in this form of RWA as follows. Observe that there can only be a maximum of L partitions of the lightpath set τ with each partition being assigned the same wavelength. It is easy to see that we need to perform L iterations of the main loop of Greedy_EDP_RWA. The pseudo code for this algorithm called as Greedy_EDP_RWA_Limλ is shown in Fig. 3. The carried traffic τ_{carried} is the set of lightpaths for which route and wavelength is assigned whereas blocked traffic τ_{blocked} is the set of unassigned lightpaths which is blocked due to limited number of wavelengths available in the network. Use of *desired* set of routes makes more sense in this case as it efficiently packs disjoint routes for each wavelength. This increases carried traffic reducing the blocking probability further.

Algorithm Greedy_EDP_RWA_Lim$\lambda(G,\tau)$
Begin
    $\lambda = 0$
    $\tau_{carried} = \phi$; $\tau_{blocked} = \phi$
    For $i = 1$ to L
        $\lambda = \lambda + 1$
        SGAforEDP(G,τ)
        Assign λ to all paths $P_i \in \mathcal{P}$
        $\tau_{carried} = \tau_{carried} \bigcup \alpha(G,\tau)$
    EndFor
    $\tau_{blocked} = \tau$
End

Fig. 3. Pseudo code for the algorithm Greedy_EDP_RWA_Limλ

4   Approximation Bounds

The maximum EDP is NP-hard and is also hard to approximate and good heuristics are not well known [5]. The best known approximation guarantee for arbitrary graphs is $\mathcal{O}(\max \sqrt{m}, \mathtt{diam}(G))$, where m is the number of edges in set E and $\mathtt{diam}(G)$ is the diameter of the graph G [5]. Srinivasan and Baveja [2] improved the approximation algorithms based on LP relaxation. Guruswami *et al* [4] have shown that simple greedy algorithms and randomized rounding algorithms can yield approximations similar to those given in [2]. Greedy algorithms have been extensively studied in combinatorial optimization due to their elegance and simplicity. The approximability of the maximum edge disjoint paths problem (EDP) in directed graphs is seemingly settled by the $\Omega(m^{1/2-\epsilon})$-hardness result of Guruswami *et al* [4] and the $\mathcal{O}(\sqrt{m})$ approximation achievable for the greedy algorithm [5]. Let OPT denote the integral optimum value for the optimal solution for a given instance. Kolliopoulos and Stein [6] show similar bounds for the shortest first greedy algorithm obtaining
$\max(OPT/\sqrt{m_0}, OPT^2/m_0, OPT/d_0)$ paths where m_0 and d_0 are the number of edges and average length of paths respectively, in some optimum solution. The underlying idea in SGA is that if one keeps routing commodities along sufficiently short paths the final number of commodities routed is lowerbounded with respect to the optimum. It is clear that the approximation ratio of SGA with construction path conflict graph is at least as good as that of BGA. As $E_0 \subseteq E$ is the set of edges used by the paths in an optimal solution and $m_0 = |E_0|$, SGA rather provides the better approximation bound than that of simple BGA. The approximation existentially improves when $|E_0| = o(|E|)$, i.e., when the optimal solution uses a small portion of the edges of the graph.

5   Conclusion

In this paper, we have presented shortest first greedy edge disjoint path based solution for the routing of lightpaths of the virtual topology of an optical network and the assignment of wavelengths to them. This method is appealing in its simplicity and employs fairness aspect providing better approximations than that of RWA with simple greedy for maximum EDP. Further, this algorithm can also be used to solve the RWA in which the total number of available wavelengths is upper bounded and the carried traffic can be improved with the construction of path conflict graph.

References

1. Banerjee D. and B. Mukherjee, "A Practical Approach for Routing and Wavelength Assignment in Large Wavelength Routed Optical Networks," *IEEE Journal on Selected Areas in Communications*, vol 14, no 5, pp. 903-908, June 1996.
2. Baveja A. and A. Srinivasan, "Approximation Algorithms for Disjoint Paths and Related Routing and Packing Problems," *Mathematics of Operations Research*, vol 25, pp. 255-280, May 2000.
3. Choi J., N. Golmie, F. Lapeyrere, F. Mouveaux, and D. Su, "A Functional Classification of Routing and Wavelength Assignment Schemes in DWDM Networks : Static Case," *Proceedings of VII International Conference on Optical Communication and Networks*, Jan 2000.
4. Guruswami V., S. Khanna, R. Rajaraman, B. Shepherd, M. Yannakakis, "Near-Optimal Hardness Results and Approximation Algorithms for Edge-Disjoint Paths and Related Problems," *Proc. of ACM Symp. on Theory Comput.*, pp. 19-28, 1999.
5. Kleinberg J., "Approximation Algorithms for Disjoint Paths Problems," *PhD Thesis*, MIT, Cambridge, MA, May 1996.
6. Kolliopoulos S., C. Stein, "Approximating Disjoint-Path Problems Using Greedy Algorithms and Packing Integer Programs," *Lecture Notes in Computer Science*, vol 1412, 1998.
7. Li G. and R. Simha, "The Partition Coloring Problem and its Application to Wavelength Routing and Assignment," *First Workshop on Optical Networks*, Dallas, TX, 2000.
8. Manohar P., D. Manjunath and R. K. Shevgaonkar, "Routing and Wavelength Assignment in Optical Networks from Edge Disjoint Path Algorithms," *IEEE Communications Letters*, vol 6, pp. 211-213, May 2002,.
9. Ramaswami R., K. Sivarajan, "Routing and Wavelength Assignment in All-Optical Networks," *IEEE/ACM Transactions on Networking*, vol 3, no 5, pp. 489-500, October 1995.
10. Rouskas G., "Routing and Wavelength Assignment in Optical WDM Networks," *Wiley Encyclopedia of Telecommunications*, 2001.
11. Zang H., J. Jue, and B. Mukherjee, "A Review of Routing and Wavelength Assignment Approaches for Wavelength-Routed Optical WDM Networks," *SPIE/Baltzer Optical Networks Magazine*, vol. 1, no. 1, Jan. 2000.

Route Optimization Technique to Support Multicast in Mobile Networks[*]

Kiyong Park[1], Sunyong Han[1], Bok-gyu Joo[2], Keecheon Kim[1], and Jinpyo Hong[3]

[1] Kon-kuk University,
[2] Hongik University,
[3] Hankuk University of Foreign Studies,
Seoul, Korea
kypark@cclab.konkuk.ac.kr, {syhan,kckim}@konkuk.ac.kr,
bkjoo@hongik.ac.kr, jphong@hufs.ac.kr

Abstract. The technology called Network Mobility (NEMO) is proposed recently by IETF to support the network mobility. Meanwhile, as the needs of group communications increase in the internet, supporting multicast services became a challenging issue on NEMO as on Mobile IP. This paper explores issues and problems of providing multicast service on mobile networks and presents a route optimization technique along with an architecture that enables optimal and efficient multicast service on NEMO environments.

1 Introduction

Since IETF proposed Mobile IP technology [1] as a technique to provide host mobility, many research works have been made to support host mobility in wireless environments. Nowadays, people want to enjoy surfing the internet even in the moving vehicles such as buses and planes, even if their equipments do not support mobility. Meanwhile, the importance of group communications has been recognized in various applications such as remote lectures, on-line games, etc. This is true even in wireless and mobile environments.

IETF recently proposed a new architecture called Network Mobility(NEMO) basic support architecture in order to answer the requirements for network mobility [2][3]. In early Mobile-IP(MIP) technologies, basic architectures such as MIP-BT, MIP-RS [7] and Multicast Proxy Server [8], were suggested to provide multicast service for mobile hosts. Because the NEMO basic support architecture is based on Mobile IP, it inherits from Mobile IP the same fundamental problems such as route optimization problem [9], tunnel convergence problem, mass of multicast problem, when it is used to support multicast for mobile networks. Therefore, we have to reinvestigate multicast support mechanism suitable for NEMO environments.

In this paper, we propose new schemes for supporting multicast on NEMO, by extending the functions of mobile router and adding new parts of router advertisement messages in MIPv6 network. We adopted Prefix Delegation mechanism suggested for route optimization in MIPv6 in order to provide route optimization on multicasting.

[*] This research is supported by University IT Research Center Project.

M. Bubak et al. (Eds.): ICCS 2004, LNCS 3038, pp. 1185–1192, 2004.
© Springer-Verlag Berlin Heidelberg 2004

For seamless multicast service even when a mobile network changes its location, we extended mobile router functions to manage the creation and coordination of dynamic tunneling with other multicast routers. Our approach provides an efficient and optimal multicasting and will become a fundamental architecture for supporting multicast on NEMO.

2 Related Researches

2.1 Network Mobility (NEMO) Basic Support Architecture

To support network mobility, some basic requirements are to be satisfied [4] on existing networks. Those requirements are session continuity, connectivity, reachability, and nested mobility support. Most of the requirements are met by employing a mobile router(MR), which is in charge of its own network mobility. Because the MR acts not only as a mobile node(MN) but also as a router, it provides a route path to on-link nodes in its network and guarantees network connectivity to the internet. To support those requirements basically in NEMO basic support architecture, bi-directional tunneling is necessary between the MR and its Home Agent (HA) so that on-link nodes in the MR's network can be reachable in the internet

A Correspondent Node (CN) may not know where the target node is placed at the first time during communication after a mobile network finished its movement. Therefore, the HA must find out the location of the node and forward data to it. The location information of the node with which the CN wants to communicate, however, does not exist in the binding list of the HA. This is because the node didn't make any registration request message for movement notification. In spite of that, the HA can find out the location information of the MR containing the node, because the MR makes a registration request process to the HA when it moves. So the HA tunnels to MR's CoA and the MR forward the data from the HA to the nodes in its network.

2.2 MIP-BT (Bi-directional Tunneling) and MIP-RS (Remote Subscription)

In MIP-BT, a mobile node in a foreign network receives multicast data packets from the mobility agent in the home network by bi-directional tunneling. This approach assumes that the home agent has multicast router functions or there is a multicast router in the home network. The home agent intercepts multicast packets that the mobile node used to receiving, encapsulates and transmits them to the mobility agent in the foreign network (called 'foreign agent'). When the foreign agent receives these packets, it decapsulates them and sends to the local network. In MIP-RS, when a mobile node moves to a new network, it sends IGMP messages to the local network in order to rejoin the multicast group [6], so that it can receive multicast data packets from the local multicast router of the foreign network.

The major advantage of MIP-BT approach is it can minimize data packets loss and there is no delay caused by reconstructing a multicast route tree. The major advantage of MIP-RS approach is that the mobile node is served directly by the local multicast router, so that multicast data packet route is optimized

2.3 Prefix Delegation (PD) Mechanism for Route Optimization on NEMO

NEMO basic support architecture preserves session continuity by employing bi-directional tunnel between a MR and its HA. This scheme is reasonable only for a small-scale mobile network because MR has to encapsulate and decapsulate all packets for its MNs.

The purpose of PD mechanism [5] is to enable MNs behind the MR to perform Mobile IPv6 route optimization. In this mechanism, every MNs under a mobile network is allocated native routable internet address as its CoA from the network prefixes in Route Advertisement (RA) messages. That means a MR does not advertise the home network prefixes but newly-acquired network prefixes from the active router (AR)'s RA message. This process is repeated recursively on nested mobile networks, and eventually all MNs and MRs under the AR can be native routable.

2.4 Summary of Current Problems

Because NEMO basic support architecture has inherited Mobile IP and related techniques, many serious problems in supporting multicast in Mobile IP still exist in NEMO. Major problems are tunnel convergence problem, mass of multicast problem, and route optimization problem. Furthermore, because NEMO basic support is using bi-directional tunnel, multicasting on NEMO has the serious weakness called the pinball route problem [10], which can be classified into a route optimization problem.

If MIP-BT is adopted on NEMO, the tunnel convergence problem occurs and it gives much overhead not only on MRs but also on AR because they must process many tunnels, while the problem gives load only to AR or mobility agent on Mobile IP. Furthermore, the more networks are nested, the more damages occur in the entire mobile network. If MIP-RS is adopted on NEMO, MNs in the mobile network can not join a multicast group, because they don't know where multicast routers are located. Moreover, their IGMP [6] messages cannot be routed to a multicast router, because their IGMP messages may not be routed by the MR's upper router for each different mobile network prefixes.

3 Architecture

3.1 Basic Concepts

Firstly, we must handle the multicast support on a mobile network itself, otherwise no MNs in mobile network are sure of the reception of multicast data. Furthermore, lots of tunnels will generate network congestions in the mobile network and its ascendants. This problem, called tunnel convergence and mass of multicast problem, cannot be solved until the mobile network has a multicast router or the MR has multicast router functions. Even though there is a multicast router in the mobile network or MR has multicast router functions, we still encounter another problem that the MR or multicast router cannot have multicast route tree statically with any fixed multicast router because of network mobility.

For above reasons, we assume that in our architecture, multicast router functions are built in the MR, and the MR and the AR must be in charge of notifying default multicast router (DmR) information to its descendant MRs to create multicast route tree. This DmR Information is an internet address of a fixed multicast router in the internet. The information can be set manually for AR and dynamically in run time for MR.

In NEMO basic support architecture, bi-directional tunnel is used for a MR to communicate with its home network. But it causes a serious problem of route optimization, called the pinball route problem. It happens again in the case of multicasting whenever a mobile network is nested. To overcome this weakness, we adopt the Prefix Delegation mechanism. It achieves route optimization of tunnel between a MR and a multicast router. That is, multicast tunnels in our architecture are made from the internet address of multicast router directly to MR's CoA.

In addition, when a MR receives a RA message it must be able to distinguish the source of the message, whether it is from the AR or an upper MR. We use a specialized bit, named 'M', in the RA message. If the 'M' bit is checked, the RA message is from an upper MR and the network is nested. We designated for 'M' bit one of five reserved bits in RA message format.

3.2 Operation Overview

Fig. 1 shows how network components interact in NEMO for multicasting and how the tunnels are established to each router.

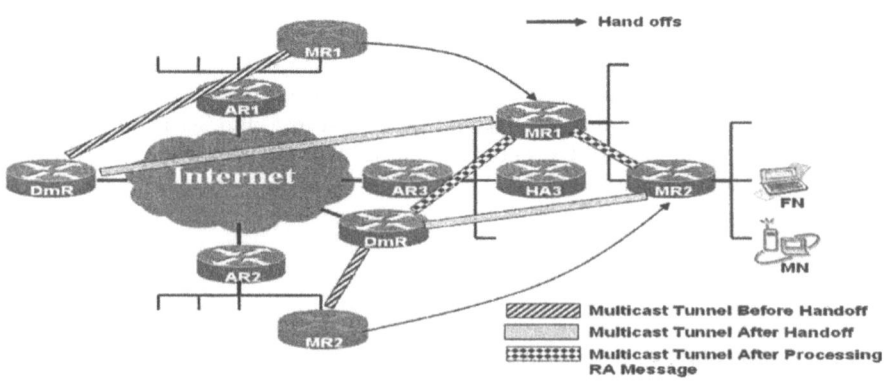

Fig. 1. Operation Flow

When a mobile network acts for the first time, the MR in the network must establish a multicast tunnel with its upper multicast router. To do that, the cases are either the MR find a multicast router by itself, or upper routers(the AR or upper MR) must inform the MR of DmR information. For the former case, if there is no local multicast router, MR cannot find one by itself, and it cannot provide multicast service on its network. For the latter case, if the MR gets DmR information, it will make a multicast tunnel to the DmR and provide successful multicasting. Therefore, in our architecture, we assume that upper routers add DmR information in RA message, and a MR receiving the information must process it. If upper routers send RA message with no DmR information, the MR must understand there is no multicast service under the router and does nothing.

Once a MR gets the information about DmR from RA message, it will establish a tunnel using its new CoA for multicast route tree construction. If the MR is attached under an AR directly, it will receive a RA message with 'M' bit not checked; then the MR must establish a tunnel to the DmR specified in RA message. On the other hand, if the mobile network is nested under another mobile network, the MR will receive a RA message with 'M' bit set. In this case the (nested) MR has to make a tunnel to its upper MR, because the DmR information with 'M' bit set in RA message means that upper MR provides multicast service.

Once the MR establishes a multicast tunnel, it changes old DmR information to newly-acquired one. The new DmR is used for its own RA message, and will be used later when the mobile network moves. We will explain about this later in detail. At this point, all IGMP messages from MNs are handled properly by MR and MNs will receive multicast data.

When a mobile network that has a multicast tunnel moves away, its multicast tunnel using the MR's previous CoA is of no use. Thus, after the MR finishes its registration request process, it has to establish a multicast tunnel to its DmR again using its new CoA and destroy previous one. There is no tree reconstruction cost or delay for this, because the DmR already has provided the multicast group it joined.

After that, the MR starts finding a DmR in the new network by examining RA messages for new DmR information. As stated above, if the MR is attached under an AR, it establishes another multicast tunnel with new DmR; if it is nested it establishes a tunnel with upper MR. Once a new multicast tunnel is established, the previous tunnel with old DmR is destroyed.

Here, we presented a mechanism to provide optimal routing for multicast data in mobile environments. Multicast packets from sender will be routed optimally to the designated MR because of PD mechanism; they will be routed efficiently because of using an appropriate multicast router. The transition is seamless because, when a mobile network changes its location, the MR keeps an old tunnel to the previous DmR until a better tunnel is established. Our approach provides a basic multicast support mechanism on NEMO environments.

3.3 Extensions

We extended each part of components in mobile network to support multicasting. Firstly, because the DmR information and 'M' bit must be specified in RA message, we extended some of the RA message format and stated those below
(1) Default Multicast Router Information Field
- Extended part of Valid Options field to store DmR address information
(2) RA Message Bit, named 'M'
- One of the reserved bits in RA message
- Because the AR may notice the DmR information to its nodes using the RA message, the AR must be extended like below
(1) Default Multicast Router(DmR) Information Configuration
- AR may have DmR information in its configure file

- AR must add DmR information on DmR field of RA message, only if it has the configure

The most important part of the extension is the functions of the MR, because it is in charge of the mobility. Many parts to extend in our architecture are in it. We stated those below.

(1) Default Multicast Router Information Process
- MR must be able to detect 'M' bit in RA message
- MR must be able to examine DmR Information field in RA message and advertise it on its own network
(2) Multicast Router Functions
(3) Multicast Tunnel Establishment
- MR must be able to establish multicast tunnel with multicast router specified in DmR Information Field
- After handoff, if RA message has 'M' bit, the MR must re-establish multicast tunnel with its upper MR
- If a MR gets the two multicast tunnels, it must destroy DmR's as soon as it receives multicast data from upper MR
(4) Prefix Delegation Mechanism for Route Optimization
- A MR must set its CoA using PD mechanism when it is nested

3.4 Characteristics

As presented in the previous sections, our mechanism eliminates the complexities of multicast data flows on NEMO environments. Specifically, in our method,

A. There is no Pinball Route problem of multicast data
B. There is no bi-directional tunnel for multicast data between MR and its HA
C. Once a MR has provided multicast service in home network, it can permanently provide multicast service till it is shut down
D. MRs in mobile networks have native route path for multicast data
E. Our mechanism minimizes multicast tree reconstruction delay

However, if nested MR supports multicast service and upper MR does not, multicast service is possible for descendants of nested MR.

4 Performance Evaluation

We compared cost variables of providing multicast among MIP-BT, MIP-RS, native-routed multicast, and our mechanism. Parameters needed to compute cost variables are stated below:

L1 : mean time of latency between nested MRs under the designated mobile network
L2 : mean time of latency between a AR and its DmR
L3 : mean time of latency between a AR and all HA that their MR is nested under the AR
L4 : mean time of latency between all home networks that one of their MR is nested under the designated AR

m :   multicast packets between CN and on-link node in the designated mobile network

n :   nesting level of MR counting from AR

r :   total number of nodes joining a multicast group in a designated mobile network

s :   mean time for multicast tree reconstruction

α :   cost constant for latency rate under AR's network

β :   cost constant for latency rate outside of AR's network.

ω :   weight constant for processing a tunnel in a router

θ :   cost constant for multicast tree reconstruction per second

We assumed an environment where a mobile router is moving deeper from root router AR in the designated network. And then, we can write down cost values (1), (2), (3) and (4)

$$\text{Adopting Native Route Path} = (\alpha L2 + \beta n\, L1\,)\, m \tag{1}$$

$$\text{MIP-BT on NEMO basic support} = (\alpha L4\, n + L3 + L1\, n\, \alpha\beta + n\, \omega)\, r\, m \tag{2}$$

$$\text{MIP-RS on NEMO basic support} = (\alpha L4\, n + L3 + L1\, n\, \alpha\beta + n\, \omega)\, m + \theta s \tag{3}$$

$$\text{our architecture} = (\alpha L3 + \beta n\, L1)\, m \tag{4}$$

We assume that α is 0.2 and β is 2, because α is the value of the local network and the local network latency may be closer to 0. And we assume that ω is 50 and m is 10. With these variables, we can draw a graph of fig 2. Note that (1) cannot occur on NEMO basic support architecture, because of the Pinball Route problem; L1 ~ L4 are generated by supposing that L4 > L3 > L2 >= L1. In Conclusion, Fig. 2 shows our architecture is the most efficient mechanism in multicasting on mobile network.

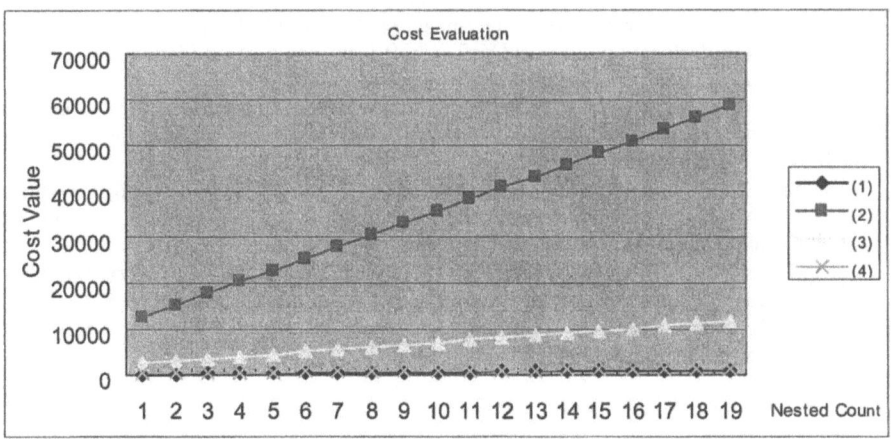

Fig. 2. Cost Evaluation Graph

5 Conclusion

In this paper, we examined and analyzed problems of providing multicast service on NEMO basic support architecture, and proposed new one that provides efficient and optimized route path for multicasting in NEMO. With our architecture, we achieved optimal routing for multicasting by adopting PD mechanism; efficient multicasting by using DmR; and seamless multicasting by keeping an old tunnel to the previous DmR as needed while a mobile network changes its location.

We extended some functions of MR and AR in order to support this. Because this extension is based on the Mobile IP and the IPv6 standards in terms of the functions and protocols, it does not result in much modification or loads to the network components. We also developed a cost analytic model to evaluate the performance of our approach. It showed much lower total cost for on-link nodes in a mobile network, compared to other techniques such as MIP-BT and MIP-RS on NEMO basic support architecture.

Since our approach provides optimal and efficient multicast service and solves the pinball route problem in mobile networks environments, this architecture can become a basic multicast support mechanism on NEMO environments.

References

1. D. Johnson, C. Perkins, J. Arkko, Mobility Support in IPv6, Internet Draft, <draft-ietf-mobileip-ipv6-24.txt>, Jun 2003.
2. Thierry Ernst, Network Mobility Support Goals and Requirements, Internet Draft, <draft-ietf-nemo-requirements-01.txt>, May 2003.
3. Thierry Earnst, Hong-You Lach, Network Mobility Support Terminology, Internet Draft, < draft-ietf-nemo-terminology-00.txt>, May 2003.
4. Vijay Devarapalli, Ryuji Wakikawa, NEMO Basic Support Protocol, Internet Draft, < draft-ietf-nemo-basic-support-01.txt>, Sep, 2003.
5. Kyeong-Jin Lee, J H Jeong, Route Optimization for Mobile Nodes in Mobile Network based on Prefix Delegation, Internet Draft, <draft-leekj-nemo-ro-pd-00.txt>, Jun 2003.
6. W. Fenne, Internet Group Management Protocol, version 2, RFC2236.
7. V. Chikarmane et al, Multicast Support for Mobile Hosts Using Mobile IP: Design Issues and Proposed Architecture, ACM/Baltzer Mobile Networks and Applications, vol 3, no 4, pp 365-379, January 1999.
8. Hee-Sook Shin, Young-Joo Suh, and Dong-Hee Kwon, Multicast Routing Protocol by Multicast Agent in Mobile Networks, Proceedings of the Proceedings of the 2000 International Conference on Parallel Processing, pp 271-278.
9. Kuang-Hwei Chi, Chien-Chao Tseng and Ting-Lu Huang, A Framework for Mobile Multicast using Dynamic Multicast Route, The computer journal, vol 42, no 6, 1999.
10. Thubert, P., and Molteni, M., Taxonomy of Route Optimization Models in the NEMO Context, Internet Draft: draft-thubert-nemo-ro-taxonomy-00, Oct 2002

PRED: Prediction-Enabled RED

Min Gyo Chung and Euinam Huh

Dept. of Computer Science, Seoul Women's University, Seoul, Korea
{mchung,huh}@swu.ac.kr

Abstract. This paper proposes a router congestion control mechanism called PRED (Prediction-enabled RED), a more adaptive and proactive version of RED (Random Early Detection). In essence, PRED predicts its queue length for an early detection of possible congestion alerts in the near future and operates adaptively to the predicted changes in traffic patterns. Typically, PRED does this by first making prediction about average queue length and then using the predicted average queue length to adjust three classic RED parameters max_{th}, min_{th}, and max_p. The incoming packets after the adjustment are now being dropped with the new probability defined by updated parameters. Due to its adaptability and proactive reaction to network traffic changes, PRED can be considered as a novel solution to dynamically configure RED. Extensive simulation results from NS-2 simulator are presented to verify the performance and characteristics of PRED.

1 Introduction

DT (Drop Tail) and RED (Random Early Detection) [1] are two well-known router congestion control algorithms. DT discards the packets when the queue becomes full, but RED drops the packets randomly before the queue is completely full. However, in order that RED operates at the maximal fairness without incurring major network disruption, RED should be properly configured. RED configuration is a problem to find out the optimal set of RED parameters given some dynamic traffic conditions such as number of active flows, connection bandwidth, congestion level, etc. The solution to the configuration problem can't be obtained easily because of the dynamic nature of the network conditions.

A number of approaches to this problem have appeared in the literature. The authors in [2,3] addressed the configuration problem and described many difficulties in the deployment of RED routers into the real world networks. The methods in [4,5] have been focused on the direct control of a packet drop probability, but Padhye et al. [6] shows that the fast and frequent fluctuation of the drop probability rather leads to lower overall throughput.

In this work, we propose PRED (Prediction-enabled RED), which is an adaptive and proactive version of RED (Random Early Detection) designed to alleviate the RED configuration difficulties. Basically, PRED keeps monitoring the current queue status for an early detection of possible congestion alerts in the near future and operates adaptively to the anticipated congestion, thereby leading to a more proactive reaction to changes in traffic patterns. Specifically, PRED

M. Bubak et al. (Eds.): ICCS 2004, LNCS 3038, pp. 1193–1200, 2004.
© Springer-Verlag Berlin Heidelberg 2004

does this by first making prediction about average queue length and then using the predicted queue size to adjust three classic RED parameters max_{th}, min_{th}, and max_p. The incoming packets after the adjustment are now dropped with the new probability defined by updated parameters. Due to its inherent characteristics of adaptability and proactive reaction, PRED can be a novel solution to the above configuration problem.

In addition, PRED has some other features. It still allows fair bandwidth sharing among TCP flows, and yields better network utilization than RED. As with standard RED, PRED can be also added to an existing FIFO-based router without any problems. To verify the performance and characteristics of PRED, we have conducted extensive simulation experiments using NS-2 simulator.

The rest of this paper is organized as follows. In Sec. 2, DT and RED congestion control algorithms are briefly introduced. In Sec. 3, we give a detailed description of PRED. In Sec. 4, detailed simulation results and analyses are presented to show the performance of PRED in many different aspects. Lastly, some concluding remarks are given in Sec. 5.

2   DT and RED

DT, a simplest form of congestion control over a router queue, accepts packets for the queue until the queue is full and then discards new incoming packets until the queue gets its room back again. However, DT occasionally allows a small number of flows to monopolize the queue space, leading to the disproportionate distribution of packet losses among flows. It also has a bias against bursty traffic and possibly causes a global synchronization problem.

RED is another congestion control scheme, specifically designed to eliminate the problems encountered in DT. RED reacts to a congestion signal proactively by dropping packets before the queue is completely full. Further, the packet drop is done at random so that all competing flows will get treated fairly in terms of packet loss rate. The drop probability used by RED, $P_d(\cdot)$, is a linearly increasing function of the average queue length and is defined as follows:

$$P_d(q) = \begin{cases} 0, & q < min_{th} \\ \frac{q - min_{th}}{max_{th} - min_{th}} max_p, & min_{th} \leq q \leq max_{th} \\ 1, & max_{th} < q, \end{cases}$$

where q denotes the (average) queue size, max_{th} a maximum queue length threshold, min_{th} a minimum queue length threshold, and max_p a maximum drop probability.

3   PRED

3.1   Motivation

RED has been used for congestion control in the Internet for many years, but it has one major drawback, namely RED configuration problem. If RED is not configured properly, it can bring about traffic disruption and lower network utilization. The author in [7] showed that under the false configuration, the fairness of RED gets even worse than the fairness of DT as the number of TCP flows increases. Observing that RED yields the best result when the average queue length comes between max_{th} and min_{th}, Hasegawa et al. [8] proposed dt-RED. Although dt-RED exhibited some improvement over its predecessors, it is still difficult to find out max_{th} and min_{th} optimal for the current queue status. Overall, finding an optimal solution to RED configuration problem is tough, and requires many resources and efforts. PRED mainly aims to relieve such configuration difficulties.

3.2 Algorithm

As shown in Fig. 1, PRED consists of two functional modules: prediction module (PM) and congestion control module (CCM). PM continuously monitors the queue to collect its current and past statistical data into a database. Based on the accumulated information, it detects possible congestion signals ahead in the near future by analytically making predictions for some unknown variables such as average queue length. Using the predicted values generated by PM, CCM is responsible for updating three classic RED parameters max_{th}, min_{th}, and max_p to be optimal for the current network conditions. The packet drop probability is now redefined by updated parameters and subsequent incoming packets will be discarded with updated probability.

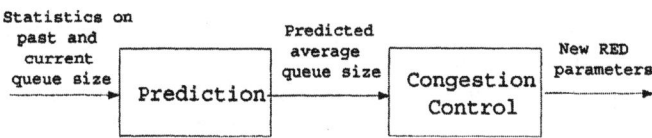

Fig. 1. PRED modules

With the average queue lengths in the current and past cycles, PRED is able to predict the variation of the average queue length in the next cycle. RED computes the average queue length in the current cycle to obtain the drop probability. On the other hand, PRED executes a built-in prediction algorithm to forecast the average queue length in the next cycle and compares the predicted average queue length with max_{th} and min_{th} to update those parameters with proper values. For example, if the predicted average queue length is greater than max_{th}, the threshold max_{th} is increased in advance so that the average queue length will not exceed max_{th} in the next cycle. Similarly, if the predicted average queue length is less than min_{th}, the threshold min_{th} is decreased before the next cycle so that the average queue length will come between max_{th} and min_{th}. In this way, PRED figures out the adequate values for thresholds max_{th} and min_{th}, adaptively to the dynamic changes in traffic patterns.

Prediction Module. Numerous prediction methods have already been available in the literature, including MACD (Moving Average Convergence/Divergence), AR (Autoregressive), and LR-Line (Linear Regression-Lines). The MACD model is based on two moving average series. When the shorter term moving average crosses over the longer term moving average, a rising tendency is predicted. Likewise, when the shorter term moving average falls below the longer term moving average, a declining signal is generated. The AR model, also known as IIR(Infinite Impulse Response) filter in engineering field, is one of linear prediction formulas that attempts to predict an output of a system using the current/past inputs and the past outputs. The LR-Lines is a statistical tool to make prediction about an unknown variable by discovering an equation for a line that most nearly fits the given data.

We tested the above three prediction approaches and discovered that all but the LR-Lines don't forecast the average queue length very well. The LR-Lines quite correctly predicts the average queue length when the queue is long enough, but as the queue gets shorter, it reports some discrepancies, that is, the predicted average queue length tends to be greater than the one measured. Nevertheless, it doesn't incur any big problems in applying the LR-Lines to PRED, because the throughput of RED-type routers doesn't get seriously deteriorated even if the average queue length goes below min_{th} due to the false prediction.

Congestion Control Module. Shortly after the predicted average queue length, denoted by avg_p, is available from PM, CCM attempts to modify three RED parameters max_{th}, min_{th} and max_p by referring to avg_p. Here, avg_p is calculated by $\frac{avg_{p_1}+avg_{p_2}+avg_{p_3}}{3}$, where avg_{p_i} is the predicted average queue length next i-cycles ahead of the current cycle. Depending on the relationship between avg_p and two thresholds (max_{th} and min_{th}), the modification can be done in three distinct ways. We will next discuss three possible situations and how CCM works in those situations.

First, consider the situation where avg_p is greater than max_{th}. This situation, for example, can take place when the amount of incoming traffic rises dramatically in a short period of time. In preparation for such a sudden traffic increase, CCM shifts max_{th} and min_{th} to the right such that avg_p is positioned between new max_{th} and min_{th} (See Fig. 2 (a)). More precisely, to calculate max_{th}, min_{th} and max_p, we use the following procedure:

$$max_{th} = \min\{avg_p(1 + i_{max}), max_{lim}\}$$
$$min_{th} = \min\{min_{th,old}(1 + i_{min}), avg_p\}$$
$$max_p = \frac{max_{p,lim}}{max_{lim}-min_{lim}}(max_{th} - min_{lim})$$

where max_{lim} and min_{lim} represent maximum and minumum values, respectively, that the thresholds max_{th} or min_{th} can possibly take; i_{max} and i_{min} represent the rate of increase for max_{th} and min_{th}, respectively; $max_{p,lim}$ is a maximum possible value for max_p; and $min_{th,old}$ is the previous value of min_{th}. Note that the equation of the straight line connecting two points ($min_{lim}, 0$) and

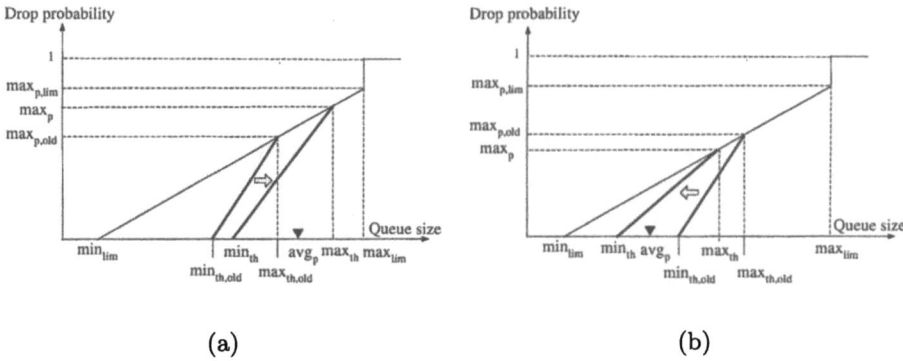

Fig. 2. PRED congestion control. (a) Situation where predicted average queue length is expected to increase, and (b) Situation where predicted average queue length is expected to decrease

$(max_{\text{lim}}, max_{\text{p,lim}})$ is $y = \frac{max_{\text{p,lim}}}{max_{\text{lim}}-min_{\text{lim}}}(x - min_{\text{lim}})$, and that max_{p} is derived from this equation by substituing $x = max_{\text{th}}$.

Second, consider the situation where avg_{p} is less than min_{th}, which is usually caused by persistent low traffic. Motivated by the fact that RED-type algorithms show the best result when avg_{p} is bounded by thresholds max_{th} and min_{th}, CCM will set up the new values for those thresholds by shifting them to the left. Detailed procedure to calculate max_{th}, min_{th} and max_{p} is shown below:

$$min_{\text{th}} = \max\{avg_{\text{p}}(1 - d_{\min}), min_{\text{lim}}\}$$
$$max_{\text{th}} = \max\{max_{\text{th,old}}(1 - d_{\max}), avg_{\text{p}}\}$$
$$max_{\text{p}} = \frac{max_{\text{p,lim}}}{max_{\text{lim}}-min_{\text{lim}}}(max_{\text{th}} - min_{\text{lim}})$$

where $d_{\max}$ and $d_{\min}$ represent the rate of decrease for max_{th} and min_{th}, respectively; $max_{\text{th,old}}$ is the previous value of max_{th}; and other names carry the same meaning as the above first situation. Note that the formula to compute max_{p} is same.

The last situation can happen when $min_{\text{th}} \leq avg_{\text{p}} \leq max_{\text{th}}$. Generally, in this circumstance, each of incoming flows sends packets at its sustained rate, so the traffic does not show any remarkable fluctuations. Since avg_{p} is already bounded by thresholds max_{th} and min_{th}, CCM does not execute any particular operations to change max_{th}, min_{th} and max_{p}.

In brief, PRED gets a control over the packet drop probability in more proactive sense than RED by predicting the average queue length ahead of time and updating max_{th}, min_{th} and max_{p} based on the predicted average queue length. Ultimately, this proactive and adaptive behavior of RED to traffic dynamics helps to facilitate RED configuration process.

3.3   Features

In this section, we will describe some other notable features that PRED retains, especially compared with standard RED. The average queue size is used by RED to determine its drop rate. Instead of instantaneous queue size, using the average queue size makes RED tolerant toward brief queue changes or bursts. However, the use of the average queue size can also bring undesirable detrimental effects to RED's overall throughput, because of its delayed reaction to dynamically changing network conditions. PRED is able to minimize this sort of detriment by reacting a few cycles earlier based on predicted quantities by LR-Lines.

In RED, the parameters max_{th}, min_{th} and max_p are fixed, but PRED dynamically adjusts their values according to the level of traffic congestion. Persistent high traffic will allow PRED to keep moving max_{th} and min_{th} toward max_{lim}, so the distance between max_{th} and min_{th} gets shorter and max_p gets bigger gradually. Similarly, as the traffic runs low, max_{th} and min_{th} come closer to min_{lim}. As a result, the distance between max_{th} and min_{th} as well as max_p becomes smaller and smaller.

In comparison with RED, PRED still allows fair bandwidth sharing among TCP flows and yields better network utilization. Further, as with standard RED, PRED can be added to an existing FIFO-based router without any big modification.

4   Simulation and Analyses

4.1   Simulation Environment

To demonstrate the performance and characteristics of PRED, we used NS-2 (Network Simulator Version 2) simulator and the topology of simulated network as shown in Fig. 3. NS is a discrete event simulator targeted at networking research [9]. The topology consists of N source nodes, one sink node and a single PRED router shared by the source and sink nodes. Each connection between a source node and the PRED router is a TCP link with 100 Mbps capacity and 10 ms propagation delay, while the connection between the sink node and the router is a TCP link with 1 Mbps and 10 ms propagation delay. TCP New-Reno was applied to those links.

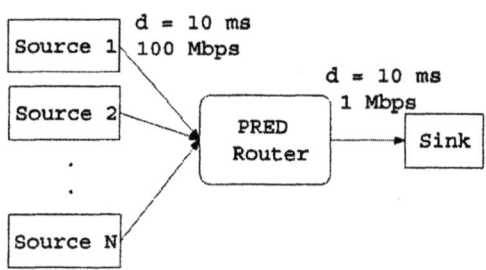

Fig. 3. Simulation Topology

4.2   Simulation Results

We consider the following simulation scenario: the number of source nodes, N, has some distinct values of 21, 100, 150, and 200, and each source node will send out packets at different bit rate every 10 second over a 70-second interval. This simulation scenario is applied to each of DT, RED, and PRED algorithms to measure the performance metrics such as fairness and the amount of dropped packets for congestion control.

The first simulation result as shown in Fig. 4 (a) supports that PRED performs well rather than the conventional TD and RED in terms of total delivered input from source nodes. This implies that PRED adjusts the drop probability to smaller than the RED as the declined traffic curve is forecasted. The second experiment as shown in Fig. 4 (b) is done under the increased traffic trend but the amount of packet delivered to the router is smaller than the peak transmission rate. PRED drops more packets than the RED as we expected. This implies the LR-Lines forecasts packet trend accurately. The fairness metric (as TCP is used in each node) is also measured from the third experiment as shown in Fig. 4 (c). PRED provides better fairness than others when the number of nodes is increased.

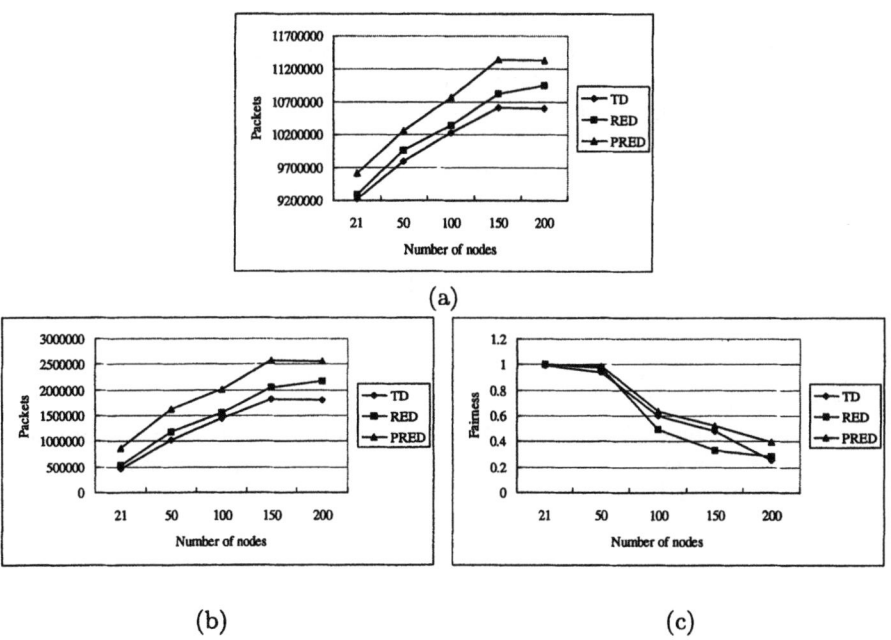

(a)

(b)   (c)

Fig. 4. Performance comparison: (a) Total amount of input data packets delivered over the link between the source nodes and the router, (b) Amount of data being dropped for congestion control when avg_p increases, and (c) Fairness index

5   Conclusion

This paper proposes a new congestion avoidance scheme called PRED, a more adaptive and proactive version of RED. PRED consists of two functional modules: prediction module (PM) and congestion control module (CCM). PM continuously examines the queue to collect its current and past statistical data into a database. Based on the accumulated information, it detects possible congestion signals ahead of time by making predictions about average queue length. Using the predicted values generated by PM, CCM is responsible for updating three classic RED parameters max_{th}, min_{th}, and max_p to be optimal for the current network conditions.

PRED provides more adaptability and proactive reaction to network traffic changes than RED, thus can be effectively used to dynamically configure RED parameters. Further, PRED allows fair bandwidth sharing among TCP flows, yields better network utilization, and can be added to an existing FIFO-based router without any big modification.

References

1. S. Floyd and V. Jacobson, "Random early detection gateways for congestion avoidance," IEEE/ACM Transactions on Networking, vol. 1, pp. 397-413, Aug. 1993
2. Martin May, Jean Bolot, Christophe Diot, and Bryan Lyles, "Reasons not to deploy RED," in Proceedings of IWQoS '99, June 1999
3. Mikkel Christiansen, Kevin Jeffay, David Ott, F. Donelson Smith, "Tuning RED for web traffic," in Proceedings of ACM SIGCOMM 2000, August 2000
4. Haining Wang and Kang G. Shin, "Refined design of random early detection gateways," in Proceedings of Globecom '99, pp. 769-775, December 1999
5. Wu-chang Feng and Dilip D.Kandlur and Debanjan Saha and Kang G. Shin, "A self-configuring RED gateway," in Proceedings of IEEE INFOCOM '99, March 1999
6. J. Padhye, V. Firoiu, D. Towsley, and J. Kurose, "Modeling TCP throughput: a simple model and itsempirical validation," in Proceedings of ACM SIGCOMM '98, pp. 303-314, Aug. 1998
7. Raj Jain, "Throughput fairness index: An explanation," ATM Forum Contribution 99-0045, February 1999
8. Go Hasegawa, Kouichi Tokuda and Masayuki Murata, "Analysis and Improvement of fairness among many TCP connections sharing Tail-Drop and RED Routers" in Proceedings of INET 2002
9. LBNL, LBNL Network Simulator-ns version 1, http://www-nrg.ee.lbl.gov/ns/

An Efficient Aggregation and Routing Algorithm Using Multi-hop Clustering in Sensor Networks*

Bo-Hyeong Lee, Hyung-Wook Yoon, Tae-Jin Lee, and Min Young Chung

School of Information and Communication Engineering
Sungkyunkwan University, Suwon, KOREA
{shaak,hwyoon,tjlee,mychung}@ece.skku.ac.kr

Abstract. Sensor networks consist of sensor nodes with small-size, low-cost, low-power, and multi-functions to sense, to process and to communicate. Minimizing power consumption of sensors is an important issue in sensor networks due to limited power in sensors. Clustering is an efficient way to reduce data flow in sensor networks and to maintain less routing information. In this paper, we propose a multi-hop clustering mechanism using global and local ID to reduce transmission power consumption and an efficient routing method to improve performance of data fusion and transmission. Simulation results show that our algorithm can increase life time and disperse power consumption among the nodes.

1 Introduction

Sensor networks consist of small size, low cost and low power sensor nodes with multi-functions to sense, to process and to communicate [1]. Minimizing power consumption of sensors is one of the most important issues in sensor networks due to limited power of sensor nodes. If a sensor node consumes all of its energy, the sensor network may not guarantee reliable data transmission due to possible partition of the network. Thus, algorithms to minimize nodes' energy consumption and to maximize network life time have been actively studied [1], [2], [3], [4].

Since each sensor node senses and transmits local proximity information, sensed data in neighbor nodes tend to have similar characteristics to one another. In order to reduce the amount of data for transmission, data with similar information of the nodes need to be aggregated, which is called data fusion or aggregation [2]. The mechanism with data fusion can lead to less energy consumption than the mechanism without data fusion because data fusion reduces total data flow in the overall network. In general, it is very difficult to optimize data fusion in sensor networks. So, the method to group sensor nodes into a cluster and aggregate data in the cluster has been proposed in [3].

Clustering is able to reduce total routing messages as well as sensed information data. However, the larger the size of a sensor network, the less the merit is. Hierarchical clustering can increase scalability, but this is more energy-consuming because data transmission can concentrate on specific nodes or some routes, and total length of ID in the header and routing table grows as the number of levels in the hierarchy increases.

---

* This paper was partially supported by BK21 program. Tae-Jin Lee is the corresponding author.

M. Bubak et al. (Eds.): ICCS 2004, LNCS 3038, pp. 1201–1208, 2004.
© Springer-Verlag Berlin Heidelberg 2004

In this paper, we propose a multi-hop clustering and algorithm for efficient data fusion and routing using global ID, local ID and cluster scheduling instead of using hierarchical clustering in order to reduce the number of cluster head (CH) nodes and concentrated traffic on specific nodes.

Our algorithm has the following advantages:

1) **Efficient data aggregation in a multi-hop cluster:** In each aggregation procedure, each node in a multi-hop cluster is allowed to transmit data only once by the cluster schedule, which is more efficient than the aggregation in a hierarchical cluster because some nodes in a hierarchical cluster may transmit data more than once.

2) **Distribution of energy consumption when aggregating:** In a hierarchical cluster, fixed aggregating nodes exist (e.g., CH nodes) and energy consumption concentrates on these nodes. Our algorithm extends single hop clustering and constructs multi-hop clusters to distribute energy consumption over many aggregation nodes using cost of local IDs.

3) **Load balancing by construction of multipaths :** Our algorithm constructs multipaths from CH nodes to sink node during the multi-hop clustering stage. When aggregated data are sent to the sink node, multipaths are used instead of the fixed single shortest path to extend network life time.

The remainder of the paper is organized as follows. In Section 2, we discuss some of the related works. In Section 3, we propose the mechanism with multi-hop clustering, ID construction, routing path construction, and data aggregation. In Section 4, simulation results and analysis are presented. Section 5 concludes the paper.

2   Related Works

Conventional routing protocols are not suitable to aggregate data and to maximize network life time in sensor networks due to sensor nodes' limited properties (e.g., bandwidth, transmission range, thousands of nodes in a network, etc.). So, Low-Energy Adaptive Clustering Hierarchy (LEACH), which is a clustering-based protocol for sensor networks, has been proposed [3]. In LEACH, formation of clusters among sensor nodes is conducted locally, and CH nodes compress local data of the nodes in their clusters to reduce the amount of communications. And the role of CH nodes is periodically rotated for distributing energy consumption. LEACH assumes that the base station (sink node) is located far from the other nodes and each node can transmit data to the base station directly. This assumption may not be suitable because of the limited capacity of sensor nodes (especially, small energy). In this paper, we assume that the base station is located at radio transmission distance as other normal sensor nodes. So, data transmission from a normal node to the sink node traverses node along a routing path.

Power-Efficient GAthering in Sensor Information Systems (PEGASIS) is a chain-based protocol that minimizes power consumption [5]. In PEGASIS, each node communicates only with close neighbor nodes and data are relayed to the sink node, thus reducing the amount of energy spent. However, all nodes are assumed to have global knowledge of the network, which is very difficult in real networks.

The hierarchical ad hoc routing is combined with clustering in sensor networks [10], [11]. If the number of levels the in hierarchical clustering is large, less routing information

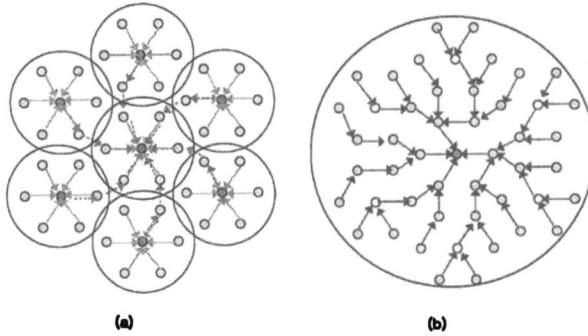

Fig. 1. (a) Hierarchical clustering and (b) multi-hop clustering.

is required [6]. However, high levels of the hierarchy is more burdensome and leads to long delay. Also, routing paths might be overlapped to transmit data as hierarchical level increases and data are concentrated in CH nodes when aggregating data (see Fig. 1). Hence, we propose and use multi-hop clustering structure, in which the distance is D hop from CH node to the nodes in a cluster, to distribute energy consumption and reduce routing overhead.

3   Proposed Multi-hop Clustering and Aggregate Routing

3.1   Multi-hop Clustering and Assignment of Node ID

Network life time, the time until the battery of any node drains-out for the first time, depends on how to determine CH nodes, i.e., clustering algorithm. Since the purpose of this paper is to realize efficient data aggregation and routing by use of local IDs for sensor nodes in a cluster, we assume that CH node is determined according to each node's current residual energy.

When the power of randomly placed sensor nodes is on, each node can become a CH node based on residual energy. If the remaining energy of a node is high, the node set short random time to become a candidate for a CH node. The node that first claims to be a CH node after random time becomes a CH node. A new CH node broadcasts CH broadcast signal with the CH's global ID, node ID, and the cluster's schedule to its neighbor nodes. The schedule of the cluster is based on the CH node's start time (see Fig. 2). Each node can know the transmission time at each level. Level i denotes the number of hops from a CH to a node in the cluster.

Neighbor nodes, which receive the CH signal, write CH ID in their information table (see Fig. 4 - (a)) and abandon being a CH node. These neighbor nodes may receive other CH broadcast signals. Then, the nodes examine their own information table, and select the CH with the smallest hop count and distance. Each node constructs a node ID based on the selected CH node's information and the node itself's (see Fig. 3). If a node determines its ID is within MAX hops from the CH, the node broadcasts CH broadcast signal again to other neighbor nodes after specified time. Nodes receiving this CH broadcast signal repeat the construction of information table with node ID.

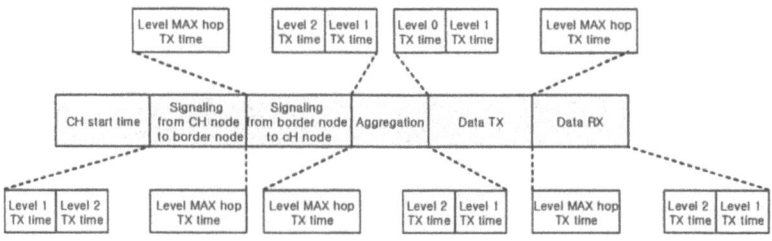

Fig. 2. Schedule in a CH broadcast signal.

CH ID	Residual energy	Hop count from CH node (level)	Shortest path cost to CH node	Link count

Fig. 3. Proposed information element (node ID).

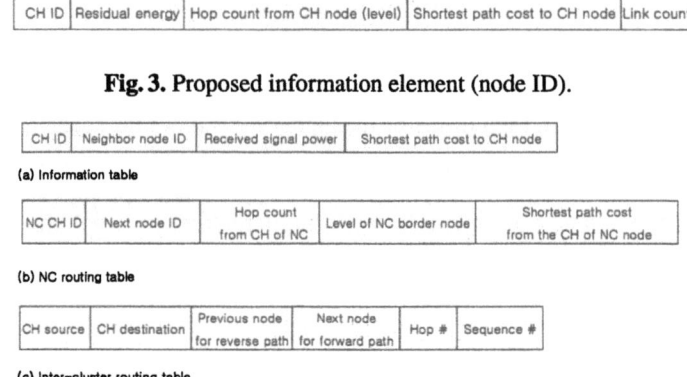

(a) Information table

(b) NC routing table

(c) Inter-cluster routing table

Fig. 4. Information and intra- and inter-cluster routing table structure.

3.2   Construction of Routing Paths

Intra-cluster routing from an intra-cluster node to the CH node: When each node construct its information table with node ID, a node records the cost of the shortest path to the CH and decides the routing path to the CH node. The construction of the shortest path is similar to the Distributed Bellman-Ford (DBF) method. The difference between DBF and our proposed intra-cluster routing method is that the proposed routing mechanism can reduce the computational load of routing path (see Fig. 5).

Routing from a CH node to a neighbor cluster(NC): The border node, connected node to a neighbor cluster, plays an important role in inter-cluster routing among CH nodes. If a border node receives neighbor cluster's CH broadcast signal, the border node transmits information of the NC to its CH node. Nodes, which relay the information from the border node to the CH node, record the hop count and the other information in the NC routing table (see Fig. 4-(b)) and construct routing paths needed when the CH node transmits data to other clusters' CH nodes. Construction of the routing path from the border node having information of NC to the CH node is similar to section 3.1 (see Fig. 6).

Inter-cluster routing from a CH to the sink node: Inter-cluster routing, which transmits aggregate data from a CH of a cluster to the sink node along the borer node of NCs, employs the same routing mechanism as in conventional ad hoc networks except replacing a node with a cluster. Routing paths determined from the previous step (see

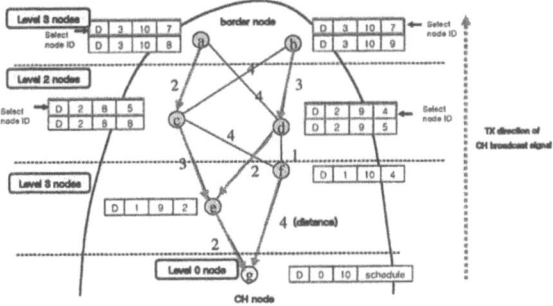

Fig. 5. Construction of intra-cluster routing path from an intra-cluster node to the CH node.

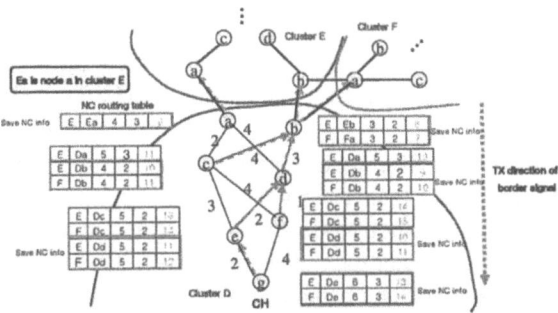

Fig. 6. Construction of routing path from a CH node to NC.

Fig. 6) are used for inter-cluster routing between CHs. Inter cluster routing table is shown in Fig. 4-(c). CH nodes manage inter-cluster routing tables which save the next cluster information to transmit aggregated data to the sink node.

3.3   Data Fusion Procedure

Nodes in a cluster periodically sense their environment and sensed transmit information to the CH node using intra-cluster routing paths. In the multi-hop clustering, the nodes at MAX hop distance from the CH node (local MAX) start to transmit in the beginning of aggregation, and the nodes at the next level transmit their own generated data and the data received from other nodes. Each node in the cluster must transmit only once to a neighbor node along the shortest path. However, data flow may be concentrated on some nodes along the shortest path, which causes unnecessary energy consumption. So, we propose the followings.

Level scheduling: When nodes start data aggregation, the nodes at MAX hop distance (level D) to the CH node transmit data. For example, when nodes in level D are able to transmit by the cluster schedule, which is informed in the clustering procedure, they transmit data to one of the nodes in level D-1. Nodes in level D-1 combine their own generated data with the received data, and transmit data again to the nodes in level D-2 when the nodes in D-1 can transmit by the cluster schedule. In this way, nodes are

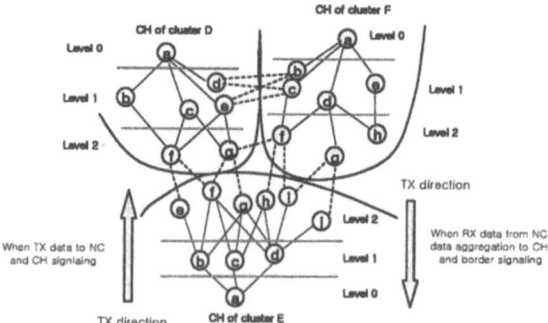

Fig. 7. Selection of receiving node from the viewpoint of cluster E.

allowed to transmit to the nodes in lower level during their level period. Finally, the CH node receives all of the nodes' data in the cluster while each node is guaranteed to transmit data only once.

Receiving node distribution: The nodes on the routing paths or receiving from many sensor nodes rapidly consumes their energy. So, it may not minimize total power consumption if we select the next hop node according to the shortest path routing. We propose two routing methods other than the shortest path. One is the random selection method and the other one is the load balancing method. In random selection method, a node transmitting data selects a random node among the nodes within the transmission distance. This realizes the random distribution of receiving nodes simply, to reduce the concentration of data to some nodes. The load balancing method selects receiving nodes based on the probability determined from cost (e.g., energy, hop count). That is a node can select a receiving node by considering the condition of both the node itself and candidate receiving nodes. For example,a node without enough energy to transmit data to a node at the long distance, increases the probability of selecting a close receiving node.

In both methods, a receiving node is selected according to the level of a transmitting node and the direction of transmission. In case of transmiting to the CH node, one of the nodes at lower level is selected. In case of transmiting to an NC, one of the nodes at higher level or in NC is selected (see Fig. 7).

4   Simulation Results

We have conducted simulations to evaluate the performance of the proposed multi-hop clustering and routing. The sensor network for the simulation consists of 50 ~ 400 nodes distributed randomly in the square area of 100 x 100 meters. The sink node is located at the position (10,10). The proposed algorithm is used for inter-cluster routing and the AODV [7] algorithm is used for intra-cluster routing. We use the same radio model as in [8]. Signals regarding clustering and routing are assumed to be broadcast with the maximum transmission range of 20m. Data are transmitted using modified RTS-CTS-data [9] as in wireless local area network. We also assume that there are no collisions and bit errors by the cluster schedule when nodes transmit and receive data. Simulations are

Table 1. Parameters used in the simulation.

Parameter	Value
MAX transmission range	20 m
MAX hop D	3
Initial energy	0.01 J
RX energy consumption	50 nJ/bit
TX energy consumption	50nj/bit + 0.1 pJ/bit/m^2 × distance^2
Aggregation period	80 times slots
Number of simulations	1000

executed until the network life time. The parameters in the simulation are summarized in Table 1.

In Fig. 8, we observe that network life time of the proposed multi-hop clustering and routing is almost the same regardless of routing methods to balance receiving nodes. The reason is that all the data packets in a cluster must be eventually collected and transmitted through the CH node to go to the sink node. So, in order to increase the network life time, periodic reclustering may be required to switch CH nodes with those having larger residual battery power. The shortest path routing, however, shows larger variance than the other receiving node selection methods as shown in Fig 9. Basically, routing paths constructed as in Section 3 is the shortest paths. Therefore, transmitting data in a cluster tends to use the same routing paths and thus energy consumption is concentrated on the nodes along the routing paths. In case of random and load balanced routing methods, energy consumption is scattered to other nodes because transmitting nodes select the next receiving node according to the remaining energy or the probability. We also note that as the number of nodes increase, random selection surpasses the load balance since the receiving nodes tend to be distributed evenly.

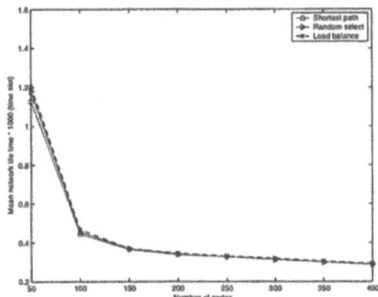
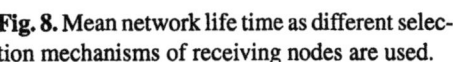

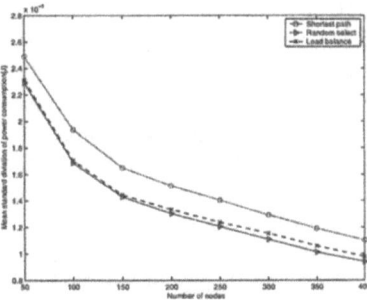

Fig. 8. Mean network life time as different selection mechanisms of receiving nodes are used.

Fig. 9. Mean standard deviation of energy consumption in sensor nodes.

5 Conclusion

In this paper, we have proposed an efficient aggregation and routing mechanism using local ID in multi-hop cluster sensor networks. The proposed algorithm can reduce the routing computational overhead and simplify routing paths since it employs level scheduling and intra- and inter-cluster routing. In addition, each node transmits data at once in the aggregation procedure, preventing from unnecessary energy consumption which has been in the previous works. We have compared the network life time and energy consumption to evaluate the performance via simulations. The proposed load balanced routing has been shown to disperse energy consumption by selecting the next receiving node at random or by the load-balancing regime.

References

1. I. F. Akyildiz and W. Su. A Survey on Sensor Network. In *IEEE communications Magazine*, volume 40, pages 102–114, Aug. 2002.
2. W. R. Heinzelman, J. Kulik, and H. Balakrishnan. Adaptive Protocols for Information Dissemination in Wireless Sensor Networks. In *Proc. of ACM MobiCom*, pages 174–185, 1999.
3. W. R. Heinzelman, A. Chandrakasan, and H.Balakrishnan. Energy-Efficient Communication Protocol for Wireless Microsensor Networks. In *Proc. of IEEE Sys. Sci.*, pages 1–10, Jan. 2000.
4. S. Singh, M. Woo, and C. S. Raghavendra. Power-Aware Routing in Mobile Ad Hoc Networks In *Proc. of ACM MobiCom*, pages 181–190, Nov. 1998
5. S. Lindsey and C. S. Raghavendra. PEGASIS: Power-efficinet GAthering in Sensor Information System. In *Proc. of IEEE Aerospace Conference*, volume 3, pages 1125–1130, Mar. 2002.
6. J. Sucec and I. Marsic. Clustering Overhead for Hierarchical Routing in Mobile Ad-hoc Networks. In *Proc. of IEEE INFOCOM*, volume 3, pages 1698–1706, Jun. 2002.
7. C. E. Perkins and E. M. Royer. Ad-hoc On Demand Distance Vector Routing. In *Proc. of IEEE WMCSA*, pages 90–100, Feb. 1999.
8. W. Heinzelman, et. al. Energy-scalable Algorithms and Protocols for Wireless Microsensor Network. In *Proc. of International Conference on Acoustics, Speech and Signal Processing*, volume 6, pages 3722–3725, Jun. 2000.
9. P. Karn. MACA - A New Channel Access Method for Packet Radio. *ARRL/CRRL Amateur*, volume 6, pages 134–140, 2000.
10. A. Iwata, C. C.Chiang, G. Pei, M. Gerla and T. W. Chen. Scalable Routing Strategies for Ad-hoc Wireless Networks. *IEEE Journal on Selected Area in Communication*, volume 17, pages 1369–1379, Aug. 1999.
11. R. Ramanathan and M. Steenstrup. Hierarchically-organized, Multi-hop Mobile Wireless Networks for Quality-of-Service Support. *ACM/Baltzer Mobile Networks and Applications*, volume 3, pages 101–119, 1998.

Explicit Routing for Traffic Engineering in Labeled Optical Burst-Switched WDM Networks*

Jing Zhang[1], Hae-Joung Lee[2], Shuang Wang[2], Xue Qiu[2], Keyao Zhu[3],
Yurong Huang[1], Debasish Datta[4], Young-Chon Kim[2], and
Biswanath Mukherjee[1]

[1] University of California, Davis, CA 95616, USA,
zhangj@cs.ucdavis.edu
[2] Chonbuk National University, Jeonju 561-756, Korea,
yckim@moak.chonbuk.ac.kr
[3] Research and Innovation Center, Alcatel Shanghai Bell, Shanghai, China
[4] Indian Institute of Technology, Kharagpur 721302, India

Abstract. Optical burst switching (OBS) is a promising technique for
supporting high-capacity bursty data traffic over optical wavelength-
division-multiplexed (WDM) networks. A label-switched path can be
established to forward a burst control packet (BCP) if each OBS node
is augmented with an IP/MPLS controller. Such a network is called a
labeled OBS (LOBS) network, and it can exploit the explicit routing and
constraint-based routing properties supported in the MPLS framework
to perform traffic and resource engineering. In this paper, we consider the
traffic-engineering problem in a LOBS network, using explicit routing to
balance network load and to reduce the burst-loss probability. We model
the traffic-engineering problem in a LOBS network as an optimization
problem and propose two new objective functions to minimize network
congestion. Our illustrative numerical examples show that the burst-loss
probability can be significantly reduced using optimization technique,
when compared with shortest-path routing.

1 Introduction

Optical burst switching (OBS) technology has been proposed and studied as a
promising technique to support high-speed data traffic over optical wavelength-
division-multiplexed (WDM) networks [1]-[3]. The basic idea of OBS is to assemble data packets into large-size bursts at the ingress node, and then to send a
burst into the network shortly after a corresponding control message is sent out,
without waiting for any acknowledgement message. The control message, usually called burst control packet (BCP), typically uses a separate control channel
and is processed electronically at each node through which it traverses. Thus, the

---

* This work has been supported in parts by NSF Grant Nos. ANI-02-07864, and INT-
03-23384, KOSEF, OIRC, and KIPA Professorship Program.

M. Bubak et al. (Eds.): ICCS 2004, LNCS 3038, pp. 1209–1216, 2004.
© Springer-Verlag Berlin Heidelberg 2004

node can configure its switch for the incoming burst according to the information carried by the BCP (such as when the burst will arrive and how long it will hold). The data burst does not need to be converted into electronic domain for processing. It can optically cut through the switching nodes and reach the destination. Through this approach, the electronic bottleneck of the optical-electric-optical (OEO) conversion occurring at the switching nodes can be eliminated.

However, data bursts from different traffic flows can share the same bandwidth on a link using statistical multiplexing. And a data burst will be dropped if contention occurs at a switching node in an OBS network with no or limited fiber delay lines (FDLs). Actually, the probability of contention for two bursts at an output link will increase if the load on the link is heavy. Thus, if the traffic can be evenly distributed or balanced over all links when it is routed, the burst blocking (or loss) probability can be greatly reduced. Hence, we investigate the traffic-engineering problem (i.e., *"put the traffic where the bandwidth is"*) in OBS networks and design intelligent routing algorithms to achieve such an objective.

In today's Internet, IP routing generally employs destination-based shortest-path routing, in which a packet is routed or forwarded according to its destination prefix at a router. Even though this routing approach can scale gracefully when network size increases, destination-based routing usually creates congested links and unbalanced traffic distribution [4]. This problem can be solved if the traffic is properly engineered using explicit-routed paths. In a MultiProtocol Label Switching (MPLS) controlled network, explicit routes are supported by establishing label-switched path (LSPs). If an OBS node is augmented with an IP/MPLS controller, the node will become a labeled OBS (LOBS) node, and it functions as a label-switched router (LSR). In such a LOBS network, a BCP will be attached a label before it is sent out on the network, and the label is swapped when the BCP is processed at each intermediate node. Both explicit routing and constraint-based routing supported in the MPLS framework can be extended to provision and engineer traffic to better utilize network resource in a LOBS network [1].

An explicit route simply performs as a point-to-point logical connection in a LOBS network. The collection of such logical connections between the various pairs of nodes essentially form a virtual network on top of the physical fiber network topology. Each explicit route is a "virtual link" of the "virtual topology". All BCPs sent out on an explicit route will follow the path through to the destination. Theoretically, any arbitrary virtual topology can be pre-planned and set up based on the traffic intensity (including full-mesh topology and topologies with more than one virtual links between a node pair). One of the main issues that needs to be addressed in a LOBS network is how to compute the explicit routes so that traffic is balanced in the virtual topology and the burst-loss probability is minimized by reducing the probability of burst contention at the various nodes.

An explicit route can be calculated either dynamically (based on "current" network resource availability) or statically. Dynamic routing can achieve better load balancing if the network resource information is accurate. However, to support dynamic routing, the network may need to be flooded by frequent network-resource-update messages, and additional signaling protocols or extensions of current signaling protocols may need to be developed. Pre-planned (static) routing can avoid such signaling overhead. And most importantly, optimization techniques can be applied in pre-planned routing such that a virtual topology is established based on a given traffic intensity matrix. Given that the physical topology of an optical network is relatively stable (i.e., this topology changes less frequently than that in an IP network), periodic or threshold-triggered re-routing of the virtual topology using optimization techniques can be performed easily to adapt to changing traffic intensities.

In this paper, we investigate the problem of pre-planned explicit routing for traffic engineering in a LOBS network. As mentioned in [1], for traffic engineering in a LOBS network, we need to address various issues unique to LOBS such as potentially excessive burst dropping due to the absence of or limited use of FDLs at the LOBS nodes when contention occurs. To tackle this situation, we model the problem of explicit routing for traffic engineering as an optimization problem with the objective of minimizing the link congestion.

The remainder of the paper is organized as follows. Section II reviews previous work, discusses the shortcomings of the objective function presented in previous work, and proposes two new objective functions to minimize congestion. In Section III, we apply pre-planned explicit routing to LOBS networks and study the performance of different objective functions using illustrative numerical examples. Section IV concludes this study and discusses future research directions.

2   Optimal Routing for Load Balancing in a LOBS Network

The problem of how to set up explicit routes between pairs of nodes so that network congestion can be minimized has been studied in [5]. Even though the authors in [5] considered explicit-routing algorithms for Internet traffic engineering instead of LOBS networks, their objective is essentially the same as our objective: namely, set up explicit routes and minimize network congestion. Therefore, we first briefly review the optimization approach proposed in [5], discuss our approaches, and then compare the two methods.

The authors in [5] formulated the problem as a linear program. To minimize network congestion or balance traffic distribution, the optimization objective they proposed is to minimize the maximum of all link utilizations. The Integer Linear Programming (ILP) formulation used in [5] (denoted as ILP-A here) is summarized below.

ILP-A:

- *Notations:*
    - i and j denote end points of a physical fiber link.
    - s and d denote source and destination of a given traffic flow.
- *Given:*
    - $G = (V, E)$: network topology with node set V and edge set E.
    - C_{ij}: capacity of link (i, j).
    - D_{sd}: bandwidth requirement of traffic flow from node s to node d.

- *Variables:*
  - X_{ij}^{sd}: $X_{ij}^{sd} = 1$ if traffic flow (s,d) is routed through fiber link (i,j); otherwise, $X_{ij}^{sd} = 0$.
  - α: maximal link utilization across the entire network. $\alpha \geq 0$.
- *Objective A:* Minimize the maximum link utilization.

$$Minimize : \alpha + r \sum_{s,d \in V} \sum_{(i,j) \in E} X_{ij}^{sd} \qquad (1)$$

where r is a positive number which is assigned a small value such that minimizing α will maintain higher priority.
- *Constraints:*
  - On physical route flow-conservation constraints:

$$\sum_{(k,j) \in E} X_{kj}^{sd} - \sum_{(i,k) \in E} X_{ik}^{sd} = 0 \ \ if \ \ k \neq s, d \ \ \forall k \in V, \forall s, d \in V \qquad (2)$$

$$\sum_{(s,j) \in E} X_{sj}^{sd} - \sum_{(i,s) \in E} X_{is}^{sd} = 1 \ \ \forall s, d \in V \qquad (3)$$

  - On link-capacity constraints:

$$\sum_{s,d \in V} D_{sd} X_{ij}^{sd} \leq C_{ij} \alpha \ \ \forall (i,j) \in E \qquad (4)$$

Note that the second term $r \sum_{s,d \in V} \sum_{(i,j) \in E} X_{ij}^{sd}$ in the objective function (i.e., Objective A) tries to avoid loops and unnecessarily long paths in routing.

Through Objective A, the authors in [5] expect that the traffic is moved away from the congested links, and traffic load is balanced across the network. However, we can see that minimizing network congestion or balancing load can not be strictly quantified. It is straightforward to see that minimizing the maximum link utilization can help reduce congestion but it cannot guarantee that the burst-loss probability is the minimum for all possible routings. For example, given two routings (routing 1 and 2) with the same α, load distribution on all the links is another parameter to evaluate how well the load is balanced. If more number of links in routing 1 have load close to α than that in routing 2, we can expect that a network with routing 1 will have larger burst-loss probability than that in a network with routing 2.

Therefore, improvements need to be made to further balance the load given the optimized maximum link utilization (i.e., α) in the network. Based on this observation, we propose two new objectives:

- Objective B: Minimize the number of links whose utilization is larger than a Watermark.
- Objective C: Minimize the sum of consumed bandwidth (or load) on links whose utilization is larger than a Watermark.

The Watermark is a positive integer, and it is smaller than α. In Objectives B and C, α provides a "ceiling" on the link load, which means that no link can have a load larger than α. Watermark provides a threshold, and a link will be marked as "(relatively) highly loaded" if its load is larger than the Watermark. With Objective B, we can ensure that the load on a link can exceed the threshold only when necessary. With Objective C, we can ensure that load on a link can only exceed the threshold by the minimum amount. With these two objectives, traffic will be shifted from heavily-loaded links to lightly-loaded links even when α is fixed.

We develop the ILP formulations for Objective B and C as follows (denoted as ILP-B and ILP-C, respectively).

ILP-B:

- *Objective B:* Minimize the number of links whose utilization is larger than Watermark.

$$Minimize : \sum_{(i,j) \in E} Y_{ij} + r \sum_{s,d \in V} \sum_{(i,j) \in E} X_{ij}^{sd} \tag{5}$$

- *Constraints:*
  - On physical route flow-conservation constraints:

$$\sum_{(k,j) \in E} X_{kj}^{sd} - \sum_{(i,k) \in E} X_{ik}^{sd} = 0 \ \ if \ \ k \neq s,d \ \ \forall k \in V, \forall s,d \in V \tag{6}$$

$$\sum_{(s,j) \in E} X_{sj}^{sd} - \sum_{(i,s) \in E} X_{is}^{sd} = 1 \ \ \forall s,d \in V \tag{7}$$

  - On link-capacity constraints:

$$\sum_{s,d \in V} D_{sd} X_{ij}^{sd} \leq C_{ij} \alpha \ \ \forall (i,j) \in E \tag{8}$$

  - On load-balancing constraints:

$$\sum_{s,d \in V} D_{sd} X_{ij}^{sd} - Watermark \leq p Y_{ij} \ \ \forall (i,j) \in E \tag{9}$$

ILP-C:

- *Objective C:* Minimize the sum of consumed bandwidth (or load) on links whose utilization is larger than a Watermark.

$$Minimize : \sum_{(i,j) \in E} Z_{ij} + r \sum_{s,d \in V} \sum_{(i,j) \in E} X_{ij}^{sd} \tag{10}$$

- *Constraints:*
  - On physical route flow-conservation constraints:

$$\sum_{(k,j) \in E} X_{kj}^{sd} - \sum_{(i,k) \in E} X_{ik}^{sd} = 0 \ \ if \ \ k \neq s,d \ \ \forall k \in V, \forall s,d \in V \tag{11}$$

$$\sum_{(s,j) \in E} X_{sj}^{sd} - \sum_{(i,s) \in E} X_{is}^{sd} = 1 \ \ \forall s,d \in V \tag{12}$$

  - On link-capacity constraints:

$$\sum_{s,d \in V} D_{sd} X_{ij}^{sd} \leq C_{ij} \alpha \ \ \forall (i,j) \in E \tag{13}$$

  - On load-balancing constraints:

$$\sum_{s,d \in V} D_{sd} X_{ij}^{sd} - Watermark \leq Z_{ij} \ \ \forall (i,j) \in E \tag{14}$$

Note that, in both ILP-B and ILP-C, α is used as a constant and its value is optimized using ILP-A. In ILP-B, Y_{ij} is a new variable defined as follows: $Y_{ij} \in \{0,1\}$; $Y_{ij} = 1$ if load on link (i,j) is larger than Watermark; otherwise, $Y_{ij} = 0$. In Eqn. (9), p is a positive integer and is larger than $\alpha - Watermark$. p is introduced to ensure that $Y_{ij} = 1$ only when load on link (i,j) is larger than Watermark. In ILP-C, Z_{ij} is a new variable defined as follows: $Z_{ij} \geq 0$; $Z_{ij} = \sum_{s,d \in V} D_{sd} X_{ij}^{sd}$ if load on link (i,j) is larger than Watermark; otherwise, $Z_{ij} = 0$.

3   Illustrative Numerical Examples

We study the performance of pre-planned routing in LOBS networks without any FDLs. As we have discussed in Section 1, each node is equipped with an IP/MPLS controller, which could route a BCP according to the label attached to it. All bursts between the same node pair will follow an explicit route (i.e., a label-switched path). In LOBS networks, burst-loss probability is mainly determined by the load of the links the burst traverses. It is also affected by the number of hops in the explicit route. If a burst needs to travel more hops, the burst will encounter higher loss probability if we assume the same load, and thus same loss probability, on each link.

Figure 1 shows the network topology we used in this study, which is a representative US nationwide network with 24 nodes and 86 unidirectional links. Each edge in Fig. 1 is composed of two unidirectional fiber links, one in each direction. Each fiber link is assumed to have 8 data channels and 1 control channel. Each channel operates at 10 Gbps. We simulate Poisson traffic at each node and the destination of a burst is uniformly distributed among all other nodes. The length of each burst follows a negative exponential distribution with mean value 10^5 bits. Contention on the control channel is ignored as the size of a BCP is usually small. The offset time for a burst is assigned as $H * t_p$, where H is the hop distance of the explicit route the burst will take and t_p is the average processing time for a BCP at a node, which is assumed to be $1/10$ of the mean burst length in this study. In addition, we assume that each node has full wavelength-conversion capability. The measured performance metrics include burst-loss probability and average hop distance.

We first pre-plan four different routings for uniform traffic: shortest-path routing, as well as optimized routing using ILP-A, ILP-B, and ILP-C. We assume one unit traffic flow for each source-destination node pair. Table 1 shows the characteristics of the four routing tables. Comparing shortest-path routing with ILP-A, we can see that 53 flows traverse the most congested link (i.e., $\alpha = 53$) in shortest-path routing but only 32 flows use the most congested link (i.e., $\alpha = 32$) in ILP-A, while the total hop distance (sum of the hop distance of all the flows) only increases by 6. This indicates that Objective A (minimize the maximum of link utilization) is effective in moving the traffic away from the congested links without affecting the hop distance too much.

In ILP-B, Watermark is fixed at 29, so the number of links with load larger than 29 (these link are denoted as heavily-loaded links) is minimized. We find

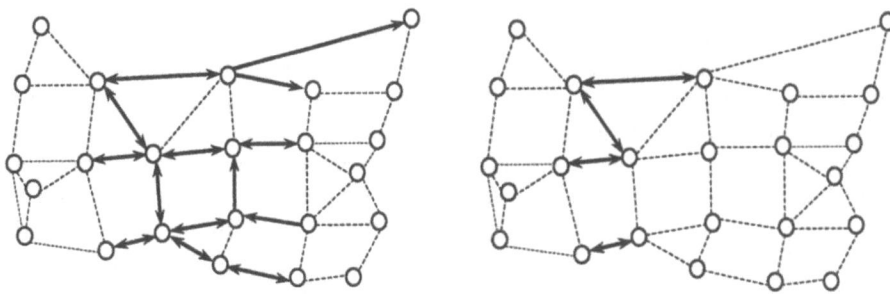

(a) After routing is optimized using ILP-A (b) After routing is optimized using ILP-B

Fig. 1. A sample network topology. The dark lines are links with more than 29 flows traversing on them (while arrows indicate in which direction)

Table 1. Characteristics of pre-planned routing tables

	Watermark	α	Total hop distance
Shortest-path routing	N/A	53	1652
ILP-A	N/A	32	1658
ILP-B	29	32	1672
ILP-C	20	32	1682

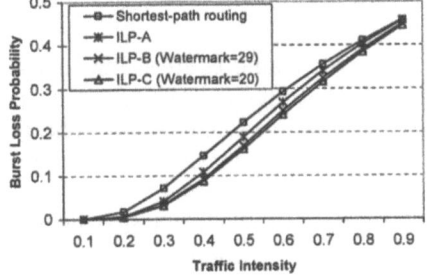

Fig. 2. Burst-loss probability

Fig. 3. Average hop distance

that only 8 links are heavily-loaded links and each of them have 32 flows in ILP-B, while in ILP-A, 24 links have load more than 29. In Fig. 1(a), we identify these 24 links in ILP-A with dark lines and in Fig. 1(b), we show the 8 heavily-loaded links in ILP-B with dark lines. We notice that these 8 links create two cuts (one from east to west and the other from west to east) in the network. One can see that, when one of the cuts is fulfilled, α cannot be reduced below the maximal link load on the cut. However, the load on other links in the network can be reduced, as shown in Fig. 1(b).

In ILP-C, Watermark is fixed as 20, which is the average number of flows (i.e., load) on a link. We choose the average link load as Watermark so the total load that exceeds the average load is minimized using ILP-C.

Figure 2 shows the burst-loss probability when we plug in the routing tables into the LOBS network. It shows that the burst-loss probability is reduced by 15%-80% (when load is less than 0.5) when routing is optimized using ILP-A compared with shortest-path routing. This is because congestion is removed in optimized routing, so the burst-loss probability is dramatically reduced. The burst-loss probability is further reduced by 15%-30% (when load is less than 0.5) using ILP-B and ILP-C, compared with ILP-A. This is because load on the "heavily-loaded" links can be further balanced using ILP-B and ILP-C, but ILP-A stops when α is optimized. ILP-C achieves the best performance because it directly minimizes the load on heavily-loaded links instead of minimizing the number of heavily-loaded links as in ILP-B; and, as we have mentioned before, the load determines the burst-loss probability. In Fig. 2, the performance gain reduces when the load increases. This is due to the fact that more short-hop bursts will succeed when the load is heavy.

Figure 3 shows the average hop distance for the pre-planned routings. As opposite to the burst-loss probability, shortest-path routing has the minimum average hop distance, which is because every burst takes the shortest path in shortest-path routing. For optimized routings, ILP-A performs best, ILP-C performs worst, and ILP-B is in between, which is comparable to the total hop distance shown in Table 1.

4   Conclusion

We considered the problem of explicit routing for traffic engineering in LOBS networks. We formulated the traffic-engineering problem as a linear program and proposed two novel objective functions to minimize congestion. The illustrative numerical results showed that the proposed optimization approaches can significantly reduce the burst-loss probability without too much sacrifice in the average hop distance for a burst. Traffic bifurcation will be studied using the optimization approach, which is our future research direction.

References

1. C. Qiao, "Labeled optical burst switching for IP-over-WDM integration," *IEEE Commun. Mag.*, vol. 38, pp. 104–114, Sep. 2000.
2. S. Junghans and C. M. Gauger, "Resource reservation in optical burst switching: architectures and realizations for reservation modules," in *Proc. OptiComm'2003*, Dallas, TX, Oct. 2003.
3. W. So, H. Lee, and Y. Kim, "QoS supporting algorithms for optical Internet based on optical burst switching," *Photonic Network Communications*, vol. 5, pp. 147–162, 2003.
4. D. Awduche, J. Malcolm, M. O'Dell, and J. McManus, "Requirements for traffic engineering over MPLS," *Internet draft, IETF*, April 1998, draft-awduche-mpls-traffic-eng-00.txt.
5. Y. Wang and Z. Wang, "Explicit routing algorithms for Internet traffic engineering," in *Proc. of ICCCN'99*, Sep. 1999.

A Mutual Authentication and Route Optimization Method between MN and CN Using AAA in Mobile IPv6*

Miyoung Kim, Hyewon K. Lee, and Youngsong Mun

School of Computing Science, Soongsil University
1-1, Sando 5Dong, Dongjak-Gu, Seoul, 156-743, Korea
{mizero31,kerenlee}@sunny.ssu.ac.kr,
mun@computing.ssu.ac.kr

Abstract. In the basic routing scheme of the mobile IP, the packets sent by the correspondent node(CN) are delevered to the mobile node(MN) via MN's home network which causes the triangle routing problem. To solve this problem, the route optimization(RO) scheme for mobile IPv6 has been defined by the mobileip working group in IETF. The return routability(RR) is devised in the mobile IPv6 to provide the authentication capability between CN and MN. However, it is inherently weak in the security capability the message exchange procedure is relatively simple. In this paper, we propose a new authentication scheme based on AAA infrastructure while support RO. We represent the performance analysis results depending on the traffic pattern and the terminal movement speed. Our proposed scheme saves the cost up in compared to 20 percent to RR.

1 Introduction

This paper describes the authentication scheme for a mobile node to perform the RO with secure manner after entering to new subnet and completing the Home Registration. To eliminate the routing delay from triangle routing, the RO enabling direct routing should be performed. However, the various attacks such as DoS, DDoS and MITM can be carried out by an attacker with a little effort which may result in packets to be intercepted and rerouted to the attacker. Thus, the security function should be added to the RO. The mobileip working group has defined the RR as an authentication scheme for making RO secure. Focusing on the operation and message exchange overheads, the RR procedure is relatively simple but not able to provide strong security function. This paper proposes an integrated mechanism with AAA for authentication and the RO to CN and shows the results of performance analysis relative to the RR in terms of the cost. The traffic and mobility property where the MN moves with the mean rate of movement and receives packets with mean rate of length for a session are assumed. We have verified the effectiveness of the cost using our proposed model that shows the cost reduction up to 20 percent.

---

* This work was done as a part of Information & Communication fundamental Technology Research Program supported by Ministry of Information & Communication in republic of Korea.

M. Bubak et al. (Eds.): ICCS 2004, LNCS 3038, pp. 1217–1223, 2004.
© Springer-Verlag Berlin Heidelberg 2004

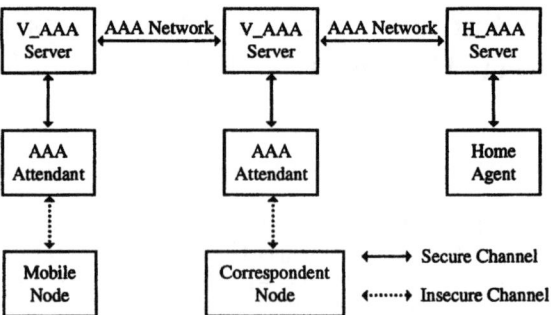

Fig. 1. The AAA Authentication Model

2 AAA Authentication Model and Entities

Fig.1 shows the communication model to exchange the session key between MN and attendant and to register the current location of the MN to its home agent(HA) based on AAA(Diameter) infrastructure[1],[2].

The attendant is the first entity the MN encounters when it moves into a subnet and it take the policy to pass, drop and block the incoming and outgoing packets using pre-configured parameters. When the authentication from the AAA server is successful, the attendant acquires the session key from it to make the wireless link between MN and attendant secure and exchanges packets with MN. When the MN is about to carry out the RO after completing the home registration without session drop, V_AAA as a AAA(Diameter) server in visited link identifies the home subnetwork of CN by the address of CN. It them sends the authentication/binding key request message to the home AAA server of CN to verify the CN and perform the RO with CN in secure manner. H_AAA carries out the authentication using information of the MN(NAI of MN[4], Message Authenticator, Home Address of MN, Address of CN, etc) and returns the reply message to V_AAA.

3 An Authentication Architecture for Route Optimization

3.1 Proposed AAA Architecture for Authentication and Route Optimization

The six messages are defined for the authentication using AAA in care of RO. In the first step, the MN sends the authentication request message(EAPoL) to attendant to authenticate itself and get the key materials where the LC, NAI of the MN, RPI, HoA, HaA, an address of CN, aaa_key, attendant_key, SecureParam_I and CR are included in the request message. Upon receiving the message, the attendant relays the message to its AAA server after making a conversion of message syntax from EAPoL to Diameter. The two groups of AVP are defined such as Address AVP (Home-Address-Option, Home-Agent-Address-Option, CN-Address-Option) and Security AVP (Nonce-Option, AAA-Key-Option, Attendant-Key-Option, Security-Parameter-Option and Authenticator-Option). After referencing the NAI option in Diameter

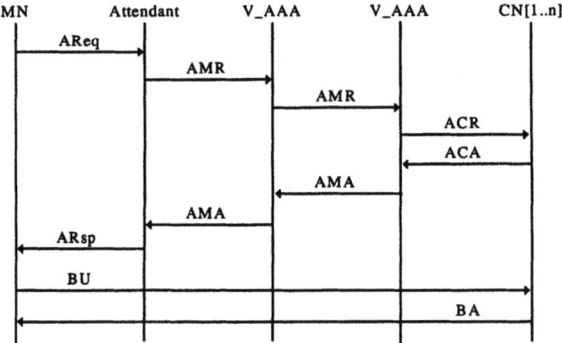

Fig. 2. Authentication and biding key exchanges between nodes in different domains

message, the V_AAA finds out the home AAA server of the MN and forwards the message to it without modification. In turn, the home AAA server sends the keying materials sent from the MN to CN after finding the address of the CN by using the Address AVP. If the destination of the message is multiple CNs, the CN-Address-Option for two or more CNs are included in the message. H_AAA copies the message and sends the same message (ACR) to multiple destination after converting the Diameter message to private message format known to CN where the address of MN and SecureParam_I are included in it. Receiving the messages, the CN stores the SecureParam_I and returns the ACR where the SecureParam_R containing the security association generated from the CN is included. H_AAA receives the ACR and sends it to V_AAA after making a conversion to Diameter format where the SecureParam_R and result code are rebuilt as a AVP form, Security AVP(Security-Parameter-Option) and Action AVP(Result-Code-Option). Upon receiving the messages, the CN stores the SecureParam_I and returns the ACR where the SecureParam_R containing the security association generated from the CN is included. H_AAA receives the ACR and sends it to V_AAA after making a conversion to Diameter format where the SecureParam_R and result code are rebuilt as a AVP form, Security AVP(Security-Parameter-Option) and Action AVP(Result-Code-Option). If the MN and CN belong to different domains, the message exchanges between them is represented in Fig.2.

4   Performance Evaluation

4.1   System Model

The performance evaluation is based on the costs during the MN is completing the registration to CN for the RO after successful home registration. This paper refers to the approach described in [9] and [10]. We assume that the CN sends the packets to MN in the ratio of λ and the MN moves into another subnet in the ratio of μ. By considering the average number of packets delivered from the CN to MN for every movement, the Packet to Mobile Ratio (PMR) is defined as $p = \lambda / \mu$.

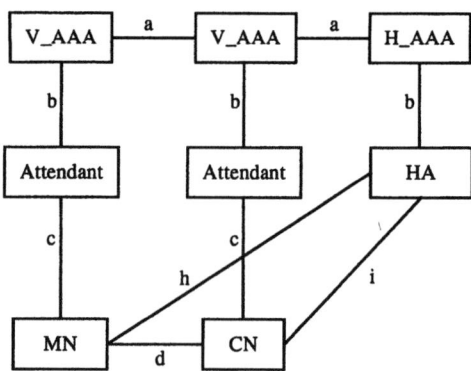

Fig. 3. A System Model for Cost Analysis

In this model, the internal and external link is connected each other and the weight is given to 1 and 2 for internal and external link, respectively. Thus a=2, b=1, c=1, d=2, i=2 and h=2 in Fig. 3.

4.2 Analysis of AAA Authentication Cost in Mobile IPv6

In the RR exchange[3], the total cost during the time where the MN moves and registers it's location to CN is denoted as C_{RR}. It is the sum of the cost s such as the cost of the MN completing the home registration($C_{BUHA - RR}$), the cost of packet loss due to the delay of home registration($C_{loss - RR}$), the cost of registering the MN's location to CN for route optimization($C_{BU - RR}$) and the tunneling cost of packets sent from CN to MN via HA during this time($C_{dt - RR}$) as follows.

$$C_{RR} = \left(C_{BUHA-RR} + C_{loss-RR}\right) + \left(C_{BU-RR} + C_{dt-RR}\right) \tag{1}$$

The $C_{BUHA - RR}$ is $(2h + 3r)$ as the home registration cost and $C_{loss - RR}$ is the cost of packets generated and sent to MN from CN during the registration delay time that are treated as the packet loss. The C_{data} the tunneling cost of single packet during the registration delay in Mobile IP is denoted as $\ell(i + h) + 3r$ which means the cost of a packet traveling from CN to MN through the HA. Thus, $C_{loss - RR}$ is obtained as follow.

$$C_{loss-RR} = \lambda \times t_{BUHA-RR} \times C_{data} = \lambda(2t_h + 3t_r) \times \ell\left((i+h) + 3r\right) \tag{2}$$

The cost of completing the binding registration for RO using the RR after home registration is defined as $C_{BU - RR}$ which is denoted by the sum of costs depending on the distance and packet processing cost at each node.

$$C_{BU-RR} = 2(i + h) + 4d + 9r = 2(2d + i + h) + 9r \tag{3}$$

During the delay in the RR process, the packets destined to MN are tunneled to HA where the tunneling cost is defined as follows.

$$C_{dt-RR} = \lambda \times t_{BU-RR} \times C_{data} \qquad (4)$$

The RR processing delay time is obtained as the sum of the transmission time of each link involved in packet traveling and the packet processing time at each node.

$$t_{BU-RR} = 2(2t_d + t_i + t_h) + 9t_r \qquad (5)$$

Therefore, putting all together, we can obtain the total cost of the RR as follows.

$$C_{RR} = 2(2d + i + 2h) + 12r +$$
$$\lambda(2(2t_d + t_i + 2t_h) + 12t_r) \times (\ell(i + h) + 3r) \qquad (6)$$

For the proposed scheme, C_{AAA} the total cost of completing the RO starting from the MN's movement is represented as formula 7 in the same manner of RR where the $C_{BUHA-AAA}$ means the home registration cost, $(2h + 3r)$ of MN registering its CoA to HA and $C_{loss-AAA}$ is the cost of the packet lost during the home registration delay since the packets are lost during that time($t_{BUHA-AAA}$) before completing the home registration. .

$$C_{AAA} = (C_{BUHA-AAA} + C_{loss-AAA}) + (C_{BU-AAA} + C_{dt-AAA}) \qquad (7)$$

$(2h + 3r)$ is obtained as follows where the $C_{data} = l(i + h) + 3r$ means the tunneling cost for a single data packet traveling from CN to MN via HA during the registration delay.

$$C_{loss-AAA} = \lambda \times t_{BUHA-AAA} \times C_{data} = \lambda((2t_h + 3t_r) \times \ell((i + h) + 3r)) \qquad (8)$$

C_{BU-AAA}, the cost of authenticating MN and registering its location is denoted as follows.

$$C_{BU-AAA} = 4(b + c) + 2(a + d) + 13r \qquad (9)$$

During the delay for authenticating peers and registering to CN, the packets from CN to MN traverse via HA. It implies the cost of tunneling packets as follows.

$$C_{dt-AAA} = \lambda \times t_{BU-AAA} \times C_{data} \qquad (10)$$

The tunneling cost of the single packet (C_{data}) is denoted by $l(i + h) + 3r$ which means the cost of packet from CN to MN via HA. In our scheme, the delay time of authentication and registration is as follows.

$$t_{BU-AAA} = 4(t_b + t_c) + 2(t_c + t_d) + 13t_r \qquad (11)$$

Therefore, the total cost of the proposed scheme is obtained as follows.

$$C_{AAA} = (4(b + c) + 2(a + d + h) + 16r) +$$
$$\lambda(4(t_b + t_c) + 2(t_a + t_d + t_h) + 16t_r) \times (\ell(i + h) + 3r) \qquad (12)$$

The cost ratio of the proposed scheme to the RR (C_{AAA} / C_{RR}) is obtained as equation 13 where the λ is replaced with $p \cdot \mu$ according to the definition of

PMR($p = \lambda / \mu$). This is for the comparison of the cost variation of C_{RR} and C_{AAA} in accordance with the variation of the PMR.

$$C_{AAA} / C_{RR} = \frac{4(b+c) + 2(a+d+h) + 16r + \rho \times \mu \times t_{BU-AAA} \times C_{data}}{2(2d+i+2h) + 12r + \rho \times \mu \times t_{BU-RR} \times C_{data}} \qquad (13)$$

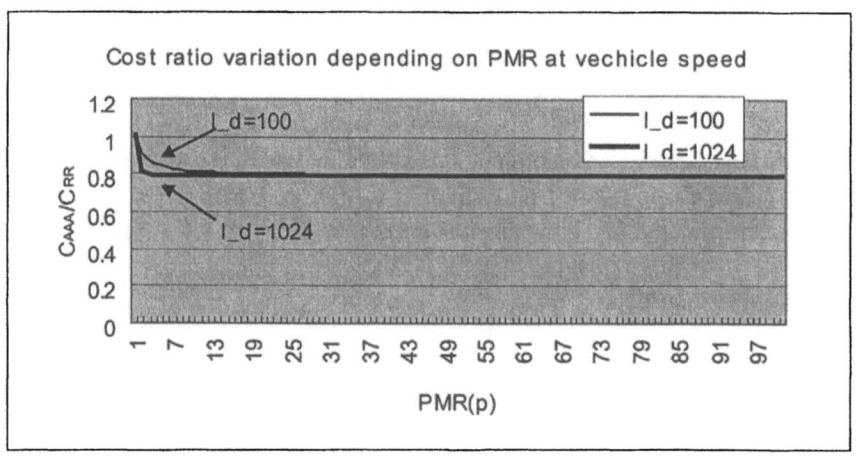

Fig. 4. The Cost ratio of C_{AAA}/C_{RR} when a mobile node moves at vehicle speed.

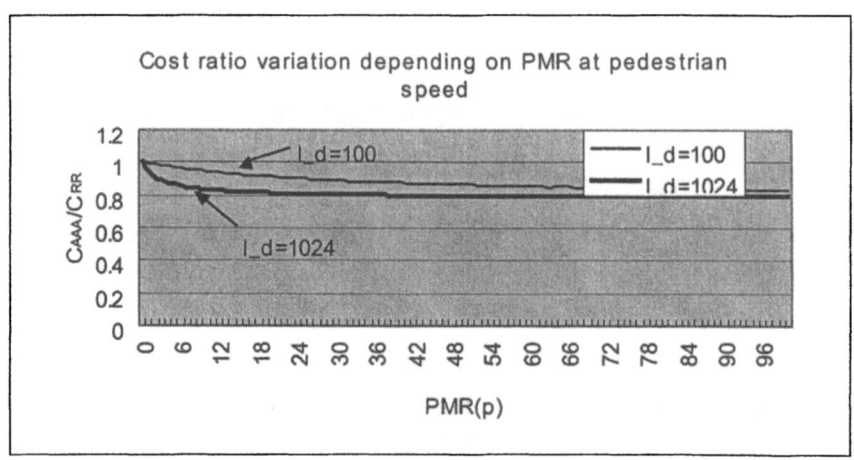

Fig. 5. The Cost ratio of C_{AAA}/C_{RR} when a mobile node moves at pedestrian

4.3 Performance Evaluation

We assume the cost of processing a message at each node is the same($r = 1$). Also according the assumption that the cost weight of the link within the domain is 1(b=c=1) and the link of entities in different domains is 2, the result of the formula 13 shows the graph in Fig. 4 and Fig. 5 depending on the variation of PMR value for the case of pedestrian speed and vehicle speed, respectively. As the MN moves faster, the traffic between MN and CN is increased and the cost ratio (C_{AAA}/C_{RR}) is significantly reduced.

In Fig. 5, at the PMR value of 43, the relative cost becoms down to 0.84 and 0.8 when the average packet length is 100 bytes and 1024 bytes, respectively.

5 Conclusions

This paper proposes a scheme to authentication the system model to analyze the cost and defines the Packet to Mobility Ratio(PMR) as average number of packets per movement. The cost the proposed scheme is compared to the RR and analyzed the variation of the cost ratio depend on PMR value. As the result of analysis, the more the PMR value is increasing, the more the cost of the proposed scheme is going down significantly in comparing with RR. The proposed scheme saves the cost about 16 and 20 percent than RR while the MN moves at the pedestrian speed where the average length of data packet is 100 and 1024 bytes, respectively. At vehicle speed, it saves the cost about 20 percent. we expect that the result of the proposed scheme is applicable to the new coming RO solution integrated with existing security infrastructure such as AAA.

References

1. F.Dupont, J.Bournelle: AAA for Mobile IPv6, draft-dupont-mipv6-aaa-01.txt, Internet Draft,IETF, Nov,2001.
2. Pat R. Calhoun, Charels E. Perkins: Diameter Mobile IPv4 Application, Intener Draft, Internet Engineerig Task Force, November 2001.
3. David B. Johnson, Charles E. Perkins, Jari Arkko: Mobility Support in IPv6, draft-ietf-mobileip-ipv6-18.txt, Internet Draft, IETF, June, 2002.
4. P.Calhoun, C.Perkins: Mobile IP Network Access Identifier Extension for IPv4, RFC 2794, IETF, March, 2000.
5. Franck Le, Basavaraj Patil, Charles E. Perkins: Diameter Mobile IPv6 Application, draft-le-aaa-diameter-mobileipv6-01.txt, Internet Draft, IETF, November, 2001.
6. Allison Mankin, Basavaraj Patil, Dan Harkins, Erik Nordmark, Pekka Nikander, Phil Roberts, Thomas Narten: Threat Model introduced by Mobil IPv6 and Requirements for Security in Mobile IPv6, draft-ietf-mobileip-ipv6-scrty-reqts-02.txt, Internet Draft, IETF, May, 2001.
7. IEEE Std. 802.1X-2001: Port-Based Network Access, Jun, 2001
8. Pat R. Calhoun, Erik Guttman, Jari Arkko: Diameter Base Protocol, draft-ietf-aaa-diameter-12.txt, Internet Draft, IETF, July, 2002.
9. R. Jain, T.Raleigh, C. Graff and M. Bereschinsky: Mobile Interner Access and QoS Guarantees Using Mobile IP and RSVP with Location Registers, in Proc. ICC'98 Conf., pp.1690-1695, Atlanta.
10. Thomas, R., H. Gilbert and G.Mazzioto: Infulence of the mobile station on the performance of a radio mobile cellualar network, Proc. 3rd Nordc Sem., paper 9.4, Copenhagen, Denmark, Sep, 1988.

Studies on a Class of AWG-Based Node Architectures for Optical Burst-Switched Networks*

Yurong (Grace) Huang[1], Debasish Datta[2], Xue Qiu[3], Jing Zhang[1],
Hyuk-Kyu Park[3], Young-Chon Kim[3], Jonathan P. Heritage[1],
and Biswanath Mukherjee[1]

[1] University of California, Davis, CA 95616, yrhuang@ece.ucdavis.edu
[2] Indian Institute of Technology, Kharagpur, West Bengal 721302, India.
[3] School of Electronics and Information, Chonbuk National University, Jeonju, Korea.
yckim@chonbuk.ac.kr

Abstract. We investigate a class of novel node architectures based on the static arrayed-waveguide gratings (AWGs) and tunable waveband converters (TWBCs) for optical burst-switched (OBS) networks. As compared to the other AWG-based architectures using tunable wavelength converters (TWCs) for switching, our design needs much fewer wavelength-converting devices, TWBCs, instead of a large number of TWCs (operating only on one wavelength at a time). Notwithstanding the inherent simplicity, AWG-based architectures, due to the static routing properties of AWGs, exhibit some internal blocking as compared to the strictly nonblocking OBS nodes employing SOA/TWC-based architectures. We address this issue in our design methodology to arrive at different candidate node architectures using multiple layers of AWGs. Our simulation results indicate that the proposed class of node architectures using TWBCs and multiple layers of AWG can offer acceptable blocking performance with a simple and cost-effective optical hardware for OBS networks.

1 Introduction

A wavelength-division-multiplexing (WDM)-based network can alleviate the problem of electronic processing bottleneck by performing some of the switching/routing functionalities. Generally, there are three categories of switching technique for all-optical networks: optical circuit switching (OCS), optical packet switching (OPS) and optical burst switching (OBS) [1]-[2]. The implementation of OCS using dedicated lightpaths between source-destination pairs may result in inefficient utilization of bandwidth resources. To improve bandwidth utilization, one might employ OPS/OBS, which can provide better resource utilization by employing statistical multiplexing and traffic engineering in optical domain. In OBS, several packets are assembled into a longer packet, called *burst*. The header of the burst is transmitted prior to its data burst with an offset time, which enables intermediate nodes to reconfigure their switches before the burst arrives. The header is processed electronically at every node

* This work has been supported in parts by NSF Grant Nos. ANI-02-07864, INT-03-23384, KOSEF, OIRC, and KIPA professorship program.

M. Bubak et al. (Eds.): ICCS 2004, LNCS 3038, pp. 1224–1232, 2004.
© Springer-Verlag Berlin Heidelberg 2004

and a modified header is transmitted again in optical domain for onward reservation, while the data burst propagates entirely in the optical domain from ingress to egress nodes.

As network bandwidth keeps increasing, performing routing in electronic domain through opto-electronic/electro-optic conversions become complicated and expensive as well. Hence, the implementation of transparent routing functionality in optical domain plays an important role for the evolution of optical networks. In general, for realizing routing functionalities in optical domain, OCS and OBS networks have different requirements for switch reconfiguration time. A slow switch might suffice OCS operations, since a lightpath is normally setup for a long period as compared to the switch reconfiguration time at nodes. However, for bursty traffic with finer traffic granularity, the switch reconfiguration time needs to be smaller for better network efficiency, thus requiring fast optical switching for OBS networks. In this paper, we explore some candidate switching schemes for OBS nodes, leading to a class of fast and cost-efficient node architectures for OBS routers.

Designing node architectures with fast and non-blocking switch configurations has been a major issue in OBS network implementation. Some of the recent efforts in this direction [3] have been to implement strictly non-blocking node architectures by using tunable wavelength converters (TWCs) in conjunction with semiconductor optical amplifiers (SOAs). However, the hardware complexity in such configurations has motivated other research groups to explore alternative node architectures that can offer simple and fast operation, however with a limited blocking in the switch fabric in some configurations. Most of these node architectures make use of TWCs along with arrayed-waveguide grating (AWG) as the basic switching hardware, the latter offering a static wavelength-selective routing functionality between its input and output ports [4]-[6]. In particular, AWG is a passive static device which is wavelength sensitive. When a lightwave enters at an input port of an AWG at a given wavelength, the output port wherefrom it will exit, depends on the wavelength it carries and the position of the input port. Employing this feature of AWG along with fast TWCs preceding each input port of AWG, a transparent switching unit can be realized for OBS networks. This type of switching scheme is attractive because it reduces network cost and complexity, and simplifies network management, as compared to the network architectures employing a large number of SOAs for switching purpose. Furthermore, because of its passive nature, AWG is a highly-reliable device without any power consumption and also offers lower insertion loss than a normal space switch [6].

In view of the above, we propose in this paper, a novel class of AWG-based OBS node architectures, wherein we reduce the optical hardware requirement furthermore by using the concept of waveband conversion (discussed later in details). We examine the performance of proposed node architectures in terms of burst-rejection probability and node throughput, and compare with the performance of OBS nodes with strictly non-blocking architecture. We also provide a methodology to improve the blocking performance with larger number of AWGs.

2 Strictly-Nonblocking OBS Node Architectures

Generally, any switching unit in a network node can be categorized in terms of its blocking performance in presence of contentions between incoming packets/bursts/calls (bursts in case of OBS networks). In particular, in a switching unit, contention might occur when two or more bursts want to exit the switch though the same output port at the same time. Indeed, one of the contending bursts can only win the contention, while the others are lost if the node does not have any buffer to store them during the period of contention. However, besides the natural blocking resulting from this output-port contention, a switching unit might exhibit another type of blocking, called *internal blocking*, when a burst loss occurs even in absence of output-port contention. Such losses occur due to the lack of alternative paths between input and out pout ports due to limitation in physical connectivity within the switch. However, the nodes that offer strictly non-blocking (i.e., no internal blocking) must employ an exhaustive connectivity between input and output ports with a complex hardware. In OBS nodes, similar issues become relevant, and the switching scheme proposed earlier in [3] using wavelength converters and SOAs offer strictly-nonblocking performance but with a high hardware complexity, and also suffer from lower reliability due to the use of large number of active devices such as tunable wavelength converters (TWCs) and SOAs. More recent switching schemes for OBS nodes [4]-[6] adopt simpler optical hardware using AWGs along with TWCs, however with limited internal blocking. In the following section, we propose a class of AWG-based node architecture, albeit with novel improvisations, which reduce internal blocking but with a significantly less optical hardware for OBS routing.

3 OBS Node Architectures Using Arrayed Waveguide Grating (AWG)

As mentioned earlier in Section 1, AWGs can be used in conjunction with TWCs to realize switching functionality in an OBS node, however, with a limited blocking. Such node architecture has been employed in [5] as shown in Fig. 1 (a). Although TWCs preceding the AWG help in choosing alternate paths through AWG by changing the incoming wavelengths to other suitable wavelengths, the static connectivity within AWG leads to some internal blocking. To minimize this internal blocking, one can use (as shown in Fig. 1 (b)) fixed wavelength converters (FWCs) at the output stage of the switch (i.e., following AWG) along with the TWCs placed at the input stage [7]. The cost and hardware complexity of such switching fabric becomes high due to the large number of TWCs and FWCs at the input and the output stages, respectively. This motivates us to employ *sharing* of the wavelength conversion process among multiple wavelengths (channels) over a given band of wavelengths, which we call as *waveband conversion* in our proposed switching schemes. In this section, we first discuss the basic operation of TWBC as well as typical AWG routing functionalities, followed by the node architectures utilizing agile TWBCs along with static AWGs.

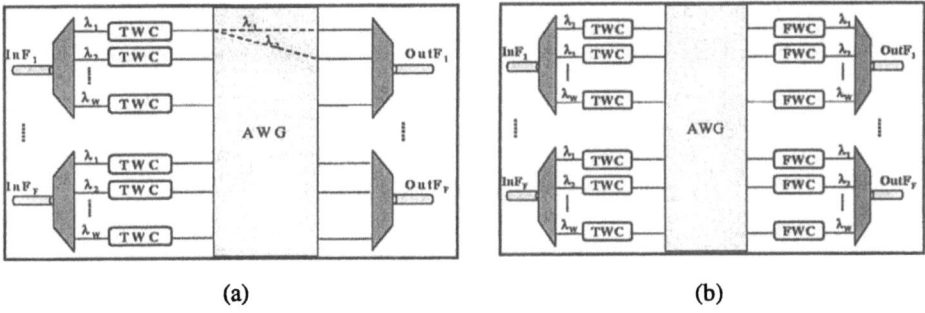

Fig. 1. (a) TWC/AWG based switching node with internal blocking; (b) TWC/AWG/FWC based nonblocking switching node

3.1 Tunable Waveband Converter (TWBC): Basic Features

Many efforts have been reported on the wavelength conversion techniques for one single wavelength channel [8]. Amongst them, the technique of parametric wavelength conversion offers also multi-wavelength conversion capability (i.e., concurrent conversion of a given multiple wavelengths to another set of multiple wavelengths) [8]. In this conversion scheme, the nonlinear interaction between pump lightwave (f_p) and signal lightwave (f_s) results in the converted wave (f_c) at the frequency of $f_c = f_p - f_s$. As shown in Fig. 2, while the pump laser converts signals between f_1 and f_4, centered around $^{f_p}/_2$, the conversion between f_2 and f_3 can be achieved concurrently as well. Thus, by employing a tunable pump laser, a TWBC could provide a wide range of waveband conversion. Having such conversion scheme, a given waveband could be converted to another specified waveband with a *single* optical device. We make use of these TWBCs along with AWGs for arriving at some viable switching schemes for OBS nodes architecture.

In the proposed switching schemes, we first divide the entire range of wavelengths (say, W wavelengths) into a number (K, say) of groups called wavebands, each consisting of an equal number of contiguous wavelengths ($M = W/K$). Next, each data burst is split into M segments at the ingress node, and all the burst segments are transmitted simultaneously using M wavelengths from anyone of the K wavebands with a duration, that is shrunk by a factor of M with respect to the original burst duration with single-wavelength transmission. Once these bursts arrive at the AWG input of a node, switching functionality to route them to different possible output ports can be realized by using TWBCs for each waveband instead of TWCs for each incoming wavelength. Thus, TWBCs being comparable to TWCs in respect of hardware implementation, the number of equivalent wavelength-converting devices for each input fiber port is decreased from W (number of TWCs) to K (number of TWBCs) with $K = W/M$, thereby leading to a significant reduction in hardware.

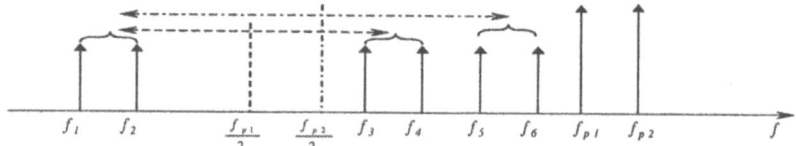

Fig. 2. A simple example for waveband conversion.

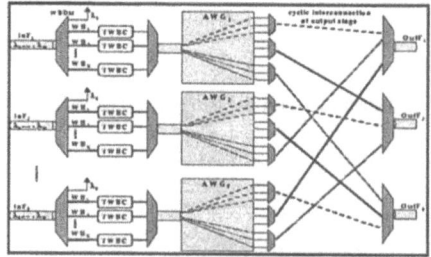

Fig. 3. A sample of WBA-SL architecture

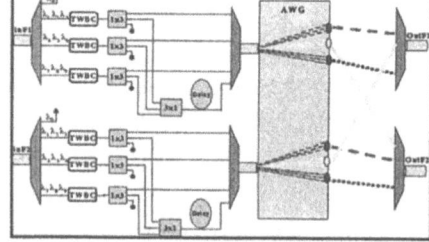

Fig. 4. A design of WBA-SL with FDL architecture

3.2   Proposed Node Architectures Using TWBC/AWG-Based Switching Schemes

In this subsection, we explore some novel switching schemes by combining capabilities of TWBCs and AWGs, which we call hereafter as *waveband-AWG* (WBA) *switching*. In particular, we present two new switching schemes based on WBA, namely *WBA with single layer of AWG* (WBA-SL) and *WBA with multiple layers of AWG* (WBA-ML).

A node is assumed to have F input fiber ports and F output fiber ports, with each input as well as output fiber port carrying W data wavelengths (or, more precisely, $K = W/M$ wavebands). First we consider the WBA-SL architecture. The proposed WBA-SL architecture employs, for each input fiber port, one waveband demultiplxer (WBDM), followed by K TWBCs for the input stage, and one AWG at the core stage, thus requiring F AWGs, F WBDMs and FK TWBCs in total (see Fig. 3). The output ports of K TWBCs for each input fiber port are combined together again (following appropriate waveband conversion) and connected to one of the appropriate input ports of the AWG, assigned for the corresponding input fiber port. Since each incoming burst is received on one of the K wavebands, they are all routed (by demultiplexing) to K distinct output ports of WBDM, which are subsequently combined (or multiplexed) again and fed into the relevant AWG. A data burst arriving on a certain waveband at a given input fiber port, is converted, if necessary, to such a required waveband, that the incoming data burst can be routed by the respective AWG to the desired output fiber port. This requires, in turn, a judicious interconnection between each of the F AWGs with all the output fiber ports. In other words, on the output side of AWGs, each AWG (corresponding to each input fiber port) needs to be interconnected to the F output fiber ports, such that at least one waveband from each input fiber port can be routed to every fiber output port. This leads to a cyclic interconnection pattern between F AWGs and F output port fibers, as shown in Fig. 3.

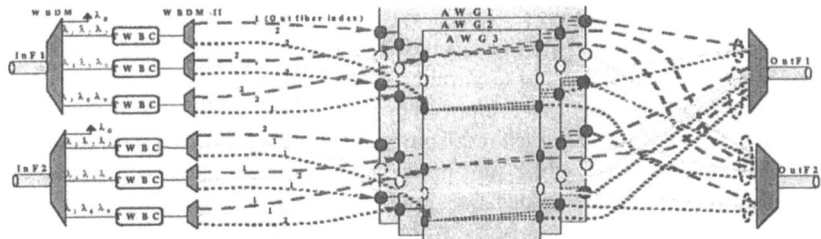

Fig. 5. The design of WBA-ML for OBS node structure

In a given input fiber port, any two or more incoming bursts destined to the same output fiber port, need to be converted to different wavebands, for being routed to the same output port without collision. However from each input fiber port, all data bursts may not be successful to reach the same output fiber port, as on the output stage, each fiber output port need to reserve at least one waveband from all the input fiber ports. In such cases, limited by the physical properties of an AWG and the output stage interconnection pattern (between AWGs and output fiber ports), burst may have to be dropped due to the internal blocking. However, such burst losses can be reduced if a delay element (e.g., FDL) can be used for each input fiber port to mitigate the contention, with the help of a (1×3) space switch for routing the waveband to the core stage, or to the delay element, or in the worst case to an optical termination to absorb the incoming light of the burst that lost the contention process. It may be noted that relatively shorter delay would be needed with the waveband-based transmission as compared to the single-wavelength transmission scheme. The WBA-SL architecture with FDLs for contention resolution is shown in Fig. 4 through an example of a node with 2 input/output fibers and three wavebands (i.e., $F = 2$, $K = 3$).

As compared to both the nonblocking OBS nodes as shown in Fig. 1 (b), and the OBS nodes employing TWC/AWG with limited internal blocking shown in Fig. 1 (a), WBA-SL architecture uses much fewer optical components, but indeed with some internal blocking. The blocking is more when one needs to avoid the (1×3) space switches and FDLs. However, fast space-switching of a waveband (i.e., multiple wavelengths at a time) may demand more complex optical hardware. To alleviate these limitations (use of switches and FDLs) of WBA-SL scheme, we next propose the WBA-ML architecture wherein the core stage is *dilated* by employing multiple layers of AWGs for each input/output fiber port (Fig. 5). In this scheme, every waveband entering an input fiber port and following TWBC can be routed to multiple AWGs, thereby utilizing the benefit of alternative paths towards output fiber ports.

In realizing the WBA-ML scheme, one can selectively route the TWBC outputs to different planes of AWGs (say, N planes) using a (1×N) space switch following each TWBC, in a manner similar to that in Fig. 4. However, space switching for waveband being difficult to realize, we propose a novel means using another stage of WBDMs following each TWBC, as shown in Fig. 5. This second or additional set of WBDMs provide a *waveband-selective routing functionality* for accessing different AWG planes with different wavebands, thus obviating the need of more complex (1×N) space switching devices. For example, consider one such additional WBDM (call it WBDM-II) with one input and two output ports (following a given TWBC, assigned for a given input fiber port), corresponding to a given transmission scheme operating with two wavebands – say, red waveband and blue waveband, coming out from the output port 1 and the output port 2 of the WBDM-II, respectively. Also, consider that the output port 1 of WBDM-II is connected to an AWG in one plane, and the output port 2 of WBDM-II is connected to another AWG in another plane, with both AWGs being allocated for the same given input fiber port. With this setting, if the input burst to WBDM-II arrives on red waveband from the preceding TWBC, it will come out of the WBDM-II from port 1 while the input on a blue waveband will come out from port 2 of the WBDM-II. By tuning the input waveband in preceding TWBC, a burst can be routed to different output ports of the WBDM-II and hence to different planes of AWG. Thus such additional stage of WBDMs provides a *switchless routing* of wavebands to different planes of AWG.

It is worthwhile to note that, the connectivity between the output ports of WBDM-IIs and the input ports of AWGs in different planes and the connectivity between output ports of AWGs and the fiber output ports of the node are closely related. Moreover, the connectivity on the output side would necessarily exhibit some cyclic pattern, which needs to be optimized for achieving best possible node throughput. This is one of the issues for our future study. Here, we show one possible intuitive solution for illustration, as shown in Fig. 5, wherein we have also *integrated* all the constituent AWGs in a given plane into *one single* AWG. For example, in Fig.5, each of the three AWG planes would have used two separate AWGs without the proposed integration.

As mentioned above, each plane in the proposed architecture (without AWG integration) would have F AWGs, thus requiring $N \times F$ AWGs for the entire node, employing N planes of AWG. However, using some inherent cyclic property of AWGs, one can integrate all the AWGs in a given plane into one composite AWG, thereby reducing the number of AWG devices to F only, although the individual AWGs will become larger in size following the proposed. In our node design, we consider an AWG with a size of $FW \times FW$ using F input/output fiber ports and supporting W equally-spaced wavelengths. Such an AWG would offer a fixed (i.e., static) routing matrix, such that an optical signal on $\lambda_i (i = 0, 1, …, W\text{-}1)$, at the input port $INP_j (j = 0, 1,…, FW\text{-}1)$ is routed to the output port $OP_x, x = (i + j)$ mod FW. By using the input ports $j = fW, f = 0, 1,…, F\text{-}1$, for input fiber f, each output port only forwards a predetermined wavelength from one input port. This allows one to reuse (spatially) the same waveband at the different input/output ports leading to the desired integration process.

4 Illustrative Numerical Examples

In this section, we examine our WBA-ML design to evaluate its performance in terms of node throughput and burst rejection probability due to internal blocking. The node throughput is evaluated as the ratio of the total number of bursts successfully routed to the total number of bursts arrived at the node input during a given simulation time.

The node load is defined as *node load* $= \dfrac{\alpha \times L}{W \times F \times R}$, where α represents burst arrival rate at the node, L is average burst length, W is number of wavelengths per fiber, F is node degree, and R is per-channel transmission rate. The burst internal-rejection probability is evaluated as a ratio of the total number of rejected bursts due to an

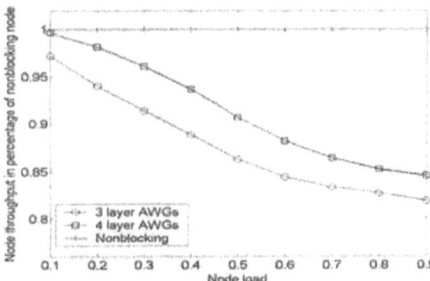

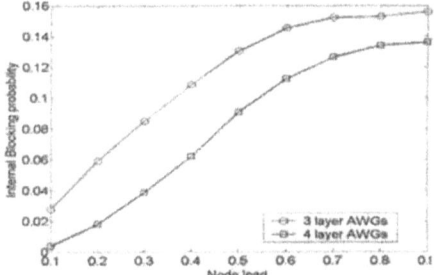

Fig. 6. Node throughput for WBA-ML architecture as percentage of the nonblocking node.

Fig. 7. Burst rejection probability due to internal blocking.

internal blocking in the switching unit to the total number of incoming bursts during a given simulation time.

For illustration purpose, we simulate a node with two input/output fiber ports ($F = 2$) each of which carries 12 wavelengths ($W = 12$) for payload transmission. In the switching fabric, the number of AWG layers is same as the number of wavebands, K. A waveband in an input or output fiber where a burst arrives or departs is uniformly distributed. Both, burst arrival and burst length are assumed to be Poisson distributed.

As discussed in Section 3, the number of TWCs is significantly reduced by using TWBCs in our design for an OBS node. For example, 24 (i.e., 12 wavelengths/fiber × 2 input fiber ports) TWCs and FWCs are used in the node as shown in Fig. 1 (b), whereas in our design, 6 (i.e., 3 wavebands/fiber × 2 fibers) TWBC are required. Thus, the optical hardware could be reduced significantly (in this example, by a factor of 4[1(for TWC) + 1(for FWC)] = 8). In our design, while saving cost from using TWBCs and AWGs, we might sacrifice the node performance with lower throughput and high burst rejection probability, because of the static properties of an AWG. However, as mentioned earlier, this limitation in node performance can be overcome by increasing the number of AWG layers, i.e., K.

Fig. 6 shows the node throughput (relative) in percentage of that of a nonblocking node (i.e., {throughput of the proposed node / throughput of a nonblocking node} × 100%) as a function of the node load. We find that the proposed OBS node can achieve 89% of the throughput offered by a nonblocking node at a load of 0.4 when employing 3 layers of AWGs. The throughput can be further improved to 93.7% of a nonblocking node by using one more layer of AWG, i.e., 4 layers of AWGs. Thus, the results indicate that the node throughput can be appropriately improved by increasing the number of AWG layers. In this example, by increasing K from 3 to 4, the throughput increases by 5.4% at a load of 0.4. It is expected that, one can approach close to the nonblocking performance with a reasonable increase in optical hardware by adding a required number of AWG layers.

The results of burst rejection probability vs. node load are plotted in Fig. 7, wherein we again observe a similar performance improvement (reduced burst rejection) with larger number of AWG layers. Thus by increasing the number of AWG layers, one can achieve a specified node performance, without needing significant increase in optical hardware. It may be noted that, the additional cost and complexity from a few more layers of AWGs are much less than the cost of large number of TWCs, replaced by much fewer TWBCs in our design. Thus, the proposed class of architectures offers a novel tradeoff towards engineering a cost-effective as well as efficient design of OBS nodes.

5 Conclusion

We have studied a class of novel architectures for OBS nodes, by using TWBCs along with passive AWGs with static crossconnect functionality. In comparison with other AWG-based node architectures reported so far in the literature, our node architectures offer simpler implementation with fewer optical devices, which are achieved by harnessing the novel concept of waveband conversion with TWBCs as a replacement of single-wavelength-based TWCs. Although the passive AWGs with planar integrated-optics implementation offer simpler optical hardware, their inherent static routing characteristics along with coarser granularity of TWBCs result in internal blocking in the basic WBA-SL scheme (TWBCs with single layer of AWG), which we proposed as the first and generic design step. In the next step, we improvised our basic scheme with WBA-ML scheme, wherein we employed multiple layers of AWGs along with a switchless dynamic routing scheme realized by additional stage of waveband demultiplexers. By utilizing the physical properties of AWGs, we also proposed to integrate all the AWGs in a given layer into one single AWG, without any additional degradation in blocking performance. Simulation results indicate that the proposed class of WBA-based node architectures using TWBC/AWG combination can offer acceptable blocking performance with a simple and cost-effective hardware for OBS networks.

References

1. S. Verma, H. Chaskar, and R. Ravikanth, "Optical burst switching: a viable solution for terabit IP backbone," *IEEE Network*, Nov./Dec. 2000.
2. M. Yoo, C. Qiao, and S. Dixit, "Optical burst switching for service differentiation in the next-generation optical internet," *IEEE Comm. Magazine*, Feb. 2001.
3. C. M. Gauger, H. Buchta, E. Patzak, and J. Saniter, "Performance meets technology – an integrated evaluation of OBS nodes with FDL buffers," *Proc. First International Workshop on Optical Burst Switching (WOBS 2003)*, Dallas, Texas October 2003.
4. E. Zouganeli et al., "Performance evaluation of a mesh optical burst switched network with tunable lasers and tunable wavelength converters," *ConTEL* Jun. 2003.
5. J. Cheyns, et al., "Performance improvement of an internally blocking optical pack/burst switch," *IEEE ICC*, 2003.
6. S. Bregni, A. Pattavina, and G. Vegetti, "Architectures and performance of AWG-based optical switching nodes for IP networks," *IEEE JSAC*, Sept. 2003.
7. S. J. B. Yoo et al., "High-performance optical-label switching packet routers and smart edge routers for the next-generation internet," *IEEE JLT*, Sept. 2003.
8. S. J. B. Yoo, "Wavelength conversion technologies for WDM network applications", *IEEE JLT*, Jun. 1996.

Self-Organizing Sensor Networks

Doina Bein and Ajoy K. Datta

School of Computer Science, University of Nevada Las Vegas, NV 89154
{siona,datta}@cs.unlv.edu

Abstract. We propose a self-organizing protocol in a sensor network. The algorithm starting from an arbitrary state establishes a reliable communication (based on the directed diffusion strategy) in the network in finite number of steps. In directed diffusion protocol [1], a request for data from a an initiator is broadcast in the network, and the positive answers from the sensors are forwarded back to the initiator.

Keywords: Directed diffusion, routing, self-organizing, self-stabilization, sensor networks

1 Introduction

The process of sensing, data processing, and information communication is the basis of sensor networks ([2], [3]). Due to the large number of nodes and thus, the amount of overhead, the sensor nodes may not have any global identification (ID). In some cases, they may carry a global positioning system (GPS). Sensor nodes are equipped with a processor, but they have limited memory. They can carry out simple tasks and perform simple computations. The communication is wireless: radio, infrared, or optical media, and the chosen transmission medium must be available worldwide. Recent developments in wireless communications have produced low-power, low-cost, and multifunctional sensor nodes which can communicate with each other unhindered within small distances.

The nodes in sensor networks are usually deployed for specific tasks: surveillance, reconnaissance, disaster relief operations, medical assistance, etc. Increasing computing and wireless communication capabilities will expand the role of the sensors from mere information dissemination to more demanding tasks as sensor fusion, classification, collaborative target tracking. They may be deployed in an hostile environment, inaccessible terrains, and through a cooperative effort, proper information has to be passed to a higher level. Their positions are not predetermined, i.e., the network can start in an arbitrary topology.

Contributions. The goal of this paper is to design a self-organizing sensor network using self-stabilization. Both the sensors and the sensor network infrastructure are prone to failures, insufficient power supply, high error rate, disconnection, and little or no network support. Many protocols and algorithms have been proposed for traditional wireless ad-hoc networks, but they do not take into consideration frequent topology changes, sensors failures, and possible nonexistent global ID. A distributed self-configuring and self-healing algorithm for

M. Bubak et al. (Eds.): ICCS 2004, LNCS 3038, pp. 1233–1240, 2004.
© Springer-Verlag Berlin Heidelberg 2004

multi-hop wireless networks is proposed in [4]. Being self-stabilizing guarantees that the system will converge to the intended behavior in finite time, regardless of the system starting state (initial state of the sensor nodes and the initial messages on the links).

In this paper, we deal with the communication reliability of the network, and we present a self-organizing protocol in a sensor network. The protocol constructs in finite number of steps a reliable communication in the sensor network in which requests for data sensed by the sensors are answered back to the initiator on the shortest path, which can send it to a higher level network (e.g. Internet, satellite etc) or application level.

Related Work. Given a general asynchronous network with at most n nodes, a similar self-stabilizing protocol is proposed in [5]. Its idea of maintaining the correct distances in the routing table regarding the closest nodes can be used for our algorithm in a sensor network, where the criterion is not the shortest distance among neighbors, but correct collected data for each request.

We use directed diffusion protocol ([1]) for implementing all the distributed services and for retrieving data through dynamically changing ad-hoc sensor networks. In directed diffusion (see Subsection 2.1), the nodes are not addressed by their IP addresses but by the data they generate. In order to be able to distinguish between neighbors, nodes may have local unique IDs. Examples of such identifiers are 802.11 MAC addresses ([6]), Bluetooth cluster addresses ([7]).

The most general technique of designing a system to tolerate arbitrary transient faults is self-stabilization ([8]). A *self-stabilizing* system is guaranteed to converge to the intended behavior in finite time, regardless of the initial state of the nodes and the initial messages on the links ([9]). In a distributed self-organizing protocol, with no initialization code and with only local information, the global objective to be achieved is to construct a reliable communication in the network in which requests for data sensed by the sensors are answered back through the shortest path with the correct existing data up to the current moment, meanwhile taking care of topology changes as well.

Paper Organization. In Section 2 we present several aspects regarding sensor networks, directed diffusion protocol, and self-stabilization as a particular case of fault tolerance. In Section 3 we present the self-stabilizing directed diffusion protocol, and we make some concluding remarks and future work in Section 4.

2   Sensor Networks as Distributed Systems

A sensor node is made up of four basic components: a *sensing unit*, a *processing unit*, a *power unit*, and a *transceiver unit*. Additional components as *power generator*, *global positioning system*, *location finding unit* can be also attached, depending on the application. The *processing unit*, equipped with a small memory, manages the work of other components, it is responsible for information communication, and also supervises an eventual collaboration of the node with other nodes in accomplishing assigned sensing tasks. So it is important for our routing algorithm to have a low computation time and memory requirements. The *transceiver unit* connects the node to the network.

The sensor nodes are generally scattered in a *sensor field*, and there are one or more nodes called *initiators*, capable to communicate with higher level networks (Internet, satellite etc) or applications. Each sensor is responsible for collecting data appropriate to its type and specifications, and maintaining a shortest path communication with the initiator nodes.

Whenever a new data is acquired regarding some past request not expired yet, a sensor node communicates it to the initiator, and facilitates the shortest path forwarding. A sensor node is able to do these tasks concurrently for any number of initiators and any number of request per initiator. Some sensor nodes may fail due to hard conditions, lack of power, environmental interference. The failure of a node should not obstruct the network functionality, and its overall task, as long as the network still remains connected and does not get disconnected.

2.1 Directed Diffusion Protocol

Directed diffusion ([1]) is a data-centric protocol, where the nodes are not identified by their ID, but by the data they generate as result of detecting (sensing). The data is organized as *attribute-value* pairs. An initiator node makes a request for a certain data by broadcasting an *interest* for a certain data throughout the sensor network. Different nodes may matched the request on different degrees, and gradients are kept in order to point out the neighbor (or the neighbors) toward the initiator (or initiators) of that particular request. The possible answers are forwarded back to the initiator, and intermediate nodes may perform a pre-filtering of the answers.

In our sensor network, we consider simple attribute-value scheme, with an upper bound of K in the number of such pairs. Each attribute has an fixed length associated value range, but not a fixed range. We impose these restrictions as the memory in a sensor unit has limited capacity. When broadcasting an interest, an initiator will specify the exact attribute name, and a much larger interval attribute. The message contains also a distance field (initially 0), and some control information: a timestamp, and an expiration value, which added to the timestamp specifies the moment in the future when the data is not important anymore (it has expired).

Every node maintains a so called *interest cache*, that contains for each existent *attribute-values* pair, a *gradient* field, which specifies the neighboring node through which an **interest** request has been received, the newest *match* up to the current moment, some control fields (*timestamp* and *expiration*), a *newest* field (identifies the node from which the newest matching data has been received) and a *distance* field (keeps track of the shortest distance toward *initiator*). An attribute exists in the cache if, at some not far moment in the past, an initiator has manifested an interest for that attribute, so the content of the cache is *interest driven*. The variables used are:
- IC = the interest cache data structure, which keeps all the data
- *no_entries_IC*: the current number of entries in the IC

- *sensed*: it is set to *true* by the sensing unit whenever new data is collected
- *SU*: the sensing unit data structure which contains the complete collected data, and it has the following fields: *attr* (attribute name), *value* (value of the attribute detected by the sensing unit), and *other* (other data). The sensing unit sets the Boolean variable *sensed* to *true* whenever new data is collected, and stores in the data structure *SU* the complete data.

There are three macros that operate over the variable *IC*:
- macro *add_to_IC(msg, nbr)* adds a new entry based on the field values of the message *msg* of type *INTEREST*, sets the gradient to be *nbr*, and also increments *no_entries_IC*. If *K* (the upper bound in the number of entries) is reached then the oldest message, breaking ties by expiration date, is removed.
- macro *modify_IC(e, msg, nbr)* modifies the entry *e* of *IC* based on the field values of the message *msg* of type *INTEREST*, and sets the gradient to be *nbr*
- macro *delete_from_IC(e)* deletes the entry *e*, and decrements *no_entries_IC*.

The *interest cache* is comparable with a routing cache, where instead of node IDs, we keep attribute-values pairs. Whenever an interest request is received, the node checks to see if it has an entry in its interest cache. If not, it creates an entry, using the parameters received in the message. For any node, an initiator is identified by a *gradient* field, which specifies the neighboring node through which the request has been received. The newest match in the data for an interest is kept, together with the node local ID (in the field *newest*) from which this match comes. The shortest distance toward the initiator is kept in the *distance* field. If a node receives data from its sensing unit or from the neighbors which have no existent interest, the data is simply dropped. Also, if the data received is older than the existent one, it is also dropped.

We prevent the case in which, as a result of an arbitrary initialization, for an interest received, wrong matches are stored initially in the cache with good (newer) timestamps. Whenever a new data corresponding to some pair attribute-values is received by a node (which can be either from the node itself sensing unit or from one of its neighbors) and that node ID is stored in the interest cache as the *newest*, that data will override the entry, independent of the values stored currently, even if the new value does not have a better timestamp for that interest. Later on the values received from the node sensing unit or from its neighbors will correct this.

Because the node has to self-organize, the interest cache is checked periodically for expired interests, and those entries are zapped. It is not necessary to send messages to the neighbors to do the same, because they already have the expiration date stored in their own cache, so they can do the same. Also, periodically, the entries in the cache are compared with the one of the gradients, to make sure that fictive interests are not stored as a result of an arbitrary initialization.

In [1], the model proposed forces the initiator to periodically broadcast an interest to the rest of the network, with the purpose to maintain the robustness and reliability of the network. In our algorithm, this is not necessary, because each node is responsible for keeping track of changes in its neighborhood, so an initiator will re-broadcast its interest only if, within a certain timeout, he did not receive an answer or a proper answer back. An addition to the [1], we allow any number of initiators and any number of requests per initiator, with the condition that we have an upper bound of K in the number of entries in the interest cache, as the memory in a sensor unit has limited capacity. Another difference from [1] is that we save space by requiring only one gradient per entry, while in [1] there is a gradient for each neighbor.

2.2 Self-Stabilizing Distributed Systems

In a sensor network, the nodes communicate by messages, with two actions: *send(message)*, and *receive(message)*. The messages have variable length and they are sent through wireless links. Our algorithm is asynchronous, so it is guaranteed to run correctly in networks with arbitrary timing guarantees. A common assumption is to bound the interval of time for transmitting a message, called *timeout*, after which the message is considered lost.

There are some assumptions we make in our algorithm: independent of the node/link failure, the network never becomes disconnected, and that we have FIFO ordering among the messages on the channel, which means that the messages are delivered in a node in the order they have been sent on the channel by a neighboring node. The messages used:

- *CHECK, CLEAR* : fields *id* (sender local ID), *attr* (attribute name), *interval* (interval of values for an interest), *time* (initiator sending time), *expir* (expiration time).
- *DATA* : fields *id* (sender local ID), *attr* (attribute name), *value* (attribute value), *other_data* (other data), *time* (sending time).
- *INTEREST* : fields *attr* (attribute name), *interval* (interval of values for the attribute), *time* (initiator sending time), *expir* (expiration time of the interest), *dist* (total length of the crossed path)

Each node v has a unique local ID, LID_v, and knows only its direct neighbors (variable N_v), so it can distinguish among its adjacent wireless links. We assume that N_v as well as the *current_time* are maintained by an underlying local topology maintenance protocol, which can modify N_v value during the lifetime of the node because of adding/crashing of neighboring sensors.

Each sensor node has a local state, identified by its current interest cache, and its variables. The global state of the system is the union of the local state of its nodes, as well as the messages on the links. The distributed program consists of a finite set of actions. Each action is uniquely identified by a label and it is part of a *guarded command*: $< label >::< guard >\rightarrow< action >$

The action can be executed only if its guard, a Boolean expression involving the variables, evaluates to *true*. An action is atomically executed: the evaluation of the guard and the execution of the corresponding action are done in one atomic step. In the *distributed daemon* system, one or more nodes execute an action, and a node may take at most one action at any moment.

A self-stabilizing system S guarantees that, starting from an arbitrary global state, it reaches a legal global state within a finite number of state transitions, and remains in a legal state unless a change occurs. In a non-self-stabilizing system, the system designer needs to enumerate the accepted kinds of faults, such as node/link failures, and he must add special mechanisms for recovery. Ideally, a system should continue its work by correctly restoring the system state whenever a fault occurs.

3   Self-Stabilizing Routing Algorithm in Sensor Networks

The macros/functions used by the algorithm are:
- *restart* sets the variables IC, SU, *no_entries_IC*, *turn*, *sensed* to their default values
- *check_entry* returns *true* if an $INTEREST$ message has a matching entry in IC on the data fields, but the control fields show a newer request, *false* otherwise
- *check_data* returns *true* if the data collected by the sensing unit or received in a DATA emssagematches an interest entry in IC, *false* otherwise
- *match_entry* returns *true* if the data and control fields from a CLEAR/CHECK message are matching some entry in the IC, *false* otherwise

The purpose of the algorithm is to construct a reliable communication in the sensor network in which requests for data sensed by the sensors are answered back through the shortest path with the correct existing data.

The guard 1.01 has the role to check for errors as a result of an arbitrary initialization of the network, and to keep up with topology changes in the immediate neighborhood. Periodically, each entry in the interest cache IC is checked for expiration and consistency with the gradient node. If it is not expired, the data and the control fields of the entry is sent to the gradient node (message $CHECK$), to make sure that no changes have occurred in the interest requested in the past. If a change has occur or wrong interests are stored in IC, the gradient node answers back (message $CLEAR$), and the entry is cleared out of the interest cache.

Whenever a data is collected through the sensing unit, its content is stored in the data structure SU, and the variable *sensed* is set to *true*. Regulary the data collected is checked to see if it matches a previous, and not yet expired interest. If it does, then the data is sent (message $DATA$) to the corresponding gradient. In any case, *sensed* is set back to *false*. When a $DATA$ message is received, similar tests are done. If the message matches an interest, it is forwarded to the gradient, but having in the *id* field the current node (IDs can be checked only locally, they have no meaning further.)

When an initiator manifests an **interest** for some particular data, it sends an $INTEREST$ message, with data fields (attribute name and interval of values), control fields (timestamp and expiration), and a *distance* field, initially set by the initiator to be 0. Whenever an $INTEREST$ message is received by a node from a neighbor, the length of the link to the neighbor is added to the *distance* field. Among identical $INTEREST$ messages received, the one with the smallest distance value is selected, in order for the node to be always oriented using the gradient toward the initiator node through the shortest path.

Algorithm 1 $\mathcal{SOSN}$ Self-Stabilizing Directed Diffusion Protocol

---

$error \equiv \ \neg(0 \leq turn < no\_entries\_IC) \vee \neg(IC[turn].time > current\_time) \vee$
$(IC[turn].gradient \in N_v) \vee \neg(IC[turn].newest \in (N_v \cup LID_v))$

1.01 $error \quad \longrightarrow \quad restart$

1.02 **timeout** $\wedge (0 \leq turn < no\_entries\_IC) \quad \longrightarrow$
$e \leftarrow IC[turn]$
if $(e.time + e.expir < current\_time)$ **then** delete_from_IC(e)
else send $CHECK(LID_v, e(attr, values, time, expir))$ TO $e.gradient$
$turn := (turn + 1)$ mod $no\_entries\_IC$

1.03 Upon receipt of msg $CHECK$ from neighbor $nbr \quad \longrightarrow$
if $(msg\_id \neq nbr) \wedge (msg.id \notin N_v) \vee (msg.id \in N_v \wedge (\exists$ an entry e in IC:
$match\_entry(e, msg)))$ **then** discard message msg
else send $CLEAR(LID_v, msg(attr, interval, time, expir))$ TO nbr

1.04 Upon receipt of msg $CLEAR$ from neighbor $nbr \quad \longrightarrow$
if $(msg\_id \neq nbr) \wedge (msg.id \notin N_v) \vee (msg.id \in N_v \wedge \neg(\exists$ an entry e in IC:
$match\_entry(e, msg)))$ **then** discard message msg
else $delete\_from\_IC(e)$

1.05 **timeout** $\wedge$ **sensed** $\quad \longrightarrow$
if $(\exists$ an entry e in IC: $check\_data(LID_v, e, SU(attr, value), current\_time))$
then send $DATA(LID_v, SU(attr, value, other), current\_time)$ TO $e.gradient$
$sensed := false$

1.06 Upon receipt of msg $INTEREST$ from $nbr \quad \longrightarrow$
if $\neg(msg.id \in N_v \wedge msg.id == nbr)$ **then** discard message msg
$msg.dist = msg.dist + length\_path\_to(nbr)$
if $(\exists$ an entry e in IC: $match\_entry(e, msg))$ **then**
if $(msg.dist < e.dist \vee nbr == e.gradient)$ **then**
$modify\_IC(e, msg, nbr)$
send msg TO all nodes in $N_v \setminus nbr$
else discard message msg
else
if $(\exists$ an entry e in IC: $check\_entry(e, msg))$ **then**
$modify\_IC(e, msg, nbr)$
else $add\_to\_IC(msg, nbr)$
send msg TO all nodes in $N_v \setminus nbr$

1.07 Upon receipt of msg $DATA$ from $nbr \quad \longrightarrow$
if $(msg.id \in N_v \wedge msg.id == nbr) \wedge (\exists$ an entry e in IC:$check\_data(msg.id,$
$e, msg(attr, value, time)))$ **then** send $DATA(LID_v, msg(attr, value, other,$
$time))$ TO $e.gradient$

---

4 Conclusion

We presented a self-organizing protocol that guarantees that starting in an arbitrary state, and having only local information, in finite number of steps builds a reliable communication in the network based on directed diffusion method. An interesting open problem is a comparative study between self-stabilizing directed diffusion and snap-stabilization. Snap-stabilization was first introduced in [10], and guarantees that a system will always behave according to its specifications ([11]). Snap-stabilizing propagation of information with feedback algorithm has been presented in [12] and is used extensively in distributed computing to solve problems like spanning tree construction, termination detection, synchronization.

References

1. Chalermek Intanagonwiwat, Ramesh Govindan, and Deborah Estrin. Directed diffusion: a scalable and robust communication paradigm for sensor networks. *Proceedings of the 6th annual international conference on Mobile computing and networking, Boston, Massachusetts, United States*, pages 56 – 67, 2000.
2. Ian F. Akyildiz, Weilian Su, Yogesh Sankarasubramanian, and Erdal Cayirci. A survey on sensor networks. *IEEE Communications Magazine August 2002*, pages 102–114, 2002.
3. G. Hoblos, M. Staroswiecki, and A. Aitouche. Optimal design of fault tolerant sensor networks. *IEEE International Conference Cont. Apps. Anchorage, AK*, pages 467–472, 2000.
4. H. Zhang and A. Arora. GS^3: Scalable self-configuration and self-healing in wireless networks. In *21st ACM Symposium on Principles of Distributed Computing*, July 2002.
5. D. Bein, A. K. Datta, and V. Villain. Self-stabilizing routing protocol for general networks. *Second edition of RoEduNet International Conference In Networking, Iasi, Romania*, pages 15–22, 2003.
6. IEEE Computer Society LAN MAN Standards Committee. Wireless lan medium access control (mac) and physical layer (phy) specifications. *Technical Report 802.11-1997, Institute of Electrical and Electronics Engineers New York, NY*, 1997.
7. The Bluetooth Special Interest Group. Bluetooth v1.0b specification. *http://www.bluetooth.com*, 1999.
8. E. W. Dijkstra. Self stabilizing systems in spite of distributed control. *Communications of the ACM*, 17:643–644, 1974.
9. M. G. Gouda. *Elements of network protocol design*. John Wiley & Sons, Inc., 1998.
10. A. Bui, AK Datta, F Petit, and V Villain. Space optimal snap-stabilizing pif in tree networks. *Proceedings of the Fourth Workshop on Self-Stabilizing Systems*, pages 78–85, 1999.
11. A Cournier, AK Datta, F Petit, and V Villain. Optimal snap-stabilizing pif in un-oriented trees. *5th International Conference On Principles Of Distributed Systems (OPODIS 2001)*, pages 71–90, 2001.
12. A Cournier, AK Datta, F Petit, and V Villain. Snap-stabilizing PIF algorithm in arbitrary networks. In *IEEE 22nd International Conference on Distributed Computing Systems (ICDCS 02)*, pages 199–206. IEEE Computer Society Press, 2002.

The Application of GLS Algorithm to 2 Dimension Irregular-Shape Cutting Problem

Luiza Budzyńska and Paweł Kominek*

Institute of Computing Science, Poznan University of Technology,
Piotrowo 3a, 60-965 Poznan
{Luiza.Budzynska,Pawel.Kominek}@cs.put.poznan.pl

Abstract. This paper describes the application of the Genetics Local Search Algorithm (GLS) to the 2 Dimension irregular-shape Cutting Problem. We describe different recombination operators used to generate feasible solutions, as well as an algorithm that can be used to rotate figures in this problem. In our case, each offspring resulting from a recombination is a starting point for local optimization. The resulting solutions are permutations of the figure labels (i.e. sequence-coding). One of the distinctive features of our method is a specific representation of figures, in which a 32 bit binary vector is transformed into an integer value. Figures obtained in this way are then placed on the strip (encoded in the same way) in order to obtain the final solution. The paper includes results of computational experiments demonstrating the efficiency of the proposed approach.

1 Introduction

The Genetic Local Search (GLS) is a metaheuristic algorithm that combines genetic (evolutionary) algorithms with local optimization. Other frequently used names are Memetic Algorithms or Hybrid Genetic Algorithms [11]. It is well known that the way in which the evolutionary algorithms are adapted to a particular problem may have a crucial influence on its performance. The paper considers 2-Dimension Cutting (2-DC) problem [1] of minimizing the stripe length. Because of high computational complexity [7] of the problem the use of evolutionary algorithm with local search (GLS) based on significant features is proposed [4, 9]. Heuristics based on the GLS scheme often prove to be extremely efficient in combinatorial optimization [6, 9].

It is quite difficult to track the single origin of GLS. An important element to adopt GLS to a particular problem is the representation of data that we use to create its solution. Frequently the representations come from intuition [10] but a successful representation should not only be natural, but it should also be efficient from the point of view of local search. Traditional genetic algorithms use binary coding of solutions [8] but for many problems, such coding is not natural. Recently, dataset are often encoded with some specialized data structures [5].

---

* This work was supported by grant no 3 T11F 004 26 from the State Committee for Scientific Research.

M. Bubak et al. (Eds.): ICCS 2004, LNCS 3038, pp. 1241–1248, 2004.
© Springer-Verlag Berlin Heidelberg 2004

In this paper, we propose a dynamic structure to remember free pixels on the strip and we propose to use an array of integer values to minimize the amount of used memory and improve efficiency of combinatorial optimization. After this introduction the 2-DC problem to which we apply our approach is described in the second section. In the third section, we describe in detail the encoded solution and the figure representations. Next, we describe the adaptation of the dynamic structure to the GLS algorithm to solve the 2-DC problem. Computational experiments are described in the fifth section. In the sixth section, conclusions are presented.

2   Problem Definition

The 2-Dimension Cutting (2-DC) problem [1] considered in this paper consists in minimization of used strip length. The strip is a rectangle with unlimited length and determined width. To find a solution we use a procedure called Bottom-Up-Left-Justified, in which the first figure to be put is placed in left-up corner of the strip. The figures are moved pixel-by-pixel across the strip, first down and then if it not possible to the right, until a free location is found. None of the figures can overlap. The figures are any polygons, with possible holes made of other polygons. Depending on the instance of the problem, the rotations of figures are allowed or not.

3   Coding for the 2-DC Problem

3.1   Solution Coding

In our method, every solution is represented as a sequence of numbers, called a sequence list. These sequences are the identifiers of the figures, which are ordered according to procedure Bottom-Up-Left-Justified described above (Fig. 1 shows an example of 20 figures).

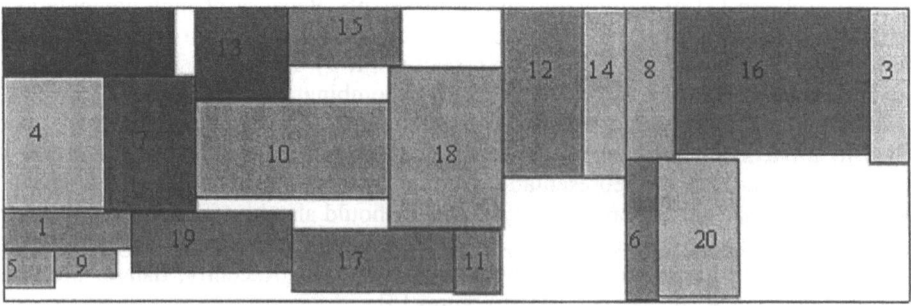

Fig. 1. Example of sequence coding: [2-4-1-5-9-7-9-13-10-15-17-18-11-12-14-8-6-20-16-3]

3.2 Figure Representation

The first implementation of this problem using binary and integer maps was created for the requirements of a vector machine [12]. It seemed that such encoding would not be effective on machines without vector registers. In spite of all we decided to verify this methodology on PC computers. In the considered problem all figures are polygons. Possible holes in the polygons are also represented as polygons. During the conversion of the representation from vectors (vertices) to binary map representation we approximate sides of figures with segments of line. The polygons obtained in such way are greater than or equal to the input figures. This results from the resolution of the binary map. Next we fill the polygons' interiors using the following algorithm:

```
ALGORITHM_1 FILL_THE_FIGURE
BEGIN
    For the considered polygon take all pairs of
    vertices and compute a - the slope of the line of
    its sides, and b - the intercept (y=ax+b);
    From all identified sides remove those, which are
    parallel to the x-axis (a=0);
    Sort all sides preserved in the preceding step
    according to the x-coordinate and then y-coordinate
    of their vertices;
    Take each pair of the preserved sides and
    horizontally fill the area between them;
END.
```

This algorithm is one of many we have tested for filling the interiors of polygons. It proved fast and accurate in that it always correctly filled polygons, even those having holes. We used the binary map representation for the figures, but the computational times achieved were not fully satisfying. Therefore, we converted the binary map into an integer map [4].

4 Implemented Algorithm and Data Structures

To solve the 2-DC problem we used the Genetic Local Search algorithm in the following version:

```
ALGORITHM_2 GENETIC_LOCAL_SEARCH
BEGIN
  P:=0;
  FOR i:=1 TO N DO
          BEGIN
              Construct a new feasible solution x;
              Apply local search to x obtaining x';
              Add x' to P;
          END
      REPEAT
```

```
        Draw at random two solutions x and y from P ;
        Recombine x and y obtaining z;
        Apply local search to z obtaining z';
        if z' is better than the worst solution in P and
        different to all solutions in P then Add z' to P
        and delete the worst solution from P;
        P:=P+1;
    UNTIL stopping criterion is met
    END.
```

where,

```
    P    - population id,
    N    - size of population,
```

4.1 Proposed Recombine Operators

Several operators take advantage of the sequence lists. One of the proposed recombination operators finds pairs of figures adjacent to each other and common to two parent solutions and places them in their offspring, and completes the offspring with randomly selected figures. Another proposed operator finds pairs or triplets of figures (not necessarily adjacent) following the same order in parent solutions and places them in the offspring, completing it with randomly selected figures. The last proposed operator constructs an offspring that has the same features in common with its parents as the two parents have in common with each other.

All operators were tested on different instances [3]. The results indicate that the use of GLS with the last recombination operator indeed improves both the quality of the new local optima and reduces the CPU time needed to reach them.

4.2 Identification of Free Pixels on the Stripe

Initially, the strip represented by array of values contains only empty values. Whenever an insertion of a figure is evaluated, the appropriate bits on the strip are tested. Each time the figure-placing procedure starts from the first row and the first column of the array. To speed up the procedure we proposed to additionally define a vector of columns including indices of first free row in each column (denoted as FFR) or an additional array including all indexes of free row [3] (denoted as AFR). However, this approach did not reduce the computational time of GLS.

4.3 Rotation of Figures

To rotate figures we perform the rotation of coordinates at an a priori given angle φ (fig. 2 shows an example).

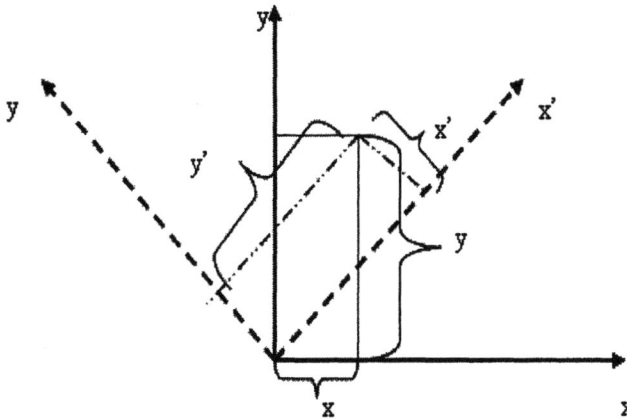

Fig. 2. Rotation of coordinates

To calculate new coordinates of figure vertices x' and y' we use the following formula (1)

$$x = x'\cos(\varphi) - y'\sin(\varphi), \, y = x'\sin(\varphi) + y'\cos(\varphi)$$
$$x' = x\cos(\varphi) + y\sin(\varphi), \, y' = -x\sin(\varphi) + y\cos(\varphi)$$

(1)

5   Computational Experiments

At first our implementation used arrays to encode data, then we applied vector containers from STL library. During the first experiment on PC, when array structures were used, computational time was satisfying. This encouraged us to perform further research aiming at finding the best appropriate structures. The proposed vector container resembles a C++ array in that it holds zero or more objects of the same type. The vector container can contain elements of any types. The STL library implements different methods in vector container, e.g.: insert, add, clear, resize, which are used in our implementation. In the experiments, we used seven different instances of the problem. The instances differed in the sets of figures to be put on the strip and the angle of rotation. For the instances denoted as "43a shape" and "10a pros" the rotation of 90° and 180° was allowed for each figure. In [5] Oliveira at all presents the best known layout for "43 shape" and "43a shape".

The experiments were performed on a PC with Pentium 733 MHz processor. We have noticed that an increase of the running time did not influence the quality of results significantly. Table 1 presents comparison of creation times of one randomly selected solution for binary (BM) and integer map (IM) representation of the figures.

Table 2 contains experimental results obtained with a number of different methods described in section 4.2. In all cases, populations of size 40 were used.

The proposed methods are also compared to a standard GLS algorithm without any additional structures (see Table 3). In addition, we test the best kind of GLS (i.e. the one with FFR structure) on several instances (see Table 4). All the algorithms shared a common code, for the most part.

Table 1. Creation time of solution

Name of instance	Fitness function [pixel]	Time [s]
10 pros (BM)	720	2,6
10 pros (IM)	720	1,6
23 pros (BM)	650	271,9
23 pros (IM)	650	4,8
45 pros (BM)	1490	4,9
45 pros (IM)	1490	0,6

Table 2. Average of 100 solutions for GLS algorithm with FFR and AFR structure for different instances

Name of instance	AFR structure				FFR structure			
	Fitness function [pixel]		Time [s]		Fitness function [pixel]		Time [s]	
10 pros (BM)	809	±97,7	5,5	±4,0	804,4	±73,3	2,3	±0,87
23 pros (BM)	713,6	±76,4	200,2	±36,4	672	±38,8	165,4	±59,7
45 pros (BM)	1582,4	±48,6	2,2	±1,4	1596,9	±49,6	2,2	±0,9

Table 3. Average of 100 solutions for GLS algorithm without any additional structure for different instances

Name of instance	Fitness function [pixel]		Time [s]	
10 pros (BM)	792,4	±60,1	3,7	±3,7
23 pros (BM)	671,6	±79,7	201,3	±46,5
45 pros (BM)	1572,4	±44,9	2	±0,8

Table 4. Average of 100 solutions for GLS algorithm without any additional structure for different instances

Name of instance	Fitness function [pixel]		Time [s]	
43 shape	67,6	± 1,1	70,6	±24,6
43a shape	63,6	±0,5	165,7	±12,8
10 pros	682	±17,9	33,5	±10,1
10a pros	626	±15,8	50,9	±5,9
23 pros	571,7	±10,7	176,44	±11,3
45 pros	1473,57	±19,8	31,3	±1,3
20 polygon	1001,4	±8,3	123,7	±5,4

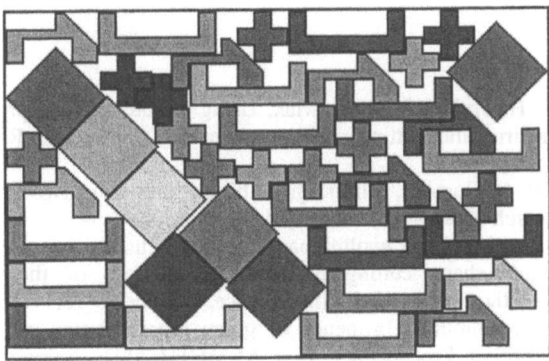

Fig. 3. The best known layout for instance "43 shape" obtained with GLS

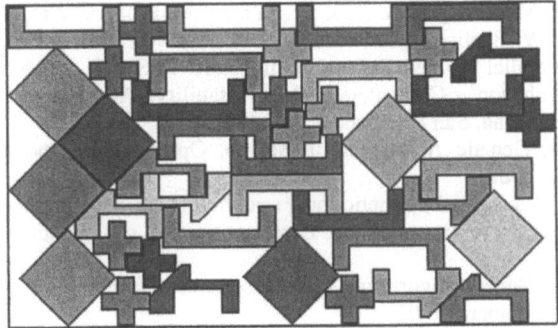

Fig. 4. The best known layout for instance "43a shape" obtained with GLS

Figures 3 and 4 present the best layout for "43 shape" and "43a shape" obtained by GLS algorithm solving 2-DC problem where figures are transformed to the integers map and the first bits on the strip are known.

6 Conclusion

In this research we have tested two figure representations for the 2-DC problem using GLS algorithm. The first of them is a binary map representation of the figure denoted as (BM) and the second is an integer map denoted as (IM).

The results of the experiments prove that the GLS based on integer map representation outperforms the GLS based on binary map representation.

Another interesting issue to point out is the use of a dynamic structure to remember free pixels on the strip. The results indicate that the GLS algorithm using additional dynamic structure does not give better results than GLS without such structure.

Software for 2-DC problem is allowed under location:
http://www-idss.cs.put.poznan.pl/~pkominek

References

1. J. Blazewicz, P. Hawryluk, R. Walkowiak, Using a tabu search approach for solving the two-dimensional irregular cutting problem, Annals of Operations Research, Vol 41, pp 313-327, 1993.
2. P. C. Gilmore, R. E. Gomory, Multistage Cutting Problems of two and more dimensions, Operations Research 13:94--119, 1965.
3. L. Dolata, P. Kominek: An evolutionary algorithm using expert's knowledge for 2-dimension irregular-shape cutting problem, Proceedings of the AI-METH 2002 - Symposium on Methods of Artificial Intelligence, Gliwice 13-15.11.2002, pp. 147-150.
4. L. Budzyńska, P. Kominek, Influence of given representation on Performance of an evolutionary algorithm, Proceedings of the AI-METH 2003 - Symposium on Methods of Artificial Intelligence, Gliwice 5-7.11.2003, pp. 135-138.
5. J.F. Oliveira, A.M. Gomes, J.S. Ferreira, TOPOS - a new constructive algorithm for nesting problems, OR Spektrum 22, 263–284, 2000.
6. B. Freisleben, P. Merz, A genetic local search algorithm for travelling salesman problem., In H.-M. Voigt, W. Ebeling, I.-Rechenberg, H.-P. Schwefel (eds.), Proceedings of the 4th Conference on Parallel Problem Solving fram Nature- PPSN IV, pp. 890-900, 1996.
7. M. Garey, D. Johnson , Computers and intractability: A guide to the theory of NP-completeness, Freeman, San Francisco, Calif, 1979.
8. D. E. Goldberg, Genetic Algorithms in Search, Optimization, and Machine Learning, Addison Wesley, 1988.
9. A. Jaszkiewicz, P. Kominek, Genetic local search with distance preserving recombination operator for a vehicle routing problem. European Journal of Operational Research, 151/2, 352-364, 2003.
10. Z. Michalewicz, Genetic algorithms + Data structures = Evolution programs, Springer Verlag, Berlin Heidelberg, 1992.
11. N. J Radcliffe., P. D. Surry, Formal memetic algorithms, in: T. Fogarty (Ed.), Evolutionary Computing: AISB Workshop, Springer-Verlag), 1994.
12. L. Dolata, Przetwarzanie wektorowe dla problemu rozkroju, Praca magisterska, Politechnika Poznańska, Poznań, 2000.

Biologically-Inspired: A Rule-Based Self-Reconfiguration of a Virtex Chip

Gunnar Tufte and Pauline C. Haddow

The Norwegian University of Science and Technology
Department of Computer and Information Science
Sem Selandsvei 7-9, 7491 Trondheim, Norway
{gunnart,pauline}@idi.ntnu.no

Abstract. To be able to evolve digital circuits with complex structure and/or complex functionality we propose an artificial development process as the genotype-phenotype mapping. To realistically evolve such circuits a hardware implementation of the development process together with high-speed reconfiguration logic for phenotype implementation is presented. The hardware implementation of the development process is a programmable reconfiguration processor. The high-speed reconfiguration logic for evaluation of the phenotype is capable of exploiting the advantage of massive parallel processing due to the cellular automata like structure.

1 Introduction

In Evolvable Hardware (EHW), evolutionary algorithms can be used to evolve electronic circuits. In general, a one-to-one mapping for the genotype-phenotype transition is assumed.

Introducing inspiration from biological development, where a genotype represents a building plan of how to assemble an organism rather than a blueprint of the assembled phenotype may be a way of artificially developing complex phenotypes from relatively simple genotypes [1,2]. Combining artificial development with evolution thus aims to improve the complexity of electronic circuits that may be evolved [3,4,5].

In this work we continue to use the knowledge-rich development for circuit design on a Virtual Sblock FPGA platform [5]. The main goal herein is to define a hardware solution combining artificial development with intrinsic evolution to be able to realistic evolve complex structure and/or complex functionality.

The hardware used in this project is a cPCI host computer and a Nallatech BenERA [6] cPCI FPGA card featuring a Xilinx XCV1000E-6 [7].

The article is laid out as follows: Section 2 introduces the Sblock and the Virtual FPGA concept, which is the platform for our development process. Section 3 describes the mapping of Sblocks to our target technology. Section 4 describes our development process. Section 5 explains the implementation of our hardware solution combining the development process and the developed Sblock phenotype on an FPGA. Finaly a conclusion is presented in Section 6.

M. Bubak et al. (Eds.): ICCS 2004, LNCS 3038, pp. 1249–1256, 2004.
© Springer-Verlag Berlin Heidelberg 2004

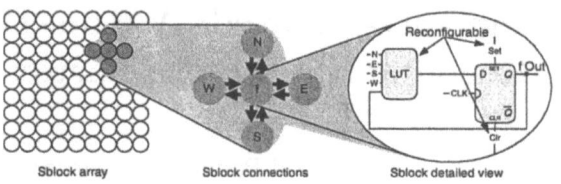

Fig. 1. Sblocks in a Virtual Sblock FPGA

2   Virtual Sblock FPGA

The Virtual Sblock FPGA is a technology independent platform for evolvable hardware. The key feature is a more evolution and development friendly hardware platform for evolving digital circuits than what is offered from commercial FPGAs. Figure 1 illustrates the Virtual Sblock FPGA: the grid of Sblocks; local connections between neighbouring Sblocks and the internal logic of an Sblock.

The Virtual Sblock FPGA [8] contains blocks — *Sblocks* — laid out as a symmetric grid where neighbouring blocks are connected. Each Sblock neighbours onto Sblocks on its four sides (north N, east E, south S and west W). The output value of an Sblock is synchronously updated and sent to all its four neighbours and as a feedback to itself.

An Sblock may be configured as either a logic or memory element with direct connections to its four neighbours or it may be configured as a routing element. Functionality of an Sblock (Sblock type) is defined by the content of its look-up table (LUT), a function generator capable of generating all possible five-input boolean functions. The five inputs to the LUT consist of inputs from the 4 neighbours and its own output value (Sblock state).

The Virtual Sblock FPGA may also be viewed at as a two-dimensional cellular automata (CA). The cellular automata may be uniform, if all Sblocks have the same LUT configuration, or non-uniform if different Sblocks have different LUT configurations.

In this work, the cellular automata interpretation is chosen. The CA i.e. the complete circuit, starts from some initial condition (interpreted as the input) and runs for a number of interactions to some final condition (interpreted as the output). This is the most common interpretation of CA computation [9]. This interpretation is used in [10,5].

A detailed description of Sblocks and the Virtual Sblock FPGA is presented in [8]. Evolution and development friendly features of the platform may be found in [11]. The term development friendly is used to reflect the fact that properties of the architecture are more suitable to developmental techniques than those found in today's technology.

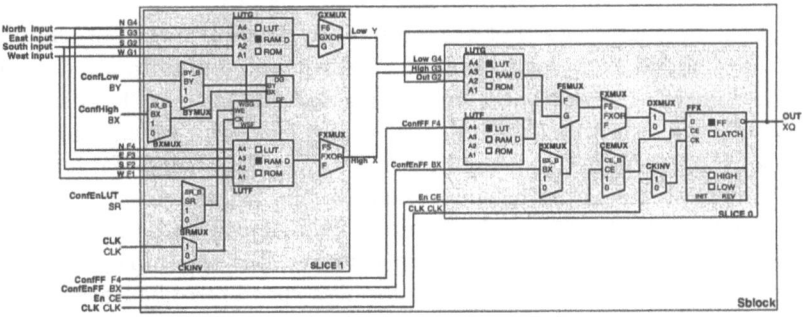

Fig. 2. Sblocks hard macro implementation for on-chip reconfiguration

3   Sblock Architecture on a Virtex-E FPGA

The mapping of the Sblock architecture to an FPGA aims to utilize the underlying technology of the target device. The Virtex-E [7] series of FPGAs from Xilinx consists of an array of configurable logic blocks (CLBs) and fast internal BlockRam (BRAM) memory. BRAM modules are placed in-between CLB columns at fixed intervals. Exploiting the 2-d array of CLBs in the device, we form the 2-d grid of Sblocks where an Sblock fits into one CLB. This adaptation gives us the possibility to constrain the placement of Sblocks to specific CLB locations in the target device. Knowledge of placement gives us two advantages. First, knowledge of Sblock placement is crucial for partial reconfiguration, since it is essential to know where the desired Sblock is located to be able to update the LUT. Second, the constrained placement of Sblocks ensures adjacent Sblocks to be placed in CLBs according to their neighbourhood. This implies short and conform routing between Sblocks, giving approximately the same propagation delay for all Sblock connections. The Sblock implementation is shown in Figure 2.

Configuration of an Sblock FPGA is a two stage process. The first stage configures fixed logic, assigns and configures Sblocks into their respective CLBs, and sets up the routing resources to connect Sblocks. In the second stage reconfiguration manipulates LUTs in desired Sblocks, utilizing partial dynamic reconfiguration of Sblocks. The first configuration stage uses standard configuration mode provided by the device. The second configuration stage may use the standard partial reconfiguration modes i.e. frame based reconfiguration [12], or custom reconfiguration as explained in Section 5.

The LUTs are used as both function generators and as shift registers. An Sblock requires two slices, while one Sblock fits in to one CLB using both slices. Bold letters indicate input and output ports on the slices. The LUTs inputs are connected to form the five input LUT, due to the use of the *BXMUX* in the shift register implementation the five input LUT here consists of *LUTG* and *LUTF* in slice 1 and *LUTG* in slice 0. The shown Sblock implementation gives a maximum number of Sblocks on an FPGA that equals the number of CLBs.

Extra signals in Figure 2 are data and control signals for the custom reconfiguration of the Sblocks. Instead of using frames to alter LUT content and flip-flop state we have added extra logic on the FPGA to shift configuration data in and out of the Sblocks. The data port for shifting data into the LUTs is split in two ports: *ConfLow* for shifting data into *LUTG* and *ConfHigh* for shifting data into *LUTF*. *ConfEnLut* enables reconfiguration of LUTs. The *ConfFF* is a data input used to shift in states to the flip-flop when *ConfEnFF* is active.

The function of the Sblock in Figure 2 can of course be reconfigured by changing LUT bits by reading and writing frames, but the implementation also supports a custom reconfiguration mode for high-speed reconfiguration. The custom reconfiguration mode exploits the possibility of using LUTs as both function generators and shift registers. Instead of altering LUT contents by writing frames, on-chip reconfiguration logic can address desired Sblocks and alter their LUTs by shifting in new data.

4   Development on an Sblock Platform Inspired by Biology

As stated in Section 1, instead of using a genotype representation as a blueprint for a circuit a development process uses a building plan of how to assemble the circuits. In this work the building plan consists of rules.

The rules are based on two types of rules i.e. change and growth rules. These rules are restricted to expressions consisting of the type of the target cell and the types of the cells in its neighbourhood. Firing of a rule can cause the target cell to change type, die (implemented as a change of type) or cause another cell to grow into it.

Figure 3 illustrates the process of applying growth and change rules. A growth rule targets the **C** Sblock which is an empty Sblock. This means that the condition of this change rule, including types of the four neighbours and the target itself match. The rule triggered grows a new Sblock into *C* by copying the Sblock type from the *E* neighbour. A change rule then targets the Sblock in position *C*. The change rule result is that the targeted Sblock *C* is changed.

The direction of growth is built into the rules themselves [5]. Growth is not triggered by an internal cell in the growing organism but in fact by a free cell — empty Sblock, neighbouring the growing organism. When a growth rule triggers, it matches to a free cell and its given neighbourhood. Four different growth rules

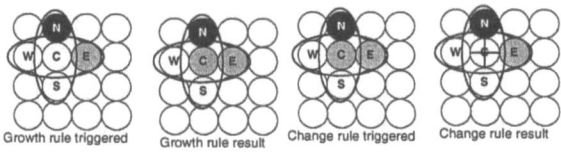

Growth rule triggered   Growth rule result   Change rule triggered   Change rule result

Fig. 3. Change and Growth on an Sblock Grid

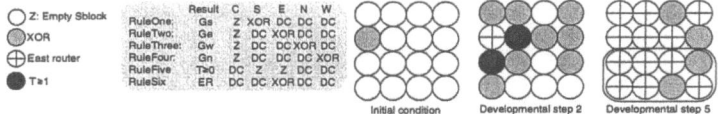

Fig. 4. Example of development to a symmetrical organism

control growth from the North (G_N), East (G_E), South (G_S) and West (G_W) respectively. This means that if, for example, a north growth rule targets a free cell, then the type of the cell to the North is copied to the free cell.

In Figure 4 development of an organism is shown. The example shows development of an organism with structural properties, no functionality is considered. Three different Sblock types are possible: XOR, East Router (ER), and a threshold element ($T \geq 1$).

The organism develops from a single cell (axiom) shown in the initial condition. After two development steps, the organism has grown from the axiom to eleven Sblocks and differentiated to consist of all three Sblock types. At the fifth and final development step, the organism has developed into an organism with two symmetrical patterns (one of the two patterns is indicated). The genome (rules) used is as shown. The rules are ranked, rule one has the highest priority and the last rule lowest priority. Details concerning development of structural organisms can be found in [13] and for functionality in [5].

5   Rule Based Reconfiguration Inspired by Biology

The reconfiguration speed depends on both the time spent on downloading configuration data and on the time spent to do the reconfiguration by changing the functionality of the device. Inspired by the building plan idea we propose a small on-chip processor [14] for handling the reconfiguration. The FPGA is partitioned in two: a reconfigurable and a fixed logic partition. The Processor is the fixed partition. The reconfigurable partition is an Sblock array.

Following the idea of using a building plan as a genome the only configuration data needed may be the instructions of how to build the phenotype i.e. the rules. The on-chip reconfiguration logic is designed to use the rules to develop a phenotype in the on-chip Sblock array. The functionality of the phenotype may be the interpretation of the states of the Sblocks in the developing organism [5].

The use of a processor design as reconfiguration logic has several advantages. The two most important features may be: First the speed of downloading reconfiguration data. The compact rule representation requires only a small amount of data to be transferred to the chip. Second, the processor design allows an instruction set. Using an instruction set is flexible, making it possible to use the processor to configure Sblocks by addressing target Sblocks and downloading Sblock types to the chip. A more powerful method is to use the processor to configure the Sblock array using developmental rules. Another flexibility is the possibility to write small programs. Such programs may be setting a number of

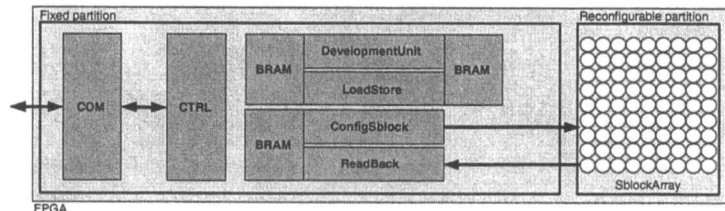

Fig. 5. Implementation of Sblock on-chip reconfiguration

development steps and/or clock cycles to run the Sblock array. Programs can also be used to monitor the Sblock array during the development process.

The possibility a hardware implementation gives for massive parallel processing in CA like structures is exploited for running the Sblock array, the development process where parallel processing includes implementing the rules in every cell, was considered to resource consuming. Therefore the development process is sequential.

Figure 5 shows a block drawing of our reconfiguration processor together with the Sblock array. The Sblocks is implemented as shown in Figure 2, where reconfiguration of the Sblocks is done by shifting data into the LUTs. Running state steps is done in parallel i.e. all Sblocks output values are updated in parallel each cycle. The processor is divided into seven modules:

- COM: Interface for external communication with the host on the cPCI bus.
- CTRL: This is the control unit of the processor. This module also includes the instruction memory and manages communication with the *COM* module.
- BRAM: Intermediate storage of type and state of the Sblock array. Split in two BRAM modules, storing current development steps state and type and a second for the developing organism.
- DevelopmentUnit: Hardware implementation of the development process in Section 4. To speedup the process the development process works on a copy of the actual Sblock array and the copies is stored in BRAM. The current Sblock array configuration is stored in one *BRAM* module. The development process reads the current Sblock array and updates the second *BRAM* module with new Sblock types for those that triggers a rule. The *DevelopmentUnit* is a six stage pipeline, processing two Sblocks each clock cycle.
- ConfigSblock: Reconfiguration of the Sblock array. The module reads out Sblock types and states from *BRAM* and reconfigures the Sblocks with the Sblock array under development stored in *BRAM*. Implemented as a seven stage pipeline, reconfigures two Sblocks every cycle.
- LoadStore: Handles reading and writing of Sblock types and states to *BRAM* and communicates with external devices through the *COM* interface.
- ReadBack: Used to monitor Sblock types and states. Read back of Sblock states is done from the Sblock array by the *Readback* module and stored as state information in *BRAM*. Sblock types are already present in *BRAM* from the latest development step.

A typical experiment for the system will be to run a genetic algorithm (GA) on the host computer and the development process with the developing phenotype in hardware. As stated in Section 3 the configuration process is a two stage process, where the first stage configures the FPGA with the processor design and the Sblock array. The configuration in the first stage is done after synthesis of the processor design with an Sblock array of desired size. After the first configuration stage, the FPGA contains the reconfiguration processor and the Sblock array.

Before the experiments starts, the development process must be defined i.e. defining the number of development steps and number of state steps for each evaluation. Criteria for read back of information from the evaluation can also be defined. These definitions are done by writing and downloading a program to the reconfiguration processor. The downloaded program is executed to evaluate each downloaded rule genome from the GA in the host.

To evaluate our hardware solution we first compared reconfiguration time using the configuration processor with Virtex frame based reconfiguration using the JTAG interface. The task was to reconfigure a 32 by 32 Sblock array. Since the goal is to evaluate configuration speed we wanted to include both download and reconfiguration time in both cases. To manage this the reconfiguration processor was only used to receive configuration data from the host and configure the addressed Sblock i.e. all reconfiguration data was in full downloaded from the host for both cases. The test was done by reconfiguring all Sblocks in the array a 1000 times. The frame based reconfiguration used 170,38 seconds to accomplish the task. The configuration processor accomplished the same task in 0.55 seconds, giving a speedup of more then 300.

A comparison of performance of the system with a software simulation was also tried. A software simulator simulated both the development process and the Sblock states on an 8 by 8 Sblock array. The experiment included both development steps and state steps. The development process was 10 000 steps with 50 000 state steps at each development step. For each development step all Sblock types in the array was stored together with the last state step information. Storing information implies writing back data from the FPGA to a file on the host over the cPCI bus while for the software simulation storing is writing to a file. The software simulation accomplished the task in 18.06 minutes, the runtime in hardware was measured to 6.2 seconds giving a speedup of 175 times. The software simulation was performed on a 2.3 GHz Pentium 4 PC.

6   Conclusion

The presented hardware system consisting of a hardware platform including a development process and high-speed reconfigurable logic shows a promising speedup both for intrinsic evolution and for evolution using development as genotype-phenotype mapping. The implementation of a reconfiguration processor is capable of both running intrinsic EHW experiments requiring high-speed reconfiguration and to speedup the genotype-phenotype mapping.

References

1. Kitano, H.: Building complex systems using development process: An engineering approach. In: Evolvable Systems: from Biology to Hardware, ICES. (1998) 218–229
2. Bentley, P.J., Kumar, S.: Three ways to grow designs: A comparison of embryogenies for an evolutionary design problem. In: Genetic and Evolutionary Computation Conference (GECCO '99). (1999) 35–43
3. Gordon, T.G.W., Bentley, P.J.: Towards development in evolvable hardware. In: the 2002 NASA/DOD Conference on Evolvable Hardware (EH'02). (2002) 241 –250
4. Miller, J.F., Thomson, P.: A developmental method for growing graphs and circuits. In: 5th International Conference on Evolvable Systems (ICES03). (2003) 93–104
5. Tufte, G., Haddow, P.C.: Identification of functionality during development on a virtual sblock fpga. In: Congress on Evolutionary Computation. (2003)
6. Nallatech: BenERA User Guide. Nt107-0072 (issue 3) 09-04-2002 edn. (2002)
7. Xilinx: Xilinx Virtex-E 1.8 V Field Programmable Gate Arrays Production Product Specification. Ds022-1 (v2.3) july 17, 2002 edn. (2002)
8. Haddow, P.C., Tufte, G.: Bridging the genotype-phenotype mapping for digital FPGAs. In: the 3rd NASA/DoD Workshop on Evolvable Hardware. (2001) 109–115
9. Mitchell, M., Hraber, P.T., Crutchfield, J.P.: revisiting the egde of chaos: Evolving cellular automata to perform computations. Complex Systems 7 (1993) 89–130 Santa Fe Institute Working Paper 93-03-014.
10. van Remortel, P., Lenaerts, T., Manderick, B.: Lineage and induction in the Development of evolved genotypes for non-uniform 2d cas. In: 15th Australian Joint Conference on Artificial Intelligence. (2002) 321–332
11. Haddow, P.C., Tufte, G., van Remortel, P.: Evolvable hardware: pumping life into dead silicon. In Kumar, S., ed.: On Growth, Form and Computers. Elsevier Limited Oxford UK (2003) 404–422
12. Xilinx: Xilinx XAPP 151 Virtex Configuration Architecture Advanced Users' Guide. 1.1 edn. (1999)
13. Tufte, G., Haddow, P.C.: Building knowledg into developmental rules for circuite design. In: 5th International Conference on Evolvable Systems ICES. (2003) 69–80
14. Djupdal, A.: Design and Implementation of Hardware Suitable for Sblock Based Experiments, Masters Thesis. The University of Science and technology, Norway (2003)

Designing Digital Circuits for the Knapsack Problem

Mihai Oltean[1], Crina Groşan[1], and Mihaela Oltean[2]

[1] Department of Computer Science,
Faculty of Mathematics and Computer Science,
Babeş-Bolyai University, Kogălniceanu 1
Cluj-Napoca, 3400, Romania.
{moltean,cgrosan}@cs.ubbcluj.ro
[2] David Prodan College, Cugir, 2566, Romania.
olteanmihaelaelena@yahoo.com

Abstract. Multi Expression Programming (MEP) is a Genetic Programming variant that uses linear chromosomes for solution encoding. A unique feature of MEP is its ability of encoding multiple solutions of a problem in a single chromosome. In this paper we use Multi Expression Programming for evolving digital circuits for a well-known NP-Complete problem: the knapsack (subset sum) problem. Numerical experiments show that Multi Expression Programming performs well on the considered test problems.

1 Introduction

The problem of evolving digital circuits has been deeply analyzed in the recent past [4]. A considerable effort has been spent on evolving very efficient (regarding the number of gates) digital circuits. J. Miller, one of the pioneers in the field of the evolvable digital circuits, used a special technique called Cartesian Genetic Programming (CGP) [4] for evolving digital circuits. CGP architecture consists of a network of gates (placed in a grid structure) and a set of wires connecting them. The results [4] show that CGP is able to evolve digital circuits competitive to those designed by human experts.

In this paper, we use Multi Expression Programming (MEP)[1] [5] for evolving digital circuits. MEP is a Genetic Programming (GP) [3] variant that uses linear chromosomes of fixed length. A unique feature of MEP is its ability of storing multiple solutions of a problem in a single chromosome. Note that this feature does not increase the complexity of the MEP decoding process when compared to other techniques that store a single solution in a chromosome.

In this paper we present the way in which MEP may be efficiently applied for evolving digital circuits. We describe the way in which multiple digital circuits may be stored in a single MEP chromosome and the way in which the fitness

---

[1] MEP source code is available from www.mep.cs.ubbcluj.ro

M. Bubak et al. (Eds.): ICCS 2004, LNCS 3038, pp. 1257–1264, 2004.
© Springer-Verlag Berlin Heidelberg 2004

of this chromosome may be computed by traversing the MEP chromosome only once.

In this paper MEP is used for evolving digital circuits for a well-known NP-Complete [2] problem: the knapsack (subset sum) problem. Since this problem is NP-Complete we cannot realistically expect to find a polynomial-time algorithm for it. Instead, we have to speed-up the existing techniques in order to reduce the time needed to obtain a solution. A possibility for speeding-up the algorithms for this problem is to implement them in assembly language. This could lead sometimes to improvements of over two orders of magnitude. Another possibility is to design and build a special hardware dedicated to that problem. This approach could lead to significant improvements of the running time. Due to this reason we have chosen to design, by the means of evolution, digital circuits for several instances of the knapsack problem.

The knapsack problem may also be used as benchmarking problem for the evolutionary techniques which design electronic circuits. The main advantage of the knapsack problem is its scalability: increasing the number of inputs leads to more and more complicated circuits. The results show that MEP performs very well for all the considered test problems.

The paper is organized as follows. In section 2, the problem of designing digital circuits and the knapsack problem are shortly described. The Multi Expression Programming technique is presented in section 3. Several numerical experiments are performed in section 4.

2   Problem Statement and Representation

The problem that we are trying to solve in this paper may be briefly stated as follows:

Find a digital circuit that implements a function given by its truth table.

The gates that are usually used in the design of digital circuits along with their description are given in Table 1.

Table 1. Function set (gates) used in numerical experiments. These functions are taken from [4]

#	Function	#	Function
0	$a \cdot b$	5	$a \oplus b$
1	$a \cdot \bar{b}$	6	$a + b$
2	$\bar{a} \cdot b$	7	$a + \bar{b}$
3	$\bar{a} \cdot \bar{b}$	8	$\bar{a} + b$
4	$a \oplus b$	9	$\bar{a} + \bar{b}$

The knapsack problem (or the subset sum problem) may be stated as follows:

Let M be a set of numbers and a target sum k. Is there a subset $S \subseteq M$ which has the sum k?

The knapsack problem is a well-known NP-Complete problem [2]. No polynomial-time algorithm is known for this problem.

Instead of designing a heuristic for this problem we will try to evolve digital circuits which will provide the answer for a given input.

In the experiments performed in this paper the set M consists of several integer numbers from the set of consecutive integers starting with 1. For instance if the base set is $\{1, 2, 3, 4, 5, 6, 7\}$ then M may be $\{2, 5, 6\}$. We will try to evolve a digital circuit that is able to provide the correct answer for all subsets M of the base set.

The input for this problem is a sequence of bits. A value of 1 in position k means that the integer number k belongs to the set M, otherwise the number k does not belong to the set M.

For instance consider the consecutive integer numbers starting with 1 and ending with 7. The string 0100110 encodes the set $M = \{2, 5, 6\}$. The numbers 1, 3, 4 and 7 do not belong to M since the corresponding positions are 0. The possible subsets of M instance have the sum 2, 5, 6, 7, 8, 11 or 13. In our approach, the target sum is fixed and we are asking if is there a subset of given sum.

The number of training instances for this problem depends on the number of consecutive integers used as base for M. If we use numbers 1, 2 and 3, we have $2^3 = 8$ training instances. If we use numbers 1, 2, 3, 4, 5, 6 and 7, we have $2^7 = 128$ training instances. In this case, whichever subset M of $\{1, \ldots, 7\}$ will be presented to the evolved circuit we have to obtain a binary answer whether the target sum k may or not be obtained from a subset of M.

3   Multi Expression Programming

In this section, *Multi Expression Programming* (MEP) [5] is briefly described.

3.1   MEP Representation

MEP genes are represented by substrings of a variable length. The number of genes per chromosome is constant. This number defines the length of the chromosome. Each gene encodes a terminal or a function symbol. A gene encoding a function includes pointers towards the function arguments. Function arguments always have indices of lower values than the position of that function in the chromosome.

This representation is similar to the way in which *C* and **Pascal** compilers translate mathematical expressions into machine code [1].

The proposed representation ensures that no cycle arises while the chromosome is decoded (phenotypically transcripted). According to the proposed representation scheme the first symbol of the chromosome must be a terminal symbol. In this way only syntactically correct programs (MEP individuals) are obtained.

Example. A representation where the numbers on the left positions stand for gene labels is employed here. Labels do not belong to the chromosome, they are being provided only for explanation purposes.

For this example we use the set of functions $F = \{+, *\}$, and the set of terminals $T = \{a, b, c, d\}$. An example of chromosome using the sets F and T is given below:

1: a
2: b
3: $+ 1, 2$
4: c
5: d
6: $+ 4, 5$
7: $* 3, 6$

3.2   Decoding MEP Chromosomes and Fitness Assignment Process

In this section it is described the way in which MEP individuals are translated into computer programs and the way in which the fitness of these programs is computed.

This translation is achieved by reading the chromosome top-down. A terminal symbol specifies a simple expression. A function symbol specifies a complex expression obtained by connecting the operands specified by the argument positions with the current function symbol.

For instance, genes 1, 2, 4 and 5 in the previous example encode simple expressions formed by a single terminal symbol. These expressions are:

$$E_1 = a,$$
$$E_2 = b,$$
$$E_4 = c,$$
$$E_5 = d,$$

Gene 3 indicates the operation $+$ on the operands located at positions 1 and 2 of the chromosome. Therefore gene 3 encodes the expression:

$$E_3 = a + b.$$

Gene 6 indicates the operation $+$ on the operands located at positions 4 and 5. Therefore gene 6 encodes the expression:

$$E_6 = c + d.$$

Gene 7 indicates the operation $*$ on the operands located at position 3 and 6. Therefore gene 7 encodes the expression:

$$E_7 = (a + b) * (c + d).$$

E_7 is the expression encoded by the whole chromosome.

There is neither practical nor theoretical evidence that one of these expressions is better than the others. Moreover, Wolpert and McReady [7] proved that we cannot use the search algorithm's behavior so far for a particular test function to predict its future behavior on that function. This is why each MEP chromosome is allowed to encode a number of expressions equal to the chromosome length. Each of these expressions is considered as being a potential solution of the problem.

The value of these expressions may be computed by reading the chromosome top down. Partial results are computed by dynamic programming and are stored in a conventional manner.

As MEP chromosome encodes more than one problem solution, it is interesting to see how the fitness is assigned.

Usually the chromosome fitness is defined as the fitness of the best expression encoded by that chromosome.

For instance, if we want to solve symbolic regression problems the fitness of each sub-expression E_i may be computed using the formula:

$$f(E_i) = \sum_{k=1}^{n} |o_{k,i} - w_k|, \tag{1}$$

where $o_{k,i}$ is the obtained result by the expression E_i for the fitness case k and w_k is the targeted result for the fitness case k. In this case the fitness needs to be minimized.

The fitness of an individual is set to be equal to the lowest fitness of the expressions encoded in chromosome:

$$f(C) = \min_i f(E_i). \tag{2}$$

When we have to deal with other problems we compute the fitness of each sub-expression encoded in the MEP chromosome and the fitness of the entire individual is given by the fitness of the best expression encoded in that chromosome.

3.3   Search Operators

Search operators used within MEP algorithm are crossover and mutation. Considered search operators preserve the chromosome structure. All offspring are syntactically correct expressions.

Crossover. By crossover two parents are selected and are recombined. For instance, within the uniform recombination the offspring genes are taken randomly from one parent or another.

Example. Let us consider the two parents C_1 and C_2 given in Table 2. The two offspring O_1 and O_2 are obtained by uniform recombination as shown in Table 2.

Table 2. MEP uniform recombination

Parents		Offspring	
C_1	C_2	O_1	O_2
1: **b**	1: *a*	1: *a*	1: **b**
2: * **1, 1**	2: *b*	2: * **1, 1**	2: *b*
3: + **2, 1**	3: + 1, 2	3: + **2, 1**	3: + 1, 2
4: *a*	4: *c*	4: *c*	4: *a*
5: * **3, 2**	5: *d*	5: * **3, 2**	5: *d*
6: *a*	6: + 4, 5	6: + 4, 5	6: *a*
7: - **1, 4**	7: * 3, 6	7: - **1, 4**	7: * 3, 6

Mutation. Each symbol (terminal, function of function pointer) in the chromosome may be target of mutation operator. By mutation some symbols in the chromosome are changed. To preserve the consistency of the chromosome its first gene must encode a terminal symbol.

Example. Consider the chromosome C given in Table 3. If the boldfaced symbols are selected for mutation an offspring O is obtained as shown in Table 3.

Table 3. MEP mutation

C	O
1: *a*	1: *a*
2: * 1, 1	2: * 1, 1
3: **b**	3: + **1, 2**
4: * 2, 2	4: * 2, 2
5: *b*	5: *b*
6: + **3, 5**	6: + **1, 5**
7: *a*	7: *a*

3.4   MEP Algorithm

In this paper we use a steady-state [6] as underlying mechanism for Multi Expression Programming. The algorithm starts by creating a random population of individuals. The following steps are repeated until a stop condition is reached. Two parents are selected using a selection procedure. The parents are recombined in order to obtain two offspring. The offspring are considered for mutation. The best offspring replaces the worst individual in the current population if the offspring is better than the worst individual.

The algorithm returns as its answer the best expression evolved along a fixed number of generations.

4   Numerical Experiments

In this section several numerical experiments for evolving digital circuits for the knapsack problem are performed. The general parameters of the MEP algorithm are given in Table 4. Since different instances of the problem being solved will have different degrees of difficulty we will use different population sizes, number of genes in a chromosome and number of generations for each instance. Particular parameters are given in Table 5.

Experimental results are given in Table 6. We are interested in computing the number of successful runs and the number of gates in the shortest evolved circuit.

Table 6 shows that MEP successfully found at least a solution for the considered test problems. The difficulty of evolving a digital circuit for this problem increases with the number of inputs of the problem. Only 20 individuals are required to obtain 39 solutions (out of 100 runs) for the instance with 4 inputs. In return, 1000 individuals (50 times more) are required to obtain 10 perfect solu-

Table 4. General parameters of the MEP algorithm for evolving digital circuits

Parameter	Value
Crossover probability	0.9
Crossover type	Uniform
Mutations	5 / chromosome
Function set	Gates 0 to 9 (see Table 1)
Terminal set	Problem inputs
Selection	Binary Tournament

Table 5. Particular parameters of the MEP algorithm for different instances of the knapsack problem. In the second column the base set of numbers is given for each instance. In the third column the target sum is given.

#	Set of numbers	Sum	Number of fitness cases	Population size	Number of genes	Number of generations
1	$\{1\ldots4\}$	5	16	20	10	51
2	$\{1\ldots5\}$	7	32	100	30	101
3	$\{1\ldots6\}$	10	64	500	50	101
4	$\{1\ldots7\}$	14	128	1000	100	201

Table 6. Results obtained by MEP for the considered test problems. 100 independent runs have been performed for all problems

#	Set of numbers	Sum	Successful runs	Number of gates in the shortest circuit
1	$\{1\ldots4\}$	5	39 out of 100	3
2	$\{1\ldots5\}$	7	31 out of 100	5
3	$\{1\ldots6\}$	10	10 out of 100	11
4	$\{1\ldots7\}$	14	7 out of 100	21

tions (out of 100 independent runs) for the instance with 7 inputs. Also the size of the evolved circuits increases with the number of problem inputs. However, due to the reduced number of runs we cannot be sure that we have obtained the optimal circuits. Additional experiments are required in this respect.

Due to the NP-Completeness of the problem it is expected that the number of gates in the shortest circuit to increase exponentially with the number of inputs.

References

1. Aho A., Sethi R., and Ullman J.: Compilers: Principles, Techniques, and Tools, Addison Wesley, (1986)
2. Garey M.R., Johnson D.S.: Computers and Intractability: A Guide to NP-completeness, Freeman & Co, San Francisco, (1979)
3. Koza J. R.: Genetic Programming: On the Programming of Computers by Means of Natural Selection, MIT Press, Cambridge, MA, (1992)
4. Miller J. F., Job D. and Vassilev V.K.: Principles in the Evolutionary Design of Digital Circuits - Part I, Genetic Programming and Evolvable Machines, Vol. 1(1), Kluwer Academic Publishers, (1999) 7-35
5. Oltean M.: Solving Even-Parity Problems using Multi Expression Programming, in Proceedings of the the 7th Joint Conference on Information Sciences, Research Triangle Park, North Carolina, Edited by Ken Chen (et. al), (2003) 315-318
6. Syswerda G.: Uniform Crossover in Genetic Algorithms, In Proceedings of the 3rd International Conference on Genetic Algorithms, J.D. Schaffer (eds), Morgan Kaufmann Publishers, CA, (1989) 2-9
7. Wolpert D.H. and McReady W.G.: No Free Lunch Theorems for Search, Technical Report, SFI-TR-05-010, Santa Fe Institute, (1995)

Improvements in FSM Evolutions from Partial Input/Output Sequences

Sérgio G. Araújo, A. Mesquita, and Aloysio C.P. Pedroza

Electrical Engineering Dept.
Federal University of Rio de Janeiro
C.P. 68504 - CEP 21945-970 - Rio de Janeiro - RJ - Brazil
Tel: +55 21 2260-5010 - Fax: +55 21 2290-6626
{granato,aloysio}@gta.ufrj.br, mesquita@coe.ufrj.br

Abstract. This work focuses on the synthesis of finite-state machines (FSMs) by observing its input/output behaviors. Evolutionary approaches that have been proposed to solve this problem do not include strategies to escape from local optima, a typical problem found in simple evolutionary algorithms, particularly in the evolution of sequential machines. Simulations show that the proposed approach improves significantly the state space search.

1 Introduction and Related Works

The task of modeling existing systems by exclusively observing their input/output (I/O) behavior is of interest when the purpose is (1) to uncover the states of natural systems, (2) to model synthetic systems implemented from scratch, in an ad hoc manner or for which the documentation has been lost and (3) to generate automata from scenario (trace) specifications, a fruitful area in the telecommunications domain. At present, most of these problems are outside the scope of conventional techniques.

Evolutionary Algorithms (EAs), and variations, have been used in the synthesis of sequential machines to explore the solutions space. Early attempts dating back to the 60's [1] used some of the today EAs concepts such as population, random initialization, generation, mutation and reproduction (cloning) to evolve an automaton that predicts outputs based on known input sequences. However, these early approaches have shown poor performance by lack of the crossover operator [7].

More recent attempts have been successful in synthesizing sequential systems with the aid of Genetic Algorithms (GAs) [2, 3, 4]. In [1] an approach to synthesize synchronous sequential logic circuits from partial I/O sequences by using *technology-based* representation (a netlist of gates and flip-flops) is presented. The method is able to synthesize a variety of small FSMs, such as serial adders and 4-bit sequence detectors. With the same purpose other works [3, 4] used technology-independent *state-based* representation in which the next-state and the output, corresponding to each current-state/input pair of the state-transition table, are coded in a binary string defined as the *chromosome*. Because, in this case, the number of states of the finite-state automaton is unknown, a large number of states must be used at the start.

The proposed approach differs from the previous ones by the use of the fitness gradient to improve the system performance in the presence of evolutionary

M. Bubak et al. (Eds.): ICCS 2004, LNCS 3038, pp. 1265–1272, 2004.
© Springer-Verlag Berlin Heidelberg 2004

stagnation phenomenon. Stagnation problems refer to a situation in which the optimum seeking process stagnates before finding a global optimal solution. Stagnation at local optima where all the neighboring solutions are non-improving is frequently found in simple EAs [13], particularly in the evolution of complex systems such as state-machines. This gap in the automata evolution may persevere for thousandths of generations due to an "unproductive" population. To overcome this problem, the proposed approach penalizes the best individual and its variants and a new population emerges in order to surpass the previous best fitness value. Simulations show that the proposed approach improves significantly the state space search.

2 Definitions

2.1 FSMs

Finite-state machines (FSMs) are commonly used for specifying reactive systems due to its simplicity and precise definition of the temporal ordering of interactions [5]. From the two traditional Moore/Mealy FSMs models it is found that in the Mealy machines, since the outputs are associated with the transitions, some behaviors can be implemented with fewer states than Moore machines.

The Mealy FSM model M is formally defined as a 7-tuple $\{Q,V,t,q_0,V',o,D\}$ where $Q \neq \emptyset$ is a finite set of states of M, V is a finite input alphabet, t is the state transition function, $q_0 \in Q$ is the initial state, V' is a finite output alphabet, o is the output function and D is the specification domain, which is a subset of $Q \times V$. t and o together characterize the behavior of the FSM, i.e., $t(q,v)$: $Q \times V \rightarrow Q$ and $o(q,v)$: $Q \times V \rightarrow V'$. If $D = Q \times V$, then t and o are defined for all possible state/input combinations and therefore the FSM is said to be *completely specified*. Deterministic and completely specified FSMs are preferred strategies for automata modeling since they tell it exactly how to react in every situation it perceives [11].

2.2 Evolutionary Computation

Evolutionary Computation (EC) concerns the design and analysis of probabilistic algorithms inspired by the principles of Darwinian natural selection. The Genetic Algorithm (GA) and the Genetic Programming (GP) are the most familiar instances of EC algorithms.

The *chromosome* is the basic component of the GA [6]. It represents a point (an individual) in the problem solution space. The fitness value measures how close the individual is to the solution. By iteratively applying the genetic operators (fitness evaluation, fitness-based selection, reproduction, crossover and mutation) to a randomly started population of individuals, it evolves in order to breed at least one offspring with a given target behavior. GAs usually employ *fixed length* chromosome strings in which one or more parameters are coded. GA solutions are best applied to poorly understood or highly non-linear problems for which deterministic solutions are not available. GP [7] is a branch of the GA, the solutions representation being the main difference. Unlike GA, GP can easily code chromosomes of *variable length*, which increases the flexibility in structure production.

3 Methodology

The proposed methodology is aimed to solve the problem of finding a completely specified deterministic FSM consistent with a given sample of I/O sequences called *training sequences* (*TSs*). A *TS* is defined as a finite sequence of *correct* input-output pairs $< v_1/v_1', v_2/v_2' ... v_L/v_L' >$, where $v \in V$, $v' \in V'$ and L is the length of the sequence (L inference is discussed in subsection 3.1). The methodology execution flow is shown in Fig. 1a. The population of FSMs supplied by the GP is evaluated for the fitness by means of the *TSs*; if at least one individual reproduces the I/O behavior specified by the *TSs*, the algorithm stops. The resulting FSM describes the observed system.

The *black-box* approach depicted in Fig. 1b is adopted for fitness evaluation. In this case the automaton to be evolved interacts with the environment through an event-driven interface. In each evolutionary step (generation) the system probes a population of FSMs with input sequences and records the corresponding output sequences. These output sequences are compared with the *correct* (desired) output sequences (the outputs of the *TSs*) and a fitness value is assigned to each candidate solution.

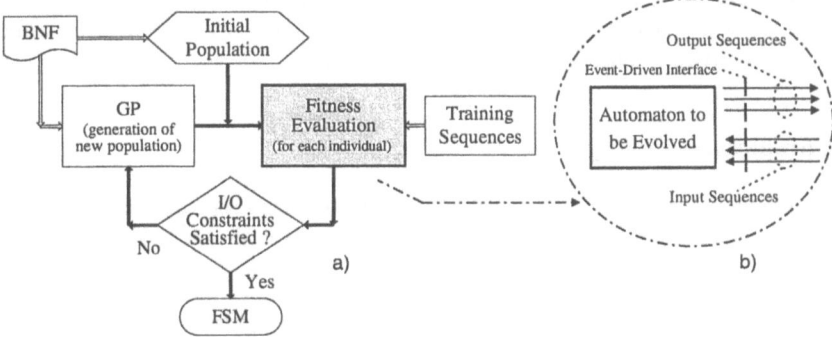

Fig. 1. a) Execution flow of the methodology; b) Model for fitness evaluation

Unlike other approaches [3, 4], the proposed methodology includes a heuristic to help the algorithm to evade from local optima. One of the main causes of premature convergence is loss of genetic diversity due to selection pressure [13]. Moreover, population diversity tends to get worst at local optima. To support the emergence of new populations, carrying out alternative solutions, the heuristic penalizes the fitness of the best individual and of its variants each time a local-optimum condition is detected. The penalty applied to outstanding fitness FSMs is defined by a *penalty factor Pf* ($0.0 \leq Pf < 1.0$) and variants of the best individual are defined by a *similarity factor* based on the Hamming distance concept. The classic binary Hamming distance was extended to a more adequate notation system, being redefined as the number of components (genes) by which two vectors (chromosomes) differ. Given a reference chromosome with K-genes, C_R, the similarity factor Sf of any chromosome with respect to C_R is defined as $Sf = (K-H)/K$, where H is the Hamming distance, in K-dimensional Hamming space, between them. For instance, the similarity factor between 130212 012001 140120 (C_R coding a 3-state FSM) and 130102 012023 140122 is $Sf = (18-5)/18 = 0.72$.

The intention of the heuristic is to eliminate the building blocks (or schemas), which push the individuals to the local optimum, from the current population. Pf is triggered as the gradient of the best individual fitness (GR) drops below a pre-defined threshold. At this time, the system reduces diversity-decreasing operators (i.e., crossover and cloning - see [12]) rate and raises the mutation probability (p_m) to increase the population diversity. Pf remains active for sufficient time to break the "bad" building blocks. However, the genetic operators only return to their original values as the system breeds a solution improving the previous best fitness value.

3.1 Training Sequence (TS) Length

The TSs must be long enough to exercise all paths of the FSM that describes the observed system. Ref. [2] gives an approximation formula to estimate the length of the TSs that yields a correct FSM, based on the *waiting times in sampling* problem solution [10]. This formula defines the length of the input sequence as $L = E(S) \times E(I)$, where $E(S)$ and $E(I)$ are the *expected number of state transitions* and *expected number of inputs*, respectively. As example, $E(N)$ can be computed using $E(N) = N (1 + \frac{1}{2} + \dots + \frac{1}{N})$. However, since the number of states S required to describe the system is unknown, it must be overestimated a priori.

3.2 Chromosome Coding

The chromosome, which encodes a FSM, uses state-based representation (SBR). The resulting string (chromosome) with S states and I inputs is shown in Fig. 2.

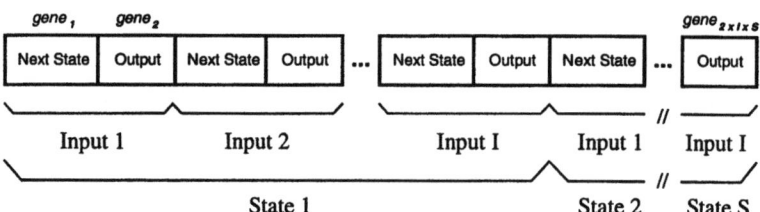

Fig. 2. Chromosome coding using SBR

The GP Kernel (GPK) [8], a complex G^3P (Grammar-Guided GP) system [9], was used in the present work to evolve automata encoded as syntactic trees. GPK system requires a Backus-Naur form (BNF) for chromosome structuring. The BNF that allows the generation of chromosomes representing six-state FSMs is defined as:

```
S          := <stat>;
<stat>     := <in_even><in_even>><in_even>><in_even>><in_even>><in_even>;
<in_even>  := <next_st><out> <next_st><out> <next_st><out> <next_st><out>
              <next_st><out> <next_st><out>;
<next_st>  := "0"|"1"|"2"|"3"|"4"|"5";
<out>      := "0"|"1"|"2"|"3"|"4"|"5"|"6";
```

Note that *<in_even>* fixes the number of input events (six, in the above BNF, each one yielding a next-state/output pair, i.e., *<next_st> <out>*). In contrast to the classic GA, where genes are simple bit-sequences and crossover may be performed at any point, GPK only permits crossover by swapping sub-trees (of a couple) starting with the same non-terminal symbols (i.e., symbols between corner brackets). Moreover, the mutation operation is implemented as a crossover operation between the selected individual and a temporary random-derived tree.

3.3 Fitness Evaluation

The fitness value assigned to a given FSM behavior, here evaluated through an input/output sequence perspective, is defined by the fitness function F in the form:

$$F = \sum_{i=1}^{N} w_i H_i \qquad (1)$$

Where w_i is a weighting factor for fitness case i, N is the number of fitness cases (*TSs*) and H_i is the number of *output hits* due to the fitness case i (*TS$_i$*). H_i is evaluated as follows. Initially, the FSM must be in the reset (idle) state. In the sequence, for each input of the *TS$_i$*, its output is compared with the *correct* output and an output hit is signed in case of a match.

4 Experiment

The goal is to generate a protocol entity specification, in FSM model, for the sender entity of a connection-oriented protocol (*PS$_{SND}$*) from given I/O event samples. The *TS* length was evaluated using six inputs (see coding table of Fig. 4) and an estimated value of five for *S*, leading to a 168-input/output *TS* (see subsection 3.1). Moreover, it is desirable to have multiple *TSs* to improve the performance of the learning system (see [4]). In fact, eighteen *TSs* ($N = 18$) were used, each with 32 bits in length, which corresponds to more than three 168-length sequences. w_i was set to 1 for all i since the *TSs* have the same length.

The population size (M) was set to 200 and the maximum number of generations (G_{MAX}), 5,000. The crossover (two-point shaped), the mutation and the reproduction probabilities were set to $p_c = 0.65$, $p_m = 0.05$ and $p_r = 0.30$, respectively. Linear rank selection was used considering the elitist strategy. The population of FSMs was shaped using the BNF described in subsection 3.2, which defines six-states FSMs.

Individuals with *Sf* ≥ 0.38 were defined as the variants of the best individual and, therefore, will be penalized at local optima. *Sf* comprised only next-state genes (see

Fig. 2). In fact, the heuristic did not consider the similarity factor with respect to output genes, since they have no influence in the state-machine graph. *GR* was evaluated over 50 generations and *GR* threshold for heuristic activation was set to 0.1. In case of heuristic activation, the system works with $p_m = 0.15$, $p_c = 0.60$ and $p_r = 0.25$. At generation corresponding to 95% of the maximum fitness value the heuristic was disabled since at this evolution stage *GR* is naturally close to zero. *Pf* was defined as 0.5 (fitness value penalized in 50%), remaining active throughout the next 50 generations.

Fig. 3 depicts the fitness curves of the best FSM of a typical run for three setups: <u>1: full heuristic</u>, <u>2: change genetic operators only</u> at local optima (no penalty applied) and <u>3: without heuristic</u>. As noted, until G = 266 the three curves are the same, as the heuristic isn't yet activated. For setup 1 (full heuristic), from $G = 290$ to $G = 1028$, as GR drops below 0.1 five times, the fitness value of the best FSM "backtracks" five times (at G = 268, G = 560, G = 760, G = 904 and G = 1028) and the system finally converges to global optimum ($F_{MAX} = 576$) at G = 1230. For setup 2, the system only converges to global optimum at G = 2944. For setup 3, the system did not escape from local optimum ($F = 554$) considering $G_{MAX} = 5,000$.

The resulting FSM using the full heuristic, which successfully describes the PS_{SND}, is given using state-transition graph (STG) in Fig. 4, after 576 output hits at G = 1230 (note: label v/v' represents an edge e_{ij} between two states q_i and q_j iff $o(q_i,v)=q_j$ and $t(q_i,v)=v'$). This FSM has one redundant state (state 4) and one unreachable state (state 1). Conventional methods may be used for state minimization. Table 1 compares the performance among the three setups. It shows that the full heuristic, which penalizes the best individual and its variants, yields 38 global optimum convergences (using *Sf* > 0.27) among 50 runs, each evolving up to a maximum of 1,500 generations. This result improves setups 2 and 3 in 46% and 280%, respectively. This table also indicates that the heuristic is quite sensitive to *Sf*.

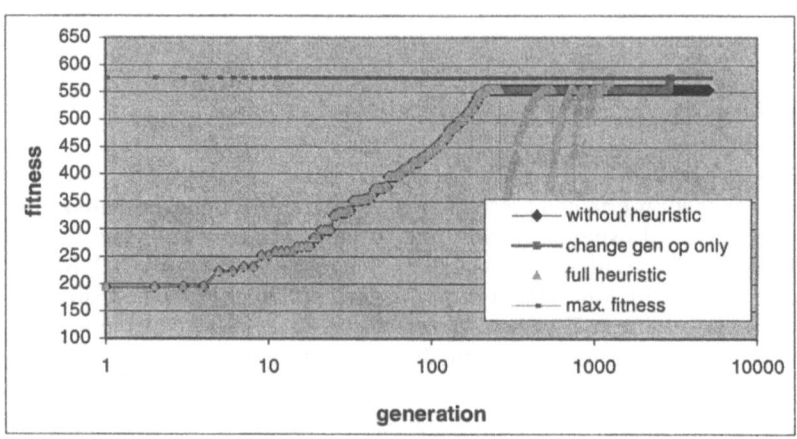

Fig. 3. Fitness curves of the best individual for three setups

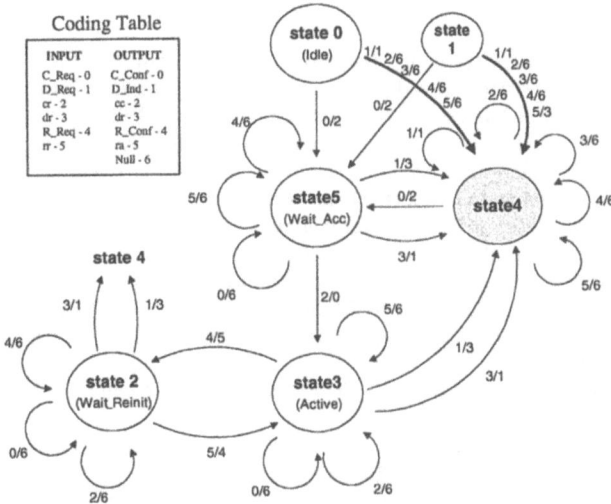

Fig. 4. PS_{SND}, using STG, of the fittest individual

Table 1. Comparison among three different setups

	Setup 3: Without heuristic	Setup 2: Change genetic operators only	Setup 1: Full heuristic Penalty ($Pf = 0.5$) applied to:			
			$Sf > 0.16$	$Sf > 0.27$	$Sf > 0.38$	$Sf > 0.55$
Convergence to Global Optimum for $G_{MAX} = 1,500$ (50 runs)	10	26	32	38	32	27

5 Conclusion

A methodology for generating state machines from given input/output sequences was proposed. Application fields range from reverse engineering approaches, focusing on creating representations for already implemented systems, to forward engineering ones, by moving from high-level abstractions (e.g., specifications by means of user cases or traces) to physical implementations of the system.

The proposed approach has as distinctive feature a specific strategy for escaping from local optima. Preliminaries results show improvements of about 46% in the convergence to the global optimum, considering a maximum number of generations. However, the right most values for the penalty and similarity factors, as well as for the number of generations the penalty factor is maintained active, still has to be chosen upon experiments. Our main focus on the future is to quantify these parameters regarding to the problem size. In addition, improvements that may be done on the presented search algorithm contemplate a heuristic with memory cells to retain all previous penalized schemas, guiding new populations to explore a refined state space.

References

[1]   L. Fogel, *Autonomous Automata*, Industrial Research, 4:14-19, 1962.
[2]   C. Manovit, C. Aporntewan and P. Chongstitvatana, *Synthesis of Synchronous Sequential Logic Circuits from Partial Input/Output Sequences*, ICES'98, pp. 98-105, 1998.
[3]   R. Collins and D. Jefferson, *Representation for Artificial Organisms*, in Proc. of the 1ˢᵗ Int. Conf. on Simulation of Adaptive Behavior, MIT Press, 1991.
[4]   P. Chongstitvatana and C. Aporntewan, *Improving Correctness of Finite-State Machine Synthesis from Multiple Partial Input/Output Sequences*, in Proc. of the 1ˢᵗ NASA/DoD Workshop on Evolvable Hardware, pp. 262-266, 1999.
[5]   G. Bockmann and A. Petrenko, *Protocol Testing: A Review of Methods and Relevance for Software Testing*, ISSTA'94, ACM, Seattle, U.S.A., pp. 109-124, 1994.
[6]   J. Holland, *Adaptation in Natural and Artificial Systems*, The University of Michigan, 1ˢᵗ Edition, 1975.
[7]   J. Koza, *Genetic Programming: On the Programming of Computers by Means of Natural Selection*, MIT Press, 1992.
[8]   H. Hörner, *A C++ Class Library for Genetic Programming*, Release 1.0 Operating Instructions, Viena University of Economy, 1996.
[9]   P. Whigham, *Grammatically-Based Genetic Programming*, in Proc. of the Workshop on G.P.: From the Theory to Real-World Applications, pp. 33-41, Morgan Kaufman, 1995.
[10] W. Feller, *An Introduction to Probability Theory and its Applications*, Vol. I, Wiley, pp. 224-225, 1968.
[11] W. Spears and D. Gordon, *Evolving FSM Strategies for Protecting Resources*, in Proceedings of the, 2000.
[12] W. Langdon, *Evolution of Genetic Programming Populations*, University College London Technical Report RN/96/125, 1996.
[13] R. Ursem, *Diversity-Guided Evolutionary Algorithms*. In Proceedings of Parallel Problem Solving from Nature VII (PPSN-2002), p. 462-471, 2002.

Intrinsic Evolution of Analog Circuits on a Programmable Analog Multiplexer Array

José Franco M. Amaral[1], Jorge Luís M. Amaral[1], Cristina C. Santini[2],
Marco A.C. Pacheco[2], Ricardo Tanscheit[2], and Moisés H. Szwarcman[2]

[1] UERJ - Rio de Janeiro State University
Rua São Francisco Xavier, 524, Maracanã, Rio de Janeiro, 20550-013, RJ, Brazil
{franco,jamaral}@uerj.br
[2] ICA: Applied Computational Intelligence Laboratory,
Department of Electrical Engineering, PUC-Rio
R. Marquês de S. Vicente 225, Gávea, Rio de Janeiro, CEP 22453-900, RJ, Brazil
{santini,marco,ricardo,moises}@ele.puc-rio.br

Abstract. This work discusses an Evolvable Hardware (EHW) platform for the intrinsic evolution of analog electronic circuits. The EHW analog platform, named PAMA-NG (Programmable Analog Multiplexer Array-Next Generation), is a reconfigurable platform that consists of integrated circuits whose internal connections can be programmed by Evolutionary Computation techniques, such as Genetic Algorithms (GAs), to synthesize circuits. The PAMA-NG is classified as Field Programmable Analog Array (FPAA). FPAAs are reconfigurable devices that will become the building blocks of a forthcoming class of hardware, with the important features of self-adaptation and self-repairing, through automatic reconfiguration. The PAMA-NG platform architectural details, concepts and characteristics are discussed. Two case studies, with promising results, are described: a logarithmic amplifier and an *S* membership function circuit of a fuzzy logic controller.

1 Introduction

This work focuses on evolutionary design of analog circuits on a platform named PAMA-NG (Programmable Analog Multiplexer Array – Next Generation). Evolutionary Computation is applied to carry out the process of automatic circuit synthesis through a set of search algorithms, called Evolutionary Algorithms (EAs), which borrow from biological evolution their main principles. This particular area of research, where Evolutionary Algorithms are applied to electronic circuit synthesis, receives the name of Evolvable Hardware (EHW)[1].

The PAMA-NG platform is based on PAMA (Programmable Analog Multiplexer Array)[2]. PAMA-NG is classified as an FPAA [2][3][4] and supplies an environment to evolve generic analog circuits based on discrete components, without the need of simulators, i.e. intrinsic evolution. PAMA is a flexible platform with respect to granularity: its building blocks can be chosen by the designer, from transistors to high level analog components, such as operational amplifiers. PAMA-NG introduces new *important* features, such as protection against configurations that may damage

M. Bubak et al. (Eds.): ICCS 2004, LNCS 3038, pp. 1273–1280, 2004.
© Springer-Verlag Berlin Heidelberg 2004

electronic components and the possibility of analyzing circuits that have been evolved, since it provides access to individual circuit elements.

A traditional Genetic Algorithm (GA) [5] drives PAMA's intrinsic evolution. In the evolutionary process, a population of chromosomes is randomly generated to represent a pool of circuit architectures. The chromosomes (control bit strings) are uploaded to the reconfigurable hardware. Circuit responses are compared to specifications of a target response and individuals are ranked in accordance with how close they satisfy the specifications. A new iteration loop involves generation of a new population from the pool of the individuals in the previous generation. Some individuals are taken as they are and some are modified by genetic operators, such as crossover and mutation. The process is repeated for a number of generations, resulting in increasingly better individuals. This process ends after a given number of generations or when the obtained result is close enough to the target response. In practice, one or several solutions may be found among the individuals of the last generation.

This paper is divided into three additional sections. Section 2 describes how PAMA-NG: internal interfaces and data flow are shown and the Reconfigurable Circuit implementation and purpose are described. Section 3 presents two case studies and section 4 concludes the paper.

2 PAMA-NG

The PAMA-NG evolutionary platform is an analog platform based on analog multiplexers, which are responsible for the interconnections of the different discrete components that can be plugged into the board. Figure 1 depicts PAMA-NG block diagram.

The platform performs intrinsic evolution of analog electronic circuits through a traditional GA. The chromosome configures the connection of the discrete components and each gene drives the select input signals of a particular analog multiplexer. As shown in Figure 1, a multifunction I/O board connected to the PC bus is responsible for A/D conversion and for chromosome upload. An eight-channel D/A converter allows for the application of different input values to the circuits being evolved, in order to perform the input/output mapping for fitness evaluation.

2.1 Reconfigurable Circuit

The Analog Reconfigurable Circuit (ARC) is divided into three layers: discrete components, analog multiplexers and analog bus (Figure 1). The designer chooses the components, from low-level discrete ones such as transistors, resistors and capacitors, to higher level circuits, such as operational amplifiers and comparators, or even more complex building blocks, such as multipliers and root mean-square circuits.

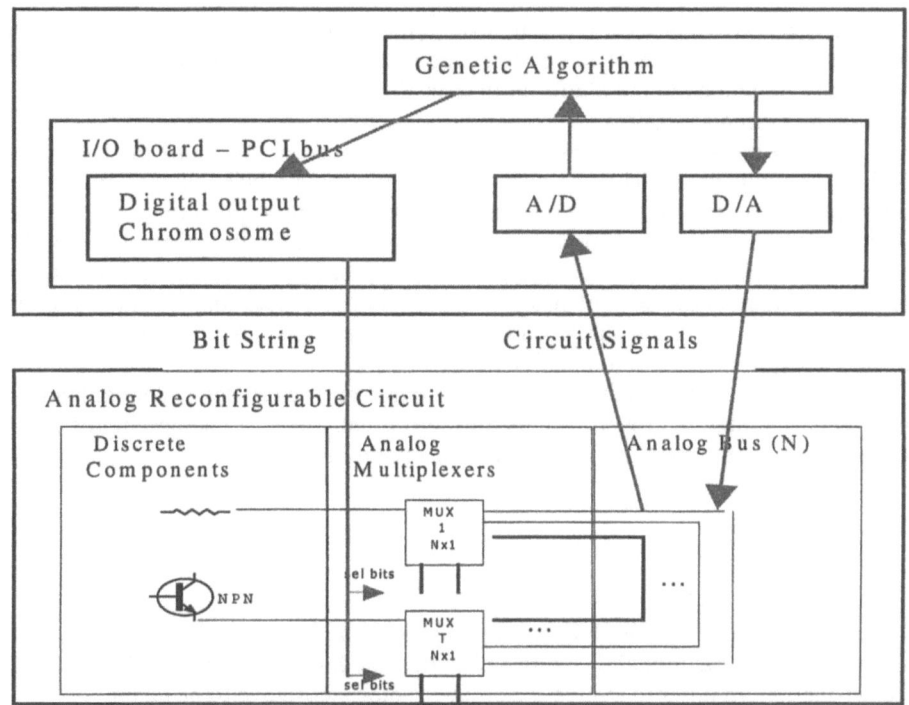

Fig. 1. The PAMA-NG platform block diagram

Each component terminal is connected to a specific line of an analog bus through an analog multiplexer. The chromosome bits, which are set by the GA running on the PC, select for each analog multiplexer a line of the analog bus to be connected to the component's terminal. Each line of the analog bus corresponds to one interconnection point of the circuit (node); some of them can be associated with external points, such as an input signal, power supply, ground and circuit output, while others can correspond to with internal points of interconnection.

The purpose of the ARC is to be generic, flexible and scalable. Discrete components can be of any type and in any number to fulfill application requirements. These components may be on sockets or on protoboard, making it easy to change them.

One ARC circuit module implements a 32 16-to-1 analog multiplexer. The low on-resistance (Ron) of 100 ohms of the analog multiplexers prevents the ARC from damage caused by potentially hazardous configurations during evolution. The search space associated to the platform is $16^{32} \sim 10^{38}$ (number of different possible solutions).

2.2 Important Issues

There are some important issues to discuss when analyzing an evolutionary platform, such as the circuit evaluation time, the platform's granularity and architecture and its protection against damaging configurations.

PAMA-NG presents a fast circuit evaluation time. Its multifunction I/O board starts the A/D conversion of the circuit's input and output signals immediately after the circuit's download to the platform. In tests, the period between two digital samples has been measured as 0.08ms.

The flexibility is another important feature. The optimal choice of the elementary block type and granularity is task dependent. PAMA-NG allows all discrete components to be selected by the designer, according to the application.

There are some other important characteristics that directly affect the evolution, such as the platform's protection against damaging configurations. In PAMA-NG, the parasitic resistances of the switches (Ron) provide this protection. Another desirable characteristic is that there are no architecture constraints, so that the evolution is allowed to exploit any interconnection pattern.

In view of the above, PAMA presents desirable characteristics when compared to other reconfigurable platforms designed to intrinsically evolve analog circuits [3][4]. Those characteristics have been contrasted in a previous work [6].

3 Case Studies

In the case studies, PAMA-NG is used to perform intrinsic evolution of analog circuits through a traditional GA with a chromosome length of 128 bits (32 mux times 4 select-inputs). In order to evolve the circuits, some steps have to be taken.

First, the discrete components have to be chosen. The platform has to be provided with a sufficient number of discrete components to be capable of achieving the desired circuit. The type of these components must be adequate. Since PAMA has originally a limited number of component terminals (32), its flexibility can be exploited to obtain higher level circuits, by using higher level building blocks as discrete components.

Next, the number of input and output signals must be assigned. The inputs and outputs of the circuit have to be connected to the multifunction I/O board.

Finally, a fitness measure has to be created. A precise function translates the output of the desired circuit. Experience has shown that the choice of an adequate fitness function plays a significant role in the evolution of an analog circuit.

The Genetic Algorithm used in the experiments has the following characteristics:

    Binary Representation: 128 bits
    Population Initialization: Random
    Reproduction Technique: Steady State
    Operators: One-point Crossover an Mutation
    Fitness Technique: Exponential Normalization

In the fitness technique used – exponential normalization – probabilities of the ranked individuals are exponentially weighted [6] and are given by Equation (1).

$$p_i = \frac{c-1}{c^n-1} c^{n-i} \tag{1}$$

The closer c is to 1, the lower is the "exponentiality" of the selection method.

3.1 Logarithmic Amplifier

In a logarithmic amplifier the output is proportional to the logarithm of the input signal. In [7] the intrinsic evolution of a logarithmic amplifier with no gain has been reported. In that experiment the objective was to demonstrate the platform's flexibility with respect to granularity; coarse-grained components, such as operational amplifiers, were used. The discrete components were then chosen based on conventional configurations for logarithmic amplifiers and some connections were fixed.

Here a logarithmic amplifier with an amplification of two is presented. This evolution has no restrictions concerning the discrete components connections.

The following GA parameters are used:

    Generations: 300
    Population: 400
    Crossover: 0.7
    Mutation: 0.15
    Steady State: 10
    Exponential Normalization c parameter: 0.9
    Evolution Time: 46 minutes

The discrete components are:

    2 amplifiers from IC TL075 (supply voltages: +5V and -5V)
    matched transistors from IC3086
    1 PNP transistor
    1 diode
    1 zener 3.3V
    resistors (1K, 22K, 47K, 3K3, 2K2, 330K, 100K)

Six of the 16 analog channels are fixed as external:

    2 power supplies (+5V)
    2 ground reference (0V)
    input
    output.

The other 10 channels (circuit nodes) are left as internal interconnection points of the circuit.

The input signal is a saw-tooth waveform of 100Hz with an amplitude range of 150mV to 500mV.

The fitness function used for evaluation of the configured circuits is:

$$\text{desired output} = -2 * \ln (Vin). \tag{2}$$

$$\text{Fitness} = (\, |\text{desired output} - \text{actual output}| \,)^5 \tag{3}$$

Figure 2 shows the schematic diagram of the best circuit evolved. About 200 discrete measurements were made to evaluate each circuit configured in PAMA-NG.

Figure 3 presents the input samples and the actual and desired outputs for the best circuit. This result shows that the evolved circuit performs well within expectations since only a small mapping error takes place.

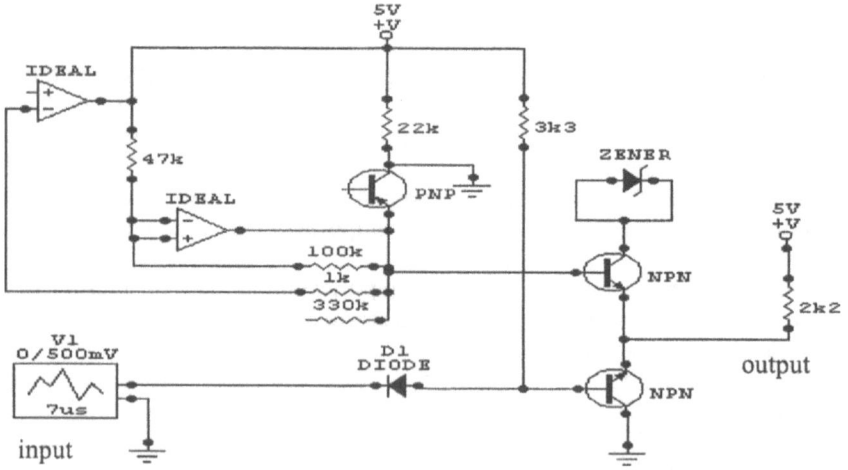

Fig. 2. Schematic of the best circuit evolved (without multiplexer's Ron)

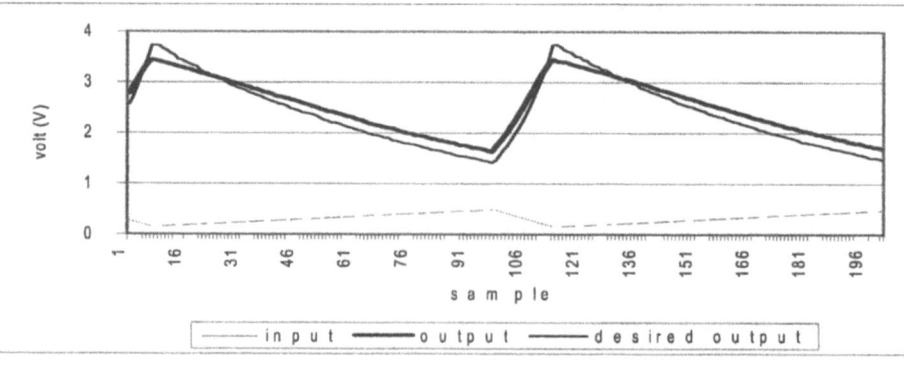

Fig. 3. Response of the best circuit evolved: a Logarithmic Amplifier with gain of two

3.2 *S* Membership Function Circuit

This experiment focuses on the development of a type *S* function – frequently used as a membership function in fuzzy inference systems – with amplitude of 3 volts and break points at 1 and 2 volts. The GA parameters are:

Generations: 100
Population: 400
Crossover: 0.7
Mutation: 0.1
Steady State: 50
Exponential Normalization *c* parameter : 0.9

The fitness function corresponds to the Mean Square Error of the difference between the desired output (*S* membership function) and the actual output. The transfer function and the schematic of the evolved circuit are illustrated in Figure 4 and Figure 5, respectively. Figure 4 shows the performance of the evolved circuit; it

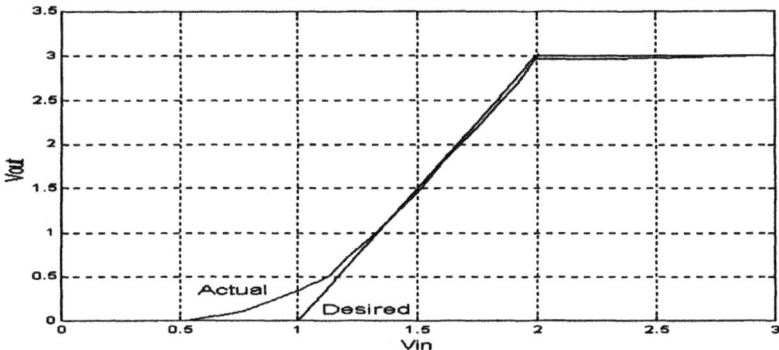

Fig. 4. Transfer function of type S membership function

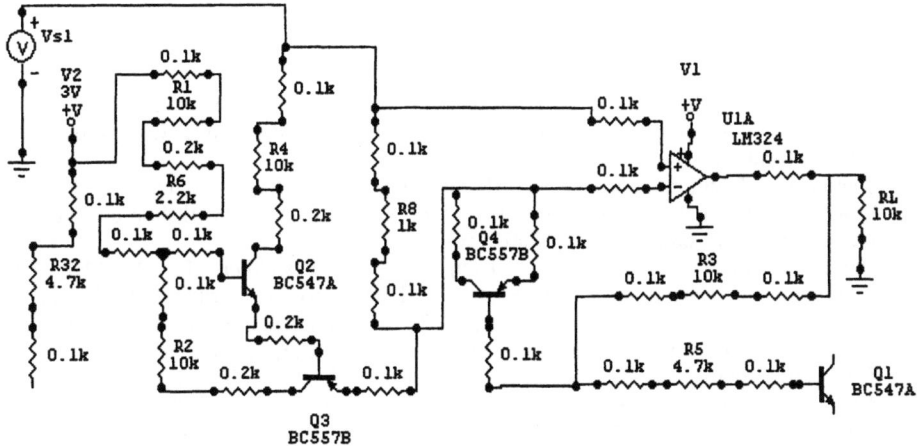

Fig. 5. Schematic of the evolved S membership function circuit

can be observed that only a small mapping error takes place at the break point of 1 volt. Figure 5 shows an unusual transistor configuration (Q4). This occurs quite often with EHW tools and demonstrates their ability to search for different circuit topologies

4 Conclusions

This work presented the PAMA-NG (Programmable Platform Multiplexer Array – Next Generation), a FPAA class platform that makes use of analog multiplexers, and some recent electronic designs carried out on it.

The PAMA-NG platform itself is being constantly updated. The latest version presents some advantages concerning the previous prototype platform which was first used to intrinsically evolve circuits [2]. The new ARC (Analog Reconfigurable Circuit) allows more component terminals (32 instead of 9), allowing the evolution of a great number of circuits. In the new platform, the chromosome (configuration bits) is downloaded in parallel, in contrast with the serial interface used in previous versions. In PAMA-NG the converters were updated and the A/D conversion is performed by a multifunction I/O board connected to the PC bus. The circuit evaluation speed has increased considerably. The most recent enhancement is a D/A converter.

Further experiments shall consider the use of multi-objective criteria in circuit synthesis. The main difficulty to overcome is the access to different measures such as power dissipation and frequency responses. The synthesis of higher level circuits is also planned, particularly those related to analog fuzzy logic controllers [8]. The flexible PAMA-NG platform shall then be used to intrinsically evolve circuits based on coarse-grained building blocks. PAMA may also be used as a basic cell in a scalable and parallel system.

References

1. Zebulum, R.S., Pacheco, M.A., Vellasco, M.M.: Evolutionary Electronics: Automatic Design of Electronic Circuits and Systems by Genetic Algorithms. CRC Press, Boca Raton, Florida (2001)
2. Santini, C., Zebulum, R.S., Pacheco, M.A., Vellasco, M.M., Szwarcman, M.: PAMA - Programmable Analog Multiplexer Array. Proc. 3rd NASA DoD Workshop on Evolvable Hardware. IEEE Computer Press (2001) 36-43
3. Hamilton, A., Papathanasiou, K., Tamplin, M.R., Brandtner, T.: PALMO: Field Programmable Analog and Mixed-Signal VLSI for Evolvable Hardware. In: Sipper, M., Mange, D., Perez-Uribe, A. (eds.): Proc. 2nd International Conference on Evolvable Systems. Lecture Notes in Computer Science, Vol. 1478. Springer-Verlag (1998) 335-344
4. Stoica, A., Keymeulen, D., Zebulum, R.S., Thakoor, A., Daud, T., Klimeck, G., Jin, Y., Tawel, R., Duong, V.: Evolution of Analog Circuits on Field Programmable Transistor Arrays. Proc. 2nd NASA DoD Workshop on Evolvable Hardware. IEEE Computer Press (2000) 99-108
5. Goldberg, D.: Genetic Algorithms in Search, Optimization and Machine Learning. Addison-Wesley Publishing Compan, Reading, Massachusetts (1989)
6. Zebulum, R.S., Santini, C., Sinohara, H., Pacheco, M.A., Vellasco, M.M., Szwarcman, M.: A Reconfigurable Plataform for the Automatic Synthesis of Analog Circuits. Proc. 2nd NASA/DoD Workshop on Evolvable Hardware. IEEE Computer Press (2000) 91-98
7. Santini, C., Amaral, J.F.M., Pacheco, M.A., Vellasco, M.M., Szwarcman, M.: Evolutionary Analog Circuit Design on a Programmable Analog Multiplexer Array. Proc. IEEE Int. Conf. on Field Programmable Technology, Hong Kong (2002) 189-196
8. Amaral, J. F. M., Amaral, J. L. M., Santini, C., Tanscheit, R., Vellasco, M.M., Pacheco, M.A.: Towards Evolvable Analog Fuzzy Logic Controllers. Proc. 2002 NASA/DoD Conference on Evolvable Hardware. IEEE Computer Press (2002) 123-128

Encoding Multiple Solutions in a Linear Genetic Programming Chromosome

Mihai Oltean[1], Crina Groşan[1], and Mihaela Oltean[2]

[1] Department of Computer Science,
Faculty of Mathematics and Computer Science,
Babeş-Bolyai University, Kogălniceanu 1
Cluj-Napoca, 3400, Romania.
{moltean,cgrosan}@cs.ubbcluj.ro
[2] David Prodan College, Cugir, 2566, Romania.
olteanmihaelaelena@yahoo.com

Abstract. Linear Genetic Programming (LGP) is a Genetic Programming variant that uses linear chromosomes for solution encoding. Each LGP chromosome is a sequence of C language instructions. Each instruction has a destination variable and several source variables. One of the variables is usually chosen to provide the output of the program. In this paper, we enrich the LGP technique by allowing it to encode multiple solutions for a problem in the same chromosome. Numerical experiments show that the proposed Multi-Solution LGP significantly outperforms the standard Single-Solution LGP on the considered test problems.

1 Introduction

Linear Genetic Programming (LGP) [1] is a Genetic Programming [2] variant that uses linear chromosomes for solution encoding. Each LGP chromosome is a sequence of C language instructions. Each instruction has a destination variable and several source variables. One of the variables is usually chosen to provide the output of the program.

In this paper an improved variant of Linear Genetic Programming is proposed. The obtained technique is called Multi-Solution Linear Genetic Programming (MS-LGP). In the proposed variant each chromosome stores multiple solutions of the problem being solved. All the solutions represented in a MS-LGP individual are decoded by traversing the chromosome only once. Partial results are stored by using Dynamic Programming. The best solution encoded in a MS-LGP chromosome will represent (will provide the fitness of) that individual.

Several numerical experiments with MS-LGP and with the standard Single-Solution Linear Genetic Programming (SS-LGP) are performed by using 4 test functions. For each test problem the relationships between the success rate and the population size and the code length are analyzed. Results show that MS-LGP significantly outperforms SS-LGP for all the considered test problems.

The paper is organized as follows. In section 2 Linear Genetic Programming is described. In sub-section 2.3 is described the way in which multiple solutions

M. Bubak et al. (Eds.): ICCS 2004, LNCS 3038, pp. 1281–1288, 2004.
© Springer-Verlag Berlin Heidelberg 2004

are encoded in a LGP chromosome. Several numerical experiments are performed in section 3.

2   Linear Genetic Programming

Linear Genetic Programming (LGP) [1] uses a specific linear representation of computer programs. Instead of the tree-based GP expressions [2] of a functional programming language (like *LISP*), programs of an imperative language (like *C*) are evolved.

A LGP individual is represented by a variable-length sequence of simple *C* language instructions. Instructions operate on one or two indexed variables (registers) r or on constants c from predefined sets. The result is assigned to a destination register, e.g. $r_i = r_j * c$.

An example of the LGP program is the following one:

```
void LGP(double r[8])
{
  r[0] = r[5] + 73;
  r[7] = r[3] - 59;
  r[2] = r[5] + r[4];
  r[6] = r[7] * 25;
  r[1] = r[4] - 4;
  r[7] = r[6] * 2;
}
```

2.1   Decoding LGP Individuals

A linear genetic program can be turned into a functional representation by successive replacements of variables starting with the last effective instruction [1].

Usually one of the variables ($r[0]$) is chosen as the output of the program. This choice is made at the beginning of the program and is not changed during the search process. In what follows we will denote this LGP variant as Single-Solution Linear Genetic Programming (SS-LGP).

2.2   Genetic Operators

The variation operators used in conjunction with Linear Genetic Programming are crossover and mutation. Standard LGP crossover works by exchanging continuous sequences of instructions between parents [1].

Two types of standard LGP mutations are usually used: micro mutation and macro mutation. By micro mutation an operand or an operator of an instruction is changed [1].

Macro mutation inserts or deletes a random instruction [1].

Since we are interested more in multi-solutions paradigm rather than in variable length chromosomes we will use fixed length chromosomes in all experiments performed in this paper. Genetic operators used in numerical experiments are uniform crossover and micro mutation.

LGP uniform crossover. LGP uniform crossover works between instructions. The offspring's genes (instructions) are taken with a 50% probability from the parents.

Example. Let us consider the two parents C_1 and C_2 given in Table 1. The two offspring O_1 and O_2 are obtained by uniform recombination as shown in Table 1.

Table 1. LGP uniform recombination

Parents		Offspring	
C_1	C_2	O_1	O_2
$r[5] = r[3] * r[2];$	$r[2] = r[0] + r[3];$	$r[5] = r[3] * r[2];$	$r[2] = r[0] + r[3];$
$r[3] = r[1] + 6;$	$r[1] = r[2] * r[6];$	$r[1] = r[2] * r[6];$	$r[3] = r[1] + 6;$
$r[0] = r[4] * r[7];$	$r[4] = r[6] - 4;$	$r[0] = r[4] * r[7];$	$r[4] = r[6] - 4;$
$r[5] = r[4] - r[1];$	$r[6] = r[5] / r[2];$	$r[5] = r[4] - r[1];$	$r[6] = r[5] / r[2];$
$r[1] = r[6] * 7;$	$r[2] = r[1] + 7;$	$r[2] = r[1] + 7;$	$r[1] = r[6] * 7;$
$r[0] = r[0] + r[4];$	$r[1] = r[2] + r[4];$	$r[1] = r[2] + r[4];$	$r[0] = r[0] + r[4];$
$r[2] = r[3] / r[4];$	$r[0] = r[4] * 3;$	$r[0] = r[4] * 3;$	$r[2] = r[3] / r[4];$

LGP Mutation. LGP mutation works inside of a LGP instruction. By mutation each operand (source or destination) or operator is affected with a fixed mutation probability.

Example. Consider an individual C which is affected by mutation. An offspring O is obtained as shown in Table 2 (modified variables are written in boldface):

Table 2. LGP mutation

C	O
$r[5] = r[3] * r[2];$	$r[5] = r[3] * r[2];$
$r[3] = r[1] + 6;$	$r[3] = r[6] + r[0];$
$r[0] = r[4] * r[7];$	$r[0] = r[4] + r[7];$
$r[5] = r[4] - r[1];$	$r[4] = r[4] - r[1];$
$r[1] = r[6] * 7;$	$r[1] = r[6] * 2;$
$r[0] = r[0] + r[4];$	$r[0] = r[0] + r[4];$
$r[2] = r[3] / r[4];$	$r[0] = r[3] / r[4];$

2.3   Multi Solutions Linear Genetic Programming

We enrich the LGP structure in two ways:

(i) We allow as each destination variable to represent the output of the program. In the standard LGP only one variable is chosen to provide the output.

(ii) We check for the program output after each instruction in chromosome. This is again different from the standard LGP where the output was checked after the execution of all instructions in a chromosome.

After each instruction, the value stored in the destination variable is considered as a potential solution of the problem. The best value stored in one of the destination variables is considered for fitness assignment purposes.

Example. Consider the chromosome C given below:

```
void LGP(double r[8])
{
  r[5] = r[3] * r[2];
  r[3] = r [1] + 6;
  r[0] = r[4] * r[7];
  r[6] = r[4] - r[1];
  r[1] = r[6] * 7;
  r[2] = r[3] / r[4];
}
```

Instead of encoding the output of the problem in a single variable (as in SS-LGP) we allow that each of the destination variables ($r[5]$, $r[3]$, $r[0]$, $r[6]$, $r[1]$ or $r[2]$) to store the program output. The best output stored in these variables will provide the fitness of the chromosome.

For instance, if we want to solve symbolic regression problems, the fitness of each destination variable $r[i]$ may be computed using the formula:

$$f(r[i]) = \sum_{k=1}^{n} |o_{k,i} - w_k|,$$

where $o_{k,i}$ is the result obtained in variable $r[i]$ for the fitness case k, w_k is the targeted result for the fitness case k and n is the number of fitness cases. For this problem the fitness needs to be minimized.

The fitness of an individual is set to be equal to the lowest fitness of the destination variables encoded in the chromosome:

$$f(C) = \min_i f(r[i]).$$

Thus, we have a Multi-Solution program at two levels: first level is given by the possibility that each variable to represent the output of the program and the second level is given by the possibility of checking for the output at each instruction in the chromosome.

Our choice was mainly motivated by the No Free Lunch Theorems for Search [4]. There is neither practical nor theoretical evidence that one of the variables employed by the LGP is better than the others. More than that, Wolpert and McReady [4] proved that we cannot use the search algorithm's behavior so far for a particular test function to predict its future behavior on that function.

The Multi-Solution ability has been tested within other evolutionary model such as Multi Expression Programming [3]. For these methods it has been shown [3] that encoding multiple solutions in a single chromosome leads to significant improvements.

3   Numerical Experiments

In this section several experiments with SS-LGP and MS-LGP are performed. For this purpose we use several well-known symbolic regression problems. The problems used for assessing the performance of the compared algorithms are:

$f_1(x) = x^4 + x^3 + x^2 + x,$
$f_2(x) = x^6 - 2x^4 + x^2,$
$f_3(x) = sin(x^4 + x^2),$
$f_4(x) = sin(x^4) + sin(x^2).$

For each function 20 fitness cases have been randomly generated with a uniform distribution over the [0, 1] interval.

The general parameters of the LGP algorithms are given in Table 3. The same settings are used for Multi Solution LGP and for Single-Solution LGP.

Table 3. The parameters of the LGP algorithm for symbolic regression problems

Parameter	Value
Number of generations	51
Crossover probability	0.9
Mutations	2 / chromosome
Function set	$F = \{+, -, *, /, sin\}$
Terminal set	Problem inputs + 4 supplementary registers
Selection	Binary Tournament
Algorithm	Steady State

For all problems the relationship between the success rate and the chromosome length and the population size is analyzed. The success rate is computed as the number of successful runs over the total number of runs.

3.1   Experiment 1

In this experiment the relationship between the success rate and the chromosome length is analyzed. The population size was set to 50 individuals. Other parameters of the LGP are given in Table 3. Results are depicted in Figure 1.

Figure 1 shows that Multi-Solution LGP significantly outperforms Single-Solution LGP for all the considered test problems and for all the considered parameter setting. More than that, large chromosomes are better for MS-LGP than short chromosomes. This is due to the multi-solution ability: increasing the chromosome length leads to more solutions encoded in the same individual. The

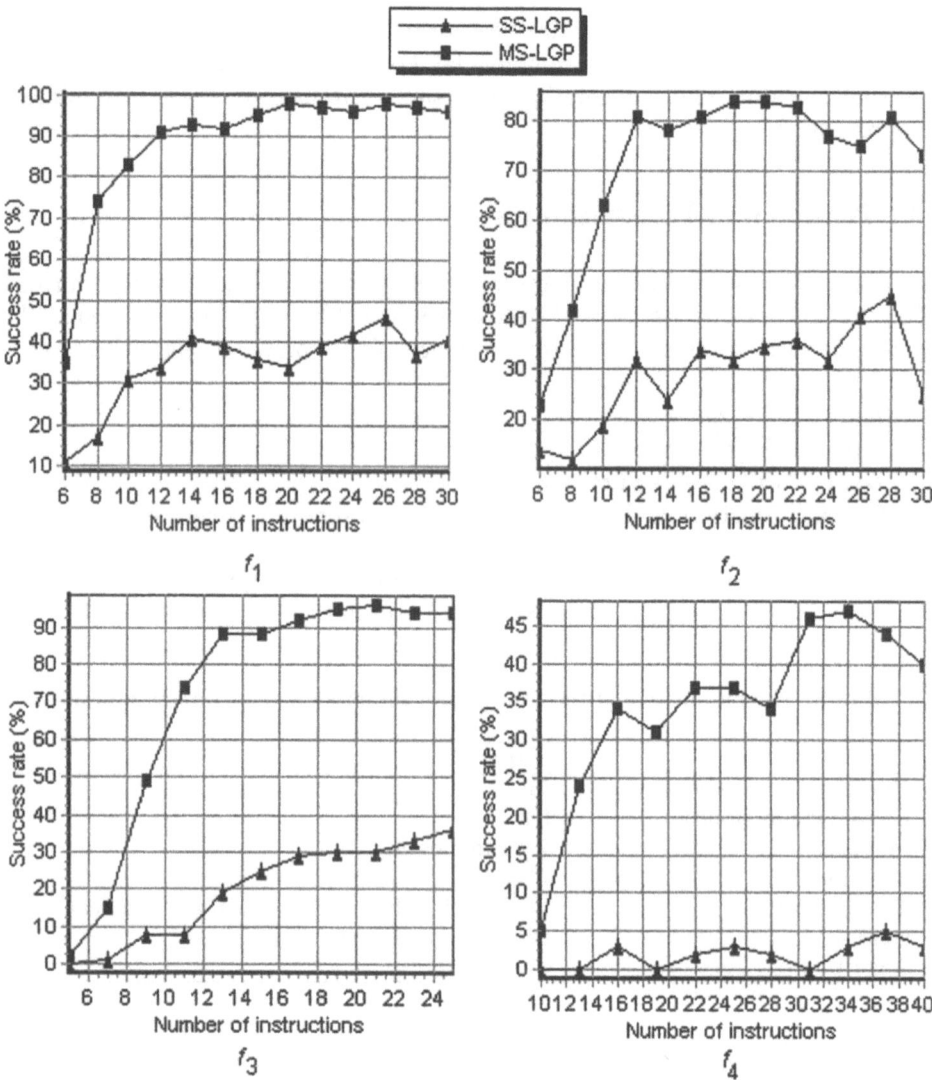

Fig. 1. The relationship between the success rate and the number of instructions in a chromosome. Results are averaged over 100 runs.

easiest problem is f_1. MS-LGP success rate for this problem is over 90% when the number of instructions in a chromosome is larger than 12. The most difficult problem is f_4. For this problem and with the parameters given in Table 3, the success rate of the MS-LGP algorithm never increases over 47%. However, these results are very good compared to those obtained by SS-LGP (the success rate never increases over 5%).

3.2   Experiment 2

In this experiment the relationship between the success rate and the population size is analyzed. The number of instructions in a LGP chromosome was set to 12. Other parameters for the LGP are given in Table 3. Results are depicted in Figure 2.

Figure 2 shows that Multi-Solution LGP performs better than Single-Solution LGP. Problem f_1 is the easiest one and problem f_4 is the most difficult one.

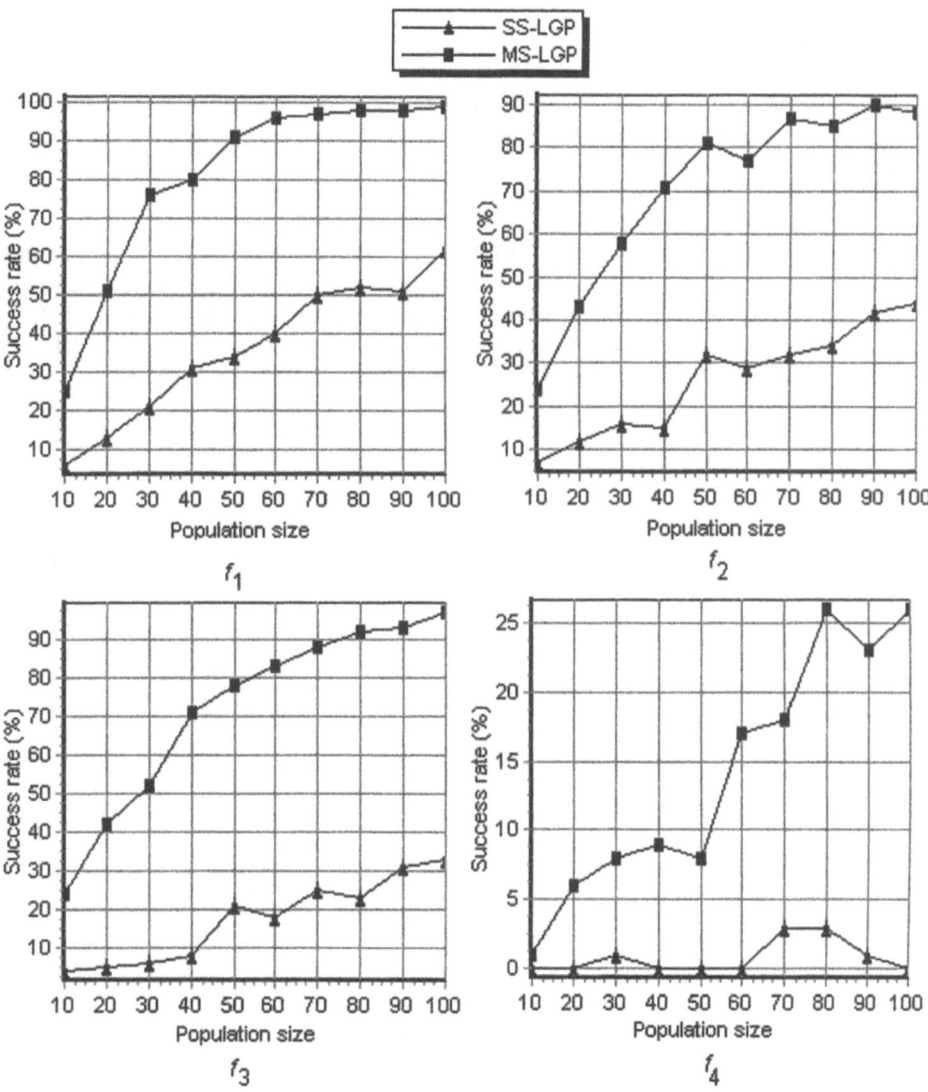

Fig. 2. The relationship between the population size and the success rate. Population size varies between 10 and 100. Results are averaged over 100 runs.

4   Conclusions

In this paper an improved variant of the Linear Genetic Programming technique has been proposed. The improvement consists in encoding multiple solutions of a problem in a single chromosome. It has been show how to efficiently decode this chromosome by traversing it only once. Numerical experiments have shown that Multi-Solution LGP significantly outperforms Standard Single-Solution LGP for all the considered test problems.

References

1. Brameier M. and Banzhaf W., A Comparison of Linear Genetic Programming and Neural Networks in Medical Data Mining, IEEE Transactions on Evolutionary Computation, 5, 17-26, 2001.
2. Koza J. R., Genetic Programming: On the Programming of Computers by Means of Natural Selection, MIT Press, Cambridge, MA, 1992.
3. Oltean M., Solving Even-Parity Problems using Multi Expression Programming, in Proceedings of the the 7th Joint Conference on Information Sciences, Research Triangle Park, North Carolina, Edited by Ken Chen (et. al), pp. 315-318, 2003.
4. Wolpert D.H. and McReady W.G., No Free Lunch Theorems for Search, Technical Report, SFI-TR-05-010, Santa Fe Institute, 1995.

Evolutionary State Assignment for Synchronous Finite State Machines

Nadia Nedjah and Luiza de Macedo Mourelle

Department of Systems Engineering and Computation, Faculty of Engineering,
State University of Rio de Janeiro,
Rio de Janeiro, Brazil
{nadia,ldmm}@eng.uerj.br
http://www.eng.uerj.br/~ldmm

Abstract. Among other important aspects, finite state machines represent a powerful way for synchronising hardware components so that these components may cooperate in the fulfilment of the main objective of the hardware design. In this paper, we propose to optimally solve the state assignment *NP*-complete problem, which is inherent to designing any synchronous finite state machines using evolutionary computations. This is motivated by to reasons: first, finite state machines are very important in digital design and second, with an optimal state assignment one can physically implement the state machine in question using a minimal hardware area and reduce the propagation delay of the machine output signals.

1 Introduction

Sequential digital systems or simply finite state machines have two main characteristics: *(i)* there is at least one feedback path from the system output signal to the system input signals; and *(ii)* there is a memory capability that allows the system to determine current and future output signal values based on the previous input and output signal values [1].

Traditionally, the design process of a state machine passes through five main steps: *(i)* the specification of the sequential system, which should determine the next states and outputs of every present state of the machine. This is done using state tables and state diagrams; *(ii)* the state reduction, which should reduce the number of present states using equivalence and output class grouping; *(iii)* the state assignment, which should assign a distinct combination to every present state. This may be done using Armstrong-Humphrey heuristics [1]; *(iv)* the minimisation of the control combinational logic using K-maps and transition maps; *(v)* and finally, the implementation of the state machine, using gates and flip-flops.

In this paper, we concentrate on the third step of the design process, i.e. the state assignment problem. We present a genetic algorithm designed for finding a state assignment of a given synchronous finite state machine, which attempts to minimise the cost related to the state transitions.

The remainder of this paper is organised into five sections. In Section 2, we explain thoroughly the state assignment problem and show that a better assignment improves

M. Bubak et al. (Eds.): ICCS 2004, LNCS 3038, pp. 1289–1296, 2004.
© Springer-Verlag Berlin Heidelberg 2004

considerably the cost related to state transitions. In Section 3, we give an overview on evolutionary computations and genetic algorithms and their application to solve *NP*-problems. In Section 4, we design a genetic algorithm for evolving best state assignment for a given state machine specification. We describe the genetic operators used as well as the fitness function, which determines whether a state assignment is better that another and how much. In Section 5, we present results evolved through our genetic algorithm for some well-known benchmarks. Then we compare the obtained results with those obtained by another genetic algorithm described in [2] as well as with NOVA, which is uses well established but non-evolutionary method [3].

2 State Assignment Problem

Once the specification and the state reduction step has been completed, the next step is then to assign a code to each state present in the machine. It is clear that if the machine has N present states then, one needs N distinct combinations of 0s and 1s. So one needs K flip-flops to store the machine present state, wherein K is the smallest positive integer such that $2^K \geq N$. The state assignment problem under consideration consists of finding the *best* assignment of the flip-flop combinations to the machine states. Since a machine state is nothing but a counting device, combinational control logic is necessary to activate the flip-flops in the desired sequence. Given a state transition function, it is expected that the complexity (area and time) as well as cost of the control logic will vary for different assignments of flip-flop combinations to allowed states. Consequently, the designer should seek the assignment that minimises the complexity and the cost of the combinational logic required to control the state transitions. For instance, consider the state machine of one input signal and 4 states whose state transition function is given in tabular form in Table 1 and we are using JK-flip-flops to store the machine current state. Then the state assignment $\{s_0 \equiv 00, s_1 \equiv 10, s_2 \equiv 01, s_3 \equiv 11\}$ requires a control logic that consists of 4 NOT gates, 3 AND gates and 1 OR gate while the assignments $\{s_0 \equiv 00, s_1 \equiv 11, s_2 \equiv 01, s_3 \equiv 10\}$, $\{s_0 \equiv 10, s_1 \equiv 01, s_2 \equiv 11, s_3 \equiv 00\}$ and $\{s_0 \equiv 01, s_1 \equiv 10, s_2 \equiv 00, s_3 \equiv 11\}$ require a control logic that consists of only 2 NOT gates and 1 OR gate.

Table 1. State transition function

Present State	Next Sate	
	$I = 0$	$I = 1$
q_0	q_0	q_1
q_1	q_2	q_1
q_2	q_0	q_3
q_3	q_2	q_1

3 Evolutionary Computations

Evolutionary algorithms are computer-based solving systems, which use evolutionary computational models as key element in their design and implementation. A variety of evolutionary algorithms have been proposed. The most popular ones are *genetic algorithms* [4]. They have a conceptual base of simulating the evolution of individual structures via the Darwinian natural selection process. The process depends on the adherence of the individual structures as defined by its environment to the problem pre-determined constraints. Genetic algorithms are well suited to provide an efficient solution of *NP*-hard problems [5].

Genetic algorithms maintain a *population* of *individuals* that evolve according to *selection* rules and other *genetic operators*, such as *mutation* and *recombination*. Each individual receives a measure of *fitness*. *Selection* focuses on individuals, which shows high fitness. *Mutation* and *crossover* provide general heuristics that simulate the *recombination* process. Those operators attempt to perturb the characteristics of the parent individuals as to generate *distinct* offspring individuals.

Genetic algorithms are implemented through the following generic algorithm described by Algorithm 1, wherein parameters *ps, f* and *gn* are the population size, the fitness of the expected individual and the number of generation allowed respectively.

Algorithm 1. GA(ps, f, gn):individual;
```
1: generation  := 0;
2: population  := initialPopulation();
3: fitness := evaluate(population);
4: do parents   := select(population);
5:    population := reproduce(parents);
6:    fitness   := evaluate(population);
7:    generation := generation + 1;
8: while(fitness[i] < f, ∀ i ∈ population) & (generation < gn);
9: return fittestIndividual(population);
End
```

In Algorithm 1, function *intialPopulation* returns a valid random set of individuals that compose the population of the first generation, function *evaluate* returns the fitness of a given population. Function *select* chooses according to some criterion that privileges fitter individuals, the individuals that will be used to generate the population of the next generation and function *reproduction* implements the crossover and mutation process to yield the new population.

4 Application to the State Assignment Problem

The identification of a good state assignment has been thoroughly studied over the years. In particular, Armstrong [6] and Humphrey [7] have pointed out that an assignment is good if it respects two rules, which consist of the following: *(i)* two or more states that have the same next state should be given adjacent assignments; *(ii)* two or more states that are the next states of the same state should be given adjacent assignment. State adjacency means that the states appear next to each other in the mapped representation. In other terms, the combination assigned to the states should differ in only one position; and *(iii)* the first rule should be given more important the second. For instance, state codes 0101 and 1101 are adjacent while state codes 1100 and 1111 are not adjacent.

Now we concentrate on the assignment encoding, genetic operators as well as the fitness function, which given two different assignment allows one to decide which is fitter.

4.1 Assignment Encoding

Encoding of individuals is one of the implementation decisions one has to make in order to use genetic algorithms. It very depends on the nature of the problem to be solved. There are several representations that have been used with success [4].

In our implementation, an individual represents a state assignment. We use the *integer encoding*. Each chromosome consists of an array of N entries, wherein entry i is the code assigned to i^{th} machine state. For instance, chromosome in Fig. 1 represents a possible assignment for a machine with 6 states:

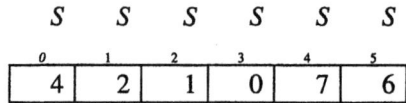

Fig. 1. State assignment encoding

4.2 The Individual Reproduction

Besides the parameters, which represent the population size, the fitness of the expected result and the maximum number of generation allowed, the genetic algorithm has several other parameters, which can be adjust by the user so that the result is up to his or her expectation. The selection is performed using some *selection probabilities* and the reproduction, as it is subdivided into *crossover* and *mutation* processes, depends on the kind of crossover and the mutation rate and degree to be used.

Given the parents populations, the reproduction proceeds using replacement as a reproduction scheme, i.e. offspring replace their parents in the next generation. Obtaining offspring that share some traits with their corresponding parents is performed by the *crossover* function. There are several *types* of crossover schemes [4]. The newly obtained population can then suffer some mutation, i.e. some of the genes of some of the individuals. The crossover type, the number of individuals that should be mutated and how far these individuals should be altered are set up during the initialisation process of the genetic algorithm.

There are many ways how to perform crossover and these may depend on the individual encoding used [4]. We present crossover techniques used with binary, permutation and value representations. *Single-point crossover* consists of choosing randomly one *crossover point*, then, the part of the bit or integer sequence from beginning of offspring till the crossover point is copied from one parent, the rest is copied from the second parent. *Double-points crossover* consists of selecting randomly two *crossover points*, the part of the bit or integer sequence from beginning of offspring to the first crossover point is copied from one parent, the part from the first to the second crossover point is copied from the second parent and the rest is copied from the first parent. *Uniform crossover* copies integers randomly from the first or from the second parent. Finally, *arithmetic crossover* consists of applying some arithmetic operation to yield a new offspring.

The single point and two points crossover use randomly selected crossover points to allow variation in the generated offspring and to avoid premature convergence on a local optimum [4]. In our implementation, we tested single-point and double-point crossover techniques.

Mutation consists of changing some genes of some individuals of the current population. The number of individuals that should be mutated is given by the parameter *mutation rate* while the parameter *mutation degree* states how many genes of a selected individual should be altered. The mutation parameters have to be chosen carefully as if mutation occurs very often then the genetic algorithm would in fact change to *random search* [4]. Fig. 2 illustrates the genetic operators.

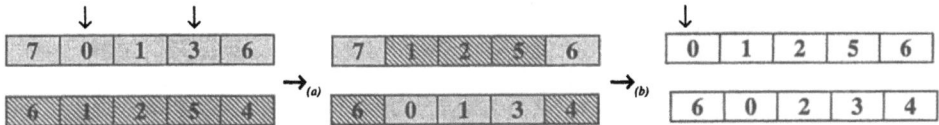

Fig. 2. State assignment genetic operators: *(a)* double-point crossover and *(b)* mutation

4.3 The Fitness Evaluation

This step of the genetic algorithm allows us to classify the individuals of a population so that fitter individuals are selected more often to contribute in the constitution of a new population. The fitness evaluation of state assignments is performed with respect to two rules of Armstrong [6] and Humphrey [7]: *(i)* how much a given state assignment adheres to the first rule, i.e. how many states in the assignment, which have the same next state, have no adjacent state codes; *(ii)* how much a given state in the assignment adheres to the second rule, i.e. how many states in the assignment, which are the next states of the same state, have no adjacent state codes.

In order to efficiently compute the fitness of a given state assignment, we use an $N{\times}N$ *adjacency matrix*, wherein N is the number of the machine states. The triangular bottom part of the matrix holds the expected adjacency of the states with respect to the first rule while the triangular top part of it holds the expected adjacency of the states with respect to the second rule. The matrix entries are calculated as in Equation (1), wherein *AM* stands for the adjacency matrix, functions *next(s)* and *prev(s)* yield the set of states that are next and previous to state *s* respectively. For instance, for the state machine in Table 1, we get the 4×4 adjacency matrix in Fig 3.

$$AM_{i,j} = \begin{cases} \#\left(next(q_i) \cap next(q_j)\right) & i > j \\ \#\left(prev(q_i) \cap prev(q_j)\right) & i < j \\ 0 & i = j \end{cases} \qquad (1)$$

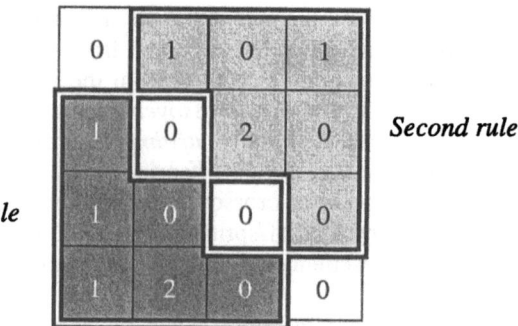

Fig. 3. Adjacency matrix for the machine state in Table 1

Using the adjacency matrix *AM*, the fitness function applies a penalty of 2, respectively 1, every time the first rule, respectively the second rule, is broke. Equation (2) states the details of the *fitness* function applied to a state assignment *SA*, wherein function *nadjacent* (q, p) returns 0 if the codes representing states q and p are adjacent and 1 otherwise.

$$fitness(SA) = \sum_{i=0}^{N-1} \sum_{j=i+1}^{N-1} \left(AM_{i,j} + 2 \times AM_{j,i} \right) \times nadjacent(SA_i, SA_j) \quad (2)$$

For instance, considering the state machine whose state transition function is described in Table 1, the state assignment $\{s_0 \equiv 00, s_1 \equiv 10, s_2 \equiv 01, s_3 \equiv 11\}$ has a fitness of 5 as the codes of states s_0 and s_3 are not adjacent but $AM_{0,3} = 1$ and $AM_{3,0} = 1$ and the codes of states s_1 and s_2 are not adjacent but $AM_{1,2} = 2$ while the assignments $\{s_0 \equiv 00, s_1 \equiv 11, s_2 \equiv 01, s_3 \equiv 10\}$ has a fitness of 3 as the codes of states s_0 and s_1 are not adjacent but $AM_{0,1} = 1$ and $AM_{,1} = 1$.

The objective of the genetic algorithm is to find the assignment that minimise the fitness function as described in Equation (2). Assignments with fitness 0 satisfy all the adjacency constraints. Such an assignment does not always exist.

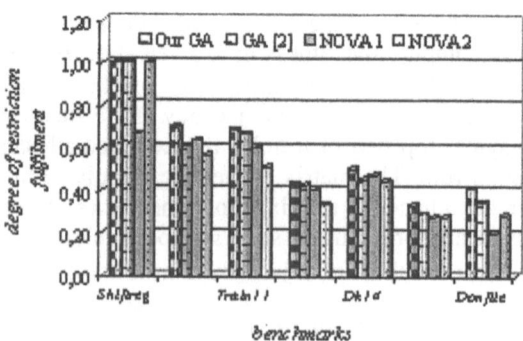

Fig. 4. Graphical comparison of the degree of fulfilment of rule 1 and 2 reached by the systems

Table 2. Best state assignment yield by the compared systems for the benchmarks

FSM	Size	System	State Assignment
Shiftreg	8/16	GA [2]	[0,2,5,7,4,6,1,3]
		NOVA1	[0,4,2,6,3,7,1,5]
		NOVA2	[0,2,4,6,1,3,5,7]
		Our GA	[5,7,4,6,1,3,0,2]
Lion9	9/25	GA [2]	[0,4,12,13,15,1,3,7,5]
		NOVA1	[2,0,4,6,7,5,3,1,11]
		NOVA2	[0,4,12,14,6,11,15,13,7
		Our GA	[10,8,12,9,13,15,7,3,11]
Train11	11/25	GA [2]	[0,8,2,9,13,12,4,7,5,3,1]
		NOVA1	[0,8,2,9,1,10,4,6,5,3,7]
		NOVA2	0,13,11,5,4,7,6,10,14,15,12]
		Our GA	[2,6,1,4,0,14,10,9,8,11,3]
Bbarra	10/60	GA [2]	[0,6,2,14,4,5,13,7,3,1]
		NOVA1	[4,0,2,3,1,13,12,7,6,5]
		NOVA2	[9,0,2,13,3,8,15,5,4,1]
		Our GA	[3,0,8,12,1,9,13,11,10,2]
Dk14	7/56	GA [2]	[0,4,2,1,5,7,3]
		NOVA1	[5,7,1,4,3,2,0]
		NOVA2	[7,2,6,3,0,5,4]
		Our GA	[3,7,1,0,5,6,2]
Bbsse	16/56	GA [2]	[0,4,10,5,12,13,11,14,15,8,9,2,6,7,3,1]
		NOVA1	[12,0,6,1,7,3,5,4,11,10,2,13,9,8,15,14]
		NOVA2	[[2,3,6,15,1,13,7,8,12,4,9,0,5,10,11,14]
		Our GA	[15,14,9,12,1,4,3,7,6,10,2,11,13,0,5,8]
Donfile	24/96	GA [2]	[0,12,9,1,6,7,2,14,11,17,20,23,8,15,10,16,21,19,4,5,22,18,13,3]
		NOVA1	[12,14,13,5,23,7,15,31,10,8,29,25,28,6,3,2,4,0,30,21,9,17,12,1]
		NOVA2	[6,30,11,28,25,19,0,26,1,2,14,10,31,24,27,15,12,8,29,23,13,9,7,3]
		Our GA	[2,18,17,1,29,21,6,22,7,0,4,20,19,3,23,16,9,8,13,5,12,28,25,24]

Table 3. Fitness of best assignments yield by the compared systems

State machine	#AdjRes	Our GA	GA [2]	NOVA1	NOVA2
Shiftreg	24	0	0	8	0
Lion9	69	21	27	25	30
Train11	57	18	19	23	28
Bbara	225	127	130	135	149
Dk14	137	68	75	72	76
Bbsse	305	203	215	220	220
Donfile	408	241	267	326	291

5   Comparative Results

In this section, we compare the assignment evolved by our genetic algorithm to those yield by another genetic algorithm [2] and to those obtained using the non-evolutionary assignment system called NOVA [3]. The examples are well-known benchmarks for testing synchronous finite state machines [8]. Table 2 shows the best state assignment generated by the compared systems. The size column shows the total number of states/transitions of the machine.

Table 3 gives the fitness of the best state assignment produced by our genetic algorithm, the genetic algorithm from [2] and the two versions of NOVA system [3]. The *#AdjRes* stands for the number of expected adjacency restrictions. Each adjacency according to rule 1 is counted twice and that with respect to rule 2 is counted just once. For instance, in the case of the *Shiftreg* state machine, all 24 expected restrictions were fulfilled in the state assignment yielded by the compared systems. However, the state assignment obtained the first version of the NOVA system does not fulfil 8 of the expected adjacency restrictions of the state machine.

The chart of Fig 4 compares graphically the degree of fulfilment of the adjacency restrictions expected in the other state machines used as benchmarks. The chart shows clearly that our genetic algorithm always evolves a better state assignment.

6   Conclusion

In this paper, we exploited evolutionary computation to solve the *NP*-complete problem of state encoding in the design process of asynchronous finite state machines. We compared the state assignment evolved by our genetic algorithm for machine of different sizes evolved to existing systems. Our genetic algorithm always obtains better assignments (see Table 3 of Section 5).

References

1.   V.T. Rhyne, *Fundamentals of digital systems design*, Prentice-Hall, Electrical Engineering Series. 1973.
2.   J.N. Amaral, K. Tumer and J. Gosh, *Designing genetic algorithms for the State Assignment problem*, IEEE Transactions on Systems Man and Cybernetics, vol., no. 1999.
3.   T. Villa and A. Sangiovanni-Vincentelli, *Nova: state assignment of finite state machine for optimal two-level logic implementation*, IEEE Transactions on Computer-Aided Design, vol. 9, pp. 905-924, September 1990.
4.   Z. Michalewics, *Genetic algorithms + data structures = evolution program*, Springer-Verlag, USA, third edition, 1996.
5.   K. DeJong and W.M. Spears, *Using genetic algorithms to solve NP-complete problems*, Proceedings of the Third International Conference on Genetic Algorithms, pp. 124-132, Morgan Kaufmann, 1989.
6.   D.B. Armstrong, *A programmed algorithm for assigning internal codes to sequential machines*, IRE Transactions on Electronic Computers, EC 11, no. 4, pp. 466-472, August 1962.
7.   W.S. Humphrey, *Switching circuits with computer applications*, New York: McGraw-Hill, 1958.
8.   Collaborative Benchmarking Laboratory, North Carolina State University, http://www.cbl.ncsu.edu/pub/Benchmark_dirs/LGSynth89/fsmexamples/, November 27th, 2003.

Author Index

Lecture Notes in Computer Science

For information about Vols. 1–2970

please contact your bookseller or Springer-Verlag

Vol. 3022: T. Pajdla, J. Matas (Eds.), Computer Vision - ECCV 2004. XXVIII, 621 pages. 2004.

Vol. 3021: T. Pajdla, J. Matas (Eds.), Computer Vision - ECCV 2004. XXVIII, 633 pages. 2004.

Vol. 3019: R. Wyrzykowski, J. Dongarra, M. Paprzycki, J. Wasniewski (Eds.), Parallel Processing and Applied Mathematics. XIX, 1174 pages. 2004.

Vol. 3016: C. Lengauer, D. Batory, C. Consel, M. Odersky (Eds.), Domain-Specific Program Generation. XII, 325 pages. 2004.

Vol. 3015: C. Barakat, I. Pratt (Eds.), Passive and Active Network Measurement. XI, 300 pages. 2004.

Vol. 3014: F. van der Linden (Ed.), Software Product-Family Engineering. IX, 486 pages. 2004.

Vol. 3012: K. Kurumatani, S.-H. Chen, A. Ohuchi (Eds.), Multi-Agnets for Mass User Support. X, 217 pages. 2004. (Subseries LNAI).

Vol. 3011: J.-C. Régin, M. Rueher (Eds.), Integration of AI and OR Techniques in Constraint Programming for Combinatorial Optimization Problems. XI, 415 pages. 2004.

Vol. 3010: K.R. Apt, F. Fages, F. Rossi, P. Szeredi, J. Váncza (Eds.), Recent Advances in Constraints. VIII, 285 pages. 2004. (Subseries LNAI).

Vol. 3009: F. Bomarius, H. Iida (Eds.), Product Focused Software Process Improvement. XIV, 584 pages. 2004.

Vol. 3008: S. Heuel, Uncertain Projective Geometry. XVII, 205 pages. 2004.

Vol. 3007: J.X. Yu, X. Lin, H. Lu, Y. Zhang (Eds.), Advanced Web Technologies and Applications. XXII, 936 pages. 2004.

Vol. 3006: M. Matsui, R. Zuccherato (Eds.), Selected Areas in Cryptography. XI, 361 pages. 2004.

Vol. 3005: G.R. Raidl, S. Cagnoni, J. Branke, D.W. Corne, R. Drechsler, Y. Jin, C.G. Johnson, P. Machado, E. Marchiori, F. Rothlauf, G.D. Smith, G. Squillero (Eds.), Applications of Evolutionary Computing. XVII, 562 pages. 2004.

Vol. 3004: J. Gottlieb, G.R. Raidl (Eds.), Evolutionary Computation in Combinatorial Optimization. X, 241 pages. 2004.

Vol. 3003: M. Keijzer, U.-M. O'Reilly, S.M. Lucas, E. Costa, T. Soule (Eds.), Genetic Programming. XI, 410 pages. 2004.

Vol. 3002: D.L. Hicks (Ed.), Metainformatics. X, 213 pages. 2004.

Vol. 3001: A. Ferscha, F. Mattern (Eds.), Pervasive Computing. XVII, 358 pages. 2004.

Vol. 2999: E.A. Boiten, J. Derrick, G. Smith (Eds.), Integrated Formal Methods. XI, 541 pages. 2004.

Vol. 2998: Y. Kameyama, P.J. Stuckey (Eds.), Functional and Logic Programming. X, 307 pages. 2004.

Vol. 2997: S. McDonald, J. Tait (Eds.), Advances in Information Retrieval. XIII, 427 pages. 2004.

Vol. 2996: V. Diekert, M. Habib (Eds.), STACS 2004. XVI, 658 pages. 2004.

Vol. 2995: C. Jensen, S. Poslad, T. Dimitrakos (Eds.), Trust Management. XIII, 377 pages. 2004.

Vol. 2994: E. Rahm (Ed.), Data Integration in the Life Sciences. X, 221 pages. 2004. (Subseries LNBI).

Vol. 2993: R. Alur, G.J. Pappas (Eds.), Hybrid Systems: Computation and Control. XII, 674 pages. 2004.

Vol. 2992: E. Bertino, S. Christodoulakis, D. Plexousakis, V. Christophides, M. Koubarakis, K. Böhm, E. Ferrari (Eds.), Advances in Database Technology - EDBT 2004. XVIII, 877 pages. 2004.

Vol. 2991: R. Alt, A. Frommer, R.B. Kearfott, W. Luther (Eds.), Numerical Software with Result Verification. X, 315 pages. 2004.

Vol. 2990: J. Leite, A. Omicini, L. Sterling, P. Torroni (Eds.), Declarative Agent Languages and Techniques. XII, 281 pages. 2004. (Subseries LNAI).

Vol. 2989: S. Graf, L. Mounier (Eds.), Model Checking Software. X, 309 pages. 2004.

Vol. 2988: K. Jensen, A. Podelski (Eds.), Tools and Algorithms for the Construction and Analysis of Systems. XIV, 608 pages. 2004.

Vol. 2987: I. Walukiewicz (Ed.), Foundations of Software Science and Computation Structures. XIII, 529 pages. 2004.

Vol. 2986: D. Schmidt (Ed.), Programming Languages and Systems. XII, 417 pages. 2004.

Vol. 2985: E. Duesterwald (Ed.), Compiler Construction. X, 313 pages. 2004.

Vol. 2984: M. Wermelinger, T. Margaria-Steffen (Eds.), Fundamental Approaches to Software Engineering. XII, 389 pages. 2004.

Vol. 2983: S. Istrail, M.S. Waterman, A. Clark (Eds.), Computational Methods for SNPs and Haplotype Inference. IX, 153 pages. 2004. (Subseries LNBI).

Vol. 2982: N. Wakamiya, M. Solarski, J. Sterbenz (Eds.), Active Networks. XI, 308 pages. 2004.

Vol. 2981: C. Müller-Schloer, T. Ungerer, B. Bauer (Eds.), Organic and Pervasive Computing – ARCS 2004. XI, 339 pages. 2004.

Vol. 2980: A. Blackwell, K. Marriott, A. Shimojima (Eds.), Diagrammatic Representation and Inference. XV, 448 pages. 2004. (Subseries LNAI).

Vol. 2979: I. Stoica, Stateless Core: A Scalable Approach for Quality of Service in the Internet. XVI, 219 pages. 2004.

Vol. 2978: R. Groz, R.M. Hierons (Eds.), Testing of Communicating Systems. XII, 225 pages. 2004.

Vol. 2977: G. Di Marzo Serugendo, A. Karageorgos, O.F. Rana, F. Zambonelli (Eds.), Engineering Self-Organising Systems. X, 299 pages. 2004. (Subseries LNAI).

Vol. 2976: M. Farach-Colton (Ed.), LATIN 2004: Theoretical Informatics. XV, 626 pages. 2004.

Vol. 2973: Y. Lee, J. Li, K.-Y. Whang, D. Lee (Eds.), Database Systems for Advanced Applications. XXIV, 925 pages. 2004.

Vol. 2972: R. Monroy, G. Arroyo-Figueroa, L.E. Sucar, H. Sossa (Eds.), MICAI 2004: Advances in Artificial Intelligence. XVII, 923 pages. 2004. (Subseries LNAI).

Vol. 2971: J.I. Lim, D.H. Lee (Eds.), Information Security and Cryptology -ICISC 2003. XI, 458 pages. 2004.

MIX
Papier aus verantwortungsvollen Quellen
Paper from responsible sources
FSC® C105338

If you have any concerns about our products,
you can contact us on
ProductSafety@springernature.com

In case Publisher is established outside the EU,
the EU authorized representative is:
**Springer Nature Customer Service Center GmbH
Europaplatz 3, 69115 Heidelberg, Germany**

Printed by Libri Plureos GmbH
in Hamburg, Germany